Handbook of Research on Algae as a Sustainable Solution for Food, Energy, and the Environment

Mostafa M. El-Sheekh
Tanta University, Egypt

Norhayati Abdullah
Universiti Teknologi Malaysia, Malaysia

Imran Ahmad
Universiti Teknologi Malaysia, Malaysia

IGI Global
PUBLISHER of TIMELY KNOWLEDGE

A volume in the Practice, Progress, and Proficiency in Sustainability (PPPS) Book Series

Published in the United States of America by
IGI Global
Engineering Science Reference (an imprint of IGI Global)
701 E. Chocolate Avenue
Hershey PA, USA 17033
Tel: 717-533-8845
Fax: 717-533-8661
E-mail: cust@igi-global.com
Web site: http://www.igi-global.com

Library of Congress Cataloging-in-Publication Data

Names: El-Sheekh, Mostafa, editor. | Abdullah, Norhayati, editor. | Ahmad, Imran, 1982- editor.
Title: Handbook of research on algae as a sustainable solution for food, energy, and the environment / Mostafa El-Sheekh, Norhayati Abdullah, Imran Ahmad.
Description: Hershey : Engineering Science Reference, [2022] | Includes bibliographical references and index. | Summary: "This book provides an opportunity to the readers to understand about the biological and biotechnological processes going on in the cultivation and harvesting of algae, DNA sequencing, and genomics of algae, giving detailed insight into various applications of algal biomass in agriculture, food, feed, and medicines"-- Provided by publisher.
Identifiers: LCCN 2021054771 (print) | LCCN 2021054772 (ebook) | ISBN 9781668424384 (hardback) | ISBN 9781668424407 (ebook)
Subjects: LCSH: Algae. | Algae as food. | Algae as feed. | Algal biofuels.
Classification: LCC QK566 .E93 2022 (print) | LCC QK566 (ebook) | DDC 579.8--dc23/eng/20211123
LC record available at https://lccn.loc.gov/2021054771
LC ebook record available at https://lccn.loc.gov/2021054772

This book is published in the IGI Global book series Practice, Progress, and Proficiency in Sustainability (PPPS) (ISSN: 2330-3271; eISSN: 2330-328X)

British Cataloguing in Publication Data
A Cataloguing in Publication record for this book is available from the British Library.

For electronic access to this publication, please contact: eresources@igi-global.com.

Practice, Progress, and Proficiency in Sustainability (PPPS) Book Series

Ayman Batisha
International Sustainability Institute, Egypt

ISSN:2330-3271
EISSN:2330-328X

Mission

In a world where traditional business practices are reconsidered and economic activity is performed in a global context, new areas of economic developments are recognized as the key enablers of wealth and income production. This knowledge of information technologies provides infrastructures, systems, and services towards sustainable development.

The **Practices, Progress, and Proficiency in Sustainability (PPPS) Book Series** focuses on the local and global challenges, business opportunities, and societal needs surrounding international collaboration and sustainable development of technology. This series brings together academics, researchers, entrepreneurs, policy makers and government officers aiming to contribute to the progress and proficiency in sustainability.

Coverage

- Socio-Economic
- Outsourcing
- Knowledge clusters
- Green Technology
- Innovation Networks
- Intellectual Capital
- Global Content and Knowledge Repositories
- E-Development
- Strategic Management of IT
- Sustainable Development

IGI Global is currently accepting manuscripts for publication within this series. To submit a proposal for a volume in this series, please contact our Acquisition Editors at Acquisitions@igi-global.com or visit: http://www.igi-global.com/publish/.

The Practice, Progress, and Proficiency in Sustainability (PPPS) Book Series (ISSN 2330-3271) is published by IGI Global, 701 E. Chocolate Avenue, Hershey, PA 17033-1240, USA, www.igi-global.com. This series is composed of titles available for purchase individually; each title is edited to be contextually exclusive from any other title within the series. For pricing and ordering information please visit http://www.igi-global.com/book-series/practice-progress-proficiency-sustainability/73810. Postmaster: Send all address changes to above address.

Titles in this Series

For a list of additional titles in this series, please visit: www.igi-global.com/book-series/practice-progress-proficiency-sustainability/73810

Principles, Practices, and Future Outlooks for Orchards Mangement
Waleed Fouad Abobatta (Horticulture Research Institute, Agriculture Research Center, Egypt) Ahmed Farag (Central Laboratory for Agricultural Climate, Agriculture Research Center, Egypt) and Mohamed Abdel-Raheem (National Research Center, Egypt)
Engineering Science Reference • © 2022 • 335pp • H/C (ISBN: 9781668424230) • US $215.00

Socio-Economic Sustainability in the Post-Pandemic Era
Jozef Oleński (High School Technology and Economics, Jaroslaw, Poland) Jeffrey Sachs (Columbia University, USA) Masayuki Susai (Nagasaki University, Japan) Yannis Tsekouras (University of Macedonia, Greece) and Arjan Gjonça (London School of Economics and Political Science, UK)
Information Science Reference • © 2022 • 315pp • H/C (ISBN: 9781799897606) • US $215.00

Climate Change, World Consequences, and the Sustainable Development Goals for 2030
Ana Pego (Nova University of Lisbon, Portugal)
Engineering Science Reference • © 2022 • 335pp • H/C (ISBN: 9781668448298) • US $250.00

Innovative Economic, Social, and Environmental Practices for Progressing Future Sustainability
Chai Lee Goi (Curtin University, Malaysia)
Engineering Science Reference • © 2022 • 369pp • H/C (ISBN: 9781799895909) • US $215.00

Impact of Artificial Reefs on the Environment and Communities
Jorge H. P. Ramos (CinTurs, University of Algarve, Portugal)
Engineering Science Reference • © 2022 • 304pp • H/C (ISBN: 9781668423448) • US $240.00

Wetland Biodiversity, Ecosystem Services, and the Impact of Climate Change
Abdelkrim Ben Salem (Mohammed V University, Morocco) Laila Rhazi (Mohammed V University, Morocco) and Ahmed Karmaoui (University of Moulay Ismail, Morocco & Southern Center of Culture and Science, Morocco)
Engineering Science Reference • © 2022 • 310pp • H/C (ISBN: 9781799892892) • US $250.00

Handbook of Research on Organizational Sustainability in Turbulent Economies
Rafael Ignacio Perez-Uribe (Santo Tomas University, Bogota, Colombia) Carlos Salcedo-Perez (Politecnico GranColombiano University Institución, Colombia) and Andres Carvajal-Contreras (EAN University, Colombia)
Business Science Reference • © 2022 • 428pp • H/C (ISBN: 9781799893011) • US $325.00

IGI Global
PUBLISHER of TIMELY KNOWLEDGE

701 East Chocolate Avenue, Hershey, PA 17033, USA
Tel: 717-533-8845 x100 • Fax: 717-533-8661
E-Mail: cust@igi-global.com • www.igi-global.com

Dedicated to our parents and mentors, especially the late Prof. Dr Qamar-e-Hasan, who was an eminent academician, an able administrator, and a role model father.

Editorial Advisory Board

List of Contributors

Table of Contents

Detailed Table of Contents

Algae encompass a phylogenetically diverse group; its members are placed in the domains Bacteria and Eukarya, in the latter dispersed in five supergroups. Contrary to common assumption, algae are found almost everywhere from aquatic to terrestrial habitats, some of them in symbiotic relationship, others thriving in harsh environments such as deserts, hot springs, or on ice. Around 45% of atmospheric oxygen is released by algae, which underlines their pivotal role in global element cycles. This high diversity is also reflected in different morphologies from microscopic single cells to parenchymatous organisms measuring several meters in size. The total species number is estimated from 70,000 to more than 10,000,000. Algae produce various valuable compounds, but this high natural potential is only marginally being used. To date, only a few taxa are commercially exploited.

"Algae" represent a diverse group of photosynthetic organisms ranging from single cell to massive kelp that are found in all ecosystems on our planet. Algae play an essential role for life; they account for almost 50% of the photosynthesis activity on earth. The biochemical composition of algae includes a variety of high-value products such as pigments, lipids, carbohydrates, amino acids, and many other bioactive compounds. Algal biochemical compounds are variable among the different species and are highly dependent on the algal culture conditions such as temperature, nutrients, light, etc. In the recent years, algal biotechnological applications are on the spot. Algae are exploited for several biotechnological uses such as for biofuels, bio fertilizers, pharmaceuticals, nutraceuticals, bioremediators, and others. This chapter discusses the biochemistry and biotechnology of algae with emphasizing on the high-value biochemical algal compounds and trending algal biotechnological applications.

Intensive research efforts are aimed at increasing and modifying the algal biomass production and selection for different purposes. Financial aspects for biomass production techniques always remain a challenge that needs to be addressed. Using cost-effective media for the growth and choosing high lipid content strain is always aimed to reduce the cost of yield of algal biomass. With each passing day, there is advancement in the use of algae for the vested interests. Different species are expected to function well at different niche and environmental conditions. Therefore, adaptation of robust method and selection of algal trait is most relevant for yielding large scale algal biomass. The overarching significance of producer strain has driven research in recent years towards genetically modified species. This chapter particularly focusses on the selection and breeding of algae like different cultivation aspects in open pond, photobioreactor, bio flocculation, and advantages and disadvantages thereof.

Algae, including cyanobacteria, dominate aquatic habitats. They are the principal producers of aquatic environments. On the other hand, microbes, are essential algal helpers and are known as holobionts. Holobionts are algae-associated microbes that include bacteria, fungi, and viruses. Over millions of years, many interaction mechanisms between algal cells and their halobiont have evolved. These interactions include mutualism, commensalism, and parasitism. These interactions are critical for ecosystem resistance and resilience. Microbes, for example, regulate algal cell proliferation by producing toxic metabolites that control the algal growth. Alternatively, the production of vitamins and growth factors by microbes might promote algal cell proliferation. Moreover, in biotechnological applications, the algae and bacteria co-cultures are very promising as a sustainable application to persistent environmental issues and green energy solutions. Various mechanisms of intracellular and extracellular algae microbe interactions were discussed in this chapter. This is an endeavor to get knowledge about algae-microbe interactions for biomass-based energy solutions and other environmental applications.

Algae importance is spectacularly increasing in many biotechnological applications, such as human food, animal feed, biofuels, bioplastics, bioremediation, pharmaceuticals, and cosmetics. With the widespread use of "omics" technologies over the past two decades, recent advanced research attempts to understand the pathways of the promising algae species by whole genomes sequencing (genomics) and revealing lipid pathways (lipidomics), microarray to study all RNA transcripts (transcriptomics), all protein sets produced by the algal cell (proteomics). DNA alteration as classical mutagenesis caused a random mutation such as ethyl methane-sulfonate as chemical mutagenic and ultraviolet radiation as a physical mutagenic. On the other hand, the CRISPR-Cas9 modern technique is used to genetically engineer a protein with maximum editing efficiency. Incorporating omics and mutations techniques helps to thoroughly understand the systems biology of algae in the new era called integrated omics.

Algae are a fascinatingly diverse group of photosynthetic organisms existing in diverse environments (ranging from oceans, rivers, lakes, ponds, and brackish waters). Comprising the base of the aquatic food ecosystem, algae have pivotal ecological functions as oxygen producers. Ranging in size from unicellular microalgae to the giant kelp, they have a wide range of (food, pharmaceutical, and industrial) applications. Physiology of algae comprises the study of algal function and behaviour. It encompasses all the dynamic processes of growth, metabolism, reproduction, defence, communication of algae (that account for algae being alive), and the processes underlying large biogeographical patterns of algae. Several biotic and abiotic environmental variables such as nutrients, light, temperature stress, salinity stress, desiccation, global warming, and ocean acidification affect algal growth and occurrence. This chapter provides a rudimentary insight regarding the growth, reproduction, and biochemistry of algae under varying environmental conditions.

During the past decades, algae have attracted worldwide attention as a sustainable bioresource to produce various biochemicals and biofuels. However, the prohibitive cost of algal biomass production and processing casts doubt on the industrial applications of algae. Hence, many efforts have been made to enhance the viability of these species. One serious challenge is maximizing algal biomass production. Since algal

extraordinary uses because of its large yields, short growth time, and variable culture conditions. Algal polymers, blends, and combinations of biomass or algal biomolecules with other polymeric materials are examples of algae-based materials. Polysaccharides and sulfate polysaccharides such as agar, carrageenan, alginates, and polyhydroxyalkanoates are essential algal polymers. Algal and biomass polymers have improved mechanical characteristics, biocompatibility, and biodegradability when included into synthetic polymer systems. Algal-based biomaterials are interesting contenders to replace existing, non-renewable polymer materials derived from fossil fuels. This chapter discusses the numerous applications of biomaterials obtained from algae. Furthermore, as biotechnology advances, algae-based polymers, blends, and composites have found many applications in a variety of domains of human existence, ranging from medicinal applications to sophisticated technological applications.

Chapter 13
A Sustainable Supply Chain Model for the Development of Green Fuel Production From Microalgae 332
Nima Norouzi, Bournemouth University, UK

This study presents designing and managing a green fuel supply chain based on algae to investigate the development of such fuels in the country. On this basis, a definitive model is first developed to model all the activities of the green fuel supply chain, which includes the supply of raw materials for the growth of algae, the cultivation of algae and their conversion into fuel, and finally, the supply of fuel in the country. This deterministic model is extended to a robust network model to secure supply chain decisions against uncertainty. Using the proposed model for the development of algal fuels in Iran shows that the green fuel production cost is currently 27 cents/liter. The current cost of producing fuel from algae cannot compete with fossil fuels, but this cost can be greatly reduced in the future by slightly increasing the growth rate of algae and their oil content.

Chapter 14
Application of Algae for Hydrogen Generation and Utilization 354
Eman R. Sallam, National Institute of Oceanography and Fisheries, Egypt
Hanan M. Khairy, National Institute of Oceanography and Fisheries, Egypt
Mostafa Elshobary, Department of Botany, Faculty of Science, Tanta University, Egypt
Howida A. Fetouh, Faculty of Science, Alexandria University, Egypt

Hydrogen is a promising future fuel with high energy content for both heat and electrical energy without emission of any hazardous gases such as carbon dioxide or ozone-harming substances. Bio-hydrogen driven from microalgae has recently gained considerable attention. As it is more sustainable than other sources, further developments in such systems are still in their early stages and require improving efficiency and achieving a real-world application on a large scale. This chapter focuses on assessing the potential of microalgae applied sciences for the industrial manufacturing of hydrogen from algae using solar energy. It summarizes the principle key of hydrogen production, the viable and theoretical limits of microalga hydrogen manufacturing systems, and the rising techniques to engineer next-generation structures and

how these fitting into a developing hydrogen economy. In addition, it discusses the utilization of a solar collector system for long-range storage of thermal solar energy in dark conditions or winter.

Rare earth metals (REMs) are some of the most expensive materials due to global demand with a constant rise due to their critical role in advancements of technology. This drives us to find eco-friendly approaches to their efficient recovery and reuse. Many microorganism groups can develop various mechanisms to chelate metals, among these the algae that are considered to be promising an emerging solution for collecting rare metals and substances from the environments due to their high bioremediation ability through different mechanisms such as bioaccumulation, biodegradation, or biosorption inside their biomass. Hence, it is easy to recover these accumulated metals using different methodologies and reuse them in different technological aspects; on the other hand, through the bioremediation process, other substances can be produced as secondary metabolites that can be utilized as useful materials. The chapter will discuss the importance of rare metals and the effective biotechnological role of algae in order to recover them for their economic reuse in different approaches.

The commercial viability of various algal species increases with the growing demand of natural food supplements. Utilisation of algal biomass as health supplements is paving new ways in the fields of nutraceutical and pharmaceutical industries. They also play a crucial nutritional role in livestock feed and aquaculture. Various algal species are rich sources of bioactive compounds like fatty acids, essential minerals, bioactive peptides, carotenoids, vitamins, etc. and thus has the potential to compete with their synthetic counterparts in the market. The increase in demand for high value health supplements and market trends has motivated researchers and industries in developing algal novel products containing functional ingredients. Some important algae that are used as human food, antioxidants, and nutritional supplements are reviewed in this chapter. This chapter also summarizes the role of algae in animal feed industry and aquaculture. Major challenges in the application of algae as nutraceuticals and food are also discussed along with possible future directions.

Nanotechnology has been a catchphrase in recent years. Its expansion into a new field has been phenomenal. Because of their various shapes and sizes, nanoparticles differ from their conventional material. They have a larger surface area, which is necessary for many chemical methods. One of the possible solutions to the above-mentioned limitations is algae-mediated nanoparticle production. This chapter focuses on the use of algae to synthesis nanoparticles and the possible benefits of this technology over traditional methods. The creation of nanoparticles by cyanobacteria, microalgae, and macroalgae is taken into account. Metal nanoparticles derived from algae, such as gold, silver, and iron, have a wide range of

Biomass from algae, which is rich in proteins, carbohydrates, and lipids, could be used for the production of biofuels and chemicals. Because algal cultivation and harvesting require high energy and costs, algae-based fuel production is a challenging commercial application. At the pilot scale, this is a common bottleneck problem in algae processing for fuels or chemicals. By implementing an integrated algae biorefinery concept, the need for energy and costs can be reduced. Biopolymers, biochemicals, biofuels, and biofertilizers can all be recovered with higher economic efficiency than conventional methods. A green economy based on algae will also be more viable by reducing production costs. The purpose of this mini-review is to give information about the development of integrated biorefineries for recovery of algal-based bioproducts and their potential applications. The authors discuss the lifecycle assessment and the economic aspects of an integrated algal biorefinery. A discussion of the challenges and future directions of integrated algal biorefinery is concluded.

Chapter 22
Towards Sustainable Use of Algae as Adsorbents for Wastewater Treatment.................................... 547
Victor Odhiambo Shikuku, Kaimosi Friends University College, Kenya
George Oindo Achieng', Maseno University, Kenya
Patrick Ssebugere, Makerere University, Uganda

The occurrence of heavy metals, dyes, micronutrients, phenols, pharmaceuticals, and personal care products (PPCPs) in water resources continue to raise environmental concerns since they are known to cause detrimental effects on living organisms even at low concentrations. Conventional wastewater treatment plants have also been pointed out as point sources of loading these pollutants into the recipient surface waters. Because of the non-biodegradable nature of heavy metals and the stable structure of dyes and PPCPs, these pollutants are persistent in the environment. Studies have shown that algae (micro and macro) present an alternative source of low-cost, efficient, and sustainable biosorbent for the removal of various pollutants from water either singly or in synergy with other wastewater treatment processes. This chapter is a brief review of recent studies on the use of algae-based biosorbents for the sequestration of heavy metals, dyes, and PPCPs from wastewater. Microalgae and macroalgae are shown to be promising and sustainable materials for the biosorption of water pollutants.

Chapter 23
Role of Algae in Cancer.. 562
Fazrena Nadia Md Akhir, Malaysia-Japan International Institute of Technology, Universiti Teknologi Malaysia, Malaysia
Farhah Hanani Mohd Nazaruddin, Malaysia-Japan International Institute of Technology, Universiti Teknologi Malaysia, Malaysia
Nor'azizi Othman, Malaysia-Japan International Institute of Technology, Universiti Teknologi Malaysia, Malaysia
Hirofumi Hara, The University of Tokyo, Japan
Imran Ahmad, Malaysia-Japan International Institute of Technology, Universiti Teknologi Malaysia, Malaysia

Cancer is ranked as the second most common cause of death worldwide and searching new therapeutic agents for cancer treatment remains a major challenge. Despite the remarkable developments in cancer therapy in past decades, there is still an insistent necessity for innovative drugs in cancer biology, particularly in the under-explored area of marine anticancer compounds. Algae are photosynthetic organisms consisting of a total of 30,000 species that thrive in a watery environment. The identification

Foreword

I am extremely happy to write an introductory note on this timely and scholarly volume titled *Handbook of Research on Algae as a Sustainable Solution for Food, Energy, and the Environment* on Algal science and technology catering to a large audience encompassing students, teachers, researchers, environmentalists, food specialists, medicinal & health professionals, industrialists, and entrepreneurs.

Photosynthetic organisms are centric to biomass production, with multifarious applications. Algae being an efficient harvester of solar energy and recycler of inorganic molecules of the ecosystem, are highly desirable organisms contributing to the sustainable production of biomass. The studies on the bioprospecting of the algal population have led to the identification of more than 75,000 species occurring in various habitats in marine, aquatic and terrestrial systems. They exhibit a wide diversity which is classified based on the pigment production namely, green, blue-green, red, and brown forms. Being a natural member of the ecosystem from over 1.8- 1.5 billion years, it fulfills a primary role as a food producer providing nourishment to the organisms in the food chain. The utility of algae in producing goods and services for mankind is being exploited sustainably.

This volume addresses various issues related to sustainable development goals (SDGs) of the UN through algal Science and Technologies. It deals with the breeding of algae for cultivation in various ecological regions in industrial and semi-industrial levels, processing of biomass for food, feed, health, and bioenergy applications. It also describes the sustainable production of algal biomass for pharmaceuticals. The augmentation of the utility of algal forms is facilitated by studies on functional genomics, leading to our enhanced understanding of the genetic basis of fundamental aspects of growth and metabolism, which are elegantly described. Aspects of large-scale production of biomass coupled with biorefinery approaches to downstream processing are superbly described for profitable exploitation. Aspects of the circular economy for sustainability is a unique feature of this book for complete utilization of the algal biomass and by-products for sustainable exploitation in an economical manner.

Sustainable energy production through algae by fixation of atmospheric carbon dioxide is the need of the hour . It also serves as an alternative to fossil fuels providing solutions without the generation of greenhouse gases. This has implications for reducing global warming, and gaining carbon credits for industrial production purposes. The aspects of production of lipids for biodiesel; hydrocarbons for high-grade fuels; and hydrogen as clean energy have been explicitly presented.

The utility of algae for food purposes is of paramount importance as they produce novel metabolites such as PUFA, DHA, EPA etc which are exploited as nutraceuticals. They produce high-quality protein, vitamins, minerals, pigments, antioxidants, and health-promoting substances of use by humans. Genetic engineering of algae has empowered us to produce value-added products which are of high utility.

Bioremediation of industrial wastes adopting algal consortia has been dealt with. This not only provides solutions for the mitigation of pollution but also facilitates the production of biomass for feeds, fertilizers, and chemicals of utility in an economical manner.

All the above-mentioned information are elegantly presented in the form of 24 chapters. Each of these chapters is contributed by experts in the respective fields rendering a volume that is unique and authoritative. It provides the readers with comprehensively presented information on algal technologies meeting their demands for obtaining all the information in one platform. I congratulate the editors for fulfilling the requirements of all stake holders desirous of reading a comprehensive volume of with authoritative narration on the topic of algal resources and technologies for sustainable production and commercial exploitation for the benefit of mankind.

I sincerely hope that this volume which is elegantly brought out by IGI publishers will be well-received by all the stakeholders who are interested in this topic. I wish all the best to the editors and contributors.

Gokare A. Ravishankar
Dayananda Sagar Institutions, Bengaluru, India & Central Food Technological Research Institute (CSIR), Mysore, India

Preface

The world population is projected to reach 9.7 billion people by 2050. One out of nine people in the world is suffering from hunger/malnutrition. Expanding food production and economic growth have often come at a heavy cost to the natural environment. The Earth can fulfil the demands of food security; however, the agricultural sector will need innovative technologies and research to retrieve its full capacity for production. Therefore, easy to produce and cost-effective sources that can rapidly produce substantial amounts of nutritional value are needed. Algae can become a future food supplement and nutraceutical compound because of its unique composition. Algae (eukaryotic and prokaryotic) are capable of bio-converting CO_2 (using light) into Algae biomass which in turn is the potential source of biofuels, nutritional supplements, and pharmaceuticals. Algae are having many advantages like high biomass yields per unit area, and the ability to be grown on non-arable land or in the photobioreactors (closed or open), they can be grown in saline water and wastewater. Algae are an extremely diverse group with an estimated number of species ranging from 200,000 to 800,000, out of which about 30,000 are archived. Some of the advantages are as follows: Algae are having the capability to fix CO_2 in the atmosphere and its biomass can be efficient in the bio-fixation of waste CO_2, thereby playing a vital role in the reduction of the major contributor to global warming and greenhouse effect. Algae can be grown around the year. The harvesting of algae is a cost incentive as compared to other crops and they can be grown independently without affecting the human food chain thus eradicating the food versus fuel feud. Algae can be grown in several environments that are unsuited for growing other crops, such as fresh, brackish, saltwater, or non-arable lands. Moreover, they can be cultivated in open ponds and photobioreactors (PBRs). Therefore because of this significant and non-selective growth, microalgae can be produced at a superior rate (yield/hectare) without disturbing the ecology. Algae can be cultivated using wastewater providing a dual advantage of wastewater treatment and biomass production. Seawater is a viable alternative for the bulk cultivation of algae, which will reduce the cost of production and reduce the global consumption of freshwater. Since seawater is a good source of nutrients and may enhance the productivity of microalgae biomass. Algae form a stable and efficient biological system to utilize solar energy and produce organic compounds. Algae have the potential of transforming 9–10% of sunlight into biomass with a theoretical yield of about 280 ton/ha/year. Algae biomass is used to obtain many beneficial products (co/by) like biofuels, bio-pesticides, biopolymers, carbohydrates, proteins, and biofertilizers, etc. Generally, algae double their biomass within a day but exponential growth results in a doubling of biomass in just 3-4 hours. On average alga have oils in the range of 20 to 50% by weight of dry biomass, but sometimes even higher productivities can be obtained by varying the growth factors.

The increasing depletion of fossil-based fuels and their deleterious impacts on the environment caused by the emissions of greenhouse gases like CO_2 have led to the shift in generating energy from renewable sources. It will reduce the dependency on fossil fuels and the severe problems associated with environmental pollution. This critical juncture led to the advent of biofuels which are renewable, nontoxic, and ecofriendly fuel produced from various feedstocks. Amid these feedstocks Algae is having immense potential to replace fossil fuels and play a vital role in energy security and obviate global environmental problems, ultimately leading to their sustainability. Algae for their cultivation need light as an energy source to convert the absorbed water and CO_2 into biomass through photosynthesis. The requirement of major nutrients is fulfilled by nitrogen and phosphorus, which are 10-20% of the algae biomass. The growth is supplemented by the presence of some macro (Na, Mg, Ca, and K) and micronutrients (Mo, Mn, B, Co, Fe, and Zn). The oils contained in Algae have similar physicochemical properties as that of vegetable oils hence they are rich raw materials in the production of biofuels.

Today's planet faces several critical problems such as resource depletion, environmental destruction, and climate change that affect all areas of life as we know it. Figuring out how to address these issues and prioritizing Earth's health has been at the forefront of study as it is a key issue that affects us all. One element that requires further investigation is algae regarding its potential for creating a more sustainable future across the food, energy, and environmental sectors. It came to our mind to write a book on algae covering all the domains of cultivation and applications.

Handbook of Research on Algae as a Sustainable Solution for Food, Energy, and the Environment provides insight into the biotechnological and biorefinery aspects of algae together with their unique applications in the agriculture and pharmaceutical industry. Furthermore, this book considers the biological and biotechnological processes happening in the cultivation and harvesting of algae, DNA sequencing, and genomics of algae. Moreover, it examines the bio-remediation aspects of algae and its utilization to produce biofuels, methane, hydrogen, and other useful renewable sources of energy, thereby contributing to environmental sustainability. Covering topics such as cell biology and food science, this reference work is ideal for academicians, researchers, industry professionals, scholars, practitioners, instructors, and students

Algae were once the driving force behind evolution, and they remain so now. In terms of appearance, physiology, and ecology, algae are very diverse and versatile. This versatility is evident in the large spectrum of high-quality chemicals they manufacture, many of which are still awaiting commercialization. The estimated number of algal species is between 70,000 and more than 10,000,000. Algae continue to play an important part in biogeochemical cycles today. This group is responsible for about 45 percent of all oxygen generated (every second breath you take). Chapter 1, "Algae: Their World Explored," focuses on the classification, lineage, morphology, and phylogeny of Algae giving readers an understanding of their possible categorization.

Chapter 2, "Biochemistry and Biotechnology of Algae," provides insight into these photosynthetic organisms, known as "algae," that can be found in both marine and freshwater habitats and have a wide range of biotechnological applications. They use many ways to fix atmospheric carbon dioxide and convert it to biomass efficiently by utilizing nutrients. Algae are regarded as a bio factory for many chemicals with various applications due to their biochemical composition. Algae biotechnology usage are growing, and algae are being used in a variety of industries. Scientists are always arguing the obstacles to algae's promotion as a viable biotechnological resource.

Chapter 3, “Selection of Strains and Breeding of Algae,” gives an acumen about diverse algal strains, their cultivation methods, generated biomass, and phenotype variation for various biotechnological applications. The biggest challenge in generating the requisite biomass for any strain, according to the study, is the production protocol for large-scale synthesis. At large-scale production, a strain may face a variety of circumstances, contamination, or suitability constraints that are not present in a controlled environment. As a result, production may suffer or fail to reach laboratory-scale levels. Light intensities, temperature, alternating light, and dark, and probable exterior environmental contaminants may all cause significant changes. Bioprospecting or synthetic biology can help tackle the problem. While humanity may profit from quick growth that harvests energy from photosynthesis, low-cost production, and simple scalable cultivation of algae.

Chapter 4, “Cell Biology and Microbial Interactions in Algal Cells,” explains the story that exploitation of algae and bacteria co-cultures is very crucial nowadays as a solution for environmental, energy issues besides their contribution to the enhancement of the nutritional values of feed and food by extracting valuable compounds from their growth compared to those of monocultures. In wastewater treatment, the algae-microbe relationship boosted nutrient uptake while also contributing to the system’s stability during the remediation phase, making it more sustainable. Furthermore, this symbiosis aids in cost-effective aeration, greenhouse gas emissions sequestration, and the creation of flocs, all of which make biomass management easier. Furthermore, the related microorganisms aid algal hydrogenase in hydrogen generation by facilitating anaerobic micromovement for effective bio-hydrogen production in photo-fermentation. As a result, breakthroughs in our understanding of microbe interaction tactics will be extremely useful in addressing the world’s most pressing problems.

The purpose of Chapter 6, “Physiology of Algae: An Insight,” is to summarise current knowledge of algal adaptation and acclimatization processes in response to stable and unstable environmental conditions, including biotic interactions between algae, abiotic factors (temperature, light, UV irradiation), hydrodynamic characteristics, salinity, desiccation, and ocean acidification (pH). This chapter clearly shows that optimal environmental conditions are essential for algae to grow, mature, and function physiologically. Because of changes in climatic and anthropogenic activity in the environment over the last decade, these processes have received a lot of attention. These factors are responsible for causing changes in the physiology of algal species, such as changes in the rate of physiological and biochemical processes (reduction in photosynthesis rate, respiration rate, etc.), changes in bio-membrane structure/composition, and thus changes in their fluidity (extremes of low and high temperature, desiccation, salt, and pH, etc.). Algal acclimation and adaptation may help to reduce the risk of extinction for algae in a changing environment.

Although there have been significant advancements in algal bioprocessing, commercialization is still a long way off. As a result, improvements in both upstream and downstream processing of algal biomass are necessary. Therefore, Chapter 7, “Cultivation of Algae and Its Biorefinery Approach,” provides an overview of the bioprocessing of algae, focusing on the cultivation and biorefinery of algal biomass. When it comes to algal species culture, the focus should be on increasing biomass productivity and upgrading target chemicals. Furthermore, due to algae’s potential to sequester CO_2 and perform bioremediation, algal production employing industrial flue gases and wastewaters should be investigated. Furthermore, the major problem in algal biorefinery is the lack of mild methods for the separation of all algae components, which necessitates increased efforts to improve cell disruption and extraction procedures.

The understanding of the algal growth parameters and the ability to monitor and control these factors are basic requirements for efficient scaling up from flask scale to commercial scales. Hence, Chapter 8, "Scaling Up and Harvesting of Algae," elaborates on adequate studies and technologies for qualitative and quantitative monitoring of the algal cultivation. Some considerations should be made while designing culture systems, such as providing optimal development parameters such as temperature, enough illumination, an oxygen removal system, a CO_2 supply system, and enhanced stirring and mixing mechanisms. It is the responsibility of researchers to develop these designs to give a cost-effective way for algae cultivation. High yields of Algae are sought by researchers using culture optimization settings and more efficient harvesting procedures. Bio-flocculation systems that involve algal-algal, algal-fungi, and algal-bacteria interaction are more efficient, eco-friendly, and low-cost, but harvesting efficiency varies depending on the species utilized and the conditions.

Chapter 9, "In Silico Models on Algal Cultivation and Processing: An Approach for Engineered Optimization," expounds on the mathematical models and computational approaches. The development of optimal Algae growing systems for mass-scale biomass production for biorefinery purposes, which is sometimes difficult, can be accomplished using optimization frameworks that integrate both experimental and computational approaches. In the case of Algae processes, this necessitates an understanding of how cells and their intracellular components react to their surroundings (e.g., nutrients, light intensity, or temperature at which they are grown). Even though various modeling frameworks can describe the complex dynamics of Algae growth, a growing body of literature indicates that efforts are being made to develop models that can simulate both cellular growth processes as well as carbohydrate and lipid accumulation in response to a variety of growth-limiting factors. The use of strong modeling frameworks can lead to the discovery of optimal cultivation situations, which has significant implications for the creation of custom Algae-based biorefineries.

The manufacturing and marketing of algae-based products are accomplished by identifying the roadblocks and restrictions in the entire process, from start to finish, as determined by the Life Cycle Analysis (LCA). Chapter 10, "Algal Life Cycle Analysis and Its Contribution to the Circular Economy," will put the flash on the life cycle analysis of different products obtained from algae. To assist with the LCA, there is a variety of software available, each having a predetermined set of modules for simplicity of use. There are a few unique Algae products on the market right now. The demand for these products as nutraceuticals related to human health and well-being contributes significantly to the economy. The LCA also reveals process bottlenecks and aids in identifying areas where additional study is required to make the manufacturing chain more environmentally sustainable.

Modern agricultural practices led to soil and environmental degradation. Increasing needs for food and plant products necessitates sustainable agriculture and higher crops yields. Chapter 11, "Role of Algae in Agriculture," provides an insight into the exploitation of algae as a green chemical for sustainable agriculture. Since algae have the potential to produce huge biomass by fixing atmospheric CO_2. This algal biomass is being used in various algal-based products as agricultural inputs. It is a need of time to integrate these bio-based products into modern agriculture. Furthermore, technical advancement, awareness among the farming community, and policy regulation will bring an epoch-making advancement in algae-related industries that will ultimately lead toward environmentally benign, highly proactive, resource-efficient agricultural systems.

Chapter 12, "Role of Algae in the Production of Biomaterials," gives a comprehensive account of the application of algae to produce materials. The ever-increasing world population and their living standards result in the production of synthetic polymer products in a variety of applications, leading to the release of toxic materials such as dioxins into the environment. Sustainable Algae-based bioplastics, which can degrade without releasing any harmful chemicals, are easy to recycle, renewable, do not use fossil fuels, are cost-effective, require less energy to produce, can be used to encounter the needs of the population without harming the environment. Algae have the potential to produce biomaterials since the polymers needed for biomaterials are generated when Algae are thriving in harsh climates. For growth, they use light energy and inorganic nutrients such as carbon dioxide, phosphorus, and nitrogen, and they can synthesize valuable molecules from biomass. The core focus of development should be on cutting the cost of synthesizing algae-based biodegradable polymers.

Chapter 13 is "A Sustainable Supply Chain Model for the Development of Green Fuel Production From Algae." Increasing demand for fossil fuels, limited oil resources, and environmental pollution are among the most important motivations for developing green fuels. Algae, as one of the most valuable raw materials for the manufacturing of green fuel, has gotten a lot of attention in recent years around the world. To analyze the growth of such fuels in the country, this study proposes an algal supply chain network design model that includes all activities related to fuel production, from raw material supply to fuel production and supply. In this case, a solid optimization method has been used to deal with the supply chain uncertainty. The results show that a slight increase in the cost of the entire supply chain can improve the supply chain algae stability. More precisely, the cost of stability for the supply chain in question is small for limited oscillation intervals. The production costs show that the cost of producing each liter of green fuel from algae is higher than fossil fuels, but this cost is significantly reduced with a slight improvement in algal growth and conversion of dry algae to biodiesel.

Chapter 14, "Application of Algae for Hydrogen Generation and Utilization," gives a detailed account of the possible routes of algal conversion to hydrogen. Economic growth, energy use, the environment, and global warming are linked, yet "securing energy and the environment at the lowest cost" is becoming more challenging. Although natural and manmade factors contribute to climate change, global warming caused by greenhouse gases, particularly CO_2, is the primary cause of anthropogenic climate change. Switching to a hydrogen economy and employing renewable energy sources would be the best option for maintaining economic growth. In the long run, hydrogen created from algal biomass and solar energy will be the most renewable energy currencies. To achieve timely implementation of educational, financial, legislative, social, and technical initiatives is required.

Chapter 15, "Recovery of Rare Metals and Substance Production by Algae," is all about the necessity for finding out an eco-friendly approach for the recovery and reuse of rare earth metals (REMs). Among different Microorganisms, groups that can develop various mechanisms to chelate metals are algae the algal biomass (living or nonliving) is considered to be promising an emerging approach for re-collecting REMs from the aquatic environments due to their high ability for bioremediation through different mechanisms was achieved by bioaccumulation, or biosorption inside the algal biomass, on the other hand through the recovery process induce a byproduct metabolites substances that can be utilized as useful materials.

Chapter 16 is "Application of Algae in Food Science, Antioxidants, Animal Feed, and Aquaculture." Algae are considered the richest sources of renewable, sustainable, and economically feasible bioactive medicinal products, and food and feed supplements. The lack of incentives for microalgal-based foods production and little awareness about its health benefits are major hurdles to the success of micro-algae-derived foods. Therefore, extra work should be done on the genetic manipulation of microalgal species for increasing the production of target compounds. Further effort involving biotechnological treatment of the macroalgae by enzymatic degradation of algal fibres could improve protein digestibility and, therefore, will increase their nutritional value.

Chapter 17, "Algal Nanobiotechnology and Its Applications," is all about the studies of phyco-nanotechnology. Several scientists have looked at the biological potential of algae to produce nanoparticles of various shapes and sizes under varied settings. It was discovered that algae biomolecules act as a reducing and capping agent without the use of toxic compounds, resulting in stabilized nanoparticles (NPs). This chapter attempts to highlight the work that has been done in the field of phyco-nanotechnology. The biological approach to the synthesis of metallic NPs utilizing algae is discussed. AgNPs, AuNPs, and FeNPs were given special attention because of their unusual properties, which have piqued scientific and technological curiosity. For this reason, the major parameters influencing a steady and effective biosynthesis, as well as the most used algae, were detailed. Some metallic NP uses in nanomedicine, as well as their underlying primary molecular mechanism-induced cytotoxicity were discussed. Algal-mediated NPs synthesis is a quick, cost-effective, and efficient technique that has allowed nanotechnologists to synthesize desirable nanomaterials using renewable energy processes.

Chapter 18, "Utilization of Algae in Crop Improvement and Crop Protection for a Better Agricultural System," discusses the wider applications of algae-based various applications in agriculture and other sectors along with their tentative remedies. To produce green energy and protein-rich foods in the future, algae are very promising for urban agriculture, as they can play a fundamental role in carbon reduction and the management of greenhouse gases. Environmental sustainability and the inadequate availability of fossil fuels have prompted the research for innovations and expansion of algal science. However, algal research only mitigated the several issues related to fossil fuel production, enhancement in agricultural production, and environmental protection but also has unique properties for a wider range of applications. Furthermore, algal liquid extract can improve crop yield in adverse conditions by reducing toxic stress. Further, there are certain hindrances in the commercialization of products that resulted from the algae.

Chapter 19, "Microalgae as a Renewable Resource for Bioplastic Production," encompasses the feasibility of bioplastics production from algae. Despite the outstanding in-vitro successes of algae-based bioplastics, industrialization and large-scale production of algae-based plastics are still restricted by certain challenges. It is possible to boost biopolymer synthesis through stress-induced procedures, although this typically comes at the sacrifice of productivity. To that end, genetic engineering techniques such as gene editing, recombinant DNA, or blending microalgae with other plastics polymers might be advantageous to enable changed species to be commercialized. Thus, the problems of algal bioplastics must be solved as there is already an attraction factor from the market, which finally drives the producers/users to replace the fossil-based plastics. It is critical to work inside and outside technical domains to generate a product at the forefront of innovation, both in terms of sustainability and acceptability, to close the material loop for bioplastics.

Chapter 20 is "Role of Microalgae to Mitigate Fossil-Based Plastic Wastes Impacts." The chapter describes that microalgae can be used as a viable solution for the management and valorization of plastic wastes. Prior knowledge of the components of various plastic polymers might help to build the potential strategies like pretreatment, size reduction, sterilization, or addition of supplements, etc. for bioremediation through microalgae. Various bioreactors like a biofilm-based membrane, bioelectrical and MFC-based systems, and multistep treatment strategic technologies could improve microalgal growth on polymers. For future perspectives, microalgae cultivation with the 'sterile' polymers and cultivation in batch mode are promising to produce value-added biomass. Moreover, the biorefinery will maximize the value creation from microalgal-based valorization platforms of plastics. Using different types of plastics as a carbon source, microalgae cultivation on various support materials also requires advanced research in the future. Either using the single strain of microalgae, a consortium of multiple microalgae might give extraordinary biomass productivity with effective removal of various components of plastics waste from the environment.

Chapter 21 is "Algae Biomass Conversion Technologies." Due to the variety of biochemical compositions available in algae biomass, algae are extremely suitable for multi-product biorefineries. The most challenging aspect remained, however, the integration of various technologies for biomass conversion in a complete algae biorefinery. For algae biorefineries to be more sustainable and economically viable, more research is needed. According to the current LCA evaluation at a laboratory scale, it is still feasible to realize multiple products through algae biorefinery with reduced environmental impacts. Algae biorefineries, however, can only be realized at the commercial level soon by counterbalancing biofuels costs with profits from bioproducts.

Chapter 22 is "Towards Sustainable Use of Algae as Adsorbents for Wastewater Treatment." Microalgae-based wastewater treatment processes are less energy-intensive, relative to conventional and advanced oxidation processes since the energy required for algal growth is supplied through photosynthesis. The results show that microalgae offer an environmentally safe, low-cost, and efficient means for heavy metals, pharmaceuticals, and personal care products (PPCPs), dyes, and nutrient removal from water and can be used singly or amalgamated with other treatment methods. Additionally, the resultant byproduct, algal biomass, is a possible feedstock for biofuel production.

Chapter 23, "Role of Algae in the Treatment of Cancer," outlines cancer, which is a significant issue, ranking second only to cardiac problems in terms of danger. Microalgae have emerged as a vast, mostly unexplored repository of common substances. It is used not just like a functional food, but it also has a historical past of use in cancer therapy in countries in Asia. According to the literature analysis, microalgae act as a useful and effective finding point with a promising outcome. It includes details about microalgae and their bioactive components, which have the possibility to treat cancer. Further investigation of isolated local algae species is required to properly study this amazing resource and emphasize the relevance of employing this organic compound to supplement current medications and offer comparable outcomes with reduced or no adverse consequences. Microalgae bioactive will offer the most likely natural alternative for chemical medicines in the future, with greater recovery rates and fewer side effects, once practical and economically possible production of marine microalga biomass is achieved.

Chapter 24 is "Polyhydroxyalkanoates (PHAs) Production From Microalgae Cultivated in Wastewater." Microalgae have been proven capable of synthesizing bioplastics like polyhydroxy butyrate (PHB) as a form of energy and carbon storage, under stress. This has been validated by researchers even when microalgae were cultivated in wastewater, providing an eco-friendly solution for the necessary nutrients. This is expected to significantly reduce the carbon footprint of bioplastics and provide an eco-friendly alternative to petrochemical plastics. Research is still necessary to optimize and improve the bioplastic procedure followed in the scope of minimizing the use of organic solvents and the overall environmental footprint of the process. The downstream part of the bioplastic production process can also have a strong influence on the properties of the final product and as a result the applications for which it can be used. Through the research carried out in this field, the gap between the properties of bioplastics and fossil-based plastics has been significantly reduced, with bioplastics becoming suitable for more applications.

This book contains the perspectives of researchers and academics with extensive scientific understanding and practical expertise. We are confident that the book will improve the foresight of present and future researchers working on algae as a future source of food, fuel, and Environment. This book offers an all-in-one resource for researchers, graduate students, and industry professionals working in the areas of agriculture, biofuels, biomedicine, bioremediation, biotechnology, and chemical engineering. Furthermore, this book includes structured foundational content on algae and its applications for undergraduate and graduate students working in biology and life sciences.

Mostafa M. El-Sheekh
Tanta University, Egypt

Norhayati Abdullah
Universiti Teknologi Malaysia, Malaysia

Imran Ahmad
Universiti Teknologi Malaysia, Malaysia

Acknowledgment

Praise, glory, and thanks to Allah, who gave us strength and helped us to clear all the stages from the preparation of the proposal to the publication of the book. The world is a better place thanks to people who want to spread knowledge and unify humans without discrimination. What makes it even better are the people who share the gift of their time to mentor the aspirants so that they can serve the people who are suffering, irrespective of their race. Thanks to everyone who strives to grow and help others grow.

Humanity is the body, and humans living in different countries are like body parts. If a part suffers, others cannot ignore or overlook it, but rather feel the pain and stand with it. This was the drive for this book to contribute to humanity in terms of food, fuel, and the environment, which are the inevitable and vital requirements for the sustainable survival of humanity.

Having some ideas and turning them into a book is as hard as it sounds. The experience is both internally challenging and rewarding. We especially want to thank all the individuals that helped make this happen. Heartfelt gratitude to all the authors who contributed their valuable and scientifically relevant chapters. Thank you to the Editorial Advisory Board, which helped a lot during the important process of reviewing the chapters. Without the help and advice of our peers and the powerful and unique platform provided by IGI Publications, this book would not be here.

Sometimes the scripts are written on the heart turn to become the blueprint of real black and white. We are grateful to our home universities, Tanta University, Universiti Teknologi Malaysia, and Aligarh Muslim University, where the seeds are sown and nurtured. Writing a book is harder than we thought and more rewarding than we could ever have imagined. None of this would have been possible without the continuous support and encouragement of the staff members of the home universities.

Lastly, a warm thanks to the families and friends who stood with us at all the times of struggle and consolidated us to move on.

Chapter 1
Algae:
Their World Explored

Michael Schagerl
University of Vienna, Austria

ABSTRACT

Algae encompass a phylogenetically diverse group; its members are placed in the domains Bacteria and Eukarya, in the latter dispersed in five supergroups. Contrary to common assumption, algae are found almost everywhere from aquatic to terrestrial habitats, some of them in symbiotic relationship, others thriving in harsh environments such as deserts, hot springs, or on ice. Around 45% of atmospheric oxygen is released by algae, which underlines their pivotal role in global element cycles. This high diversity is also reflected in different morphologies from microscopic single cells to parenchymatous organisms measuring several meters in size. The total species number is estimated from 70,000 to more than 10,000,000. Algae produce various valuable compounds, but this high natural potential is only marginally being used. To date, only a few taxa are commercially exploited.

INTRODUCTION

Algae are an essential component of our world and human existence, but rarely noticed by non-biologists. In daily newspapers, algae are largely neglected. If they are mentioned, new press releases mostly focus on negative headlines such as problems associated with eutrophication or toxic algae (Sha et al., 2021; Zingone et al., 2021). This group was, however, responsible for the Great Oxidation Event approximately 2.4 billion years ago, which was the key driver of biological evolution (Martin & Allen, 2018; Sánchez-Baracaldo et al., 2022; Schirrmeister et al., 2015) and the ancestor of land plants (Donoghue et al., 2021), which evolved around 500 million years ago (Yoon et al., 2004). Even today, algae play a pivotal role in biogeochemical cycles. Around 45% of the produced oxygen originates from this group (every second breath you take)(Field et al., 1998), although they constitute less than 1% of the global biomass. This high net productivity is explained by their fast growth rates, which surpass those of vascular plants by a factor of ten.

DOI: 10.4018/978-1-6684-2438-4.ch001

The group shows an exceptional morphological spectrum. Even a single drop of water already contains single cells measuring around 1 µm in diameter, but also more complex algae > 1 mm. Importantly, the size scale extends to organisms several tens of meters in length known as submarine kelp forests. Contrary to what is expected by non-biologists, algae settle not only aquatic habitats. They inhabit snow and glaciers (cryo-vegetation), deserts, tree bark, rocks, house facades, and many other biotopes, some of them exhibiting extreme environmental conditions. Such a life on the edge requires physiological adaptations, for example the synthesis of compatible solutes to cope with salt concentrations at the limit of solubility (Schagerl & Burian, 2016). Algae are also mutualistic symbionts in lichens, protists and larger animals such as corals or sloths (Grube et al., 2017; Kaup et al., 2021).

The estimated number of algae species is between 70,000 and more than 10,000,000 (De Clerck et al., 2013; Guiry, 2012; Norton et al., 1996). This big discrepancy reflects our limited knowledge (lots of species still await discovery), but also problems in species definition. The biological species definition, i.e. a natural reproductive community, is often not applicable. In such cases, alternative approaches such as the traditional morphological species concept, or the modern phylogenetic species concept, are applied.

The extraordinary biodiversity and the global occurrence of algae including many extreme habitats is still a hidden treasure. Only a handful taxa are currently commercially exploited, many others still wait to be studied.

DELIMITATION OF ALGAE

The term algae comprises organisms that release oxygen as a by-product of photosynthesis. Moreover, algae do not show the characteristic features of vascular plants – they are not differentiated into leaves, roots and shoots and do not develop vascular tissues.

More closely examining this definition reveals that algae do not have a common, unique feature. Algae are polyphyletic, they show diverse evolutionary lineages. Based on this definition, cyanobacteria are included in this group, which is understandable in that chloroplasts in all eukaryotic phyla originated from cyanobacteria. Moreover, cyanobacteria resemble eukaryotic algae; they share ecological niches and show similarities in morphology and physiology. Nonetheless, opinions differ in this respect because sometimes solely the eukaryotic forms are considered as algae. Another point deserves mention here: many heterotrophic taxa are included in this group, as they are closely related to their photoautotrophic counterparts. Prime examples for phagotrophy and osmotrophy are found in the phyla Euglenophyta and Dinophyta. Mixotrophic organisms are also represented, among them the chrysophyte genus *Dinobryon*, which switches between phagotrophy and autotrophy depending on the irradiance level (Caron et al., 1993). Mixotrophic cultivation is a promising method for generating high value products (Patel et al., 2021).

MORPHOLOGICAL FEATURES

Algae show a great morphological diversity from single-celled organisms to giant kelps with distinct tissues. Multicellular algae, together with fungi, lichens and mosses, are called thallophytes. They are separated from cormophytes, comprising vascular plants and ferns, based on their lack of the characteristic differentiation into shoots, leaves and roots along with the absence of vascular tissues. Pascher (1931) was the first to categorise algae bodies according to their organisational level (Fig. 1). This artificial system is still applied within phyla, mainly for the taxonomic level of orders. The **monadal** organisation level comprises flagellates. Flagellates can be single-celled, but colony formation and coenobia also exist. Algal flagellates are characterised by a polar cell organisation, with flagellae usually inserting at the anterior pole. This is where the contractile vacuoles and eye-spots are also located. **Amoeboid** (rhizopodial) algae are able to move by means of cytoplasmic extensions; they constantly change their shape. **Capsal** forms have features of flagellates and show cell polarity, but they are no longer able to move. They are commonly enveloped by mucilage. **Coccal** organisms are unable to move and are usually surrounded by a distinct cell wall; they no longer show polarity of cell structures. **Trichal** (filamentous) forms can be unbranched or branched, uni- to multiseriate, homo- or heterotrichal. **Siphonous** forms lack cross walls, and the thalli – often in form of bladders or tubes – are multinucleate single compartments. **Siphonocladal** forms have multinucleate cells.

Sometimes, filaments can adhere to one another and form macroscopic thalli. As they do not resemble true tissues, but are based on filaments, this level is referred to as **pseudoparenchymatous**. Examples are the so-called plectenchyma of Rhodophyta. In contrast, true **parenchymatous** thalli develop from apical cells and meristems. This organisation level is realised in giant kelps (Phaeophyceae), which are often organised in cauloids, phylloids and rhizoids resembling the organs of cormophytes.

SPECIES CONCEPTS AND THE PHYLOGENY OF ALGAE

The vast diversity of algae calls for an approach that brings order to the system. An example for categorising an organism down to species level is provided in Fig. 2. Exact identification down to species level and assignment of unique strain numbers is crucial both for research for industrial exploitation because individual strains within species have unique physiological properties. In many cases, large-scale cultivation is performed without providing these essential details. Two taxa serve as case studies here. "*Chlorella* sp." is commercially cultivated, but genomics revealed that the name summarizes various taxa with comparable morphology that are not necessarily related to each other (Krienitz et al., 2015). "*Spirulina platensis*" can be taken as a trade name for cyanobacterial taxa with unbranched coiled filaments (Nowicka-Krawczyk et al., 2019; Sili et al., 2012).

Books have been published dealing with different species concepts (McCoy et al., 2020; Simpson, 2010), whereby the taxonomic category "species" is usually treated as the basic rank, and all other taxonomic ranks are derived from this natural unit. In phycology, the **morphological species** concept is most commonly applied. Species are described based on specific morphological traits such as size, shape, arrangement of cells, cell protusions, and the fine structure of cell walls. This traditional concept clearly does not reflect true natural units. Especially modern methods based on sequencing have provided new insights. In the past, hidden species with identical morphology (convergent evolution) have sometimes been summarized as a single taxon, but the opposite outcome also occurred, i.e. the same species was described as different species. This can be explained by heteromorphic life cycles or morphological acclimations to specific environmental conditions. The **biological species** concept is based on interbreeding between organisms resulting in viable offspring. For algae, its application is challenging and time-consuming and therefore used only occasionally. Moreover, it can be applied only to groups with sexual reproduction. Finally, horizontal gene transfer between different species complicates this concept. The **phylogenetic species** concept is based on a common ancestor. The species shows a unique molecular, morphological or biochemical feature. Problems arise with setting species boundaries. Ad-

Figure 1a. Organisation levels of algae with examples: amoeboid Chlamydomyxa (Chrysophyceae), monadoid Phacus (Euglenophyceae), capsal Hydrurus (Chrysophyceae), coccal Cosmarium (Zygnematophyceae), trichal Draparnaldia (Chloropyhceae)
read area = eye-spot, dotted circle = contractile vacuole, green areas = chloroplasts, grey areas = nuclei

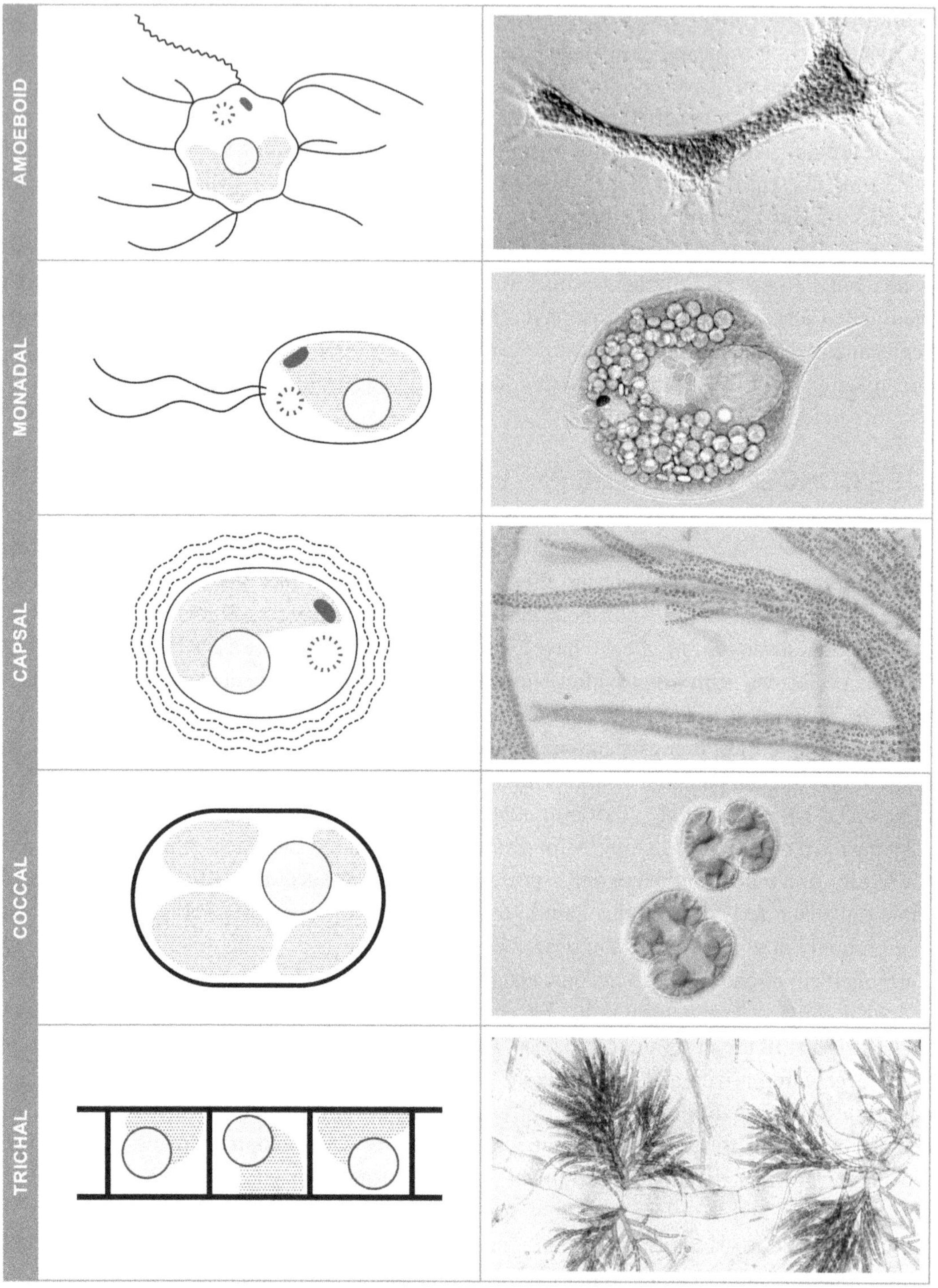

Figure 1b. Organisation levels of algae with examples: siphonocladal Cladophora (Ulvophyceae), siphonaceous Acetabularia (Ulvophyceae), pseudoparenchymatous Lemanea (Florideophyceae), parechymatous Dictyota (Phaeophyceae)
green areas = chloroplasts, grey areas = nuclei

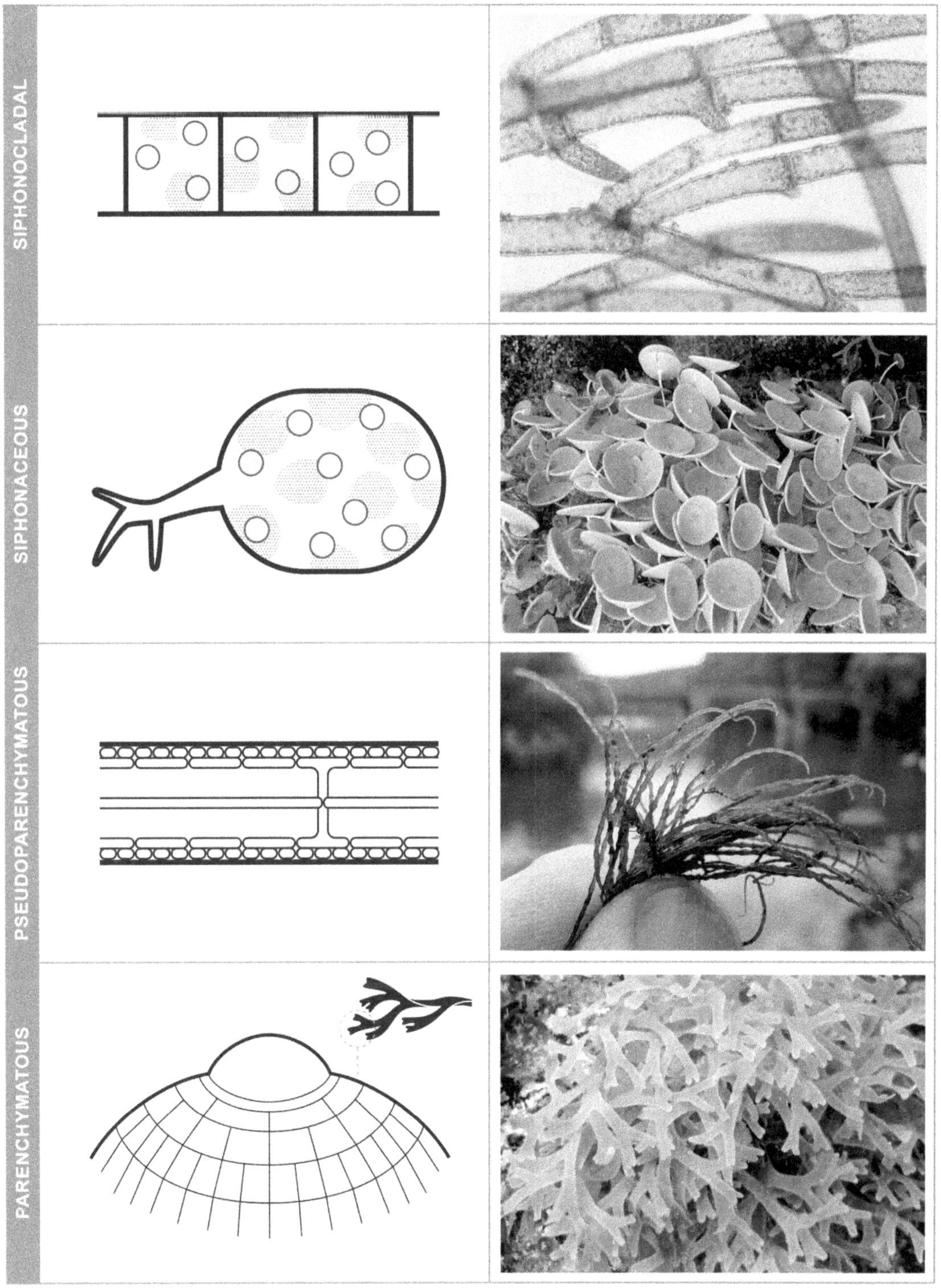

Figure 2. Classification of the marine macroalga Anadyomene stellata (Wulfen) C.Agardh 1823. Top - natural habitat, bottom - close-up of of the little beauty.

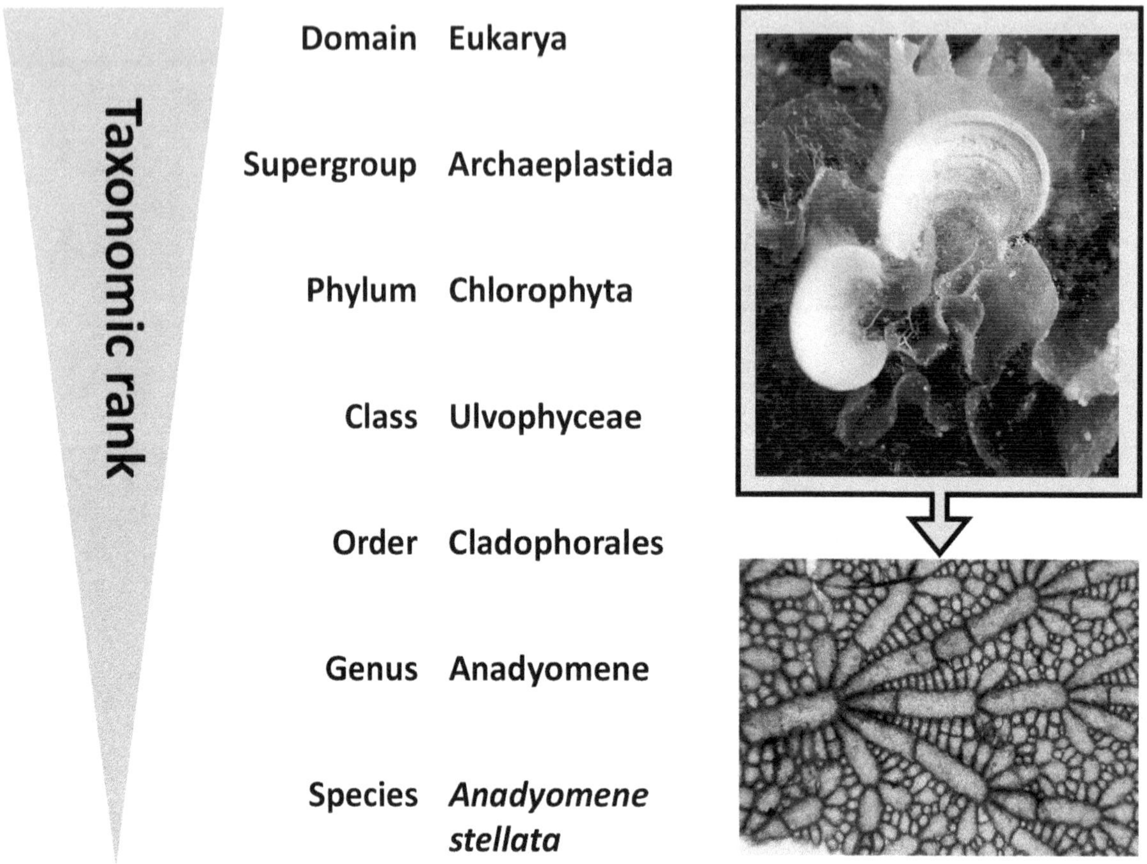

ditionally, species delineations can change if additional data are included. In summary, each concept has advantages, but also drawbacks.

The modern tree of life comprises three domains: Archea, Bacteria and Eukaryotes, with the latter sometimes also referred to as Eukarya (Hug et al., 2016). It must be noted that the tree of life is still a model with different versions and interpretations, depending on data taken into account. It is however obvious that algae do not have a common ancestor. Cyanobacteria are placed in the domain Bacteria, all other phyla are spread wide across domain Eukaryotes. Within the latter domain, five to nine supergroups are delimited, depending on the characteristics used to resolve the groups (Burki et al., 2020). A recent consensus tree based on molecular data resulted in a distinct rearrangement of nine supergroups (Burki et al., 2020), whereas another model that also considered biochemistry and ultrastructure of group members resolved six supergroups and two more groups with uncertain affiliation (Boenigk, 2021). Algal phyla are allocated in all supergroups except Amorphea (Fig. 3).

Oxygenic photosynthesis evolved around 2.35 x 10^9 years ago (Fig. 4)(Fischer et al., 2016). The wide dispersal of algal phyla in various evolutionary lineages is explained by repeated **endosymbiosis** by heterotrophs from different, unrelated supergroups (Fig. 5). Both molecular data and cellular traits provide strong evidence for this evolutionary process. In Cryptophyta and Chlorarachniophyta, a remnant of the

Figure 3. Tree of life. Algal phyla are allocated in most supergroups of Eukarya and in the Bacteria

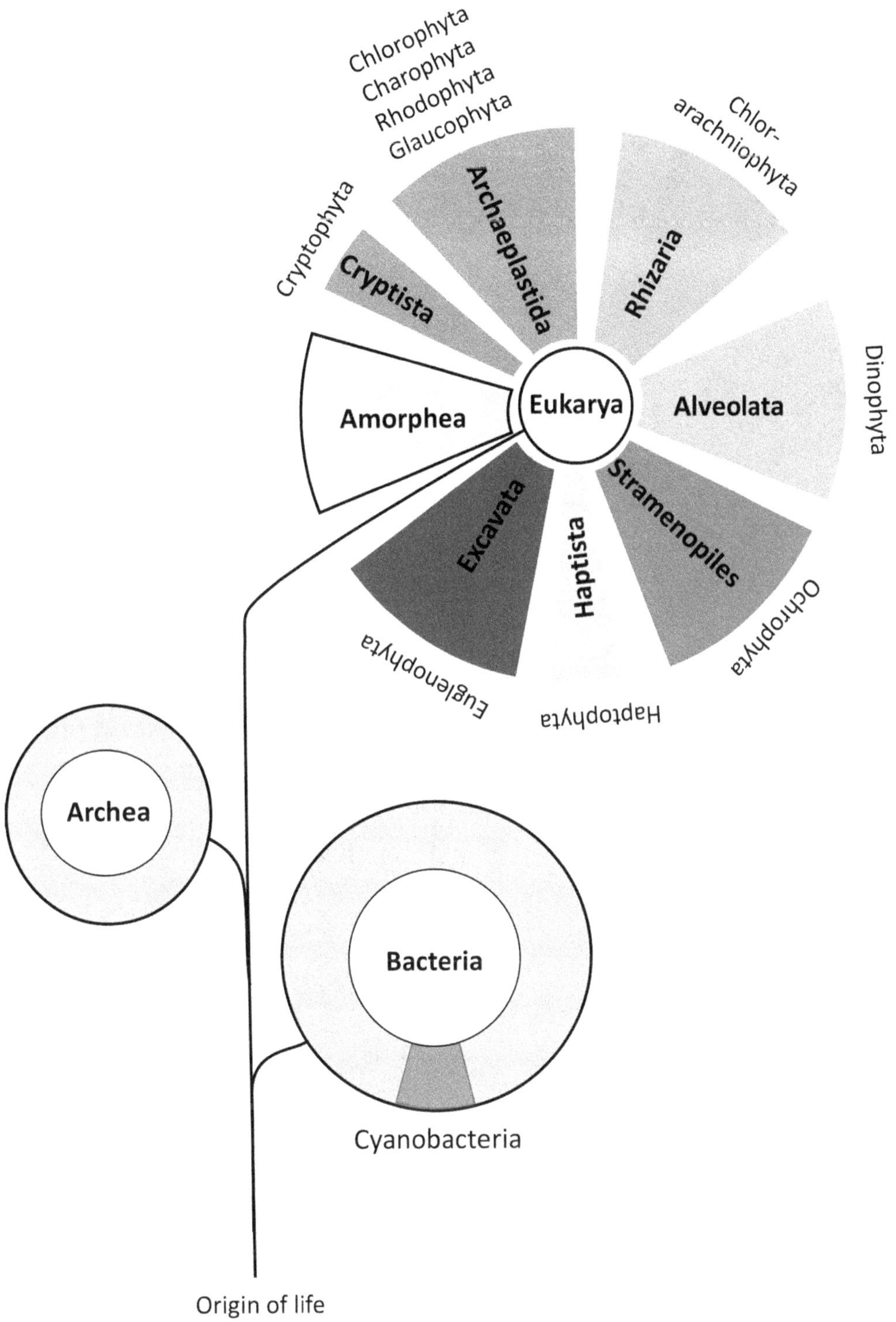

former nucleus of the endosymbiont (nucleomorph) still remains in the host cell in addition to the host cell's nucleus. Moreover, depending on the type of endosymbiosis, between two and four membranes

Figure 4. Timeline of algal evolution. For an easier comparison, the formation of the Earth 4.6 billion years (Giga annuum Ga) ago until today is converted to a dial. Data combined from Janouškovec et al. (2017), Yoon et al. (2004)

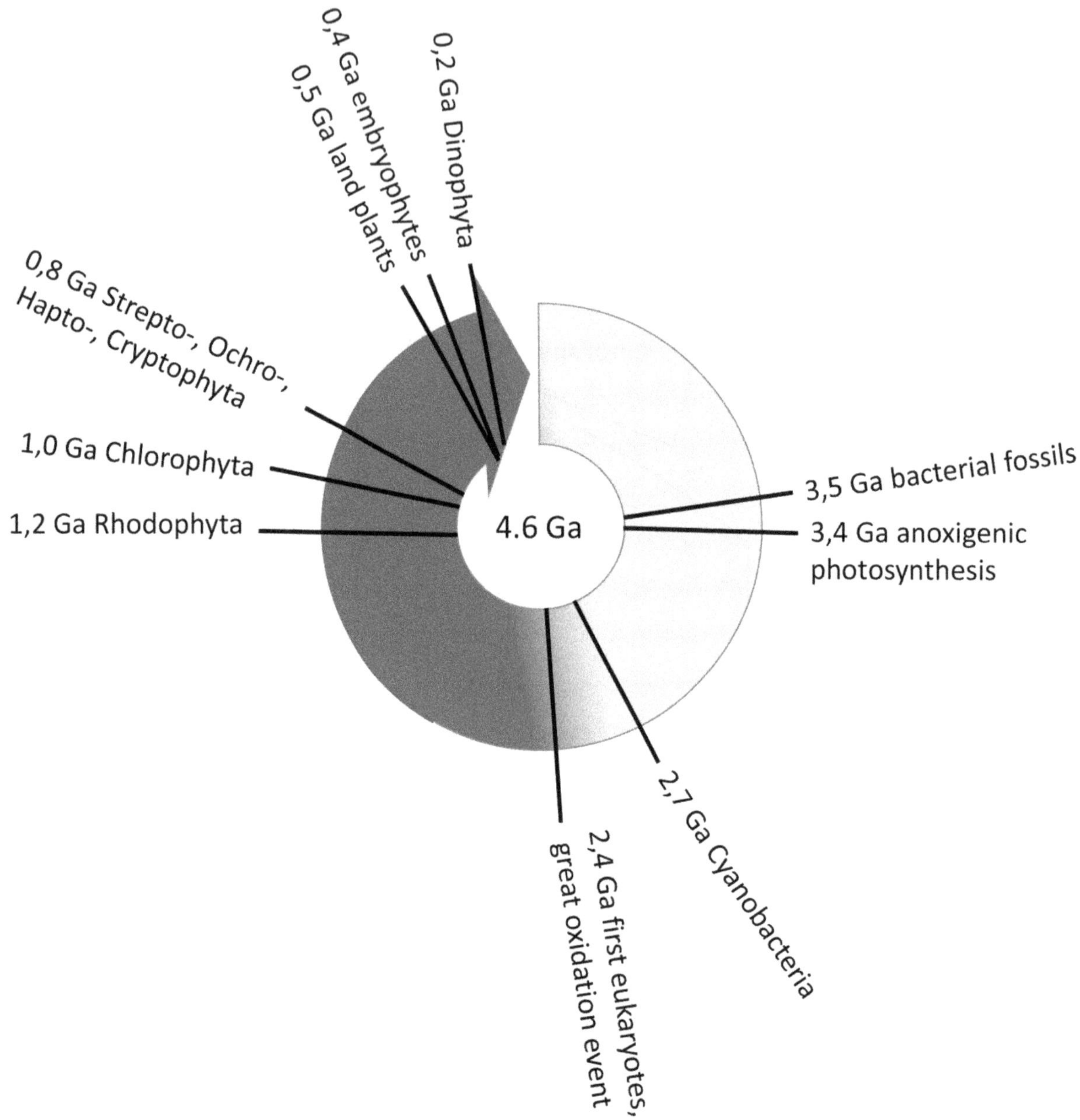

envelop plastids. Plastids still share some genes with cyanobacteria, but most of the coding genes have been transferred to the nucleus of the host cell (Qiu et al., 2013; Sibbald & Archibald, 2020). Plastids are prime examples of horizontal gene transfer between different species.

Genomic studies confirmed that all plastids stem from a single cyanobacterium that was engulfed by a heterotrophic eukaryote (Sibbald & Archibald, 2020). This so-called primary endosymbiosis occurred just once around 10^9 years ago: plastids are therefore monophyletic. Primary endosymbiosis is

Figure 5. The evolution of algae took place by primary, secondary and tertiary endosymbiosis

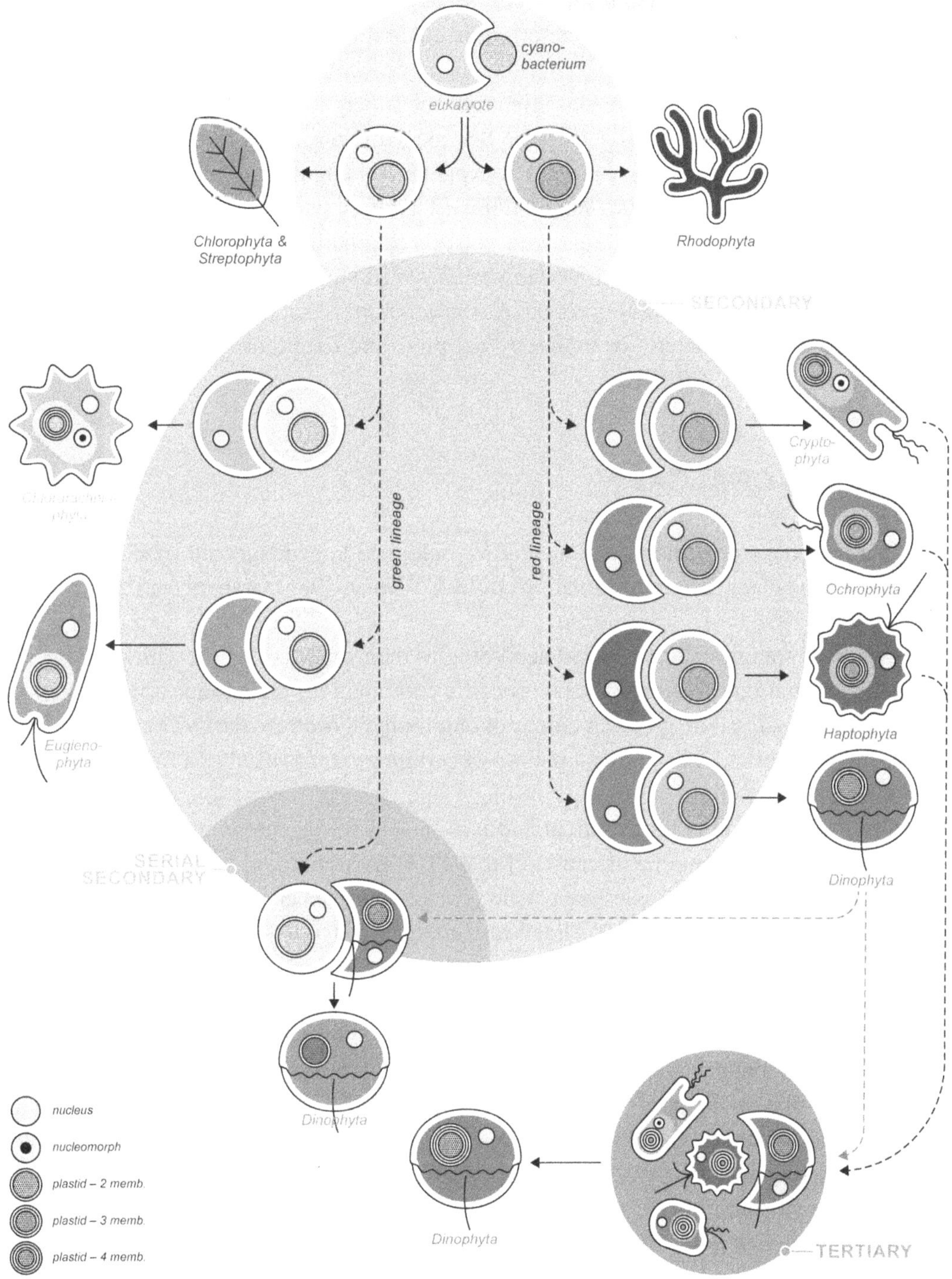

characteristic of the supergroup Archaeplastida, i.e., Rhodophyta, Glaucophyta and Viridiplantae, the latter encompassing the green lineages Strepto- and Chlorophyta. A group of streptophytic algae finally conquered terrestrial habitats and developed to vascular plants (Martin & Allen, 2018).

From primary endosymbiosis, two lineages (red and green) evolved to all other algal phyla by the process of secondary endosymbiosis. In secondary endosymbiosis, a photoautotrophic organism of the Archaeplastida was ingested by another heterotrophic eukaryotic host cell. Secondary endosymbiosis took place at least twice (Fig. 5).

In the phylum Dinophyta, nowadays classified into the supergroup Alveolata, serial secondary and tertiary endosymbiosis occurs in addition to the archetypical dinophytes hosting plastids from a red alga. The process of serial secondary endosymbiosis is explained by engulfment of an additional green alga by an archetypical dinoflagellate (genus *Lepidodinium*). Tertiary endosymbiosis – also realized in the Dinophyta – describes ingestion of an alga with plastids originating from secondary endosymbiosis. Examples are *Karenia* (haptophyte) and *Kryptoperidinium* (diatom). In some dinoflagellates, kleptoplastidy is also observed, e.g., in *Dinophysis* and *Nusuttodinium* (Takano et al., 2014). These flagellates ingest cryptomonads, either directly or indirectly via prey, and temporarily host their plastids before digestion (Hehenberger et al., 2019).

CLASSIFICATION OF ALGAE

Some excellent textbooks providing fundamental phycological knowledge and specific traits of phyla are highly recommended for further information (Bold & Wynne, 1985; Graham et al., 2016; Lee, 2018; van den Hoek et al., 1996).

Algae phyla were traditionally classified according to their colour (Fig. 6), which is also reflected in their names. Colouration is mainly based on specific combinations and unique absorption properties of photosynthetic pigments. The greenish colour of chlorophylls is often masked by carotenoids or/and in some cases by phycobilins. The pigment pattern of certain groups is used as a fingerprint for phylum identification.

Pigments together with other **biochemical features** such as storage products and the composition of the cell wall support the current classification (Table 1). **Ultrastructural features** are also taken into account to categorize groups. Examples are plastid ultrastructure depending on the endosymbiotic process (number of membranes, pyrenoids). The ultrastructure of flagella (isokont in Viridiplantae versus heterokont in Ochrophyta), basal bodies of flagella (especially in Viridiplantae) and eyespots are additional important features to characterize groups. **Life history features**, both sexual and vegetative, are also considered for group definitions. Sexual generation cycles alternate and – depending on zygote and gamete formation – haplontic, haplodiplontic or diplontic life cycles can be distinguished (Graham et al., 2016). The mode of gamete fusion is either isogamy, anisogamy or oogamy. Oogamy is mainly found in highly evolved taxa. Female and male sexual organs may develop on the same plant (homothallic) or on different plants (heterothallic); both types sometimes occur within the same genus. Generations either show identical morphology (isomorphic, e.g., *Cladophora*) or differ in form (heteromorphic, e.g., *Derbesia*). In many cases, vegetative reproduction patterns are present, from simple binary fission to simultaneous spore formation, the latter for dispersal. Last but not least, **genetic features** provide fundamental insights into algal phylogeny and evolution. Sequence comparisons of nuclear and/or plastid regions or whole genomes are taken to construct phylogenetic trees that reflect relationships. The application of such highly sophisticated methods, including complicated algorithms, enormously increased our knowledge and led to a re-organisation of taxa ranks from supergroups down to species level.

Figure 6. Some examples of major algal lineages. The importance of specific pigment patterns is reflected by their scientific and trivial names: (a) Nostoc – Cyanobacteria – blue-green alga, (b) Hildenbrandia – Rhodophyta – red alga, (c) Synura – Chrysophceae – golden alga, (d) Dictyota – Phaeophyceae – brown alga, (e) Vaucheria – Xanthophyceae – yellow-green alga, (f) Micrasterias – green alga

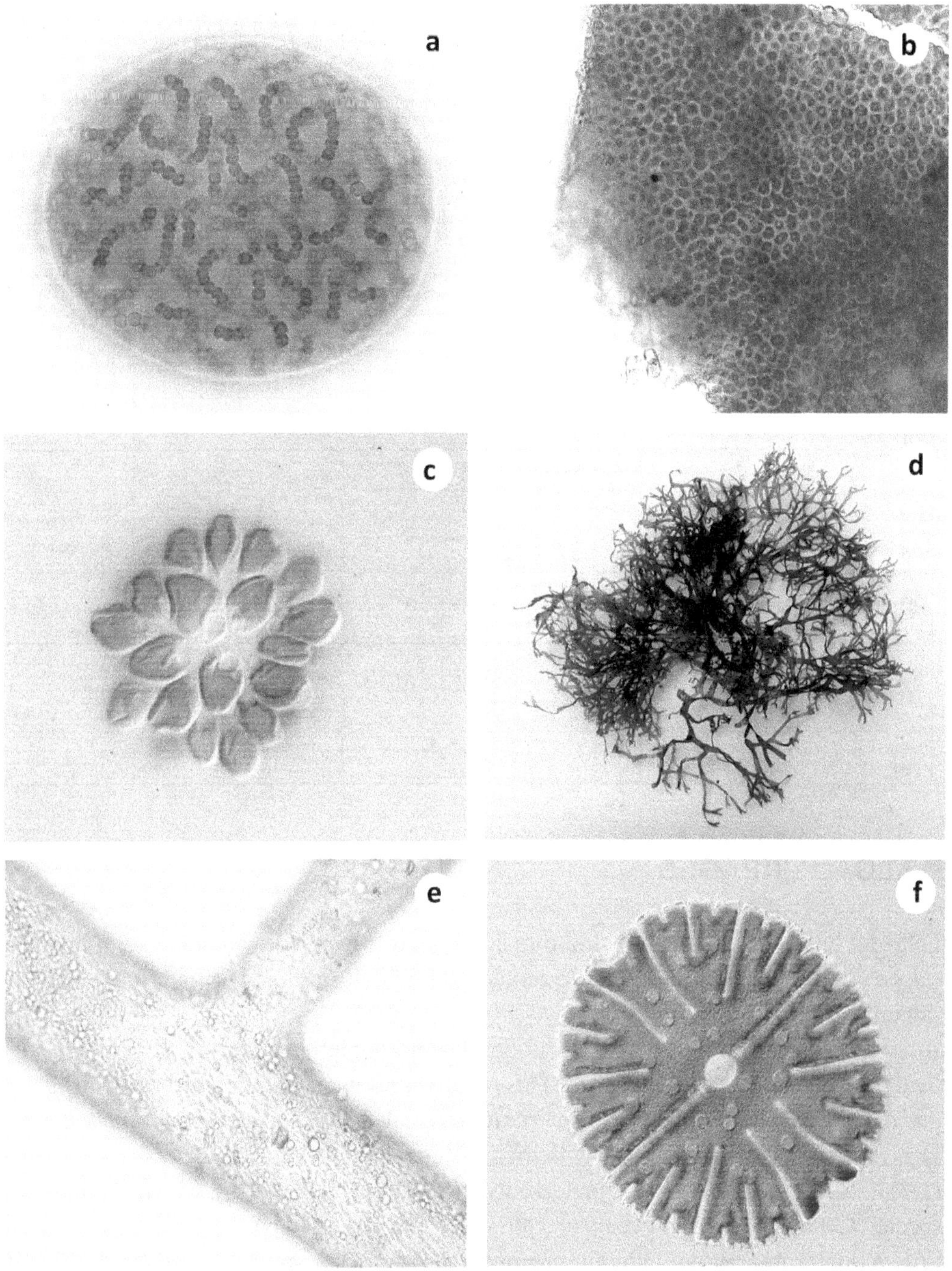

The different approaches provide detailed information from the molecular level to life history traits.

Although genomic data are the foundation of all other traits, the derived features must not be neglected. A perfect match of the different methodological approaches highly increases the informative value.

Table 1. Classification of algal lineages and specific traits. Chl = Chlorophyll, SAR-Clade = the three supergroups Stramenopiles, Alveolata, Rhizaria are combined to a clade because of their common ancestor Telonemia

Domain	Supergroup	Phylum	Major Photosynthetic Pigments	Cell Envelope	Storage Products
Bacteria		Cyanobacteria	Chl-a, phycobilins in phycobilisomes, carotenoids (echinenone, myxoxanthophyll, β-carotene)	Peptidoglycan (murein)	Cyanophytan starch resembling glycogen, cyanophycin (N-storage), volutin (P-storage)
Eukarya	Archaeplastida	Glaucophyta	Same as Cyanobacteria (plastids called cyanelles are enveloped by murein)	Naked or cellulose	Starch stored in cytosol
Eukarya	Archaeplastida	Rhodophyta	Chl-a, phycobilins in phycobilisomes, carotenoids (α-, β-carotene, zeaxanthin)	Cellulose, carrageenan, galactan, agar, some calcified	Floridean starch resembling to glycogen
Eukarya	Archaeplastida	Chlorophyta & Streptophyta	Chls-a, b, carotenoids (lutein, neoxanthin, violaxanthin, β-carotene)	Naked, sometimes organic scales, often cellulose, some calcifed	Starch
Eukarya	Excavata	Euglenophyta	Chls-a, b, carotenoids (neoxanthin, diadinopxanthin, ß-carotene)	Proteinaceous envelop called pellicle	Paramylon in characteristic bodies
Eukarya	Cryptista	Cryptophyta	Chls-a, c, phycobilins dissolved in plastid lumen, carotenoids (alloxanthin, α- and β-carotene)	Proteinaceous envelop calle periplast	Starch
Eukarya	Haptista	Haptophyta	Chls-a, c, carotenoids (diadinoxanthin, fucoxanthin, β-carotene)	Often carbonate scales	Chrysolaminarin
Eukarya	Stramenopiles (SAR-Clade)	Ochrophyta	Chls-a, c, carotenoids (diadinoxanthin, fucoxanthin, β-carotene)	Some naked, some with silica scales, some with cellulose and alginates	Chrysolaminarin, lipids
Eukarya	Alveolata (SAR-Clade)	Dinophyta	Chls-a, c, various carotenoids depending on endosymbiosis type (the archetypical taxa contain peridinin)	Envelop called amphiesma consists of vesicles often containing cellulose plates	Starch
Eukarya	Rhizaria (SAR-Clade)	Chlorarachniophyta	Chls-a, b, carotenoids (lutein, loroxanthin, neoxanthin, violaxanthin, β-carotene)	Naked	Carbohydrate (β-1,3-glucan)

MAJOR ALGAL LINEAGES

Based in current knowledge, there are eleven major algal lineages, with cyanobacteria included under the term algae (Table 1). Textbooks provide detailed summaries on the specific traits (Archibald et al., 2017; Graham et al., 2016; Lee, 2018).

Cyanobacteria are at the base of algal evolution. Beyond single-celled forms, this group also contains colonies and filaments, sometimes branched. This group is present in aquatic and terrestrial habitats, with some representatives thriving in extreme environments such as hot springs (*Mastigocladus*), alkaline-saline lakes (*Limnospira*) and rock faces with high UV-insolation (*Gloeocapsa*). Some taxa such as the terrestrial *Nostoc* fully recover after complete desiccation (Dodds et al., 1995), which also classifies them pioneer species. Cyanobacteria show the characteristic cell structures of prokaryotes with a gram-negative cell wall (peptidoglycane), often secreting mucopolysaccharides. Especially in terrestrial habitats, the mucilaginous sheath is pigmented to protect cells from damaging UV-radiation. These natural sunscreens, such as the substance scytonemin, are of interest for industrial exploitation. Many species are able to fix molecular nitrogen. The enzyme responsible for incorporation, nitrogenase, is oxygen-sensitive and must be protected from oxygen, which is released as a by-product of photosynthesis. Protection is realised either by spatial separation (heterocyte formation) or by temporal separation (photosynthesis during the day and dinitrogen fixation at night). In addition to chlorophyll-a and carotenoids, Cyanobacteria are capable of synthesising red (phycoerithrin) and blue pigments (phycocyanin and allo-phycocyanin), which are located in phycobilisomes. Cyanobacteria use these pigments to acclimate to the actual light supply for efficient light harvesting – a process known as complementary chromatic adaptation. Phycocyanins are effective antioxidants and are already being commercially used. The genus *Limnospira* (different species are sold under the trade name Spirulina platensis) is one of the few algae that is cultivated in large-scale outdoor systems and traded all over the world. Some cyanobacteria synthesise strong toxins called cyanotoxins, which can pose a threat to human health (Chorus & Welker, 2021). Especially in highly eutrophicated systems, cyanobacteria can develop long-lasting blooms, sometimes also producing cyanotoxins (e.g., *Microcystis, Aphanizomenon, Anabaena*)(Huisman et al., 2018). Blooms are also facilitated by carbon-concentrating mechanisms found in cyanobacteria, along with internal nitrogen (cyanophycin) and phosphorus (volutin) storage compounds. Eutrophication is often accompanied by a shift from phosphorus to nitrogen limitation, which creates an ecological niche for dinitrogen-fixing species. Some planktonic forms are specially adapted to turbid systems: they develop gas vesicles to improve buoyancy.

Glaucophyta is a small phylum within the Archaeplastida and sometimes found in swampy freshwaters. The photosynthetic organelles called cyanelles differ from the typical plastids in certain aspects. They are surrounded by a peptidoglycan envelope and show also other features of cyanobacteria, e.g., pigmentation, phycobilisomes and carboxysome-like bodies. The cyanelles are interpreted as an early evolutionary step towards the plastids of other algal phyla. They cannot survive outside the cells and, compared to free-living cyanobacteria, they have lost about 90% of their genes (Löffelhardt et al., 1997).

Rhodophyta are mostly found in marine waters, but a few have conquered freshwaters (e.g., *Batrachospermum, Lemanea*) and even terrestrial habitats (*Porphyridium*). Genera such as *Hildenbrandia* occur in both marine and freshwater habitats. Some taxa are able to use hydrogen carbonate for carbon fixation, others are limited to carbon dioxide and therefore restricted to habitats with high atmospheric exchange (air-water interface). The order Corallinales contains calcifying red algae: calcification is linked to hydrogen carbonate uptake. Rhodophyta played a central role in geological history, and they formed the primary mass of mountains. Many taxa have a complex, triphasic life cycle with special sexual organs (Graham et al., 2016). Red algae can be single-celled, but also grow as complex, macroscopic thalli. Interestingly, no flagellated forms exist, which is explained by a secondary loss of monadoid stages (Yoon et al., 2017). Plastids still show characteristics of cyanobacteria, e.g., phycobilins located in phycobilisomes. These bodies are located at the plastid membranes and act as light-harvesting complexes that highly efficiently transfer energy to photosystem II. Phycobilins bridge the green absorption gap of chlorophyll-a, which is advantageous at dim light conditions. Red algae have been recorded at less than 0.1% surface irradiance in more than 200 m depth (Littler et al., 1985). Some marine taxa (e.g., *Gelidium, Palmaria, Chondrus, Kappaphycus*) are economically important because the gelling substances agar and carrageenan are extracted. These polygalactan sulphates are an amorphous component of the extracellular matrix and are embedded between cellulose microfibrils of the cell walls. *Porphyra/Pyropia/Neopyropia* is dried (nori sheets) and used to wrap sushi rolls. Farming of seaweeds has a long tradition in East Asia (Chopin & Tacon, 2021; Radulovich et al., 2015).

In the 1970s, growing evidence was emerging that **green algae** comprise more than one group. Based on mitochondria ultrastructure, Cavalier-Smith (1981) presented a new system of eukaryotic kingdoms.

The term Viridiplantae was applied for green plants containing plastids with starch as a storage product. Based on cladistics analysis with various characters considered, e.g., biochemistry, morphology, life history, but not genomics, Bremer (1985) proposed two major monophyletic lineages Streptophyta and Chlorophyta within the Viridiplantae. The term streptophytes (streptos = twisted) was originally introduced for plants with twisted sperms (Jeffrey, 1967) and at that time comprised only embryophytes and charophytes in the strict sense (stoneworts). The two phyla **Chlorophyta** and Streptophyta are still valid (Leliaert et al., 2012). Today, Charophyceae, Conjugatophyceae and four smaller classes are counted as streptophytic algae; these groups are summarised under the umbrella term **Charophyta** and thus delimited from Embryophyta (vascular plants). From the phycologist's perspective, a small group of streptophytes finally conquered terrestrial habitats and developed to vascular plants; there is no deep phylogenetic incision between higher plants and charophytes. All other classes of green algae are Chlorophyta. Some groups such as prasinophytes still remain to be classified correctly. Although green algae also inhabit marine systems, their biodiversity is much higher in freshwaters. Chlorophytes are also prominent in the littoral of the sea, but Charophyta are almost exclusively restricted to freshwaters and terrestrial habitats. Their morphology is extremely diverse, ranging from single flagellates such as the model organism *Chlamydomonas* to macroalgae resembling higher plants (Characeae). Different life histories, biochemistry, cell wall construction, division patterns and structural features also reflect the two lineages. The ultrastructure of flagellar basal bodies, which anchor the flagella in the cell, is still a central argument for differentiation into the phyla and also into classes. Four types of basal bodies are represented in Viridiplantae (Pröschold & Leliaert, 2007). Nonetheless, the two major lineages still share some common features, which justifies their entity in the Viridiplantae. Flagellates have an isokont flagellation, the thylakoids are surrounded by two membranes, the main storage product is starch, and both chlorophyll a and b are synthesised (Table 1). Some green algae are of high commercial interest and already exploited on a large scale. *Haematococcus lacustris* is cultivated in closed systems for astaxanthin production and sold for aquaculture, as a food additive and for cosmetics. *Dunaliella salina* is grown in salterns at the limit of salt solubility and used for ß-carotene production. *Chlorella* is cultivated in indoor-systems and sold as a food additive.

Chlorarachniophyta are a small phylum of about 15 species. They inhabit shallow marine waters and resemble coloured amoebas because they contain green plastids. Cells additionally contain a nucleomorph from the endosymbiont.

Euglenophyta comprise unicellular flagellates, many of them heterotrophic forms (osmo-, phagotrophic, in some cases bacterivory is the main feeding mode). An epizooic life style has also been recorded, e.g., on copepods (*Colacium*). They occupy mainly freshwater habitats, such as littoral zones and shallow waters enriched in organic substances. Some taxa are found in marine environments (e.g., *Eutreptia, Eutreptiella, Klebsiella*). Sexual life cycles are unknown; reproduction involves cell division. The cells are enveloped by an organic proteinaceous layer (the pellicula), which is arranged in stripes around the cells. The plates are connected to each other like roof-tiles. Some taxa have a rigid cell envelope and a defined shape, e.g., the leaf-like cells of *Phacus*. *Trachelomonas* and *Strombomonas* are encased by an additional resistant mucilage coating (lorica), which is enriched with minerals and therefore appears brownish. Some taxa show metabolic movement. The group is closely related to the Kinetoplastida, whose representatives *Leishmania* and *Trypanosoma* are known to cause serious health problems. Some euglenophytes are known to produce toxins (Triemer et al., 2004), which can pose a challenge for for aquaculture. *Euglena sanguinea* accumulates large amounts of astaxanthin and is reddish (Laza-Martínez et al., 2019), similar to the chlorophyte *Haematococcus*. *Euglena sanguinea*, in contrast, also synthesises euglenophycin, which has herbicidal activity but is also toxic for mammals (Zimba et al., 2017). *Euglena gracilis* powder is already available on the market and sold as a food additive and for cosmetics.

Cryptophyta is a small group of asexual freshwater flagellates. Their shape is asymmetrical, flattened and somewhat resemble a coffee bean. The ingested endosymbiont was a red alga, which is confirmed from the biochemistry and ultrastructure of the plastid: phycobilins are present, which explains their blueish, reddish to brownish colour. The pigments are, however, dispersed between the thylakoids. Colourless representatives such as the genus *Chilomonas* are also known. In addition to the nucleus of the host, cells contain a nucleomorph from the endosymbiont. The cells are enveloped by small protein plates in a layer termed the periplast. Cryptophyta are a high-quality food for herbivorous zooplankton and are preferably ingested. As a defence strategy, Cryptophyta developed special structures, so-called ejectosomes. They are suddenly released like a spring from a gullet near the anterior pole to escape from predators.

Haptophyta (Prymnesiophyta) are marine flagellates, with a few exceptions inhabiting freshwater. They are important primary producers in the sea and sometimes form blooms. In most taxa, a flagellum-like organelle known as a haptonema is present. The haptonema is involved in escape reactions but also in catching particles (Bloodgood, 2020). Particles agglutinate to the haptonema and are then transferred to its tip. The haptonema then bends to the cell surface, where the particles are ingested through phagocytosis (Eikrem et al., 2017). Many haptophytes are covered by organic scales, and some also develop calcium carbonate scales called coccoliths (e.g., *Emiliana*). After dying, the coccoliths sink to the bottom and are deposited on the sea floor, ultimately forming thick layers. This carbon sequestration over long periods is thought to influence the climate. The group is phylogenetically related to Stramenopiles – they have storage products and a photosynthetic pigment pattern comparable to Ochrophyta – but they are equipped with isokont flagella. They reproduce vegetatively and sexually, in some cases with generations having a different morphology. Taxa such as *Chrysochromulina* and *Prymnesium* synthesise ichthyotoxins, which cause serious problems including massive fish kills. *Phaeocystis* is very common and occurs in polymorphous stages. During blooms, cells aggregate to large colonies and produce foam, often floating on the sea surface. The slime clogs fish gills and also has a negative impact on tourism due to its unpleasant appearance. Other taxa such as *Isochrysis* and *Pavlova* are cultivated on a large scale and serve as high-quality food for bivalve aquaculture.

Detailed relationships within the monophyletic phylum **Ochrophyta** (photosynthetic Stramenopiles) remain to be resolved (depending on the respective author, between 9 and 13 classes are distinguished). The term Ochrophyta was introduced by Cavalier-Smith & Chao (1996) and comprises several prominent classes such as Phaeo-, Bacillario-, Chryso- and Xanthophyceae. Formerly, the term Heterokontophyta was common, but this name was abandoned because it also included certain fungi. The phylum is present in aquatic habitats, but Xanthophycaeae, although less studied, are also a component of terrestrial habitats. Group characteristics include the heterokont flagellation, the storage product chrysolaminaran (starch is never found in the Ochrophyta), and the plastids are additionally enveloped by a fold of the endoplasmic reticulum. In most classes, the carotenoid fucoxanthin is a dominant carotenoid of the light-harvesting complexes. This pigment is responsible for the yellowish to brownish colouration of the organisms. Life cycles include haplonts, haplodiplonts and diplonts, as well as isomorphic and heteromorphic generations. In some classes, silicate cell walls (Bacillariophyceae and cysts of other taxa) or silica scales (Synurophyceae) cover the cells. The growth mode and morphology are quite diverse. Single flagellates, colonies, capsal and coccal forms, filaments, but also true parenchymatous thalli with apical meristems have been described, the latter developing to organisms several meters in length (phaeophyceae giant kelps). Siphonaceous taxa such as *Vaucheria* and *Botrydium* (Xanthophyceae) also occur. A very prominent class inhabiting both marine and freshwater systems is the Bacillariophyceae (diatoms), which are always single-celled, coccal forms, but sometimes aggregate to filament-like structures. Cells walls of diatoms are silicated and consist of two halves constructed like a box with lid. During cell division, always the smaller bottom called hypotheca is reproduced, which reduces the size of one of the two daughter cells. After several divisions, sexual reproduction needs to be performed, yielding a large zygote – otherwise the population will die out. Precisely this step is critical for cultivation, which explains the lack of available strains in culture collections. Diatoms are used for water quality assessment (biomonitoring) as well as for industrial purposes. Diatomaceous earth (diatomite consisting of deposits of fossil cells) is mined and used as a filter material to clarify wine and other beverages. Diatomaceous earth is contained in tooth paste, cosmetics and food additives. Another commercially exploited phylum is Phaeophyceae. These brown algae play a central role in marine littoral communities. Kelps thrive along nutrient-rich coastal stretches and are directly consumed or harvested and used for extracting gelling substances (alginates). In former times, iodine was also obtained from macroscopic brown algae. Kelps can be annual to perennial and have high growth rates. *Macrocystis pyrifera* is listed amongst the fastest growing species, with a length increase of several cm per day. Beyond seaweeds, also microalgae of the phylum Ochrophyta have received attention for industrial exploitation. One promising genus is the Eustigmatophyeae *Nannochloropsis*, which synthesises valuable lipid compounds such as Ω-3fatty acids (Brennan & Regan, 2020). Some taxa placed in the Chrysophyceae (*Uroglena, Dinobryon*) have a characteristic rancid smell, pointing to high amounts of polyunsaturated fatty acids. The Ochrophyta also contain toxic species. One example causing massive fish kills is *Chattonella,* placed in the Raphidophyceae (Imai & Yamaguchi, 2012).

Dinophyta are usually found in their flagellated stage and are very abundant in both marine and freshwater systems. Modern dinoflagellates developed relatively short time ago, about 200×10^6 million years ago (Janouškovec et al., 2017). The cells are surrounded by a complex cell envelope, the amphiesma, including polygonal, flattened vesicles at the cell periphery. In many taxa, the vesicles are filled with cellulose and formed to structured plates called thecal plates. Plate number and arrangement are conserved and used for species delimitation. Usually two flagella emerge on the ventral side of the cell. The transverse flagellum is located in a transverse furrow and responsible for turning the cell on its own axis. The longitudinal flagellum is responsible for forward movement. This phylum features different types of endosymbiosis, resulting in different plastid structures and pigmentation. The archetypical Dinophyta is assumed to contain a red alga symbiont and appear brownish due to the dominant carotenoid peridinin. Serial secondary and tertiary endosymbiosis is also realized in this phylum. Such cells have a different pigmentation. Chromosomes are condensed throughout the whole cell cycle, giving the nucleus a unique appearance (dinokaryon). In many cases, a tubular, sac-like structure (the pusule) is connected to the furrow and thought to have an osmoregulatory function. Moreover, defense structures to escape predators such as trichocysts (Bouck & Sweeney, 1966) and mucocysts are located at the cell periphery. Dinophyta have a haploid life cycle, and in some members the zygotes remain as resting spores (hypnozygotes, also called dinocysts) at the floor of water bodies to survive unfavourable living conditions. Noteworthy is the presence of mixotrophic taxa. Importantly, around 50% of the species are heterotrophic, more precisely micropredators or parasites. Mixo- and heterotrophic forms developed different mechanisms for food capture and ingestion. In some taxa, a pseudopodium-like structure (pallium) is expanded outside the amphiesma. The prey is taken up into the pallium and digested outside the cell envelope (*Diplopsalis, Protoperidinium*). In other forms, a feeding tube (the peduncle) consisting of bundles of microtubules is protuded (*Tyrannodinium*). Simple engulfment of particles in the sulcus-region also occurs (*Ceratium*). Finally, ballistic, highly advanced mechanisms exist. These nematocysts are fired against the prey (Gavelis Gregory et al., 2017). Interestingly, many members of the sister group Apicomplexa – the pathogens *Toxoplasma* and *Plasmodium* are placed in this group – were once photoautotrophs, but they lost that ability during evolution. In analogy to nucleomorphs, their cells contain so-called apicoplasts, which are remnants of plastids (McFadden, 2011). Apicoplasts are used to generate fatty acids. Some marine Dinophyta, such as *Pyrocystis, Lingulodinium* and *Noctiluca*, show bioluminescence, which functions as a defence mechanism against predators (Haddock et al., 2009). Either the predator is warned by the flashes (startle hypothesis), or animals two trophic levels higher up the food web are attracted and feed on the direct predators of dinophytes (burglar alarm hypothesis). Bioluminescence takes place in spherical particles known as scintillons, which contain the enzyme luciferase, the bioluminescent substance luciferin and proteins. The bioluminescence occurs as very short flashes of blue light and is most prominent during the night. Dinophytes also produce major toxins that pose a threat even to humans either via aerosols or through direct ingestion of food, e.g., bivalves. This problem becomes serious during bloom formation, which is known under the name red tide because the mass development of dinoflagellates causes the water to turn a reddish colour. Although the syndrome shellfish poising is sometimes also caused by other algal phyla, dinoflagellates are the main contributors in the sea (e.g., *Alexandrium, Karenia, Pfiesteria*). Amnesic (mainly caused by diatoms), diarrheal, neurotoxic and paralytic shellfish poisoning are known and in some cases fatal. Dinophyta are important symbiotic partners (endosymbiotic zooxanthellae), with corals serving as a prime example (*Symbodinium*). In addition, other animals including sea anemones, jellyfish and foraminifera host this group for sugar generation.

DIPPING INTO THE ECOLOGY OF ALGAE

Non-biologists regard algae as water organisms, but algae exist everywhere. In aquatic habitats, they play a pivotal role as biomass generators at the base of food webs. They are present in both the pelagic and littoral, which requires different adaptation strategies. Phytoplankton faces the problem of sinking to dark regions. One strategy to minimise sinking velocity is minimising size: phytoplankton therefore tend to be small (picoplankton). In addition, the form resistance is increased in many taxa, e.g., by spines, protuberances and shapes that deviate from a sphere (Padisák et al., 2003). Another adaptation is reducing the density, for example by forming gas vesicles (cyanobacteria) and synthesising lipids.

Benthic algae on hard, exposed substrates develop holdfast and rhizoids to prevent them from being washed away. Certain diatoms and desmids have stalk systems for attachment to the substrate. Other diatom communities inhabiting mudflats migrate to maximise their fitness (Consalvey et al., 2004). Algae of the intertidal are additionally adapted to periodic desiccation, e.g., by synthesising osmolytes (Kirst, 1990). Species exposed to excessive light and high UV-radiation synthesise substances protecting them against damage (Karsten, 2008), and some of them are able to move to avoid excess radiation.

High nutrient input into water bodies may cause mass developments of algae called blooms. If toxic species are involved, the massive occurrence is termed a harmful algal bloom (Shumway et al., 2018). However, non-toxic blooms can also create serious problems, for example through oxygen depletion during the night or decaying biomass.

Algae inhabiting extreme environments are known as extremophiles. Such systems include highly saline-alkaline lakes (Krienitz & Schagerl, 2016), but also acidic waters (Gross, 2000) and hot springs (Shu & Huang, 2022; Ward & Castenholz, 2000). Specifically such extremophiles are of prime interest for industrial exploitation because they have developed strategies to cope with extreme conditions. One example is the use of algal enzymes as washing agents.

Algae are, however, not restricted to aquatic habitats; they are also an important component of terrestrial ecosystems (Hoffmann, 1989). A prominent example for a terrestrial species is the cyanobacterium *Chroococcidiopsis*, which is present even in salt deserts (Casero et al., 2021). Other cyanobacterial species inhabit rocks, both as epilithic and endolithic forms (Büdel & Friedl, 2021), with some of them contributing significantly to system metabolism because of their dinitrogen fixing ability (Singh, 2021). Another special habitat is snow and ice. The so-called cryovegetation here can already be recognized with the naked eye because of the red colouration (Fig. 7). Cryovegetation consists mostly of green algae, but other phyla can also be present. The red colour is caused by the pigments purpurogallin and astaxanthin and is interpreted as a protective measure against excessive irradiation. In addition, these organisms synthesise substances to avoid the formation of ice-crystals, which would otherwise damage cells (Hoham & Remias, 2020). Finally, polyunsaturated fatty acids are increasingly synthesised at low temperatures to maintain membrane fluidity. All these specific adaptations offer high potential for industrial exploitation (Hulatt et al., 2017).

Algae are also present in caves (Falasco et al., 2014), cause skin diseases (Schöfer, 2018) and irritations of the respiratory system (Sharma et al., 2007), grow on bark, house facades, rocks and soil (Metting, 1981), and are frequent in air (airborne algae)(Tesson et al., 2016). They are found almost everywhere.

A FEW THOUGHTS ON APPLIED PHYCOLOGY

The high potential of algae exploitation has attracted investors and companies. Although many do not realize it, we all use products containing algae or algal substances in our daily lives (Fig. 8). Due to their gelling properties, alginates and carrageenan are backbones of the food industry, partly replacing gelatine. Algae and their products are found in cosmetics, pharmaceutical products and food supplements, and some species are treated as novel foods.

The different applications of algae and algae-derived products call for a unique labelling and conscientious quality control, which is, however, not realised in many cases (Fig. 9). In several countries, the law does not regulate dietary supplements. Moreover, the sloppy use of scientific names often creates major uncertainty for both basic and applied research. A prime example is 'Spirulina culture', which is a collective name for various *Limnospira*, *Arthrospira* and maybe even true *Spirulina* strains. Most likely, 'Spirulina' powder on the market has originated from *Limnospira fusiformis* or *Limnospira maxima*, from which several strains are available from culture collections. The scientific name *Limnospira* was introduced by Nowicka-Krawczyk et al. (2019), but since then more than 15,000 papers have been published with 'Spirulina' mentioned in the title (source Google Scholar). Considering that species-specific traits are the norm here, merely mentioning a trade name instead of the specific strain borders on negligence. As species names can change, providing strain numbers with specifications such as their origin is highly recommended.

The use of seaweeds for human purposes has a long tradition. Seaweeds are harvested from their natural habitat and also cultivated (Pereira & Yarish, 2008). In contrast, the high potential of microalgae

Figure 7. Algae exist everywhere: (a) surface scums of Euglena sanguinea in a fish pond. Notice the colour change from the shaded to the sunny area, (b) endosymbiotic zoochlorellae in a ciliate, (c) intertidal zone during low tide, (d) epilithic algae community inhabing rock faces called 'Tintenstriche', (e) hot springs, (f) cryovegetation called 'bloody snow' covering a snow field

still awaits commercial use; only a handful of species is being cultivated at large scale. From time to time, algae come into the fore for biofuel production, which, however, is not economical (Mu et al., 2020). This brings up a basic but largely neglected fact, namely the conversion of available light into chemical energy through photosynthesis. This so-called photosynthetic efficiency ranges from a theoretical

Figure 8. Products containing algae or algal substances in our daily lives. (a) food and dietary supplements, (b) personal care products and pharmaceutical products, (c) animal food

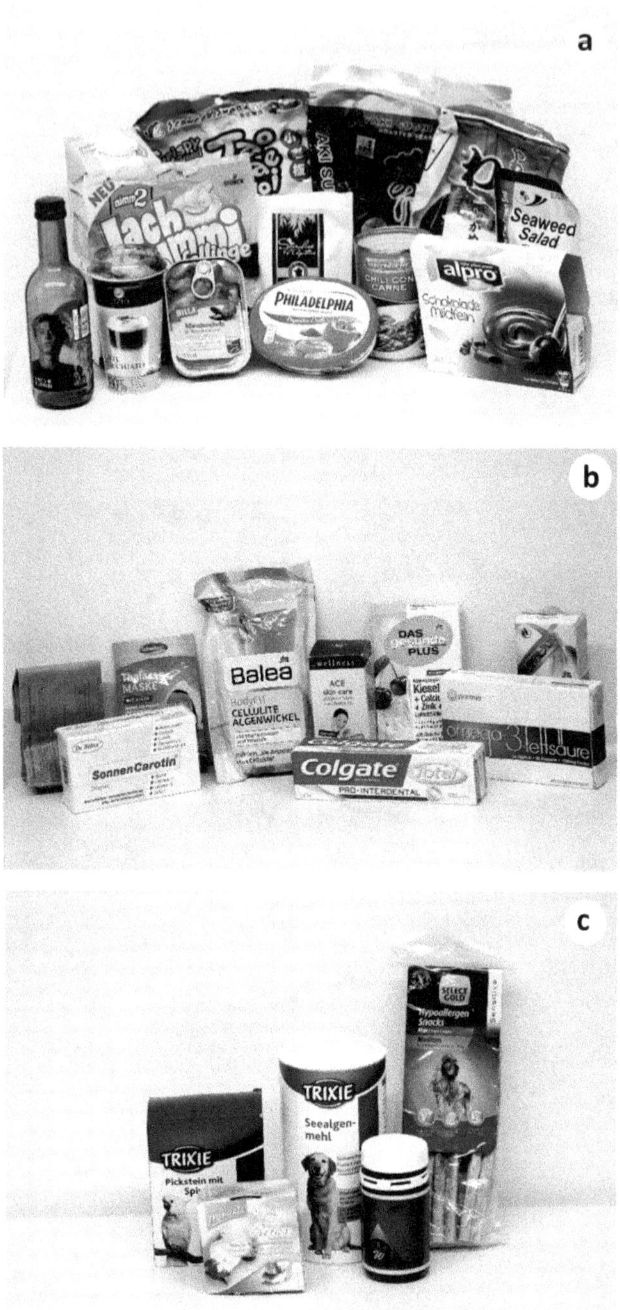

maximum of 10% to real values of < 1% in natural systems (Ooms et al., 2016). The photosynthetic efficiency can be increased to about 3 to 5% in photobioreactors with an optimal irradiance supply (Vecchi et al., 2020). These values, however, are valid only for simple organic hydrocarbons. The synthesis of complex macromolecules results in reduced net productivity (Wilhelm & Jakob, 2011) because energy

Figure 9. Quality check of algal products. (a, b) diatomaceous earth with a few fossil diatoms and many other inorganic particles, (c, d) 'Spirulina' fish food with high purity, (e, f) 'Chlorella' pills as dietary supplement, high impurities with 'Spirulina'

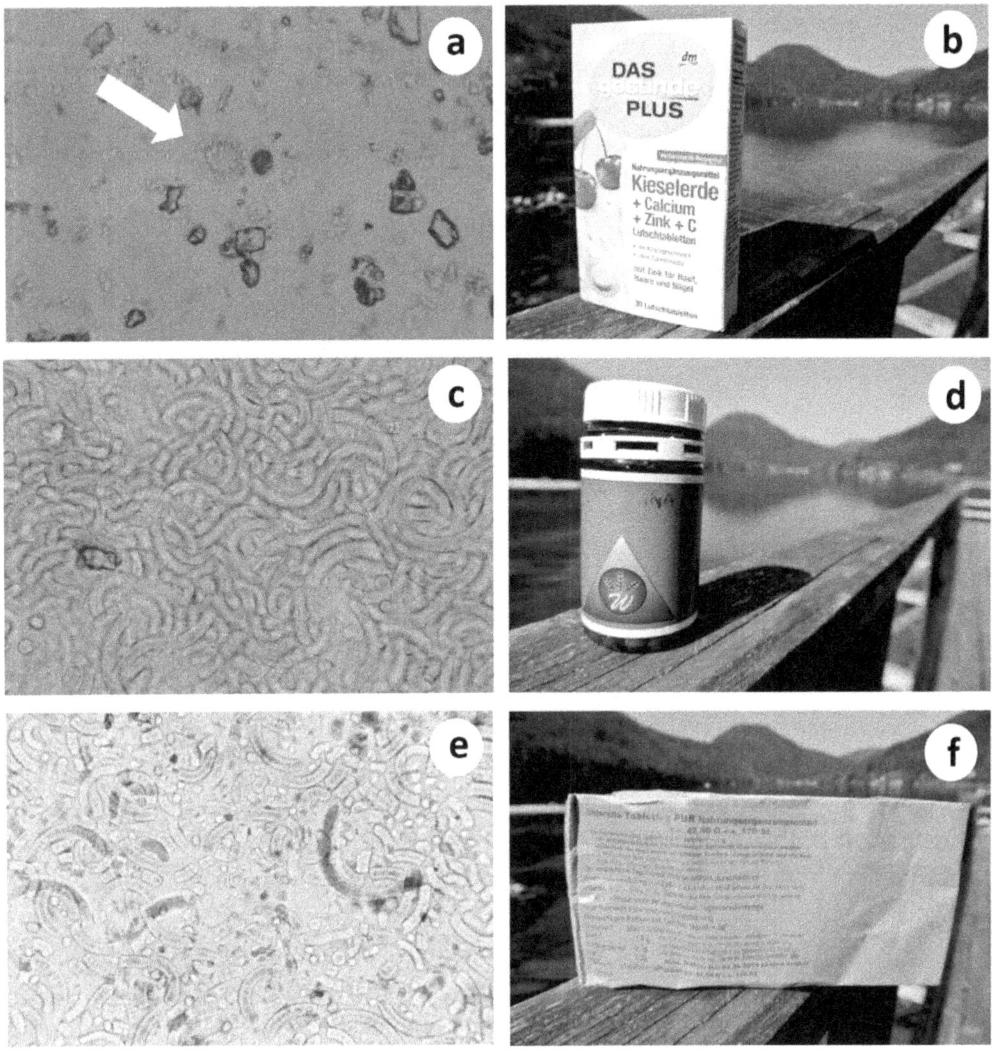

is needed for alternative elaborate metabolic pathways. Although the efficiency of valuable product generation can be increased by specific modifications of the cultivation mode, genetic engineering and other sophisticated techniques (Khan & Fu, 2020), this increase goes hand in hand with a reduced biomass yield. It is not possible to yield more energy equivalents than the generated ones (Fig. 10). A typical example is approaches to exploit the green alga *Botryococcus*, which synthesises high-quality lipids resembling kerosene, but at the expense of extremely slow growth. Many papers have been published and millions of dollars invested in the commercial exploitation of the taxon, which has still failed to reach commercial viability.

This raises another important fact, one which is largely neglected by companies. Algae as living organisms need special care and a basic understanding of algal ecology and physiology. Scaling up from

Figure 10. You cannot make something out of nothing. Biochemical energy generated from photosynthesis is either (a) mainly transformed to high-quality compounds resulting in low biomass yield (left) or (b) mainly converted to high biomass containing low amounts of high-quality products (right).

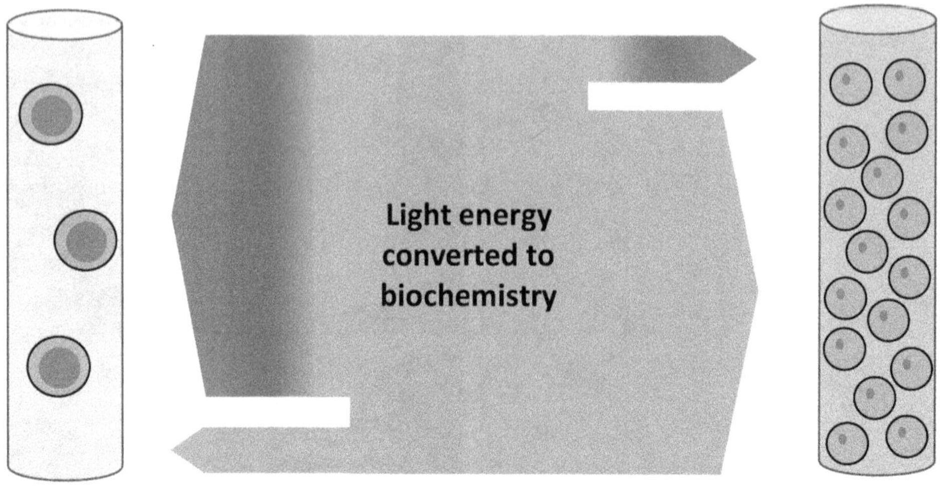

laboratory experiments to industrial facilities, and maintaining algae cultures at large scale, is challenging and requires the services not only of technicians, but also of phycologists. Importantly, providing specific culture conditions can boost the production of the substances of interest. A prime example is the two-stage cultivation of *Haematococcus lacustris*: the first stage focuses on maximum biomass generation, followed by a second step for astaxanthin accumulation (Oslan et al., 2021).

A promising alternative to the traditional approach of focusing solely on a single bioproduct is multipurpose exploitation of algal biomass. This biorefinery approach reduces costs because the harvested biomass is used for more than a single compound (Khoo et al., 2019). The crucial and costly step is the extraction of compounds; therefore, the whole biomass containing the valuable compounds is commonly sold as a powder.

Different cultivation systems are in use depending on the algae of interest. The cheapest systems (although also with the lowest biomass yield) are open systems such as raceway ponds and salterns. These systems are currently restricted to extremophile cultivation: extreme conditions are provided, which are tolerated or even preferred by the cultivated organism, but potential competing organisms will not survive. Cultivation of algae other than extremophiles requires protective measures against such competitors. These closed systems are known as photobioreactors and are constructed in different ways, e.g., flat panels, tubes and columns. Their common feature is that they are manufactured individually, with very specific specifications, which makes them expensive. For some products, an alternative approach might be feasible: even fully phototrophic algae such as *Chlorella* can be cultivated heterotrophically in fermenters for maximum biomass yield (Hu et al., 2018). Applying heterotrophic cultivation, however, depends on the product of interest. It makes little sense to produce sugar for bioethanol production via heterotrophic algae cultivation, which is based on organic media. A largely neglected but profoundly important aspect for profitable algal cultivation is downstream processing. About 20 to 30% of the overall production costs are caused by the harvest procedure (Barros et al., 2015). Afterwards, the algal biomass needs immediate processing to minimise bacterial decomposition; the necessary speed of processing entails additional costs.

Besides the frequently mentioned valuable compounds – biofuels, polyunsaturated fatty acids, gelling substances – algal biomass can also be used as biofertilizer (Chatterjee et al., 2017), which makes it interesting for organic farming. The cultivation can be implemented as an additional step of waste water treatment (Li et al., 2018). Another unconventional application is the use of algae as an additive of animal feed. Astaxanthin is used for aquaculture because it strengthens the immune system of fish and causes the special orange colour of salmons. Seaweeds have potential for use as methane reducers for cattle farming (Lean et al., 2021). Finally, the interest in algae goes far beyond the products originating from algae biomass. Algae are also used for bioremediation, for biomonitoring and in the field of paleolimnology (e.g., to help reconstruct past climates).

CONCLUSION

The exact definition of algae depends on the respective scientist's viewpoint, as this group is an artificial, functional term that does not provide any information on relationships. Algae drove evolution in the past, and they remain indispensable for element cycles today. Algae are extremely diverse and adaptable in terms of morphology, physiology and ecology. This diversity is also reflected in the wide range of high-quality compounds they produce, many of them waiting to be exploited.

ACKNOWLEDGMENT

This chapter is dedicated to Jörg Ott and Michael Stachowitsch, who introduced the author of this chapter to marine science many years ago. Jörg and Mischka are always open for fruitful discussions and cooperation. Their enthusiasm and selfless support have inspired many people, including the author of this chapter. Many thanks to Hannah Zinöcker for investing her valuable time in designing the figures and her patience.

REFERENCES

Archibald, J. M., Simpson, A. G. B., & Slamovits, C. H. (Eds.). (2017). *Handbook of the Protists* (2nd ed.). Springer.

Barros, A. I., Gonçalves, A. L., Simões, M., & Pires, J. C. M. (2015). Harvesting techniques applied to microalgae: A review. *Renewable & Sustainable Energy Reviews*, *41*, 1489–1500. doi:10.1016/j.rser.2014.09.037

Bloodgood, R. A. (2020). Prey capture in protists utilizing microtubule filled processes and surface motility. *Cytoskeleton*, *77*(11), 500–514. doi:10.1002/cm.21644 PMID:33190423

Boenigk, J. (2021). Evolution und Systematik der Protisten. In J. Boenigk (Ed.), *Boenigk, Biologie* (pp. 815–858). Springer Spektrum., doi:10.1007/978-3-662-61270-5_33

Bold, H. C., & Wynne, M. J. (1985). *Introduction to the Algae. Structure and reproduction*. Prentice Hall, Inc.

Bouck, G. B., & Sweeney, B. M. (1966). The fine structure and ontogeny of trichocysts in marine dinoflagellates. *Protoplasma, 61*(1), 205–223. doi:10.1007/BF01247920 PMID:5920039

Bremer, K. (1985). Summary of green plant phylogeny and classification. *Cladistics, 1*(4), 369–385. doi:10.1111/j.1096-0031.1985.tb00434.x PMID:34965683

Brennan, B., & Regan, F. (2020). In-situ lipid and fatty acid extraction methods to recover viable products from Nannochloropsis sp. *The Science of the Total Environment, 748*, 142464. doi:10.1016/j.scitotenv.2020.142464 PMID:33113682

Büdel, B., & Friedl, T. (Eds.). (2021). *Life at Rock Surfaces. Challenged by Extreme Light, Temperature and Hydration Fluctuations*. De Gruyter. doi:10.1515/9783110646467

Burki, F., Roger, A. J., Brown, M. W., & Simpson, A. G. B. (2020). The New Tree of Eukaryotes. *Trends in Ecology & Evolution, 35*(1), 43–55. doi:10.1016/j.tree.2019.08.008 PMID:31606140

Caron, D. A., Sanders, R. W., Lim, E. L., Marrasé, C., Amaral, L. A., Whitney, S., Aoki, R. B., & Porters, K. G. (1993). Light-dependent phagotrophy in the freshwater mixotrophic chrysophyte Dinobryon cylindricum. *Microbial Ecology, 25*(1), 93–111. doi:10.1007/BF00182132 PMID:24189708

Casero, M. C., Ascaso, C., Quesada, A., Mazur-Marzec, H., & Wierzchos, J. (2021). Response of Endolithic Chroococcidiopsis Strains From the Polyextreme Atacama Desert to Light Radiation. *Frontiers in Microbiology, 11*, 3607. doi:10.3389/fmicb.2020.614875 PMID:33537015

Cavalier-Smith, T. (1981). Eukaryote kingdoms: Seven or nine? *Bio Systems, 14*(3), 461–481. doi:10.1016/0303-2647(81)90050-2 PMID:7337818

Cavalier-Smith, T., & Chao, E. E. (1996). 18S rRNA sequence of Heterosigma carterae (Raphidophyceae), and the phylogeny of heterokont algae (Ochrophyta). *Phycologia, 35*(6), 500–510. doi:10.2216/i0031-8884-35-6-500.1

Chatterjee, A., Singh, S., Agrawal, C., Yadav, S., Rai, R., & Rai, L. C. (2017). Role of Algae as a Biofertilizer. In R. P. Rastogi, D. Madamwar, & A. Pandey (Eds.), *Algal Green Chemistry* (pp. 189–200). Elsevier. doi:10.1016/B978-0-444-63784-0.00010-2

Chopin, T., & Tacon, A. G. J. (2021). 2021/04/03). Importance of Seaweeds and Extractive Species in Global Aquaculture Production. *Reviews in Fisheries Science & Aquaculture, 29*(2), 139–148. doi:10.1080/23308249.2020.1810626

Chorus, I., & Welker, M. (2021). Toxic Cyanobacteria in Water: A Guide to Their Public Health Consequences, Monitoring and Management (2nd ed.). CRC Press. doi:10.1201/9781003081449

Consalvey, M., Paterson, D. M., & Underwood, G. J. C. (2004). The ups and downs of life in a benthic biofilm: Migration of benthic diatoms. *Diatom Research, 19*(2), 181–202. doi:10.1080/0269249X.2004.9705870

De Clerck, O., Guiry, M. D., Leliaert, F., Samyn, Y., & Verbruggen, H. (2013). Algal taxonomy: A road to nowhere? *Journal of Phycology*, *49*(2), 215–225. doi:10.1111/jpy.12020 PMID:27008509

Dodds, W. K., Gudder, D. A., & Mollenhauer, D. (1995). The ecology of Nostoc. *Journal of Phycology*, *31*(1), 2–18. doi:10.1111/j.0022-3646.1995.00002.x

Donoghue, P. C. J., Harrison, C. J., Paps, J., & Schneider, H. (2021). 2021/10/11/). The evolutionary emergence of land plants. *Current Biology*, *31*(19), R1281–R1298. doi:10.1016/j.cub.2021.07.038 PMID:34637740

Eikrem, W., Medlin, L. K., Henderiks, J., Rokitta, S., Rost, B., Probert, I., Throndsen, J., & Edvardsen, B. (2017). Haptophyta. In J. M. Archibald, A. G. B. Simpson, & C. H. Slamovits (Eds.), *Handbook of the Protists* (pp. 893–953). Springer International Publishing. doi:10.1007/978-3-319-28149-0_38

Falasco, E., Ector, L., Isaia, M., Wetzel, C. E., Hoffmann, L., & Bona, F. (2014). Diatom flora in subterranean ecosystems: A review. *International Journal of Speleology*, *43*(3), 231–251. doi:10.5038/1827-806X.43.3.1

Field, C. B., Behrenfeld, M. J., Randerson, J. T., & Falkowski, P. (1998). Primary production of the biosphere: Integrating terrestrial and oceanic components. *Science*, *281*(5374), 237–240. doi:10.1126cience.281.5374.237 PMID:9657713

Fischer, W. W., Hemp, J., & Johnson, J. E. (2016). Evolution of Oxygenic Photosynthesis. *Annual Review of Earth and Planetary Sciences*, *44*(1), 647–683. doi:10.1146/annurev-earth-060313-054810

Gavelis Gregory, S., Wakeman Kevin, C., Tillmann, U., Ripken, C., Mitarai, S., Herranz, M., Özbek, S., Holstein, T., Keeling Patrick, J., & Leander Brian, S. (2017). Microbial arms race: Ballistic "nematocysts" in dinoflagellates represent a new extreme in organelle complexity. *Science Advances*, *3*(3), e1602552. doi:10.1126ciadv.1602552 PMID:28435864

Graham, L. E., Graham, J. M., Wilcox, L. W. W., & Cook, M. E. (2016). *Algae (Version* (3.3 ed.). LJLM Press.

Gross, W. (2000). Ecophysiology of algae living in highly acidic environments. *Hydrobiologia*, *433*(1), 31–37. doi:10.1023/A:1004054317446

Grube, M., Seckbach, J., & Muggia, L. (Eds.). (2017). Algal and Cyanobacteria Symbioses. World Scientific. doi:10.1142/q0017

Guiry, M. D. (2012). How many species of algae are there? *Journal of Phycology*, *48*(5), 1057–1063. doi:10.1111/j.1529-8817.2012.01222.x PMID:27011267

Haddock, S. H. D., Moline, M. A., & Case, J. F. (2009). Bioluminescence in the Sea. *Annual Review of Marine Science*, *2*(1), 443–493. doi:10.1146/annurev-marine-120308-081028 PMID:21141672

Hehenberger, E., Gast, R. J., & Keeling, P. J. (2019). A kleptoplastidic dinoflagellate and the tipping point between transient and fully integrated plastid endosymbiosis. *Proceedings of the National Academy of Sciences of the United States of America*, *116*(36), 17934–17942. doi:10.1073/pnas.1910121116 PMID:31427512

Hoffmann, L. (1989). Algae of terrestrial habitats. *Botanical Review*, *55*(2), 77–105. doi:10.1007/BF02858529

Hoham, R. W., & Remias, D. (2020). Snow and Glacial Algae: A Review1. *Journal of Phycology*, *56*(2), 264–282. doi:10.1111/jpy.12952 PMID:31825096

Hu, J., Nagarajan, D., Zhang, Q., Chang, J.-S., & Lee, D.-J. (2018). Heterotrophic cultivation of microalgae for pigment production: A review. *Biotechnology Advances*, *36*(1), 54–67. doi:10.1016/j.biotechadv.2017.09.009 PMID:28947090

Hug, L. A., Baker, B. J., Anantharaman, K., Brown, C. T., Probst, A. J., Castelle, C. J., Butterfield, C. N., Hernsdorf, A. W., Amano, Y., Ise, K., Suzuki, Y., Dudek, N., Relman, D. A., Finstad, K. M., Amundson, R., Thomas, B. C., & Banfield, J. F. (2016). A new view of the tree of life. *Nature Microbiology*, *1*(5), 16048. doi:10.1038/nmicrobiol.2016.48 PMID:27572647

Huisman, J., Codd, G. A., Paerl, H. W., Ibelings, B. W., Verspagen, J. M. H., & Visser, P. M. (2018). Cyanobacterial blooms. *Nature Reviews. Microbiology*, *16*(8), 471–483. doi:10.103841579-018-0040-1 PMID:29946124

Hulatt, C. J., Berecz, O., Egeland, E. S., Wijffels, R. H., & Kiron, V. (2017). Polar snow algae as a valuable source of lipids? *Bioresource Technology*, *235*, 338–347. doi:10.1016/j.biortech.2017.03.130 PMID:28384586

Imai, I., & Yamaguchi, M. (2012). Life cycle, physiology, ecology and red tide occurrences of the fish-killing raphidophyte Chattonella. *Harmful Algae*, *14*, 46–70. doi:10.1016/j.hal.2011.10.014

Janouškovec, J., Gavelis, G. S., Burki, F., Dinh, D., Bachvaroff, T. R., Gornik, S. G., Bright, K. J., Imanian, B., Strom, S. L., Delwiche, C. F., Waller, R. F., Fensome, R. A., Leander, B. S., Rohwer, F. L., & Saldarriaga, J. F. (2017). Major transitions in dinoflagellate evolution unveiled by phylotranscriptomics. *Proceedings of the National Academy of Sciences of the United States of America*, *114*(2), E171. doi:10.1073/pnas.1614842114 PMID:28028238

Jeffrey, C. (1967). The Origin and Differentiation of the Archegoniate Land Plants: A Second Contribution. *Kew Bulletin*, *21*(2), 335–349. doi:10.2307/4108533

Karsten, U. (2008). Defense Strategies of Algae and Cyanobacteria Against Solar Ultraviolet Radiation. In C. D. Amsler (Ed.), *Algal Chemical Ecology* (p. 313). Springer. doi:10.1007/978-3-540-74181-7_13

Kaup, M., Trull, S., & Hom, E. F. Y. (2021). On the move: Sloths and their epibionts as model mobile ecosystems. *Biological Reviews of the Cambridge Philosophical Society*, *96*(6), 2638–2660. doi:10.1111/brv.12773 PMID:34309191

Khan, S., & Fu, P. (2020). Biotechnological perspectives on algae: A viable option for next generation biofuels. *Current Opinion in Biotechnology*, *62*, 146–152. doi:10.1016/j.copbio.2019.09.020 PMID:31683048

Khoo, C. G., Dasan, Y. K., Lam, M. K., & Lee, K. T. (2019). Algae biorefinery: Review on a broad spectrum of downstream processes and products. *Bioresource Technology*, *292*, 121964. doi:10.1016/j.biortech.2019.121964 PMID:31451339

Kirst, G. O. (1990). Salinity Tolerance of Eukaryotic Marine Algae. *Annual Review of Plant Physiology and Plant Molecular Biology*, *41*(1), 21–53. doi:10.1146/annurev.pp.41.060190.000321

Krienitz, L., Huss, V. A. R., & Bock, C. (2015). Chlorella: 125 years of the green survivalist. *Trends in Plant Science*, *20*(2), 67–69. doi:10.1016/j.tplants.2014.11.005 PMID:25500553

Krienitz, L., & Schagerl, M. (2016). Tiny and Tough: Microphytes of East African Soda Lakes. In M. Schagerl (Ed.), *Soda Lakes of East Africa* (pp. 149–177). Springer. doi:10.1007/978-3-319-28622-8_6

Laza-Martínez, A., Fernández-Marín, B., & García-Plazaola, J. I. (2019). Rapid colour changes in Euglena sanguinea (Euglenophyceae) caused by internal lipid globule migration. *European Journal of Phycology*, *54*(1), 91–101. doi:10.1080/09670262.2018.1513571

Lean, I. J., Golder, H. M., Grant, T. M. D., & Moate, P. J. (2021). A meta-analysis of effects of dietary seaweed on beef and dairy cattle performance and methane yield. *PLoS One*, *16*(7), e0249053. doi:10.1371/journal.pone.0249053 PMID:34252114

Lee, R. E. (2018). Phycology (5th ed.). Cambridge University Press. doi:10.1017/9781316407219

Leliaert, F., Smith, D. R., Moreau, H., Herron, M. D., Verbruggen, H., Delwiche, C. F., & De Clerck, O. (2012). Phylogeny and Molecular Evolution of the Green Algae. *Critical Reviews in Plant Sciences*, *31*(1), 1–46. doi:10.1080/07352689.2011.615705

Li, Y., Tarpeh, W. A., Nelson, K. L., & Strathmann, T. J. (2018). Quantitative Evaluation of an Integrated System for Valorization of Wastewater Algae as Bio-oil, Fuel Gas, and Fertilizer Products. *Environmental Science & Technology*, *52*(21), 12717–12727. doi:10.1021/acs.est.8b04035 PMID:30256626

Littler, M. M., Littler, D. S., Blair, S. M., & Norris, J. N. (1985). Deepest Known Plant Life Discovered on an Uncharted Seamount. *Science*, *227*(4682), 57–59. doi:10.1126cience.227.4682.57 PMID:17810025

Löffelhardt, W., Bohnert, H. J., Bryant, D. A., & Hagemann, R. (1997). The Cyanelles of Cyanophora Paradoxa. *Critical Reviews in Plant Sciences*, *16*(4), 393–413. doi:10.1080/07352689709701955

Martin, W. F., & Allen, J. F. (2018). An Algal Greening of Land. *Cell*, *174*(2), 256–258. doi:10.1016/j.cell.2018.06.034 PMID:30007415

McCoy, S. J., Krueger-Hadfield, S. A., & Mieszkowska, N. (2020). Evolutionary Phycology: Toward a Macroalgal Species Conceptual Framework. *Journal of Phycology*, *56*(6), 1404–1413. doi:10.1111/jpy.13059 PMID:32726874

McFadden, G. I. (2011). The apicoplast. *Protoplasma*, *248*(4), 641–650. doi:10.100700709-010-0250-5 PMID:21165662

Metting, B. (1981). The systematics and ecology of soil algae. *Botanical Review*, *47*(2), 195–312. doi:10.1007/BF02868854

Mu, D., Xin, C., & Zhou, W. (2020). Life Cycle Assessment and Techno-Economic Analysis of Algal Biofuel Production. In A. Yousuf (Ed.), *Microalgae Cultivation for Biofuels Production* (pp. 281–292). Academic Press. doi:10.1016/B978-0-12-817536-1.00018-7

Norton, T. A., Melkonian, M., & Andersen, R. A. (1996). Algal biodiversity. *Phycologia*, *35*(4), 308–326. doi:10.2216/i0031-8884-35-4-308.1

Nowicka-Krawczyk, P., Mühlsteinová, R., & Hauer, T. (2019). Detailed characterization of the Arthrospira type species separating commercially grown taxa into the new genus Limnospira (Cyanobacteria). *Scientific Reports*, *9*(1), 694. doi:10.103841598-018-36831-0 PMID:30679537

Ooms, M. D., Dinh, C. T., Sargent, E. H., & Sinton, D. (2016). Photon management for augmented photosynthesis. *Nature Communications*, *7*(1), 12699. doi:10.1038/ncomms12699 PMID:27581187

Oslan, S. N., Shoparwe, N. F., Yusoff, A. H., Rahim, A. A., Chang, C. S., Tan, J. S., Oslan, S. N., Arumugam, K., Ariff, A. B., Sulaiman, A. Z., & Mohamed, M. S. (2021). A Review on Haematococcus pluvialis Bioprocess Optimization of Green and Red Stage Culture Conditions for the Production of Natural Astaxanthin. *Biomolecules*, *11*(2), 256. Advance online publication. doi:10.3390/biom11020256 PMID:33578851

Padisák, J., Soróczki-Pintér, É., & Rezner, Z. (2003). Sinking properties of some phytoplankton shapes and the relation of form resistance to morphological diversity of plankton – an experimental study. *Hydrobiologia*, *500*(1), 243–257. doi:10.1023/A:1024613001147

Pascher, A. (1931). Systematische Übersicht über die mit Flagellaten in Zusammenhang stehenden Algenreihen und Versuch einer Einreihung dieser Algenstämme in die Stämme des Pflanzenreiches. *Botanisches Zentralblat*. *Beihefte*, *48*, 317–332.

Patel, A. K., Singhania, R. R., Dong, C.-D., Obulisami, P. K., & Sim, S. J. (2021). 2021/12/01/). Mixotrophic biorefinery: A promising algal platform for sustainable biofuels and high value coproducts. *Renewable & Sustainable Energy Reviews*, *152*, 111669. doi:10.1016/j.rser.2021.111669

Pereira, R., & Yarish, C. (2008). Mass Production of Marine Macroalgae. In S. E. Jørgensen & B. D. Fath (Eds.), *Encyclopedia of Ecology* (pp. 2236–2247). Academic Press. doi:10.1016/B978-008045405-4.00066-5

Pröschold, T., & Leliaert, F. (2007). *Systematics of the green algae: Conflict of classic and modern approaches*. doi:10.1201/9780849379901.ch7

Qiu, H., Yoon, H. S., & Bhattacharya, D. (2013). Algal endosymbionts as vectors of horizontal gene transfer in photosynthetic eukaryotes. *Frontiers in Plant Science*, *4*, 366. doi:10.3389/fpls.2013.00366 PMID:24065973

Radulovich, R., Neori, A., Valderrama, D., Reddy, C. R. K., Cronin, H., & Forster, J. (2015). Farming of seaweeds. In B. K. Tiwari & D. J. Troy (Eds.), *Seaweed Sustainability* (pp. 27–59). Academic Press. doi:10.1016/B978-0-12-418697-2.00003-9

Sánchez-Baracaldo, P., Bianchini, G., Wilson, J. D., & Knoll, A. H. (2022). Cyanobacteria and biogeochemical cycles through Earth history. *Trends in Microbiology*, *30*(2), 143–157. doi:10.1016/j.tim.2021.05.008 PMID:34229911

Schagerl, M., & Burian, A. (2016). The Ecology of African Soda Lakes: Driven by Variable and Extreme Conditions. In Soda Lakes of East Africa (pp. 295-320). Springer. doi:10.1007/978-3-319-28622-8_12

Schirrmeister, B. E., Gugger, M., & Donoghue, P. C. J. (2015). Cyanobacteria and the Great Oxidation Event: Evidence from genes and fossils. *Palaeontology*, *58*(5), 769–785. doi:10.1111/pala.12178 PMID:26924853

Schöfer, H. (2018). Cutaneous infection – would you think of algae? *Journal of the European Academy of Dermatology and Venereology*, *32*(9), 1401–1402. doi:10.1111/jdv.15202 PMID:30136764

Sha, J., Xiong, H., Li, C., Lu, Z., Zhang, J., Zhong, H., Zhang, W., & Yan, B. (2021). 2021/07/01/). Harmful algal blooms and their eco-environmental indication. *Chemosphere*, *274*, 129912. doi:10.1016/j.chemosphere.2021.129912 PMID:33979937

Sharma, N. K., Rai, A. K., Singh, S., & Brown, R. M. Jr. (2007). Airborne algae: Their present status and relevance. *Journal of Phycology*, *43*(4), 615–627. doi:10.1111/j.1529-8817.2007.00373.x

Shu, W.-S., & Huang, L.-N. (2022). 2022/04/01). Microbial diversity in extreme environments. *Nature Reviews. Microbiology*, *20*(4), 219–235. doi:10.103841579-021-00648-y PMID:34754082

Shumway, S. E., Burkholder, J. M., & Morton, S. L. (Eds.). (2018). *Harmful Algal Blooms: A Compendium Desk Reference*. John Wiley & Sons. doi:10.1002/9781118994672

Sibbald, S. J., & Archibald, J. M. (2020). Genomic Insights into Plastid Evolution. *Genome Biology and Evolution*, *12*(7), 978–990. doi:10.1093/gbe/evaa096 PMID:32402068

Sili, C., Torzillo, G., & Vonshak, A. (2012). Arthrospira (Spirulina). In B. A. Whitton (Ed.), *Ecology of Cyanobacteria II: Their Diversity in Space and Time* (pp. 677–705). Springer Netherlands. doi:10.1007/978-94-007-3855-3_25

Simpson, M. (2010). Species and Conservation in Systematics. In M. Simpson (Ed.), *Plant Systematics* (pp. 649–668). Academic Press. doi:10.1016/B978-0-12-374380-0.50019-1

Singh, A. (2021). Role of Algae in Soil Nitrogen Fixation. In C. Cruz, K. Vishwakarma, D. K. Choudhary, & A. Varma (Eds.), *Soil Nitrogen Ecology* (pp. 483–496). Springer International Publishing. doi:10.1007/978-3-030-71206-8_24

Takano, Y., Yamaguchi, H., Inouye, I., Moestrup, Ø., & Horiguchi, T. (2014). Phylogeny of Five Species of Nusuttodinium gen. nov. (Dinophyceae), a Genus of Unarmoured Kleptoplastidic Dinoflagellates. *Protist*, *165*(6), 759–778. doi:10.1016/j.protis.2014.09.001 PMID:25460229

Tesson, S. V. M., Skjøth, C. A., Šantl-Temkiv, T., Löndahl, J., & Voordouw, G. (2016). Airborne Microalgae: Insights, Opportunities, and Challenges. *Applied and Environmental Microbiology*, *82*(7), 1978–1991. doi:10.1128/AEM.03333-15 PMID:26801574

Triemer, R., Zimba, P., & Rowan, M. (2004). 163 Identification of Euglenoids That Produce Ichthyotoxin(S) (Euglenophyta). *Journal of Phycology*, *39*(s1), 56–57. doi:10.1111/j.0022-3646.2003.03906001_163.x

van den Hoek, C., Mann, D., & Jahns, H. M. (1996). *Algae. An Introduction to Phycology*. Cambridge University Press.

Vecchi, V., Barera, S., Bassi, R., & Dall'Osto, L. (2020). Potential and Challenges of Improving Photosynthesis in Algae. *Plants (Basel, Switzerland)*, *9*(1), 67. doi:10.3390/plants9010067 PMID:31947868

Ward, D. M., & Castenholz, R. W. (2000). Cyanobacteria in geothermal habitats. In B. A. Whitton & M. Potts (Eds.), *The ecology of cyanobacteria*. Kluwer Academic Publishers.

Wilhelm, C., & Jakob, T. (2011, December). From photons to biomass and biofuels: Evaluation of different strategies for the improvement of algal biotechnology based on comparative energy balances. *Applied Microbiology and Biotechnology*, *92*(5), 909–919. doi:10.100700253-011-3627-2 PMID:22005740

Yoon, H. S., Hackett, J. D., Ciniglia, C., Pinto, G., & Bhattacharya, D. (2004). A Molecular Timeline for the Origin of Photosynthetic Eukaryotes. *Molecular Biology and Evolution*, *21*(5), 809–818. doi:10.1093/molbev/msh075 PMID:14963099

Yoon, H. S., Nelson, W., Lindstrom, S. C., Boo, S. M., Pueschel, C., Qiu, H., & Bhattacharya, D. (2017). Rhodophyta. In J. M. Archibald, A. G. B. Simpson, & C. H. Slamovits (Eds.), *Handbook of the Protists* (pp. 89–133). Springer International Publishing. doi:10.1007/978-3-319-28149-0_33

Zimba, P. V., Huang, I. S., Gutierrez, D., Shin, W., Bennett, M. S., & Triemer, R. E. (2017). Euglenophycin is produced in at least six species of euglenoid algae and six of seven strains of Euglena sanguinea. *Harmful Algae*, *63*, 79–84. doi:10.1016/j.hal.2017.01.010 PMID:28366403

Zingone, A., Escalera, L., Aligizaki, K., Fernández-Tejedor, M., Ismael, A., Montresor, M., Mozetič, P., Taş, S., & Totti, C. (2021). 2021/02/01/). Toxic marine microalgae and noxious blooms in the Mediterranean Sea: A contribution to the Global HAB Status Report. *Harmful Algae*, *102*, 101843. doi:10.1016/j.hal.2020.101843 PMID:33875177

KEY TERMS AND DEFINITIONS

Anisogamy: Fusion of two gametes of different shape and size. Both gametes are motile. The larger gamete is treated as female, the smaller as male.

Coenobnium: Group of cells held together by cell wall structures or mucilage. One generation, 2^n cell number and genetically defined shape.

Colony: Group of cells held together by cell wall structures or mucilage. Usually more than one generation, variable cell number and no defined shape.

Diplont: The diploid generation of the life cycle is multicellular, the only haploid stage is the gametes after meiosis.

Form Resistance: Empirical variable to describe the frictional resistance of a body. A sphere has a form resistance of 1, the bigger the deviation from this shape, the larger the form resistance.

Haplo-Diplont: A haploid and a diploid multicellular generation alternates.

Haplont: The haploid generation is multicellular, the only diploid stage is the zygote.

Heterocyte: Special type of cyanobacterial cell for dinitrogen fixation with unique features.

Heterokont Flagellation: The ultrastructure of flagella inserting in a flagellate differs from each other. In the Ochrophyta, an acronematic (naked) flagellum inserts next to a pleuronematic flagellum (covered by fine hairs = mastigonema).

Heteromorphic Life Cycle: Alternate generations show different morphology.

Heterotrichal: Filaments – sometimes branched – have different morphology (e.g., erect and creeping filaments, main axis and side branches, the latter with terminal hairs).

Homotrichal: Filaments – sometimes branched – have the same morphology.

Isogamy: Fusion of two gametes which do not differ in size.

Isokont Flagellation: The ultrastructure of flagella inserting in a flagellate does not differ from each other (although they may differ in length).

Isomorphic Life Cycle: Alternate generations show same morphology.

Multinucleate: Also called coenocytic. More than one nucleus per cell.

Multiseriate: Arranged in more than a single row. Usually developing from uniseriate filaments.

Oogamy: Fusion of two gametes of different shape and size. The small gamete is mobile and treated as male. The large gamete is immobile and female.

Osmotrophy: Uptake of dissolved organic substances as energy source.

Phagotrophy: Engulfment of particulate food.

Seagrasses: Embryophytes (flowering plants) growing in marine habitats.

Seaweeds: General term for macroalgae growing in marine habitats.

Uniseriate: Arranged in a single row.

Chapter 2
Biochemistry and Biotechnology of Algae

Ghadir Aly El-Chaghaby
Regional Center for Food and Feed, Agricultural Research Center, Egypt

Sayed Rashad
Regional Center for Food and Feed, Agricultural Research Center, Egypt

ABSTRACT

"Algae" represent a diverse group of photosynthetic organisms ranging from single cell to massive kelp that are found in all ecosystems on our planet. Algae play an essential role for life; they account for almost 50% of the photosynthesis activity on earth. The biochemical composition of algae includes a variety of high-value products such as pigments, lipids, carbohydrates, amino acids, and many other bioactive compounds. Algal biochemical compounds are variable among the different species and are highly dependent on the algal culture conditions such as temperature, nutrients, light, etc. In the recent years, algal biotechnological applications are on the spot. Algae are exploited for several biotechnological uses such as for biofuels, bio fertilizers, pharmaceuticals, nutraceuticals, bioremediators, and others. This chapter discusses the biochemistry and biotechnology of algae with emphasizing on the high-value biochemical algal compounds and trending algal biotechnological applications.

INTRODUCTION

"Algae" is the term that describes a group of aquatic photosynthetic microorganisms. Generally, algae are eukaryotic organisms that have a nucleus, a number of chromosomes in which DNA is organized as nucleoproteins, and a number of membrane-bound organelles (Cabej, 2013). However, blue- green algae (cyanobacteria) are prokaryotic organisms that are single-celled and has neither a distinct nucleus with a membrane nor other specialized organelles (Cole, 2016). Based on their size, algae are classified into (1) micro-algae that are unicellular and microscopic; and (2) macro-algae that are multicellular and visible to naked human's eye (J. Liu et al., 2019). Microalgae are microscopic species that use the same photosynthetic pathway as higher plants to survive (Chai et al., 2021). They are found in both marine

DOI: 10.4018/978-1-6684-2438-4.ch002

and freshwater environments. They constitute a polyphyletic and highly diverse group of prokaryotic and eukaryotic organisms. The most abundant microalgal classes are Cyanophyceae (blue-green algae), Chlorophyceae (green algae), Bacillariophyceae (including the diatoms), and Chrysophyceae (including golden algae) (Carlsson and others 2007) (Kumar, 2021). Macroalgae (also known as seaweeds) are one of the most important biological commodities in aquatic environments (Mofeed & Deyab, 2015). Based on their pigments, macroalgae are divided into three groups based on the appearance of various: Chlorophyceae (green algae), Phaeophyceae (brown algae), and Rhodophyceae (red algae) (Barbosa et al., 2014). Algae are diverse source of important biochemical compounds (Mutanda et al., 2020) wide range of biotechnological applications. Algae biotechnology is "phycoprospecting" as its applications are countless starting from food and pharmaceuticals to biofuels(Pfeiffer, Camagakevac, et al., 2017) (Mohanty et al., 2020). This chapter discusses the biochemistry and biotechnology of algae with emphasizing on the high-value biochemical algal compounds and trending algal biotechnological applications.

BIOCHEMISTRY OF ALGAE

Photosynthesis in Algae

Back to 1893, Charles Barnes introduced the word 'photosynthesis' as the biological mechanism that involves "synthesis of complex carbon compounds out of carbonic acid, in the presence of chlorophyll, under the influence of light" (Gest et al., 2002).

Algae, like plants, have the ability to transform photosynthetic active radiation energy into biologically useful energy (Pfeiffer, Ivna, et al., 2017). All organisms capable of oxygenic photosynthesis have a relatively conserved and tightly regulated biosynthetic pathway to produce these important light harvesting and energy transducing molecules, and chlorophylls and their derivatives are essential components of the photosynthetic apparatus (Cahoon & Timko, 2003).

The mechanism of oxygenic photosynthesis can be separated into light and dark phases in algae as in higher plants (Vecchi et al., 2020). During the light photosynthetic reaction, the light is captured to provide reducing power in the form of nicotinamide–adenine dinucleotide phosphate (NADPH) and metabolic energy in the form of adenosine triphosphate (ATP); whereas during the dark reactions, NADPH and ATP are then used to synthesize carbohydrates (St. Onge, 2018).

In the biochemical reaction that takes place in the chloroplast, algae use the chlorophyll to absorb light, split the water molecule, and release oxygen gas, as well as the energy storage compounds NADPH and ATP, during in a light-dependent photosynthesis process (Petsas & Vagi, 2017).

Several algal biochemical parameters related to photosynthesis processes, such as ATP formation, CO_2 fixation, O_2 evolution, carbon absorption and chlorophyll quality, have been adopted as standard and classical markers for the assessment of environmental stress induced by several groups of different pollutants in photosynthetic algal species (Petsas & Vagi, 2017).

Figure 1. Algal photosynthesis process

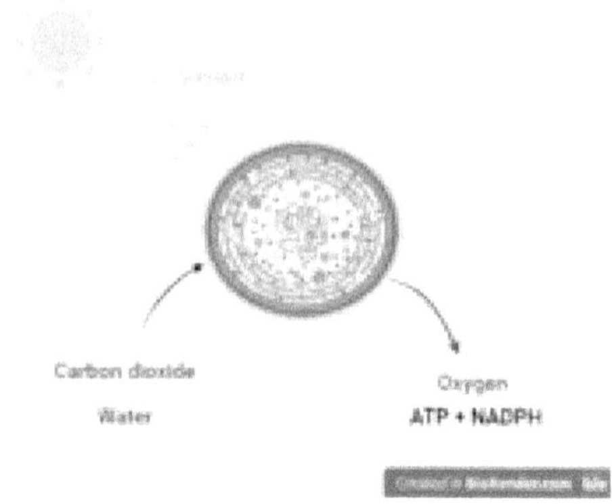

ENVIRONMENTAL FACTORS EFFECTS ON BIOCHEMICAL COMPOSITION OF ALGAE

Algae biochemical compositions are strongly dependent on their cultivation environmental factors. These factors including lighting, temperature, pH, and nutrients not only affect algal growth rate and photosynthesis, but also influence their metabolism and composition (Juneja et al., 2013).

Many algae species, in response to unfavorable environmental conditions, synthesize large amounts of storage lipids such as triacylglycerols (Goncalves et al., 2016). The chief role of fatty acids in algae is related to functions of cell membrane and metabolic processes. The degree of unsaturation of membrane fatty acids is also a significant parameter in adaptation of algae to environmental conditions(Los et al., 2013). Changes in the lipid fatty acid profile in response to elevated salinity of the medium are necessary to keep the membrane fluid and prevent its destruction(Zhila et al., 2011).

The growth of microalgae is a complicated process that involves both chemical and physical "substrates." In terms of chemical substrates, light and temperature have long been recognized as the primary determinants of algal biochemical composition(Fabris et al., 2020). The impacts of temperature on cell cultures are primarily two: one is the temperature dependency of cell component structure (particularly proteins and lipids), and the other is the temperature dependence of reaction rates, which are in turn dependent on the activation energies of those reactions(Carvalho et al., 2009).

Light intensity, is one of the most essential culture parameters in algae growth and biomass production. Algal biochemical composition depends on light intensity, in addition to regulating biological processes (Metsoviti et al., 2020). Increases in light intensity generally enhance algal growth up to a photoinhibitory threshold, although this effect's strength and the threshold vary across different species. Light intensity also affects the formation of lipids by algae(Wells et al., 2015). The increases in the intensity of light, though, diminish certain species' lipid content, while encourage or have no influence on lipid synthesis in others. That is why it is very important to perform examinations, on a species-by-species basis, the impacts of light intensity on lipid synthesis (Nzayisenga et al., 2020). During the photoautotrophic growth stage, light is the source of energy that organisms employ to transform carbon dioxide into organic molecules, particularly sugar.(Juneja et al., 2013).

Temperature is another important factor affecting the growth and composition of algae. Temperature affects all metabolic processes and the optimal temperature for a certain strain of algae will have a marked

Figure 2. Some factors affecting the growth and composition of algae

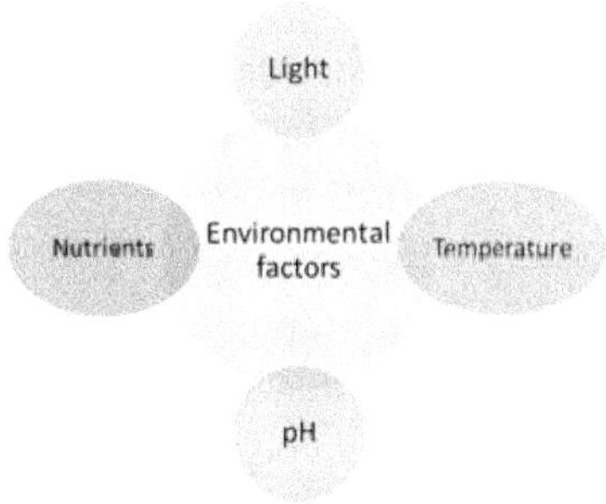

impact on crop growth(Juneja et al., 2013). All microalgae photosynthesis, breathing and growth rates are reduced when optimal temperatures are exceeded by imbalances in production and energy demand between adenosine triphosphate (ATP), inactivation or denature of necessary photosynthesis proteins. The influence of temperature also changes the content of fatty acids. Total unsaturated fatty acids have been found to decrease at high temperature (Chaisutyakorn et al., 2017). The temperature directly influences the fatty acid and lipid content in the organisms, although their type and quantity vary between the species, and is not well-known. Cells typically change their cell membrane at lower temperatures to include more unsaturated fatty acids, giving more fluidity to the membrane, whereas at higher temperatures the cell membrane becomes saturated, and therefore increasing rigidity (Nalley et al., 2018).

Nutrients and composition of the medium largely determine the amount of development and eventual concentration of the algae(Panahi et al., 2019). All critical elements in the algae growth media are required for the correct development of biomass. For algal growth, carbon, nitrogen and phosphorus are the most important nutrients. Carbon is the primary and most crucial component in the formation of algae(Yaakob et al., 2021). Nitrogen accounts for more than 10% of biomass, and its deficiency leads to a decrease in chlorophyll, increase of carotenoids, accumulation of polysaccharides and an increased production of lipids. Nitrogen deficiency is being progressively transformed into triglyceride-rich lipids by the free-fatty acid-rich compositions. Phosphorus is the third important element in the medium used for growing cell activities, such as anabolism of DNA and energy transfer (Panahi et al., 2019).

Medium pH is another important factor affecting algal composition and growth. At elevated pH, algal growth decreases which is likely to be related to inorganic carbon (HCO_3^- and CO_2) restrictions, largely owing to reduced organic cells carbon content due to abounding carbonate ions (CO_3^{2-})(Rost et al., 2006). In addition, the excessive energy consumption for internal acid-base balance resulting in a decrease in ATP synthesis available energy and hence growth is decreased. Also, the enzymes involved in intracellular metabolism are adversely affected by pH changes (Samanta et al., 2019).

ALGAL BIOCHEMICAL COMPOUNDS

Algae during their growth convert the sun's energy into a variety of metabolites known as bioactive chemicals (biochemicals). Algae represent a good source of fiber, proteins, minerals, pigments, vitamins, polyphenols, steroids, alkaloids, alginic acid, terpenoids, carotenoids, lectins, halogenated compounds, diterpenes tocopherols, tocotrienols, peptides, carbohydrates, polysaccharides, polyunsaturated fatty acids, and other lipids, many algae chemicals and products are economically important and widely used (Aditya et al., 2016; Mutanda et al., 2020).

Blue-green algae are known for their polysaccharide sheaths and extracellular polymeric secretions. Brown seaweeds are the most abundant source of fucoidans and alginates. Brown seaweed cell walls contain alginate, which is a mixed salt. It comprises major cations such as sodium, calcium, magnesium, and potassium, as well as smaller metal ions(Baweja et al., 2019).

In algae, there are three types of pigments: chlorophylls, carotenoids, and phycobiliproteins. Chlorophyll pigments are greenish fat-soluble pigments that serve an important role in photosynthesis in terrestrial plants, algae, and cyanobacteria. Carotenoid pigments include carotenes, lycopene, astaxanthin, neoxanthin, and lutein. Phycoerythrins are red pigments, while phyococyanins and allophycocyanins are blue and light blue pigments (Hentati et al., 2020).

Algal phenolic compounds include phenolic acids, phlorotannins, flavonoids, tannins, and catechins. Brown seaweeds (Pheophyceae) are noted for having high phlorolatannin levels. Both red (Rhodophycea) and green (Chlorophycea) contain a lot of flavonoids, phenolic acid, and bromophenols (Freile-Pelegrín and Robledo, 2013). Polyphenols are classified as phenolic carbonic acids (hydroxybenzoic acids and hydroxycinnamic acids), flavonoids (flavanols, anthocyanins flavones, flavonols, flavanones and flavanonols), isoflavones, stilbene, lignans, and phenolic polymers (condensed, hydrolyzable tannins and proanthocyanidins) (Zolotareva et al., 2019).

Algae can be defined as a true and natural biofactory of bioactive chemicals and vitamins for dietary consumption in biotechnology. Some vitamins (E, A, and C) have been the subject of numerous studies, vitamins A, D, E, K, C, B1, B2, B3, B5, B6, B7, B9, and B12. C, niacin, folic acid, pantothenic acid, and riboflavin make up this group (Del Mondo et al., 2020).

ALGAL BIOTECHNOLOGY

Biotechnology can be defined as: any technological application using biosystems, organisms, or derivatives thereof, to manufacture or modify bioproducts or to develop and engineer processes for specific application (Pfeifer et al., 2020). About 100 years ago, Karl Erkey defined biotechnology as, translated into English, "all the lines of work by which products are produced from raw materials with the aid of living organisms" (Keener et al., 2000). Biotechnology has had a growing impact on the environmental, agricultural, pharmaceutical, energy and industrial sectors, and has provided innovations in genetic engineering, diagnostic and tissue engineering and culture engineering (Aghmiuni et al., 2020).

Algae biotechnology continues to attract interest in sustainable bioenergy production and emission reduction(Varela Villarreal et al., 2020). Many biotechnological studies regarding microalgal strain improvement for biofuels (biodiesel, bioethanol, biogas, and biohydrogen) production have focused on either mutagenesis of single target gene or engineering of multiple genes including complete pathways involved in lipid, starch or pigment biosynthesis (Sikandar Khan & Fu, 2020).

Algal biotechnology applications comprise a wide range of activities including algal strain selection, cultivation systems, biomass harvesting, metabolites extraction, etc. (Mutanda et al., 2020). Mainly two types of algal cultivation systems are employed, both of which were extensively studied and a brief comparison of the two systems is shown in Figure 3 (Mutanda et al., 2020)(Chaumont, 1993; Y. Shen et al., 2009).

Figure 3. Comparison of the two systems is cultivation algae

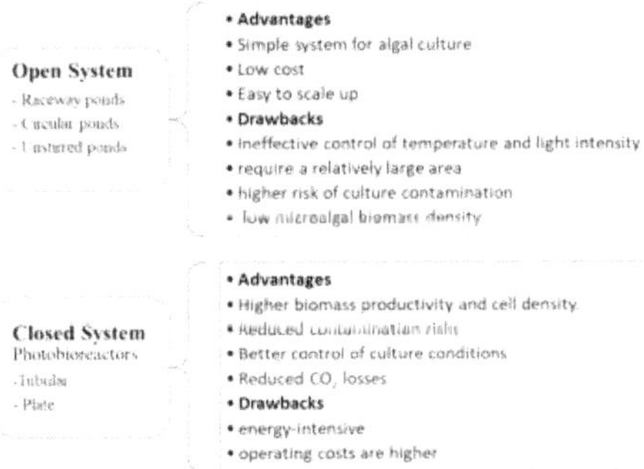

BIOFUELS AND BIOENERGY APPLICATIONS

One of the 17 Sustainable Development Goals (SDGs) set by the United Nations to address various concerns surrounding sustainable energy is affordable and clean energy (United Nations 2020). In this respect, biofuels are used as renewable fuels for on-road engines, automotive boilers, electro-generators, agricultural or architectural power sources, commercial vessels, and other applications that have properties comparable to petroleum-derived fuels such as diesel, gasoline, or fuel oil (Lin & Lu, 2021). Biofuels comprise bioethanol, biogass, biohydrogen, biomethane, biodiesel, etc. (Raheem et al., 2015). Despite the fact that various forms of feedstock are being used to generate renewable energy, algae-based bioenergy is highly attractive due to their strong photosynthesis capacity, higher rate of carbon dioxide fixation, higher biomass productivity, high lipid content, and faster growth rate(Sun et al., 2018).

Algae contain approximately 80% of the energy contained in petroleum. Algal cells contain a lipid concentration of 30%, which is higher than other sources such as soybeans and palm oils. By dry weight, microalgae have a lipid content of 30–40%, and this figure rises to 85 percent. *Botryococcus braunii* is a microalga with a hydrocarbon concentration of 30–40% that may be extracted readily(Khan et al., 2017).

Algae owing to their different biochemical composition represent an important source for biofuels production. For example, bioethanol could be produced from algal carbohydrates, whereas biodiesel could be obtained from algal oils and the leftover biomass is converted into methane or fuel oil (Saad et al., 2019).

Different pathways were proposed throughout the literature for the production of biofuels from algae. According to Raheem et al., (2018), four different pathways could be suggested for algal biofuel production. These include: (1) **Thermal breakdown** of biomass is followed by organic chemical reformation into biofuels by pyrolysis, gasification, combustion, or hydrothermal liquefaction in thermochemical conversion processing. (2) **The biochemical pathway** of adaptation involves hydrolysis of cell walls by bacteria into fermentable sugars. Fermentation mentions to the anaerobic digestion of sugars into biogas, bioethanol, or biohydrogen. (3) **Transesterification i**s a chemical reaction that produces biodiesel and glycerol by combining triglycerides with an alcohol (usually methanol or ethanol) in the presence of an acidic or basic catalyst. (4) **Photosynthetic microbial fuel cell (MFCs)** were developed in response to the oncoming energy crisis. Also, Thúy et al., (2021) According to specialists, biochemical, mechanical, and thermochemical processes are employed to produce biofuel from microalgae. Chemical pathways include torrefaction, pyrolysis, hydrothermal carbonization (HTC), hydrothermal gasification, and hydrothermal liquefaction; biochemical paths include anaerobic digestion and fermentation; and thermochemical

Figure 4. Schematic diagram of algae biofuels

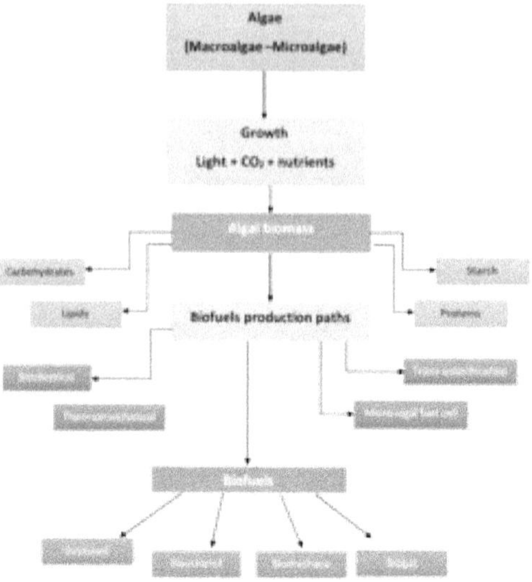

pathways include torrefaction, pyrolysis, hydrothermal carbonization (HTC), hydrothermal gasification, and hydrothermal liquefaction. Figure 4 shows a schematic diagram for algae biofuels production.

Separated lipids from microalgae can be utilized as a vital re source for generating biodiesel whereas the carbohydrate portions can be transformed to bioethanol in brewing industries (Bhattacharya & Goswami, 2020).

The yield and quality of biofuels produced from algae biomass are strongly influenced by a number of factors, including algae properties (type, chemical composition, moisture content, and particle size), process conditions (temperature, reaction time, heating rate, and carrier gas flow rate), catalyst type and dosage, and reactor configuration (Pourkarimi et al., 2019).

AGRICULTURAL APPLICATIONS

Algae biotechnology research in the field of agriculture has increased in recent years. The agricultural applications of algae as biofertilizers, plant growth biostimultants, bio-pesticides have been investigated.

Algal biofertilizers in the soil act as a slow- release fertilizer that continuously supplies nutrients to the plants. Also, algae extracts containing amino acids and minerals are applied to the surface of plant leaves liquid algal biofertilizers (Zou et al., 2021). Algal biofertilizers have several benefits: they increase the pore volume of soil, nutrients transfer, microoarganisms in soil, available nitrogen in soil, they also stabilize soil aggregates and most important algal biofertilizers fix the carbon dioxide and reduce its release to the atmosphere(Saadaoui et al., 2019; Uysal et al., 2015). Algal biofertilizers have a great advantage over chemical fertilizers, those later usually dissolve in water then settle deep in soil which makes nutrients unavailable to plants(Nosheen et al., 2021). In contrast alga biofertilizers can hold nutrient inside it and supply to plants by capillary action for a longer duration increase the water holding capacity of soil and increase the inter molecular space between soil molecule due to which proper aeration is supplied to root system (Kishimoto & Sugiura, 2014). Algal extracts act also as biostimulants of plant

Table 1. Summarizes some algae species used for the production of biofuels.

Algae	Type	Biofuel	Reference
Chlorella sp.	Microalgae	biogas	(Jehlee et al., 2017)
Nannochloropsis salina	Microalgae	biomethane	(Ma et al., 2021)
Botryococcus braunii	Microalgae	biofuel	(Tasić et al., 2016)
Chlamydomonas debaryana	Microalgae	Bio-oil	(Ansah et al., 2018)
Spirulina platensis	Microalgae	Bio-oil	(Aramkitphotha et al., 2019)
Arthrospira platensis	Microalgae	biofuel	(Serrà et al., 2020)
Porphyridium cruentum	Microalgae	Bio-ethanol	(H. M. Kim et al., 2017)
Scenedesmus sp.	Microalgae	Bio-ethanol	(H. M. Kim et al., 2017)
Nannochloropsis oculata	Microalgae	Bio-ethanol	(Reyimu & Özçimen, 2017)
Tetraselmis suecica	Microalgae	Bio-ethanol	(Reyimu & Özçimen, 2017)
Spirulina platensis	Microalgae	Bio-ethanol	(Rempel et al., 2019)
Chlorella pyrenoidosa	Microalgae	Biogas	(Prajapati et al., 2014)
Chlorella zofingiensis	Microalgae	Biogas	(Zhou et al., 2018)
Chlamydomonas reinhardtii	Microalgae	biomethane	(Klassen et al., 2017)
Cladophora sp.	Microalgae	biodiesel	(Mureed et al., 2018)
Microcystis aeruginosa	Microalgae	bioethanol	(M. I. Khan et al., 2017)
Kirchneriella lunaris	Microalgae	biodiesel	(Nascimento et al., 2013)
Ankistrodesmus fusiformis,	Microalgae	biodiesel	(Nascimento et al., 2013)
Chlamydocapsa bacillus	Microalgae	biodiesel	(Nascimento et al., 2013)
Ankistrodesmus falcatus	Microalgae	biodiesel	(Nascimento et al., 2013)
Schizocytrium *sp.*	Microalgae	bioethanol	(J. K. Kim et al., 2012)
Spirogyra Sp.	Macroalgae	biodiesel	(Ge et al., 2018)
Chara vulgaris	Macroalgae	biodiesel	(Siddiqua et al., 2015)
Macrocystis pyrifera	Macroalgae	biogas	(Fan et al., 2015)
Laminaria sp.	Macroalgae	biogas	(Montingelli et al., 2016)
Gracilaria manilaensis	Macroalgae	biomethane	(Hessami et al., 2019)
Gracilariopsis persica	Macroalgae	biomethane	(Hessami et al., 2019)
Ulva lactuca	Macroalgae	biogas	(Bruhn et al., 2011)
Ulva rigida	Macroalgae	bioethanol	(Fan et al., 2015)
Undaria pinnatifida	Macroalgae	Bio-oil	(Bae et al., 2011)
Laminaria japonica	Macroalgae	Bio-oil	(Bae et al., 2011)
Porphyra tenera	Macroalgae	Bio-oil	(Bae et al., 2011)

growth by increasing seed germination, plant growth, yield, flower set, and fruit development, as well as their post-harvest shelf life, (Michalak et al., 2016b). As bio-pesticides, algae produce biologically active secondary metabolites with proven biocidal activity. These compounds act through different forms as structural and functional modifications, disruption of the cytoplasmic membrane, enzymes inactivation and inhibition of protein synthesis in the targeted microorganisms(J. A. V. Costa et al., 2019).

The diverse positive effects reported for algal biofertilizers on soil properties is associated with the algal biochemical composition. Algal polysaccharides bind soil particles on their surface by forming a sticking mesh(Baweja et al., 2019). Moreover, the blue-green algae excrete complex organic carbon compounds that bind to the soil particles and improve soil aggregation, hence improve soil structure, soil permeability, and water-holding capacity of soil (Mobin & Chowdhury, 2019).

Many algae were used as biofertilizers, biostimulants and biopesticides around the world. Both macro- and microalgae contain numerous compounds to promote germination, leaf or stem growth, flowering and can also be used as a biological protectants agent against plant diseases (Bhattacharjee, 2016). The extracts of *Spirulina plantensis*, brown seaweed *Ascophyllum nodosum* and Baltic green macroalgae were used as biostimulants for wheat (Michalak et al., 2016a). Microalgal consortium of the two algae *Chlorella* sp. and *Scenedesmus* sp. was used as biobertlizer for tomato plants (Silambarasan et al., 2021); the seaweed *Ulva* sp. was used as an organic fertilizer for the growth of *Vigna radiata* (Akila et al., 2019); the application of *Chlorella sorokiniana* filtrates as biostimulants increased the carbon, nitrogen and phosphorus availability of calcareous soil (Marks et al., 2019); applying a mixture of microalgae consisting of *Chlorella sp., Scenedesmus sp., Spirulina sp., and Synechocystis sp.* as biostimulant showed positive effects on tomato growth(K.V. et al., 2020); *Azolla* biofertilizer achieved better nitrogen use efficiency in paddy rice fields (Yao et al., 2018); A cost-effective biopesticide for the control of *Aedes aegypti* mosquitoes was obtained from *Chlorella* sp. extract (Sigamani et al., 2020); *Amphora ovalis* algal extract spraying of onion grown under water stress increased the chlorophyll synthesis and plant growth (Sigamani et al., 2020); application of C. vulgaris as a source of macroelements resulted in improving soil nutrients and *Hibiscus sp.* Agwa et al. (2017); *Chlorella vulgaris and Chlorella* pyrenoidosa living cells were successfully used as biofertilizers for lettuce seedlings, rice, cucumber, and eggplant. (Elhafiz et al., 2015); a formulation composed of Chlorella vulgaris and plant growth-promoting bacteria showed to positive effects on lettuce (Kopta et al., 2018); also the extract from *Scenedesmus quadricauda* showed a biostimulant effect on lettuce seedlings (Puglisi et al., 2020).

MEDICINAL/PHARMACEUTICAL APPLICATIONS

Natural products originating from aquatic organisms, particularly algae (both macroalgae and microalgae), have sparked a surge in economic and scientific interest in recent years, especially as a source of food, feed and medicinal precursors (Abidizadegan et al., 2021; Pagarete et al., 2021). In this regard, a combination of green technologies and environmentally safe solvents is needed in order to enable effective extraction of biocompounds while retaining their biological qualities and allowing their use in cosmetics, and pharmaceutical industries(Chemat et al., 2020).

Today's potential pharmaceutical research is increasingly focused on the development of potent bioactive chemicals from algae. As a result, algae's unrealized potential in the field of pharmaceuticals must be further studied in order to expand and capitalize on enormous global marketing opportunities(Hamidi et al., 2020). Algae are a good source of protein, fiber, fats, and carbohydrate, as well as micronutrients (salts, minerals, and vitamins) that are important for the body's proper and healthy functioning(Leandro et al., 2020). On other hand, algae are frequently regarded as a gold mine of high-value pharmaceutically significant metabolites such as carotenoids, polyphenols, fatty acids, phycobiliproteins (Ferdous & Yusof, 2021). In Algae biological properties, both primary and secondary metabolites could be involved. Primary metabolites include proteins, polysaccharides, and lipids that have a function in physiological processes

Secondary metabolites, on the other hand, are tiny molecules such phenolic compounds, halogenated chemicals, sterols, terpenes, and short peptides(Ganesan et al., 2019). Several of these substances (oleic acid, linolenic acid, palmitoleic acid, cyanovirin, vitamin E, B12, -carotene, zeaxanthin, and lutein) have been shown to have antioxidant, anti-diabetic, antihypertensive, anti-inflammatory, antibacterial, anticancer, antiviral, fat-lowering, and neuroprotective benefits due to the bioactive substances found in them. These applications could involve both primary and secondary metabolites(Hamidi et al., 2020; Peñalver et al., 2020).

In both macro- and microalgal biotechnology, new techniques and products have been introduced, Bioactive compounds from well-studied algal forms like *Arthrospira (Spirulina), Botryococcus braunii, Chlorella vulgaris, Dunaliella salina, Haematococcus pluvialis*, and *Nostoc* have been found to have antimicrobial, antiviral, anticoagulant, antioxidant, antifungal, anti-inflammatory, and anticancer activity(Hamidi et al., 2020). Phycobiliproteins have recently been discovered to offer a variety of bioactivities, including antioxidant and radical scavenging properties, as well as anti-inflammatory and anti-cancer properties(Abidizadegan et al., 2021). The majority of phenolics are pharmacologically significant, with potent antioxidant and anti-inflammatory activities(Ruiz-Ruiz et al., 2017). Polyphenols are also found in algae, which makes them a natural supply. Pharmaceutically, polyphenolic compounds identified in macroalgae are significant for the prevention and treatment of neurodegenerative disorders. Phenolic compounds have the potential to have a favorable impact on human lifestyles when used in the creation of medicinal medications. Phenolic acids, tannic acids, flavonoids, isoflavones, cinnamic acid, benzoic acid, quercetin, and lignans are the most prevalent polyphenols present in algae and processing a variety of pharmaceutical applications (Lomartire et al., 2021).

NANOTECHNOLOGICAL APPLICATIONS

In the fields of food production, bioactive molecule synthesis, and environmental studies, algae have gathered a lot of attention (Ho et al., 2020).The utilization of bio-based systems to manufacture green metallic nanoparticles for environmental remediation has grown urgently, fueled by the growing push of green chemistry and environmental protection. It is suggested that using algae as a living cell factory or algal extract as a natural reducing agent is a green and clean technique to produce various metallic nanomaterials efficiently (S. N. Li et al., 2021). Green synthesis of nanoparticles (NPs) is an environmental friendly way for developing and producing nanomaterials with unique biological, physical, and chemical features all over the world. Algae, and cyanobacteria are among the natural sources employed in green fabrication methods (Hamida et al., 2020).

Although the mechanism for phycosynthesis of NPs is unknown, biomolecules abundant in algal membranes such as enzymes, proteins, and polysaccharides are likely to play a role. Such molecules decrease metal salts to zero valent metal. Furthermore, these compounds operate as capping agents, raising the surface charge density (zeta potential) on the nanomaterial and therefore enhancing dispersibility, due to their large structure and amphiphilic character (Gwala et al., 2021).

Algae-capped and stabilised NPs have gained widespread acceptance as a less hazardous, easy-to-handle, cost-effective, eco-friendly, and safer-to-use method in a variety of research domains in nano size. In the conversion of metal salts to metal, metal oxide, or bimetallic NPs, a natural material from algae serves as a capping, reducing, and stabilising agent. Algal biosynthesized NPs have also been studied for their biomedical applications, which include antibacterial, antioxidant, free radical scaveng-

Table 2. Bioactive compounds from algae

Algae	Bioactive compounds	Applications	Reference
Spirulina sp	Phenolic molecules, peptides	Brightness and shining effect on skin and hair	(Stirk et al., 2020)
Porphyra haitanensis	Polysaccharide, protein and essential amino acids	Protect the skin	(Geddie & Hall, 2020)
L. ochroleuca	Phenolic contents	Friendly source of antioxidant	(Otero et al., 2019)
H. pluvialis	Astaxanthin	Anti-aging natural compound activity	(Butler & Golan, 2020)
U. lactuca	Phenolic compound and pigments	Reduced destructive oxidizing agent	(Hidayati et al., 2020)
G. arcuate	Phenolic compound and pigments	Reduced destructive oxidizing agent	(Hidayati et al., 2020)
S. platensis	Phenolic compound and pigments	Reduced destructive oxidizing agent	(Hidayati et al., 2020)
Ulva pertusa	Ulvans	Antioxidant, antihyperlipidemic	(Silva et al., 2012)
Ulva lactuca	Ulvans	Antioxidant, anticoagulant,	(Silva et al., 2012)
Ulva rigida	Ulvans	antihyperlipidemic, antiviral	(Silva et al., 2012)
Ulva fasciata	Ulvans	Immunostimulating Anticoagulant	(Silva et al., 2012)
Macrocystis pyrifera	Alginates Laminarans Fucoidans	Antiallergic	(Vo et al., 2015)
Sargassum vulgare	Alginates Laminarans Fucoidans	Anticancer	(Vo et al., 2015) (Pereira & Costa-Lotufo, 2012)
Saccharina japonica	Alginates Laminarans Fucoidans	Anticoagulant, anti-inflammatory, antiangiogenic, antiviral, antithrombotic, antioxidant, antitumor	(Mayakrishnan et al., 2013)

ing, antifungal, anticancer, antiticancer, and biocomplexity properties, due to their biocompatibility and good and remarkable physicochemical properties of NPs. The rationale for algal-mediated production of diverse NPs from various algae sources has been investigated in this survey (AlNadhari et al., 2021).

Algae can be used in the production of nanomaterials (NMs) in a variety of methods, both at the molecular and cellular levels (Sharma et al., 2016). Algae offer a promising platform for the manufacture of a wide range of NMs, owing to the availability of bioactive chemicals including pigments and antioxidants in their cell extracts, which operate as biocompatible reductants (Khanna et al., 2019). Algae diversification of the nanomaterials described, such as additional oxide- and chalcogenide-based NPs, as well as new metallic and alloy-based NPs (Sharma et al., 2016).

Researchers have characterized their findings "intracellular" when NP production happens inside cells and "extracellular" when it occurs outside algal cells, even though their experimental setups do not directly include cells but rather isolated biomass or biomolecules (Dahoumane et al., 2016). Table 3 lists some algae species that were used for the synthesis of different types of nanoparticles.

Table 3. Some algae species used for the synthesis of nanoparticles

Algae	Type	Nanoparticles	Reference
Oscillatoria sancta	microalgae	Silver	(Elumalai et al., 2021)
Portieria hornemannii	Red algae	Silver	(Fatima et al., 2020)
Laurencia catarinensis	Red algae	Silver	(Abdel-Raouf *et al.*, 2018)
Spirogyra hyalina	Green algae	Silver	(Abdullah et al., 2021)
Saccharina cichorioides	Brown algae	Silver	(Yugay et al., 2020)
Fucus evanescens	Brown algae	Silver	(Yugay et al., 2020)
Gelidium corneum	Red algae	Silver	(Yılmaz Öztürk et al., 2020)
Sargassum muticum	Brown algae	Silver	(Azizi et al., 2013)
Cystoseira baccata	Brown algae	Gold	(González-Ballesteros et al., 2017)
Halymenia dilatata	Red algae	Platinum	(Sathiyaraj et al., 2021)
Sargassum ilicifolium	Brown algae	Aluminum oxide	(Koopi & Buazar, 2018)
Colpomenia sinuosa	Brown algae	Iron oxide	(Salem et al., 2019)
Pterocladia capillacea	Red algae	Iron oxide	(Salem et al., 2019)
Sargassum muticum	Brown algae	Iron oxide	(Mahdavi et al., 2013)
Padina pavonica	Brown algae	Iron oxide	(El-Kassas et al., 2016)
Sargassum acinarium	Brown algae	Iron oxide	(El-Kassas et al., 2016)
Cyanobacteria	microalgae	Silver oxide	(El-Sheekh et al., 2021)
Cyanobacteria	microalgae	Gold	(El-Sheekh et al., 2021)
Rhizoclonium fontinale	Green algae	Gold	(Parial & Pal, 2015)
Sargassum cymosum	*Brown algae*	Gold	Costa et al., 2020)
Sargassum bovinum	Brown algae	Palladium	(Momeni & Nabipour, 2015)
Chlorella vulgaris	Green algae	Palladium	(Arsiya et al., 2017)

ENVIRONMENTAL APPLICATIONS

Algae have gotten a lot of interest from evolutionists because of their fast replicability, ecological sustainability, ease of production, and ability to adapt to changing environmental conditions(Hamidi et al., 2020). Algal communities can help to create a cleaner environment, a more vibrant economy, and a better quality of life. Natural resources depletion can be slowed, and pollution can be reduced using algae(Chew et al., 2021). Algae biotechnology has recently been lauded as a feasible method to reduce climate change and regulate pollution by fixing CO_2 in the atmosphere and recovering organics from saltwater (J. Li et al., 2020). Algae-based wastewater treatment can both purify wastewater and alleviate pollution issues. Algae have been demonstrated to be effective in the treatment of municipal, agricultural, and industrial wastewater, as well as in the conversion of nutrients to biomass. Scientists are interested in the ability of microalgae to absorb nutrients and pollutants from wastewater (X. ya Liu & Hong, 2021).

The use of algal systems to recover resources from wastewater is considered particularly useful, and it has been recommended as a way to build a circular bioeconomy (Ummalyma et al., 2021). Several researches have quantified the size at which algae can contribute to carbon intake from the atmosphere, and they have discovered that the rate of uptake differs amongst organisms(Usher et al., 2014). Algae-based nutrient removal processes help with wastewater treatment by recycling nutrients, reducing the environmental impact of traditional wastewater treatment, eutrophication, CO_2 sequestration, and the production of useful biomass, which is a storehouse of energy-rich compounds with applications in biofuels, aquaculture, and agriculture (Mehariya et al., 2021).

Several researches have looked at the biological removal of carbonaceous, nitrogenous, and phosphorus substances from wastewater effluents using microalgae. This has been done with a variety of microalgal species on a variety of wastewater types, including municipal, agricultural, brewery, refinery, and industrial effluents, with different treatment efficacy and microalgae growth efficiency (Fatemeh et al., 2021).

Growing microalgae on wastewater for heavy metal removal has various benefits over previous methods, including quick metal absorption, energy savings, environmentally benign use, minimal implementation costs, and the capacity to absorb both high and low concentration levels of metals. Microalgae have particular benefits over typical treatment approaches, making it a good option for bioremediation of contaminated wastewater. Various microalgae species (*Chlorella vulgaris*, *Spirulina maxima, Spirulina platensis, Scenedesmus quadricauda, Coelastrum* sp., etc.) have showed the potential to remove heavy metals from the environment (Li et al., 2020).

Microalgae-based wastewater treatment has been extensively researched and has been shown to be more efficient, cost-effective, and environmentally friendly than standard techniques for eliminating organic contaminants, particularly nitrogen and phosphorus. In wastewater from the food processing sector, carpet mill, farm, and municipal effluent treatment facilities, microalgae species such as *Chlorella, Scenedesmus, Botrycoccus,* and *Chlamydomonas* accumulated 25-55 percent lipid (Hu et al., 2019).

The use of micro and macroalgae to remove metal ions from aqueous solutions has benefits over other approaches. Algae are more promising and novel option for the removal of pollutants due to their low cost raw material and cultivation, environmental friendliness, high adsorbing capacity and metal ion uptake, high metal selectivity, no secondary pollution, and special mechanical properties for large scale production (Ameri et al., 2020).

CONCLUSION

"Algae", these photosynthetic organisms found in both marine and freshwater environments are valuable biological resources with numerous biotechnological applications. They have different mechanisms for fixing atmospheric carbon dioxide and efficiently converting it into biomass by utilizing nutrients. Owing to their biochemical composition, algae are considered as a biofactory for a huge number of compounds with diverse applications. Algae contain a variety of bioactive chemicals and secondary metabolites that can be used in a variety of industries, including pharmaceuticals, nutraceuticals, cosmetics, fertilizers, and food. Algae are rich in dietary fibers, antioxidants, essential amino acids, phytochemicals, vitamins, polyunsaturated fatty acids, and minerals. The biotechnological applications of algae are evolving and algae are almost employed in different sectors. Scientists are continuously debating the challenges for promoting algae as a relevant biotechnological resource. This chapter focused on the biochemistry and biotechnology aspects of algae and their environmental, agricultural, pharmacological, and industrial applications for human welfare.

REFERENCES

Abdel-Raouf, N., Alharbi, R. M., Al-Enazi, N. M., Alkhulaifi, M. M., & Ibraheem, I. B. M. (2018). Rapid biosynthesis of silver nanoparticles using the marine red alga Laurencia catarinensis and their characterization. *Beni-Suef University Journal of Basic and Applied Sciences*, *7*(1), 150–157. doi:10.1016/j.bjbas.2017.10.003

Abdullah, A.-R., Al-Radadi, N. S., Hussain, T., Faisal, S., & Ali Raza Shah, S. (2022, January). Novel biosynthesis, characterization and bio-catalytic potential of green algae (Spirogyra hyalina) mediated silver nanomaterials. *Saudi Journal of Biological Sciences*, *29*(1), 411–419. Advance online publication. doi:10.1016/j.sjbs.2021.09.013 PMID:35002436

Abidizadegan, M., Peltomaa, E., & Blomster, J. (2021). The Potential of Cryptophyte Algae in Biomedical and Pharmaceutical Applications. *Frontiers in Pharmacology*, *11*(February), 1–17. doi:10.3389/fphar.2020.618836 PMID:33603668

Aditya, T., & Bitu, G., & G, M. E. (2016). The Role of Algae in Pharmaceutical Development. *Special Issue: Reviews on Pharmaceutics and Nanotechnology Research & Reviews: Journal of Pharmaceutics and Nanotechnology*, *4*, 82–89.

Aghmiuni, S. K., Siyal, S., Wang, Q., & Duan, Y. (2020). Assessment of factors affecting innovation policy in biotechnology. *Journal of Innovation and Knowledge*, *5*(3), 180–190. doi:10.1016/j.jik.2019.10.002

Akila, V., Manikandan, A., Sahaya Sukeetha, D., Balakrishnan, S., Ayyasamy, P. M., & Rajakumar, S. (2019). Biogas and biofertilizer production of marine macroalgae: An effective anaerobic digestion of Ulva sp. *Biocatalysis and Agricultural Biotechnology*, *18*, 101035. doi:10.1016/j.bcab.2019.101035

AlNadhari, S., Al-Enazi, N. M., Alshehrei, F., & Ameen, F. (2021). A review on biogenic synthesis of metal nanoparticles using marine algae and its applications. *Environmental Research, 194*(October), 110672. doi:10.1016/j.envres.2020.110672

Ameri, A., Tamjidi, S., Dehghankhalili, F., Farhadi, A., & Saati, M. A. (2020). Application of algae as low cost and effective bio-adsorbent for removal of heavy metals from wastewater: A review study. *Environmental Technology Reviews*, *9*(1), 85–110. doi:10.1080/21622515.2020.1831619

Ansah, E., Wang, L., Zhang, B., & Shahbazi, A. (2018). Catalytic pyrolysis of raw and hydrothermally carbonized Chlamydomonas debaryana microalgae for denitrogenation and production of aromatic hydrocarbons. *Fuel*, *228*, 234–242. doi:10.1016/j.fuel.2018.04.163

Aramkitphotha, S., Tanatavikorn, H., Yenyuak, C., & Vitidsant, T. (2019). Low sulfur fuel oil from blends of microalgae pyrolysis oil and used lubricating oil: Properties and economic evaluation. *Sustainable Energy Technologies and Assessments*, *31*, 339–346. doi:10.1016/j.seta.2018.12.019

Arsiya, F., Sayadi, M. H., & Sobhani, S. (2017). Green synthesis of palladium nanoparticles using Chlorella vulgaris. *Materials Letters*, *186*, 113–115. doi:10.1016/j.matlet.2016.09.101

Azizi, S., Namvar, F., Mahdavi, M., Ahmad, M., & Mohamad, R. (2013). Biosynthesis of silver nanoparticles using brown marine macroalga, Sargassum muticum aqueous extract. *Materials (Basel)*, *6*(12), 5942–5950. doi:10.3390/ma6125942 PMID:28788431

Bae, Y. J., Ryu, C., Jeon, J. K., Park, J., Suh, D. J., Suh, Y. W., Chang, D., & Park, Y. K. (2011). The characteristics of bio-oil produced from the pyrolysis of three marine macroalgae. *Bioresource Technology*, *102*(3), 3512–3520. doi:10.1016/j.biortech.2010.11.023 PMID:21129955

Barbosa, M., Valentão, P., & Andrade, P. B. (2014). Bioactive compounds from macroalgae in the new millennium: Implications for neurodegenerative diseases. *Marine Drugs*, *12*(9), 4934–4972. doi:10.3390/md12094934 PMID:25257784

Baweja, P., Kumar, S., & Kumar, G. (2019). Organic Fertilizer from Algae: A Novel Approach Towards Sustainable Agriculture. In B. Giri, R. Prasad, Q.-S. Wu, & A. Varma (Eds.), *Biofertilizers for Sustainable Agriculture and Environment* (pp. 353–370). Springer International Publishing. doi:10.1007/978-3-030-18933-4_16

Bhattacharjee, M. (2016). Pharmaceutically valuable bioactive compounds of algae. *Asian Journal of Pharmaceutical and Clinical Research*, *9*(6), 43–47. doi:10.22159/ajpcr.2016.v9i6.14507

Bhattacharya, M., & Goswami, S. (2020). Microalgae – A green multi-product biorefinery for future industrial prospects. *Biocatalysis and Agricultural Biotechnology*, *25*(March), 101580. doi:10.1016/j.bcab.2020.101580

Bruhn, A., Dahl, J., Nielsen, H. B., Nikolaisen, L., Rasmussen, M. B., Markager, S., Olesen, B., Arias, C., & Jensen, P. D. (2011). Bioenergy potential of Ulva lactuca: Biomass yield, methane production and combustion. *Bioresource Technology*, *102*(3), 2595–2604. doi:10.1016/j.biortech.2010.10.010 PMID:21044839

Butler, T., & Golan, Y. (2020). Astaxanthin Production from Microalgae. In M. A. Alam, J.-L. Xu, & Z. Wang (Eds.), *Microalgae Biotechnology for Food, Health and High Value Products* (pp. 175–242). Springer Singapore. doi:10.1007/978-981-15-0169-2_6

Cabej, N. R. (2013). Epigenetics of Reproduction in Animals. In *Building the Most Complex Structure on Earth* (pp. 59–120). Elsevier. doi:10.1016/B978-0-12-401667-5.00002-X

Cahoon, A. B., & Timko, M. P. (2003). *Biochemistry and Regulation of Chlorophyll Biosynthesis*. Springer. doi:10.1007/978-94-007-1038-2_6

Carvalho, A. P., Monteiro, C. M., & Malcata, F. X. (2009). Simultaneous effect of irradiance and temperature on biochemical composition of the microalga Pavlova lutheri. *Journal of Applied Phycology*, *21*(5), 543–552. doi:10.100710811-009-9415-z

Chai, W. S., Tan, W. G., Halimatul Munawaroh, H. S., Gupta, V. K., Ho, S. H., & Show, P. L. (2021). Multifaceted roles of microalgae in the application of wastewater biotreatment: A review. *Environmental Pollution*, *269*, 116236. doi:10.1016/j.envpol.2020.116236 PMID:33333449

Chaisutyakorn, P., Praiboon, J., & Kaewsuralikhit, C. (2017). The effect of temperature on growth and lipid and fatty acid composition on marine microalgae used for biodiesel production. *Journal of Applied Phycology, 30*(1), 37–45. doi:10.1007/s10811-017-1186-3

Chaumont, D. (1993). Biotechnology of algal biomass production: A review of systems for outdoor mass culture. *Journal of Applied Phycology*, *5*(6), 593–604. doi:10.1007/BF02184638

Chemat, F., Abert Vian, M., Fabiano-Tixier, A. S., Nutrizio, M., Režek Jambrak, A., Munekata, P. E. S., Lorenzo, J. M., Barba, F. J., Binello, A., & Cravotto, G. (2020). A review of sustainable and intensified techniques for extraction of food and natural products. *Green Chemistry*, *22*(8), 2325–2353. doi:10.1039/C9GC03878G

Chew, K. W., Khoo, K. S., Foo, H. T., Chia, S. R., Walvekar, R., & Lim, S. S. (2021). Algae utilization and its role in the development of green cities. *Chemosphere*, *268*, 129322. Advance online publication. doi:10.1016/j.chemosphere.2020.129322 PMID:33359993

Cole, L. A. (2016). Evolution of Chemical, Prokaryotic, and Eukaryotic Life. In *Biology of Life* (pp. 93–99). Elsevier. doi:10.1016/B978-0-12-809685-7.00013-7

Costa, J. A. V., Freitas, B. C. B., Cruz, C. G., Silveira, J., & Morais, M. G. (2019). Potential of microalgae as biopesticides to contribute to sustainable agriculture and environmental development. *Journal of Environmental Science and Health. Part B, Pesticides, Food Contaminants, and Agricultural Wastes*, *54*(5), 366–375. doi:10.1080/03601234.2019.1571366 PMID:30729858

Costa, L., Hemmer, J., Hillmann Wanderlind, E., Gerlach, O., Santos, A., Tamanaha, M., Bella-Cruz, A., Correa, R., Bazani, H., Radetski, C., & Almerindo, G. (2020). Green Synthesis of Gold Nanoparticles Obtained from Algae Sargassum cymosum: Optimization, Characterization and Stability. *BioNanoScience*, *10*(4), 1049–1062. Advance online publication. doi:10.100712668-020-00776-4

Dahoumane, S. A., Mechouet, M., Alvarez, F. J., Agathos, S. N., & Jeffryes, C. (2016). Microalgae: An outstanding tool in nanotechnology. *Revista Bionatura*, *1*(4), 196–201. doi:10.21931/RB/2016.01.04.7

Del Mondo, A., Smerilli, A., Sané, E., Sansone, C., & Brunet, C. (2020). Challenging microalgal vitamins for human health. *Microbial Cell Factories*, *19*(1), 1–23. doi:10.118612934-020-01459-1 PMID:33138823

El-Kassas, H. Y., Aly-Eldeen, M. A., & Gharib, S. M. (2016). Green synthesis of iron oxide (Fe3O4) nanoparticles using two selected brown seaweeds: Characterization and application for lead bioremediation. *Acta Oceanologica Sinica*, *35*(8), 89–98. doi:10.100713131-016-0880-3

El-Sheekh, M. M., Hassan, L. H. S., & Morsi, H. H. (2021). Assessment of the in vitro anticancer activities of cyanobacteria mediated silver oxide and gold nanoparticles in human colon CaCo-2 and cervical HeLa cells. *Environmental Nanotechnology, Monitoring & Management*, *16*, 100556. doi:10.1016/j.enmm.2021.100556

Elhafiz, A. A., Elhafiz, A. A., Gaur, S. S., & Hamdany, N. (2015). *Biofertilizer to grow rice, lettuce, cucumber*. doi:10.19071/rrst.2015.v7.2919

Elumalai, D., Hemavathi, M., Rekha, G. S., Pushphalatha, M., Leelavathy, R., Vignesh, A., Ashok, K., & Babu, M. (2021). Photochemical synthesizes of silver nanoparticles using Oscillatoria sancta micro algae against mosquito vectors Aedes aegypti and Anopheles stephensi. *Sensing and Bio-Sensing Research*, *100457*, 100457. Advance online publication. doi:10.1016/j.sbsr.2021.100457

Fabris, M., Abbriano, R. M., Pernice, M., Sutherland, D. L., Commault, A. S., Hall, C. C., Labeeuw, L., McCauley, J. I., Kuzhiuparambil, U., Ray, P., Kahlke, T., & Ralph, P. J. (2020). Emerging Technologies in Algal Biotechnology: Toward the Establishment of a Sustainable, Algae-Based Bioeconomy. *Frontiers in Plant Science*, *11*, 279. doi:10.3389/fpls.2020.00279 PMID:32256509

Fan, Y., Guo, R., Yuan, X., Qiu, Y., Yang, Z., Wang, F., Sun, M., & Zhao, X. (2015). Biogas production from Macrocystis pyrifera biomass in seawater system. *Bioresource Technology*, *197*, 339–347. doi:10.1016/j.biortech.2015.08.128 PMID:26344241

Fatemeh, S., Hennige, S., Willoughby, N., Adeloye, A., & Gutierrez, T. (2021). Integrating micro-algae-micro-algae into wastewater treatment : A review. *Science of the Total Environment, 752*(September), 142168. doi:10.1016/j.scitotenv.2020.142168

Fatima, R., Priya, M., Indurthi, L., Radhakrishnan, V., & Sudhakaran, R. (2020). Biosynthesis of silver nanoparticles using red algae Portieria hornemannii and its antibacterial activity against fish pathogens. *Microbial Pathogenesis*, *138*, 103780. doi:10.1016/j.micpath.2019.103780 PMID:31622663

Ferdous, U. T., & Yusof, Z. N. B. (2021). Medicinal Prospects of Antioxidants From Algal Sources in Cancer Therapy. *Frontiers in Pharmacology*, *12*(March), 1–22. doi:10.3389/fphar.2021.593116 PMID:33746748

Freile-Pelegrín, Y., & Robledo, D. (2013). Bioactive Phenolic Compounds from Algae. *Bioactive Compounds from Marine Foods: Plant and Animal Sources*, (May), 113–129. doi:10.1002/9781118412893.ch6

Ganesan, A. R., Tiwari, U., & Rajauria, G. (2019). Seaweed nutraceuticals and their therapeutic role in disease prevention. *Food Science and Human Wellness*, *8*(3), 252–263. doi:10.1016/j.fshw.2019.08.001

Ge, S., Madill, M., & Champagne, P. (2018). Use of freshwater macroalgae Spirogyra sp. for the treatment of municipal wastewaters and biomass production for biofuel applications. *Biomass and Bioenergy*, *111*, 213–223. doi:10.1016/j.biombioe.2017.03.014

Geddie, A. W., & Hall, S. G. (2020). Development of a suitability assessment model for the cultivation of intertidal macroalgae in the United States. *The Science of the Total Environment*, *699*, 134327. doi:10.1016/j.scitotenv.2019.134327 PMID:31522045

Gest, H., Engelmann, T., Kamen, M., Mcmillan, C., & Barnes, C. R. (2002). History of the word photosynthesis and evolution of its definition. In Photosynthesis Research (Vol. 73). Academic Press.

Goncalves, E. C., Wilkie, A. C., Kirst, M., & Rathinasabapathi, B. (2016). Metabolic regulation of triacylglycerol accumulation in the green algae: Identification of potential targets for engineering to improve oil yield. *Plant Biotechnology Journal*, *14*(8), 1649–1660. doi:10.1111/pbi.12523 PMID:26801206

González-Ballesteros, N., Prado-López, S., Rodríguez-González, J. B., Lastra, M., & Rodríguez-Argüelles, M. C. (2017). Green synthesis of gold nanoparticles using brown algae Cystoseira baccata: Its activity in colon cancer cells. *Colloids and Surfaces. B, Biointerfaces*, *153*, 190–198. doi:10.1016/j.colsurfb.2017.02.020 PMID:28242372

Gwala, M., Dutta, S., & Ghosh Chaudhuri, R. (2021). Microalgae Mediated Nanomaterials Synthesis. In S. K. Mandotra, A. K. Upadhyay, & A. S. Ahluwalia (Eds.), *Algae: Multifarious Applications for a Sustainable World* (pp. 295–324). Springer Singapore. doi:10.1007/978-981-15-7518-1_13

Hamida, R. S., Ali, M. A., Redhwan, A., & Bin-Meferij, M. M. (2020). Cyanobacteria – a promising platform in green nanotechnology: A review on nanoparticles fabrication and their prospective applications. *International Journal of Nanomedicine*, *15*, 6033–6066. doi:10.2147/IJN.S256134 PMID:32884261

Hamidi, M., Safarzadeh Kozani, P., Safarzadeh Kozani, P., Pierre, G., Michaud, P., & Delattre, C. (2020). Marine bacteria versus microalgae: Who is the best for biotechnological production of bioactive compounds with antioxidant properties and other biological applications? *Marine Drugs*, *18*(1), 1–38. doi:10.3390/md18010028 PMID:31905716

Hentati, F., Tounsi, L., Djomdi, D., Pierre, G., Delattre, C., Ursu, A. V., Fendri, I., Abdelkafi, S., & Michaud, P. (2020). Bioactive polysaccharides from seaweeds. *Molecules (Basel, Switzerland)*, *25*(14), 1–29. doi:10.3390/molecules25143152 PMID:32660153

Hessami, M. J., Phang, S. M., Sohrabipoor, J., Zafar, F. F., & Aslanzadeh, S. (2019). The bio-methane potential of whole plant and solid residues of two species of red seaweeds: Gracilaria manilaensis and Gracilariopsis persica. *Algal Research*, *42*, 101581. doi:10.1016/j.algal.2019.101581

Hidayati, J. R., Yudiati, E., Pringgenies, D., Oktaviyanti, D. T., & Kusuma, A. P. (2020). Comparative Study on Antioxidant Activities, Total Phenolic Compound and Pigment Contents of Tropical Spirulina platensis, Gracilaria arcuata and Ulva lactuca Extracted in Different Solvents Polarity. *E3S Web of Conferences, 147*. doi:10.1051/e3sconf/202014703012

Ho, S.-H., Zhang, C., Tao, F., Zhang, C., & Chen, W.-H. (2020). Microalgal Torrefaction for Solid Biofuel Production. *Trends in Biotechnology*, *38*(9), 1023–1033. doi:10.1016/j.tibtech.2020.02.009 PMID:32818442

Hu, X., Meneses, Y. E., Stratton, J., & Wang, B. (2019). Acclimation of consortium of micro-algae help removal of organic pollutants from meat processing wastewater. *Journal of Cleaner Production*, *214*, 95–102. doi:10.1016/j.jclepro.2018.12.255

Jehlee, A., Khongkliang, P., & O-Thong, S. (2017). Biogas Production from Chlorella sp. TISTR 8411 Biomass Cultivated on Biogas Effluent of Seafood Processing Wastewater. *Energy Procedia*, *138*, 853–857. doi:10.1016/j.egypro.2017.10.095

Juneja, A., Ceballos, R. M., & Murthy, G. S. (2013). Effects of environmental factors and nutrient availability on the biochemical composition of algae for biofuels production: A review. *Energies*, *6*(9), 4607–4638. doi:10.3390/en6094607

Keener, K., Hoban, T., & Balasubramanian, R. (2000). Biotechnology and its applications. *The North Carolina Cooperative Extension Service, D*, 1–13. https://www.ces.ncsu.edu/depts/foodsci/ext/pubs/bioapp.html

Khan, M. I., Lee, M. G., Shin, J. H., & Kim, J. D. (2017). Pretreatment optimization of the biomass of Microcystis aeruginosa for efficient bioethanol production. *AMB Express*, *7*(1), 1–9. doi:10.118613568-016-0320-y PMID:28063146

Khan, S. Siddique, R., Sajjad, W., Nabi, G., Hayat, K. M., Duan, P., & Yao, L. (2017). Biodiesel Production From Algae to Overcome the Energy Crisis. *HAYATI Journal of Biosciences, 24*(4), 163–167.

Khan, S., & Fu, P. (2020). Biotechnological perspectives on algae: a viable option for next generation biofuels. In *Current Opinion in Biotechnology* (Vol. 62, pp. 146–152). Elsevier Ltd. doi:10.1016/j.copbio.2019.09.020

Khanna, P., Kaur, A., & Goyal, D. (2019). Algae-based metallic nanoparticles: Synthesis, characterization and applications. *Journal of Microbiological Methods*, *163*(June), 105656. doi:10.1016/j.mimet.2019.105656 PMID:31220512

Kim, H. M., Oh, C. H., & Bae, H. J. (2017). Comparison of red microalgae (Porphyridium cruentum) culture conditions for bioethanol production. *Bioresource Technology*, *233*, 44–50. doi:10.1016/j.biortech.2017.02.040 PMID:28258995

Kim, J. K., Um, B. H., & Kim, T. H. (2012). Bioethanol production from micro-algae, Schizocytrium sp., using hydrothermal treatment and biological conversion. *Korean Journal of Chemical Engineering*, *29*(2), 209–214. doi:10.100711814-011-0169-3

Kishimoto, S., & Sugiura, G. (2014). Algae as a soil conditioner. *Int Achieve Future*, *5*(4), 12–23.

Klassen, V., Blifernez-Klassen, O., Wibberg, D., Winkler, A., Kalinowski, J., Posten, C., & Kruse, O. (2017). Highly efficient methane generation from untreated microalgae biomass. *Biotechnology for Biofuels*, *10*(1), 186. doi:10.118613068-017-0871-4 PMID:28725266

Koopi, H., & Buazar, F. (2018). A novel one-pot biosynthesis of pure alpha aluminum oxide nanoparticles using the macroalgae Sargassum ilicifolium: A green marine approach. *Ceramics International*, *44*(8), 8940–8945. doi:10.1016/j.ceramint.2018.02.091

Kopta, T., Pavlíková, M., Sękara, A., Pokluda, R., & Maršálek, B. (2018). Effect of Bacterial-algal Biostimulant on the Yield and Internal Quality of Lettuce (Lactuca sativa L.). *Produced for Spring and Summer Crop.*, *46*(2), 615–621. doi:10.15835/nbha46211110

Kumar, A. (2021). Current and Future Perspective of Microalgae for Simultaneous Wastewater Treatment and Feedstock for Biofuels Production. *Chemistry Africa*, (2), 249–275. Advance online publication. doi:10.100742250-020-00221-9

K.V., S., Behera, B., & P., B. (2020). Efficacy of microalgal extracts as biostimulants through seed treatment and foliar spray for tomato cultivation. *Industrial Crops and Products*, *151*, 112453. doi:10.1016/j.indcrop.2020.112453

Leandro, A., Pacheco, D., Cotas, J., Marques, J. C., Pereira, L., & Gonçalves, A. M. M. (2020). Seaweed's bioactive candidate compounds to food industry and global food security. In Life (Vol. 10, Issue 8). doi:10.3390/life10080140

Li, H., Watson, J., Zhang, Y., Lu, H., & Liu, Z. (2020). Environment-enhancing process for algal wastewater treatment, heavy metal control and hydrothermal biofuel production: A critical review. *Bioresource Technology, 298*(September), 122421. doi:10.1016/j.biortech.2019.122421

Li, J., Wu, X., Zhong, Y., Lu, Q., & Zhou, W. (2020). The 10th Asia-Pacific Conference on Algal Biotechnology: Thoughts and comments. *Journal of Cleaner Production, 264*. 10.1016/j.jclepro.2020.121626

Li, S. N., Wang, R., & Ho, S. H. (2021). Algae-mediated biosystems for metallic nanoparticle production: From synthetic mechanisms to aquatic environmental applications. *Journal of Hazardous Materials*, *420*(January), 126625. doi:10.1016/j.jhazmat.2021.126625 PMID:34329084

Lin, C. Y., & Lu, C. (2021). Development perspectives of promising lignocellulose feedstocks for production of advanced generation biofuels: A review. *Renewable and Sustainable Energy Reviews, 136*(September), 110445. doi:10.1016/j.rser.2020.110445

Liu, J., Pemberton, B., Lewis, J., Scales, P. J., & Martin, G. J. O. (2019). Wastewater treatment using filamentous algae – A review. In *Bioresource Technology* (Vol. 298, p. 122556). Elsevier Ltd. doi:10.1016/j.biortech.2019.122556

Liu, X., & Hong, Y. (2021). Microalgae-Based Wastewater Treatment and Recovery with Biomass and Value-Added Products: A Brief Review. *Current Pollution Reports*, *7*(2), 227–245. doi:10.100740726-021-00184-6

Lomartire, S., Cotas, J., Pacheco, D., Marques, J. C., Pereira, L., & Gonçalves, A. M. M. (2021). Environmental impact on seaweed phenolic production and activity: An important step for compound exploitation. *Marine Drugs*, *19*(5), 1–20. doi:10.3390/md19050245 PMID:33926129

Los, D. A., Mironov, K. S., & Allakhverdiev, S. I. (2013). Regulatory role of membrane fluidity in gene expression and physiological functions. *Photosynthesis Research*, *116*(2–3), 489–509. doi:10.100711120-013-9823-4 PMID:23605242

Ma, J., Li, L., Zhao, Q., Yu, L., & Frear, C. (2021). Biomethane production from whole and extracted algae biomass: Long-term performance evaluation and microbial community dynamics. *Renewable Energy*, *170*, 38–48. doi:10.1016/j.renene.2021.01.113

Mahdavi, M., Namvar, F., Ahmad, M., & Mohamad, R. (2013). Green biosynthesis and characterization of magnetic iron oxide (Fe 3O4) nanoparticles using seaweed (Sargassum muticum) aqueous extract. *Molecules (Basel, Switzerland)*, *18*(5), 5954–5964. doi:10.3390/molecules18055954 PMID:23698048

Marks, E. A. N., Montero, O., & Rad, C. (2019). The biostimulating effects of viable microalgal cells applied to a calcareous soil: Increases in bacterial biomass, phosphorus scavenging, and precipitation of carbonates. *The Science of the Total Environment*, *692*, 784–790. doi:10.1016/j.scitotenv.2019.07.289 PMID:31539985

Mayakrishnan, V., Kannappan, P., Abdullah, N., & Ahmed, A. B. A. (2013). Cardioprotective activity of polysaccharides derived from marine algae: An overview. *Trends in Food Science & Technology*, *30*(2), 98–104. doi:10.1016/j.tifs.2013.01.007

Mehariya, S., Goswami, R. K., Verma, P., Lavecchia, R., & Zuorro, A. (2021). Integrated approach for wastewater treatment and biofuel production in microalgae biorefineries. *Energies*, *14*(8), 2282. Advance online publication. doi:10.3390/en14082282

Metsoviti, M. N., Papapolymerou, G., Karapanagiotidis, I. T., & Katsoulas, N. (2020). Effect of Light Intensity and Quality on Growth Rate and Composition of Chlorella vulgaris. *Plants*, *9*(1), 31. Advance online publication. doi:10.3390/plants9010031 PMID:31878279

Michalak, I., Chojnacka, K., Dmytryk, A., Wilk, R., Gramza, M., & Rój, E. (2016). Evaluation of supercritical extracts of algae as biostimulants of plant growth in field trials. *Frontiers in Plant Science, 7*. doi:10.3389/fpls.2016.01591

Mohin, A., & Chowdhury, H. (2019). ScienceDirect ScienceDirect ScienceDirect ScienceDirect Commercially important bioproducts from microalgae and their International bioproducts Commercially important microalgae and their current applications – from A review current applications – A review Assessing the. *Energy Procedia, 160*(2018), 752–760. doi:10.1016/j.egypro.2019.02.183

Mofeed, J., & Deyab, M. (2015). Monitoring for the abundance and distribution of macroalgae along Suez Canal, Egypt 1. *CATRINA*, *11*(1), 81–91.

Mohanty, S., Pradhan, B., Patra, S., Behera, C., Nayak, R., & Jena, M. (2020). Screening for nutritive bioactive compounds in some algal strains isolated from coastal Odisha. *Journal of Advanced Plant Sciences*, *2020*(2), 1–8.

Momeni, S., & Nabipour, I. (2015). A Simple Green Synthesis of Palladium Nanoparticles with Sargassum Alga and Their Electrocatalytic Activities Towards Hydrogen Peroxide. *Applied Biochemistry and Biotechnology*, *176*(7), 1937–1949. doi:10.100712010-015-1690-3 PMID:26041058

Montingelli, M. E., Benyounis, K. Y., Quilty, B., Stokes, J., & Olabi, A. G. (2016). Optimisation of biogas production from the macroalgae Laminaria sp. at different periods of harvesting in Ireland. *Applied Energy*, *177*, 671–682. doi:10.1016/j.apenergy.2016.05.150

Mureed, K., Kanwal, S., Hussain, A., Noureen, S., Hussain, S., Ahmad, S., Ahmad, M., & Waqas, R. (2018). Biodiesel production from algae grown on food industry wastewater. *Environmental Monitoring and Assessment*, *190*(5), 1–11. doi:10.100710661-018-6641-3 PMID:29633020

Mutanda, T., Naidoo, D., Bwapwa, J. K., & Anandraj, A. (2020). Biotechnological Applications of Microalgal Oleaginous Compounds: Current Trends on Microalgal Bioprocessing of Products. *Frontiers in Energy Research*, *8*(October), 598803. Advance online publication. doi:10.3389/fenrg.2020.598803

Nalley, J. O., O'Donnell, D. R., & Litchman, E. (2018). Temperature effects on growth rates and fatty acid content in freshwater algae and cyanobacteria. *Algal Research*, *35*, 500–507. doi:10.1016/j.algal.2018.09.018

Nascimento, I. A., Marques, S. S. I., Cabanelas, I. T. D., Pereira, S. A., Druzian, J. I., de Souza, C. O., Vich, D. V., de Carvalho, G. C., & Nascimento, M. A. (2013). Screening Microalgae Strains for Biodiesel Production: Lipid Productivity and Estimation of Fuel Quality Based on Fatty Acids Profiles as Selective Criteria. *BioEnergy Research*, *6*(1), 1–13. doi:10.100712155-012-9222-2

Nosheen, S., Ajmal, I., & Song, Y. (2021). Microbes as biofertilizers, a potential approach for sustainable crop production. *Sustainability (Switzerland)*, *13*(4), 1–20. doi:10.3390u13041868

Nzayisenga, J. C., Farge, X., Groll, S. L., & Sellstedt, A. (2020). Effects of light intensity on growth and lipid production in microalgae grown in wastewater. *Biotechnology for Biofuels, 13*(1), 1–8. doi:10.1186/s13068-019-1646-x

Otero, P., López-Martínez, M. I., & García-Risco, M. R. (2019). Application of pressurized liquid extraction (PLE) to obtain bioactive fatty acids and phenols from Laminaria ochroleuca collected in Galicia (NW Spain). *Journal of Pharmaceutical and Biomedical Analysis*, *164*, 86–92. doi:10.1016/j.jpba.2018.09.057 PMID:30359842

Pagarete, A., Ramos, A. S., Puntervoll, P., Allen, M. J., & Verdelho, V. (2021). Antiviral Potential of Algal Metabolites-A Comprehensive Review. *Marine Drugs*, *19*(2), 1–23. doi:10.3390/md19020094 PMID:33562153

Panahi, Y., Khosroushahi, A. Y., Sahebkar, A., & Heidari, H. R. (2019). Impact of Cultivation Condition and Media Content onChlorella vulgaris Composition. *Advanced Pharmaceutical Bulletin*, *9*(2), 182–194. doi:10.15171/apb.2019.022 PMID:31380244

Parial, D., & Pal, R. (2015). Biosynthesis of monodisperse gold nanoparticles by green alga Rhizoclonium and associated biochemical changes. *Journal of Applied Phycology*, *27*(2), 975–984. doi:10.100710811-014-0355-x

Peñalver, R., Lorenzo, J. M., Ros, G., Amarowicz, R., Pateiro, M., & Nieto, G. (2020). Seaweeds as a functional ingredient for a healthy diet. *Marine Drugs*, *18*(6), 1–27. doi:10.3390/md18060301 PMID:32517092

Pereira, R. C., & Costa-Lotufo, L. V. (2012). Bioprospecting for bioactives from seaweeds:Potential, obstacles and alternatives. *Revista Brasileira de Farmacognosia*, *22*(4), 894–905. doi:10.1590/S0102-695X2012005000077

Petsas, A. S., & Vagi, M. C. (2017). Effects on the Photosynthetic Activity of Algae after Exposure to Various Organic and Inorganic Pollutants: Review. In *Chlorophyll*. InTech., doi:10.5772/67991

Pfeifer, K., Ergal, İ., Koller, M., Basen, M., Schuster, B., & Rittmann, S. K. M. R. (2020, November). Archaea Biotechnology. *Biotechnology Advances*, *47*, 107668. Advance online publication. doi:10.1016/j.biotechadv.2020.107668 PMID:33271237

Pfeiffer, T. Z., Camagakevac, I. S., Maronic, D. S., & Maksimovic, I. (2017). Regulation of Photosynthesis in Algae Under Metal Stress. *Environment and Photosynthesis a Future Prospect*, (December), 261–286.

Pfeiffer, T. Z., Ivna, Š., Maronić, D. Š., & Maksimovic, I. (2017). *Regulation of Photosynthesis in Algae Under Metal Stress "Development of new technologies for modern and sustainable production of vegetables" (TR 31036) View project*. https://www.researchgate.net/publication/322096970

Pourkarimi, S., Hallajisani, A., Alizadehdakhel, A., & Nouralishahi, A. (2019). Biofuel production through micro- and macroalgae pyrolysis – A review of pyrolysis methods and process parameters. In *Journal of Analytical and Applied Pyrolysis* (Vol. 142, p. 104599). Elsevier B.V. doi:10.1016/j.jaap.2019.04.015

Prajapati, S. K., Malik, A., & Vijay, V. K. (2014). Comparative evaluation of biomass production and bioenergy generation potential of Chlorella spp. through anaerobic digestion. *Applied Energy*, *114*, 790–797. doi:10.1016/j.apenergy.2013.08.021

Puglisi, I., Bella, E. La, Rovetto, E. I., Roberta, A., & Piero, L. (2020). *Biostimulant Effect and Biochemical Response in Lettuce Seedlings Treated with A Scenedesmus quadricauda Extract*. Academic Press.

Raheem, A., Prinsen, P., Vuppaladadiyam, A. K., Zhao, M., & Luque, R. (2018). A review on sustainable microalgae based biofuel and bioenergy production: Recent developments. *Journal of Cleaner Production*, *181*, 42–59. doi:10.1016/j.jclepro.2018.01.125

Raheem, A., Wan Azlina, W. A. K. G., Taufiq Yap, Y. H., Danquah, M. K., & Harun, R. (2015). Thermochemical conversion of microalgal biomass for biofuel production. In *Renewable and Sustainable Energy Reviews* (Vol. 49, pp. 990–999). Elsevier Ltd. doi:10.1016/j.rser.2015.04.186

Rempel, A., de Souza Sossella, F., Margarites, A. C., Astolfi, A. L., Steinmetz, R. L. R., Kunz, A., Treichel, H., & Colla, L. M. (2019). Bioethanol from Spirulina platensis biomass and the use of residuals to produce biomethane: An energy efficient approach. *Bioresource Technology*, *288*, 121588. doi:10.1016/j.biortech.2019.121588 PMID:31176943

Reyimu, Z., & Özçimen, D. (2017). Batch cultivation of marine microalgae Nannochloropsis oculata and Tetraselmis suecica in treated municipal wastewater toward bioethanol production. *Journal of Cleaner Production*, *150*, 40–46. doi:10.1016/j.jclepro.2017.02.189

Rost, B., Richter, K. U., Riebesell, U., & Hansen, P. J. (2006). Inorganic carbon acquisition in red tide dinoflagellates. *Plant, Cell & Environment*, *29*(5), 810–822. doi:10.1111/j.1365-3040.2005.01450.x PMID:17087465

Ruiz-Ruiz, J. C., Matus-Basto, A. J., Acereto-Escoffié, P., & Segura-Campos, M. R. (2017). Antioxidant and anti-inflammatory activities of phenolic compounds isolated from Melipona beecheii honey. *Food and Agricultural Immunology*, *28*(6), 1424–1437. doi:10.1080/09540105.2017.1347148

Saad, M. G., Dosoky, N. S., Zoromba, M. S., & Shafik, H. M. (2019). Algal biofuels: Current status and key challenges. *Energies*, *12*(10), 1920. Advance online publication. doi:10.3390/en12101920

Saadaoui, I., Sedky, R., Rasheed, R., Bounnit, T., Almahmoud, A., Elshekh, A., Dalgamouni, T., al Jmal, K., Das, P., & Al Jabri, H. (2019). Assessment of the algae-based biofertilizer influence on date palm (Phoenix dactylifera L.) cultivation. *Journal of Applied Phycology*, *31*(1), 457–463. doi:10.100710811-018-1539-6

Salem, D. M. S. A., Ismail, M. M., & Aly-Eldeen, M. A. (2019). Biogenic synthesis and antimicrobial potency of iron oxide (Fe3O4) nanoparticles using algae harvested from the Mediterranean Sea, Egypt. *Egyptian Journal of Aquatic Research*, *45*(3), 197–204. doi:10.1016/j.ejar.2019.07.002

Samanta, P., Jang, S., Shin, S., & Kim, J. K. (2019). Effects of pH on growth and biochemical responses in Agarophyton vermiculophyllum under different temperature conditions. *Journal of Applied Phycology, 32*(1), 499–509. doi:10.1007/s10811-019-01933-3

Sathiyaraj, G., Vinosha, M., Sangeetha, D., Manikandakrishnan, M., Palanisamy, S., Sonaimuthu, M., Manikandan, R., You, S. G., & Prabhu, N. M. (2021). Bio-directed synthesis of Pt-nanoparticles from aqueous extract of red algae Halymenia dilatata and their biomedical applications. *Colloids and Surfaces. A, Physicochemical and Engineering Aspects*, *618*, 126434. doi:10.1016/j.colsurfa.2021.126434

Serrà, A., Artal, R., García-Amorós, J., Gómez, E., & Philippe, L. (2020). Circular zero-residue process using microalgae for efficient water decontamination, biofuel production, and carbon dioxide fixation. *Chemical Engineering Journal*, *388*, 124278. doi:10.1016/j.cej.2020.124278

Sharma, A., Sharma, S., Sharma, K., Chetri, S. P. K., Vashishtha, A., Singh, P., Kumar, R., Rathi, B., & Agrawal, V. (2016). Algae as crucial organisms in advancing nanotechnology: a systematic review. In Journal of Applied Phycology (Vol. 28, Issue 3, pp. 1759–1774). Springer Netherlands. doi:10.100710811-015-0715-1

Shen, Y., Yuan, W., Pei, Z. J., Wu, Q., & Mao, E. (2009). Microalgae Mass Production Methods. *Transactions of the ASABE*, *52*(4), 1275–1287. doi:10.13031/2013.27771

Siddiqua, S., Al Mamun, A., & Enayetul Babar, S. M. (2015). Production of biodiesel from coastal macroalgae (Chara vulgaris) and optimization of process parameters using Box-Behnken design. *SpringerPlus*, *4*(1), 1–11. doi:10.118640064-015-1518-1 PMID:26636008

Sigamani, S., Chinnasamy, R., Dharmaraj, R. K., Ramamurthy, D., Devarajan, N., Narayanasamy, M., & Natarajan, H. (2020). Larvicidal potency of the extracts from Chlorella sp. against Aedes aegypti. *Biocatalysis and Agricultural Biotechnology*, *27*, 101663. doi:10.1016/j.bcab.2020.101663

Silambarasan, S., Logeswari, P., Sivaramakrishnan, R., Incharoensakdi, A., Cornejo, P., Kamaraj, B., & Chi, N. T. L. (2021). Removal of nutrients from domestic wastewater by microalgae coupled to lipid augmentation for biodiesel production and influence of deoiled algal biomass as biofertilizer for Solanum lycopersicum cultivation. *Chemosphere*, *268*, 129323. doi:10.1016/j.chemosphere.2020.129323 PMID:33359999

Silva, T. H., Alves, A., Popa, E. G., Reys, L. L., Gomes, M. E., Sousa, R. A., Silva, S. S., Mano, J. F., & Reis, R. L. (2012). Marine algae sulfated polysaccharides for tissue engineering and drug delivery approaches. *Biomatter*, *2*(4), 278–289. doi:10.4161/biom.22947 PMID:23507892

St. Onge, K. R. (2018). Photosynthesis: Light Reactions. In eLS (pp. 1–8). John Wiley & Sons, Ltd. doi:10.1002/9780470015902.a0001311.pub2

Stirk, W. A., Bálint, P., Vambe, M., Lovász, C., Molnár, Z., van Staden, J., & Ördög, V. (2020). Effect of cell disruption methods on the extraction of bioactive metabolites from microalgal biomass. *Journal of Biotechnology*, *307*, 35–43. doi:10.1016/j.jbiotec.2019.10.012 PMID:31678206

Sun, H., Zhao, W., Mao, X., Li, Y., Wu, T., & Chen, F. (2018). High-value biomass from microalgae production platforms: Strategies and progress based on carbon metabolism and energy conversion. *Biotechnology for Biofuels*, *11*(1), 1–23. doi:10.118613068-018-1225-6 PMID:30151055

Tasić, M. B., Pinto, L. F. R., Klein, B. C., Veljković, V. B., & Filho, R. M. (2016). Botryococcus braunii for biodiesel production. In *Renewable and Sustainable Energy Reviews* (Vol. 64, pp. 260–270). Elsevier Ltd. doi:10.1016/j.rser.2016.06.009

Ummalyma, S. B., Sahoo, D., & Pandey, A. (2021). Resource recovery through bioremediation of wastewaters and waste carbon by microalgae: A circular bioeconomy approach. *Environmental Science and Pollution Research International*, *28*(42), 58837–58856. Advance online publication. doi:10.100711356-020-11645-8 PMID:33527238

Usher, P. K., Ross, A. B., Camargo-Valero, M. A., Tomlin, A. S., & Gale, W. F. (2014). An overview of the potential environmental impacts of large-scale microalgae cultivation. *Biofuels*, *5*(3), 331–349. doi :10.1080/17597269.2014.913925

Uysal, O., Uysal, F. O., & Ekinci, K. (2015). Evaluation of Microalgae as Microbial Fertilizer. *European Journal of Sustainable Development*, *4*(2), 77–82. doi:10.14207/ejsd.2015.v4n2p77

Varela Villarreal, J., Burgués, C., & Rösch, C. (2020). Acceptability of genetically engineered algae biofuels in Europe: Opinions of experts and stakeholders. *Biotechnology for Biofuels*, *13*(1), 1–21. doi:10.118613068-020-01730-y PMID:32489422

Vecchi, V., Barera, S., Bassi, R., & Dall'osto, L. (2020). Potential and challenges of improving photosynthesis in algae. In Plants (Vol. 9, Issue 1). MDPI AG. doi:10.3390/plants9010067

Vo, T. S., Ngo, D. H., Kang, K. H., Jung, W. K., & Kim, S. K. (2015). The beneficial properties of marine polysaccharides in alleviation of allergic responses. *Molecular Nutrition & Food Research*, *59*(1), 129–138. doi:10.1002/mnfr.201400412 PMID:25379652

Wells, M. L., Trainer, V. L., Smayda, T. J., Karlson, B. S. O., Trick, C. G., Kudela, R. M., Ishikawa, A., Bernard, S., Wulff, A., Anderson, D. M., & Cochlan, W. P. (2015). Harmful algal blooms and climate change: Learning from the past and present to forecast the future. *Harmful Algae*, *49*, 68–93. doi:10.1016/j.hal.2015.07.009 PMID:27011761

Yaakob, M. A., Mohamed, R. M. S. R., Al-Gheethi, A., Aswathnarayana Gokare, R., & Ambati, R. R. (2021). Influence of Nitrogen and Phosphorus on Microalgal Growth, Biomass, Lipid, and Fatty Acid Production: An Overview. *Cells*, *10*(2), 10–20. doi:10.3390/cells10020393 PMID:33673015

Yao, Y., Zhang, M., Tian, Y., Zhao, M., Zeng, K., Zhang, B., Zhao, M., & Yin, B. (2018). Azolla biofertilizer for improving low nitrogen use efficiency in an intensive rice cropping system. *Field Crops Research*, *216*, 158–164. doi:10.1016/j.fcr.2017.11.020

Yılmaz Öztürk, B., Yenice Gürsu, B., & Dağ, İ. (2020). Antibiofilm and antimicrobial activities of green synthesized silver nanoparticles using marine red algae Gelidium corneum. *Process Biochemistry*, *89*, 208–219. doi:10.1016/j.procbio.2019.10.027

Yugay, Y. A., Usoltseva, R. V., Silant'ev, V. E., Egorova, A. E., Karabtsov, A. A., Kumeiko, V. V., Ermakova, S. P., Bulgakov, V. P., & Shkryl, Y. N. (2020). Synthesis of bioactive silver nanoparticles using alginate, fucoidan and laminaran from brown algae as a reducing and stabilizing agent. *Carbohydrate Polymers*, *245*, 116547. doi:10.1016/j.carbpol.2020.116547 PMID:32718640

Zhila, N. O., Kalacheva, G. S., & Volova, T. G. (2011). Effect of salinity on the biochemical composition of the alga Botryococcus braunii Kütz IPPAS H-252. *Journal of Applied Phycology*, *23*(1), 47–52. doi:10.100710811-010-9532-8

Zhou, W., Wang, Z., Xu, J., & Ma, L. (2018). Cultivation of microalgae Chlorella zofingiensis on municipal wastewater and biogas slurry towards bioenergy. *Journal of Bioscience and Bioengineering*, *126*(5), 644–648. doi:10.1016/j.jbiosc.2018.05.006 PMID:29801764

Zolotareva, E. K., Mokrosnop, V. M., & Stepanov, S. S. (2019). Polyphenol compounds of macroscopic and microscopic algae. *Algologia*, *29*(1), 3–29. doi:10.15407/alg29.01.003

Zou, Y., Zeng, Q., Li, H., Liu, H., & Lu, Q. (2021). Emerging technologies of algae-based wastewater remediation for bio-fertilizer production: A promising pathway to sustainable agriculture. *Journal of Chemical Technology and Biotechnology (Oxford, Oxfordshire)*, *96*(3), 551–563. doi:10.1002/jctb.6602

Chapter 3
Selection of Strains and Breeding of Algae

Mehwish Khalid
Institute of Environmental Sciences and Engineering (IESE), School of Civil and Environmental Engineering (SCEE), National University of Sciences and Technology (NUST), Islamabad, Pakistan

Imran Hashmi
Institute of Environmental Sciences and Engineering (IESE), School of Civil and Environmental Engineering (SCEE), National University of Sciences and Technology (NUST), Islamabad, Pakistan

ABSTRACT

Intensive research efforts are aimed at increasing and modifying the algal biomass production and selection for different purposes. Financial aspects for biomass production techniques always remain a challenge that needs to be addressed. Using cost-effective media for the growth and choosing high lipid content strain is always aimed to reduce the cost of yield of algal biomass. With each passing day, there is advancement in the use of algae for the vested interests. Different species are expected to function well at different niche and environmental conditions. Therefore, adaptation of robust method and selection of algal trait is most relevant for yielding large scale algal biomass. The overarching significance of producer strain has driven research in recent years towards genetically modified species. This chapter particularly focusses on the selection and breeding of algae like different cultivation aspects in open pond, photobioreactor, bio flocculation, and advantages and disadvantages thereof.

INTRODUCTION

Microalgae enjoy significant upper hands over other land plants for reproducing. first, they have a day to day existence pattern of not many hours or once in a while days rather than complete occasional cycles. second, their unicellular nature fundamentally helps in the scaling down of rearing methods, that lessens the expense. third, eukaryotic green growth are competent to replicate both, physically just as agamically, which accelerate the age of hereditary assortment when contrasted with the prokaryotic living beings. fourth, microalgae may be picked and evaluated for explicit aggregates applying stream cytometry or some other high throughput draws near. fifth, Ultraviolet and substance mutagenesis may

DOI: 10.4018/978-1-6684-2438-4.ch003

be further precisely utilized to microalgae than higher plants and the use of specific explicit chose tensions may incredibly upgrade the pace of strain advancement (Larkum et al., 2012).

The necessity for the examination of microalgae development boundaries under fastidious states of pH, light power, CO2 supply and blending orchestrated with speed and trustworthiness has laid the improvement of numerous sorts of photobioreactors; these are rapidly heightening in the variety of their fundamental designing plan and may make a significant impact to future algal bio-energy creation clarifications. However, a definitive field evaluation will be in enormous scope frameworks like raceways, lakes, and existing shut photobioreactors. This survey will completely explain the microalgae development, biofuel creation, importance as biofuel cells and difficulties in rearing of green growth (Larkum et al., 2012).

SIGNIFICANCE OF MICROALGAE IN HARNESSING SOLAR ENERGY

Algae are that classification of organic entities which have commonly been characterized as photograph autotrophic, shortsighted minute or plainly visible, single celled to plants consisting of many cells and are similarly profoundly power plant of energy that transform energy coming from daylight to significant biological side-effects. E.g., oil (Schenk et al., 2008). Algae are essentially water possessing life forms missing the convoluted morphological association (Barsanti and Gualtieri, 2006). They have incredible potential for creation of biofuel. Among the gathering of green growth, these microalgae are a particular gathering of unicellular creatures sufficiently capable to offer a wide scope of way outs for our energy exigences across a few pathways (Bala, 2006). They viably use CO_2 from the environment and are thus answerable for more than half of the world's all out worldwide carbon assimilation (Feng, 2011). Algae are expected as third partner of biofuels. They may assist with expanding the biomass quickly and a portion of the algal species may twofold its biomass fixation in however low as 5 hrs while some different species may be uncovering two doublings practically every day (Huber, 2009). In general, algae have the extraordinary ability to ensue oil yield and large amount of microalgae contain oil packets as their dry weight (Munir et al., 2015).

TYPES OF CULTURE COLLECTIONS

Axenicity, likewise named as sterility, is a principal state of microalgal species that uncovers a culture which is neither tainted nor developed for certain different species (Anderson, 2005). Biochemical, physiological, or atomic and natural trials for examining microalgae ought to be done in axenic conditions in order to stay away from any limitless effects impacting the result of tests. Phycologists have created a wide range of strategies for decontamination and sequestration of microalgal societies during the most recent couple of years (Rippka, 1988; Rippka et al., 1981; Waterbury and Willey, 1988; Vu et al., 2018, Waterbury and Stanier, 1981;). Under a minuscule assessment, unicellular may be extricated out and moved to some clean medium aseptically. Plate extending or splashing of arranged societies are some different methodologies utilized for such purposes, with the choices of mixes of centrifugation, filtration, or sequential weakening's, ahead of time. Treatments with anti-infection agents, higher pH or hotness shock may likewise be performed, which are particularly species explicit and require abundant exact information. Sequential weakening's for certain various blends of streaking and plate spreading is conceivably the most consistently utilized confinement strategy (Vu et al., 2018, Anderson, 2005). In any case, the accomplishment rate is enormously reliant upon the sum and measure of impurities comparable to microalgal cells. This load of strategies to keep separated axenic microalgal societies are monotonous and extremely difficult because of a few resulting cells move and hatching steps (Doppler et al., 2021).

Physical Extraction

Algal biomass is a rich wellspring of specific organically dynamic mixtures. A few methodologies have been gainfully applied in separating compounds from biomass, among various techniques dissolvable and heat/temperature treatment are renowned ones. Thinking about original strategies, handling under high tension (compressed fluid and supercritical liquid extraction) and ultrasound, microwave, and catalyst helped extraction have been considered as proficient ones. In modern production processes, phycocolloids, carotenoids, and essential unsaturated fats are of unmistakable significance. Moreover, biofuel make from green growth has likewise been completely confirmed, though phycocolloids have been effectively removed utilizing antacid hydrolysis, with respect to different parts, novel strategies demonstrated as far as adequacy, dissolvable treatment, cost and time effective usage are created. Amidist various methodologies, ultrasound-helped and SC-CO_2 extraction are reflected vowing for modern level execution (Dmytryk et al., 2018).

Fluorescence Activated Cell Sorting (FACS)

One encouraging technique to truncate the time in order to prepare axenic cultures, is fluorescence activated cell sorting (FACS). High quantities of chlorophyll within microalgal cells make them instinctively fluorescent, also known as "autofluorescence" (Pereira et al., 2018). Once the excitation of chlorophyl has been done by means of laser beam, a red color fluorescent signal is radiated which may be treated as automated cell sorting initiation (Syed et al., 2018). Downsides in gathering occasion of cell conceivably fragile ones are exorbitant shear powers, invigorated turbulences, high-recurrence throbs and electrostatic charges (Vu et al., 2018; Rutten et al., 2005). Nonetheless, these may enhance accuracy by eliminating supporter cells on microalgae planes.

Hypothetically, all open lasers that produce photons in the noticeable spectra somewhere in the range of 380 to 700 nm may be utilized. Response communities of photosystem I and II are pointed by light energy and radiated fluorescent sign normally shows a similar limit, free excitation frequency (Remelli and Santabarbara, 2018). Therefore, it is pertinent to assemble as significant FACS data related to suitable species as may really be expected. Information may be found in the writing and for this situation, axenic strains of a similar kind are accessible, which may be utilized for entryway determination for the ensuing arranging process. Since most FACS gadgets are provided with a 488 nm laser in particular (Doppler et al., 2021)

IMPORTANT PARAMETERS FOR ALGAL GROWTH

Algae are very simple to develop anyplace on the planet with less energy prerequisites and using not many of the accessible supplements. They are fit to change their shading, shape, sythesis, design, and method of endurance subject to the conditions (Mullick et al., 2017; Edwards 2010). The ideal and most

Figure 1. (a) State-of-the-art system with X = number of repetitive cycles; and (b) the developed FACS single-cell sorting worflow to make up microalgal cultures axenic (Adopted from Doppler et al., 2021)

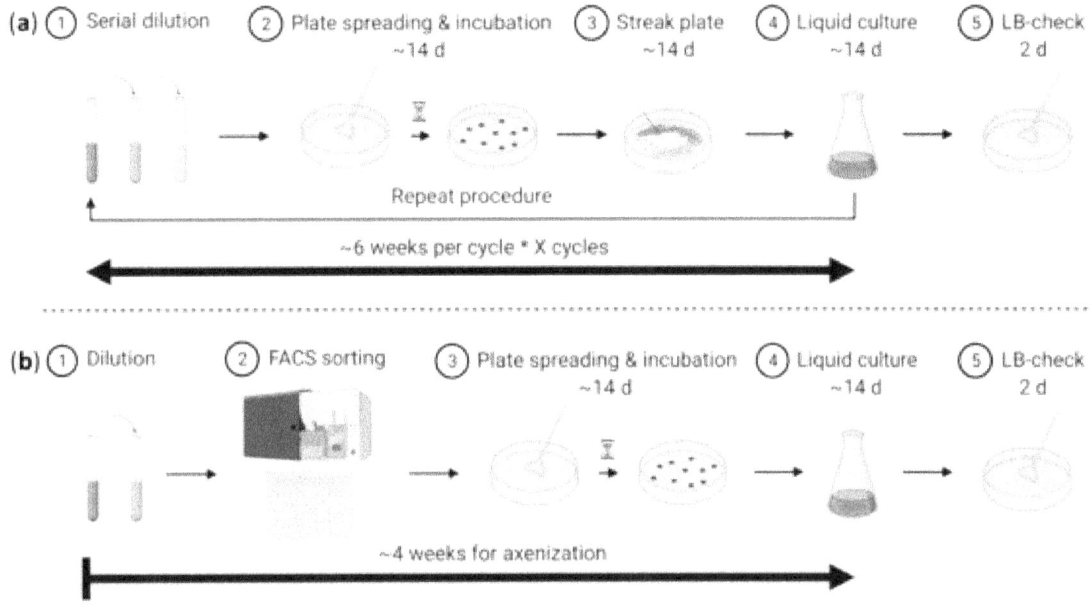

reasonable development conditions for microalgal societies are typically strain reliant and the efficiency of biomass relies upon different abiotic factors like pH, water quality, minerals, light energy, force, carbon dioxide and temperature. Biotic components contain cell thickness and delicacy. However, mechanical components incorporate steady blending, size of gas air pocket, dispersion and mass exchange, every one of these referenced are of explicit interest in photograph bioreactors (Schenk et al., 2008). Some algal species also require ions like Na^{+}, Mg^{+2}, Cl^{-}, and Ca^{+2} for their optimal growth (Mullick et al., 2017; Luo et al. 2007). It is imperative to get focussed upon inborn algal strains for their powerful utilization as a premise of biodiesel and to pick ones that have a higher development rate as well as have a more noteworthy lipid content also (Munir et al., 2015).

Light Intensity and Temperature

Light and temperature are the main elements which influence algal biomass usefulness. The necessary energy for developing green growth is given from light through photosynthesis. Adequate light energy ought to be estimably used to get more prominent biomass efficiency. Temperature influences the paces of each synthetic response that are worried about algal development and its related digestion. Indeed, even a slight difference in temperature influences the lipids and protein content in the cells biochemical design. The ideal temperature for most extreme development as a rule differs all through the season and different other ecological factors additionally assume imperative intelligent parts in deciding a definitive development rate. The ideal improvement temperature for the majority of the cyanobacteria species lies in the middle of 25°C to 35°C. Fluorescent light caused in higher development pace of green growth contrasted with the daylight or when algae situated close to windows and when presented to coordinate/

[illegible] daylight. Thus, temperature along with the light have a critical impact upon digestion, catalyst exercises and generation of algal cells (Munir et al., 2015).

CO_2 Supply

Green algae and cynobacteria also need CO_2 and dark conditions for biochemical type production, and ample quantity of biomass yield. For development of product and manufacturing of various bioprocesses there are several chokepoints like microalgal genetics, its conditions of culture and the cells retrieval (Kumar et al., 2021)

pH

Cultivation of algae depends on pH as it influences the availability of carbon, metabolic activities and biochemical structure of the cells (Richmond, 2000). In the study of Munir et al., 2015 it was inspected that for useful algal development optimum range of pH is required such as pH 7.5 is fit to develop loads of 4.89±0.091g and 4.79±0.021g for *Spirogyra sp.* and *Oedogonium sp.* respectively.

CULTIVATION OF ALGAE

There are two distinct sorts of development frameworks that are for the most part followed for the development of microalgae, raceway lakes and photobioreactors. To get most extreme biomass, open air frameworks for development including open and shut raceway lakes, lab-scale development utilizing photobioreactor should be created. On financial level, most development frameworks that are utilized till date are passed at raised expenses, which isn't reasonable for the savvy biodiesel creation and further advancement towards advertising. Along these lines, it is truly important to choose an ideal development framework for microalgae. For the great development of green growth, there are modified natural just as inorganic carbon sources utilized. Natural carbon incorporate glycerol, glucose, sodium acetic acid derivation, and sucrose and inorganic carbon incorporate CO2 and bicarbonates produce different high energy materials from microalgae like oils, lipids and and omega-3 (Kumar et al., 2021). Cultivation systems of microalgae requires numerous auxiliary nutrients and conditions as discussed in section above.

Photo-Bioreactor

Mass transfer procedures and lack of deep-rooted knowledge of hydrodynamics is a significant breakthrough in the cultivation of microalgae via photobioreactor (Kumar et al., 2021). For the most appropriate design of photobioreactors, mass transfer of nutrients and maximum deployment of photons are of major concern. Media optimization along with system biology approaches are trailed to amplify the process efficiency and lessen production costs (Ramasamy et al., 2020). Photobioreactors either flat or tubular, manifold or serpentine, helical, or made of acrylic glass. (Kumar et al., 2021).

Tubular

Tubular photobioreactors are flat, cone shaped, close to level, upward, slanted sort comprised of glass or plastic in which algal societies are made to recycle either with a siphon or some airdrop framework (Kumar et al., 2021). An arrangement of tubes positioned in a tubular reactor offers a huge surface area to volume ratio for efficacious utilization of light. These types of reactors have an external component incorporated somehow into the main element for the exchange of gas and heat, which sustains the carbon dioxide and oxygen concentration and the temperature for optimum growth. Close to even bioreactors are set at 45^{0}, past which the development expenses would increment. A variety of clear lines are associated with a base line to give gas, which builds the liquid speed, gas hold up, and supplement mass exchange among fluid and gas (Ramasamy et al., 2020).

Helical

Helical rounded bioreactors are a combination of both flat and vertical reactors and comprises of a fluorescent light in the middle as a light source. Outside cooling and degassing units are utilized to work such reactors effectively and air is given from the lower part of the reactor with the assistance of a stream meter. Degassing is accomplished at the highest point of the reactor to eliminate abundance oxygen. Photoloss and restraint are the principle worries that happen in a cylindrical reactor of little width and with a less-thick culture. The variety of cylinders in such kind of reactors give insurance from overabundance light, subsequently expanding algal development. Expanding the breadth may likewise assist with defeating the issue because of the subsequent high temperature and light, somewhat, however this may be trying for supplement mass exchange in the framework (Ramasamy et al., 2020).

Column Photobioreactor

Column reactors, for example, bubble segment and transport reactors are comprehensively utilized for algal development. These photobioreactors have a most extreme range and stature of up to 20 cm and 400 cm, separately. To get a high surface region, it is expected to keep the proportion of the measurement to stature of the reactor vessel little. Notwithstanding, expanding the segment stature past 400 cm may prompt supplement inclination in the reactor, which prompts starvation of the living beings. Besides, a long home season of oxygen may be hindering to green growth in these reactors. These reactors offer high mass exchange coefficients of supplements in the fluid gas framework (Ramasamy et al., 2020). A comparison table provided at Appendix 1

Bubble Column Reactor

A bubble column reactor is cylindrical reactor with a sparger introduced at the base, which supplies gas for mixing the substance without the requirement for any interior blending gadgets (e.g., impeller). A high mass exchange pace of CO_2 and fundamental supplements may be accomplished with an expansion in maintenance time and number of little measured air pockets. The stream qualities of the liquid and the math of the reactor unit may influence the light accessibility. In such kind of reactor, a distance across surpassing 20cm lessens the vulnerability of light, though the stature is restricted to 400cm to escape conceal in a consecutive reactor framework. Air pockets may create a cloud result in the reactor that upgrades the way length of the light inside. The cloud impact is delivered by a gathering of air pockets that shrivels the force of light coming to the microalgae. Putting punctured plates along the reactor tallness heightens bubble scattering for an upgraded blending productivity. These reactors are profoundly reasonable for high-thickness or densified cultures (Ramasamy et al., 2020).

Airlift Photobioreactor

These reactors comprise of a rounded vessel with more than one interrelating zone including a riser, where the gas combination is sparged, and a downcomer denied of any sparger (Ramasamy et al., 2020).

Flat Plate

The flat/level plate bioreactor is created with level sheet straightforward materials for engrossing extreme light force. Benefits of these are high photosynthetic effectiveness, suitability for immobilization of green growth, improved biomass creation, easy cleaning, promptly tempered, less hydrodynamic weight on cells (Kumar et al., 2021). These type pf photobioreactors surfaced as potential bioreactors to create fucoxanthin which is found plentiful in brown macroalgae (such as *Fucus, Sargassum* and *Dictyota*), diatoms, haptophytes and *Chrysophyceae species* as brown colour xanthophyll carotenoid (Sahin et al., 2019; Mao et al., 2020). As reported by Gao et al., 2017, due to high content of fucoxanthin, diatoms and *Chrysophyceae sp.* exhibit golden brown shade. For photosynthesis fucoxanthin plays an important role as the light-reaping shades in the protein-colour organization of fucoxanthin chlorophyll restricting proteins edifices (FCPs) and is accomplishing a lot of consideration from businesses and scientists these days because of its various potential medical advantages, including against oxidant, hostile to heftiness capacities, against disease, etc (Leong et al., 2021).

Open Ponds or Open Culture Systems

Shallow lakes, tanks, circular lakes and raceway lakes are the broadly utilized open frameworks (Ugwu et al., 2008). Raceway lake is used for mass development of microalgae especially at commercial grade. Positive outline of raceway lake development related to biomass surrender was depicted with some microalgal strains for instance *Spirulina* and *Dunaliella* (Kumar et al., 2021). Table 2 at Appendix 2 presents the productivity of the algal strains, grown in different cultivation systems.

The productivity of biomass of microalgae that are grown in raceway type ponds lies between 73 and 109,000 kg/ha/yr (Shen et al., 2009), and in other ponds like highly propogating raceway type, 127,000 kg/ha/yr may be attained because of the active photon flux using data for radiation and insolation (Bharathiraja et al., 2015). The most regularly grown eukaryotic as well as prokaryotic algae in raceway ponds includes *Chlorella sp.*, *Nannochloropsis sp.*, *Dunaliella salina*, *Arthrospira platensis*, *Tetraselmis sp.*, *Scenedesmus sp.*, *Haematococcus pluvialis*, *Micractinium sp.*, *Anabaena sp.*, *Phaeodactylum tricornotum*, *Actinastrum sp.* etc (Jorquera et al., 2010, Park et al., 2013, Kumar et al., 2021).

Figure 2. Schematic showing different photobioreactors used for algal cultivation (Adopted from Ramasamy et al., 2020).

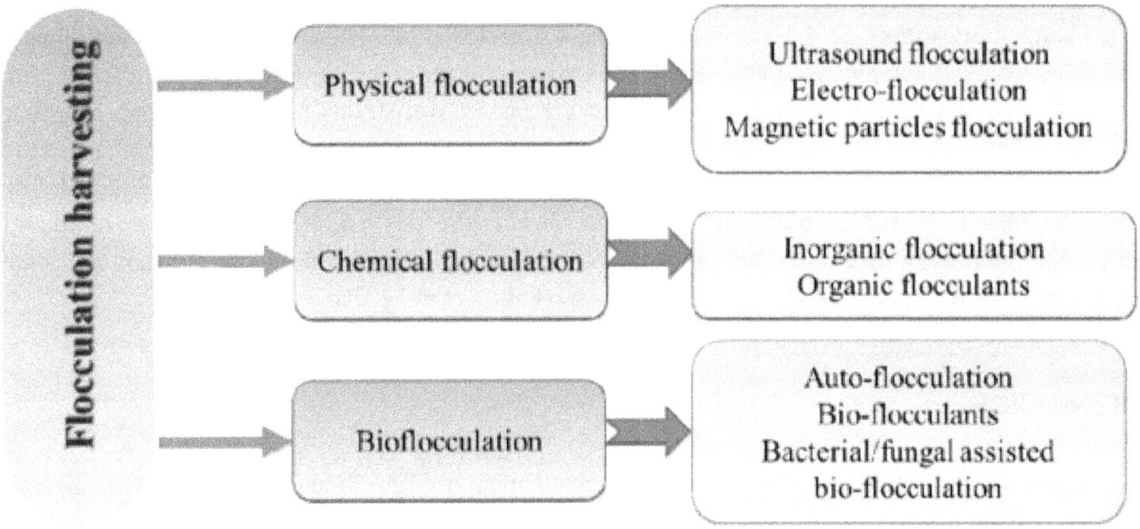

DIFFERENT TYPES OF ALGAL STRAINS AND THEIR CULTIVATION STYLES

Photosynthetic microalgae are miniature living beings that have colonized each natural surroundings on the planet and show remarkable organic variety, assessed to be more noteworthy than 200k species (Guiry, 2012), that reveals a gigantic range of ecological alterations. Dissimilar to different organisms frequently used for bio-based assembling, like yeast and microorganisms, phototrophic green growth has the benefit to utilize daylight to fix air carbon, henceforth decreasing their reliance on sugars for aging. Normally flourishing in conditions with irregular and scant supplement accessibility, numerous types of microalgae have progressed productive metabolic transformations to develop quickly under positive conditions. This results in high photosynthetic proficiency of algae than plants, which renders them into a much greater capability to generate biomass. When developed at bigger scopes –photobioreactor or lake – microalgae are proven to be more water-productive than crops. Non arable land filled with seawater or wastewater may be used for their growth with negligible utilization of freshwater. Accordingly, numerous topographical regions that are not proper or adequately fruitful for crop development may be proficiently utilized for enormous scope algal growth.

For centuries crop plants reproduced and decided to segregate explicit qualities and to get profoundly useful strains, all the current microalgal strains are viably natural secludes. Therefore, it is paramount to streamline both the life form and related climate which upholds its development in order to upgrade usefulness and increment the modern capability of microalgae.

As per the observation of Rawat et al., 2013 the general water necessity is still not exactly conventional plant-based harvests leading algae as the most demanded substitutions for cultivation. Laurens et al., 2017 reported that the possibility of mass scale algal generation has surveyed by different technical and economical calculations. In addition, the capital use of an open lake may run from ~$US 6/m^2$ to US$ $50/m^2$ (Huntley et al., 2015), though encased photobioreactors frameworks may take a toll up to 3 to 30 times higher than open ones (Punlo and Carroon, 2016). For different applications and systems operating cost may shift from US$ 0.8/kg dry weight to up to $8/kg dry weight. Moreover, Biofuels have specifically gotten more consideration and presently they are not cost competitive, with production costs at ~ US$ 3/L in contrast to <US$ 1/L of producing fuels from fossil oil. The area/volumetric usefulness is typically probably the biggest vulnerability just as drivers for progress. Strain determination is a prime factor to enhance yield for the end result. This might need novel strains planning through hereditary designing or manufactured science that is totally profiled and chosen via phenomics approach. However, while considering the expense side, dewatering is taken as the critical costs during algal handling. Algal suspensions are for the most part extremely weaken; thus, expanding the content of biomass in the development stage may observably lessen costs, which is a significant benefit for shut and joined development PBRs when contrasted with open lakes (Fabris et al., 2020).

PHENOMICS

Phenomics is by and large characterized as "the achievement of high-dimensional phenotypic information in an organism-wide scale" (Houle et al., 2010). Phenomics of algae is at present in exceptionally untimely phases of advancements, nonetheless, it holds an incredible strength in microalgal horticulture. Through creation of a database of G*E = P [where G is genome, E stands for environment and P means phenotype(s)] interactions for some given algal species, specialists may evaluate out regular and fake variety for the blend of quality alleles that will cluster together fundamental aggregates (Furbank and Tester, 2011), like increased product yield and rapid growth. Ongoing development from the area of plant phenomics features the likely effect of phenomics method and advancements in microalgae. For instance, a fresh study on phenomics- targeted on *Arabidopsis thaliana* produced a mutant exhibiting showing both expanded microorganism guard and photosynthetic development, breaking the thought trade-off between defence and growth (Fabris et al., 2020).

GENETICALLY MODIFIED STRAINS

Hereditary adjustments has been accomplished by versatile research facility advancement (ALE), irregular mutagenesis and direct hereditary designing. Lager and actual mutagens like UV light, X-beams and gamma radiation just like substance mutagens, for example, N¢-nitro-N-nitrosoguanidine (NTG) and ethyl methanesulfonate (EMS) have been adequately applied on microalgae for presentation of irregular changes. Notwithstanding, productive hereditary designing methodologies that may create explicit additions, cancellations or replacements in the host genome are relied upon to make designated adjustments while keeping away from flighty outcomes. Headways in sequencing innovation, advancement of quick, precise and proficient DNA conveyance frameworks and the advancement of high throughput genome altering apparatuses have turned into a fundamental part for the age of hereditarily better microalgal specie. The primary vital development in microalgal biotechnology was the atomic makeover of *Chlamydomonas reinhardtii* in 1990. Kindle (1990) performed tumult of cells within the sight of glass dabs covered with DNA which permitted the making of micropores in the cell film due to rubbing, causing the section of DNA atoms into the cell. Notwithstanding, mentioned advancement was done thirty years prior, the DNA conveyance into microalgal cells is as yet a significant bottleneck because of the variety in cell sizes, cell divider designs and creation among the few genera of microalgae. A few change strategies have been set up for the conveyance of exogenous DNA into various microalgal strains. The most widely recognized and fruitful strategies are electroporation, Agrobacterium tumefaciens-intervened change, molecule barrage and fomentation with glass beads (Apt et al., 1996) The greater part of these referenced strategies are confined to a predetermined number of animal categories and require difficult streamlining when carried over to other microalgae (Zhang et al., 2019). for foreseeing metabolic designing methodologies, genome-scale metabolic models are useful assets, they may be more enhanced by exploratory portrayal of proteins, motor examinations, and recognizable proof of quality guideline designs. An expansion in hereditary information accessibility and the approval of metabolic models will permit more compelling and vital plan of hereditary methodologies and will assist with working on the precision and unwavering quality of bio-informatic instruments. Despite the fact that genomics information for microalgae is in a pre-mature level, they have permitted the distinguishing proof of administrative components essential for the improvement of hereditary designing methodologies. Identification of various promoters, intertwining signals, eradicators, selection markers and reporter genes have smoothed the expression of heterologous genes in microalgae. This has hurried the improvement of different atomic devices counting state-of-the-art genome altering frameworks such as Translation Activator-Like Effector Nuclease (TALEN), RNA impedances (RNAi), Clustered Frequently Inter-Spaced Palindromic Rehashes and related proteins (CRISPR- Cas) and Zinc Finger Nulease (ZFN) (Munoz et al., 2021). Appendix 3 presents metabolism showing enhanced lipid production

Transcription Factors

Transcription Factors (TFs) are DNA-restricting proteins what work as fundamental controllers of quality articulation. Different kinds of TFs exist that are unmistakable in their system of perceiving and restricting DNA and affecting record. A solitary TF may handle the declaration of numerous qualities all the while, that makes them supreme objective to move metabolic transitions towards biosynthesis of lipid. In microalgae, a few TFs are recognized as likely focuses for hereditary designing, particularly in the species of the genera *Nannochloropsis* and *Chlamydomonas* (Munoz et al., 2021).

Genome Shuffling Technologies

Genetic engineering as of late accomplished more considerations on the grounds that new and incredible hereditary apparatuses are progressively accessible, and the genome may be adjusted to our need more precisely than previously. The metabolic methodology or engineered science has gotten more consideration, as the plan of a metabolic pathway in a framework where it doesn't exist opens up new measurements for microalgal industrialization (Ng et al., 2017). The basic gene manipulation procedure remains the same in synthetic biology, together with host choice, quality objective, plasmid development, change instruments, determination framework, and DNA altering devices. In microalgae hereditary designing, the change and determination techniques are required strides to achieve the objective quality alteration. Four primary strategies that are being utilized for change of microalgae:

1. Thorough mixing with glass beads
2. Electro-poration
3. Particle's bombardment

Figure 3. Biosynthesis pathways of microalgal lipid. Cellular organelles, plastid, endoplasmic reticulum, peroxisome, mitochondria, lipid droplets (Adopted from Munoz et al., 2021)

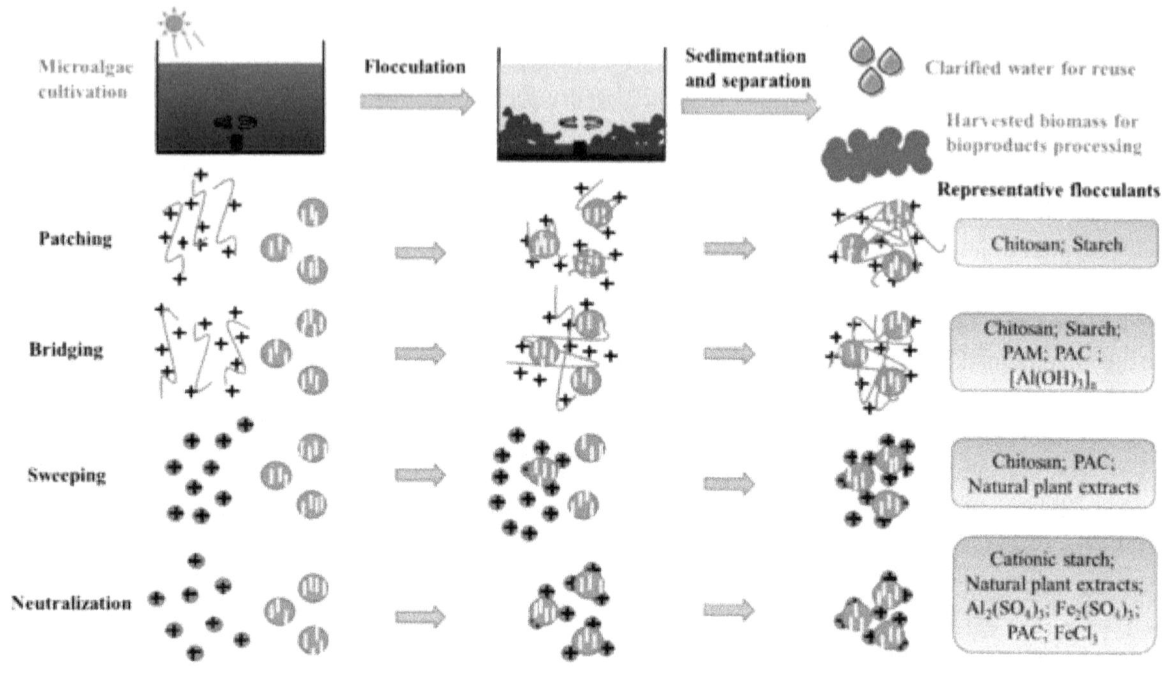

4. Agrobacte-rium-construction.

Every technique enjoys its benefits and impediments dependent on the effectiveness, incorporation, or security of the transgene (Ng et al., 2017). Alternatively, the selection system may be founded on anti-microbial opposition or correspondent quality choice. Distinctive host and advertiser strength would affect the determination effectiveness. In addition, in metabolic designing, genome-altering and quality meddling devices are vital for proficiently focusing on the quality. As of late, Clustered Regularly Interspaced Short Palindromic Repeats – CRISPR related protein 9 (CRISPR-Cas9) and Transcription Activator-Like (TAL) Effector Nucleases (TALEN) which are the latest genome-altering apparatuses, just as Zinc-Finger Nucleases (ZFN) are utilized in quality change. Wanted microalgae aggregates without quality change may be accomplished by quality meddling devices, like CRISPR-dCas9, miniature RNA (miRNA), and quiet RNA (siRNA), to initiate the quality articulation. The Omics approach is likewise an integrative procedure for the research of microalgaeae in maintainable turn of events (Ng et al., 2017).

Nuclear transformation generally happens as an arbitrary inclusion of the transgene in the atomic genome by disturbance with glass dabs, where the DNA and algal cells are unsettled in the accessibility of 0.5 mm beads and this is the most established technique announced for microalgal changes. The transgenes would be embedded aimlessly in the atomic genome and chose by anti-microbial opposition or phenotypic variety. This procedure is very straightforward and a high proportion of changed cells may be accomplished, however utilizing a cell divider less strain is needed. Electroporation demands the utilization of electrical motivations to convey exogenous DNA into cells. (Ng et al., 2017)

Figure 4. Genetic transformation procedure of microalgae. A) Nuclear transformation by glass beads agitation, electroporation and agrobacterium transfection, and random insertion. B) Chloroplasts transformation by biolistic and homologous recombination (Adopted from Ng et al., 2017)

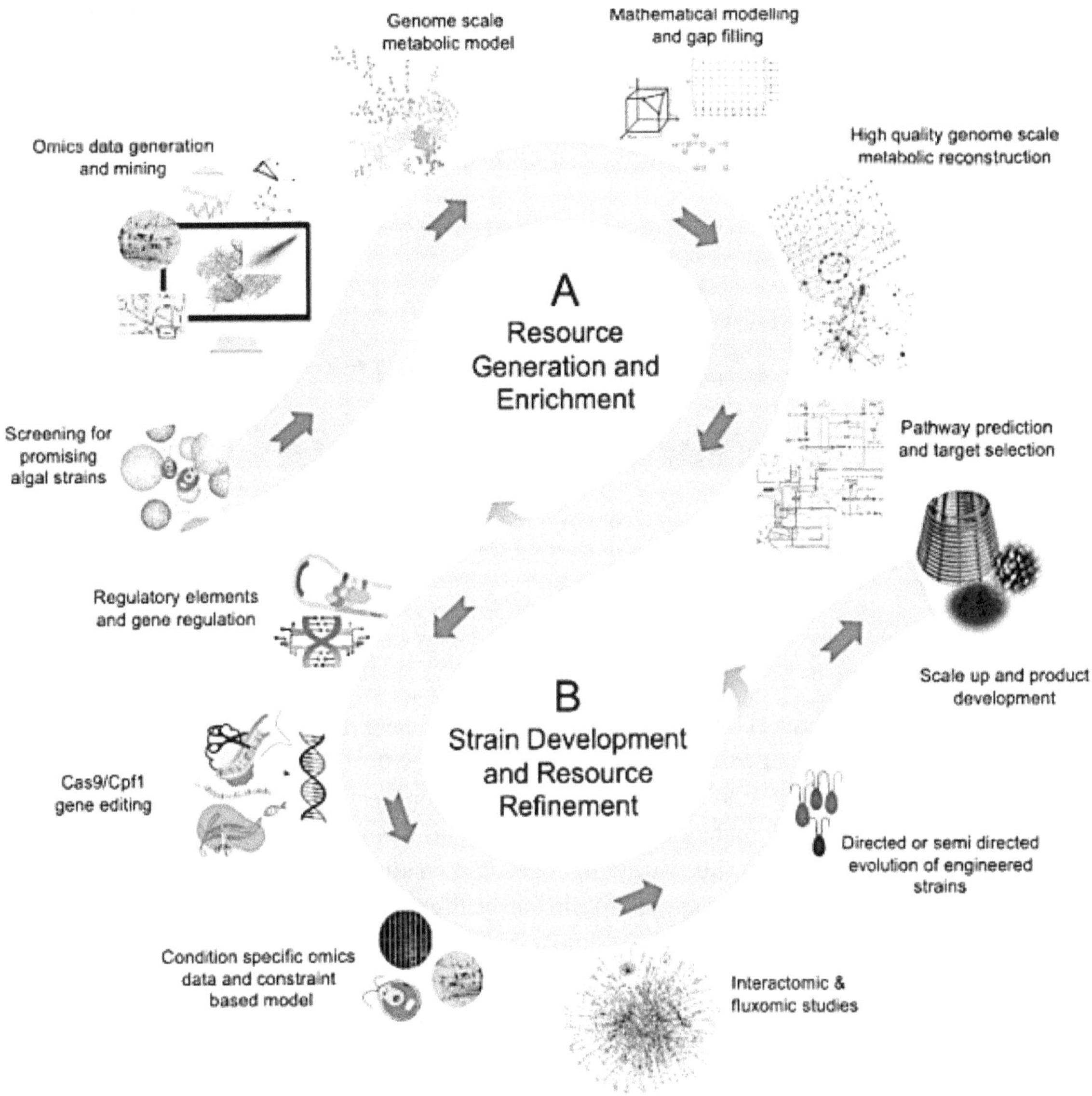

Genome sequencing initiatives tends to the huge holes in algal biodiversity and transformative history will without a doubt uncover incipient hypotheses of algal science and development, lead to the revelation of novel proteins, biochemical pathways, and undiscovered normal items (Hanschen et al., 2020)

Such a large number of transformations may bring about the deficiency of wellness because of aggregation of unbiased or pernicious changes that stifle good transformations under certain conditions (Ng et al., 2017). Such transformations may bring about proficient microalgae which during performing admirably under some ecological situation, drop their reasonable benefit in a substitute climate. For instance, Kato et al. adapted *C. reinhardtii* for growth utilising saline conditions. Developed specie exhibited predominant

Figure 5. Microalgal bioengineering data and resource driven strategy (Adopted from Kumar et al., 2020)

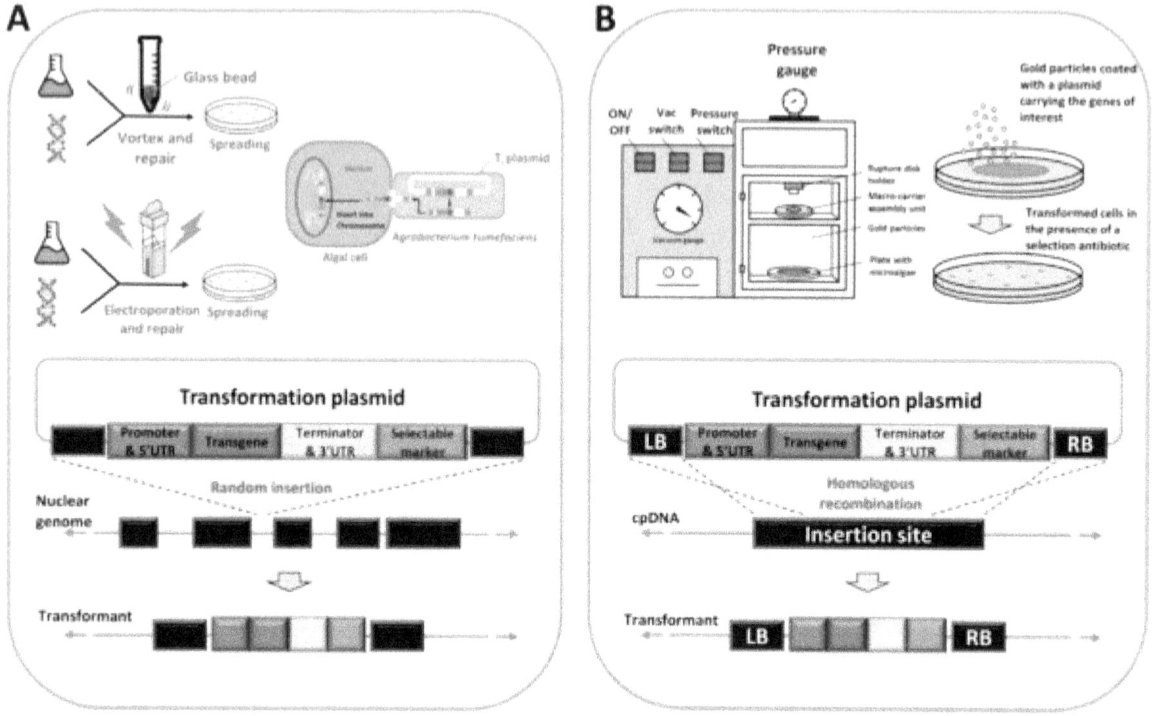

development at high saltiness levels, however showed a development imperfection in freshwater media. Such specialization might be adverse from an outside green growth strain designing viewpoint, where development situation might have to variate in space or time for ideal efficiency. Dynamic varieties in ecological conditions may limit mutational compromises. This is epitomized in a study using *C. reinhardtii*, where cultures exposed to substituting light and dim conditions exhibit predominant execution in the two conditions. Using a comparable approach, Sun et al. exposed *Schizochytrium sp.* progressively to both low temperature and high saltiness in a similar ALE cycle. The developed strain had both a more prominent biomass and a higher docosahexaenoic corrosive (DHA) part aggregate. Then again, when the provided parent strain was adjusted to just one of the two conditions, the advanced strains exhibit either an expansion in DHA part or in biomass, respectively (Lapanse et al., 2020)

HARVESTING ALGAE FROM EFFLUENT STREAMS

The low energy necessities to develop microalgae alongside the extremely quick propagation gives microalgae special capacities to use them for modern and ecological applications. Be that as it may, gathering this green growth actually remains energy burning-through and, as a result, costly (Parmentier et al., 2020). The process of gathering algal cells from their medium with no damage to their water content is known as Harvesting. A few methods are used in reaping algal biomass together with filtration, flocculation, buoyancy, centrifugation, sonication, and precipitation procedures. Sometimes, two reaping techniques may be coupled to further develop the biomass creation. In different cases, dewatering may follow gathering. Dewatering is the method involved with eliminating water content from cells to get dry mass (Saad et al., 2019)

Flocculation might be an unrivaled strategy when considering gathering effectiveness, monetary expense, energy utilization and specialized possibility. Actual flocculation procedures are energy-concentrated and need uncommon gear, making the expense obstruction for microalgal biomass collecting. Attractive molecule flocculation is significantly more productive and is additionally recyclable. Conversely, compound flocculation that includes the use of natural and inorganic flocculants, is currently at the center of attention (Li et al., 2020)

Appendix 4 presents a comparison of different harvesting methods

Centrifugation

Centrifugation is utilized as a microalgal collecting technique. This interaction is very fast and has high recuperation proficiency, however it is undeniably more energy-escalated and the shear power will harm the cell (Li et al., 2020)

Flotation

Flotation has many advantages, like a more modest region necessity, adaptability and conceivable modern versatility. Be that as it may, it requires high venture and is energy-concentrated (Li et al., 2020).

Flocculation

Flocculation is the generally utilized strategy for isolating colloidal and coarse scattering substances during water treatment, and mostly it has been utilized for division and assortment of microalgae since the 1980s. Following flocculation, the amount suspended green growth or algal fluid may accomplish effective gravity-based sedimentation, and the isolated green growth cells may be straightforwardly. caught in the reactor, consequently keeping up with high biomass and guaranteeing emanating water quality. According to the point of view of unadulterated green growth cell reaping, flocculation is the most affordable and practical strategy for treating gigantic measures of weakened green growth. Although the green algal growth cells may not straightforwardly meet the necessities of modern applications after flocculation (floc formation) and sedimentation, they may altogether decrease the energy utilization and cost of the resulting fixation procedure. Flocculation, as a propitious strategy for gathering microalgae at lower costs, has been investigated as of late, and different novel flocculation strategies have been created (Li et al., 2020).

Chemical Flocculation

As per the properties of synthetic flocculants, compound flocculation is additionally partitioned into two kinds: inorganic flocculants and natural flocculants. As of late, synthetic flocculation has been widely read for microalgae gathering. Inorganic flocculants principally incorporate decidedly charged metal particles, while natural flocculants incorporate normal natural flocculants, yet additionally manufactured natural flocculants. The principle instrument mindful in utilizing inorganic and natural flocculants in the reaping system is the charge balance, adsorption bridging and net-cleaning. Through charge balance,

Figure 6: Demonstration on various types of flocculation mechanisms in addition to the representative flocculants (Adopted from Shuangxi et al., 2020)

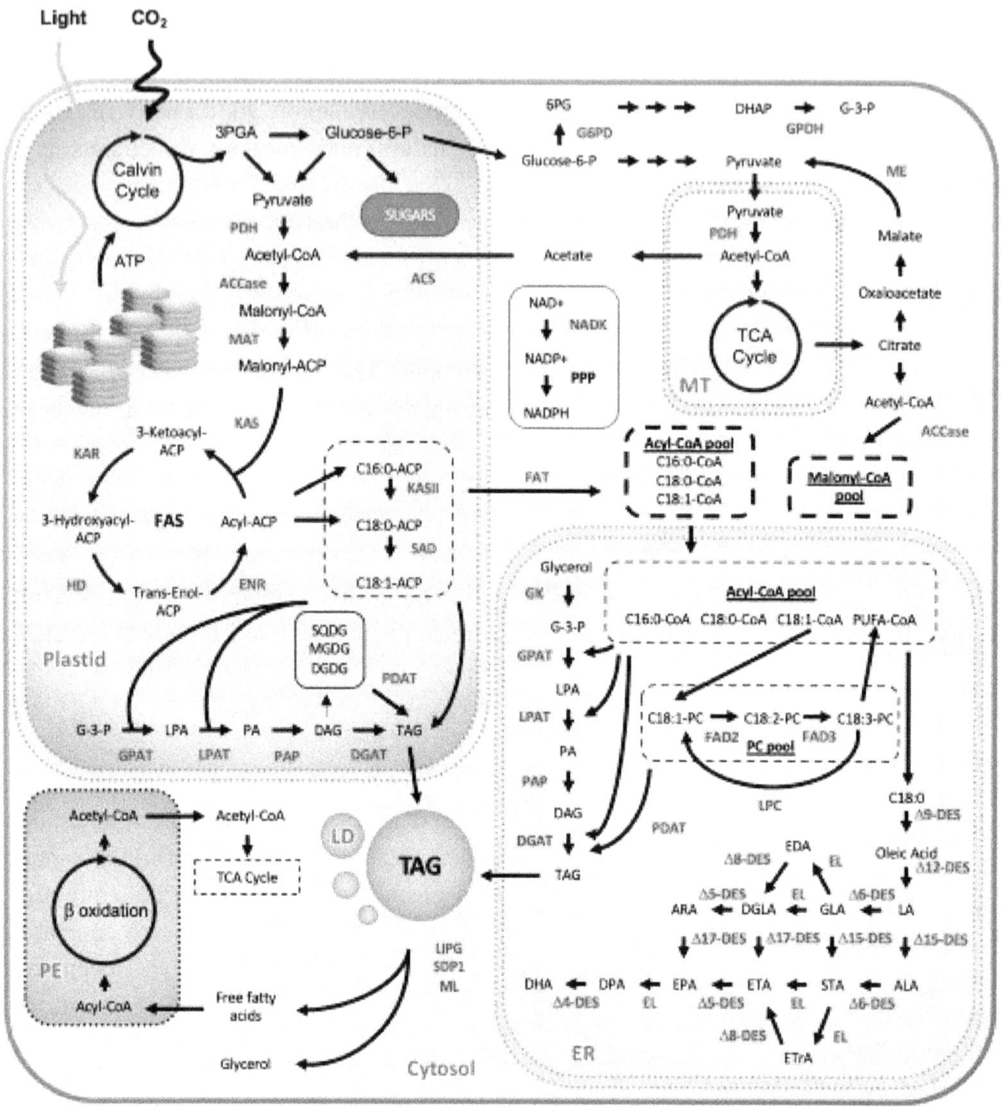

inorganic flocculants adsorb equivalent measures of adversely charge containing microalgal cells, while natural flocculants structure actual linkage with algal cells (Li et al., 2020).

Bio-Flocculation

Bioflocculation has as of late achieved a great deal of consideration as a microalgal reaping technique. It is plausible by and by with low energy input necessities and the possibility to be ecological cordial, particularly when it is contrasted and substance strategies for reaping. To be more explicit, bioflocculation frequently alludes to the flocculation incited by different microorganisms, for example, self-flocculating green growth, microscopic organisms, parasites and yeast, or the bioflocculants (for instance extracellular polymer substances (EPS) delivered by the microorganisms. Right now, many explores have been completed to make bioflocculation techniques more proficient and appropriate. In any case, most systems are just approved at a lab scale, thus far nobody has effectively applied these strategies to enormous scope microalgal collecting. Likewise, bioflocculation is very tedious and conflicting over strains (Li et al., 2020).

Physical Flocculation

The physical flocculation strategy is to accomplish flocculation and recuperation of microalgae by utilizing actual power, like power, attraction, or ultrasound. Without the expansion of compound flocculants, actual flocculation is a proper technique to stay away from contamination. Contrasted and ultrasonic flocculation, there are more examinations on attractive molecule flocculation and electro-flocculation, likely as a result of their high expulsion efficiencies and huge scope application possibilities (Li et al., 2020).

Sedimentation

Gravity based sedimentation and filtration are practical strategies, basic activity, and their usefulness is too low to even consider isolating biomass from the mass culture (Li et al., 2020).

CHALLENGES FACED IN BREEDING OF ALGAE

Enhancement of agricultural productivity in past few decades due to increase population mass rely on the maximum crop plant domestication. In this aspect, algae are at the beginning of this as no extensive information related to its cultivation and harvesting may exists. In past same crops may not be able to yield high biomass, current has been changed due to numerous technological advancements. But this challenge of mass yield remain prevails in case of algal biomass. This challenge may be overcome by using genetic modifications as done in case of higher plants. Algae being polyphyletic class of the oldest and most diverse organisms may present diverse array of genetic variations on which domestication may rely. Targets of genetic and metabolic organization which may comprise genes and pathways part of ensuing metabolites, carbon fixation, and light capture for instance, via metabolic engineering gene expression may alter both composition and content of lipids which is an essential energy packet for both fossil fuels and micro algae. Nevertheless, regulatory networks which control the rate of formation of carbon storage molecules (starches and lipids), and the composition of fatty acids, remain poorly explained.

MICROALGAE VS. MACROALGAE

Macroalgae, generally termed as seaweeds, photosynthetically consume CO2 for the formation of oils, carbohydrates, and proteins. Macroalgal strains contains lamina of leaf like structure and a floating organ known as focus. Its growth rate is usually higher compared to higher plants but lower than microalgae. Macroalgae usually store carbohydrates as energy source. This is the reason that these are not used for lipids content like microalgal strains. Macroalgal strains are used in the manufacturing process of sugars

Figure 7. Flocculation techniques for microalgae harvesting (Adopted from Shuangxi et al., 2020)

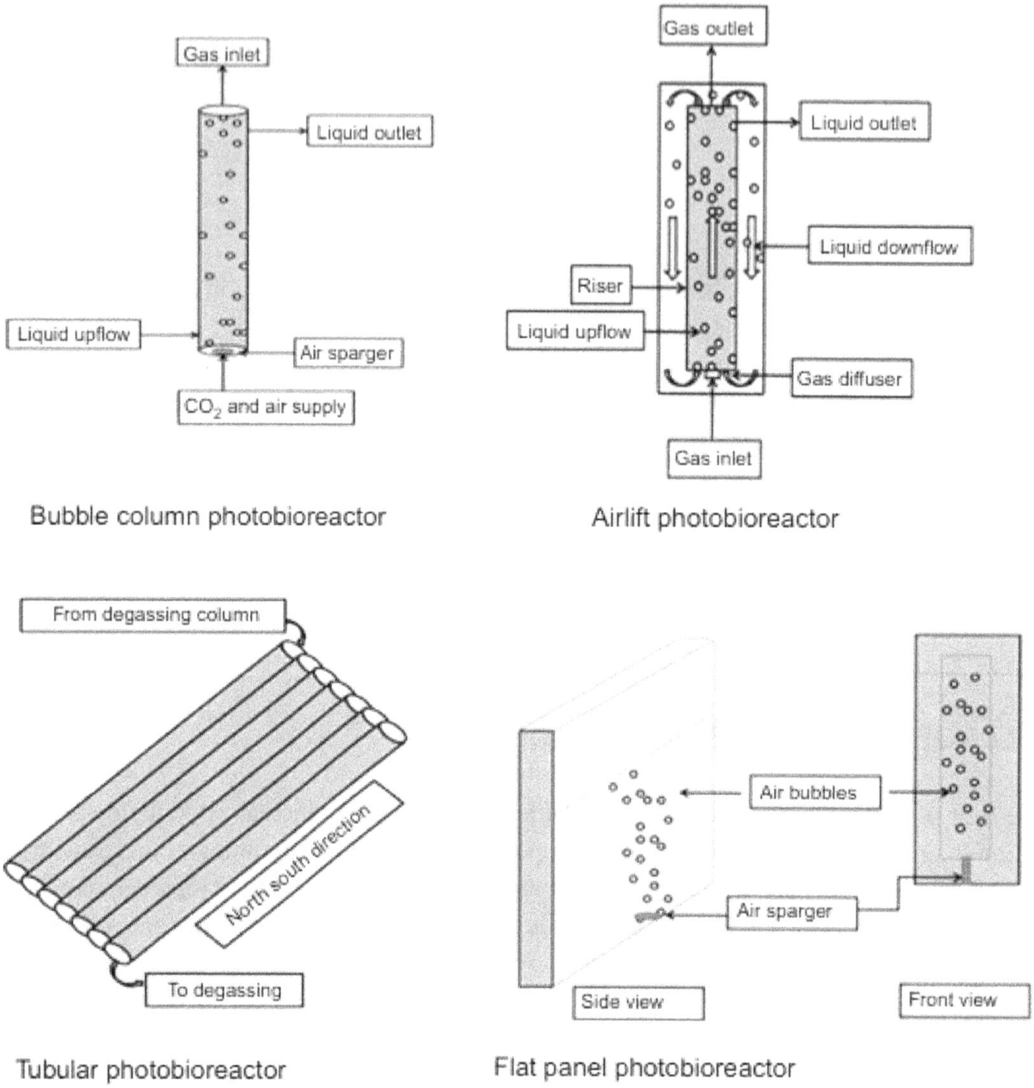

and Laminaria spp due to their stored 60% starch. Other examples of Kelp species are Laminaria digitata, L. hyperborea, Saccharina latissima, Sacchorhiza polyschides, and Alaria esculenta.

Laminaria spp and Ulva spp are most valuable macroalgal strains with respect to energy. Due to higher water content in macroalgal strains nearly about 80–90% of biomass its transformation to biodiesel via transesterification more complex. therefore, it is significant to select more effectual procedure such as use of most problem creating CO2, which may be utilized for transesterification using wet biomass. Macroalgae are mostly considered for their use in ensuing of biofuels such as bioethanol, biogas, etc. Nevertheless, larger arena exists in case of macroalgae to spread research scope to readdress its metabolic pathways towards the lipids production. This would be more exciting as production of lipids may be exploited due to their greater lengths as sea kelps.

Additionally, macroalgae store huge reserves of biomass that may be used for sequestration of carbon reserves via sea farming. In this aspect genetic tools may be of great interest as genetic mutation of macroalgae for expansion of lipid production may be done. The omics data may help in future bioprospecting in macroalgal strains. Nevertheless, this would be employed to producee lipids, carbohydrates, and biogas from macroalgal strains besides the production of additional value-oriented products (Gautam et al., 2015).

Appendix 5 presents a comparison table for micro vs macro algae

BIOFUELS PRODUCTION FROM ALGAE

Algae have high oil fillings depending on their different lipids content therefore its capacity to produce biofuel varies with specie type. Some algal strains were recognized as good fatty acid assets and some strains have additional content of fatty acids stored as dry masses. Micro algal strains may survive in varied conditions in fact with scarcer available nutrient thus these are considered as best to cultivate. Moreover, the sample collection demanded greatest care to produce whole biofuel yield with specialized instruments. The growth proportion of algae is also affected by varied ecological parameters that are not constant or identified for every region, thus method needs careful handling. The simplest process to extract fatty acids and production of biofuel is the blending method on small or experimental scale. In addition to this, selection of cultivation unit is also important as the growth rate along with specie also varied in open chamber or in closed bioreactors. The whole process either batch or continuous is established depending on the conditions available i-e pH, temperature, algal strains and the amount of algal biomass. Harvesting techniques depends on the on the geographic location and conditions. Most suggestable harvesting technique is done in settling pond or sedimentation tank. Moreover, adjustment of density and moisture content is prerequisite for the production of biofuel. Usually, spray drying is used as drying technique to remove water content. In addition to this, mechanical handlings are considered the most favourable one for disruption process. Additional requirements include the use of solvents like hexane and ethanol for active processes. In the absence of other sources, Ultrasound and microwave-based extraction methods may also be of paramount importance (Khan et al., 2017).

Utilization of microalgae for biofuel production is unquestionably required in the entire world. Although this approach is energy rich and less polluting as compared to fossil fuels, researchers are still searching for the novelty to boost up the algal biomass production from initial stage to end products in order to shift the whole process to most economical. One of the innovations is addition of nano additives. Implementation of different forms of nano-additives in various phases on microalgal growth to biofuel exhibited an excellent outcome which may project revolutionary advancement of commercial microalgal biofuels (Hossain et al., 2019).

ALGAE BASED BIOFUELS

Biohydrogen

Bio-hydrogen considered as the imminent, yet an unexplored region of biofuels production. It is a renewable, environmentally friendly and very efficient source of energy. Due to its practical limitations, its use is not feasible although it possesses high packets of energy compared to fossil fuels. Because of its high economical cost, bio hydrogen production is not feasible as it is very low. Currently, hydrogen is ensued by techniques like gasification of coal or electrolysis of water, but the use of algal biomass is a novel technique and is gaining hype nowadays. In algal biomass, bio-hydrogen is mostly produced by two different procedures namely photosynthesis and fermentation. In fermentation, biohydrogen is ensued through photo fermentation and dark fermentation, whereas photosynthetic production of biohydrogen occurs through direct bio photolysis and indirect bio photolysis (Anto et al., 2020).

Bioethanol

Extraction of ethanol (ethyl alcohol) from a living source is termed as bioethanol. It may be used as an alternative to petrol. Bioethanol having less sulphur content compared to ordinary fossil fuels therefore these are less toxic to environment. It possesses around 66% of the energy confined by petrol of the same quantity. Seeing its renewable nature, bioethanol scope is quite high. This biofuel is produced by break down of starch or other sugars such as corn, wheat, ligno-cellulosic biomass (sugarcane waste), etc mostly from first and second generation of feedstocks. Algae may overcome these disadvantages of second-generation sources. The discussion of 'food versus fuel', use of agricultural land, and water use are few of these. Algae, being the third-generation renewable source, is one of the greatest hunted after sources for the production of bioethanol. It has an ability to use industrial or residential wastewater as source therefore consume CO2 and helps in bioremediation thus results in treating of wastewater. Bio-ethanol may be ensued mainly by starch or cellulose through the process of fermentation. Species like Chlorella vulgaris may accumulate starch upto 37% of its total dry weight. Blue-green algae including Spirogyra species and Chlorococum sp. have increased levels of stored polysaccharides in their cell walls. Diverse species stored their food differently, some may store in the form of constituents like mannitol, alginate, glucan, galactan and laminarin. Algae are measured as fruitful source for bioethanol as they possess a good amount of carbohydrates. Commonly consumed algae for the production of bioethanol are *Sargassum, Glacilaria, Euglena gracilis, Prymnesium parvum, Porphyridium, Chlorella, Scenedesmus, Dunaliella, Chlamydomonas,* and *Spirulina* (Anto et al., 2020)

Biochar

Biochar is a carbonaceous material manufactured by thermal treatment of algal strains at a modest temperature and by supplying controlled O2. The manufacture of algal biochar from wet algal biomass takes place at a modest temperature for a short span through the process of HTL. Compared to ligno-cellulosic biochar, biochar generated from algae shows a lesser carbon content and reduced surface area but much higher cation exchange. Due to higher pH of biochar produced from algal strains, it may be beneficial to recover acidic soil. For maintaining healthy soil diverse inorganic elemental composition with high nitrogen content is beneficial. (Kumar et al., 2020)

Thermo-chemical conversion at 100 to 300 °C in an anaerobic chamber is called torrefaction, which may generate biochar with quite high calorific value, better hydrophobicity and lesser content of ash. hydrothermal carbonization (HTC) is used to convert Hydro-char or charred biomass by utilizing wet biomass such as agrarian residues, algal biomass etc. and may be widely applied to treat wastewater, use as adsorbent, soil amendment, activated charcoal production and hydrogen storage. It showed that hydrochar obtained from *Scenedesmus sp.* by HTC at 220 °C contained an exceptional physico-chemical and fuel property, and its large heating value reached 26.64 MJ/kg that was correspondent to medium-high calorific coal, like lignite (Zhang et al., 2021).

Biodiesel

Biodiesel is also a high-end option to fossil fuels. Biodiesel is formed by the trans-esterification of lipids attained from algae, to produce methyl esters of long chain fatty acids. Chain of fatty acid depends on the lipid content source. Mostly, oils from soybeans, palms, canola, sunflower, rapeseed, etc are the sources of biodiesel which are economically very high than fossil fuels. However, these may lead to the fuel against food conflict, high usage of agricultural land and weak infrastructure, hence denoting algae as the most convenient option. Moreover, from the ecological aspect, biodiesel from algal strains is more desirable because of slighter emission of CO2, NOx and other greenhouse gases. Both micro and macro algal strains may be utilized for its mass yeild. Few most frequently observed species for the production of biodiesel are *Chlorella sp., Chlamydomonas reinhardtii, Botryococcus braunii, Phaeodactylum tricornutum, Dunaliella salina, Thalassiosira pseudonana, Nannochloropsis and Isochrysis sp.* (Anto et al., 2020)

CONCLUSION

The current study focuses on different algal strains, their means of cultivation, generated biomass and variation of phenotype for various biotechnological applications. Study revealed that major issue in generation of required biomass for any strain lies in production protocol for large scale synthesis. Strain may experience diverse conditions, contamination or suitability constrains at large scale production which are missing in controlled chamber. Resultantly productivity may affect or it may not reach to the level of laboratory scale. Major variations may come in temperature, light intensities, alternating light and dark and possible outdoor environmental contamination. The issue may be resolved by bioprospecting or synthetic biology and humanity may take benefit of fast growth that extracts energy from photosynthesis, low-cost manufacture and modest scalable cultivation, that have been a characteristic for algae-based biofuels and represent an easy and cost-effective production method. In this way, the potential of algal based products will widen and algae may become accessible to both industry and pharma.

Way Forward and future aspects

Generation fetched of algal fuel is still tall compared to non-renewable assets in spite of the utilization of microalgae and macroalgae as potential life form for biotechnological and mechanical application. In spite of the fact that wastewater and pipe gasses have possibly being utilized to cut the supplement fetched but the innovation and gear fetched is still tall. To put it in a nutshell, essential steps utilized for algal biofuel generation and commercialization still remains a major ruin. Additionally, to produce temperate biofuel from microlage chemicals, innovation, power and labor gets to be a major deterrent. Additionally, to produce prudent biofuel from microlage synthetic substances, innovation, power and labor turns into a significant hindrance. In any case, adjustments in the development boundaries, supplement accessibility, and actual changes would foster a superior technique to improve the accumulate of interest in the microalgae. Other than this, creation of significant worth added items is likewise restricted because of the low efficiency of these particles and metabolites in microalgae. Hence, ways of expanding the delivered esteemed items from green growth should be guaranteed. As of late, green union of metallic and non-metallic nanoparticles has been achieved utilizing the cells of green growth. By utilizing physical and substance techniques, the yield of incorporated nanoparticles using green growth is low. Hence, more victory in microalgae development and generation may be accomplished by utilizing genome information of microalgae. In any case, screening the concentration of anti-microbials utilized for change is time-consuming and strain particular. For molecular advancement gene editing tools CRISPR-Cas9, TALEN, and ZFN 17 are being used to edit the genomes of mitochondria, nucleus, and chloroplast of microalgae. Omics methods will be a revolution for developing high end and cutting edge techniques. But for today, CRISPR technology holds the promising solution to change the future of microalgae as biofuel and production of value-added products in more resourceful and commercial way.

REFERENCES

Anderson, R. A. (2005). Traditional microalgae isolation techniques. In Algal Cult. Tech. Elsevier Academic Press. doi:10.1016/B978-012088426-1/50007-X

Anto, S., Mukherjee, S. S., Muthappa, R., Mathimani, T., Deviram, G., Kumar, S. S., Verma, T. N., & Pugazhendhi, A. (2020). Algae as green energy reserve: Technological outlook on biofuel production. *Chemosphere*, *242*, 125079. doi:10.1016/j.chemosphere.2019.125079 PMID:31678847

Apt, K. E., Grossman, A. R., & Kroth-Pancic, P. G. (1996). Stable nuclear transformation of the diatom Phaeodactylum tricornutum. *Mol. Gen. Genet. MGG*, *252*, 572–579. PMID:8914518

Bala, B. K. (2006). Studies on biodiesel from transformation of vegetable oils for diesel engines. *Energy Education Science and Technology*, *15*, 1–45.

Barsanti, L., & Gualtieri, P. (2006). Algae biotechnology. In Algae: Anatomy, Biochemistry, and Biotechnology. CRC Press.

Bharathiraja, B., Chakravarthy, M., Ranjith Kumar, R., Yogendran, D., Yuvaraj, D., Jayamuthunagai, J., Praveen Kumar, R., & Palani, S. (2015). Aquatic biomass (algae) as a future feed stock for bio- refineries: A review on cultivation, processing and products. *Renewable & Sustainable Energy Reviews*, *47*, 635–653. doi:10.1016/j.rser.2015.03.047

Dmytryk, A., Chojnacka, K., & Rój, E. (2018). The methods of algal biomass extraction: toward the application. In *Algae Biomass: Characteristics and applications* (pp. 49–56). Springer. doi:10.1007/978-3-319-74703-3_5

Doppler, P., Kriechbaum, R., Singer, B., & Spadiut, O. (2021). Make microalgal cultures axenic again– a fast and simple workflow utilizing fluorescence-activated cell sorting. *Journal of Microbiological Methods*, *186*, 106256. doi:10.1016/j.mimet.2021.106256 PMID:34082050

Edwards, M. (2010). *Algae 101 Part six: Algal classification.* Algae Industry Magazine.com.

Fabris, M., Abbriano, R. M., Pernice, M., Sutherland, D. L., Commault, A. S., Hall, C. C., Labeeuw, L., McCauley, J. I., Kuzhiuparambil, U., Ray, P., Kahlke, T., & Ralph, P. J. (2020). Emerging technologies in algal biotechnology: Toward the establishment of a sustainable, algae-based bioeconomy. *Frontiers in Plant Science*, *11*, 279. doi:10.3389/fpls.2020.00279 PMID:32256509

Feng, P. Z., Deng, Z., Hu, Z., & Fan, L. (2011). Lipid accumulation and growth of Chlorella zofingiensis in flat plate photobioreactors outdoors. *Bioresource Technology*, *102*(22), 10577–10584. doi:10.1016/j.biortech.2011.08.109 PMID:21955881

Furbank, R. T., & Tester, M. (2011). Phenomics–technologies to relieve the phenotyping bottleneck. *Trends in Plant Science*, *16*(12), 635–644. doi:10.1016/j.tplants.2011.09.005 PMID:22074787

Gao, B., Chen, A., Zhang, W., Li, A., & Zhang, C. (2017). Co-production of lipids, eicosapentaenoic acid, fucoxanthin, and chrysolaminarin by Phaeodactylum tricornutum cultured in a flat-plate photobioreactor under varying nitrogen conditions. *Journal of Ocean University of China*, *16*(5), 916–924. doi:10.100711802-017-3174-2

Gautam, K., Pareek, A., & Sharma, D. K. (2015). Exploiting microalgae and macroalgae for production of biofuels and biosequestration of carbon dioxide—A review. *International Journal of Green Energy*, *12*(11), 1122–1143. doi:10.1080/15435075.2014.893239

Guiry, M. D. (2012). How many species of algae are there? *Journal of Phycology*, *48*(5), 1057–1063. doi:10.1111/j.1529-8817.2012.01222.x PMID:27011267

Hanschen, E. R., & Starkenburg, S. R. (2020). The state of algal genome quality and diversity. *Algal Research*, *50*, 101968. doi:10.1016/j.algal.2020.101968

Hossain, N., Mahlia, T. M. I., & Saidur, R. (2019). Latest development in microalgae-biofuel production with nano-additives. *Biotechnology for Biofuels*, *12*(1), 1–16. doi:10.118613068-019-1465-0 PMID:31139255

Houle, D., Govindaraju, D. R., & Omholt, S. (2010). Phenomics: The next challenge. *Nature Reviews. Genetics*, *11*(12), 855–866. doi:10.1038/nrg2897 PMID:21085204

Huber & Dale. (2009). Biofuels: Grassoline at the Pump. *Sci American.*, *301*(1), 52-59.

Huntley, M. E., Johnson, Z. I., Brown, S. L., Sills, D. L., Gerber, L., Archibald, I., Machesky, S. C., Granados, J., Beal, C., & Greene, C. H. (2015). Demonstrated large-scale production of marine microalgae for fuels and feed. *Algal Research*, *10*, 249–265. doi:10.1016/j.algal.2015.04.016

Jorquera, O., Kiperstok, A., Sales, E. A., Embirucu, M., & Ghirardi, M. L. (2010). Comparative energy life-cycle analyses of microalgal biomass production in open ponds and photobioreactors. *Bioresource Technology*, *101*(4), 1406–1413. doi:10.1016/j.biortech.2009.09.038 PMID:19800784

Khan, S., Siddique, R., Sajjad, W., Nabi, G., Hayat, K. M., Duan, P., & Yao, L. (2017). Biodiesel production from algae to overcome the energy crisis. *Hayati Journal of Biosciences*, *24*(4), 163–167. doi:10.1016/j.hjb.2017.10.003

Kindle, K. L. (1990). High-frequency nuclear transformation of Chlamydomonas reinhardtii. *Proceedings of the National Academy of Sciences of the United States of America*, *87*(3), 1228–1232. doi:10.1073/pnas.87.3.1228 PMID:2105499

Kumar, B. R., Mathimani, T., Sudhakar, M. P., Rajendran, K., Nizami, A. S., Brindhadevi, K., & Pugazhendhi, A. (2021). A state of the art review on the cultivation of algae for energy and other valuable products: Application, challenges, and opportunities. *Renewable & Sustainable Energy Reviews*, *138*, 110649. doi:10.1016/j.rser.2020.110649

Kumar, G., Shekh, A., Jakhu, S., Sharma, Y., Kapoor, R., & Sharma, T. R. (2020). Bioengineering of microalgae: Recent advances, perspectives, and regulatory challenges for industrial application. *Frontiers in Bioengineering and Biotechnology*, *8*, 914. doi:10.3389/fbioe.2020.00914

Kumar, M., Sun, Y., Rathour, R., Pandey, A., Thakur, I. S., & Tsang, D. C. (2020). Algae as potential feedstock for the production of biofuels and value-added products: Opportunities and challenges. *The Science of the Total Environment*, *716*, 137116. doi:10.1016/j.scitotenv.2020.137116 PMID:32059310

LaPanse, A. J., Krishnan, A., & Posewitz, M. C. (2020). Adaptive Laboratory Evolution for algal strain improvement: Methodologies and applications. *Algal Research*, 102122.

Larkum, A. W., Ross, I. L., Kruse, O., & Hankamer, B. (2012). Selection, breeding and engineering of microalgae for bioenergy and biofuel production. *Trends in Biotechnology*, *30*(4), 198–205. doi:10.1016/j.tibtech.2011.11.003 PMID:22178650

Laurens, L. M. L., Chen-Glasser, M., & Mcmillan, J. D. (2017). A perspective on renewable bioenergy from photosynthetic algae as feedstock for biofuels and bioproducts. *Algal Research*, *24*, 261–264. doi:10.1016/j.algal.2017.04.002

Leong, Y. K., Chen, C. Y., Varjani, S., & Chang, J. S. (2021). Producing fucoxanthin from algae–Recent advances in cultivation strategies and downstream processing. *Bioresource Technology*, 126170. PMID:34678455

Li, S., Hu, T., Xu, Y., Wang, J., Chu, R., Yin, Z., Mo, F., & Zhu, L. (2020). A review on flocculation as an efficient method to harvest energy microalgae: Mechanisms, performances, influencing factors and perspectives. *Renewable & Sustainable Energy Reviews*, *131*, 110005. doi:10.1016/j.rser.2020.110005

Luo, L., Qin, B., Song, Y., & Yang, L. (2007). Seasonal and regional variations in precipitation chemistry in the Lake Taihu Basin, China. *Atmospheric Environment*, *41*(12), 2674–2679. doi:10.1016/j.atmosenv.2006.11.037

Mao, X., Chen, S. H. Y., Lu, X., Yu, J., & Liu, B. (2020). High silicate concentration facilitates fucoxanthin and eicosapentaenoic acid (EPA) production under heterotrophic condition in the marine diatom Nitzschia laevis. *Algal Research*, *52*, 102086. doi:10.1016/j.algal.2020.102086

Mullick, A., & Neogi, S. (2017). A review on acoustic methods of algal growth control by ultrasonication through existing and novel emerging technologies. *Reviews in Chemical Engineering*, *33*(5), 469–490. doi:10.1515/revce-2016-0010

Munir, N., Imtiaz, A., Sharif, N., & Naz, S. (2015). Optimization of growth conditions of different algal strains and determination of their lipid contents. *J Anim Plant Sci*, *25*(2), 546–553.

Munoz, ~. C. F., de Jaeger, L., Sturme, M. H. J., Lip, K. Y. F., Olijslager, J. W. J., Springer, J., Wolbert, E. J. H., Martens, D. E., Eggink, G., Weusthuis, R. A., & Wijffels, R. H. (2018). Improved DNA/protein delivery in microalgae – A simple and reliable method for the prediction of optimal electroporation settings. *Algal Research*, *33*, 448–455. doi:10.1016/j.algal.2018.06.021

Muñoz, C. F., Südfeld, C., Naduthodi, M. I., Weusthuis, R. A., Barbosa, M. J., Wijffels, R. H., & D'Adamo, S. (2021). Genetic engineering of microalgae for enhanced lipid production. *Biotechnology Advances*, *52*, 107836. doi:10.1016/j.biotechadv.2021.107836 PMID:34534633

Ng, I. S., Tan, S. I., Kao, P. H., Chang, Y. K., & Chang, J. S. (2017). Recent developments on genetic engineering of microalgae for biofuels and bio-based chemicals. *Biotechnology Journal*, *12*(10), 1600644. doi:10.1002/biot.201600644 PMID:28786539

Panis, G., & Carreon, J. R. (2016). Commercial astaxanthin production derived by green alga Haematococcus pluvialis: A microalgae process model and a techno- economic assessment all through production line. *Algal Research*, *18*, 175–190. doi:10.1016/j.algal.2016.06.007

Park, J. B. K., Craggs, R. J., & Shilton, A. N. (2013). Enhancing biomass energy yield from pilot-scale high rate algal ponds with recycling. *Water Research*, *47*(13), 4422–4432. doi:10.1016/j.watres.2013.04.001 PMID:23764593

Parmentier, D., Manhaeghe, D., Baccini, L., Van Meirhaeghe, R., Rousseau, D. P., & Van Hulle, S. (2020). A new reactor design for harvesting algae through electrocoagulation-flotation in a continuous mode. *Algal Research*, *47*, 101828. doi:10.1016/j.algal.2020.101828

Pereira, H., Schulze, P. S. C., Schüler, L. M., Santos, T., Barreira, L., & Varela, J. (2018). Fluorescence activated cell-sorting principles and applications in microalgal biotechnology. *Algal Research*, *30*, 113–120. doi:10.1016/j.algal.2017.12.013

) Ramasamy, S., Arun, S., & Pakshirajan, K. (2020). An overview of algal photobioreactors for resource recovery from waste. *Bioreactors*, 215-248.

Rawat, I., Kumar, R. R., Mutanda, T., & Bux, F. (2013). Biodiesel from microalgae:A critical evaluation from laboratory to large scale production. *Applied Energy*, *103*, 444–467. doi:10.1016/j.apenergy.2012.10.004

Remelli, W., & Santabarbara, S. (2018). Excitation and emission wavelength dependence of fluorescence spectra in whole cells of the cyanobacterium Synechocystis sp. PPC6803: Influence on the estimation of Photosystem II maximal quantum efficiency. *Biochimica et Biophysica Acta. Bioenergetics*, *1859*(11), 1207–1222. doi:10.1016/j.bbabio.2018.09.366 PMID:30297025

Richmond, A. (2000). Microalgal biotechnology at the turn of the millenium: A personal view. *Journal of Applied Phycology*, *12*(3/5), 441–451. doi:10.1023/A:1008123131307

Rippka, R. (1988). Isolation and purification of cyanobacteria. *Methods in Enzymology*, *167*, 3–27. doi:10.1016/0076-6879(88)67004-2 PMID:3148836

Rippka, R., Waterbury, J. B., & Stanier, R. Y. (1981). Isolation and purification of cyanobacteria: some general principles. The Prokaryotes, 212–220. doi:10.1007/978-3-662-13187-9_8

Rutten, T. P. A., Sandee, B., & Hofman, A. R. T. (2005). Phytoplankton monitoring by high performance flow cytometry: A successful approach? *Cytom. A.*, *64*(1), 16–26. doi:10.1002/cyto.a.20106 PMID:15688354

Saad, M. G., Dosoky, N. S., Zoromba, M. S., & Shafik, H. M. (2019). Algal biofuels: Current status and key challenges. *Energies*, *12*(10), 1920. doi:10.3390/en12101920

Sahin, M. S., Khazi, M. I., Demirel, Z., & Dalay, M. C. (2019). Variation in growth, fucoxanthin, fatty acids profile and lipid content of marine diatoms Nitzschia sp. and Nanofrustulum shiloi in response to nitrogen and iron. *Biocatalysis and Agricultural Biotechnology*, *17*, 390–398. doi:10.1016/j.bcab.2018.12.023

Schenk, P. M., Thomas-Hall, S. R., Stephens, E., Marx, U. C., Mussgnug, J. H., Posten, C., Kruse, O., & Hankamer, B. (2008). Second generation biofuels: High-efficiency microalgae for biodiesel production. *BioEnergy Research*, *1*(1), 20–43. doi:10.100712155-008-9008-8

Shen, Y., Yuan, W., Pei, Z. J., Wu, Q., & Mao, E. (2009). Microalgae mass production methods. *Transactions of the ASABE*, *52*, 1275–1287. doi:10.13031/2013.27771

Shurin, J. B., Burkart, M. D., Mayfield, S. P., & Smith, V. H. (2016). Recent progress and future challenges in algal biofuel production. *F1000 Research*, *5*, 5. doi:10.12688/f1000research.9217.1 PMID:27781084

Syed, M. S., Rafeie, M., Vandamme, D., Asadnia, M., Henderson, R., Taylor, R. A., & Warkiani, M. E. (2018). Selective separation of microalgae cells using inertial microfluidics. *Bioresource Technology*, *252*, 91–99. doi:10.1016/j.biortech.2017.12.065 PMID:29306136

Tsuzuki, M., Okada, K., Isoda, H., Hirano, M., Odaka, T., Saijo, H., Aruga, R., Miyauchi, H., & Fujiwara, S. (2019). Physiological properties of photoautotrophic microalgae and cyanobacteria relevant to industrial biomass production. *Marine Biotechnology (New York, N.Y.)*, *21*(3), 406–415. doi:10.100710126-019-09890-1 PMID:30927152

Ugwu, C. U., Aoyagi, H., & Uchiyama, H. (2008). Photobioreactors for mass cultivation of algae. *Bioresource Technology*, *99*(10), 4021–4028. doi:10.1016/j.biortech.2007.01.046 PMID:17379512

Vu, C. H. T., Lee, H. G., Chang, Y. K., & Oh, H. M. (2018). Axenic cultures for microalgal biotechnology: Establishment, assessment, maintenance, and applications. *Biotechnology Advances*, *36*(2), 380–396. doi:10.1016/j.biotechadv.2017.12.018 PMID:29292155

Waterbury, J. B., & Stanier, R. Y. (1981). Isolation and growth of cyanobacteria from marine and hypersaline environments. The Prokaryotes, 221–223. doi:10.1007/978-3-662-13187-9_9

Waterbury, J. B., & Willey, J. M. (1988). Isolation and growth of marine planktonic cyanobacteria. *Methods in Enzymology*, *167*, 100–105. doi:10.1016/0076-6879(88)67009-1

Zhang, K., Zhang, F., & Wu, Y. R. (2021). Emerging technologies for conversion of sustainable algal biomass into value-added products: A state-of-the-art review. *The Science of the Total Environment*, *784*, 147024. doi:10.1016/j.scitotenv.2021.147024 PMID:33895504

Zhang, Y. T., Jiang, J. Y., Shi, T. Q., Sun, X. M., Zhao, Q. Y., Huang, H., & Ren, L. J. (2019). Application of the CRISPR/Cas system for genome editing in microalgae. *Applied Microbiology and Biotechnology*, *103*(8), 3239–3248. doi:10.100700253-019-09726-x PMID:30877356

Chapter 4
Cell Biology and Microbial Interactions in Algal Cells

Nairouz H. Mahmoud
https://orcid.org/0000-0002-8561-0128
Alexandria University, Egypt

Jihan H. Mohamed
https://orcid.org/0000-0002-8469-8009
Alexandria University, Egypt

Basma A. Elsherbiny
https://orcid.org/0000-0003-1663-9565
National Institute of Oceanography and Fisheries (NIOF), Egypt

Samia S. Abouelkheir
https://orcid.org/0000-0001-8559-6787
National Institute of Oceanography and Fisheries (NIOF), Egypt

Sabrina Mohamed Moffit
https://orcid.org/0000-0002-7454-2027
Universiti Kebangsaan Malaysia, Malaysia

Bahaa Abdella
https://orcid.org/0000-0001-5237-263X
Kafrelsheikh University, Egypt

ABSTRACT

Algae, including cyanobacteria, dominate aquatic habitats. They are the principal producers of aquatic environments. On the other hand, microbes, are essential algal helpers and are known as holobionts. Holobionts are algae-associated microbes that include bacteria, fungi, and viruses. Over millions of years, many interaction mechanisms between algal cells and their halobiont have evolved. These interactions include mutualism, commensalism, and parasitism. These interactions are critical for ecosystem resistance and resilience. Microbes, for example, regulate algal cell proliferation by producing toxic metabolites that control the algal growth. Alternatively, the production of vitamins and growth factors by microbes might promote algal cell proliferation. Moreover, in biotechnological applications, the algae and bacteria co-cultures are very promising as a sustainable application to persistent environmental issues and green energy solutions. Various mechanisms of intracellular and extracellular algae microbe interactions were discussed in this chapter. This is an endeavor to get knowledge about algae-microbe interactions for biomass-based energy solutions and other environmental applications.

DOI: 10.4018/978-1-6684-2438-4.ch004

INTRODUCTION

In the aquatic environment, algae are the photosynthetic primary producers. Based on the size they are called microalgae for microscopic and macroalgae for macroscopic one, while prokaryotic algae are identified as blue-green algae or cyanobacteria (Kouzuma & Watanabe, 2015). The surface of the algal body and the micro-environment adjacent to the algal surface is a rich place with oxygen produced, making it a perfect habitat for other microorganisms. Considering that, microorganisms are ubiquitous, it is not surprising to find a variety of microbes inhabiting the phycosphere which is the microscale mucus region on the surface of the algal body. Despite the remarkable proximity of algae (primary producer) and microbe (decomposer) in nature, the classical idea of a primary producer and a decomposer located at the two ends of a food web is still accurate. Previously, the presence of bacteria in algal culture was considered as contamination but now it has been found that the associated microbes have a profound influence on agal development, growth, physiology, and disease susceptibility.

Different interaction relationships exist between the agal cells and their associated microbes. These interactions could be harmful or beneficial for one or both interacted parties. Generally, commensalism, predation (parasitism), and mutualism are the different interactions between the associated microbes and the algal host. Both microbes and hosts affect each other synergistically making it difficult to understand each of them independently from each other (Egan et al., 2013; Fuentes et al., 2016).

Algae and bacteria are abundant in the aquatic environment and both are vital for the sustainability of the aquatic ecosystem (González-Olalla et al., 2018). The evolved mechanisms of interactions between them are extraordinarily complex and still not clearly understood. Consequently, the study of the algal-microbe interactions has changed in the last two decades, it has been transitioned from basic research to biotechnological applications such as improving algae metabolism, bioactive compound production, biomass, and biofuel production. On the other hand, the use of microbes as a biocontrol agent against harmful algal blooms (HABs).

Furthermore, it is thought that the related bacteria diminish phototrophic microalgal photosynthetic oxygen tension by using O_2 as an electron acceptor. Many algae are auxotrophic for cobalamin (vitamin B12), hence the algae-bacteria symbiosis is thought to benefit algae by supplying critical nutrients including vitamins and other substances (Croft et al., 2005). Under iron-deficient environments, siderophores generated by some bacteria can enhance microalgal development (Amin et al., 2009). In addition, some algae have an extracellular sheath (made up of a variety of sugars) that provides attachment sites as well as organic carbon sources for bacterial development and photosynthetic oxygen as an electron acceptor for aerobic respiration (Park et al., 2008).

The interaction between microbes and the algal host is a dynamic process that involves both. For instance, mutualism happened when the interacted species got benefited from each other, this could illustrate by the exchange of nutrients, minerals, and vitamins (Fuentes et al., 2016). On the other hand, parasitism happened when one party benefits while the host is harmed in the case of algicidal microbes (Du et al., 2020).

This book chapter aimed to elucidate the cell biology and algae-microbe interactions in various biotechnological applications that have been leveraged for the development of sustainable energy, environment, and food.

WASTEWATER BIOREMEDIATION/PHYCOREMEDIATION

Aquatic ecosystems receive a diverse range of nutrients, pollutants, and toxins through mixing with untreated wastewaters (Rawat et al., 2011). These pollutants and toxins are extremely persistent in the environment, and often resistant to chemical or bacterial degradation (Avagyan, 2010). As a result of that, phycoremediation has gained attention to overcome the ecological crisis. Phycoremediation could be defined as a cost-effective and eco-friendly process with no secondary pollution produced. It uses macroalgae and microalgae for sequestration or the bioaccumulating ability of these bio-systems.

Currently, different strategies and culturing techniques are being used. The most used cultures are algal polycultures, which are constituted exclusively by photosynthetic microorganisms (eukaryotic and/or prokaryotic), algae-bacteria/algae-fungi consortia, and biofilms/mats which are constituted by photosynthetic microorganisms and heterotrophic bacteria. These heterogeneous systems utilize the excessive presence of phosphorus, nitrate, and ammonia, leading to the generation of valuable by-products through their application either in wastewater treatment, CO_2 sequestration, or enhancing biomass production (Emparan et al., 2019; Krug et al., 2020). The produced biomass could be reused as a renewable source of energy such as biodiesel, biofuel, biogas, hydrogen gas, and other by-products such as glycerol and bio-fertilizers (Divya & Santhanam, 2018; Scognamiglio et al., 2021). Various researches have dealt with the algae-bacteria consortium for metal bioremediation and organic pollutant degradation (Tang et al., 2010). It is also established that algal-bacterial interactions are effective in the breakdown of organophosphate pesticides such as monocrotophos, quinalphos, and methyl parathion (Subashchandrabose et al., 2013). Furthermore, the degradation of several hazardous pesticides, including Dichlorodiphenyltrichloroethane (DDT), atrazine, and α-endosulfan, has been reported (Subashchandrabose et al., 2013).

The association between algal bacteria helps in system stability by sharing nutrients, increasing resistance against intruders, and improving the overall uptake of nutrients (Zhang et al., 2017). Microalgae, on the other hand, have a negative impact on bacterial development by changing variables including pH, dissolved oxygen, temperature, and secondary metabolites, and vice versa.

Consenting bacterial symbiont, *Azospirillum brasilense,* a microalgae growth-promoting bacterium has helped *Chlorella vulgaris* in excessive removal of NH_4 -N by 93% & phosphorus removal by 83%. However, *Chlorella. vulgaris* alone was less efficient, as it has only removed 85% of the ammonium and 33% of the phosphorus (De-Bashan et al., 2002). In another study, co-culturing of *Planktothrix isothrix* and *Chlorella vulgaris* could only remove 80% of nitrogen from municipal wastewater (Silva-Benavides & Torzillo, 2012). On the other hand, *Chlorella sorokiniana* cultivated in pretreated swine slurry with activated sludge bacterial consortium had successfully removed 94-100% of the ammonium and 70-90% of the phosphorus (De Godos et al., 2009).

In an outdoor cultivation system, a co-culture of *Chlorella* sp. and *Proteobacteria* removed 72% of total nitrogen(TN), whereas an indoor cultivation system employing piggery wastewater removed 100% total phosphorus (TP) and 83% of zinc. (García et al., 2017). Bacteria can effectively oxidize the chemical oxygen demand (COD), and release CO_2 as by-products. On the other hand, algae can consume the produced CO_2 and convert it to biomass employing photosynthesis, and then releasing O_2 promotes bacterial growth (Sial et al., 2020). Furthermore, a consortium based on the bacteria *Bacillus licheniformis* and the alga *Chlorella vulgaris* removed 86.5% of COD, 88.9% of TN, and 80.2% of TP from synthetic wastewater with outstanding efficiency (Ji et al., 2018).

In wastewater treatment, the interaction of algae and fungus proved beneficial. For example, a mixed culture of *C. vulgaris* and *Mucor indicus* lowered phosphate, total ammonia, and nitrogen levels in

Figure 1. Algae-microbe symbiosis interaction and their potential influence on wastewater treatment and energy production.

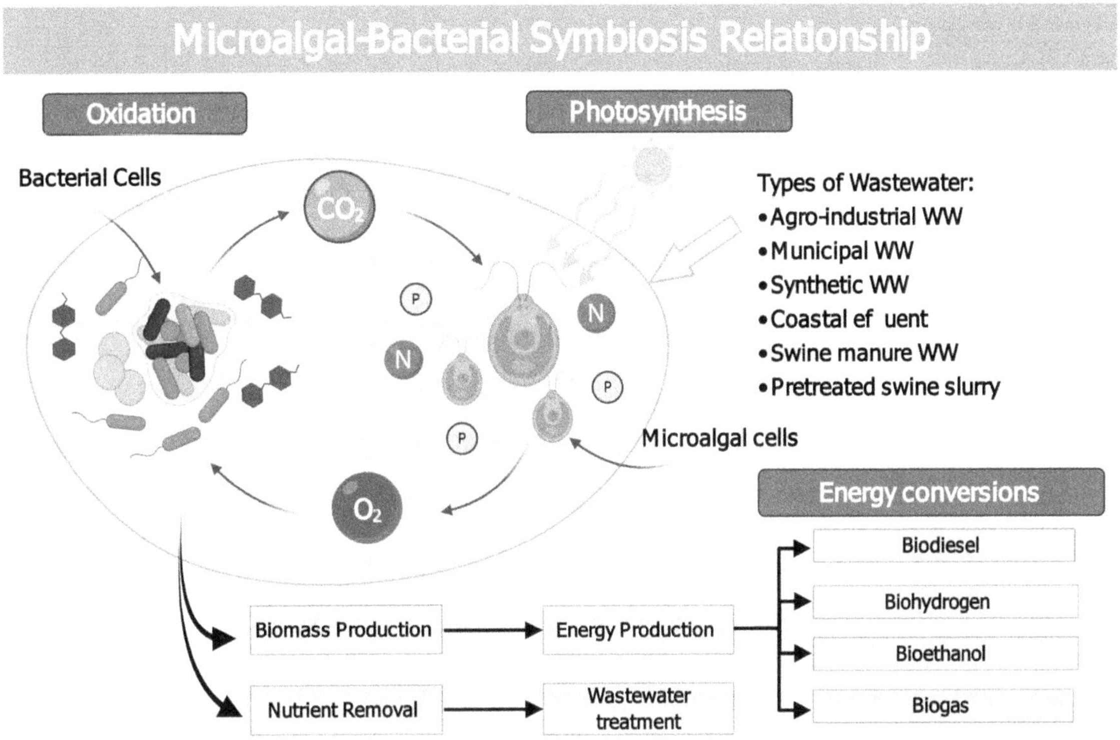

aquaculture effluent to virtually undetectable levels (Barnharst et al., 2018). Furthermore, algal-fungal co-cultivation was used to degrade biogas and treat household sewage effluent, removing 81% of TP and COD (Xu et al., 2017). Similarly, co-cultivation of *C. vulgaris* with *Ganoderma lucidum* reduced TP, TN, and COD contents from swine wastewater treatment by 84%, 74%, and 79%, respectively (Shahid et al., 2020). Furthermore, for biogas degradation, a combination of *Pseudokirchneriella subcapitata* and G. lucidum successfully eliminated CO_2 with an efficiency of 84% (Guo et al., 2017).

Figure 1 describes the cooperation in microalgal/bacterial systems in the treatment of wastewater in a symbiotic relationship, in which the microalgae produce oxygen during photosynthesis that is used by bacteria metabolism and releases CO_2 that is being consumed by the alga.

Table 1 summarizes examples of algae-microbe interactions that have been shown successful and promising results in wastewater treatment.

BIOMASS PRODUCTIVITY

Microalgal biomass is rich in carbohydrates, lipids, and proteins that can be used as a feedstock for microbial fermentation to produce other commercially important products (fatty acids, carotenoids, polyhydroxy alkenoates, etc.) (Arora et al., 2021). Bacteria in wastewater, on the other hand, might stifle microalgal development by competing for space and nutrients. Bacteria grow quicker than microalgae.

Table 1. Algae-microbe symbiosis in phosphorous and nitrogenous compounds removal from waste water

Algae	Associated microbe	Wastewater	Nutrient removal (%)	References
Algae-bacteria association				
Nannochloropsis gaditana	Spontaneous municipal wastewater microorganisms	Municipal wastewater	NH_4^+ -N (99.4%) PO_4 -P (67.7%)	Mhedhbi et al., (2020)
Chlorella vulgaris	*Azospirillum brasilense*	Synthetic wastewater	NH_4^+ -N (93%) P (75%)	De-Bashan et al., (2002)
Chlorella vulgaris	*Azospirillum brasilense*	Synthetic wastewater	P (31.5%) N (22%)	Perez-García et al., (2010)
Chlorella sorokiniana	*Activated sludge consortium*	Pretreated swine slurry	N (94-100%) P (70-90%)	De Godos et al., (2009)
Filamentous blue-green algae	*Flavobacteria, Bacteroidia, Gammaproteobacteria,* and *Betaproteobacteria*	Municipal wastewater	N (88.3%) PO_4 -P (64.8%)	Su et al., (2011)
Algae-fungi association				
Chlorella vulgaris UMN235	*Aspergillus* spp. UMN F01 and UMN F02	Swine manure wastewater	NH_4^+-N (23.23%) N (44.68%) P (84.70%)	Zhou et al., (2012)

An appropriate pretreatment procedure should be used to eliminate competing microorganisms for the effective production of algal biomass (Santiago et al., 2013).

It has been found that the consortium works better than a single organism, in terms of survival by exchanging micronutrients (vitamins) and macronutrients which help to improve biomass productivity and quality, as well as nutrient removal in wastewater bioremediation (Renuka et al., 2013). Table 2 lists examples of microalgae and bacteria-mediated substances. Furthermore, high-value metabolites such as omega-3 fatty acids, pigments, amino acids, and high sugar content may be present in algal biomass. As a result, the remaining biomass may either be immediately transformed into energy through thermo-chemical transformation or utilized to make biofuels by biological fermentation or transesterification following the extraction of high-value components (Shahid et al., 2020).

Chinnasamy et al. (2010) investigated a group of native microalgal strains capable of producing 1.47 g/L biomass when grown in sterilized carpet mill effluent with elevated CO_2 levels (6%), at a temperature of 25 °C. In another study, Riaño et al. (2011) have found that microalgae consortia can produce up to 0.555 g/L of biomass when cultivated in fish processing wastewater. Thus, the development of biorefineries for renewable biomass feedstocks production is crucial nowadays and requires a constant supply of algal biomass. Based on this, using co-cultures is a very promising application in the production of sustainable fuels such as biohydrogen, bioethanol, and biogas. It is worth noting that the biomass, growth rate, and lipid production of *Chlorella vulgaris* and *Stenotrophomona smaltophilia* co-cultures all increased by 21.9%, 20.4%, and 18%, respectively. (Scognamiglio et al., 2021). Meanwhile, co-culturing of *Chlorella vulgaris* with the bacterium *Mesorhizobium sangaii* (starting ratio of 40:1) resulted in 1.5, 2.2, and 3.3 times greater biomass, lipid content, and algal productivity than pure algaculture (Wei et al., 2020). Besides, the biomass produced through wastewater treatment usually contains 30%–50% dry weight (ash), which decreases the conversion efficiency in biofuel production. Furthermore, excessive biomass could cause secondary environmental complications, such as the release of odors and groundwater contamination, if not treated properly (Carpio et al., 2018).

Table 2. Microalgae and bacteria-mediated substances in symbiosis.

Bacteria	Microalgae	Mediated substance		Reference
		Algal origin	**Bacterial origin**	
Reugeria pomeroyi DSS-3	*Thalassiosira pseudonana CCMP1335*	2,3-Dihydroxy-propane-1-sulphonate	Vitamin B12	Girard, (2019)
Roseobacter sp. and *Marinobacter* sp.	*S. trochoidea*	Organic molecules	Vibrioferrin	Santos & Reis, (2014)
Alteromonas sp.	Diatoms	ND*	Antibiotic, 2-*n*-pentyl-4-quindinol	Kazamia et al., (2012)
Bacillus sp.	*Phaeocystis globosa*	ND*	Prolyl-methionine and hypoxanthine	Kazamia et al., (2012)
Azospirillum brasilense	*Chlorella sorokiniana*	ND*	Riboflavin /Lumichrome	Wang et al., (2020)
Azospirillum brasilense	*Chlorella vulgaris*	ND*	Siderophore-mediated N_2 fixation	Leyva et al., (2014)
ND: Not defined				

GREENHOUSE GAS REDUCTION

The production of greenhouse gases and the challenge of global warming coupled with the foreseen shortage of fossil fuel has turned the algae to be the potential successful alternative for these two challenging issues (Rawat et al., 2011). Major ongoing thrust areas of algal biotechnology are concerned with screening the suitable microalgae strains and optimizing their growth requirements as sustainable sources of biomass for a sustainable source of biofuels and at the same time reduction of CO_2 emission (Rawat et al., 2011). This could be accomplished by employing a mixed microalgae-bacteria in a photobioreactor as a downstream technology for anaerobic processes and methanotrophs-based biotransformation to effectively remove CH_4 and CO_2 through microalgae photosynthesis.

BIOMASS CONVERSIONS-BASED ENERGY

Biodiesel Production

Biodiesel is a renewable fuel produced from microbial biomass produced through transesterification separation of triacylglycerol and other lipids, by methanol and catalyst to produce fatty acid methyl esters. The maximization of the lipid content of the cultivated algae is the priority in efficient biodiesel production. It appeared that the associated microbes synthesize self-inhibitors and trace elements, which impact the deposition of lipids and fatty acids content in algal cells along with growth promotion (Sial et al., 2020). Co-culturing the green microalga *Auxenochlorella protothecoides* associated with *E. coli* enhanced algal growth up to six-fold and has doubled the neutral lipid content composition also fostering the saturated fatty acids and oleic acids production in comparison with growth and lipid content in axenic cultures (Higgins & VanderGheynst, 2014; Scognamiglio et al., 2021). Furthermore, the marine microalga *Tetraselmis striata* was identified as a biodiesel candidate due to its high lipid content and rapid growth in the presence of two microalgae symbionts, *Pelagibaca bermudensis*, and *Stappia* sp., both of which showed growth-promoting activities (Park et al., 2017). Furthermore, the microbial association with algae was extended to a fungal symbiont, it has been shown between the alga *Nannochloropsis oceanica* and the fungus *Mortierella elongate* for increasing lipids and biofuel production by the algal species (Du et al., 2019). All these findings point to the intriguing potential of microalgal-microbial consortia in improving biodiesel quality and quantity.

Biohydrogen Production

Hydrogen is one of the promising alternatives to fossil fuels derivatives and one of the cleanest sources of energy. The only by-product of its burning is water vapor, with no CO_2 released to the environment (Lam et al., 2019). Currently, the primary challenge in H_2 production by microbial anaerobic fermentation is the poor yield of hydrogen (Jiang et al., 2017). Thus, hydrogen production by other means is necessary. Microalgae generate hydrogen gas from what is known as photo-fermentation (Wang & Yin, 2018). In photo-fermentation, the algal cell produces hydrogen gas from the hydrogenase enzyme under the lack of oxygen. However, this enzyme is extremely sensitive to O_2 (Melis & Happe, 2001). Theoretically, the immediate removal of O_2 generated from photosynthesis will be favored by the hydrogenase enzyme to function. Thereby, heterotrophic bacteria could provide a localized anaerobic environment ideal for algal H_2 production through immediate consumption of the O_2 without affecting the photosynthetic process (Fuentes et al., 2016).

For instance, bacteria from the genera *Rhodococcus* sp., *Brevundimonas* sp., and *Leifsonia* sp. were discovered to boost H_2 generation from the microalga *Chlamydomonas* sp. by eliminating the O_2 through effective respiration, which is required for the activation of iron-dependent hydrogenase from *Chlamydomonas sp.* (Lakatos et al., 2014). This phenomenon is not limited to natural interactions between microalgae and bacteria; instead, artificial algal-bacterial communities can produce the same results. When a hydrogenase-deficient *Escherichia coli* was employed as an artificial symbiotic bacterium to *Chlamydomonas*, the greatest hydrogen production was achieved.

Bioethanol Production

Another valuable product that could be obtained from the algal biomass is bioethanol. After collecting or extracting useful products from the biomass, the remaining biomass might be processed into ethanol or transformed into biogas (Yao et al., 2019).

Microalgae may yield up to 40% dry weight of starch granules, which can be utilized in ethanol production (Ramanan et al., 2016). Amylase from the marine bacteria *Pseudoalterimonas undina* was used in the saccharification process of starch from microalgae to produce ethanol (Matsumoto et al., 2003). The usage of bacterial enzymes, as well as yeast fermentation, is potentially very promising for bioethanol production from algal biomass. For instance, the starch formed in the microalga *Chlamydomonas reinhardtii* was hydrolyzed using amylase from the bacterium *Bacillus licheniformis* and was fermented by the brewer's yeast *Saccharomyces cerevisiae* for ethanol production (de Farias Silva & Bertucco, 2016). Table 3 summarizes some of the symbiosis relationships that have been studied between different algal and microbial species in the production of different high-energy compounds.

Table 3. Symbiotic microbes in microalgae biofuel production.

Biofuel	Microalgae	Microorganism	References
Microalgae/ bacteria			
Biohydrogen	*Chlamydomonas* CC124	*Brevundimonas* sp., *Rhodococcus* sp., *Leifsonia* sp.	Lakatos et al., (2014)
	Chlamydomonas	*Escherichia coli*	Lakatos et al., (2014)
Biogas	*Botryococcus braunii* & *Nannochloropsis gaditana*	*Aeromonas* sp., *Raoultella* sp. & *Raoultella ornithinolytica*	Muñoz et al., (2014)
Bioethanol	Green microalgae NKG 120701	*Pseudoalteromonas undina*	Matsumoto et al., (2003)
	Chlamydomonas reinhardtii	*Bacillus licheniformis*	de Farias Silva & Bertucco, (2016)
Biodiesel	*Chlorella vulgaris*	*Pseudomonas* sp.	Bell et al., (2016)
	Tetraselmis striata	*Pelagibaca bermudensis*	Park J. et al., (2017)
	Auxenochlorella protothecoides	*Escherichia coli*	Higgins & VanderGheynst, (2014)
Microalgae/ cyanobacteria			
Biodiesel	*Chlorella vulgaris*	*Leptolyngbya* sp.	Lutzu & Dunford, (2018)
Microalgae/yeasts			
Biodiesel	*Spirulina platensis*	*Rhodotorula glutinis*	Xue et al., (2010)
	Chlorella vulgaris	*R. glutinis*	Cheirsilp et al., (2011)
	Chlorella sp. KKU-S2	*Thorulaspora globosa* YU5/2, *T. malee* YU5/2	Papone et al., (2012)
	Chlorella prototechoides	*Rhodosproridium turoloides*	Santos et al., (2013)

Biogas Production

Just like livestock and food wastes, microalgal biomass can serve as a candidate for biogas production in a process called anaerobic digestion (AD) or fermentation. Consisting of mainly four steps, the machinery of AD is significantly governed by a diverse group of microbes. While these microbes pose similar needs for macronutrients carbon, nitrogen, phosphorus, and Sulfur, their needs for trace elements are tailored more accordingly to their performance (Demirel & Scherer, 2011; Hullebusch et al., 2019). As reported by Wintsche et al., (2016), lacking a supply of trace elements such as cobalt may cause acidification and malfunction of biomass long-term mono-digestion, even at low organic rates. Besides, lacking both cobalt and nickel had been linked to decreased efficacy of biogas production. Other trace elements also crucial for AD efficiency and optimal methane production were molybdenum, tungsten, and selenium.

Degradation of algal cell walls via the cooperation of different microbial species is required before AD. As reported by Muñoz et al., (2014), the cell walls of microalgae *Botryococcus braunii* and *Nannochloropsis gaditana* have been degraded by nine bacterial strains via hydrolases activity. However, bioaugmentation with *Clostridium thermocellum*, a cellulolytic and hydrogenogenic bacteria, upon *Chlorella vulgaris* biomass, was shown to generate more methane and H_2, enhancing the overall biogas yield and efficiency (Lü et al., 2013)

Theoretically, microalgal biogas is a mixture composition of 60% methane and 40% carbon dioxide generated from AD that consists of four steps: hydrolysis, acidogenesis, acetogenesis, and methanogenesis.

Table 4. Microbes in biogas production from algal biomass.

Step	Classification	Microbe	References
Hydrolysis	Bacteria	*Cellulomonas, Clostridium, Bacteroides, Bifidobacterium, Butyrivibrio, Thermomonospora, Ruminococcus, Erwinia, Acetovibrio, Microbispora, Proteobacteria, Pseudomonas, Bacillus, Streptococcus*, and *Eubacterium*	Xu et al., (2018) Jain et al., (2015)
Acidogenesis			
Acetogenesis	Bacteria	*Acetobacterium, Sporomusa, Clostridium, Ruminococcus*, and *Eubacterium*	Jain et al., (2015)
Methanogenesis	Archaea	*Methanosaeta, Methanosarcina, Methanobacterium, Methanothermobacter, Methanospirillum, Methanobrevibacter*, and *Methanoculleus*.	Xu et al., (2018) Shin et al., (2010) Ziganshin et al., (2013)

In hydrolysis, the organic polymers protein, lipid, and carbohydrate are broken down into amino acids, glycerol, and long-chain fatty acids, as well as oligo- and monosaccharides, respectively (Adnan et al., 2019; Karuppiah & Azariah, 2019). During acidogenesis, the compounds are converted by fermentative bacteria to a mixture of short-chain volatile fatty acids as well as other minor products like CO_2, H_2, and CH_3COOH. Thereafter, in acetogenesis, the organic acids were then converted by acetogenic bacteria to acetate, CO_2 and H_2. These substrates are then taken up by methanogenic bacteria to generate CH_4 in methanogenesis (Dang et al., 2016; Zawieja & Worwąg, 2021).

Biogas production parameters like microbial growth kinetic and nutritional requirements as well as the microbial community mediate environmental conditions (temperature and pH). This presents varying nutritional demands and therefore confers slight nuances in the AD procedures. For successful conversion of the agal biomass into biogas, a diversity of microbes from eubacteria and archaea should work in harmony as summarized in Table 4.

To enhance sustainability solutions, coupling microalgal biogas production with sugarcane biorefinery seems to bring a promising outcome. In sugarcane biorefinery, sugars and bioethanol are generated as the final products along with other wastes (bagasse and vinasse) and by-products (e.g., CO_2). Combustion of bagasse can further yield CO_2 and heat. The overall CO_2 generated can be fed to microalgae. Meanwhile, the heat generated can facilitate biogas production by subjecting the microalgal biomass in AD either through mono- or co-digestion with vinasse (Abdalla et al., 2018; Klein et al., 2018; Zewdie & Ali, 2020)

BIOGAS CONVERSION TO VALUABLES

Notably, methanotrophic bacteria have been gaining attention to further leverage biogas conversion into various valuables for other potential applications. As reported by Henard et al., (2018), the cultivation of certain methanotrophic bacteria, including *Methylomicrobium alcaliphilum* $20Z^R$ (a rifampicin-resistant derivative of the strain 20Z), *Methylosinus trichosporium* OB3b, and *Methylococcus capsulatus,* within biogas confinement is proved viable. Among other strains used, *M. alcaliphilum* $20Z^R$ conferred the highest growth and biogas conversion efficacy due to having a metabolism that can properly utilize biogas to alter its intracellular redox state that may potentially be leveraged as a catalyst in biogas production. Furthermore, it was believed that the engineered *M. alcaliphilum* $20Z^R$ produced lactate from

Figure 2. Schematic representation of microbial fermentation and algae-microbe co-culture for energy production.

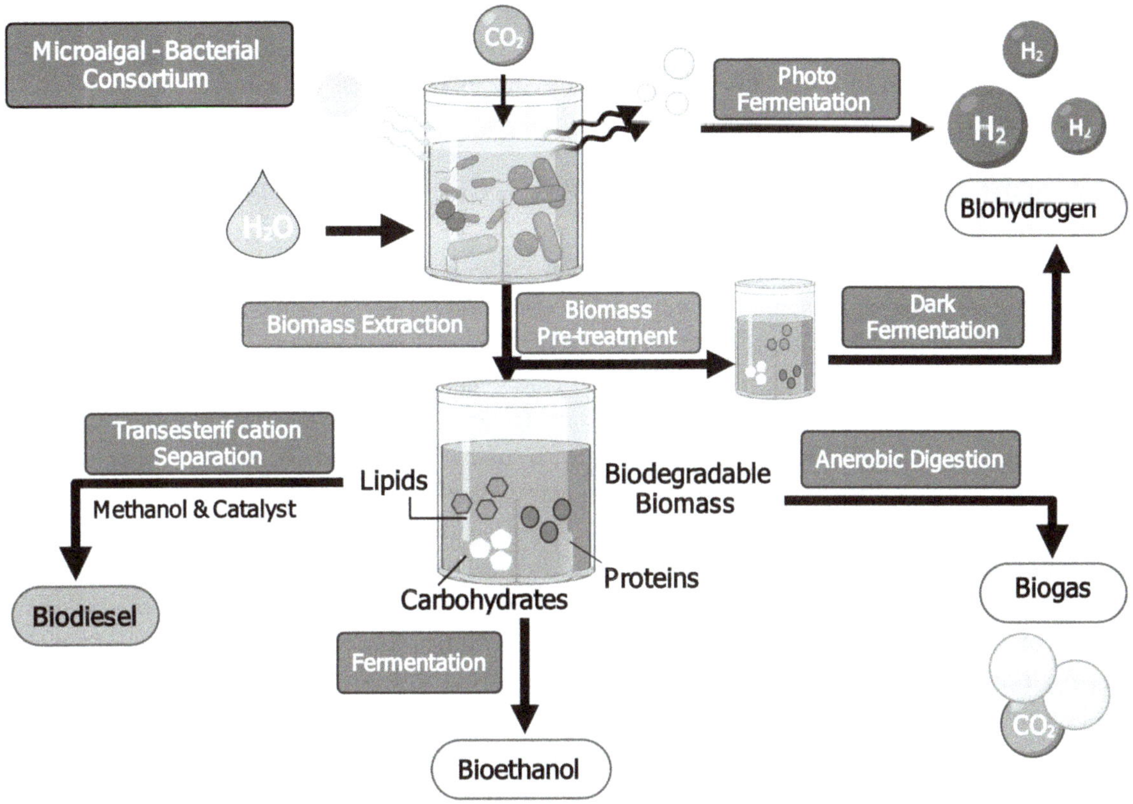

pyruvate, in a continuous gas flow bioreactor, by using lactate dehydrogenase. (Eiteman & Ramalingam, 2015). Lactate might serve as a chemical precursor for bioplastic production and can also be leveraged to generate other chemical constituents like acrylic acid, propylene glycol, and pentanol (Eiteman & Ramalingam, 2015).

Biogas can also be converted into microbial biomass via co-culture. As reported by Hill et al., (2017), cyanobacterium, *Synechococcus* PCC 7002, and methanotrophic bacterium, *Methylomicrobium alcaliphilum* 20Z were co-cultured in a steady-state continuous process. In this process, light-limited *Synechococcus* PCC 7002 produced O_2 via oxygenic photosynthesis which is necessary for both CH_4 oxidation and growth of methanotroph *M. alcaliphilum* 20Z. The steady-state metabolic coupling has been achieved as the O_2 consumption of *M. alcaliphilum* 20Z and the photosynthetic rate of *Synechococcus* PCC 7002 were both limited. Since this co-culture technique is scalable and flexible, customization of bioprocesses development with simultaneous greenhouse gases remediation can be achieved, especially those greenhouse gases generated from energy production.

In another study, the coupling of wastewater treatment and biogas valorization was optimized for recovering nutrients from AD effluent while converting biogas into microbial biomass. In producing microbial biomass, *Chlorella sorokiniana–Methylococcus capsulatus* (Bath) co-culture grown on diluted AD effluent was efficiently used to recover energy (from CH_4) and carbon (from both CH_4 and CO_2), while nutrients from wastewater were eliminated to produce treated clean water. From biomass composition analysis, it was revealed that the wastewater derived co-culture biomass can serve as an ideal candidate for bioplastic production and aquaculture feed (Roberts et al., 2020).

According to Oliver et al., (2020), lipid from algae *Schizochytrium* sp. is enriched with Omega-3 fatty acid which is suitable for feeding marine fish. In conclusion, different mechanisms of the production of high-energy compounds by algae and/or bacteria from the algal biomass are summarized in Figure 2.

ALGACIDAL MICROBES

HABs is a natural phenomenon defined as an overgrowth in water bodies caused in a brief period due to the increase of nitrogen and phosphorus. This phenomenon is distributed worldwide and influences the aquaculture and fish populations (Inaba et al., 2017). These algal blooms cause a reduction in water clarity, lead to the development of bad smells and tastes, plus clogging problems in water treatment systems. Famous species representatives of HAB are *Heterosigma akashiwo* and genus *Alexandrium* of Raphidophyceae and Dinoflagellates, respectively. *H. akashiwo* has been reported worldwide and its bloom is reported to be the cause of fish-killing in many places around the world, and also reported in Tasmania and New Zealand (Engesmo et al., 2016). Dinoflagellates *Alexandrium* produces a neurotoxin that causes paralytic shellfish poisoning in humans after the consumption of filter feeder shellfish which fed on these Dinoflagellates (Inaba et al., 2017). One promising solution for the problem of algal bloom is the use of biocontrol bacteria that have the potential to inhibit and control the overgrowth of the algae These bacteria are known as algicidal bacteria.

It is worth mentioning that *Phaeobacter gallaeciensis* is an algal opportunistic pathogen that can alter its metabolism to produce an algicidal compound known as roseobacticides. The roseobacticides are being produced when p-coumaric acid is being produced by the breakdown of algal lignin as the algae get old (Seyedsayamdost et al., 2011). It has been found that the algicide compound production by bacteria might be produced constitutively or upregulated by factors that might be related to the target algal species (Doucette et al., 1999). Furthermore, factors that affect the severity of the algal cell lysis are the density of the algicidal microorganism, the concentration of the algicidal compound, and the target sensitivity (Fuentes et al., 2016). The production of hydrolytic enzymes is another mechanism that can control algal bloom and result in bloom termination. The enzyme glycoside hydrolase in *Alteromonas* FDHY-03 is being upregulated when it is co-cultured with the dinoflagellate *Prorocentrum donghaiense,* which resulted in the breaking down of the algal cell (Shi et al., 2018). Despite the number of algicidal compounds producing bacteria is continuing to increase, less is known about the mechanism of actions of these compounds on the algal cell. In an attempt to study the mode of action, a study was conducted on prodigiosin, which is being produced by the Gram-negative bacteria *Hahella* sp. KA22 showed an increase of reactive oxygen species in *Phaeocystis globose.* The cells have lost most of their chlorophyll-producing capacity. The photosynthetic process was stopped, resulting in the algal cell's demise (Zhang et al., 2017). Table 5 lists some of the algicidal bacteria strains as well as the susceptible algal species.

Additionally, another biotechnological application of using algicidal bacteria is to use it for the lysis of algal cells for releasing the cell contents unaffected, for instance, the alga *Nannochloropsis oceanica* was treated with the bacterium *Sagittula stellata* and achieved 80.1%. of algicidal rate with 23.3% re-coverable lipid content (Wang et al., 2021)

Table 5. Algicidal bacteria and their targets.

Bacterial name	Active compound	Target harmful algal bloom species	Reference
Lactobacillus paraplantarum KCTC 5045	ND*	*Anabaena flosaquae, Anabaena crassa, Stephanodiscus hantzschii, and Peridinium bipes*	Kang et al., (2016)
Bacillus nitratireducens.	ND*	*Gymnodinium catenatum*	Prasath et al., (2021)
Brachybacterium YS-3	1-acetyl-β-carboline	*Alexandrium catenella, Prorocentrum micans, P. minimu, Cochlodinium polykrikoides, Akashiwo sanguinea, Gymnodinium impudicum, Fibriocapsa japonica, Heterosigma akashiwo, Scrippsiella trochoidea, and Chattonella marina.*	Kim et al., (2015)
Phaeobacter sp. ZM101	ND*	*Chattonella antiqua*	Inaba et al., (2019)
Phaeobacter gallaeciensis	Roseobacticides	*Emiliania huxleyi*	Seyedsayamdost et al., (2011)
Alteromonas FDHY-03	glycoside hydrolase	*Prorocentrum donghaiense*	Shi et al., (2018)
Hahella sp. KA22	Prodigiosin	*Phaeocystis globose*	Zhang et al., (2017)
Alteromonas sp. KNS-16	2-undecen-1'-yl-4-quinolone (1), 2-undecyl-4-quinolone (2), 3-hexyl-6-pentyl-4-hydroxyl-2H-pyran-2-one (3), and 6-heptyl-3-hexyl-4-hydroxyl-2H-pyran-2-one (4)	*Heterosigma akashiwo, Cochlodinium polykrikoides,* and *Alexandrium tamarense*	Cho, (2012)
Mangrovimonas yunxiaonensis LY01	ND*	*Alexandrium tamarense*	Li et al., (2014)
Deinococcus xianganensis Y35	Deinoxanthin	*Alexandrium tamarense*	Li et al., (2015)
Bacillus sp. SY-1	Mycosubtilins	*Cochlodinium polykrikoides*	Jeong & Son, (2021)
ND: Not defined.			

ALGAE-MICROBE METABOLOMICS AND QUORUM SENSING

The microbiome structure on the algal surface is affected by many factors. It is found to be species-specific and would change according to seasonal change (Parrot et al., 2019). The change is a consequence of alga producing chemical compounds to control the composition of the halobiont that inhibits the algal surface, by inhibiting the bacterial quorum sensing or directly killing the associated microbes (Kouzuma & Watanabe, 2015). Bacteria communicate with each other using autoinducer (AI) signaling molecules which are released by themselves in quorum sensing (QS) (Rémy et al., 2018). While the concentration of AI is dependent on the bacterial cell density, with a higher AI concentration found in a denser bacterial pack, AI would form a complex with a specific receptor protein once its concentration reaches the threshold. The complex would in turn activate the transcription of certain genes and modulate some physiological changes that include biofilm formation, bioluminescence production, virulence factor secretions and sporulation (Chi et al., 2017).

Apart from intraspecies use, QS signals generally have provided communication means between prokaryotes of different species as well as between prokaryotes and eukaryotes. The symbiotic interaction

Figure 3. Schematic representation of different results of QS.

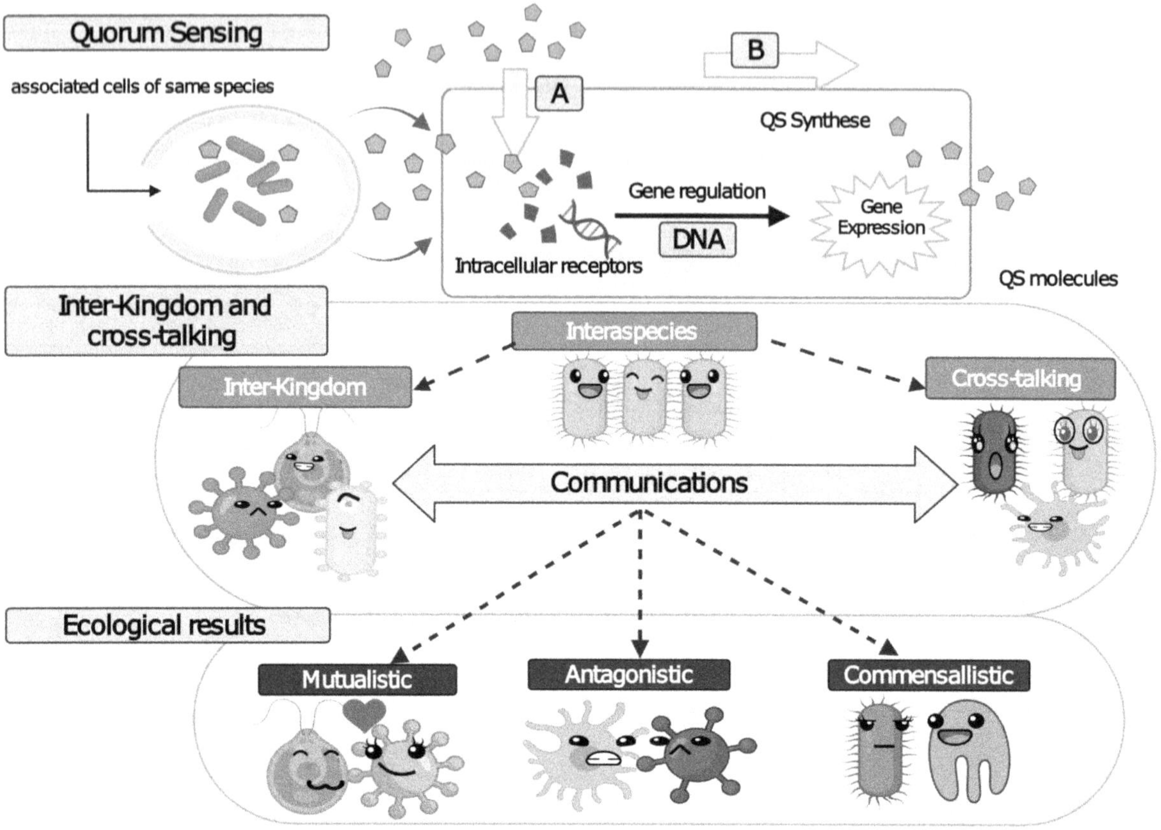

between microalgae and bacteria has been reported in a variety of other symbiotic species (Zhou et al., 2016). The ecological roles of QS signals are significant in microalgal-bacterial symbionts. While QS signals come in various classes, N-acyl-homoserine lactones and their degradation-derived tetramic acids, as well as 2-alkyl-4-quinolones, are particularly known for playing a key role in planktonic-microalgal-bacterial communication. These signaling molecules have a direct impact on planktonic microalgae development, as well as an indirect influence on algae-bacteria interactions and the microbial loop. In short, QS signals function as microalgal growth and metabolism regulator, direct photosynthesis effectors, cell cycle regulators, and diverse algicidal system regulators. On the other hand, microalgae can also exert influence upon these microscopic molecules (Dow, 2021). Figure 3, illustrates how the QS is working between different organisms and the possible results of this communication system.

Interference mechanism with QS using QS inhibitors (QSIs) stands for an anti-virulence strategy with significant promise for the treatment of bacterial infections in aquaculture and medical applications. Over the past two decades, a series of novel studies have been reported that other mechanisms such as quorum quenching, produce enzymes that destroy QS signaling molecules (Dow, 2021; Stock et al., 2020) and discovered that QSIs prevent the synthesis of QS signals, causing a significant impact.

N-acyl-homoserine Lactones (N-AHLs) are reported to be the most important signaling molecules released by Gram-negative bacteria, are very important in the virulence of marine pathogens of *Vibrio* species, and also control the pathogenicity (Abdella et al., 2017; Liu et al., 2018). Furthermore, the QS stress exerted by molecules produced in the activated sludge stage of wastewater treatment was found to reduce the growth of *Chlorophyta* sp. but has increased the lipid content by ~84%, similar results were observed when the alga was treated with pure N-hexanoyl-L-Homoserine lactone (C_6-HSL) (Zhang et al., 2018).

It transpired that the QS molecules controlled the genes responsible for lipid production in algae. Despite bacterial HSL acting as a growth control for algae, the macroalgae, in turn, are found to produce molecules that control the growth of bacteria. For instance, halogenated furanones produced by the red alga *Delisea pulchra* appeared to have suppressed AHL-dependent gene expression (Carvalho et al., 2017). The inhibition of the QS in bacterial aquaculture pathogens is a promising application as a bio-control against infectious diseases among cultivated fish and shrimps. Due to the technical difficulties of identifying the signaling molecules from bacterial strains, therefore the study of the effect of commercial homologs of the QS molecules might be an alternative for exploration of the impact of bacterial signaling molecules on algal targets. Moreover, QSIs have been mainly reported to be produced by macroalgae. For example, compared to extracts absent of symbiotic microorganisms, *Canistrocarpus cervicornis* extracts containing associated microbes showed greater QS suppression from *Chromobacterium violaceum* CV017. In another study, red macroalgae are shown to have an important role in QSI;2-dodecanoyloxye production than sulfonate of *Asparagopsis taxiformis* (Jha et al., 2013), however, brown alga *Laminaria digitata* produced oxidized halogen compounds that can suppress bacterial QS. QS suppression exerted by eleven macroalgal extracts that contain and do not contain microsymbiont studied by Batista et al. (2014) revealed that algal extracts containing microorganisms showed higher QSI activity, including species *of P. capillacea* and *Spyridia sp.* (Batista et al., 2014; Carvalho et al., 2017).

In a study involving fungi, it transpired that interactions between macroalgae, microalgal fungal microbiota, and protistan pathogens are tightly regulated by the metabolites produced by endophytic fungi, as determined by the molecular isolation and characterization of various parts of some cultivable brown algae which include *Ascophyllum nodosum*, *Pelvetia canaliculate*, *Laminaria digitata*, and *Saccharina latissima* (Vallet et al., 2018). Table 6 summarizes examples of different QS and QSI that are happening between algae and different microorganisms

In summary, QS is a huge and complex network that precisely controls biological and physiological functions in the aquatic environment. Technical problems stand against understanding and QSI discovery, including the identification of the QSI molecular target, the molecular targeting delivery, and the QSI cytotoxicity evaluation at the organism, cellular, and molecular levels. These queries on the above points can certainly provide promising trails for further efforts to discover QSIs and diversify their applications in aquaculture disease prevention.

CONCLUSION

The exploitation of algae and bacteria co-cultures is very crucial nowadays as a solution for environmental, and energy issues besides their contribution to the enhancement of the nutritional values of feed and food by extracting valuable compounds from their growth compared to those of monocultures. In wastewater treatment, the algae-microbe association has improved the nutrient uptake as well as it has contributed to the stability of the system during the remediation process, making it more sustainable. Moreover, this symbiosis helps in cost-effective aeration, the sequestration of greenhouse gas emissions, and the production of flocs, which means easier biomass management. Furthermore, the microbes associated

Table 6. Inter-kingdom signaling in the algae-microbe association.

Source organism	Active molecules	Target organism	Mode of action	References
Algae-Bacteria interaction				
Delisea pulchra	halogenated furanones	*Vibrio fischeri*	QSI	Carvalho et al., (2017)
Chlorella saccharophila CCAP211/48	Algal extract	*Vibrio harveyi, Chromobacterium violaceum* CV026, *E. coli JB523*	QSI	Natrah et al., (2011)
Canistrocarpus cervicornis	Algal extract	*Chromobacterium violaceum CV017*	QSI	Carvalho et al., (2017)
Asparagopsis taxiformis	2-dodecanoyloxyethanesulfonate	*Chromobacterium violaceum* CV026, *Serratia liquefaciens* MG44	QSI	Jha et al., (2013)
Laminaria digitata	oxidized halogen compounds	*Pseudomonas aeruginosa, Chromobacterium violaceum* CV026	QSI and BI	Carvalho et al., (2017)
Macroalgae	20 polar algal extracts	*Chromobacterium violaceum* CV017	QSI	Batista et al., (2014)
Sargassum vulgare, Colpomenia sinuosa, Padina sp., Canistrocarpus cervicornis, Spyridia aculeata, and *Pterocladiella capillacea*	Extracts of macroalgae	*C. violaceum* CV017	PI	Carvalho et al., (2017)
		Pseudomonas aeroginosa PA01	BI	
		Shewanella putrefaciens	CI	
		(*Vibrio aestuarianus, Pseudoalteromonas elyakovii, Polaribacter irgensii,* and *Pseudomonas fluorescens*)	BFI	
Bacteria-algae interaction				
NA	N-hexanoyl-L-Homoserine lactone	*Chlorophyta* sp.	GI	Zhang et al., (2018)
Pseudomonas sp. and *Pseudoalteromonas* sp.	2-heptyl-4-quinolone	*Emiliania huxleyi*	GI	Pollara et al., (2021)
Loktanella sp. II 4.36, *Roseovarius* sp. D12-1	N-9-hexadecenoylalanine methyl ester (9-C16:1-NAME)	*Skeletonema costatum*	GI	Ziesche et al., (2015)
Synthesized	C14-HSL	*Seminavis robusta*	GP	Stock et al., (2020)
Synthesized	OH-C14-HSL, oxo-C14-HSL, and tetramic acid	*Seminavis robusta*	GI	Stock et al., (2020)
Synthesized	oxo-C12-AHL	*Phaeodactylum tricornutum*	GI	Stock et al., (2020)
Fungi-algae interaction				
Endophytic fungi	Metabolites (pyrenocines)	*Ascophyllum nodosum, Pelvetia canaliculata, Laminaria digitata,* and *Saccharina latissima*	Protection against infection	Vallet et al., (2018)
GI= Growth Inhibition, GP= Growth Promotion, QSI= Quorum Sensing Inhibitor, PI= Pigment Inhibition, BI= Biofilm Inhibition, CI= Corrosion Inhibition, BFI= Biofouling Inhibition				

help the algal hydrogenase in hydrogen production, by making anaerobic micromovement for efficient production of biohydrogen in photo-fermentation. In addition, the algal-bacterial biomass produced could be leveraged for a downstream application like by increasing the lipid contents for biodiesel production and biogas production through anaerobic fermentation. Next, cultivating methanotrophic bacteria in biogas can be optimized for wastewater treatment and greenhouse gas remediation, while resulting in aquaculture feed and bioplastics as the end products. In sustainable aquaculture, the algicidal microbe was found to be a promising biocontrol agent for the HABs, which affect the water quality, farmed fish, and consumers. Likewise, the understating of the signaling molecules that changed between the microbes and their host will help in disease prevention in aquaculture through QS interferences. Therefore, the advances in the understanding of microbe interaction strategies will be of great application to the challenging issues that face humankind.

ACKNOWLEDGMENT

This book chapter received no specific grant from any funding agency in the public, commercial, or not-for-profit sectors.

REFERENCES

Abdalla, A. M., Hassan, T. H., & Mansour, M. E. (2018). Performance of Wet and Dry Bagasse Combustion in Assalaya Sugar Factory - Sudan. *Innovative Energy & Research*, *7*(1). Advance online publication. doi:10.4172/2576-1463.1000179

Abdella, B., El-Wazzan, E., El-Sersy, N. A., Sabry, S. A., & El-Helow, E. R. (2017). Pathogenicity and antibiotic susceptibility of two bacterial pathogens associated with the clam Tapes decussatus in some Egyptian fisheries. *Ege Journal of Fisheries and Aquatic Sciences*, *34*(4), 383–389. doi:10.12714/egejfas.2017.34.4.04

Amin, S. A., Green, D. H., Hart, M. C., Kupper, F. C., Sunda, W. G., & Carrano, C. J. (2009). Photolysis of iron-siderophore chelates promotes bacterial-algal mutualism. *Proceedings of the National Academy of Sciences of the United States of America*, *106*(40), 17071–17076. doi:10.1073/pnas.0905512106 PMID:19805106

Arora, K., Kaur, P., Kumar, P., Singh, A., Patel, S. K. S., Li, X., Yang, Y.-H., Bhatia, S. K., & Kulshrestha, S. (2021). Valorization of Wastewater Resources Into Biofuel and Value-Added Products Using Microalgal System. *Frontiers in Energy Research*, *9*, 646571. Advance online publication. doi:10.3389/fenrg.2021.646571

Avagyan, A. B. (2010). New Design & Build Biological System through the Use of Microalgae Addressed to Sustainable Development. *Journal of Environmental Protection*, *01*(02), 183–200. doi:10.4236/jep.2010.12023

Barnharst, T., Rajendran, A., & Hu, B. (2018). Bioremediation of synthetic intensive aquaculture wastewater by a novel feed-grade composite biofilm. *International Biodeterioration & Biodegradation*, *126*, 131–142. doi:10.1016/j.ibiod.2017.10.007

Batista, D., Carvalho, A. P., Costa, R., Coutinho, R., & Dobretsov, S. (2014). Extracts of macroalgae from the Brazilian coast inhibit bacterial quorum sensing. *Botanica Marina*, *57*(6), 441–447. Advance online publication. doi:10.1515/bot-2014-0031

Bell, T. A. S., Prithiviraj, B., Wahlen, B. D., Fields, M. W., & Peyton, B. M. (2016). A Lipid-Accumulating Alga Maintains Growth in Outdoor, Alkaliphilic Raceway Pond with Mixed Microbial Communities. *Frontiers in Microbiology*, *6*. Advance online publication. doi:10.3389/fmicb.2015.01480 PMID:26779138

Adnan, Ong, Nomanbhay, Chew, & Show. (2019). Technologies for Biogas Upgrading to Biomethane: A Review. *Bioengineering (Basel, Switzerland)*, *6*(4), 92. doi:10.3390/bioengineering6040092

Carpio, R., Kuo, C.-T., de Leon, R., Schideman, L. C., & Zhang, Y. (2018). Hydrothermal liquefaction of demineralized wastewater algal biomass. *International Journal of Smart Grid and Clean Energy*, *7*(1), 13–23. doi:10.12720gce.7.1.13-23

Carvalho, A. P., Batista, D., Dobretsov, S., & Coutinho, R. (2017). Extracts of seaweeds as potential inhibitors of quorum sensing and bacterial growth. *Journal of Applied Phycology*, *29*(2), 789–797. doi:10.100710811-016-1014-1

Cheirsilp, B., Suwannarat, W., & Niyomdecha, R. (2011). Mixed culture of oleaginous yeast Rhodotorula glutinis and microalga Chlorella vulgaris for lipid production from industrial wastes and its use as biodiesel feedstock. *New Biotechnology*, *28*(4), 362–368. doi:10.1016/j.nbt.2011.01.004 PMID:21255692

Chi, W., Zheng, L., He, C., Han, B., Zheng, M., Gao, W., Sun, C., Zhou, G., & Gao, X. (2017). Quorum sensing of microalgae associated marine Ponticoccus sp. PD-2 and its algicidal function regulation. *AMB Express*, *7*(1), 59. doi:10.118613568-017-0357-6 PMID:28281272

Chinnasamy, S., Bhatnagar, A., Hunt, R. W., & Das, K. C. (2010). Microalgae cultivation in a wastewater dominated by carpet mill effluents for biofuel applications. *Bioresource Technology*, *101*(9), 3097–3105. doi:10.1016/j.biortech.2009.12.026 PMID:20053551

Cho, J. Y. (2012). Algicidal Activity of Marine *Alteromonas* sp. KNS-16 and Isolation of Active Compounds. *Bioscience, Biotechnology, and Biochemistry*, *76*(8), 1452–1458. doi:10.1271/bbb.120102 PMID:22878186

Croft, M. T., Lawrence, A. D., Raux-Deery, E., Warren, M. J., & Smith, A. G. (2005). Algae acquire vitamin B12 through a symbiotic relationship with bacteria. *Nature*, *438*(7064), 90–93. doi:10.1038/nature04056 PMID:16267554

Dang, Y., Holmes, D. E., Zhao, Z., Woodard, T. L., Zhang, Y., Sun, D., Wang, L.-Y., Nevin, K. P., & Lovley, D. R. (2016). Enhancing anaerobic digestion of complex organic waste with carbon-based conductive materials. *Bioresource Technology*, *220*, 516–522. doi:10.1016/j.biortech.2016.08.114 PMID:27611035

De-Bashan, L. E., Moreno, M., Hernandez, J. P., & Bashan, Y. (2002). Removal of ammonium and phosphorus ions from synthetic wastewater by the microalgae Chlorella vulgaris coimmobilized in alginate beads with the microalgae growth-promoting bacterium Azospirillum brasilense. *Water Research*, *36*(12), 2941–2948. doi:10.1016/S0043-1354(01)00522-X PMID:12171390

de Farias Silva, C. E., & Bertucco, A. (2016). Bioethanol from microalgae and cyanobacteria: A review and technological outlook. *Process Biochemistry*, *51*(11), 1833–1842. doi:10.1016/j.procbio.2016.02.016

De Godos, I., González, C., Becares, E., García-Encina, P. A., & Muñoz, R. (2009). Simultaneous nutrients and carbon removal during pretreated swine slurry degradation in a tubular biofilm photobioreactor. *Applied Microbiology and Biotechnology*, *82*(1), 187–194. doi:10.100700253-008-1825-3 PMID:19122998

Demirel, B., & Scherer, P. (2011). Trace element requirements of agricultural biogas digesters during biological conversion of renewable biomass to methane. *Biomass and Bioenergy*, *35*(3), 992–998. doi:10.1016/j.biombioe.2010.12.022

Divya, M., & Santhanam, P. (2018). Bioremediation of wastewater using a novel method of microalgae immobilized on Twin-Layer Recirculation System (TLRS). In Basic and Applied Phytoplankton Biology (pp. 177–190). doi:10.1007/978-981-10-7938-2_8

Doucette, G. J., McGovern, E. R., & Babinchak, J. A. (1999). Algicidal bacteria active against Gymnodinium breve (Dinophyceae). I. Bacterial isolation and characterization of killing activity. *Journal of Phycology*, *35*(6, SUPPL.), 1447–1454. doi:10.1046/j.1529-8817.1999.3561447.x

Dow, L. (2021). How Do Quorum Sensing Signals Mediate Algae–Bacteria Interactions? *Microorganisms*, *9*(7), 1391. doi:10.3390/microorganisms9071391 PMID:34199114

Du, X.-P., Cai, Z.-H., Zuo, P., Meng, F.-X., Zhu, J.-M., & Zhou, J. (2020). Temporal variability of virioplankton during a Gymnodinium catenatum algal bloom. *Microorganisms*, *8*(1), 1–15. doi:10.3390/microorganisms8010107 PMID:31940944

Du, Z.-Y., Zienkiewicz, K., Vande Pol, N., Ostrom, N. E., Benning, C., & Bonito, G. M. (2019). Algal-fungal symbiosis leads to photosynthetic mycelium. *eLife*, *8*, e47815. Advance online publication. doi:10.7554/eLife.47815 PMID:31307571

Egan, S., Harder, T., Burke, C., Steinberg, P., Kjelleberg, S., & Thomas, T. (2013). The seaweed holobiont: Understanding seaweed-bacteria interactions. *FEMS Microbiology Reviews*, *37*(3), 462–476. doi:10.1111/1574-6976.12011 PMID:23157386

Eiteman, M. A., & Ramalingam, S. (2015). Microbial production of lactic acid. *Biotechnology Letters*, *37*(5), 955–972. doi:10.100710529-015-1769-5 PMID:25604523

Emparan, Q., Harun, R., & Danquah, M. K. (2019). Role of phycoremediation for nutrient removal from wastewaters: A review. *Applied Ecology and Environmental Research*, *17*(1), 889–915. doi:10.15666/aeer/1701_889915

Engesmo, A., Eikrem, W., Seoane, S., Smith, K., Edvardsen, B., Hofgaard, A., & Tomas, C. R. (2016). New insights into the morphology and phylogeny of Heterosigma akashiwo (Raphidophyceae), with the description of Heterosigma minor sp. nov. *Phycologia*, *55*(3), 279–294. doi:10.2216/15-115.1

Fuentes, J. L., Garbayo, I., Cuaresma, M., Montero, Z., González-del-Valle, M., & Vílchez, C. (2016). Impact of microalgae-bacteria interactions on the production of algal biomass and associated compounds. *Marine Drugs*, *14*(5), 100. doi:10.3390/md14050100 PMID:27213407

García, D., Posadas, E., Grajeda, C., Blanco, S., Martínez-Páramo, S., Acién, G., García-Encina, P., Bolado, S., & Muñoz, R. (2017). Comparative evaluation of piggery wastewater treatment in algal-bacterial photobioreactors under indoor and outdoor conditions. *Bioresource Technology*, *245*, 483–490. doi:10.1016/j.biortech.2017.08.135 PMID:28898848

Girard, L. (2019). Quorum sensing in *Vibrio* spp.: The complexity of multiple signalling molecules in marine and aquatic environments. *Critical Reviews in Microbiology*, *45*(4), 451–471. doi:10.1080/1040841X.2019.1624499 PMID:31241379

González-Olalla, J. M., Medina-Sánchez, J. M., Lozano, I. L., Villar-Argaiz, M., & Carrillo, P. (2018). Climate-driven shifts in algal-bacterial interaction of high-mountain lakes in two years spanning a decade. *Scientific Reports*, *8*(1), 1–12. doi:10.103841598-018-28543-2 PMID:29980756

Guo, G., Cao, W., Sun, S., Zhao, Y., & Hu, C. (2017). Nutrient removal and biogas upgrading by integrating fungal–microalgal cultivation with anaerobically digested swine wastewater treatment. *Journal of Applied Phycology*, *29*(6), 2857–2866. doi:10.100710811-017-1207-2

Henard, C. A., Franklin, T. G., Youhenna, B., But, S., Alexander, D., Kalyuzhnaya, M. G., & Guarnieri, M. T. (2018). Biogas Biocatalysis: Methanotrophic Bacterial Cultivation, Metabolite Profiling, and Bioconversion to Lactic Acid. *Frontiers in Microbiology*, *9*, 2610. Advance online publication. doi:10.3389/fmicb.2018.02610 PMID:30429839

Higgins, B. T., & VanderGheynst, J. S. (2014). Effects of *Escherichia coli* on Mixotrophic Growth of Chlorella minutissima and Production of Biofuel Precursors. *PLoS One*, *9*(5), e96807. doi:10.1371/journal.pone.0096807 PMID:24805253

Hill, E. A., Chrisler, W. B., Beliaev, A. S., & Bernstein, H. C. (2017). A flexible microbial co-culture platform for simultaneous utilization of methane and carbon dioxide from gas feedstocks. *Bioresource Technology*, *228*, 250–256. doi:10.1016/j.biortech.2016.12.111 PMID:28092828

Inaba, N., Trainer, V. L., Nagai, S., Kojima, S., Sakami, T., Takagi, S., & Imai, I. (2019). Dynamics of seagrass bed microbial communities in artificial Chattonella blooms: A laboratory microcosm study. *Harmful Algae*, *84*(April), 139–150. doi:10.1016/j.hal.2018.12.004 PMID:31128798

Inaba, N., Trainer, V. L., Onishi, Y., Ishii, K. I., Wyllie-Echeverria, S., & Imai, I. (2017). Algicidal and growth-inhibiting bacteria associated with seagrass and macroalgae beds in Puget Sound, WA, USA. *Harmful Algae*, *62*, 136–147. doi:10.1016/j.hal.2016.04.004 PMID:28118888

Jain, S., Jain, S., Wolf, I. T., Lee, J., & Tong, Y. W. (2015). A comprehensive review on operating parameters and different pretreatment methodologies for anaerobic digestion of municipal solid waste. *Renewable & Sustainable Energy Reviews*, *52*, 142–154. doi:10.1016/j.rser.2015.07.091

Jeong, S.-Y., & Son, H.-J. (2021). Effects of mycosubtilin homolog algicides from a marine bacterium, *Bacillus* sp. SY-1, against the harmful algal bloom species Cochlodinium polykrikoides. *Journal of Microbiology (Seoul, Korea)*, *59*(4), 389–400. doi:10.100712275-021-1086-8 PMID:33779952

Jha, B., Kavita, K., Westphal, J., Hartmann, A., & Schmitt-Kopplin, P. (2013). Quorum sensing inhibition by asparagopsis taxiformis, a marine macro alga: Separation of the compound that interrupts bacterial communication. *Marine Drugs*, *11*(1), 253–265. doi:10.3390/md11010253 PMID:23344114

Ji, X., Jiang, M., Zhang, J., Jiang, X., & Zheng, Z. (2018). The interactions of algae-bacteria symbiotic system and its effects on nutrients removal from synthetic wastewater. *Bioresource Technology*, *247*, 44–50. doi:10.1016/j.biortech.2017.09.074 PMID:28946093

Jiang, L.-L., Zhou, J.-J., Quan, C.-S., & Xiu, Z.-L. (2017). Advances in industrial microbiome based on microbial consortium for biorefinery. *Bioresources and Bioprocessing*, *4*(1), 11. doi:10.118640643-017-0141-0 PMID:28251041

Kang, Y. H., Kang, S. K., Park, C. S., Joo, J. H., Lee, J. W., & Han, M. S. (2016). Use of lactic acid bacteria as a biological agent against the cyanobacterium Anabaena flos-aquae. *Journal of Applied Phycology*, *28*(3), 1747–1757. doi:10.100710811-015-0701-7

Karuppiah, T., & Azariah, V. E. (2019). Biomass Pretreatment for Enhancement of Biogas Production. In *Anaerobic Digestion*. IntechOpen. doi:10.5772/intechopen.82088

Kazamia, E., Czesnick, H., Van Nguyen, T. T., Croft, M. T., Sherwood, E., Sasso, S., Hodson, S. J., Warren, M. J., & Smith, A. G. (2012). Mutualistic interactions between vitamin B12-dependent algae and heterotrophic bacteria exhibit regulation. *Environmental Microbiology*, *14*(6), 1466–1476. doi:10.1111/j.1462-2920.2012.02733.x PMID:22463064

Kim, Y. S., Son, H. J., & Jeong, S. Y. (2015). Isolation of an algicide from a marine bacterium and its effects against the toxic dinoflagellate Alexandrium catenella and other harmful algal bloom species. *Journal of Microbiology (Seoul, Korea)*, *53*(8), 511–517. doi:10.100712275-015-5303-1 PMID:26224453

Klein, C. B., Bonomi, A., & Filho, M. R. (2018). Integration of microalgae production with industrial biofuel facilities: A critical review. *Renewable & Sustainable Energy Reviews*, *82*, 1376–1392. doi:10.1016/j.rser.2017.04.063

Kouzuma, A., & Watanabe, K. (2015). Exploring the potential of algae/bacteria interactions. *Current Opinion in Biotechnology*, *33*, 125–129. doi:10.1016/j.copbio.2015.02.007 PMID:25744715

Krug, L., Erlacher, A., Markut, K., Berg, G., & Cernava, T. (2020). The microbiome of alpine snow algae shows a specific inter-kingdom connectivity and algae-bacteria interactions with supportive capacities. *The ISME Journal*, *14*(9), 2197–2210. doi:10.103841396-020-0677-4 PMID:32424246

Lakatos, G., Deák, Z., Vass, I., Rétfalvi, T., Rozgonyi, S., Rákhely, G., Ördög, V., Kondorosi, É., & Maróti, G. (2014). Bacterial symbionts enhance photo-fermentative hydrogen evolution of Chlamydomonas algae. *Green Chemistry*, *16*(11), 4716–4727. doi:10.1039/C4GC00745J

Lam, M. K., Loy, A. C. M., Yusup, S., & Lee, K. T. (2019). Biohydrogen Production From Algae. In *Biohydrogen* (pp. 219–245). Elsevier. doi:10.1016/B978-0-444-64203-5.00009-5

Leyva, L. A., Bashan, Y., Mendoza, A., & De-Bashan, L. E. (2014). Accumulation fatty acids of in Chlorella vulgaris under heterotrophic conditions in relation to activity of acetyl-CoA carboxylase, temperature, and co-immobilization with Azospirillum brasilense. *Naturwissenschaften*, *101*(10), 819–830. doi:10.100700114-014-1223-x PMID:25129521

Li, Y., Zhu, H., Lei, X., Zhang, H., Guan, C., Chen, Z., Zheng, W., Xu, H., Tian, Y., Yu, Z., & Zheng, T. (2015). The first evidence of deinoxanthin from Deinococcus sp. Y35 with strong algicidal effect on the toxic dinoflagellate Alexandrium tamarense. *Journal of Hazardous Materials*, *290*, 87–95. doi:10.1016/j.jhazmat.2015.02.070 PMID:25746568

Li, Y., Zhu, H., Zhang, H., Chen, Z., Tian, Y., Xu, H., Zheng, T., & Zheng, W. (2014). Toxicity of algicidal extracts from Mangrovimonas yunxiaonensis strain LY01 on a HAB causing Alexandrium tamarense. *Journal of Hazardous Materials*, *278*, 372–381. doi:10.1016/j.jhazmat.2014.06.032 PMID:24997253

Liu, J., Fu, K., Wu, C., Qin, K., Li, F., & Zhou, L. (2018). "In-Group" communication in marine *Vibrio*: A review of N-Acyl Homoserine Lactones-Driven Quorum Sensing. *Frontiers in Cellular and Infection Microbiology*, *8*(MAY), 1–17. doi:10.3389/fcimb.2018.00139 PMID:29868495

Lü, F., Ji, J., Shao, L., & He, P. (2013). Bacterial bioaugmentation for improving methane and hydrogen production from microalgae. *Biotechnology for Biofuels*, *6*(1), 92. doi:10.1186/1754-6834-6-92 PMID:23815806

Lutzu, G. A., & Dunford, N. T. (2018). Interactions of microalgae and other microorganisms for enhanced production of high-value compounds. *Frontiers in Bioscience*, *23*(8), 4656. doi:10.2741/4656 PMID:29293446

Matsumoto, M., Yokouchi, H., Suzuki, N., Ohata, H., & Matsunaga, T. (2003). Saccharification of Marine Microalgae Using Marine Bacteria for Ethanol Production. *Applied Biochemistry and Biotechnology*, *105*(1–3), 247–254. doi:10.1385/ABAB:105:1-3:247 PMID:12721490

Melis, A., & Happe, T. (2001). Hydrogen Production. Green Algae as a Source of Energy. *Plant Physiology*, *127*(3), 740–748. doi:10.1104/pp.010498 PMID:11706159

Mhedhbi, E., Khelifi, N., Foladori, P., & Smaali, I. (2020). Real-Time Behavior of a Microalgae–Bacteria Consortium Treating Wastewater in a Sequencing Batch Reactor in Response to Feeding Time and Agitation Mode. *Water (Basel)*, *12*(7), 1893. doi:10.3390/w12071893

Muñoz, C., Hidalgo, C., Zapata, M., Jeison, D., Riquelme, C., & Rivas, M. (2014). Use of Cellulolytic Marine Bacteria for Enzymatic Pretreatment in Microalgal Biogas Production. *Applied and Environmental Microbiology*, *80*(14), 4199–4206. doi:10.1128/AEM.00827-14 PMID:24795376

Natrah, F. M. I., Kenmegne, M. M., Wiyoto, W., Sorgeloos, P., Bossier, P., & Defoirdt, T. (2011). Effects of micro-algae commonly used in aquaculture on acyl-homoserine lactone quorum sensing. *Aquaculture (Amsterdam, Netherlands)*, *317*(1–4), 53–57. doi:10.1016/j.aquaculture.2011.04.038

Oliver, L., Dietrich, T., Marañón, I., Villarán, M. C., & Barrio, R. J. (2020). Producing Omega-3 Polyunsaturated Fatty Acids: A Review of Sustainable Sources and Future Trends for the EPA and DHA Market. *Resources*, *9*(12), 148. doi:10.3390/resources9120148

Papone, T., Kookkhunthod, S., & Leesing, R. (2012). Microbial Oil Production by Monoculture and Mixed Cultures of Microalgae and Oleaginous Yeasts using Sugarcane Juice as Substrate. *International Journal of Biological. Life and Agricultural Sciences*, *5*(4). Advance online publication. doi:10.5281/zenodo.1332532

Park, J., Park, B. S., Wang, P., Patidar, S. K., Kim, J. H., Kim, S.-H., & Han, M.-S. (2017). Phycospheric Native Bacteria Pelagibaca bermudensis and Stappia sp. Ameliorate Biomass Productivity of Tetraselmis striata (KCTC1432BP) in Co-cultivation System through Mutualistic Interaction. *Frontiers in Plant Science*, *8*. Advance online publication. doi:10.3389/fpls.2017.00289 PMID:28321229

Park, Y., Je, K.-W., Lee, K., Jung, S.-E., & Choi, T.-J. (2008). Growth promotion of Chlorella ellipsoidea by co-inoculation with Brevundimonas sp. isolated from the microalga. *Hydrobiologia*, *598*(1), 219–228. doi:10.100710750-007-9152-8

Parrot, D., Blümel, M., Utermann, C., Chianese, G., Krause, S., Kovalev, A., Gorb, S. N., & Tasdemir, D. (2019). Mapping the Surface Microbiome and Metabolome of Brown Seaweed Fucus vesiculosus by Amplicon Sequencing, Integrated Metabolomics and Imaging Techniques. *Scientific Reports*, *9*(1), 1061. doi:10.103841598-018-37914-8 PMID:30705420

Perez-Garcia, O., de-Bashan, L. E., Hernandez, J. P., & Bashan, Y. (2010). Efficiency of growth and nutrient uptake from wastewater by heterotrophic, autotrophic, and mixotrophic cultivation of chlorella vulgaris immobilized with azospirillum brasilense. *Journal of Phycology*, *46*(4), 800–812. doi:10.1111/j.1529-8817.2010.00862.x

Pollara, S. B., Becker, J. W., Nunn, B. L., Boiteau, R., Repeta, D., Mudge, M. C., Downing, G., Chase, D., Harvey, E. L., & Whalen, K. E. (2021). Bacterial Quorum-Sensing Signal Arrests Phytoplankton Cell Division and Impacts Virus-Induced Mortality. *MSphere*, *6*(3), e00009-21. Advance online publication. doi:10.1128/mSphere.00009-21 PMID:33980670

Prasath, B. B., Wang, Y., Su, Y., Zheng, W., Lin, H., & Yang, H. (2021). Coagulant plus *bacillus nitratireducens* fermentation broth technique provides a rapid algicidal effect of toxic red tide dinoflagellate. *Journal of Marine Science and Engineering*, *9*(4), 395. Advance online publication. doi:10.3390/jmse9040395

Ramanan, R., Kim, B. H., Cho, D. H., Oh, H. M., & Kim, H. S. (2016). Algae-bacteria interactions: Evolution, ecology and emerging applications. *Biotechnology Advances*, *34*(1), 14–29. doi:10.1016/j.biotechadv.2015.12.003 PMID:26657897

Rawat, I., Ranjith Kumar, R., Mutanda, T., & Bux, F. (2011). Dual role of microalgae: Phycoremediation of domestic wastewater and biomass production for sustainable biofuels production. *Applied Energy*, *88*(10), 3411–3424. doi:10.1016/j.apenergy.2010.11.025

Rémy, B., Mion, S., Plener, L., Elias, M., Chabrière, E., & Daudé, D. (2018). Interference in Bacterial Quorum Sensing: A Biopharmaceutical Perspective. *Frontiers in Pharmacology*, *9*, 203. Advance online publication. doi:10.3389/fphar.2018.00203 PMID:29563876

Renuka, N., Sood, A., Ratha, S. K., Prasanna, R., & Ahluwalia, A. S. (2013). Evaluation of microalgal consortia for treatment of primary treated sewage effluent and biomass production. *Journal of Applied Phycology*, *25*(5), 1529–1537. doi:10.100710811-013-9982-x

Riaño, B., Molinuevo, B., & García-González, M. C. (2011). Treatment of fish processing wastewater with microalgae-containing microbiota. *Bioresource Technology*, *102*(23), 10829–10833. doi:10.1016/j.biortech.2011.09.022 PMID:21983413

Roberts, N., Hilliard, M., He, Q. P., & Wang, J. (2020). A Microalgae-Methanotroph Coculture is a Promising Platform for Fuels and Chemical Production From Wastewater. *Frontiers in Energy Research*, *8*, 563352. Advance online publication. doi:10.3389/fenrg.2020.563352

Santiago, A. F., Calijuri, M. L., Assemany, P. P., Calijuri, M. D. C., & Dos Reis, A. J. D. (2013). Algal biomass production and wastewater treatment in high rate algal ponds receiving disinfected effluent. *Environmental Technology (United Kingdom)*, *34*(13–14), 1877–1885. doi:10.1080/09593330.2013.812670 PMID:24350441

Santos, C. A., Caldeira, M. L., Lopes da Silva, T., Novais, J. M., & Reis, A. (2013). Enhanced lipidic algae biomass production using gas transfer from a fermentative Rhodosporidium toruloides culture to an autotrophic Chlorella protothecoides culture. *Bioresource Technology*, *138*, 48–54. doi:10.1016/j.biortech.2013.03.135 PMID:23612161

Santos, C. A., & Reis, A. (2014). Microalgal symbiosis in biotechnology. *Applied Microbiology and Biotechnology*, *98*(13), 5839–5846. doi:10.100700253-014-5764-x PMID:24816618

Scognamiglio, V., Giardi, M. T., Zappi, D., Touloupakis, E., & Antonacci, A. (2021). Photoautotrophs–Bacteria Co-Cultures: Advances, Challenges and Applications. *Materials (Basel)*, *14*(11), 3027. doi:10.3390/ma14113027 PMID:34199583

Seyedsayamdost, M. R., Case, R. J., Kolter, R., & Clardy, J. (2011). The Jekyll-and-Hyde chemistry of Phaeobacter gallaeciensis. *Nature Chemistry*, *3*(4), 331–335. doi:10.1038/nchem.1002 PMID:21430694

Shahid, A., Malik, S., Zhu, H., Xu, J., Nawaz, M. Z., Nawaz, S., Asraful Alam, M., & Mehmood, M. A. (2020). Cultivating microalgae in wastewater for biomass production, pollutant removal, and atmospheric carbon mitigation; a review. *The Science of the Total Environment*, *704*, 135303. doi:10.1016/j.scitotenv.2019.135303 PMID:31818584

Shi, X., Liu, L., Li, Y., Xiao, Y., Ding, G., Lin, S., & Chen, J. (2018). Isolation of an algicidal bacterium and its effects against the harmful-algal- bloom dinoflagellate Prorocentrum donghaiense (Dinophyceae). *Harmful Algae*, *80*(September), 72–79. doi:10.1016/j.hal.2018.09.003 PMID:30502814

Shin, S. G., Lee, S., Lee, C., Hwang, K., & Hwang, S. (2010). Qualitative and quantitative assessment of microbial community in batch anaerobic digestion of secondary sludge. *Bioresource Technology*, *101*(24), 9461–9470. doi:10.1016/j.biortech.2010.07.081 PMID:20705457

Sial, A., Zhang, B., Zhang, A., Liu, K., Imtiaz, S. A., & Yashir, N. (2020). Microalgal–Bacterial Synergistic Interactions and Their Potential Influence in Wastewater Treatment: A Review. *BioEnergy Research*. Advance online publication. doi:10.100712155-020-10213-9

Silva-Benavides, A. M., & Torzillo, G. (2012). Nitrogen and phosphorus removal through laboratory batch cultures of microalga Chlorella vulgaris and cyanobacterium Planktothrix isothrix grown as monoalgal and as co-cultures. *Journal of Applied Phycology*, *24*(2), 267–276. doi:10.100710811-011-9675-2

Stock, F., Bilcke, G., De Decker, S., Osuna-Cruz, C. M., Van den Berge, K., Vancaester, E., De Veylder, L., Vandepoele, K., Mangelinckx, S., & Vyverman, W. (2020). Distinctive Growth and Transcriptional Changes of the Diatom Seminavis robusta in Response to Quorum Sensing Related Compounds. *Frontiers in Microbiology*, *11*(June), 1–14. doi:10.3389/fmicb.2020.01240 PMID:32582129

Su, Y., Mennerich, A., & Urban, B. (2011). Municipal wastewater treatment and biomass accumulation with a wastewater-born and settleable algal-bacterial culture. *Water Research*, *45*(11), 3351–3358. doi:10.1016/j.watres.2011.03.046 PMID:21513965

Subashchandrabose, S. R., Ramakrishnan, B., Megharaj, M., Venkateswarlu, K., & Naidu, R. (2013). Mixotrophic cyanobacteria and microalgae as distinctive biological agents for organic pollutant degradation. *Environment International*, *51*, 59–72. doi:10.1016/j.envint.2012.10.007 PMID:23201778

Tang, X., He, L. Y., Tao, X. Q., Dang, Z., Guo, C. L., Lu, G. N., & Yi, X. Y. (2010). Construction of an artificial microalgal-bacterial consortium that efficiently degrades crude oil. *Journal of Hazardous Materials*, *181*(1–3), 1158–1162. doi:10.1016/j.jhazmat.2010.05.033 PMID:20638971

Vallet, M., Strittmatter, M., Murúa, P., Lacoste, S., Dupont, J., Hubas, C., Genta-Jouve, G., Gachon, C. M. M., Kim, G. H., & Prado, S. (2018). Chemically-Mediated Interactions Between Macroalgae, Their Fungal Endophytes, and Protistan Pathogens. *Frontiers in Microbiology*, *9*, 3161. Advance online publication. doi:10.3389/fmicb.2018.03161 PMID:30627120

van Hullebusch, E. D., Yekta, S. S., Fermoso, B. C., & Fermoso, F. G. (2019). Biogeochemistry of major elements in anaerobic digesters: carbon, nitrogen, phosphorus, sulfur and iron. In *Trace Elements in Anaerobic Biotechnologies* (pp. 1–22). IWA Publishing. doi:10.2166/9781789060225_0001

Wang, J., & Yin, Y. (2018). Fermentative hydrogen production using pretreated microalgal biomass as feedstock. *Microbial Cell Factories*, *17*(1), 22. doi:10.118612934-018-0871-5 PMID:29444681

Wang, M., Chen, S., Zhou, W., Yuan, W., & Wang, D. (2020). Algal cell lysis by bacteria: A review and comparison to conventional methods. *Algal Research*, *46*, 101794. doi:10.1016/j.algal.2020.101794

Wang, M., Yuan, W., Chen, S., Wang, L., Zhao, S., & Li, S. (2021). Algal Lysis by Sagittula stellata for the Production of Intracellular Valuables. *Applied Biochemistry and Biotechnology*, *193*(8), 2516–2533. doi:10.100712010-021-03502-2 PMID:33779932

Wei, Z., Wang, H., Li, X., Zhao, Q., Yin, Y., Xi, L., Ge, B., & Qin, S. (2020). Enhanced biomass and lipid production by co-cultivation of Chlorella vulgaris with Mesorhizobium sangaii under nitrogen limitation. *Journal of Applied Phycology*, *32*(1), 233–242. doi:10.100710811-019-01924-4

Wintsche, B., Glaser, K., Sträuber, H., Centler, F., Liebetrau, J., Harms, H., & Kleinsteuber, S. (2016). Trace Elements Induce Predominance among Methanogenic Activity in Anaerobic Digestion. *Frontiers in Microbiology*, *7*. Advance online publication. doi:10.3389/fmicb.2016.02034 PMID:28018337

Xu, J., Wang, X., Sun, S., Zhao, Y., & Hu, C. (2017). Effects of influent C/N ratios and treatment technologies on integral biogas upgrading and pollutants removal from synthetic domestic sewage. *Scientific Reports*, *7*(1), 10897. doi:10.103841598-017-11207-y PMID:28883448

Xu, R., Zhang, K., Liu, P., Khan, A., Xiong, J., Tian, F., & Li, X. (2018). A critical review on the interaction of substrate nutrient balance and microbial community structure and function in anaerobic co-digestion. *Bioresource Technology*, *247*, 1119–1127. doi:10.1016/j.biortech.2017.09.095 PMID:28958888

Xue, F., Miao, J., Zhang, X., & Tan, T. (2010). A New Strategy for Lipid Production by Mix Cultivation of Spirulina platensis and Rhodotorula glutinis. *Applied Biochemistry and Biotechnology*, *160*(2), 498–503. doi:10.100712010-008-8376-z PMID:18931954

Yao, S., Lyu, S., An, Y., Lu, J., Gjermansen, C., & Schramm, A. (2019). Microalgae-bacteria symbiosis in microalgal growth and biofuel production: A review. *Journal of Applied Microbiology*, *126*(2), 359–368. doi:10.1111/jam.14095 PMID:30168644

Zawieja, I., & Worwąg, M. (2021). Biogas Production from Excess Sludge Oxidized with Peracetic Acid (PAA). *Energies*, *14*(12), 3434. doi:10.3390/en14123434

Zewdie, D. T., & Ali, A. Y. (2020). Cultivation of microalgae for biofuel production: Coupling with sugarcane-processing factories. *Energy, Sustainability and Society*, *10*(1), 27. doi:10.118613705-020-00262-5

Zhang, C., Li, Q., Fu, L., Zhou, D., & Crittenden, J. C. (2018). Quorum sensing molecules in activated sludge could trigger microalgae lipid synthesis. *Bioresource Technology*, *263*(March), 576–582. doi:10.1016/j.biortech.2018.05.045 PMID:29783193

Zhang, H., Wang, H., Zheng, W., Yao, Z., Peng, Y., Zhang, S., Hu, Z., Tao, Z., & Zheng, T. (2017). Toxic effects of prodigiosin secreted by Hahella sp. KA22 on harmful alga Phaeocystis globosa. *Frontiers in Microbiology*, *8*(JUN), 1–12. doi:10.3389/fmicb.2017.00999 PMID:28634473

Zhou, J., Lyu, Y., Richlen, M. L., Anderson, D. M., & Cai, Z. (2016). Quorum Sensing Is a Language of Chemical Signals and Plays an Ecological Role in Algal-Bacterial Interactions. *Critical Reviews in Plant Sciences*, *35*(2), 81–105. doi:10.1080/07352689.2016.1172461 PMID:28966438

Zhou, W., Cheng, Y., Li, Y., Wan, Y., Liu, Y., Lin, X., & Ruan, R. (2012). Novel fungal pelletization-assisted technology for algae harvesting and wastewater treatment. *Applied Biochemistry and Biotechnology*, *167*(2), 214–228. doi:10.100712010-012-9667-y PMID:22538982

Ziesche, L., Bruns, H., Dogs, M., Wolter, L., Mann, F., Wagner-Döbler, I., Brinkhoff, T., & Schulz, S. (2015). Homoserine Lactones, Methyl Oligohydroxybutyrates, and Other Extracellular Metabolites of Macroalgae-Associated Bacteria of the Roseobacter Clade: Identification and Functions. *ChemBioChem*, *16*(14), 2094–2107. doi:10.1002/cbic.201500189 PMID:26212108

Ziganshin, A. M., Liebetrau, J., Pröter, J., & Kleinsteuber, S. (2013). Microbial community structure and dynamics during anaerobic digestion of various agricultural waste materials. *Applied Microbiology and Biotechnology*, *97*(11), 5161–5174. doi:10.100700253-013-4867-0 PMID:23624683

Chapter 5
Integrated Omics and Mutation in Algae

Eladl Eltanahy
https://orcid.org/0000-0003-0894-0438
Algae Biotechnology Lab, Faculty of Science, Mansoura University, Mansoura, Egypt

Aya Torky
Algae Biotechnology Lab, Faculty of Science, Mansoura University, Mansoura, Egypt

ABSTRACT

Algae importance is spectacularly increasing in many biotechnological applications, such as human food, animal feed, biofuels, bioplastics, bioremediation, pharmaceuticals, and cosmetics. With the widespread use of "omics" technologies over the past two decades, recent advanced research attempts to understand the pathways of the promising algae species by whole genomes sequencing (genomics) and revealing lipid pathways (lipidomics), microarray to study all RNA transcripts (transcriptomics), all protein sets produced by the algal cell (proteomics). DNA alteration as classical mutagenesis caused a random mutation such as ethyl methane-sulfonate as chemical mutagenic and ultraviolet radiation as a physical mutagenic. On the other hand, the CRISPR-Cas9 modern technique is used to genetically engineer a protein with maximum editing efficiency. Incorporating omics and mutations techniques helps to thoroughly understand the systems biology of algae in the new era called integrated omics.

INTRODUCTION

Algae are photosynthetic cell factories capable of converting light energy into chemical energy in the form of biomass and numerous natural products, and therefore, they occupy the main base level in food chains as primary producers. For this reason, algae are considered the main food for fish, aquatic organisms, and protozoa. Algae contain high-value compounds such as carotenoids, proteins, fatty acids, carbohydrates, and solidifying agents such as carrageenan and alginate. These bioproducts used in different applications as biofuels, biofertilizers, antimicrobial compounds and biomedical applications

DOI: 10.4018/978-1-6684-2438-4.ch005

(AlPod & Elkatuy, 2022, Alsenani et al., 2020; Beaumont et al., 2021; Eltanahy & Torky, 2021; Lina M. González-González, Eltanahy, & Schenk, 2019; Lina María González-González et al., 2018).

Algae, whether micro or macro, are considered one of the oldest autotrophic organisms present on the Earth's surface, and they are credited with forming oxygen resulting from the photosynthesis process to maintain the ecosystem and prevent global warming and the increase in carbon dioxide in the atmosphere. Furthermore, the interest in using algae began a long time ago in different cultures as a source of food and alternative medicine in ancient civilizations. With the increase in technology and development in scientific research, discoveries related to these algae increased with the isolation of nova species of algae from promising and virgin wild environments and the possibility of benefiting in the manufacturing of food for human or animal feed and pharmaceutical industries as well as biofuels production and wastewater treatment and bioremediation.

It is indisputable that the field of algal biotechnology has some interesting challenge points from the scientific community and despite of the presence of many algal products in the market, full potential of algae is hardly tapped due to some limitations in scaling up algal cultivation for industrial production including the difficulty of obtaining high biomass accompanied with high desirable metabolites production, issues in harvesting and downstream processing and lack of the ease of adaptation to both biotic and abiotic stress factors by algal strain in addition to bioprospecting for more promising strains with distinctive traits, raising the specific growth rate of isolated species, or inducing the algal cell productivity of high-value biological components through changes in the culture nutritional and environmental conditions (Mosey, Douchi, Knoshaug, & Laurens, 2021; Pierobon et al., 2018) while with the emerge of the omics era, other solutions such as breeding, sexual hybridization, mutation and genetic and metabolic engineering present a more controlled promising alternatives (Gan, Lim, & Phang, 2016).

INTEGRATED OMICS

Integrated-omics, multi-omics, pan-omics, poly-omics and trans-omics are different names for the study of two or more omics data sets in order to support the analysis of the data and the huge results produced by different omics analyzes (Krassowski, Das, Sahu, & Misra, 2020). Also, how to use them to explain metabolic pathways and visualize the biological mechanism of algae growth and an accurate understanding of how different physical and chemical factors affect the environment algae influence its various metabolic processes, as well as changing the bioactive metabolic compounds. Although omics techniques have been considered somewhat modern techniques during the past two decades, most of the early research was mainly concerned with the medical fields and pathogenic bacteria because of their impact on human health and the possibility of using them in the pharmaceutical industries (Jamers, Blust, & De Coen, 2009) in addition to the very high costs of the analysis, which were not available in many research laboratories at that time. Over time, these omics techniques became cheaper in price, and with the development of next-generation sequencing techniques and the issuance of many types of equipment that could perform these analyzes, the use of omics techniques blowout to start the era of phycology omics.

The use of omics techniques such as genomics, transcriptomics, proteomics, and metabolomics in recent years has provided a huge amount of information about the biological systems inside the algal cell to reach a complete understanding of systems biology, which is known as the interaction between components of biological systems. Unfortunately, these data individually depict an effect on only one level inside the cell, which does not give a comprehensive picture of what happens more accurately in the metabolic pathways. Therefore, it was necessary to develop more comprehensive systems that depend on the correct understanding of the biology of living cells, integrating mathematical modelling and theoretical hypotheses while relying on some tools and algorithms for integrating the different results of omics with complementary biological analyzes (Ma, Lam, & Zaidah, 2015). The flow of integrated omics analysis and its applications is summarized in Figure.1.

In order to understand more clearly the importance, main obstacles, and how to effectively utilize the omics output results by combining more than one dataset to proceed with the integrated omics analysis, the following part will focus on each branch of omics in further detail.

Genomics

The most critical challenge in algae is the extraction of the complete genome due to the distinctive nature of the algal cell wall, which makes it challenging to break easily, in addition to the large and complex amounts of polysaccharides that are secreted by most of the red and brown macroalgae. There are many traditional methods for extracting DNA from microalgae and seaweeds (Coyer, Robertson, & Alberte, 1994; Fain, Druehl, & Baillie, 1988; Goff & Coleman, 1988; Murray & Thompson, 1980). In general, DNA extraction protocols are based on organic extraction by using the phenol and chloroform, or nonorganic extraction method, which depends on salting out in addition to proteinase K treatment, and adsorption method by using silica–gel membrane (Gupta, 2019). For microalgae, the most common methods are based on the precipitation of DNA with ethanol, and sometimes CsCl density-gradient purification is used to obtain DNA with high purity. But for seaweeds, it is found that the algae contain many compounds that hinder the extraction or amplification process by inhibiting the action of Taq polymerase enzyme (Jin et al., 1997). The most important of these compounds are the polysaccharides in the composition of the alga thallus, which is the biggest obstacle to extracting DNA from cells. Therefore, it requires the use of additional extraction steps to be able to extract the DNA properly, the most important of which is the cationic detergent cetyltrimethyl ammonium bromide (CTAB), which is used with salinity concentrations higher than 0.5M at room temperature, to precipitate the polysaccharides after the formation of complexes. Next are the polyphenolic compounds, which can also be eliminated by using polyvinylpolypyrrolidone (PVPP) alkaline buffer and beta-mercaptoethanol, which are added to inactivate polyphenols and form complex to get rid of them during DNA extraction (Murray & Thompson, 1980).

Some studies have developed these methods to become more accessible and practical and showed positive results with *Porphyra yezoensis*. The extracted DNA concentration was 1.5 μg 100 mg^{-1} from the algal tissue. By measuring the quality of DNA and the impurities in the extract, the results showed a ratio of 1.8 and 0.4 for the calculations of wavelengths A260/A280 and A230/ A260, respectively. These specifications are sufficient for PCR, genomic library preparation, and restriction digestion (Nakajima, Kitade, Iitsuka, Fukuda, & Saga, 2000).

In a similar study, Wattier, Prodöhl, and Maggs (2000) tested a new protocol based on the method used by Dellaporta, Wood, and Hicks (1983) with some modifications to get rid of the polysaccharides that cause extraction problems. The procedure was tested on twelve species of red algae, one green alga, and a wild plant to verify the efficiency of the modified method. As a result of following this protocol, the DNA concentration reached 5 μg/10 mg algae dry weight without containing RNA or degrading the DNA even after storage of DNA extracts at four °C, and the electrophoreses results showed intact bands without any smears or DNA degradation. In another study on 15 species of red algae using 1 mg

Figure 1. Flow chart of integrated omics analysis from algae

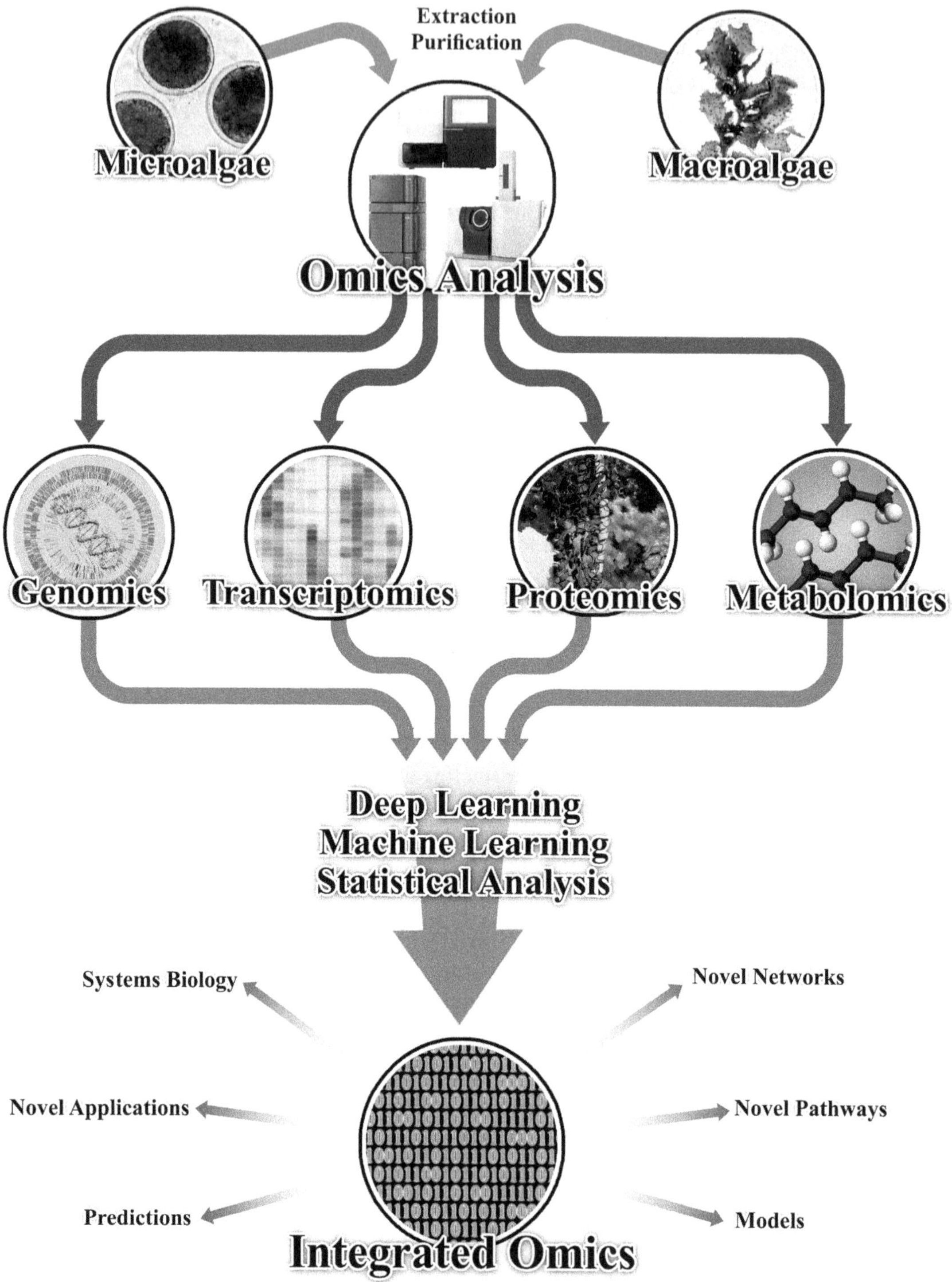

fresh algal thallus, it yielded 0.1 g of high-quality DNA with an A260/A280 ratio of 1.68–1.90 (Z. Hu, Zeng, Wang, Shi, & Duan, 2004).

Joubert and Fleurence (2005) found that the main issues of co-precipitation of polysaccharides with DNA during the extraction process can be overcome by performing the extraction process in two steps extraction process. The first step depends on enzymatic analysis using specific enzymes cocktail consisting of β-glucuronidase, β-glucanase, xylanase, cellulase and agarase, which convert the polysaccharides into simpler form that does not precipitate later. Then the second step is to do centrifugal precipitation of the extract and then purify the DNA from the pellet. This method was tested on 50 mg of *Palmaria palmata* and *Gracilaria verrucosa*, and the resulting RNA was 40 μg and 18 μg, respectively.

Transcriptomics

As in DNA extraction, extracting the essential RNA for the transcriptome analysis that one of the biggest obstacles to extraction is the presence of polysaccharides in the cell walls, in addition to polyphenolic compounds. Therefore, many scientists are developing their methods for extracting RNA and adding extractive factors, such as adding sand particles during the milling process to extract RNA from *Gracilaria changii* with the use of phenol in the initial stages, which increases the yield of the RNA to 0.65–1.14 $\mu g\ g^{-1}$ fresh weight with the resulting RNA having a high quality of up to, ratio 1.80–2.05 at A260:280 (Chan, Teo, Ho, Othman, & Phang, 2004).

Many studies accomplished analyzing the transcriptome of the algae. For example, *Porphyra yezoensis* leafy gametophyte, 10,154 5¢-end expressed sequence tags (EST) were read from an RNA sample after it was converted to cDNA then normalized and found that the samples contained 2,140 were unique species (Nikaido et al., 2000) while in *Griffithsia okiensis* just 1,104 cDNA clones were sequenced using the cDNA library (H. Lee, Lee, An, & Lee, 2007)

Similar studies on *Chondrus crispus* showed the presence of 2,002 ESTs, while 2,052 were found in the thallus. Clustering's analysis found 2,291 unique ESTs, and 50% of the ESTs showed similarity with known sequences ($e<10^{-4}$). The research also showed that ESTs related to stress-related biological processes in the protoplasts were five times higher than thallus ESTs (Collen et al., 2006).

As for *Sargassum binderi* alga, it has gained interest from researchers because it contains many critical metabolic substances such as alginate, alginic acid and mannitol, in addition to unsaturated and saturated oils and photosynthetic pigments. In a study on *S. binderi* (Wong, Ho, Lee, Rahim, & Phang, 2007), 1876 ESTs were obtained, including 1,270 unique genes, out of a total of 2,304 cDNA clones, of which approximately 40% of the genes are identical with those in the databases. Accordingly, they were divided into nine subgroups, the most important of which are the genes of metabolic pathways (25%), the transcription and translation genes (22.5%), and finally, the protein biosynthesis, folding, and degradation genes with only 5.6%.

According to the *Laminaria* alga, J. Yao, Fu, Wang, and Duan (2009) developed a specific protocol for the genus *L. japonica* by experimenting with four different methods for extracting RNA. It showed decent results with RT-PCR tests, and the harvest rate was high, reaching up to 68 $\mu g\ g^{-1}$ fresh weight, and also the RNA purity reached an A260/280 ratio of 1.96 ± 0.05.

Proteomics

Proteomics is the large-scale study of proteins and proteome is the entire set of proteins (Anderson & Anderson, 1998). Proteomics provides the complementary information needed so that complex biochemical processes within the cell can be detected and understood following the genomics and transcriptomics research.

Before developing high throughput proteomic analysis tools, it was challenging to predict cell behaviour using metabolic and cellular networks. Moreover, the same gene could produce more than one protein, or the same RNA sequence was sliced into pieces, and each part makes a specific protein. Also, after the protein formation, it may undergo a modification after the translation process and thus may perform a different function. This confirms that it was impossible to know all these alterations and changes in the protein produced in the cell without directly analyzing the protein using proteomic techniques (Chakdar, Hasan, Pabbi, Nevalainen, & Shukla, 2021).

Metabolomics

Analogous to the terms' genomics, 'transcriptomics,' and 'proteomics', metabolomics refers to the study of small molecules, commonly known as metabolites set of low molecular weight metabolites, but its size varies greatly depending on the organism (Fiehn, 2002). The metabolites are directly differentiated due to the difference in the cell genome, and because algae mostly live in an aquatic environment, the metabolites are affected by the ultimate variations in the environmental factors in which the algae grow, whether temperature, pressure, light intensity, pH, and even environmental pollution in the oceans and seas, such as oil spills for example (Sogin, Puskás, Dubilier, Liebeke, & Huber, 2019).

Most studies in algae are concerned with the study of biologically active secondary metabolites that can be used in applications for the food, feed, medicine, cosmetics and biofuel industries, based on the quantitative measurement of protein, fatty acids, polysaccharides, carotenoids, astaxanthin and antioxidants (Richmond, 2004).

While environmental studies, which are rather few, are concerned with measuring secondary metabolites in algae and the qualitative change of the metabolites, the most important of these studies was on the alga *Chlamydomonas*. It was found that when the algae were exposed to nitrogen, phosphorous, sulfur and iron-depleted growth conditions, the cells secreted more than 800 new metabolites compared to the alga when grown under optimal growth conditions (Bölling & Fiehn, 2005) and in a similar mutant strain of *Chlamydomonas reinhardtii*, BAF-J5, which is a starch-less mutant it was found that, if the algae exposed to nitrogen depleted growth condition the cells will store lipids up to 65% of algal dry weight more than control treatment (James et al., 2011).

APPLICATIONS OF INTEGRATED OMICS

Recent studies in phycology have been greatly affected by the availability of advanced investigates of integrated omics by maximizing the applications of algae in various pragmatic fields using different technologies such as algal transgenics to utilize algae as green factories of recombinant products such as the production of toxins, vaccines, pharmaceuticals, and the increase in biofuel productivity (Cardozo et al., 2007; Dhandayuthapani, Malathy, Mulla, & Gupta, 2021; S. Kumar, Singh, Kumar, & Shukla, 2021; Siripornadulsil, Dabrowski, & Sayre, 2007).

In a recent study on the desert alga *Chlorella ohadii*, integrated omics including transcripts, metabolites, and lipid response interactions revealed how this alga can withstand extreme irradiation levels of up to 3,000 µmol photons $m^{-2} s^{-1}$, which is equivalent to six times the energy required for saturation in the process of photosynthesis (Treves et al., 2020). Its cell doubling rate is three times that of *C. reinhardtii,* and its growth is characterized by different phases. The first phase is a slow growth rate and a high rate of photosynthesis, and the second is when the cells are in a rapid growth rate associated with respiration at its peak. Furthermore, the extreme irradiation levels cause a decrease in sugar metabolites (sucrose and glucose), respiratory intermediates (citrate, malate, and pyruvate) and major amino acids (aspartate, glycine, and glutamate).

Using integrated omics analyses on the microalga *Chromochloris zofingiensis*, which grows in fresh water and produces astaxanthin, as well as accumulates lipids, the effect of salinity was studied, and it was found that a concentration of 0.2 M NaCl is the optimal concentration for the production of both astaxanthin and triacylglycerol (TAG). The effect of salinity was associated with a decrease in photosynthesis and CO_2 fixation but with a significant increase in reactive oxygen species (ROS) accumulation, protein turnover, keratin synthesis and oleaginousness. Also, a harmonious increase was observed in the metabolic pathways of carbon formation from starch, production of fatty acids and acetyl-CoA, which is the central component of most of the biological processes and formation of TAG and the oil droplets that collect inside it. As for astaxanthin, it was found that the metabolic pathway of carotenoids is converted into the production of astaxanthin through the lutein pathway by lycopene beta cyclase (LCYb) enzymes and diacylglycerol acyltransferase (DGAT), which is responsible for the assembly of TAG inside algal cells (Mao, Zhang, Wang, & Liu, 2020).

Integrated Omics study on seaweed contributed to understanding the molecular mechanisms of adaptation to various environmental stressors and the influence of the surrounding microbial environment (microbiome) of the alga on its internal metabolism and the substances secreted from its cells in its surrounding environment. Furthermore, one of the most important environmental influences on algae is the thermal effect, as it affects the cell growth, internal components, and rearranges the cell's priorities in depleting energy. In that case, the cell's metabolism becomes toward the repair and protection processes that can be remedied (Harley et al., 2012). Eventually, complete death of its cells may occur (Davison & Pearson, 1996).

The seaweed *Saccharina japonica* is of economic importance, so it is used in many applications through direct harvesting or cultivation. However, for reasons that are not clear, the algae die due to the presence of copper leaking from the pipes of poor-quality cooling systems in the water. It is known that copper in low concentrations causes an increase in algae growth in general, but it may reach high concentrations with a toxic effect on algae, while the mechanism is not precise. In the study of proteomics for this effect, the algae were exposed to different concentrations of copper ranging from 10 to 200 µg/L, and then differentially expressed genes (DEGs) were analyzed. The results demonstrated changes in four critical pathways related to algal metabolism such as protein synthesis, photosynthesis, redox processes and their reduction to levels that lead to the phenomenon of cell bleaching and then the death of the alga (Zhang, Wang, Shan, Pang, & Xu, 2019).

In a systematic review of omics research on macroalgae over the past three decades, it was found that the largest interest of research was on the red algae reaching about 45%. Furthermore, more than half of these studies were on the economic algae *Pyropia* (family Bangiaceae) and *Gracilaria* (family Gracilariaceae) due to their importance as an antiviral food source in fish farms because they contain antioxidants and antiviral compounds (Lozano, Wacyk, Carrasco, & Cortez-San Martín, 2016). Also, *Gracilariopsis* (family Gracilariaceae) and *Porphyra* (family Bangiaceae) but with fewer priorities. *Gracilaria* and Gracilariopsis have been deemed the primary source of agar industries, while *Pyropia* and *Porphyra* are used for nori in sushi rolls products. In the second place are brown algae (35%), and most research was limited to the genera *Saccharina* and *Sargassum* in addition to *Ectocarpus*. Thirdly, omics research on green algae mainly (18%) to *Ulva* alga (Patwary et al., 2021).

Marine macroalgae in the Mediterranean Sea and its epiphytic microorganisms that live on the surfaces of the algal thallus, especially the alga *Taonia atomaria*, were studied using integrated omics methods. The analysis focused on metabolomics and microbiomes (epiphytic cells densities) by combining the data of multiplatform mass spectrometry-based metabolomics analysis based on the molecular networking to identify the differentiation in the metabolites and highlighted a clear chemical differentiation at the algal surface along the thallus with similar clustering as for microbial communities (Paix et al., 2020). In comparison with the diversity found on the rocks in the same area, one of the most important results of this study was that the alga has a role in selecting the epiphytic microbial diversity on its surface for apical parts where it was higher amounts of sesquiterpenes, phosphatidylcholines (PCs), and diacylglycerylhydroxymethyl-N,N,N-trimethyl-β-alanines (DGTAs). In contrast, the carotenoids and dimethylsulfoniopropionate (DMSP) were found to be accumulated in the algal basal parts acting as an antioxidant, anti-adhesion agent, defence against herbivory, and cryoprotection (Lyons, Scheibling, & Van Alstyne, 2010).

The researchers Roach et al. (2020) explored the effect of the interference of corals and algae in the marine environment using metagenomic sequencing techniques in order to obtain sufficient data to know the sequence of RNA and genes and their role in gene expression with metabolomic profiling technology to identify the compounds that are formed from organisms in This ecosystem and the use of epifluorescence microscopy technology for sample examination and characterization and the results indicated an increase in the abundance of microbes and an increase in the size of cells within the holobionts.

Macroalgae that grow on the shores of oceans and seas are greatly affected by low tides, and the alga thallus is exposed to repetitive desiccation effects. Therefore, this area of shores is greatly affected by tidal factors, and one of the most important marine algae present in these areas is the algae *Pyropia haitanensis*, which is used as food (nori products) and characterized by its strong resistance to the effects of desiccation and thallus still alive despite its loss of approximately 90% of its water content. Scientists have attributed this property to the ability of the alga to accumulate fluoridoside as an osmoregulator inside its cells by upregulation of *PhNHO1* and *PhGPDH* genes, which encodes glycerokinase, and glycerol-3-phosphate dehydrogenase, respectively, in addition to upregulation of *PhLOX1* and *PhLOX2,* which are both involved in the production of volatile short-chain compounds to protect the desiccated thallus against bacteria (Qian et al., 2015). In an advanced study, Huang, Peng, and Yan (2021) studied the same alga using integrated omics (transcriptome, proteome, and metabolome) analyses to compare the effect of desiccation alone and the effect of three physical environmental factors, namely desiccation, high temperature and high light intensity. The study proved that gyoxylate and dicarboxylate metabolism (pathway ko00630), which is related to osmoregulator biosynthesis and its accumulation is upregulated. Not only that, but also increased gene expression of all genes related to lipid metabolism in the alga cells as a result of exposure to desiccation stress, such as *sPLA2*, *PLD1/2*, *LPCAT*, and *ETP* in an association of phospholipid metabolites (GPE: glycerol phosphatidylglycolamine and GPC: glycerol phosphatidylcholine). It was also found that the metabolic pathway of photosynthesis and carbon fixation (pathway ko00710) is partially restricted to reduce the production of oxidizing agents and free oxygen atoms. In contrast, the Photosynthesis-antenna proteins (*Lhca1*) were overproduced to prevent damage from the light intensity.

INTEGRATED OMICS TOOLS AND RESOURCES

Recently, many different sources of libraries and data sets have been published on the Internet containing many metabolic pathways and whole-genome sequences for some species of algae of economic importance, in addition to massive data on algae resulting from mass spectrometry data analyses.

There are also many free or paid tools and programs for different omics analyses, with the availability of many methods of visualization that show the results of data set analyzes. In the following table (Table.1), the most popular programs and various resources for algae and omics analyzes are sorted alphabetically

MUTATION IN ALGAE

Mutation, which is defined as an alteration in the genetic material, can occur in different forms in algae; spontaneous mutation occurs in the absence of mutagens, the random mutation which takes place in response to chemical and physical mutagens, the adaptive mutation which occurs as a response to different stress conditions and finally, the specific mutation which occurs deliberately using genetic engineering approaches (figure 2).

Spontaneous Mutation

Mutations are the origin of genetic variations and biodiversity that ultimately lead to evolutionary and adaptive changes within species. Quantifying mutation rate and studying its effect on fitness and other traits in mutant species is fundamental to understanding the adaptive and evolutionary potentials in a species (Wu et al., 2017). Mutation rates can be estimated usually through mutation accumulation experiments (MA) in which the presence of reference genome sequences is of high quality (Halligan & Keightley, 2009; W. Wei et al., 2014). The development of next-generation sequencing and other high throughput technologies improves the accessibility of total genome sequences of numerous algal species (Vandepoele et al., 2013). MA experiments' principle is to monitor genetic variations in MA lines kept at a low population size. As the strength of selections is inversely proportional to population size, at very low population size, the selection is limited, which allows spontaneous mutations to become fixed and estimated (Elena & Lenski, 2003). The spontaneous mutation rate is identified by comparing the MA genome sequences to those of their ancestors while monitoring the number of cell divisions to estimate the mutation rate per cell division (Lynch et al., 2016).

Mutation rate has been estimated in several species through MA techniques, and this includes *Picochlorum costavermella*, Trebouxiophycean green alga with different biotechnological potentials, whose mutation rate ($\mu = 10.12 \times 10^{-10}$) was found to be one of the highest reported among unicellular microalgae (Krasovec, Sanchez-Brosseau, Grimsley, & Piganeau, 2018). Another study evaluated the spontaneous mutation rate in four haploid marine microalgae; *Bathycoccus prasinos*, *Ostreococcus tauri*, *Ostreococcus mediterraneus*, and *Micromonas pusilla*, it varied from $\mu = 4.4 \times 10^{-10}$ to 9.8×10^{-10} mutations per nucleotide per generation (Krasovec, Eyre-Walker, Sanchez-Ferandin, & Piganeau, 2017).

As in terms of fitness, a negative correlation between the total number of mutations in *C. reinhardtii* and the change in fitness was observed, MA lines were found to be less fit than their ancestors and the mutational effects of mutations located in coding regions were found to fit two categories according to fitness date: one deleterious and one neutral (Kraemer, Böndel, Ness, Keightley, & Colegrave, 2017).

Table 1. Common tools and resources used in integrated omics analysis

Alga-PrAS (Algal Protein Annotation Suite)	Database of physicochemical and structural properties and post-translational modifications in algal proteomes	http://alga-pras.riken.jp/
Global Natural Products Social Molecular Networking (GNPS)	Web-based mass spectrometry ecosystem that aims to be an open-access knowledge base for community-wide organization and sharing of raw, processed, or annotated fragmentation mass spectrometry data (MS/MS)	https://gnps.ucsd.edu/
ALCOdbCyano	Gene co-expression for blue-green algae *Synechocystis* sp. PCC 6803, *Synechococcus elongatus* PCC 7942, and *Anabaena* sp. PCC 7120	http://alcodb.jp/cyano/
Algae Gene Coexpression database (AlCODb)	Gene co-expression for *C. reinhardtii* and *Cyanidioschyzon merolae* as a green and red algae, respectively	http://alcodb.jp/
AlgaePath	Web-based database that integrates gene information, biological pathways, and NGS datasets for the green algae *C. reinhardtii* and Neodesmus sp. UTEX 2219–4	http://algaepath.itps.ncku.edu.tw/
Comet	Tandem mass spectrometry data analysis program used for protein identification	http://comet-ms.sourceforge.net/
CrGDB	CrGDB displays high quality spliced alignments for EST, cDNA, PUT, and model species proteins, including *C. reinhardtii*	https://www.plantgdb.org/CrGDB/
dEMBF	Database of enzymes of microalgal biofuel feedstock	http://bbprof.immt.res.in/embf/
Ensembl	Different omics tools for data analysis	https://www.ensembl.org/info/docs/tools/index.html
Escher	Web-based tool for building, viewing, and sharing visualizations of biological pathways	https://escher.github.io/
Expasy	An extensible and integrative portal which provides access to over 160 databases and software tools	https://www.expasy.org/
GIM(3)E	Models of cellular metabolism developed from metabolomics and expression data	http://opencobra.sourceforge.net/
Greenhouse	Whole genomes database of 42 algae	https://greenhouse.lanl.gov/greenhouse/
JIC Euglena	Dedicated non-redundant protein database *Euglena gracilis*	https://jicbio.nbi.ac.uk/euglena/
KaPPA-View4	Web-based database which can represent user-uploaded '-omics' data on the metabolic pathway maps	http://kpv.kazusa.or.jp/
Kyoto Encyclopedia of Genes and Genomes (KEGG)	Database resource for understanding high-level functions and utilities of the biological system	https://www.genome.jp/kegg/
MADMAX	Management and analysis database for multiple omics experiments	http://madmax2.bioinformatics.nl
MarVis-Suite	Toolbox for interactive ranking, filtering, combination, clustering, visualization, and functional analysis of data sets containing intensity-based profile vectors (data set features or marker candidates), as obtained e.g. from mass spectrometry (MS), microarray, or RNA-seq experiments	http://marvis.gobics.de/
MASCOT	Search engines for the identification of algal peptides and proteins	https://www.matrixscience.com/
MassTRIX	Annotates metabolites in high precision mass spectrometry data. It marks the identified chemical compounds on KEGG pathway maps using the KEGG/API	http://masstrix.org
MaxQuant	Quantitative proteomics software package designed for analyzing large mass-spectrometric data sets	https://www.maxquant.org/
MeltDB 2.0	Software platform for the Analysis and Integration of Data from Metabolomics Experiments	https://meltdb.cebitec.uni-bielefeld.de
MetaCyc	Databases for three microalgae *C. reinhardtii*, *Auxenochlorella protothecoides* and *Ostreococcus taurii*	https://metacyc.org/
mixOmics	The data analyzed with mixOmics may come from high throughput sequencing technologies, such as 'omics data (transcriptomics, metabolomics, proteomics, microbiome/metagenomics …) but also beyond the realm of omics (e.g. spectral imaging)	http://mixomics.org/
NCBInr and NCBIprot databases	Collection of sequences from several sources, including translations from annotated coding regions in GenBank, RefSeq and TPA	https://www.ncbi.nlm.nih.gov/protein/
Paintomics	Integrative visualization of multiple omic datasets onto KEGG pathways	https://www.paintomics.org/
PathVisio	Free open-source pathway analysis and drawing software which allows drawing, editing, and analyzing biological pathways	https://pathvisio.github.io/
PhycoCosm	Algal multi-omics portal, developed by the US Department of Energy containing genomic information of over 50 microalgal species	https://phycocosm.jgi.doe.gov
Pico-PLAZA 3.0	Integration of structural and functional annotation of 39 plant species, including algae	https://bioinformatics.psb.ugent.be/plaza/versions/plaza_pico_03/
PLAZA diatoms 1.0	Access point for integration of structural and functional annotation of 26 species of diatoms	https://bioinformatics.psb.ugent.be/plaza/versions/plaza_diatoms_01/
Reactome	Free, open-source, curated and peer-reviewed pathway database	https://reactome.org/
realDB	Database of genome and transcriptome data for the less represented red algae or the phylum Rhodophyta	http://realDB.eplant.org
Sequence Read Archive (SRA)	The largest publicly available repository of high throughput sequencing data. The archive accepts data from all branches of life as well as metagenomic and environmental surveys	https://www.ncbi.nlm.nih.gov/sra
SIMCA®	Multivariate data analysis software	https://www.sartorius.com/en/products/process-analytical-technology/data-analytics-software/mvda-software/simca
The Diatom EST Database	Search for expressed sequence tag (EST) data from two eukaryotic microalgae of the class Bacillariophyceae, *Phaeodactylum tricornutum* and *Thalassiosira pseudonana*	http://avesthagen.sznbowler.com/
The European Bioinformatics Institute (EMBL-EBI)	Maintains the world's most comprehensive range of freely available and up-to-date molecular data resources	https://www.ebi.ac.uk/services
UniProt	A scientific community with a comprehensive, high-quality and freely accessible resource of protein sequence and functional information	https://www.uniprot.org/
Vanted	Visualization and analysis of networks containing experimental Data	https://www.cls.uni-konstanz.de/software/vanted/

In 2019, the first spontaneous mutation rate was estimated in diatoms, particularly the diatom *Phaeo-*

Figure 2. Types of mutations in algae, EMS: ethyl methane sulphonate, NTG: N-methyl-N-nitro-N-nitrosoguanidine, UV: Ultraviolet, RNAi: RNA interference, ZFNs: Zinc finger nucleases, TALENs: Transcription Activator-Like Effector Nucleases and CRISPR/Cas 9: Clustered, regularly interspaced, short palindromic repeats (CRISPR) and associated genes (Cas genes)

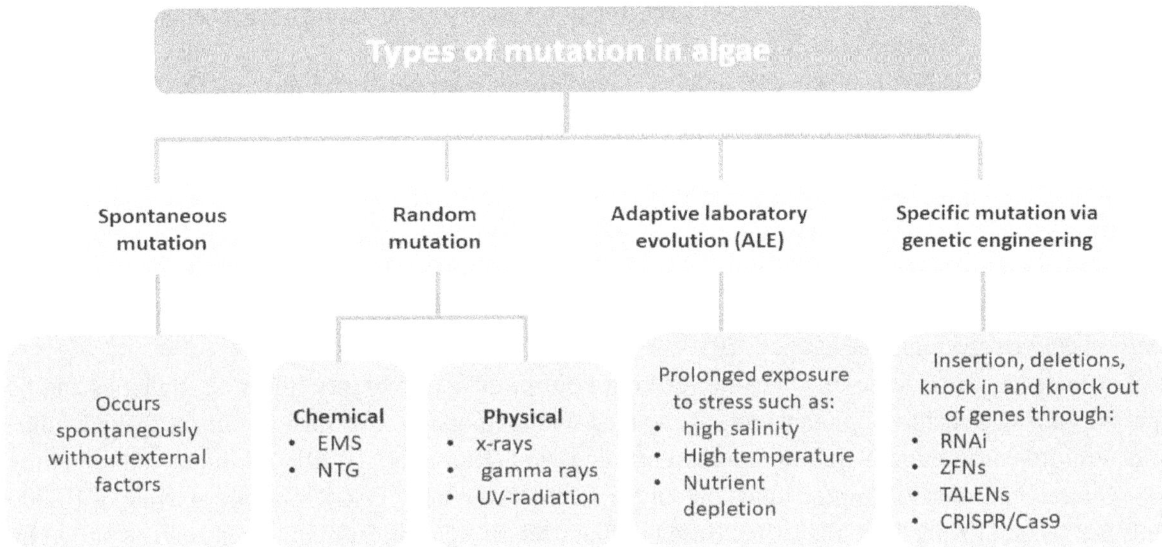

dactylum tricornutum (strain RCC2967). Thirty-six MA lines were sequenced, and 156 *de novo* mutations were identified. The base substitution mutation rate per site per generation was $\mu_{bs}= 4.77 \times 10^{-10}$, while the insertion-deletion mutation rate was $\mu_{id}= 1.58 \times 10^{-11}$. Consistently with previous observations in other species, the mutation rate was biased towards excessive mutations from GC to AT. Also, a higher mutation rate was observed in mitochondria as mutation rates differ between the genomes of the nucleus and other organelles (chloroplasts and mitochondria) (Krasovec, Sanchez-Brosseau, & Piganeau, 2019).

Spontaneous mutation has also been reported in macroalgae; a green-coloured *Pyropia kinositae* was found in a natural population. Studies showed that compared to the wild type, the growth of the conchocelis (a haploid stage preceding the gametophyte) and gametophytic blade in the *Pyropia kinositae* mutant was slower. It also had a significantly lower phycoerythrin content. Surprisingly, the green mutant's colour phenotype was genetically stable in both the gametophytic and the sporophytic conchocelis phases, which suggests its use in different breeding techniques as new breeding material for *Pyropia* (Fumina, Koki, & Kyosuke, 2020). Katju and Bergthorsson (2019) have extensively studied this research point in their article published in 2019.

Random Mutation

Random mutation involves applying external factors known as mutagens to make random changes in the genetic makeup of organisms. It is important to produce algae with improved characteristics that their ancestral strains did not have. There are two main classes of random mutagenesis: (1) physical mutagenesis through x-rays, gamma rays and ultraviolet (UV) radiation. Gamma radiation possesses strong penetration power that leads to the formation of reactive oxygen species such as H_2O_2, $O_2^{\ -}$ and OH^- eventually resulting in cellular DNA damage (Beyaz & Yildiz, 2017). In comparison, UV radiation induces the deletion of adenine (A) thymine (T) base pairs or the formation of thymine dimers that cause the transition of cytosine (C) and guanine (G) to adenine and thymine (Muthuraj, Selvaraj, Palabhanvi, Kumar, & Das, 2019), (2) chemical mutagenesis using chemicals such as ethyl methane sulphonate (EMS) and N-methyl-N-nitro-N-nitrosoguanidine (NTG) where EMS causes C/G to A/T transitions by removing C and G base pairs while NTG targets the replication fork inducing a wide range of mutations (Kuo et al., 2017; Sarayloo et al., 2017).

Random mutagenesis is a two-step process. First, mutants are generated through the application of mutagens. Secondly, a screening strategy is applied to select and isolate positive mutants (Arora, Yen, & Philippidis, 2020), making efficient screening strategies crucial for the success of random mutagenesis. Traditional screening methods by cultivating each mutant colony individually and then analyzing it to test the acquisition of the desired properties is inefficient and labour and time-consuming. Therefore, large scale screening methods such as fluorescence-activated cell sorting (FACS) and chemical inhibitors of certain products, in addition to percoll density gradient centrifugation, have been used for a faster and more selective screening (Yu et al., 2020).

For instance, in the case of screening for high lipid producing mutants, all three strategies can be applied. In FACS methods, mutant cells are stained with a lipophilic dye such as Nile red which binds to neutral lipids resulting in yellow fluorescence emission (Rumin et al., 2015). Mutant cells showing high fluorescence due to elevated lipid content are sorted through the FACS machine at a rate of 10,000 cells per second (Yamada et al., 2016). Based on their relative densities, mutant cells can be sorted by percoll centrifugation, where cells with high lipid content will have lower density as lipids have a lower density than other metabolites such as proteins and carbohydrates (Sung, Choi, Lee, Hong, & Sim, 2019). The application of lipid synthesis inhibitors is considered an indirect method for selecting high lipid producing mutants. Cells showing higher tolerance to the chemical either enhance the synthesis or the activity of the inhibited enzyme to lighten the inhibitor's effect or activate alternative pathways suggesting an overall increased lipid synthesis (Arora et al., 2020).

Just as in the case with lipid synthesis inhibitors, amino acid inhibitors can be used to select single cell protein-rich mutants, which will form larger colonies on selective plates. However, there are no universal inhibitors that collectively inhibit amino acid production. Therefore, individual or combined amino acids inhibitors should be used not only to select mutants with high protein content but also with high essential amino acids concentrations (Spalvins, Raita, Valters, & Blumberga, 2021).

Another technique used for selecting mutants is the use of selective media that only mutants will grow on or show different characteristic behaviour. For example, the addition of glufosinate, a herbicide, to a growth medium induces carotenoid accumulation by impairing ammonia assimilation by inhibiting the crucial enzyme glutamine synthetase, which results in a change of colony colour to orange. In a study evaluating the effect of chemical mutation by EMS on the growth and astaxanthin production in *Coelastrum* sp., a mutant G1-C1 selected using glufosinate showed the highest total carotenoids content of 45.48 mg/L and astaxanthin content of 28.32mg/L, which was twice higher than that of the wild type (Tharek et al., 2021).

EMS has been extensively used to induce random mutations in different microalgal species resulting in mutants with new beneficial traits. *Chlorella minutissima* was subjected to 2M EMS for induction of lipid accumulation. Among three different mutants (CM2, CM5, CM7), CM7 showed increased biomass by 1.6-fold and lipid content by 1.55-fold. It also showed an increase in monounsaturated fatty acids coupled with a decrease in saturated and polyunsaturated fatty acids suggesting its possible application for good quality biodiesel production (Mehtani et al., 2017).

Also, EMS induced mutagenesis was applied on a *Dunaliella tertiolecta* strain to improve zeaxanthin content, a xanthophyll pigment that helps treat and prevent degenerative diseases. Mutant strain with 10–15% higher zeaxanthin content was obtained, then optimization of culture conditions for this mutant, high daily zeaxanthin productivity (8 mg·L^{-1}) was achieved by growth at 0.6 M NaCl and 140–160 µmol photons·m^{-2}·s^{-1} (M. Kim, Ahn, Jeon, & Jin, 2017).

Photobioreactors cultivation faces some constraints, such as the photoinhibition in surface-exposed cells as well as uneven light distribution due to elevated optical density of the culture, so the use of strains with improved light use efficiency and tolerance to high irradiation is critical for industrial application. Carotenoids play an important role in preventing photooxidative damage. A study implemented EMS induced chemical mutagenesis for obtaining high carotenoids producing, photooxidative stress-tolerant *Chlorella vulgaris* strain. About 60,000 mutagenized lines were screened by evaluating their resistance to the carotenoid biosynthesis inhibitor norflurazon, and only 12 lines were selected for further analysis, among which NFR-3 and NFR-13 mutants showed higher carotenoids content compared to the wild type strain (25% and 14% increase, respectively) (Guardini et al., 2021).

Moreover, chemical mutagenesis through treatment with EMS and NTG was applied to the model diatom *Phaeodactylum tricornutum* where a significant correlation between carotenoids content and neutral lipids was found during the exponential growth phase. The correlative metabolic pathways between fucoxanthin synthesis and lipid synthesis were analyzed. Among 1000 strains screened through the FACS method, only four mutants exhibited long term stability and the highest increase of fucoxanthin content by 69.3% (Zhiqian Yi et al., 2018).

Mutant *Pyropia yezoensis* macroalga exhibiting improved growth rate was also obtained through EMS induced mutagenesis. Thallus was treated with 2mM EMS and physiological analysis showed that the mutant strain PyE2 had a faster growth rate, increased phycoerythrin content and improved photochemical efficiency than the wild type strain. Proteomic analysis of the blades grown under culture conditions revealed that 19 protein species were expressed differently between the mutant and wild type strains. The increased growth rate can be owing to the up-regulation of proteins involved in photosynthesis and protein synthesis. Random mutation techniques can be employed to develop new cultivars used in the nori industry (H.-J. Lee & Choi, 2018).

Physical mutagenesis was also used in several microalgal species; *Scenedesmus obliquus* mutant strain, namely SDEC-1M, was obtained through exposure to UV radiation. SDEC-1M showed improved characteristics such as high light conversion efficiency, tolerance to high CO_2 levels (15% v/v) and increased carbohydrates and lipids contents (37.26% and 24.80%, respectively) compared to the wild strain making it a better candidate for biofuel production and CO_2 bio-fixation (Qi, Wu, Mu, Zhang, & Xu, 2018).

With the purpose of developing high temperatures tolerant strains, *Pyropia tenera* was subjected to gamma irradiation of 1 kGy and a mutant (Pt1k) showing a change in blade colour from red to dark green was chosen for further analysis under different temperatures. A four-fold increase in biomass and growth rate accompanied by high phycocyanin content were obtained in Pt1k at 12 °C. When cultured at 25 °C for three weeks, chlorophyll-a and phycocyanin contents were higher in the Pt1k mutant. Furthermore, antioxidant enzymes; ascorbate peroxidase and superoxide dismutase activities increased in both strains during the first week of exposure then decreased rapidly in the wild strain while kept constant in the Pt1k mutant indicating a possible involvement of antioxidant enzyme activity in the tolerance to high temperatures (H.-J. Lee & Choi, 2019).

Random mutagenesis has several advantages over other strain improvement techniques, including its simplicity and possible application on a large scale, avoiding issues with outdoor cultivation of genetically modified organisms (GMOs), in addition to the little knowledge required about the metabolic pathways of the targeted products. On the other hand, random mutation can cause multiple mutations, which makes identifying the gene(s) responsible for developing the new traits a very hard and time-consuming process (Perin et al., 2015).

Adaptive Laboratory Evolution

Adaptive laboratory evolution (ALE) is another approach applied for strain improvement where the flexibility of microorganisms to adapt to diverse environmental conditions is exploited. In ALE, microorganisms are cultivated under definite stress conditions such as nutrient depletion, high salinity and high temperature for a prolonged time ranging from weeks to years. This adaptation leads to the accumulation of several DNA mutations, including limited insertions and deletions and single nucleotide polymorphisms (SNPs), resulting in the formation of evolved strains exhibiting improved features such as high growth rate and tolerance to extreme growth conditions and the presence of toxic compounds (Arora et al., 2020).

Although the ALE approach has been employed in bacteria and yeasts for more than two decades, its microalgae application has only emerged in recent years. Two cultivation techniques have been used for ALE; batch cultivation and continuous cultivation. In batch cultivation, adaption is performed in deep-well plates or shake flasks where culture is serially transferred to a new medium at regular intervals. Despite this method's simplicity and low cost, it has multiple drawbacks, such as the variable population density and instability of factors affecting growth rates such as nutrient availability and pH. On the other hand, continuous cultivation in bioreactors offers tight regulation of nutrient availability, dissolved oxygen, and pH. While bioreactors result in a steady growth rate and population density, their main disadvantage is their high cost (Dragosits & Mattanovich, 2013).

In a study conducted to evolve a microalgal strain with high CO_2 fixation ability and tolerance to high CO_2 concentrations, two adapted *Chlorella* sp. strains, AE10 and AE20, were obtained after applying ALE under 10% and 20% CO_2, respectively, for 31 cycles. Both strains grew in 30% CO_2 and maximum biomass was observed in the AE10 strain and was found to be 1.22 and 2.94 times higher than those of AE20 and the original strain, respectively. The chlorophyll content of both adapted strains was found to be significantly higher than the original strain as well (D. Li, Wang, Zhao, Wei, & Sun, 2015).

In another study, ALE was performed in order to enhance salinity tolerance in a freshwater *Chlorella* sp. AE10. An evolved strain (*Chlorella* sp. S30) with the ability to withstand 30 g/L salts was obtained after 138 days. Possible tolerance mechanism was investigated through comparative transcriptomic analysis, and results showed up-regulation in genes involved in central carbon metabolism, amino acid biosynthesis and antioxidant enzymes to combat oxidative phosphorylation and decreased photosynthesis caused by high salinity(X. Li et al., 2018).

Schizochytrium species were studied to employ the high salinity adaptive laboratory evolution to improve lipid content and antioxidant activity. After 150 days, an adapted strain (ALE150) exhibiting a 32.7% increase in dry biomass, 53.31% increase in lipid content and higher antioxidant activity was obtained. Transcriptomic analysis of the adapted strain showed up-regulation of genes involved in central carbon metabolism and antioxidant enzymes in addition to changes in metabolic fluxes towards polyketide synthase and fatty acid synthase pathways (Sun et al., 2018).

Phenol is an extreme environmental contaminant that microalgae also show little tolerance and a low degradation rate. Many trials were applied to improve the effectiveness of algae in phenolic pollution treatment; for instance, ALE was applied on *Chlorella* sp., resulting in an adapted strain which was obtained after treatment with 500 mg/L phenol for 95 days and the strain tolerated up to 700 mg/L phenol without substantial inhibition. Maximum biomass of 3.40 g/L and 2.70 g/L were obtained under 500 and 700 mg/L phenol, respectively, which were two folds higher than that of the original strain. Also, at a cell density of 0.6 g/L, the adapted strain removed 500 mg/L phenol in 7 days (Wang et al., 2016).

In a recent study, co-cultivation of twelve strains of *Tisochrysis lutea* for six months under cumulative thermal stress was employed as an ALE technique to enhance the polyunsaturated fatty acids (PUFAs) content in polar lipids. Only two strains survived, and results showed a 2-fold increase in lipid content affecting different lipid classes; saturated fatty acids decreased while monounsaturated and polyunsaturated fatty acids increased, which suggests their possible role in regulating membrane fluidity to survive temperature changes. Among important PUFAs, DHA content was found to be triple that of the original strain (Gachelin et al., 2021).

Yi *et al.* used semi-continuous cultivation under red and blue light illumination to improve pigments and neutral lipids accumulation in the diatom *Phaeodactylum tricornutum.* After 11 ALE cycles, the growth rate was doubled, but there was a slight decrease (5.3%) in neutral lipids content. Also, pigments content was estimated, and both chlorophyll-a and β-carotene contents remained unchanged, while fucoxanthin content showed a 2.1-fold increase compared to the original strain (Z. Yi et al., 2015).

Specific Mutation via Genetic Engineering Approaches

One more form of mutation (genetic modification) that can occur in algae is the specific mutation through genetic engineering, which has been almost exclusively investigated in microalgae. Genetic engineering approaches involve modifying the algal genome through the insertion, deletion, knock-in and knock-out of genes resulting in tailored strains acquiring new desirable traits. Although the successful transformation of plant and animal cells has been reported for decades, the transformation of the model alga *C. reinhardtii* only came in the late 1980s (Rochaix & van Dillewijn, 1982). This delay was mainly caused by problems developing successful transformation and genome editing techniques in microalgae. So far, only a handful of genome-editing techniques have been developed for *Chlamydomonas* and a small number of other microalgal species.

Multiple editing techniques have been employed in microalgal engineering, including the use of RNA interference, nucleases such as Zinc Finger nucleases (ZFNs) and Transcription activator-like effector nucleases (TALENs), in addition to Clustered Regularly Interspaced Short Palindromic Repeats – CRISPR associated protein 9 (CRISPR-Cas9) system (Jeon et al., 2017).

RNA interference (RNAi) system is a reverse genetics approach that identifies genes functions through targeting complementary sequences on mRNA, thus silencing any sequence of interest (Ketting, 2011). In 2015, RNAi was used to study the effect of nitrogen depletion on lipids synthesis in *Phaeodactylum tricornutum*. Results showed an increase in lipids content upon the knock-down of a gene encoding for nitrate reductase (Levitan et al., 2015).

Despite being an effective gene silencing technique, off-target changes such as silencing desired phenotypes and incomplete suppression of target genes set a major drawback of RNAi (Boettcher & McManus, 2015). This was the drive for searching for other targeted mutagenesis approaches. One of such approaches is the use of programmable sequence-specific nucleases such as zinc finger nucleases (ZFNs) and transcription activator-like effector nuclease (TALENs).

ZFNs are protein based nucleases. In this system, endonuclease of *Flavobacterium okeanokoites* (FokI), which has a DNA binding domain at the N-terminal and a non-specific DNA binding domain at the C-terminal, is used where the zinc finger proteins which are responsible for DNA recognition are fused with the cleavage domain of the FokI endonuclease at the C terminal. Together, they bind to three nucleotides in the target sequence. FokI endonuclease is a dimeric protein that cleaves DNA resulting in double-strand breaks, and then the DNA is repaired either by homologous recombination, which allows insertion and deletion modifications or non-homologous end joints, which leads to small mutations(Fayyaz et al., 2020a).

When introduced to a mutant cell of *Chlamydomonas* that was designed with a non-functional aphVIII marker through the insertion of the *cop3* gene, the zinc finger nucleases cleaved the gene resulting in restoring the function of the aphVIII marker (Sizova, Greiner, Awasthi, Kateriya, & Hegemann, 2013). ZFNs are a promising gene-editing tool. However, it has a few limitations as the ZFNs must be individually programmed for each target. Their construction is complex and expensive, and they can cause some off-target changes (Ng, Tan, Kao, Chang, & Chang, 2017).

TALENs were discovered in phytopathogenic bacteria of the genus *Xanthomonas*. They can affect plants by the secretion of effector proteins that their ability was found to mimic transcription factors in eukaryotes in activating target genes(Nemudryi, Valetdinova, Medvedev, & Zakian, 2014). these proteins contain three structural domains: DNA binding domain, nuclear localization signal and transcriptional activation domain and are made up of arrays of protein repeats (12-27) with different 34 amino acids in each domain. Like ZFNs, TALENs are engineered by fusion of the DNA binding domain with the cleavage domain of the FokI endonuclease (Joung & Sander, 2013). However, they possess several advantages over ZFNs for being simpler in engineering and more specific as each TALEN domain recognizes only one nucleotide while ZFNs recognize three, which reduces the chances of off-target modification (Abdallah, Prakash, & McHughen, 2015).

In 2014, it was reported that in the first targeted and stable genome editing of the diatom *Phaeodactylum tricornutum* using TALENs, 56% of colonies showed targeted mutagenesis and a 45-fold increase in triacylglycerol content was obtained upon the interruption of the UDP-glucose pyrophosphorylase gene proving that genetic engineering is a very powerful tool to obtain strains with high lipids content for biofuel production (Daboussi et al., 2014).

ZFNs and TALENs were frequently used in genome editing till the discovery of Clustered, regularly interspaced, short palindromic repeats (CRISPR) and associated genes (Cas genes) system. This system is constituted of the Cas9 protein, which displays the endonuclease activity and a guide RNA molecule (gRNA or crRNA) to guide Cas9 to the target sequence leading to site-specific double-strand breaks(Gasiunas, Barrangou, Horvath, & Siksnys, 2012). As RNA guidance is based on the base-pairing of complementary sequences, multiple genes can be modified by adding gRNA for each target. The CRISPR/Cas9 system is the most appreciated genome editing technique for its efficiency, precision and ease of handling (H. Kim & Kim, 2014; Piatek et al., 2015).

The genetic engineering process takes place in primarily four steps: host selection, choice of a genetic engineering technique, transformation and finally, screening and selection (Ng et al., 2017). The transformation methods represent an integral part of the genetic engineering process, and they are mainly dependent on the organelle to be transformed. While biolistics is the technique used to transform chloroplasts, it can be used alongside electroporation and conjugative bacteria for nuclear transformation. In electroporation, an electrical field is applied to target cells making micropores in the biological membrane, thus facilitating the transfer of foreign DNA molecules. Biolistics is a method that delivers

DNA to cells through high-speed particle bombardment where nucleic acid-coated particles are propelled by a pressurized gun (gene gun) to transform cells or organelles. Transformation by conjugative bacteria is accomplished by manipulating the cellular machinery of *Agrobacterium tumefaciens* which enables it to transfer bacterial T-DNA into the plant genome. More recently, a successful chemical transformation has been reported in *Pleurochrysis carterae* where protoplasts were generated before incubation in 20% polyethylene glycol (PEG) to allow transformation (Mosey et al., 2021).

Nuclear transformation is the most studied as proteins encoded in the nucleus are involved in various functions such as cell division, cell maintenance, central metabolism and nutrient transport. On the other hand, chloroplasts present a more targeted organelle for transformation as gene expression rates are higher than that of nuclear transformation and the fact that most energy metabolism reactions are localized within the chloroplast makes it a great candidate for genetic modification in order to enhance biomass production (Mosey et al., 2021).

Transformation is typically performed in cell densities of 10^5 to 10^9 as only a very small proportion will successfully transform and express the exogenous DNA. Hence, comes the need for robust screening and selection techniques. The most common technique is by expressing antibiotic or herbicide resistance in the transformed cells. However, this can be challenging for some algal species that exhibit a natural resistance to existing antibiotics, so selecting the proper antibiotic is crucial for the screening process. Zeocin and hygromycin are two generally used antibiotics, although typically at high concentrations, which might be a problem. There is no guarantee that the inserted resistance gene will promote resistance at such high concentrations. This issue could be worsened in algae in wastewater bodies as they might be subjected to repeated low doses of antibiotics, therefore, acquiring resistance (Barancheshme & Munir, 2019).

Also, herbicides such as atrazine, norflurazon and glufosinate represent suitable herbicides for the effective selection of algal mutants. Choice of the selectable marker is an extremely important step in validating successful transformation (Sinning, Michel, Mathis, & Rutherford, 1989).

Efficient genome editing tools and next-generation sequencing technologies made a remarkable shift in the availability of functional-genomes profiling in an unprecedented manner. Genome editing mediated by CRISPR-Cas can make libraries, large collections of mutant strains, that link genotype to phenotype.

Libraries can be built in two different formats: arrayed and pooled. In arrayed libraries, each mutant strain is individually validated by sequencing. It is considered the standard method for genome profiling as mutants are studied as pure cultures, allowing direct phenotype measurement. The disadvantages of this method are the time and cost required to screen thousands of mutants. Conversely, pooled libraries can be constructed while all library members are present in a single reaction vessel. Each mutant is barcoded by a short genetic sequence then their abundance is measured when grown under a definite condition by sequencing barcodes via next-generation sequencing. Although pooled libraries allow rapid profiling of mutants, mutants of interest remain individually validated afterwards (Fenster & Eckert, 2021).

In a study to investigate the genotype-phenotype relationship in *Synechocystis* sp alga, PCC 6803, an inducible CRISPRi gene repression library was developed where mutants were grown under different conditions, and gene fitness was estimated. The results showed a 49% increase in growth rate upon the repression of the peroxiredoxin *bcp2* gene. In contrast, other mutants with improved growth rates exhibited upregulation of genes related to cyclic electron flow (L. Yao et al., 2020).

Engineering Calvin cycle for the optimized photosynthetic activity of different microalgal species has been addressed in multiple studies. In one study, biomass production of *Nannochloropsis oceanica* was found to significantly increase with the overexpression of the RuBisCO activase enzyme (L. Wei,

Wang, Xin, Lu, & Xu, 2017). Other manipulation targets are the low abundant enzymes sedoheptulose 1,7- bisphosphatase, fructose–1,6-bisphosphatase and fructose 1,6-bisphosphate aldolase. The introduction of cyanobacterial fructose 1,6-bisphosphate aldolase to the green alga *C. vulgaris* enhanced its photosynthetic capacity by 1.2-fold (Yang et al., 2017).

In addition, research employing knock-down strategies has shown great results. Lipid content was enhanced by two folds upon the knock-down of the transcription regulator ZnCys in *Nannochloropsis gaditana* (Ajjawi et al., 2017), while the knock-down of zeaxanthin epoxidase in *C. reinhardtii* resulted in a 47-fold increase in productivity and a 56-fold increase in zeaxanthin content (Baek et al., 2018). Table 2. shows different examples of successful CRISPR-Cas mediated genetic engineering in microalgae.

Even though most chloroplast transformations were performed in the model species *C. reinhardtii*, successful attempts were obtained in other microalgal species reviewed (Siddiqui, Wei, Boehm, & Ahmad, 2020). More than a hundred different recombinant proteins, including antibodies, vaccines, and therapeutic proteins, have been reported to be successfully expressed in algal chloroplasts (Dyo & Purton, 2018).

Genetic modification of algae for the enhanced production of biomass and products of interest is a powerful technology that has proven its significance. However, it faces major constraints, including the lack of genetic information needed to select and modify target genes, which are now only available for the model *species C. reinhardtii* and a handful of other microalgal species. Also, GMOs, in general, are considered dangerous to humans as the survival of genetically modified algae in nature is still unknown, and there is the risk of undesired gene flow creating new strains that may possess harmful traits. Therefore, careful monitoring and risk assessment are crucial before transferring them from the lab to the natural habitat. Omics (genomics, transcriptomics, proteomics, lipidomics and metabolomics) can help in the prediction and studying of the effect of such genetic modifications, and possible interaction with other species in the environment is a step towards approved GMOs (Fayyaz et al., 2020a; Spicer & Molnar, 2018).

FUTURE PERSPECTIVE

Despite the great advances in mutation and genetic engineering techniques of microalgae, they still face major constraints in the way of exploitation of their full potential in microalgae biotechnology. For example, current transformation methods exhibit low efficiency, which is mainly due to the robustness and rigidity of the cell wall, which requires the development of improved transformation methods and tools. Also, the lack of sufficient genomic and transcriptomic data poses an obstacle to the design of target sites and understanding of intact metabolic pathways. Moreover, increasing the efficiency of screening markers and the genetic stability of engineered strains are other constraints to overcome. However, the implementation of multi-omics and genome editing tools boosted the progress of biorefinery of important microalgae for the production of valuable metabolites and increased tolerance to various stress conditions via multiple gene activation and inactivation. A combination of gene editing tools with advanced multi-omics can lead to obtaining improved microalgae, maintaining a high rate of both growth and bioactive compounds production (Lin, Tan, Hsiang, Sung, & Ng, 2019).

Incomplete reference datasets, lack of specialized software and high cost are among the main challenges. More investment in the production of data on microalgae and their metabolism is required, along with the provision of bioinformatics tools for their integration. In 2018, Maes and co-workers reported the first integrated pipeline to manage the biological data of *C. reinhardtii*, including genes, transcripts and proteins, namely MinOmics (Methods for Integrated analysis of Multiple Omics datasets). This tool will be of great importance when used on omics data of other beneficial microalgal species. Artificial intelligence, more specifically deep learning, has been used to find correlations among large unstructured data sets in multiple biological areas and can be very useful in microalgal -omics for novel metabolites production and high photosynthetic efficiency (Lauritano, Ferrante, & Rogato, 2019; Maes et al., 2018).

Once genetically engineered improved microalgal strains are developed, strict monitoring and regulations are required to assure their safety for both humans and the environment, especially when cultivated outdoors. Recently, the non-transgenic and marker-free CRISPR has revolutionized microalgae engineering by the production of non-GMO algal products. The addition of a non-GMO tag to engineered microalgae is expected to enhance their acceptance and solve the dispute in both the academic and industrial community regarding the use of genetically modified algal strains. Also, more attention and work should be given to non-model microalgae species as they have several advantages in biotechnological and industrial applications compared to model species (G. Kumar et al., 2020).

The implementation of genetically engineered microalgae can be helpful for the production of economic and sustainable algal products. For instance, it was found to reduce the biofuels production by 15–20%, while the combination of natural and transgenic strains reduced the cost by 85%. Omics approaches can provide deep insights into the genomic and metabolic behaviour of microalgae which can assist researchers in their experimental designs and predict possible results. However, the currently available information on algal metabolic pathways and response at the molecular level is inadequate. Hence, there is an urgent need for whole-genome sequencing of promising algal species with commercial value. 10KP (10,000 Plants) is a genome sequencing project aiming at sequencing the genome of 10,000 species, including 3,000 autotrophic and heterotrophic protists and 1,000 green algae by 2023. Similar projects are highly encouraged for the provision of a better understanding of microalgae and tailoring species with desired characteristics unlocking their full potential (Fayyaz et al., 2020b).

CONCLUSION

Microalgae are excellent sources of various industrially valuable compounds, and there has been extensive work and research on the improvement of the characteristics of these microalgae to meet commercial application requirements. One of the most promising techniques is the alteration of their genetic make-up to exhibit more beneficial properties such as high growth rate and high production of desired compounds in addition to tolerance to biotic and abiotic stress conditions. Advances in bioinformatics and sequencing tools opened the doors for genomics, transcriptomics, proteomics, and integrated omics for the exploration of metabolic regulation of microalgae. Further research, including the development of efficient transformation and genome editing techniques, large data sets and analysis tools are required for a better understanding of the potential of genetic engineering in microalgae. The application of strict regulations for the controlled production of genetically modified algae, their commercial production and their possible release to the environment is of great importance as it may lead to the development of new species with unknown characteristics.

Table 2. Examples of CRISPR/Cas mediated genetic engineering in algae for enhanced productivity

Transformation technique	Microalgal strain	Modification	Result	Reference
Biolistic particle bombardment	*Fistulifera solaris*	Overexpression of the endogenous glycerol kinase (GK) gene	An increase in lipid productivity by 12% was obtained	Muto et al. (2015)
	Tetraselmis sp.	Knock-out of ADP-glucose pyrophosphorylase gene	Lipid content was enhanced by 2.7 and 3.1 fold in two different mutant strains	Chang et al. (2020)
Electroporation	*Chlamydomonas reinhardtii*	Knock-out of the *ZEP* gene	Zeaxanthin content was doubled in the mutant strain	Baek et al. (2016)
		Knock-out of the phospholipase A2 gene	A significant increase of 64.25% in lipids accumulation was obtained	Shin et al. (2019)
		Knock-out of the Cre01. g000300 gene involved in fatty acid degradation	A 5% increase in total lipids content was accompanied by a 27.2% increase in the C18:1 proportion	Nguyen et al. (2020)
	Nannochloropsis oceanica	Development of a transcriptional activation system based on CRISPR/ dCas9	growth and (Fv/Fm) of mutants increased by 23% and 12%, respectively	L. Wei, Jiang, and Liu (2022)
	Nannochloropsis salina	Knock-in of the *FAD12* gene, encoding Δ12-fatty acid desaturase	Linoleic acid (LA) production increased by four folds while eicosapentaenoic acid (EPA) increased by 1.5-fold	Ryu et al. (2021)
	Synechococcus elongatus	Knock-in of *gltA/ppc* gene and knock-out of *glgc* gene	Succinate content was enhanced by 11 folds compared to the wildtype	H. Li et al. (2016)
Glass beads	*Chlamydomonas reinhardtii*	Repression of *PEPC1* gene, which controls the carbon flux	Higher biomass and lipid accumulation rates were obtained	Kao and Ng (2017)
Salt gradient method	*Dunaliella salina*	Knock-down of the β-carotene hydroxylase (*Dschyb*) gene	β-carotene levels in mutants significantly increased to up to 1.4 μg/ml compared to the wildtype	L. Hu et al. (2021)

REFERENCES

Abdallah, N. A., Prakash, C. S., & McHughen, A. G. (2015). Genome editing for crop improvement: Challenges and opportunities. *GM Crops and Food: Biotechnology in Agriculture and the Food Chain*, *6*(4), 183–205. doi:10.1080/21645698.2015.1129937 PMID:26930114

Ajjawi, I., Verruto, J., Aqui, M., Soriaga, L. B., Coppersmith, J., Kwok, K., ... Xu, W. (2017). Lipid production in *Nannochloropsis gaditana* is doubled by decreasing expression of a single transcriptional regulator. *Nature Biotechnology*, *35*(7), 647–652. doi:10.1038/nbt.3865 PMID:28628130

AlProl, A. E., & Elkatory, M. R. (2022). The current status of various algal industries. In Handbook of Algal Biofuels (pp. 123-147). doi:10.1016/B978-0-12-823764-9.00010-8

Alsenani, F., Tupally, K. R., Chua, E. T., Eltanahy, E., Alsufyani, H., Parekh, H. S., & Schenk, P. M. (2020). Evaluation of microalgae and cyanobacteria as potential sources of antimicrobial compounds. *Saudi Pharmaceutical Journal*, *28*(12), 1834–1841. doi:10.1016/j.jsps.2020.11.010 PMID:33424272

Anderson, N. L., & Anderson, N. G. (1998). Proteome and proteomics: New technologies, new concepts, and new words. *Electrophoresis*, *19*(11), 1853–1861. doi:10.1002/elps.1150191103 PMID:9740045

Arora, N., Yen, H.-W., & Philippidis, G. P. (2020). Harnessing the power of mutagenesis and adaptive laboratory evolution for high lipid production by oleaginous microalgae and yeasts. *Sustainability*, *12*(12), 5125. doi:10.3390u12125125

Baek, K., Kim, D. H., Jeong, J., Sim, S. J., Melis, A., Kim, J.-S., Jin, E. S., & Bae, S. (2016). DNA-free two-gene knockout in *Chlamydomonas reinhardtii* via CRISPR-Cas9 ribonucleoproteins. *Scientific Reports*, *6*(1), 30620. doi:10.1038rep30620 PMID:27466170

Baek, K., Yu, J., Jeong, J., Sim, S. J., Bae, S., & Jin, E. (2018). Photoautotrophic production of macular pigment in a Chlamydomonas reinhardtii strain generated by using DNA-free CRISPR-Cas9 RNP-mediated mutagenesis. *Biotechnology and Bioengineering*, *115*(3), 719–728. doi:10.1002/bit.26499 PMID:29150930

Barancheshme, F., & Munir, M. (2019). Development of antibiotic resistance in wastewater treatment plants. In Y. Kumar (Ed.), *Antimicrobial Resistance-A Global Threat*. IntechOpen. doi:10.5772/intechopen.81538

Beaumont, M., Tran, R., Vera, G., Niedrist, D., Rousset, A., Pierre, R., Shastri, V. P., & Forget, A. (2021). Hydrogel-Forming Algae Polysaccharides: From Seaweed to Biomedical Applications. *Biomacromolecules*, *22*(3), 1027–1052. doi:10.1021/acs.biomac.0c01406 PMID:33577286

Beyaz, R., & Yildiz, M. (2017). The use of gamma irradiation in plant mutation breeding. In S. Jurić (Ed.), *Plant Engineering* (pp. 33–46). doi:10.5772/intechopen.69974

Boettcher, M., & McManus, M. T. (2015). Choosing the Right Tool for the Job: RNAi, TALEN, or CRISPR. *Molecular Cell*, *58*(4), 575–585. doi:10.1016/j.molcel.2015.04.028 PMID:26000843

Bölling, C., & Fiehn, O. (2005). Metabolite Profiling of *Chlamydomonas reinhardtii* under Nutrient Deprivation. *Plant Physiology*, *139*(4), 1995–2005. doi:10.1104/pp.105.071589 PMID:16306140

Cardozo, K. H. M., Guaratini, T., Barros, M. P., Falcão, V. R., Tonon, A. P., Lopes, N. P., Campos, S., Torres, M. A., Souza, A. O., Colepicolo, P., & Pinto, E. (2007). Metabolites from algae with economical impact. *Comparative Biochemistry and Physiology. Toxicology & Pharmacology: CBP*, *146*(1), 60–78. doi:10.1016/j.cbpc.2006.05.007 PMID:16901759

Chakdar, H., Hasan, M., Pabbi, S., Nevalainen, H., & Shukla, P. (2021). High-throughput proteomics and metabolomic studies guide re-engineering of metabolic pathways in eukaryotic microalgae: A review. *Bioresource Technology*, *321*, 124495. doi:10.1016/j.biortech.2020.124495 PMID:33307484

Chan, C.-X., Teo, S.-S., Ho, C.-L., Othman, R. Y., & Phang, S.-M. (2004). Optimisation of RNA extraction from *Gracilaria changii* (Gracilariales, Rhodophyta). *Journal of Applied Phycology*, *16*(4), 297–301. doi:10.1023/B:JAPH.0000047782.20940.de

Chang, K. S., Kim, J., Park, H., Hong, S. J., Lee, C. G., & Jin, E. (2020). Enhanced lipid productivity in AGP knockout marine microalga *Tetraselmis* sp. using a DNA-free CRISPR-Cas9 RNP method. *Bioresource Technology*, *303*, 122932. doi:10.1016/j.biortech.2020.122932 PMID:32058903

Collen, J., Roeder, V., Rousvoal, S., Collin, O., Kloareg, B., & Boyen, C. (2006). An Expressed Sequence Tag Analysis of Thallus and Regenerating Protoplasts of *Chondrus Crispus* (Gigartinales, Rhodophyceae). *Journal of Phycology*, *42*(1), 104–112. doi:10.1111/j.1529-8817.2006.00171.x

Coyer, J. A., Robertson, D. L., & Alberte, R. S. (1994). Genetic variability within a population and between diploid/haploid tissue of *Macrocystis pyrifera* (Phaeophyceae). *Journal of Phycology*, *30*(3), 545–552. doi:10.1111/j.0022-3646.1994.00545.x

Daboussi, F., Leduc, S., Maréchal, A., Dubois, G., Guyot, V., Perez-Michaut, C., Amato, A., Falciatore, A., Juillerat, A., Beurdeley, M., Voytas, D. F., Cavarec, L., & Duchateau, P. (2014). Genome engineering empowers the diatom *Phaeodactylum tricornutum* for biotechnology. *Nature Communications*, *5*(1), 3831. doi:10.1038/ncomms4831 PMID:24871200

Davison, I. R., & Pearson, G. A. (1996). Stress Tolerance in Intertidal Seaweeds. *Journal of Phycology*, *32*(2), 197–211. doi:10.1111/j.0022-3646.1996.00197.x

Dellaporta, S. L., Wood, J., & Hicks, J. B. (1983). A plant DNA minipreparation: Version II. *Plant Molecular Biology Reporter*, *1*(4), 19–21. doi:10.1007/BF02712670

Dhandayuthapani, K., Malathy, S., Mulla, S. I., & Gupta, S. K. (2021). An Insight into the Potential Application of Microalgae in Pharmaceutical and Nutraceutical Production. In *Algae* (pp. 135–179). Springer. doi:10.1007/978-981-15-7518-1_7

Dragosits, M., & Mattanovich, D. (2013). Adaptive laboratory evolution – principles and applications for biotechnology. *Microbial Cell Factories*, *12*(1), 64. doi:10.1186/1475-2859-12-64 PMID:23815749

Dyo, Y. M., & Purton, S. (2018). The algal chloroplast as a synthetic biology platform for production of therapeutic proteins. *Microbiology*, *164*(2), 113–121. doi:10.1099/mic.0.000599 PMID:29297850

Elena, S. F., & Lenski, R. E. (2003). Evolution experiments with microorganisms: The dynamics and genetic bases of adaptation. *Nature Reviews. Genetics*, *4*(6), 457–469. doi:10.1038/nrg1088 PMID:12776215

Eltanahy, E., & Torky, A. (2021). *Microalgae as Cell Factories: Food and Feed-grade High-value Metabolites*. Academic Press.

Fain, S. R., Druehl, L. D., & Baillie, D. L. (1988). Repeat and single copy sequences are differentially conserved in the evolution of kelp chloroplast DNA. *Journal of Phycology*, *24*(3), 292–302. doi:10.1111/j.1529-8817.1988.tb04469.x

Fayyaz, M., Chew, K. W., Show, P. L., Ling, T. C., Ng, I.-S., & Chang, J.-S. (2020a). Genetic engineering of microalgae for enhanced biorefinery capabilities. *Biotechnology Advances*, *43*, 107554. doi:10.1016/j.biotechadv.2020.107554 PMID:32437732

Fayyaz, M., Chew, K. W., Show, P. L., Ling, T. C., Ng, I. S., & Chang, J.-S. (2020b). Genetic engineering of microalgae for enhanced biorefinery capabilities. *Biotechnology Advances*, *43*, 107554. doi:10.1016/j.biotechadv.2020.107554 PMID:32437732

Fenster, J. A., & Eckert, C. A. (2021). High-Throughput Functional Genomics for Energy Production. *Current Opinion in Biotechnology*, *67*, 7–14. doi:10.1016/j.copbio.2020.09.010 PMID:33152605

Fiehn, O. (2002). Metabolomics — the link between genotypes and phenotypes. In C. Town (Ed.), *Functional genomics* (pp. 155–171). Springer Netherlands. doi:10.1007/978-94-010-0448-0_11

Fumina, S., Koki, M., & Kyosuke, N. (2020). Identification, growth, and pigment content of a spontaneous green mutant of *Pyropia kinositae* (Bangiales, Rhodophyta). *Journal of Applied Phycology*, *32*(3), 1983–1994. doi:10.100710811-020-02085-5

Gachelin, M., Boutoute, M., Carrier, G., Talec, A., Pruvost, E., Guihéneuf, F., Bernard, O., & Sciandra, A. (2021). Enhancing PUFA-rich polar lipids in *Tisochrysis lutea* using adaptive laboratory evolution (ALE) with oscillating thermal stress. *Applied Microbiology and Biotechnology*, *105*(1), 301–312. doi:10.100700253-020-11000-4 PMID:33201276

Gan, S.-Y., Lim, P.-E., & Phang, S.-M. (2016). Genetic and Metabolic Engineering of Microalgae. In F. Bux & Y. Chisti (Eds.), *Algae Biotechnology: Products and Processes* (pp. 317–344). Springer International Publishing. doi:10.1007/978-3-319-12334-9_16

Gasiunas, G., Barrangou, R., Horvath, P., & Siksnys, V. (2012). Cas9–crRNA ribonucleoprotein complex mediates specific DNA cleavage for adaptive immunity in bacteria. *Proceedings of the National Academy of Sciences of the United States of America*, *109*(39), E2579–E2586. doi:10.1073/pnas.1208507109 PMID:22949671

Goff, L. J., & Coleman, A. W. (1988). The use of plastid DNA restriction endonuclease patterns in delineating red algal species and populations. *Journal of Phycology*, *24*(3), 357–368. doi:10.1111/j.1529-8817.1988.tb04478.x

González-González, L. M., Eltanahy, E., & Schenk, P. M. (2019). Assessing the fertilizing potential of microalgal digestates using the marine diatom *Chaetoceros muelleri*. *Algal Research*, *41*, 101534. doi:10.1016/j.algal.2019.101534

González-González, L. M., Zhou, L., Astals, S., Thomas-Hall, S. R., Eltanahy, E., Pratt, S., Jensen, P. D., & Schenk, P. M. (2018). Biogas production coupled to repeat microalgae cultivation using a closed nutrient loop. *Bioresource Technology*, *263*, 625–630. doi:10.1016/j.biortech.2018.05.039 PMID:29800924

Guardini, Z., Dall'Osto, L., Barera, S., Jaberi, M., Cazzaniga, S., Vitulo, N., & Bassi, R. (2021). High Carotenoid Mutants of *Chlorella vulgaris* Show Enhanced Biomass Yield under High Irradiance. *Plants*, *10*(5), 911. doi:10.3390/plants10050911 PMID:34062906

Gupta, N. (2019). DNA Extraction and Polymerase Chain Reaction. *Journal of Cytology / Indian Academy of Cytologists*, *36*(2), 116–117. doi:10.4103/JOC.JOC_110_18 PMID:30992648

Halligan, D. L., & Keightley, P. D. (2009). Spontaneous Mutation Accumulation Studies in Evolutionary Genetics. *Annual Review of Ecology, Evolution, and Systematics*, *40*(1), 151–172. doi:10.1146/annurev.ecolsys.39.110707.173437

Harley, C. D. G., Anderson, K. M., Demes, K. W., Jorve, J. P., Kordas, R. L., Coyle, T. A., & Graham, M. H. (2012). Effects of climate change on global seaweed communities. *Journal of Phycology*, *48*(5), 1064–1078. doi:10.1111/j.1529-8817.2012.01224.x PMID:27011268

Hu, L., Feng, S., Liang, G., Du, J., Li, A., & Niu, C. (2021). CRISPR/Cas9-induced β-carotene hydroxylase mutation in *Dunaliella salina* CCAP19/18. *AMB Express*, *11*(1), 83. doi:10.118613568-021-01242-4 PMID:34097133

Hu, Z., Zeng, X., Wang, A., Shi, C., & Duan, D. (2004). An efficient method for DNA isolation from red algae. *Journal of Applied Phycology*, *16*(3), 161–166. doi:10.1023/B:JAPH.0000048456.26639.1a

Huang, L., Peng, L., & Yan, X. (2021). Multi-omics responses of red algae *Pyropia haitanensis* to intertidal desiccation during low tides. *Algal Research*, *58*, 102376. doi:10.1016/j.algal.2021.102376

Jamers, A., Blust, R., & De Coen, W. (2009). Omics in algae: Paving the way for a systems biological understanding of algal stress phenomena? *Aquatic Toxicology (Amsterdam, Netherlands)*, *92*(3), 114–121. doi:10.1016/j.aquatox.2009.02.012 PMID:19304329

James, G. O., Hocart, C. H., Hillier, W., Chen, H., Kordbacheh, F., Price, G. D., & Djordjevic, M. A. (2011). Fatty acid profiling of *Chlamydomonas reinhardtii* under nitrogen deprivation. *Bioresource Technology*, *102*(3), 3343–3351. doi:10.1016/j.biortech.2010.11.051 PMID:21146403

Jeon, S., Lim, J.-M., Lee, H.-G., Shin, S.-E., Kang, N. K., Park, Y.-I., Oh, H.-M., Jeong, W.-J., Jeong, B., & Chang, Y. K. (2017). Current status and perspectives of genome editing technology for microalgae. *Biotechnology for Biofuels*, *10*(1), 267. doi:10.118613068-017-0957-z PMID:29163669

Jin, H.-J., Kim, J. H., Sohn, C. H., DeWreede, R. E., Choi, T. J., Towers, G. H. N., Hudson, J. B., & Hong, Y. K. (1997). Inhibition of Taq DNA polymerase by seaweed extracts from British Columbia, Canada and Korea. *Journal of Applied Phycology*, *9*(4), 383–388. doi:10.1023/A:1007925202219

Joubert, Y., & Fleurence, J. (2005). DNA isolation protocol for seaweeds. *Plant Molecular Biology Reporter*, *23*(2), 197–197. doi:10.1007/BF02772712

Joung, J. K., & Sander, J. D. (2013). TALENs: A widely applicable technology for targeted genome editing. *Nature Reviews. Molecular Cell Biology*, *14*(1), 49–55. doi:10.1038/nrm3486 PMID:23169466

Kao, P. H., & Ng, I. S. (2017). CRISPRi mediated phosphoenolpyruvate carboxylase regulation to enhance the production of lipid in *Chlamydomonas reinhardtii. Bioresoure Technology, 245*(Pt B), 1527-1537. doi:10.1016/j.biortech.2017.04.111

Katju, V., & Bergthorsson, U. (2019). Old Trade, New Tricks: Insights into the Spontaneous Mutation Process from the Partnering of Classical Mutation Accumulation Experiments with High-Throughput Genomic Approaches. *Genome Biology and Evolution*, *11*(1), 136–165. doi:10.1093/gbe/evy252 PMID:30476040

Ketting, R. F. (2011). The Many Faces of RNAi. *Developmental Cell*, *20*(2), 148–161. doi:10.1016/j.devcel.2011.01.012 PMID:21316584

Kim, H., & Kim, J.-S. (2014). A guide to genome engineering with programmable nucleases. *Nature Reviews. Genetics*, *15*(5), 321–334. doi:10.1038/nrg3686 PMID:24690881

Kim, M., Ahn, J., Jeon, H., & Jin, E. (2017). Development of a *Dunaliella tertiolecta* Strain with Increased Zeaxanthin Content Using Random Mutagenesis. *Marine Drugs*, *15*(6), 189. doi:10.3390/md15060189 PMID:28635662

Kraemer, S. A., Böndel, K. B., Ness, R. W., Keightley, P. D., & Colegrave, N. (2017). Fitness change in relation to mutation number in spontaneous mutation accumulation lines of *Chlamydomonas reinhardtii*. *Evolution*, *71*(12), 2918–2929. doi:10.1111/evo.13360 PMID:28884790

Krasovec, M., Eyre-Walker, A., Sanchez-Ferandin, S., & Piganeau, G. (2017). Spontaneous Mutation Rate in the Smallest Photosynthetic Eukaryotes. *Molecular Biology and Evolution*, *34*(7), 1770–1779. doi:10.1093/molbev/msx119 PMID:28379581

Krasovec, M., Sanchez-Brosseau, S., Grimsley, N., & Piganeau, G. (2018). Spontaneous mutation rate as a source of diversity for improving desirable traits in cultured microalgae. *Algal Research*, *35*, 85–90. doi:10.1016/j.algal.2018.08.003

Krasovec, M., Sanchez-Brosseau, S., & Piganeau, G. (2019). First Estimation of the Spontaneous Mutation Rate in Diatoms. *Genome Biology and Evolution*, *11*(7), 1829–1837. doi:10.1093/gbe/evz130 PMID:31218358

Krassowski, M., Das, V., Sahu, S. K., & Misra, B. B. (2020). State of the Field in Multi-Omics Research: From Computational Needs to Data Mining and Sharing. *Frontiers in Genetics*, *11*(1598), 610798. Advance online publication. doi:10.3389/fgene.2020.610798 PMID:33362867

Kumar, G., Shekh, A., Jakhu, S., Sharma, Y., Kapoor, R., & Sharma, T. R. (2020). Bioengineering of Microalgae: Recent Advances, Perspectives, and Regulatory Challenges for Industrial Application. *Frontiers in Bioengineering and Biotechnology*, *8*, 914. doi:10.3389/fbioe.2020.00914 PMID:33014997

Kumar, S., Singh, N. K., Kumar, A., & Shukla, P. (2021). Next Generation Biofuel Production in the Omics Era: Potential and Prospects. In Omics Technologies for Sustainable Agriculture and Global Food Security (vol. 2, pp. 293-311). Springer.

Kuo, C.-M., Lin, T.-H., Yang, Y.-C., Zhang, W.-X., Lai, J.-T., Wu, H.-T., Chang, J.-S., & Lin, C.-S. (2017). Ability of an alkali-tolerant mutant strain of the microalga *Chlorella* sp. AT1 to capture carbon dioxide for increasing carbon dioxide utilization efficiency. *Bioresource Technology*, *244*(Pt 1), 243–251. doi:10.1016/j.biortech.2017.07.096 PMID:28780257

Lauritano, C., Ferrante, M. I., & Rogato, A. (2019). Marine Natural Products from Microalgae: An -Omics Overview. *Marine Drugs*, *17*(5), 269. doi:10.3390/md17050269 PMID:31067655

Lee, H., Lee, H. K., An, G., & Lee, Y. K. (2007). Analysis of expressed sequence tags from the red alga *Griffithsia okiensis*. *Journal of Microbiology (Seoul, Korea)*, *45*(6), 541–546. PMID:18176538

Lee, H.-J., & Choi, J.-i. (2018). Isolation and characterization of a high-growth-rate strain in *Pyropia yezoensis* induced by ethyl methane sulfonate. *Journal of Applied Phycology*, *30*(4), 2513–2522. doi:10.100710811-018-1426-1 PMID:30147236

Lee, H.-J., & Choi, J.-i. (2019). Enhancing temperature tolerance of *Pyropia tenera* (Bangiales) by inducing mutation. *Phycologia*, *58*(5), 496–503. doi:10.1080/00318884.2019.1623547

Levitan, O., Dinamarca, J., Zelzion, E., Lun, D. S., Guerra, L. T., Kim, M. K., Kim, J., Van Mooy, B. A. S., Bhattacharya, D., & Falkowski, P. G. (2015). Remodeling of intermediate metabolism in the diatom *Phaeodactylum tricornutum* under nitrogen stress. *Proceedings of the National Academy of Sciences of the United States of America*, *112*(2), 412–417. doi:10.1073/pnas.1419818112 PMID:25548193

Li, D., Wang, L., Zhao, Q., Wei, W., & Sun, Y. (2015). Improving high carbon dioxide tolerance and carbon dioxide fixation capability of *Chlorella* sp. by adaptive laboratory evolution. *Bioresource Technology*, *185*, 269–275. doi:10.1016/j.biortech.2015.03.011 PMID:25776894

Li, H., Shen, C. R., Huang, C.-H., Sung, L.-Y., Wu, M.-Y., & Hu, Y.-C. (2016). CRISPR-Cas9 for the genome engineering of cyanobacteria and succinate production. *Metabolic Engineering*, *38*, 293–302. doi:10.1016/j.ymben.2016.09.006 PMID:27693320

Li, X., Yuan, Y., Cheng, D., Gao, J., Kong, L., Zhao, Q., Wei, W., & Sun, Y. (2018). Exploring stress tolerance mechanism of evolved freshwater strain *Chlorella* sp. S30 under 30 g/L salt. *Bioresource Technology*, *250*, 495–504. doi:10.1016/j.biortech.2017.11.072 PMID:29197772

Lin, W.-R., Tan, S.-I., Hsiang, C.-C., Sung, P.-K., & Ng, I.-S. (2019). Challenges and opportunity of recent genome editing and multi-omics in cyanobacteria and microalgae for biorefinery. *Bioresource Technology*, *291*, 121932. doi:10.1016/j.biortech.2019.121932 PMID:31387837

Lozano, I., Wacyk, J. M., Carrasco, J., & Cortez-San Martín, M. A. (2016). Red macroalgae *Pyropia columbina* and *Gracilaria chilensis*: Sustainable feed additive in the Salmo salar diet and the evaluation of potential antiviral activity against infectious salmon anemia virus. *Journal of Applied Phycology*, *28*(2), 1343–1351. doi:10.100710811-015-0648-8

Lynch, M., Ackerman, M. S., Gout, J.-F., Long, H., Sung, W., Thomas, W. K., & Foster, P. L. (2016). Genetic drift, selection and the evolution of the mutation rate. *Nature Reviews. Genetics*, *17*(11), 704–714. doi:10.1038/nrg.2016.104 PMID:27739533

Lyons, D., Scheibling, R. E., & Van Alstyne, K. (2010). Spatial and temporal variation in DMSP content in the invasive seaweed *Codium fragile* ssp. fragile: Effects of temperature, light and grazing. *Marine Ecology Progress Series*, *417*, 51–61. doi:10.3354/meps08818

Ma, N. L., Lam, S. S., & Zaidah, R. (2015). The application of algae for cosmeceuticals in the Omics Age. *Genomics, Proteomics, Metabolomics in Nutraceuticals Functional Foods*, 476-488. doi:10.1002/9781118930458.ch37

Maes, A., Martinez, X., Druart, K., Laurent, B., Guégan, S., Marchand, C. H., Lemaire, S. D., & Baaden, M. (2018). MinOmics, an integrative and immersive tool for multi-omics analysis. *Journal of Integrative Bioinformatics*, *15*(2), 20180006. doi:10.1515/jib-2018-0006 PMID:29927748

Mao, X., Zhang, Y., Wang, X., & Liu, J. (2020). Novel insights into salinity-induced lipogenesis and carotenogenesis in the oleaginous astaxanthin-producing alga *Chromochloris zofingiensis*: A multi-omics study. *Biotechnology for Biofuels*, *13*(1), 73. doi:10.118613068-020-01714-y PMID:32322303

Mehtani, J., Arora, N., Patel, A., Jain, P., Pruthi, P. A., Poluri, K. M., & Pruthi, V. (2017). Augmented lipid accumulation in ethyl methyl sulphonate mutants of oleaginous microalga for biodiesel production. *Bioresource Technology*, *242*, 121–127. doi:10.1016/j.biortech.2017.03.108 PMID:28366694

Mosey, M., Douchi, D., Knoshaug, E. P., & Laurens, L. M. (2021). Methodological review of genetic engineering approaches for non-model algae. *Algal Research*, *54*, 102221. doi:10.1016/j.algal.2021.102221

Murray, M. G., & Thompson, W. F. (1980). Rapid isolation of high molecular weight plant DNA. *Nucleic Acids Research*, *8*(19), 4321–4326. doi:10.1093/nar/8.19.4321 PMID:7433111

Muthuraj, M., Selvaraj, B., Palabhanvi, B., Kumar, V., & Das, D. (2019). Enhanced lipid content in *Chlorella* sp. FC2 IITG via high energy irradiation mutagenesis. *Korean Journal of Chemical Engineering*, *36*(1), 63–70. doi:10.100711814-018-0180-z

Muto, M., Tanaka, M., Liang, Y., Yoshino, T., Matsumoto, M., & Tanaka, T. (2015). Enhancement of glycerol metabolism in the oleaginous marine diatom *Fistulifera solaris* JPCC DA0580 to improve triacylglycerol productivity. *Biotechnology for Biofuels*, *8*(1), 4. doi:10.118613068-014-0184-9 PMID:25632299

Nakajima, M., Kitade, Y., Iitsuka, O., Fukuda, S., & Saga, N. (2000). Rapid extraction of high-quality genomic DNA from *Porphyra yezoensis* (Bangiales, Rhodophyta). *Phycological Research*, *48*(1), 15–17. doi:10.1046/j.1440-1835.2000.00179.x

Nemudryi, A. A., Valetdinova, K. R., Medvedev, S. P., & Zakian, S. M. (2014). TALEN and CRISPR/Cas Genome Editing Systems: Tools of Discovery. *Acta Naturae*, *6*(3), 19–40. doi:10.32607/20758251-2014-6-3-19-40 PMID:25349712

Ng, I.-S., Tan, S.-I., Kao, P.-H., Chang, Y.-K., & Chang, J.-S. (2017). Recent Developments on Genetic Engineering of Microalgae for Biofuels and Bio-Based Chemicals. *Biotechnology Journal*, *12*(10), 1600644. doi:10.1002/biot.201600644 PMID:28786539

Nguyen, T. H. T., Park, S., Jeong, J., Shin, Y. S., Sim, S. J., & Jin, E. (2020). Enhancing lipid productivity by modulating lipid catabolism using the CRISPR-Cas9 system in *Chlamydomonas*. *Journal of Applied Phycology*, *32*(5), 2829–2840. doi:10.100710811-020-02172-7

Nikaido, I., Asamizu, E., Nakajima, M., Nakamura, Y., Saga, N., & Tabata, S. (2000). Generation of 10,154 Expressed Sequence Tags from a Leafy Gametophyte of a Marine Red Alga, *Porphyra yezoensis*. *DNA Research*, *7*(3), 223–227. doi:10.1093/dnares/7.3.223 PMID:10907854

Paix, B., Carriot, N., Barry-Martinet, R., Greff, S., Misson, B., Briand, J.-F., & Culioli, G. (2020). A Multi-Omics Analysis Suggests Links Between the Differentiated Surface Metabolome and Epiphytic Microbiota Along the Thallus of a Mediterranean Seaweed Holobiont. *Frontiers in Microbiology*, *11*(494), 494. doi:10.3389/fmicb.2020.00494 PMID:32269559

Patwary, Z. P., Paul, N. A., Nishitsuji, K., Campbell, A. H., Shoguchi, E., Zhao, M., & Cummins, S. F. (2021). Application of omics research in seaweeds with a focus on red seaweeds. *Briefings in Functional Genomics*, *20*(3), 148–161. doi:10.1093/bfgp/elab023 PMID:33907795

Perin, G., Bellan, A., Segalla, A., Meneghesso, A., Alboresi, A., & Morosinotto, T. (2015). Generation of random mutants to improve light-use efficiency of *Nannochloropsis gaditana* cultures for biofuel production. *Biotechnology for Biofuels*, *8*(1), 161. doi:10.118613068-015-0337-5 PMID:26413160

Piatek, A., Ali, Z., Baazim, H., Li, L., Abulfaraj, A., Al Shareef, S., Aouida, M., & Mahfouz, M. M. (2015). RNA-guided transcriptional regulation in planta via synthetic dCas9-based transcription factors. *Plant Biotechnology Journal*, *13*(4), 578–589. doi:10.1111/pbi.12284 PMID:25400128

Pierobon, S., Cheng, X., Graham, P., Nguyen, B., Karakolis, E., & Sinton, D. (2018). Emerging microalgae technology: A review. *Sustainable Energy & Fuels*, *2*(1), 13–38. doi:10.1039/C7SE00236J

Qi, F., Wu, D., Mu, R., Zhang, S., & Xu, X. (2018). Characterization of a Microalgal UV Mutant for CO_2 Biofixation and Biomass Production. *BioMed Research International*, *4375170*, 1–8. Advance online publication. doi:10.1155/2018/4375170 PMID:30671452

Qian, F., Luo, Q., Yang, R., Zhu, Z., Chen, H., & Yan, X. (2015). The littoral red alga *Pyropia haitanensis* uses rapid accumulation of floridoside as the desiccation acclimation strategy. *Journal of Applied Phycology*, *27*(1), 621–632. doi:10.100710811-014-0336-0

Richmond, A. (2004). Part III. Economic applications of micro algae. In Handbook of Microalgal Culture. Biotechnology Applied Phycology (pp. 255-353): Blackwell Publishing.

Roach, T. N., Little, M., Arts, M. G., Huckeba, J., Haas, A. F., George, E. E., ... Silveira, C. B. (2020). A Multiomic Analysis of in Situ Coral-Turf Algal Interactions. *Proceedings of the National Academy of Sciences of the United States of America*, *117*(24), 13588–13595. doi:10.1073/pnas.1915455117 PMID:32482859

Rochaix, J. D., & van Dillewijn, J. (1982). Transformation of the green alga *Chlamydomonas reinhardii* with yeast DNA. *Nature*, *296*(5852), 70–72. doi:10.1038/296070a0 PMID:7038510

Rumin, J., Bonnefond, H., Saint-Jean, B., Rouxel, C., Sciandra, A., Bernard, O., Cadoret, J.-P., & Bougaran, G. (2015). The use of fluorescent Nile red and BODIPY for lipid measurement in microalgae. *Biotechnology for Biofuels*, *8*(1), 42. doi:10.118613068-015-0220-4 PMID:25788982

Ryu, A. J., Jeong, B., Kang, N. K., Jeon, S., Sohn, M. G., Yun, H. J., Lim, J. M., Jeong, S. W., Park, Y.-I., Jeong, W. J., Park, S., Chang, Y. K., & Jeong, K. J. (2021). Safe-Harboring based novel genetic toolkit for *Nannochloropsis salina* CCMP1776: Efficient overexpression of transgene via CRISPR/Cas9-Mediated Knock-in at the transcriptional hotspot. *Bioresource Technology*, *340*, 125676. doi:10.1016/j.biortech.2021.125676 PMID:34365302

Sarayloo, E., Tardu, M., Unlu, Y. S., Simsek, S., Cevahir, G., Erkey, C., & Kavakli, I. H. (2017). Understanding lipid metabolism in high-lipid-producing *Chlorella vulgaris* mutants at the genome-wide level. *Algal Research*, *28*, 244–252. doi:10.1016/j.algal.2017.11.009

Shin, Y. S., Jeong, J., Nguyen, T. H. T., Kim, J. Y. H., Jin, E., & Sim, S. J. (2019). Targeted knockout of phospholipase A(2) to increase lipid productivity in *Chlamydomonas reinhardtii* for biodiesel production. *Bioresource Technology*, *271*, 368–374. doi:10.1016/j.biortech.2018.09.121 PMID:30293032

Siddiqui, A., Wei, Z., Boehm, M., & Ahmad, N. (2020). Engineering microalgae through chloroplast transformation to produce high-value industrial products. *Biotechnology and Applied Biochemistry*, *67*(1), 30–40. doi:10.1002/bab.1823 PMID:31538673

Sinning, I., Michel, H., Mathis, P., & Rutherford, A. W. (1989). Characterization of four herbicide-resistant mutants of Rhodopseudomonas viridis by genetic analysis, electron paramagnetic resonance, and optical spectroscopy. *Biochemistry*, *28*(13), 5544–5553. doi:10.1021/bi00439a031 PMID:2550055

Siripornadulsil, S., Dabrowski, K., & Sayre, R. (2007). Microalgal Vaccines. In R. León, A. Galván, & E. Fernández (Eds.), *Transgenic microalgae as green cell factories* (pp. 122–128). Springer New York. doi:10.1007/978-0-387-75532-8_11

Sizova, I., Greiner, A., Awasthi, M., Kateriya, S., & Hegemann, P. (2013). Nuclear gene targeting in *Chlamydomonas* using engineered zinc-finger nucleases. *The Plant Journal*, *73*(5), 873–882. doi:10.1111/tpj.12066 PMID:23137232

Sogin, E. M., Puskás, E., Dubilier, N., Liebeke, M., & Huber, J. A. (2019). Marine Metabolomics: A Method for Nontargeted Measurement of Metabolites in Seawater by Gas Chromatography-Mass Spectrometry. *mSystems*, *4*(6), e00638–e00619. doi:10.1128/mSystems.00638-19 PMID:31822601

Spalvins, K., Raita, S., Valters, K., & Blumberga, D. (2021). Improving single cell protein yields and amino acid profile via mutagenesis: Review of applicable amino acid inhibitors for mutant selection. *Agronomy Research (Tartu)*, *19*(S3). Advance online publication. doi:10.15159/ar.21.083

Spicer, A., & Molnar, A. (2018). Gene Editing of Microalgae: Scientific Progress and Regulatory Challenges in Europe. *Biology (Basel)*, *7*(1), 21. doi:10.3390/biology7010021 PMID:29509719

Sun, X.-M., Ren, L.-J., Bi, Z.-Q., Ji, X.-J., Zhao, Q.-Y., & Huang, H. (2018). Adaptive evolution of microalgae *Schizochytrium* sp. under high salinity stress to alleviate oxidative damage and improve lipid biosynthesis. *Bioresource Technology*, *267*, 438–444. doi:10.1016/j.biortech.2018.07.079 PMID:30032058

Sung, Y. J., Choi, H. I., Lee, J. S., Hong, M. E., & Sim, S. J. (2019). Screening of oleaginous algal strains from *Chlamydomonas reinhardtii* mutant libraries via density gradient centrifugation. *Biotechnology and Bioengineering*, *116*(12), 3179–3188. doi:10.1002/bit.27149 PMID:31429923

Tharek, A., Yahya, A., Salleh, M. M., Jamaluddin, H., Yoshizaki, S., Hara, H., Iwamoto, K., Suzuki, I., & Mohamad, S. E. (2021). Improvement and screening of astaxanthin producing mutants of newly isolated *Coelastrum* sp. using ethyl methane sulfonate induced mutagenesis technique. *Biotechnology Reports (Amsterdam, Netherlands)*, *32*, e00673. doi:10.1016/j.btre.2021.e00673 PMID:34621628

Treves, H., Siemiatkowska, B., Luzarowska, U., Murik, O., Fernandez-Pozo, N., Moraes, T. A., Erban, A., Armbruster, U., Brotman, Y., Kopka, J., Rensing, S. A., Szymanski, J., & Stitt, M. (2020). Multi-omics reveals mechanisms of total resistance to extreme illumination of a desert alga. *Nature Plants*, *6*(8), 1031–1043. doi:10.103841477-020-0729-9 PMID:32719473

Vandepoele, K., Van Bel, M., Richard, G., Van Landeghem, S., Verhelst, B., Moreau, H., Van de Peer, Y., Grimsley, N., & Piganeau, G. (2013). pico-PLAZA, a genome database of microbial photosynthetic eukaryotes. *Environmental Microbiology*, *15*(8), 2147–2153. doi:10.1111/1462-2920.12174 PMID:23826978

Wang, L., Xue, C., Wang, L., Zhao, Q., Wei, W., & Sun, Y. (2016). Strain improvement of *Chlorella* sp. for phenol biodegradation by adaptive laboratory evolution. *Bioresource Technology*, *205*, 264–268. doi:10.1016/j.biortech.2016.01.022 PMID:26803904

Wattier, R. A., Prodöhl, P. A., & Maggs, C. A. (2000). DNA isolation protocol for red seaweed (rhodophyta). *Plant Molecular Biology Reporter*, *18*(3), 275–281. doi:10.1007/BF02823999

Wei, L., Jiang, Z., & Liu, B. (2022). A CRISPR/dCas9-based transcription activated system developed in marine microalga *Nannochloropsis oceanica*. *Aquaculture (Amsterdam, Netherlands)*, *546*, 737064. doi:10.1016/j.aquaculture.2021.737064

Wei, L., Wang, Q., Xin, Y., Lu, Y., & Xu, J. (2017). Enhancing photosynthetic biomass productivity of industrial oleaginous microalgae by overexpression of RuBisCO activase. *Algal Research*, *27*, 366–375. doi:10.1016/j.algal.2017.07.023

Wei, W., Ning, L.-W., Ye, Y.-N., Li, S.-J., Zhou, H.-Q., Huang, J., & Guo, F.-B. (2014). SMAL: A Resource of Spontaneous Mutation Accumulation Lines. *Molecular Biology and Evolution*, *31*(5), 1302–1308. doi:10.1093/molbev/msu073 PMID:24531082

Wong, T. K.-M., Ho, C.-L., Lee, W.-W., Rahim, R. A., & Phang, S.-M. (2007). Analyses of expressed sequence tags from *Sargassum binderi* (phaeophyta). *Journal of Phycology*, *43*(3), 528–534. doi:10.1111/j.1529-8817.2007.00349.x

Wu, W., Liu, X., Wang, M., Meyer, R. S., Luo, X., Ndjiondjop, M.-N., Tan, L., Zhang, J., Wu, J., Cai, H., Sun, C., Wang, X., Wing, R. A., & Zhu, Z. (2017). A single-nucleotide polymorphism causes smaller grain size and loss of seed shattering during African rice domestication. *Nature Plants*, *3*(6), 17064. doi:10.1038/nplants.2017.64 PMID:28481332

Yamada, K., Suzuki, H., Takeuchi, T., Kazama, Y., Mitra, S., Abe, T., Goda, K., Suzuki, K., & Iwata, O. (2016). Efficient selective breeding of live oil-rich *Euglena gracilis* with fluorescence-activated cell sorting. *Scientific Reports*, *6*(1), 26327. doi:10.1038rep26327 PMID:27212384

Yang, B., Liu, J., Ma, X., Guo, B., Liu, B., Wu, T., Jiang, Y., & Chen, F. (2017). Genetic engineering of the Calvin cycle toward enhanced photosynthetic CO 2 fixation in microalgae. *Biotechnology for Biofuels*, *10*(1), 1–13. doi:10.118613068-017-0916-8 PMID:29034004

Yao, J., Fu, W., Wang, X., & Duan, D. (2009). Improved RNA isolation from *Laminaria japonica* Aresch (Laminariaceae, Phaeophyta). *Journal of Applied Phycology*, *21*(2), 233–238. doi:10.100710811-008-9354-0

Yao, L., Shabestary, K., Björk, S. M., Asplund-Samuelsson, J., Joensson, H. N., Jahn, M., & Hudson, E. P. (2020). Pooled CRISPRi screening of the cyanobacterium *Synechocystis* sp PCC 6803 for enhanced industrial phenotypes. *Nature Communications*, *11*(1), 1666. doi:10.103841467-020-15491-7 PMID:32245970

Yi, Z., Su, Y., Xu, M., Bergmann, A., Ingthorsson, S., Rolfsson, O., Salehi-Ashtiani, K., Brynjolfsson, S., & Fu, W. (2018). Chemical Mutagenesis and Fluorescence-Based High-Throughput Screening for Enhanced Accumulation of Carotenoids in a Model Marine Diatom *Phaeodactylum tricornutum*. *Marine Drugs*, *16*(8), 272. doi:10.3390/md16080272 PMID:30081564

Yi, Z., Xu, M., Magnusdottir, M., Zhang, Y., Brynjolfsson, S., & Fu, W. (2015). Photo-Oxidative Stress-Driven Mutagenesis and Adaptive Evolution on the Marine Diatom *Phaeodactylum tricornutum* for Enhanced Carotenoid Accumulation. *Marine Drugs*, *13*(10), 6138–6151. doi:10.3390/md13106138 PMID:26426027

Yu, Q., Li, Y., Wu, B., Hu, W., He, M., & Hu, G. (2020). Novel mutagenesis and screening technologies for food microorganisms: Advances and prospects. *Applied Microbiology and Biotechnology*, *104*(4), 1517–1531. doi:10.100700253-019-10341-z PMID:31919586

Zhang, Y., Wang, X., Shan, T., Pang, S., & Xu, N. (2019). Transcriptome profiling of the meristem tissue of *Saccharina japonica* (Phaeophyceae, Laminariales) under severe stress of copper. *Marine Genomics*, *47*, 100671. doi:10.1016/j.margen.2019.03.006

Chapter 6
Physiology of Algae:
An Insight

Harshita Mishra
University of Allahabad, India

Ashutosh Pathak
https://orcid.org/0000-0001-5404-1690
Forest Research Institute, India

P. V. Subba Rao
Aquaculture Foundation of India, India

K. Suresh Kumar
University of Allahabad, India

ABSTRACT

Algae are a fascinatingly diverse group of photosynthetic organisms existing in diverse environments (ranging from oceans, rivers, lakes, ponds, and brackish waters). Comprising the base of the aquatic food ecosystem, algae have pivotal ecological functions as oxygen producers. Ranging in size from unicellular microalgae to the giant kelp, they have a wide range of (food, pharmaceutical, and industrial) applications. Physiology of algae comprises the study of algal function and behaviour. It encompasses all the dynamic processes of growth, metabolism, reproduction, defence, communication of algae (that account for algae being alive), and the processes underlying large biogeographical patterns of algae. Several biotic and abiotic environmental variables such as nutrients, light, temperature stress, salinity stress, desiccation, global warming, and ocean acidification affect algal growth and occurrence. This chapter provides a rudimentary insight regarding the growth, reproduction, and biochemistry of algae under varying environmental conditions.

DOI: 10.4018/978-1-6684-2438-4.ch006

INTRODUCTION

Algae are the key primary producers present in aquatic and terrestrial environments and are a representative of several emerging genetic model systems (Armbrust et al., 2004; Hopes et al., 2016; Nymark et al., 2016). They also play progressively more significant role in human nutrition (FAO, 2014). Algal photosynthesis provides about one-half of the total oxygen that we breathe, and their genomes revealed the story of a tangled past that traverses the tree of life through processes of endosymbiosis and the horizontal gene transfer (HGT) (Price et al., 2012; Cenci et al., 2017).

They are known to demonstrate a wide range of morphological diversity and inhabit the diverse aquatic systems. Algae frequently occur in extreme environments, and this reflects their remarkable ability to endure a variety of natural as well anthropogenic stresses. They need to survive with low concentrations of few essential nutrients (particularly carbon, nitrogen, phosphorus and trace elements) in natural water and, low availability of light and temperature. Besides withstanding such limiting factors, they have to survive and grow in the adverse habitats that are enriched with salts, toxic metals and pesticides, elevated temperature, pH, UV–B radiation and light intensity. Algae are known to be the principal primary producers of water bodies – from a rain pond to the oceans. Therefore, the tolerance range of algae to diverse stress factors assumes tremendous significance from an ecological standpoint. The basic metabolic mechanism of algae is similar to that of higher plants.

Cyanobacteria and microalgae present in the atmosphere, are exposed to a range of stress factors such as, humidity, temperature, oxidative stress, nitrogen starvation, radiation, and osmotic stress (Fröhlich–Nowoisky et al., 2016, Després et al., 2012, Tesson et al., 2016, Wiśniewska et al., 2019). There is one theory suggesting that this stress is a result of evolutionary force that causes selection pressure and hence affects spread and the evolution of organisms (Fröhlich–Nowoisky et al.,2016). Their atmospheric survival capability depends totally on the adaptability of the cyanobacteria and microalgae towards the ever changing environmental condition (Fröhlich–Nowoisky et al., 2016; Tesson et al., 2016; Wiśniewska et al., 2019). The stress response generated is of great importance to their capability to colonize the new habitats and also to survive. They also endure the ever demanding environmental conditions such as irradiance, temperature, and humidity gradients. According to Tesson et al., (2012), there may be a possibility that microalgae may change the stage of life e.g., into dormant cell stage, during their stay in the unfavourable environmental conditions. Therefore, the further investigation of physiological modifications that affects microalgae as they disperse is quintessential. The diversity of algae includes such genotypes that can, through evolution, grow over a large range of mean and extreme value of flux in photo-synthetically active radiation (PAR) of both low and high concentration of dissolved nutrients and of extensive ranges in temperature (Raven & Geider, 1988).

Nearly 50% of the primary biomass production on the earth is based upon the aquatic ecosystems (Siegenthaler & Sarmiento 1993, Woodwell et al., 1998). Most of the aquatic yield is due to phytoplankton, where the algae and cyanobacteria play a core role, especially in the coastal areas. In addition to the biomass productivity, "forests of algae" serve as the breeding grounds for fish, molluscs, and crustaceans. Both algae and cyanobacteria are ecologically and economically paramount group of living organisms in the world and they form an important constituent of the nitrogen-fixing microbial population both in the aquatic and the terrestrial ecosystems especially in the tropical rice fields (Sinha & Hader, 1996). Algae also have considerable economic potentials, being used in the production of food for the humans and animals, fertilizers, and cosmetics. Light is probably the most decisive factor in the coastal environment, which regulates the vertical distribution of algae (Luning & Dring, 1979). Algae also faces a serious stress when it is exposed to unfiltered solar irradiation and have been found to respond to this with the reversible photoinhibition. (Franklin and Forster 1997; Hader, 1998).

Most of the marine macroalgae live attached to the substrate in the coastal areas and are characterized by a variable environment mainly due to the tide cycles which more or less depends on the geographical situation. Macroalgae are spread over coastal zones according to their resistance towards several abiotic and biotic factors. Some species are found on the intertidal zones and are outside the water for a more or less long time every day; on the other hand, some macroalgae are constantly found underwater if they are living in sub-tidal zone or in pools in which the abiotic parameters can strongly be modified during a tidal cycle. During these alternating phases of immersion/emersion, macroalgae deal with important modifications of numerous abiotic parameters, such as light, temperature, salinity, and biotic parameters, such as the grazing, fouling, pathogenic interactions (Falkowski & Raven, 2013). In order to counteract these important variations, macroalgae from intertidal zone develops a resistance mechanism in which they synthesize some special compounds, like photo-protectors and osmolytes. In this chapter, after the presentation of the major environmental parameters facing intertidal seaweeds in the temperate environment condition, some of the adaptation mechanisms of these algae too are presented with an emphasis on both microalgae including cyanobacteria and macroalgae particularly the brown and red seaweeds.

EFFECT OF BIOTIC FACTORS IN THE PHYSIOLOGY OF ALGAE

Biotic interactions between the algae and the other eukaryotes (Worden et al., 2015) are found extremely widespread in aquatic and in terrestrial ecosystems. The extent to which nature has experimented with these relationships is in broad range, including the interactions among the organisms that maintain few functional associations, to those that have evolved to a highly integrated suite of functions. Algae are also found engaged in the extracellular or surface interactions in the phycosphere, i.e. ecologically and physiologically integrated neighbourhood inhabited by algae (Bell & Mitchell, 1972). Epi–biosis known as the surface colonization of one organism, the basibiont. Epi–biotic interaction between alga–alga, alga–bacterium and alga– virus, plays a dominant role in the acquisition of nutrient and recycling, metabolic flux, energy flow and as well in other developmental processes. Similarly, like herbivory, epi–biosis represents one of the most significant interactions that can determine the fate of algae and has been well known to shape the entire communities of marine ecosystem (Korpinen et al., 2007).

Biotic Interaction with its Plastid Origin–Endosymbiosis in Archaeplastida

Algae have known to be originated as a result of the primary plastid endosymbiosis. This is a process in which a mitochondria-containing (single–celled eukaryote) engulfs and retains a cyanobacterium that eventually becomes a photosynthetic organelle or plastid (Cavalier–Smith, 1982; Bhattacharya et al., 2004). This product of 1.6 billion–yr–old endosymbiotic event (Yoon et al., 2004) ultimately splits into the three primary plastid lineages, the red algae, the glaucophyte algae and the green algae. (Price et al., 2012; Adl et al., 2012). Algae from these groups themselves are frequently engulfed by other protists, giving rise to a rainbow of the serially derived plastids that distributed throughout the tree of life (Palmer, 2003; Gould et al., 2008).

In fact, there are only two bona fide primary endosymbiosis known that give rise to widespread organelles over the long history of eukaryotes; the event from which all plastids originated, led to the evolution of mitochondria. Other more taxonomically limited cases of organelle origin are associated with the photosynthetic amoeba lineage *Paulinella*, the non–photosynthetic organelle of the trypanosomatids (Kostygov et al., 2016; Morales et al., 2016) and nitrogen–fixing spheroid bodies in the rhopalodiacean diatoms (Nakayama et al., 2014; Zehr et al., 2016). The rarity of primary endosymbiosis has awestruck the scientists for several years and is usually attributed to the extensive innovations required for organelle establishment.

Mutualistic Interaction between Free–Living Algae and Fungi

A well-known mutualistic relationship is known to exist about 480 million years ago between the fungi and algae known as lichen (Lutzoni et al., 2018). Lichen symbiosis is much adaptive wherein it allows the fungal partner mycobiont and algal partner photobiont symbionts to survive in such habitats and environments that is almost uninhabitable by either species growing alone, such as those growing on a rock outcrop or in a desert crust. Lichenized fungi have been known to have multiple independent origins in the Ascomycota and Basidiomycota, and are *per se* the meta–organisms that includes communities of chlorophyte algae, cyanobacteria, in addition to the basidiomyceteous yeasts (Spribille et al., 2016). Nutrient exchange often underlines mutualisms between both the photobionts and mycobionts. The reciprocal transfer of the carbon and nitrogen was known for synthetic consortia which is composed of the one or two photobiont and a diverse panel of fungi belonging to the class Ascomycete that demonstrates a latent capacity of ascomycetous yeasts and the filamentous fungi to interact with algae (Hom & Murray, 2014). In a study, the filamentous ascomycetous fungi *Alternaria infectoria* was demonstrated to provide nitrogen to *Chlamydomonas reinhardtii* in a long–lived bipartite system, whereby the nitrogen–starved alga responded favourably to the growing fungus (Simon et al., 2017). A non-lichen algal–fungal mutualistic relationship was described which involves the chytrid fungus *Rhizidium phycophilum* and the green alga *Bracteacoccus* providing the evidence that early diverging fungi have evolved the mutualistic rapport with algae based on the solute exchange (Picard et al., 2013). However, in every known examples of fungal–algal symbiosis, algal cells remain external to the fungal hyphae and are not known to enter into the living fungal cells.

In an algal–fungal mutual interaction *Nannochloropsis oceanica* algal cells become internalized within the hyphae of the fungus *Mortierella elongata* (Zhi –Yan Du et al., 2019). This obvious symbiosis begins with the close physical contact and nutrient exchange, including the carbon and nitrogen transfer between the fungal and algal cells as it has been reported by isotope tracer experiments. This mutualistic interaction appears to be much stable, as both the partners remain physiologically active over months of co-cultivation, leading to eventual internalization of photosynthetic algal cells, which persist to function, grow and divide within the fungal hyphae.

While the clear fungal–algal symbiosis may conjure the concept of a lichen formation, it may differ in many respects. The fungus *Mortierella* generally lacks the distinct tissue differentiation or the hyphal structures such as thallus and haustoria. Furthermore, the hyphae of *Mortierella* harbours an algal cell intracellularly while lichens maintain the algae in a matrix of fungi, but external to their cells. However, *Geosiphon pyriformis*, which is an early diverged fungal relative of *Mortierella* and arbuscular mycorrhizal fungi, shows the formation of a unique intracellular association with the photosynthetic cyanobacterium *Nostoc punctiforme* (Mollenhauer et al., 1996). In this the fungal–bacterial photobiont symbiosis, the fungus envelops the photosynthetic cyantobacterium *Nostoc* within a special swollen structure multinucleate fungal 'bladder' that is morphologically distinct from rest of the fungal mycelium (Schu¨ßler et al.,

1996). Within this bladder-like specialized structure, the cyanobacterium (*Nostoc*) are surrounded by the host–derived symbiosome membrane specialized for acquiring the photosynthate (Brenner et al., 2008).

Nannochloropsis and *Mortierella* are known to be biotechnologically important species for the production of lipids and biofuel, with the available genomes and the molecular tool kits. The recent observations provide us with exclusive opportunities for studying the algal–fungal mutualistic relationship including the mechanisms leading to the endosymbiosis.

Algal Bacterial Interactions that Balances the Planktonic Microbiome

Plankton comprises of all the free-floating organisms in the open water, including the microalgae that are the basis of marine and the fresh water food webs. Early interpretation of single metabolites from the microalgae that act as regulators of the growth of competitors or can act as defence metabolites, has paved the way for the concept of chemically mediated phytoplankton interactions (Gross, 2003). This concept predicts that the interactions by the means of production, storage and the release of chemical mediators can influence the growth as well as the prevalence of phytoplankton and also their associated microorganisms (Pohnert, 2004). Some of the selected compound classes were extensively studied, which revealed that single metabolites can exhibit a huge number of cascading effects. As for example, oxylipins produced by the marine diatoms are used as an activated chemical defence of these algae against the grazer population, for the antibacterial effects can even be used for the regulation of cell death within diatom blooms. The major chemicals, metabolites and their functions are discussed in Table 1.

Compact association between the microalgae and bacteria often have resulted in the evolution of a complex network of cross–kingdom interactions and a fine specialization of different organisms. Such interactions are mediated by diverse molecules and through some recognition mechanisms. Bacteria belonging to *Roseobacter* and *Halomonas* species are able to exchange the vitamins or vitamin precursor with algae that cannot synthesize them *de novo* (Croft et al., 2005; weinhausen et al., 2017). *Phaeobacter inhibens* and *Sulfitobacter* can provide auxins, as well as ammonium ions, to the diatoms and coccolithophores such as *Pseudonitzschia multiseries* and *Emiliania huxleyi* in exchange for the amino acid tryptophan (Amin et al.,2015; Sergev et al., 2016). *Rugeria pomeroyi* is able to detect the different sulphuric compounds released by the microalgae and react both by realizing auxins that can sustain algal growth and quorum sensing molecules that promotes the bacterial proliferation (Jhonson et al., 2016; Durham et al., 2017). The beneficial interaction between *Marinobacter* and the dinoflagellates *Scrippsiella trochoidea* is based on the exchange of Fe^{2+}, made bioavailable through the production of a light–labile siderophore (viberoferrin) by the bacterium. (Amin et al., 2009)

Metabolic Hotspots around the Algal Cells

Marine pelagic bacteria have evolved several strategies to utilize the resources released from the living and decaying primary producing microalgae. Some of the most prominent algicidal species are found capable of controlling the entire algal blooms by inducing algal lysis, and also facilitating the uptake of metabolites released during the process (Meyer et al., 2017). One such example is seen in the interaction between *Kordia algicida* and *Skeletonema costatum* (Paul & Pohnert, 2011) an enzyme known as protease is produced by the bacterium causes the alga to lyse and also the concomitant release of organic some compounds that sustain growth of bacteria. Such kind of interactions can go beyond the random associations as it is strikingly shown by antagonistic bacteria that have evolved the 'wolf pack' hunting

Table 1. The major chemicals, metabolites and their functions in the interaction with bacteria

S.no.	Molecules	Function	Refrences
1.	Amino acids	Nutrients and signaling molecules; tryptophan produced by microalgae serves as precursors for the biosynthesis of auxins in bacteria and promotes mutualistic association	Amin et al.,(2015) Segev et al.,(2016)
2.	Vibrioferrin	Binding and uptake of iron ions; the Fe complex with vibrioferrin can be photolysed and releases Fe^{2+} that can be taken up by algae.	Amin et al.,(2009)
3.	Oxylipins	Defence metabolites, antibiotics and regulators, oxylipins are polyunsaturated aldehydes which are metabolites derived from the oxidative transformation of polyunsaturated fatty acids .	Lanora et al.,(2011)
4.	Orfamide A	Demobilization and killing; this lipopeptide is produced by the bacterium *P. protegens* and induces deflagellation and disruption of Ca^{2+} homeostasis in the motile microalga *C. reinhardtrii*	Aiyar et al.,(2017)
5.	DMSP	Osmolyte, nutrient, regulator, signaling molecule DMSP stimulates chemotaxis in ceratin marine bacteria, which can metabolise it as sulfur, carbon and energy source.	Seymoura et al(2010), Amin et al.(2015) Smriga et al(2016)
6.	Auxins	Growth promotion molecules in plants and algae; IAA is produced by different bacteria and shows hormone effects; at low concentration promotes algal growth, at high concentration reduces growth and leads to cell death.	Amin et al,(2015) Segev et al,(2016)
7.	Vitamins	Nutrients and growth promoters; certain microalgae cannot synthesise some vitamins de novo synthesis, so they use vitamins or their precursor produced by bacteria.	Croft et al.(2005) Weinhausen et al.,(2017)
8.	Organic acids	Signaling molecules; some algae release organic acids, such as p-couramic acid during senescence that can be recognized by bacteria.	Segev et al.,(2016)
9.	Algicidal compounds	Induce algal death or inhibit algal growth; *Phaeobacter* releases Roseobacticides when *E.huxleyi* reaches senescence and promotes algal lysis. Other algicidal compounds inhibit algal growth.	Seyeddsayamdost et al,(2011)

strategies (Aiyar et al., 2017). In this scenario the bacterium *Pseudomonas protegens* assembles around *Chlamydomonas reinhardtii* and immobilize the alga by the process of deflagellation, this process is triggered by lipopeptide orfamide A, and also disruption of algal Ca^{2+} homeostasis (Aiyar et al., 2017). Therefore, the bacterium shapes its immediate environment and benefits from the resources released upon by following process of cell lysis. The phycosphere can be considered like the marketplace where the cross kingdom communications are mediated through the release and the uptake of the organic compounds (Wienhausen et al., 2017). Gradients present within this sphere guide the chemo–attraction of bacteria in micro-scale interactions (Sonnenschein et al., 2012; Smriga et al., 2016; Seymour & Raina, 2018). Till date, no spatially resolved chemical imaging of phycosphere around a microalga is available. However, Raman microscopy on the macroalgal surface suggests that this zone ramifies to some hundreds of micrometres into the water and a steep gradient with up to millimolar concentrations of the chemical mediators directly above the algal surface can be seen (Grosser et al., 2012).

Recently, the concept of the chemical communication was extended beyond the mere diffusion which has limited the exchange of metabolites. Schatz et al. (2017) found that the extracellular vesicles, produced by the virus–infected microalgae *Emilinia huxleyi* enable the cell–cell communication. Vesicles usually carry small RNAs that target metabolism of sphingolipid and also the cell-cycle pathways. Their internalization by *E. huxleyi* promotes spontaneous infection of the algal community. The mechanism of vesicle mediated exchange might extend beyond this example and more advanced study might extend it to other microorganisms, as evidenced by bacteria that also produces the extracellular vesicles to deliver the different substances such as DNA, antimicrobial peptides, quorum sensing molecules (Schatz & Vardi, 2018).

Notwithstanding the steep local chemical gradients and the temporally fluctuating composition of organic molecules in the oceans, the overall species composition in plankton seems to be surprisingly balanced (Teeling et al., 2016). The annually recurring patterns of the prevalent groups indicates that a pronounced local shaping might be translated in the cascading network to inter annual the stable succession patterns in communities of planktons.

Untargeted exo–metabolome survey of two strains of the most widely distributed clade Roseobacter revealed that the bacteria supply some essential vitamins, the plant growth promoters, as well as some of the amino acids that might serve as the source of nitrogen to the algae (Wienhausen et al., 2017). In turn, the heterotrophic bacteria receive the nutrients and can consume up to 82% of all the algal–derived organic matter (Hornak et al., 2017). The release of vitamins and the vitamin precursors by the bacteria is one of the most common and well–studied means for mediating the algae–bacteria interactions (Croft et al., 2005; Paerl et al., 2017). Vitamin transfers the support to algal growth and can be considered as bacterial farming of algae as suppliers of all the organic resources. Intensively studied examples of a metabolite that forms structures of the marine environment is the ubiquitous sulfur containing dimethylsulfoniopropionate (DMSP). DMSP compound is produced by the most of the phytoplankton species, but also certain bacteria contribute to the DMSP pool (Thume et al., 2018). DMSP along with its degradation products, also do some essential physiological functions as osmo–regulators and antioxidants in the producing algae (Johnston et al., 2016). Leaked DMSP is a kind of chemo attractant, that guides the bacteria and predators towards the phytoplankton cells (Seymour et al., 2010) and that stimulates the production of the quorum-sensing molecules among bacteria (Johnson et al., 2016). This metabolite therefore promotes mutualistic relationship. In recent times the concept of DMSP as an essential metabolite in the sulfur cycling has been expanded by description of the structurally related oxidized dimethylsulfoxonium propionate (DMSOP). This metabolite, that is formed by algae and bacteria can serve as the precursor for oceanic dimethylsulfoxide (DMSO) and is expanding the oceanic sulfur cycle that is driven by the associated metabolism of the algae and bacteria (Thume et al., 2018). With ongoing discovery of the other widely distributed organic osmolytes containing reduced sulfur, including dimethylsulfonioacetate and 2,3–dihydroxypropane–1–sulfonate (DHPS), the field of algal–bacterial sulphur shuttling is expanding (Gebser & Pohnert, 2013; Durham et al., 2015). Recent studies investigating the coordinated regulation of these metabolites underosmotic stress already suggests a novel degree of complexity (Gebser & Pohnert, 2013). By untangled interaction networks, it can be well recognized that sulfur cycling is also connected to the nitrogen cycling as evidenced by the interaction of *Phaeobacter inhibens* with a cosmopolitan microalga. *Phaeobacter inhibens* attaches to the cells of the coccolithophore *E. huxleyi* and exploits the released DMSP as a source of carbon and sulfur (Segev et al., 2016). After association, the bacteria initially increase the algal growth by releasing growth promoting indole–3–acetic acid (IAA). Production of IAA is further increased by the supply of the biosynthetic precursor known as tryptophan an amino acid by *E. huxleyi*. During the later stages of the interaction, a switch in bacterial strategy is observed, and the bacteria trigger algal cell death by the activation of pathways of oxidative stress. Such a shift from this mutualistic to pathogenic relationship seems to be more frequent as this was also observed in the *Phaeobacter gallaeciensis –E. huxleyi* interactions (Seyedsayamdost et al., 2011) in which initial production of auxins happens.

ABIOTIC STRESS FACED BY ALGAE

There are wide range of stresses faced by algae and can be divided (Davison & Pearson, 1996) into two kinds: limitation stress which is caused by insufficient supply of resources such as low light or nutrient deficiency and disruptive stress resulting from the damage caused by the adverse conditions or by allocation of resources in order to prevent the damage. Larger-sized plankton algae are dependent on the turbulence for the transport of the nutrients but are not stressed mechanically by it to any significant extent. Nonetheless, the movement of water has a most important mechanical impact on the benthic macroalgae attached to a substratum. Water currents exert a drag on such algae, the magnitude of which varies according to pattern of flow and shape and size of the algal body. The most macroscopic algae are pliant and respond to the tension, shear, bending, and twisting imposed by moving water. This suppleness reduces the drag. The hydrodynamic forces can also be reduced by the mucilage secretion, which increases the viscosity locally and so it favours laminar flow rather than turbulence. Mechanical stresses imposed on the thallus may be powerful enough to tear the tissues altogether (Koehl, 1986). Environmental conditions in which the algal communities are stressed by the water movements are usually complicated in both the pattern and variations with time. Stream periphyton is subject to shear stress, which for much of time may not vary to any great extent, but the spates may impose some damaging stress. Communities may differ in their resistance, e.g. non–filamentous diatom assemblages are even more resistant than filamentous communities. Spates have important effects on the longer–term composition and the dynamics of periphyton growths (Biggs & Thomsen, 1995).

The structure and the flexibility of large intertidal algae are such that they minimize these stressful conditions. Their cell walls are composed of a wide variety of different long–chain polysaccharides, the bonding of which may be different during synthesis to produce gels or fibres, making different degrees of tensile strength or the flexibility possible. However, a structural feature which improves the resistance towards mechanical stress may reduce the performance of some other (Koehl, 1986). A very simple example of this is the density of plants per unit area; the higher it is, the greater will be its resistance to the stress of water movement, but lesser will be the photosynthetic efficiency. The stability will depend more or less on the conditions under which the algae have been grown. Undoubtedly, the resistance to mechanical stress is an important factor in determining the location and success of a seaweed on a rocky shore but the interactions with other stresses, all are compounded by the daily vicissitudes of the tide, makes it difficult to interpret its role in establishing the observed patterns of distribution of littoral and sublittoral algae.

OTHER ABIOTIC FACTORS AFFECTING THE PHYSIOLOGY OF ALGAE

Enhanced Effects of Ultraviolet (UV)-B Radiation on Algae and Cyanobacteria

Effect of UV–B radiation is seen on the growth and development, biomass production, sensitivity, photosynthetic pigments, UV–B absorbing compounds, photosynthesis, protein and DNA damage, enzyme activity, nitrogen fixation and assimilation of nitrogen, protective mechanism of algae and cyanobacteria, accommodation of algae and cyanobacteria to wide range of environmental stress and its effects on ecological system.

(Clendennen et al. 1996; Dring et al. 1996a; Dring et al.1996b; Hanelt et al. 1997a, 1997b; Bischof et al. 1998b) exhibited the amazing effects of UV–B radiation; but typically these studies were conducted under the conditions with supplemental UV–B irradiance that was higher than it would ever occur outside the experimental conditions or natural conditions. Enhanced UV–B usually decreased the chlorophyll content, whereas it increased UV–B absorbing compounds in several algae. Decrease in the photosynthetic activity particularly at the higher UV–B doses, was due to the both direct, i.e. effect on photosystem, and indirect, i.e. decrease in pigments, effects. The decrease in the amount of chlorophyll pigments and photosynthesis has resulted into lower biomass production. Although algae and the cyanobacteria have evolved various avoidance and repair mechanisms to protect themselves against the damaging effects caused by UV radiation to adapt to the enhanced UV–B radiation.

Enhanced UV–B may have two kinds of effects on the algae and cyanobacteria; on the one hand, the algae and cyanobacteria are damaged by the radiation of UV–B; on the other hand, algae and the cyanobacteria show protection or defence mechanisms to accommodate the exposure to enhanced UV–B radiation.

Cyanobacteria and some other organisms of aquatic ecosystem under prolonged UVR exposure develops UV absorbing/screening compounds such as (MAA) Mycosporine–like Amino Acid, scytonemin, melanin and fluorescent pigments that protects them against the UV–A and UV–B radiation. If these defence mechanisms fail, UVR starts damaging the biomolecules such as DNA and proteins. (Fuentes–Tristan et al. 2019)

MAAs production depends on the availability of nitrogen and is induced by exposing the organisms to the UV light (Sathasivam et.al, 2019), being UV-B the most influential most (Sundararaman et al., 2007). A study was done on *Lyngya purpurea* determination that the abundant availability of nitrogen, oxidative stress, and UV-B presented an elevation in MAAs' production (Chandra et al. 2019).

MAAs act as sunscreens and may provide some additional protection as an antioxidants (Sundararaman et al. 2007). They are capable of dissipating the absorbed radiation as the harmless heat without producing the ROS (reactive oxygen species) and have been found to prevent the three out of ten photons from reaching to sensitive cellular targets in the cytoplasm of cyanobacteria (Sinha, 2015 and Conde et al., 2007). Other protective roles that MAAs exhibit include an action against osmotic, thermal, and desiccation stress as well.

Another UV–A and UV–B, and even UV–C screening compound is Scytonemin, one of the highly stable, yellow-brown, lipid-soluble inducible pigment found exclusively in the polysaccharide sheaths of some of the cyanobacteria (Fuentes–Tristan et al. 2019) as a protective mechanism against a short UV wavelength (Rastogi et al.2010 and Rastogi et al., 2015). During the periods of desiccation, Scytonemin becomes even more important as a UV protector due to the inactivation of the other screening mechanisms (Sharma et al 2013). It has been found in about 300 cyanobacterial species living in the soil crusts, intertidal mats, and the biofilms. Some of the genera that accumulate this screening compound in their sheaths surrounding cells includes, *Lyngbya, Calothrix, Chlorogloeopsis, Scytonema, Nostoc* and *Chroococcidiopsis* .*Lyngbya aestuarii* is characterized by having a sheath with Scytonemin as a photoprotective mechanism (Gröniger et al., 2000)

The study on the effect of UV–B on biomass yield of cyanobacteria has shown that although the low exposure dose of UVB (2J at 0.4 mW/cm^2) resulted in only reduction of 11% in the biomass yield of *Anabaena*, an increase in UVB intensity (0.5 mW/cm^2) resulted in a sharp decline of 27% in the biomass production. Under similar conditions (2J at 0.5 mW/cm^2), *Nostoc* and *Scytonema* produced nearly 12% and 10% loss of the biomass, respectively. At higher exposure dose (4J at 0.6mW/cm^2), *Nostoc*'s biomass production was depleted by 38%. (Xue et al.,2005)

They have mostly revealed a adverse effect of UV–B upon the algal metabolism. Bischof et al. (1998, 2002) exhibited that among all the photosynthetic pigments, lutein concentration was substantially found elevated under UV exclusion. In addition, marked UV effects on the xanthophyll cycle were also found, exclusion of solar UV radiation (and particularly UV–B) resulted in an increased ratio of zeaxanthin concentration to the total xanthophyll content, indicating the adverse effects of UV–B on efficiency of the photoprotection under high irradiances of photosynthetically-active radiation(PAR).

Eswaran et al. (2002) has reported that UV–B inhibited the accumulation of chlorophyll and phycobiliproteins, and also lowered the yield of agar by 23–43% and its gel strength by 22–36% under 12–72 hrs of exposure. The longer the exposure to UV-B radiation, the more significant will be the impact on the pigments and phycocolloids such as alginates, agars, and carrageenans in *Gracilaria edulis* was observed.

Till date, the information on UV–B effects on marine macroalgae is available from studies concerning photosynthetic activity (Xu et al., 2005). Measurements of chlorophyll, fluorescence of photosystem II (PSII), have revealed that the maximal quantum yield (*Fv/Fm*) and also the rate of electron transport (ETR_{max}) of macroalgae are negatively affected by UV–B radiation (Bischof et al. 1998b). Whereas $^{14}CO_2$ uptake was drastically inhibited (Tyagi et al. 2003) (78% inhibition) by 30 min exposure to the UV–B, about 76% activity remained when the UV–B exposure was given to the cultures in the presence of ascorbic acid.

Studies have shown that the damaging effects of UV–B are substantially minimized by certain reducing agents, the protective effect being particularly strong on the O_2 sensitive enzyme, nitrogenase. The presence of these agents or chemicals in their natural habitat or inside the cells of living organisms may protect/repair the damaging effects of UV–B radiation. Tyagi et al. (2003) considered that the protection of $^{14}CO_2$ uptake activity could be due to the role of ascorbic acid in maintaining the integrity of photosynthetic membranes, reaction centre of PS II and probably the activity of RUBISCO, even after the UV–B exposure. Callieri et al. (2001) studied the photosynthetic response of pico–and nanoplanktonic algae to UVB, UVA, and PAR in a high mountain lake; it was found that the UV–A plus UV–B significantly affects the algal assemblage in lakes; however, most of the effect could be attributed to UVA. The natural assemblages were found more inhibited by the UV than the autochthonous assemblages. The photosynthetic rate of picoplankton under full UV exposure was reduced by 73% and 55%, respectively, relative to PAR only. During the events of photo inhibition, the photosynthetic electron transport chain in the photosystem II is actively declined (Tevini et al. 1991) to protect the photosynthetic apparatus from excessive irradiance produced by solar energy. The underlying biochemical mechanism is radiation induced damage to the D1 protein of PS–II. During the period of recovery, the lost protein is replaced by the newly synthesized D1. Photo-inhibition is considered as an active regulatory process (Montero et al. 2002) by which the photosynthetic electron transport is reduced.

Table 2. Temperature tolerance and temperature optima of few microalgae isolated from different habitats

Species	Origin	Temperature optima	Temperature tolerance	References
Cyanobacterial assemblages	Antarctica	20 °C	0– 45°C	Fritsen & Priscu (1998)
Phormidium subfuscum	Antarctica	15°C	5– 20°C	Tang & Vincent (1999)
Nannochloropsis oceanica	Temperate	25 – 29°C	14.5 – 35.7°C	Sandnes et al. (2005
Amphora coffeaeformis	Tropical	28 – 33°C	28 – 40°C	Rajadurai et al. (2005)
*Micromonas*sp.	Arctic	6 – 8°C	0 – 15°C	Lovejoy et al. (2007)
Amphidinium sp.	Okinawa, Japan	24 – 29°C	21 – 35°C	Kitaya et al. (2008)
Symbiodinium californium	Santa Barbara, California	15 – 28°C	5 – 30°C	Teoh et al. (2010)

RESPONSE OF ALGAE TO INCREASED TEMPERATURE

Temperature is one of the fundamental environmental factors that strongly regulate growth of algae (Eppley, 1972; Raven & Geider, 1988). The liaison between the temperature and a given biological rate, photosynthesis has often been described by the temperature coefficient Q10 or Arrhenius functions (the factor by which a biological rate is increased by a 10 °C rise in the temperature) (Teoh et al., 2010).

Temperature is the most important ecological parameter that affects almost each and every aspect of aquatic life. The effects may fluctuate from increase in the metabolic rate of organisms to the displacement or even mortality of the some sensitive organisms (Rajadurai et al., 2005). Temperature-growth range of an alga is important ecologically because it defines the range over which an alga can be metabolically active and it determines the distribution of algae.

The productivity rate of algae and its survival are strongly affected by the change in their physiological and biochemical processes, as well as the biotic and abiotic factors in the environment. In the recent years, global climate change such as increase in temperature and elevated ultraviolet radiation (UVR) due to ozone depletion has hugely impacted the organisms particularly the ones in marine ecosystem. In the upcoming years, climate model projections summarized by the Intergovernmental Panel on Climate Change (IPCC) indicat that average global surface temperature might rise further from 0.5 to 1.6 °C by the year 2030, and keep rising to 1.1 to 6.4°C by the year 2100. According to Qin et al. (2007) algae serve as the primary producer of the food chain in both the marine and in the terrestrial ecosystems. (Teoh et al .2010)

Different species of algae have different ranges of temperature tolerance (Table 2) and physiological responses towards temperature changes. According to Li (1980), some of the algae can survive at extreme habitats with the temperatures ranging from –2°C in the Arctic and Antarctic to 75°C in the thermophilic hot springs.

Temperature can enforce a significant effect on the specific growth rate of algae. At temperature below the optimum for the growth, μ (μ optima: Specific growth rate (s) at optimal temperature(s)) increases with the increasing temperature but it declines distinctly at above the optimal temperature. For example, the growth rates of the three tropical algae from Australia *Cryptomonas* sp., *Rhodomonas* sp. and prymnesiophyte NT19 increase with the temperature within 25 to 30°C, but decline at temperature above 30 °C (Renaud et al., 2002).

In the same way, *Nannochloropsis oceanica* showed increasing μ as function of the temperature, from 14.5°C, with a peak at 25°C – 29°C. Above 30°C, the cultures showed the dramatic reduction in μ, with no cultures growing at temperatures over 35°C (Sandnes et al., 2005). In marine diatom *Chaetoceros calcitrans*, the growth rate increased with the temperature from 0.3 d^{-1} at 6°C to 1.0 d^{-1} at 15°C, and 1.4 d^{-1} at 25°C (Anning et al., 2001). Similar trends were evidenced in two mesophilic microalgae, *Microcystis aeruginosa* and *Scenedesmus acutus* (Staehr & Birkeland, 2006), Arctic cyanobacterium, *Schizothrix calcicola* (Tang & Vincent, 2000) and *Phormidium sp.* from a high Arctic lake where μ increased with increasing temperature from 5°C to 25°C.

DESSICATION AND STRESS TOLERANCE IN ALGAE

Most of the green algae are typically found occurring in the aquatic ecosystem, and many of the algal species also live partly or permanently under aero–terrestrial conditions where the cells are exposed regularly to the atmosphere and experience dehydration this capability of an algal cells to survive under air dried state is termed to be as desiccation tolerance. The mechanisms involved in the desiccation tolerance of the green algae are still not well understood. Beginning from its, physiological, structural changes and biochemical consequences of desiccation will be addressed in different algal lineages of green algae (Holzinger & Karsten, 2013).

This helps in maintaining the structural integrity in the dried state and allows cell to maintain the turgor pressure for a long period of time during the process of dehydration. The physiological strategies in an aero terrestrial green algae generally includes a rapid reduction of photosynthetic rate during the desiccation, but also shows the quick recovery after the rewetting, whereas an aquatic species are very sensitive to the drying. The underlying mechanisms such as the affected molecular components of the photosynthetic machinery are not well understood in the case of green algae. Therefore, the modern approaches based on the transcriptomics, proteomics, and/ or metabolomics are urgently required to better understand the molecular mechanisms involved in the desiccation stress physiology of these organisms living in the aquatic environment.

Water is the indispensable requirement for all the organisms on the Earth, and thus the elimination of water from algal cells represents a severe more often lethal stress in those organisms. The structure of the intracellular biomolecules and the membranes is maintained by molecules of water, and thus the dehydration leads to an often irreversible aggregation of the macromolecules and disintegration of the organelles.

The different life forms may strongly affect the desiccation tolerance in these organisms, because they can be distinguished based on their size into micro and macro–algae, found in the ocean. While many of the green micro algae live in the ephemeral freshwater ponds, marine green macroalgae such as the seaweeds occur in the intertidal or supra–littoral zone along the rocky coasts, where they preferentially grow as sessile organisms attached to the hard substrata such as gravel, rocks, or coral reefs, or as an epiphytes on salt-marsh plants, mangroves, and mussel colonies. (Holzinger & Karsten, 2013). Here they are chiefly confronted with the dehydration when it is exposed at low tides. In contrast to this ephemeral ponds may experience the desiccation at regular and for much longer intervals and for much longer periods (Evans, 1958).

Consequently, dehydration during exposure at the low tide in the marine environment is usually a local and "short–term" rather than global and "long–term" factor, even though it can be highly variable in the coastal regions due to the neap tides and meteorological conditions. In addition to this dehydration in the marine green algae can also be mediated by the hyper-saline conditions, due to the tidal flows, wind, hydrology, and evaporation, or by freezing. Stress due to salinity and desiccation is the different types of water scarcity. Whereas under the condition of hyper-salinity, seaweed cells are still in full contact with the liquid water of decreased water potential, desiccation leads to even more intense cellular dehydration. Therefore, the salinity stress is often defined as the "physiological drought (Kirst, 1990). Both hyper-salinity and desiccation truly affect the internal osmotic potential, which are necessary for maintaining the turgor pressure as the driving force for growth. The acclimation responses of marine green algae are comparable (Karsten, 2012).

Although most of the green algae typically inhabits aquatic environment, many taxa occur on land, where they participate in the symbiotic relationship with fungi, forming the lichens (EttlandGärtner,1995; Friedl,1997; Friedl & Rybalka, 2012)), or grow on man–made surfaces such as roof tiles or on natural surfaces such as tree bark (Rindi & Guiry, 2004; Karsten et al., 2007). In addition, many green algal taxa can grow under rocks (hypolithic) or on the surface (epidaphic, or just below the surface (endedaphic) of the soil (Singh, 1941; Friedmannetal., 1967; Bell, 1993). Green algae are also typical components of the so-called biological soil crusts, which are found on all the continents on earth, in arid and semi-arid regions (Flechtner, 2007) as well as from the tropical to polar regions, in the high alpine zone, i.e., inhabitants where soil moisture is limiting and the vascular plant cover is meagre (Fritsch & Haines, 1923; Belnap & Lange, 2001). Aero–terrestrial algae form an close association with the soil particles, which exists within, or directly on top of the upper most millimeters of soil (Büdel, 2005; Cardonetal., 2008). There, the biological crusts form the water–stable aggregates that have significant ecological roles in the primary production, water retention, nutrient cycling, and in stabilization of soils (Lewis, 2007).

Compared to the aquatic environments, aero–terrestrial green algae are exposed to much harsher environmental conditions, such as colossal differences in the water potential between the terrestrial habitat (e.g., soil) and the atmosphere, resulting in normal desiccation leading to dehydration of the cells. In the alpine regions of Europe, for example, water availability frequently fluctuates, from fluid droplets after rain or snow, to extended periods of dryness or freezing. Availability of water, which includes precipitation, condensation, and water vapour is therefore the key ecological prerequisite for the long–term survival of the aero-terrestrial algae in such habitats, because only fully or well-hydrated and ultra-structurally intact cells will be able to function physiologically (Gray et al., 2007; Karsten et al., 2010; Karsten and Holzinger, 2012). In the terrestrial ecology, drought is defined as an extended period of months or years when a particular region undergoes a shortage in its water supply, such as lack of precipitation. Although drought is normal, recurring feature of the climate in the many parts of the world, global-change scenarios envisage intensified desertification (Brito et al., 2013), which of course will show substantial impact on the terrestrial ecosystems and organisms of the that affected region.

Effects of Dessication on Photosynthesis and Respiration

In the dehydrated state, photosynthesis is generally completely blocked no more excitation energy absorption can be used for the electron transport, and hence it may result in photo-inhibition or even in the photo–damage (Wieners et al., 2012). For cyanobacteria, lichens, green algae, and bryophytes(mosses), various sites sensitive to desiccation in the photosynthetic apparatus have been reported in the photosystems(PS I and PS II), particularly the photosystemII (PSII) with its oxygen–evolving complex(OEC), the ATP–generation, and carbon assimilation processes (Allakhverdiev et al.,2008). Photoinhibition is generally caused by damage to the D1 protein of PSII. In the green algae, photoinhibition, in principle, occurs continuously while the cells are exposed to the light, and the damage is repaired continuously, which consists of degradation and *de novo* synthesis of the D1 protein, followed by the activation of the reaction centre. Because of this quick repair, most of PSII reaction centres are constantly in fully functional state, even under the high light conditions. However, desiccation may also constrain the supply of carbon dioxide for use in the carbon fixation, which decreases the rate of repair of the D1 protein in PSII by direct inactivation of translation machinery (Takahashi and Murata, 2008). Consequently, not the photo–damage, but rather the loss of the repair capacity is main mechanism for depression of photosynthesis in green algae during the desiccation. Interrupting in carbon fixation results in the creation of reactive oxygen species (ROS), which in turn not only blocks the biosynthesis of PSII proteins but also biochemically affects the nucleic acids and polyunsaturated fatty acids (PUFA) (lipid peroxidation). A proteomic study done on a desiccation–tolerant grass established that as the photosynthetic activity is blocked during the dehydration, photosynthetic proteins also generally decreases to prevent the formation of reactive oxygen species (ROS) (Oliver et al., 2011).

Protective Strategies

One of the strategy of aero-terrestrial and the aquatic green algae against the desiccation is to elude dehydration by avoidance or self–protection. Additional factors that contribute to avoidance strategy or at least retardation of the water loss includes a low algal surface–to–volume ratio and the morphological features such as the thick cell walls and mucilage layers (*Prasiola*; Jacob et al.,1992). Extracellular polysaccharides (EPS) are critical in desiccation tolerance of cyanobacteria and have beneficial effects on desiccation tolerance in the green alga *Chlorella sp.*(Knowles & Castenholz, 2008) under the exposed conditions, macroalgal canopies of *Ulva sp.* In the upper intertidal zone in southern Spain form sheet–like, multiple–layered structures in which the top player usually bleaches due to strong solar radiation, desiccation, and other abiotic stresses, thereby providing photo-protection and moisture for sub canopy thalli (Bischof et al., 2002).

SALINITY STRESS

Increasing salinity is a challenging environmental stress for organisms which needs to be tackled. Unicellular photosynthetic microalgae are especially more vulnerable as they have to struggle not only with the osmotic stress and ionic imbalance but also with the reactive oxygen species (ROS) generated interfering with photosynthesis.

Effect of High Salt Concentration on Green Algae

There are several morphological and molecular changes that occur simultaneously which improve the survival of salt sensitive algae under elevated salt stress. These are described in greater detail below.

Basic Morphological Changes

Stress caused by the high salt concentration retards the cell division, decreases size, stops motility, and triggers the formation of palmelloid in *Chlamydomonas* species (Nakamura et al., 1975; Hema et al., 2007; Neelam and Subramanyam, 2013; Khona et al., 2016). Growth rate is directly impacted which can be easily detected. *Chlamydomonas reinhardtii* cells under the high salt stress condition have shown lower growth rate compared to the untreated cells. Untreated (control) salt sensitive algal cells reached to an optical density (OD) of 1 at 750 nm in four days while in the salt treated cells it took six days to achieve the same OD as of control. The cells were smaller in the dividing stage across every salt concentration (Neelam & Subramanyam, 2013). Upsurge in the concentration of salt has negatively affected the growth of some other freshwater algae species such as *Chlorella vulgaris, Chlorella salina, Chlorella emersonii* (Talebi, et al. 2013), and *Scenedesmus opoliensis* (Demetriou et al. 2007).

Some algal genera such as *Dunaliella* and *Chlorella* do not form palmelloids and show diverse mechanisms to tackle stress caused by salt. *Chlorella* cells have a rigid cell wall, thus limiting its ability to change volume of cell. Hence, osmoregulation through production of the organic solutes and accumulation of inorganic ions are used to maintain the osmotic homeostasis. Furthermore, cations present in the media are bound within the intracellular spaces reducing the osmotic activity (Ahmad & Hellebust, 1984). In contrast, *Dunaliella* lacks a rigid cell wall which allows the cells to change the volume of cells rapidly under salt stress condition by adjusting with intracellular ion and glycerol concentration eventually restoring the cells turgor pressure (Kaçka & Dönmez, 2008).

The quaternary ammonium solute glycine betaine is involved in the osmoacclimitisation of a number of the halotolerant cyanobacteria (Reed et al., 1984). Glycine betaine preserves the thylakoid and plasma membrane integrity when exposed to the saline conditions (Rhodes and Hanson 1993). Drastic increase in the proline content at all the concentrations of NaCl (up to 0.4M) and the *Chlorella vulgaris*. Glycine betaine is the cardinal solute in the highly salt tolerant halophilic forms. It was also augmented with increase in the NaCl concentrations up to 0.3 M and thereafter it declined. (Hiremath and Mathad et al. 2010). In the higher plants, proline is considered to play an crucial role in the defence mechanism of the stressed cells providing carbon, nitrogen and energy source after stress by the degradation. (Szekely, 2004)

According to Hong et al. (2000) increased resistance to the oxidative stress is due to the accumulation of proline and the other metabolites. He also made an observation that proline also increased the salt tolerance of microorganisms. Therefore in microalgae and other plants, proline acts as a free radical scavenger and increases the salt tolerance of microorganisms. It is likely that proline accumulation may be one of the major mechanism of salinity tolerance by the alga. An increasing (0.12 M) NaCl concentration inhibited the Biomass growth rate of the unicellular organisms of algae *Pseudokirchneriella subcapitata* (Sikorski, 2021)

HYDRODYNAMISM

Hydrodynamism refers to motion of water, due to the tidal cycle, waves or the ocean currents resulting from winds and differences in the water mass densities (Hurd et al., 2014). This factor influences distribution and the vertical zonation of algae on the shore, because of following reasons: (1) the currents and waves affect other environmental factors such as light penetration, temperature, and the availability of nutrients and all elements required for the development of algae; (2) hydrodynamism influences the fauna distribution and ultimately affects algae through spatial competition and the potential impact of herbivores; (3)Moreover, to survive on sites exposed, macroalgae must be resistant enough to remain attached to their substrate and to withstand wave action, (Hurd, 2000; Luning, 1990). Consequently, from the hydrodynamism, macroalgal species are often acclimatized to a type of wave exposure, either sheltered (low hydrodynamism), exposed (high hydrodynamism) or the intermediate, semi–exposed environment.

Hydrodynamism can affect the growth, the morphology or the phenology of seaweeds. In this sense, Kraemer and Chapman (1991) have revealed that juveniles of the brown alga *Egregia menziesii* grown under high hydrodynamic conditions were about two times more stronger and stiffer than those grown under low hydrodynamic conditions to resist the hydrodynamism and not being swept away by the waves and tide. Furthermore, seaweeds can adapt their size, shape and direction to thrive and disperse with the flow and increase their resistance towards currents (Boller & Carrington, 2006). It has also been established that the red alga *Chondrus crispus* can realign its stipe and compact its thallus in response to flow (Boller & Carrington, 2006). Likewise, nutrient uptake by macroalgae is influenced by the motion of water, impacting its growth and photosynthesis (Gerard, 1982; Hurd, 2000).Indeed, with the increasing current, the nutrient absorption rate increases due to the decrease in the thickness of the diffusion boundary layer (Luning, 1990). Photosynthesis rate increases with hydrodynamism until enzymes saturation, as studied in the brown alga *Macrocystis pyrifera* (Wheeler, 1980). In order to compensate for low nutrient and gas absorption rates, an alga in low hydrodynamic environments tends to have a higher surface–to–volume ratio (Stewart & Carpenter, 2003). The reproduction can also be influenced by the hydrodynamism. Water motion mainly affects gametes and spores release, and their dispersal distance Gordon and Brawley (2004). Thus, they have demonstrated an increase in the zoospores released from the brown alga *Alaria esculenta* under the shaken conditions, and on the contrary, there were more gametes released by antheridia under calm conditions. The presence of few molecules in seaweeds could be influenced by the hydrodynamism, especially polysaccharides in brown algae (Phaeophyceae). Craigie et al. (1984) found that the alginic acid composition in the brown alga *Saccharina longicruris* was related to the degree of exposure of wave, although this hypothesis is still contentious (Kraemer & Chapman, 1991). Moreover, Le Lann et al. (2014) showed that the brown alga *Bifurcaria bifurcata*, living in a semi–exposed/sheltered environment, synthesized more terpenes, including bifurcanone which is specific to the sheltered environment. According to Rönnberg and Ruokolahti (1986), phenolic content of *F. vesiculosus* has been shown to depend on the degree of wave exposure of the habitat, with minor fluctuations of phenolic content at exposed site compared to sheltered site, as it was demonstrated in the brown alga *Sargassum muticum* by Plouguerné et al. (2006b). It is sometimes challenging to interpret the significance of the difference in the composition or morphology with the exposure mode, as many other factors can interact, including desiccation and herbivory. In this intellect, Kraemer and Chapman (1991) did not find any variation in alginic acid content in the Laminariales *E. menziesii* exposed to different wave velocities; and Fuller and Mathieson (1972) showed no major difference in carrageenan content of *C. crispus* according to the wave exposure.

LIGHT

As algae are recognised to be photosynthetic organisms, light is the important factor for their growth and for their distribution. Chiefly, algae need to absorb enough Photosynthetically Active Radiation (PAR) in their thylakoids to transform the light energy into chemical energy through photosynthesis, crucial for their development. In this sense, growth of the red macroalgae such as *Gracilaria domingensis* (Ramlov et al., 2011), *C. crispus* and *Palmaria palmate* (Manríquez–Hernandez, Duston, & Garbary, 2016) have been enhanced under the high photon flux intensity, indicating that the light can be a restraining factor. There is a photosynthetic light compensation point, which corresponds to photon fluence rate, for which the amount of oxygen produced by the photosynthesis accurately compensates for the oxygen absorption related to the respiration (Hurd et al., 2014).

Light is the energy source during the photoautotrophic growth phase and organisms use light energy to transform carbon dioxide to organic compounds i.e. sugars. Light is the source of energy during photoautotrophic growth phase and organisms use light energy to convert carbon dioxide into an organic compound–especially, sugars. Intensity of light effects the growth of algae through its impact on photosynthesis (Stockenreiter et al., 2013). Although growth rate under an increasing light intensity is a function of strain and temperature of culture, the growth rate of algae is found to be maximal at saturation intensity and decreases with both increase or decrease in the light intensity (Plouguerné et al., 2006b). The photo–adaptation and photo–acclimation processes in algae leads to changes in cell properties according to the light availability and an increase in the photosynthetic efficiency (Dubinsky et al., 1995). Adaptation can take place through several mechanisms such as changes in the types and quantities of pigments, growth rate, rate of dark respiration or the availability of essential fatty acids (Fábregas et al., 2004) Morphological photo-acclimation is accompanied by changes in the volume of cell and the number and density of thylakoid membranes (Berner et al.,1989). Algae overcome the light limitation by desaturation of chloroplast membranes (Mock et al., 2002). Light intensity when increases above the saturating limits causes photo–inhibition (You et al., 2004). This occurs due to the disruption in the lamellae of chloroplast caused by the high light intensity (Brody et al., 1959) and inactivation of enzymes involved in the carbon dioxide fixation (Iqbal & Zafar, 1993). Growth rate of *Dunaliella viridis* was observed to decreased to 63% with increase in the light intensity from 700 to 1500 $\mu mol \cdot m^{-2} \cdot s^{-1}$ (Gordillo et al., 1998) Light intensity also impacts the cellular composition of algae. *Dunaliela tertiolecta* exhibits a decrease in protein content and increase in lipid fraction with the increasing light intensities up to saturation (Cuhelet al., 1984). Low light intensity has been observed to end result in higher protein content while high photon flux density (PFD) results in an increased extracellular polysaccharide content (Iqbal & Zafar, 1993). In the absence of light, it was observed that the total lipid content increased in the *D. virdis* but it reduces triglycerides, free fatty acids, free alcohols and sterols (Smithet al., 1993). In *Nannochloropsis* sp. grown under the low light conditions (35 $\mu E \cdot m^{-2} \cdot s^{-1}$), 40% of the total lipids were found to be galactolipids and 26% were found to be triacylglycerols. In the same system, high light (550 $\mu E \cdot m^{-2} \cdot s^{-1}$) conditions resulted in an increased synthesis of triacylglycerol with a reduction in galactolipid synthesis (Sukenik et al., 1989).

EFFECT OF OCEAN ACIDIFICATION ON ALGAL METABOLISM

Ocean acidification is defined as the reduction in pH of the ocean over a comprehensive period of time, caused primarily by the uptake of carbon dioxide (CO_2) from the atmosphere and by the anthropogenic sources of carbon dioxide (CO_2), in turn the gradual change occurs in the carbonate chemistry of the ocean. Oceans are one of the leading sinks of this increasing CO_2 as they can absorb over 25 million tons of anthropogenic sources of CO_2 daily, causing unprecedented variations to the ocean chemistry (IPCC, 2007). Raised CO_2 concentrations in the ocean modify the speciation of dissolved inorganic carbon in seawater and reduces the pH by carbonate buffer system, along with the varying abilities of some of the macrophytes to use CO_2 and HCO_3. By the end of millennium, pH of the ocean surface from a pre–industrial value 8.2 to 7.4 (Caldeira and Wickett, 2003). When pH is altered, the carbon speciation in seawater is also changed, which has the strong consequences on the photosynthetic activity, respiration, and calcification metabolism.

Algal physiologists and seawater carbonate chemists pool their knowledge in order to provide the fundamental information on the carbon physiology and seawater carbonate chemistry required to understand the complexities of how the ocean acidification can affect the metabolism of algae.

Seaweeds might get benefit from the rising CO_2 concentration through increased photosynthesis and the carbon acquisition, with subsequent and superior growth rates (Mackey et al., 2015, Aires et al., 2018; Cornwall & Hurd, 2020). Unlike the photosynthesis, some other metabolic processes, such as ion homeostasis, respiration, nutrient uptake and enzyme activity, are apparently suppressed by the ocean acidification conditions (Hofmann et al., 2013; Gutow et al., 2014; Fernández et al., 2015). This change in algal metabolism may promote the modifications in chemistry of seaweed and can also change the dietary quality of tissue for the grazers. In addition, association microbiota can also be influenced by the environmental changes, with feedback results (Aires et al., 2018).

In marine ecosystems, enrichment of different levels of CO_2 in seawater consequently affects the acidification of seawater, which greatly influences the marine algae metabolism. The overgrowth of *Ulva sp.* consequent to elevated pCO_2 in eutrophic estuaries can be directly promoted through acidification (Young & Gobler, 2016a). However, Reidenbach et al., (2017) detected no changes in *U. australis* growth by the decreasing pCO_2, which influenced carbon and nitrogen metabolisms. Axiomativelly, both lowered and increased seawater pH exerts significant physiological stress on *U. lactuca* germlings (Chen et al., 2017).

Young and Gobler (2016) have reported that *Gracilaria* and *Ulva's* growth rates were significantly boosted by an average of about 70% and 30%, respectively, beyond the controlled treatment when exposed to the raised levels of pCO_2. Moreover, *Gracilaria* and *Ulva* show a physiological shift from near-exclusive use of HCO_3^- to mainly CO_2 use when subjected to the elevated level of pCO_2 via detecting $\delta^{13}C$ isotopes. This shift in carbon dependence coupled with growth rate increased in response to increased pCO_2, proposed that these seaweeds' photosynthesis depended on their inorganic carbon source. Young and Gobler, (2016a) and Semesi et al., (2009) demonstrated that increasing dissolved CO_2 concentration to a specific level of ~ 26 mmol kg^{-1} caused significant enhancement in the photosynthetic rates of *Hydrolithon* sp. by 13%. Furthermore, the negative impact of the higher pCO_2 on growth of *U. fasciata* in the present study may contribute to the lowered pH. Elevated pCO_2 reduces pH at the surface of the cell, which could modify both extracellular and intracellular acid–base balance (Flynn et al., 2012). Intracellular metabolic activities including the photosynthesis and development, may be impaired by the disrupted homeostasis (Xu et al., 2017). A reduced growth rate caused by the decreased pH was also found in diverse seaweeds such as *Pyropia yezoensis* (Gao et al., 2019), *P. haitanensis* (Xu et al., 2017) and *U. lactuca* (Olischläger et al., 2013).

The marine macroalga *Ulva sp.* is an apt candidate raw biomass with a high growth rate and high protein, lipid, carbohydrate yield suitable for food application (Khairy & El–Shafay, 2013; Kazir et al., 2019). Recent study has amply established that the pH-shift of *Ulva sp.* improves the nutrient contents and gives a superior grade of food applications (Barakat et al., 2021)

CONCLUSION

The chapter aims to briefly summarize our present knowledge regarding algal adaptation, and acclimatization process towards stable and unstable environmental conditions such as biotic interactions of algae, abiotic factors (temperature, light (UV irradiation), hydrodyanism salinity, desiccation and ocean acidification (pH). The present study clearly indicates that, the optimum environmental conditions are required for the enhanced growth, development and physiological activity of algae. The conditions required are the optimum temperature pH, salinity and light intensity and biotic interaction of algal species. There are some algal species that showed better growth under all biotic and abiotic and factors compared to some other that do not show growth. During the last decade, these processes were extensively studied because of the change in the climatic and anthropogenic activities in the ecosystem. The members of algae are often used as the model organisms for such investigations owing to their experimental simplicity. These factors are responsible to bring about the change in the physiology of algal species. Following changes were observed e.g. changes in the rate of physiological and biochemical processes (reduction of photosynthesis rate, respiration rate etc.), changes in the biomembrane structure/composition and consequently brings change in their fluidity (extremes of low and high temperature, desiccation, salt and pH, etc.). Both the acclimatization and adaptation of algae may eradicate the risk of extinction for algal species in a changing environment.

ACKNOWLEDGMENT

This study was supported by UGC–BSR Start–Up–Grand (Ref. No. F.30-373/2017 (BSR)) New Delhi for financial and instrumental assistance in connection with this research works.

REFERENCES

Adl, S. M., Simpson, A. G. B., Lane, C. E., Lukeš, J., Bass, D., Bowser, S. S., ... Spiegel, F. W. (2013). Erratum: The revised classification of eukaryotes. *The Journal of Eukaryotic Microbiology*, *60*(3).

Ahmad, I., & Hellebust, J. A. (1984). Osmoregulation in the extremely euryhaline marine micro-alga Chlorella autotrophica. *Plant Physiology*, *74*(4), 1010–1015. doi:10.1104/pp.74.4.1010 PMID:16663495

Aires, T., Serebryakova, A., Viard, F., Serrão, E. A., & Engelen, A. H. (2018). Acidification increases abundances of Vibrionales and Planctomycetia associated to a seaweed-grazer system: Potential consequences for disease and prey digestion efficiency. *PeerJ*, *6*, e4377. doi:10.7717/peerj.4377 PMID:29610702

Aiyar, P., Schaeme, D., García-Altares, M., Carrasco Flores, D., Dathe, H., Hertweck, C., Sasso, S., & Mittag, M. (2017). Antagonistic bacteria disrupt calcium homeostasis and immobilize algal cells. *Nature Communications*, *8*(1), 1–13. doi:10.103841467-017-01547-8 PMID:29170415

Aiyar, P., Schaeme, D., Garcia-Altares, M., Flores, D. C., Dathe, H., Hertweck, C., Sasso, S., & Mittag, M. (2017). Antagonistic bacteria disrupt calcium homeostasis and immobilize algal cells. *Nature Communications*, *8*(1), 1756. doi:10.103841467-017-01547-8 PMID:29170415

Allakhverdiev, S. I., Kreslavski, V. D., Klimov, V. V., Los, D. A., Carpentier, R., & Mohanty, P. (2008). Heat stress: An overview of molecular responses in photosynthesis. *Photosynthesis Research*, *98*(1), 541–550. doi:10.100711120-008-9331-0 PMID:18649006

Amin, S. A., Green, D. H., Hart, M. C., Küpper, F. C., Sunda, W. G., & Carrano, C. J. (2009). Photolysis of iron–siderophore chelates promotes bacterial–algal mutualism. *Proceedings of the National Academy of Sciences of the United States of America*, *106*(40), 17071–17076. doi:10.1073/pnas.0905512106 PMID:19805106

Amin, S. A., Hmelo, L. R., Van Tol, H. M., Durham, B. P., Carlson, L. T., Heal, K. R., Morales, R. L., Berthiaume, C. T., Parker, M. S., Djunaedi, B., Ingalls, A. E., Parsek, M. R., Moran, M. A., & Armbrust, E. V. (2015). Interaction and signalling between a cosmopolitan phytoplankton and associated bacteria. *Nature*, *522*(7554), 98–101. doi:10.1038/nature14488 PMID:26017307

Anning, T., Harris, G., & Geider, R. J. (2001). Thermal acclimation in the marine diatom *Chaetoceros calcitrans* (Bacillariophyceae). *European Journal of Phycology*, *36*(3), 233–241. doi:10.1080/09670260110001735388

Armbrust, E. V., Berges, J. A., Bowler, C., Green, B. R., Martinez, D., Putnam, N. H., Zhou, S., Allen, A. E., Apt, K. E., Bechner, M., Brzezinski, M. A., Chaal, B. K., Chiovitti, A., Davis, A. K., Demarest, M. S., Detter, J. C., Glavina, T., Goodstein, D., Hadi, M. Z., ... Rokhsar, D. S. (2004). The genome of the diatom Thalassiosira pseudonana: Ecology, evolution, and metabolism. *Science*, *306*(5693), 79–86. doi:10.1126cience.1101156 PMID:15459382

Barakat, K. M., El–Sayed, H. S., Khairy, H. M., El–Sheikh, M. A., Al–Rashed, S. A., Arif, I. A., & Elshobary, M. E. (2021). Effects of ocean acidification on the growth and biochemical composition of a green alga (*Ulva fasciata*) and its associated microbiota. *Saudi Journal of Biological Sciences*, *28*(9), 5106–5114. doi:10.1016/j.sjbs.2021.05.029 PMID:34466088

Behringer, G., Ochsenkühn, M. A., Fei, C., Fanning, J., Koester, J. A., & Amin, S. A. (2018). Bacterial communities of diatoms display strong conservation across strains and time. *Frontiers in Microbiology*, *9*, 659. doi:10.3389/fmicb.2018.00659 PMID:29681892

Bell, R. A. (1993). Cryptoendolithic algae of hot semiarid lands and deserts. *Journal of Phycology*, *29*(2), 133–139. doi:10.1111/j.0022-3646.1993.00133.x

Bell, W., & Mitchell, R. (1972). Chemotactic and growth responses of marine bacteria to algal extracellular products. *The Biological Bulletin*, *143*(2), 265–277. doi:10.2307/1540052

Belnap, J., & Lange, O. L. (2003). Biological Soil Crusts: Structure, Function, and Management. In I. T. Baldwin, M. M. Caldwell, G. Heldmaier, O. L. Lange, H. A. Mooney, E.-D. Schulze, & U. Sommer (Eds.), *Ecological Studies Series 150* (Vol. 150, p. 503). Springer- Verlag. doi:10.1007/978-3-642-56475-8

Berner, T., Dubinsky, Z., Wyman, K., & Falkowski, P. G. (1989). Photoadaptation and the "package" effect in Dunaliella tertiolecta (Chlorophyceae) 1. *Journal of Phycology*, *25*(1), 70–78. doi:10.1111/j.0022-3646.1989.00070.x

Bhattacharya, D., Yoon, H. S., & Hackett, J. D. (2004). Photosynthetic eukaryotes unite: Endosymbiosis connects the dots. *BioEssays*, *26*(1), 50–60. doi:10.1002/bies.10376 PMID:14696040

Biggs, B. J. F., & Thomsen, H. A. (1995). Disturbance of stream periphyton by perturbations in shear stress: Time to structural failure and differences in community resistance. *Journal of Phycology*, *31*(2), 233–241. doi:10.1111/j.0022-3646.1995.00233.x

Bischof, K., Hanelt, D., Tug, H., Karsten, U., Brower, P. E. M., & Wiencke, C. (1998). Acclimation of brown algal photosynthesis to ultraviolet radiation in Arctic coastal waters. (Spitsbergen, Norway). *Polar Biology*, *20*(6), 388–395. doi:10.1007003000050319

Bischof, K., Krabs, G., Wiencke, C., & Hanelt, D. (2002). Solar ultraviolet radiation affects the activity of ribulose-1, 5-bisphosphate carboxylase-oxygenase and the composition of photosynthetic and xanthophyll cycle pigments in the intertidal green alga *Ulva lactucaL. Planta*, *215*(3), 502–509. doi:10.100700425-002-0774-9 PMID:12111234

Bischof, K., & Peralta, G. (2002). Effects of solar UV-B radiation on canopy structure of *Ulva* communities from southern Spain. *Journal of Experimental Botany*, *53*(379), 2411–2421. doi:10.1093/jxb/erf091 PMID:12432033

Boller, M. L., & Carrington, E. (2006). The hydrodynamic effects of shape and size change during reconfiguration of a flexible macroalga. *The Journal of Experimental Biology*, *209*(10), 1894–1903. doi:10.1242/jeb.02225 PMID:16651555

Brenner, K., You, L., & Arnold, F. H. (2008). Engineering microbial consortia: A new frontier in synthetic biology. *Trends in Biotechnology*, *26*(9), 483–489. doi:10.1016/j.tibtech.2008.05.004 PMID:18675483

Brito, J. C., Godinho, R., Martínez-Freiría, F., Pleguezuelos, J. M., Rebelo, H., Santos, X., Vale, C. G., Velo-Antón, G., Boratyński, Z., Carvalho, S. B., Ferreira, S., Gonçalves, D. V., Silva, T. L., Tarroso, P., Campos, J. C., Leite, J. V., Nogueira, J., Álvares, F., Sillero, N., ... Carranza, S. (2014). Unravelling biodiversity, evolution and threats to conservation in the Sahara-Sahel. *Biological Reviews of the Cambridge Philosophical Society*, *89*(1), 215–231. doi:10.1111/brv.12049 PMID:23848599

Brody, M., & Vatter, A. E. (1959). Observations on cellular structures of Porphyridium cruentum. *The Journal of Biophysical and Biochemical Cytology*, *5*(2), 289–294. doi:10.1083/jcb.5.2.289 PMID:13654450

Büdel, B. (2005). Microorganisms of biological crusts on soil surfaces. In *Microorganisms in soils: roles in genesis and functions* (pp. 307–323). Springer. doi:10.1007/3-540-26609-7_15

Buschmann, A. H. (2016). *Seaweed Ecology and Physiology*. Cambridge University Press.

Caldeira, K., & Wickett, M. E. (2003). Anthropogenic carbon and ocean pH. *Nature*, *425*(6956), 365–365. doi:10.1038/425365a PMID:14508477

Callieri, C., Morabito, G., Huot, Y., Neale, P. J., & Litchman, E. (2001). Photosynthetic response of pico- and nanoplanktonic algae to UVB, UVA and PAR in a high mountain lake. *Aquatic Sciences*, *63*(3), 286–293. doi:10.1007/PL00001355

Cardon, Z. G., Gray, D. W., & Lewis, L. A. (2008). The green algal underground: Evolutionary secrets of desert cells. *Bioscience*, *58*(2), 114–122. doi:10.1641/B580206

Cavalier-Smith, T. (1982). The origins of plastids. *Biological Journal of the Linnean Society. Linnean Society of London*, *17*(3), 289–306. doi:10.1111/j.1095-8312.1982.tb02023.x

Cenci, U., Bhattacharya, D., Weber, A. P., Colleoni, C., Subtil, A., & Ball, S. G. (2017). Biotic host–pathogen interactions as major drivers of plastid endosymbiosis. *Trends in Plant Science*, *22*(4), 316–328. doi:10.1016/j.tplants.2016.12.007 PMID:28089380

Chandra, R., Das, P., Vishal, G., & Nagra, S. (2019). Factors affecting the induction of UV protectant and lipid productivity in *Lyngbya* for sequential biorefinery product recovery. *Bioresource Technology*, *278*, 303–310. doi:10.1016/j.biortech.2019.01.084 PMID:30708334

Chen, B., Lin, L., Ma, Z., Zhang, T., Chen, W., & Zou, D. (2019). Carbon and nitrogen accumulation and interspecific competition in two algae species, *Pyropia haitanensis and Ulva lactuca*, under ocean acidification conditions. *Aquaculture International*, *27*(3), 721–733. doi:10.100710499-019-00360-y

Clendennen, S. K., Zimmerman, R. C., Powers, D. A., & Alberte, R. S. (1996). Photosynthetic responses of the giant kelp Macrocystis pyrifera. Phaeophyceae to ultraviolet radiation. *Journal of Phycology*, *32*(4), 614–620. doi:10.1111/j.0022-3646.1996.00614.x

Conde, F. R., Churio, M. S., & Previtali, C. M. (2007). Experimental study of the excited-state properties and photostability of the mycosporine-like amino acid palythine in aqueous solution. *Photochemical & Photobiological Sciences*, *6*(6), 669–674. doi:10.1039/b618314j PMID:17549269

Cornwall, C. E., & Hurd, C. L. (2019). Variability in the benefits of ocean acidification to photosynthetic rates of macroalgae without CO_2 concentrating mechanisms. *Marine and Freshwater Research*, *71*(3), 275–280. doi:10.1071/MF19134

Craigie, J. S., Morris, E. R., Rees, D. A., & Thom, D. (1984). Alginate block structure in Phaeophyceae from Nova Scotia: Variation with species, environment and tissue-type. *Carbohydrate Polymers*, *4*(4), 237–252. doi:10.1016/0144-8617(84)90001-8

Croft, M. T., Lawrence, A. D., Raux-Deery, E., Warren, M. J., & Smith, A. G. (2005). Algae acquire vitamin B12 through a symbiotic relationship with bacteria. *Nature*, *438*(7064), 90–93. doi:10.1038/nature04056 PMID:16267554

Cuhel, R. L., Ortner, P. B., & Lean, D. R. (1984). Night synthesis of protein by algae 1. *Limnology and Oceanography*, *29*(4), 731–744. doi:10.4319/lo.1984.29.4.0731

Davison, I. R., & Pearson, G. A. (1996). Stress tolerance in intertidal seaweeds. *Journal of Phycology*, *32*(2), 197–211. doi:10.1111/j.0022-3646.1996.00197.x

Demetriou, G., Neonaki, C., Navakoudis, E., & Kotzabasis, K. (2007). Salt stress impact on the molecular structure and function of the photosynthetic apparatus—The protective role of polyamines. *Biochimica et Biophysica Acta (BBA)- Bioenergetics*, *1767*(4), 272–280. doi:10.1016/j.bbabio.2007.02.020 PMID:17408588

Després, V. R., Huffman, J. A., Burrows, S. M., Hoose, C., Safatov, A. S., Buryak, G., Fröhlich-Nowoisky, J., Elbert, W., Andreae, M. O., Pöschl, U., & Jaenicke, R. (2012). Primary biological aerosol particles in the atmosphere: A review. *Tellus. Series B, Chemical and Physical Meteorology*, *64*(1), 15598–15656. doi:10.3402/tellusb.v64i0.15598

Dring, M. J., Wagner, A., Boeskov, J., & Luning, K. (1996b). Sensitivity of intertidal and subtidal red algae to UVA and UVB radiation, as monitored by chlorophyll. uorescence measurements: Influence of collection depth and season, and length of irradiation. *European Journal of Phycology*, *31*(4), 293–302. doi:10.1080/09670269600651511

Dring, M. J., Makarov, V., Schoschina, E., Lorenz, M., & Luning, K. (1996a). Influence of ultraviolet–radiation on chlorophyll fluorescence and growth in different life-history stages of three species of Laminaria, Phaeophyta. *Marine Biology*, *126*, 183–191. doi:10.1007/BF00347443

Du, Z. Y., Zienkiewicz, K., Pol, N. V., Ostrom, N. E., Benning, C., & Bonito, G. M. (2019). Algal-fungal symbiosis leads to photosynthetic mycelium. *eLife*, *8*, e47815. doi:10.7554/eLife.47815 PMID:31307571

Dubinsky, Z., Matsukawa, R., & Karube, I. (1995). Photobiological aspects of algal mass culture. *Journal of Marine Biotechnology*, *2*, 61–65.

Durham, B. P., Dearth, S. P., Sharma, S., Amin, S. A., Smith, C. B., Campagna, S. R., Armbrust, E. V., & Moran, M. A. (2017). Recognition cascade and metabolite transfer in a marine bacteria-phytoplankton model system. *Environmental Microbiology*, *19*(9), 3500–3513. doi:10.1111/1462-2920.13834 PMID:28631440

Durham, B. P., Sharma, S., Luo, H., Smith, C. B., Amin, S. A., Bender, S. J., Dearth, S. P., Van Mooy, B. A., Campagna, S. R., Kujawinski, E. B., Armburst, E. V., & Moran, M. A. (2015). Cryptic carbon and sulfur cycling between surface ocean plankton. *Proceedings of the National Academy of Sciences of the United States of America*, *112*(2), 453–457. doi:10.1073/pnas.1413137112 PMID:25548163

Eppley, R. W. (1972). Temperature and phytoplankton growth in the sea. *Fish Bulletin*, *70*, 1063–1085.

Eswaran, K., Mairh, O. P., & Subba Rao, P. V. (2002). Inhibition of pigments and phycocolloid in a marine red alga Gracilaria edulis by ultraviolet–B radiation. *Biologia.Plantarum. Prague*, *45*(1), 157–159. doi:10.1023/A:1015158204181

Evans, J. H. (1958). The survival of freshwater algae during dry periods: Part I. An investigation of the algae of five small ponds. *Journal of Ecology*, *46*(1), 149–167. doi:10.2307/2256910

Fábregas, J., Maseda, A., Domínguez, A., & Otero, A. (2004). The cell composition of Nannochloropsis sp. changes under different irradiances in semicontinuous culture. *World Journal of Microbiology & Biotechnology*, *20*(1), 31–35. doi:10.1023/B:WIBI.0000013288.67536.ed

Falkowski, P. G., & Raven, J. A. (2013). *Aquatic photosynthesis*. Princeton University Press.

FAO. (2014). *The state of world fisheries and aquaculture*. FAO.

Fernández, P. A., Roleda, M. Y., & Hurd, C. L. (2015). Effects of ocean acidification on the photosynthetic performance, carbonic anhydrase activity and growth of the giant kelp Macrocystis pyrifera. *Photosynthesis Research*, *124*(3), 293–304. doi:10.100711120-015-0138-5 PMID:25869634

Figueroa, F. L., Salles, S., Aguilera, J., Jimenez, C., Mercado, J., Vin. Egla, B., Flores-Moya, A., & Altamirano, M. (1997). Effects of solar radiation on photoinhibition and pigmentation in the red alga Porphyra leucostica. *Marine Ecology Progress Series*, *151*, 81–90. doi:10.3354/meps151081

Flechtner, V. R. (2007). North American desert microbiotic soil crust communities. In *Algae and cyanobacteria in extreme environments* (pp. 537–551). Springer. doi:10.1007/978-1-4020-6112-7_29

Flynn, K. J., Blackford, J. C., Baird, M. E., Raven, J. A., Clark, D. R., Beardall, J., Brownlee, C., Fabian, H., & Wheeler, G. L. (2012). Changes in pH at the exterior surface of plankton with ocean acidification. *Nature Climate Change*, *2*(7), 510–513. doi:10.1038/nclimate1489

Franklin, L. A., & Forster, R. M. (1997). The changing irradiance environment: Consequences for marine macrophyte physiology, productivity and ecology. *European Journal of Phycology*, *32*(3), 207–232. doi:10.1080/09670269710001737149

Friedl, T. (1997). The evolution of the green algae. In *Origins of Algae and their Plastids* (pp. 87–101). Springer. doi:10.1007/978-3-7091-6542-3_4

Friedl, T., & Rybalka, N. (2012). Systematics of the green algae: A brief introduction to the current status. *Progress in Botany*, *73*, 259–280. doi:10.1007/978-3-642-22746-2_10

Friedmann, I., Lipkin, Y., & Ocampo-Paus, R. (1967). Desert algae of the Negev (Israel). *Phycologia*, *6*(4), 185–200. doi:10.2216/i0031-8884-6-4-185.1

Fritsch, F. E., & Haines, F. M. (1923). The moisture-relations of terrestrial algae. II. The changes during exposure to drought and treatment with hypertonie solutions. *Annals of Botany*, *37*(148), 683–728. doi:10.1093/oxfordjournals.aob.a089874

Fritsen, C. H., & Priscu, J. C. (1998). Cyanobacterial assemblages in permanent ice covers on Antarctic lakes: Distribution, growth rate, and temperature response of photosynthesis. *Journal of Phycology*, *34*(4), 587–597. doi:10.1046/j.1529-8817.1998.340587.x

Fröhlich-Nowoisky, J., Kampf, C. J., Weber, B., Huffman, J. A., Pöhlker, C., Andreae, M. O., Lang-Yona, N., Burrows, S. M., Gunthe, S. S., Elbert, W., Su, H., Hoor, P., Thines, E., Hoffmann, T., Després, V. R., & Pöschla, U. (2016). Bioaerosols in the earth system: Climate, health, and ecosystem interactions. *Atmospheric Research*, *182*(15), 346–376. doi:10.1016/j.atmosres.2016.07.018

Fuentes-Tristan, S., Parra-Saldivar, R., & Iqbal, H. M. N. (2019). Bioinspired biomolecules: Mycosporine-like amino acids and scytonemin from Lyngbya sp. with UV-protection potentialities. *Journal of Photochemistry and Photobiology. B, Biology*.

Fujita, R. M., Wheeler, P. A., & Edwards, R. L. (1989). Assessment of macroalgal nitrogen limitation in a seasonal upwelling region. *Marine Ecology Progress Series*, *53*, 53. doi:10.3354/meps053293

Fuller, S. W., & Mathieson, A. C. (1972). Ecological studies of economic red algae. IV. Variations of carrageenan concentration and properties in *Chondrus crispus* Stackhouse. *Journal of Experimental Marine Biology and Ecology*, *10*(1), 49–58. doi:10.1016/0022-0981(72)90092-5

Gao, G., Gao, Q., Bao, M., Xu, J., & Li, X. (2019a). Nitrogen availability modulates the Gebser, B., & Pohnert, G. (2013).Synchronized regulation of different zwitterionic metabolites in theosmoadaption of phytoplankton. *Marine Drugs*, *11*, 2168–2182.

Gerard, V. A. (1982). In situ water motion and nutrient uptake by the giant kelp Macrocystis pyrifera. *Marine Biology*, *69*(1), 51–54. doi:10.1007/BF00396960

Gordillo, F. J., Goutx, M., Figueroa, F. L., & Niell, F. X. (1998). Effects of light intensity, CO2 and nitrogen supply on lipid class composition of Dunaliella viridis. *Journal of Applied Phycology*, *10*(2), 135–144. doi:10.1023/A:1008067022973

Gordon, R., & Brawley, S. H. (2004). Effects of water motion on propagule release from algae with complex life histories. *Marine Biology*, *145*(1), 21–29. doi:10.100700227-004-1305-y

Gould, S. B., Waller, R. F., & McFadden, G. I. (2008). Plastid evolution. *Annual Review of Plant Biology*, *59*(1), 491–517. doi:10.1146/annurev.arplant.59.032607.092915 PMID:18315522

Gray, D. W., Lewis, L. A., & Cardon, Z. G. (2007). Photosynthetic recovery following desiccation of desert green algae (Chlorophyta) and their aquatic relatives. *Plant, Cell & Environment*, *30*(10), 1240–1255. doi:10.1111/j.1365-3040.2007.01704.x PMID:17727415

Gröniger, A., Sinha, R. P., Klisch, M., & Häder, D. P. (2000). Photoprotective compounds in cyanobacteria, phytoplankton and macroalgae—A database. *Journal of Photochemistry and Photobiology. B, Biology*, *58*(2-3), 115–122. doi:10.1016/S1011-1344(00)00112-3 PMID:11233638

Gross, E. M. (2003). Allelopathy of aquatic autotrophs. *Critical Reviews in Plant Sciences*, *22*(3-4), 313–339. doi:10.1080/713610859

Grosser, K., Zedler, L., Schmitt, M., Dietzek, B., Popp, J., & Pohnert, G. (2012). Disruptionfree imaging by Raman spectroscopy reveals a chemical sphere with antifouling metabolites around macroalgae. *Biofouling*, *28*(7), 687–696. doi:10.1080/08927014.2012.700306 PMID:22783962

Gutow, L., Rahman, M. M., Bartl, K., Saborowski, R., Bartsch, I., & Wiencke, C. (2014). Ocean acidification affects growth but not nutritional quality of the seaweed *Fucus vesiculosus* (Phaeophyceae, Fucales). *Journal of Experimental Marine Biology and Ecology*, *453*, 84–90. doi:10.1016/j.jembe.2014.01.005

Hader, D. P., Kumar, H. D., Smith, R. C., & Worrest, R. C. (1998). Effects on aquatic ecosystems. *Journal of Photochemistry and Photobiology. B, Biology*, *46*(1–3), 53–68. doi:10.1016/S1011-1344(98)00185-7

Han, T. (1996a). Effect of ultraviolet-B radiation on Ulva perusa Kjellman, Chlorophyta. I. Growth and pigment content. *Algae. Kor. J. Phycol.*, *11*, 155–159.

Hanelt, D., Melchersmann, B., Wiencke, C., & Nultsch, W. (1997a). Effects of high light stress on photosynthesis of polar macroalgae in relation to depth distribution. *Marine Ecology Progress Series*, *149*, 255–266. doi:10.3354/meps149255

Hema, R., Senthil-Kumar, M., Shivakumar, S., Chandrasekhara Reddy, P., & Udayakumar, M. (2007). Chlamydomonas reinhardtii, a model system for functional validation of abiotic stress responsive genes. *Planta*, *226*(3), 655–670. doi:10.100700425-007-0514-2 PMID:17431668

Hiremath, S., & Mathad, P. (2010). Impact of Salinity on the Physiological and Biochemical Traits of Chlorella vulgaris Beijerinck. *Journal of Algal Biomass Utilization*, *1*(2), 51–59.

Hofmann, L. C., Straub, S., & Bischof, K. (2013). Elevated CO2 levels affect the activity of nitrate reductase and carbonic anhydrase in the calcifying rhodophyte *Corallina officinalis*. *Journal of Experimental Botany*, *64*(4), 899–908. doi:10.1093/jxb/ers369 PMID:23314813

Holzinger, A., & Karsten, U. (2013). Desiccation stress and tolerance in green algae: Consequences for ultrastructure, physiological and molecular mechanisms. *Frontiers in Plant Science*, *4*, 327. doi:10.3389/fpls.2013.00327 PMID:23986769

Hom, E. F., & Murray, A. W. (2014). Niche engineering demonstrates a latent capacity for fungal-algal mutualism. *Science*, *345*(6192), 94–98. doi:10.1126cience.1253320 PMID:24994654

Hong, Z., Lakkineni, K., Zhang, Z., & Verma, D. P. S. (2000). Removal of feedback inhibition of A1 – pyrroline - 5- carboxylate synthetase results in increased proline accumulation and protection of plants from osmotic stress. *Plant Physiology*, *122*(4), 1129–1136. doi:10.1104/pp.122.4.1129 PMID:10759508

Hopes, A., Nekrasov, V., Kamoun, S., & Mock, T. (2016). Editing of the urease gene by CRISPR-Cas in the diatom Thalassiosira pseudonana. *Plant Methods*, *12*(1), 49. doi:10.118613007-016-0148-0 PMID:27904648

Hornak, K., Kasalicky, V., Simek, K., & Grossart, H. P. (2017). Strain-specific consumption and transformation of alga-derived dissolved organic matter by members of the Limnohabitans-C and Polynucleobacter-B clusters of Betaproteobacteria. *Environmental Microbiology*, *19*(11), 4519–4535. doi:10.1111/1462-2920.13900 PMID:28856804

Hurd, C. L. (2000). Water motion, marine macroalgal physiology, and production. *Journal of Phycology*, *36*(3), 453–472. doi:10.1046/j.1529-8817.2000.99139.x PMID:29544017

Hurd, C. L., Harrison, P. J., Bischof, K., & Lobban, C. S. (2014). *Seaweed ecology and physiology*. Cambridge University Press. doi:10.1017/CBO9781139192637

Ianora, A., Bentley, M. G., Caldwell, G. S., Casotti, R., Cembella, A. D., Engström-Öst, J., Halsband, C., Sonnenschein, E., Legrand, C., Llewellyn, C. A., Paldavičienë, A., Pilkaityte, R., Pohnert, G., Razinkovas, A., Romano, G., Tillmann, U., & Vaiciute, D. (2011). The relevance of marine chemical ecology to plankton and ecosystem function: An emerging field. *Marine Drugs*, *9*(9), 1625–1648. doi:10.3390/md9091625 PMID:22131962

Iqbal, M., & Zafar, S. I. (1993). Effects of photon flux density, CO2, aeration rate, and inoculum density on growth and extracellular polysaccharide production by Porphyridium cruentum. *Folia Microbiologica*, *38*(6), 509–514. doi:10.1007/BF02814405

Jacob, A., Wiencke, C., Lehmann, H., & Kirst, G. O. (1992). *Physiology and ultrastructure of desiccation in the green alga Prasiola crispa from Antarctica*. Academic Press.

Johnson, W. M., Kido Soule, M. C., & Kujawinski, E. B. (2016). Evidence for quorum sensing and differential metabolite production by a marine bacterium in response to DMSP. *The ISME Journal*, *10*(9), 2304–2316. doi:10.1038/ismej.2016.6 PMID:26882264

Kaçka, A., & Dönmez, G. (2008). Isolation of Dunaliella spp. from a hypersaline lake and their ability to accumulate glycerol. *Bioresource Technology*, *99*(17), 8348–8352. doi:10.1016/j.biortech.2008.02.042 PMID:18406610

Karsten, U., Franklin, L. A., Luning, K., & Wiencke, C. (1998a). Natural ultraviolet radiation and photosynthetic active radiation induce formation of mycosporine-like amino acids in the marine macroalga *Chondrus crispus* Rhodophyta. *Planta*, *205*(2), 257–262. doi:10.1007004250050319

Karsten, U., & Holzinger, A. (2012). Light, temperature, and desiccation effects on photosynthetic activity, and drought-induced ultrastructural changes in the green alga Klebsormidium dissectum (Streptophyta) from a high alpine soil crust. *Microbial Ecology*, *63*(1), 51–63. doi:10.100700248-011-9924-6 PMID:21811791

Karsten, U., Lütz, C., & Holzinger, A. (2010). Ecophysiological performance of the aeroterrestrial green alga Klebsormidium crenulatum (Charophyceae, Streptophyta) isolated from an alpine soil crust with an emphasis on desiccation stress 1. *Journal of Phycology*, *46*(6), 1187–1197. doi:10.1111/j.1529-8817.2010.00921.x

Karsten, U., Sawall, T., Hanelt, D., Bischof, K., Figueroa, F. L., Flores Moya, A., & Wiencke, C. (1998b). An inventory of UV-absorbing mycosporine-like amino acids in macroalgae from polar to warm-temperate regions. *Botanica Marina*, *41*(1-6), 443–453. doi:10.1515/botm.1998.41.1-6.443

Karsten, U., Schumann, R., & Mostaert, A. (2007). Aeroterrestrial algae growing on man-made surfaces. In *Algae and cyanobacteria in extreme environments* (pp. 583–597). Springer. doi:10.1007/978-1-4020-6112-7_32

Karsten, U., & Wiencke, C. (1999). Factors controlling the formation of UVabsorbing mycosporine-like amino acids in the marine red alga *Palmaria palmate* from Spitzbergen (Norway). *Journal of Plant Physiology*, *155*(3), 407–415. doi:10.1016/S0176-1617(99)80124-2

Kazir, M., Abuhassira, Y., Robin, A., Nahor, O., Luo, J., Israel, A., Golberg, A., & Livney, Y. D. (2019). Extraction of proteins from two marine macroalgae, Ulva sp. and Gracilaria sp., for food application, and evaluating digestibility, amino acid composition and antioxidant properties of the protein concentrates. *Food Hydrocolloids*, *87*, 194–203. doi:10.1016/j.foodhyd.2018.07.047

Khairy, H. M., & El-Shafay, S. M. (2013). Seasonal variations in the biochemical composition of some common seaweed species from the coast of Abu Qir Bay, Alexandria, Egypt. *Oceanologia*, *55*(2), 435–452. doi:10.5697/oc.55-2.435

Khona, D. K., Shirolikar, S. M., Gawde, K. K., Hom, E., Deodhar, M. A., & D'Souza, J. S. (2016). Characterization of salt stress-induced palmelloids in the green alga, Chlamydomonas reinhardtii. *Algal Research*, *16*, 434–448. doi:10.1016/j.algal.2016.03.035

Kirst, G. O. (1990). Salinity tolerance of eukaryotic marine algae. *Annual Review of Plant Biology*, *41*(1), 21–53. doi:10.1146/annurev.pp.41.060190.000321

Kitaya, Y., Xiao, L., Masuda, A., Ozawa, T., Tsuda, M., & Omasa, K. (2008).Effects of temperature, photosynthetic photon flux density, photoperiod and O2 and CO2 concentrations on growth rates of the symbiotic dinoflagellate, Amphidinium sp. In *Nineteenth International Seaweed Symposium* (pp. 287–292). Springer. 10.100710811-008-9331-7

Knowles, E. J., & Castenholz, R. W. (2008). Effect of exogenous extracellular polysaccharides on the desiccation and freezing tolerance of rock-inhabiting phototrophic microorganisms. *FEMS Microbiology Ecology*, *66*(2), 261–270. doi:10.1111/j.1574-6941.2008.00568.x PMID:18710394

Koehl, M. A. R. (1986). Seaweeds in moving water: form and mechanical function. In T. J. Givnish (Ed.), *On the economy of plant form and function* (pp. 603–634). Cambridge University Press.

Korpinen, S., Honkanen, T., Vesakoski, O., Hemmi, A., Koivikko, R., Loponen, J., & Jormalainen, V. (2007). Macroalgal communities face the challenge of changing biotic interactions: Review with focus on the Baltic Sea. *Ambio*, *36*(2/3), 203–211. doi:10.1579/0044-7447(2007)36[203:MCFTCO]2.0.CO;2 PMID:17520935

Kraemer, G. P., & Chapman, D. J. (1991). biomechanics and alginic acid composition during hydrodynamic adaptation by egregia menziesii (phaeophyta) juveniles. *Journal of Phycology*, *27*(1), 47–53. doi:10.1111/j.0022-3646.1991.00047.x

Krienitz, L. (1996). Syllabus der Boden-Luft-und Flechtenalge. Gustav Fischer Verlag.

Lawrence, K. P., Long, P. F., & Young, A. R. (2018). Mycosporine-like amino acids for skin photoprotection. *Current Medicinal Chemistry*, *25*(40), 5512–5527. doi:10.2174/0929867324666170529124237 PMID:28554325

Le Lann, K., Rumin, J., Cérantola, S., Culioli, G., & Stiger-Pouvreau, V. (2014). Spatiotemporal variations of diterpene production in the brown macroalga *Bifurcaria bifurcata* from the western coasts of Brittany (France). *Journal of Applied Phycology*, *26*(2), 1207–1214. doi:10.100710811-013-0148-7

Lewis, L. A. (2007). *Chlorophytaon land: independent lineages of green eukaryotes from arid lands.* In J. Seckbach (Ed.), *Algae and Cyanobacteria in extreme Environments* (pp. 571–582). Springer.

Li, W. K. W. (1980). Temperature adaptation in phytoplankton: cellular and photosynthetic characteristics. In P. G. Falkowski (Ed.), *Primary productivity in the sea* (pp. 259–279). Plenum Press. doi:10.1007/978-1-4684-3890-1_15

Lovejoy, C., Vincent, W. F., Bonilla, S., Roy, S., Martineau, M. J., Terrado, R., Potvin, M., Massana, R., & Pedrós-Alió, C. (2007). distribution, phylogeny, and growth of cold-adapted picoprasinophytes in arctic seas. *Journal of Phycology*, *43*(1), 78–89. doi:10.1111/j.1529-8817.2006.00310.x

Luning, K., & Dring, M. J. (1979). Continuous underwater light measurement near Helgoland.North Sea and its significance for characteristic light limits in the sublittoral region. *Helgoländer Wissenschaftliche Meeresuntersuchungen*, *32*(4), 403–424. doi:10.1007/BF02277985

Lüning, K., & Tom Dieck, I. (1989). *Environmental triggers in algal seasonality.* Academic Press.

Lutzoni, F., Nowak, M. D., Alfaro, M. E., Reeb, V., Miadlikowska, J., Krug, M., Arnold, A. E., Lewis, L. A., Swofford, D. L., Hibbett, D., Hilu, K., James, T. Y., Quandt, D., & Magallón, S. (2018). Contemporaneous radiations of fungi and plants linked to symbiosis. *Nature Communications*, *9*(1), 5451. doi:10.103841467-018-07849-9 PMID:30575731

Mackey, K. R., Morris, J. J., Morel, F. M., & Kranz, S. A. (2015). Response of photosynthesis to ocean acidification. *Oceanography (Washington, D.C.)*, *28*(2), 74–91. doi:10.5670/oceanog.2015.33

Manríquez-Hernández, J. A., Duston, J., & Garbary, D. J. (2016). Effect of irradiance on bioremediation capacity of the red algae *Chondrus crispus* and *Palmaria palmata*. *Aquaculture International*, *24*(1), 39–55. doi:10.100710499-015-9907-6

Meyer, N., Bigalke, A., Kaulfuss, A., & Pohnert, G. (2017). Strategies and ecological roles of algicidal bacteria. *FEMS Microbiology Reviews*, *41*(6), 880–899. doi:10.1093/femsre/fux029 PMID:28961821

Mock, T., & Kroon, B. M. (2002). Photosynthetic energy conversion under extreme conditions—II: The significance of lipids under light limited growth in Antarctic sea ice diatoms. *Phytochemistry*, *61*(1), 53–60. doi:10.1016/S0031-9422(02)00215-7 PMID:12165302

Mollenhauer, D., Mollenhauer, R., & Kluge, M. (1996). Studies on initiation and development of the partner association in *Geosiphon pyriforme* (Kutz) v. Wettstein, a unique endocytobiotic system of a fungus (Glomales) and the cyanobacterium *Nostoc punctiforme* (Kutz) Hariot. *Protoplasma*, *193*(1-4), 3–9. doi:10.1007/BF01276630

Montero, O., Sobrino, C., Pares, G., & Lubian, L. M. (2002). Photoinhibition and recovery after selective short-term exposure to solar radiation of five chlorophyll c-containing marine microalgae. *Ciencias Marinas*, *28*(3), 223–236. doi:10.7773/cm.v28i3.229

Nakamura, K. A. Z. U. O., Bray, D. F., & Wagenaar, E. B. (1975). Ultrastructure of Chlamydomonas eugametos palmelloids induced by chloroplatinic acid treatment. *Journal of Bacteriology*, *121*(1), 338–343. doi:10.1128/jb.121.1.338-343.1975 PMID:1116991

Neelam, S., & Subramanyam, R. (2013). Alteration of photochemistry and protein degradation of photosystem II from Chlamydomonas reinhardtii under high salt grown cells. *Journal of Photochemistry and Photobiology. B, Biology*, *124*, 63–70. doi:10.1016/j.jphotobiol.2013.04.007 PMID:23685462

Nymark, M., Sharma, A. K., Sparstad, T., Bones, A. M., & Winge, P. (2016). A CRISPR/Cas9 system adapted for gene editing in marine algae. *Scientific Reports*, *6*(1), 24951. doi:10.1038rep24951 PMID:27108533

Oliver, M. J., Jain, R., Balbuena, T. S., Agrawal, G., Gasulla, F., & Thelen, J. J. (2011). Proteome analysis of leaves of the desiccation-tolerant grass, Sporobolus stapfianus, in response to dehydration. *Phytochemistry*, *72*(10), 1273–1284. doi:10.1016/j.phytochem.2010.10.020 PMID:21109273

Paerl, R. W., Bouget, F. Y., Lozano, J. C., Verge, V., Schatt, P., Allen, E. E., Palenik, B., & Azam, F. (2017). Use of plankton-derived vitamin B1 precursors, especially thiazole-related precursor, by key marine picoeukaryotic phytoplankton. *The ISME Journal*, *11*(3), 753–765. doi:10.1038/ismej.2016.145 PMID:27935586

Palmer, J. D. (2003). The symbiotic birth and spread of plastids: How many times and whodunit? *Journal of Phycology*, *39*(1), 4–12. doi:10.1046/j.1529-8817.2003.02185.x

Paul, C., & Pohnert, G. (2011). Interactions of the algicidal bacterium *Kordia algicida* with diatoms: Regulated protease excretion for specific algal lysis. *PLoS One*, *6*(6), e21032. doi:10.1371/journal.pone.0021032 PMID:21695044

Perez-Rodriguez, E., Gomez, I., Karsten, U., & Figueroa, F. L. (1998). Effects of UV-radiation on photosynthesis and excretion of UV-absorbing compounds of *Dasycladus vermicularis* Dasycladales, Chlorophyta from southern Spain. *Phycologia*, *37*(5), 379–387. doi:10.2216/i0031-8884-37-5-379.1

Picard, K. T., Letcher, P. M., & Powell, M. J. (2013). Evidence for a facultative mutualist nutritional relationship between the green coccoid alga *Bracteacoccus* sp. (Chlorophyceae) and the zoosporic fungus *Rhizidium phycophilum* (Chytridiomycota). *Fungal Biology*, *117*(5), 319–328. doi:10.1016/j.funbio.2013.03.003 PMID:23719219

Plouguerné, E., Le Lann, K., Connan, S., Jechoux, G., Deslandes, E., & Stiger-Pouvreau, V. (2006). Spatial and seasonal variation in density, reproductive status, length and phenolic content of the invasive brown macroalga Sargassum muticum (Yendo) Fensholt along the coast of Western Brittany (France). *Aquatic Botany*, *85*(4), 337–344. doi:10.1016/j.aquabot.2006.06.011

Pohnert, G. (2004). Chemical defense strategies of marine organisms. *Topics in Current Chemistry*, *239*, 179–219. doi:10.1007/b95453 PMID:22160234

Post, A., & Larkum, A. W. D. (1993). UV-absorbing pigments, photosynthesis and UV exposure in Antarctica: Comparison of terrestrial and marine algae. *Aquatic Botany*, *45*(2-3), 231–243. doi:10.1016/0304-3770(93)90023-P

Price, D. C. (2012). Cyanophora paradoxa genome elucidates origin of photosynthesis in algae and plants (Science (843)). *Science*, *335*(6074), 1302.

Price, D. C., Chan, C. X., Yoon, H. S., Yang, E. C., Qiu, H., Weber, A. P., Schwacke, R., Gross, J., Blouin, N. A., Lane, C., Reyes-prietodion, A., Durnfordjonathan, D. G., Neilsonb, J. A. D., Franz Lang, B., Burger, G., Steiner, J. M., Löffelhardt, W., Meuser, J. E., Posewitz, M. C., ... Bhattacharya, D. (2012). *Cyanophora paradoxa* genome elucidates origin of photosynthesis in algae and plants. *Science*, *335*(6070), 843–847. doi:10.1126cience.1213561 PMID:22344442

Qin, D., Chen, Z., Averyt, K. B., Miller, H. L., Solomon, S., Manning, M., & Tignor, M. (2007). *Ipcc, 2007: Summary for policymakers*. Academic Press.

Rai, L. C., & Gaur, J. P. (Eds.). (2001). Algal Adaptation to Environmental Stresses Physiological, Biochemical and Molecular Mechanisms. Springer Berlin Heidelberg. doi:10.1007/978-3-642-59491-5

Rajadurai, M., Poornima, E. H., Narasimhan, S. V., Rao, V. N. R., & Venugopalan, V. P. (2005). Phytoplankton growth under temperature stress: Laboratory studies using two diatoms from a tropical coastal power station site. *Journal of Thermal Biology*, *30*(4), 299–305. doi:10.1016/j.jtherbio.2005.01.003

Rajadurai, M., Poornima, E. H., Narasimhan, S. V., Rao, V. N. R., & Venugopalan, V. P. (2005). Phytoplankton growth under temperature stress: Laboratory studies using two diatoms from a tropical coastal power station site. *Journal of Thermal Biology*, *30*(4), 299–305. doi:10.1016/j.jtherbio.2005.01.003

Ramlov, F., Souza, J., Faria, A. V., Maraschin, M., Horta, P. A., & Yokoya, N. S. (2011). Growth and accumulation of carotenoids and nitrogen compounds in Gracilaria domingensis (Kütz.) Sonder ex Dickie (Gracilariales, Rhodophyta) cultured under different irradiance and nutrient levels. *Revista Brasileira de Farmacognosia*, *21*(2), 255–261. doi:10.1590/S0102-695X2011005000081

Rastogi, R. P., Sinha, R. P., Singh, S. P., & Häder, D. P. (2010). Photoprotective compounds from marine organisms. *Journal of Industrial Microbiology & Biotechnology*, *37*(6), 537–558. doi:10.100710295-010-0718-5 PMID:20401734

Rastogi, R. P., Sonani, R. R., & Madamwar, D. (2015). Cyanobacterial sunscreen scytonemin: Role in photoprotection and biomedical research. *Applied Biochemistry and Biotechnology*, *176*(6), 1551–1563. doi:10.100712010-015-1676-1 PMID:26013282

Raven, J. A., & Geider, R. J. (1988). Temperature and algal growth. *The New Phytologist*, *110*(4), 441–461. doi:10.1111/j.1469-8137.1988.tb00282.x

Reed, R. H. (1984). Transient breakdown in the selective permeability of the plasma membrane of Chlorella emersonii in response to hyperosmotic shock: Implications for cell water relations and osmotic adjustment. *The Journal of Membrane Biology*, *82*(1), 83–88. doi:10.1007/BF01870734

Reidenbach, L. B., Fernandez, P. A., Leal, P. P., Noisette, F., McGraw, C. M., Revill, A. T., Hurd, C. L., & Kübler, J. E. (2017). Growth, ammonium metabolism, and photosynthetic properties of *Ulva australis* (Chlorophyta) under decreasing pH and ammonium enrichment. *PLoS One*, *12*(11), e0188389. doi:10.1371/journal.pone.0188389 PMID:29176815

Renaud, S. M., Thinh, L. V., Lambrinidis, G., & Parry, D. L. (2002). Effect of temperature on growth, chemical composition and fatty acid composition of tropical Australian microalgae grown in batch cultures. *Aquaculture (Amsterdam, Netherlands)*, *211*(1-4), 195–214. doi:10.1016/S0044-8486(01)00875-4

Rhoades, D., & Hanson, A. D. (1993). Quaternary ammonium and tertiary sulphonium compounds in higher plants. *Annual Review of Plant Physiology and Plant Molecular Biology*, *44*(1), 357–384. doi:10.1146/annurev.pp.44.060193.002041

Rindi, F., & Guiry, M. D. (2004). Composition and spatial variability of terrestrial algal assemblages occurring at the bases of urban walls in Europe. *Phycologia*, *43*(3), 225–235. doi:10.2216/i0031-8884-43-3-225.1

Rönnberg, O., & Ruokolahti, C. (1986, January). Seasonal variation of algal epiphytes and phenolic content of Fucus vesiculosus in a northern Baltic archipelago. In *Annales Botanici Fennici* (pp. 317–323). The Finnish Botanical Publishing Board.

Sandnes, J. M., Källqvist, T., Wenner, D., & Gislerød, H. R. (2005). Combined influence of light and temperature on growth rates of Nannochloropsis oceanica: Linking cellular responses to large-scale biomass production. *Journal of Applied Phycology*, *17*(6), 515–525. doi:10.100710811-005-9002-x

Sathasivam, R., Radhakrishnan, R., & Hashem, A., & Abd Allah, E. F. (2019). Microalgae metabolites: A rich source for food and medicine. *Saudi Journal of Biological Sciences*, *26*(4), 709–722. doi:10.1016/j.sjbs.2017.11.003 PMID:31048995

Schatz, D., & Vardi, A. (2018). Extracellular vesicles—New players in cell–cell communication in aquatic environments. *Current Opinion in Microbiology*, *43*, 148–154. doi:10.1016/j.mib.2018.01.014 PMID:29448174

Schussler, A., Bonfante, P., Schnepf, E., Mollenhauer, D., & Kluge, M. (1996). Characterization of the Geosiphon pyriforme symbiosome by affinity techniques: Confocal laser scanning microscopy (CLSM) and electron microscopy. *Protoplasma*, *190*(1/2), 53–67. doi:10.1007/BF01281194

Segev, E., Wyche, T. P., Kim, K. H., Petersen, J., Ellebrandt, C., Vlamakis, H., ... Kolter, R. (2016). Dynamic metabolic exchange governs a marine algal-bacterial interaction. *elife, 5*, e17473.

Segev, E., Wyche, T. P., Kim, K. H., Petersen, J., Ellebrandt, C., Vlamakis, H., Barteneva, N., Paulson, J. N., Chai, L., Clardy, J., & Kolter, R. (2016). Dynamic metabolic exchange governs a marine algal-bacterial interaction. *eLife*, *5*, e17473. doi:10.7554/eLife.17473 PMID:27855786

Semesi, I. S., Kangwe, J., & Björk, M. (2009). Alterations in seawater pH and CO2 affect calcification and photosynthesis in the tropical coralline alga, Hydrolithon sp.(Rhodophyta). *Estuarine, Coastal and Shelf Science*, *84*(3), 337–341. doi:10.1016/j.ecss.2009.03.038

Seyedsayamdost, M. R., Case, R. J., Kolter, R., & Clardy, J. (2011). The Jekyll-and-Hyde chemistry of *Phaeobacter gallaeciensis*. *Nature Chemistry*, *3*(4), 331–335. doi:10.1038/nchem.1002 PMID:21430694

Seymour, J. R., & Raina, J. B. (2018). Swimming in the sea: Chemotaxis by marine bacteria. *Microbiology Australia*, *3*(1), 12–16. doi:10.1071/MA18005

Seymour, J. R., Simo, R., Ahmed, T., & Stocker, R. (2010). Chemoattraction to dimethylsulfoniopropionate throughout the marine microbial food web. *Science*, *329*(5989), 342–345. doi:10.1126cience.1188418 PMID:20647471

Sharma, N. K., Rai, A. K., & Stal, L. J. (2013). *Cyanobacteria: an economic perspective*. John Wiley & Sons.

Siegenthaler, U., & Sarmiento, J. L. (1993). Atmospheric carbon dioxide and the ocean. *Nature*, *365*(6442), 119–259. doi:10.1038/365119a0

Sikorski, L. (2021). Effects of Sodium Chloride on Algae and Crustaceans—The Neighbouring Links of the Water Trophic Chain. *Water (Basel)*, *13*(18), 2493. doi:10.3390/w13182493

Simon, J., Kósa, A., Bóka, K., Vági, P., Simon-Sarkadi, L., Mednyánszky, Z., Horváth, Á. N., Nyitrai, P., Böddi, B., & Preininger, É. (2017). Self-supporting artificial system of the green alga *Chlamydomonas reinhardtii* and the ascomycetous fungus *Alternaria infectoria*. *Symbiosis*, *71*(3), 199–209. doi:10.100713199-016-0430-y

Singh, R. N. (1941). On some phases in the life history of the terrestrial alga, Fritschiella tuberosa Iyeng., and its autecology. *The New Phytologist*, *40*(3), 170–182. doi:10.1111/j.1469-8137.1941.tb07041.x

Sinha, R. P. (2015). Biochemical characterization of sunscreening mycosporine-like amino acids from two Nostoc species inhabiting diverse habitats. *Protoplasma*, *252*(1), 199–208. doi:10.100700709-014-0674-4 PMID:24997122

Sinha, R. P., & Hader, D. P. (1996). Photobiology and ecophysiology of rice field cyanobacteria. *Photochemistry and Photobiology*, *64*(6), 887–896. doi:10.1111/j.1751-1097.1996.tb01852.x

Smith, R. E., Cavaletto, J. F., Eadie, B. J., & Gardner, W. S. (1993). Growth and lipid composition of high Arctic ice algae during the spring bloom at Resolute, Northwest Territories, Canada. *Marine Ecology Progress Series*, *97*, 19–29. doi:10.3354/meps097019

Smriga, S., Fernandez, V. I., Mitchell, J. G., & Stocker, R. (2016). Chemotaxis toward phytoplankton drives organic matter partitioning among marine bacteria. *Proceedings of the National Academy of Sciences of the United States of America*, *113*(6), 1576–1581. doi:10.1073/pnas.1512307113 PMID:26802122

Sonnenschein, E. C., Syit, D. A., Grossart, H. P., & Ullrich, M. S. (2012). Chemotaxis of *Marinobacter adhaerens* and its impact on attachment to the diatom *Thalassiosira weissflogii. Applied and Environmental Microbiology*, *78*(19), 6900–6907. doi:10.1128/AEM.01790-12 PMID:22820333

Sorokin, C., & Krauss, R. W. (1958). The Effects of Light Intensity on the Growth Rates of Green Algae. *Plant Physiology*, *33*(2), 109–113. doi:10.1104/pp.33.2.109 PMID:16655087

Spribille, T., Tuovinen, V., Resl, P., Vanderpool, D., Wolinski, H., Aime, M. C., Schneider, K., Stabentheiner, E., Toome-Heller, M., Thor, G., Mayrhofer, H., Johannesson, H., & McCutcheon, J. P. (2016). Basidiomycete yeasts in the cortex of ascomycete macrolichens. *Science*, *353*(6298), 488–492. doi:10.1126cience.aaf8287 PMID:27445309

Staehr, P. A., & Birkeland, M. J. (2006). Temperature acclimation of growth, photosynthesis and respiration in two mesophilic phytoplankton species. *Phycologia*, *45*(6), 648–656. doi:10.2216/06-04.1

Stewart, H. L., & Carpenter, R. C. (2003). The effects of morphology and water flow on photosynthesis of marine macroalgae. *Ecology*, *84*(11), 2999–3012. doi:10.1890/02-0092

Stockenreiter, M., Haupt, F., Graber, A. K., Seppälä, J., Spilling, K., Tamminen, T., & Stibor, H. (2013). Functional group richness: Implications of biodiversity for light use and lipid yield in microalgae. *Journal of Phycology*, *49*(5), 838–847. doi:10.1111/jpy.12092 PMID:27007310

Sukenik, A., Carmeli, Y., & Berner, T. (1989). Regulation of fatty acid composition by irradiance level in the eustigmatophyte Nannochloropsis sp. 1. *Journal of Phycology*, *25*(4), 686–692. doi:10.1111/j.0022-3646.1989.00686.x

Sundararaman, M., Boopathi, T., & Gopinath, S. (2007). Status of Mangrove Ecosystem. In *Algae and Cyanobacteria in Extreme Environments* (pp. 209–224). Springer. doi:10.1007/978-1-4020-6112-7_11

Szekely, G. (2004). The role of proline in Arabidopsis thaliana osmotic stress response. *Acta Biologica Szegediensis*, *48*, 81.

Takahashi, S., & Murata, N. (2008). How do environmental stresses accelerate photoinhibition? *Trends in Plant Science*, *13*(4), 178–182. doi:10.1016/j.tplants.2008.01.005 PMID:18328775

Talebi, A. F., Tabatabaei, M., Mohtashami, S. K., Tohidfar, M., & Moradi, F. (2013). Comparative salt stress study on intracellular ion concentration in marine and salt-adapted freshwater strains of microalgae. *Notulae Scientia Biologicae*, *5*(3), 309–315. doi:10.15835/nsb539114

Tang, E. P. Y., Tremblay, R., & Vincent, W. F. (1997a). Cyanobacterial dominance of polar freshwater ecosystems: Are high-latitude mat-formers adapted to low temperature? *Journal of Phycology*, *33*(2), 171–181. doi:10.1111/j.0022-3646.1997.00171.x

Tang, E. P. Y., & Vincent, W. F. (2000). Effects of daylength and temperature on the growth and photosynthesis of an Arctic cyanobacterium, *Schizothrix calcicola* (Oscillatoriaceae). *European Journal of Phycology*, *35*(3), 263–272. doi:10.1080/09670260010001735861

Teeling, H., Fuchs, B. M., Bennke, C. M., Kruger, K., Chafee, M., Kappelmann, L., Reintjes, G., Waldmann, J., Quast, C., Glockner, F. O., Lucas, J., Wichels, A., Gerdts, G., Wiltshire, K. H., & Amann, R. I. (2016). Recurring patterns in bacterioplankton dynamics during coastal spring algae blooms. *eLife*, *5*, e11888. doi:10.7554/eLife.11888 PMID:27054497

Teoh, M. L., Chu, W. L., & Phang, S. M. (2010). Effect of temperature change on physiology and biochemistry of algae: A review. *Malaysian Journal of Science*, *29*(2), 82–97. doi:10.22452/mjs.vol29no2.1

Tesson, S. V. M., Skjøth, C. A., Santl-Temkiv, T., & Londahl, J. (2016). Airborne microalgae: Insights, opportunities, and challenges. *Applied and Environmental Microbiology*, *82*(7), 1978–1991. doi:10.1128/AEM.03333-15 PMID:26801574

Tevini, M., Braun, J., & Fieser, G. (1991). The protective function of the epidermal layer of rye seedlings against ultraviolet-B radiation. *Photochemistry and Photobiology*, *53*(3), 229–333. doi:10.1111/j.1751-1097.1991.tb03636.x

Thume, K., Gebser, B., Chen, L., Meyer, N., Kieber, D., & Pohnert, G. (2018). The metabolite dimethylsulfoxonium propionate extends the marine organosulfur cycle. *Nature*, *563*(7731), 412–415. doi:10.103841586-018-0675-0 PMID:30429546

Tyagi, R., Kumar, A., Tyagi, M. B., Jha, P. N., Kumar, H. D., Sinha, R. P., & Hader, D. P. (2003). Protective role of certain chemicals against UV-B-induced damage in the nitrogen-fixing cyanobacterium, *Nostoc muscorum*. *Journal of Basic Microbiology*, *43*(2), 137–147. doi:10.1002/jobm.200390014 PMID:12746856

Wieners, P. C., Mudimu, O., & Bilger, W. (2012). Desiccation-induced non-radiative dissipation in isolated green lichen algae. *Photosynthesis Research*, *113*(1), 239–247. doi:10.100711120-012-9771-4 PMID:22833109

Wienhausen, G., Noriega-Ortega, B. E., Niggemann, J., Dittmar, T., & Simon, M. (2017). The exometabolome of two model strains of the Roseobacter group: A marketplace of microbial metabolites. *Frontiers in Microbiology*, *8*, 1985. doi:10.3389/fmicb.2017.01985 PMID:29075248

Wiśniewska, K., Lewandowska, A. U., & Śliwińska-Wilczewska, S. (2019). The importance of cyanobacteria and microalgae present in aerosols to human health and the environment–Review study. *Environment International*, *131*(104964), 1–11. doi:10.1016/j.envint.2019.104964 PMID:31351382

Wood, W. F. (1987). Effect of solar ultra-violet radiation on the kelp *Ecklonia radiata*. *Marine Biology*, *96*(1), 143–150. doi:10.1007/BF00394848

Woodwell, G. M., Mackenzie, F. T., Houghton, R. A., Apps, M. J., Gorham, E., & Davidson, E. A. (1998). Biotic feedbacks in the warming of the Earth. *Climatic Change*, *40*(3-4), 495–518. doi:10.1023/A:1005345429236

Worden, A. Z., Follows, M. J., Giovannoni, S. J., Wilken, S., Zimmerman, A. E., & Keeling, P. J. (2015). Rethinking the marine carbon cycle: Factoring in the multifarious lifestyles of microbes. *Science*, *347*(6223), 1257594. doi:10.1126cience.1257594 PMID:25678667

Xu, K., Chen, H., Wang, W., Xu, Y., Ji, D., Chen, C., & Xie, C. (2017). Responses of photosynthesis and CO2 concentrating mechanisms of marine crop *Pyropia haitanensis* thalli to large pH variations at different time scales. *Algal Research*, *28*, 200–210. doi:10.1016/j.algal.2017.10.023

Xue, L., Zhang, Y., Zhang, T., An, L., & Wang, X. (2005). Effects of enhanced ultraviolet-B radiation on algae and cyanobacteria. *Critical Reviews in Microbiology*, *31*(2), 79–89. doi:10.1080/10408410590921727 PMID:15986833

Yoon, H. S., Hackett, J. D., Ciniglia, C., Pinto, G., & Bhattacharya, D. (2004). A molecular timeline for the origin of photosynthetic eukaryotes. *Molecular Biology and Evolution*, *21*(5), 809–818. doi:10.1093/molbev/msh075 PMID:14963099

You, T., & Barnett, S. M. (2004). Effect of light quality on production of extracellular polysaccharides and growth rate of Porphyridium cruentum. *Biochemical Engineering Journal*, *19*(3), 251–258. doi:10.1016/j.bej.2004.02.004

Young, C. S., & Gobler, C. J. (2016). Ocean acidification accelerates the growth of two bloom-forming macroalgae. *PLoS One*, *11*(5), e0155152. doi:10.1371/journal.pone.0155152 PMID:27176637

Chapter 7
Cultivation of Algae and Its Biorefinery Approach

Mohammadhosein Rahimi
School of Biosystems and Food Enginering, University College Dublin, Ireland

Mina Tajmirriahi
Independent Researcher, UK

Ronald Halim
University College Dublin, Ireland

ABSTRACT

During the past decades, algae have attracted worldwide attention as a sustainable bioresource to produce various biochemicals and biofuels. However, the prohibitive cost of algal biomass production and processing casts doubt on the industrial applications of algae. Hence, many efforts have been made to enhance the viability of these species. One serious challenge is maximizing algal biomass production. Since algal growth is strain-specific, the optimization of cultivation conditions (pH, illumination, temperature, and nutrients) can significantly tackle the problem of algal biomass production. Another way of reducing the production costs and enhancing the viability of algal biotechnology is the fractionation of all major components, known as a multi-product biorefinery. Various upstream and downstream processes are involved in an algae biorefinery. Therefore, having detailed knowledge about these bioprocesses and how to optimize them is a milestone for the commercialization of algae. Consequently, this chapter aims to provide an overview of algae cultivation methods and parameters affecting algae growth as well as different microalgae cultivation systems. Besides, it describes the bioprocesses involved in an algae biorefinery and their bioproducts.

INTRODUCTION

In today's populated and industrialized world, algae are considered a sustainable resource for producing various products to address societies' demands (Bhattacharya and Goswami 2020). Algal biomass produc-

DOI: 10.4018/978-1-6684-2438-4.ch007

tion is advantageous in comparison with terrestrial plants. For instance, algae do not require freshwater for cultivation and can grow in various harsh conditions such as wastewater and saline water. They also do not need fertile land, require much less area than terrestrial crops, and are well known for their high photosynthetic rate and the ability for high CO_2 fixation (Xu et al. 2019).

Algae are photosynthetic species that use sunlight, CO_2, nutrients, and water to produce carbohydrates, lipids, and proteins as their main products (Rahimi and Jazini 2021). These species are applicable in several industry sectors such as food, energy, and the environment. For instance, due to the presence of high-value antioxidants, carotenoids, minerals, and vitamins, algae are potential choices for food and feed (Chandra et al. 2019; M. P. Sudhakar et al. 2019). In the energy sector, producing various biofuels such as biodiesel, bioethanol, and biogas using algal species is possible (Chew et al. 2017). Besides, from an environmental point of view, algal species have been considered potential organisms because they consume CO_2, grow on several industrial wastewaters, and simultaneously consume hazardous components (Javed et al. 2019). However, due to the excessive costs of algal biomass processing, their industrial utilization is still doubtful. There are still challenges in both upstream and downstream processing of algal biomass; They require several technological enhancements and optimizations. Upstream processing is related to cultivation and maximizing the biomass concentration, while downstream processing refers to harvesting, cell disruption and extraction, and purification of algal biomass (Alavijeh et al. 2020).

Since algae can produce a wide range of valuable products, one conceivable way for their commercialization could be the utilization of their whole fractions in which no waste remains (Alavijeh et al. 2020). In this regard, they are considered potential bioresources in the biorefinery concept. As defined by the International Energy Agency (IEA), the sustainable processing of biomass into a spectrum of valuable products (biochemicals and biofuels) is known as biorefinery (Chandra et al. 2019). Algae are advantageous over other generations of biorefineries in many ways. For instance, compared to the first-generation biorefineries, they do not compete with the global food supply (del Río et al. 2020). Besides, compared with the second-generation biorefineries, the algal biomass structure is not complex, though the extraction of biomolecules is more convenient (Xu et al. 2019). Because of the high growth rate of algae compared to terrestrial crops, obtaining considerable amounts of algal biomass is faster than crops or lignocellulosic biomass. It is crucial since feedstock availability is essential for a biorefinery process.

This chapter aims to give a broad overview of algal bioprocessing, focusing on biorefinery. Various algal biomass products are discussed in this respect. General information on algal bioprocessing is then provided, emphasizing cultivation and cell disruption. Furthermore, available information on algal biomass harvesting and conversion is presented. Finally, future difficulties and outlooks on algal biorefineries are discussed.

BIOPROCESSES INVOLVED IN BIOREFINERY

Varied upstream and downstream processes are involved in the bioprocessing of algal biomass (Javed et al. 2019). In the case of microalgae, these processes include cultivation, harvest, cell disruption and extraction, and conversion (Figure 1). The bioprocessing of macroalgae is remarkably similar to microalgae. To obtain seaweed biomass, utilization of cultivated or naturally occurring seaweed is possible. There are two methods for macroalgae harvesting: manual and mechanical (i.e., dredge, moving boat, and mesh conveyor). After harvesting, an extra process, pretreatment, is required, consisting of several steps, including foreign objects removal, drying, and grinding. After performing suitable methods for cell wall

Figure 1. Various processes that should be considered in the bioprocessing of microalgae species to reach a profitable biorefinery.

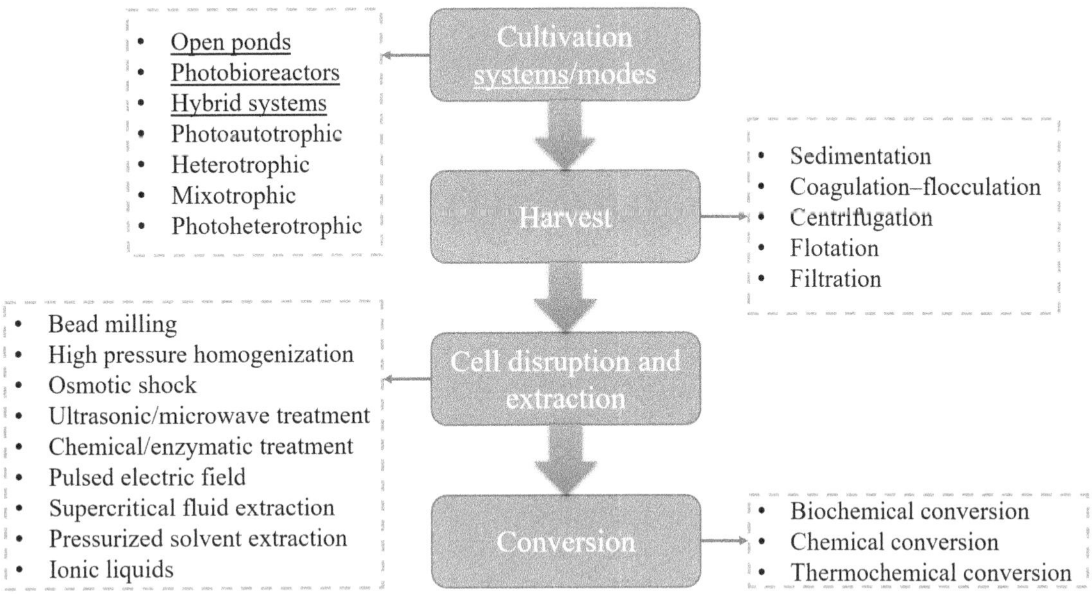

disruption, the macroalgal biomass is converted into several products through biochemical, chemical, or thermochemical processes (K. Sudhakar et al. 2018). Accordingly, a successful approach combines suitable technologies related to cultivation, harvesting, cell disruption and extraction of the products, and conversion of algal biomass to convert all fractions most efficiently by lowering the operating costs and improving the products recovery and quality (Günerken et al. 2015).

Algae have the potential to generate a wide range of bioproducts, such as antioxidants, pigments, minerals, and vitamins. These valuable products can be used in several sectors, including food, feed, cosmetics, pharmaceuticals, and fertilizers (Mohan et al. 2020). However, the biochemical composition of algal biomass is related to several factors, such as the species and cultivation conditions. Besides, the cultivation parameters must be optimized by considering the target products. For instance, offshore cultivation of macroalgae provides controlled conditions of growth, and therefore higher efficiency and productivity of biomass can be obtained compared to natural growth (K. Sudhakar et al. 2018). As another example, when the target product is biodiesel, strategies to enhance lipid fraction and biomass productivity of certain species during the cultivation should be the priority. Then, other downstream processes should be evaluated to enhance the viability of the entire process (Abomohra et al. 2020).

Following growth and the production of significant volumes of algal biomass, a succession of appropriate technologies is required to use all fractions of algal biomass in order to achieve the most efficient biorefinery (Gerardo et al. 2015). In this case, the segregation of various algal products without damaging other fractions is important. Therefore, developing enhanced mild techniques is needed for efficient extraction of the target products without damaging other cell components. Besides, the proposed technology should be cost and energy effective.

Figure 2. A simplified schematic of an algal biorefinery that shows the various algae products and their application.

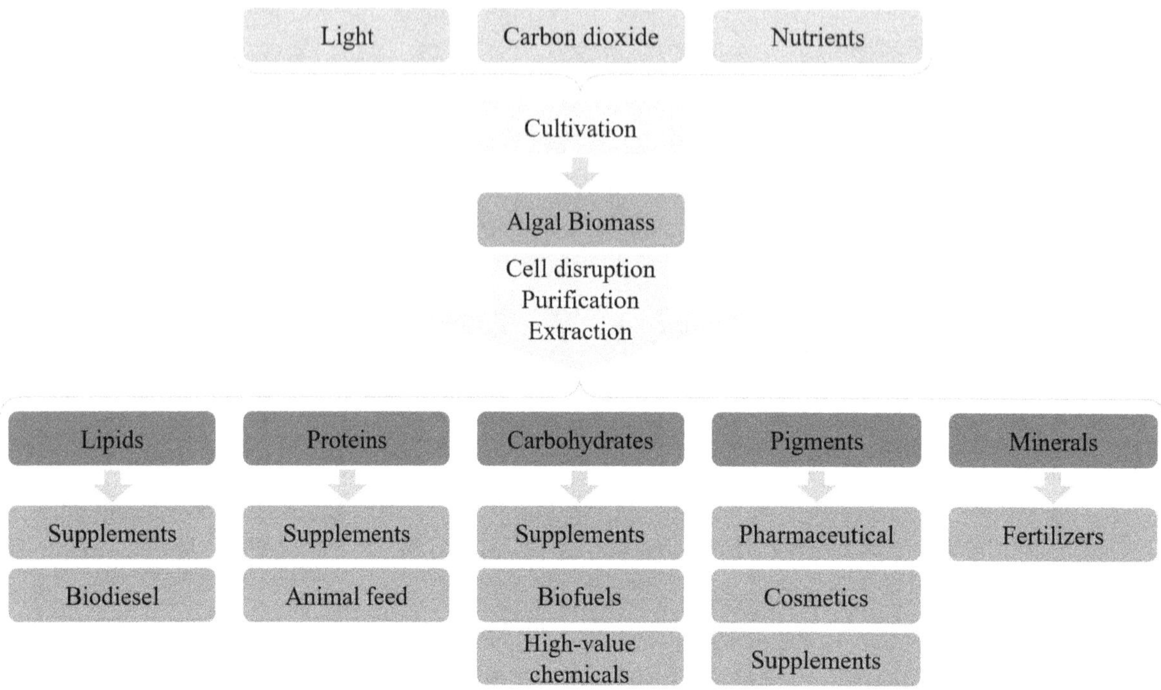

VARIOUS PRODUCTS OF ALGAE

Due to the energy crisis and the growing need for fuel, algae have attracted attention. Algae is a sustainable feedstock for producing various products, for instance, biofuels and other bioproducts, since it contains vast quantities of lipids, carbohydrates, and proteins (Figure 2) (Chew et al. 2017). Producing algal biofuel is not economically feasible because of the high costs of capital investments and operations. Hence, the production of co-products is a helpful strategy for improving the economy of the algal biorefinery. There are four main processes for biofuels production from Algal biomass: thermochemical conversion, biochemical conversion, transesterification, and photosynthetic microbial fuel cell, which are elaborated in Table 2. In the following, some critical products of these processes are presented (Chew et al. 2017).

Biofuels

Biofuel refers to a solid, liquid, or gaseous material derived from living or dead biomass to generate energy. Bio-oil, biodiesel, biohydrogen, biogas, bioethanol, and biobutanol are distinct types of biofuels (Bhattacharya and Goswami 2020).

Bio-Oil

Bio-oil, also called biocrude oil or pyrolysis oil, is an alternative to petroleum-derived oils in energy production. Bio-oil is a clean fuel compared to fossil fuels. It does not emit SO_X, and its NO_2 emissions are 50 percent less than diesel oil. The conversion of biomass to bio-oil involves two main stages: pyrolysis and hydrothermal liquefaction. During pyrolysis, heat decomposes the organic compounds in the absence of oxygen and converts them into liquid (bio-oil), solid (charcoal), and gaseous substances (Xiu and Shahbazi 2012). There are many studies on bio-oil extraction from microalgae and macroalgae. Anand et al. extracted bio-oil from a microalga, *Schizochytrium limacinum*, using the fast pyrolysis method (Anand, Gautam, and Vinu 2017). Rahbari et al. investigated the bio-oil production from a brown macroalga, *Sargassum* species, and analyzed the produced bio-oil (Rahbari et al. 2019).

Biodiesel

Algal biodiesel can be used as a substitute for petroleum diesel because it does not require any significant changes in the engine design. This biofuel generates much less CO, SO_2, and unburned hydrocarbons compared to petroleum diesel fuel (Chandra et al. 2019).

Biohydrogen

Cyanobacteria and green algae are capable of biohydrogen production. Hydrogen is a clean fuel since its only major by-product is water, and its combustion emits no CO_2 into the atmosphere. There are several barriers to biohydrogen production, such as sophisticated machinery and high production costs. Therefore, using algae for biohydrogen production is beneficial because it decreases the environmental impacts and production costs (Lam and Lee 2013).

Biogas

Biogas, a promising biofuel, is mainly composed of methane and carbon dioxide. This biofuel is generated during the anaerobic digestion process that converts organic substances into biogas (Chew et al. 2017). Anaerobic digestion has four stages: hydrolysis, acidogenesis, acetogenesis, and methanogenesis. Biomethane can be used as fuel or chemical feedstock. There are various substances for biomethane production, such as algae, wood, grass, and solid waste. Among these materials, green algal biomass is suitable for biomethane production (Bhattacharya and Goswami 2020).

Bioethanol

Bioethanol production from algal biomass is increasing significantly because of some advantages that algae have, including high biomass efficiency and high starch, glucose, and cellulose contents that make algae a potential feedstock for bioethanol production (Chandra et al. 2019). Algal waste utilization is a valuable strategy for the economical production of bioethanol. Several microalgae are used for bioethanol production, such as *Dunaliella*, *Chlorella*, and *Spirulina*, to name a few. There are two different processes for bioethanol production from algae, fermentation and gasification (conversion of syngas to ethanol through indirect routes)(Bhattacharya and Goswami 2020). Among macroalgae, brown macroalgae (such as *Laminaria hyperborean*, *Sargassum sagamianum*, and *Undaria pinnatifida*), green macroalgae (i.e., *Enteromorpha intestinalis*, *Ulva fasciata*, and *Ulva lactuca*), and red macroalgae (e.g., *Gelidium elegans*, *Kappaphycusalvarezii*, and *Gracilaria* sp.) can be used for bioethanol production (Ramachandra and Hebbale 2020).

Biobutanol

In some countries, algal culture development is utilized for biogas and bioethanol production. However, in the USA, they are used for biobutanol production. This biofuel has been used in the transportation sector for many years. It has low vapor pressure and high energy density, making it a suitable replacement for bioethanol as a petroleum additive material. The feedstocks for biobutanol production are starch, sugars, and cellulose present in algal biomass, making it economically profitable. Acetone-butanol-ethanol (ABE) fermentation is a process for biobutanol production using algae such as *Clostridium sp.* (Bhattacharya and Goswami 2020).

Lipid and Polysaccharide

Algae have the capacity of accumulating 30 to 50% of lipid under specific culture conditions. Algae can produce valuable lipids, such as polyunsaturated fatty acids (PUFAs). There are two distinct types of these essential fatty acids: (1) omega-3 PUFAs including Eicosapentaenoic acid (EPA), docosahexaenoic acid (DHA), α-linolenic acid, and (2) omega-6 PUFAs, i.e., linoleic acid. PUFAs are synthesized in various algae, including *Arthrospira platensis*, *Crypthecodinium cohnii*, *Schizochytrium* spp., *Chlorella vulgaris*, and *Nannochloropsis oculata* (Eltanahy and Torky 2021). Omega-3 fatty acids, DHA, and EPA, are among PUFAs that are vital for human health and nutrition. Omega-3 fatty acids are beneficial lipids since they reduce the risk of cardiovascular disease. DHA and EPA serve as health food supplements (Chandra et al. 2019).

Algal species are rich sources of various polysaccharides, including starch, sucrose, mannitol, mannose, sulfated polysaccharides, and carrageenan. Some microalgae with high carbohydrate contents are *Chlorella*, *Nannochloropsis*, and *Isochrysis* (Eltanahy and Torky 2021). Red seaweed contains agar, cellulose, and carrageenan. Brown seaweed has several polysaccharides, including alginates, fucoidans, and cellulose (del Río et al. 2020). These polysaccharides have various functions and applications. Sulfated polysaccharides have cosmetic, nutritional, and pharmaceutical functions. Carrageenan is utilized as a gelling agent. Alginates are used for drug delivery and have antitumor properties. Fucoidan exhibits various biological functions, including antioxidant, antiviral, anti-inflammatory, and antitumor activities (M. P. Sudhakar et al. 2019).

Proteins

The growing population of the world has increased the need for food production. Protein is one of the major nutrients, and seeking alternative sources and new production methods for its production is necessary in order to meet the global need and prevent its shortage. Micro/macroalgae are potential protein sources and are advantageous compared to protein-rich crops. Their protein yield per unit area is higher than terrestrial crops. For instance, the protein content of algae can be up to 78% of their dry weight, while the protein contents of soybean, rice, and pea are 38%, 10%, and 2.8%, respectively (Eltanahy and Torky 2021). *Chlorella*, *Arthrospira platensis*, and *Scenedesmus* are three microalgae with high protein contents. Unlike plants, algae do not require potable water and fertile lands for cultivation. Different algae have different amino acids contents and types, with most demonstrated to have balanced profiles containing all essential amino acids for food/feed applications. Red algae contain leucine and isoleucine, while brown algae contain methionine, cysteine, and lysine. Besides, green algae often contain arginine, threonine, phenylalanine aspartic acid, glutamic acid, glycine, alanine, serine, and valine (Bleakley and Hayes 2017).

Pigments and Vitamins

Different algae can produce valuable pigments, for example, chlorophylls, phycobiliproteins, and carotenoids. Chlorophyll is not only a natural food dye, but it is also convertible to chlorophyllin, a potent antimutagen (Chandra et al. 2019). Phycobiliproteins are highly fluorescent with antioxidant properties and food coloring application. Antioxidants protect us from disease (M. P. Sudhakar et al. 2019). Many carotenoids, including β-carotene and astaxanthin, are produced by microalgae and macroalgae (Chandra et al. 2019). Carotenoids are one of the most extensive pigments available in red, orange, and yellow colors. Carotenoids have several practical applications for plants and humans. They attract pollinators to flowers and protect the plant from light-induced damage to cells. Carotenoids also play a vital role in human health. For instance, lutein and zeaxanthin are responsible for protecting our retina by reducing oxidative stresses. Carotenoids also have antioxidant and food colorant functions (Ngamwonglumlert and Devahastin 2018). β-carotene, a carotenoid, is a crucial nutrient and food colorant agent which is highly demanded in the food and pharmaceutical industries. Astaxanthin, another carotenoid, prevents us from certain eye diseases (Chandra et al. 2019).

Microalgae can synthesize different vitamins. *Chlorella* is a rich source of vitamin A, B1, B2, B3, and C. *Microchloropsis* has the highest vitamin E content compared to other microalgae. This vitamin is helpful in Alzheimer's and heart disease treatment (Eltanahy and Torky 2021). Several macroalgae are capable of producing vitamin C, including *Porphyrau mbilicalis*, *Himanthalia elongata*, *Gracilaria changii*, and *Eisenia arborea*. Besides, β-carotene contents of some macroalgae, such as *Codium fragile* and *Gracilaria chilensis*, can be higher than carrots (Wells et al. 2017).

Bioplastics and Biocomposites

Algal biomass has the potential to be used for producing various bioplastics, such as hybrid plastics, cellulose-based plastics, and polylactic acid (PLA). Algal biomass is used as filler material for hybrid plastics synthesis. After oil extraction from algae, the residues are rich in cellulose, protein, and carbohydrates (Bhattacharya and Goswami 2020). The residual cellulose in the algal remains can be used for cellulose-based plastics. Algal biomass is a valuable feedstock for the lactic acid production of bacteria. PLA is the product of the polymerization of lactic acid (Bhattacharya and Goswami 2020). Biocomposites, natural fiber-reinforced biopolymers, are blends of varied materials used for several applications because of their biocompatibility, mechanical properties, and biodegradability. Biocomposites are practical products that can be obtained using algal biomass. Algal residues after oil extraction can be used as filler for biocomposites production (G. M. Kim, Chang, and Kim 2021).

Nanocellulose

Nanocellulose is an influential biopolymer in various industries, including food, pharmaceutical, and agriculture. Nanocellulose is renewable and environmentally friendly, and it enjoys specific mechanical properties. According to previous studies, nanocellulose fibers could be extracted from green filamentous algae. Algae contain insignificant lignin content compared to woody biomass. Delignification, a necessary step for producing nanocellulose from lignocellulosic materials, is a time- and energy-consuming process. So, the main advantage of utilizing algae for nanocellulose production is that it requires a mild chemical treatment for the delignification process (Bhattacharya and Goswami 2020).

Phenolic Compounds

Previous studies report that microalgae and macroalgae are valuable sources of bioactive metabolites. These compounds are synthesized by algae and have antioxidant, antitumoral, or antimicrobial characteristics that make them useful for utilization in different industrial sectors. Some identified phenolic compounds are benzoic acid, gallic acid, syringic acid, vanillic acid, sinapic acid, ferulic acid, caffeic acid, and protocatechuic acid (M. P. Sudhakar et al. 2019).

CULTIVATION OF ALGAE

Algae have a high photosynthesis yield and can store a substantial amount of bioproducts inside their cells. Therefore, they are a good option for biotechnological purposes. Compared to other agricultural products, algae cultivation does not need fertile soil, an enormous freshwater volume, and different pesticides and herbicides. Besides, wastewater and milling effluents can be used for cultivation. By consuming the atmospheric carbon dioxide through photosynthesis, algae cultivation can also reduce the concentration of this greenhouse gas, which slows down global warming (Tan et al. 2020).

The three most widely used microalgae cultivation systems are open ponds, photobioreactors (PBRs), and hybrid systems, which are currently used in both research and industrial scale (Jankowska, Sahu, and Oleskowicz-Popiel 2017). The most common systems for microalgae cultivation are open ponds that provide more than 80 percent of algal biomass worldwide (Moreno-Garcia et al. 2017). Macroalgal cultivation is a necessary step for utilizing these organisms on a sustainable basis. Knowledge of macroalgae genetics can help select more productive species to cultivate. Macroalgae cultivation can be performed in natural waters (offshore or nearshore) or open ponds. The cultivation techniques for macroalgae are included but are not limited to line cultivation, net cultivation, floating raft cultivation, and tank/pond cultivation (Radulovich et al. 2015). Since the nearshore zone has various uses, including oil extraction and shipping, offshore and land-based operations are more practical and appropriate for cultivation (Liang et al. 2015). Further information about the scaling up of algal biomass will be explained in another chapter of the present book.

Different Cultivation Modes

Algae are cultured in four different modes: photoautotrophy, heterotrophy, mixotrophy, and photoheterotrophy (Zhan, Rong, and Wang 2016). In photoautotrophic cultivation, sunlight serves as the energy source, and the carbon dioxide content of the air is the inorganic carbon source. This type of cultivation is advantageous since it is performed outdoors and can utilize atmospheric carbon dioxide (Moreno-Garcia et al. 2017). Besides, it is possible to utilize non-drinking water and non-agricultural land for culture in this way, so it does not interfere with raising food crops (Khan et al. 2016). Nevertheless, this method is for particular areas with constant and intense solar radiation. This technique of algae cultivation is suitable for commercial-scale because it is both technically and economically viable (Moreno-Garcia et al. 2017). As it is clear, sunlight is a limiting factor in this method, and this problem can be solved by using another culture method called heterotrophic cultivation (Khan et al. 2016).

Some algae can grow heterotrophically since they can use organic substrates (i.e., glycerol, glucose, and acetate) as the source of both carbon and energy. This type of cultivation has many advantages compared to autotrophic, including higher biomass and lipid productivity, a more straightforward scaling-up process, inexpensive bioreactor design, and the capability to remove carbon, nitrogen, phosphorus, and other elements from wastewater. However, there are some limitations associated with this culture method. It is applicable to just a few algal species. Besides, the addition of organic substrates is expensive and increases the risk of contamination. In addition, light-induced metabolites are not produced under this condition (Khan et al. 2016).

In the third mode, mixotrophic cultivation, algae can use sunlight and organic substances as energy sources and carbon dioxide and organic matter as carbon sources. This culture method improves biomass production and lipid accumulation. The growth rate of algae under this condition is faster than in the autotrophy condition. Among different cultivation conditions, mixotrophic cultivation is more favorable since it combines the advantages of autotrophic and heterotrophic cultivations while overcoming their disadvantages (Zhan, Rong, and Wang 2016).

In photoheterotrophic conditions, algae can use solar energy as an energy source, but they cannot consume carbon dioxide as a carbon source. So, it requires organic compounds as its carbon source. The main difference between this method and mixotrophic cultivation is that in the mixotrophic condition, the energy source must be organic compounds, not sunlight (Zhang 2013). The appropriate method should be chosen based on the species, environmental conditions, and the aim of the biorefinery process. For instance, if the synthesis of carotenoids is the aim of the process, performing an appropriate cultivation method to enhance the carotenoids content during the cultivation is important (Rahimi and Jazini 2021).

Factors Affecting the Algal Growth

There are a number of factors that influence algal growth and its biomass productivity, including carbon source, light, nutrients, pH, salinity, temperature, and mixing (Gatamaneni, Orsat, and Lefsrud 2018).

Whether organic or inorganic, the carbon source is an essential factor that influences biomass and lipid production. The effect of organic carbon source depends on its type (i.e., glucose, acetate, and xylose), its concentration, and the algae species. Among different carbon sources, such as glucose, galactose, and xylose, glucose is an ideal substrate because other substances are more complicated to convert during metabolic pathways (Gatamaneni, Orsat, and Lefsrud 2018). Previous reports show that fat synthesis declines sharply by increasing the initial glucose concentration in the medium (Zhan, Rong, and Wang 2016). Burning fossil fuels to meet energy needs emit greenhouse gases, including carbon dioxide. The increasing emission of this gas is responsible for global warming. Algae are more effective than terrestrial plants in fixing carbon dioxide. This gas is the key factor for algal growth. As carbon dioxide concentration increases, the growth of algae increases as well. However, some algae are not carbon dioxide tolerant as excessive CO_2 leads to a pH reduction. There is a limit to the concentration of this gas in the culture (Gatamaneni, Orsat, and Lefsrud 2018).

As it is obvious, lighting has a substantial impact on algal growth in phototrophic and mixotrophic modes. Light intensity and duration (photoperiod) enhance cell densities, lipid accumulation, biomass productivity, and pigment production (Moreno-Garcia et al. 2017). There are three distinct conditions of lighting: light limitation, light saturation, and photoinhibition. In light limiting conditions, as the light intensity increases, the algal growth also increases. Photoinhibition can happen when the light intensity is more than the saturation limit, diminishing the photosynthesis efficiency and declining the algal growth rate (Gatamaneni, Orsat, and Lefsrud 2018). Since different algae need specific light wavelengths to grow, the operational parameters of cultivation should be scrutinized in advance (Jankowska, Sahu, and Oleskowicz-Popiel 2017).

The two most important nutrients for algal cultivation are nitrogen and phosphorus. Nitrogen is the structural component of amino acids. Nucleotides, building blocks of nucleic acids, consist of a nitrogen-based part. Besides, nucleic acids and phospholipids contain phosphorus. Previous studies suggest that the insufficiency of these two essential macronutrients in the culture media can shift the metabolic pathways in algae (Gatamaneni, Orsat, and Lefsrud 2018). Nitrogen depletion shifts the metabolism from protein production to lipid production, which increases lipid productivity. Phosphorus depletion inhibits growth and also increases lipid accumulation (Moreno-Garcia et al. 2017).

In addition to nitrogen and phosphorus, trace metals, including calcium, copper, iron, magnesium, nickel, and zinc, are necessary for the metabolic functions of algae and their limitation can hinder algal growth (Gatamaneni, Orsat, and Lefsrud 2018).

The pH affects cell metabolism and biomass productivity. The bicarbonate in the culture medium is converted to carbon dioxide and hydroxyl ions by carbonic anhydrase. Algae consume carbon dioxide for the photosynthesis process. The produced hydroxyl ions are capable of increasing pH. The optimum pH range for algal growth is between 6 and 10. Whether high pH or low pH, it decreases the rate of photosynthesis. High pH can alter the trends of nutrients or trace metals absorption. Low pH increases the risk of contamination by other microorganisms because of enzyme inhibition (Gatamaneni, Orsat, and Lefsrud 2018).

Previous investigations have shown that increasing salt concentration in the culture medium reduces cell growth and increases lipid production. Different strains of algae are capable of tolerating different concentrations of salt. Marine algae species are more salt-tolerant than freshwater ones (Gatamaneni, Orsat, and Lefsrud 2018).

Temperature is an essential factor in the growth and lipid accumulation of algae. Hence, in order to control the growth of algae, this parameter should be optimized. Arrhenius' equation expresses that a 10-degree Celsius increase in temperature doubles the growth rate until reaching an optimal temperature. After this, an increase in temperature leads to a decrease in growth. It is because there are a few enzymes responsible for carbon fixation and lipid synthesis. Increasing the temperature above the optimal results in enzymes deactivation (Gatamaneni, Orsat, and Lefsrud 2018).

As mentioned above, light and temperature are two principal factors that are expensive to control. An inexpensive way to distribute light and nutrients to all cells uniformly is by mixing. It also speeds up heat transfer and prevents sedimentation. There are several types of mixing, such as aeration, pumping, or mechanical agitation. The selection of mixing type depends on some factors, including microalgal strain, type of culture system, and scale of culture systems (Moreno-Garcia et al. 2017).

HARVESTING OF ALGAE

Large-scale algae production is commonly proposed to be economically viable. Because dilute quantities of algae are conceivable in this situation, an appropriate method for harvesting algal biomass is required. When harvesting algal biomass, two crucial factors to consider are the quantity and quality of the bio-

Table 1. Various harvesting methods for microalgae and the advantages and disadvantages of the harvesting procedures.

Method	Description	Advantages	Disadvantages
Sedimentation	The cells left to settle by gravitational settling Can be used in combination with other methods	Requires low capital and operating costs Does not damage the cells Does not require chemicals or energy	Slow Requires high land area Low solid concentration compared to other methods Species-specific
Centrifugation	Using centrifugal force to enhance the sedimentation rate	Fast Applicable to all species Chemical-free biomass High recovery rate	Species-specific (may damage some species with fragile cell walls and their content)
Coagulation-floccultaion	Aggregation of single cells to form bigger flocs through various methods (i.e., autoflocculation, biological, chemical, physical, and physicochemical flocculation)	Easy Fast No contamination (Physical method)	Biomass contamination (Chemical/biological methods) Requires high land area Might cause cells death Sometimes cost a lot and requires high energy (e.g., physical method) High cost of flocculants/ coagulants
Flotation	Low-density cells float upward rapidly using surfactants or coagulants	Relatively less expensive	High cost of surfactants/ coagulants Depends on air bubbles distribution Species-specific
Filtration	Using a filtration system to harvest algal biomass	Fast	Fouling and clogging of membranes

mass. However, in order to perform a proper procedure, numerous parameters must be considered. In this scenario, an appropriate approach should be chosen based on the algae species and growth conditions. Either wild or cultivated macroalgae harvesting is done by hand or mechanically using a dredge, moving boat, and mesh conveyor (K. Sudhakar et al. 2018). Table 1 provides the most performed methods for harvesting microalgal biomass (Gerardo et al. 2015). In some cases, two methods can be combined to provide better efficiency. It is noteworthy that extensive information on harvesting algal biomass is available in another chapter of the present book.

METHODS FOR CELL DISRUPTION AND EXTRACTION OF ALGAL COMPONENTS

Biorefinery is highly dependent on the procedure of components removal from algal biomass. These technologies consist of two steps. First, a suitable method is required to disrupt the cells and provide the cell contents available for extraction. Second, the released products should be purified through appropriate procedures. While different algal species possess specific characteristics, it is necessary to consider the algal structure when choosing an extraction method.

Up to now, several methods have been proposed for cell disruption of algal species and the extraction of algal components, including bead milling, high-pressure homogenization, ultrasonication, pulsed electric field, osmotic shock, microwaves, enzymatic treatment, ionic liquids, and chemical hydrolysis (Günerken et al. 2015; Halim 2020). As discussed above, the chosen method for cell disruption and extraction in the biorefinery process should be mild and provide suitable conditions for the extraction of various components. For instance, the main concentration of the appropriate method is to perforate the outer cell wall/membrane while they do not damage the contents inside the cells, resulting in easier extraction of the cell compounds. Moreover, the cell disruption method selection depends on cell wall thickness and product localization in the cytoplasm (Alavijeh et al. 2020). The following paragraphs will explain the present methods used for the cell disruption and extraction of valuable algal compounds.

Bead Milling

Bead milling is a method in which the algal cells disrupt due to the grinding of cells in contact with the solid beads. Numerous factors affect the performance of this method. For example, the diameter of beads above 0.5 mm (about 0.02 in) negatively affects the cell disruption, while the smaller diameters (than 0.5 mm) have a positive effect. Besides, the density of beads is important (Günerken et al. 2015). For instance, low-density beads such as glass beads are more efficient in low viscosity media. On the contrary, high viscosity beads such as zirconium are beneficial in high viscosity media. Other parameters also, such as dry cell weight and retention time, should be considered when using this method. Disadvantages of bead milling include high energy requirements and high-temperature rise leading to potential thermo-degradation of target components (Günerken et al. 2015; Vanthoor-Koopmans et al. 2013).

High-Pressure Homogenization

High-pressure homogenization has been reported to be efficient in recovering components such as high-value proteins, lipids, and pigments. In this mechanical method, the cell cultures are subjected to high-pressure conditions resulting in the disruption of the cells. In other words, a suspension containing cells pumps through a valve collides with an impact ring, and comes out from the valve (Halim 2020). The hydrodynamic cavitation, shear forces provided by the high pressure on the impact ring and valve seat, and the pressure the cells tolerate when passing from the valve seat are important parameters in this method. However, using this method could be highly species-specific, in which its efficiency differs according to cell wall rigidity (Dong et al. 2016). However, as with bead milling, this method is not useful in algal biorefineries due to high energy requirements for downstream processing.

Chemical Cell Disruption

Chemical methods have been proposed to be efficient for cell walls disruption by breaking the bonds of walls (Dong et al. 2016). Various chemicals such as acids, antibiotics, bases, and solvents could be used for cell disruption, and the suitable chemical should be selected according to the algal species structure (Günerken et al. 2015). These methods require low energy and prevent the formation of an emulsion. However, chemical methods are not suitable for biorefinery processes while these methods are not mild and are approximately slow compared to novel methods. Moreover, chemical methods may damage other cell components during the extraction. Besides, Chemical processes performed at extreme pH also have a high capital cost due to extensive corrosion.

Enzymatic Cell Disruption

The algal cell walls could be hydrolyzed when subjected to enzymatic treatment (Demuez et al. 2015). In this case, knowing the cell wall structure is vital. Enzymes could be extremely effective, compared to mechanical cell disruption methods, by consuming less energy. Moreover, enzymes reduce the extraction time, minimize solvent usage, preserve biological activity, and increase the extract yield (Terme et al. 2020). As well discussed in the literature, the composition of different algal cell walls varies widely, requiring specific enzymes for cell wall disruption. In this regard, one or more enzymes (e.g., lysozyme, cellulase, Macerozyme, etc.) could be performed in the process (Vanthoor-Koopmans et al. 2013). Although enzymes utilization has been considered a mild method for cell wall disruption, performing them in the industry is doubtful due to their high costs. Accordingly, to be advantageous in a biorefinery process, they should be reused or immobilized to reduce the costs. Additionally, tailored enzymatic cocktails specifically designed based on the cell-wall architecture of the specific species, can be deployed in order to reduce the cost of unnecessary enzymes and lower enzyme concentrations (Vanthoor-Koopmans et al. 2013).

Osmotic Shock

As a result of high osmotic pressure followed by sudden dilution, osmotic shock weakens cells by causing lysis. The reaction is caused by the rapid entry of water into cells with increased internal pressure. Osmotic shock is another method for algal cell disruption, considered low-cost and requires lower energy than other methods. However, this method is highly species-specific, which should be tuned for the specific species (Naghdi et al. 2016). The osmotic shock disrupts the cells due to the high-pressure differences between the media and cells due to the abrupt change in the medium's solute concentration (Halim 2020). This method also requires longer times in comparison with other cell disruption methods.

Ultrasound

In ultrasonication, cell disruption happens due to extensive shear forces in the medium. This method has been utilized to recover various algal components such as lipids, chlorophylls, and carotenoids (e.g., astaxanthin) (Rahimi and Jazini 2021). This method possesses some disadvantages. For instance, the sample temperature may increase during the process, which destroys the cell component (e.g., proteins) (Günerken et al. 2015). Besides, this method's efficiency is highly dependent on the structure of the cells and may not be suitable for all species. On the contrary, ultrasonic treatment is simple, cost-effective (when used on concentrated slurries rather than dilute cultures), and could be developed for specific species to provide high yield in short durations. Moreover, various solvents could be used in the process, and compared to other extraction methods, the equipment cost is lower. Furthermore, it is an exceptionally low energy-intensive method, and the combination of this method with other available disruption methods or solvents is proposed to reduce further energy demands (Vanthoor-Koopmans et al. 2013).

Microwave

Performing microwave energy is beneficial to recovery and increasing the yield of algal components such as lipids and pigments from algal species. When a suspension is subjected to microwave energy, the interaction between the waves and liquid molecules results in increasing local heat. Therefore, the provided pressure on the cells disrupts the cells (Günerken et al. 2015). The extraction time, solvents reduction, and extraction yields increment are among the advantages of microwave extraction. However, there are some limitations in using this method in algal biorefineries, which make this method complicated in an industrial-scale setup. For instance, this method is not suitable for volatile target components. Besides, emulsion formation is possible when using this procedure, which prevents solvent recovery (Halim 2020).

Pulsed Electric Field

The pulsed electric field applies short electric treatments to provide electroporation, resulting in the permeability of algal cells though the cell contents could be extracted easily (Martínez et al. 2017). For instance, this method resulted in the potential extraction of pigments and c-phycocyanin from *Chlorella vulgaris* and *Artrosphira platensis*, respectively (Luengo et al. 2014; Martínez et al. 2017). This method is nonthermal, and its intensity is determined by the provided electric pulses in the media (100-300 V/cm to 20-80 kV/cm) (Barba et al. 2015). This method is advantageous as it is environmentally friendly and reduces the whole energy of the process as the energy input is not high, does not require solvents for protein recovery, and elevated temperature. On the contrary, as it is an almost new method, efforts are needed to provide extensive information on the operation process as well as the effect of this method on various algal species.

Ionic Liquids

Ionic liquids are anionic and cationic salts that are useful for extracting high-value compounds from algal biomass (Vermuë et al. 2018). Up to now, imidazolium-based ionic liquids that are commercial have been used for bioproducts extraction from microalgae. Moreover, ionic liquids from cholinium, known as toxic, were performed in algal biotechnology, but the information about using pyridinium- and phosphonium-based ionic liquids is still low (Tan et al. 2020). The low melting point is the advantage of ionic liquids, which helps the extraction reaction perform at low temperatures. Moreover, they are not pollutants and can be recycled easily compared to organic solvents. According to the presence of functional groups, ionic liquids can be used explicitly for selective extraction of the target compounds. However, the high cost of ionic liquids is a significant drawback in their utilization (Niemi and Gentili 2021).

Supercritical Fluid Extraction

The separation of one component from others using supercritical fluids (e.g., carbon dioxide) as the solvent is called supercritical fluid extraction. This method is promising for extracting algal components. Supercritical fluid extraction is reported to be efficient in extracting several components such as bioactive compounds, lipids, and pigments (Niemi and Gentili 2021). It is also advantageous when it is low cost, environmentally friendly, and reduces the extraction duration compared to conventional fluids (Shanmugam, Ganesan, and Rajauria 2021). Moreover, by optimizing this method, several components could be extracted. However, the equipment costs a lot, and this method has limited effectiveness when applied to wet microalgae slurry (rather than dry biomass). Besides, the technique requires more investigation due to the limited knowledge of the mixtures and solvents.

Pressurized Solvent Extraction

Pressurized solvent extraction is an emerging technology in algal biorefineries. In this method, desorption of the compounds of interest from the matrix is followed by diffusion of the molecules through the sample pores and collection of the molecules in the solvent (Garcia-Vaquero et al. 2021). Different solvents could be used during this process, but as the dielectric constant of water in high temperatures (e.g., 250 °C) is comparable to some organic solvents such as ethanol or methanol, using water as a solvent could be the best alternative (Gallego et al. 2018). Pressure and temperature are two essential parameters that need to be controlled during this process. Reduction of the extraction duration and low concentrations of organic solvents are among the advantages of this method. On the contrary, this method could not easily be scaled up in order to the prohibitive costs of the equipment and difficulties in large-scale maintaining of extraction conditions.

CONVERSION OF ALGAL BIOMASS

As discussed in previous sections, algae have been considered as potential candidates for energy production. Besides, according to the biorefinery concept, the simultaneous production of bulk chemicals and biofuels is essential. Moreover, a suitable method is required to reach the maximum energy yield. In this regard, conversion technologies are divided into biochemical, chemical, thermochemical, and photosynthetic conversion processes. Table 2 provides an overview of different conversion methods, descriptions, and products (Sankaran et al. 2018). Moreover, another chapter of the present book focused on the biomass conversion processes.

ALGAL BIOREFINERY CONCEPT AND CHALLENGES

However, algal biomass has been identified as a potential resource for a variety of biochemicals and biofuels, the production of a single product such as proteins for food purposes or even biofuels is not yet economically feasible. The reason is high total production and processing costs (Vanthoor-Koopmans et al. 2013). As a result, multi-product biorefineries should replace single-product ones in order to increase the viability of algal biotechnology. In this scenario, the procedures associated with harvesting, cell disruption, and extraction of target products are critical. However, the processes related to algal biorefineries are expensive due to the more unit operations necessary for the separation and purification of various products (Vermuë et al. 2018). One cost-cutting strategy is to produce higher-value goods, such as pigments, which cannot be relied on much. However, the most significant shortcoming is the absence of mild technologies for obtaining various fractions of algal biomass. Consequently, more emphasis should be placed on cell disruption, extraction, and purification procedures.

Downstream Processing of Algal Biomass

Although harvesting-related technologies are prospering, more work needs to be done to improve the process. Centrifugation and filtration are two potential procedures that significantly increase biomass harvesting. Moreover, these processes provide high-quality biomass as there is no need for chemicals.

Table 2. Various processes for the conversion of algal biomass to bioenergy. Biochemical, chemical, thermochemical, and photosynthetic conversion methods could be performed to produce different products.

Conversion method	Description/conditions	Products
Biochemical conversion		
Fermentation	Conversion of carbohydrates using various microorganisms (e.g., *Clostridium acetobutylicum*, and *Saccharomyces cerevisiae*)	Biobutanol Bioethanol
Anaerobic digestion	Conversion of organic matter to using bacteria in the absence of oxygen	Biogas
Photobiological hydrogen generation	Conversion of water to H_2 during the cultivation of algae	Biohydrogen
Chemical Conversion		
Transesterification	Conversion of algal oil to glycerol and biodiesel in the presence of an alcohol (and catalyst)	Biodiesel
Thermochemical conversion		
Gasification	Incomplete oxidation of algal biomass at high temperatures (700-1300 °C)	Syngas
Hydrothermal liquefaction	Performing water in average temperature (250-350 °C) and high pressure (5-20 MPa) for algal biomass conversion	Bio-oil
Pyrolysis	Thermal decomposition of algal biomass in: Atmospheric pressure 200-750 °C Inert atmosphere	Biochar Bio-oil Syngas
Torrefaction	Thermochemical conversion of algal biomass in the absence of oxygen in 200-300 °C for direct combustion.	Biochar
Photosynthetic conversion		
Microbial fuel cell	Conversion of chemical energy to electricity by microorganisms' action	Bioelectricity

These approaches, however, are prohibitively expensive in terms of operational and capital expenses (Chandra et al. 2019). Other techniques for harvesting algal biomass, such as coagulation, flocculation, and flotation, have been developed effectively. Although these approaches are less expensive, they have certain drawbacks. For example, harvested biomass may contain flocculant/coagulant/surfactant leftovers (e.g., chemicals, bioflocculants, and nanoparticles) (Chandra et al. 2019; Gerardo et al. 2015). If these components are not of food quality, their presence may diminish the value of collected biomass (Vermuë et al. 2018). Furthermore, there are numerous ways for their separation that may raise the overall operating costs. Future research is still essential to enhance knowledge of algal physiology, technological advancements, and harvesting materials regarding harvesting technologies.

Cell disruption is an essential step in a biorefinery. A cell disruption method should be mild to preserve all cell components. As discussed in previous sections, several chemical, mechanical, and biological procedures have been proposed and developed over the last few decades (Alavijeh et al. 2020; Buchmann et al. 2019; Rahimi and Jazini 2021). Changes in pH and high temperatures in chemical techniques severely influence the cell structure (Schwenzfeier, Wierenga, and Gruppen 2011). Although mechanical procedures are milder than biochemical and chemical approaches because they do not need pH or temperature increases, they require high amounts of energy (Coons et al. 2014). Furthermore, using these procedures (e.g., bead milling) disrupts the entire cell, causing difficult separation and purification steps.

In this case, developing physical cell disruption technologies (e.g., pulsed electric field) may be beneficial (Haberkorn et al. 2021). However, these methods are species-specific and may not be applicable when the cell wall is thick. In this case, additional pretreatments are needed to reduce the cell wall resistance. Therefore, in addition to the technology enhancement requirements in this field, understanding the cell wall microstructure of individual algae species is also important.

When executing a technique for the extraction and/or purification of the products, considering the separation of all fractions of algal biomass without losing any product is essential. Three different recent extraction techniques were described in this chapter. However, there are still doubts about performing the suitable procedure, requiring the in-depth study of these techniques.

Integrated Bioprocessing of Algae

In addition to improving and optimizing individual downstream steps, the entire process should be considered well. As shown in Figure 3, four routes could be proposed in the bioprocessing of algal biomass. As usual, the first procedure is conducting each step individually (i.e., cultivation, harvest, cell disruption, and extraction), which optimization of these processes is required to lower the overall process cost. According to procedures 2, 3, and 4, the combination of various steps is possible.

The combination of cultivation and algal harvest (Procedure 2, Figure 3) has been reported previously. In this regard, the biomass of *Chlorella vulgaris* was successfully dewatered utilizing recyclable flocculants (Morrissey et al. 2015). Regarding the third process (Figure 3), integrated harvesting and disruption of cells have been reported several times. Ferric ions, for example, were used to capture and disturb wet microalgal biomass (*Chlorella* sp. KR-1) at the same time (D.-Y. Kim et al. 2015). However, extreme conditions (e.g., temperatures between 50 and 70 °C) are frequently used in these procedures, which might destroy algal proteins. Although the integration of harvesting and cell disruption can make the process more straightforward, combining cell disruption and extraction (procedure 4, Figure 3) could be beneficial. The use of ionic liquids has been described multiple times in this regard. Ionic liquids, for example, were used to extract high-value astaxanthin from the microalgae *Haematococcus pluvialis* (Desai et al. 2016).

CONCLUSION

The major products of algae are carbohydrates, lipids, and proteins which can be used for several purposes, such as biofuels, nutraceuticals, and pharmaceuticals. Although many advances have been made in the bioprocessing of various algal species, their utilization on an industrial scale is not feasible. Therefore, technological enhancements are required in the upstream and downstream processing of algal biomass. This chapter provides an overview of the bioprocessing of algae, focusing on the cultivation and biorefinery of algal biomass. Regarding the cultivation of algal species, the concentration should be on improving biomass productivity and upgrading the target compounds. Besides, due to the ability of algae for CO_2 sequestration and bioremediation, the possibility of algae cultivation using industrial flue gases and wastewaters should be considered. Moreover, in terms of algal biorefinery, the most significant challenge is the lack of mild technologies for the fractionation of all algal components, which requires more effort to enhance cell disruption and extraction techniques. However, to have a successful biorefinery, the whole process should be considered, in which significant knowledge of algae structure and biochemical composition is essential.

Figure 3. An overview of integrated unit processes in algal bioprocessing. 1) Simple procedure, 2) Combination of cultivation and harvest, 3) Combination of harvest and cell disruption, and 4) Combination of cell disruption and extraction.

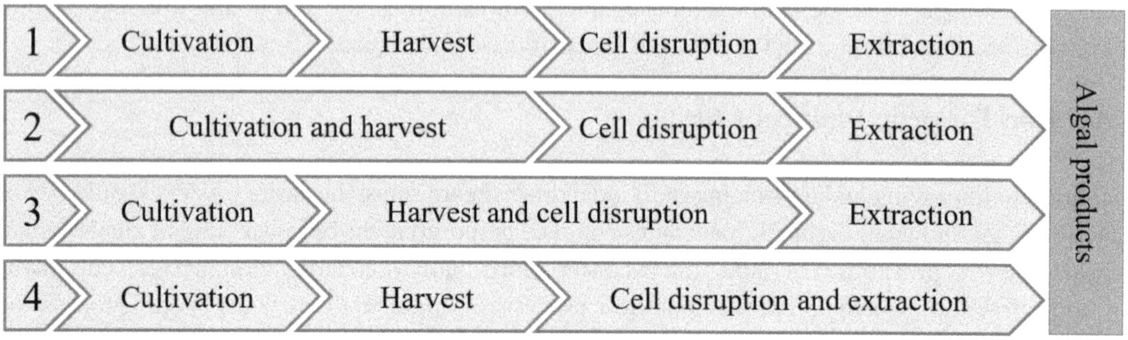

REFERENCES

Abomohra, A. E.-F., El-Naggar, A. H., Alaswad, S. O., Elsayed, M., Li, M., & Li, W. (2020). Enhancement of Biodiesel Yield from a Halophilic Green Microalga Isolated under Extreme Hypersaline Conditions through Stepwise Salinity Adaptation Strategy. *Bioresource Technology*, *310*, 123462. doi:10.1016/j.biortech.2020.123462 PMID:32388355

Alavijeh, R. S., Karimi, K., Wijffels, R. H., van den Berg, C., & Eppink, M. (2020). Combined Bead Milling and Enzymatic Hydrolysis for Efficient Fractionation of Lipids, Proteins, and Carbohydrates of Chlorella Vulgaris Microalgae. *Bioresource Technology*, *309*, 123321. doi:10.1016/j.biortech.2020.123321 PMID:32305840

Anand, V., Gautam, R., & Vinu, R. (2017). Non-Catalytic and Catalytic Fast Pyrolysis of Schizochytrium Limacinum Microalga. *Fuel*, *205*, 1–10. doi:10.1016/j.fuel.2017.05.049

Barba, F. J. (2015). Current Applications and New Opportunities for the Use of Pulsed Electric Fields in Food Science and Industry. *Food Research International, 77*, 773–98.

Bhattacharya, M., & Goswami, S. (2020). Microalgae – A Green Multi-Product Biorefinery for Future Industrial Prospects. *Biocatalysis and Agricultural Biotechnology*, *25*, 101580. doi:10.1016/j.bcab.2020.101580

Bleakley, S., & Hayes, M. (2017). Algal Proteins: Extraction, Application, and Challenges Concerning Production. *Foods*, *6*(5), 1–34. doi:10.3390/foods6050033 PMID:28445408

Buchmann, L., Brändle, I., Haberkorn, I., Hiestand, M., & Mathys, A. (2019). Pulsed Electric Field Based Cyclic Protein Extraction of Microalgae towards Closed-Loop Biorefinery Concepts. *Bioresource Technology*, *291*, 121870. doi:10.1016/j.biortech.2019.121870 PMID:31382092

Chandra, R., Iqbal, H. M. N., Vishal, G., Lee, H.-S., & Nagra, S. (2019). Algal Biorefinery: A Sustainable Approach to Valorize Algal-Based Biomass towards Multiple Product Recovery. *Bioresource Technology, 278*, 346–359. doi:10.1016/j.biortech.2019.01.104 PMID:30718075

Chew, K. W., Yap, J. Y., Show, P. L., Suan, N. H., Juan, J. C., Ling, T. C., Lee, D.-J., & Chang, J.-S. (2017). Microalgae Biorefinery: High Value Products Perspectives. *Bioresource Technology, 229*, 53–62. doi:10.1016/j.biortech.2017.01.006 PMID:28107722

Coons, J. E., Kalb, D. M., Dale, T., & Marrone, B. L. (2014). Getting to Low-Cost Algal Biofuels: A Monograph on Conventional and Cutting-Edge Harvesting and Extraction Technologies. *Algal Research, 6*, 250–270. doi:10.1016/j.algal.2014.08.005

del Río, P. G., Gomes-Dias, J. S., Rocha, C. M. R., Romaní, A., Garrote, G., & Domingues, L. (2020). Recent Trends on Seaweed Fractionation for Liquid Biofuels Production. *Bioresource Technology, 299*, 122613. doi:10.1016/j.biortech.2019.122613 PMID:31870706

Demuez, M., Mahdy, A., Tomás-Pejó, E., González-Fernández, C., & Ballesteros, M. (2015). Enzymatic Cell Disruption of Microalgae Biomass in Biorefinery Processes. *Biotechnology and Bioengineering, 112*(10), 1955–1966. doi:10.1002/bit.25644 PMID:25976593

Desai, R. K., Streefland, M., Wijffels, R. H., & Eppink, M. H. M. (2016). Novel Astaxanthin Extraction from Haematococcus Pluvialis Using Cell Permeabilising Ionic Liquids. *Green Chemistry, 18*(5), 1261–1267. doi:10.1039/C5GC01301A

Dong, T., Knoshaug, E. P., Pienkos, P. T., & Laurens, L. M. L. (2016). Lipid Recovery from Wet Oleaginous Microbial Biomass for Biofuel Production: A Critical Review. *Applied Energy, 177*, 879–895. doi:10.1016/j.apenergy.2016.06.002

Eltanahy, E., & Torky, A. (2021). Microalgae as Cell Factories: Food and Feed-Grade High-Value Metabolites. In *Microalgal Biotechnology: Recent Advances, Market Potential, and Sustainability*. Royal Society of Chemistry. doi:10.1039/9781839162473-00001

Gallego, R., Montero, L., Cifuentes, A., Ibáñez, E., & Herrero, M. (2018). Green Extraction of Bioactive Compounds from Microalgae. *Journal of Analysis and Testing, 2*(2), 109–123. doi:10.100741664-018-0061-9

Garcia-Vaquero, M., Sweeney, T., O'Doherty, J., & Rajauria, G. (2021). Recent Advances in the Use of Greener Extraction Technologies for the Recovery of Valuable Bioactive Compounds from Algae. In *Recent Advances in Micro and Macroalgal Processing: Food and Health Perspectives* (pp. 96–122). Wiley Online Library. doi:10.1002/9781119542650.ch4

Gatamaneni, B. L., Orsat, V., & Lefsrud, M. (2018). Factors Affecting Growth of Various Microalgal Species. *Environmental Engineering Science, 35*(10), 1037–1048. doi:10.1089/ees.2017.0521

Gerardo, M. L. (2015). Harvesting of Microalgae within a Biorefinery Approach: A Review of the Developments and Case Studies from Pilot-Plants. *Algal Research, 11*, 248–62.

Günerken, E., D'Hondt, E., Eppink, M. H. M., Garcia-Gonzalez, L., Elst, K., & Wijffels, R. H. (2015). Cell Disruption for Microalgae Biorefineries. *Biotechnology Advances*, *33*(2), 243–260. doi:10.1016/j.biotechadv.2015.01.008 PMID:25656098

Haberkorn, I., Buchmann, L., Häusermann, I., & Mathys, A. (2021). Nanosecond Pulsed Electric Field Processing of Microalgae Based Biorefineries Governs Growth Promotion or Selective Inactivation Based on Underlying Microbial Ecosystems. *Bioresource Technology*, *319*, 124173. doi:10.1016/j.biortech.2020.124173 PMID:33017777

Halim, R. (2020). Industrial Extraction of Microalgal Pigments. In *Pigments from Microalgae Handbook* (pp. 265–308). Springer. doi:10.1007/978-3-030-50971-2_12

Jankowska, E., Sahu, A. K., & Oleskowicz-Popiel, P. (2017). Biogas from Microalgae: Review on Microalgae's Cultivation, Harvesting and Pretreatment for Anaerobic Digestion. *Renewable & Sustainable Energy Reviews*, *75*, 692–709. doi:10.1016/j.rser.2016.11.045

Javed, F., Aslam, M., Rashid, N., Shamair, Z., Khan, A. L., Yasin, M., Fazal, T., Hafeez, A., Rehman, F., Rehman, M. S. U., Khan, Z., Iqbal, J., & Bazmi, A. A. (2019). Microalgae-Based Biofuels, Resource Recovery and Wastewater Treatment: A Pathway towards Sustainable Biorefinery. *Fuel*, *255*, 115826. doi:10.1016/j.fuel.2019.115826

Khan, M. (2016). Heterotrophic Growth of Micro Algae. Recent Advances in Microalgal Biotechnology.

Kim, D.-Y., Oh, Y.-K., Park, J.-Y., Kim, B., Choi, S.-A., & Han, J.-I. (2015). An Integrated Process for Microalgae Harvesting and Cell Disruption by the Use of Ferric Ions. *Bioresource Technology*, *191*, 469–474. doi:10.1016/j.biortech.2015.03.020 PMID:25817422

Kim, G. M., Chang, W.-S., & Kim, Y.-K. (2021). Biocomposites Using Whole or Valuable Component-Extracted Microalgae Blended with Polymers: A Review. *Catalysts*, *12*(1), 25. doi:10.3390/catal12010025

Lam, M. K., & Lee, K. T. (2013). Biohydrogen Production from Algae. In Biohydrogen. Elsevier B.V. doi:10.1016/B978-0-444-59555-3.00008-8

Liang, Y. (2015). *Industrial Biorefineries and White Biotechnology. In Algal Biorefineries*. Elsevier B.V.

Luengo, E., Condón-Abanto, S., Álvarez, I., & Raso, J. (2014). Effect of Pulsed Electric Field Treatments on Permeabilization and Extraction of Pigments from Chlorella Vulgaris. *The Journal of Membrane Biology*, *247*(12), 1269–1277. doi:10.100700232-014-9688-2 PMID:24880235

Martínez, J. M., Luengo, E., Saldaña, G., Álvarez, I., & Raso, J. (2017). C-Phycocyanin Extraction Assisted by Pulsed Electric Field from Artrosphira Platensis. *Food Research International*, *99*, 1042–1047. doi:10.1016/j.foodres.2016.09.029 PMID:28865615

Mohan, S. V. (2020). Algal Biorefinery Models with Self-Sustainable Closed Loop Approach: Trends and Prospective for Blue-Bioeconomy. *Bioresource Technology, 295*, 122128.

Moreno-Garcia, L., Adjallé, K., Barnabé, S., & Raghavan, G. S. V. (2017). Microalgae Biomass Production for a Biorefinery System: Recent Advances and the Way towards Sustainability. *Renewable & Sustainable Energy Reviews*, *76*(March), 493–506. doi:10.1016/j.rser.2017.03.024

Morrissey, K. L. (2015). Charge-tunable Polymers as Reversible and Recyclable Flocculants for the Dewatering of Microalgae. *Biotechnology and Bioengineering, 112*(1), 74–83.

Naghdi, F. G., Bai, X., Thomas-Hall, S. R., Sharma, K., & Schenk, P. M. (2016). Lipid Extraction from Wet Chaetoceros Muelleri Culture and Evaluation of Remaining Defatted Biomass. *Algal Research*, *20*, 205–212. doi:10.1016/j.algal.2016.10.011

Ngamwonglumlert, L., & Devahastin, S. (2018). Carotenoids. Encyclopedia of Food Chemistry, 40–52.

Niemi & Gentili. (2021). Extraction Technologies for Functional Lipids. In *Recent Advances in Micro and Macroalgal Processing: Food and Health Perspectives*. Wiley Online Library.

Radulovich, R. (2015). Farming of Seaweeds. In *Seaweed Sustainability* (pp. 27–59). Elsevier. doi:10.1016/B978-0-12-418697-2.00003-9

Rahbari, H., Akram, A., Pazoki, M., & Aghbashlo, M. (2019). Bio-Oil Production from Sargassum Macroalgae: A Green and Healthy Source of Energy. *Jundishapur Journal of Health Sciences*. Advance online publication. doi:10.5812/jjhs.84301

Rahimi, M., & Jazini, M. (2021). Mixotrophic Cultivation of Chromochloris Zofingiensis on Glycerol, Acetate, and Vinasse. *Journal of Applied Phycology*, *33*(6), 3579–3590. Advance online publication. doi:10.100710811-021-02568-z

Ramachandra, T. V., & Hebbale, D. (2019, February). Bioethanol from Macroalgae: Prospects and Challenges. *Renewable & Sustainable Energy Reviews*, *117*, 109479. doi:10.1016/j.rser.2019.109479

Sankaran, R., Show, P. L., Nagarajan, D., & Chang, J. S. (2018). Exploitation and Biorefinery of Microalgae. In *Waste Biorefinery: Potential and Perspectives* (pp. 571–601). Elsevier. doi:10.1016/B978-0-444-63992-9.00019-7

Schwenzfeier, A., Wierenga, P. A., & Gruppen, H. (2011). Isolation and Characterization of Soluble Protein from the Green Microalgae Tetraselmis Sp. *Bioresource Technology*, *102*(19), 9121–9127. doi:10.1016/j.biortech.2011.07.046 PMID:21831634

Shanmugam, Ganesan, & Rajauria. (2021). Extraction Technologies to Recover Dietary Polyphenols from Macro-and Microalgae" In Recent Advances in Micro and Macroalgal Processing: Food and Health Perspectives. Wiley Online Library.

Sudhakar, K., Mamat, R., Samykano, M., Azmi, W. H., Ishak, W. F. W., & Yusaf, T. (2018). An Overview of Marine Macroalgae as Bioresource. *Renewable & Sustainable Energy Reviews*, *91*, 165–179. doi:10.1016/j.rser.2018.03.100

Sudhakar, M. P., & Ramesh Kumar, B. (2019). A Review on Bioenergy and Bioactive Compounds from Microalgae and Macroalgae-Sustainable Energy Perspective. *Journal of Cleaner Production*, *228*, 1320–1333. doi:10.1016/j.jclepro.2019.04.287

Tan, J. S., Lee, S. Y., Chew, K. W., Lam, M. K., Lim, J. W., Ho, S.-H., & Show, P. L. (2020). A Review on Microalgae Cultivation and Harvesting, and Their Biomass Extraction Processing Using Ionic Liquids. *Bioengineered*, *11*(1), 116–129. doi:10.1080/21655979.2020.1711626 PMID:31909681

Terme, N. (2020). Emerging Seaweed Extraction Techniques: Enzyme-Assisted Extraction a Key Step of Seaweed Biorefinery? In *Sustainable Seaweed Technologies* (pp. 225–256). Elsevier. doi:10.1016/B978-0-12-817943-7.00009-3

Vanthoor-Koopmans, M., Wijffels, R. H., Barbosa, M. J., & Eppink, M. H. M. (2013). Biorefinery of Microalgae for Food and Fuel. *Bioresource Technology*, *135*, 142–149. doi:10.1016/j.biortech.2012.10.135 PMID:23186688

Vermuë, M. H., Eppink, M. H. M., Wijffels, R. H., & Van Den Berg, C. (2018). Multi-Product Microalgae Biorefineries: From Concept towards Reality. *Trends in Biotechnology*, *36*(2), 216–227. doi:10.1016/j.tibtech.2017.10.011 PMID:29132753

Wells, M. L., Potin, P., Craigie, J. S., Raven, J. A., Merchant, S. S., Helliwell, K. E., Smith, A. G., Camire, M. E., & Brawley, S. H. (2017). Algae as Nutritional and Functional Food Sources: Revisiting Our Understanding. *Journal of Applied Phycology*, *29*(2), 949–982. doi:10.100710811-016-0974-5 PMID:28458464

Xiu, S., & Shahbazi, A. (2012). Bio-Oil Production and Upgrading Research : A Review. *Renewable & Sustainable Energy Reviews*, *16*(7), 4406–4414. doi:10.1016/j.rser.2012.04.028

Xu, S., Elsayed, M., Ismail, G. A., Li, C., Wang, S., & Abomohra, A. E.-F. (2019). Evaluation of Bioethanol and Biodiesel Production from Scenedesmus Obliquus Grown in Biodiesel Waste Glycerol: A Sequential Integrated Route for Enhanced Energy Recovery. *Energy Conversion and Management*, *197*, 111907. doi:10.1016/j.enconman.2019.111907

Zhan, J., Rong, J., & Wang, Q. (2016). Mixotrophic Cultivation, a Preferable Microalgae Cultivation Mode for Biomass / Bioenergy Production, and Bioremediation, Advances and Prospect. *International Journal of Hydrogen Energy*, 1–13.

Zhang, B. (2013). New Golden Age of the Algal Technology. *Journal of Petroleum & Environmental Biotechnology*, *04*(06), 7463. doi:10.4172/2157-7463.1000e120

KEY TERMS AND DEFINITIONS

Algae Bioprocessing: Employing different processes to produce high-value biochemicals and biofuels from algae.

Algae Biorefinery: The process in which all fractions of algae are valorized without any remaining wastes.

Cell Disruption: Breaking the algae's membrane and/or cell wall to release their intracellular components.

Cultivation: The action of producing algal biomass.

Downstream Processing: The processes related to harvesting of algal biomass, as well as cell disruption and extraction of algae components.

Harvest: The separation of algal biomass from water and other nutrients.

Mild Conditions: Processes used to valorize algae components without damaging other compounds.

Chapter 8
Scaling Up and Harvesting of Algae

Ragaa A. Hamouda
University of Sadat City, Egypt

Nada M. Doleib
University of Jeddah, Saudi Arabia

Muhammad A. Abuelmagd
Mansoura University, Egypt

ABSTRACT

The scaling up and increment of the algal cultures cultivation process is a complex task that requires experienced staff. Some parameters such as biomass yield, biomass productivity, and specific growth should be calculated using the findings of laboratory scale that might be relevant for large-scale production as it provides a baseline to visualize and to verify production balance-related problems in the algal production system. The main goal of scale-up is to increase the production quantities with comparable or higher productivity and product quality. The harvesting process of the algal biomass represents a major hindrance in microalgae industry as it is approximately ranged from 20 to 30% of the total cost of the cultivation. There are many harvesting techniques such as physical, chemical, biological methods, and magnetic particle facilitated separation. This chapter has summarized the research progress in algal scaling up by optimizing different parameters such as light, temperature, nutrients, and strain selection.

INTRODUCTION

Scaling up of the algal cultures to the large scales depends mainly on environmental conditions, such as temperature and solar radiation, though this complex process requires skilled and experienced personnel, and also necessary to find the optimum concentrations of each element of the process to achieve the maximum productivity and minimize the required cost and saving the time (Borowitzka & Vonshak, 2017). Nowadays, the maintenance of a suitable cultivation technologies which achieves a stable, long-

DOI: 10.4018/978-1-6684-2438-4.ch008

term large-scale cultures, high-productivity faces constraints such as highly risk of contamination, mass transfer rate, poor gas-liquid, low mixing, and biomass concentration, enormous evaporation of water, poor temperature management and insufficient light penetration (Das, 2015). It is necessary to choose the most appropriate photobioreactor design and suitable conditions to perform algae cultivation on a high scale (Guler et al., 2019). The prospective of algal scaling up is the production of a valuable bioproducts and for other applications such as oil production, biohydrogen, low-cost fertilizer, feed application and wastewater treatment (Couto et al., 2021; Emblemsvåg et al., 2020; Khosravitabar, 2020; Morillas-España et al., 2021; Norsker et al., 2021). There is a number of microalgal species being cultured on a large-scale commercially; the major one is *Arthrospira platensis* and *Chlorella* species, which used basically as a nutritional supplements, in addition to *Dunaliella salina* as sources of β-carotene. Microalgal species such as *Porphyridium cruentum* are being cultivated for produce polysaccharides and phycobilin pigments on small commercial scale (Arad Malis and van Moppers 2013), furthermore, the extensive production of algal to be used as feed in hatcheries for live aquaculture involving species such as *Isochrysis galbana*, *Chaetoceros* spp., *Tisochrysis lutea Nannochloropsis* spp., *Skeletonema costatum*, *Tetraselmis* spp., *Pavlova lutheril; Thalassiosira pseudonana* etc. (Borowitzka 1998; Muller-Fuega 2004). This study concentrated on the scale-up and development of the optimum production conditions for biomass and/or different bio-chemicals in the large scale level. Furthermore, this chapter insight the algal conventional and advanced harvesting methods, as well as clarify their advantage and disadvantage.

SCALING UP OF ALGAL CULTURES

Problems in Scaling Up

Scaling up of algal cultures is usually done by a factor of 10 per step, i.e. raising from 1 L to 10 L and so on (Borowitzka & Vonshak, 2017). Scale-up is actually a complex process whereas transition from flask (laboratory) scale to pond or photobioreactor (industrial) scale is the foremost bottleneck stage so a various factors must be considered to avoid the collapse in the algal scaling up process;

1. Achieving maximum algal growth rate and/or accumulate the target algal product, using the available conditions with cost effective techniques.
2. Control in the chemical and physical parameters affects algal growth and accumulation target products.
3. Robustness of the culture process as the algal productivity affected by the modification in nutrient feeding process, temperature, and light.
4. Competition with other algal species and/ or bacterial contaminants.
5. Nutrient availability and its predilection whereas during growth, CO_2, nitrogen and phosphorus utilization differs from species to another.
6. Efficiency of harvesting with low cell lysed percentage, extraction and purification of bioproducts should be also considered. For algal and cyanobacterial cultures the main factors affecting growth and productivity are temperature variations, light intensity and nutrients availability, most commercial-scale algal biomass production is performed out outdoors in sun light, exposing the algal culture to daily variations in light as well as shorter term variations owing to clouds (Das,

2015; Riveros et al., 2018). A lot of recent commercial large-scale cultures used open air culture systems, particularly paddle- wheel driven raceway systems as well as closed photobioreactors.

Algal Cultivation Systems

Generally, algae grow photoautotrophically which means that they use light as energy source and carbon dioxide as carbon source. Thus, Algal cultivation can be performed in open ponds or closed photobioreactors.

Open Ponds System

Open pond system is a large shallow pond that provides an economic and simple method for algal cultivation, as well as it provides algal culture with maximum light penetration and free solar radiation. There are different designs for open ponds system such as tanks, wide shallow ponds, circular ponds as well as raceway ponds. Race-way pond design is the first design and highest pond for algal cultivation that proposed by Oswald and used for majority of algal cultivation (about 95%) as a result of its ease of mixing and high biomass production (Das, 2015; Mendoza et al., 2013). The raceway pond usually consists of a rectangular shallow canal with a central middle wall dividing the mother pond in two channels arranged in a race track closed loop, current flows from a supply end to an exit one (Chisti, 2007; Rayen et al., 2019). The raceway pond may be drilled into the land or built above it using different materials such as cement, fiber-glass, epoxy or poly vinyl chloride (Grobbelaar, 2013). The raceway pond could be agitated by different tools such as air-lifts, low shear force pumps, and paddle wheels, meanwhile the paddle wheel is the most common tool (Soeder, 1981). The dimensions of raceway pond are very critical parameters such as the length/width ratio and pond depth and its designing as larger width may causes weakness in the current speed, which is not favourable for mixing media and its transfer, while the depth influence the light penetration percentage, variation in the temperature of the culture, and its mixing and associated energy consumption by the wheel (Das, 2015 ; Rayen et al., 2019). However, open ponds also have different disadvantages such as the algal culture is subjected to contamination risk with other microbes, that limits algal biomass production and make algal biomass impure for the extraction of high valuable products. In addition, just a few algal species such as *Chlorella*, *Dunaliella*, *Spirulina*, etc. displayed a successful growing capability in the open system (Das, 2015). Meanwhile, the inefficient stirring may resulted in low mass transfer, hence causes lower in biomass productivity. The current development in open culture system technology includes the development of mixing tools to avoid sedimentation and to enhance light penetration percentage (Kumar & Das, 2012). But the major problem is that, the physical and chemical factors are difficult to be under control such as fluctuation in light availability and temperature during diurnal cycles and seasonal variations (Gupta et al., 2015). As a result of these technical and biological obstacles of these open systems the concerns have given rise to enhancement of closed photobioreactors. Figure 1 represents the schematic diagram of the raceway pond.

Closed Systems (Photobioreactors)

Photobioreactors are designed to control the difficulties related to open pond cultivation methods and to achieve the optimum factors for algal growth temperature, mixing, light penetration, pH as well as to obtain an axenic culture and elimination the risk of contamination for a prolonged duration (Chisti,

Figure 1. Raceway pond schematic design (arrows represent the flow direction)

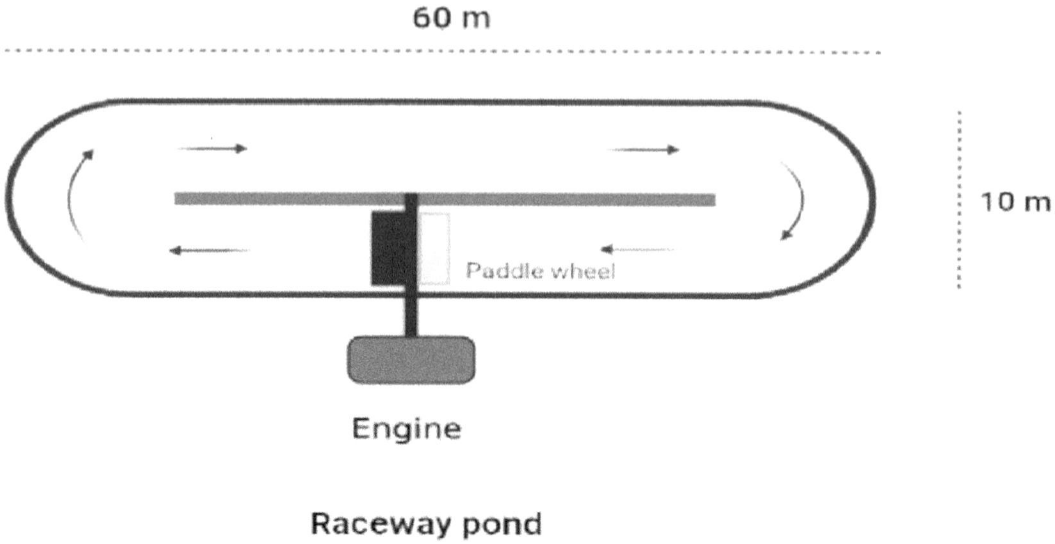

2007; Das, 2015). A photobioreactor is a closed and under control system which consist of culture vessel, motor for agitation the culture and media component, system for gas collection, and control system (Yan & Hino, 2016). Consequently, photobioreactors are able to achieve maximum biomass productivity per unit reactor volume. An uniform illumination is supplied in photobioreactor system via electric lights in order to minimize mutual shading (Andersen, 2005; Gupta et al., 2015). Previous studies reported the use of two fluorescent tubes to create a solar cycle via switching on tubes every 1.5 h to enhance irradiance starting with light intensity quantified as 290 µmol photons / m^2 s1 that elevated to 800 and 1400 µmol photons / m^2 s1 which lasted for 3 h, finally decreased in the same manner (Roncero-Ramos et al., 2019). Agitation system should also attain a fast CO_2 and O_2 transfer (Gupta et al., 2015). Four phases are included in typical photobioreactor; solid (algal biomass), liquid (medium) gaseous (carbon dioxide and oxygen), finally superimposed phase (light/radiation) (Posten, 2009). There are three type of photobioreactors; airlift, flat plate and helical tubular (Yan & Hino, 2016). As a result of closed and controlled surface nature of the photobioreactor, gaseous CO_2 can be pumped using sparing in an organized manner which increases the contact time between gaseous CO_2 and liquid phase hence improves the media stirring, CO_2 sequestration, removal of generated O_2, finally the mass productivity will be increased (Das, 2015). There are different types of photobioreactors such as vertical tubular, horizontal tubular, helical tubular and flat panel. In vertical tubular photobioreactor height/diameter ratio is mostly more than 2.0, CO_2-rich gas is pumped through the bottom via sparger, in case of horizontal tubular photobioreactor a long tubes located horizontally and have a small diameter, the CO_2-rich gas is pumped from one side of the tube through the dedicated gas exchange unit. this type of reactor is suitable for algal cultivation in the outdoor environment (Das, 2015). Helical tubular photobioreactor have coiled shape structure made from flexible, transparent materials and the gas (high CO_2 percentage) is pumped from the bottom opening of the helical reactor. The irradiance source can be positioned co-axially at the centre of the photobioreactor to reduce the loss of light energy and to achieve the maximum biomass productivity (Das, 2015). Finally, the flat panel photobioreactor has flat surface at two sides for receiving the maximum amount of light, so it has a high surface/volume ratio, the stirring is achieved using

Figure 2. A schematic designs for flat panel photobioreactor.

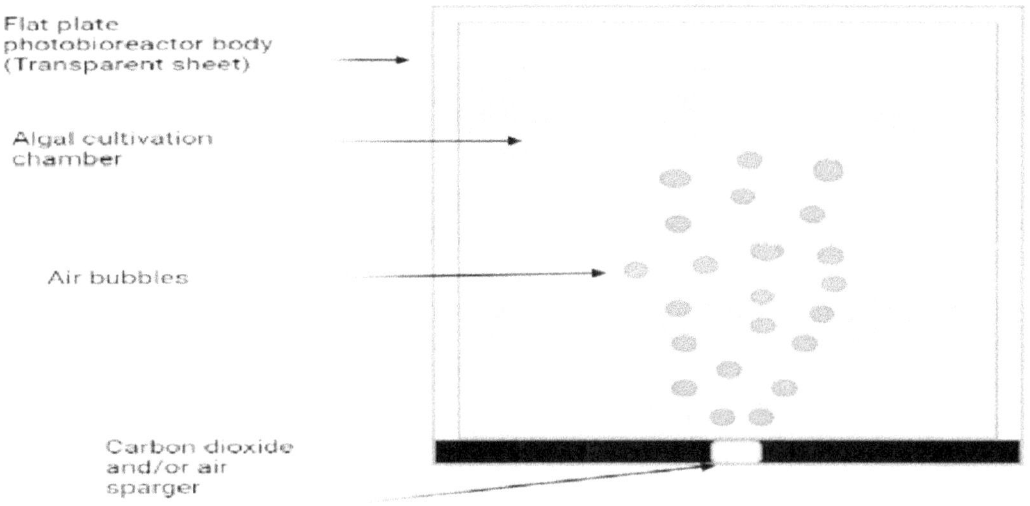

sprager with high percentage of CO_2 gas, which is a fixed tube at the bottom of photobioreactor (Das, 2015). Some flat panel photobioreactors (dimensions; 48 cm height, 26.5 cm width, and 4 cm thickness) were designed to use sun light as natural source for energy and surrounding temperature to minimize cost production as can be, by placing their wider side to East-West direction to make best use of sun light (Rajvanshi et al., 2019). The pilot-scale photobioreactor displayed a good efficiency in the production of high oil content about 30% of dry biomass in outdoor environment (average daily irradiation 10-30 MJ/m2/day, throughout the year, and temperature mostly, less than 35°C) (Norsker et al., 2021). Roncero-Ramos et al. (2019) using sterilized air (filtered with 0.22 µm Millipore filter) to perform a continuous aeration and stirring processes with rate quantified as 0.1 (air flow volume / volume of culture / minute). Flat panel photobioreactor could be stirred using tube sparger fixed at the bottom of the photobioreactor (Rajvanshi et al., 2019). Figure 2 represents a schematic designs for flat panel photobioreactor, and, figure 3 represents diagram for tubular photobioreactor

Light

Basically, a great effect has been produced as a result of the night / day cycle and seasonal variations in irradiance, for any site. The micro-algal cells allocated at the surface are bared to the serious light effect on the surface, however a various factors have a great effect on the quantity of photon that strikes the photosynthetic apparatuses of the microalgae, including the culture density, optical features of the algal cells, stirring rate, optical path of the photobioreactor, culture depth (Guler et al., 2019). Lights are the most important factors affecting the structure and composition of phytoplankton diversity in aquatic environments (Hill 1996; Wellnitz and Ward 2002). Recently many reports clarified that the quality of irradiance significantly affecting a series of physiological processes in algae, involving photosynthesis (Luimstra et al. 2019), gene expression (Wei et al. 2020), growth and development (Tan et al. 2020). The light availability and its quality are recognized as the main factors limiting the algal production in commercial scale (Burlew 1953b; Richmond 1996; Borowitzka, 1998). Incident light that penetrate the

Figure 3. A schematic diagram for tubular photobioreactor.

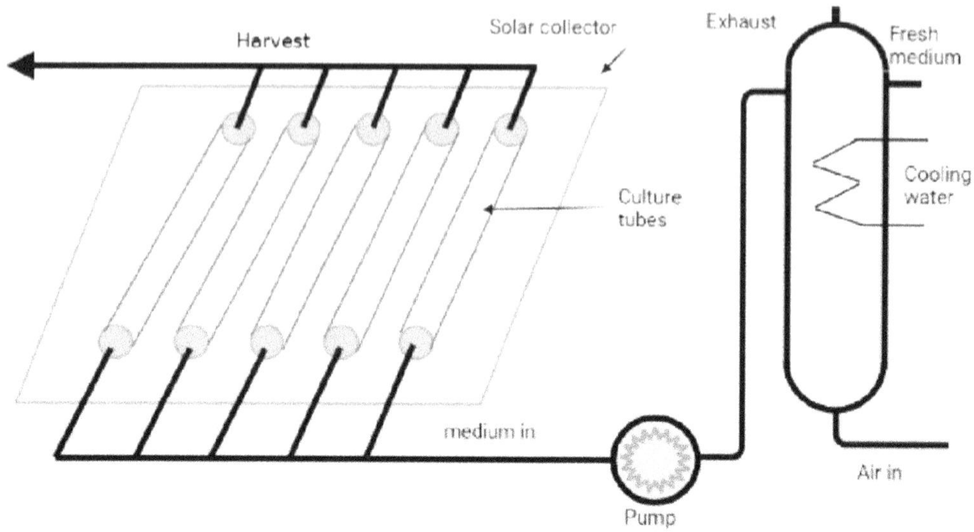

sea surface is partially absorbed by water, particulates such as organic and mineral suspends,, dissolved materials such as yellow substances, detritus, and phytoplankton. Consequence, the quantity of light will be decreases below water surface (Kirk 1994; Gallegos and Jordan 2002), as well as the variations in light quality as a result of penetration the water column (Altas and Bannister 1980; Kirk 1994; Schwarz and Markager 1999; Stomp et al. 2007). However the influence of light on algal production is not straightforward, nor is the interaction of light with other surrounding factors such as variation in temperature. There is a variation among phytoplankton species in their response to the quality of light. As an example, rapid growth and vertical migration of marine diatoms could be stimulated by shorter wavelengths such as purple and blue light (Shikata et al. 2009; Wenderoth and Rhiel 2004), while in case of *Anabaena circinalis*, blue light inhibits their germination and akinete formation (Thompson et al. 2009). In compare, red and yellow light are useful to the growth and production of dry biomass in Spirulina (Chen et al. 2011; Markou 2014). Moreover, many reports have revealed the value of using light quality to adjust or enhance the output of biological products of algal metabolite, such as oil, protein or medicines, generally, all these studies express that light quality have a great influence on physiological responses in phytoplankton culture (Gatamaneni Loganathan et al. 2020; Satthong et al. 2019). Growing outdoors culturing of algae in closed photobioreactors or in open ponds will be irradiated with very high light quantity which fluctuate during the day and each season (Borowitzka et al. 2016). Besides the daily variation in light, another shorter term of variation resulting from clouds which causes reflection and focusing of the irradiance. Cell density of algal culture has a significant effect on light environment experienced by the algal culture. In shallow ponds and photobioreactors in which dense cultures are generally used the irradiance decreases speedily from the surface of the culture, thus cells deeper within the culture might actually stay in the dark (Oswald 1988a; Janssen et al. 2003). Mixing systems rotates the microalgal cells from the low light sides to the layer of high light surface allowing the cells to expose to an alternating light/dark cycles whose frequency is detected by the design of photobioreactor (flat plate, open pond, vertical or horizontal tubular and so on.). Prussi et al. (2014) reported that the vertical stirring in raceway ponds leads to poor except, in the vicinity of the paddle, however the addition of extra foils or wings in the channels could enhance the vertical mixing (Voleti, 2012).

Temperature

Temperature is one of the most significant factors affecting all the living cells, also different algal species. Each algal species has own optimum temperature at which they achieve maximum growth. The temperature has a great effect on the cellular enzymatic reactions. The optimum temperature of most of the algal species ranges from 25-35 °C. Algal cells are more sensitive to high temperature than lower one. In fact, a little increment over the optimal temperature may seriously affect the algal cells (Das, 2015). Temperature impacts all metabolic processes and the optimum temperature of a certain algae strain will have a clear effect on the achieving culture production. Scaling up outdoor production of microalgae can be subject to temperature instabilities (Ras et al., 2013). Variation in seasonal temperature in addition to rapid daily fluctuations is responsible of modifying the growth conditions of microalgae and therefore affects productivity (Ras et al., 2013). Different strategies were organized depending on the species to counteract the effect of exceeding optimal temperatures such as cell shrinking and energy re-balancing. Furthermore, long term adaptation of specific species over generation cycles has also been verified efficiently to increase optimal temperatures (Ras et al., 2013). Cultivation of algae for commercial scale need a variety of important energy-related inputs to attain the planned yields required for economic practicability (Ras et al., 2013). One important properties of the temperature adaptability of algal cells that is the deadly growth temperature is generally only slightly above the optimum temperature (Bernard and Rémond 2012; Butterwick et al. 2005). When algal cells at their optimum temperature, they achieve a good performance to capture the available light and are less possible to be photoinhibited (Borowitzka 1998; Torzillo and Vonshak 1994). The productivity of growing season could be extended for several months by selecting strain that better adapted to lower temperatures (Borowitzka et al., 2016). Likewise, *Dunaliella salina* which have broad temperature tolerance (Borowitzka, 1988). *Dunaliella salina* species could be grown the whole year in Australia at the production plants (Borowitzka, 2013). *Neochloris oleoabundans* generating improved growth rates under stringent temperature control conditions at 22°C, while *Scenedesmus dimorphus* yielded slightly higher growth under conditions of temperature ñuctuation. Most microalgae species tend to perform photosynthesis and cellular division over a wide range of temperatures generally fall between 15 and 30 °C, with optimal conditions between 20 and 25 °C (Li, 1980). The best growth temperatures permit the cell to undergo photosynthesis without any changing concern inherent biochemical or physiological parameters. Temperature of maximum growth rates for mesophilic species are reported to be between 20 - 25 °C, but it could increase more than 40 °C for some thermophilic strains such as *Anacystis nidulans*, or decline below than 17 °C for psychrophilic strains such as *Asterionella formosa*. It is well known that *Chlorella* species have a wide range of optimal growth temperatures (Borowdszma et al, 2016). Growth rates versus optimal temperatures were studied for seventeen different Chlorella strains it is demonstrated that Chlorella species grew successfully about 26 °C for *C. vulgaris* and *C. prothotecoides*, also about 36 °C for *C. fusca* and *C. kessleri* (Kessler, 1985). Accordingly Researches highlighting the ability of Chlorella species to colonize a wide range of diverse natural habitats and then recommended this genus as an exciting candidate for high rate large scale productivity. Environmental conditions could change the optimal temperature cursor either way. *Tetraselmis sp.* exhibited successful growth in media conditions close to the natural waters compared to an enriched nutrient medium it revealed lower optimum temperatures (Maddux and Jones, 1964).

Moreover the optimum growth temperature of *Dunaliella tertiolecta* increased by 6 °C when NaCl concentration increased from 0.125 to 1.5 M (Eppley 1972). Optimum temperatures could therefore be related to the environmental conditions form which they have been isolated (Fawley 1984). Kudo et al. (2000) stated the optimum growth temperatures for *Phaeodactylum tricornutum* ranges from 20 to 23. Such differences in optimum temperature might not only be owing to the variation in environmental conditions but may result from the procedure used to assessment this optimum temperature from the current data. *Chlorella vulgaris*, was successfully grown at 5°C give lower chlorophyll content compared to cultures at 27°C (Maxwell et al., 1994). Cells grown at 5 °C were able to adapt their photosynthetic machinery in response to the excitation pressure on the photosystem II in a result to the excess of light. Similar adaptation has been recorded for *Dunaliella salina* (Krol et al. 1997), and *Dunalliella tertiolecta* (Levasseur et al. 1990). Meanwhile lowering of the growth temperature increased chlorophyll content of *Skeletonema costatu*. This was explained by the acquired carboxylase activity which ensures the consumption of over productive energy (Mortain-Bertrand et al. 1988).

Selection of Algal Strain and Genetic Modification

A valuable commercial products can be produced using algal large scale such as β-carotene, food, feeds, biofuels and fertilizers (Borowitzka, 2013). There are a different factors should be considered in the choice of algal strain such as growth medium, native environmental conditions, the scale of operation, and target product, some harmful algal species are able to form blooms and generate toxins in the natural environments, these toxins are able to cause a severe problems such as killing fish, birds and animals (Beacham et al., 2017). For any propose algae process the selected strain must displayed high productivity of the desired product when grown at large scale (Borowitzka, 2013; Borowitzka & Moheimani, 2013). For large scale of phycocyanin and protein production *Arthrospira platensis* is preferred (Yu et al., 2019). Meanwhile the oil production is the main target a high producing lipid algae is selected such as *Botryococcus braunii* (Prathima & Karthikeyan, 2017). The manipulation in genetic material of algal strains can be improves metabolic pathways, consequently final metabolic products and unlimited biotechnological application (Beacham et al., 2017). Different technologies were included such as homologous recombination and CRISPR/Cas9 genome editing unlike higher plants the development of genetically modified algae is still in its early steps and limited to a minor algal species, also little is acknowledged about the stability of these modified algal species in large-scale cultures or if they able to have any environmental risks (Beacham et al., 2017; Borowitzka, 2013).

Oxygen

The oxygen produced by the algal cells during photosynthetic process has a negative impact on the physiological and metabolic process, so the removal of the resulted oxygen from the culture system is essential (Morillas-España et al., 2021). Accumulation of high percentage of O_2 causes photo-oxidative stress to the algal cells, the maximum tolerable dissolved O_2 percentage was reported not to be more than 400% (Kumar & Das, 2012). Accumulation of oxygen is considered a limiting factor for the algae productivity in the photobioreactor design (Janssen et al., 2003). The amount of dissolved O_2 was measured at 2 points along pilot scale photobioreactor, the 1st one before the degassing tank which sparged with atmospheric air meanwhile the 2nd one between circulation pump and buffer tank (Norsker et al., 2021).

Mixing and Flow Rate

By increasing the culture density system, the hydrodynamic environment faces an enormous variation, hence it's essential to achieve a mixing system which provides a uniform chemical and light environment for the algal culture that prevents settling of the cells. Ponds are provided by a group of baffles, and water is stirred through the ponds to achieve proper mixing of nutrients and oxygen which produce uniform algal growth. In the small scale systems it's easy to perform mixing process meanwhile, in the large photobioreactors and ponds uniform mixing becomes more difficult that may cause the reduction in productivity and increase the risk of culture collapse (Borowitzka & Vonshak, 2017). Mixing in large-scale system vary significantly from small laboratory-scale ones. As in large ponds the flow rate differs along the channels while vertical mixing occurs in the vicinity of the paddle wheel where water flow is turbulent, whereas in the long channels flow is predominantly laminar (Prussi et al., 2014).

Nutrient Source and Supply

Algae have different nutrient requirements in both composition and concentration of the nutrients supplied which is basically composed of both micro- and macro-nutrients, trace metals as well as several vitamins (Rawat et al., 2013). There is no general growth media which works for all algal species, as it grouped depending on fresh water or salt water species, so researchers are obligatory to find great care on how growth media is composed and used (Das, 2015). Macronutrients including

C, N, P as well as some major ions such as Mg, Na, K, Ca, SO_4 and Cl as a base media meanwhile micro or trace nutrients include iron, manganese, zinc, cobalt, copper, molybdenum and a small amount of metalloid selenium (He et al., 2013).

Recycling of the Culture Medium

The extracted solid material and other media from recycling culture media at the end of the culture process could be used as an additional source of nutrients such as phosphorus and nitrogen (Beacham et al., 2017; El Asri et al., 2020). A filter can be installed above the pond in order to recycle the filtrate. The culture to be harvested should be passed through a sieve (about 200 µm) or fine weave cloth to remove any foreign particles such as larvae, insects, leaves and clumps of exopolysaccharide (Das, 2015).

Monitoring of Cultivation Process

Physical, chemical and biological information should be gathered and analysed for the surrounding production site such as climate conditions, native flora prior to the production process, this data will be effective for monitoring algal culture for a long time (Beacham et al., 2017). Monitoring of culture media was performed each day to estimate algal dry biomass, pH of algal culture, photosynthetic activity, as well as nutrient absorption rate, also, the assessment of main metabolic products such as carbohydrates, lipids, proteins, exopolysaccharides were performed periodically (Pezzolesi et al., 2019). Previous studies reported daily manual monitoring for cultivation process, Roncero-Ramos et al. (2019) perform a manual shaking for 12 column reactors with a hemi-spherical base in sterile condition and replaced the evaporated water with autoclaved distilled quantities, different growth parameters such as dry biomass and EPS production were also estimated.

Contaminants and Disease

All large-scale microalgal cultures whether they are cultured in open systems or cultured using closed systems are subject to contamination risk by either algal species, fungi, bacteria or even predatory protozoa (Borowitzka & Vonshak, 2017). The relation between cultured algal species and microbial contaminants in pond, photobioreactors, inoculated media, or even surrounding medium is still under research (Höger et al, 2021). Algal species are closely associated with the microbial communities in native environments, and their interactions have been fashioned in a long coexisting evolutionary history of coexistence. The relationship between phytoplankton and bacteria represents an essential role in ecosystem functioning and has an enormous impact on biogeochemical cycling in marine environments (Azam 1998; Buchan et al. 2014). The microbial environment of the phycosphere permits the exchange of chemical compounds and manifold metabolites that could result in a wide spectrum of communalistic, competitive, antagonistic, mutualisti, or parasitic relationships (Cirri and Pohnert, 2019). The relationships between algal strains and different bacterial species might be modified by various abiotic factors such as light intensity and/or fluctuations in temperature (Piwosz et al. 2020). The association between algal and bacterial species thus considered to be a dynamic scale of subsequent states of extremely complex networks affected by the variations in environmental conditions (Cirri and Pohnert 2019). Culturing of microalgal species with special selective medium, such as high salinity medium (*D. salina*), high pH value about 10 (*Arthrospira platensis*) or high available nutrients such as *Chlorella*, is an important strategy to minimize the risk of contamination. It also needs to be realized that all commercial-scale algal cultures will not be axenic cultures and an associated bacterial flora will be exist in the culture medium (Borowitzka & Vonshak, 2017). Functional Bacterial traits could be obtained from 16S rRNA gene–based approaches by FAPROTAX which was created for ocean environments and then recently developed and extended to other environment. Long time ago, bacterial communities were considered to be mere contaminants that could inhibit productivity of algae or terminate its populations. But recently it found that there is a positive effect of algae–bacterial interactions which considered as a promising biotechnological technique. These techniques could improve biomass productivity, harvesting, bioremediation for sustainable aquaculture and wastewater treatment (Fuentes et al. 2016). Algae also could affected by fungus infection, that may cause a high threat for plant designer as they leads to massive and sudden collapse of algal culture in industrial systems as well as natural environments (Carney and lane, 2014). Such zoosporic parasites, impact efficiently on ecological aquatic food chain dynamics by controlling phytoplankton population size and keeping access of nutrients to higher trophic levels (Frenken et al. 2017; Jephcott et al. 2017). It has been observed in some cases (De Bruin et al., 2004) that different susceptibility of different strains of the same species, suggested by judicious strain selection, that this property might use to minimize the risk of contamination. Fungal infection, represent extra problematic management in both closed photobioreactors and open pond systems (Gutman et al., 2009). Algal cultures become more susceptible to infection and predation if they do not grow under their optimum conditions (Borowitzka and Vonsha 2017). Therefore, the important factor for reducing the risks and effects of these contaminants is to provide optimum conditions in the variable environment they are exposed to (Flynn et al., 2017). This emphasizes the significance of selecting strains with optimum temperature related to the temperatures prevailing at production site. In addition to high energy and financial expenditures for conditions of cultivation, harvesting and final operations, one of the main challenge to the sustainable production of algal biomass is the maintaining of biological steadiness in large-scale production plants (Scott et al. 2010; Carney and Lane 2014).

MICROALGAE HARVESTING TECHNIQUE

Physical Harvesting Methods

Centrifugation

Centrifugation method is noted the separation of algae based on densities (Mannweiler and Hoare 1992). Heasman et al. (2000) reported that the harvesting efficiency of nine microalgae using a disk stack centrifuge were greater than 95% at a force of 13,000g, and the recovery efficiency elevated with increase in the gravitational force. Japar et al., (2017) used four different centrifugations speed (1000, 3000, 5000, and 7000 rpm) for five minutes, the highest harvesting efficiency was achieved by 7000 rpm. The harvesting efficiency by centrifugation methods close to 100%, but demands high cost more than yields, so it can be reducing cost by coupling centrifugation methods with other harvesting methods such as flocculation (Najjar et al., 2020).

Gravity Sedimentation

In this method, the solids are separated from liquids by gravitational forces (Salim et al., 2011). Japar et al., (2017) reported the sedimentation carried by transferred 500 ml of cultured microalgae into separation funnel, and the alga was sediment in the bottom of the funnel, another method was the sedimentation with flocculent by adding 5% (v/v) of 1 M alum ($Al_2(SO_4)_3$), then stirrer for one minute, and the transfer to separation funnel, the sedimentation occurred after 8 hours. Grima et al. (2003), reported sedimentation with flocculants is the most efficient method that attain 80% harvesting efficiency. There are many factors that influence the rate of deposition and harvesting processes for examples light intensity, temperature, time, size and density of microalgae (Mariam et al. 2015). The difference in cells density and size of microalgae are influence in harvesting processes (Wu, 2010). Morphology and shape of Arthrospira trichomes has a vital role in consequences on harvesting of biomass (Cuellar-Bermudez et al., 2020). The density of fresh water is various from the density of marine water and also the densities of fresh algae are various from the densities of marine algae, but the densities of fresh and salt water are the same the densities of microalgae so the depositions of algae are low (Granados et al., 2012; Murphy and Allen, 2011). The rate of deposition is dependent on the type of microalgae, but no relation between the size of cells and deposition rate (Cole and Buchak 1995; Peperzak et al., 2003). Many studies recorded that the deposition rate of microalgae varied from 0 .4 to 2.2 m/d (Peperzak et al., 2003), 0.1-0.3 m/d (Yang et al. 2000). The deposition rate of small algae were less than 1.0 cm/h, and in case of large algae were 2.6 cm/h (Choi et al., 2006). The results of some studies that related to the effects of temperature on deposition rate were differing. When pH was adjusted in range of 8-8.5, and the temperature was 4°C, the deposition rate of microalgae were increased (Knuckey et al., 2006). Meanwhile, Davis et al. (1995) recorded that the deposition rate of microalgae in colder water was lower due to increase of viscosity. The highest deposition rate of *Chaetoceros calcitrans* was obtained at pH 8, and 27°C, at day 8 in dark conditions (Harith et al., 2009). Meanwhile, Greenwell et al. (2010) reported the microalgae were deteriorating when harvesting by sedimentation process in high temperature. The algal growth phase is influence in the sedimentation rate. The sedimentation rate of microalgae extensively increased in the stationary growth phase (Choi et al. 2006). The highest sedimentation rate was observed in stationary phase with *Scenedesmus* sp. but in log phase in case of *C. vulgaris* (Manheim and Nelson, 2013). The

Harvesting process by sedimentation is low cost, but the using of this method is not widely used in industries (Uduman et al.,2010; Shelef et al., 1984).The harvesting microalgae by only sedimentation is not efficient method since the cell recovery rates of 60-65% are little (Mata et al., 2010; Ras et al., 2010) Gonzalez-Fernandez and Ballesteros, (2013) reported in sedimentation methods the time is consumed and the composition of algae changed.

Filtration

This type of harvesting methods use permeable membrane to separate algae biomass from culture filtrates, this method require pressure various according to the types of permeable membranes, which can be driven by vacuum, pressure or gravity (Al Hattab et al., 2015). And also the permeable membrane filter can be classified depending on the size of the pores, macro filtration, micro-filtration, ultrafiltration and reverse osmosis (Brennan and Owende, 2010).Vacuum filtration separates solids from solution by catching the solid particles onto a filter while drawing the liquid through by pressure from the filter (Sparks T., 2012; Koline-Sanderson, 2014).There are many types of filter membranes that can be consumed in vacuum filtration, such as starch pre-coated drum filter, suction filter, vacuum drum filter, filter thickener, and belt filter (Mohn, 1980).Cuellar-Bermudez et al., (2020) used nylon membranes with nominal pore size1, 5, 10, 25, and 40 µm for harvesting Arthrospira strains, and determined the harvesting efficiency by balance and filtration rate of distilled water. Microgreen algae *Chlamydomonas reinhardtii,* micro sized *Dunaliella* species were harvested by pre-coated drum filter) diatoacmeous earth filter((Gudin and Chaumont, 1991; Uduman et al., 2010). Microalgae Chlorella zofingiensis, Scenedesmus acuminatus,, C. pyrenoidosa, Nannochloropsis sp., Dictyosphaerium sp., Dunaliella salina, Pavlova lutheri and Microcystis have been successfully harvested using membrane-based operations, the efficiency of harvesting was 100% (Castro-Muñoz and García-Depraect, 2021).

Flotation

In this method the gas pass through a liquid-solid suspension, resulting the microalgae to float to the medium surface by holding to the gaseous bubbles (Singh et al., 2011). The results of air flotation methods were 98% of microalgae harvesting (Yan and Jameson, 2004). The efficiency of dispersed air flotation methods were greater than 93% when harvesting *Botryococcus braunii*, *Chlorella vulgaris* and *Scenedesmus obliquus* (Xu et al., 2010; Kurniawati et al., 2014), and greater than 95 in case of *Microcystis* and filamentous *Cylindrospermospsis* (Milledge and Heaven, 2013)

Chemical Methods

Flocculation

Flocculation is a significant process in water treatment with aimed to aggregate particles due to the addition of a clarifying agent. So the dispersed cells microalgae aggregate and form flocs and sedimentation by addition of some agents (Knuckey et al., Yang et al., 2011). During flocculation algae (negative charge) naturalize with positive charges of cationic flocculants forming larger particles that aggregate and settle via gravitational forces (Zhu et al., 2018). Flocculation are the best physical conditions for harvesting microalgae and used in harvesting large volume of algal cultures (Borges et al., 2011). Flocculation is

Figure 4. Schematic diagram for different type of flocculations

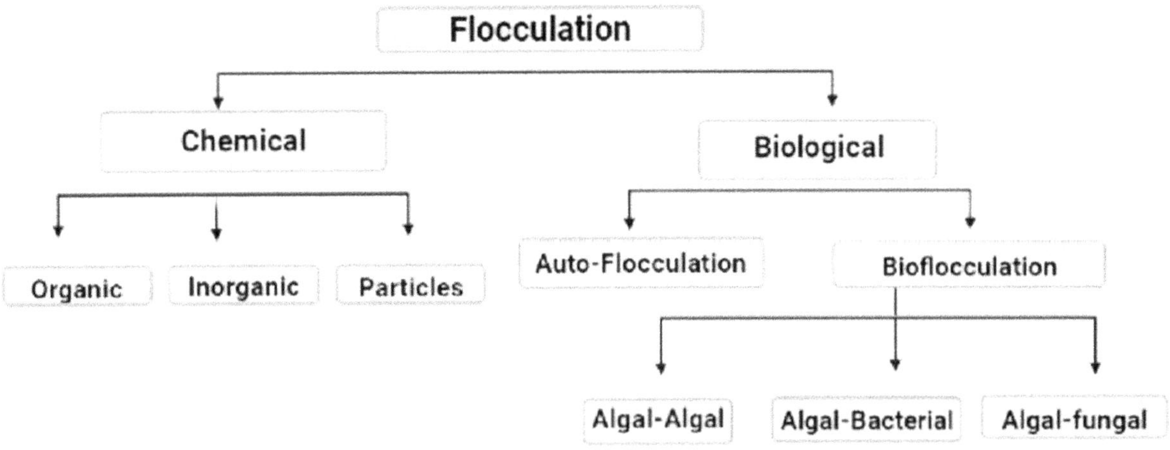

deliberated as an efficient, convenient and preferable process for the harvesting of microalgal biomass (Gupta et al., 2016). The chemical methods of harvesting microalgae are obtained by flocculation are shown in figure (4).

Inorganic Flocculants

Many positively charged metals are used as inorganic flocculants in aqueous solution, that destabilize negatively charged algal cells, causing algae flocculation and hence precipitation, examples of inorganic flocculants like, aluminium sulphate ($Al_2(SO_4)_3$, Ferric sulfate ($Fe_2(SO_4)_3$) and ferric chloride ($FeCl_3$), calcium oxide (CaO), calcium carbonate $CaCO_3$ and magnesium hydroxide($Mg(OH)_2$) (Kim et al., 2017; Surendhiran and Vijay, 2013; Ma et al., 2016; Choi, 2015 ; Vandamme et al., 2015). Inorganic flocculants have been commonly consumed in production of algae biofuel, but the liberation of such chemicals into the environment that is a critical environmental and health concern (Matter et al., 2019). The efficiency of inorganic flocculants shows in table (1).

There are many factors that affecting in flocculation such as pH, algal cell size and Cationic Inducer (Surendhiran and Vijay, 2014). The results obtained by Vandamme et al., (2012) proved that pH 10 is more efficiency in the flocculation processes, due to when pH increase, the negative charge of algal cells increase. Adding of divalent cationic salts in the medium improved the flocculation at high pH by interlocking the cells, forming dense flocs (Kim et al., 2011). But flocculation had disadvantage in case of addition chemical substances due to be highly toxic, nondegradable and also intermediate products are harmful to the environment (Tao and Salihon, 2011; Yokoi et al., 1996; Kumar et al., 2004).

Organic Flocculants

Despite of the inorganic flocculants have efficacy and cost effectiveness, but cause major toxicity to human health and environmental pollution (Okaiyeto et al., 2016). So the researchers seek for alternative materials that eco-friendly and do not affects in human health. Some cationic polymers have been used for algae harvesting, due to positive charges and negative charges of algal cells, these polymers are

Table 1. Efficiency of inorganic flocculants

Algae	flocculants	pH	Time (min)	Efficiency	References
N. oculata	$Fe_2(SO_4)_3$ (0.4 g/L)	-	180	93.8%	Kim et al., (2017)
N. oculata	$FeCl_3$ (0.6 g/L)	-	-	87.3%	Surendhiran and Vijay, (2013)
Chlorella	$Fe_2(SO_4)_3$	low-pH	-	~98%	Kim et al., (2017)
C. vulgaris	CaO (60 mg/L)	-	-	85%	Ma et al., (2016)
C. vulgaris	$CaCO_3$ (80 mg/L)	6	20	99%	Choi, (2015)
Chlorella	$Mg(OH)_2$ (145 mg/L)	10.7	-	90%	Vandamme et al., (2015)
Chlorella	$Mg(OH)_2$ (145 mg/L)	8	-	95%	Vandamme et al., (2015)
Chlorella zofingiensis	$FeCl_3$	4.0±0.3		<90%	Wyatt et al., (2012)
Algal bloom	20 ppm $FeCl_3$	8.2	30	98%	Loganathan et al., (2018)
Chlorococcum sp	50 mg/L ammonium	-	-	84%	Lv et al.,(2018)
Nannochloropsis sp	$Al2(SO_4)_3$ (87.5 ppm)			95.2% (±1.1)	Chua et al., (2019)
Nannochloropsis sp	$FeCl_3$ (87.5 ppm)			95.6% (±1.2)	Chua et al., (2019)
Tetraselmis sp. KCTC12236BP	1.2 g/L $Al_2(SO_4)_3$	-	30	86%	Kwon et al., (2014)

less toxic eco-friendly (Peng et al., 2017; Fast et al., 2014). Organic polymers such as chitosan (cationic polyelectrolytic) are used in algal harvesting (Sanyano et al., 2013). The mechanism of chitosan flocculation related to algal cells is adsorption, and charge neutralization (Toh et al.,2017; Fast et al., 2014). There are many factors are affecting in chitosan flocculation such as dosage of chitosan, ionic strength, cell density, growth stage and also algal species (Matter et al., 2016; Xu et al., 2013).Cationic starch has often been used for algae harvesting, and can be also used in blend with other polymers such as chitosan, and blend with inorganic flocculent such as aluminum salts to increase the harvesting efficiency (Peng et al., 2017; Choy et al., 2018). The best efficiency of algae harvesting by chitosan at low pH 7, but in case of cationic starch was under alkaline conditions (Vandamme et al., 2013; Tork et al., 2017). Alginate is an anionic polysaccharide that is derived from brown algae (Facchi et al., 2017). The improvement of algal harvesting processes might be attributable in the combination of alginate and chitosan (Matter et al., 2019). Soluble poly-L-lysine (PLL) is a cationic, linear biopolymer, that is secreted by various *Streptomycetaceae* bacteria Tannins that extracted from some plants such as *Acacia mearnsii,* inulin (polysaccharide extracted from some plant) are also used as flocculants for harvesting microalgae (Shima et al., 1984; Wang et al., 2013). Meanwhile the high efficiency of using organic polymers in harvesting of some microalgae, there are some obstacles such as high cost, low efficiency of some marine microalgae and narrow operating pH range that limits commercial technologies for algae harvesting (Sanyano et al., 2013). The efficiency of organic flocculants is shown in table 2.

Table 2. The efficiency of organic flocculants on algal harvesting.

Algae	Flocculants	pH	Efficiency	Referencres
Algal bloom	10 ppm alum +1pp chitosan	8.2	98%	Loganathan et al., (2018)
C. sorokiniana	chitosan (10mg)	below 7	99%	Tork et al., (2017)
S. obliquus	10 mg/L chitosan	6	Less 70%	Matter et al., (2018)
S. obliquus	10 mg/L alginate +10 mg/L chitosan	6	85%	Matter et al., (2018)
N. oculata	10 mg/L of Tanfloc	6	99%	Roselet et al.,(2016)
Scenedesmus sp.	Tanfloc 210 mg/L	7.8	96.7%	Selesu et al., (2016)
Botryococcus sp.	60 mg/L. Inulin	-	88.6%	Rahul et al., (2015)
Scenedesmus sp.	8 mg/L polyamine polymer	-	90%	Gupta et al., (2014)

Particles-Based Flocculation

Many studies are focused of using nanoparticles as a flocculants to harvesting microalgae, due to nanoparticles causing cell disruption and lipid extraction and hence more effective in biofuel productions by microalgae (Lee et al., 2015). Aminoclay nanoparticles are blend with cationic heavy metals and used as flocculants to harvesting microalgae, such as magnesium aminoclay (Mg-AC) can used in to flocculate the alga C. vulgaris (1.0 g/L) at pH ranged from 5 to 12 with harvesting efficiency more than 90% (Farooq et al., 2013).The harvesting efficiency of cyanobacteria was reached 100%, when mixing cerium aminoclay (Ce-AC) and magnesium aminoclay (Mg-AC) after 1h (Ji et al., 2016). Magnetic particles have also been used in harvesting microalgae, due to easier separation of algae and reuse of culture media, which cause the decrease in the harvesting cost, and applied in production of biofuel by algae (Hu et al., 2013; Wang et al., 2015). Fe_3O_4-embedded carbon microparticles could be consumed for fresh algae flocculation and hydrophobic-interaction that causing of lipid extraction, 2 g/L of Fe3O4-embedded carbon microparticles was caused of harvesting 99% of Chlorella within 1 min, the high doses of Fe_3O_4-embedded carbon microparticles caused lipid extraction after ten minutes of sonication (Seo et al., 2015). Almomani et al., (2020) synthesized iron –nanoparticles and used in harvesting *Spirulina platensis* and *Chlorella vulgaris,* each and mixed, the results proved that the iron –nanoparticles is more efficiency in adsorption *Spirulina platensis* more than *Chlorella vulgaris* and mixed culture. The efficiency of *Chlorella zofingiensis* was more than 95% in one min with pH ranged from 4 to 11 when used magnetic nanoparticles (Li et al., 2021).

Biological Methods

Auto-Flocculation

Auto-Flocculation is promising techniques because algae harvesting without addition any flocculant (Rashid et al., 2018).The changing in acidic or alkaline conditions and culture aging cause the suspended algal cells self-aggregate, creating large flocs, which make their gravitational sedimentation, that due to reduce intensities of negative charges of algae (Liu et al., 2013). When pH increases, algae secret exopolysaccharide, to protective cell wall, and when pH decrease, the surface charge of the cell wall charge due to the dissociations of carboxyl and amine groups on the cell wall (Shen et al., 2013). The harvesting of *Chlorella vulgaris* by auto-flocculation strategy, increased when added adequate amount of glycine into the medium, that stimulating exo-polysaccharide secretion by alga and causes alga precipitation (Shen et al.,2015). Under pH 4 the harvesting efficiency of *Chlorella. ellipsoideum*, *Chlorella. nivale* and *Scenedesmus* sp. were 95, 94,98% at 15 minutes respectively (Liu et al., 2013). Meanwhile at alkaline conditions, the harvesting efficiency of *Chlorella muelleri,* was 100% at pH 11.5 for 30 mins (Huo et al., 2014) *Chlorella. vulgaris,* 95%, pH 11 with 60 mins (Vandamme et al., 2012) *Chlorococcum sp,* 94% pH 12 with 10 mins (Vandamme et al., 2012). The composition and of algal medium and concentrations of nutrients can influence in algal auto-flocculation efficiency, Ca^{2+} and Mg^{2+} that originally demands in algal media causes algal auto-flocculation, due to its positive charges interact with negative charges algal cell wall (Nguyen et al., 2014). The nutrients residue in the media after the stopping of algal growth can behave as flocculants and increase the harvesting efficiency (Kim et al., 2020). The harvesting algae by auto-flocculation techniques is eco-friendly biomass, low cost, but slow and highly species-specific (Branyikova et al., 2018; Matter et al., 2015; Ummalyma et al., 2017; Yoo et al., 2015). Auto-flocculation microalgae Synechocystis sp and Tribonema sp. have potential to reduce harvesting cost (Cheng et al., 2020).

Bioflocculation

Bioflocculants are based on using exopolysaccharides produced by microorganisms during their growth (Zhu et al., 2012). Recent studies are deal with bioflocculation agent that is valuable over chemical flocculants due to their high efficiency, ecofriendliness, biodegrading nature, and nontoxicity (Zheng et al., 2008; Pan et al., 2009). The bioflocculation mechanism between microalgal and bacterial cells is occurred as chains of inter-bridging between cells, that cause neutralization of charges and enhancement precipitation (Zheng et al., 2012). The bioflocculation enhanced by addition cationic inducer such as Zn^{+2} which act as linker between functional groups (Zheng et al., 2009). But the higher concentration of the inducer led to flocculation efficiency became lesser due to destruction of cells (Surendhiran and Vijay, 2014). The optimum conditions that cause bio flocculation efficiency of *Nannochloropsis oculata*, by using bacterium *Bacillus subtilis* as natural flocculant were at 30.78 °C, pH 10.8, flocculation time 6.7 h, bioflocculant size 0.38 mL, and cationic inducer concentration 0.036 mM (Surendhiran and Vijay, 2014).

Interactions between Microorganisms in Bioflocculation

Bioflocculation caused when interaction between algal-algal, algal-bacterial, algal-fungal (Ummalyma et al. 2017).

Algal-Algal Interactions

The algal–algal interaction is an excellent method for algae harvesting in biofuel industry but still limited used (Chen et al., 2018). The algal–algal interaction method reduced cost because does not need different cultivation conditions and the contamination is reduced (Salim et al. 2011). The bioflocculation may occur between two algae when algae producing flocculating agents like exopolysaccharide or glycoproteins that adhere to neighbouring cells or by change of medium charge that stimulate bioflocculation

(Ummalyma et al. 2017). The harvesting efficiency of *C. vulgaris and N. oleoabundans* was 68% and 46% when amended by microalgae *Teraselmis suecica* each (Salim et al. 2011; Alam et al. 2014).

Algal–Bacterial Interactions

Bacteria and microalgae have negative charges (Arau´jo et al. 2010). The main mechanisms for bioflocculation causes between bacteria and microalgae are extracellular polymeric substances (polysaccharides, proteins or other bioflocculant agents) that increase the microalgae sedimentation, and also charge neutralization in the presence of ions (Alam et al., 2016). Bioflocculation causes between bacteria and microalgae by addition cations to neutralize the negative charges such as Ca^{+2} and Mg^{+2} ions (Wang et al., 2012). The harvesting efficiency of axenic culture Chlorella was 2%, meanwhile when addition bacteria the harvesting efficiency was 94% (Lee et al., 2013).

Algal-Fungal Interactions

An important method using in harvesting algae is bio-flocculation by algal-fungi interaction, due to microalgae (negative charge) have great ability to attract to fungi (positive charge) (Chen et al. 2018a; Rashid et al. 2018; Alam et al., 2016). Sometimes bio-flocculations are obtained by simultaneously cultivation of algae and fungi, this culture may be positive or negative effects on algal growth according to their species or algal culture conditions)Rashid et al. 2018(. Microalgae give oxygen to fungi, meanwhile fungi protect algae from pathogenic bacteria (Alam et al. 2016; Dao and Beardall 2016; Rashid et al. 2018). And also co-cultivation of algae- fungi can enhancement biomass and lipid production increase and positive effective in removal heavy metals and nutrients when grow in wastewater (Wrede et al., 2014).

CONCLUSION AND FUTURE PROSPECS

The understanding of the algal growth parameters and the ability to monitor and control these factors are basic requirements for the efficient scaling up from flask scale to commercial scales. This chapter was fulfilled by adequate studies and technologies for qualitative and quantitative monitoring of the algal cultivation. Some factors should be followed in the design of the cultivation systems i.e it must provide the growth optimum parameters such as temperature, adequate illumination, oxygen elimination system, CO_2 provide system, stirring and mixing mechanism should be improved. Researchers have a duty to develop these designs to provide a cost effective method for algal cultivation. Researchers seek for the high yields of microalgae by applied culture optimization conditions, and more efficiency of harvesting processes. Bioflocculation methods are efficient methods for harvesting microalgae. The bioflocculation methods using algal-algal, algal-fungi and algal-bacteria interaction are more efficiency and eco-friendly methods and low cost, but harvesting efficiency various according to species used and condition. So the authors must be concentrate studies the optimization conditions of algal-algal, algal-fungi and algal-bacteria interaction. Also studies should be focus in using nanoparticles in harvesting microalgae.

REFERENCES

Al Hattab, M., Ghaly, A., & Hammouda, A. (2015). Microalgae harvesting methods for industrial production of biodiesel: Critical review and comparative analysis. *J Fundam Renew Energy Appl*, *5*(2), 1000154. doi:10.4172/2090-4541.1000154

Alam, M. A., Vandamme, D., Chun, W., Zhao, X., Foubert, I., Wang, Z., & Yuan, Z. (2016). Bioflocculation as an innovative harvesting strategy for microalgae. *Reviews in Environmental Science and Biotechnology*, *15*(4), 573–583. doi:10.100711157-016-9408-8

Alam, M. A., Wan, C., Guo, S. L., Zhao, X. Q., Huang, Z. Y., Yang, Y. L., & Bai, F. W. (2014). Characterization of the flocculating agent from the spontaneously flocculating microalga *Chlorella vulgaris* JSC-7. *Journal of Bioscience and Bioengineering*, *118*(1), 29–33. doi:10.1016/j.jbiosc.2013.12.021 PMID:24507901

Almomani, F. (2020). Algal cells harvesting using cost-effective magnetic nano-particles. *The Science of the Total Environment*, *720*, 137621. doi:10.1016/j.scitotenv.2020.137621 PMID:32146403

Altas, D., & Bannister, T. T. (1980). Dependence of mean spectral extinction coefficient of phytoplankton on depth, water color, and species. *Limnology and Oceanography*, *25*(1), 157–159. doi:10.4319/lo.1980.25.1.0157

Andersen, R. A. (2005). Recipes for freshwater and seawater media. *Algal Culturing Techniques*.

Arad, S., & van Moppers, D. (2013). Novel sulphated polysaccharides of red microalgae; basics and applications. In Handbook of microalgal culture: applied phycology and biotechnology. Wiley-Blackwell.

Arau'jo, E. A., Andrade, N. J., Carvalho, A. F., Ramos, A. M., Silva, C. D. S., & Silva, L. H. M. (2010). Aspectos coloidais da adesa˜o de microorganismos. *Quimica Nova*, *33*(9), 1940–1948. doi:10.1590/S0100-40422010000900022

Azam, F. (1998). Microbial control of oceanic carbon flux: The plot thickens. *Science*, *280*(5364), 694–696. doi:10.1126cience.280.5364.694

Beacham, T. A., Sweet, J. B., & Allen, M. J. (2017). Large scale cultivation of genetically modified microalgae: A new era for environmental risk assessment. *Algal Research*, *25*, 90–100.

Bernard, O., & Rémond, B. (2012). Validation of a simple model accounting for light and temperature effect on microalgal growth. *Bioresource Technology*, *123*, 520–527.

Borges, L., Moron-Villarreyes, J. A., D'Oca, M. G. M., & Abreu, P. C. (2011). Effects of flocculants on lipid extraction and fatty acid composition of the microalgae Nannochloropsis oculata and Thalassiosira weissflogii. *Biomass and Bioenergy*, *35*(10), 4449–4454.

Borowitzka, M. A. (1988a). Algal growth media and sources of cultures. In M. A. Borowitzka & L. J. Borowitzka (Eds.), *Micro-algal biotechnology* (pp. 456–465). Cambridge University Press.

Borowitzka, M. A. (1997). Algae for aquaculture: Opportunities and constraints. *Journal of Applied Phycology*, *9*, 393–401.

Borowitzka, M. A. (1998). Limits to growth. In Y. S. Wong & N. F. Y. Tam (Eds.), *Wastewater treatment with algae* (pp. 203–226). Springer-Verlag.

Borowitzka, M. A. (1998). Limits to growth. In Y. S. Wong & N. F. Y. Tam (Eds.), *Wastewater treatment with algae* (pp. 203–226). Springer-Verlag.

Borowitzka, M. A. (2013). *Dunaliella*: biology, production, and markets. In A. Richmond & Q. Hu (Eds.), *Handbook of microalgal culture* (pp. 359–368). Wiley-Blackwell.

Borowitzka, M. A. (2013). High-value products from microalgae—Their development and commercialisation. *Journal of Applied Phycology*, *25*(3), 743–756.

Borowitzka, M. A., Beardall, J., & Raven, J. R. (2016). *The physiology of Microalgae*. Springer.

Borowitzka, M. A., & Moheimani, N. R. (2013). *Algae for biofuels and energy* (Vol. 5). Springer.

Borowitzka, M. A., & Vonshak, A. (2017). Scaling up microalgal cultures to commercial scale. *European Journal of Phycology*, *52*(4), 407–418.

Branyikova, I., Filipenska, M., Urbanova, K., Ruzicka, M. C., Pivokonsky, M., & Branyik, T. (2011). Physicochemical approach to alkaline flocculation of Chlorella vulgaris induced by calcium phosphate precipitates. *Colloids and Surfaces. B, Biointerfaces*, *2018*(166), 54–60.

Brennan, L., & Owende, P. (2010). Biofuels from microalgae a review of technologies for production, processing, and extractions of biofuels and co-products. *Renewable & Sustainable Energy Reviews*, *14*, 557–577.

Buchan, A., LeCleir, G. R., Gulvik, C. A., & González, J. M. (2014). Master recyclers: Features and functions of bacteria associated with phytoplankton blooms. *Nature Reviews. Microbiology*, *12*, 686–698.

Burlew, J. S. (1953). Current status of large-scale culture of algae. In J. S. Burlew (Ed.), *Algal culture from laboratory to pilot plant* (pp. 3–23). Carnegie Institution.

Butterwick, C., Heaney, S. I., & Talling, J. F. (2005). Diversity in the influence of temperature on the growth rates of freshwater algae, and its ecological relevance. *Freshwater Biology*, *50*(2), 291–300.

Carney, L. T., & Lane, T. W. (2014). Parasites in algae mass culture. *Frontiers in Microbiology*, *5*, 278.

Castro-Muñoz, R., & García-Depraect, O. (2021). Membrane-Based Harvesting Processes for Microalgae and Their Valuable-Related Molecules: A Review. *Membranes*, *11*(8), 585.

Chen, C. Y., Yeh, K. L., Aisyah, R., Lee, D. J., & Chang, J. S. (2011). Cultivation, photobioreactor design and harvesting of microalgae for biodiesel production: A critical review. *Bioresource Technology*, *102*, 71–81. https://doi.org/10.1016/j.biortech.2010.06.159

Chen, C. Y., Yeh, K. L., Aisyah, R., Lee, D. J., & Chang, J. S. (2011). Cultivation,photobioreactor design and harvesting of microalgae for biodiesel production: A critical review. *Bioresource Technology*, *102*, 71–81. https://doi.org/10.1016/j.biortech.2010.06.159

Chen, J., Leng, L., Ye, C., Lu, Q., Addy, M., Wang, J., & Zhou, W. (2018). A comparative study between fungal pellet-and spore-assisted microalgae harvesting methods for algae bioflocculation. *Bioresour Technol, 259*, 181-190. .1016/j.biortech.2018.03.040 doi:https://doi.org/10

Cheng, P., Chen, D., Liu, W., Cobb, K., Zhou, N., Liu, Y., & Ruan, R. (2020). Auto-flocculation microalgae species Tribonema sp. and Synechocystis sp. with T-IPL pretreatment to improve swine wastewater nutrient removal. *The Science of the Total Environment*, *725*, 138263.

Chisti, Y. (2007). Biodiesel from microalgae. *Biotechnology Advances*, *25*(3), 294–306.

Choi, H. J. (2015). Effect of eggshells for the harvesting of microalgae species. *Biotechnology, Biotechnological Equipment*, *29*, 666–672.

Choi, S. K., Lee, J. Y., Kwon, D. Y., & Cho, K. J. (2006). Settling characteristics of problem algae in the water treatment process. *Water sci Technol: A j Int Association. Water Research*, *53*, 113–119.

Choy, S. Y., Prasad, K. M. N., Wu, T. Y., Raghunandan, M. E., Phang, S. M., Juan, J. C., & Ramanan, R. N. (2018). Separation of *Chlorella* biomass from culture medium by flocculation with rice starch. *Algal Research*, *30*, 162–172.

Chua, E. T., Eltanahy, E., Jung, H., Uy, M., Thomas-Hall, S. R., & Schenk, P. M. (2019). Efficient Harvesting of Nannochloropsis Microalgae via Optimized Chitosan-Mediated Flocculation. *Global Challenges*, *3*(1), 1800038.

Cirri, E., & Pohnert, G. (2019). Algae-bacteria interactions that balance the planktonic microbiome. *The New Phytologist*, *223*, 100–106.

Cole, T. M., & Buchak, E. M. (1995). *CE-QUAL-W2: A two dimensional, laterally averaged, hydrodynamic and water quality model, version 2.0: User manual*. Instruction Report EL-95-1, US Army Engineer Waterways Experiment Station.

Couto, C., Hernández, C. P., Alves Sobrinho, R. C. M., Mendes, C. R. B., Roselet, F., & Abreu, P. C. (2021). Optimization of a low-cost fertilizer-based medium for large-scale cultivation of the coastal diatom Conticribra weissflogii using response surface methodology and its effects on biomass composition. *Journal of Applied Phycology*, *33*(5), 2767–2781.

Cuellar-Bermudez, S. P., Kilimtzidi, E., Devaere, J., Goiris, K., Gonzalez-Fernandez, C., Wattiez, R., & Muylaert, K. (2020). Harvesting of Arthrospira platensis with helicoidal and straight trichomes using filtration and centrifugation. *Separation Science and Technology*, *55*(13), 2381–2390.

Dao, L. H., & Beardall, J. (2016). Effects of lead on growth, photosynthetic characteristics and production of reactive oxygen species of two freshwater green algae. *Chemosphere*, *147*, 420–429. .2015.12.117 doi:10.1016/j.chemosphere

Das, D. (2015). *Algal Biore nery: An Integrated Approach*. Springer International Publishing AG.

Davis, E. M., Downs, T. D., Shi, Y., & Ajgaonkar, A. A. (1995). Recycle as an alternative to algal TSS and BOD removal from an industrial waste stabilization pond system. *Industrial Waste Conference*, *50*, 65-73.

De Bruin, A., Ibelings, B. W., Rijkeboer, M., Brehm, M., & Van Donk, E. (2004). Genetic variation in Asterionella formosa (Bacillariophyceae): Is it linked to frequent epidemics of host-specific parasitic fungi? *Journal of Phycology*, *40*, 823–830.

El Asri, O., Afilal, M. E., Laiche, H., & Elfarh, A. (2020). Evaluation of physicochemical, microbiological, and energetic characteristics of four agricultural wastes for use in the production of green energy in Moroccan farms. *Chemical and Biological Technologies in Agriculture*, *7*(1), 1–11.

Emblemsvåg, J., Kvadsheim, N. P., Halfdanarson, J., Koesling, M., Nystrand, B. T., Sunde, J., & Rebours, C. (2020). Strategic considerations for establishing a large-scale seaweed industry based on fish feed application: A Norwegian case study. *Journal of Applied Phycology*, *32*(6), 4159–4169.

Eppley, R. W. (1972). Temperature and phytoplankton growth in the sea. *Fish Bulletin*, *70*, 1063–1085.

Facchi, D. P., Lima, A. C., de Oliveira, J. H., Lazarin-Bidóia, D., Nakamura, C. V., Canesin, E. A., Bonafé, E. G., Monteiro, J. P., Visentainer, J. V., Muniz, E. C., & … . (2017). Polyelectrolyte complexes based on alginate/tanfloc: Optimization, characterization and medical application. *International Journal of Biological Macromolecules*, *103*, 129–138.

Farooq, W., Lee, Y. C., Han, J. I., Darpito, C. H., Choi, M., & Yang, J. W. (2013). Efficient microalgae harvesting by organo-building blocks of nanoclays. *Green Chemistry*, *15*, 749–755.

Fast, S. A., Kokabian, B., & Gude, V. G. (2014). Chitosan enhanced coagulation of algal turbid waters–comparison between rapid mix and ultrasound coagulation methods. *Chemical Engineering Journal*, *244*, 403–410.

Fawley, M. W. (1984). Effects of light intensity and temperature interactions on growth characteristics of Phaeodactylum tricornutum (Bacillariophyceae). *Journal of Phycology*, *20*, 67–72.

Flynn, K. J., Kenny, P., & Mitra, A. (2017). Minimising losses to predation during microalgae cultivation. *Journal of Applied Phycology*, *29*, 1829–1840.

Frenken, T., Alacid, E., Berger, S. A., Bourne, E. C., Gerphagnon, M., Grossart, H.-P., Gsell, A. S., Ibelings, B. W., & … . (2017). Integrating chytrid fungal parasites into plankton ecology: Research gaps and needs. *Environmental Microbiology*, *19*, 3802–3822.

Fuentes, J. L., Garbayo, I., Cuaresma, M., Montero, Z., González-del-Valle, M., & Vílchez, C. (2016). Impact of microalgae-bacteria interactions on the production of algal biomass and associated compounds. *Marine Drugs*, *14*, 100.

Gallegos, C. L., & Jordan, T. E. (2002). Impact of the spring 2000 phytoplankton bloom in Chesapeake Bay on optical properties and light penetration in the Rhode River, Maryland. *Estuaries*, *25*, 508–518. doi:10.1007/BF02804886

Gonzalez-Fernandez, C., & Ballesteros, M. (2013). Microalgae Autoflocculation: An alternative to high-energy consuming harvesting methods. *Journal of Applied Phycology*, *25*, 991–999.

Granados, M. R., Acién, F. G., Gómez, C., Fernández-Sevilla, J. M., & Grima, E. M. (2012). Evaluation of flocculants for the recovery of freshwater microalgae. *Bioresource Technology*, *118*, 102–110.

Greenwell, H. C., Laurens, L. M. L., Shields, R. J., Lovitt, R. W., & Flynn, K. J. (2010). Placing microalgae on the biofuels priority list: A review of the technological challenges. *Journal of the Royal Society, Interface*, *7*, 703–726.

Grima, E. M., Acie, F. G., Medina, A. R., & Chisti, Y. (2003). Recovery of microalgal biomass and metabolites: Process options and economics. *Biotechnology Advances*, *20*, 491–515.

Grobbelaar, J. U. (2013). Mass production of microalgae at optimal photosynthetic rates. *Photosynthesis*, 357-371.

Gudin, C., & Chaumont, D. (1991). Cell fragility, the key problem of microalgae mass production in closed photobioreactors. *Bioresource Technology*, *38*, 141–151.

Guler, B. A., Deniz, I., Demirel, Z., Oncel, S. S., & Imamoglu, E. (2019). Transition from start-up to scale-up for fucoxanthin production in flat plate photobioreactor. *Journal of Applied Phycology*, *31*(3), 1525–1533.

Gupta, P. L., Lee, S.-M., & Choi, H.-J. (2015). A mini review: Photobioreactors for large scale algal cultivation. *World Journal of Microbiology & Biotechnology*, *31*(9), 1409–1417.

Gupta, P. L., Lee, S. M., & Choi, H. J. (2016). Integration of microalgal cultivation system for wastewater remediation and sustainable biomass production. *World Journal of Microbiology & Biotechnology*, *32*, 139–149.

Gupta, S. K., Kumar, M., Guldhe, A., Ansari, F. A., Rawat, I., Kanney, K., & Bux, F. (2014). Design and development of polyamine polymer for harvesting microalgae for biofuels production. *Energy Conversion and Management*, *85*, 537–544.

Gutman, J., Zarka, A., & Boussiba, S. (2009). The host-range of *Paraphysoderma sedebokerensis*, a chytrid that infects *Haematococcus pluvialis*. *European Journal of Phycology*, *44*, 509–514.

Harith, Z. T., Yusoff, F. M., Mohammed, M. S., Shariff, M., & Din, M. (2009). Effect of different flocculants on the flocculation performance of microalgae, Chaetoceros calcitrans, cells. *African Journal of Biotechnology*, *8*, 5971–5978.

He, P. J., Mao, B., Lü, F., Shao, L. M., Lee, D. J., & Chang, J. S. (2013). The combined effect of bacteria and Chlorella vulgaris on the treatment of municipal wastewaters. *Bioresource Technology*, *146*, 562–568.

He, P. J., Mao, B., Lü, F., Shao, L. M., Lee, D. J., & Chang, J. S. (2013). The combined effect of bacteria and Chlorella vulgaris on the treatment of municipal wastewaters. *Bioresource Technology*, *146*, 562–568.

Heasman, M., Diemar, J., O'Connor, W., Sushames, T., & Foulkes, L. (2000). Development of extended shelf-life microalgae concentrate diets harvested by centrifugation for bivalve molluscs-a summary. *Aquaculture Research*, *31*, 637–659.

Hill, W. (1996). Effects of light. In Algal ecology: freshwater benthic ecosystems, Aquatic Ecology Series. Academic Press. https://doi.org/10.1016/B978-012668450-6/50034-5.

Höger, A., Griehl, C., & Noll, M. (2021). Infection with intracellular parasite *Amoeboaphelidium protococcarum* induces shifts in associated bacterial communities in microalgae cultures. *Journal of Applied Phycology*, *33*, 2863–2873. https://doi.org/10.1007/s10811-021-02542-9

Hu, Y. R., Wang, F., Wang, S. K., Liu, C. Z., & Guo, C. (2013). Efficient harvesting of marine microalgae *Nannochloropsis maritima* using magnetic nanoparticles. *Bioresource Technology*, *138*, 387–390.

Huo, S., Wang, Z., Zhu, S., Cui, F., Zou, B., You, W., Yuan, Z., & Dong, R. (2014). Optimization of alkaline flocculation for harvesting of *Scenedesmus quadricauda* 507 and *Chaetoceros muelleri* 862. *Energies*, *7*, 6186–6195.

Janssen, M., Tramper, J., Mur, L. R., & Wijffels, R. H. (2003). Enclosed outdoor photobioreactors: Light regime, photosynthetic efficiency, scale-up, and future prospects. *Biotechnology and Bioengineering*, *81*(2), 193–210.

Jephcott, T. G., Ogtrop, F.F., Gleason, F.H., Macarthur, D.J., & Scholz, B. (2017). The ecology of chytrid and aphelid parasites of phytoplankton. In The Fungal Community. CRC Press.

Ji, H. M., Lee, H. U., Kim, E. J., Seo, S., Kim, B., Lee, G. W., Oh, Y.-K., Kim, J. Y., Huh, Y. S., & Song, H. A. (2016). Efficient harvesting of wet blue-green microalgal biomass by two-aminoclay [AC]-mixture systems. *Bioresource Technology*, *15*, 313–318.

Kessler, E. (1985). Upper limits of temperature for growth in *Chlorella. Plant Systematics and Evolution*, *151*, 67–71.

Khosravitabar, F. (2020). Microalgal biohydrogen photoproduction: Scaling up challenges and the ways forward. *Journal of Applied Phycology*, *32*(1), 277–289.

Kim, D. G., La, H. J., Ahn, C. Y., Park, Y. H., & Oh, H. M. (2011). Harvest of *Scenedesmus* sp. with bioflocculant and reuse of culture medium for subsequent high-density cultures. *Bioresource Technology*, *102*(3), 3163–3168.

Kim, D. Y., Lee, K., Lee, J., Lee, Y. H., Han, J. I., Park, J. Y., & Oh, Y.-K. (2017). Acidified-flocculation process for harvesting of microalgae: Coagulant reutilization and metal-free-microalgae recovery. *Bioresource Technology*, *239*, 190–196.

Kim, M. S., & Kwak, D. H. (2020). Auto/bio-flocculation conditions to separate algal particles without chemical coagulants for flotation and sedimentation processes. *Separation Science and Technology*, *55*(6), 1185–1196.

Kirk, J. T. O. (1994). *Light and photosynthesis in aquatic ecosystems* (2nd ed.). Cambridge University Press.

Knuckey, R. M., Brown, M. R., Robert, R., & Frampton, D. M. F. (2006). Production of microalgal concentrates by flocculation and their assessment as aquaculture feeds. *Aquacultural Engineering*, *35*, 300–313.

Koline-Sanderson. (2014). *Rotary drum vacuum filters*. Accessed on October 23, 2014 from http://www.komline.com/docs/rotary_ drum_vacuum_filter.html

Krol, M., Maxwell, D. P., & Huner, N. P. A. (1997). Exposure of *Dunaliella salina* to low temperature mimics the high lightinduced accumulation of carotenoids and the carotenoid binding protein (Cbr). *The Plant Cell*, *38*(2), 213–216.

Kudo, I., Miyamoto, M., Noiri, Y., & Maita, Y. (2000). Combined effects of temperature and iron on the growth and physiology of the marine diatom *Phaeodactylum tricornutum* (Bacillariophyceae). *Journal of Phycology*, *36*, 1096–1102.

Kumar, C. G., Joo, H. S., Kavali, R., Choi, J. W., & Chang, C. S. (2004). Characterization of an extracellular biopolymer flocculant from a haloalkalophilic Bacillus isolate. *World Journal of Microbiology & Biotechnology*, *20*(8), 837–843.

Kumar, K., & Das, D. (2012). Growth characteristics of Chlorella sorokiniana in airlift and bubble column photobioreactors. *Bioresource Technology*, *116*, 307–313.

Kurniawati, H. A., Ismadji, S., & Liu, J. C. (2014). Microalgae harvesting using natural saponin and chitosan. *Bioresource Technology*, *166*, 429–434.

Kwon, H., Lu, M., Lee, E. Y., & Lee, J. (2014). Harvesting of microalgae using flocculation combined with dissolved air flotation. *Biotechnology and Bioprocess Engineering; BBE*, *19*, 143–149.

Lee, J., Cho, D. H., Ramanan, R., Kim, B. H., Oh, H. M., & Kim, H. S. (2013). Microalgae-associated bacteria play a key role in the flocculation of *Chlorella vulgaris*. *Bioresource Technology*, *131*, 195–201.

Lee, Y.-C., Lee, K., & Oh, Y.-K. (2015). Recent nanoparticle engineering advances in microalgal cultivation and harvesting processes of biodiesel production: A review. *Bioresource Technology*, *184*, 63–72.

Levasseur, M. E., Morissette, J. C., Popovic, R., & Harrison, P. (1990). Effects of long term exposure to low temperature on the photosynthetic apparatus of *Dunaliella tertiolecta* (Chlorophyceae). *Journal of Phycology*, *26*, 479–484.

Li, W. K. W. (1980). Temperature adaptation in phytoplankton: cellular and photosynthetic characteristics. In P. G. Falkowski (Ed.), *Primary productivity in the sea* (pp. 259–279). Plenum Press.

Li, X., Liu, B., Lao, Y., Wan, P., Mao, X., & Chen, F. (2021). Efficient magnetic harvesting of microalgae enabled by surface-initiated formation of iron nanoparticles. *Chemical Engineering Journal*, *408*, 127252.

Liu, J., Zhu, Y., Tao, Y., Zhang, Y., Li, A., Li, T., Sang, M., & Zhang, C. (2013). Freshwater microalgae harvested via flocculation induced by pH decrease. *Biotechnology for Biofuels*, *6*, 98.

Loganathan, G. B., Orsat, V., Lefsrud, M., & Wu, B. S. (2020). A comprehensive study on the effect of light quality imparted by lightemitting diodes (LEDs) on the physiological and biochemical properties of the microalgal consortia of *Chlorella variabilis* and *Scenedesmus obliquus* cultivated in dairy wastewater. *Bioprocess and Biosystems Engineering*, *43*, 1445–1455. https://doi.org/10.1007/s00449-020-02338-0

Loganathan, K., Saththasivam, J., & Sarp, S. (2018). Removal of microalgae from seawater using chitosan-alum/ferric chloride dual coagulations. *Desalination*, *433*, 25–32.

Luimstra, V. M., Schuurmans, J. M., de Carvalho, C. F. M., Matthijs, H. C. P., Hellingwerf, K. J., & Huisman, J. (2019). Exploring the low photosynthetic efficiency of cyanobacteria in blue light using a mutant lacking phycobilisomes. *Photosynthesis Research, 141*, 291–301. https://doi.org/10.1007/s11120-019-00630-z

Lv, J., Wang, X., Liu, W., Feng, J., Liu, Q., Nan, F., Jiao, X., & Xie, S. (2018). The performance of a self-flocculating microalga *Chlorococcum* sp. GD in wastewater with different ammonia concentrations. *International Journal of Environmental Research and Public Health, 15*, 434.

Ma, X., Zheng, H., Zhou, W., Liu, Y., Chen, P., & Ruan, R. (2016). Enhanced harvesting of *Chlorella vulgaris* using combined flocculants. *Applied Biochemistry and Biotechnology, 180*, 791–804.

Maddux, W. J., & Jones, R. F. (1964). Some interactions of temperature, light intensity, and nutrient concentration during the continuous culture of *Nitzschia closterium* and *Tetraselmis* sp. *Limnology and Oceanography, 9*, 79–86.

Manheim, D., & Nelson, Y. (2013). Settling and bioflocculation of two species of algae used in wastewater treatment and algae biomass production. *Environmental Progress & Sustainable Energy, 32*, 946–954.

Mannweiler, K., & Hoare, M. (1992). The scale-down of an industrial disk stack centrifuge. *Bioprocess Engineering, 8*, 19–25.

Mariam, A. H., Ghaly, A., & Amal, H. (2015). Microalgae harvesting methods for industrial production of biodiesel: Critical review and comparative analysis. *Fundamentals of Renewable Energy and Applications, 5*(2), 1–26.

Markou, G. (2014). Effect of various colors of light-emitting diodes (LEDs) on the biovolume composition of *Arthrospira platensis* cultivated in semi-continuous mode. *Appl Biochem Biotechnol, 172*, 2758-2768. doi:10.1007/s12010-014-0727-3

Mata, T. M., Martins, A. A., & Caetano, N. S. (2010). Microalgae for biodiesel production and other applications: A review. *Renewable & Sustainable Energy Reviews, 14*, 217–232.

Matter, I. A., Bui, V. K. H., Jung, M., Seo, J. Y., Kim, Y. E., Lee, Y. C., & Oh, Y. K. (2019). Flocculation harvesting techniques for microalgae: *A review. Applied Sciences (Basel, Switzerland), 9*(15), 3069.

Matter, I. A., Darwesh, O. M., & Eida, M. F. (2018). Harvesting of microalgae *Scenedesmus obliquus* using chitosan-alginate dual flocculation system. *Biosci. Res, 15*, 540–548.

Matter, I. A., Darwesh, O. M., & El-baz, F. K. (2016). Using the natural polymer chitosan in harvesting *Scenedesmus* species under di_erent concentrations and cultural pH values. *International Journal of Pharma and Bio Sciences, 7*, 254–260.

Maxwell, D. P., Fal, S., Trick, C. G., & Huner, N. (1994). Growth at low temperature mimics high-light acclimation in Chlorella vulgaris. *Plant Physiology, 105*(2), 535–543.

Mendoza, J. L., Granados, M. R., De Godos, I., Acién, F. G., Molina, E., Banks, C., & Heaven, S. (2013). Fluid-dynamic characterization of real-scale raceway reactors for microalgae production. *Biomass and Bioenergy, 54*, 267–275.

Milledge, J. J., & Heaven, S. (2013). A review of the harvesting of micro-algae for biofuel production. *Reviews in Environmental Science and Biotechnology*, *12*, 165–178.

Mohn, F. H. (1980). Experiences and strategies in the recovery of biomass from mass cultures of microalgae. In G. Schelef & C. J. Soeder (Eds.), *Algae Biomass* (pp. 547–571). Elsevier.

Morillas-España, A., Lafarga, T., Acién-Fernández, F. G., Gómez-Serrano, C., & González-López, C. V. (2021). Annual production of microalgae in wastewater using pilot-scale thin-layer cascade photobioreactors. *Journal of Applied Phycology*, 1–11.

Mortain-Bertrand, A., Descolas-Gros, C., & Jupin, H. (1988). Pathway of dark inorganic carbon fixation in two species of diatoms: Influence of light regime and regulator factors on diel variations. *Journal of Plankton Research*, *10*, 199–217.

Muller-Fuega, A. (2004). Microalgae for aquaculture. The current global situation and future trends. In A. Richmond (Ed.), *Microalgal culture: biotechnology and applied phycology* (pp. 352–364). Blackwell Science.

Murphy, C. F., & Allen, D. T. (2011). Energy-water nexus for mass cultivation of algae. *Environmental Science & Technology*, *45*, 5861–5868.

Najjar, Y. S., & Abu-Shamleh, A. (2020). Harvesting of microalgae by centrifugation for biodiesel production: A review. *Algal Research*, *51*, 102046.

Nguyen, T. D. P., Frappart, M., Jaouen, P., Pruvost, J., & Bourseau, P. (2014). Harvesting *Chlorella vulgaris* by natural increase in pH: E_ect of medium composition. *Environmental Technology*, *35*, 1378–1388.

Norsker, N.-H., Cuaresma, M., de Vree, J., Ruiz-Domínguez, M. C., García, M. C. M., Uronen, P., & Wijffels, R. (2021). *Neochloris oleoabundans* oil production in an outdoor tubular photobioreactor at pilot scale. *Journal of Applied Phycology*, *33*(3), 1327–1339.

Okaiyeto, K., Nwodo, U. U., Okoli, S. A., Mabinya, L. V., & Okoh, A. I. (2016). Implications for public health demands alternatives to inorganic and synthetic flocculants: Bioflocculants as important candidates. *MicrobiologyOpen*, *5*(2), 177–211.

Oswald, W. J. (1967). *Current status of microalgae from wastes (Water purification through biological oxidation of organic wastes using controlled photosynthesis, discussing algae production and harvesting)*. Academic Press.

Oswald, W. J. (1988). Large-scale algal culture systems (engineering aspects). In M. A. Borowitzka & L. J. Borowitzka (Eds.), *Micro-algal biotechnology* (pp. 357–394). Cambridge University Press.

Pan, Y., Shi, B., & Zhang, Y. (2009). Research on flocculation property of bioflocculant PG.a21 Ca. *Modern Applied Science*, *3*(6), 106–112.

Peng, C., Li, S., Zheng, J., Huang, S., & Li, D. (2017). Harvesting microalgae with di_erent sources of starch-based cationic flocculants. *Applied Biochemistry and Biotechnology*, *181*, 112–124.

Peperzak, L. F., Koeman, C. R., Gieskes, W. W. C., & Joordens, J. C. A. (2003). Phytoplankton sinking rates in the Rhine region of freshwater influence. *Journal of Plankton Research*, *25*, 365–368.

Pezzolesi, L., Mazzotti, M., Vanucci, S., & Pistocchi, R. (2019). Assimilation of inorganic nitrogen for scaling up Desmodesmus communis (Scenedesmaceae) biomass production. *Journal of Applied Phycology, 31*(5), 2833–2844.

Piwosz, K., Vrdoljak, A., Frenken, T., González-Olalla, J. M., Šantić, D., & McKay, R. M. (2020). Light and primary production shape bacterial activity and community composition of aerobic anoxygenic phototrophic bacteria in a microcosm experiment. *mSphere 5*. doi:10.1128/mSphere.00354-20

Posten, C. (2009). Design principles of photo-biorcactors for cultivation of microalgae. *Engineering in Life Sciences*, *9*(3), 165–177.

Prathima, A., & Karthikeyan, S. (2017). Characteristics of micro-algal biofuel from Botryococcus braunii. *Energy Sources. Part A, Recovery, Utilization, and Environmental Effects*, *39*(2), 206–212.

Prussi, M., Buffi, M., Casini, D., Chiaramonti, D., Martelli, F., Carnevale, M., & Rodolfi, L. (2014). Experimental and numerical investigations of mixing in raceway ponds for algae cultivation. *Biomass and Bioenergy*, *67*, 390–400.

Rahul, R., Kumar, S., Jha, U., & Sen, G. (2015). Cationic inulin: A plant based natural biopolymer for algal biomass harvesting. *International Journal of Biological Macromolecules*, *72*, 868–874.

Rajvanshi, M., Gautam, K., Manjre, S., Kumar, G. R. K., Kumar, C., Govindachary, S., & Dasgupta, S. (2019). Stoichiometrically balanced nutrient management using a newly designed nutrient medium for large scale cultivation of Cyanobacterium aponinum. *Journal of Applied Phycology*, *31*(5), 2779–2789.

Ras, M., Lardon, L., Bruno, S., Bernet, N., & Steyer, J. P. (2011). Experimental study on a coupled process of production and anaerobic digestion of *Chlorella vulgaris*. *Bioresource Technology*, *102*, 200–206.

Ras, M., Steyer, J., & Bernard, O. (2013). Temperature effect on microalgae: A crucial factor for outdoor production (A review). *Reviews in Environmental Science and Biotechnology*, *12*, 153–164. doi:10.100711157-013-9310-6

Rashid, N., & Nayak, M., & Suh, W.I. (2019). Efficient microalgae removal from aqueous medium through auto-flocculation: Investigating growth-dependent role of organic matter. *Environmental Science and Pollution Research International*, *26*, 27396–27406.

Rashid, N., Park, W. K., & Selvaratnam, T. (2018). Binary culture of microalgae as an integrated approach for enhanced biomass and metabolites productivity, wastewater treatment, and bioflocculation. *Chemosphere*, *194*, 67–75.

Rawat, I., Kumar, R. R., Mutanda, T., & Bux, F. (2013). Biodiesel from microalgae: A critical evaluation from laboratory to large scale production. *Applied Energy*, *103*, 444–467.

Rayen, F., Behnam, T., & Dominique, P. (2019). Optimization of a raceway pond system for wastewater treatment: A review. *Critical Reviews in Biotechnology*, *39*(3), 422–435.

Richmond, A. (1996). Efficient utilization of high irradiance for production of photoautotrophic cell mass: A survey. *Journal of Applied Phycology*, *8*, 381–387.

Riveros, K., Sepúlveda, C., Bazaes, J., Marticorena, P., Riquelme, C., & Acién, G. (2018). Overall development of a bioprocess for the outdoor production of Nannochloropsis gaditana for aquaculture. *Aquaculture Research*, *49*(1), 165–176.

Roncero-Ramos, B., Román, J. R., Gómez-Serrano, C., Cantón, Y., & Acién, F. G. (2019). Production of a biocrust-cyanobacteria strain (Nostoc commune) for large-scale restoration of dryland soils. *Journal of Applied Phycology*, *31*(4), 2217–2230.

Roselet, F., Burkert, J., & Abreu, P. C. (2016). Flocculation of nannochloropsis oculata using a tannin-based polymer: Bench scale optimization and pilot scale reproducibility. *Biomass and Bioenergy*, *87*, 55–60.

Salim, S., Bosma, R., Vermuë, M. H., & Wijffels, R. H. (2011). Harvesting of microalgae by bio-flocculation. *Journal of Applied Phycology*, *23*, 849–855.

Sanyano, N., Chetpattananondh, P., & Chongkhong, S. (2013). Coagulation–flocculation of marine *Chlorella* sp. for biodiesel production. *Bioresource Technology*, *147*, 471–476.

Satthong, S., Saego, K., Kitrungloadjanaporn, P., Nuttavut, N., Amornsamankul, S., & Triampo, W. (2019). Modeling the effects of light sources on the growth of algae. *Adv Differ Equ,* 170. doi:10.1186/s13662-019-2112-6

Schwarz, A.-M., & Markager, S. (1999). Light absorption and photosynthesis of a benthic moss community: Importance of spectral quality of light and implications of changing light attenuation in the water column. *Freshwater Biology*, *42*, 609–623. doi:10.1046/j.1365-2427.1999.00496.x

Scott, S. A., Davey, M. P., Dennis, J. S., Horst, I., Howe, C. J., Lea-Smith, D. J., & Smith, A. G. (2010). Biodiesel from algae: Challenges and prospects. *Current Opinion in Biotechnology*, *21*, 277–286.

Selesu, N. F., de Oliveira, V. T., Corrêa, D. O., Miyawaki, B., Mariano, A. B., Vargas, J. V., & Vieira, R. B. (2016). Maximum microalgae biomass harvesting via flocculation in large scale photobioreactor cultivation. *Canadian Journal of Chemical Engineering*, *94*, 304–309.

Seo, J. Y., Lee, K., Praveenkumar, R., Kim, B., Lee, S. Y., Oh, Y.-K., & Park, S. B. (2015). Tri-functionality of Fe_3O_4-embedded carbon microparticles in microalgae harvesting. *Chemical Engineering Journal*, *2015*, 206–214.

Shelef, G., Sukenik, A., & Green, M. (1984). *Microalgae harvesting and processing: a literature review*. Solar Energy Res Inst Golden.

Shen, Y., Cui, Y., & Yuan, W. (2013). Flocculation optimization of microalga *Nannochloropsis oculata. Applied Biochemistry and Biotechnology*, *169*, 2049–2063.

Shen, Y., Fan, Z., Chen, C., & Xu, X. (2015). An auto-flocculation strategy for Chlorella vulgaris. *Biotechnology Letters*, *37*(1), 75–80.

Shikata, T., Nukata, A., Yoshikawa, S., Matsubara, T., Yamasaki, Y., Shimasaki, Y., Oshima, Y., & Honjo, T. (2009). Effects of light quality on initiation and development of meroplanktonic diatom blooms in a eutrophic shallow sea. *Marine Biology*, *156*, 875–889.

Shima, S., Matsuoka, H., Iwamoto, T., & Sakai, H. (1984). Antimicrobial action of -poly-l-lysine. *The Journal of Antibiotics*, *37*, 1449–1455.

Singh, A., Nigam, P. S., & Murphy, J. D. (2011). Mechanism and challenges in commercialisation of algal biofuels. *Bioresource Technology*, *102*, 26–34.

Soeder, C. J. (1981). Types of algal ponds. Wastewater for aquaculture. UFS Publication Series C, 131-135.

Sparks, T. (2012). Solid-liquid filtration: Understanding filter presses and belt filters. *Filtration & Separation*, *49*, 20–24.

Stomp, M., Huisman, J., Stal, L. J., & Matthijs, H. C. P. (2007). Colorful niches of phototrophic microorganisms shaped by vibrations of the water molecule. *The ISME Journal*, *1*, 271–282.

Surendhiran, D., & Vijay, M. (2013). Study on flocculation efficiency for harvesting *Nannochloropsis oculata* for biodiesel production. *International Journal of Chemtech Research*, *5*, 1761–1769.

Surendhiran, D., & Vijay, M. (2014). Exploration on bioflocculation of *Nannochloropsis oculata* using response surface methodology for biodiesel production. *TheScientificWorldJournal*.

Tan, X., Zhang, D., Duan, Z., Parajuli, K., & Hu, J. (2020). Effects of light color on interspecific competition between *Microcystis aeruginosa* and *Chlorella pyrenoidosa* in batch experiment. *Environmental Science and Pollution Research International*, *27*, 344–352. https://doi.org/10.1007/s11356-019-06650-5

Tao, D. G., & Salihon, J. (2011). The Optimisation of levels of the variables pH and Concentration of Ferric chloride for Harvesting marine microalgae by flocculation. In *Proceedings of the International Conference on Food Engineering and Biotechnology (IPCBEE)*. IACSIT Press.

Thompson, P. A., Jameson, I., & Blackburn, S. I. (2009). The influence of light quality on akinete formation and germination in the toxic cyanobacterium *Anabaena circinalis*. *Harmful Algae*, *8*, 504–512. https://doi.org/10.1016/j.hal.2008.10.004

Toh, P. Y., Azenan, N. F., Wong, L., Ng, Y. S., Chng, L. M., Lim, J., & Chan, D. J. C. (2017). The role of cationic coagulant-to-cell interaction in dictating the flocculation-aided sedimentation of freshwater microalgae. *Arabian Journal for Science and Engineering*, *43*, 1–9.

Tork, M. B., Khalilzadeh, R., & Kouchakzadeh, H. (2017). Efficient harvesting of marine Chlorella vulgaris microalgae utilizing cationic starch nanoparticles by response surface methodology. *Bioresource Technology*, *243*, 583–588.

Torzillo, G., & Vonshak, A. (1994). Effect of light and temperature on the photosynthetic activity of the cyanobacterium *Spirulina plantensis*. *Biomass and Bioenergy*, *6*, 399–403.

Uduman, N., Qi, Y., Danquah, M. K., Forde, G. M. A., & Hoadley, A. (2010). Dewatering of microalgal cultures: A major bottleneck to algae-based fuels. *Journal of Renewable and Sustainable Energy*, *2*, 012701–012715.

Ummalyma, S. B., Gnansounou, E., Sukumaran, R. K., Sindhu, R., Pandey, A., & Sahoo, D. (2017). Bioflocculation: An alternative strategy for harvesting of microalgae—An overview. *Bioresource Technology*, *242*, 227–235.

Vandamme, D., Beuckels, A., Markou, G., Foubert, I., & Muylaert, K. (2015). Reversible flocculation of microalgae using magnesium hydroxide. *BioEnergy Research*, *8*, 716–725.

Vandamme, D., Foubert, I., Fraeye, I., Meesschaert, B., & Muylaert, K. (2012). Flocculation of *Chlorella vulgaris* induced by high pH: Role of magnesium and calcium and practical implications. *Bioresource Technology*, *105*, 114–119.

Vandamme, D., Foubert, I., & Muylaert, K. (2013). Flocculation as a low-cost method for harvesting microalgae for bulk biomass production. *Trends in Biotechnology*, *31*, 233–239.

Voleti, R. S. (2012). Experimental studies of vertical mixing in an open channel raceway for algae biofuel production (M.Sc. thesis). Utah State University.

Wang, H., Laughinghouse, H. D., Anderson, M. A., Chen, F., Willliams, E., Place, A. R., & Hill, R. T. (2012). Novel bacterial isolate from Permian groundwater, capable of aggregating potential biofuel-producing microalga *Nannochloropsis oceanica* IMET1. *Applied and Environmental Microbiology*, *78*, 1445–1453. https:// doi.org/10.1128/AEM.06474-11

Wang, L., Liang, W., Yu, J., Liang, Z., Ruan, L., & Zhang, Y. (2013). Flocculation of Microcystis aeruginosa using modified larch tannin. *Environmental Science & Technology*, *47*, 5771–5777.

Wang, Y., Guo, W., Yen, H. W., Ho, S. H., Lo, Y. C., Cheng, C. L., Ren, N., & Chang, J. S. (2015). Cultivation of Chlorella vulgaris JSC-6 with swine wastewater for simultaneous nutrient/COD removal and carbohydrate production. *Bioresource Technology*, *198*, 619–625.

Wei, L., You, W., Gong, Y., El Hajjami, M., Liang, W., Xu, J., & Poetsch, A. (2020). Transcriptomic and proteomic choreography in response to light quality variation reveals key adaption mechanisms in marine *Nannochloropsis oceanica. The Science of the Total Environment*, *720*, 137667. https:// doi.org/10.1016/j.scitotenv.2020.137667

Wellnitz, T. A., & Ward, J. V. (2002). Does light intensity modify the effect mayfly grazers have on periphyton? *Freshwater Biology*, *39*, 135–149. https://doi.org/10.1046/j.1365-2427.1998.00270.x

Wenderoth, K., & Rhiel, E. (2004). Influence of light quality and gassing on the vertical migration of diatoms inhabiting the Wadden Sea. *Helgoland Marine Research*, *58*, 211–215. https://doi.org/10.1007/s10152-004-0187-1

Wrede, D., Taha, M., Miranda, A. F., Kadali, K., Stevenson, T., Ball, A. S., & Mouradov, A. (2014). Co-cultivation of fungal and microalgal cells as an efficient system for harvesting microalgal cells, lipid production and wastewater treatment. *PLoS One*, *9*, e113497.

Wu, G. (2010). *Assay Development: Fundamentals and Practices*. Wiley.

Wyatt, N. B., Gloe, L. M., Brady, P. V., Hewson, J. C., Grillet, A. M., Hankins, M. G., & Pohl, P. I. (2012). Critical conditions for ferric chloride-induced flocculation of freshwater algae. *Biotechnology and Bioengineering*, *109*(2), 493–501.

Xu, L., Wang, F., Li, H. Z., Hu, Z. M., & Guo, C. (2010). Development of an efficient electroflocculation technology integrated with dispersed-air flotation for harvesting microalgae. *Journal of Chemical Technology and Biotechnology (Oxford, Oxfordshire)*, *85*, 1504–1507.

Xu, Y., Purton, S., & Baganz, F. (2013). Chitosan flocculation to aid the harvesting of the microalga *Chlorella sorokiniana. Bioresource Technology, 129*, 296–301.

Yan, X. L., & Hino, R. (2016). *Nuclear hydrogen production handbook*. CRC press.

Yan, Y. D., & Jameson, G. J. (2004). Application of the Jameson Cell technology for algae and phosphorus removal from maturation ponds. *International Journal of Mineral Processing, 73*, 23–28.

Yang, M. D., Sykes, R. M., & Merry, C. J. (2000). Estimation of algal biological parameters using water quality modeling and SPOT satellite data. *Ecological Modelling, 125*, 1–13.

Yang, Z., Shang, Y., & Lu, Y. (2011). Flocculation properties of biodegradable amphoteric chitosan-based flocculants. *Chemical Engineering Journal, 172*(1), 287–295.

Yokoi, H., Arima, T., Hirose, J., Hayashi, S., & Takasaki, Y. (1996). Flocculation properties of poly(γ-glutamic acid) produced by Bacillus subtilis. *Journal of Fermentation and Bioengineering, 82(*1), 84–87.

Yoo, C., La, H. J., Kim, S. C., & Oh, H. M. (2015). Simple processes for optimized growth and harvest of Ettlia sp. by pH control using CO_2 and light irradiation. *Biotechnology and Bioengineering, 112*, 288–296.

Yu, J., Hu, H., Wu, X., Wang, C., Zhou, T., Liu, Y., & Zheng, H. (2019). Continuous cultivation of Arthrospira platensis for phycocyanin production in large-scale outdoor raceway ponds using microfiltered culture medium. *Bioresource Technology, 287*, 121420.

Zheng, Gao, Yin, Tang, Ji, & Huang. (2012). Harvesting of microalgae by flocculation with poly (γ-glutamic acid). *Bioresource Technology, 112*, 212–220.

Zheng, Y., Ye, Z., Fang, X. L., Li, Y. H., & Cai, W. M. (2008). Production and characteristics of a bioflocculant produced by Bacillus sp. F19. *Bioresource Technology, 99*(16), 7686–7691.

Zheng, Y., Ye, Z., Fang, X. L., Li, Y. H., & Cai, W. M. (2008). Production and characteristics of a bioflocculant produced by Bacillus sp. F19. *Bioresource Technology, 99*(16), 7686–7691.

Zhu, C., Chen, C., & Zhao, L. (2012). Bioflocculant produced by Chlamydomonas reinhardtii. *Journal of Applied Phycology, 24*(5), 1245–1251.

Zhu, L., Li, Z., & Hiltunen, E. (2018). Microalgae *Chlorella vulgaris* biomass harvesting by natural flocculant: Effects on biomass sedimentation, spent medium recycling and lipid extraction. *Biotechnology for Biofuels, 11*(1), 1–10.

Chapter 9
In Silico Models on Algal Cultivation and Processing:
An Approach for Engineered Optimization

Lamiaa H. Hassan
https://orcid.org/0000-0001-7127-6723
Menoufia University, Egypt

Imran Ahmad
https://orcid.org/0000-0001-9774-9111
Universiti Teknologi Malaysia, Malaysia

Mostafa El Sheekh
Tanta University, Egypt

Norhayati Abdullah
Universiti Teknologi Malaysia, Malaysia

ABSTRACT

In modern system-level metabolic engineering, genome-wide metabolic reconstructions are used as a systems-based framework for integrating and analyzing large "omics" data sets as well as for assessing cell design molecular and bioinformatics approach "in silico". Microalgae growth processes are based on the concurrent interaction of micronutrients (Mg, Fe, Zn, etc.), macronutrients (N, C, P), and environmental parameters (temperature and light). Blackbox models or macroscopic models give the reliable interrelationship amidst the growth kinetics of microalgae and its potential of lipid and starch accumulation in response to any of the growth restraining factors. This chapter provides an insight into the different in silico models for the growth and cultivation of microalgae. Various factors such as light intensity/distribution, the temperature during cultivation, and nutrient concentration are considered. The chapter also summarises the role of different photobioreactors (PBRs) in optimising algae-based products using genome-scale models.

DOI: 10.4018/978-1-6684-2438-4.ch009

INTRODUCTION

The attention given to algae in different commercial applications for biofuels, vital nature components production, etc. make it necessary to develop system-level metabolic engineering. The bioinformatics tools in understanding and guidance of the metabolic pathway and algal bioactive components regulation, as microalgae are considered as bio-cell factories. Therefore, a synergistic approach to *in silico* analysis and then conditioning *in vivo* will improve robustness and enhance the marketability of carbon-neutral fuels from algae (Banerjee *et al*., 2020).

In silico analysis and tools have been developed in the field of genome scale metabolic reconstructions (GEMs) and microbial metabolic engineering, which give a clear prediction of algal genes, sequences and allow driving the pathway to produce high contents of a certain compounds such as triacyl-glycerides (TAGs), the precursor for biodiesel (http://genome.jgi-psf.org/Chlre4/Chlre4.home.html). Despite, there is no available genome scale reconstruction (GEMs) for algae, several studies aimed the improving of *in silico* analysis and 'omics' databases (Dal'Molin *et al*., 2011). *Chlamydomonas reinhardtii* was used as a first trial remodel a large metabolic reconstruction of algae, which demonstrated 484 reactions and 458 metabolites sited in the chloroplast, cytosol and mitochondria (Boyle and Morgan, 2009; Herrera-Valencia *et al*., 2012).

This chapter provides an insight into the different *in silico* models for the growth and cultivation of microalgae. Various factors such as light intensity/distribution, the temperature during cultivation and nutrient concentration are considered. The chapter also summarises the role of different Photobioreactors (PBRs) in optimising algae-based products using genome scale models.

Background

In 2007, the first nuclear genome of *C. reinhardtii* was published, followed by *Thalassiosira pseudonana*, while *Nannochloropis gaditana* was in 2012 (Salehi-Ashtiani *et al*., 2015). Modern approaches progress the microalgae metabolic modeling, which has been called genome-scale metabolic models. This makes a clear focus on special genetic targets to enhance certain compounds productivity such as lipids and carbohydrates for biofuel production. In addition, phenotypes behavior can be elucidated under the dynamic growth conditions. The bioinformatics approaches help in detection of the functional annotations and associating genes, or multiple genes with specific biochemical reactions, which generate the gene-protein-reaction (Salehi-Ashtiani *et al*., 2015).

Algae are an active source of biofuel precursor, such as TAGs for biodiesel and sugars for bioethanol production. The biofuel production from algae require optimization and cultivation under certain conditions to increase the synthesis of TAG and polysaccharides. So this needs a deep understanding of algal metabolomics, fluxomics, genomic, and proteomics. Algae can produce valuable secondary metabolites which have a wide interest in food, pharmaceutical, and cosmetics industries (Reijnders *et al*., 2014). Nowadays, some *in silico* tools are available to help in understanding of algal metabolic pathways, transcriptomics, genomes, and proteomics.

Algal diversity faces a great variance in habitat types and climatic changes. The classical identification methods by using cell morphology is not sufficient, other data on biochemical and genomic sequences must be included. In addition, the survival rate of species may be affected by sampling time and their transferring to the laboratory. A technique of using environmental DNA and genomic fragments that released by broken cells depends on "soil memory effect" (Foucher *et al*., 2020). By using the *in silico* tools let it available to test the hard abiotic conditions such as, high radiation, desiccation, extreme winds heavy rain and snow, extreme changes in temperature, and chronic nutrient scarcity (Stewart *et al.*, 2021). On another hand, using bioinformatics approaches help in finding the relation between different species and their ancestors to illustrate their origin (Fernándeza *et al.*, 2010).

During the toxicity investigations, *in silico* modeling is very helpful instead of *in vivo* and *in vitro* approaches (Das *et al.*, 2015). It allows access to valuable data related to the eco-toxicity effect of algae on the surrounding ecosystem, as well as the effect of different environmental biohazard chemical compounds and dyes on algal growth (Bakire *et al.*, 2018; Ding *et al.*, 2018).

Algal Diversity and Origin

A team of researchers studied green algae biodiversity and distribution in France Alps mountain environments from 1,250 to 3,000 m high from a various habitat including forests grasslands and rocky areas. They tested the environmental variables of these sits, including elevation from sea level, pH and organic matter of the soil, Carbon, Nitrogen levels and C/N ratio. Two new DNA meta-barcoding studies, targeted Chlorophyta "Chlo01" designed in the V7 region of the 18S ribosomal RNA (Taberlet *et al.*, 2012), and a Chlorophyceae marker "Chlo02" designed in the 23S ribosomal RNA chloroplast sequence. NCBI database (https://www.ncbi.nlm.nih.gov) was used for 1,628 complete chloroplast sequences, plastid genomes, and 23 Chlorophyceae genomes. *In silico* tools were used, the corresponding primer sequences were refined by ecoPCR (http://metabarcoding.org/ecopcr) and the ROBIBarcodes R package (https://git.metabarcoding.org/obitools/ROBIBarcodes), and OBITools (Boyer *et al.*, 2016) was used to filter the results. The taxonomic resolution was calculated using the method suggested by Ficetola *et al.* (2010). EMBL sequence database for taxonomic assignment used to realize those tests (Amid *et al.*, 2020). OligoCalc (http://biotools.nubic.northwestern.edu/OligoCalc.html) was used in detecting hybridization temperature (Taberlet *et al.*, 2018). Chlo02 was designed using the ecoPrimers software (https://www.genome.jp/kegg/pathway.html) (Riaz *et al.*, 2011). The DNA of each species constituting the mock community was extracted using the Macherey Nagel NucleoSpin Plant II extraction kit according to the instruction manual (http://biotools.nubic.northwestern.edu/OligoCalc.html). The results exhibited the Chlo01 database represents 295 genera and 62 families belonging the Chlorophyta. The Chlo02 database represents 42 genera and 19 families belonging the Chlorophyceae. The Euka03 database represents 5179 genera and 2488 families belonging the Eukaryota.

Fernándeza *et al.* (2010) used the bioinformatics applications to evidence that green algae are ancestral to land plant nsHbs, depending on genomes sequenced from diverse green algae is a valuable source of information to search for the globin ancestor of land plant nsHbs. They reported the high similarity to land plant nsHbs from the genome of the prasinophyceae algae *Micromonas pusilla*, *M.* sp. RCC299, and *Ostreococcus* sp. RCC809. Land plants Hemoglobins (Hbs) are O_2-binding proteins, which classified into three types: symbiotic Hbs, nonsymbiotic Hbs (nsHbs), and truncated Hbs (tHbs), depending on sequence similarity, synthesis in plant organs, and postulated function plant Hbs. This study worked on different *in silico* analysis by using various tools such as, the Neighbor-Joining method of the Clustal X program (Thompson *et al.*, 1997) which detected the multiple sequence alignment and cluster analysis of *Micromonas* and *Ostreococcus* nsHb-like globins with the selected land plant and bacterial Hbs. Other *in silico* analysis were performed using the BLASTP (https://blast.ncbi.nlm.nih.gov/Blast.cgi?PAGE=Proteins) from the Genbank database to detect sequence similarity and identity values between the *Micromonas* and *Ostreococcus* nsHb-like globins and individual Hbs. Predicting for chloroplast, mitochondrial, and signal peptide targeting sequences (ChloroP 1.1, MitoProt II and SignalP-NN/HMM tools, respectively) were obtained from ExPaSy web site (https://web.expasy.org/protparam/). The automated mode of Swiss Model (http://swissmodel.expasy.org) is another *in silico* analysis tool used for modelling of the tertiary structure of the *Micromonas* and *Ostreococcus* nsHb-like globins. Model reliability performed using Verify3D from the Swiss Model workspace (http://swissmodel.expasy.org). The VMD program (Humphrey *et al.*, 1996) and Adobe Photoshop® software were used for models editing. The results exhibited that *Micromonas* and *Ostreococcus* nsHb like globins are closely related to bacterial Hbs, while no nsHb-like sequences were identified in the genome of other green algae, but from the diatom *Thalassiosira pseudonana*.

Algal Genome and Metabolism

Protein database tools enable the combining data on enzymes, genes, proteins, metabolic pathways, and substrates. Table 2 show the most common protein databases. Pathway/Genome Databases (PGDBs) allow forming a visualize data on the metabolic analysis, biochemical reactions, and genomic evidence for the enzymatic reactions, or gene-protein-reactions. BioCyc (http://biocyc.org/) database contain about 3,530 PGDBs, only 7 are on algae (*Acaryochloris marina*, *Anabaena cylindrica*, *Anabaena variabilis*, *Chlamydomonas reinhardtii*, *Nannochloropsis gaditana*, *Synechococcus elongatus,* and *Thalassiosira pseudonana*). MetaCyc (http://metacyc.org/), Kbase (http://kbase.us), and Biomart (http://www.biomart.org/index.html) are other PGDBs

Chlamydomonas reinhardtii genome was used to develop a genome-scale metabolic network model (AlgaGEM) (Dal'Molin *et al.*, 2011). This study output AlgaGEM with 866 functions of unique ORFs, 1862 metabolites, 1725 unique reaction, and 2249 gene-enzyme-reaction-association entries of the cytoplasm, mitochondrion, plastid and microbody. The results gave significant data on *C. reinhardtii* behaviours, catabolism, and recycling of phosphoglycolate in photorespiration. *C. reinhardtii* genome scale model, reaction database, and compound IDs were gained from Kyoto Encyclopedia of Genes and Genomes (KEGG) (https://www.genome.jp/kegg/pathway.html) (Kanehisa *et al.*, 2016), ChlamyCyc database (http://genome.jgi-psf.org/Chlre4/Chlre4.home.html), Metacyc database (http://metacyc.org/), and ExPASy Biochemical Pathways database (http://chlamyto.mpimp-golm.mpg.de/chlamycyc/index.jsp). In addition, they used DOE Joint Genome Institute (JGI); *Chlamydomonas reinhardtii* v4.0 (http://metacyc.org/) and online Informatics Resource for *Chlamydomonas* (Chlamy Center) (http://www.expasy.ch/cgi-bin/search-biochem-index) to collect the *C. reinhardtii* genome data. ExPASy Enzyme Database (Ehttps://enzyme.expasy.org), AraPerox (Arabidopsis Protein from Plant Peroxisomes) (http://www.araperox.uni-goettingen.de/), SUBA (Arabidopsis subcellular database) (https://www.plantenergy.uwa.edu.au/applications/suba2/index.php), PPDB (Plant proteome database) (http://ppdb.tc.cornell.edu/default.aspx), UniproKB/SwissProt (http://ca.expasy.org/sprot/relnotes/relstat.html), and Transport DB (http://www.membranetransport.org/) were used for Enzyme/Protein Localization and others. The resulted data gave an over view on the ability of *C. reinhardtii* increase in hydrogen production when cyclic electron flow is disrupted as seen in high production spikes derived from mutagenesis studies, which predicted the physiological pathway for H_2 production.

Glycerol-3-phosphate dehydrogenase (GPDH) is a catalytic enzyme which play role in glycerol and glycerolipid synthesis in microalgae. Herrera-Valencia *et al.* (2012) reported that biochemical and genetic studies of the GPDH could enhance glycerol and TAGs synthesis for biodiesel production from microalgae, by using *in silico* characterization of three putative GPDH genes from *C. reinhardtii*: *CrGPDH*1, *CrG*-

PDH2, and *CrGPDH3*. These sequences were similar to GPDH gene of *Dunaliella salina* and *Dunaliella viridis*. BLASTP search in the GenBank (https://blast.ncbi.nlm.nih.gov/Blast.cgi?PAGE=Proteins) was used to get the protein sequences of *D. salina* (DsGPDH2) and *D. viridis* (DvGPDH2) protein which helped in the search on retrieve GPDH-like sequences from the *C. reinhardtii* genome sequence v4.3 at Phytozome (http://www.phytozome.net/). The sequences were edited by Lasergene software package version 7.2 (DNASTAR, Madison, WI). Meg-Align program of the Lasergene software package version 7.2 (DNASTAR, Madison, WI) and SMART database (http://smart.embl-heidelberg.de) were used to calculate the amino acid identity between sequences and detect the conserved domains, respectively. Chloroplast transit peptides were detected by the ChloroP 1.1 server (http://www.cbs.dtu.dk/services/ChloroP). The three-dimensional structure and models visualizing of the GPDH domain was carried out by the SWISS-MODEL workspace (http://swissmodel.expasy.org) and Jmol program version 12.0 (http://www.jmol.org), respectively. The alignment of the predicted protein sequences were performed using ClustalX program version 2.0 (Larkin *et al.*, 2007). BoxShade program helped in shading the identical amino acids in the alignment with black color and conservative substitutions with gray color (http://www.ch.embnet.org). A phylogenetic tree was performed according to the neighborjoining (NJ) method by using the Molecular Evolutionary Genetic Analysis (MEGA) software version 5 (Tamura *et al.*, 2011).

Genome-scale model (GSM) for the diatom *Phaeodactylum tricornutum* (iLB1025) (Levering *et al.*, 2016; Broddrick *et al.*, 2019) was used as a template for the metabolic model of the polar diatom *Fragilariopsis cylindrus* (CCMP1102) depending on MATLAB/Python workflow to analyze the sensitivity and robustness of the model (Pianosi *et al.*, 2015). The web-based software (ChloroP, MitoProt, and DeepLoc-1.0) (Lavoie *et al.*, 2020) was predicted 96.6% of homologous proteins between *F. cylindrus* and *P. tricornutum*. The protein IDs of *F. cylindrus* were edited according to manual protein annotations stated by the Joint Genome Institute website (https://genome.jgi.doe.gov/pages/search-for-genes.jsf?organism=Fracy1) and BLASTP (https://blast.ncbi.nlm.nih.gov/Blast.cgi?PAGE=Proteins) analyses. The divergent allelic variants were resolved and validated using independent sequencing approaches and genome assembly approaches (Paajanen *et al.*, 2017). Flux Balance Analysis (FBA) mathematical equation system was used, which gave a unique flux distribution that helped the cellular metabolism and maximized the growth (Lavoie *et al.*, 2020). MATLAB 2018 (The MathWorks Inc., Natick, MA, USA) tool was used in the estimation of model development, sensitivity analyses, and flux distribution, while the custom codes were gained from GitHub (https://github.com/michellavoie4/Resilience-Fcylindrus). Scripts for analyzing local and global sensitivity to changes in 69 parameters of the model are detailed in the files localsens_script.m and globalSA_script.m using the localsensFBA.m and FBAlight.m functions. Morris method was used to perform the global sensitivity analysis by using the Sensitivity Analysis For Everyone (SAFE) toolbox in MATLAB considering the Latin cube resampling strategy and evaluating 6500 models (Lavoie *et al.*, 2020). FBA and quadratic minimization of the metabolic adjustment (MOMA) were performed by the Gurobi Optimizer Version 7.5.2 (Gurobi Optimization Inc., Houston, TX, USA) solver (http://www.gurobi.com), which helped in the calculations of the modified structure (MMS) and the full initial model (IM) with the Constraint-Based Reconstruction and Analysis COBRA toolbox v3.0 (Heirendt *et al.*, 2019).

Rapid Annotation of Photosynthetic Systems (RAPS) algorithm is based on a number of protein-gene interaction relationships derived from previously published and manually annotated algal genome scale metabolic network reconstructions (Metcalf *et al.*, 2020), depending on the genome scale metabolic models of *Chlamydomonas reinhardtii* (Imam *et al.*, 2015) and *Nannochloropsis gaditana* (Shah *et al.*, 2017). In RAPS database, each metabolite was referenced to its associated InChIKey (Heller *et al.*,

2015) as possible, which obtained from KEGG (https://www.genome.jp/kegg/pathway.html), PubChem (https://pubchem.ncbi.nlm.nih.gov/), MetaCyc (http://metacyc.org/), MetaNetX (Ganter *et al.*, 2013), and BIGG databases (http://bigg.ucsd.edu/). Different models were involved in RAPS algorithm such as COBRApy (Ebrahim *et al.*, 2013), Biopython (Cock *et al.*, 2009), LibSBML (Bornstein *et al.*, 2008), Matplotlib (Hunter, 2007), BLAST (https://blast.ncbi.nlm.nih.gov/Blast.cgi?PAGE=Proteins), and Open Babel (O'Boyle *et al.*, 2011). COBRApy (Ebrahim *et al.*, 2013), and SBML (Hucka *et al.*, 2003) format tools were used for modifying and creating the data, respectively. The authors checked the quality of RAPS' metabolic network reconstruction by comparing with RAPS and predicted protein for *Chlorella variabilis* model (Blanc *et al.*, 2010; Juneja *et al.*, 2016). Metcalf *et al.* (2020) created a new genome scale metabolic network depending on RAPS, which include *Chlorella variabilis*, *Coccomyxa subellipsoidea*, *Dunaliella salina*, *Micromonas commoda* RCC299, *Micromonas pusilla* CCMP1545, and *Volvox carteri* by the aid of Phytozome (Goodstein *et al.*, 2012) or Phycocosm (Nordberg *et al.*, 2014) protein database.

In green algae, the cellular chromophores absorb light by using photoreceptors. Mahdavi *et al.* (2020) worked on *Volvox* species and to detect the encoding nucleotide sequences that affected by light. DASH gene transcripts (*Drosophila*, *Arabidopsis*, *Synechocystis*, *Homo*) cryptochromes (DASH-Crys) were assessed depending on their light absorption, and DNA binding and repairing, which considered Light-Inducible DNA-Repair (LIDR) tools in the optogenetics investigations. This study worked on two DASH-type proteins known as VcCPF5 and VcCPF8 on 10 strain of *V. carteri* f. *nagariensis*. NCBI (https://www.ncbi.nlm.nih.gov) and Phytozome (https://phytozome.jgi.doe.gov/pz/portal.html) databases were the sources of *V. carteri* complete genome, ESTs (Expressed Sequence Tags), and RNAseq. DASH- cryptochrome related sequences was identified using the BLAST (Camacho *et al.*, 2009), while the suitable DASH-related proteins in *V. carteri* were obtained by BLASTP (https://blast.ncbi.nlm.nih.gov/Blast.cgi?PAGE=Proteins) depending on two known DASH-cryptochrome amino acid sequences, called Cry3-*A. thaliana* and SynCryDASH (Uniprot codes: Q84KJ5 and P77967, respectively). The putative exons were checked by tBLASTn by executing the EST and genome database of VcCPF5 and VcCPF8 proteins. The Photolyase Homology Region (PHR) domain was obtained by exploring the protein sequences against entries in the Pfam, PROSITE and NCBI-CDD databases (El-Gebali *et al.*, 2019). PHR domain was used to build multiple sequence alignments using Clustal-W method. The alignment output was optimized using BioLign software (version 4.0.6) (http://en.bio-soft.net). A likelihood-based phylogenetic trees was performed by PHYML server (http://www.atgc-montpellier.fr/phyml), and Phylogeny.fr server (http://www.phylogeny.fr/one_task.cgi?task_type=mrbayes). The two trees were edited using iTOL web server (http://www.itol.embl.de/). The frequency of the cis-regulatory motifs in the promoter and calculate the physiochemical features of VcCPF5 and VcCPF8 sequences was performed using the PlantCARE algorithm (http://bioinformatics.psb.ugent.be/webtools/plantcare/html/), and ProtParam server (https://web.expasy.org/protparam/), respectively. The protein solubility was analyzed using PROSO II server (http://mbiljj45.bio.med.uni-muenchen.de:8888/prosoII/prosoII.seam). The 3D models were performed by Chimera software (Pettersen *et al.*, 2004), MODELLER 9v21 software (Alva *et al.*, 2016), Swiss-Model (http://swissmodel.expasy.org), RaptorX (Källberg *et al.*, 2012), Phyre2 (http://www.sbg.bio.ic.ac.uk/phyre2/) and Robetta (Kim, *et al.*, 2004) tools. ModEval11 (Shen and Sali, 2006), ModFOLD15 (Maghrabi, and McGuffin 2017), PROSESS16 (Berjanskii *et al.*, 2010), ProSA17 (Wiederstei and Sippl, 2007), QMEAN4 QMEANDisCo13 (Studer *et al.*, 2020), and VERIFY-3D14 (Eisenberg *et al.*, 1997) servers were run the model quality. The protein ligand quality models were performed using AutoDock-Vina (https://vina.scripps.edu/) algorithm, Pubchem (https://pubchem.ncbi.nlm.nih.gov/) or PDB (http://ppdb.tc.cornell.edu/default.aspx) databases, and OpenBabel program (O'Boyle *et al.*, 2011). FunFOLD2 18 (Roche *et al.*, 2013) was applied to get a prediction of more reliable protein-ligand binding sites. The surface electrostatic potential of two modeled proteins was calculated by in Chimera software plugin (Pettersen *et al.*, 2004) depending on two algorithms PDB2PQR and APBS. The pH field and strength parameters were set to 7.

Schierenbeck *et al.* (2015) investigated the high light tolerance of microalgae to allow their outdoor cultivation on a large scale. They worked on *C. reinhardtii* using *in vivo* oxygen evolution activity on UV mutagenesis of *Chlamydomonas reinhardtii* WT, CC124 strain, then they performed the DNA and genome sequencing and identification of the mutations. They encoded the generation sequencing data supporting the results of this article is available in the NCBI Sequence Read Archive (SRA) (https://www.ncbi.nlm.nih.gov) under BioProject accession number SRP037721 (PRJNA23 8037) with each sequence file under numbers SRS557198 (WT), SRS558641 (hit1) and SRS558642 (hit2). The availability of reference genome sequences and removal of common background variants in the wild-type strain and each mutant, enabled the identification of two single nucleotide polymorphisms within the same gene Cre02.g085050, later named LRS1 (putative Light Response Signaling protein 1). The depended on the *in silico* analysis to identify genes involved in high light adaptation in untagged mutants. SDSC biology workbench (http://workbench.sdsc.edu/) was the tool used for Protein alignments, I-TASSER (Roy *et al.*, 2012) performed the amino acid sequences of the three-dimensional structure of LRS1 (Cre02.g085050), DeepView-Swiss-PdbViewer software (Guex and Peitsch, 1997) was used in modelling, and CLC sequence viewer (CLC bio, a QIAGEN Company) was used for protein alignments.

The study of Raj-Kumar *et al.* (2017) aimed to reassess the prevalence of alternative splicing (AS) and to explore the sequence features responsible for intron retention in *C.reinhardtii*. The authors used GMAP (Wu and Watanabe, 2005), BLAT (Kent, 2002) and the Program to Assemble Spliced Alignment (PASA) (http://pasa.sourceforge.net/) (Raj-Kumar *et al.*, 2017) to identify splice isoforms, and incorporating existing Phytozome v9.0 (http://www.phytozome.net/). Phytozome Assembly v5.3. (https://phytozome.jgi.doe.gov/pz/portal.html) was used to get *C. reinharditii* genome sequence and annotation model. We used Pyro sequencing cDNA reads (6,317,641) from JGI (http://genome.jgipsf.org/chlamy/chlamy.info.html) and Genoscope cDNAs (689,548) from NCBI SRA (https://www.ncbi.nlm.nih.gov) were used. Sequence contaminants were processed by Sanger ESTs and pyro sequencing separately, by screening for sequencing adapters, vector sequence, regions of low read-quality, short sequence fragments and regions of low complexity/repeats using DUST (Campbell *et al.*, 2006). Genome-guided assembly of the dataset was performed using CLC genomics (http://www.clcbio.com/). The genomic coordinates of the introns were used to ensure that the classes of introns were not redundant. Alignment gaps (i.e, introns) were determined by the presence of N in the CIGAR coordination, when Illumina RNA-Seq data sets retrieved from SRA were aligned using BWA aln (Li *et al.*, 2009), and visualized on an IGV browser (Robinson *et al.*, 2011).

Models for Microalgae Cultivation

One of the methods to optimize the processes is to develop sturdy mathematical models which provide accurate representation of vital parameters of algal cultivation in response to the factors limiting the growth (Yuan *et al.*, 2020). Though the modelling of the algal cultivation systems is a complicated task as there are many growth limiting factors responsible to control the growth. The growth of algae depends on the synergetic interaction of various micronutrients(i.e., Zn, Fe, Mg etc.), macronutrients (i.e., C, N, P) and environmental factors (temperature and light intensity) (Murwanashyaka *et al.*, 2020). Macroscopic

Figure 1. Extracellular and Intracellular elements controlling microalgae growth

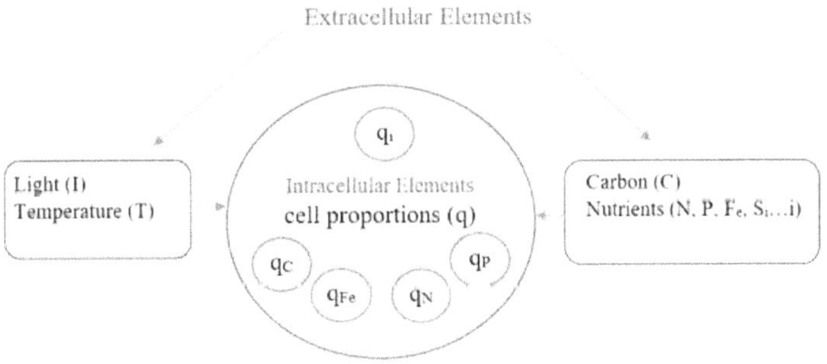

models (Black box) that accurately establishes the relation between algal growth kinetics and lipid/starch accumulation (output of models) in response to a growth limiting factor (model input) (Theodoropoulos & Sun, 2019). This can contribute to the prediction of the outcome for a specific cultivation process, without carrying time and cost consuming experiments. Therefore, models can be used as optimization tools for the bioprocesses in the identification of conditions leading to the maximum productivity in terms of biomass for sustainable and economically viable large scale cultivation system (Bekirogullari *et al.*, 2018). Till date many kinetic models are developed and implemented to algal cultivation systems yeah to optimize and control biomass, starch/lipids productivity under the effect of single/multiple substrates and environmental parameters like light intensity, micronutrients, process inhibitors, and dissolved O_2 and CO_2 concentrations (Ryu *et al.*, 2019).

The growth of algae depends on the interaction of cells with the nutrients and environmental conditions. Cell concentration increase with time as the nutrients are consumed from the medium and the environmental factors affect the cell Physiology assisting them to carry out photosynthesis. Precisely mathematical models are formulated based on μ (sp. growth rate) which is rate of cell growth with time in response to a growth limiting factor. Algal growth limiting factors are categorized as intracellular and extracellular elements (Bekirogullari *et al.*, 2020). Broadly two types of single substrate models are present in the literature which are classified as Monod's model (depending on extracellular elements) and Droop's model (depending on intracellular elements) (Droop, 1968; Monod, 1949).

Monod Model

The application of Monod model in the algal biomass growth concept assumes that a particular substrate (S) is exclusively responsible for the growth of the cells and at the same time variation in other substrates will cause a negligible effect. The specific growth rate of algae hyperbolically increases with the increase in substrate concentration until a point of saturation is attained and the growth rate becomes constant (Monod, 1949).

$$\mu = \mu_{max} \frac{S}{K_s + S}$$

Where K_s = Saturation constant; μ_{max} = Maximum Specific growth rate; and Ks is also termed as half velocity constant, since its value is equivalent to S, when $\mu = 1/2\mu_{max}$.

Monod's model is used by various researchers to express the growth of different microalgal species (*Chlorella sp., Chlamydomonas sp.* and *Scenedesmus sp.*) under the effect of single growth limiting factor like carbon, phosphorus, nitrogen and light irradiance (Pérez *et al.*, 2008).

Andrews Model

It is revealed that the substrates at high concentrations pose a risk of inhibition for microalgal growth. Quite a few models are being proposed to explain the inhibition due to substrate. However, the most common model employed for substrate inhibited growth kinetics was developed by Andrews (1968) (Andrews, 1968). It was corresponding to the model proposed by Haldane (1930). This model differs from the Monod model by the induction of an inhibitory constant K*is* that makes the value of specific growth rate to decrease at the high substrate concentration. The equation can be expressed as

$$\mu = \mu_{max} \frac{S}{S + K_s + \frac{S^2}{K_{is}}}$$

Where,

$$\mu_{max} = \frac{\mu}{1 + 2\sqrt{\frac{k_s}{k_i}}}$$

and

$$S = \sqrt{k_s . k_i}$$

Inhibitory models are effectively utilized for the optimized initial growth conditions which are necessary in achieving maximum cell concentration and productivity. This model is widely employed to define the relationship between biomass growth and substrate concentration in the species like *Chlorella sp., C. reinhardtii, and Nannochloropsis sp* (Fouchard *et al.*, 2009).

Droop Model

It is considered as one of the most accepted and applied model for the growth of microalgal species. During the investigation of *Monochrysis lutheri* by virtue of Vitamin B_{12} growth limited kinetics; Droop found that the plot for the specific growth rate vs the concentration of external vitamin does not follow the standard hyperbolic pattern as per the prediction of Monod equation. Rather Droop explained that the growth rate is the sigmoid function depending on the intrinsic quota of vitamin (Q). Q can be defined as vitamin weight/unit cell (Droop, 1968). Droop model is used on a large scale for modelling growth as a function of intrinsic concentration of a limiting nutrient with respect to biomass containing carbon. The model can be expressed as

$$\mu = \bar{\mu} \cdot \left(1 - \frac{q_{s,0}}{q_s}\right)$$

Where, μ= maximum value of specific growth rate obtained at a hypothetical infinite cell quota and $q_{s,0}$= the minimum value cell quota or subsistence quota.

It is important to mention that since Droop model depends on internal concentration it is supported by an uptake expression as a saturation function of the extrinsic concentration. After that Droops model has been used extensively to explain the growth of various microalgae species (Adesanya *et al.*, 2014; Packer *et al.*, 2011). As compared to Monod equation it is having upper edge because it can express the microalgal growth even when the concentration of extrinsic nutrient is finished. Other models for the single substrate growth kinetic models are Caperon-Meyer model, Molina-Grima model, Geider model, and Martinez-Sancho model.

Modelling the Effect of Light and Temperature

Photosynthesis is the main cellular process of microalgae which is a metabolic route converting CO2 and solar energy into ATP and O_2. One of the most critical criteria in microalgae production is light intensity (Brennan & Owende, 2010). Individual cells in dense cultures receive varying amounts of light based on their location within the PBR.

Effect of Light

Nonetheless, in well-mixed systems, cells move from high light zones to near dark zones in short light cycles. Light intensity must be controlled to account for light gradients and ultimately enhance algal biomass growth and bioproduct productivity while minimizing capital and operating costs (Béchet *et al.*, 2013).

There are three types of light intensity models: a) models that consider incident or average light intensity reaching the cultures (Type I); b) models that consider short light cycles (Type II); and c) models that account for both light intensity and short light cycles (Type III) (Type III).

Type I – Light inhibition model

$$P(I) = P_s \frac{\mathrm{I}}{K_I + I + \frac{I^2}{K'_I}}$$

Type II – Monod-like model

$$\mu = \rho_m \varphi E_a \frac{K_I I}{K_I + I}$$

Type III – Indoor airlift reactor model

$$I(l) = I_0 \exp(-(k_1 + k_2 X_p) X l)$$

Here *P*(*I*) is the photosynthetic rate, *I* is the light intensity, *Ps*, *KI*, *k*1, *k*2 and *KI*¢ are constants, *ρm* is the maximum energetic yield for photon conversion, φ the mass quantum yield, *Ea* the spectral absorption coefficient and *XP* the pigment content (Cornet & Dussap, 2009).

Effect of Temperature

Following the Van't Hoff thermodynamic equation for chemical processes, Van't Hoff and Arrhenius (so-called Arrhenius equation) proposed the following equation to characterize the influence of temperature on the specific rate constant of chemical reactions.

$$\frac{dlnK}{dT} = \frac{E}{RT^2}$$

Where k represents the specific reaction rate constant, R represents the universal gas constant, T represents the temperature, and E represents the activation energy.

The Arrhenius equation, on the other hand, fails to accurately represent how temperature affects microbial development. Researchers have since sought to offer a modified version of this law that is suitable for microbial growth by substituting the growth rate constant for the reaction rate constant (Ratkowsky *et al.*, 1982).

$$\mu = A_0 \, exp\left(-\frac{E}{RT}\right)$$

Various modified Arrhenius formulas have been reported in the literature to simulate the effect of temperature on biomass growth rate. A modified version of the Arrhenius equation was proposed by Pérez et al. (2008):

$$\mu = A_0 \, exp\left[\frac{-E_a}{R}\left(\frac{T - T_0}{T_0}\right)\right] - B_0 \, exp\left[\frac{E_b}{R}\left(\frac{T - T_0}{T_0}\right)\right]$$

where, the first and second halves of the equation represent the stimulation and inhibition of microalgal growth rate by temperature, respectively. T0 = 293 K was utilised as the reference temperature in this study, and the following values were found for the model's main kinetic parameters: A0 = 0.26 h1, B0

= 0.18 h1, Ea = 28 kcal/mol, and Eb = 39 kcal/mol. Using these kinetic parameter values, a temperature of 20.4 ° C was estimated as the ideal temperature for *P. tricornutum* growth (Pérez *et al.*, 2008).

Roels (1983) proposed a modified version of the Arrhenius equation for the rate of deactivation of microbial growth systems as a function of temperature, which goes like this:

$$\mu(I) = \mu_{m,0}(I)\frac{exp\left(-\frac{E_a}{kT}\right)}{1 + K \cdot exp\left(-\frac{E_a'}{kT}\right)}$$

Where (I) denotes the specific growth rate (d^{-1}) at the specified light intensity. T is the temperature (K), E is activation energy (j/K), and k is a dimensionless constant. I, m, 0(I) is the maximum specific growth rate (d^{-1}), K is the Boltzmann constant (j/K), R is the universal gas constant, T is the temperature (K), E is activation energy (j/K), and k is a dimensionless constant (Roels, 1983).

Biofuel Production

Omics witness available applications that can be investigated on genome sequences for multiple algal strains to enhance the metabolite production of developed mutant strains, genome sequence, genetic transformation, gene targeting, different promoters, selection markers and molecular biology techniques. Depending on this, major lipid droplet protein (MLDP) in *C. reinhardtii* was demonstrated as a marker gene for lipid accumulation by using RNA interference strategy, which led to increase the lipid droplets size (Banerjee *et al.*, 2020). CRISPR-Cas9 system is a helpful application that can modify DNA and provide a stable mutation by using RNP mediated gene transfer. This improved the efficiency, specificity and robustness of multi-gene in the micro-algae system (Tanwar and Kumar, 2020).

The process of lipid synthesizing in microalgae depends on two main enzymes, type II fatty acid synthase (FAS) and acetyl-CoA carboxylase (ACCase). ACCase enzyme could improve lipids accumulation by converting acetyl-CoA to malonyl-CoA, which form TAGs (Behera *et al.*, 2019). Behera *et al.* (2019) worked on the ACCase enzyme of the microalga *Chlamydomonas reinhardtii.* The protein sequence, and the template structure data of ACCase in FASTA format were obtained from NCBI (https://www.ncbi.nlm.nih.gov) database, and BLASTP (https://blast.ncbi.nlm.nih.gov/Blast.cgi?PAGE=Proteins), respectively. Then, MODELER (http://www.salilab.org/modeller/) software was used to draw the 3D structure of the protein by using python scripts. The protein chain, folds, family, and domain analysis were retrieved by PHYRE (http://www.sbg.bio.ic.ac.uk/phyre2/) software. The annotations were done using BLAST link by Amigo-gene (http://www.amigo.geneontology.org/) ontology tool. The structural validation was obtained by RAMPAGE (http://mordred.bioc.cam.ac.uk/~rapper/rampage.php) tool. The ligands protein structure, and docking site were done by PubChem (https://pubchem.ncbi.nlm.nih.gov/), and metaPocket (http://metapocket.eml.org/), while the docking was done by AutoDockVina (https://vina.scripps.edu/). PYMOL (https://pymol.org/) software showed the binding of the docked structures, which helped to detect the ligand–protein stability. The resulted data illustrated that the docking analysis confirmed the stability of interaction of the selected ligands (ACP, AMP, biotin, and glycine), which was evidence on enhancing the enzyme activity for lipid synthesis.

Table 1. Comparison between different single substrate models with their functions, advantages, and disadvantages

Model Name	Mechanism of growth Kinetics	Growth Inhibiting factor/ No. of parameters	Model functions	Advantages	Disadvantages
Monod	Saturation of substrate	Extracellular/2	To estimate O_2 production and CO_2 uptake for better PBR control (Eriksen *et al.*, 2007). Evaluate light effect on the growth of biomass in PBRs (Sasi *et al.*, 2011). To estimate the growth dynamics of microalgae coupled with wastewater treatment process (Xin *et al.*, 2010).	It has been widely used to study the dynamics of microbial growth at low and moderate substrate concentrations.	It is incapable of describing growth inhibition at large concentrations of external substrate. It is unable to foresee biomass growth in the absence of an external substrate.
Andrews	Inhibition of substrate	Extracellular/3	Determination of organic carbon's toxicity and determination of the appropriate concentration in heterotrophic microalgal cultures (Chen & Johns, 1994). Optimisation of the biomass productivity of microalgae in PBRs (Pfaffinger *et al.*, 2016).	The model can be used to depict non-interactive algae growth suppression.	It is unable to forecast biomass growth in the absence of an external substrate. In comparison to the Monod model, one additional parameter must be evaluated.
Droop	Threshold for quota	Intracellular/2	Estimation of biomass and lipid productivities in nitrogen-limited Cultures (Mairet *et al.*, 2011). Estimation of microalgal growth dynamics for phytoremediation Purposes (Kwon *et al.*, 2013).	Even when an external substrate is depleted, it can characterize biomass growth.	The theoretical maximum specific growth rate lacks a biological interpretation (μ=μ¯ when q®¥). Internal nutrient concentrations, or quotas, are difficult to quantify accurately, thus they are usually calculated via indirect methods.
Molina-Grima	Saturation of light	Extracellular/3	Estimation of light-dependent microalgal productivities, simulation of light attenuation by cells, and prediction of transition periods in chemostat cultures while adjusting dilution rates (Grima *et al.*, 1994). Light-gradient simulation inside PBRs, as well as evaluation and optimization of cultivation parameters (Barbosa *et al.*, 2003).	It uses simple Monod-like kinetics to describe the effect of light on the growth rate of microalgae. The shape-parameter, *n*, makes it possible to depict growth kinetics at low light intensities more accurately.	In comparison to the Monod model, one more parameter must be evaluated.
Caperon-Meyer	Threshold for quota	Intracellular/3	Co-nutrient limitation in microalgal growth estimation and nutrient competition between species evaluation (Davidson & Gurney, 1999). Understanding the mechanics of phosphate transfer in microalgal cultures that are subjected to nutrient transient regimes (John & Flynn, 2000). Calculation of the best N/C ratios for nutrient-limited agriculture (Liu *et al.*, 2001).	Even when an external substrate is depleted, it can indicate biomass growth. The saturation-type kinetics used have a quota threshold that limits the specific growth rate.	Internal nutrient concentrations, or quotas, are difficult to quantify accurately and are usually inferred using indirect methods. In comparison to the Droop model, one more parameter must be evaluated.
Geider	Saturation of light	Extracellular/4	During phytoplankton growth dynamics, it is important to understand the link between light and fluctuating chlorophyll: carbon content (Geider *et al.*, 1997).	It has been used to describe the pace of photosynthesis as a function of chlorophyll for a long time.	Because chlorophyll is a diminute and changeable cell component, reliable measurement of chlorophyll: carbon ratios may be problematic.
Martinez-Sancho	Saturation of substrate	Extracellular/3	Simulation of microalgal growth kinetics in the absence or high extracellular phosphorous (Sancho *et al.*, 1997).	Even after an external substrate is depleted, it can characterize biomass growth.	Compared to the Monod model, one more parameter must be estimated.

The genome-scale metabolic model (GSMM) of the microalga *Nannochloropsis salina* was assessed, due to its ability to produce polyunsaturated fatty acids and triacylglycerols, which can be used as food supplements, like Omega 3 and in biofuel biosynthesis. Loira *et al.* (2017) generated iNS934, a functional

GSMM tool for metabolic improvement of the alga *N. salina*. During this work, RNA extraction of *N. salina* cells and *de novo* transcriptome were processed. He genome assembly was done using GMAP by NCBI BioProject, the coding sites were transcripts using TransDecoder (http://transdecoder.github.io). BLAST, and InterProScan were used for annotation of coding sites, and to identify protein domains, respectively. The metabolic model (iRC1080) of *Chlamydomonas reinhardtii* was used as a reference for the reconstruction of *N. salina* metabolic network. The resulted iNS934 model showed that *Nannochloropsis* could have 50% lipids of its biomass.

The production of H_2 as biofuel was found to be linked with [FeFe]-hydrogenases (HYDA) in anaerobic metabolism in many green algae. Meuser *et al.* (2011) worked on model of H_2 metabolism in *Chlorella variabilis* NC64A relative to *Chlamydomonas*, depending on the presence of HYDA genes encoding accessory FeS cluster binding domains (F-cluster) in *C. variabilis* NC64A (Meyer, 2007). *In silico* models were used, cDNA sequencing on *C. variabilis* NC64A was detected by *C. variabilis* NC64A draft genome v 1.0, 2008 (http://genome.jgi-psf.org/ChlNC64A_1/) (Blanc *et al.*, 2010), which predicted H_2 metabolism during anaerobic pathways by BLAST comparisons to *C. reinhardtii* orthologs. While the F-cluster HYDA1 mRNA transcripts was confirmed by BLASTP, which used to gather HYDA-homolog-deduced amino-acid sequences from genomic sequences using the DOE-IMG and the NCBI Genome Blast servers in July of 2009 and again in September 2009. ClustalX (version 2.0.8) (Larkin *et al.*, 2007) was used to aligned the resulted data. Phylogenetic analysis was performed using PhyML server (http://www.atgc-montpellier.fr/phyml), which representing the primary HYDA strains were experimentally selected for further phylogenetic analysis. This indicated that algal HYDA H-cluster domains are evolved from a single ancestral F-cluster HYDA. A Bayesian inferred likelihood tree, tree topologies, and a composite phylogram were generated with MrBayes (version 3.1) (http://smart.embl-heidelberg.de), WAG substitution model ProtTest (version 2.4) (Abascal *et al.*, 2005), and FigTree (version 1.2.2) (http://tree.bio.ed.ac.uk/software/figtree/), respectively.

Toxicity

Some researches assessed the ionic liquids (ILs) toxicity towards algae as ecological indicator organisms by using *in silico* modeling. Villain *et al.* (2014) assessed the sensitive toxicity to *Selenastrum capricornutum* alga using an initial dataset including more than 401 eco-toxicological data of pesticides. The toxic ratio (TR) was calculated depending on the predicted and experimental values, while quantile regression (QR) and quantile support vector machine regression (QSVMR) was clearly noticed. The eco-toxicological data were downloaded from the OECD website (https://www.env.go.jp/chemi/sesaku/02e.pdf). The biological data about algae was obtained from ECB website (http://esis.jrc.ec.europa.eu/index.php?PGM=hpv), and AQUatic Information REtrieval (AQUIRE) (https://www.envirosource.com/domino/thielen/envrsrc.nsf/BookSearch/71714B1ED9A65DED86256621007E3CCC?OpenDocument). Two *in silico* tools were used to detect the logarithmic octanol-water partition coefficient (LogKOW), logP (open-source software KOWWIN) (https://www.ecetoc.org/report/estimated-partitioning-property-data/computational-methods/log-kow/), and a second one named A LogP, an atom-based method (https://www.molinspiration.com/services/logp.html).

Another study predicted the growth inhibition of freshwater alga *Pseudokirchneriella subcapitata* (log EC_{50}) after 72 h exposure to 348 chemical compounds, which were classified into five models according to toxic modes (Bakire *et al.*, 2018). The experimental data and the classification of the chemicals were performed by Chemical Risk Information Platform (CHRIP) (https://www.env.go.jp/chemi/sesaku/02e.

pdf) and Toxtree software (http://ecb.jrc.it/qsar/qsar-tools/index.php?c¼TOXTREE). LogKOW, and the molecular structures of all compounds were estimated by EPI Suite 4.11 (US EPA, 2012), and Gaussian 09 program package (https://gaussian.com/). They used GsGrid software package (Version 1.7) to calculate the atomic charge model (Mulliken charge), while the 3D geometric, topological, and molecular properties were obtained by Dragon software 6.0 (2012) (Bakire *et al.*, 2018). The statistical analysis of the QSAR model was implemented by Weka 3.8.0 data mining suite (https://www.cs.waikato.ac.nz/ml/weka/). Another study assessed the effect of chronic aquatic chemical toxicity on *P. subcapitata* (Dal'Molin *et al.*, 2011). The data was collected from the Ministry of the Environment in Japan website (http://www.env.go.jp/en/chemi/sesaku/aquatic_Mar_2016.pdf). The resulted data exhibited that 552 chemicals with growth inhibition toxicity data (72 h NOEC) to *P. subcapitata*. MOPAC 2016 software (http://OpenMOPAC.net, 2016) and Dragon software (version 6.0) (http://www.Talete.Mi.It/, 2012) were used in the molecular structures modeling, and on the optimized geometric structures, respectively. Log KOW were calculated from the US EPA EPI Suite 4.1TM (U.S. EPA, 2012). AMBIT Discover (version 0.04) (http://ambit.sourceforge.net/download_ambitdiscovery.html) was used in the classification models applicability domain depending on the Euclidean distance-based method.

A similar work investigated the toxicity of a large dataset of 352 organic chemicals on *P. subcapitata* (Kusk *et al.*, 2018), they designed a linear discriminant analysis (LDA) model for organic chemicals classification based on their acute toxicity on algae. The molecular structures of the chemicals were obtained from Chemicalbook–chemical (https://www.chemicalbook.com/) search engine, crosschecked (http://www.lookchem.com; https://webbook.nist.gov/chemistry/casser.html), and drawn by Marvin sketch ChemAxon (http://www.chemaxon.com) (version 14.10.27) software (Khan and Ojha, 2020). The molecular descriptors were estimated using Dragon (http://www.talete.mi.it/products/dragondescription.htm) (version 7) and PaDEL-descriptor (http://www.yapcwsoft.com/dd/padeldescriptor) (version 2.20) software. Data-division ver. 1.2 software (http://teqip.jdvu.ac.in/QSAR_Tools/DTCLab) was used for dataset division. The LDA model was developed by STATASTICA software (Hilbe, 2007) (version 13.4), and Discovery Studio (http://accelrys.com/products/discovery-studio/) (version 4.1) was used in pharmacophore model developing.

The effect on dyes was investigated on the microalga *Raphidocelis subcapitata* depending on an *in silico* approach (Croce *et al.*, 2017). The results showed the algal sensitivity to dye toxicity. The in-house standalone application istSimilarity (v.1.0.5) was used in determining the chemical structures, and similarity of the selected dyes. The similarity index (SI) was calculated by VEGA platform of QSARs (http://www.vega-qsar.eu). The chemical structures were retrieved and verified by using ChemIDplus Advanced (http://chem.sis.nlm.nih.gov/chemidplus/), ChemAgora Portal web databases (https://chemagora.jrc.ec.europa.eu/chemagora/), and PubChem Compound (https://www.ncbi.nlm.nih.gov/pccompound). An *in vitro* investigation was tested the dyes toxicity on *R. subcapitata,* by the traditional cultivation method in BG-11 medium with 100 mg/L dyes for 72 h. The statistical analyses were done by Prism5 (GraphPad Software, Inc.) (Croce *et al.*, 2017).

Talukdar and Pal (2017) studied the effect of the cyanobacterium *Mycrocystis* phytocompounds toxins on algae, daphnid and fish using *in silico* approach regarding to Quantitative Structure Activity Relationship (QSAR) modelling and toxicokinetics (ADMET). They used anatoxin-a, saxitoxin, neosaxitoxin, microcystin-LR, microcystin-LA and microcystin-RR. The *in silico* modeling of QSAR was done using Ecological Structure Activity Relationship (ECOSAR) tool (Version, 1.11) (Mayo-Bean *et al.*, 2012). The predictive results showed the ecotoxicological risk of anatoxin-a, microcystin-LR and LA on to three

Table 2. Features of commonly used resources

Database		Description
NCPI (https://www.ncbi.nlm.nih.gov) (The National Center for Biotechnology Information)		NCPI is part of the United States National Library of Medicine, a branch of the National Institutes of Health. It is now a leading source for public biomedical databases, software tools for analyzing molecular and genomic data, and research in computational biology. NCBI creates and maintains over 40 integrated databases for the medical and scientific communities as well as the general public.
Protein database	ExPASy (Expert Protein Analysis System) (http://chlamyto.mpimp-golm.mpg.de/chlamycyc/index.jsp)	ExPASy acts as a proteomics server to analyze protein sequences and structures and two dimensional gel electrophoresis (2-D Page electrophoresis).
	UniProt (Universal Protein Resource) (http://metacyc.org/)	UniProt contains a large amount of information about the biological function of proteins, protein sequence and functional information, many entries being derived from genome sequencing projects.
	AraPerox (Arabidopsis Protein from Plant Peroxisomes) (http://www.araperox.uni-goettingen.de/)	AraPerox is bioinformatically predicted peroxisome proteins, which give extracted microarray data for transcripts of genes involved in peroxisomal pathways.
	PPDB (Plant proteome database) (http://ppdb.tc.cornell.edu/default.aspx)	PPDB is dedicated to plant plastids, but has now expanded to the whole plant proteome. The PPDB stores experimental data from in-house proteome and mass spectrometry analysis, curated information about protein function, properties and subcellular localization. Importantly, proteins are particularly curated for possible (intra) plastid location and their plastid function.
	UniProtKB/Swiss-Pro (http://ca.expasy.org/sprot/relnotes/relstat.html)	UniProtKB/Swiss-Prot is the expertly curated component of UniProtKB (produced by the UniProt consortium). It contains hundreds of thousands of protein descriptions, including function, domain structure, subcellular location, post-translational modifications and functionally characterized variants.
	SMART (Simple Modular Architecture Research Tool) (http://smart.embl-heidelberg.de)	SMART allows the identification and annotation of genetically mobile domains and the analysis of domain architectures. Each domain found in a non-redundant protein database as well as search parameters and taxonomic information are stored in a relational database system. User interfaces to this database allow searches for proteins containing specific combinations of domains in defined taxa.
	BLAST (Basic Local Alignment Search Tool) (https://blast.ncbi.nlm.nih.gov/Blast.cgi?PAGE=Proteins)	BLAST finds regions of similarity between biological sequences. The program compares nucleotide or protein sequences to sequence databases and calculates the statistical significance
	metaPocket (http://metapocket.eml.org/)	metaPocket is a server to identify pockets on protein surface to predict ligand-binding sites, which helps to detect protein function annotation and structure-based drug design.
	Alga-PrAS (http://alga-pras.riken.jp/)	Alga-PrAS (Algal Protein Annotation Suite) is a database of physicochemical and structural properties and post-translational modifications in algal proteomes.
Enzyme portal	BRENDA (BRaunschweig ENzyme DAtabase) (http://www.chlamy.org/chlamydb.html)	BRENDA is the enzyme information system, which integrates information from seven databases to provide functional molecular and biochemical information on enzymes that have been classified by the IUBMB.
	ExPASy Enzyme Database (Ehttps://enzyme.expasy.org)	ENZYME is a repository of information relative to the nomenclature of enzymes. It describes each type of characterized enzyme for which an EC (Enzyme Commission) number has been provided.
DNA and genome tools	ecoPCR (http://metabarcoding.org/ecopcr)	ecoPCR is an electronic PCR software developed by the LECA. It helps to estimate Barcode primers quality. In conjunction with OBITools, you can postprocess ecoPCR output to compute barcode coverage and barcode specificity. New barcode primers can be developed using the ecoPrimers software.
	ROBIBarcodes (https://git.metabarcoding.org/obitools/ROBIBarcodes)	The ROBIBarcodes package provides a set a function producing statistics and graphics helping in the objective analysis of a barcode quality from the result of an ecoPCR.
	OligoCalc (http://biotools.nubic.northwestern.edu/OligoCalc.html)	OligoCalc provides a convenient web interface for calculating the physical properties of DNA and RNA oligonucleotides including melting temperature, molecular weight, %GC content and absorbance coefficient for a given oligonucleotide sequence.
Metabolic pathways	KEGG (Kyoto Encyclopedia of Genes and Genomes) (https://www.genome.jp/kegg/pathway.html)	KEGG is a collection of manually drawn pathway maps representing our knowledge of the molecular interaction, reaction and relation networks for Metabolism (carbohydrate, lipid, nucleotide, and amino acid).
	MetaCyc (http://metacyc.org/)	MetaCyc contains pathways involved in both primary and secondary metabolism of all domains of life, in addition to associated metabolites, reactions, enzymes, and genes. It contains 2937 pathways from 3295 different organisms.
	Expasy (ExPASy Biochemical Pathways) (http://www.expasy.ch/cgi-bin/search-biochem-index)	A Biochemical Pathways consists of a graphical representation of the main metabolic pathways. Each enzyme mentioned in the chart is linked to its corresponding entry in ENZYME.
	BiGG (http://bigg.ucsd.edu/)	BiGG model helps to predict metabolic pathway usage and growth phenotypes by redesigning Biochemical, Genetic and Genomic knowledge base. It contains more than 75 high-quality, manually-curated genome-scale metabolic models, which can generate and test hypotheses when integrated with experimental data.
	BioCyc (http://biocyc.org/)	BioCyc includes metabolic reconstructions, regulatory networks, protein features, orthologs, gene essentiality, and atom mappings. It contains 19534 Pathway/Genome Databases (PGDBs) for model eukaryotes and for thousands of microbes, plus software tools for exploring them.
Chemistry databases	Pubchem (https://pubchem.ncbi.nlm.nih.gov/)	Pubchem is a popular chemical information resource that serves the scientific community. It contains information on chemical substances and their biological activities.

test models and the ADMET. As anatoxin-a was neurotoxin and P-glycoprotein inhibitor microcystin-

LB and LA inhibited the metabolism but all compounds were non mutagenic and non carcinogenic.

CONCLUSION

The development of optimal microalgae growing systems for mass-scale biomass production for biorefinery purposes, which is sometimes difficult, can be accomplished using optimization frameworks that integrate both experimental and computational approaches. Predictive models have been found to be effective tools for evaluating and optimizing bioprocesses involving many growth-limiting parameters, such as microalgal cultivation. Several mathematical models have been constructed to predict how microalgal cells grow and collect carbohydrates and lipids in response to various growth-limiting conditions such as nutrition, light, and temperature. The most appropriate model structure for simulation or optimization will be determined by the model's intended application, thus great thought should be made to maintaining a positive balance between a model's predictive capacity and its mathematical complexity. In the case of microalgae processes, this necessitates an understanding of how cells and their intracellular components react to their surroundings (e.g., nutrients, light intensity, or temperature at which they are grown). Even though various modelling frameworks are capable of describing the complex dynamics of microalgae growth, a growing body of literature indicates that efforts are being made to develop models that can simulate both cellular growth processes as well as carbohydrate and lipid accumulation in response to a variety of growth-limiting factors. The use of strong modelling frameworks can lead to the discovery of optimal cultivation situations, which has significant implications for the creation of custom microalgae-based biorefineries.

If models are to be used in this way, they should be evaluated for their ability to predict fed-batch or continuous cultivation dynamics so that large-scale nutrient feeding strategies and operating conditions can be identified and implemented to enable microalgae biomass to transition from a promising biorefinery substrate to a commercial reality.

ACKNOWLEDGMENT

The Authors want to thank IGI Publications for providing us with such a good opportunity.

Funding: This research received no specific grant from any funding agency in the public, commercial, or not-for-profit sectors.

REFERENCES

Abascal, F., Zardoya, R., & Posada, D. (2005). ProtTest: Selection of best-fit models of protein evolution. *Bioinformatics (Oxford, England)*, *21*(9), 2104–2105. doi:10.1093/bioinformatics/bti263 PMID:15647292

Adesanya, V. O., Davey, M. P., Scott, S. A., & Smith, A. G. (2014). Kinetic modelling of growth and storage molecule production in microalgae under mixotrophic and autotrophic conditions. *Bioresource Technology*, *157*, 293–304. doi:10.1016/j.biortech.2014.01.032 PMID:24576922

Adesanya, V. O., Davey, M. P., Scott, S. A., & Smith, A. G. (2014). Kinetic modelling of growth and storage molecule production in microalgae under mixotrophic and autotrophic conditions. *Bioresource Technology*, *157*, 293–304. doi:10.1016/j.biortech.2014.01.032 PMID:24576922

Alva, V., Nam, S.-Z., Söding, J., & Lupas, A. N. (2016). The MPI bioinformatics Toolkit as an integrative platform for advanced protein sequence and structure analysis. *Nucleic Acids Research*, *44*(1), 410–416. doi:10.1093/nar/gkw348 PMID:27131380

Amid, C., Alako, B. T. F., Balavenkataraman, K. V., Burdett, T., Burgin, J., Fan, J., Harrison, P. W., Holt, S., Hussein, A., Ivanov, E., Jayathilaka, S., Kay, S., Keane, T., Leinonen, R., Liu, X., Martinez-Villacorta, J., Milano, A., Pakseresht, A., Rahman, N., ... Cochrane, G. (2020). The European Nucleotide Archive in 2019. *Nucleic Acids Research*, *48*, 70–76. doi:10.1093/nar/gkz1063 PMID:31722421

Andrews, J. F. (1968). A mathematical model for the continuous culture of microorganisms utilizing inhibitory substrates. *Biotechnology and Bioengineering*, *10*(6), 707–723. doi:10.1002/bit.260100602

Andrews, J. F. (1968). A mathematical model for the continuous culture of microorganisms utilizing inhibitory substrates. *Biotechnology and Bioengineering*, *10*(6), 707–723. doi:10.1002/bit.260100602

Bakire, S., Yang, X., Ma, G., Wei, X., Yu, H., Chen, J., & Lin, H. (2018). Developing predictive models for toxicity of organic chemicals to green algae based on mode of action. *Chemosphere*, *190*, 463–470. doi:10.1016/j.chemosphere.2017.10.028 PMID:29028601

Banerjee, S., Banerjee, S., Ghosh, A. K., & Das, D. (2020). Maneuvering the genetic and metabolic pathway for improving biofuel production in algae: Present status and future prospective. *Renewable & Sustainable Energy Reviews*, *133*, 1–18. doi:10.1016/j.rser.2020.110155

Barbosa, M. J., Hoogakker, J., & Wijffels, R. H. (2003). Optimisation of cultivation parameters in photobioreactors for microalgae cultivation using the A-stat technique. *Biomolecular Engineering*, *20*(4-6), 115–123. doi:10.1016/S1389-0344(03)00033-9 PMID:12919788

Béchet, Q., Shilton, A., & Guieysse, B. (2013). Modeling the effects of light and temperature on algae growth: State of the art and critical assessment for productivity prediction during outdoor cultivation. *Biotechnology Advances*, *31*(8), 1648–1663. doi:10.1016/j.biotechadv.2013.08.014 PMID:23981914

Béchet, Q., Shilton, A., & Guieysse, B. (2013). Modeling the effects of light and temperature on algae growth: State of the art and critical assessment for productivity prediction during outdoor cultivation. *Biotechnology Advances*, *31*(8), 1648–1663. doi:10.1016/j.biotechadv.2013.08.014 PMID:23981914

Behera, B., Selvanayaki, S., Jayabalan, R., & Balasubramanian, P. (2019). An in-silico Approach for enhancing the lipid productivity in microalgae by manipulating the fatty acid biosynthesis. In Soft Computing for Problem Solving, Advances in Intelligent Systems and Computing 816 (pp. 877-889). Springer Nature Singapore Pte Ltd. doi:10.1007/978-981-13-1592-3_70

Bekirogullari, M., Figueroa-Torres, G. M., Pittman, J. K., & Theodoropoulos, C. (2020). Models of microalgal cultivation for added-value products-A review. *Biotechnology Advances*, *44*, 107609. doi:10.1016/j.biotechadv.2020.107609 PMID:32781245

Bekirogullari, M., Figueroa-Torres, G. M., Pittman, J. K., & Theodoropoulos, C. (2020). Models of micro algal cultivation for added-value products-A review. *Biotechnology Advances*, *44*, 107609. doi:10.1016/j.biotechadv.2020.107609 PMID:32781245

Bekirogullari, M., Pittman, J., & Theodoropoulos, C. (2018). Multi-factor kinetic modelling of microalgal biomass cultivation for optimised lipid production. *Bioresource Technology*, *269*, 417–425. doi:10.1016/j.biortech.2018.07.121 PMID:30265993

Bekirogullari, M., Pittman, J., & Theodoropoulos, C. (2018). Multi-factor kinetic modelling of microalgal biomass cultivation for optimised lipid production. *Bioresource Technology*, *269*, 417–425. doi:10.1016/j.biortech.2018.07.121 PMID:30265993

Berjanskii, M., Liang, Y., Zhou, J., Tang, P., Stothard, P., Zhou, Y., Cruz, J., MacDonell, C., Lin, G., Lu, P., & Wishart, D. S. (2010). PROSESS: A protein structure evaluation suite and server. *Nucleic Acids Research, 38,* 633–640. . doi:10.1093/nar/gkq375

Blanc, G., Duncan, G., Agarkova, I., Borodovsky, M., Gurnon, J., Kuo, A., Lindquist, E., Lucas, S., Pangilinan, J., Polle, J., Salamov, A., Terry, A., Yamada, T., Dunigan, D. D., Grigoriev, I. V., Claverie, J.-M., & Van Etten, J. L. (2010). The *Chlorella variabilis* NC64A genome reveals adaptation to photosymbiosis, coevolution with viruses, and cryptic sex. *The Plant Cell*, *22*(9), 2943–2955. doi:10.1105/tpc.110.076406 PMID:20852019

Bornstein, B. J., Keating, S. M., Jouraku, A., & Hucka, M. (2008). LibSBML: An API library for SBML. *Bioinformatics (Oxford, England)*, *24*(6), 880–881. doi:10.1093/bioinformatics/btn051 PMID:18252737

Boyer, F., Mercier, C., Bonin, A., Le Bras, Y., Taberlet, P., & Coissac, E. (2016). OBITools: A unix-inspired software package for DNA metabarcoding. *Molecular Ecology Resources*, *16*(1), 176–182. doi:10.1111/1755-0998.12428 PMID:25959493

Boyle, N. R., & Morgan, J. (2009). Flux balance analysis of primary metabolism in *Chlamydomonas reinhardtii. BMC Systems Biology*, *3*(4), 32–45. doi:10.1186/1752-0509-3-4 PMID:19128495

Brennan, L., & Owende, P. (2010). Biofuels from microalgae—A review of technologies for production, processing, and extractions of biofuels and co-products. *Renewable & Sustainable Energy Reviews*, *14*(2), 557–577. doi:10.1016/j.rser.2009.10.009

Brennan, L., & Owende, P. (2010). Biofuels from microalgae—A review of technologies for production, processing, and extractions of biofuels and co-products. *Renewable & Sustainable Energy Reviews*, *14*(2), 557–577. doi:10.1016/j.rser.2009.10.009

Broddrick, J. T., Du, N., Smith, S. R., Tsuji, Y., Jallet, D., Ware, M. A., Peers, G., Matsuda, Y., Dupont, C. L., Mitchell, B. G., Palsson, B. O., & Allen, A. E. (2019). Cross-compartment metabolic coupling enables flexible photoprotective mechanisms in the diatom *Phaeodactylum tricornutum. The New Phytologist*, *222*(3), 1364–1379. doi:10.1111/nph.15685 PMID:30636322

Camacho, C., Coulouris, G., Avagyan, V., Ma, N., Papadopoulos, J., Bealer, K., & Madden, T. L. (2009). BLAST+: Architecture and applications. *BMC Bioinformatics*, *10*(1), 421. doi:10.1186/1471-2105-10-421 PMID:20003500

Campbell, M. A., Haas, B. J., Hamilton, J. P., Mount, S., & Buell, C. (2006). Comprehensive analysis of alternative splicing in rice and comparative analyses with Arabidopsis. *BMC Genomics*, *7*(1), 327–339. doi:10.1186/1471-2164-7-327 PMID:17194304

Chen, F., & Johns, M. R. (1994). Substrate inhibition of Chlamydomonas reinhardtii by acetate in heterotrophic culture. *Process Biochemistry*, *29*(4), 245–252. doi:10.1016/0032-9592(94)80064-2

Cock, P. J. A., Antao, T., Chang, J. T., Chapman, B. A., Cox, C. J., Dalke, A., Friedberg, I., Hamelryck, T., Kauff, F., Wilczynski, B., & de Hoon, M. J. L. (2009). Biopython: Freely available Python tools for computational molecular biology and bioinformatics. *Bioinformatics (Oxford, England)*, *25*(11), 1422–1423. doi:10.1093/bioinformatics/btp163 PMID:19304878

Cornet, J. F., & Dussap, C. G. (2009). A simple and reliable formula for assessment of maximum volumetric productivities in photobioreactors. *Biotechnology Progress*, *25*(2), 424–435. doi:10.1002/btpr.138 PMID:19340890

Cornet, J. F., & Dussap, C. G. (2009). A simple and reliable formula for assessment of maximum volumetric productivities in photobioreactors. *Biotechnology Progress*, *25*(2), 424–435. doi:10.1002/btpr.138 PMID:19340890

Croce, R., Cinà, F., Lombardo, A., Crispeyn, G., Cappelli, C. I., Vian, M., Maiorana, S., Benfenati, E., & Baderna, D. (2017). Aquatic toxicity of several textile dye formulations: Acute and chronic assays with *Daphnia magna* and *Raphidocelis subcapitata*. *Ecotoxicology and Environmental Safety*, *144*, 79–87. doi:10.1016/j.ecoenv.2017.05.046 PMID:28601520

Dal'Molin, C. G. O., Quek, L., Palfreyman, R. W., & Nielsen, L. K. (2011). AlgaGEM – a genome-scale metabolic reconstruction of algae based on the *Chlamydomonas reinhardtii* genome. *BMC Genomics*, *12*(S4), S5. doi:10.1186/1471-2164-12-S4-S5 PMID:22369158

Das, R. N., Roy, K., & Popelier, P. L. A. (2015). Interspecies quantitative structure–toxicity–toxicity (QSTTR) relationship modeling of ionic liquids, Toxicity of ionic liquids to *V. fischeri, D. magna* and *S. vacuolatus*. *Ecotoxicology and Environmental Safety*, *122*, 497–520. doi:10.1016/j.ecoenv.2015.09.014 PMID:26414597

Davidson, K., & Gurney, W. (1999). An investigation of non-steady-state algal growth. II. Mathematical modelling of co-nutrient-limited algal growth. *Journal of Plankton Research*, *21*(5), 839–858. doi:10.1093/plankt/21.5.839

Ding, F., Wang, Z., Yang, X., Shi, L., Liu, J., & Chen, G. (2018). Development of classification models for predicting chronic toxicity of chemicals to *Daphnia magna* and *Pseudokirchneriella subcapitata*. *SAR and QSAR in Environmental Research*, *30*(1), 39–50. doi:10.1080/1062936X.2018.1545694 PMID:30477347

Droop, M. R. (1968). Vitamin B12 and marine ecology. IV. The kinetics of uptake, growth and inhibition in Monochrysis lutheri. *Journal of the Marine Biological Association of the United Kingdom*, *48*(3), 689–733. doi:10.1017/S0025315400019238

Droop, M. R. (1968). Vitamin B12 and marine ecology. IV. The kinetics of uptake, growth and inhibition in Monochrysis lutheri. *Journal of the Marine Biological Association of the United Kingdom*, *48*(3), 689–733. doi:10.1017/S0025315400019238

Ebrahim, A., Lerman, J. A., Palsson, B. O., & Hyduke, D. R. (2013). COBRApy: COnstraints-based reconstruction and analysis for python. *BMC Systems Biology*, *7*(1), 74–79. doi:10.1186/1752-0509-7-74 PMID:23927696

Eisenberg, D., Lüthy, R., & Bowie, J. U. (1997). VERIFY3D: Assessment of protein models with three-dimensional profiles. *Methods in Enzymology*, *277*, 396–404. doi:10.1016/S0076-6879(97)77022-8 PMID:9379925

El-Gebali, S., Mistry, J., Bateman, A., Eddy, S. R., Luciani, A., Potter, S. C., Qureshi, M., Richardson, L. J., Salazar, G. A., Smart, A., Sonnhammer, E. L. L., Hirsh, L., Paladin, L., Piovesan, D., Tosatto, S. C. E., & Finn, R. D. (2019). The Pfam protein families database in 2019. *Nucleic Acids Research*, *47*(1), 427–432. doi:10.1093/nar/gky995 PMID:30357350

Eriksen, N. T., Riisgård, F. K., Gunther, W. S., & Iversen, J. J. L. (2007). On-line estimation of O 2 production, CO 2 uptake, and growth kinetics of microalgal cultures in a gas-tight photobioreactor. *Journal of Applied Phycology*, *19*(2), 161–174. doi:10.100710811-006-9122-y PMID:19396354

Fernándeza, I., Vinogradovb, S. N., & Arredondo-Petera, R. (2010). Identification and *in silico* characterization of a putative ancestor to land plant non-symbiotic hemoglobins from the prasinophyceae algae *Micromonas* and *Ostreococcus. Global Journal of Biochemistry*, *1*(1), 18–30.

Ficetola, G. F., Coissac, E., Zundel, S., Riaz, T., Shehzad, W., Bessiere, J., Taberlet, P., & Pompanon, F. (2010). An *in silico* approach for the evaluation of DNA barcodes. *BMC Genomics*, *11*(1), 434–443. doi:10.1186/1471-2164-11-434 PMID:20637073

Fouchard, S., Pruvost, J., Degrenne, B., Titica, M., & Legrand, J. (2009). Kinetic modeling of light limitation and sulfur deprivation effects in the induction of hydrogen production with Chlamydomonas reinhardtii: Part I. Model development and parameter identification. *Biotechnology and Bioengineering*, *102*(1), 232–245. doi:10.1002/bit.22034 PMID:18688816

Fouchard, S., Pruvost, J., Degrenne, B., Titica, M., & Legrand, J. (2009). Kinetic modeling of light limitation and sulfur deprivation effects in the induction of hydrogen production with Chlamydomonas reinhardtii: Part I. Model development and parameter identification. *Biotechnology and Bioengineering*, *102*(1), 232–245. doi:10.1002/bit.22034 PMID:18688816

Foucher, A., Evrard, O., Ficetola, G. F., Gielly, L., Poulain, J., Giguet-Covex, C., Laceby, J. P., Salvador-Blanes, S., Cerdan, O., & Poulenard, J. (2020). Persistence of environmental DNA in cultivated soils: implication of this memory effect for reconstructing the dynamics of land use and cover changes. *Scientific Reports, 10*(1). doi:10.1038/s41598-020-67452-1

Ganter, M., Bernard, T., Moretti, S., Stelling, J., & Pagni, M. (2013). MetaNetX.org: A website and repository for accessing, analysing and manipulating metabolic networks. *Bioinformatics (Oxford, England)*, *29*(6), 815–816. doi:10.1093/bioinformatics/btt036 PMID:23357920

Geider, R., MacIntyre, H., & Kana, T. (1997). Dynamic model of phytoplankton growth and acclimation: responses of the balanced growth rate and the chlorophyll a: carbon ratio to light, nutrient-limitation and temperature. *Marine Ecology Progress Series*, *148*, 187–200. doi:10.3354/meps148187

Goodstein, D. M., Shu, S., Howson, R., Neupane, R., Hayes, R. D., Fazo, J., Mitros, T., Dirks, W., Hellsten, U., Putnam, N., & Rokhsar, D. S. (2012). Phytozome: A comparative platform for green plant genomics. *Nucleic Acids Research, 40*(D1), 1178–1186. doi:10.1093/nar/gkr944 PMID:22110026

Grima, E. M., Camacho, F. G., Perez, J. S., Sevilla, J. F., Fernández, F. A., & Gomez, A. C. (1994). A mathematical model of microalgal growth in light-limited chemostat culture. *Journal of Chemical Technology & Biotechnology: International Research in Process. Environmental AND Clean Technology, 61*(2), 167–173.

Guex, N., & Peitsch, M. C. (1997). SWISS-MODEL and the Swiss-PdbViewer: An environment for comparative protein modeling. *Electrophoresis, 18*(15), 2714–2723. doi:10.1002/elps.1150181505 PMID:9504803

Heirendt, L., Arreckx, S., Pfau, T., Mendoza, S. N., Richelle, A., Heinken, A., Haraldsdóttir, H. S., Wachowiak, J., Keating, S. M., Vlasov, V., Magnusdóttir, S., Ng, C. Y., Preciat, G., Žagare, A., Chan, S. H. J., Aurich, M. K., Clancy, C. M., Modamio, J., Sauls, J. T., ... Fleming, R. M. T. (2019). Creation and analysis of biochemical constraint-based models using the COBRA Toolbox v.3.0. *Nature Protocols, 14*(3), 639–702. doi:10.103841596-018-0098-2 PMID:30787451

Heller, S. R., McNaught, A., Pletnev, I., Stein, S., & Tchekhovskoi, D. (2015). InChI, the IUPAC international chemical identifier. *Journal of Cheminformatics, 7*(1), 23–56. doi:10.118613321-015-0068-4 PMID:26136848

Herrera-Valencia, V. A., Macario-Gonza'lez, L. A., Casais-Molina, M. L., Beltran-Aguilar, A. G., & Peraza-Echeverrı'a, S. (2012). *In Silico* cloning and characterization of the Glycerol-3-Phosphate Dehydrogenase (GPDH) gene family in the green microalga *Chlamydomonas reinhardtii. Current Microbiology, 64*(5), 477–485. doi:10.100700284-012-0095-6 PMID:22358185

Hilbe, J. M. (2007). STATISTICA 7: An overview. *The American Statistician, 61*(1), 91–94. doi:10.1198/000313007X172998

Hucka, A. F. M., Sauro, H. M., Bolouri, H., Doyle, J. C., Kitano, H., Arkin, A. P., Bornstein, B. J., Bray, D., Cornish-Bowden, A., Cuellar, A. A., Dronov, S., Gilles, E. D., Ginkel, M., Gor, V., Goryanin, I. I., Hedley, W. J., Hodgman, T. C., Hofmeyr, J.-H., Hunter, P. J., ... Wang, J. (2003). The systems biology markup language (SBML): A medium for representation and exchange of biochemical network models. *Bioinformatics (Oxford, England), 19*(4), 524–531. doi:10.1093/bioinformatics/btg015 PMID:12611808

Humphrey, W., Dalke, A., & Schulten, K. (1996). VMD-Visual molecular dynamics. *Journal of Molecular Graphics, 14*(1), 33–38. doi:10.1016/0263-7855(96)00018-5 PMID:8744570

Hunter, J. D. (2007). Matplotlib: A 2D graphics environment. *Computing in Science & Engineering, 9*(3), 90–95. doi:10.1109/MCSE.2007.55

Imam, S., Schäuble, S., Valenzuela, J., López García de Lomana, A., Carter, W., Price, N. D., & Baliga, N. S. (2015). A refined genome-scale reconstruction of *Chlamydomonas* metabolism provides a platform for systems-level analyses. *The Plant Journal, 84*(6), 1239–1256. doi:10.1111/tpj.13059 PMID:26485611

John, E. H., & Flynn, K. J. (2000). Modelling phosphate transport and assimilation in microalgae; how much complexity is warranted? *Ecological Modelling*, *125*(2-3), 145-157. doi:10.1016/S0304-3800(99)00178-7

Juneja, A., Chaplen, F. W. R., & Murthy, G. S. (2016). Genome scale metabolic reconstruction of *Chlorella variabilis* for exploring its metabolic potential for biofuels. *Bioresource Technology*, *213*, 103–110. doi:10.1016/j.biortech.2016.02.118 PMID:26995318

Källberg, M., Wang, H., Wang, S., Peng, J., Wang, Z., Lu, H., & Xu, J. (2012). Template-based protein structure modeling using the RaptorX web server. *Nature Protocols*, *7*(8), 1511–1522. doi:10.1038/nprot.2012.085 PMID:22814390

Kanehisa, M., Sato, Y., Kawashima, M., Furumichi, M., & Tanabe, M. (2016). KEGG as a reference resource for gene and protein annotation. *Nucleic Acids Research*, *44*(1), 457–462. doi:10.1093/nar/gkv1070 PMID:26476454

Kent, W. J. (2002). BLAT---The BLAST-Like Alignment Tool. *Genome Research*, *12*(4), 656–664. doi:10.1101/gr.229202 PMID:11932250

Khan, K., & Ojha, P. K. (2020). Ecotoxicological modeling of organic chemicals for their acute toxicity in algae using classification and toxicophore-based approaches. *International Journal of Quantitative Structure-Property Relationships*, *5*(2), 17–72. doi:10.4018/IJQSPR.2020040102

Kim, D. E., Chivian, D., & Baker, D. (2004). Protein structure prediction and analysis using the Robetta server. *Nucleic Acids Research*, *32*, 526–531. doi:10.1093/nar/gkh468 PMID:15215442

Kusk, K. O., Christensen, A. M., & Nyholm, N. (2018). Algal growth inhibition test results of 425 organic chemical substances. *Chemosphere*, *204*, 405–412. doi:10.1016/j.chemosphere.2018.04.047 PMID:29677648

Kwon, H. K., Oh, S. J., & Yang, H.-S. (2013). Growth and uptake kinetics of nitrate and phosphate by benthic microalgae for phytoremediation of eutrophic coastal sediments. *Bioresource Technology*, *129*, 387–395. doi:10.1016/j.biortech.2012.11.078 PMID:23262016

Larkin, M. A., Blackshields, G., Brown, N. P., Chenna, R., McGettigan, P. A., McWilliam, H., Valentin, F., Wallace, I. M., Wilm, A., Lopez, R., Thompson, J. D., Gibson, T. J., & Higgins, D. G. (2007). Clustal W and Clustal X version 2.0. *Bioinformatics (Oxford, England)*, *23*(21), 2947–2948. doi:10.1093/bioinformatics/btm404 PMID:17846036

Lavoie, M., Saint-Béat, B., Strauss, J., Guérin, S., Allard, A., Hardy, S. V., Falciatore, A., & Lavaud, J. (2020). Genome-Scale Metabolic Reconstruction and *in Silico* Perturbation Analysis of the Polar Diatom *Fragilariopsis cylindrus* Predicts High Metabolic Robustness. *Biology (Basel)*, *9*(2), 30–49. doi:10.3390/biology9020030 PMID:32079178

Levering, J., Broddrick, J., Dupont, C. L., Peers, G., Beeri, K., Mayers, J., Gallina, A. A., Allen, A. E., Palsson, B. O., & Zengler, K. (2016). Genome-scale model reveals metabolic basis of biomass partitioning in a model diatom. *PLoS One*, *11*(5), e0155038. doi:10.1371/journal.pone.0155038 PMID:27152931

Li, H., Handsaker, B., Wysoker, A., Fennell, T., Ruan, J., Homer, N., Marth, G., Abecasis, G., & Durbin, R. (2009). The sequence alignment/ map format and SAMtools. *Bioinformatics (Oxford, England)*, *25*(16), 2078–2079. doi:10.1093/bioinformatics/btp352 PMID:19505943

Liu, H., Laws, E. A., Villareal, T. A., & Buskey, E. J. (2001). Nutrient-limited growth of Aureoumbra lagunensis (Pelagophyceae), with implications for its capability to outgrow other phytoplankton species in phosphate-limited environments. *Journal of Phycology*, *37*(4), 500–508. doi:10.1046/j.1529-8817.2001.037004500.x

Loira, N., Mendoza, S., Cortés, M. P., Rojas, N., Travisany, D., Di-Genova, A., Gajardo, N., Ehrenfeld, N., & Maass, A. (2017). Reconstruction of the microalga *Nannochloropsis salina* genome-scale metabolic model with applications to lipid production. *BMC Systems Biology*, *11*(1), 66–82. doi:10.118612918-017-0441-1 PMID:28676050

Maghrabi, A. H. A., & McGuffin, L. J. (2017). ModFOLD6: An accurate web server for the global and local quality estimation of 3D protein models. *Nucleic Acids Research*, *45*(1), 416–421. doi:10.1093/nar/gkx332 PMID:28460136

Mahdavi, S., Razeghi, J., Pazhouhandeh, M., Movafeghi, M., Kosari-Nasab, M., & Kianianmomeni, A. (2020). Characterization of two predicted DASH-related proteins from the green alga *Volvox carteri* provides new insights into their light-mediated transcript regulation and DNA repair activity. *Algal Research*, *52*, 102116. doi:10.1016/j.algal.2020.102116

Mairet, F., Bernard, O., Masci, P., Lacour, T., & Sciandra, A. (2011). Modelling neutral lipid production by the microalga Isochrysis aff. galbana under nitrogen limitation. *Bioresource Technology*, *102*(1), 142–149. doi:10.1016/j.biortech.2010.06.138 PMID:20656476

Mayo-Bean, K. E., Moran-Bruce, K., Nabholz, J. V., Meylan, W. M., & Howard, P. H. (2012). Operation manual for the ecological structure-activit,y relationship model (ECOSAR) class program estimating toxicity of industrial chemicals to aquatic organisms using the ecosar (ecological structure activity relationship) class program. ms-windows version 1.11. Office of Pollution Prevention and Toxics, US Environmental Protection Agency Syracuse Research Corporation Consortium for Environmental Risk Management.

Metcalf, A. J., Nagygyor, A., & Boyle, N. R. (2020). Rapid Annotation of Photosynthetic Systems (RAPS): Automated algorithm to generate genome-scale metabolic networks from algal genomes. *Algal Research*, *50*(101967), 1–7. doi:10.1016/j.algal.2020.101967

Meuser, J. E., Boyd, E. S., Ananyev, G., Karns, D., Radakovits, R., Murthy, U. M. N., Ghirardi, M. L., Dismukes, G. C., Peters, J. W., & Posewitz, M. C. (2011). Evolutionary significance of an algal gene encoding an [FeFe]-hydrogenase with F-domain homology and hydrogenase activity in *Chlorella variabilis* NC64A. *Planta*, *234*(4), 829–843. doi:10.100700425-011-1431-y PMID:21643991

Meyer, J. (2007). [FeFe] hydrogenases and their evolution: A genomic perspective. *Cellular and Molecular Life Sciences*, *64*(9), 1063–1084. doi:10.100700018-007-6477-4 PMID:17353991

Monod, J. (1949). The growth of bacterial cultures. *Annual Review of Microbiology*, *3*(1), 371–394. doi:10.1146/annurev.mi.03.100149.002103

Monod, J. (1949). The growth of bacterial cultures. *Annual Review of Microbiology*, *3*(1), 371–394. doi:10.1146/annurev.mi.03.100149.002103

Murwanashyaka, T., Shen, L., Yang, Z., Chang, J.-S., Manirafasha, E., Ndikubwimana, T., Chen, C., & Lu, Y. (2020). Kinetic modelling of heterotrophic microalgae culture in wastewater: Storage molecule generation and pollutants mitigation. *Biochemical Engineering Journal*, *157*, 107523. doi:10.1016/j.bej.2020.107523

Murwanashyaka, T., Shen, L., Yang, Z., Chang, J.-S., Manirafasha, E., Ndikubwimana, T., Chen, C., & Lu, Y. (2020). Kinetic modelling of heterotrophic microalgae culture in wastewater: Storage molecule generation and pollutants mitigation. *Biochemical Engineering Journal*, *157*, 107523. doi:10.1016/j.bej.2020.107523

Nordberg, H. C. M., Dusheyko, S., Hua, S., Poliakov, A., Shabalov, I., Smirnova, T., Grigoriev, I. V., & Dubchak, I. (2014). The genome portal of the Department of Energy Joint Genome Institute: 2014 updates. *Nucleic Acids Research*, *42*(1), 26–31. doi:10.1093/nar/gkt1069 PMID:24225321

O'Boyle, N. M., Banck, M., James, C. A., Morley, C., Vandermeersch, T., & Hutchison, G. R. (2011). Open Babel: An open chemical toolbox. *Journal of Cheminformatics*, *3*(1), 33. doi:10.1186/1758-2946-3-33 PMID:21982300

Paajanen, P., Strauss, J., van Oosterhout, C., McMullan, M., Clark, M. D., & Mock, T. (2017). Building a locally diploid genome and transcriptome of the diatom *Fragilariopsis cylindrus. Scientific Data*, *4*(1), 170149. doi:10.1038data.2017.149 PMID:28994819

Packer, A., Li, Y., Andersen, T., Hu, Q., Kuang, Y., & Sommerfeld, M. (2011). Growth and neutral lipid synthesis in green microalgae: A mathematical model. *Bioresource Technology*, *102*(1), 111–117. doi:10.1016/j.biortech.2010.06.029 PMID:20619638

Packer, A., Li, Y., Andersen, T., Hu, Q., Kuang, Y., & Sommerfeld, M. (2011). Growth and neutral lipid synthesis in green microalgae: A mathematical model. *Bioresource Technology*, *102*(1), 111–117. doi:10.1016/j.biortech.2010.06.029 PMID:20619638

Pérez, E. B., Pina, I. C., & Rodríguez, L. P. (2008). Kinetic model for growth of Phaeodactylum tricornutum in intensive culture photobioreactor. *Biochemical Engineering Journal*, *40*(3), 520–525. doi:10.1016/j.bej.2008.02.007

Pérez, E. B., Pina, I. C., & Rodríguez, L. P. (2008). Kinetic model for growth of Phaeodactylum tricornutum in intensive culture photobioreactor. *Biochemical Engineering Journal*, *40*(3), 520–525. doi:10.1016/j.bej.2008.02.007

Pettersen, E. F., Goddard, T. D., Huang, C. C., Couch, G. S., Greenblatt, D. M., Meng, E. C., & Ferrin, T. E. (2004). UCSF Chimera—A Visualization System for Exploratory Research and Analysis. *Journal of Computational Chemistry*, *25*(13), 1605–1612. doi:10.1002/jcc.20084 PMID:15264254

Pfaffinger, C. E., Schöne, D., Trunz, S., Löwe, H., & Weuster-Botz, D. (2016). Model-based optimization of microalgae areal productivity in flat-plate gas-lift photobioreactors. *Algal Research*, *20*, 153–163. doi:10.1016/j.algal.2016.10.002

Pianosi, F., Sarrazin, F., & Wagener, T. A. (2015). A Matlab toolbox for global sensitivity analysis. *Environmental Modelling & Software*, *70*, 80–85. doi:10.1016/j.envsoft.2015.04.009

Raj-Kumar, P.-K., Vallon, O., & Liang, C. (2017). *In silico* analysis of the sequence features responsible for alternatively spliced introns in the model green alga *Chlamydomonas reinhardtii. Plant Molecular Biology*, *94*(3), 253–265. doi:10.100711103-017-0605-9 PMID:28364390

Ratkowsky, D. A., Olley, J., McMeekin, T., & Ball, A. (1982). Relationship between temperature and growth rate of bacterial cultures. *Journal of Bacteriology*, *149*(1), 1–5. doi:10.1128/jb.149.1.1-5.1982 PMID:7054139

Ratkowsky, D. A., Olley, J., McMeekin, T., & Ball, A. (1982). Relationship between temperature and growth rate of bacterial cultures. *Journal of Bacteriology*, *149*(1), 1–5. doi:10.1128/jb.149.1.1-5.1982 PMID:7054139

Reijnders, M. J. M. F., Van-Heck, R. G. A., Lam, C. M. C., Scaife, M. A., Dos-Santos, V. A. P. M., Smith, A. G., & Schaap, P. J. (2014). Green genes: Bioinformatics and systems-biology innovations drive algal biotechnology. *Trends in Biotechnology*, *32*(12), 617–626. doi:10.1016/j.tibtech.2014.10.003 PMID:25457388

Riaz, T., Shehzad, W., Viari, A., Pompanon, F., Taberlet, P., & Coissac, E. (2011). ecoPrimers: Inference of new DNA barcode markers from whole genome sequence analysis. *Nucleic Acids Research*, *39*(21), 145–155. doi:10.1093/nar/gkr732 PMID:21930509

Robinson, J. T., Thorvaldsdóttir, H., Winckler, W., Guttman, M., Lander, E. S., Getz, G., & Mesirov, J. P. (2011). Integrative genomics viewer. *Nature Biotechnology*, *29*(1), 24–26. doi:10.1038/nbt.1754 PMID:21221095

Roche, D. B., Buenavista, M. T., & McGuffin, L. J. (2013). The FunFOLD2 server for the prediction of protein-ligand interactions. *Nucleic Acids Research*, *41*(1), 303–307. doi:10.1093/nar/gkt498 PMID:23761453

Roels, J. A. (1983). *Energetics and kinetics in biotechnology*. Elsevier Biomedical Press.

Roels, J. A. (1983). *Energetics and kinetics in biotechnology*. Elsevier Biomedical Press.

Roy, A., Yang, J., & Zhang, Y. (2012). COFACTOR an accurate comparative algorithm for structure-based protein function annotation. *Nucleic Acids Research*, *40*(1), 471–477. doi:10.1093/nar/gks372 PMID:22570420

Ryu, K. H., Kim, B., & Lee, J. H. (2019). A model-based optimization of microalgal cultivation strategies for lipid production under photoautotrophic condition. *Computers & Chemical Engineering*, *121*, 57–66. doi:10.1016/j.compchemeng.2018.10.004

Ryu, K. H., Kim, B., & Lee, J. H. (2019). A model-based optimization of microalgal cultivation strategies for lipid production under photoautotrophic condition. *Computers & Chemical Engineering*, *121*, 57–66. doi:10.1016/j.compchemeng.2018.10.004

Salehi-Ashtiani, K., Koussa, J., Dohai, B. S., Chaiboonchoe, A., Cai, H., Dougherty, K. A. D., Nelson, D. R., Jijakli, K., & Khraiwesh, B. (2015). Toward applications of genomics and metabolic modeling to improve algal biomass productivity. In Biomass and Biofuels from Microalgae (pp. 173-189). Springer International Publishing Switzerland. doi:10.1007/978-3-319-16640-7_10

Sancho, M. M., Castillo, J. J., & El Yousfi, F. (1997). Influence of phosphorus concentration on the growth kinetics and stoichiometry of the microalga Scenedesmus obliquus. *Process Biochemistry*, *32*(8), 657–664. doi:10.1016/S0032-9592(97)00017-4

Sasi, D., Mitra, P., Vigueras, A., & Hill, G. A. (2011). Growth kinetics and lipid production using Chlorella vulgaris in a circulating loop photobioreactor. *Journal of Chemical Technology and Biotechnology (Oxford, Oxfordshire)*, *86*(6), 875–880. doi:10.1002/jctb.2603

Schierenbeck, L., Ries, D., Rogge, K., Grewe, S., Weisshaar, B., & Kruse, O. (2015). Fast forward genetics to identify mutations causing a high light tolerant phenotype in *Chlamydomonas reinhardtii* by whole-genome-sequencing. *BMC Genomics, 16*(1), 57, 1-15. . doi:10.1186/s12864-015-1232-y

Shah, A. R., Ahmad, A., Srivastava, S., & Jaffar Ali, B. M. (2017). Reconstruction and analysis of a genome-scale metabolic model of *Nannochloropsis gaditana*. *Algal Research*, *26*, 354–364. doi:10.1016/j.algal.2017.08.014

Shen, M. Y., & Sali, A. (2006). Statistical potential for assessment and prediction of protein structures. *Protein Science*, *15*(11), 2507–2524. doi:10.1110/ps.062416606 PMID:17075131

Stewart, A., Rioux, D., Boyer, F., Gielly, L., Pompanon, F., Saillard, A., Thuiller, W., Valay, J.-G., Maréchal, E., & Coissac, E. (2021). Altitudinal Zonation of Green Algae Biodiversity in the French Alps. *Frontiers in Plant Science*, *12*, 679428. doi:10.3389/fpls.2021.679428 PMID:34163510

Studer, G., Rempfer, C., Waterhouse, A. M., Gumienny, R., Haas, J., & Schwede, T. (2020). QMEAN-DisCo – Distance Constraints Applied on Model Quality Estimation. *Bioinformatics (Oxford, England)*, *36*(6), 1765–1771. doi:10.1093/bioinformatics/btz828 PMID:31697312

Taberlet, P., Bonin, A., Zinger, L., & Coissac, E. (2018). *Environmental DNA: For Biodiversity Research and Monitoring*. Oxford University Press. doi:10.1093/oso/9780198767220.001.0001

Taberlet, P., Prud'Homme, S. M., Campione, E., Roy, J., Miquel, C., Shehzad, W., Gielly, L., Rioux, D., Choler, P., Clément, J.-C., Melodelima, C., Pompanon, F., & Coissac, E. (2012). Soil sampling and isolation of extracellular DNA from large amount of starting material suitable for metabarcoding studies. *Molecular Ecology*, *21*(8), 1816–1820. doi:10.1111/j.1365-294X.2011.05317.x PMID:22300434

Talukdar, P., & Pal, R. (2017). Predictive toxicity assessment of toxins released from *Microcystis aeruginosa* kutz: An *in silico* approach. *World Journal of Pharmaceutical Research*, *6*(14), 411–423.

Tamura, K., Peterson, D., Peterson, N., Stecher, G., Nei, M., & Kumar, S. (2011). MEGA5: Molecular evolutionary genetics analysis using maximum likelihood, evolutionary distance, and maximum parsimony methods. *Molecular Biology and Evolution*, *28*(10), 2731–2739. doi:10.1093/molbev/msr121 PMID:21546353

Tanwar, A., & Kumar, S. (2020). *Genome editing of algal species by CRISPR Cas9 for biofuels. Genome engineering via CRISPR-cas9 system*. Academic Press.

Theodoropoulos, C., & Sun, C. (2019). *Bioreactor Models and Modeling Approaches. In Comprehensive Biotechnology*. Elsevier BV.

Theodoropoulos, C., & Sun, C. (2019). Bioreactor Models and Modeling Approaches. In *Comprehensive Biotechnology* (pp. 663–680). Elsevier BV. doi:10.1016/B978-0-444-64046-8.00144-0

Thompson, J. D., Gibson, T. J., Plewniak, F., Jeanmougin, F., & Higgins, D. G. (1997). The clustal X windows interface: Flexible strategies for multiple sequence alignment aided by quality analysis tools. *Nucleic Acids Research*, *24*(24), 4876–4882. doi:10.1093/nar/25.24.4876 PMID:9396791

U.S. EPA. (2012). *Estimation Programs Interface Suite™ for Microsoft Windows, v 4.10*. United States Environmental Protection Agency.

Villain, J., Lozano, S., Halm-Lemeille, M.-P., Durrieu, G., & Bureau, R. (2014). Quantile regression model for a diverse set of chemicals: Application to acute toxicity for green algae. *Journal of Molecular Modeling*, *20*(12), 2508. Advance online publication. doi:10.100700894-014-2508-x PMID:25431186

Wiederstein, M., & Sippl, M. J. (2007). ProSA-web: Interactive web service for the recognition of errors in three-dimensional structures of proteins. *Nucleic Acids Research*, *35*(Web Server), 407–410. doi:10.1093/nar/gkm290 PMID:17517781

Wu, T. D., & Watanabe, C. K. (2005). GMAP: A genomic mapping and alignment program for mRNA and EST sequences. *Bioinformatics (Oxford, England)*, *21*(9), 1859–1875. doi:10.1093/bioinformatics/bti310 PMID:15728110

Xin, L., Hong-Ying, H., Ke, G., & Ying-Xue, S. (2010). Effects of different nitrogen and phosphorus concentrations on the growth, nutrient uptake, and lipid accumulation of a freshwater microalga Scenedesmus sp. *Bioresource Technology*, *101*(14), 5494–5500. doi:10.1016/j.biortech.2010.02.016 PMID:20202827

Yuan, S., Hu, J., Liu, Z., Hong, Y., & Wang, X. (2020). Modeling microalgae growth in continuous culture: Parameters analysis and temperature dependence. *Energy*, *195*, 116961. doi:10.1016/j.energy.2020.116961

Yuan, S., Hu, J., Liu, Z., Hong, Y., & Wang, X. (2020). Modeling microalgae growth in continuous culture: Parameters analysis and temperature dependence. *Energy*, *195*, 116961. doi:10.1016/j.energy.2020.116961

Chapter 10
Algal Life Cycle Analysis and Its Contribution to the Circular Economy

Wafaa Fathi Zohir
https://orcid.org/0000-0002-7635-6965
International Centre for Genetic Engineering and Biotechnology, New Delhi, India & Department of Botany and Microbiology, Faculty of science, Damanhour University, Egypt

Vikas U. Kapase
International Centre for Genetic Engineering and Biotechnology, India

Prachi Nawkarkar
International Centre for Genetic Engineering and Biotechnology, India

Shashi Kumar
International Centre for Genetic Engineering and Biotechnology, India

ABSTRACT

Microalgae are a diverse group of microscopic and highly efficient photosynthetic organisms with rapid growth that contains a wide range of biochemical components such as pigments, lipids, carbohydrates, and proteins, making them a viable feedstock for various commercial applications in biofuel production, nutraceutical, pharmaceutical, and environmental sectors. Life-cycle assessment (LCA) is one of the most appealing and attractive tools used nowadays by the scientific and decision-makers communities to ensure environmentally sustainable production/consumption of various products. It is a systematic and standardized methodology that has turned into a crucial communication tool for the projects, the target markets, and the general public. Recently, LCA has been applied to quantify algal biofuels' environmental sustainability. As a result, the importance of algal life cycle analysis, its consequences, and contribution to the circular economy will be discussed in this chapter in terms of developing industrial applications.

DOI: 10.4018/978-1-6684-2438-4.ch010

INTRODUCTION

In a world of growing human population, interest in the use of photosynthetic organisms for industrial purposes is becoming more urgent and popular to reduce the generation of synthetic products. The photosynthetic organisms called algae have been identified as key elements for use in various industries, including pharmaceutical, cosmetic, and food industries. Algae are natural renewable biofactories that are rapidly gaining prominence due to their long-term sustainable and economical sources of biofuels production, bioactive medicinal products, food, and feed. Also, microalgae can contribute to atmospheric CO_2 mitigation and wastewater treatment (Khan et al., 2018; Ratnapuram et al., 2018). Upgradation of algal fuel and bioproducts technology from pilot scale to commercial level is possible by overcoming the associated challenges and limitations. Considering the potential applications of microalgae, an algal biorefinery could be a promising option for the commercial production of biofuel and bioproducts from microalgae. In an attempt to establish sustainable environmental benefits, it is necessary to investigate the life cycle assessment (LCA) of algae biorefinery systems. The LCA provides a way to evaluate and assess the environmental impact of the global warming potential (GWP) of the algal biorefinery system before any implementation on a large scale (Khan et al., 2018).

BACKGROUND

Algae are prokaryotic or eukaryotic photosynthetic organisms that can grow in a variety of sizes, shapes, and habitats i.e., fresh and saltwater under a wide range of environmental conditions. Mostly, Microalgae can grow in all existing earth ecosystems - aquatic habitats, including lakes, ponds, rivers, oceans, thermophilic and hypersaline environments, and even wastewater (Ratnapuram et al., 2018), and terrestrial - desert crusts, tree trunks, and animal fur. Also, these can tolerate and resist a wide range of conditioners like temperatures, salinities, pH values, light intensities; and conditions in reservoirs or deserts and can grow alone or in symbiosis with other organisms (Khan et al., 2018). Microalgae are important primary producers and form the basis of the food chain in aquatic environments for several bioproducts essential for human diets/nutrition. Cyanobacteria are blue-green photosynthetic prokaryotic microorganisms that are included in this broad circumscription. It is estimated that more than 200,000 species exist, but only a limited number, of around 30,000, have been studied and analyzed. Hence, there is a huge potential for this untapped and unexplored broad bioresource for potential biotechnological applications making these a major source for commercial and scientific interest (Mutanda et al., 2020).

Microalgae have rapid growth and produce a significant quantity of carbohydrates, proteins, lipids, and bioactive compounds due to the photosynthesis process. These have also been known to produce a range of commercially important secondary metabolites that are currently sourced from conventional agriculture (Kaparapu and Geddada, 2020). However, there are numerous advantages to using microalgae over land plants. Microalgae have the ability to fix CO_2 using solar energy with efficiency 10 times greater than that of terrestrial plants. Microalgae are efficient at consuming harmful pollutants, have minimal resource requirements and do not compete with food or agriculture for land & water resources, and have a higher capacity to generate biomass (Mutanda et al., 2020). Furthermore, many geographical areas that are not suitable for crop cultivation could be effectively used for large-scale algal cultivation; leading to a broad range of bio value-add products. In this regard, microalgal biofactories have the potential to be the most cost effective, economical, and sustainable platforms that are naturally predisposed to produce particular plant-derived products (Kaparapu and Geddada, 2020).

Microalgal Classification

Algae have been classified into three major groups based on the pigments and evolutionary origins as (i) Rhodophyta (red algae), (ii) Phaeophyta (brown algae), and (iii) Chlorophyta (green algae) and classified by size as macroalgae or microalgae. Macroalgae (seaweed) are multicellular, large-size algae while microalgae are unicellular microscopic cells and maybe (i) prokaryotic, similar to cyanobacteria (Chloroxybacteria), or (ii) eukaryotic, similar to green algae (Chlorophyta) (Khan et al., 2018; Mutanda et al., 2020). The phylum Chlorophyta includes green algae related to plants (Embryophyta), the Phaeophyta called kelps, and Rhodophyceae, brown and red algae respectively, containing large groups of seaweeds. Other groups of microalgae are known as dinoflagellates, cryptophytes, haptophytes, and others (Lenntech). Algae have different shapes and sizes, many species are microscopic, uni- and multicellular; others, like leafy kelp, have plant-like bodies. The large-bodied algae lack the more complex structural features in the plants like roots, stems, leaves, and a vascular system to transport nutrients and water. Furthermore, algae have the ability to produce a variety of metabolites under different cultivation conditions viz. photoautotrophic, heterotrophic, and mixotrophic growth conditions. The growth kinetics of microalgae vary due to prevailing key growth controlling factors and are dependent upon the nutrient supply (Suganya et al., 2016). These algae are considered a promising sustainable alternative source to fossil fuels, because of their ability to store and accumulate a large quantity of carbon in the form of lipids which is the base for biofuel production.

APPLICATIONS OF MICROALGAE

Microalgae are a rich source for a wide variety of bioproducts or bio-compounds, which include polysaccharides, lipids, pigments, proteins, vitamins, bioactive compounds, and antioxidants. These have multiple applications in biofuels, nutraceuticals (health supplements), pharmaceuticals, and cosmetic industries (Khan et al., 2018). Also, they have been used in aquaculture, animal feeds, wastewater remediation, and atmospheric CO_2 mitigation. Nowadays, algae are produced in large quantities and sold directly as food and nutrient supplements, while their processed products or extracts are used in biopharmaceuticals and cosmetics (Kaparapu and Geddada, 2020). Because of the benefits of microalgae application in biofuel and bioproducts, their industrial cultivation has increased dramatically over the last few decades. The possible industrial-scale cultivation, along with the renewable and sustainable nature of microalgae has given rise to a biorefinery concept whereby two bio-products can be obtained from algae biomass by simultaneous or gradual processes. Further, techniques for enhancement in growth or product yield and genetic engineering can be employed to improve their potential as a future source of renewable and sustainable bioproducts (Khan et al., 2018).

Microalgae: Application in Biofuels

Microalgae have been identified as viable sources for biofuel feedstock due to their ability to synthesize high content of oil/lipid along with high biomass production. Oleaginous species of microalgae cultured in growth-optimized conditions in photobioreactors have the potential to yield 5,000-15,000 gal of microalgal oil per acre per year (Mulumba and Farag, 2012). Algal biomass can be utilized directly as food/feed or processed into liquid and gas fuels by applying different biochemical or thermochemical conversion processes. Algal biomass can be used as a feedstock for biodiesel production as the extracted oil is similar in properties to the standard biodiesel.

While considering algal biofuel properties, neutral lipids with suitable carbon chain length and degree of saturation play an important role. Algal oil contains saturated and monounsaturated fatty acids. The high proportion of saturated and monounsaturated fatty acids is considered optimal from a fuel quality standpoint, it increases the calorific value of the fuel and reduces polymerization during combustion than what would occur with polyunsaturated fatty acid-derived fuel (El-Sheekh et al., 2020). However, the presence of polyunsaturated fatty acids helps to maintain the fluidity of the fuel at low temperatures. After oil extraction from algae, the remaining biomass fraction can be used as a high-protein feed for livestock, adding value to this process and reducing waste (Kaparapu and Geddada, 2020). Macroalgae is a plant-like structure and easier to harvest than microalgae. The macroalgae Rhodophyte- *Gracilaria vermiculophylla, G. multipartite, Chlorophyta- Ulva lactuca, U. intestines,* and the phaeophyte- *Padina boryana* are widely used in biofuel extraction due to high biomass production and relatively higher lipid content (Abomohra et al., 2018).

The most abundantly synthesized fatty acids by microalgae are C14:0, C16:0, C18:1, C18:2, and C18:3, while others such as C16:2, C16:3, C16:4, C18:4, C20:5, C22:6 are reported to be strain-specific. Commonly, microalgae store lipids in the form of triglycerides, and environmental conditions such as nutrient availability, temperature, and irradiation, can affect the microalgal fatty acids profile (Ratnapuram et al., 2018). However, under physiological stress such as temperature, light intensity, pH, CO_2, nitrogen starvation, and phosphate limitation, the lipids are synthesized abundantly and accumulated at the expense of other macromolecules. Thereby nutrient limitation/starvation is a widely used strategy for enhancing lipid yield for biofuel/biodiesel production. But its compromised biomass productivity poses a grave problem. Optimized cultivation modes such as photoautotrophic, heterotrophic, and mixotrophic; have been found appropriate for achieving both higher biomass and lipid productivity for biodiesel applications (El-Sheekh et al., 2020). A two-stage cultivation strategy has been proposed as a lucrative option to balance cell growth and lipid production (Ratnapuram et al., 2018). Among all microalgal strains, the ability to grow fast and accumulate upto 58% of triglycerides of the total lipid content makes *Chlorella vulgaris* as the leading candidate for commercial algal lipid production (El-Mohsnawy et al., 2020). The lipids produced are primarily composed of palmitic acid C16:0, stearic acid C18:0, palmitoleic acid C16:1, and oleic acid C18:1, all of which are suitable constituents for biodiesel production (Kaparapu and Geddada, 2020).

Microalgae: As Food and Feed

Microalgae produce a variety of valuable products that elevate their importance as a nutritional food for humans and animals (Khan et al., 2018). Microalgae are known to be a suitable source of protein (produce essential amino acids) and carbohydrates which make them an important food source. Also, they produce different types of essential fatty acids (e.g., eicosapentaenoic acid [EPA], docosahexaenoic acid [DHA]), pigments (carotenoids including antioxidant substances), carbohydrates, and proteins for medicinal and commercial purposes (Suganya et al., 2016; Hu et al., 2019). Microalgae accumulate large amounts of lipids, present in the form of glycerol then esterified to different types of fatty acids that have nutritional and medicinal applications.

Due to microalgae therapeutic applications; several research on extracting & identifying the bioproducts from microalgae to determine their biological and medicinal activities is on the escalation. Microalgae are becoming economical sources of natural bioproducts for use in different applications like food, feed, cosmetics, pharmaceuticals, and nutraceuticals. For example, Blue-green algae (cyanobacteria) are rich in phycobiliproteins that are used as nutrients and food coloring agents. Many microalgae strains, including *Aphanizomenon flosaquae*, *Chlorella* sp., and *Arthrospira*, are commercially farmed for their high protein content and other health-promoting bioproducts (Fabris et al., 2020).

Microalgae: As Nutraceuticals

Ingredients and products developed from microalgal biomass can be found in a variety of commercial food markets. Various metabolites identified in microalgal biomass have found substantial applications in the nutraceutical and pharmaceutical industries (Mutanda et al., 2020). Different microalgal strains synthesize varying amounts of intracellular bio-molecules based on the growing parameters, such as nutrient availability, light quality, intensity, and so on. (Fabris et al., 2020). There are some of the multi-application natural bioproducts, are produced by microalgae and described below:

Polyunsaturated Fatty Acids (PUFAs)

Microalgae are considered primary producers and more sustainable sources of most essential PUFAs in nature. Essential long-chain PUFAs (LC-PUFAs) such as alpha-linolenic acid (ALA), arachidonic acid (AA), eicosapentaenoic acid (EPA), docosahexaenoic acid (DHA) are reported to play a vital role in human health applications such as physical, mental, and visual developments in infants (Suganya et al., 2016), and treatment of various diseases (cardiovascular, arthritis, cancer, etc.). However, the human body cannot synthesize these vital fatty acids and so is dependent upon fish and fish oils as sources for LC-PUFAs. Due to the toxicity issues and fishy smell, attention has been diverted to marine microalgae which are the dominant primary producers. DHA is used in newborn formulae, fortified foods, beverages, and dietary supplements since it is required for neurological development (Hu et al., 2019). Microalgae can accumulate DHA with concentrations of up to 50% of total biomass and it can be produced by the most suited microalgae *Crypthecodinium cohnii*, *Schizochytrium* sp., *Ulkenia* sp. (Suganya et al., 2016), and *Pavlova lutheri* (Chang et al., 2014). EPA is found in marine microalgae and produced with high quality across several microalgal classes like Bacillariophyceae (diatoms), Chlorophyceae, Chrysophyceae, Cryptophyceae, Eustigamatophyceae, and Prasinophyceae (Wen and Chen, 2003). Moreover, Omega-3 and omega-6 fatty acids are important for tissue integrity and have beneficial effects on human health. So, several microalgae have been studied for their ability to synthesize these long-chain PUFAs including *Porphyridium cruentum, Arthrospira platensis, Odontella, Isochrysis galbana* (Khan et al.,2018; El-Mohsnawy et al., 2020).

Microalgal Pigments

In microalgae, the pigments play a major role in light-harvesting, CO_2 fixation, non-photochemical quenching, etc. In algae, the pigments are responsible for light-harvesting, CO_2 fixation, protection against damage by excessive irradiation, and the coloration of the algal culture (Khanra et al., 2018; Kiran & Venkata Mohan, 2021). Three major groups of pigments are found in microalgae, chlorophylls (green coloration), carotenoids (carotenes for orange and xanthophylls for yellowish shade), phycobilins (red or blue coloration) (Bhalam-urugan et al., 2018). The pigment fraction from microalgae have nutraceutical, pharmaceutical, veterinary, and medical applications such as anti-inflammatory effects, anti-oxidative effect, cancer prevention, as well as in the cosmetic industry, aquaculture, and food technology (Kiran & Venkata Mohan, 2021; Khanra et al., 2018; Bhalam-Murugan et al., 2018). Due to their wide applications, they are being considered the algal products with the highest commercial attention along with the high market value. These facts and applications are collected in **Table 1**, which presents the most important algal pigments.

Microalgae: As Pharmaceutical and Biomedical Applications

Over-the-counter pharmaceuticals mainly consist of tablets or liquid forms of artificial or herbal therapeutic products. However, the current trend sees the availability of highly nutritional microalgal products, as pure extracts in the form of tablets, capsules, and additives to several food products. Because of increasing awareness and demand for nutritional bioproducts, the microalgal market is growing in a good direction. From a pharmaceutical point of view, the primary and secondary metabolites from microalgae can be used as active pharmaceutical ingredients as antiviral, anti-inflammatory, immunological modulators, and immune adjuvants (Kaparapu and Geddada, 2020). Currently, microalgal species like *Arthrospira (Spirulina), Chlorella, Dunaliella, Haematococcus,* and *Nostoc* sp. are commonly cultivated for the production of bioactive products of pharmaceutical interest. Furthermore, microalgae can be used to make vaccines and treat the world's most dangerous viral infections, such as coronaviruses and the human immunodeficiency virus (HIV) (Talukdar et al., 2020).

According to the developments in microalgal biotechnology and the role of microalgae in combating the human viruses, *Arthrospira platensis* is highly used as a common natural source to produce sulphated polysaccharides named Spirulan (existing as an ionic form as calcium or sodium) that have proved to be effective against enveloped viruses, including HSV-1, mumps virus, measles virus, human cytomegalovirus, influenza A virus and HIV-1. A recent study has demonstrated the great potential of this substance against herpesviruses both in vitro and in a clinical trial (Mader et al. 2016; Pagarete et al. 2021). *Chlamydomonas reinhardtii* is the most exploited microalgae that have been used in the production of therapeutic proteins (erythropoietin, interferon β insulin, and immunoglobulin A) and Glycerol when the alga was stressed by sulfur (Yan et al., 2016). Moreover, Naviculan from the diatom, Navicula *directa,* revealed a potential activity against-HSV-1 and HSV-2 (Dewi et al., 2018) and polysaccharides from dinoflagellates (*Gyrodinium impudicum* and *Cochlodinium polykrikoides*) were found to inhibit infections of the encephalomyocarditis virus (EMCV), respiratory syncytial virus-A and -B (RSV-A; RSV-B), influenza A and B viruses (Rodrigues et al., 2017). Nostaflan isolated from *Nostoc flagelliforme* inhibits influenza A virus, HSV-1, HSV-2, and cytomegalovirus (CMV) (Ahmadi et al., 2015; Elaya Perumal and Sundarara, 2020).

Moreover, Cyanovirin, a bioactive molecule derived from Nostoc, was found to be effective in treating HIV and influenza-A symptoms (H1N1) (Bhattacharjee, 2016). Also, Vitamin B complexes such as B12, β-carotene, lutein, ascorbic acid, and α-tocopherol are reported to be abundant in C*hlorella* biomass. These bioactive compounds are used to prevent macular degeneration and reduce the risk of

Table 1. Microalgal pigments and potential fields of application — an overview (modified from Koller et al., 2014; Kiran & Venkata Mohan, 2021).

Pigment group	The dominant color of algal culture	Examples of the pigment	Common Microalgal	Color pigments	Application of the pigment	References
Chlorophylls	Green	Chlorophyll a,	All phototrophic oxygenic algae	Green	Pharmaceutical, cosmetic (deodorant), Food additive (E140), antioxidant antimutagenic activities activity, immune activators, cytotoxic towards tumoral cells,	(Khanra et al., 2018; Odjadjare et al., 2017)
Carotenoids (Carotenes)	Red, Brown orange	ß-Carotene	*Dunaliella salina* (up to 14%), *D. bardawil, Botryococcus braunii; Scenedesmus almeriensis, Coelastrella striolata* var. multistriata	Yellow	Pro-vitamin supplements, Antioxidant, food additive E160a; coloration of egg yolk, possess antiaging and anticancer properties.	(Bhalam-urugan et al. 2018; Galasso et al., 2019)
Carotenoids (Xanthophylls)		Astaxanthin	*Haematococcus pluvialis, B.braunii, S. obliquus,Chlorella zofingiensis*	Reddish- salmon	Food additive (E161); antioxidant; Coloration, farming of salmon and trout, immune-response; Cancer defense, inflammation, metabolic syndrome, diabetes, neurodegenerative and ocular diseases, lung injury, repressed alveolar wall swelling, and myeloperoxidase activity	(Galasso et al., 2019; Lu et al., 2021)
Carotenoids (Xanthophylls)		Lutein	*Chlorella protothecoides, C. zofingiensis, B. braunii, Chlorococcum citriforme, D. salina, Galdieria sulphuraria. Muriellopsis sp., S. almeriensis, Neospongiococcus gelatinosum*	Yellow-orange	Food additive E161b; feed additive; coloration of egg yolk, pigmentation of animal tissues. Pharmaceutical: anti-macular degeneration, anti-colon cancer. Cosmetics; Antioxidant and anticancer activity, Cataract, atherosclerosis, diabetic retinopathy, and age-related retinal degeneration	(Liu et al., 2017)
Carotenoids (Xanthophylls)		Zeaxanthin	*B. braunii, D. salina, Nannochloropsis oculata, N. gaditana*	Orange-yellow	Food additive E 161 h, animal feed; Pharmaceutical: anti-colon cancer, eye health	(Gong and Bassi, 2016; Liu et al., 2017)
Carotenoids (Xanthophylls)		Canthaxanthin	*N. oculata, N. salina, N. gaditana*	Golden-orange	Food additive E 161 g, farming of salmonids and chicken Tanning pills	(Novoveská et al., 2019)
Carotenoids (Xanthophylls)		Violaxanthin	*B.braunii, D. tertiolecta, Nannochloropsis sp*	Orange	Food additive E161e; anti-cancer; Anti-proliferative activity	(Koller et al., 2014)
Phycobilins	Red	Phycocyanin	*Arthrospira, Spirulina* (cyanobacteria)	Blue-green	Food colorant (as beverages, ice cream, sweets); cosmetics; Immunofluorescence techniques; antibody labels, receptors, and other biological molecules; anti-inflammatory	(Parmar et al., 2011; Koller et al., 2014)
		Phycoerythrin	Cyanobacteria, *Porphyridium*	Red	Immunofluorescence techniques; labels for antibodies, receptors, and other biological molecules; Cosmetics	(Parmar et al., 2011; Markou and Nerantzis, 2013)

cancer (Bhalam-urugan et al., 2018). While bioactive substances such as enzymes, vitamins, and anti-microbials are abundant in *Dunaliella.* (Bhattacharjee, 2016).

Recently, ocean extracts and phytoplanktonic bioactive molecules or whole-cell supplements are acquiring more and more attention. Due to, the presence of a variety of valuable therapeutic bioactive compounds such as polyunsaturated aldehydes (PUAs), chrysolaminarin polysaccharide, violaxanthin, fucoxanthin, stigmasterol, nonyl 8-acetoxy-6-methyl octanoate (NAMO) that can be used in different nutraceutical and pharmaceutical industries (Wang et al., 2021).

Microalgae as a Source of Biofertilizers, Biostimulants, and Biopesticides

Microalgae and cyanobacteria are potential sources for plant growth-promoting substances such as polysaccharides, phenolics, hormone-like substances, and proteins. Considering their roles, microalgal/cyanobacterial products, and biomass can be classified as biofertilizers, biostimulants, and biopesticides (Gonçalves, 2021).

Biofertilizers promote the productivity of the crop by improving the soil properties, enhancing fertility, and supplying the essential nutrients (e.g., N, P, K) for plants' growth. Consequently, biofertilizers are classified as: (i) plant-growth-promoting; (ii) compost; (iii) nitrogen-fixators; (iv) phosphate- and potassium solubilising biofertilizers; and (v) phosphorus-mobilisation biofertilizers. The use of biofertilizers in agriculture presents several benefits like (i) enhancement in crop productivity per unit of area and time; (ii) reduction in energetic requirements; (iii) control and maintenance of adequate soil properties and fertility; (iv) lower risks of soil and water contamination; and (v) crops' protection against pathogenic organisms (Gonçalves, 2021; Ronga et al., 2019).

Biostimulants act directly on the plants and promote crop productivity. These include a variety of bioactive substances such as polysaccharides, phenolic compounds, hormone-like compounds, vitamins, etc. These compounds are responsible for improving respiration, photosynthetic activity, nucleic acid synthesis, and ion uptake, thereby enhancing plants' metabolism and plants. These compounds can also act as protectants against stress conditions. The application of biostimulants in agriculture has advantages like (i) increase in crop productivity (ii) increase in nutrients utilization efficiencies; and (iii) enhancement in crops' quality (Ronga et al., 2019).

The biopesticides protect plants from disease infestations. Antimicrobial, antioxidant, antiviral, and antifungal chemicals, commonly included in crop-promoting formulas, help plant growth by protecting them from diseases (Gonçalves, 2021).

Environmental Benefits of Using Microalgae

Microalgae have several lucrative characteristics that make them hopeful feedstocks for the generation of biofuels and high-value products. These have the best carbon sequestration capacity making them carbon neutral and therefore serve as better candidates for climate change mitigation. Microalgal systems have a higher photon conversion efficiency, and can couple CO_2-neutral fuel production with CO_2 sequestration thereby producing non-toxic and highly biodegradable biofuels (Cobos et al., 2017). Since microalgal cultivation and metabolite extraction does not interfere with food security concerns, the microalgal food supplement industry is currently booming. According to Hildebrand et al. (2013), microalgal productivity is dependent on the efficiency of carbon fixation and the downstream cellular processes that convert it into useful precursors and final metabolites. However, the major drawback is the apparent low yield of desirable metabolites e.g., pigments and neutral lipids. The low yield of the desirable metabolites can be overcome by carefully manipulating environmental stresses and optimizing growth conditions for maximal product yield. Microalgal cultivation does not require large tracks of arable land as compared to some crops such as maize, sunflower, cotton, and soybean; neither require copious volumes of freshwater resources. Wastewater streams such as municipal, domestic, and agricultural wastewaters with sufficient concentrations of nitrates, phosphates, and trace elements can be harnessed for microalgal cultivation (Chen et al., 2017). The search for superior and unique microalgal strains is an ongoing exercise and currently bioprospecting for indigenous strains is an intense activity among research communities.

Hurdles in the Processes

Despite the varied applications of microalgae in the fields of bioenergy and nutraceuticals, microalgae technology is faced with numerous technical hurdles. The scaling up of microalgal culture needs to overcome challenges that are responsible for its low product yields (Hamilton et al. 2016). Upscaling required optimization in areas of strain selection, culture development, product induction, and extraction technologies. All operations start from the laboratory scale to define the techniques and operating conditions. Transfer of microalgae culture from the laboratory to commercial scale faces the challenge of adaptation to environmental as well as operating conditions. The effects of operating conditions on microalgal cultivation at a large scale include temperature, mixing, microbial contamination, pH, oxygen build-up, photoperiod, light intensity, biofouling, and salinity. Consequently, several large-scale operations produce lower biomass yields than those realized under laboratory conditions. Nutrient, energy, and water recycling represent the main technical and economic problems while shifting lab cultures to a large scale. Four cultivation challenges have emerged and addressed for economic and commercial-scale microalgal cultivation: 1) Culture stability; 2) Standardized metrics for system-level productivity analysis; 3) Nutrient source scaling, energy costs, sustainability, and management; and 4) Water conservation, management, and recycling (Mutanda et al., 2020; Fabris et al., 2020; Hamilton et al., 2016). These four challenges can be overcome by applying more experimental and modeling studies to predict the optimum operating conditions and parameters. It is also important to know the unique characteristics of each microalgal strain, and its behavior at the laboratory level that can be affected at a large scale. Scaling up influences the stability and productivity rate of the culture as the operating conditions are affected during cultivation due to size expansion. The size expansion has a significant impact on the operating and environmental conditions during cultivation. Furthermore, each aspect has financial implications when scaling up, hence the economics related to these challenges have to be taken into consideration and require more studies (Fabris et al., 2020; Hamilton et al., 2016).

Algal Biotechnological Advancements

The development of the most sustainable technology for the production of biofuel, food, feed, and mitigation of environmental CO_2 is necessary to meet the ecological demands of a rapidly growing population. Microalgae have received a great deal of attention as a potential resource to address urgent needs of food, feed, and fuel. Microalgae's biological diversity can be enhanced to generate a variety of useful products by genetic modification. At the moment, microalgae are only used in a limited number of industrial applications for food, feed, and high-value products. The transition from algal-based bioenergy to high-value bioproducts, as well as the model of algae-based biorefineries, have been defined in detail in recent works (Fabris et al., 2020). Still a range of certain conditions is to be fulfilled to develop a commercially and economically sustainable algae-based biorefinery (**Fig.1**). Previously, overcoming the major obstacles of taking microalgal processing systems to a commercially feasible scale has proved difficult, but the expanded research in cultivation systems, harvesting, downstream processing, and genetic modifications of microalgae leading to enhanced biomass and product yields, will pave the way to develop a new algae-based biorefinery for commercial use. The strategies of genetic engineering include gene knockout, overexpression, insertion, deletion, and mutations using RNAi, Zinc finger Nucleases (ZFNs), transcription activator-like effector nucleases (TALENs), CRISPR/Cas9, and chemical/ physical mutagens (Sproles et al., 2021).

In recent years, several genetic manipulation strategies were employed for specific objectives like enhanced growth, high TAG content, high PUFA content, and pigments. Although, algal genome engineering faces a number of challenges like low efficiency, cell wall composition and structure, genomic ploidy, and selection agent sensitivity, some microalgal species like model organism *C.reinhardtii, Nannochloropsis sp., Picochlorum sp., Synechocystis sp. PCC 6803, Cyanidioschyzon merolae, Phaeodactylum tricornutum, Tetraselmis chui, Neochloris oleoabundans, Chlorella* sp., *Dunaliella* sp., and many more were successfully genetically transformed (Mosey et al., 2021). Some recent examples of algal genetic modifications are stated in the below table.

Some efforts have been made recently to enhance the production of EPA and DHA by altering the metabolic pathways via genetic manipulation. Most recently *P. tricornutum* attracted attention as a potential source of EPA and DHA production (Koller et al., 2014). The diatom *P. tricornutum* has been genetically sequenced and modified to enhance the production of Omega-3 polyunsaturated fatty acids like EPA and DHA etc. It has been reported that the genetically engineered strain of *P. tricornutum* produces a maximum yield of 36.5 and 23.6% of DHA and EPA respectively per total fatty acid, indicating its feasibility for production at a commercial scale (Chauton et al., 2015).

LIFE CYCLE ANALYSIS

Microalgae species have received considerable interest as a third-generation feedstock for biofuel, as Neutral- and pharmaceutical products with high productivity with simultaneous mitigation of CO_2. Several private/government companies involved in R & D consider algae as the best alternative for conventional products. Before the incorporation of any new technology on a large scale, however, it is necessary to understand its environmental impacts, to avoid ecological problems. Life cycle assessment (LCA) has emerged as a critical assessment tool to understand the relative environmental performance processing technologies at the systems level. An analytical tool -Life Cycle Analysis (LCA) provides insights to understand the environmental impacts such as global warming potential (GWP) and computing the greenhouse gas footprint of biomass-based energy processes. It is a methodology that evaluates efficient resource utilization and sustainability of any products and involved processes.

LCA as an environmental management tool began to shape up in the 1960s under different names and with different forms. It was not till the 1990s that the term life cycle assessment was used to refer to studies on the environmental life cycle (De Lassio et al., 2016) that included consumption of raw materials and energy, air, water pollution, and waste production, etc. Nowadays, the LCA examines the environmental footprint of a product from the initial stages of raw material/resource extraction to the final product and its end use until its final disposal, thereby encompassing the entire life cycle of the product system. It is designed to make comprehensive comparisons of available technological options by quantifying and assessing all energy inputs and related outputs as environmental burdens of a product during its entire life cycle. This information thus enables one to make informed decisions. For example, in its application in the construction industry, the LCA can consider analyses of products in the industry, individual buildings, and groups of buildings (Khasreen et al., 2009). The LCA framework has been accepted internationally

Figure 1. Algal biorefinery spectrum: Products, processes, and culture systems involved.

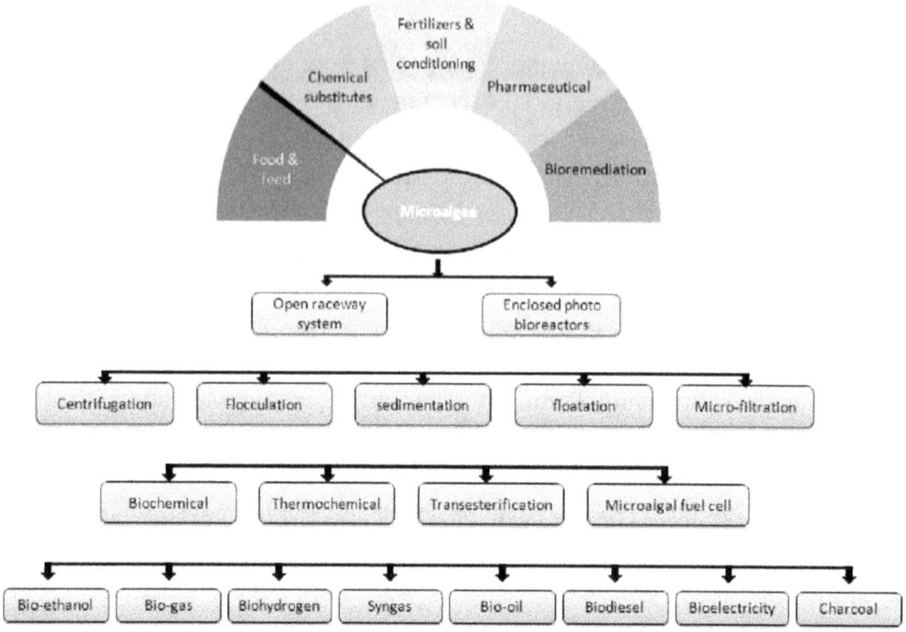

with the well-established best practices through environmental system standards - ISO 14040 and ISO 14044 for evaluating requirements and impact technologies, processes, and products.

The LCA allows us to compare the possibility of various biofuel production approaches, like open ponds and LED/solar-lit photobioreactor growth systems for the production of biomass, algae oil/ biodiesel, or high-value products. These are the key points to studying a production scenario to achieve the target and resolve existing problems. The growth system infrastructure, functioning, and downstream processing play an important role in algae-based production systems. However, due to the unavailability of commercial algae-based fuel production systems, the results of any study depend on the assumptions made to make the calculations. The LCA has the following major steps: (1) goal and scope definition, (2) life cycle inventory (LCI) analysis, (3) life cycle impact assessment (LCIA), and (4) improvement and interpretation (Adhikari & Pellegrino, 2015).

The circular economy in the EU According to European Union policies, it is defined as a broad term that encompasses a variety of processes and products. It's not just about the products, but also about the processes that are involved in its production, its value addition and activities are all taken into account. In another sense, it is a development model that aims to uphold the environmental and sustainable values by implementing principles such as the "3 Rs": reduce, reuse, and recycle. It has driven great interest in recent years to develop the sustainable technologies to reduce environmental impacts through the use of recyclable packaging, the promotion of ecological products, the reduction of emissions and waste, the assessment of renewable and alternative energies, energy conservation, the use of low-environmental-impact consumer goods, eco-design, waste recovery, and dematerialization (Dahiya et al. 2022).

LCA is a standardized (ISO 14,040–14,044:2006) and science-based technique for assessing the impacts of technology, products, or materials, which can aid in understanding the environmental implications of the circular economy. When evaluating the circular economy solutions, many concerns arise that may be

Table 2. Case studies for genetic engineering in microalgae

Microalgal strain	Genetic modification	Results	Reference
Cyanidioschyzon merolae	Overexpression of endogenous CmFAX1(nuclear-encoded chloroplast membrane protein)	2.4-fold increase in intracellular TAGs & 0.4-fold decrease in intracellular free fatty acids	(Takemura et al., 2019)
Synechocystis sp. PCC 6803	Co-overexpression of native plsX and plsC(fatty acid/phospholipid synthesis)	Increase in total lipid by 2 fold than WT and 1.2 fold than single overexpression	(Towijit et al., 2018)
Phaeodactylum tricornutum	Inactivation of the gene encoding a Hotdog-fold thioesterase involved in acyl-CoA hydrolysis (ptTES1)	1.7-fold increase in TAG content	(Hao et al., 2018)
Cyanidioschyzon merolae	Heterologous expression of a cyanobacterial acyl-ACP reductase	TAG content increased by 3.0 times	(Sumiya et al., 2015)
Chlamydomonas reinhardtii	Overexpression of Fructose-1,6-bisphosphate aldolases (FBAs)	34% increase in total fatty acids content	(Lee et al., 2020)
Tetraselmis chui	Heterologous expression of DGAT genes	TAG increases from 40% up to 115% of dry weight	(Úbeda-Mínguez et al., 2017)
Chlorella ellipsoidea	Overexpression of transcription factor *LEAFY COTYLEDON1 (LEC1) of Arabidopsis thaliana*	total fatty acid content and total lipid increased by 24.20–32.65 and 22.14–29.91% respectively	(Liu et al., 2021)
Neochloris oleoabundans	Co-overexpression of endogenous plastidial lysophosphatidic acid acyltransferase (NeoLPAAT1) and endoplasmic reticulum-located diacylglycerol acyltransferase 2 (NeoDGAT2)	Neutral lipid content increased by 2-fold	(Chungjatupornchai & Fa-aroonsawat, 2021)
Chlamydomonas reinhardtii	Overexpression of lysophosphatidic acyltransferase gene (*c-part*) and glycerol-3-phosphate dehydrogenase gene (*c-gpd1*)	increase of 44.5 and 67.5% lipid content observed after triple heat shock	(Wang et al., 2018)
Chlamydomonas reinhardtii	Overexpression of acetyl-CoA synthetase (*ACS2*)	TAG content increases by 2.4-fold in nitrogen-free medium	(Rengel et al., 2018)
Chlorella vulgaris	Overexpression of DNA binding with one finger (DOF)-type transcription factor	1.5-fold higher neutral lipid content under nitrogen-deficient conditions	(Tokunaga et al., 2019)
Nannochloropsis oceanica	overexpression of a nuclear-encoded, CBBC- homologous, candidate RuBisCO activase	The growth rate increased by ~ 32%, biomass accumulation by ~ 46%, lipid productivity by ~ 41%, and photosynthesis by 28%	(Wei et al., 2017)
Nannochloropsis salina	Overexpression of bZIP transcription factor	better growth rate, enhance biomass productivity with lipid increase under the high salt concentration	(Kwon et al., 2018)
Phaeodactylum tricornutum	Overexpression of an endogenous type 2 diacylglycerol acyltransferase	2-fold and 3.5-fold increase in TAG content under nitrogen replete and deplete conditions respectively	(Haslam et al., 2020)
Neochloris oleoabundans	Overexpression of endogenous lysophosphatidic acid acyltransferase (NeoLPAAT1)	total lipid content increased 1.8- to 1.9-fold and TAG content increased 2.1- to 2.2-fold	(Chungjatupornchai et al., 2019)

answered using LCA and related approaches, which provide insight into trade offs of impacts such as water consumption, energy, carbon, material use, and recycled content, as well as social and economic implications. In the case of the circular economy of algal biorefineries, where the system produces one major product and some secondary co-products. Along with the raw materials and products, emphasis is given to waste recovery and management, the ultimate goal being zero waste to make it more environmentally sustainable. The analysis of all the processes is made considering the involvement of algal biorefinery in the circular economy (Ubando et al. 2020).

Goal and Scope Definition

The LCA is used in a variety of contexts, including the industrial sector, community organizations, and certifying entities. In the case of private companies, the objectives can be defined as obtaining eco-labels and certifications, business marketing, legal compliance, scenario comparison, materials, and product costs. The general environmental goal of the analysis is to minimize the entire GWP associated with the speculative commercial-scale algal biofuel/value-added product production that uses different types of technologies. In this section, the purpose of the analysis needs to be confirmed and decisions on the specifics of the production system should be made. A well-established goal in turn helps to determine the study's nature and limits. How the limits of research are drawn specify the start and end of the life cycle analysis of the particular process and identify the operations included within the production system. The functional unit defines the product or process to be analyzed. It represents the system outcomes at the production unit level. It can be further shaped by study goals (Curran, 2017). The goal and scope of the study help to evaluate the possibilities for each step and design parameters which collectively indicate the most potentially sustainable commercial-scale algal production system for biofuel and value-added products.

In the case of the algae-based production system, the goal and scope can be different cultivation systems like open ponds and LED/solar-lit photobioreactors, downstream processing like hydrothermal liquefaction (HTL), anaerobic digestion (AD), nutrient recycling process with a specific functional unit (Bello et al., 2017; Stephenson et al., 2010). For example, in the case of algal biodiesel production, the system starts with microalgae cultivation, harvesting by flocculation or centrifugation and dewatering, lipid extraction, or direct conversion into biodiesel by HTL or AD, and byproduct management. These all steps have their own goal and scope and it needs to evaluate LCA for each step to develop a sustainable algal biorefinery (Brentner et al., 2011). For algal biofuel production, LCA goals should be accurately defined at the very commencement of the study. Defining the goal is the first step of LCA, whether the LCA analysis is (i) limited to a particular of algal fuels or (ii) is a comparative analysis of algal and other fuels. The following six aspects, as elaborated by Bjørn et al. (2018) and Albertı et al. (2019), should be addressed during the goal definition in the LCA study of algal fuel production systems:

1. Prospect use(s) of the outcomes/deliverables;
2. Limitations imposed because of the methodology, assumptions, and impact coverage;
3. Motivations behind the survey and decision-making
4. Intended audience of the outcomes/deliverables;
5. Comparative surveys to be revealed to the public; and
6. Commissioner of the study and other influential actors.

Table 3. Types of functional units used in the LCA study of different systems (Adapted from (Carneiro et al., 2017; Hosseinzadeh-Bandbafha et al., 2020).

Functional unit types	LCA system
Input-based unit:	This FU is determined based on a unit of input biomass like energy and mass.
Output-based unit:	This FU is determined on the basis of a unit of output like useful energy generated or a kilometer of transportation
Agricultural land-based unit	This FU exhibits the agricultural land area being cultivated to yield a given amount of algal biomass feedstock, e.g., one ha algal cultivation.
Year-based unit	This FU reports the final results on the basis of the annual average.

The objective for the LCA study should be clearly defined in the goal definition, it should be identified as closely as possible. For example, a specific algal species, a processing step, a policy option, etc., and their function in the LCA platform under investigation should be specified.

Functional Unit (FU) and Reference Flow

Functional unit refers to the exact attribute which is being studied. It is a quantified description of the product performance (Table 3). The reference flow is the measurement of the needed inputs from a process in a system that is required for a particular production system to achieve the efficiency described in the terms of the functional unit. The FU(s) of the study and/or the reference flow(s) should be determined quantitatively and qualitatively with revision according to the goal of the study. In the published literature, there is no agreement on the selection of FUs, making comprehensive analysis much more challenging. In LCA analyses of algal fuel production systems, four different classes of FUS can be identified, and the interpretation of final results is greatly influenced by the FU selected. (Carneiro et al., 2017). Therefore, it is advisable to show the results using several FUs so that the system under investigation has comparability with those reported in the literature.

Defining System Boundary

The system boundaries refer to the list of processes to be included in the LCA study. The system boundary definition is a subjective choice based on the scope of the study conducted. This initial setting should be substantially revised later, if necessary (Wolf et al., 2012). Defining boundaries permits a comparative assessment of LCA studies of competing systems. According to Jose and Archanaa (2017), a system boundary or control surface should be defined in order to determine what unit processes/operations should be considered in an LCA. For example, in algae cultivation, concerns related to harvesting, dehydration, handling, and biofuels production by different methods need to be included in the scope of the study. Algal fuels such as bioethanol, biodiesel, biomass, and oil can be investigated through three different approaches as follows (Wolf et al., 2012; Hosseinzadeh-Bandbafha et al., 2020).

1. **Well to Gate (WTG):** It includes algae biomass production, collection, pretreatment, and biofuels synthesis;

2. **Well to Tank or Pump (WtT):** It includes algae biomass production, collection, pretreatment, biofuels synthesis, and transferring the synthesized biofuels to a fuel distribution station; and
3. **Well to Wheel (WtW):** that includes algae production, collection, pretreatment, biofuels synthesis, transferring the synthesized biofuels to a fuel distribution station, and use in an engine or other power generation equipment.

The choice of system boundary primarily depends on the goal definition of the LCA. The system boundary takes into account unit processes & datasets as well as product & waste streams. This means that all products that enter (e.g., algae, material, and energy to biofuel production step) and waste or product that leaves (e.g., biofuel, by-product, and emissions from biofuel production step) the process should appear in its inventory. Moreover, all the elementary streams leaving or entering the process into the ecosphere directly or indirectly, along with all the streams crossing the system boundary should be taken into consideration (European Commission, 2010). The system boundary should be schematically represented in the life-cycle stages of the process revealing the included and excluded parts.

Defining Cut-Off

The cut-off criteria allow the LCA study to be conducted without having to model 100% of the process. The quantitative cut-off which has been targeted should be defined clearly, except for the cases where these have been identified in the goal definition previously (European Commission, 2010). This primary aim is amenable to revision later if the study is aimed at comparing different processes, or if the intended integrity cannot be achieved because of the unavailability of databases or lack of resources. The latter is applicable in rare conditions where the overall goal of the study cannot be attained and requires revisions. It is noteworthy that in algal fuel production systems, a 5% cut-off can be considered for impact categories to be included according to the ILCD handbook (Bradley et al., 2015).

Life Cycle Inventory (LCI) Analysis

A life cycle inventory (LCI) has an important role as a basis for life cycle impact assessment and economic analysis. It Includes all the input/output data of the hypothetical algal biorefinery/system as well as the additional materials and energies utilized in the process. The inventory analysis needs to be made for individual processes like infrastructure for cultivation, nutrients, power requirement, GHG emissions, water requirement, and land requirements. The LCI datasets used and the method for allocating co-products are both important parameters.

The allocation mechanism divides the upstream environmental loads out of all of the multi-output processes & co-products. When co-products are generated with the main product(s) in a certain process, it is difficult to distinguish between environmental impacts. Thus the allocation is used in the LCA to estimate the number of individual products (e.g. biodiesel) and co-products (e.g. glycerol) on the overall environmental impacts of the process. Assessment of various theoretical and practical characteristics associated with the selected allocation method is a critical step in decision-making as the choice of allocation influences the energy and environmental impacts. The allocation criteria are classified into four major classes as follows (Carneiro et al. 2017; Hosseinzadeh-Bandbafha et al., 2020):

1. **Null allocation:** describes that the main product is proportional to the entire energy requirements, whereas the co-products are referred to as the null energy requirements.
2. **Economic allocation or market value:** the economic value of each product determines the energy requirements given.
3. **Energy content or mass allocation:** the energy requirements are distributed according to the number of physical properties of the products such as mass, calorific value, etc.
4. **System substitution method:** This means that the co-products are excluded from the system and attention is kept on the main product only.

Life Cycle Impact Assessment (LCIA)

In order to compile the final results for the comparative LCA studies focused on the type of algal fuel produced, an impact assessment must be undertaken as well. The LCIA evaluates the magnitude and significance of the potential environmental impacts of the product/service by analyzing LCI results. LCIA study has to follow certain substeps (Mu et al., 2020);

a. Selection of the Impact category for analysis-as is mentioned in **Table 4**.
b. Category-based classification of LCI results
c. Characterization of potential impact indicators

In the case of the LCIA study of algal biorefinery, environmental impact categories can be assigned based on project requirements. In many studies, only major environmental impacts are covered and other impacts are ignored. Considering the goal of replacing fossil fuels with algal biofuel, it is critical to investigate whether the production of algae biofuels is environmentally and economically feasible to the production of petroleum-based fuels. It is also required to determine the GHG emissions in an attempt to prove the superiority of algae biofuel over traditional fossil fuels. Along with GHG emissions of algal biofuel, the other environmental impacts like eutrophication (N), ozone depletion (O3), acid rain (SO2), respiration (PM2.5), water use, land use, etc. need to be studied.

Normalization and weighting is an elective step and is used to characterize the LCIA results. In the case of normalization, the LCIA results can be multiplied by normalization factors representing the overall inventory of a reference. For weighting, the normalized LCIA results of algal fuels may be multiplied by a set of weighting factors. Then, the weighted LCIA results could be summed up to obtain a single value for all impact categories in LCA (Hosseinzadeh-Bandbafha et al., 2020).

There are two common purposes for normalizing/ weighting the results of LCIA for algal fuels, (Hauschild and Bjørn, 2018): (a) Application in supporting interpretation (b) Application in cutting-off quantification. A common reference should be used as the basis of the normalization of results of different impact categories. These Normalized results may later be weighted across different indicators. Also, the normalized results of algal fuels should not be summed up directly across different impact categories as this would suggest an even weighting of all impact categories (European Commission, 2010). For weighted results of algal fuels, these can be summed up across the impact categories or areas of protection, but cannot be used in comparative studies for public disclosure (Hauschild and Bjørn, 2018).

Table 4. The categories considered while assessing the environmental impact analysis in LCA studies (adapted from Mu et al., 2020).

Sr. No	Major Categories	Impacts
1	Ecosystem Impacts	1. Climate Change 2. Acid Rain 3. Eutrophication 4. Land Use Change 5. Solid Waste 6. Toxicity
2	Human Impacts	1. Ozone Depletion 2. Smog 3. Particulate Matter 4. Carcinogens 5. Toxicity
3	Resource Depletion	1. Fossil Fuel 2. Freshwater 3. Soil 4. Forest 5. Grassland 6. Minerals

Interpretation and Sensitivity Analysis

Lastly, interpretation is the final stage/phase of LCA. In this fourth phase, all the conclusions and recommendations must be drawn from the LCIA results (European Commission, 2010). The following guidelines/ purposes should be considered for interpreting algal fuels:

1. Analyse and Identify the significant bottlenecks in algal fuels production, which comprise key processes, parameters, assumptions, and elementary flows.
2. Evaluate the sensitivity/ limitation of these issues and their impact on the overall LCA results in the production of algal fuel (Davis et al., 2017). This will help to evaluate the completeness and consistency of the issues in the LCI/LCA during the production of algal fuel.
3. Provide recommendations and conclusions from the results of LCA of the production of the algal fuel. Providing additional interpretation in comparative analysis of two or more algal fuel production systems should be helpful in the development of a sustainable process (Khasreen et al., 2009).

The sensitivity analysis documented in the LCA study is seriously affected by the variability of the input and output data that are investigated as significant (or not) to the results of the study. Few guidelines should be taken into account for the analysis (Khasreen et al., 2009; Hosseinzadeh-Bandbafha et al., 2020):

1. Test the data used for algal fuels LCA;
2. Identify and test the important/key parameters which are affected by the energy of an element or LC of algae fuel e.g., maintenance and replacement rates;
3. Determines which materials are sensitive to location and the importance of life cycle embodied energy compared to operational energy for an algae fuel.

Materials and Tools Required for Performing the LCA of Algal Fuels

Following are the main software platforms used to Conduct the LCA of algal fuels (Hosseinzadeh-Bandbafha et al., 2020):

1. **SimaPro LCA Software**: It provides a professional framework for compiling, assessing, and tracking the impacts of algal fuel projects on the environment. This software platform enables systematic & transparent modeling as well as analysis of complex life cycles of algal fuel production chains in line with the recommendation of ISO 14040 and ISO 14044 series.
2. **GaBi Software:** This platform is available in various versions for educational to professional purposes. The software measures and evaluates the environmental, cost, and social aspects of algal fuels during their life cycles. This platform also offers EcoInvent data and GaBi databases that provide thorough worldwide coverage.
3. **OpenLCA Software:** This is a free open-source program that is used to model and evaluate the LCA of algae fuel systems. It comes with different import and export options, and different in-built modules to choose from. A basic LCA computation as well as two plugins, a format converter, and an uncertainty module, may be implemented and performed on this platform. The format converter can convert relevant LCA data formats to the other in a lossless manner, while the uncertainty module can diagnose, measure, display, and interpret uncertainty in LCA algal fuel systems.

LCA STUDIES CONDUCTED ON BIOFUELS IN RECENT YEARS

Based on the environmental impact like greenhouse gas emissions (GHG), eutrophication potential, and energy utilization, a life cycle analysis was conducted to evaluate various species/growth circumstances, taking into account each biomass fraction and ultimate product substitution. Two freshwater microalgal species *Neochloris oleoabundans* (Neo+) and *Chlorella sorokiniana* (Chl+) and two marine microalgal species *Nannochloropsis oculata* (Nan+) and *Tetraselmis suecica* (Tet+) were cultivated under nitrogen replete and deplete conditions for their growth analysis. The controlled growth analysis reveals distinct variances among the algae species studied, which are reflected in the LCA assessment for environmental impacts. Even though all of the growth conditions produce biodiesel with positive net GHG emissions, the N-replete freshwater scenarios Chl+ and Neo+ have the lowest net GHG emissions, with 2.4 and 0.5 kg CO_2 per kg of biodiesel, respectively. Although N-replete conditions had lower lipid contents than their equivalent N-depleted conditions, among all scenarios Chl+ showed the highest volumetric lipid productivity of 550 mg/L (Soh et al., 2014). The study done by Montazeri et al., also suggests the 0.5 kg CO_2 per kg of biodiesel with overall energy demand (1.4 MJ per Kg of biodiesel), eutrophication factors (22 g N eq per Kg of biodiesel), when Chl+ grown in N-replete conditions (Montazeri et al., 2016).

Through multi-objective optimization, Bello et al. presented a case study of a hypothetical integrated biofuel system that uses AD and HTL nutrient recycling pathways. Three distinct scenarios were evaluated, each one corresponding to a particular stage of future maturity. The findings demonstrate that AD or HTL nutrient recycling technology integration into the algal biofuel production system helps to lower the cost and environmental impact of algal biofuels. Additionally, it is discovered that the AD technology performs significantly better than the HTL process in terms of displacing excess fertilizer demand, energy recovery, GHG emission reduction, and maturity. However, HTL is a novel and viable nutrient recycling technology that exhibits cost advantages over AD. Additional pilot and demonstration-scale research are required before endorsing HTL or AD as viable nutrient recycling routes (Bello et al., 2017).

Stephenson et al explored the algal cultivation in traditional raceways and air-lift tube bioreactors along with different downstream processing methods. According to their study findings, if the future benchmark for algal lipid productivity, such as *C. vulgaris*, of 40 tons /ha/ year, cultivation in normal raceways would have much more environmental sustainability than closed air-lift tubular bioreactors. While biodiesel produced from the same cultivation strategy has 80 percent lower GWP than fossil-derived fuel (based on net energy content), and much higher for air-lift tube bioreactors (Stephenson et al., 2010).

The Combinatorial Life Cycle Assessment to Inform Process Design of Industrial Production of Algal Biodiesel was studied by Brentner et al. They divided the system into five independent process steps: (1) microalgae cultivation system, (2) harvesting and/or dewatering, (3) lipid extraction, (4) conversion (transesterification) into biodiesel, and (5) byproduct management. For each process stage, a number of technology possibilities were investigated, and potential technology combinations were evaluated for their environmental sustainability by LCA. The results indicate that the flat panel photobioreactors are best for the cultivation of microalgae, as it has the lowest environmental impact. The overall energy requirement for the flat panel PBR was >30 times, 10 times, and 20% lower than the tubular PBR, the annular PBR, and the open raceway pond respectively. While the flocculation method of harvesting and supercritical methanol for direct transesterification were found to be good for the algae-based biorefinery system (Brentner et al., 2011).

The production and testing of algae-based bio-jet fuels have increased in recent times. Fortier et al studied an LCA for a 1 GJ functional unit of bio-jet fuel produced using hydrothermal liquefaction (HTL) of wastewater-grown microalgae. A comparative study of HTL location at a Wastewater Treatment Plant (WWTP) or the refinery has been performed. Base cases were built in part for each approach utilizing primary algal production data in wastewater effluent and HTL studies at Kansas University. In these circumstances, the LC-GHG emissions were compared to conventional jet fuel along with sensitivity analysis and an analysis of Monte Carlo. The results indicate the HTL at a WWTP (35.2 kg CO_2 eq/GJ for the base case) performs better than HTL at a refinery (86.5 kg CO_2 eq/GJ for the base case) concerning LC-GHG emissions. The results from LCA were especially sensitive to the level of heat integration, the HTL heat source, and the dewatered algae's solid content. In comparison with conventional jet fuel, GHG emissions from algae-based bio-jet fuels can be minimized by 76%, and these sensitive parameters can be improved and HTL location at the WWTP (Fortier et al., 2014). While Orfield et al performed aquatic waste products hydrothermal gasification to the on-site energy recovery and *Escherichia coli* cultivation on waste products and biomass recycling back through the reactor for increased oil yields via regrowth pathway. In the case of catalytic hydrothermal gasification, it was reported that the highest net energy ratio of 1.9 and the lowest GWP of 1.0 kg CO_2/L-oil1 along with the minimum reaction temperature and reaction times, 250 °C and ~1 h respectively. On the other hand, optimal economical results were observed at a maximum temperature of 400 °C and reaction time of 5 min with the cost of USD 1.64 L-oil−1 (or USD 263 bbl^{-1}) of algal oil (Orfield et al., 2014). Whereas, Bennion et al, studied LCA of HTL and Pyrolysis conversion systems for algal biofuel production. By keeping ''well to pump'' (WTP) as a system boundary and based on the metrics of net energy ratio (NER), which is defined as the ratio of energy consumed to energy generated along with GHG emissions, the environmental impact of the above two systems were quantified. The NER for HTL and Pyrolysis was found to be 1.23 and 2.27 for WTP biofuel production, respectively. The reason for the high NER of industrial-scale pyrolysis systems is more energy consumption requirement in the drying process, which accounts for nearly half (0.97) of the overall NER. The biofuel production using HTL (-11.4 g CO_2 per MJ renewable diesel) has lower GHG emissions than through the pyrolysis pathway (210 g CO_2 per MJ renewable diesel) (Bennion et al., 2015).

Barlow et al., (2016) performed a techno-economic and life-cycle assessment of biodiesel production by HTL of biomass cultivated using a rotating algal biofilm reactor and validated the biorefinery model based on pilot-scale growth analysis and laboratory-scale HTL experiments. The techno-economic study revealed that the cost of biomass feedstock influenced the minimum selling price of the fuel (MFSP), with a base case of USD104.31 per gallon. Based on a WTP system boundary, life-cycle assessment findings suggest a base case GWP of 80 g CO_2-e MJ1 and a NER of 1.65. Algal productivity had the greatest impact on fuel selling price, according to sensitivity analysis, underscoring the necessity of improving biomass production. The MFSP, GWP, and NER of the system are reduced to USD11.90 Gal1, 44 g CO_2-e MJ1, and 0.33, respectively, after optimization. The environmental effect of combining algae farming with wastewater treatment was shown to be greatly reduced at the system level (Barlow et al., 2016)

CONTRIBUTION OF ALGAE AND ALGAL PRODUCTS IN THE ECONOMY

There are a number of economically profitable applications for algae. Besides food, it is used as feed for aquaculture, as fertilizer, as a soil binding agent, as well as an alkalinity reclaimer, and is incorporated into many commercial products. Algae are also found in a variety of household products, including toothpaste and pharmaceutical products. The high demand for plant-derived proteins has accelerated market growth.

Global Microalgae Market Scope and Market Size

The microalgae market is divided into food & beverage, dietary supplements, pharmaceuticals, cosmetics & personal care, biofuel, inks, animal feed, and others, based on the application. As infectious diseases continue to rise, demand for plant-based dietary supplements is also seen to be rising. Hence, the dietary supplements segment is expected to dominate the market in 2021. (databridgemarketresearch.com, 2021). The market is divided based on the type of microalgae, their species, their application, grades, modes, and processes. The growth of segments allows you to analyze niche pockets of growth and market-entry strategies, as well as determine your core application areas and target market differences. The *Spirulina* segment is expected to dominate the market in 2021, on account of its health benefits, including anti-inflammatory, anti-oxidant, and brain-protective functions. It is also the most widely cultivated microalgae. The microalgae market is divided into other segments such as *Dunaliella salina*, *Spirulina*, *Chlorella*, and based on product type. As *Haematococcus* species produces astaxanthin as a protective substance exploited for its high antioxidant properties, it is expected to dominate the market in 2021. For antioxidants, the other microalgae such as *Phaeodactylum tricornutum*, *Porphyridium cruentum*, *Nannochloropsis* spp. are also available options in the current market Organic nutrients are easily accessible to microalgae, and most people have been reported to prefer microalgae derived products grown on organic sources, so the organic segment is expected to dominate the market by 2021. The first report of commercial mass cultivation of microalgae was in Japan in the 1960s, where *Chlorella* was used. Now algae culturing has expanded to encompass food and feed, biofuels, and biopharmaceuticals. Natural products are in high demand for use in cosmetics and medicinal products and are being extracted from microalgae. According to one estimate, about 5000 metric tons of dry algal biomass processed for bio-products generates USD 1.25 × 109 each year (Iwamoto, 2004).

The Global Microalgae Market is expanding as a result of technological and scientific advancements. The growing interest in different industry sectors will stimulate the microalgae's market development. The global algae market was valued at USD 717.14 million in 2018 and is projected to reach USD 1365.8 million by 2027, at a compound annual growth rate (CAGR) of 5.35% between 2019 and 2027 (Boukid & Castellari, 2021). The global demand for algae biofuels, estimated at USD 6.8 billion in 2020, is expected to rise to USD 11.4 billion by 2027, rising at a CAGR of 7.5 percent between 2020- and 2027. The Algae Biofuels market in the U.S. is estimated at USD 1.8 Billion in the year 2020. China, the world's second-largest economy, is expected to reach a projected market size of USD 2.5 billion by 2027, with a CAGR of 11.4 percent from 2020 to 2027. Among the other notable geographies, Japan is expected to grow at 4.1% and Canada by 6.7% between 2020-2027, while Germany is expected to grow at a CAGR of around 4.9 percent in Europe (Algae Biofuels - Global Market Trajectory & Analytics, 2021).

Agar-Agar

Agar made from seaweed (macroalgae) is a galactose-based heterogeneous polysaccharide. Traditional sources of polysaccharides include red algae from the genera *Gracilaria*, *Ahnfeltia, Gelidium*, and *Pterocladiella*. Because of its ability to jellify, agar is widely used in many industries. There are unique properties of agar due to the differences in the jellification temperature and melting temperature. Roughly 90% of the agar is used for foodstuffs, and only 10% for industrial purposes (Titlyanov et al., 2017). It is composed of agarose and agaropectin polymers that are heterogeneous in nature. Agar is typically made up of 70% agarose and 30% agaropectin. Agarose is a polysaccharide made up of repeated galactose disaccharides that is linear in nature (no branch points). Agaropectin has a more diverse composition, containing D and L isomers of galactose with sulfate and pyruvate substituents, giving the polymer a strong negative charge. Agar is best known as a growth medium for microbial organism identification and enumeration. This global agar market is expected to expand at a compound annual growth rate of 3.9% over the next four years, from 12500 million USD to 17700 million USD, according to the forecast published figures by the 360 research reports. Whereas the data given by the KSU Sentinel newspaper it will reach from 302.9 to 399.8 million USD by the end of 2026, with a CAGR of 4.0% during 2021-2026. The overall reports indicate that the global agar market has good exponential growth in the near future.

Carotenoids

1. **Astaxanthin:** In recent years, natural astaxanthin has been widely used in human healthcare products, with anti-inflammatory, anti-oxidative, anti-atherosclerotic, and anti-aging effects (Lu et al., 2021). Astaxanthin is used in the nutraceutical, pharmaceutical, and food processing industries. It is also used to produce salmonids (salmon and trout), Red Marine, and Tai coloring additives as animal feed additives. Astaxanthin is also used to protect against skin aging in cosmeceutical applications. Astaxanthin (3.3¢-dihydroxy-β, β-carotene -4.4¢-dione), ketocarotenoid with high antioxidant activity, was found in some microalgae like *Chlorella zofingiensis*, *Haematococcus pluvialis*, *Chlorella sorokiniana*, *Tetraselmis sp*., *Chlorella protothecoides,* and *Scenedesmus sp*., etc. The astaxanthin market is divided into synthetic and natural categories, depending on the source.

Animal feed is commonly supplemented with synthetic astaxanthin. Due to the low production cost, the synthetic astaxanthin has captured 95% of the astaxanthin market while the US Food and Drug Administration (FDA) has not approved synthetic astaxanthin for human consumption due to concerns about food safety and the limited bioavailability of chemically manufactured supplements. While the natural astaxanthin can be consumed directly as it doesn't have any side effects on human consumption. By 2027, the global astaxanthin market is expected to reach 3.4 billion USD, growing at a CAGR of 16.2% due to its high commercial demand in various industries.

2. **β-carotene:** The approximate market value of ß-carotene produced by *Dunaliella bardawil* was about 0.6 USD per 1,000 mg. Further, ß-carotene is important for retinal synthesis in humans, necessary for the rhodopsin generation (Koller et al., 2014). The commercial ß-carotene production was done by cultivating the marine alga *Dunaliella salina*, which contains approx 14% of ß-carotene (Koller et al., 2014). Rising demand for naturally derived goods in the food, beverage and dietary supplement industries is projected to be a major driver in the global market in the future. The market value of β-carotene was calculated as USD 314.14 billion in 2020, with an annual growth rate of 3.9 percent and it is predicted to reach USD 380.37 billion by 2025 (Papadaki & Mantzouridou, 2021).
3. **Lutein:** Lutein is a lipid-soluble carotenoid that humans get from their diet. It helps to prevent macular degeneration, lowering the risk of stroke & heart attack, and reducing the risk of other debilitating metabolic disorders. Several microalgae species have been discovered as potential lutein producers in both marine and freshwater habitats like *Scenedesmus obliquus*, *Chlamydomonas* sp. JSC4, *Tetraselmis* sp. CTP4, *Chlorella protothecoides*, *Auxenochlorella protothecoides* SAG 211-7a, etc. Over the forecast period of 2020–2027, the global lutein market is estimated to reach EUR 409 million, growing at a compound annual growth rate (CAGR) of 6.10 percent (Saha et al., 2020). *Scenedesmus almeriensis* is known to produce up to 4.5 mg/g (of dry weight) of lutein, in outdoor culture conditions (Del Campo et al., 2007); and its content can be increased up to 5.4 mg/g (dry weight) by manipulating light intensity and temperature (Sánchez et al., 2008). The cost of lutein produced from *S. almeriensis* was approximately 2.5 USD per 1,000 mg of lutein (Bhalamurugan et al., 2018) The market value of phycocyanin alone reached between 5-10 million US$ (Odjadjare et al., 2017).

PUFA

As we know, long unsaturated hydrocarbons with more than one double bond are known as PUFAs, and they have a wide range of applications in the feed and nutraceutical industries. Long-chain PUFAs such -as linoleic acid (18:3), arachidonic acid (20:4), eicosapentaenoic acid (20:5), and docosahexaenoic acid (20:6) are abundant in microalgae Recently, the production costs related to EPA/DHA from microalgae reached to US$ 40/kg EPA + DHA, but advanced technologies could reduce this to ~USD 10/kg EPA + DHA, which is competitive if compared to fish oil (~USD 8/kg EPA + DHA) (Chauton et al., 2015). Docosahexaenoic acid (DHA) and eicosapentaenoic acid (EPA) are abundant omega-3 fatty acids with a market value of 80 to 160 USD kg1 and a global value of USD 898.7 million by 2025 (Kiran & Venkata Mohan, 2021).

CONCLUSION

Microalgae have been highlighted as the most promising and sustainable source for numerous metabolites and biofuels, because of their substantial benefits to human health and the environment. Globally, microalgae biofuels and metabolites have a wide spectrum of applications and are in great demand in different markets/ industries. Thus, there has been a gradual increase and continuous development in their technologies, application, and commercial utilization. However, the microalgal technology for the production and marketing of algae-based products is at its infancy and needs to be streamlined to make it more sustainable. This is possible by identifying the obstacles & limitations in the complete process, from origin to final market, which is correctly evaluated by its Life Cycle Analysis. LCA is a systematic and standardized tool designed to evaluate different resources in a certain system and their impact on the environment and public health. The LCA takes into account certain system boundaries, which define the scope of a particular study and determines its impact on the environment. There is multiple software available, with a predefined set of modules for ease of operation, to help conduct the LCA. There are certain unique products from microalgae that are currently available in the market. These products contribute to the economy largely because of their demands as nutraceuticals related to human health and well-being. The LCA also sheds light on the bottlenecks in the process and helps to determine areas where specific research efforts are needed, to make the production chain environmentally more sustainable.

REFERENCES

Abomohra, A. E. F., El-Naggar, A. H., & Baeshen, A. A. (2018). Potential of microalgae for biodiesel production: Screening and evaluation studies. *Journal of Bioscience and Bioengineering*, *125*(2), 231–237. doi:10.1016/j.jbiosc.2017.08.020 PMID:29037768

Adhikari, B., & Pellegrino, J. (2015). Life-cycle assessment of five microalgae-to-biofuels processes of varying complexity. *Journal of Renewable and Sustainable Energy*, *7*(4), 043136. doi:10.1063/1.4929462

Ahmadi, A., Zorofchian Moghadamtousi, S., Abubakar, S., & Zandi, K. (2015). Antiviral potential of algae polysaccharides isolated from marine sources: A review. *BioMed Research International*, *2015*, 2015. doi:10.1155/2015/825203 PMID:26484353

Albertí, J., Brodhag, C., & Fullana-i-Palmer, P. (2019). First steps in life cycle assessments of cities with a sustainability perspective: A proposal for goal, function, functional unit, and reference flow. *The Science of the Total Environment*, *646*, 1516–1527. doi:10.1016/j.scitotenv.2018.07.377 PMID:30235636

Barlow, J., Sims, R. C., & Quinn, J. C. (2016). Techno-economic and life-cycle assessment of an attached growth algal biorefinery. *Bioresource Technology*, *220*, 360–368. doi:10.1016/j.biortech.2016.08.091 PMID:27595701

Bello, M., Ranganathan, P., & Brennan, F. (2017). Life cycle optimization for sustainable algal biofuel production using integrated nutrient recycling technology. *ACS Sustainable Chemistry & Engineering*, *5*(11), 9869–9880. doi:10.1021/acssuschemeng.7b01833

Bennion, E. P., Ginosar, D. M., Moses, J., Agblevor, F., & Quinn, J. C. (2015). Lifecycle assessment of microalgae to biofuel: Comparison of thermochemical processing pathways. *Applied Energy*, *154*, 1062–1071. doi:10.1016/j.apenergy.2014.12.009

Bhalamurugan, G. L., Valerie, O., & Mark, L. (2018). Valuable bioproducts obtained from microalgal biomass and their commercial applications: A review. *Environmental Engineering Research*, *23*(3), 229–241. doi:10.4491/eer.2017.220

Bhattacharjee, M. (2016). Pharmaceutically valuable bioactive compounds of algae. *Asian Journal of Pharmaceutical and Clinical Research*, *9*(6), 43–47. doi:10.22159/ajpcr.2016.v9i6.14507

Bjørn, A., Laurent, A., Owsianiak, M., & Olsen, S. I. (2018). Goal definition. In *Life Cycle Assessment* (pp. 67–74). Springer. doi:10.1007/978-3-319-56475-3_7

Boukid, F., & Castellari, M. (2021). Food and beverages containing algae and derived ingredients launched in the market from 2015 to 2019: A front-of-pack labeling perspective with a special focus on Spain. *Foods*, *10*(1), 173. doi:10.3390/foods10010173 PMID:33467009

Bradley, T., Maga, D., & Antón, S. (2015). Unified approach to Life Cycle Assessment between three unique algae biofuel facilities. *Applied Energy*, *154*, 1052–1061. doi:10.1016/j.apenergy.2014.12.087

Brentner, L. B., Eckelman, M. J., & Zimmerman, J. B. (2011). Combinatorial life cycle assessment to inform process design of industrial production of algal biodiesel. *Environmental Science & Technology*, *45*(16), 7060–7067. doi:10.1021/es2006995 PMID:21662987

Carneiro, M. L. N., Pradelle, F., Braga, S. L., Gomes, M. S. P., Martins, A. R. F., Turkovics, F., & Pradelle, R. N. (2017). Potential of biofuels from algae: Comparison with fossil fuels, ethanol and biodiesel in Europe and Brazil through life cycle assessment (LCA). *Renewable & Sustainable Energy Reviews*, *73*, 632–653. doi:10.1016/j.rser.2017.01.152

Chang, K. J. L., Nichols, C. M., Blackburn, S. I., Dunstan, G. A., Koutoulis, A., & Nichols, P. D. (2014). Comparison of thraustochytrids Aurantiochytrium sp., Schizochytrium sp., Thraustochytrium sp., and Ulkenia sp. for production of biodiesel, long-chain omega-3 oils, and exopolysaccharide. *Marine Biotechnology (New York, N.Y.)*, *16*(4), 396–411. doi:10.100710126-014-9560-5 PMID:24463839

Chauton, M. S., Reitan, K. I., Norsker, N. H., Tveterås, R., & Kleivdal, H. T. (2015). A techno-economic analysis of industrial production of marine microalgae as a source of EPA and DHA-rich raw material for aquafeed: Research challenges and possibilities. *Aquaculture (Amsterdam, Netherlands)*, *436*, 95–103. doi:10.1016/j.aquaculture.2014.10.038

Chen, C. Y., Ho, S. H., Liu, C. C., & Chang, J. S. (2017). Enhancing lutein production with *Chlorella sorokiniana* Mb-1 by optimizing acetate and nitrate concentrations under mixotrophic growth. *Journal of the Taiwan Institute of Chemical Engineers*, *79*, 88–96. doi:10.1016/j.jtice.2017.04.020

Chomkhamsri, K., Wolf, M. A., & Pant, R. (2011). International reference life cycle data system (ILCD) handbook: Review schemes for life cycle assessment. *Towards Life Cycle Sustainability Management*, 107-117.

Chungjatupornchai, W., Areerat, K., & Fa-Aroonsawat, S. (2019). Increased triacylglycerol production in oleaginous microalga Neochloris oleoabundans by overexpression of plastidial lysophosphatidic acid acyltransferase. *Microbial Cell Factories*, *18*(1), 1–11. doi:10.118612934-019-1104-2 PMID:30866936

Chungjatupornchai, W., & Fa-Aroonsawat, S. (2021). Enhanced triacylglycerol production in oleaginous microalga *Neochloris oleoabundans* by co-overexpression of lipogenic genes: Plastidial LPAAT1 and ER-located DGAT2. *Journal of Bioscience and Bioengineering*, *131*(2), 124–130. doi:10.1016/j.jbiosc.2020.09.012 PMID:33069576

Cobos, M., Paredes, J. D., Maddox, J. D., Vargas-Arana, G., Flores, L., Aguilar, C. P., Marapara, J., & Castro, J. C. (2017). Isolation and characterization of native microalgae from the Peruvian Amazon with potential for biodiesel production. *Energies*, *10*(2), 224. doi:10.3390/en10020224

Curran, M. A. (2017). Overview of goal and scope definition in life cycle assessment. In *Goal and scope definition in life cycle assessment* (pp. 1–62). Springer. doi:10.1007/978-94-024-0855-3_1

Dahiya, D., Sharma, H., Rai, A. K., & Nigam, P. S. N. (2022). Application of biological systems and processes employing microbes and algae to Reduce, Recycle, Reuse (3Rs) for the sustainability of a circular bioeconomy. *AIMS Microbiology*, *8*(1), 83–102. doi:10.3934/microbiol.2022008 PMID:35496985

Davis, J., De Menna, F., Unger, N., Östergren, K., Loubiere, M., & Vittuari, M. (2017). *Generic strategy LCA and LCC: Guidance for LCA and LCC focused on prevention, valorisation and treatment of side flows from the food supply chain*. Academic Press.

De Lassio, J., França, J., Espirito Santo, K., & Haddad, A. (2016). Case study: LCA methodology applied to materials management in a Brazilian residential construction site. *Journal of Engineering (Stevenage, England)*.

Del Campo, J. A., García-González, M., & Guerrero, M. G. (2007). Outdoor cultivation of microalgae for carotenoid production: Current state and perspectives. *Applied Microbiology and Biotechnology*, *74*(6), 1163–1174. doi:10.100700253-007-0844-9 PMID:17277962

Dewi, I. C., Falaise, C., Hellio, C., Bourgougnon, N., & Mouget, J. L. (2018). Anticancer, antiviral, antibacterial, and antifungal properties in microalgae. In *Microalgae in health and disease prevention* (pp. 235–261). Academic Press. doi:10.1016/B978-0-12-811405-6.00012-8

El-Mohsnawy, E., El-Sheekh, M. M., Mabrouk, M. E. M., & Zohir, W. F. (2020). Enhancing accumulation of omega 3 and 9 fatty acids in Chlorella vulgaris under mixotrophic nutrition. *Journal of Animal and Plant Sciences*, *30*(2), 485–492.

El-Sheekh, M. M., El-Mohsnawy, E., Mabrouk, M. E., & Zoheir, W. F. (2020). Enhancement of biodiesel production from the green microalga Micractinium reisseri via optimization of cultivation regimes. *Journal of Taibah University for Science: JTUSCI*, *14*(1), 437–444. doi:10.1080/16583655.2020.1745505

Elaya Perumal, U., & Sundararaj, R. (2020). Algae: A potential source to prevent and cure the novel coronavirus–A review. *Int. J. Emerg. Technol*, *11*(2), 479–483.

Fabris, M., Abbriano, R. M., Pernice, M., Sutherland, D. L., Commault, A. S., Hall, C. C., Labeeuw, L., McCauley, J. I., Kuzhiuparambil, U., Ray, P., Kahlke, T., & Ralph, P. J. (2020). Emerging technologies in algal biotechnology: Toward the establishment of a sustainable, algae-based bioeconomy. *Frontiers in Plant Science*, *11*, 279. doi:10.3389/fpls.2020.00279 PMID:32256509

Fortier, M. O. P., Roberts, G. W., Stagg-Williams, S. M., & Sturm, B. S. (2014). Life cycle assessment of bio-jet fuel from hydrothermal liquefaction of microalgae. *Applied Energy*, *122*, 73–82. doi:10.1016/j.apenergy.2014.01.077

Galasso, C., Gentile, A., Orefice, I., Ianora, A., Bruno, A., Noonan, D. M., Sansone, C., Albini, A., & Brunet, C. (2019). Microalgal derivatives as potential nutraceutical and food supplements for human health: A focus on cancer prevention and interception. *Nutrients*, *11*(6), 1226. doi:10.3390/nu11061226 PMID:31146462

Gonçalves, A. L. (2021). The use of microalgae and cyanobacteria in the improvement of agricultural practices: A review on their bio fertilising, biostimulating and biopesticide roles. *Applied Sciences (Basel, Switzerland)*, *11*(2), 871. doi:10.3390/app11020871

Gong, M., & Bassi, A. (2016). Carotenoids from microalgae: A review of recent developments. *Biotechnology Advances*, *34*(8), 1396–1412. doi:10.1016/j.biotechadv.2016.10.005 PMID:27816618

Hamilton, M. L., Powers, S., Napier, J. A., & Sayanova, O. (2016). Heterotrophic production of omega-3 long-chain polyunsaturated fatty acids by trophically converted marine diatom Phaeodactylum tricornutum. *Marine Drugs*, *14*(3), 53. doi:10.3390/md14030053 PMID:27005636

Handbook, I. L. C. D. (2010). *International Reference Life Cycle Data System. General guide for Life Cycle Assessment. Detailed guidance. European Commission, Joint Research Centre*. Institute for Environment and Sustainability.

Hao, X., Luo, L., Jouhet, J., Rébeillé, F., Maréchal, E., Hu, H., Pan, Y., Tan, X., Chen, Z., You, L., Chen, H., Wei, F., & Gong, Y. (2018). Enhanced triacylglycerol production in the diatom Phaeodactylum tricornutum by inactivation of a Hotdog-fold thioesterase gene using TALEN-based targeted mutagenesis. *Biotechnology for Biofuels*, *11*(1), 1–18. doi:10.118613068-018-1309-3 PMID:30455741

Haslam, R. P., Hamilton, M. L., Economou, C. K., Smith, R., Hassall, K. L., Napier, J. A., & Sayanova, O. (2020). Overexpression of an endogenous type 2 diacylglycerol acyltransferase in the marine diatom Phaeodactylum tricornutum enhances lipid production and omega-3 long-chain polyunsaturated fatty acid content. *Biotechnology for Biofuels*, *13*(1), 1–17. doi:10.118613068-020-01726-8 PMID:32467729

Hauschild, M. Z., & Bjørn, A. (2018). LCA cookbook. In *Life Cycle Assessment* (pp. 963–1048). Springer. doi:10.1007/978-3-319-56475-3_37

Hildebrand, M., Abbriano, R. M., Polle, J. E., Traller, J. C., Trentacoste, E. M., Smith, S. R., & Davis, A. K. (2013). Metabolic and cellular organization in evolutionarily diverse microalgae as related to biofuels production. *Current Opinion in Chemical Biology*, *17*(3), 506–514. doi:10.1016/j.cbpa.2013.02.027 PMID:23538202

Hosseinzadeh-Bandbafha, H., Tabatabaei, M., Aghbashlo, M., Sulaiman, A., & Ghassemi, A. (2020). Life-Cycle Assessment (LCA) analysis of algal fuels. In *Biofuels from Algae* (pp. 121–151). Humana.

Hu, H., Li, J. Y., Pan, X. R., Zhang, F., Ma, L. L., Wang, H. J., & Zeng, R. J. (2019). Different DHA or EPA production responses to nutrient stress in the marine microalga Tisochrysis lutea and the freshwater microalga Monodus subterraneus. *The Science of the Total Environment*, *656*, 140–149. doi:10.1016/j.scitotenv.2018.11.346 PMID:30504016

Iwamoto, H. (2004). Industrial production of microalgal cell-mass and secondary products-major industrial species-Chlorella. In Handbook of Microalgal Culture: Biotechnology and Applied Phycology. Blackwell Science.

Jose, S., & Archanaa, S. (2017). Environmental and economic sustainability of algal lipid extractions: an essential approach for the commercialization of algal biofuels. In *Algal Biofuels* (pp. 281–313). Springer. doi:10.1007/978-3-319-51010-1_14

Kaparapu, J., & Geddada, M. N. R. (2020). *Commercial applications of algae in the field of biotechnology. In Applied Algal Biotechnology*. Nova Science Publishers, Inc.

Khan, M. I., Shin, J. H., & Kim, J. D. (2018). The promising future of microalgae: Current status, challenges, and optimization of a sustainable and renewable industry for biofuels, feed, and other products. *Microbial Cell Factories*, *17*(1), 1–21. doi:10.118612934-018-0879-x PMID:29506528

Khanra, S., Mondal, M., Halder, G., Tiwari, O. N., Gayen, K., & Bhowmick, T. K. (2018). Downstream processing of microalgae for pigments, protein and carbohydrate in industrial application: A review. *Food and Bioproducts Processing*, *110*, 60–84. doi:10.1016/j.fbp.2018.02.002

Khasreen, M. M., Banfill, P. F., & Menzies, G. F. (2009). Life-cycle assessment and the environmental impact of buildings: A review. *Sustainability*, *1*(3), 674–701. doi:10.3390u1030674

Kiran, B. R., & Venkata Mohan, S. (2021). Microalgal Cell Biofactory—Therapeutic, Nutraceutical and Functional Food Applications. *Plants*, *10*(5), 836. doi:10.3390/plants10050836 PMID:33919450

Koller, M., Muhr, A., & Braunegg, G. (2014). Microalgae as versatile cellular factories for valued products. *Algal Research*, *6*, 52–63. doi:10.1016/j.algal.2014.09.002

Kwon, S., Kang, N. K., Koh, H. G., Shin, S. E., Lee, B., Jeong, B. R., & Chang, Y. K. (2018). Enhancement of biomass and lipid productivity by overexpression of a bZIP transcription factor in Nannochloropsis salina. *Biotechnology and Bioengineering*, *115*(2), 331–340. doi:10.1002/bit.26465 PMID:28976541

Lee, B. S., Koo, K. M., Ryu, J., Hong, M. J., Kim, S. H., Kwon, S. J., Kim, J.-B., Choi, J., & Ahn, J. W. (2020). Overexpression of fructose-1, 6-bisphosphate aldolase 1 enhances accumulation of fatty acids in Chlamydomonas reinhardtii. *Algal Research*, *47*, 101825. doi:10.1016/j.algal.2020.101825

Liu, T., Liu, W. H., Zhao, J. S., Meng, F. Z., & Wang, H. (2017). Lutein protects against β-amyloid peptide-induced oxidative stress in cerebrovascular endothelial cells through modulation of Nrf-2 and NF-κb. *Cell Biology and Toxicology*, *33*(1), 57–67. doi:10.100710565-016-9360-y PMID:27878403

Liu, X., Zhang, D., Zhang, J., Chen, Y., Liu, X., Fan, C., Wang, R. R.-C., Hou, Y., & Hu, Z. (2021). Overexpression of the Transcription Factor AtLEC1 Significantly Improved the Lipid Content of Chlorella ellipsoidea. *Frontiers in Bioengineering and Biotechnology*, *9*, 113. doi:10.3389/fbioe.2021.626162 PMID:33681161

Lu, Q., Li, H., Zou, Y., Liu, H., & Yang, L. (2021). Astaxanthin as a microalgal metabolite for aquaculture: A review on the synthetic mechanisms, production techniques, and practical application. *Algal Research*, *54*, 102178. doi:10.1016/j.algal.2020.102178

Mader, J., Gallo, A., Schommartz, T., Handke, W., Nagel, C. H., Gunther, P., Brune, W., & Reich, K. (2016). Calcium spirulan derived from Spirulina platensis inhibits herpes simplex virus 1 attachment to human keratinocytes and protects against herpes labialis. *The Journal of Allergy and Clinical Immunology*, *137*(1), 197–203. doi:10.1016/j.jaci.2015.07.027 PMID:26341274

Markou, G., & Nerantzis, E. (2013). Microalgae for high-value compounds and biofuels production: A review with focus on cultivation under stress conditions. *Biotechnology Advances*, *31*(8), 1532–1542. doi:10.1016/j.biotechadv.2013.07.011 PMID:23928208

Montazeri, M., Soh, L., Pérez-López, P., Zimmerman, J. B., & Eckelman, M. J. (2016). Time-dependent life cycle assessment of microalgal biorefinery co-products. *Biofuels, Bioproducts & Biorefining*, *10*(4), 409–421. doi:10.1002/bbb.1649

Mosey, M., Douchi, D., Knoshaug, E. P., & Laurens, L. M. (2021). Methodological review of genetic engineering approaches for non-model algae. *Algal Research*, *54*, 102221. doi:10.1016/j.algal.2021.102221

Mu, D., Xin, C., & Zhou, W. (2020). Life cycle assessment and techno-economic analysis of algal biofuel production. *Microalgae Cultivation for Biofuels Production*, 281-292.

Mulumba, N., & Farag, I. H. (2012). Tubular photobioreactor for microalgae biodiesel production. *International Journal of Engineering Science and Technology*, *4*(2), 703–709.

Mutanda, T., Naidoo, D., Bwapwa, J. K., & Anandraj, A. (2020). Biotechnological Applications of Microalgal Oleaginous Compounds: Current Trends on Microalgal Bioprocessing of Products. *Frontiers in Energy Research*, *8*, 598803. doi:10.3389/fenrg.2020.598803

Novoveská, L., Ross, M. E., Stanley, M. S., Pradelles, R., Wasiolek, V., & Sassi, J. F. (2019). Microalgal carotenoids: A review of production, current markets, regulations, and future direction. *Marine Drugs*, *17*(11), 640. doi:10.3390/md17110640 PMID:31766228

Odjadjare, E. C., Mutanda, T., & Olaniran, A. O. (2017). Potential biotechnological application of microalgae: A critical review. *Critical Reviews in Biotechnology*, *37*(1), 37–52. doi:10.3109/07388551.2015.1108956 PMID:26594785

Orfield, N. D., Fang, A. J., Valdez, P. J., Nelson, M. C., Savage, P. E., Lin, X. N., & Keoleian, G. A. (2014). Life cycle design of an algal biorefinery featuring hydrothermal liquefaction: Effect of reaction conditions and an alternative pathway including microbial regrowth. *ACS Sustainable Chemistry & Engineering*, *2*(4), 867–874. doi:10.1021c4004983

Pagarete, A., Ramos, A. S., Puntervoll, P., Allen, M. J., & Verdelho, V. (2021). Antiviral Potential of Algal Metabolites—A Comprehensive Review. *Marine Drugs*, *19*(2), 94. doi:10.3390/md19020094 PMID:33562153

Papadaki, E., & Mantzouridou, F. T. (2021). Natural β-Carotene production by Blakeslea trispora cultivated in spanish-style green olive processing wastewaters. *Foods*, *10*(2), 327. doi:10.3390/foods10020327 PMID:33557096

Parmar, A., Singh, N. K., Kaushal, A., Sonawala, S., & Madamwar, D. (2011). Purification, characterization and comparison of phycoerythrins from three different marine cyanobacterial cultures. *Bioresource Technology*, *102*(2), 1795–1802. doi:10.1016/j.biortech.2010.09.025 PMID:20889334

Peng, Y., Xie, E., Zheng, K., Fredimoses, M., Yang, X., Zhou, X., Wang, Y., Yang, B., Lin, X., Liu, J., & Liu, Y. (2012). Nutritional and chemical composition and antiviral activity of cultivated seaweed Sargassum naozhouense Tseng et Lu. *Marine Drugs*, *11*(1), 20–32. doi:10.3390/md11010020 PMID:23271422

Ratnapuram, H. P., Vutukuru, S. S., & Yadavalli, R. (2018). Mixotrophic transition induced lipid productivity in Chlorella pyrenoidosa under stress conditions for biodiesel production. *Heliyon*, *4*(1), e00496. doi:10.1016/j.heliyon.2017.e00496 PMID:29322110

Rengel, R., Smith, R. T., Haslam, R. P., Sayanova, O., Vila, M., & Leon, R. (2018). Overexpression of acetyl-CoA synthetase (ACS) enhances the biosynthesis of neutral lipids and starch in the green microalga Chlamydomonas reinhardtii. *Algal Research*, *31*, 183–193. doi:10.1016/j.algal.2018.02.009

Rodrigues, J. A. G., Eloy, Y. R. G., Vanderlei, E. D. S. O., Cavalcante, J. F., Romanos, M. T. V., & Benevides, N. M. B. (2017). An anti-dengue and anti-herpetic polysulfated fraction isolated from the coenocytic green seaweed Caulerpa cupressoides inhibits thrombin generation in vitro. *Acta Scientiarum. Biological Sciences*, *39*(2), 149–159. doi:10.4025/actascibiolsci.v39i2.28081

Ronga, D., Biazzi, E., Parati, K., Carminati, D., Carminati, E., & Tava, A. (2019). Microalgal biostimulants and biofertilisers in crop productions. *Agronomy (Basel)*, *9*(4), 192. doi:10.3390/agronomy9040192

Saha, S. K., Ermis, H., & Murray, P. (2020). Marine microalgae for potential lutein production. *Applied Sciences (Basel, Switzerland)*, *10*(18), 6457. doi:10.3390/app10186457

Sánchez, J. F., Fernández-Sevilla, J. M., Acién, F. G., Cerón, M. C., Pérez-Parra, J., & Molina-Grima, E. (2008). Biomass and lutein productivity of Scenedesmus almeriensis: Influence of irradiance, dilution rate and temperature. *Applied Microbiology and Biotechnology*, *79*(5), 719–729. doi:10.100700253-008-1494-2 PMID:18491039

Sarayloo, E., Tardu, M., Unlu, Y. S., Simsek, S., Cevahir, G., Erkey, C., & Kavakli, I. H. (2017). Understanding lipid metabolism in high-lipid-producing Chlorella vulgaris mutants at the genome-wide level. *Algal Research*, *28*, 244–252. doi:10.1016/j.algal.2017.11.009

Soh, L., Montazeri, M., Haznedaroglu, B. Z., Kelly, C., Peccia, J., Eckelman, M. J., & Zimmerman, J. B. (2014). Evaluating microalgal integrated biorefinery schemes: Empirical controlled growth studies and life cycle assessment. *Bioresource Technology*, *151*, 19–27. doi:10.1016/j.biortech.2013.10.012 PMID:24189381

Sproles, A. E., Fields, F. J., Smalley, T. N., Le, C. H., Badary, A., & Mayfield, S. P. (2021). Recent advancements in the genetic engineering of microalgae. *Algal Research*, *53*, 102158. doi:10.1016/j.algal.2020.102158

Stephenson, A. L., Kazamia, E., Dennis, J. S., Howe, C. J., Scott, S. A., & Smith, A. G. (2010). Life-cycle assessment of potential algal biodiesel production in the United Kingdom: A comparison of raceways and air-lift tubular bioreactors. *Energy & Fuels*, *24*(7), 4062–4077. doi:10.1021/ef1003123

Suganya, T., Varman, M., Masjuki, H. H., & Renganathan, S. (2016). Macroalgae and microalgae as a potential source for commercial applications along with biofuels production: A biorefinery approach. *Renewable & Sustainable Energy Reviews*, *55*, 909–941. doi:10.1016/j.rser.2015.11.026

Sumiya, N., Kawase, Y., Hayakawa, J., Matsuda, M., Nakamura, M., Era, A., Tanaka, K., Kondo, A., Hasunuma, T., Imamura, S., & Miyagishima, S. Y. (2015). Expression of cyanobacterial acyl-ACP reductase elevates the triacylglycerol level in the red alga Cyanidioschyzon merolae. *Plant & Cell Physiology*, *56*(10), 1962–1980. doi:10.1093/pcp/pcv120 PMID:26272551

Surzycki, R., Greenham, K., Kitayama, K., Dibal, F., Wagner, R., Rochaix, J. D., Ajam, T., & Surzycki, S. (2009). Factors effecting expression of vaccines in microalgae. *Biologicals*, *37*(3), 133–138. doi:10.1016/j.biologicals.2009.02.005 PMID:19467445

Takemura, T., Imamura, S., & Tanaka, K. (2019). Identification of a chloroplast fatty acid exporter protein, CmFAX1, and triacylglycerol accumulation by its overexpression in the unicellular red alga Cyanidioschyzon merolae. *Algal Research*, *38*, 101396. doi:10.1016/j.algal.2018.101396

TalukdarJ.DasguptaS.NagleV.BhadraB. (2020). COVID-19: Potential of microalgae-derived natural astaxanthin as adjunctive supplement in alleviating cytokine storm. *Available at* SSRN 3579738. doi:10.2139/ssrn.3579738

Titlyanov, A. E., Titlyanova, V. T., Li, X., & Huang, H. (2016). *Coral reef marine plants of Hainan island*. Academic Press.

Tokunaga, S., Sanda, S., Uraguchi, Y., Nakagawa, S., & Sawayama, S. (2019). Overexpression of the DOF-type transcription factor enhances lipid synthesis in Chlorella vulgaris. *Applied Biochemistry and Biotechnology*, *189*(1), 116–128. doi:10.100712010-019-02990-7 PMID:30877635

Towijit, U., Songruk, N., Lindblad, P., Incharoensakdi, A., & Jantaro, S. (2018). Co-overexpression of native phospholipid-biosynthetic genes plsX and plsC enhances lipid production in Synechocystis sp. PCC 6803. *Scientific Reports*, *8*(1), 1–13. doi:10.103841598-018-31789-5 PMID:30201972

Ubando, A. T., Felix, C. B., & Chen, W. H. (2020). Biorefineries in circular bioeconomy: A comprehensive review. *Bioresource Technology*, *299*, 122585. doi:10.1016/j.biortech.2019.122585 PMID:31901305

Úbeda-Mínguez, P., García-Maroto, F., & Alonso, D. L. (2017). Heterologous expression of DGAT genes in the marine microalga Tetraselmis chui leads to an increase in TAG content. *Journal of Applied Phycology*, *29*(4), 1913–1926. doi:10.100710811-017-1103-9

Wang, C., Li, Y., Lu, J., Deng, X., Li, H., & Hu, Z. (2018). Effect of overexpression of LPAAT and GPD1 on lipid synthesis and composition in green microalga Chlamydomonas reinhardtii. *Journal of Applied Phycology*, *30*(3), 1711–1719. doi:10.100710811-017-1349-2 PMID:29899598

Wang, K., Xu, Z., & Liao, X. (2021). Bioactive compounds, health benefits, and functional food products of sea buckthorn: A review. *Critical Reviews in Food Science and Nutrition*, 1–22. doi:10.1080/10408398.2021.1905605 PMID:33783272

Wei, L., Wang, Q., Xin, Y., Lu, Y., & Xu, J. (2017). Enhancing photosynthetic biomass productivity of industrial oleaginous microalgae by overexpression of RuBisCO activase. *Algal Research*, *27*, 366–375. doi:10.1016/j.algal.2017.07.023

Wen, Z. Y., & Chen, F. (2003). Heterotrophic production of eicosapentaenoic acid by microalgae. *Biotechnology Advances*, *21*(4), 273–294. doi:10.1016/S0734-9750(03)00051-X PMID:14499126

Yan, N., Fan, C., Chen, Y., & Hu, Z. (2016). The potential for microalgae as bioreactors to produce pharmaceuticals. *International Journal of Molecular Sciences*, *17*(6), 962. doi:10.3390/ijms17060962 PMID:27322258

Chapter 11
Role of Algae in Agriculture

Muhammad Zia Ul Haq
https://orcid.org/0000-0001-9914-3367
Department of Agronomy, University of Agriculture, Faisalabad, Pakistan

Mohamed S. Abdel-Kareem
Alexandria University, Egypt

Iftikhar Ali
Department of Agronomy, University of Agriculture, Faisalabad, Pakistan

Alaa A. Fathy
Qassim University, Saudi Arabia

Abdul Khaliq
Department of Agronomy, University of Agriculture, Faisalabad, Pakistan

Eman T. El-Kenany
Alexandria University, Egypt

ABSTRACT

The sustainable development of modern agriculture faces many obstacles, including biodiversity loss and environment and soil degradation. Algae possess the ability to fix carbon through photosynthesis and produce enormous biomass. The potential use of algae in bio-fertilizers, nutrient recycling, crop stimulants against abiotic stresses, and bio-control agent against plant pests provides a way forward for sustainable agriculture development. This chapter summarizes the use of algae in agriculture ranging from bio-fertilizers to crop stimulants. It is expected that the integration of algae in inputs will transform modern agriculture into a more environmentally benign and resource-efficient system, hence making it more productive.

DOI: 10.4018/978-1-6684-2438-4.ch011

INTRODUCTION

Primary production from terrestrial plants is the basis of today's agriculture. Soil-based plants produce every calorie a human consumes except for seafood. Agriculture is the most significant direct and indirect cause of farmland degradation in various ways. Overcutting vegetation, unbalanced crop rotations, and overgrazing contribute to soil erosion. The overuse of chemical fertilizer to restore crop production resulted in a shift in soil nutrient ratios (Clercq et al., 2018). The traditional method of food production is currently experiencing significant changes. Using new agricultural technologies resulted in substantial improvements, thanks to modern farming methods such as modern irrigation techniques, fertilizers and pesticides, and crop varieties with high productivity (World Bank, 2008).

The United Nations 2030 Sustainable Development Agenda and the international community committed to ending hunger in 2015. Was the UN able to achieve this goal? Unfortunately, not even close. Starvation affects around 800 million people worldwide. And according to current projections, approximately 650 million people will still be undernourished by 2030 (Clercq et al., 2018). Extensive growth of the human population through the last decades resulted in an urgent need for the continuous production of food supply, primarily cultivated crops. This was achieved mainly through a highly productive and intensive agricultural system, using massive amounts of various chemical fertilizers (Schwencke & Carú, 2001). Meanwhile, the overuse of synthetic chemical fertilizers in the new agricultural systems has increased their soil salinity and inhibited the growth of soil microflora, resulting in a significant reduction in biodiversity and ecosystem functionality. Moreover, excessive use of synthetic fertilizers in the soil led it to accumulate in plant tissues and increase nitrate and phosphorus concentrations in water bodies, initiating eutrophication and hypoxia in lakes and estuaries (Vance et al., 2003).

Nowadays, awareness is changed from eating food grown with chemical fertilizers to food grown with organic fertilizers due to its hazardous effects on the body. Biofertilizers can help solve the problem of food needs of the ever-increasing global population. It is essential to realize the valuable aspects of seaweeds fertilizers to apply them in modern agricultural practice and develop new food products (Roohinejad et al., 2017). The application of biofertilizers containing beneficial microbes promotes, to a large extent, crop productivity. These potential biological fertilizers would play a key role in the productivity and sustainability of soil and protect the environment as eco-friendly and cost-effective inputs for the farmers. Using biological and organic fertilizers as a low input system can help to achieve sustainability of farming.

Algae are primarily aquatic organisms and, like terrestrial plants, possess the ability of photosynthesis by capturing the energy from sunlight and fixing carbon dioxide (CO_2) into biomass (Trentacoste et al., 2015). Algae are supposed to be the ancestor of modern plants; therefore, the lineage of their photosynthetic machinery connects to cyanobacteria (Falcón et al., 2010). Although algae and plants are different in many aspects, they share the most important fundamental process: photosynthesis. That makes algae distinguished and valuable.

The use of algae in nutraceuticals, supplements, and as fertilizer in rice fields dates back thousands of years (Kiple & Ornelas, 2000). The ability of algae to produce fuel precursor molecules was discovered in the early 1940s. And by the 1950s, large-scale cultivation and biomass production of algae had been started in the USA, Japan, Netherlands, and Germany (Borowitzka, 2013). The research focused on the beneficial effects of algae on plants was practically started in 1960-70s (Dmytryk & Chojnacka, 2018). The use of algae and algal-based products has been utilized in today's agriculture to supply nutrients and growth regulation of plants under stressful environments. The present chapter focuses on the use of algae as organic fertilizer, their role in nutrient cycling, their use as plant growth promoters, and the management of plant pests.

BIOFERTILIZERS FROM ALGAE

Agriculture is vital to our food supply; it provides jobs for almost two-thirds of the world's extremely poor people. Furthermore, food production directly depends on natural resources such as biodiversity, land, vegetation, rainfall, and sunlight. (Boliko, 2019). Higher crop yields and cultivation efficiency have been achieved by the use of chemical fertilizers. Nevertheless, severe environmental damage is caused by the excessive use of these fertilizers (Kang et al., 2021). Moreover, soil acidification, salinity, and reduction in soil microflora and fauna are some of the consequences of excessive chemical fertilizer use. To tackle the abovementioned problems, the use of organic fertilizers seems logical.

Before now, the term biofertilizers was used to include organic fertilizer. However, technically, there is a big difference between them. Vishal & Abhishek (2014) distinguished between biofertilizers and organic fertilizers and stated that "biofertilizers are microbial inoculants consisting of living cells of micro-organisms like bacteria, algae, and fungi help in increasing crop productivity". So, biological activities are markedly enhanced by microbial interactions in the rhizosphere of plants. On the other hand, organic fertilizers are produced from animal sources such as animal manure or plant sources like green manure.

Biofertilizers are formulations of live micro-organisms on the application in new agricultural systems (seed inoculation, foliar spray on crops, or to soil); they promote the growth, yield, and total productivity of cultivated crops (Van Oosten et al., 2017). Plant bio-stimulants are substances derived from natural sources that enhance crop growth and productivity when used in agricultural systems (Illera-Vives et al., 2015; Sandepogu, 2018; Van Oosten et al., 2017). They are also known to regulate many plants' physiological processes and enhance plants' adaptability against different abiotic stresses (Mansori et al., 2019).

Algal biofertilizers are natural recyclers and reservoirs of nutrients, increase plant growth, and offer all said advantages. Recently, various algae have been explored for their impact on cultivation, soil, and environment, and novel industrial processes have been developed for extensive scale cultivation of algae and production of algal biofertilizers. The chapter introduces the diverse nature of algae as a biostimulant. The biochemical compounds of algae, which impact the growth of plants, are also essential parts of the discussion, along with the functions of algae as biofertilizers (Iqbal et al., 2021).

Biofertilizers derived from algae proved better than farmyard manure since algal fertilizers have reasonable organic matter contents that help moisture-retaining and nutrient availability in soil (Baweja et al., 2019). Furthermore, algal fertilizers possess several other benefits, including enhanced plant resistance to diseases and environmental stress, soil structure stability, improved soil aeration, nutrient availability, and water holding capacity (Kumar & Sahoo, 2011). Figure 1 summarizes the algae species used in biofertilizer production. Algae play a significant role in agriculture, used as biofertilizers and soil stabilizers. Algae, mainly the seaweeds, are used as fertilizers, resulting in less nitrogen and phosphorous runoff than the one from livestock manure (Abdel-Raouf et al., 2012). The review (Abdel-Raouf et al., 2012) aims to support that algae are important constituents of arid and semi-arid ecosystems.

Moreover, their distribution and situation may specify the health of the environment. Also, the existence of algae leads to diminished erosion by controlling the water flow into soils. Likewise, they play a part in soil fertility, soil recovery, controlling agricultural pests, forming a microbiological coating, agricultural wastewater treatment, and recycling of treated water. Like other organisms, algae found in

Figure 1. Various species of algae used as biofertilizers.

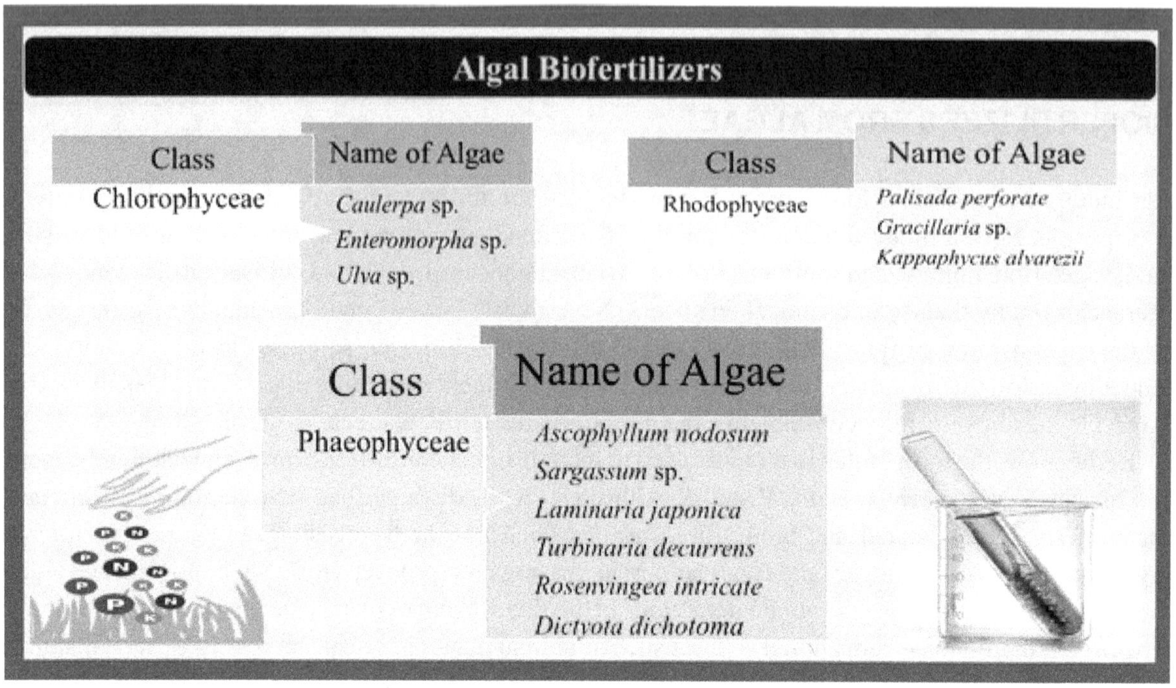

diverse soil types may help the soil recover its characteristics such as carbon content, texture, aeration (Ibraheem, 2007), and nitrogen fixation. Figure 2 elaborates on the advantage and limitations of using biofertilizers.

Algal Diversity for Fertilizer Production

Algae make the diverse community of unicellular to multicellular organisms, i.e., micro and macroalgae. Blue-green algae (BGA) are distinctive among microalgae, sharing certain characteristics with plants and bacteria. Microalgae are extensively spread across tropical, subtropical, and temperate areas. Furthermore, these have been discovered in practically every possible ecosystem from the Arctic to the Antarctic, with tropical and subtropical regions being the most prevalent. Collaborating with other organisms generates microbial mats, biofilms, and benthic communities (Zutshi & Fatima, 2015). BGA thrive in damp conditions of paddy fields, earning them the name "Algalization". Their genomic sequence and metabolism are similar to bacteria, but the existence of phycobiliprotein and chlorophyll corroborate autotrophic nourishment, similar to that of eukaryotic plant cells. BGA is important in ecosystem stabilization and agriculture because it can perform photosynthesis and nitrogen fixation. *Anabaena variabilis*, *Nostoc muscorum*, *Aulosira fertissima*, and *Tolypothrix tenuis* are the species of BGA found to be a prominent source of biofertilizer (Baweja et al., 2019).

Marine macroalgae, sometimes recognized as seaweeds, are employed as organic fertilizers in several countries. They are found in waters worldwide and are separated into three types: green, brown, and red algae. Many nations, including China, Japan, Korea, and the Philippines, produce marine algae economically. Seaweeds impact the biological and physicochemical qualities of soil, affecting plant productivity.

Figure 2. Advantages and limitations of using biofertilizers. (Adopted from Chakraborty & Akhtar 2021).

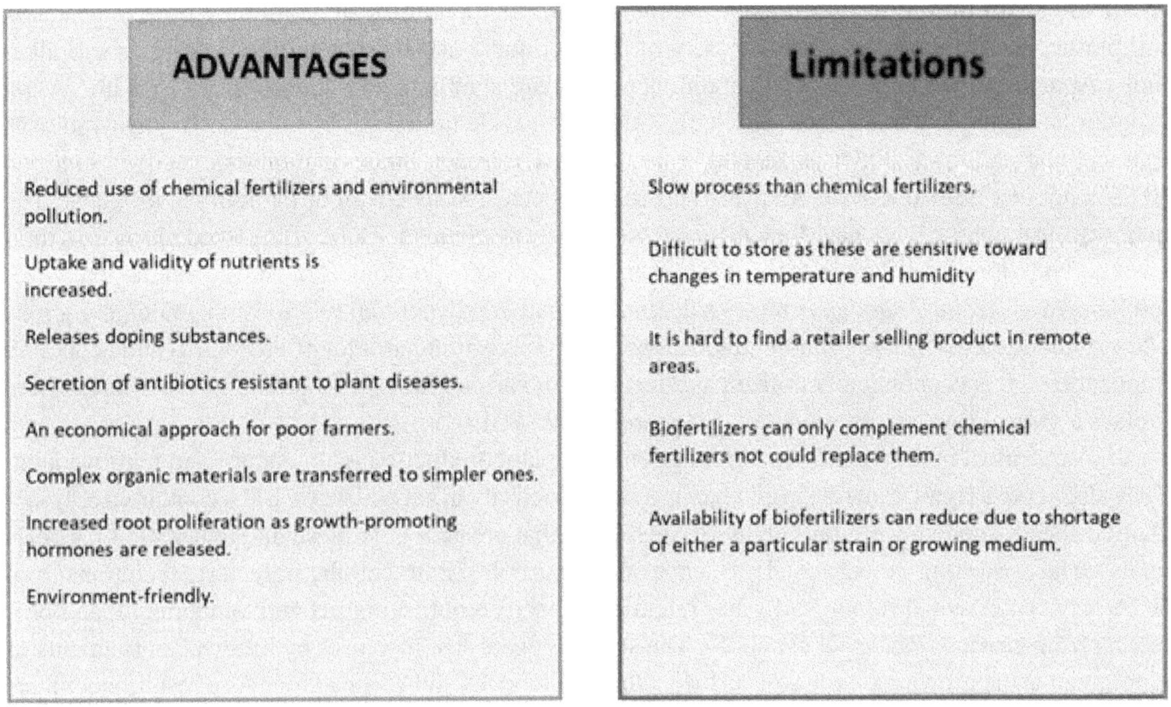

By enhancing moisture retention capacity, the seaweeds and their extracts improve soil structure and biology. Many seaweed liquid extracts include bioactive chemicals that are employed in a variety of agricultural and horticultural applications. Seaweed extracts are becoming increasingly popular for a variety of crops, including vegetables, grains, and flowers. *Ulva*, *Enteromorpha*, *Laminaria*, *Undaria*, *Sargassum*, and *Gracilaria* are among the marine algae utilized as soil conditioners (Bajwa et al., 2019).

Biofertilizers Derived from Microalgae

BGA possesses the capability to capture and fix atmospheric nitrogen. They can adapt to various soil types and environments, which has made them cosmopolitan in distribution. Species like *Nostoc linkia*, *Anabaena variabilis*, *Aulosira fertilisima*, Calothrix sp., Tolypothrix sp., and Scytonema sp. are significant nitrogen-fixers and identified from various regions and used for rice production (Prasad & Prasad, 2001). Nitrogen is the second limiting factor, after water, for plant growth in many fields, and its deficiency is replaceable by fertilizers (Malik et al., 2001). Blue-green algae are photosynthetic nitrogen fixers and are free-living and play an important role in maintaining and building soil fertility, consequently increasing rice growth and yield as natural biofertilizers (Song et al., 2005). They add growth-promoting constituents, including vitamin B12, to recover the soil's aeration and water holding capacity.

Rice growth and yield are enhanced by using BGA in rice paddy areas. BGA has been used in rice farming in several Asian nations, including China, Vietnam, and India, as an alternative to nitrogen fertilizers (Raj & Kumari, 2020). A study at IRRI demonstrated that using only BGA with no additional nitrogen fertilizer, 23 consecutive crops may be cultivated effectively for 12 years (Watanabe et al., 1977).

Besides rice crops, oat, tomato, radish, barley, cotton, sugarcane, maize, chilli, and lettuce have all been shown to benefit from BGA inoculation. On the other hand, total nitrogen fixation by BGA depends on soil physicochemical qualities and a variety of other climatic and biotic conditions, such as soil alkalinity. Apart from working as a soil conditioner, there are symbiotically competent BGA with several superior attributes that make them explicitly valuable to scale up the list of nitrogen-fixing symbiosis. Extracellular and intracellular partnerships are the most common kinds of symbiotic BGA. Symbiotic BGA is not just limited to roots; they also possess an extensive spectrum of plant tissue as a host. The majority of symbiotic BGA have their nitrogenase defence mechanism. BGA offers fixed nitrogen to their hosts and provides fixed carbon to the host's non-photosynthetic portions. The BGA's interaction with a fern known as *Azolla-Anabaena* creates a distinctive mutually beneficial relationship among numerous other symbiotic associations of blue-green algae. This connection has a lot of ecological and economic implications. It is economically utilized as a fertilizer in various sectors, including rice farming, where it plays a critical role in crop productivity (Baweja et al., 2019).

BGA is critical in nitrogen fixation for agricultural plant production maintenance and improvement. They also serve a significant role in enhancing soil productivity in several ways. BGA contains a polysaccharide sheath and extracellular polymeric secretions that create a glue mesh and cohere soil particles on its surface, exerting a mechanical influence on soil particles. Extracellular polymeric secretions from BGA serve an essential role in soil water retention capacity, controlling pH and temperature, and protecting from erosion (Pandey et al., 2005). The soil aggregate size increases by interwoven filaments of blue-green algae growing on the soil surface, which minimizes soil compaction. As a result, inoculants containing blue-green algae have been utilized to enhance soil aggregation, boost soil fertility, and repair destroyed soil crusts (Bajwa et al., 2019). Furthermore, algal inoculations provide a large quantity of organic matter to the soil. It plays a significant role in reducing the soil's oxidizable matter content by releasing oxygen during carbon fixation, significant in locations where more than one rice crop is planted each year (Pabbi, 2015).

Biofertilizers Derived from Macroalgae

Recently, some labourers in the agricultural field have been made to convert to an economy built on biological and renewable resources. This requires the manipulation of the agricultural ecosystem for alternative natural resources, for example, the use of seaweeds as a composite or biofertilizer, which appears to be auspicious biotechnology in the new agricultural systems, thus enhancing the progress and sustainability of food production (Piwowar & Harasym, 2020).

Seaweeds or marine macroalgae can also be used as a natural fertilizer. They are commonly employed in agricultural settings as bio-stimulants and soil conditioners. Seaweeds are abundant globally and obtained as drifting ones from beaches worldwide. They do not fix nitrogen from the atmosphere like microalgae, but they are an excellent supply of growth hormones and essential nutrients. Seaweeds are utilized as liquid extracts or pulp residuals following extraction. There have also been accounts of seaweed being used as fertilizers in agricultural areas in the form of dried powder. Several experiments have been carried out on several crops utilizing organic fertilizers made from various seaweeds, and considerable increases in crop output have been recorded. Even if the liquid extracts are administered at lesser doses, seaweed extract enhances seed germination percentage. The liquid extracts that are commercially available are mostly made from brown seaweeds and differ in viscosity, colour, odour, and pH.

Fucoidans and alginates are primarily found in brown seaweeds, and their chelating and gelling abilities make these polysaccharides particularly useful in crop production (Baweja et al., 2019).

Meanwhile, global seaweed aquaculture production in 2014 was estimated at 27.3 million tons wet weight, roughly a quarter of farmed fish production by weight, and has accomplished a massive productivity increase (Boliko, 2019). Fertilizers derived from seaweed as extract and dry mass are the best-recommended alternatives to synthetic chemical fertilizers (Semary, 2017). Meanwhile, the conversion of seaweed biomass towards functional agricultural fertilizers may require specific pretreatment of fresh or dried algae or physiochemical modification of raw biomass (Tuhy et al., 2020).

Seaweeds are an outstanding source of vitamins A, B1, B12, C, D and A, riboflavin, niacin, pantothenic and folic acid (Safinaz & Ragaa 2013). Thirumaran et al. (2009) stated that seaweed liquid fertilizer contained macronutrients, trace elements, organic substances like amino acids, and plant growth regulators such as auxin, cytokinin, and gibberellins. The application of seaweed liquid fertilizer enhanced the soil's water retention capacity. Seaweed extracts enhance seed germination, improve plant growth, induce resistance to frost, fungal, and insect attack and increase nutrient uptake from soil. Several countries use marine algae as manure. The potential of seaweeds in current crop cultivation has been extensively investigated recently to develop new applicable products from seaweeds in the form of liquid fertilizer and biomass (Nabti et al. 2017).

ALGAE PRODUCTS AS BIOFERTILIZERS

Extracts from Algae

Algae extracts are the most well-known and widely used algal compounds for agricultural purposes. The synthesis of algal extracts employing several physical and chemical processes and their usage as biostimulants has been thoroughly studied (Tuhy et al., 2020). Extraction with organic solvents (ethanol and methanol) appears to be the most prevalent approach (López et al., 2011). Many instances of the usage of organic and inorganic solvents in various temperature and pressure variations may be found in the scientific literature. In addition, the extraction procedure takes a variety of times. Traditional extraction procedures were previously supplemented with so-called aided techniques like microwaves or ultrasounds to increase extraction efficiency and get a favourable biochemical composition of extracts (García-Salgado et al., 2012). Various extracts derived from different seaweed species have been produced for the agricultural market in various formulations over the last two decades (Illera-Vives et al., 2015; Mansori et al., 2015; Sandepogu, 2018). Also, several vital chemical compounds to many crops were reported in extracts of many seaweeds species, such as plant growth hormones, plant pigments, antioxidants, and polysaccharides (Tuhy et al., 2020). Thirumaran et al. (2009) specified that the latest research showed that seaweed fertilizers are favoured due to their nitrogen, phosphorus, and potash content and trace elements and metabolites similar to plant growth regulators. Recently, seaweed extracts as liquid fertilizers have come on the market because they contain many growth-promoting hormones like auxin, gibberellin, trace elements, vitamins, amino acids, and micronutrients. Wells et al. (2017) showed that the essential amino acids liberated from *Palmaria palmata* (raw and boiled for 30 min) were much more than that of wheat, rice, and cornflour (Table 1). With the same respect, El-Din (2015) found promising quantities of essential amino acids in algal extracts of *Sargassum vulgare* and *Codium tomentosum* (Figure 3).

Table 1. Essential amino acids liberated from 1 g dry weight of Palmaria palmata (raw and boiled), wheat, rice, and cornflour (Adopted from Wells 2017).

Amino acids (mg)	Raw algae	Boiled algae	Wheat	Rice	Cornflour
Histidine	1.5	2.6	1.7	1	0.9
Isoleucine	3.8	7	2.4	1.6	0.9
Leucine	6.4	12.7	3.9	3.3	4
Lysine	7.7	15.5	1.2	2.3	0.9
Methionine	2.6	4	1.2	1.1	1
Phenylalanine	5.2	8.7	3.5	2.5	2.4
Threonine	5.2	8.6	2.3	2.2	2
Valine	6.4	13.5	3	2.5	1.7
Total	38.8	72.6	19.2	16.5	13.8

Algae extracts include a multitude of chemical components that are advantageous to plant culture. The most well-known plant growth hormones, pigments, antioxidants, and polysaccharides, whose composition is highly dependent on extraction conditions, are the most prominent (Chojnacka et al., 2012). Microalgae water extract was tested as a bio-stimulant in pot studies on vegetable crops. Plant growth metrics such as height, biomass, root and shoot length, and leaf count were measured forty days after planting, and the findings were statistically analyzed. The difference in growth metrics between the control plants and the plants treated with algae extract was statistically significant. The presence of plant growth regulators in the algal extracts was identified as a primary cause of the tested formulation's bio-stimulation action after HPLC analysis (Shariatmadari et al., 2013). A study demonstrated the fa-

Figure 3. Amino acid content of the seaweed liquid fertilizers from Sargassum vulgare and Codium tomentosum. (Adopted from El-Din 2015).

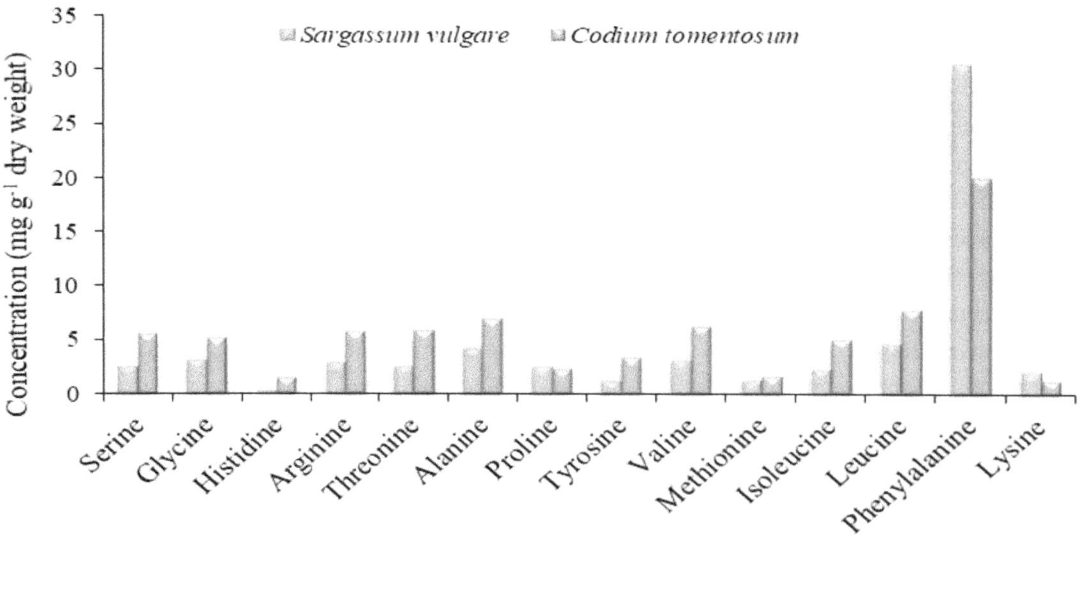

vourable impact of extracts produced from Ascophyllum nodosum on plants in a controlled environment on tomatoes grown in a hydroponic system under greenhouse conditions. In comparison to the control group, growth indices such as shoot length, shoot diameter, dry biomass, and vitamin C content were considerably more significant in plants treated with brown algae extract. It was also discovered that the greater the bio-stimulant dosage, the larger the claimed bio-stimulation effect (Araghian et al., 2015).

In a phytotron chamber, the effect of seaweed extract (a popular bio-stimulant) on the development of spinach under water-scarce conditions was investigated. Under modest drought stress, *Ascophyllum nodosum* extract enhanced spinach growth positively. Using a bio-stimulant yielded higher relative water content, plant biomass, and leaf area. Drought-induced suppression of gas exchange was also decreased (Xu & Leskovar, 2015). The response of *Phaseolus vulgaris* L. to water-limited conditions in the presence of bio-stimulant was examined using algal extracts derived from *Ulva rigida* and *Fucus spiralis*. Under water shortage, algae extracts improved growth metrics and physiological attributes. Algal extracts increased the generation of phenolics while also increasing the activities of superoxide dismutase and catalase. The researchers concluded that seaweed extracts could protect from abiotic stress (Mansori et al., 2016). It has been demonstrated that there are many studies available that describe the prospective use of algal extracts as bio-stimulants, and the data is continually rising. The choice of algae species and algal extract manufacturing techniques is determined by the extracts' desired composition and intended use.

Compost from Algae

Composting algal biomass is another approach to turning micro-and macroalgal biomass into valuable agricultural products. The presence of thermophilic bacteria allows stable bio-stimulants to be produced from algal biomass (Tuhy et al., 2020). A study examined the utilitarian qualities of algal biomass supplemented with zinc ions by biosorption on *Lepidium sativum*. Plants treated with bio-preparations had considerably larger plant mass and a much higher number of germinated seeds. In addition, excellent bioavailability of essential nutrients to plants was observed compared to standard micronutrient fertilizers such as organic chelates and inorganic minerals, resulting in zinc biofortification (Tuhy et al., 2014).

Other critical elements such as iron, copper, manganese, and sulphur were also higher in plants treated with algal bio-preparations. It was discovered that biomass provides a source of additional nutrients in plant-available forms. The use of micronutrient bio-preparations to fertilize *Lepidium sativum* increased chlorophyll content.

Composting seaweeds can be a useful technique for producing bio-stimulants that are ready to use in organic agriculture. Under a controlled environment, the impact of various composting doses on tomato and lettuce crops was evaluated. The suggested bio-stimulant was compared to mineral fertilizer and approved organic fertilizer in terms of effectiveness. There were significant variations in the yield of both tested crops between groups treated with the novel bio-stimulant and groups treated with reference materials. In addition, the application of algal compost increased the weight and diameter of tomatoes (Illera-Vives et al., 2015). Algal compost has been found to work as a soil conditioner and controlled-release fertilizer. Because of its advantageous micro and macronutrient content, algal compost may be used to fertilize effectively. It is concluded that composting algal biomass is a useful approach for using problematic material while also producing an acceptable agricultural product.

Algal Meal

A course, powdered grind from the edible part of any grain, and any ground or powdery item derived from a byproduct might be characterized as a "meal." Many beneficial formulations in the form of a meal are utilized to use renewable resources and enhance naturally occurring processes in agricultural systems. Because numerous useful components such as nitrogen, carbon, phosphorus, and micronutrients are provided, this may represent a chance to lessen the environmental effect of chemical fertilizer manufacture and usage while also improving soil fertility and biological activity. Natural seaweed has recently been used as a fertilizer, allowing for a partial substitute for synthetic fertilizer (Tuhy et al., 2020). The term "algae meal" can refer to the algal biomass without oil and de-oiled oil cake. When the oil from the algal biomass is removed, the resultant cake has little lipid content and is mostly composed of protein and carbs (Mata et al., 2010). Algal meals might be used as organic soil additives in agriculture. Algal meals, like other organic additions, may still hold considerable levels of organic carbon and a wide spectrum of nutritional components, including nitrogen and phosphorus, making their application as soil amendments viable. There are a variety of commercial algal products that proved beneficial for their utilization in agriculture and horticulture, including liquid extracts used as foliar sprays, soil drench, and granular soil conditioners and manure. Because some useful minerals, like potash, are leached from algal dung while stored wet, drying it before application makes it more effective. There are a variety of dry seaweed fertilizers on the market. Mechanical distributors can simply apply them (Tseng, 2001).

EFFECT OF ALGAE ON SOIL HEALTH

Algal proteoglycans have adhesive characteristics that allow cells to adhere to solid surfaces and soil aggregates. Soil aggregation appears to be particularly essential since it directly impacts soil warmth, aeration, and infiltration rates, which enhances the crop's physical environment (Chatterjee et al., 2017). Additionally, inoculation of *Nostoc* strains on clay soils improves soil aggregation and porosity owing to improved water infiltration.

Even though some researchers have examined the influence of algal biofertilizers on soil microflora, little is investigated about the alteration in the soil microbial community that occurs as a result of BGA inoculation. After inoculating burned soils with cyanobacteria, research revealed unit increases of larger than four logarithmic in heterotrophic bacteria, actinomycetes, algal, and fungal propagules, and three logarithmic unit increases in fungal mycelia (Acea et al., 2001). After the soil was inoculated with *Nostoc muscorum*, a study found a considerable change in the heterotrophic microbial community. These findings imply that increased carbon and energy sources provided by cyanobacterial polysaccharides may be a factor in the growth of heterotrophic microbial communities. Additionally, increasing the total nitrogen concentration of inoculated soil promotes the growth of indigenous soil microbes. The nutrient condition of the soil, particularly nitrogen and phosphorus, dictates the mineralization of accessible carbon, which affects the microbial population (Chatterjee et al., 2017).

NUTRIENT CIRCULATION MEDIATED BY ALGAE

Microalgae are primarily composed of proteins, carbohydrates, and lipids, and their composition varies according to the microalgae species. The carbohydrate proportion of microalgae biomass is 64%, and the protein and lipid fractions are up to 71 and 22, respectively, of cell dry weight (Razzak et al., 2013). Microalgae can use carbon in the form of CO_2 emitted by industrial power plants and inorganic and organic carbon (Kuenz et al., 2021). One hundred tons of algal biomass fix about 183 tons of CO_2 (Razzak et al., 2013). CO_2 exists in the nongaseous form of bicarbonate in water, which may be translocated and concentrated in algae. The bicarbonate is transformed into CO_2 inside the cell and can then be taken up by RuBisCO. Following various processes, these molecules become substrates to synthesize starch and oil (Sayre, 2010).

Biofertilizers not only fix atmospheric nitrogen and make it available to the plant but also solubilize the insoluble forms of phosphates like tricalcium, iron, and aluminium phosphates into available forms, scavenge soil's phosphate, generate hormones and anti-metabolites which endorse root growth, decaying organic matter and help in mineralization in soil and increase the availability of nutrients and improve the yield by 10 to 25% without unfavourably affecting the soil environment. In tropical countries, rice production mainly depends on biofertilizers (Vaishampayan et al., 2001). Latent cells of competent strains of N_2-fixing, phosphate solubilizing, or mobilizing micro-organisms are used for application to seed or soil with the intent to increase the number of such micro-organisms and hasten those microbial processes which enhance the accessibility of nutrients that can be assimilated by plants. They improve the soil structure, restore soil nutrients, build soil organic matter, water uptake, plant growth, and plant tolerance to abiotic and biotic factors (Akram et al., 2020).

Furthermore, microalgae provide a possibility for wastewater treatment owing to their capacity to grow biomass using inorganic nitrogen and phosphorus (Zhao et al., 2019). Algae directly absorb nitrate and ammonium ions from the surrounding water; however, ammonium is favoured for cellular nitrogen compound formation. Algae convert nitrate to ammonium by the enzyme nitrate reductase (Graham et al., 2009). Algae prefer to absorb phosphorus in the form of orthophosphate. After biological degradation (aerobic or anaerobic) of wastewater, the inorganic components (nitrate, ammonium, and phosphate ions) are present in ample concentrations to cause eutrophication of the surrounding aquatic ecosystem. Microalgae have a great capability for absorbing these nutrients and using inorganic nitrogen and phosphorus for development (Zhao et al., 2019). Iron is a cofactor for various enzymes, including ferredoxin, catalase, cytochromes, glutamate synthetase, nitrogenase, nitrate, and nitrite reductase (Graham et al. 2009). Sulfur is typically absorbed and digested as sulphate, and its incorporation into a range of sulfur-containing chemicals required for protein, lipid, polysaccharide, and signalling molecules production, is vital (Giordano et al., 2008). Apart from their trace element requirements, algae are well-known for their propensity to absorb and accumulate heavy metals and organic chemicals such as organochlorides. Additionally, certain species of microalgae secrete extracellular esterase that dissolves deltamethrin (insecticide) and has the potential to digest a variety of hydrocarbons (found in oily wastes) (Kuenz et al., 2021).

Figure 4. Benefits of using algal extract in agriculture systems. (Adopted from Anuradha & Singh 2021).

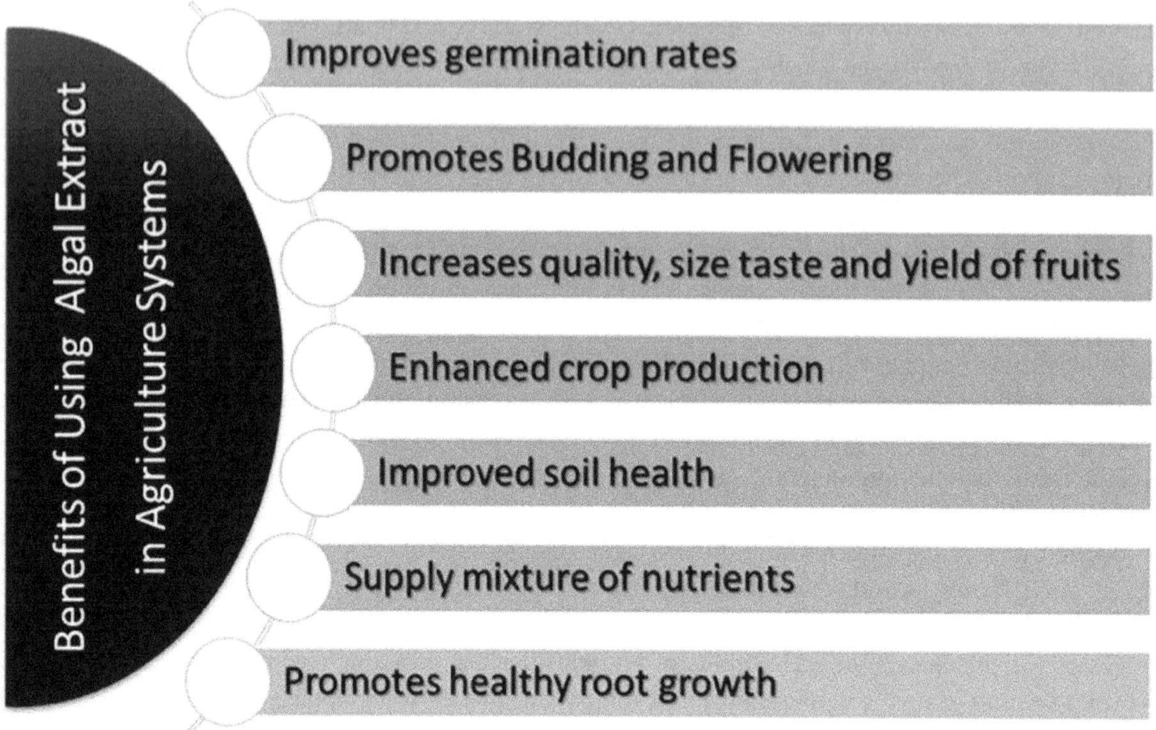

EFFECTS OF MACROALGAE ON PLANTS

Most crops are unable to grow in extreme climatic conditions, and overcoming such harsh conditions necessitates very complicated and expensive agricultural practices and techniques, which are frequently unavailable due to a lack of financial resources or/and appropriate infrastructure. As a result, innovative means and solutions are required to overcome adverse climatic stresses, increasing crop yield to meet the urgent and essential needs of food supply to keep pace with the ongoing increase in population numbers. The use of algae extracts looks to be a promising possibility to increase agriculture production efficiency and improve their tolerance to unfavourable environmental conditions (Anuradha & Singh 2021). Several studies have found that macroalgae have a positive impact on crop development and yield, notably via encouraging root growth (Mancuso et al., 2006). In general, extracts from macroalgae boost seedling emergence, plant vigour, nutrient acquisition, and biotic and abiotic stress tolerance in crop plants. Some of those benefits are represented in figure 4.

Crop Growth and Yield

Seed emergence and enhanced seedling vigour are widely established to substantially influence seedling establishment, crop growth, development, and final crop yield. The organic components of Ascophyllum nodosum extract generated amylase activity in barley may work in tandem with gibberellic acid-dependent amylase synthesis to improve germination and seedling vigour (Rayorath et al., 2008). Furthermore, due to the higher absorption capacity of available nutrients, okra seedlings treated with seaweed products like Kelpak® showed more significant root growth and seedling vigour (Papenfus et al., 2013). Plants sprayed with a commercial seaweed product like Actiwave® showed higher carboxylation activity, linked to higher chlorophyll contents and stomatal conductance, which resulted in a considerable increase in photosynthesis. In terms of the enhanced chlorophyll content, Fe is a necessary component of chlorophyll production, and its uptake could be positively influenced by seaweed extract application (Spinelli et al., 2009). Seaweed extracts promote anthesis by promoting strong plant development as well as fruit set in a variety of crop plants. A significant boost in wheat grain yield due to foliar spray of *K. alvarezii* and *G. edulis* solution was mainly attributed to an increase in yield-related components such as grains per spike and 1000-grain weight (Shah et al., 2013). Similarly, tomato plants, foliar sprayed with seaweed extracts of *Ecklonia* showed a significantly increased fruit yield.

Dhargalkar & Pereira (2005) showed that lower concentrations of seaweed extracts enhanced the rate of seed germination of chillies and turnips. Strik et al. (2004) stated that seaweed extracts are significant fertilizers for many crops. El-Barody et al. (2007) found the addition of different successive extracts of *Asparogopsis taxiformis* thallus powder to the soil as biofertilizers gave a significant increase in the growth of *Vicia faba*. Sabh et al. (2008) found that NPK in plants treated with Sargassum sp. reached four folds of the negative control. Thamida et al. (2011) showed that the agronomic characters and yield components of rice in algalized plots were statistically identical to the treatment with the recommended dose of urea-N.

Laurencia obtusa and *Jania rubens* was the best treatment to enhance the growth of maize plants during the early stages of planting and also caused a high increase in the whole plant length. Thus, using algae as biofertilizers improved the vegetative characteristics of maize plants (Safinaz & Ragaa 2013). Furthermore, the study of El-Zabalawy et al. (2015) clarified the important role of marine algae in monitoring polluted soil by reducing the amount of heavy metals available in the soil through adsorption of a large portion of these metals. At the same time, the mixing of marine algae stimulated the roots of the studied crops (*Lactuca sativa* var. capitata and *Brassica oleraceae* var. capitata Form rubra) to increase the absorption of salts and minerals.

Tolerance to Abiotic Stresses

Many abiotic stresses like slat stress, low/high-temperature stress, and drought stress increase the generation of reactive oxygen species (ROS), which can harm normal plant functions. Researchers revelated that the application of various seaweed products can effectively alleviate the effects of abiotic stress by inducing the antioxidant system of plants. For example, the amount of superoxide dismutase, which scavenges superoxide, was enhanced in turf grasses treated with an *A. nodosum* seaweed extract (Fike et al., 2001). Similarly, glutathione reductase, superoxide dismutase, and ascorbate peroxidase activity were substantially enhanced in tall feccue plants in response to Tasco, a seaweed product. Several studies, on the other hand, concluded that the growth-promoting effects of seaweed extracts against several abiotic stresses are also associated with cytokinin activity which reduces stress-induced free radicals by scavenging them directly and reducing the generation of ROS through blocking xanthine oxidation (Fike et al., 2001). However, bioactive molecules other than cytokinin present in seaweed extracts may also play their role in scavenging free radicals under abiotic stresses.

Bio stimulants

The minerals, amino acids, vitamins, cytokinins, auxins, and abscisic acid-like growth factors present in seaweed or algae extracts modify cellular metabolism in treated plants, resulting in improved growth and crop yield (Khan et al., 2009). The plant growth hormonal components included in the extracts, particularly cytokinins, are hypothesized to be linked to increased crop yield under seaweed extract application. Since seaweed products contain growth-stimulating qualities, their application as bio-stimulants in agriculture is widely known. Bio-stimulants, also known as metabolic enhancers, are compounds other than fertilizers that increase plant growth when administered in tiny doses (Khan et al., 2009). At low quantities, such as 1:1000 or greater, seaweed extracts are bioactive. Seaweed extracts may contain chelating substances like mannitol which aids plants in nutrient uptake. Seaweed extracts may also contain various plant growth regulators organic osmolytes such as betaines, amino acids, mineral nutrients, vitamins, and their derivatives (Spinelli et al., 2009). *A. nodosum* and *F. serratus* have been found to include glycine betaine, aminobutyric acid betaine, aminovaleric acid betaine, and laminine (Blunden et al., 2010). Seaweeds synthesize agars, alginates, carrageenans, fucans, phlorotannins, and other polymers that are not present in other plant species (Khan et al., 2009). The principal polymers identified in brown seaweeds include polysaccharides, phlorotannins, proteins, and nucleic acids (Papenfus et al., 2013), which play a role in nutrient acquisition and are accumulated in plants in response to stress.

Auxins, cytokinins, gibberellins, abscisic acid, and ethylene are the main phytohormones found in seaweed extracts. Auxins are involved in plant tissue elongation, apical dominance, cell division, organ movements, and plant ageing. Auxins and auxin-like substances are said to be abundant in marine algae. Stirk et al. (2013) investigated 31 species of seaweeds and found evidence that cytokinins are broadly distributed throughout them. Although residues of gibberellins were found by mass spectrometry in a solution of *A. nodosum*, there appears to be no direct chemical evidence for gibberellins in brown algae (Khan et al., 2009). Abscisic acid and ethylene modulate plant response to biotic or abiotic stress, inhibit cell growth, and accelerate ageing in plants (Tuhy et al., 2013). However, ABA's role in algae remains uncertain, despite the fact that it promotes protein synthesis, which is required for drought tolerance in terrestrial plants (Verslues et al., 2006). Although a possible ethylene precursor, 1-aminocyclopropane-1-carboxylic acid, was discovered in Kelpak 66, a commercial extract of *E. maxima*, evidence for ethylene synthesis in brown seaweeds.

BIOCONTROL POTENTIAL AGAINST PLANT PATHOGENS AND PESTS

Controlling soil-borne plant diseases is one of the most critical difficulties confronting contemporary agriculture. Naturally, plants and animals have several basic defence mechanisms to withstand pests and other insects. Natural substances like algal extracts, laminarin and carrageenans application are used to get rid of pests. They are observed in tobacco leaves to induce signalling and defence gene expression (Pardee et al., 2004). Several scientific studies prove that different chemicals, fungicides and insecticides have been replaced by microalgae (Ibraheem, 2007). Major disease-causing groups among plants are fungi and bacteria. *Sclerotina sclerotiorum* and *Rhizoctonia solani* Plant pathogens can be efficiently controlled by cyanobacterium *Nostoc mucorum*. Certain substances produced by the free-floating unicellular organism *Chlorella elliposidea* affect mosquito larvae growth and immunity.

Seaweed products benefit plant health by modifying the microbial population in the rhizosphere and affecting plant physiology and metabolism (Khan et al., 2009). Extracts of seaweed have been shown to boost a plant's resilience to pests and diseases. The suggested mechanism of action for seaweed products is that the readily degradable organic matter in seaweed offers food for hostile bacteria, allowing them to grow and colonize the root zone, or may contain specific organic compounds like alginate, which can directly inhibit disease infections (Sultana et al., 2011). Among other things, these antimicrobial compounds can kill disease-causing bacteria by disrupting the cytoplasmic membrane and blocking protein production (Swain et al., 2017).

Additionally, following twice-weekly spraying with a commercial *A. nodosum* product, red spider mite populations were significantly controlled in strawberry plants. While it is possible that seaweed extracts include chelated metals that might reduce red spider mite fertility, the mode of action is unknown (Khan et al., 2009). Algae contain a higher concentration of antioxidant polyphenols with antibacterial effects. Khan et al. (2009) proposed one possible mechanism in which seaweed extracts may block the disease incidence pathway seen in pathogenic bacteria. Finally, seaweed extracts may serve as an eco-friendly substitute for fungicides and other pesticides (Ara et al., 2005).

MICROALGAE AND ITS COMMERCIALIZATION IN AGRICULTURE

Microalgae, particularly cyanobacteria, can improve crop growth and development by mineralization and decomposition of organic compounds and enhancing the availability of major and micronutrients, and producing relatively more bioactive compounds (Prasanna et al., 2016; Stirk et al., 2013). Cyanobacteria have a long history of being used in agriculture due to their capacity to enhance crop growth and yield, nutrient uptake, and soil microbial activity (Table 2). Because of their production of hydrolytic enzymes and biocidal chemicals, cyanobacteria are regarded as promising biocontrol agents against a number of plant diseases (Gupta et al., 2013). Food, energy, biofertilizers, secondary metabolites, cosmetics, and medications may all be made from cyanobacterial biomass on a big scale. As a result, cyanobacteria are utilized in environmentally friendly, long-term agriculture practices to produce high-value biomass (Chittora et al., 2020). The use of the living culture of microalgae is another useful approach as it supplies sustained nutrient sequestration throughout the plant's life cycle, as well as other benefits such as soil erosion protection, nutrient leaching prevention, and soil structure and fertility preservation. To make this realistic, a significant amount of microalgal biomass will be needed. On the other hand, the economics of algal fertilizer production is important to their success. The utilization of waste materials to generate algal fertilizers is gaining interest as a cost-effective alternative with added environmental advantages (Renuka et al., 2018). On the other hand, the issues connected with waste-produced algal biomass require more research and field evaluation before being commercialized. Finally, because of their proven anti-stressor effect, algae-based products have a tremendous possibility to be launched, particularly as bio-stimulants and pesticides (Dmytryk & Chojnacka, 2018). A sustainable algae-based industry might be a vital component of the future bioeconomy, allowing for more resource-efficient food and fuel production as well as the development of new products, businesses, and employment opportunities (Ullmann & Grimm, 2021).

Table 2. Microalgae as potential tool for crop growth and yield enhancer

Crop	Microalgae	Effects	Reference
Oryza sativa	*Consortia, Anabaena, Nostoc* strains, *Calothrix elenkinii, Anabaena* combined with *Azotobacter*	Increased crop yields and soil fertility, improved antioxidant enzymes activity, alteration in soil microbiological diversity	Zayadan et al. 2014; Priya et al. 2015; Ranjan et al. 2016; Prasanna et al. 2015
Triticum aestivum	*Anabaena doliolum, Cylindrospermum sphaerica, Anabaena* combined with *Azotobacter, Calothrix ghosei, Hapalosiphon intricatus, Chlorella vulgaris*	Increased crop growth and yield, better mineralization of carbon and nitrogen, soil enzymatic activity improvement, improved uptake of nutrients	Nisha et al. 2007; Nain et al. 2010; Rana et al. 2015; Renuka et al. 2017
Zea mays	*Anabaena* combined with *Azotobacter, Nostoc* strains, *Chlorella vulgaris*	Plant productivity enhancement, increased nutrient uptake, improved antioxidant enzymes activity, enhanced microbial activity led to improved nitrogen uptake, initiation of plant microbiome and rhizosphere functionality	Yilmaz and Sönmez 2017; Uysal et al. 2015
Gossypium hirsutum	*Anabaena* combined with *Trichoderma and Calothrix sp.*	Improved nutrient uptake and helpful in insect pest control	Prasanna et al. 2015
Solanum lycopersicum	*Acutodesmus dimorphus, Nannochloropsis*	Fruit quality improvement, germination and plant growth improvement	Coppens et al. 2016
Cicer arietinum	*Anabaena-Mesorhizobium ciceri biofilm and Anabaena laxa*	Plant productivity enhancement, improved plant growth and metabolic activity of nodules, initiation of plant microbiome and rhizosphere functionality	Prasanna et al. 2017

FUTURE RESEARCH DIRECTIONS

The present chapter elaborates on the potential use of algae in agriculture as biofertilizers, bio-stimulants, and nutrient recycling, to name a few. Further research is required to identify low-cost sources of nutrients for mass-scale cultivation of algae. Also, identifying novel species and strains for use in the agricultural system is needed. Awareness regarding the integration of algal-based product and their facilitated commercialization is a need of time for the development of sustainable agricultural systems.

CONCLUSION

Modern agricultural practices led to soil and environmental degradation. Increasing needs for food and plant products necessitates sustainable agriculture and higher crops yields. Algae has the potential to produce huge biomass by fixing atmospheric CO_2. This algal biomass is being used in various algal-based products as agricultural inputs. It is a need of time to integrate these bio-based products into modern agriculture. Furthermore, technical advancement, awareness among the farming community, and policy regulation will bring an epoch-making advancement in algae-related industries that will ultimately lead toward environmentally benign, highly proactive, resource-efficient agricultural systems.

FUNDING

The present book chapter did not receive any grant from the funding agency.

REFERENCES

Abdel-Raouf, N., Al-Homaidan, A. A., & Ibraheem, I. B. M. (2012). Agricultural importance of algae. *African Journal of Biotechnology*, *11*(54), 11648–11658. doi:10.5897/AJB11.3983

Acea, M. J., Diz, N., & Prieto-Fernandez, A. (2001). Microbial populations in heated soils inoculated with cyanobacteria. *Biology and Fertility of Soils*, *33*(2), 118–125. doi:10.1007003740000298

Akram, M. S., Cheema, M. A., Nadeem, F., Waqas, M., Bilal, M., & Saeed, M. (2020). Role of biofertilizers in sustainable agriculture. *Mediterranean Journal of Basic and Applied Sciences*, *4*(2), 8–23.

Anuradha, & Singh, J. (2021). Organic farming by biofertilizers. In *Biofertilizers*. Scrivener Publishing LLC.

Ara, J., Sultana, V., Qasim, R., Ehteshamul-Haque, S., & Ahmad, V. U. (2005). Biological activity of Spatoglossum asperum: A brown alga. *Phytotherapy Research*, *19*(7), 618–623. doi:10.1002/ptr.1699 PMID:16161033

Araghian, S., Bagherzadeh, A., & Sadrabadi, R. (2015). Effect of brown algae and vermicompost application on some cherry tomato traits in hydroponic system. *Agroecology Journal*, *10*(4), 77–83.

Baweja, P., Kumar, S., & Kumar, G. (2019). Organic fertilizer from algae: a novel approach towards sustainable agriculture. In *Biofertilizers for Sustainable Agriculture and Environment* (pp. 353–370). Springer. doi:10.1007/978-3-030-18933-4_16

Blunden, G., Morse, P. F., Mathe, I., Hohmann, J., Critchley, A. T., & Morrell, S. (2010). Betaine yields from marine algal species utilized in the preparation of seaweed extracts used in agriculture. *Natural Product Communications, 5*(4).

Boliko, M. C. (2019). FAO and the situation of food security and nutrition in the world. *Journal of Nutritional Science and Vitaminology*, *65*(Supplement), S4–S8. doi:10.3177/jnsv.65.S4 PMID:31619643

Borowitzka, M. A., & Moheimani, N. R. (Eds.). (2013). *Algae for Biofuels and Energy* (Vol. 5, pp. 133–152). Springer. doi:10.1007/978-94-007-5479-9_8

Chakraborty, T., & Akhtar, N. (2021). Biofertilizers: Prospectives and challenges for future. In Biofertilizers. Scrivener Publishing LLC.

Chatterjee, A., Singh, S., Agrawal, C., Yadav, S., Rai, R., & Rai, L. C. (2017). Role of algae as a biofertilizer. In *Algal Green Chemistry* (pp. 189–200). Elsevier. doi:10.1016/B978-0-444-63784-0.00010-2

Chittora, D., Meena, M., Barupal, T., Swapnil, P., & Sharma, K. (2020). Cyanobacteria as a source of biofertilizers for sustainable agriculture. *Biochemistry and Biophysics Reports*, *22*, 100737. doi:10.1016/j.bbrep.2020.100737 PMID:32083191

Chojnacka, K. (2012). Biologically Active Compounds in Seaweed Extracts - the Prospects for the Application. *The Open Conference Proceedings Journal*, *3*(1), 20–28. doi:10.2174/1876326X01203020020

Coppens, J., Grunert, O., Van Den Hende, S., Vanhoutte, I., Boon, N., Haesaert, G., & De Gelder, L. (2016). The use of microalgae as a high-value organic slow-release fertilizer results in tomatoes with increased carotenoid and sugar levels. *Journal of Applied Phycology*, *28*(4), 2367–2377. doi:10.100710811-015-0775-2

De Clercq, M., Vats, A., & Biel, A. (2018). Agriculture 4.0: The Future of Farming Technology. *World Government Summit in Collaboration with Oliver Wyman*, (February), 30.

Dhargalkar, V. K., & Pereira, N. (2005). Effect of seaweed extract on the growth, yield and nutrient uptake of soybean (*Glycine max*) under rainfed conditions. *Journal of Applied Phycology*, *1*(2), 351–355.

Dmytryk, A., & Chojnacka, K. (2018). Algae as fertilizers, biostimulants, and regulators of plant growth. In *Algae Biomass: Characteristics and Applications* (pp. 115–122). Springer. doi:10.1007/978-3-319-74703-3_10

El-Barody, G. S., Moussa, M. Y., Shallan, A. M., Ali, A. M., Sabh, Z. A., & Shalaby, A. E. (2007). Contribution to the Aroma, Biological Activities, Minerals, Protein, Pigments and Lipid Contents of the Red Alga, *Asparagopsis taxiformes* (Delie) Trevisan. *Journal of Applied Sciences Research*, *3*(12), 1825–1834.

El-Din, S. M. (2015). Utilization of seaweed extracts as bio-fertilizers to stimulate the growth of wheat seedlings. *The Egyptian Journal of Experimental Biology*, *11*(1), 31–39.

El-Zabalawy, K. M., Abou-Shleel, S. M., & Abdel-Kareem, M. S. (2015). Effect of marine algae on bio-accumulation of heavy metals from polluted soil by some leafy vegetables. *Nature and Science*, *13*(3), 109–116.

Falcón, L. I., Magallón, S., & Castillo, A. (2010). Dating the cyanobacterial ancestor of the chloroplast. *The ISME Journal*, *4*(6), 777–783. doi:10.1038/ismej.2010.2 PMID:20200567

Fike, J. H., Allen, V. G., Schmidt, R. E., Zhang, X., Fontenot, J. P., Bagley, C. P., Ivy, R. L., Evans, R. R., Coelho, R. W., & Wester, D. B. (2001). Tasco-Forage: I. Influence of a seaweed extract on antioxidant activity in tall fescue and in ruminants. *Journal of Animal Science*, *79*(4), 1011–1021. doi:10.2527/2001.7941011x PMID:11325175

García-Salgado, S., Quijano, M. A., & Bonilla, M. M. (2012). Arsenic speciation in edible alga samples by microwave-assisted extraction and high performance liquid chromatography coupled to atomic fluorescence spectrometry. *Analytica Chimica Acta*, *714*, 38–46. doi:10.1016/j.aca.2011.12.001 PMID:22244135

Giordano, M., Norici, A., Ratti, S., & Raven, J. A. (2008). Role of sulfur for algae: acquisition, metabolism, ecology and evolution. *Sulfur Metabolism in Phototrophic Organisms*, 397-415.

Graham, J. E., Wilcox, L. W., & Graham, L. E. (2009). *Algae*. Pearson Education, University of Wisconsin.

Gupta, V., Ratha, S. K., Sood, A., Chaudhary, V., & Prasanna, R. (2013). New insights into the biodiversity and applications of cyanobacteria (blue-green algae)—Prospects and challenges. *Algal Research*, *2*(2), 79–97. doi:10.1016/j.algal.2013.01.006

Ibraheem, I. (2007). Cyanobacteria as alternative biological conditioners for bioremediation of barren soil. *Egyptian Journal of Phycology*, *8*(1), 99–117. doi:10.21608/egyjs.2007.114548

Illera-Vives, M., Labandeira, S. S., Brito, L. M., López-Fabal, A., & López-Mosquera, M. E. (2015). Evaluation of compost from seaweed and fish waste as a fertilizer for horticultural use. *Scientia Horticulturae*, *186*, 101–107. doi:10.1016/j.scienta.2015.02.008

Iqbal, M., Muhammad, G., Aslam, S., Hussain, M., Shafiq, Z., & Razzaq, H. (2021). Algal Biofertilizer. In Biofertilizers: Study and Impact. John Wiley & Sons, Inc. doi:10.1002/9781119724995.ch22

Kang, Y., Kim, M., Shim, C., Bae, S., & Jang, S. (2021). Potential of algae–bacteria synergistic effects on vegetable production. *Frontiers in Plant Science*, *12*, 556. doi:10.3389/fpls.2021.656662 PMID:33912211

Khan, W., Rayirath, U. P., Subramanian, S., Jithesh, M. N., Rayorath, P., Hodges, D. M., Critchley, A. T., Craigie, J. S., Norrie, J., & Prithiviraj, B. (2009). Seaweed extracts as biostimulants of plant growth and development. *Journal of Plant Growth Regulation*, *28*(4), 386–399. doi:10.100700344-009-9103-x

Kiple, K. F., & Ornelas, K. (2000). *The Cambridge World History of Food*. Cambridge University Press.

Kuenz, A., Grimm, D., & Rahmann, G. (2021). Versatility of algae—Exploring the potential of algae for nutrient circulation. *Organic Agriculture*, *11*(2), 251–260. doi:10.100713165-020-00308-0

Kumar, G., & Sahoo, D. (2011). Effect of seaweed liquid extract on growth and yield of *Triticum aestivum* var. Pusa Gold. *Journal of Applied Phycology*, *23*(2), 251–255. doi:10.100710811-011-9660-9

López, A., Rico, M., Rivero, A., & de Tangil, M. S. (2011). The effects of solvents on the phenolic contents and antioxidant activity of *Stypocaulon scoparium* algae extracts. *Food Chemistry*, *125*(3), 1104–1109. doi:10.1016/j.foodchem.2010.09.101

Malik, F. R., Ahmed, S., & Rizki, Y. M. (2001). Utilization of lignocellulosic waste for the preparation of nitrogenous biofertilizer. *Pakistan Journal of Biological Sciences*, *4*(10), 1217–1220. doi:10.3923/pjbs.2001.1217.1220

Mancuso, S., Briand, X., Mugnai, S., & Azzarello, E. (2006). Marine bioactive substances (IPA Extract) improve foliar ion uptake and water stress tolerance in potted" *Vitis vinifera*" plants. *Advances in Horticultural Science*, *20*(2), 1–6.

Mansori, M., Chernane, H., Latique, S., Benaliat, A., Hsissou, D., & El Kaoua, M. (2016). Effect of seaweed extract (*Ulva rigida*) on the water deficit tolerance of *Salvia officinalis* L. *Journal of Applied Phycology*, *28*(2), 1363–1370. doi:10.100710811-015-0671-9

Mansori, M., Farouk, I. A., Hsissou, D., & Kaoua, M. E. (2019). Seaweed extract treatment enhances vegetative growth and antioxidant parameters in water stressed *Salvia officinalis* L. *Journal of Materials and Environmental Sciences*, *10*(8), 756–766.

Mata, T. M., Martins, A. A., & Caetano, N. S. (2010). Microalgae for biodiesel production and other applications: A review. *Renewable & Sustainable Energy Reviews*, *14*(1), 217–232. doi:10.1016/j.rser.2009.07.020

Nabti, E., Jha, B., & Hartmann, A. (2017). Impact of seaweeds on agricultural crop production as biofertilizer. *International Journal of Environmental Science and Technology*, *14*(5), 1119–1134. doi:10.100713762-016-1202-1

Nain, L., Rana, A., Joshi, M., Jadhav, S. D., Kumar, D., Shivay, Y. S., Paul, S., & Prasanna, R. (2010). Evaluation of synergistic effects of bacterial and cyanobacterial strains as biofertilizers for wheat. *Plant and Soil*, *331*(1), 217–230. doi:10.100711104-009-0247-z

Nisha, R., Kaushik, A., & Kaushik, C. P. (2007). Effect of indigenous cyanobacterial application on structural stability and productivity of an organically poor semi-arid soil. *Geoderma*, *138*(1-2), 49–56. doi:10.1016/j.geoderma.2006.10.007

Pabbi, S. (2015). Blue green algae: a potential biofertilizer for rice. In *The Algae World* (pp. 449–465). Springer. doi:10.1007/978-94-017-7321-8_17

Pandey, K. D., Shukla, P. N., Giri, D. D., & Kashyap, A. K. (2005). Cyanobacteria in alkaline soil and the effect of cyanobacteria inoculation with pyrite amendments on their reclamation. *Biology and Fertility of Soils*, *41*(6), 451–457. doi:10.100700374-005-0846-7

Papenfus, H. B., Kulkarni, M. G., Stirk, W. A., Finnie, J. F., & van Staden, J. (2013). Effect of a commercial seaweed extract (Kelpak®) and polyamines on nutrient-deprived (N, P and K) okra seedlings. *Scientia Horticulturae*, *151*, 142–146. doi:10.1016/j.scienta.2012.12.022

Pardee, K. I., Ellis, P., Bouthillier, M., Towers, G. H. N., & French, C. J. (2004). Plant virus inhibitors from marine algae. *Canadian Journal of Botany*, *82*(3), 304–309. doi:10.1139/b04-002

Piwowar, A., & Harasym, J. (2020). The importance and prospects of the use of algae in agribusiness. *Sustainability*, *12*(14), 1–13. doi:10.3390u12145669

Prasad, R. C., & Prasad, B. N. (2001). Cyanobacteria as a source biofertilizer for sustainable agriculture in Nepal. *Journal in Plant Science Botanica Orientalis*, *1*, 127–133.

Prasanna, R., Babu, S., Bidyarani, N., Kumar, A., Triveni, S., Monga, D., Mukherjee, A. K., Kranthi, S., Gokte-Narkhedkar, N., Adak, A., Yadav, K., Nain, L., & Saxena, A. K. (2015). Prospecting cyanobacteria-fortified composts as plant growth promoting and biocontrol agents in cotton. *Experimental Agriculture*, *51*(1), 42–65. doi:10.1017/S0014479714000143

Prasanna, R., Kanchan, A., Ramakrishnan, B., Ranjan, K., Venkatachalam, S., Hossain, F., Shivay, Y. S., Krishnan, P., & Nain, L. (2016). Cyanobacteria-based bioinoculants influence growth and yields by modulating the microbial communities favourably in the rhizospheres of maize hybrids. *European Journal of Soil Biology*, *75*, 15–23. doi:10.1016/j.ejsobi.2016.04.001

Prasanna, R., Ramakrishnan, B., Simranjit, K., Ranjan, K., Kanchan, A., Hossain, F., & Nain, L. (2017). Cyanobacterial and rhizobial inoculation modulates the plant physiological attributes and nodule microbial communities of chickpea. *Archives of Microbiology*, *199*(9), 1311–1323. doi:10.100700203-017-1405-y PMID:28669069

Priya, H., Prasanna, R., Ramakrishnan, B., Bidyarani, N., Babu, S., Thapa, S., & Renuka, N. (2015). Influence of cyanobacterial inoculation on the culturable microbiome and growth of rice. *Microbiological Research*, *171*, 78–89. doi:10.1016/j.micres.2014.12.011 PMID:25644956

Raj, R., & Kumari, M. (2020). Blue green algae (BGA) and its application. *Journal of Pharmacognosy and Phytochemistry*, *9*(2), 287–296.

Rana, A., Kabi, S. R., Verma, S., Adak, A., Pal, M., Shivay, Y. S., Prasanna, R., & Nain, L. (2015). Prospecting plant growth promoting bacteria and cyanobacteria as options for enrichment of macro-and micronutrients in grains in rice–wheat cropping sequence. *Cogent Food & Agriculture*, *1*(1), 1037379. doi:10.1080/23311932.2015.1037379

Ranjan, K., Priya, H., Ramakrishnan, B., Prasanna, R., Venkatachalam, S., Thapa, S., Tiwari, R., Nain, L., Singh, R., & Shivay, Y. S. (2016). Cyanobacterial inoculation modifies the rhizosphere microbiome of rice planted to a tropical alluvial soil. *Applied Soil Ecology*, *108*, 195–203. doi:10.1016/j.apsoil.2016.08.010

Rayorath, P., Jithesh, M. N., Farid, A., Khan, W., Palanisamy, R., Hankins, S. D., Critchley, A. T., & Prithiviraj, B. (2008). Rapid bioassays to evaluate the plant growth promoting activity of *Ascophyllum nodosum* (L.) Le Jol. using a model plant, *Arabidopsis thaliana* (L.) Heynh. *Journal of Applied Phycology*, *20*(4), 423–429. doi:10.100710811-007-9280-6

Razzak, S. A., Hossain, M. M., Lucky, R. A., Bassi, A. S., & De Lasa, H. (2013). Integrated CO_2 capture, wastewater treatment and biofuel production by microalgae culturing—A review. *Renewable & Sustainable Energy Reviews*, *27*, 622–653. doi:10.1016/j.rser.2013.05.063

Renuka, N., Guldhe, A., Prasanna, R., Singh, P., & Bux, F. (2018). Microalgae as multi-functional options in modern agriculture: Current trends, prospects and challenges. *Biotechnology Advances*, *36*(4), 1255–1273. doi:10.1016/j.biotechadv.2018.04.004 PMID:29673972

Renuka, N., Prasanna, R., Sood, A., Bansal, R., Bidyarani, N., Singh, R., Shivay, Y. S., Nain, L., & Ahluwalia, A. S. (2017). Wastewater grown microalgal biomass as inoculants for improving micronutrient availability in wheat. *Rhizosphere*, *3*, 150–159. doi:10.1016/j.rhisph.2017.04.005

Roohinejad, S., Koubaa, M., Barba, F. J., Saljoughian, S., Amid, M., & Greiner, R. (2017). Application of seaweeds to develop new food products with enhanced shelf-life, quality and health-related beneficial properties. *Food Research International*, *99*, 1066–1083. doi:10.1016/j.foodres.2016.08.016 PMID:28865618

Sabh, A. Z., & Shallan, A. M. (2008). Effect of organic fertilization on Bean (*Vicia faba* L.) by using different marine macroalgae in relation to the morphological, anatomical characteristics and chemical constituents of the plant. *Australian Journal of Basic and Applied Sciences*, *4*, 1076–1091.

Safinaz, A. F., & Ragaa, A. H. (2013). Effect of some red marine algae as biofertilizers on growth of maize (*Zea mayz* L.) plants. *International Food Research Journal*, *20*(4), 1629–1632.

Sandepogu, M. (2018). *Seaweed Extract and Humic Acids Biostimulant to improve Growth and Post Harvest Quality of Spinach and Lettuce*. Univ.

Sayre, R. (2010). Microalgae: The Potential for Carbon Capture. *Bioscience*, *60*(9), 722–727. doi:10.1525/bio.2010.60.9.9

Schwencke, J., & Carú, M. (2001). Advances in actinorhizal symbiosis: Host plant-Frankia interactions, biology, and applications in arid land reclamation. A review. *Arid Land Research and Management*, *15*(4), 285–327. doi:10.1080/153249801753127615

Semary, N. A. E. (2017). Algae and Chain Aquaculture: An Approach Towards Sustainable Agriculture. Handbook of Environmental Chemistry, 5(Part N), 1-12.

Shah, M. T., Zodape, S. T., Chaudhary, D. R., Eswaran, K., & Chikara, J. (2013). Seaweed sap as an alternative liquid fertilizer for yield and quality improvement of wheat. *Journal of Plant Nutrition*, *36*(2), 192–200. doi:10.1080/01904167.2012.737886

Shariatmadari, Z., Riahi, H., Seyed Hashtroudi, M., Ghassempour, A., & Aghashariatmadary, Z. (2013). Plant growth promoting cyanobacteria and their distribution in terrestrial habitats of Iran. *Soil Science and Plant Nutrition*, *59*(4), 535–547. doi:10.1080/00380768.2013.782253

Song, T., Mårtensson, L., Eriksson, T., Zheng, W., & Rasmussen, U. (2005). Biodiversity and seasonal variation of the cyanobacterial assemblage in a rice paddy field in Fujian, China. *FEMS Microbiology Ecology*, *54*(1), 131–140. doi:10.1016/j.femsec.2005.03.008 PMID:16329979

Spinelli, F., Fiori, G., Noferini, M., Sprocatti, M., & Costa, G. (2009). Perspectives on the use of a seaweed extract to moderate the negative effects of alternate bearing in apple trees. *The Journal of Horticultural Science & Biotechnology*, *84*(6), 131–137. doi:10.1080/14620316.2009.11512610

Stirk, W. A., Ördög, V., Novák, O., Rolčík, J., Strnad, M., Bálint, P., & van Staden, J. (2013). Auxin and cytokinin relationships in 24 microalgal strains1. *Journal of Phycology*, *49*(3), 459–467. doi:10.1111/jpy.12061 PMID:27007035

Strik, W. A., Arthur, G. D., Lourens, A. F., Novok, O., Strand, M., & Van-Staden, J. (2004). Changes in seaweed concentrates when stores at an elevated temperature. *Journal of Applied Phycology*, *16*, 31–39. doi:10.1023/B:JAPH.0000019057.45363.f5

Sultana, V., Baloch, G. N., Ara, J., Ehteshamul-Haque, S., Tariq, R. M., & Athar, M. (2012). Seaweeds as an alternative to chemical pesticides for the management of root diseases of sunflower and tomato. *Journal of Applied Botany and Food Quality*, *84*(2), 162.

Swain, S. S., Paidesetty, S. K., & Padhy, R. N. (2017). Antibacterial, antifungal and antimycobacterial compounds from cyanobacteria. *Biomedecine & Pharmacotherapie*, *90*, 760–776. doi:10.1016/j.biopha.2017.04.030 PMID:28419973

Thamida, B. Z. N., Mandal, R., & Saiful, I. (2011). Effect of Cyanobacterial biofertilizer on the growth and yield components of two HYV of rice. *Journal of Algal Biomass Utilization*, *2*(1), 1–9.

Thirumaran, G., Arumugam, M., Arumugam, R., & Anantharaman, P. (2009). Effect of Seaweed Liquid Fertilizer on Growth and Pigment Concentration of *Abelmoschus esculentus* (l) medikus. *American-Eurasian Journal of Agronomy*, *2*(2), 57–66.

Trentacoste, E. M., Martinez, A. M., & Zenk, T. (2015). The place of algae in agriculture: Policies for algal biomass production. *Photosynthesis Research*, *123*(3), 305–315. doi:10.100711120-014-9985-8 PMID:24599393

Tseng, C. K. (2001). Algal biotechnology industries and research activities in China. *Journal of Applied Phycology*, *13*(4), 375–380. doi:10.1023/A:1017972812576

Tuhy, L., Chowañska, J., & Chojnacka, K. (2013). Seaweed extracts as biostimulants of plant growth [review]. *Chemik*, *67*(7), 636–641.

Tuhy, Ł., Saeid, A., & Chojnacka, K. (2020). Algae Fertilizers. Encyclopedia of Marine Biotechnology, 1, 105-119.

Tuhy, Ł., Samoraj, M., Michalak, I., & Chojnacka, K. (2014). The application of biosorption for production of micronutrient fertilizers based on waste biomass. *Applied Biochemistry and Biotechnology*, *174*(4), 1376–1392. doi:10.100712010-014-1074-0 PMID:25108517

Ullmann, J., & Grimm, D. (2021). Algae and their potential for a future bioeconomy, landless food production, and the socio-economic impact of an algae industry. *Organic Agriculture*, *11*(2), 261–267. doi:10.100713165-020-00337-9

Uysal, O., Uysal, F. O., & Ekinci, K. (2015). Evaluation of microalgae as microbial fertilizer. *European Journal of Sustainable Development*, *4*(2), 77–77. doi:10.14207/ejsd.2015.v4n2p77

Vaishampayan, A., Sinha, R. P., Hader, D. P., Dey, T., Gupta, A. K., Bhan, U., & Rao, A. L. (2001). Cyanobacterial biofertilizers in rice agriculture. *Botanical Review*, *67*(4), 453–516. doi:10.1007/BF02857893

Van Oosten, M. J., Pepe, O., De Pascale, S., Silletti, S., & Maggio, A. (2017). The role of biostimulants and bioeffectors as alleviators of abiotic stress in crop plants. *Chemical and Biological Technologies in Agriculture*, *4*(1), 1–12. doi:10.118640538-017-0089-5

Vance-Carroll, P., Uhde-Stone, C., & Allan, D. L. (2003). Phosphorus acquisition and use: Critical adaptations by plants for securing a nonrenewable resource. *The New Phytologist*, *157*(3), 423–447. doi:10.1046/j.1469-8137.2003.00695.x PMID:33873400

Verslues, P. E., Agarwal, M., Katiyar-Agarwal, S., Zhu, J., & Zhu, J. K. (2006). Methods and concepts in quantifying resistance to drought, salt and freezing, abiotic stresses that affect plant water status. *The Plant Journal*, *45*(4), 523–539. doi:10.1111/j.1365-313X.2005.02593.x PMID:16441347

Vishal, K. D., & Abhishek, C. (2014). Isolation and characterization of Rhizobium leguminosarum from root nodules of *Pisums sativum* L. *Journal of Academia and Industrial Research*, *2*(8), 464–467.

Watanabe, I., Lee, K. K., Alimagno, B. V., Sato, M., Del Rosario, D. C., & De Guzmán, M. R. (1977). *Biological nitrogen fixation in paddy field studied by in situ acetylene-reduction assays*. IRRI Research Paper Series (Philippines). no. 3.

Wells, M. L., Potin, P., Craigie, J. S., Raven, J. A., Merchant, S. S., Helliwell, K. E., Smith, A. G., Camire, M. E., & Brawley, S. H. (2017). Algae as nutritional and functional food sources: Revisiting our understanding. *Journal of Applied Phycology*, *29*(2), 949–982. doi:10.100710811-016-0974-5 PMID:28458464

World Bank. (2008). World Development Report 2008: Agriculture for Development. Washington, DC. © World Bank. *Journal of Chemical Information and Modeling*, *53*(9).

Xu, C., & Leskovar, D. I. (2015). Effects of *A. nodosum* seaweed extracts on spinach growth, physiology and nutrition value under drought stress. *Scientia Horticulturae*, *183*, 39–47. doi:10.1016/j.scienta.2014.12.004

Yilmaz, E., & Sönmez, M. (2017). The role of organic/bio–fertilizer amendment on aggregate stability and organic carbon content in different aggregate scales. *Soil & Tillage Research*, *168*, 118–124. doi:10.1016/j.still.2017.01.003

Zayadan, B. K., Matorin, D. N., Baimakhanova, G. B., Bolathan, K., Oraz, G. D., & Sadanov, A. K. (2014). Promising microbial consortia for producing biofertilizers for rice fields. *Microbiology*, *83*(4), 391–397. doi:10.1134/S0026261714040171 PMID:25844458

Zhao, Z., Liu, S., Yang, X., Lei, Z., Shimizu, K., Zhang, Z., Lee, D.-J., & Adachi, Y. (2019). Stability and performance of algal-bacterial granular sludge in shaking photo-sequencing batch reactors with special focus on phosphorus accumulation. *Bioresource Technology*, *280*, 497–501. doi:10.1016/j.biortech.2019.02.071 PMID:30819417

KEY TERMS AND DEFINITIONS

Algalization: Mass cultivation system of algae for biofertilizer production.

Antioxidant System: The system that inhibits the oxidation of membranes in plants by quenching and chelation of free radicals and redox metals, respectively.

Bio-Preparations: Any preparation that is derived directly from any living entity or the product of that entity.

Bio-Stimulants: A product obtained from living organisms or micro-organisms applied to plants for growth enhancement and abiotic stress tolerance.

IRRI: International Rice Research Institute located at Los Baños, Philippines.

Phytotron: The greenhouse facility to conduct an experiment under a strictly controlled environment.

Quorum Sensing Pathway: It is a communication mechanism that exists in bacteria to maintain control of specific processes.

Chapter 12
Role of Algae in the Production of Biomaterials

Aishwarya Mogal
Lovely Professional University, India

Shristy Gautam
https://orcid.org/0000-0002-5769-0562
Lovely Professional University, India

ABSTRACT

Algae have drawn a significant attention for the manufacture of biomaterials with unique features and extraordinary uses because of its large yields, short growth time, and variable culture conditions. Algal polymers, blends, and combinations of biomass or algal biomolecules with other polymeric materials are examples of algae-based materials. Polysaccharides and sulfate polysaccharides such as agar, carrageenan, alginates, and polyhydroxyalkanoates are essential algal polymers. Algal and biomass polymers have improved mechanical characteristics, biocompatibility, and biodegradability when included into synthetic polymer systems. Algal-based biomaterials are interesting contenders to replace existing, non-renewable polymer materials derived from fossil fuels. This chapter discusses the numerous applications of biomaterials obtained from algae. Furthermore, as biotechnology advances, algae-based polymers, blends, and composites have found many applications in a variety of domains of human existence, ranging from medicinal applications to sophisticated technological applications.

INTRODUCTION

Green materials that are renewable and biodegradable are aimed to produce and develop due to the shortage of non-renewable petroleum resources and their impact on the environment (Xia et al., 2017) at a rate 25 x 10^6 tonnes/year (Balaji et al., 2013). Green materials include bioplastic, biopolymers, and bio-composites which are driven from the bio-based materials like agricultural waste, feed stock, and plant proteins like corn zein, wheat protein, sunflower protein, and soy protein (Xia et al., 2017) to reduce greenhouse gases. Increasing population results in more production of these green materials, this would

DOI: 10.4018/978-1-6684-2438-4.ch012

be burden on the food, arable land, water supply, and would lead to the competition. There are three types of feedstocks: 1st generation feedstocks comprise edible biomass such as maize, sugarcane, and whey, which contributes to food competition in society; 2nd generation feedstocks contain non-edible biomass such as lignocellulosic feedstocks from wastes such as agriculture, forest, or energy crops, which leads to competing for land use and soil replenishment. To overcome the competition for food, agricultural land, and their byproducts, micro-algal biomass, a 3rd generation feedstock with high biomass and CO_2 capture potential (Coppola et al., 2021a; Hjuler & Hansen, 2018), is an excellent alternative to produce green materials. Alga, also known as water plant or seaweeds are present in marine, fresh, and waste water environment (Gautam & Mannan, 2020); are autotrophic organisms having capacity to transform CO_2 into biomass including carbohydrates, proteins, lipids, and fats. They are microscopic organisms having fast doubling time, requires less space and water, classified as macroalgae and microalgae. Both these algae are excellent source of vitamins, proteins, fats, carbohydrates, fibers, lipids, and secondary compounds. Macroalgae are multicellular organisms having chlorophycease, phaeophyceae, rhodophyceae, brown algae, and red algae while microalgae are microscopic having diatoms, blue green algae, and dinoflagellates (Sanjeewa et al., 2016). High-performance biodegradable polymers must be blended with low-cost macromolecules like natural fibers, proteins, and starch to develop green materials like bioplastics. Plastics have become the most extensively used material in today's society, with applications in medical devices, computer equipment packages, domestic appliances, and automobiles. (Balaji et al., 2013). Polymers, or solid macromolecules, on the other hand, are so linked with people in every part of life that it is difficult to fathom living today without them. They come from a range of sustainable sources, and because they are easy to sterilize, biodegradable, and have a long shelf life, they do not create any hazardous or inflammatory reactions (Azeem et al., 2017). Algae-based biopolymers have better mechanical properties than petroleum-based polymers, and they can be modified by adding plasticizers, additives, and compatibilizers to improve their durability, strength, and flexibility (Onen Cinar et al., 2020). The use of algae as a natural fiber to build bio-composite from residual biofuel process of extraction or water reservoir (Constante et al., 2015) is advantageous since it does not harm the environment. This chapter reviews important green materials like bioplastics, biopolymers, and bio composites derived from algae.

BIOPLASTICS

Plastics play an important role in everyday life as they have undoubted utility and convenience but they are synthesized from non-biodegradable petroleum which put up two major world crises-depletion of fossil fuels and degradation of environment by accumulating in marine and terrestrial region (García et al., 2021). Plastics are produced by chemical synthesis of polymers whose molecular weight is comparatively high. Synthetic plastics, derived by poly-condensation, polymerization or poly-addition of monomers can be grouped into four categories- thermosets, elastomers, synthetic fibers, and thermoplastics. The most common synthetic polymers are polyethylene (PE), polystyrene (PS), polyvinyl chloride (PVC), polycarbonate (PC), polypropylene (PP), and polyamides (PA). These plastics have slow breakdown, hence non-biodegradable (Coppola et al., 2021a). To overcome plastic waste accumulation in the world, an environmentally friendly, renewable and biodegradable substitute, bioplastics have taken much attention (García et al., 2021).

Bioplastics are similar to traditional plastics in terms of properties but are sustainable as they produce less greenhouse gases and comes under green material which lower carbon footprints on the environment (Agustin et al., 2014; Coppola et al., 2021a). The association of bioplastic industry, European Bioplastics (EUBP) defines bioplastic as biodegradable plastics produced from renewable resources. IUPAC defined bioplastic as a derivative of biomass or monomers which have plant origin (Vert et al., 2012). Bioplastics are also synthesized from carbohydrates that are starch-based proteins, soy proteins and other carbon sources like wastewater treatment byproducts. Thermoplastic starch is a most known bioplastic generated from enzymatic bio-reactions (Coppola et al., 2021a). Bioplastics from vegetable wastes are biodegradable and have alike mechanical properties with other bioplastics (Perotto et al., 2018). For the production of bioplastics, physical reinforcing methods like ultrasounds application, increase in mold temperature, and de-hydrothermal treatment are applied. Bioplastics has an extensive variety of thermal and mechanical characteristics. When soy protein-based bioplastics are treated with the thermal treatment, it enhances the mechanical properties, ultrasounds lead to the smaller pores in the structure, and when de-hydrothermal treatment is applied, it increases the superabsorbent capacity (García et al., 2021; Jiménez-Rosado et al., 2020).

Majority of the bioplastics are produced from the raw materials like carbohydrates and proteins which are mainly derived from the agricultural plant sources like corn, wheat, soybean, and potato, which have an impact on the arable land, resources, also pose a potential threat to food security (Fabra et al., 2017). Hence, microalgae biomass is considered to be an eco-friendly, innovative, cost effective, biocompatible, biodegradable resource for the production of bioplastics. Around 1% share of the 370 million tons of total global plastic produced is of bioplastic market which can go up to 30% until 2025 (García et al., 2021). The positive market growth projection and a large market size gave an insight that bioplastics can dramatically impact the global plastics supply chain. A large horizon of traditional plastics can be directly replaced with algae-based bioplastics which includes biodegradable resins, engineered resins and thermoplastics (Hunt, 2018).

ALGAE: AS THE PRODUCERS OF BIOPLASTICS

Algae are the primary producers of aquatic ecosystem which transform sunlight and carbon dioxide into metabolites like pigments, carbohydrates, lipids, and proteins. These can be converted into products for pharmaceuticals, nutraceuticals (Sathasivam et al., 2019), food, or for the production of green energy (Raheem et al., 2018). Microalgae procure nitrogen and phosphorous (Cuellar-Bermudez et al., 2017; Passero et al., 2015) required for their growth from waste streams, which can also be used as carbon sources to produce bioplastics, poly-hydroxy-alkanoates (PHAs) (Arias et al., 2020; Mannina et al., 2020). This attribute of microalgae combined with the sunlight as an energy source, their fast-doubling time, low water and space requirements, makes them commercially captivating substitute as PHA producing organisms (Tharani & Ananthasubramanian, 2020). Poly-hydroxy-alkanoates (PHA) obtained from microalgae acts as a raw material (Rumin et al., 2020) which are a promising alternative to replace plant source and suitable for a wide variety of applications (García et al., 2021).

Poly Hydroxy Alkanoates (PHAs)

PHAs are natural energy and carbon storage compounds (Coppola et al., 2021a), the only group of green plastics present in microalgae. They can be synthesized in nutrient deficient condition, only in the presence of carbon source (García et al., 2021). PHAs synthesis starts from bio-based renewable resources; are products of biosynthesizing secondary metabolism of prokaryotes; are biodegradable by the action of isolated enzymes or organisms; are compostable as they compost to water, carbon dioxide or methane and lastly, they are biocompatible as they do not exert any negative effects on living beings or the environment (Coppola et al., 2021a). The degradation of PHAs does not raise the carbon dioxide level in the environment, so it does not contribute to the global warming and change in the climate; hence their entire creation to disposal life cycle is a part of nature's closed cycle of carbon (Koller, 2019).

PHAs are linear polyesters, has an extensive variety of thermal and mechanical characteristics (Khatami et al., 2020) produced by chemical, enzymatic, or microbiological processes, hence have the wide range of applications. PHAs are used in medical applications (Zhang et al., 2018) which require a product with high purity to manufacture implants (Chen & Zhang, 2018; Narancic et al., 2020), drug delivery carriers, wound healing dressings (Elmowafy et al., 2019), post-surgical ulcer treatment, and cancer detection (Abd El-malek et al., 2020). They are used in the nanotechnology sector for nanoparticles, nanocomposites, and films and for the fabrication of compostable batteries, because of their mechanical properties and blending ability (García et al., 2021). Food industry has concerns with the packaging of the food stuffs. The need for packaging food with low environmental impact, low burden, high economic, and ease of customization can be achieved by biodegradable bioplastics (Jabeen et al., 2015). PHAs based bioplastics can be used in agricultural applications to make nets, grow bags, and mulch films. Nets are substitute to the polyethylene having high density, used to protect crops from winds, birds, and insects and also to increase crop's yield and quality. Grow bags are used for the seedling of the crops. Mulch films made of bioplastics are necessary to uphold the soil, control moisture, prevent contamination, and control weeds (Abd El-malek et al., 2020).

Various studies propose that there is an important linkage between PHAs biosynthesis pathway and glycogen pathway which is the major cellular carbohydrate forms in microalgae, degrading to a simpler compound glucose. The work accomplished by Koch et al., 2019 found out that the biosynthesis pathways for PHA and carbohydrates accumulation are interconnected and the products of glycogen degradation are the key precursors for PHA production. The PHA biosynthesis starts with utilizing two acetyl – CoA molecules to form one acetoacetyl-CoA which is catalyzed by β-ketothiolase (PhaA) enzyme in a condensation reaction. Later nicotinamide adenine dinucleotide phosphate (NADPH)-dependent acetoacetyl-CoA reductase (PhaB) enzyme reduces acetoacetyl-CoA molecule to D-3-hydroxybutyryl-CoA. Finally, the binding of D-3-hydroxybutyryl to an existing PHB molecule through an ester bond, giving CoA is catalyzed by PHB synthase (PhaC) enzyme. Acetyl-CoA concentration within the cell and free CoA plays an essential role in the synthesis of PHA, as it is regulated at the enzymatic level (Figure-1)

STRESS CONDITION FOR THE PRODUCTION OF BIOPLASTICS

For the production of bioplastics, environmental conditions, nutritional factors and promising species are the factors that influence production metabolism. Based on the literature, there are certain main factors which influence the production of PHAs, salinity, iron, phosphorous, nitrogen and glucose concentra-

Figure 1. Schematic overview of PHB (PHA) biosynthesis in microalgae. Acetyl Co-A, Acetyl Co-enzyme A; PHB, Polyhydroxybutyryl; TCA cycle, Citric acid cycle; HSCoA, Coenzyme A; PhaA, β-ketothiolase; PhaB, acetoacetyl-CoA reductase; PhaC, PHB synthase; NADPH+H⁺, Reduced nicotinamide adenine dinucleotide phosphate; NADP⁺, Nicotinamide adenine dinucleotide phosphate (modified from García et al., 2021).

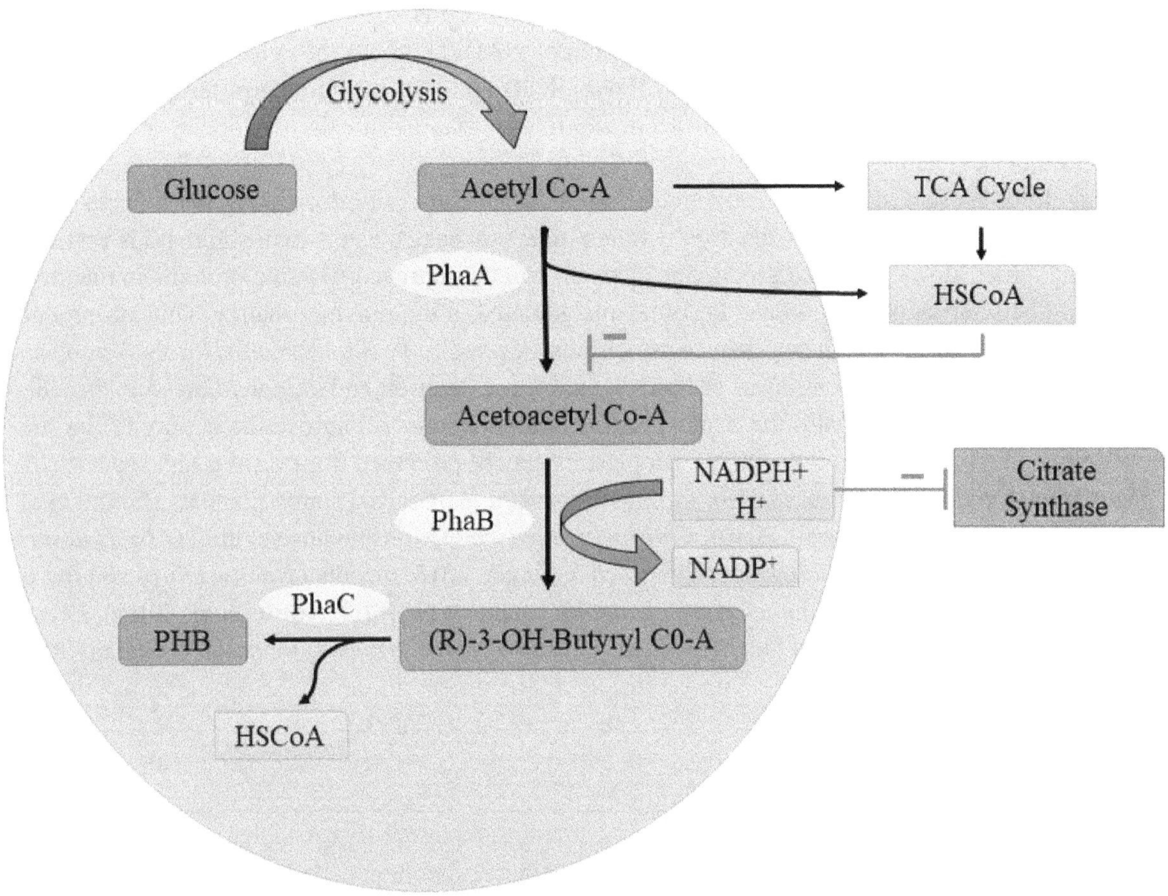

tion (García et al., 2021). Various studies stated that when microalgae are grown under normal growth conditions, concentration of PHA production was less than 3.5% w/w DW. Other studies reported 9% lower PHA concentration of the dry weight per liter. This concentration can be markedly increased when microalgae are grown under suitable culture conditions (Mendhulkar & Shetye, 2017). Microalgae culture with nutrient management is a key point linked to increase the production of PHAs. The culture medium with stress condition, especially limiting nutrient conditions (nitrogen, phosphate or sulfur), microalgae collect the biopolymers such as lipids, polysaccharides including starch, proteins or pigments. Research also gives the proof for the production of bioplastics when algae is induced to stress condition (Obruca et al., 2018). Studies have shown that when strains of *Chlamydomonas reinhardtii* are grown in sulfur deficient media, they produce up to 49% w/w of starch and demonstrated the plasticization capacity of the microalgae (Rumin et al., 2020). Other study shown that in normal growth condition, *Scenedesmus sp*. yield only 8.61% w/w DW PHA. But when they were grown in phosphate deficient media having low glucose level, the production yield increases up to 29.92%. When the *Scenedesmus sp.* were grown in iron deficient and low glucose media, the PHA accumulation was 12.2% w/w DW (García et al., 2021). *Nostoc muscorum* grown under normal growth condition and supplemented with acetate and propionate yields 31% PHA (Mallick et al., 2007) whereas *Spirulina sp.* yields 30.7% under nitrogen deficient media (Coelho et al., 2015). It has been suggested that when phosphorous is restricted, the enzymes in the PHA synthetic pathways increases the level of PHA. Phosphorous deficiency reduces NADP to yield NADPH increasing the concentration of NADPH within the cells and stimulates PHA synthesis while decreasing the size of total ATP pool. It also inhibits the citrate synthase activity in the TCA cycle, promoting PHA production and accumulations by making sure the availability of acetyl-CoA for β-ketothiolase (García et al., 2021).

Algae lipids present in the bioplastic product emits unpleasant odor decreasing the number of applications of bioplastics. Tendency of carbohydrates/sugars to make clusters in the biomass is very time consuming which slows down the production. This all can be improved with the proteins in the algae, which lengthen during production and become amalgamated in the polymer matrix. Thus, to produce higher quality bioplastics, a higher protein feedstock must be present. Two-fold benefits for algal bioplastic production is achieved by the removal of lipids by solvent extraction and carbohydrates by fermentation, which reduces or eliminates the manufacturing complications to improve the quality of the final bioplastic product (Beckstrom et al., 2020). Bioplastics may be produced from various algal species like *Chlorella pyrenoidosa* (Das et al., 2018), *Chlorella sorokiniana*-derived starch granules (Gifuni et al., 2017), *Spirulina sp.* (Zhang et al., 2020), *Chlorogloea fritschii* (Monshupanee et al., 2016), *Calothrix scytonemicola*, *Neochloris oleoabundans* (Johnsson & Steuer, 2018), residual *Nannochloropsis* after oil extraction (Yan et al., 2016), *Nannocloropsis gaditana* (Fabra et al., 2017; Torres et al., 2015), *Phaeodactylum tricornutum* (Hempel et al., 2011), and *Scenedesmus almeriensis* (Johnsson & Steuer, 2018) (Table 1 and 2).

BIOPOLYMERS

Biopolymers are the solid biomolecule, comprising distinct component called as monomers which are linked together by covalent bond (Mohan et al., 2016) in a complicated fashion to gain more complex repeating units (Rao et al., 2014). Biopolymers are derived from renewable sources which can be great alternative to fossil-based polymers which are non-renewable. Algae, the photosynthetic organisms are the fast-growing organism can be an exceptional source of biopolymers which contains long chain polysaccharides such as agar, carrageenan, alginate, (Balakrishnan et al., 2013) fucoidan, and ulvan having abundant industrial applications in pharmaceuticals, biotechnology industries, and food (Kinch et al., 2003). The ability of an algae to grow in the diverse environment and gave a high yield of the component make them an outstanding feedstock for biopolymer production (Wei et al., 2013).

ALGAE PRODUCING BIOPOLYMERS

Some of the algae which are great source of the biopolymers are brown algae like *Laminaria japonica* and *Hizikia fusiforme*, possess sulfated fucans, or fucoidans having medicinal properties to treat diseases like human immunodeficiency syndrome (AIDS) viruses, herpes, fertilization functions, cell proliferation, hepatitis, anticoagulant heparin inflammation, and adhesion (Wijesinghe & Jeon, 2012). Red algae which possess sulfated galactans and carrageenans have therapeutic properties as antiviral, antioxidant, anticoagulant, and antitumor. Green algae, Ulva lactuca consist of ulvan having anti-influenza properties to cure gastric ulcers (Azeem et al., 2017) (Table 3). Various biopolymers are produced from microalgae such as agar, Carrageenans, Alginates, Fucans and Ulvan are discussed below:

Table 1. Different algae used for the production of bioplastics

Algae	Name of the Algae	Bioplastic	Reference
Green Algae	*Chlamydomonas reinhardtii*	Bioplastic (starch)	(Rumin et al., 2020)
	Scenedesmus sp.	Bioplastic (PHA)	(García et al., 2021)
	Chlorella pyrenoidosa	Bioplastic	(Das et al., 2018)
	Chlorella sarokiniana	Bioplastic (starch granules)	(Gifuni et al., 2017)
	Neochloris oleoabundans	Bioplastic	(Johnsson & Steuer, 2018)
	Scenedesmus almeriensis	Bioplastic	(Johnsson & Steuer, 2018)
Blue Green Algae	*Nostoc muscorum*	Bioplastic (PHA)	(Mallick et al., 2007)
	Spirulina sp.	Bioplastic (PHA)	(Coelho et al., 2015)
	Chlorogloea fritschii	Bioplastic	(Monshupanee et al., 2016)
	Calothrix scytonemicola	Bioplastic	(Johnsson & Steuer, 2018)
Brown Algae	*Residual Nannochloropsis*	Bioplastic	(Yan et al., 2016)
	Nannochloropsis gaditana	Bioplastic	(Fabra et al., 2017; Torres et al., 2015)
Diatom	*Phaeodactylum tricornutum*	Bioplastic	(Hempel et al., 2011)

Table 2. Advantages and disadvantages of using algae for bioplastic production (Coppola et al., 2021b; Özçimen et al., 2017; Thomas et al., 2019)

Advantages	Disadvantages
High yield, rapid doubling time, and high biomass	Selection of the appropriate strain of algae
Easily cultured in a habitat	High cultivation and manufacturing energy costs
Capable of growing in wide range of environment, from freshwater, marine water to polluted water source	The massive scale production requires more water
Cultured on non-arable land	Ability to grow in stressful conditions, although cell growth rate slows down.
Ability to grow in stress condition	Cultivation, harvesting, and extraction on a huge scale necessitate expensive technologies.
No competition with food	-
Remediation of sewage	-

Agar

Agar is a heterogeneous compound made up of potassium, calcium, and magnesium (Rioux & Turgeon, 2015) with disaccharide 3 – linked D-galactose and 4-linked 3, 6-anhydro-L-galactose (3, 6-AG) residues in a repeating unit having methoxyl, sulfate, and/or pyruvate substituents at numerous sites. The chemical structure of agar is directly linked to the physical properties like melting temperatures, gelling, gel strength; the gelling ability of agar increases when the amount of 3, 6-anhydro-L-galactose increased. Agar gelation is conciliated with the hydrogen bonds present in helical conformation of agar polysaccharides. The gelling temperature is affected by the methoxyl group present in the agar (Azeem et al., 2017). Agar is the first additive approved by FDA and are related to carrageenans in terms of physical properties (Hernandez-Carmona et al., 2013). Agar is a polymer, divided in to two subgroups, agropectin which is an acid polymer carrying methyl, sulfate and pyruvate groups and agarose which is linear and neutral polysaccharide (Rioux & Turgeon, 2015). Extraction of agar start with acidic or alkaline pretreatment followed by hot water extraction, hot filtration, cooling, freezing and thawing, drying, milling and at last blending. Worlds total agar production that is 53% is gained by two prime species of algae, *Gracilaria* and *Gracilariopsis* (Villanueva et al., 2010). Other algal species like *Gelidiella*, *Pterocladia*, and *Gelidium* are also rich sources of agar (Azeem et al., 2017). Agar has various applications in the industries. Around 90% agar produced is greatly used in the food industries due to its high nutritional value and fiber content, in sweet, cheese (Kronberga et al., 2011), and as vegetarian gelatin and remaining 10% is used in many biotechnological and bacteriological applications (Bixler & Porse, 2011). Agar also has anti-tumor activity, anti-aggregation activity, and lower the blood glucose level (Azeem et al., 2017). Agar is also used as stabilizer and gelling agent in fish products, dairy products, water gels, sauces, and canned meat (Rioux & Turgeon, 2015).

Carrageenans

Carrageenans are made up of linear chains of D-galactose and 3, 6-anhydro-D-galactose units joined in an alternating manner with magnesium, sulfate, ammonium, potassium, and calcium. The structure of carrageenans is in three different forms, kappa (Ḱ), iota (ι), and lambda (λ). The three forms of carrageenans differ by the supfate group and (3, 6)-anhydro-d-galactose rings present in them. Kappa carrageenans consist of alternating d-galactose-4-sulfate and (3,6)-anhydro-d-galactose units while iota carrageenans differ from kappa carrageenans in sulfate group which is in position two on the (3,6)-anhydro-d-galactose units and lambda carrageenans contains notable amount of sulfate and less (3,6)-anydro-d-galactose ring. Kappa is mainly isolated from *K. alvarezii*, iota from *Eucheuma denticulatum*, and lambda from Gigartina and *Furcellaria*. Both the carrageenans, kappa and iota are hot water soluble while lambda carrageenan is cold water soluble. Carrageenans are extracted by alkaline extraction method followed by filtration and centrifugation. Carrageenans have various application in food industry like in preparing creams, milk shakes, puddings, and ice creams. Kappa carrageenans with iota carrageenans can be used in cake frostings and gel desserts. Sodium salts of kappa and lambda carrageenans are used in fruit juices (Azeem et al., 2017; Rioux & Turgeon, 2015).

Alginates

Alginate is the linear polysaccharide composed of two different monomer units, β-D-mannuronic acid (M) and α-L-guluronic acid (G) bound with β-(1,4) which are arranged randomly (Özçimen et al., 2017). Alginate is present in the cell wall providing mechanical strength and flexibility to the seaweeds (Algal-Based Biopolymers). Alginates are isolated from red algae *Gelidium*, brown algae *Laminaria saccharina*, *Ascophyllum nodosum*, *Laminaria hyperborea* (Azeem et al., 2017; Rioux & Turgeon, 2015). Alginates are obtained by pretreatment followed by the extraction method, filtration and precipitation. Alginate is used in the cell transplantation, acts as a barrier between the transplant and the immune system of host. Alginate is used as a thickening agent in in ketchup, mayonnaise, caramels, ice creams, granola bars. It is used as gelling agent in puddings, and jams. It is also used as stabilizing, and emulsifying agent. Alginate is used in encapsulating many products like probiotics, drugs, and proteins (Azeem et al., 2017).

Fucans

Fucans, sulfated polysaccharides are found in the cell wall of brown algae like *Dictyota mertensii*, *Fucus vesiculosus*, and *Spatoglossum schroederi*. The fucans have several medicinal importance having anti-microbial and antiviral properties (Queiroz et al., 2008), prevent HIV infections (Li et al., 2008). Fucans have main three types consisting of glycurona-galacatofucans, xylofucoglycuronans, and fucoidans. Out of the three types, fucoidan is an important polymer present in algae. Fucoidan is a heterogeneous polymer mainly composed of L-fucose, D-galactose, D-xylose, D-uronic acid, and L-fucose sulfate present in the extracellular matrix of brown algae (Van Weelden et al., 2019). Fucoidan is an excellent inhibitor, have anti-inflammatory, antioxidant, anti-viral, anti-coagulant, anti-tumor (Hmelkov et al., 2018), and anti-thrombosis (Azeem et al., 2017), anti-cancer activities (Etman et al., 2020), gelling and filming properties (Udani & Hesslink, 2012). Several methods to extract fucoidan from algae includes treating algal biomass with deionized water, boiled and centrifuged (Azeem et al., 2017).

Ulvan

Ulvan, the polyanionic hetero-polysaccharide are water soluble (Kim & Chojnacka, 2015) and good polymers which are extracted from cell wall of green algae, Ulva species. Ulvans are rich source of sugars like rhamnose, an important component of ulvans and also a rare sugar which is a precursor for synthesis of aroma compounds; iduronic acid which acts as an anti-coagulant substance (Azeem et al., 2017). Ulvan is mainly composed of L-rhamnose, D-glucose, D-xylose and D-glucuronic acid. The extraction of ulvan from algae is done in hot water solutions and in acidic or alkaline solution, calcium as a chelating agent (Azeem et al., 2017). Ulvan is used in tissue engineering for cartilage combined with poly D-lactic acid and also have immune-modulatory, anti-inflammatory, anti-hyperlipidemic, antioxidant, and anti-cancer activities (Cindana Mo'o et al., 2020).

BIOCOMPOSITES

Green composites that is bio-composites are low in weight and cost, gives less dependency on fossil-based composites, contain bio-resins, biopolymers, and natural fibers (Mukherjee & Kao, 2011). There

Table 3. Different algae used for the biopolymers

Algae	Name of the algae	Biopolymer	Reference
Brown Algae	*Laminaria japonica*	Sulfated fucans	(Wijesinghe & Jeon, 2012)
	Hizikia fusiforme	Sulfated fucans	(Wijesinghe & Jeon, 2012)
	Laminaria saccharina	Alginate	(Azeem et al., 2017; Rioux & Turgeon, 2015)
	Ascophyllum nodosum	Alginate	(Azeem et al., 2017; Rioux & Turgeon, 2015)
	Laminaria hyperborea	Alginate	(Azeem et al., 2017; Rioux & Turgeon, 2015)
	Dictyota mertensii	Fucans	(Queiroz et al., 2008)
	Fucus vesiculosus	Fucans	(Queiroz et al., 2008)
	Spatoglossum schroederi	Fucans	(Queiroz et al., 2008)
Green Algae	*Ulva lactuca*	Ulvan	(Azeem et al., 2017)
Red Algae	*Gracilaria sp.*	Agar	(Villanueva et al., 2010)
	Gracilariopsis sp.	Agar	(Villanueva et al., 2010)
	Gelidiella sp.	Agar	(Azeem et al., 2017)
	Pterocladia sp.	Agar	(Azeem et al., 2017)
	Gelidium sp.	Agar	(Azeem et al., 2017)
	Kappaphycus alvarezii	Kappa carrageenan	(Azeem et al., 2017; Rioux & Turgeon, 2015)
	Eucheuma denticulatum	Iota carrageenan	(Azeem et al., 2017; Rioux & Turgeon, 2015)
	Gigartina sp.	Lambda carrageenan	(Azeem et al., 2017; Rioux & Turgeon, 2015)
	Furcellaria sp.	Lambda carrageenan	(Azeem et al., 2017; Rioux & Turgeon, 2015)
	Gelidium sp.	Alginate	(Azeem et al., 2017; Rioux & Turgeon, 2015)

are various biomass-based materials used for the formulation of bio-composites, which are wheat gluten (Kunanopparat et al., 2008), cellulose (Azizi et al., 2014), bamboo fiber (Lee & Wang, 2006), agro-industrial wastes (Chiellini et al., 2001), and chitin (Davila-Rodriguez et al., 2009). But this bio-based bio-composites relies on the crops for the production which competes with the food application, feed, arable land, fertilizers, and takes time to grow to give sufficient amount of the material. To solve this problem, microalgae can be an excellent alternative for the production of bio-composites. For the production and development of bio-composites, various strains of microalgae are used such as *Spirulina sp., Nannochloropsis sp.* (Shi et al., 2012), *Chlorella sp.* (Zhang et al., 2000), *Botryococcus Braunii* (Toro et al., 2013), and red algae (*Gelidium Elegance*) (Lee et al., 2008). Bio-composite materials are generally formulated by biomass-based fillers with wide range of synthetic polymers (Tran et al., 2018), which are petroleum based synthetic thermoplastics such as polypropylene (Liu et al., 2014), low density polyethylene (Luo et al., 2013), and high-density polyethylene (Ou et al., 2014), are non-renewable, non-biodegradable and their accretion in the planet will impact the environment. The alternative to the synthetic thermoplastic which can be used with bio-based fillers is poly vinyl alcohol (PVA), poly butylene adipate-co-terephthalate (PBAT), and poly butylene succinate (PBS) having high water solubility, tensile strength, and biodegradability (Tran et al., 2018).

Table 4. Different algae used for bio-composites

Algae	Name of the algae	Reference
Blue green Algae	*Spirulina sp.*	(Shi et al., 2012)
Brown Algae	*Nannochloropsis sp.*	(Shi et al., 2012)
Green Algae	*Chlorella sp.*	(Zhang et al., 2000)
	Botryococcus braunii	(Toro et al., 2013)
Red Algae	*Gelidium elegance*	(Lee et al., 2008)

ALGAE AS SOURCE FOR BIOCOMPOSITES

Residual microalgal biomass (RMB), the residue after the production of third generation biodiesel is the best alternative to use as filler for bio-composite, reducing the cost of production, is inexpensive, and efficient to replace 20-30% of polymer matrix (Toro et al., 2013). RMB can vary between 30 – 60% of the dry biomass (Chisti, 2008), and contain negligible amount of oils, and high amount of nitrogen free extract, ashes, proteins, and carbohydrates. Biocomposites production by RMB is the best alternative to wood-based fillers and plant fibers (Ashori & Nourbakhsh, 2010; Xu et al., 2010) RMB filler can be used with the matrix of poly vinyl alcohol (PVA) (Tran et al., 2018), poly butylene adipate-co-terephthalate (PBAT), and poly butylene succinate (PBS). Bio-composites can be prepared with extrusion, the melting process which is followed by injection molding technique (Toro et al., 2013; Torres et al., 2015).

Increasing the filler content helps to improve the tensile modulus, and storage modulus of bio-composites, elucidate by a crystalline structure, physicochemical interactions, and intramolecular bonds of the bio-composites (Toro et al., 2013) while it decreases the elongation percentage and strength. Mechanical strength and tensile strength (Tran et al., 2016) of bio-composites can be achieved by giving the stress to the filler and matrix, which requires tough interface of filler-matrix to transfer stress from the matrix to filler (Toro et al., 2013). But when filler content increases, the tensile strength decreases caused due to high filler-filler interaction instead of filler-polymer interaction. The filler-filler interaction results into slippage of the polymer chains (Tran et al., 2016). To improve the mechanical properties of the bio composites, the morphology, particle size, and particle size distribution of the filler are an important variable (Toro et al., 2013). The smaller the particle size, the higher surface area of the filler resulting higher interfacial reaction of the filler-matrix (Fu et al., 2008). The application of bio-composites have not been explored much, but the use of bio-composites for the structural automotive applications such as floor covers, door trims, and interior panels offers better damping, mechanical properties, and light weight (Faruk et al., 2014; Mohanty et al., 2000) (Table 4).

BIOMATERIALS AND THEIR INTERACTION WITH THE ENVIRONMENT

Biomaterials produce from biodegradable materials have negligible impact on the environment. Biomaterials are sustainable products producing less greenhouse gases, making the environment less polluted. The biomaterials come under green materials which lower carbon footprints while degrading in the marine or terrestrial environment. Also, biomaterials are bio-based materials are biocompatible and biodegradable in the environment making it suitable to use (Agustin et al., 2014). Microalgae can generate higher biomass and can capture 816.1 gm of carbon dioxide while releasing 75% of oxygen into the atmosphere. They have an ability to grow in 40% CO_2 condition with a fixation rate of 0.73 to 2.22 g/L/day (Khan et al., 2018; Valdovinos-García et al., 2020). As a conclusion, microalgae-based bioplastics are a viable path to a circular economy. Algae-based bioplastics can be reused as peat and in subsequent biopolymer production processes, cutting biopolymer production costs and minimizing environmental impact, fostering a cyclic bio-economy (Devadas et al., 2021).

ALGAL VS. OTHER NATURAL AND SYNTHETIC FIBERS

Natural / synthetic biomaterials are produced by non-renewable petroleum resources causing depletion in the fossil fuels and also degradation of the environment. They have slow breakdown, hence considered as non-biodegradable (Coppola et al., 2021a; Xia et al., 2017). Because synthetic and natural polymers do not degrade in the environment, they collect in nature without being discarded. If this is the case, there will be more non-biodegradable items in the aquatic ecosystem than fish by 2050. To solve this, more eco-friendly and long-lasting materials created by microalgae should be used (Muneer et al., 2020). Algae is an excellent option for the production of biomaterials as they have fast doubling time, need low water and space, have ability to grow in an adverse environment and also do not compete with human food supply. They have ability to convert atmospheric carbon dioxide into biomass having carbohydrates, proteins, and lipids which are useful for the production of biomaterials (Sathasivam et al., 2019). Other benefits of algal biomaterials over natural or synthetic biomaterials: they use less fossil fuel, are non-toxic, biodegradable, and produce less energy. Natural and synthetic fiber requires arable land, 1st generation feedstocks such as maize, whey, sugarcane, and soybean, and 2nd generation lignocellulosic feedstocks, whereas algal biomaterials do not require arable land and are not reliant on edible biomass (Thomas et al., 2019). As a result, algal biomass is becoming increasingly important as a source of energy and a substitute for natural and synthetic fibers. One of the advantages of using algal biomass for biomaterials is that it does not create harmful greenhouse gases (Özçimen et al., 2017).

CHALLENGES AND FUTURE PROSPECTS

The use of algae for the production of biomaterials will have many applications in various fields like food, pharmaceuticals, biotechnology, medical, and so on (Sathasivam et al., 2019). But algal biomass harvesting and processing to make material usable needs extra cost which makes the process challenging from a sustainability point of view (Kumar et al., 2020). Identification and cultivation of suitable microalgae are critical steps in the production of biomaterials such as bioplastics, biopolymers, and bio-composites. In order to make biodegradable plastics, it is essential to choose appropriate polymers and extract those from algae, as opposed to traditional plastics. Feedstock renewability, biodegradability, polymer size, degradation rate, and moisture content should all be put into consideration while considering polymers. According to Beckstrom's research (Beckstrom et al., 2020), bioplastic generated from algal lipid does have a foul odor, which should be taken into account as plastics are utilized in a wide variety of domains like food packaging, beverage industries, appliances, textiles, and consumer products. Another challenge is large-scale algae cultivation. They can be grown in open pond or closed photo-bioreactors. Open pond systems are low-cost and simple to maintain, but their output is low, there is a substantial risk of contamination, and specific strains needed to make biomaterials are not always available. Closed photo-bioreactors have a high productivity, minimal contamination, and a high growth rate of selected microalgae strains, but they are more costly to maintain and scale up than an open pond system (Chia et al., 2020). Biomaterials derived from microalgae could contribute in the battle against waste mismanagement, as well as plastic pollution and debris in the environment (Muneer et al., 2020). With its enzyme, microalgae potentially break plastics and convert them to metabolites like carbon dioxide, water, and new biomass (Chia et al., 2020).

CONCLUSION

The ever-increasing world's population and their living standards result in the production of synthetic polymer products in a variety of applications, varying from packaging materials, beverage industries, to high-value medical devices, equipment's leads to the release of toxic materials such as dioxins into the environment, which is a major cause of global warming. Synthetic polymers also consume more time, have an influence on food chain, and provide very little biomass. Sustainable microalgae-based bioplastics, which have the ability to degrade without releasing any harmful chemicals, are easy to recycle, renewable, do not use fossil fuels, are cost effective, and require less energy to produce, can be used to encounter the needs of the population without harming the environment (Karan et al., 2019; Thiruchelvi et al., 2021). Microalgae have a simple cellular structure and the ability to dwell in harsh climates, and so they have the potential to produce biomaterials since the polymers needed for biomaterials are generated when microalgae are thriving in harsh climates. For growth, they use light energy and inorganic nutrients such as carbon dioxide, phosphorus, and nitrogen, and they can synthesize valuable molecules from biomass (Costa et al., 2019). The core focus of development should be on cutting the cost of synthesizing algae-based biodegradable polymers. At the completion of their lifecycle, algae-based biomaterials release CO_2 and water, which is eventually taken up by microalgae in the environment, releasing oxygen and becoming a third-generation feedstock for biomaterial production (Devadas et al., 2021). More improvement is still necessary for biomaterial extraction from microalgae to address economic feasibility concerns in industrial scale implementations (Costa et al., 2019).

ACKNOWLEDGMENT

This research received no specific grant from any funding agency in the public, commercial, or not-for-profit sectors.

REFERENCE

Agustin, M. B., Ahmmad, B., Alonzo, S. M. M., & Patriana, F. M. (2014). Bioplastic based on starch and cellulose nanocrystals from rice straw. *Journal of Reinforced Plastics and Composites*, *33*(24), 2205–2213. doi:10.1177/0731684414558325

Arias, D. M., García, J., & Uggetti, E. (2020). Production of polymers by cyanobacteria grown in wastewater: Current status, challenges and future perspectives. *New Biotechnology*, *55*, 46–57. doi:10.1016/j.nbt.2019.09.001 PMID:31541716

Ashori, A., & Nourbakhsh, A. (2010). Reinforced polypropylene composites: Effects of chemical compositions and particle size. *Bioresource Technology*, *101*(7), 2515–2519. doi:10.1016/j.biortech.2009.11.022 PMID:19948401

Azeem, M., Batool, F., & Iqbal, N. (2017). Algal-based biopolymers. In Algae Based Polymers, Blends, and Composites (pp. 1-31). Elsevier. doi:10.1016/B978-0-12-812360-7.00001-X

Azizi, S., Ahmad, M. B., Ibrahim, N. A., Hussein, M. Z., & Namvar, F. (2014). Preparation and properties of poly (vinyl alcohol)/chitosan blend bio-nanocomposites reinforced by cellulose nanocrystals. *Chinese Journal of Polymer Science*, *32*(12), 1620–1627. doi:10.100710118-014-1548-0

Balaji, S., Gopi, K., & Muthuvelan, B. (2013). A review on production of poly β hydroxybutyrates from cyanobacteria for the production of bio plastics. *Algal Research*, *2*(3), 278–285. doi:10.1016/j.algal.2013.03.002

Balakrishnan, C., Jenifer, P., & Esakkilingam, M. (2013). Algal documentation and phytochemical studies of red algae Gracilaria corticata of Manapad Coast, Tamil Nadu. *Journal of Pharmacognosy and Phytochemistry*, *2*(4).

Beckstrom, B. D., Wilson, M. H., Crocker, M., & Quinn, J. C. (2020). Bioplastic feedstock production from microalgae with fuel co-products: A techno-economic and life cycle impact assessment. *Algal Research*, *46*, 101769. doi:10.1016/j.algal.2019.101769

Bixler, H. J., & Porse, H. (2011). A decade of change in the seaweed hydrocolloids industry. *Journal of Applied Phycology*, *23*(3), 321–335. doi:10.100710811-010-9529-3

Chen, G.-Q., & Zhang, J. (2018). Microbial polyhydroxyalkanoates as medical implant biomaterials. *Artificial Cells, Nanomedicine, and Biotechnology*, *46*(1), 1–18. doi:10.1080/21691401.2017.1371185 PMID:28849679

Chia, W. Y., Tang, D. Y. Y., Khoo, K. S., Lup, A. N. K., & Chew, K. W. (2020). Nature's fight against plastic pollution: Algae for plastic biodegradation and bioplastics production. *Environmental Science and Ecotechnology*, *4*, 100065. doi:10.1016/j.ese.2020.100065

Chiellini, E., Cinelli, P., Imam, S. H., & Mao, L. (2001). Composite films based on biorelated agro-industrial waste and poly (vinyl alcohol). Preparation and mechanical properties characterization. *Biomacromolecules*, *2*(3), 1029–1037. doi:10.1021/bm010084j PMID:11710006

Chisti, Y. (2008). Biodiesel from microalgae beats bioethanol. *Trends in Biotechnology*, *26*(3), 126–131. doi:10.1016/j.tibtech.2007.12.002 PMID:18221809

Cindana Mo'o, F. R., Wilar, G., Devkota, H. P., & Wathoni, N. (2020). Ulvan, a polysaccharide from macroalga Ulva sp.: A review of chemistry, biological activities and potential for food and biomedical applications. *Applied Sciences (Basel, Switzerland)*, *10*(16), 5488. doi:10.3390/app10165488

Coelho, V. C., da Silva, C. K., Terra, A. L., Costa, J. A. V., & de Morais, M. G. (2015). Polyhydroxybutyrate production by Spirulina sp. LEB 18 grown under different nutrient concentrations. *African Journal of Microbiological Research*, *9*(24), 1586–1594. doi:10.5897/AJMR2015.7530

Constante, A., Pillay, S., Ning, H., & Vaidya, U. K. (2015). Utilization of algae blooms as a source of natural fibers for biocomposite materials: Study of morphology and mechanical performance of Lyngbya fibers. *Algal Research*, *12*, 412–420. doi:10.1016/j.algal.2015.10.005

Coppola, G., Gaudio, M. T., Lopresto, C. G., Calabro, V., Curcio, S., & Chakraborty, S. (2021a). Bioplastic from renewable biomass: a facile solution for a greener environment. *Earth Systems and Environment*, 1-21.

Coppola, G., Gaudio, M. T., Lopresto, C. G., Calabro, V., Curcio, S., & Chakraborty, S. (2021b). Bioplastic from renewable biomass: A facile solution for a greener environment. *Earth Systems and Environment*, *5*(2), 231–251. doi:10.100741748-021-00208-7

Costa, S. S., Miranda, A. L., de Morais, M. G., Costa, J. A. V., & Druzian, J. I. (2019). Microalgae as source of polyhydroxyalkanoates (PHAs)—A review. *International Journal of Biological Macromolecules*, *131*, 536–547. doi:10.1016/j.ijbiomac.2019.03.099 PMID:30885732

Cuellar-Bermudez, S. P., Aleman-Nava, G. S., Chandra, R., Garcia-Perez, J. S., Contreras-Angulo, J. R., Markou, G., Muylaert, K., Rittmann, B. E., & Parra-Saldivar, R. (2017). Nutrients utilization and contaminants removal. A review of two approaches of algae and cyanobacteria in wastewater. *Algal Research*, *24*, 438–449. doi:10.1016/j.algal.2016.08.018

Das, S. K., Sathish, A., & Stanley, J. (2018). Production of biofuel and bioplastic from chlorella pyrenoidosa. *Materials Today: Proceedings*, *5*(8), 16774–16781. doi:10.1016/j.matpr.2018.06.020

Davila-Rodriguez, J. L., Escobar-Barrios, V. A., Shirai, K., & Rangel-Mendez, J. R. (2009). Synthesis of a chitin-based biocomposite for water treatment: Optimization for fluoride removal. *Journal of Fluorine Chemistry*, *130*(8), 718–726. doi:10.1016/j.jfluchem.2009.05.012

Devadas, V. V., Khoo, K. S., Chia, W. Y., Chew, K. W., Munawaroh, H. S. H., Lam, M.-K., Lim, J.-W., Ho, Y.-C., Lee, K. T., & Show, P. L. (2021). Algae biopolymer towards sustainable circular economy. *Bioresource Technology*, *325*, 124702. doi:10.1016/j.biortech.2021.124702 PMID:33487515

El-malek, F. A., Khairy, H., Farag, A., & Omar, S. (2020). The sustainability of microbial bioplastics, production and applications. *International Journal of Biological Macromolecules*, *157*, 319–328. doi:10.1016/j.ijbiomac.2020.04.076 PMID:32315677

Elmowafy, E., Abdal-Hay, A., Skouras, A., Tiboni, M., Casettari, L., & Guarino, V. (2019). Polyhydroxyalkanoate (PHA): Applications in drug delivery and tissue engineering. *Expert Review of Medical Devices*, *16*(6), 467–482. doi:10.1080/17434440.2019.1615439 PMID:31058550

Etman, S. M., Elnaggar, Y. S., & Abdallah, O. Y. (2020). Fucoidan, a natural biopolymer in cancer combating: From edible algae to nanocarrier tailoring. *International Journal of Biological Macromolecules*, *147*, 799–808. doi:10.1016/j.ijbiomac.2019.11.191 PMID:31770552

Fabra, M. J., Martínez Sanz, M., Gómez Mascaraque, L. G., Coll-Marqués, J. M., Martínez, J. C., & López-Rubio, A. (2017). Development and characterization of hybrid corn starch-microalgae films: Effect of ultrasound pre-treatment on structural, barrier and mechanical performance. *Algal Research*, *28*, 80–87. doi:10.1016/j.algal.2017.10.010

Faruk, O., Bledzki, A. K., Fink, H. P., & Sain, M. (2014). Progress report on natural fiber reinforced composites. *Macromolecular Materials and Engineering*, *299*(1), 9–26. doi:10.1002/mame.201300008

Fu, S.-Y., Feng, X.-Q., Lauke, B., & Mai, Y.-W. (2008). Effects of particle size, particle/matrix interface adhesion and particle loading on mechanical properties of particulate–polymer composites. *Composites. Part B, Engineering*, *39*(6), 933–961. doi:10.1016/j.compositesb.2008.01.002

García, G., Sosa-Hernández, J. E., Rodas-Zuluaga, L. I., Castillo-Zacarías, C., Iqbal, H., & Parra-Saldívar, R. (2021). Accumulation of PHA in the microalgae Scenedesmus sp. under nutrient-deficient conditions. *Polymers*, *13*(1), 131. doi:10.3390/polym13010131 PMID:33396913

Gautam, S., & Mannan, M. A.-u. (2020). The Role of Algae in Nutraceutical and Pharmaceutical Production. In *Bioactive Natural products in Drug Discovery* (pp. 665–685). Springer. doi:10.1007/978-981-15-1394-7_24

Gifuni, I., Olivieri, G., Krauss, I. R., D'Errico, G., Pollio, A., & Marzocchella, A. (2017). Microalgae as new sources of starch: Isolation and characterization of microalgal starch granules. *Chemical Engineering Transactions*, *57*, 1423–1428.

Hempel, F., Bozarth, A. S., Lindenkamp, N., Klingl, A., Zauner, S., Linne, U., Steinbüchel, A., & Maier, U. G. (2011). Microalgae as bioreactors for bioplastic production. *Microbial Cell Factories*, *10*(1), 1–6. doi:10.1186/1475-2859-10-81 PMID:22004563

Hernandez-Carmona, G., Freile-Pelegrín, Y., & Hernández-Garibay, E. (2013). Conventional and alternative technologies for the extraction of algal polysaccharides. In *Functional ingredients from algae for foods and nutraceuticals* (pp. 475–516). Elsevier. doi:10.1533/9780857098689.3.475

Hjuler, S. V., & Hansen, S. B. (2018). LCA of Biofuels and Biomaterials. In *Life cycle assessment* (pp. 755–782). Springer. doi:10.1007/978-3-319-56475-3_30

Hmelkov, A. B., Zvyagintseva, T. N., Shevchenko, N. M., Rasin, A. B., & Ermakova, S. P. (2018). Ultrasound-assisted extraction of polysaccharides from brown alga Fucus evanescens. Structure and biological activity of the new fucoidan fractions. *Journal of Applied Phycology*, *30*(3), 2039–2046. doi:10.100710811-017-1342-9

Hunt, R. (2018). *Bloom Foam: Transforming Green Water Into Green Products*. Academic Press.

Jabeen, N., Majid, I., & Nayik, G. A. (2015). Bioplastics and food packaging: A review. *Cogent Food & Agriculture*, *1*(1), 1117749.

Jiménez-Rosado, M., Bouroudian, E., Perez-Puyana, V., Guerrero, A., & Romero, A. (2020). Evaluation of different strengthening methods in the mechanical and functional properties of soy protein-based bioplastics. *Journal of Cleaner Production*, *262*, 121517. doi:10.1016/j.jclepro.2020.121517

Karan, H., Funk, C., Grabert, M., Oey, M., & Hankamer, B. (2019). Green bioplastics as part of a circular bioeconomy. *Trends in Plant Science*, *24*(3), 237–249. doi:10.1016/j.tplants.2018.11.010 PMID:30612789

Khan, M. I., Shin, J. H., & Kim, J. D. (2018). The promising future of microalgae: Current status, challenges, and optimization of a sustainable and renewable industry for biofuels, feed, and other products. *Microbial Cell Factories*, *17*(1), 1–21. doi:10.118612934-018-0879-x PMID:29506528

Khatami, K., Perez-Zabaleta, M., Owusu-Agyeman, I., & Cetecioglu, Z. (2020). *Waste to bioplastics: How close are we to sustainable polyhydroxyalkanoates production?* Waste Management.

Kim, S.-K., & Chojnacka, K. (2015). *Marine algae extracts: processes, products, and applications*. John Wiley & Sons. doi:10.1002/9783527679577

Kinch, J., Bagita, J., & Bate, M. (2003). *Exploring the potential for seaweed farming in Milne Bay Province*. Fisheries Newsletter-South Pacific Commission.

Koller, M. (2019). Switching from petro-plastics to microbial polyhydroxyalkanoates (PHA): The biotechnological escape route of choice out of the plastic predicament. *EuroBiotech J*, *3*(1), 32–44. doi:10.2478/ebtj-2019-0004

Kronberga, M., Karklina, D., Murniece, I., & Kruma, Z. (2011). Changes of agar-agar gel properties after replacing sucrose by inulin syrup. *Proceedings of 6th Baltic Conference on Food Science and Technology FOODBALT-2011, 'Innovations for Food Science and Production'*. Latvia University of Agriculture.

Kumar, M., Sun, Y., Rathour, R., Pandey, A., Thakur, I. S., & Tsang, D. C. (2020). Algae as potential feedstock for the production of biofuels and value-added products: Opportunities and challenges. *The Science of the Total Environment*, *716*, 137116.

Kunanopparat, T., Menut, P., Morel, M.-H., & Guilbert, S. (2008). Reinforcement of plasticized wheat gluten with natural fibers: From mechanical improvement to deplasticizing effect. *Composites. Part A, Applied Science and Manufacturing*, *39*(5), 777–785. doi:10.1016/j.compositesa.2008.02.001

Lee, M. W., Han, S. O., & Seo, Y. B. (2008). Red algae fibre/poly (butylene succinate) biocomposites: The effect of fibre content on their mechanical and thermal properties. *Composites Science and Technology*, *68*(6), 1266–1272. doi:10.1016/j.compscitech.2007.12.016

Lee, S.-H., & Wang, S. (2006). Biodegradable polymers/bamboo fiber biocomposite with bio-based coupling agent. *Composites. Part A, Applied Science and Manufacturing*, *37*(1), 80–91. doi:10.1016/j.compositesa.2005.04.015

Li, B., Lu, F., Wei, X., & Zhao, R. (2008). Fucoidan: Structure and bioactivity. *Molecules (Basel, Switzerland)*, *13*(8), 1671–1695. doi:10.3390/molecules13081671 PMID:18794778

Liu, R., Peng, Y., Cao, J., & Chen, Y. (2014). Comparison on properties of lignocellulosic flour/polymer composites by using wood, cellulose, and lignin flours as fillers. *Composites Science and Technology*, *103*, 1–7. doi:10.1016/j.compscitech.2014.08.005

Luo, X., Li, J., Feng, J., Xie, S., & Lin, X. (2013). Evaluation of distillers grains as fillers for low density polyethylene: Mechanical, rheological and thermal characterization. *Composites Science and Technology*, *89*, 175–179. doi:10.1016/j.compscitech.2013.10.001

Mallick, N., Gupta, S., Panda, B., & Sen, R. (2007). Process optimization for poly (3-hydroxybutyrate-co-3-hydroxyvalerate) co-polymer production by Nostoc muscorum. *Biochemical Engineering Journal*, *37*(2), 125–130. doi:10.1016/j.bej.2007.04.002

Mannina, G., Presti, D., Montiel-Jarillo, G., Carrera, J., & Suárez-Ojeda, M. E. (2020). Recovery of polyhydroxyalkanoates (PHAs) from wastewater: A review. *Bioresource Technology*, *297*, 122478. doi:10.1016/j.biortech.2019.122478 PMID:31810735

Mendhulkar, V. D., & Shetye, L. A. (2017). Synthesis of biodegradable polymer polyhydroxyalkanoate (PHA) in cyanobacteria Synechococcus elongates under mixotrophic nitrogen-and phosphate-mediated stress conditions. *Industrial Biotechnology (New Rochelle, N.Y.)*, *13*(2), 85–93. doi:10.1089/ind.2016.0021

Mohan, S., Oluwafemi, O. S., Kalarikkal, N., Thomas, S., & Songca, S. P. (2016). Biopolymers–application in nanoscience and nanotechnology. *Recent Advances in Biopolymers, 1*(1), 47-66.

Mohanty, A., Misra, M., & Hinrichsen, G. (2000). Biofibres, biodegradable polymers and biocomposites: An overview. *Macromolecular Materials and Engineering*, *276*(1), 1–24. doi:10.1002/(SICI)1439-2054(20000301)276:1<1::AID-MAME1>3.0.CO;2-W

Monshupanee, T., Nimdach, P., & Incharoensakdi, A. (2016). Two-stage (photoautotrophy and heterotrophy) cultivation enables efficient production of bioplastic poly-3-hydroxybutyrate in auto-sedimenting cyanobacterium. *Scientific Reports*, *6*(1), 1–9. doi:10.1038rep37121 PMID:27845413

Mukherjee, T., & Kao, N. (2011). PLA based biopolymer reinforced with natural fibre: A review. *Journal of Polymers and the Environment*, *19*(3), 714–725. doi:10.100710924-011-0320-6

Muneer, F., Rasul, I., Azeem, F., Siddique, M. H., Zubair, M., & Nadeem, H. (2020). Microbial polyhydroxyalkanoates (PHAs): Efficient replacement of synthetic polymers. *Journal of Polymers and the Environment*, *28*(9), 2301–2323. doi:10.100710924-020-01772-1

Narancic, T., Cerrone, F., Beagan, N., & O'Connor, K. E. (2020). Recent advances in bioplastics: Application and biodegradation. *Polymers*, *12*(4), 920. doi:10.3390/polym12040920 PMID:32326661

Obruca, S., Sedlacek, P., Koller, M., Kucera, D., & Pernicova, I. (2018). Involvement of polyhydroxyalkanoates in stress resistance of microbial cells: Biotechnological consequences and applications. *Biotechnology Advances*, *36*(3), 856–870. doi:10.1016/j.biotechadv.2017.12.006 PMID:29248684

Onen Cinar, S., Chong, Z. K., Kucuker, M. A., Wieczorek, N., Cengiz, U., & Kuchta, K. (2020). Bioplastic production from microalgae: A review. *International Journal of Environmental Research and Public Health*, *17*(11), 3842. doi:10.3390/ijerph17113842 PMID:32481700

Ou, R., Xie, Y., Wolcott, M. P., Yuan, F., & Wang, Q. (2014). Effect of wood cell wall composition on the rheological properties of wood particle/high density polyethylene composites. *Composites Science and Technology*, *93*, 68–75. doi:10.1016/j.compscitech.2014.01.001

Özçimen, D., İnan, B., Morkoç, O., & Efe, A. (2017). A review on algal biopolymers. *J. Chem. Eng. Res. Updates*, *4*, 7–14. doi:10.15377/2409-983X.2017.04.2

Passero, M., Cragin, B., Coats, E. R., McDonald, A. G., & Feris, K. (2015). Dairy wastewaters for algae cultivation, polyhydroxyalkanote reactor effluent versus anaerobic digester effluent. *BioEnergy Research*, *8*(4), 1647–1660. doi:10.100712155-015-9619-9

Perotto, G., Ceseracciu, L., Simonutti, R., Paul, U. C., Guzman-Puyol, S., Tran, T.-N., Bayer, I. S., & Athanassiou, A. (2018). Bioplastics from vegetable waste via an eco-friendly water-based process. *Green Chemistry*, *20*(4), 894–902. doi:10.1039/C7GC03368K

Queiroz, K., Medeiros, V., Queiroz, L., Abreu, L., Rocha, H., Ferreira, C., Jucá, M., Aoyama, H., & Leite, E. (2008). Inhibition of reverse transcriptase activity of HIV by polysaccharides of brown algae. *Biomedicine and Pharmacotherapy*, *62*(5), 303–307. doi:10.1016/j.biopha.2008.03.006 PMID:18455359

Raheem, A., Prinsen, P., Vuppaladadiyam, A. K., Zhao, M., & Luque, R. (2018). A review on sustainable microalgae based biofuel and bioenergy production: Recent developments. *Journal of Cleaner Production*, *181*, 42–59. doi:10.1016/j.jclepro.2018.01.125

Rao, M. G., Bharathi, P., & Akila, R. (2014). A comprehensive review on biopolymers. *Sci. Revs. Chem. Commun*, *4*(2), 61–68.

Rioux, L.-E., & Turgeon, S. L. (2015). Seaweed carbohydrates. In *Seaweed sustainability* (pp. 141–192). Elsevier. doi:10.1016/B978-0-12-418697-2.00007-6

Rumin, J., Nicolau, E., Gonçalves de Oliveira, R. Junior, Fuentes-Grünewald, C., & Picot, L. (2020). Analysis of scientific research driving microalgae market opportunities in Europe. *Marine Drugs*, *18*(5), 264. doi:10.3390/md18050264 PMID:32443631

Sanjeewa, K. K. A., Fernando, I. P. S., Samarakoon, K. W., Lakmal, H. H. C., Kim, E.-A., Kwon, O.-N., Dilshara, M. G., Lee, J.-B., & Jeon, Y.-J. (2016). Anti-inflammatory and anti-cancer activities of sterol rich fraction of cultured marine microalga Nannochloropsis oculata. *Algae - Korean Phycological Society*, *31*(3), 277–287. doi:10.4490/algae.2016.31.6.29

Sathasivam, R., Radhakrishnan, R., & Hashem, A., & Abd Allah, E. F. (2019). Microalgae metabolites: A rich source for food and medicine. *Saudi Journal of Biological Sciences*, *26*(4), 709–722. doi:10.1016/j.sjbs.2017.11.003 PMID:31048995

Shi, B., Wideman, G., & Wang, J. H. (2012). A new approach of bioCO 2 fixation by thermoplastic processing of microalgae. *Journal of Polymers and the Environment*, *20*(1), 124–131. doi:10.100710924-011-0329-x

Tharani, D., & Ananthasubramanian, M. (2020). Microalgae as sustainable producers of bioplastic. In *Microalgae biotechnology for food, health and high value products* (pp. 373–396). Springer. doi:10.1007/978-981-15-0169-2_11

Thiruchelvi, R., Das, A., & Sikdar, E. (2021). Bioplastics as better alternative to petro plastic. *Materials Today: Proceedings*, *37*, 1634–1639. doi:10.1016/j.matpr.2020.07.176

Thomas, N. E., Balasubramanian, D. R., & Anamika, G. (2019). Microalgae as an Alternative Source for Bio-Plastic Production: A Review. *Baselius Researcher*, 116.

Toro, C., Reddy, M. M., Navia, R., Rivas, M., Misra, M., & Mohanty, A. K. (2013). Characterization and application in biocomposites of residual microalgal biomass generated in third generation biodiesel. *Journal of Polymers and the Environment*, *21*(4), 944–951. doi:10.100710924-013-0609-8

Torres, S., Navia, R., Campbell Murdy, R., Cooke, P., Misra, M., & Mohanty, A. K. (2015). Green composites from residual microalgae biomass and poly (butylene adipate-co-terephthalate): Processing and plasticization. *ACS Sustainable Chemistry & Engineering*, *3*(4), 614–624. doi:10.1021c500753h

Tran, D.-T., Lee, H. R., Jung, S., Park, M. S., & Yang, J.-W. (2018). Lipid-extracted algal biomass based biocomposites fabrication with poly (vinyl alcohol). *Algal Research*, *31*, 525–533. doi:10.1016/j.algal.2016.08.016

Tran, D. T., Lee, H. R., Jung, S., Park, M. S., Yang, J. W., & Chang, Y. K. (2016). Preparation and characterization of poly (vinyl alcohol) biocomposites with microalgae ash. *Journal of Applied Polymer Science*, *133*(26). Advance online publication. doi:10.1002/app.43599

Udani, J., & Hesslink, R. (2012). The potential use of fucoidans from brown seaweed as a dietary supplement. *Journal of Nutrition & Food Sciences*, *2*(171), 2. doi:10.4172/2155-9600.1000171

Valdovinos-García, E. M., Barajas-Fernández, J., Olán-Acosta, M. Á., Petriz-Prieto, M. A., Guzmán-López, A., & Bravo-Sánchez, M. G. (2020). Techno-economic study of CO2 capture of a thermoelectric plant using microalgae (Chlorella vulgaris) for production of feedstock for bioenergy. *Energies*, *13*(2), 413. doi:10.3390/en13020413

Van Weelden, G., Bobiński, M., Okła, K., Van Weelden, W. J., Romano, A., & Pijnenborg, J. (2019). Fucoidan structure and activity in relation to anti-cancer mechanisms. *Marine Drugs*, *17*(1), 32. doi:10.3390/md17010032 PMID:30621045

Vert, M., Doi, Y., Hellwich, K.-H., Hess, M., Hodge, P., Kubisa, P., Rinaudo, M., & Schué, F. (2012). Terminology for biorelated polymers and applications (IUPAC Recommendations 2012). *Pure and Applied Chemistry*, *84*(2), 377–410. doi:10.1351/PAC-REC-10-12-04

Villanueva, R., Sousa, A., Gonçalves, M., Nilsson, M., & Hilliou, L. (2010). Production and properties of agar from the invasive marine alga, Gracilaria vermiculophylla (Gracilariales, Rhodophyta). *Journal of Applied Phycology*, *22*(2), 211–220. doi:10.100710811-009-9444-7

Wei, N., Quarterman, J., & Jin, Y.-S. (2013). Marine macroalgae: An untapped resource for producing fuels and chemicals. *Trends in Biotechnology*, *31*(2), 70–77. doi:10.1016/j.tibtech.2012.10.009 PMID:23245657

Wijesinghe, W., & Jeon, Y.-J. (2012). Biological activities and potential industrial applications of fucose rich sulfated polysaccharides and fucoidans isolated from brown seaweeds: A review. *Carbohydrate Polymers*, *88*(1), 13–20. doi:10.1016/j.carbpol.2011.12.029

Xia, W., Zhu, N., Ni, Z., & Chen, M. (2017). Fabrication and characterization of biodegradable composites from poly (butylene succinate-co-butylene adipate) and Taihu Lake blue algae. *Advanced Composites Letters*, *26*(5), 096369351702600501. doi:10.1177/096369351702600501

Xu, Y., Wu, Q., Lei, Y., & Yao, F. (2010). Creep behavior of bagasse fiber reinforced polymer composites. *Bioresource Technology*, *101*(9), 3280–3286. doi:10.1016/j.biortech.2009.12.072 PMID:20064712

Yan, C., Wang, R., Wan, J., Zhang, Q., Xue, S., Wu, X., Zhang, J., Zhang, J., Lu, Y., & Cong, W. (2016). Cellulose/microalgae composite films prepared in ionic liquids. *Algal Research*, *20*, 135–141. doi:10.1016/j.algal.2016.09.024

Zhang, C., Wang, C., Cao, G., Wang, D., & Ho, S.-H. (2020). A sustainable solution to plastics pollution: An eco-friendly bioplastic film production from high-salt contained Spirulina sp. residues. *Journal of Hazardous Materials*, *388*, 121773. doi:10.1016/j.jhazmat.2019.121773 PMID:31836373

Zhang, F., Kabeya, H., Kitagawa, R., Hirotsu, T., Yamashita, M., & Otsuki, T. (2000). An exploratory research of PVC-Chlorella composite material (PCCM) as effective utilization of Chlorella biologically fixing CO2. *Journal of Materials Science*, *35*(10), 2603–2609. doi:10.1023/A:1004779415778

Zhang, J., Shishatskaya, E. I., Volova, T. G., da Silva, L. F., & Chen, G.-Q. (2018). Polyhydroxyalkanoates (PHA) for therapeutic applications. *Materials Science and Engineering C*, *86*, 144–150. doi:10.1016/j.msec.2017.12.035 PMID:29525089

ADDITIONAL READING

Abdelhamid, M. A., & Pack, S. P. (2021). Biomimetic and bioinspired silicifications: Recent advances for biomaterial design and applications. *Acta Biomaterialia*, *120*, 38–56. doi:10.1016/j.actbio.2020.05.017 PMID:32447061

Adiram-Filiba, N., Geiger, Y., Kumar, S., Keinan-Adamsky, K., Elbaum, R., & Goobes, G. (2020). Peptides from diatoms and grasses harness phosphate ion binding to silica to help regulate biomaterial structure. *Acta Biomaterialia*, *112*, 286–297. doi:10.1016/j.actbio.2020.05.006 PMID:32434074

Fernández, F. G. A., Reis, A., Wijffels, R. H., Barbosa, M., Verdelho, V., & Llamas, B. (2021). The role of microalgae in the bioeconomy. *New Biotechnology*, *61*, 99–107. doi:10.1016/j.nbt.2020.11.011 PMID:33249179

Orejuela-Escobar, L., Gualle, A., Ochoa-Herrera, V., & Philippidis, G. P. (2021). Prospects of Microalgae for Biomaterial Production and Environmental Applications at Biorefineries. *Sustainability*, *13*(6), 3063. doi:10.3390u13063063

Salati, M. A., Khazai, J., Tahmuri, A. M., Samadi, A., Taghizadeh, A., Taghizadeh, M., Zarrintaj, P., Ramsey, J. D., Habibzadeh, S., Seidi, F., Saeb, M. R., & Mozafari, M. (2020). Agarose-based biomaterials: Opportunities and challenges in cartilage tissue engineering. *Polymers*, *12*(5), 1150. doi:10.3390/polym12051150 PMID:32443422

Wang, Y., Xue, Y., Zhang, T., Fang, Q., Jin, M., Wang, X., Wang, Z., Hu, Y., Zhao, W., Lou, D., & Tan, W.-Q. (2021). Photosynthetic biomaterials: Applications of photosynthesis in algae as oxygenerator in biomedical therapies. *Bio-Design and Manufacturing*, *4*(3), 1–16. doi:10.100742242-021-00129-4

Chapter 13
A Sustainable Supply Chain Model for the Development of Green Fuel Production From Microalgae

Nima Norouzi
https://orcid.org/0000-0002-2546-4288
Bournemouth University, UK

ABSTRACT

This study presents designing and managing a green fuel supply chain based on algae to investigate the development of such fuels in the country. On this basis, a definitive model is first developed to model all the activities of the green fuel supply chain, which includes the supply of raw materials for the growth of algae, the cultivation of algae and their conversion into fuel, and finally, the supply of fuel in the country. This deterministic model is extended to a robust network model to secure supply chain decisions against uncertainty. Using the proposed model for the development of algal fuels in Iran shows that the green fuel production cost is currently 27 cents/liter. The current cost of producing fuel from algae cannot compete with fossil fuels, but this cost can be greatly reduced in the future by slightly increasing the growth rate of algae and their oil content.

INTRODUCTION

Environmental pollution from fossil fuel consumption, together with rapid consumption, limited oil reserves, and increasing energy demand are the most important motivations for developing new energy. Among new energy sources, biomass-based green fuels have gained considerable importance in recent years. This is because biomass fuels can replace fossil fuels without changing the transport fleet (Yue et al., 2014). In addition, the uptake of carbon dioxide during growth reduces the net amount of carbon dioxide entering the atmosphere and consequently reduces environmental hazards (Sims et al., 2010). To date, various raw materials have been used to produce green fuels. The most important raw materials

DOI: 10.4018/978-1-6684-2438-4.ch013

are corn, sugarcane, soybeans, and oilseeds (Mata et al., 2010). Although the production of fuel from these materials has been done economically in many parts of the world, it has faced severe criticism in recent years because these fuels are extracted from the edible food and adversely affect the agricultural market (Davis et al., 2011). Such problems have led to attention being paid to the use of non-food primary sources for fuel production. The most important primary non-food sources known as the second generation of fossil fuels are corn husks, grass, spruce, jatropha, and bamboo (Sims, 2010). Although this generation has fewer adverse effects on food markets, they do not have high productivity and need adequate water and fertilizers for the agricultural sector, making them unable to produce green fuel on a large scale. Among the types of raw materials without a food source, microalgae (tiny-celled species of algae) have many biological and technical properties and characteristics that have made them one of the most desirable and promising raw materials for green fuel production in coming years. The most important advantages of microalgae are high growth rate as well as high oil storage capacity (Sims, 2010), the ability to grow in brackish water and wastewater, which can help in minimizing the need for fresh water. In addition, microalgae having the capability to use the greenhouse gases emitted from the power plants and converted into fuel in their cells (Maity et al., 2014).

With increasing attention to green fuels from algae in different parts of the world, the importance of managing and planning the supply chain of green fuels based on algae is felt more than ever. Optimal and efficient supply chain network design models and determining the optimal green fuel supply chain decisions extracted from algae show which areas in the field of algae fuels need to be directed to the best conditions for growth. Provide this industry faster. In general, green fuel supply chain network design models extracted from biomass can be classified into three categories (Balaman & Selim, 2014). The first category is related to green fuel supply chain models from food raw materials. These models include the following supply chain activities: Food raw materials (for example, soybeans) are collected from farms in various locations, then transported to conversion refineries, and after the conversion of raw materials to fuel in refineries, the final fuel is going to be directed to the supply stations. Based on these activities, supply chain models in this category optimize decisions such as supplier selection (Sokhansanj et al., 2014), capacity, location, and technology of conversion refineries (An et al., 2011;Shabani et al., 2014), and method of transportation of raw materials and final fuel (Awudu & Zhang, 2013; Azadeh et al., 2014). The second category includes green fuel supply chain models from non-food raw materials like fibres of industrial crops, where the raw materials are collected from optimal places of predetermined farms (Zhang et al., 2013). In the third category are the models of the green fuel supply chain extracted from algae. The reason for classifying algal fuels into separate category is that the algae supply chain network structure is significantly different from other green fuel supply chain networks. More specifically, algal growth requires specific raw materials such as carbon dioxide, water, and nutrients, incorporating into the algae's green fuel supply chain (Mata, 2010). Although algae are one of the most important sources for green fuel production, less attention has been paid to modeling and optimizing the supply chain of algae in different parts of the world, and few studies in this area have limited parts of the activities required for green fuel production which contains algae. This study develops a comprehensive supply chain model covering all the steps required to produce fuel from algae to cover this research gap.

Uncertainty is an important issue in the green fuel supply chain derived from biomass, which strongly influences supply chain decisions (Osmani & Zhang, 2014). Despite this importance, most green fuel supply chain network design models are definitive. In recent years, as the role of uncertainty in the green fuel supply chain has become more prominent, some researchers have developed uncertain models (Maity, 2010). Probabilistic planning is the method used to deal with the uncertainty in these studies. However, this method has serious limitations in the field of supply chain network design. For example, contingency planning for uncertainty modeling requires the probability distribution function of uncertain parameters. At the same time, in the green fuel supply chain, due to the nascent green fuel industry, insufficient historical data is available, and as a result, the distribution is obtained. Probabilistic parameters are difficult (Sims, 2010). Probabilistic scenario-based optimization is considered as a popular method in supply chain network optimization to deal with uncertainty requires considering a large set of scenarios that increase the complexity of the problem and consequently increase the time. Eventually, it becomes computational processed (Balaman & Selim, 2014), which gives an alternative solution that overcomes such problems and, more importantly, provides an answer that remains feasible and robust for all realities of uncertain parameters (Sadjadi & Omrani, 2008). Accordingly, in this study, a robust supply chain network design model for the production of fuel from algae is developed. Applying a robust optimization approach in this model ensures that the strategic and tactical supply chain decisions obtained are less sensitive to uncertain parameters and do not lose their value due to small fluctuations in parameters.

BACKGROUND

The Process of Producing Fuel from Algae

In general, the cycle of green fuel production from algae consists of several stages. At the beginning of this cycle, there are ponds in which algae are grown. More precisely, these ponds are full of a mixture of water and nutrients needed for the productive growth of algae (carbon dioxide, nitrogen, and phosphorus) and are built in areas of several hectares. After algae growth in these ponds, it is necessary to separate the algae from the water around them to prevent their decay and prepare them for the next stages of fuel production. For this purpose, a mixture of water and algae enters the harvesting and drying stages, respectively. In the harvesting stage, a significant amount of water is separated from the mixture of water and algae using the flocculation process, and then in the drying neighborhood, a small amount of remaining water is evaporated by heat, and the algae are completely dried (Davis et al., 2011; Norouzi, 2021a). Dried algae can be converted to different fuels by different processes, but the production of diesel fuel from this raw material is the most well-known method, and on the other hand, biodiesel fuel is the most valuable fuel among different green fuels (Davis et al., 2011; Norouzi & Kalantari, 2020). To obtain diesel fuel, the oil in the dried algae is extracted and then converted to biodiesel fuel by chemical reactions(Norouzi, 2021b).

Supply Chain Network Design Model

The supply chain network model can be considered according to Figure 1 for fuel production from algae based on the described fuel production process. This chain, which includes all the activities required to produce fuel from the supply of raw materials to the supply of fuel to the final market, consists of the following layers:

Required Raw Materials

Optimal and productive algae growth requires an adequate and sustainable supply of materials such as water, carbon dioxide, and nutrients, so the network design model must consider these materials' supply carefully. Water is the main requirement for the growth of algae, which provides the environment needed for the growth and development of algal biomass. According to various studies, the production of each liter of fuel from algae requires 400 liters of water (Yang et al., 2011). This amount of water is a major obstacle to the large-scale production of algal fuel due to the scarcity of freshwater resources. To alleviate this concern, the ability of algae to grow in wastewater can be exploited. Therefore, in this network design model, wastewater is used and freshwater to produce algae. Carbon dioxide is another important raw material needed in the process of photosynthesis of algae like other plants. Because algae grow below the water's surface, their carbon uptake from the atmosphere is not sufficient for optimal growth due to the surface reversibility of water, so the use of a concentrated carbon dioxide source is required (Mata, 2010). Although different sources can supply carbon dioxide, carbon dioxide emitted from power plants is of particular interest. The reason for this is that carbon dioxide emitted from power plants has a sufficient concentration in all seasons, and also, by using this carbon dioxide to produce green fuel, it is possible to reduce the carbon dioxide emissions into the atmosphere (Pate et al., 2010). Nitrogen and phosphorus are the two main nutrients needed for algal growth (Mata, 2010; Norouzi, 2020). The preparation of agricultural fertilizers provides these two nutrients. In addition, wastewater may contains high amounts of the derivatives of these nutrients that stimulating the growth condition.

Production Sites

At the middle layer of the supply chain are production sites. According to the described stages of fuel production, production sites are composed of different units. In mathematical supply chain modeling, it is assumed that the fuel production process takes place in two stages. In the first stage, there are algae production ponds that produce dry algae multiplied by the growth rate of algae (amount of dry matter per hectare) multiplied by the total area of the ponds (hectares). In the second stage, the dry algae are converted to green diesel fuel with a certain conversion rate(Norouzi, 2019; Norouzi & Fani, 2021).

One of the most important problems facing green fuels is that fuel demand is the same throughout the year, but fuel production depends on weather conditions and decreases significantly in cold seasons (Norouzi et al., 2010). To reduce this problem, a fuel storage network design model must be taken into consideration. In other words, if there is demand, fuel is transferred to the demand areas directly. Otherwise, the produced fuel is stored to be consumed in later periods.

Fuel Supply Areas

In the last layer of the algae supply chain are the fuel supply areas. Since the early stages of developing green fuels extracted from algae can not be expected to distribute fuel throughout the country, several provinces are considered experimental markets, and fuel supply is done in these provinces(Wigmosta et al., 2011).

Figure 1. Structure of microalgae-based biofuel supply chain

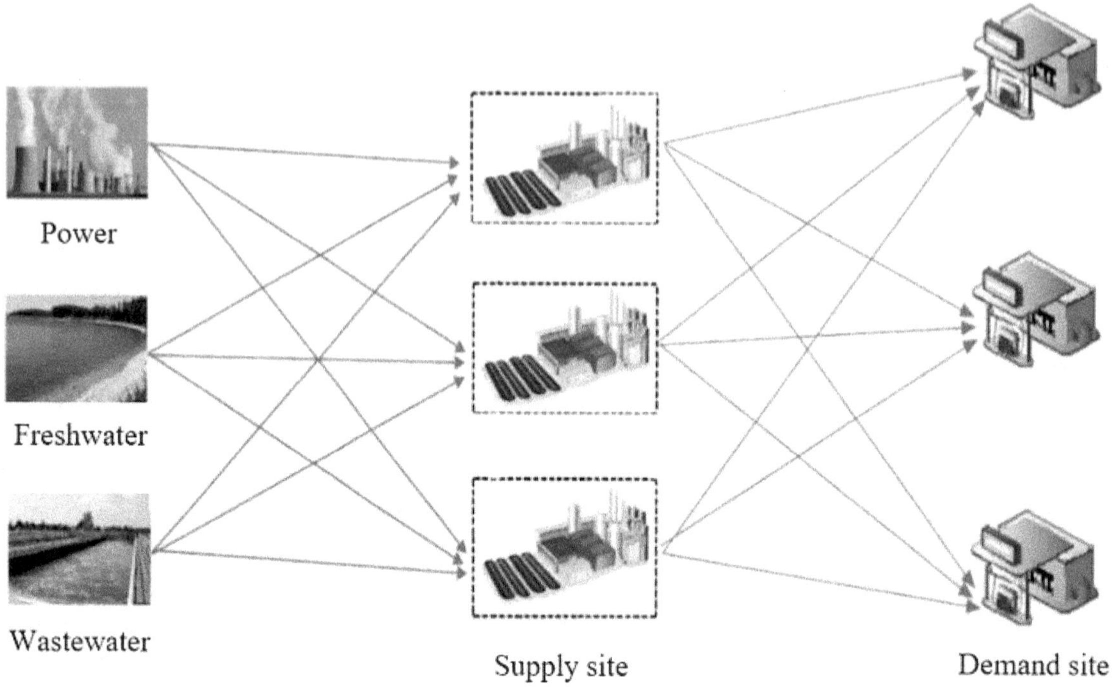

Transportation

To communicate between the mentioned supply chain layers, two methods of transportation are considered: transportation by pipeline and transportation by tanker. Since water, sewage, and carbon dioxide need to be transported to production sites without interruption, and a pipeline is considered to transport these raw materials. In addition, because transport tanks for transporting gasoline and diesel fuel have had a good performance, for transporting green fuel produced at production sites to experimental markets are considered(Norouzi et al., 2021; Yu & Cruz, 2019).

MATERIALS AND METHODS

Model Description

Before presenting the mathematical model, the sets, variables, and parameters used in the network design model are listed below. It should be noted that the indefinite parameters are shown with.

According to the parameters and variables described above, the objective function and the limitations of the mathematical model of algal network design are presented below.

Table 1. Parameters and variables in the model

Parameter	Description
F	Collection of freshwater sources, (f Î F)
W	Collection of wastewater sources, (w Î W)
C	Collection of carbon dioxide sources (power plants), (c Î C)
I	Collection of candidate locations for production sites, (i Î L)
Q	demand, (q_f, q_s, q_w, q_c Î Q)
M	Collection of places of demand, (m Î M)
T	Collection of time, (t Î T)
$capf_f^t$	Amount of freshwater available in region f over time t
$capw_w^t$	Amount of wastewater available in the region w over time t
$capc_c^t$	Amount of carbon dioxide available in region c over time t
$tf_{f \to i}$	Shipping cost per unit of freshwater from source f to production region i
$tw_{w \to i}$	Cost of transportation per unit of wastewater from source w to production region i
$tc_{c \to i}$	Cost of transportation per unit of carbon dioxide from source c to production region i
$tb_{i \to m}$	Transportation cost per biodiesel unit from the production region i to the region of demand m
$C_{i,q}$	Annual investment cost of production site i with capacity q
O_i	Cost of production of biodiesel at the production site i
$C_{qf}^{f \to i}$	Annual investment cost of capacity pipeline f to i with capacity of q_f
$C_{qw}^{w \to i}$	Annual investment cost of capacity pipeline w to i with capacity of q_w
$C_{qc}^{c \to i}$	Annual investment cost of capacity pipeline c to i with capacity of q_c
$Cap_{qf}^{f \to i,t}$	Maximum capacity of the pipeline f to i with capacity q_f in time t
$Cap_{qw}^{w \to i,t}$	Maximum capacity of the pipeline w to i with capacity q_w in time t
$Cap_{qc}^{c \to i,t}$	Maximum capacity of the pipeline c to i with capacity q_c in time t
δq	Area of production site ponds with capacity q
ϑ_i^t	Fertility (production) of dry algae per unit area in area i in period t
βw	Water required per unit of algae produced
βn	Nitrogen required per unit of algae produced
βp	Phosphorus required per unit of algae produced
βc	Carbon dioxide required per unit of algae produced
ϕn	Nitrogen available per unit of wastewater
ϕp	Phosphorus available per unit of wastewater
P_n^t	Nitrogen prices in time t
P_p^t	Phosphorus price in time t
h_i	Cost of maintaining biodiesel at the production site i
σ	The rate of conversion of dried algae to biodiesel
$d_{m,t}$	Demand in the m region in period t
$xf^t_{f \to i}$	The amount of fresh water transferred from source f to the production site i in period t
$xw^t_{w \to i}$	The amount of wastewater transferred from source w to the production site i in period t
$xc^t_{c \to i}$	The amount of carbon dioxide transferred from source c to the production site i in period t
$xb^t_{i \to m}$	The amount of biodiesel transferred from the production site i to the demand area m in time period t

continued on following page

Table 1. Continued

Parameter	Description
v^t_i	The amount of biodiesel stored at the production site i in period t
xn_i^t	The amount of nitrogen fertilizer purchased at the production site i in period t
xp_i^t	The amount of phosphorus fertilizer purchased at the production site i in period t
$Y_{i,q}$	Variable 0-1, equal to 1 in case of construction of production site with capacity q, otherwise 0
$L_{qf}^{f\to i}$	Variable 0-1, equal to 1 in case of construction of pipeline with capacity q between source f and site i, otherwise 0
$L_{qw}^{w\to i}$	Variable 0-1, equal to 1 if a pipeline with capacity between source w and site i is constructed, otherwise 0
$L_{qc}^{c\to i}$	Variable 0-1, equal to 1 if a pipeline with capacity between source c and site i is constructed, otherwise 0

Objective Function

The following is the objective function for the network design model. This objective function minimizes the costs of the entire supply chain along the planning horizon. These costs include: costs of transferring raw materials from sources to production sites (FC), fixed investment cost for piping (PC), cost of nutrients (EC), costs of production sites (operating and investment costs) (PF), costs of inventory (IC) and biodiesel transfer costs from production sites to demand areas (BC).

$$\text{Min}Z = FC + PC + EC + PF + IC + BC \tag{1}$$

$$FC = \sum_f \sum_i \sum_t tf_{f\to i} xf_{f\to i}^t + \sum_w \sum_i \sum_t tw_{w\to i} xw_{w\to i}^t + \sum_c \sum_i \sum_t tc_{c\to i} xc_{c\to i}^t \tag{2}$$

$$PC = \sum_f \sum_i \sum_{qf} C_{qf}^{f\to i} L_{qf}^{f\to i} + \sum_w \sum_i \sum_{qw} C_{qw}^{w\to i} L_{qw}^{w\to i} + \sum_c \sum_i \sum_{qc} C_{qc}^{c\to i} L_{qc}^{c\to i} \tag{3}$$

$$PF = \sum_i \sum_m \sum_t O_i xb_{i\to m}^t + \sum_i \sum_q C_{i,q} Y_{i,q} \tag{4}$$

$$EC = \sum_i \sum_t P_n^t xn_i^t + \sum_i \sum_t P_p^t xp_i^t \tag{5}$$

$$IC = \sum_i \sum_q \sum_t h_i v_i^t \tag{6}$$

$$BC = \sum_i \sum_m \sum_t tb_{i,m} xb^t_{i \to m} \tag{7}$$

Model Constraints

The amount of raw materials available (freshwater, sewage, and carbon dioxide) varies in different regions and in different seasons of the year. Constraints (8) to (10) ensure that the harvest of these raw materials from their source does not exceed the maximum available quantity.

$$\sum_i xf^t_{f \to i} \le capf^t_f \tag{8}$$

$$\sum_i xw^t_{w \to i} \le capw^t_w \tag{9}$$

$$\sum_i xc^t_{c \to i} \le capc^t_c \tag{10}$$

Constraint (11) ensures that the amount of water required at each production site, which is equal to the amount of algae produced multiplied by the amount of water required, is guaranteed by fresh water and sewage.

$$\sum_q \beta_w \vartheta_{i,t} \delta_q Y_{i,q} \le \sum_f xf^t_{f \to i} + \sum_w xw^t_{w \to i} \tag{11}$$

Constraint (12) ensures that the amount of nitrogen required is supplied through wastewater to the production site and nitrogen purchased from the open market. There are similar conditions for phosphorus shown in constraint (13). Constraint (14) ensures that the required amount of carbon dioxide is provided at each site and each period of carbon dioxide transported from the power plants.

$$\sum_q \beta_n \vartheta_{i,t} \delta_q Y_{i,q} \le \varphi_n \sum_f xw^t_{w \to i} + xn^t_i \tag{12}$$

$$\sum_q \beta_p \vartheta_{i,t} \delta_q Y_{i,q} \le \varphi_p \sum_f xw^t_{w \to i} + xp^t_i \tag{13}$$

$$\sum_{q}\beta_{c}\vartheta_{i,t}\delta_{q}Y_{i,q} \le \sum_{f}xc^{t}_{c\to i} \tag{14}$$

Constraint (15) ensures that the amount of biodiesel produced at each production site in each period in addition to the biodiesel inventory stored at the end of the previous period is greater than the biodiesel transferred from the production site to the demand markets and inventories at the end of the current period.

$$\sum_{q}\beta_{c}\vartheta_{i,t}\delta_{q}Y_{i,q} + v_{i}^{t-1} \ge \sum_{f}xb^{t}_{i\to m} + v_{i}^{t} \tag{15}$$

Constraint (16) to (18) ensure that the amount of fresh water, sewage and carbon dioxide transferred from raw material sources to production sites during each period of pipeline capacity does not exceed the capacity of the pipeline.

$$xf^{t}_{f\to i} \le \sum_{qf} cap^{f\to i}_{qf} L^{f\to i}_{q_f} \tag{16}$$

$$xw^{t}_{w\to i} \le \sum_{qw} cap^{w\to i}_{qw} L^{w\to i}_{q_w} \tag{17}$$

$$xc^{t}_{c\to i} \le \sum_{qc} cap^{c\to i}_{qc} L^{c\to i}_{qc} \tag{18}$$

Constraint (19) to (22) ensures that only one capacity level is selected for each production site and each freshwater, wastewater, and carbon dioxide pipeline.

$$\sum_{q} Y_{i,q} \le 1 \tag{19}$$

$$\sum_{qf} L^{f\to i}_{qf} \le 1 \tag{20}$$

$$\sum_{qw} L^{w\to i}_{qw} \le 1 \tag{21}$$

$$\sum_{qc} L^{c\to i}_{qc} \le 1 \tag{22}$$

Constraint (23) ensures that the amount of fuel supplied to the candidate provinces exceeds the demand in those provinces in each period. Finally, the constraints (24) ensure that the mentioned decision variables are negative.

$$\sum_{i}\sum_{m} xb^{t}_{m\rightarrow i} \geq d_{m,t} \tag{23}$$

$$xf^{t}_{f\rightarrow i}, xw^{t}_{w\rightarrow i}, xc^{t}_{c\rightarrow i}, xb^{t}_{m\rightarrow i}, v^{t}_{i}, xn^{t}_{i}, xp^{t}_{i} \geq 0 \tag{24}$$

Case Study

In this section, the performance of the developed model is evaluated and analyzed using a case study to design a supply chain network for the production of green fuel development from algae in the country. It should be noted that the present study is the first study to develop the supply chain model of algae green fuel production in the country; the whole country has been considered in a case study. With the further development of this type of fuel in the country, supply chain models can be studied separately for different provinces. Among Asian countries, Iran has different capabilities and incentives to produce green fuel from algae, making the application of the model in the country a practical case study. One of the most important motivations for the production of green algal fuel in Iran can be mentioned as follows: (1) Extensive geographical area that makes available different climatic conditions throughout the year (for example, the southern provinces of the country even in cold seasons of the year have high temperatures that allow the growth of algae and fuel production throughout the year), (2) the existence of extensive sandy and salty soils in the country that can be used to build algae growth ponds, (3) Availability of abundant saline water resources in the Caspian Sea and the Persian Gulf, which is a suitable raw material for the growth of some species of algae, and (4) the existence of power plants in suitable climatic zones that can be used as a source. Carbon dioxide should be used to grow algae and reduce the releasing of the most of the greenhouse gases into the atmosphere (Tabatabaei et al., 2011). It should be noted that the green fuel industry of algae is very fledgling, and the early phases of its development are taking shape in Australia, the United States, and Japan (Soyester, 1973). Considering that economic analyses related to the production of green fuel from algae have not been performed in the country so far, so in this case study, the required data of the model are estimated using economic analyses performed in other countries.

Model Assumptions and Data

In this case study, the following assumptions and data have been used:

(1) The planning horizon is for a year consisting of 12 months.
(2) One of the most important economic factors influencing the developed model is the investment and operating costs. Considering that the piping method to transfer water to algae production sites is very similar to urban water piping, the costs related to water piping have been calculated based

on the price list of the water distribution network in 2014. These calculations are based on piping with a diameter of 200 mm. Also, the piping price is considered for the lowest slope in this price list (less than 8 degrees).

(3) Regarding the piping required to transfer carbon dioxide from power plants to production sites, there are many similarities between this type of piping and city gas piping, so the costs associated with carbon dioxide transfer are based on The specific price list of oil and gas facilities in 2011 is estimated. These calculations are based on 12-inch steel pipes. To transfer carbon dioxide from power plants, it should be noted that it is necessary to build carbon dioxide storage units in the power plants and piping to the costs calculated based on urban piping (~20%) (Rashid et al., 2013).

(4) Investment and operating costs of different stages of fuel production (cultivation of algae, separation of algae from the pond, drying of algae, oil extraction, conversion of oil into fuel) based on the latest economic-technological studies related to the production of fuel from algae in The United States has been estimated by Landquist et al. (2010). In these economic studies, the capacity of production sites has been considered according to the capacity of the first stage of the production process (algae cultivation ponds), and the costs of different stages related to ponds with an area of 400 have been estimated.

(5) (5) In the above calculations, the investment costs of water transmission lines, carbon dioxide, and production sites have been calculated for the base capacity. The following equation is used to estimate the investment costs of transmission lines and production sites with higher capacity(Norouzi et al., 2021).

$$Cost = Cost_{ref} \times \left(\frac{Cap}{Cap_{ref}} \right)^{\beta} \tag{25}$$

In this regard, transmission lines and production sites with basic capacity, the investment cost of transmission lines, and production sites with basic capacity with higher capacity and scale factor. It should be noted that the scale coefficient in these calculations is 0.8 (Lundquist et al., 2010) and 0.9 (Eskandarpour et al., 2015) for water and carbon dioxide transmission lines and 0.85 for production sites, respectively.

(6) Other parameters of the model, including the number of raw materials required per unit of fuel production, conversion rates at different stages of fuel production, and the amount of nitrogen and phosphorus in wastewater, are shown in Table 3.

(7) For the construction of algae growth sites, 20 proposed sites have been considered in different provinces.

(8) For fuel supply (demand) areas, eight provinces of Tehran, Qazvin, Isfahan, Shiraz, Qom, Markazi, Kermanshah, and Yazd are considered pilot provinces in the early phases of green fuel development.

(9) The purpose of this study is to replace the consumption of 5% of current fuel with green fuel, and the amount of consumption in the mentioned provinces is estimated based on the information of Iran Petroleum Products Distribution Company(Bertsimas, 2004).

(10) 25 power plants in the country, 30 wastewater treatment plants, 35 freshwater sources, and 13 saltwater sources are identified based on GIS analysis and used in the model.

Table 2. Input parameters and conversion rates

Parameter	Amount	Reference
Rate of conversion of dry algae to biodiesel	25%	[23]
Nitrogen in wastewater	42 mg/liter	[26]
Phosphorus in wastewater	6.9 mg/liter	[26]
Water required for algal production	400 liters/kg of algae	[16]
Nitrogen required for algal production	1.6 kg/kg of algae	[16]
Phosphorus required for algal production	3.2 kg/kg of algae	[16]
Carbon dioxide required for algal production	6.2 kg/kg of algae	[16]

RESULTS AND DISCUSSION

Total Supply Chain Cost and Computational Performance

First, it is necessary to compare the design model of the supply chain network of algae indefinite and stable states in terms of the total cost of the supply chain and computational performance. Table 2 shows the total cost of the supply chain and the computational time of the supply chain model in a definite and stable state. In this experiment, for robust optimization models, the amplitude range of uncertain parameters is set at 10%, 20%, and 30%, and also the level of conservatism controlled by the uncertainty budget is set at 70%, 85%, and 99%. The results show that robust models lead to higher supply chain costs compared to deterministic models. Although it was expected that robust optimization models would be more expensive than deterministic models due to increased supply chain stability and the capture of supply chain environment uncertainties, the cost difference needs to be considered. For example, a highly conservative version of the solid model (degree of conservatism = 99%, amplitude of oscillation = 30%) leads to a 71% increase in the cost of the batch production model. Although the cost of stability for this sample is extremely high, conservatism levels and lower volatility that provide a balance between stability and cost increase lead to a more reasonable cost increase. For example, a degree of conservatism of 70% and a range of fluctuations of 10% increase the cost of the supply chain by 13%, which is a relatively reasonable cost. The important point is that conservatism is determined by the degree of risk-taking of the decision-maker. The risk-averse decision-maker uses higher levels of uncertainty and consequently has to pay a higher price for it. Risk-takers, on the other hand, prefer to increase reliability at a reasonable cost. One of the most important results of this table is that solid models do not create a gap concerning the model solution time and provide a suitable computational time compared to definite models.

Optimal Supply Chain Design in a Definite and Stable State

As one of the most important decisions of the algae supply chain network, the location and capacity of production sites and transmission lines are examined in this section. Figure 3 shows the capacity and position of production sites and pipelines under a definite and stable model (oscillation 20% and degree of conservatism 99%). As is obvious, solid models suggest similar candidate locations for production sites, so it can be said that the optimal location of sites is largely stable. In other words, the optimality

Table 3. Total cost and computational time of deterministic and robust models

Oscillation range	Level of conservatism					
	99%		85%		70%	
10%	617	586	579	588	559	524
20%	695	436	647	656	609	634
30%	863	501	879	598	713	651
Final Model	521			425		

of these decisions does not change under data fluctuations, and the position of sites remains optimal for almost all values of uncertain parameters. Another important result of Figure 3 can be derived by comparing the optimal capacity of production sites' indefinite and robust models. In general, robust models determine production equipment with higher capacity or more equipment with lower capacity than definitive models because they provide the ability to satisfy the worst possible values of indeterminate parameters. As the results show, higher capacity equipment has been suggested in the case of the algae supply chain. This is also the case with pipeline transmission lines. In other words, in the robust model, pipeline transmission lines with higher capacity than the definitive model have been proposed.

Figure 2. Optimal supply chain design for deterministic and robust models

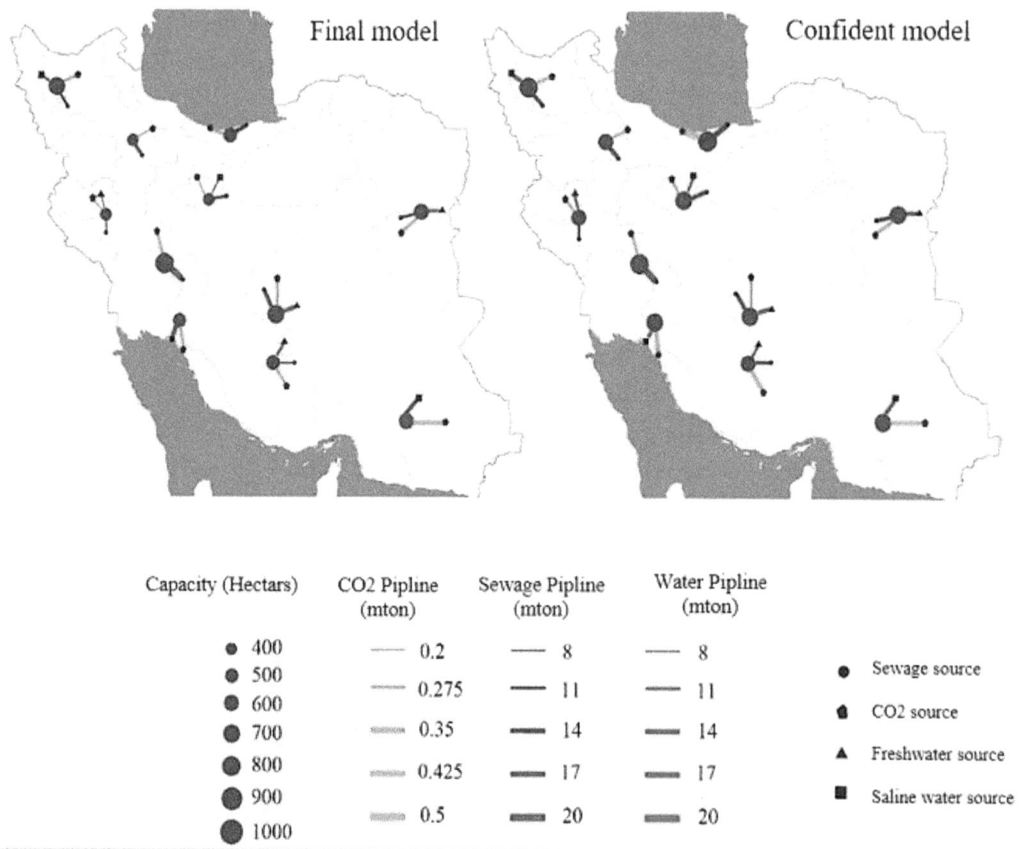

Sensitivity Analysis

There are various operational parameters and costs in the supply chain of microalgae that can play an important role in reducing costs. As we know, the sensitivity analysis method seeks to change the value of various input parameters and investigate its effect on the objective function. Performing this method for the operational parameters and costs of the algae supply chain reveals the importance of each of them to the production cost and indicates which of the following parameters can be used to achieve lower fuel costs. By identifying the most important effective parameters, it becomes clear which parameters and parts of the microalgae supply chain need to be reduced. In this section, the effects of the most important factors in the supply chain of microalgae, which include (1) the rate of algae growth, (2) the rate of conversion of dry algae to biodiesel, (3) investment cost of piping, (4) operating cost of piping, (5) cost Investment of production sites, (6) operating costs of production sites, (7) the amount of fresh water available, (8) the amount of wastewater available, (9) the amount of carbon dioxide available, and (10) transportation costs, on the cost of producing each liter of fuel, are analyzed (see table 4).

Validation of Developed Models

As observed in the results section of the models, the answers obtained from the definitive and robust models of supply chain network design are different. Therefore, it is necessary to review and compare the answers obtained from definite and solid models and evaluate the desirability (strength) of the answers obtained from them. For this purpose, the validation of robust planning models presented by Ben-Tal & Nemirovski (2000) is used. This method, which seeks to generate probable reality in the future to evaluate the results, consists of the following steps:

(1) Definite and solid models are solved under nominal data, and the answers of the definite model (x_d^*, y_d^*) and the solid model (x_r^*, y_r^*) are stored.

(2) (2) The answers obtained from the definite and independent models are replaced in the following "realism model":

$$Min\,Z = t^{real}x^* + c^{real}y^* + \sum_i \pi R_i \tag{26}$$

$$\sum_j a_{ij}^{real}x^* + R_i \le b_i \tag{27}$$

In this linear programming model, a compact form of the mathematical model of supply chain network design, the objective function coefficients and constraints are randomly generated from their corresponding probable range. For example, in an indefinite coefficient of constraints (a_{ij}), a random value between $[a_{ij} - \hat{a}_{ij}, a_{ij} + \hat{a}_{ij}]$ its probable upper and lower limit is generated randomly. It should be noted that since some constraints (x^*,y^*) may have unmodulated constraints, R_i is considered a variable, but a penalty

Table 4. The variation range of sensitivity analysis parameters

Point	Parameter	Estimation
0	Basic model	
1+	Growth of algae	+25
1-	Growth of algae	-25
2+	rate of conversion of algae to biodiesel	+25
2-	rate of conversion of algae to biodiesel	-25
3+	Piping investment cost	+25
3-	Piping investment cost	-25
4+	Operating cost of piping	+25
4-	Operating cost of piping	-25
5+	Cost of investing in production sites	+25
5-	Cost of investing in production sites	-25
6+	Operating costs of production sites	+25
6-	Operating costs of production sites	-25
7+	Availability of fresh water	+25
7-	Availability of fresh water	-25
8+	Sewage availability	+25
8-	Sewage availability	-25
9+	Availability of carbon dioxide	+25
9-	Availability of carbon dioxide	-25
10+	Transport cost	+25
10-	Transport cost	-25

Figure 3. Results of sensitivity analysis

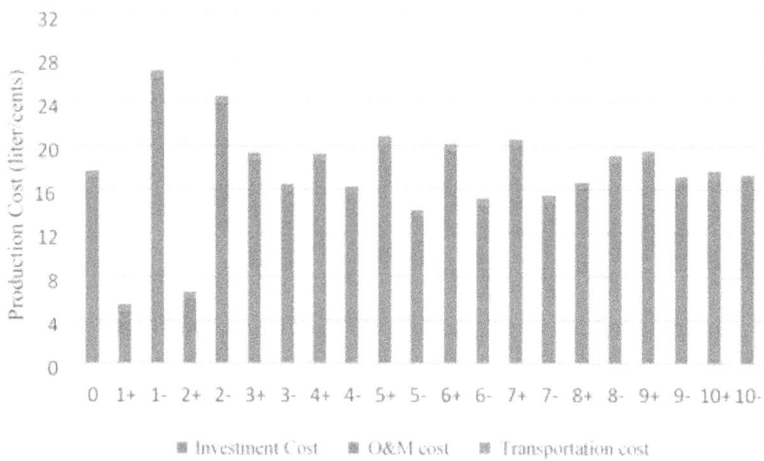

for violating this constraint is considered. As a result, R in this model is only a random variable, and other values are placed as parameters in the model.

(3) Step two (generating a random value for uncertain coefficients) is repeated 50 times, and the above model is solved for each random value generated, and the values of the obtained objective function (Z) are stored.
(4) Mean and standard deviation of objective function values (Z) are calculated as evaluation indicators to compare models(Doctor et al., 2005; Kally & Fishelson, 1993).

The results of this experiment to evaluate definite and solid answers are shown in Figure 4. Based on the results, the mean value indicates that the robust model performed better than the definitive model in all cases. The important point is that with increasing the oscillation range, the difference between the mean values of the two models increases. This suggests that solid models perform better in highly uncertain environments where it is extremely difficult to predict the exact value of the parameters compared to the definite model. On the other hand, Figure 4 shows that the standard deviation of a stable model under realistic values (simulation of reality) is much less than the standard deviation of a definite model. This result reveals that the answer obtained from the stable model remains close to the optimal value in case of deviation and fluctuation of the parameters (the concept of the stability of the answer of the stable model), while in the case of parameter fluctuation, the optimality of the definite model answer to Intensity decreases. As a result, using a solid model in uncertain environments where the probability of parameter fluctuations is extremely high is a more appropriate choice. It should be noted that although the robust model in this case study had a better performance than the definitive model, the performance of the robust model doesn't need to be better in other studies. The risk of the studied environment is an important criterion that determines the desired model(Li et al., 2011).

CONCLUSION

Increasing demand for fossil fuels, limited oil resources, and environmental pollution are among the most important motivations for developing green biofuels. Microalgae, are considered as the most profitable sources among the raw materials using for the production of green fuel which has attracted much attention in recent years in the world. This study presents an algae supply chain network design model to investigate the development of such fuels in the country. A network design model has been developed that includes all activities related to fuel production, from raw material supply to fuel production and then fuel supply. One of the most important concerns of green fuel network design models is the issue of uncertainty. This is because the production of green fuels is in the early stages of development, and it is difficult to determine the exact amount of parameters required by the supply chain model. In this case, a solid optimization method has been used to deal with the supply chain uncertainty. The results showed that a slight increase in the cost of the entire supply chain can improve the supply chain algae stability. More precisely, the cost of stability for the supply chain in question is small for limited oscillation intervals. The production costs pointed out that the cost of producing each liter of green biofuel from algae is higher than fossil fuels, but this cost is significantly reduced with a slight improvement in algal growth and conversion of dry algae to biodiesel. As a result, with the reduction of fossil fuel reserves in the future and consequently the increase in fuel prices, algae can be considered as one of

Figure 4. Mean and variance of the objective function of realization model under the solution of deterministic and robust model

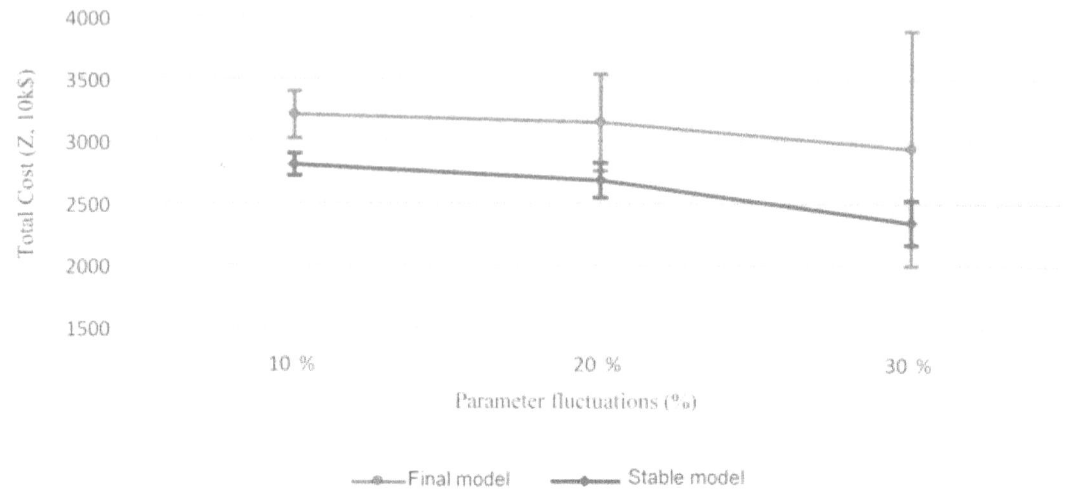

the most suitable alternatives for the production of eco-friendly biofuel other than consuming the fossil fuels in the country.

ACKNOWLEDGMENT

This research received no specific grant from any funding agency in the public, commercial, or not-for-profit sectors.

REFERENCES

An, H., Wilhelm, W. E., & Searcy, S. W. (2011). A mathematical model to design a lignocellulosic biofuel supply chain system with a case study based on a region in Central Texas. *Bioresource Technology*, *102*(17), 7860–7870. doi:10.1016/j.biortech.2011.05.060 PMID:21724388

Awudu, I., & Zhang, J. (2013). Stochastic production planning for a biofuel supply chain under demand and price uncertainties. *Applied Energy*, *103*, 189–196. doi:10.1016/j.apenergy.2012.09.025

Azadeh, A., Arani, H. V., & Dashti, H. (2014). A stochastic programming approach towards optimization of biofuel supply chain. *Energy*, *76*, 513–525. doi:10.1016/j.energy.2014.08.048

Balaman, Ş. Y., & Selim, H. (2014). A network design model for biomass to energy supply chains with anaerobic digestion systems. *Applied Energy*, *130*, 289–304. doi:10.1016/j.apenergy.2014.05.043

Ben-Tal, A., & Nemirovski, A. (2000). Robust solutions of linear programming problems contaminated with uncertain data. *Mathematical Programming*, *88*(3), 411–424. doi:10.1007/PL00011380

Bertsimas, D., & Sim, M. (2004). The price of robustness. *Operations Research*, *52*(1), 35–53. doi:10.1287/opre.1030.0065

Davis, R., Aden, A., & Pienkos, P. T. (2011). Techno-economic analysis of autotrophic microalgae for fuel production. *Applied Energy*, *88*(10), 3524–3531. doi:10.1016/j.apenergy.2011.04.018

Doctor, R., Palmer, A., Coleman, D., Davison, J., Hendriks, C., Kaarstad, O., & Austell, M. (2005). *Transport of CO2. IPCC Special Report on Carbon Dioxide Capture and Storage, R. Pichs-Madruga* (S. Timashev, Ed.). Intergovernmental Panel on Climate Change.

Eskandarpour, M., Dejax, P., Miemczyk, J., & Péton, O. (2015). Sustainable supply chain network design: An optimization-oriented review. *Omega*, *54*, 11–32. doi:10.1016/j.omega.2015.01.006

Kally, E., & Fishelson, G. (1993). *Water and peace: water resources and the Arab-Israeli peace process*. Praeger.

Li, Z., Ding, R., & Floudas, C. A. (2011). A comparative theoretical and computational study on robust counterpart optimization: I. Robust linear optimization and robust mixed integer linear optimization. *Industrial & Engineering Chemistry Research*, *50*(18), 10567–10603. doi:10.1021/ie200150p PMID:21935263

Lundquist, T. J., Woertz, I. C., Quinn, N. W. T., & Benemann, J. R. (2010). A realistic technology and engineering assessment of algae biofuel production. Energy Biosciences Institute.

Maity, J. P., Bundschuh, J., Chen, C. Y., & Bhattacharya, P. (2014). Microalgae for third generation biofuel production, mitigation of greenhouse gas emissions and wastewater treatment: Present and future perspectives–A mini review. *Energy*, *78*, 104–113. doi:10.1016/j.energy.2014.04.003

Mata, T. M., Martins, A. A., & Caetano, N. S. (2010). Microalgae for biodiesel production and other applications: A review. *Renewable & Sustainable Energy Reviews*, *14*(1), 217–232. doi:10.1016/j.rser.2009.07.020

Norouzi, N. (2019). An overview on Water, Energy & Environment by 2030. *International Journal of Management Perspective*, *8*(2), 11–19.

Norouzi, N. (2020). COVID-19 and management models in SMEs of the energy industry. *Majlesi Journal of Energy Management, 9*(2).

Norouzi, N. (2020). COVID-19 and management models in SMEs of the energy industry. *Majlesi Journal of Energy Management, 9*(2).

Norouzi, N. (2021a). The Pahlev Reliability Index: A measurement for the resilience of power generation technologies versus climate change. *Nuclear Engineering and Technology*, *53*(5), 1658–1663. doi:10.1016/j.net.2020.10.013

Norouzi, N. (2021b). An Overview on the renewable hydrogen market. *International Journal of Energy Studies*, *6*(1), 67–94.

Norouzi, N., & Fani, M. (2021). The prioritization and feasibility study over renewable technologies using fuzzy logic: A case study for Takestan plains. *Journal of Energy Management and Technology*, *5*(2), 12–22.

Norouzi, N., Fani, M., & Nasiri, Z. (2021). The development of a Nexus based green architecture ranking system in Iran. *Civil Engineering Beyond Limits (CEBEL), 2*(2).

Norouzi, N., Kalantari, G., & Talebi, S. (2020). Combination of renewable energy in the refinery, with carbon emissions approach. *Biointerface Research in Applied Chemistry, 10*(4), 5780–5786. doi:10.33263/BRIAC104.780786

Osmani, A., & Zhang, J. (2014). Economic and environmental optimization of a large scale sustainable dual feedstock lignocellulosic-based bioethanol supply chain in a stochastic environment. *Applied Energy, 114*, 572–587. doi:10.1016/j.apenergy.2013.10.024

Pate, R., Klise, G., & Wu, B. (2010). Resource demand implications for US algae biofuels production scale-up. *Applied Energy, 88*(10), 3377–3388. doi:10.1016/j.apenergy.2011.04.023

Rashid, N., Rehman, M. S. U., Memon, S., Rahman, Z. U., Lee, K., & Han, J. I. (2013). Current status, barriers and developments in biohydrogen production by microalgae. *Renewable & Sustainable Energy Reviews, 22*, 571–579. doi:10.1016/j.rser.2013.01.051

Sadjadi, S. J., & Omrani, H. (2008). Data envelopment analysis with uncertain data: An application for Iranian electricity distribution companies. *Energy Policy, 36*(11), 4247–4254. doi:10.1016/j.enpol.2008.08.004

Shabani, N., Sowlati, T., Ouhimmou, M., & Rönnqvist, M. (2014). Tactical supply chain planning for a forest biomass power plant under supply uncertainty. *Energy, 78*, 346–355. doi:10.1016/j.energy.2014.10.019

Sims, R. E., Mabee, W., Saddler, J. N., & Taylor, M. (2010). An overview of second generation biofuel technologies. *Bioresource Technology, 101*(6), 1570–1580. doi:10.1016/j.biortech.2009.11.046 PMID:19963372

Sokhansanj, S., Kumar, A., & Turhollow, A. F. (2006). Development and implementation of integrated biomass supply analysis and logistics model (IBSAL). *Biomass and Bioenergy, 30*(10), 838–847. doi:10.1016/j.biombioe.2006.04.004

Soyster, A. L. (1973). Convex programming with set-inclusive constraints and applications to inexact linear programming. *Operations Research, 21*(5), 1154–1157. doi:10.1287/opre.21.5.1154

Tabatabaei, M., Tohidfar, M., Jouzani, G. S., Safarnejad, M., & Pazouki, M. (2011). Biodiesel production from genetically engineered microalgae: Future of bioenergy in Iran. *Renewable & Sustainable Energy Reviews, 15*(4), 1918–1927. doi:10.1016/j.rser.2010.12.004

Wigmosta, M. S., Coleman, A. M., Skaggs, R. J., Huesemann, M. H., & Lane, L. J. (2011). National microalgae biofuel production potential and resource demand. *Water Resources Research, 47*(3). Advance online publication. doi:10.1029/2010WR009966

Yang, J., Xu, M., Zhang, X., Hu, Q., Sommerfeld, M., & Chen, Y. (2011). Life-cycle analysis on biodiesel production from microalgae: Water footprint and nutrients balance. *Bioresource Technology, 102*(1), 159–165. doi:10.1016/j.biortech.2010.07.017 PMID:20675125

Yu, M., Cruz, J. M., & Li, D. M. (2019). The sustainable supply chain network competition with environmental tax policies. *International Journal of Production Economics*, *217*, 218–231. doi:10.1016/j.ijpe.2018.08.005

Yue, D., You, F., & Snyder, S. W. (2014). Biomass-to-bioenergy and biofuel supply chain optimization: Overview, key issues and challenges. *Computers & Chemical Engineering*, *66*, 36–56. doi:10.1016/j.compchemeng.2013.11.016

Zhang, J., Osmani, A., Awudu, I., & Gonela, V. (2013). An integrated optimization model for switchgrass-based bioethanol supply chain. *Applied Energy*, *102*, 1205–1217. doi:10.1016/j.apenergy.2012.06.054

ADDITIONAL READING

Bullen, R. A., Arnot, T. C., Lakeman, J. B., & Walsh, F. C. (2006). Biofuel cells and their development. *Biosensors & Bioelectronics*, *21*(11), 2015–2045. doi:10.1016/j.bios.2006.01.030 PMID:16569499

Davis, R., Aden, A., & Pienkos, P. T. (2011). Techno-economic analysis of autotrophic microalgae for fuel production. *Applied Energy*, *88*(10), 3524–3531. doi:10.1016/j.apenergy.2011.04.018

Demirbas, A. (2008). Biofuels sources, biofuel policy, biofuel economy and global biofuel projections. *Energy Conversion and Management*, *49*(8), 2106–2116. doi:10.1016/j.enconman.2008.02.020

Fargione, J., Hill, J., Tilman, D., Polasky, S., & Hawthorne, P. (2008). Land clearing and the biofuel carbon debt. *Science*, *319*(5867), 1235–1238. doi:10.1126cience.1152747 PMID:18258862

George, B., Pancha, I., Desai, C., Chokshi, K., Paliwal, C., Ghosh, T., & Mishra, S. (2014). Effects of different media composition, light intensity and photoperiod on morphology and physiology of freshwater microalgae Ankistrodesmus falcatus–A potential strain for bio-fuel production. *Bioresource Technology*, *171*, 367–374. doi:10.1016/j.biortech.2014.08.086 PMID:25218209

Minteer, S. D., Liaw, B. Y., & Cooney, M. J. (2007). Enzyme-based biofuel cells. *Current Opinion in Biotechnology*, *18*(3), 228–234. doi:10.1016/j.copbio.2007.03.007 PMID:17399977

Sims, R. E., Mabee, W., Saddler, J. N., & Taylor, M. (2010). An overview of second generation biofuel technologies. *Bioresource Technology*, *101*(6), 1570–1580. doi:10.1016/j.biortech.2009.11.046 PMID:19963372

Tsukahara, K., & Sawayama, S. (2005). Liquid fuel production using microalgae. *Journal of the Japan Petroleum Institute*, *48*(5), 251–259. doi:10.1627/jpi.48.251

KEY TERMS AND DEFINITIONS

Algaculture: Algaculture is a form of aquaculture involving the farming of species of algae. The majority of algae that are intentionally cultivated fall into the category of microalgae (also referred to as phytoplankton, microphytes, or planktonic algae). Macroalgae, commonly known as seaweed, also have many commercial and industrial uses. Still, due to their size and the specific requirements of the environment in which they need to grow, they do not lend themselves as readily to cultivation (this may change, however, with the advent of newer seaweed cultivators, which are algae scrubbers using upflowing air bubbles in small containers). Commercial and industrial algae cultivation has numerous uses, including food ingredients such as omega-3 fatty acids or natural food colorants and dyes, food, fertilizer, bioplastics, chemical feedstock (raw material), pharmaceuticals, etc. algal fuel. It can also be used as a means of pollution control. Global production of farmed aquatic plants, overwhelmingly dominated by seaweeds, grew in output volume from 13.5 million tonnes in 1995 to just over 30 million tonnes in 2016.

Algae Fuel: Algae fuel, algal biofuel, or algal oil is an alternative to liquid fossil fuels that use algae to source energy-rich oils. Also, algae fuels are an alternative to commonly known biofuel sources, such as corn and sugarcane. When made from seaweed (macroalgae), it can be known as seaweed fuel or seaweed oil. Several companies and government agencies are funding efforts to reduce capital and operating costs and make algae fuel production commercially viable. Like fossil fuel, algae fuel releases CO_2 when burnt, but unlike fossil fuel, algae fuel and other biofuels only release CO_2 recently removed from the atmosphere via photosynthesis as the algae or plant grew. The energy crisis and the world food crisis have ignited interest in algaculture (farming algae) for making biodiesel and other biofuels using land unsuitable for agriculture. Among algal fuels' attractive characteristics are that they can be grown with minimal impact on freshwater resources, can be produced using saline and wastewater, have a high flash point, and are biodegradable and relatively harmless to the environment if spilled. Algae cost more per unit mass than other second-generation biofuel crops due to high capital and operating costs but are claimed to yield between 10 and 100 times more fuel per unit area. The United States Department of Energy estimates that if algae fuel replaced all the petroleum fuel in the United States, it would require 15,000 square miles (39,000 km^2), which is only 0.42% of the US map, or about half of the land area of Maine. This is less than 1¤7 the area of corn harvested in the United States in 2000.

Biofuel: Biofuel is the fuel produced through contemporary processes from biomass, rather than by the very slow geological processes involved in forming fossil fuels, such as oil. Since biomass can be used as a fuel directly (e.g., wood logs), some people use biomass and biofuel interchangeably. However, the word biomass simply denotes the biological raw material the fuel is made of or some form of thermally/chemically altered solid end product, like torrefied pellets or briquettes.

First-Generation Biofuels: First-generation biofuels are fuels made from food crops grown on arable land. The crop's sugar, starch, or oil content is converted into biodiesel or ethanol, using transesterification or yeast fermentation.

Fourth-Generation Biofuels: This class of biofuels includes electrofuels and solar fuels. Electrofuels are made by storing electrical energy in the chemical bonds of liquids and gases. The primary targets are butanol, biodiesel, and hydrogen, but include other alcohols and carbon-containing gases such as methane and butane. Solar fuel is a synthetic chemical fuel produced from solar energy. Light is converted to chemical energy, typically by reducing protons to hydrogen or carbon dioxide to organic compounds.

Second-Generation Biofuels: Second-generation biofuels are fuels made from lignocellulosic or woody biomass or agricultural residues/waste. The feedstock used to make the fuels either grow on arable land but are byproducts of the main crop or grown on marginal land. Second-generation feedstocks include straw, bagasse, perennial grasses, jatropha, waste vegetable oil, municipal solid waste, and so forth.

Solar Fuel: A solar fuel is a synthetic chemical fuel produced from solar energy. Solar fuels can be produced through photochemical (i.e., activation of certain chemical reactions by photons), photobiological (i.e., artificial photosynthesis), thermochemical (i.e., through the use of solar heat supplied by concentrated solar thermal energy to drive a chemical reaction), and electrochemical reactions (i.e., using the electricity from solar panels to drive a chemical reaction). Light is used as an energy source, with solar energy being transduced to chemical energy, typically by reducing protons to hydrogen or carbon dioxide to organic compounds. Solar fuel can be produced and stored for later use when sunlight is not available, making it an alternative to fossil fuels and batteries. Examples of such fuels are hydrogen, ammonia, and hydrazine. Diverse photocatalysts are being developed to carry these reactions in a sustainable, environmentally friendly way.

Third-Generation Biofuels: Algae can be produced in ponds or tanks on land and out at sea. Algal fuels have high yields, can be grown with minimal impact on freshwater resources, can be produced using saline water and wastewater, have a high ignition point, and are biodegradable and relatively harmless to the environment if spilled. Production requires large amounts of energy and fertilizer, the produced fuel degrades faster than other biofuels, and it does not flow well in cold temperatures. By 2017, due to economic considerations, most efforts to produce fuel from algae have been abandoned or changed to other applications.

Chapter 14
Application of Algae for Hydrogen Generation and Utilization

Eman R. Sallam
National Institute of Oceanography and Fisheries, Egypt

Hanan M. Khairy
National Institute of Oceanography and Fisheries, Egypt

Mostafa Elshobary
https://orcid.org/0000-0002-2131-0679
Department of Botany, Faculty of Science, Tanta University, Egypt

Howida A. Fetouh
Faculty of Science, Alexandria University, Egypt

ABSTRACT

Hydrogen is a promising future fuel with high energy content for both heat and electrical energy without emission of any hazardous gases such as carbon dioxide or ozone-harming substances. Bio-hydrogen driven from microalgae has recently gained considerable attention. As it is more sustainable than other sources, further developments in such systems are still in their early stages and require improving efficiency and achieving a real-world application on a large scale. This chapter focuses on assessing the potential of microalgae applied sciences for the industrial manufacturing of hydrogen from algae using solar energy. It summarizes the principle key of hydrogen production, the viable and theoretical limits of microalga hydrogen manufacturing systems, and the rising techniques to engineer next-generation structures and how these fitting into a developing hydrogen economy. In addition, it discusses the utilization of a solar collector system for long-range storage of thermal solar energy in dark conditions or winter.

DOI: 10.4018/978-1-6684-2438-4.ch014

INTRODUCTION

The ongoing electricity disaster and world warming have challenged the scientific community to boost choice electricity sources. A large variety of substances has been investigated to produce bioenergy, such as industrial or crop wastes. The scarceness of traditional fossil fuel resources and global warming resulting from CO_2(g) emission from the combustion of natural fossil fuels is a distasteful problem that threatens human health and the entire ecosystem. Furthermore, around eighty percent of initial energy provision and more than sixty percent of electricity generation depend on fossil fuels (Baykara, 2018). Frighteningly, the energy demands and populations are severely increasing, which leads to a rapid increase in fossil fuel consumption, increased $CO_{2(g)}$ emission, and what it entails of several catastrophes such as floods, storms, forest fires, and the whole ecosystem damage (Nagarajan et al., 2017b). Increasing the dependence on fossil fuels has led to increased CO_2(g) emission from 388.5 ppm in 2010 to 409.95 ppm in only 10 years (NOAA/ESRL, 2019).

Consequently, ensuring energy security, reducing CO_2 emission and developing the transition to renewable energies are real challenges to scientists and political and economic decision-makers. Clean energy technologies such as solar energy, wind, hydropower, biomass, geothermal, tidal and wave power need rapid improvement. These inexhaustible energy sources are more eco-friendly than fossil energy sources. While, the main challenge with these renewable resources is that energy produced from these sources is difficult to store, consume or transport, making them distinctively suitable to limited applications (Elshobary, et al., 2021a).

Hydrogen is the universe's lightest, simplest, and most considerable chemical element. However, it takes place blended with different elements, specifically oxygen (O) in water (H_2O) and with carbon (C), nitrogen (N) and oxygen in residing organisms and fossil fuels. Combustion of hydrogen ($H_{2(g)}$) with air oxygen generates water vapour (H_2O(v)) solely as the spinoff and, even after its entire combustion, launch zero emissions of poisonous gases or wastes to the environment. Moreover, it has a 2.7–3.09 greater electricity density levels from about a hundred and twenty LHV to one hundred forty HHV $kJ.g^{-1}$ than different hydrocarbon fuels (Brentner et al., 2010). Bio-hydrogen production has no impact on food products because it allows renewable hydrogen generation from less expensive "waste" feedstocks (Ghimire et al., 2015; Sallam et al., 2021), wastewater (Skonieczny & Yargeau, 2009), sewage sludge (Sittijunda et al., 2010), or microalgae (Rathore & Singh, 2013).

The main objective of this review is to present an insight into the generation of biohydrogen via microalgae. In this context, relevant perspectives are critically discussed, focusing on the overviews of biohydrogen production, the principle key international technology, biological challenges and constraints in microalgae hydrogen generation, and the rising techniques to engineer next-generation structures and how these fit into the biohydrogen economy.

A promising issue will be discussed in detail in this chapter: the utilization of a solar collector system such as solar pond for long-range storage of thermal solar energy to be used in dark conditions or winter. This solar energy technology would enhance the photosynthetic ability of the microalgae as the principal driving force for water splitting and energy output.

Table 1. Different hydrogen production sources with its advantage and disadvantages.

Sources	Feedstocks	Methods	Advantages	Disadvantages	References
Fossil fuel	Methane Coal	Steam reforming of methane and oxidation processes. Coal gasification.	Higher efficiency (50-75%) and lower product cost ranges	Depleted sources and emitted large quantities of CO_2	(Ashik et al., 2015; Bicer & Dincer, 2017; Damen et al., 2006; Yan & Hoekman, 2014)
Biomass	Biomass, biooil, biogas and organic wastes	Pyrolysis/gasification, Thermal methods Photochemical and photoelectrochemical processes	Renewable	Huge biomass needed according to its lower calorific value High cost thermal and pressurized steps	(Angeli et al., 2014; Kalinci et al., 2009) (Parthasarathy & Narayanan, 2014; Uddin et al., 2013)
Water	Water	Water electrolysis, direct thermal decomposition or thermolysis, thermochemical processes and photolysis	High efficiency ranged from 70 to 90%. Clean source.	• Corrosion problems. • Increasing the manufacturing costs due to high operating pressure and electrolyzers • Complex system design	(Bicer & Dincer, 2017; Genç et al., 2012) (Acar et al., 2014; Genç et al., 2012; Ismail & Bahnemann, 2014; Parthasarathy & Narayanan, 2014; Uddin et al., 2013)
Metal hydrides	$LiH+MgH_2$ $LiBH_4+NaBH_4$, $LiAlH_4+NaAlH_4$	Pyrolysis and Hydrolysis	• Clean source without CO_2 emission • A promising technique for hydrogen storage	High operating pressure and temperature	(Moussa et al., 2013; N. Patel & Miotello, 2015; S. K. S. Patel et al., 2021)
Solar	Sunlight	Solar Thermal Dissociation Electrolysis Thermal chemical	Clean source of zero-emission	Limited in some area	(Abanades et al., 2006; Baykara, 2004, 2018; D'Souza, 2013; Steinfeld, 2005; Xing et al., 2017)
Nuclear	Nuclear reaction	Electrolysis Thermal chemical	Clean source of zero-emission	Not safe	(Bicer & Dincer, 2017; Dincer & Balta, 2011)

H_2 PRODUCTION FROM NONRENEWABLE RESOURCES

Hydrogen can be generated via various methods, including fossil fuels, alkaline water electrolysis, biomass, metal hydrides, and microorganisms, including bacteria, algae and fungi (Lee et al., 2017; Singh & Wahid, 2015). Even though hydrogen created through physical and chemical pathways does not emit $CO_{2(g)}$ on the ignition, its production method imposed a high-cost energy source (Brar et al., 2022). Table 1 summarizes some hydrogen production sources with its advantage and disadvantages.

Natural gasoline steam reforming, coal gasification, alkaline water electrolysis, wind& photo voltaic energy), biomass gasification, thermochemical water splitting, and high-temperature water electrolysis are all techniques producing hydrogen gasoline from nonrenewable sources. The most commn used industrial process for hydrogen production is the reforming of fossil hydrocarbons in oil refineries by using superheated water vapor (mostly when waste heat is available), because it is more energy and cost-effective than direct hydrogen production from splitting of water. Steam catalyzed reforming of methane $CH_{4(g)}$ (SRM) at 700°C represented in eqn. (1). The partial oxidation of methane (POM) is represented in eqn. (2). The CO_2 reforming of methane $CH_{4(g)}$ (CRM) can synchronously convert $CO_{2(g)}$ and CH_4 into valuc-addcd syngas (H_2 + CO) eqn. (3). All $CO_{(g)}$ in the mixture oxidized to $CO_{2(g)}$ by steam eqn. (4) (Roblero Lucchetti, 2020).

$$CH_{4(g)} + H_2O_{(g)} = CO_{(g)} + 3\ H_{2(g)} \ldots\ldots \Delta H^0_{298} = +206\ \text{kJ mol}^{-1} \quad (1)$$

$$CH_{4(g)} + 0.5\ O_{2(g)} = 2CO_{2(g)} + 4H_{2(g)} \ldots\ldots \Delta H^0_{298} = -38\ \text{kJ mol}^{-1} \quad (2)$$

$$CH_{4(g)} + CO_{2(g)} \rightarrow 2H_{2(g)} + 2CO_{(g)} \ldots\ldots \Delta H^0_{298} = +247\ \text{kJ mol}^{-1} \quad (3)$$

$$CO_{(g)} + H_2O_{(g)} = CO_{2(g)} + H_{2(g)} \quad (4)$$

BIOPROCESSING FOR HYDROGEN GAS PRODUCTION

Renewable, low-cost biomass can be used as a substrate and a prospective feedstock for sustainable hydrogen production. Furthermore, because agricultural and urban wastes can be disposed of simultaneously, bio-hydrogen production from wastes can provide a dual benefit of pure energy generation and waste remediation. On the other hand, biomass-based hydrogen production is constrained by low hydrogen production levels and substrate degradation rates.

The predominant avenues for hydrogen manufacturing from biomass are thermochemical applied sciences (e.g., gasification and pyrolysis), fermentation, and microbial gasoline cells. Although biomass gasification, together with steam gasification and supercritical water gasification, suggests excessive workable in field-scale applications, the selectivity and affectivity of hydrogen manufacturing want enchantment to impenetrable within your means industrial purposes with excessive atom economy. All of these techniques matter on protons (H+) eqn. 5, 6 as an electron sink for two electron equivalents to structure H2 in accordance to the following response (Fetouh et al., 2019):

$$H^+ + e^- = H \ \text{Discharge step} \quad (5)$$

$$H + H = H_{2(g)} \ \text{Recombination step} \quad (6)$$

Figure 1. Mechanisms of hydrogen production by microalgae and cyanobacteria

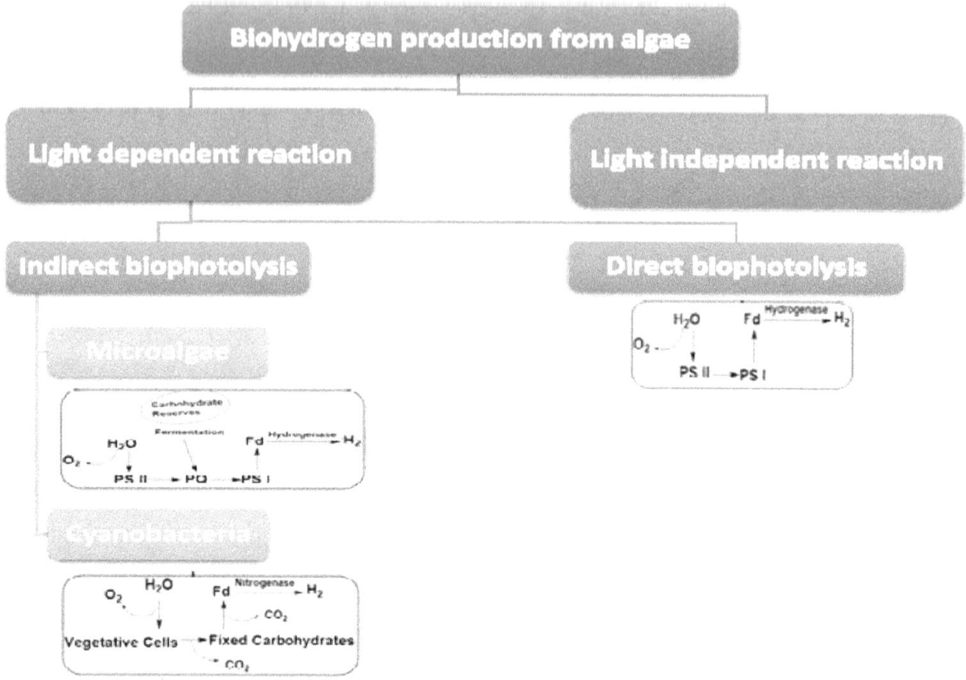

1. H_2 Production From Algae as Sustainable Bio-Resources

In this context, algae could offer various kind of biofuels, including: biodiesel (Abomohra & Elshobary, 2019; Ashour et al., 2019; Elshobary et al., 2019; Elshobary, et al., 2021a; Osman et al., 2020); methane (CH_4) (Zabed et al., 2019); bioelectricity (Elshobary, et al., 2021a) and bioethanol (Abomohra & Elshobary, 2019; Elshobary, et al., 2021b; Osman et al., 2020). Interestingly, Algae represents a viable option for bio-hydrogen production according to: its rapid growth rate, opportunities for year-round harvesting cycles, the ability of duplication of their biomass in a short time and high efficient photosynthetic and $CO_{2(g)}$ sequestration rates (Mubarak et al., 2019; Yin et al., 2020). The facility to produce hydrogen from algae has been investigated, and a variety of cultivation approaches have been used to increase hydrogen production from algae. In addition, carbohydrate-accumulated algae whether microalgae or macroalgae have been used as a feedstock for fermented bacteria to produce hydrogen. These renewable sources has the priority for energy generation as there is no pollution by emission of $CO_{2(g)}$, much less strength consumption, because they work at room temperature and atmospheric pressure (Karthic & Joseph, 2012). Bio-hydrogen production from algae can be categorized into i) bio-photolysis ii) darkish fermentation (Figure 1).

Algae are morphologically, physiologically, and biochemically diversified group of microorganisms without vascular tissue. Algae have numerous fascinating secondary metabolites and various bioactive substances due to their biodiversity and expansion in varied living environments: proteins, pigments, vitamins, oils, fats, polyunsaturated fatty acids, amino acids, lipids, minerals, fibers, carbohydrates, foodstuff, hydrocarbons, carbonyl compounds (aldehydes, ketones), alcohols, acids, esters, terpenes, sterols, volatiles and some polar compounds, alkaloids, flavonoids, steroids, phlobatannins. pigments, phenols, tannins, peroxides, polysaccharides, micronutrients, and vitamins (Elshobary et al., 2020; Elshobary, et al., 2021b; Khairy & El-Shafay, 2013).

Algae, like other plant organisms, have chlorophyll pigment, which allows performing photosynthetic processes. Algae rapidly grow and are widely distributed in freshwater, brackish water, and marine water. Many algae bioactive constituents can be extracted and processed for hydrogen production (Khairy & El-Shafay, 2013; Sallam et al., 2021), especially the polysaccharides macromolecules. Macroalgae have a higher biomass production than microalgae and can be grown without agricultural land or fresh water; they can also create unique substances that terrestrial plants do not produce.

2. Bio-Photolysis

Bio-photolysis appears to be a potential H_2-production method. Cultivating microalgae in large-scale compact photobioreactors could produce hydrogen spontaneously during the photochemical phase of photosynthesis (Vargas et al., 2014). Microalgae, such as microalgae (eukaryotic) and cyanobacteria (prokaryotic), can restore $CO_{2(g)}$ immediately from the air by using a photosynthetic apparatus comparable to greater to higher plants in the presence of sunlight that can be captured through the antenna chlorophylls and transferred to P680 of photosystem I and P700 of photosystem II. Once a sunlight-photon excite these photosystems, the electron transportation is initiated across the cell membrane to produce ATP unit energy (Srirangan et al., 2011). The liberated free electrons are transferred to ferredoxin NADP+ oxidoreductase (FNR) via the ferredoxin enzyme, generating NADPH for CO_2 fixation. A water molecule can be spitted when at the suitable electrical power proving the overpotential for hydrogen and oxygen evolution, resulting in the production of hydrogen (eqn. 7).

$$4H_2O \xrightarrow{\text{quantized photom "light energy"}} 2O_{2(g)} + 4H_{2(g)} \tag{7}$$

Both microalgae and cyanobacteria can release H_2 by bio-photolysis of water under anaerobiotic conditions. Some green algal species such *as Chlamydomonas reinhardtii, Chlorella, Chlorococcum littorale, Scenedesmus* spp., *Platymonas subcordiformis* and *Chlorella fusca* are favored in hydrogen production as they have high active hydrogenase enzyme, and ease of cultivation (Melis, 2007; Oey et al., 2016).

a. Direct Bio-Photolysis

Direct bio-photolysis produces hydrogen directly from water splitting using a microalgae photosynthesis system (Figure 2), which is presumably the simplest and most appealing approach. In anaerobic conditions, light photons (h energized PSI and PSII to transmit electrons directly from the water to reduce ferredoxin (Fd) (Kim & Kim, 2011). Reduced Fd works as an electron donor, allowing the hydrogenase enzyme to generate hydrogen from protons (H^+) on reduction (Ha et al., 2020). Direct bio-photolysis is more useful because the culture media relies only on water, sunlight, atmospheric CO_2 and the nutrients ($C_6H_{12}O_6$)as a carbon source (Eroglu & Melis, 2016).

Figure 2. Schematic representation of direct bio-photolysis in eukaryotic microalgae

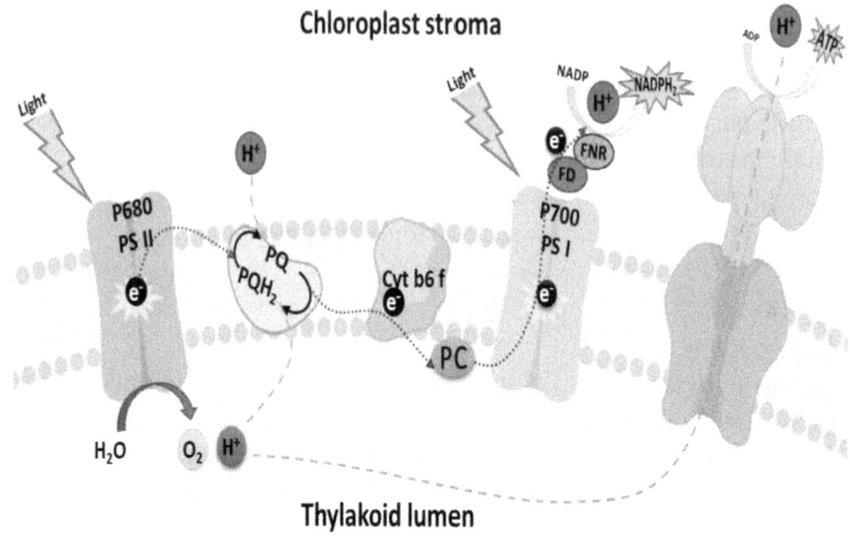

b. Indirect Bio-Photolysis

During indirect bio-photolysis, microalgae and cyanobacteria produce hydrogen from photosynthetic products such as starch and glycogen. It is summarized in two steps: growing and carbon fixation phase using light energy, and second, non-growing and hydrogen production phase via fermentation of the stored fixed carbon (Voloshin et al., 2016). In green algae, natural variation in light intensity temporally separates the oxygen evolution and hydrogen generation during the day. During the light period, photosynthesis is activated to fix CO_2 and generate oxygen. In dark periods, the fixed carbon is catabolized, generating a flow of electrons transferred from NADPH to the Plasto-Quinone (PQ) (Figure 3) as represented by (Kumari et al., 2017).

This process still working till oxygen is generated in the medium. However, under the anaerobic condition of oxygen depletion, the hydrogenase is activated to produce hydrogen from the electrons (plus H+) formed during fermentation. These electrons transfer to PSI, FDX1 and finally hydrogenase enzyme (K.-Y. Y. Show et al., 2013). The procedure is not continuous; as soon as mild length resumes, photosynthetic oxygen is released and prohibited hydrogenase. The hydrogen production can be summarized in the following chemical equations (Applying Hess's law for the chemical eqns. 8, 9:

The Overall reaction: H_2O+ light $\rightarrow H_2+O_2$

$$C_6H_{12}O_6+2H_2O \rightarrow 4H_2+2CH_3COOH+2CO_2 \quad (8)$$

$$2CH_3COOH+4H_2O+light \rightarrow 8H_2+4CO_2 \quad (9)$$

In heterocystous cyanobacteria, the O_2 sensitivity problem of hydrogenase is naturally solved by spatially separating oxygenic photosynthesis and hydrogen production. Cyanobacteria can fix atmospheric nitrogen by nitrogenase in specialized cells called heterocysts micro-anaerobic conditions with the en-

Figure 3. Schematic representation of indirect biophotolysis in eukaryotic microalgae

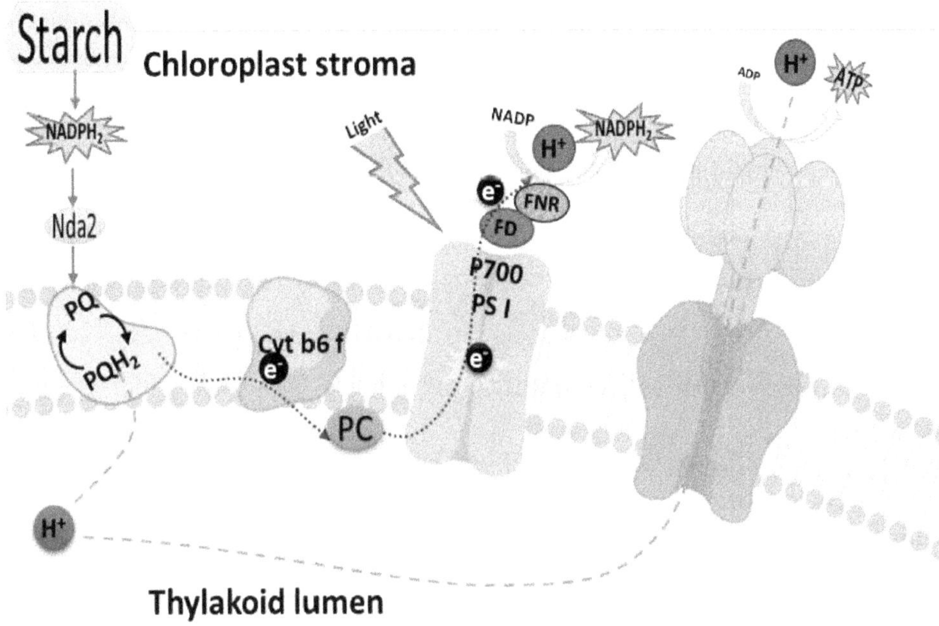

Table 3. Photo fermentation biohydrogen production using microalgal biomass as substrate

Phylum	microagal species	condition	Biohydrogen yield	Productivity	References
Chlorophyta	*Scenedesmus obliquus*	purple light	128 mL L^{-1}	204.8 mL $L^{-1}day^{-1}$	(Ruiz-Marin et al., 2020)
	Chlorella vulgaris	purple light	60.4 mL L^{-1}	39.18 mL $L^{-1}day^{-1}$	(Ruiz-Marin et al., 2020)
	Chlorella vulgaris	Photobioreactor, HM medium, Light irradiance 150 (μmol m^{-2} s^{-1})		5.3 mL L^{-1} h^{-1}	(Touloupakis et al., 2021)
	Chlamydomonas reinhardtii	Air + CO_2	86 ml L^{-1}		(Grechanik et al., 2021)
	C. reinhardtii L159I–N230Y	Photobioreactor, TAP-S medium, Light irradiance 140 (μmol m^{-2} s^{-1})		5.8 mL L^{-1} h^{-1}	(Torzillo et al., 2009)
Cyanophyta	*Spirulina Platensis*	optimized photo-fermentation factors are pH-5.95, D.F.-20.30 folds, Fe(II) sulfate-0.412 μM	337.0 ml g^{-1} DW		(Pandey et al., 2021)
	Arthospira maxima	Electrochemical Sequential Batch Reactor	15.16 μm mg^{-1}DW	2.65 m^3 m^{-3} d^{-1}	(H. Singh et al., 2022)

ergetic compounds obtained from the catabolic of stored fixed carbon, whereas CO_2 fixation performs in the vegetative cells (Benemann, 2000).

The frequent hydrogen-producing cyanobacteria are the filamentous genera, including *Nostoc, Anabaena, Calothrix* and *Oscillatoria*, whereas the unicellular and non-heterocystous filamentous cyanobacteria have a temporal separation mechanism through performing photosynthesis (oxygen releasing system) at the daylight hours and nitrogen fixation (hydrogen production) all through nighttime (Compaore & Stal, 2010). Moreover, non-nitrogen fixing species like *Synechocystis* and *Synechococcus* can additionally produce hydrogen by hydrogenase as a substitute for nitrogenase (Benemann, 2000). Cyanobacteria are capable of generating hydrogen through [NiFe] hydrogenases which perform along with a [MoFe] nitrogenase (Srirangan et al., 2011). During ordinary nitrogen fixation through [MoFe] nitrogenase, hydrogen is generated as a byproduct, represented by the following equation (Eroglu & Melis, 2016), eqn. 10:

$$N_2 + 8e^- + 8H^+ + 16ATP \rightarrow 2NH_3 + H_2 + 16ADP + 16Pi \quad (10)$$

Under nitrogen depletion, nitrogenase acts as effective hydrogenase with photosynthetic ATP to generate a lot greater rate of hydrogen. Hydrogen manufacturing by way of non-nitrogen fixing cyanobacteria is confirmed in eqn. 11 (Kim & Kim, 2011):

$$8H^+ + 8e^- + 4ATP \rightarrow 4H_2 + 4ADP + 4Pi \quad (11)$$

Indirect bio-photolysis is a competitive process due to the continued stable availability of fixed carbon sources through the photosynthetic growth phase used in bio-hydrogen production. Furthermore, most cyanobacteria can utilize alternative carbon sources to provide protons and electrons for hydrogen production. (Mathews & Wang, 2009). Interestingly, indirect bio-photolysis is supposed to be practicable for commercial hydrogen production, especially if photon conversion and photosynthetic rate can increase the fixed carbon (Nagarajan et al., 2017b).

c. Dark Fermentation

Bio-hydrogen synthesis by dark fermentation is a simple, environmentally friendly, low-cost technology that does not require light energy (Łukajtis et al., 2018). Fermentation of carbon-containing waste materials is regarded as an important method for bio-hydrogen manufacture and carbon waste treatment. In the absence of light and oxygen, fermentative bacteria can synthesize bio-hydrogen by fermenting rich carbon wastes. Protons (H^+) are produced by the breakdown of carbon wastes, which interact with electrons (e^-) to make hydrogen gas (H. Y. Lee et al., 2011). This process depends on different physiological factors such as temperature, pH and the organic substrate. Algae can play a significant role in bio-hydrogen generation through dark fermentation (Figure 4). The carbohydrate in the algal biomass is a promising feedstock in dark fermentation (Ho et al., 2013). Algal carbohydrates are present in simple polymers such as glucans and glycogen, easily converted to simple fermentable sugar such as glucose. Recent special techniques have been upgraded to hydrolysis of algae biomass, including mechanical, thermal, chemical (acid and alkaline hydrolysis) and enzymatic hydrolysis, which enhanced the hydrogen production many times (Abomohra & Elshobary, 2019; Nagarajan et al., 2017a). Table 4 summerized the utilization of various algal biomass as a feedstock for biohydrogen production through Dark fermentation.

Figure 4. Dark fermentative biohydrogen using algal biomass

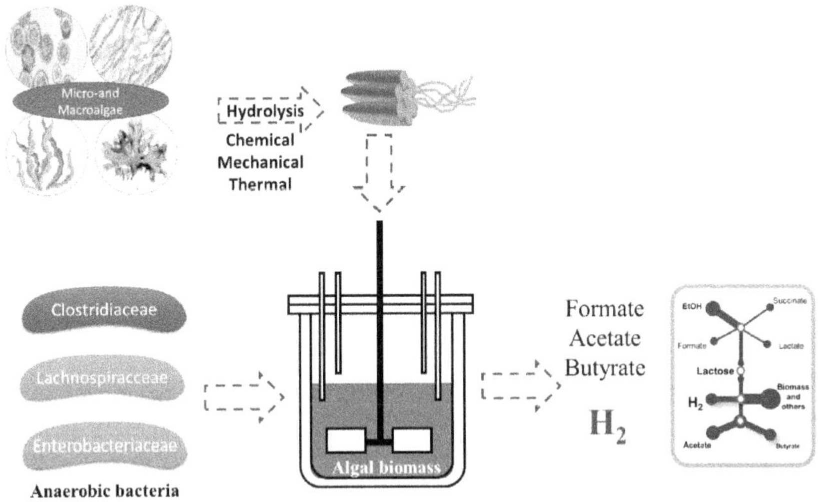

Table 4. Dark fermentative biohydrogen production using microalgal biomass as substrate

	Algal species	Conditions	Biohydrogen yield	References
Microalgae	*Scenedesmus obliquus*	Sonication	36.2 mL g^{-1} VS	(H. Singh et al., 2022)
	Chlorella vulgaris	ground, homogenized and autoclaved biomass	30.4 mL g^{-1} VS	(H. Singh et al., 2022)
	Spirulina Platensis	20.0% (v/v) and inoculum age-48 h of co-culture of *Rhodobacter sphaeroides* NMBL-01 and *Bacillus firmus* NMBL-03 under conditions pH-5.95	1510 mL L^{-1}	(Pandey et al., 2021)
	Arthospira maxima	Enzymatic	78.7mL g^{-1} TS	(Cheng et al., 2011)
	Mixed algal bloom	Hydrothermal acid	25.0 mL g^{-1}	(Cheng et al., 2019)
	Mxed microalgae biomass	pH (5.5) and heat pretreatment	9.5 mL g^{-1} VS	(G. Kumar et al., 2016)
Macroalgae	*Ulva lactuca*	35°C, anaerobic digester sludge	10.0 mLg^{-1} TS	(G. Kumar et al., 2015)
	Gelidium amansii	pH 5.5, 35°C, anaerobic digester sludge, pre-treated at 90°C, 20 min	43.1 mL g^{-1} TS	(G. Kumar et al., 2015)
	Laminaria japonica	pH 5.5, 35°C	67.0 mL g^{-1} TS	(G. Kumar et al., 2015)

BIOLOGICAL CHALLENGES AND CONSTRAINTS IN MICROALGAE HYDROGEN GENERATION

Many obstacles and challenges significantly restrict bio-hydrogen production, including the presence of toxic compounds accompanied by hydrogen production, carbon dioxide gas (CO_2)need to be separated during hydrogen production, slow and low production rate (Anwar et al., 2019).

1. Oxygen Sensitivity of the Hydrogenase

However, hydrogen production in microalgae is not a constant process as hydrogenase is extremely sensitive to the oxygenic environment as oxygen deactivates hydrogenase and prevents hydrogen production. As a result, hydrogen is only produced for a short period after illuminating till the O_2 level becomes suppressive. This suppression can be overcome by exhausting the released oxygen by stimulating the respiration rate by breaking down exogenous or endogenous substrates (Razu et al., 2019). Separation O_2 evolution from photosynthesis and hydrogen generation processes in green algae *Chlamydomonas reinhardtii* (Kosourov et al., 2005). This method depends on the elimination of sulfur sources from the cultivation medium. Sulfur depletion prevents the synthesis of photosynthetic enzymes and proteins of photosynthetic system II inhibits electron transport at PSII, while the activity of PSI is maintained. This is an efficient approach was used in g *Chlorella, Scenedesmus*, and cyanobacterium *Synechocystis*. It was discovered that mutants derived from microalgae and cyanobacteria produce more oxygen and so produce more hydrogen (Hallenbeck, 2002). Hydrogenase is extremely sensitive to oxygen during hydrogen production in microalgae, which deactivates hydrogenase and prevents hydrogen production. So, hydrogen is only produced for a short time after illuminating until the oxygen level becomes suppressive, but this suppression can be overcome by exhausting the released oxygen by stimulating the respiration rate by breaking down exogenous or endogenous substrates (Razu et al., 2019). Separation O_2 evolution from photosynthesis and hydrogen production processes in green algae *Chlamydomonas reinhardtii* (Kosourov et al., 2005).

2. Molecular Engineering of Algal H_2 Production

Many green algae produce hydrogen by biophotolysis of water by transferring electrons (charge) from water molecules to hydrogenase or nitrogenase enzymes under anaerobic conditions (Weaver et al., 1980). Hydrogenase is found in the chloroplast of eukaryotic microalgae and lacks nitrogenase, whereas hydrogenase is found in the cytoplasm of prokaryotic cyanobacteria and has nitrogenase. Specific technique conditions can influence the activation of these enzymes; for example, hydrogenase only activates at high pH and low intensity. In the lack of nitrogen, solar light is used to activate nitrogenase (N2). There are two forms of hydrogenase enzyme: iron-only (Fe-Fe-hydrogenase) and nickel-iron (Ni-Fe-hydrogenase). A binuclear iron middle is coupled to a protein with only one bond between cysteine amino acid residues in Fe-Fe-hydrogenases. All [Fe]-hydrogenases have a [2Fe-2S] cluster at the catalytic website (the H-cluster) that is linked to a [4Fe-4S] cluster by a cysteine residue, as well as unusual ligands such CO, CN, and di(thiol) methylamine. (Fan & Hall, 2001; Happe & Naber, 1993; Nicolet et al., 2000). (Happe & Naber, 1993) confirmed that H_2 reactions in *Chlamydomonas reinhardtii* are catalyzed by a monomeric, 49-kDa reversible [Fe]-hydrogenase enzyme. Nicolet et al. (2000) discovered that the H-cluster is bound by three distinct conserved cysteine residues bound to the protein matrix. In addition, most of these enzymes have supplementary iron-sulfur facilities that serve as electron relays between donor and acceptor carriers.

H_2-PRODUCTION IMPROVEMENT STRATEGIES-SUPPORTIVE TARGETS

Hydrogen manufacturing has been explored in many cyanobacterial species, considering they need the easiest dietary conditions. Furthermore, Cyanobacteria have several unique techniques to defend O2-sensitive enzymes from the photosynthetically advanced oxygen gas. In the lack of a nitrogen source, many nitrogen-fixing filamentous Cyanobacteria undergo mobile differentiation, which uses specialized heterocystous cells to separate the oxygenic photosynthesis and nitrogen-fixing enzymes, which are ordinarily interspersed among the vegetative cells. The formation of heterocysts across the vegetative filaments necessitates the incorporation of several external multiple external as well as internal signals, rules regulating genes, and a variety of cell methods and cellular processes. (R. Kumar et al., 2009).

1. Metabolomic and Recombinant DNA Technology of Microalgae for Enhanced Hydrogen Production

The life cycle alternate alternates between the diploid sporophyte and haploid gametophyte. Repeated mitotic divisions produce the haploid gametes, the zygote is formed by fusion between two gametes which then develops into diploid sporophytic thallus of Ulva. Diploid sporophytes are produced from the fusion of two gametes of opposite mating type. Both gametophyte and sporophyte developed follows the same pattern (Bourdareau et al., 2021). Algae cells show unspecialized cellular differentiation where each cell can go for photosynthesis and reproduction. Some species grows attached to rocks by a small disc-shaped holdfast (Tverskoi et al., 2018).

Cell wall external cover or extracellular matrix produced from complex biosynthetic machinery often use alga's photosynthetically fixed carbon multilayered. Each layer is composed of a highly irregular electron-dense fibrillar network with variable thickness. The thick multi-laminate fibrillar cell walls consist of polysaccharides such as cellulose, xyloglucan and glucuronan. Algae produce uncommon sugars through anabolism: generating sugar nucleotide precursors in the cytoplasm, polymerization of precursors in cellular endomembrane system, exporting, and assembling in the cell wall are all steps in the biosynthesis of matrix polysaccharides.(Wanke et al., 2021).

2. Elimination of Pathways Competing for Electrons

Photosynthetic microalgae use phycobilisomes (PBS) complexes protein to break up water into oxygen plus protons and electrons, which activate P700 and P680 chlorophyll pigments at the photosystem (PS) I and II response facilities embedded in the thylakoid membrane, respectively. The freed electrons are utilized along the photosynthetic electron transport chain, diminishing equivalents that provide energy for cell metabolism. (Luimstra et al., 2019). PSII, cytochrome (Cyt) b6f, plastocyanin (PC)/Cyt c6, and PSI transport photosynthetic electrons to ferredoxin (Fd) (distribution hub for electrons). The photosynthetic electron transfer chain (PETC) also generates electrical potential across the thylakoid membrane; the H+ motive force (pmf) drives ATP production from ADP via ATP synthase. (Malone et al., 2021).

Algae maintain photochemical reactions and metabolic needs in response to environmental conditions (light intensity, temperature, and nutrients). Microalgae developed several auxiliary electron transport (AET) pathways to dissipate the potentially dangerous oversupply of absorbed light photons. These AET pathways are electron sinks, such as flavodiiron proteins (FDPs) or different terminal oxidases, and pathways that recycle electrons around photosystem I, such as NADPH-dehydrogenase-like complexes (NDH) or the ferredoxin-plastoquinone reductase (FQR). Under controlled conditions, there is no need to pursue AET pathways that will become energetically unfavourable. These AET pathways are electron sinks, such as flavodiiron proteins (FDPs) or different terminal oxidases, and pathways that recycle electrons around photosystem I, such as NADPH-dehydrogenase-like complexes (NDH) or the ferredoxin-plastoquinone reductase (FQR). Under controlled conditions, there is no need to pursue AET pathways that will become energetically unfavourable (Nikkanen et al., 2021). Based on the environment and the metabolic state of the cell, electrons are transferred from FdviaFd-NAD(P)H-oxidoreductase (FNR), converting NADP+ to electron provider NADPH, which is primarily used in the CO2 fixation (CBB) cycle. (Nikkanen et al., 2021).

However, if the sink capability of CBB cycle is saturated, PETC may additionally grow to be excessively decreased, producing reactive oxygen species (ROS). This might also happen all through fluctuations in light intensity or nutrient availability. Over-reduction and huge ROS production can brutally injure the photosynthetic machinery, which has required the evolution of photoprotective mechanisms and auxiliary electron transport (AET) pathways. (Qiao et al., 2021) represented that photosystems are vulnerable to excessive reduction, and whilst PSII has a speedy restore cycle, restoration of PSI from injury is slow. Photosynthetic algae make investments closely in protecting PSI through multiple mechanisms to preserve PSI and its P700 reaction center chlorophyll pair in an oxidized state.

3. Photobioreactor for the Improvement of Hydrogen Production

Based on the topics illustrated in this chapter, the novel Photobioreactor for improving hydrogen production is represented (Wang et al., 2019). As shown in Figure (4), photosynthesis of algae involves reaction glucose sugar ($C_6H_{12}O_6$) with water in the presence of carbon dioxide and sunlight photon (eqn. 12):

$$C_6H_{12}O_6 + 2H_2O \rightarrow 4H_2 + 2CH_3COOH + 2CO_2 \quad (12)$$

As represented by (Daud et al., 2021), the suggested fuel cell converting the evolved hydrogen gas from algae biomass to electricity was classified according to the operating temperature as: low temperature (25-100°C); medium temperature (100-500°C); high temperature above 1000°C, and very high temperature (above 1000°C). The latter high-temperature fuel cell can be efficiently operated without the need for a catalyst. The polarization of a fuel cell caused by the slow diffusion of oxygen reduces the current. Figure (5) shows the schematic representation of a hydrogen-oxygen fuel cell with a solid acidic or alkaline conducting electrolyte, ion exchange membrane that is impermeable to the reactant's hydrogen and oxygen gases but is permeable to ions hydrogen (H^+), and OH^- which carries the current between the two electrodes. The ultra-thin ion-exchange membrane keeps the cell resistance as low as possible. To facilitate the operation of the cell at a low-temperature range from:40-60°C, the electrodes are covered with a finely divided platinum catalyst.

Water is drained out of the cell during operation. Fuel cell of this general type is too efficient to be used successfully in the space program. In hydrogen-oxygen fuel cell, the half-cell reactions are:

$$H_{2(g)} = 2H^+ + 2e^-$$

$$½O_{2(g)} + 2H^+ + 2e^- = H_2O_{(l)}$$

Figure 5. Schematic representation of a modified photoreactor for hydrogen production from algal biomass (Wang et al., 2019).

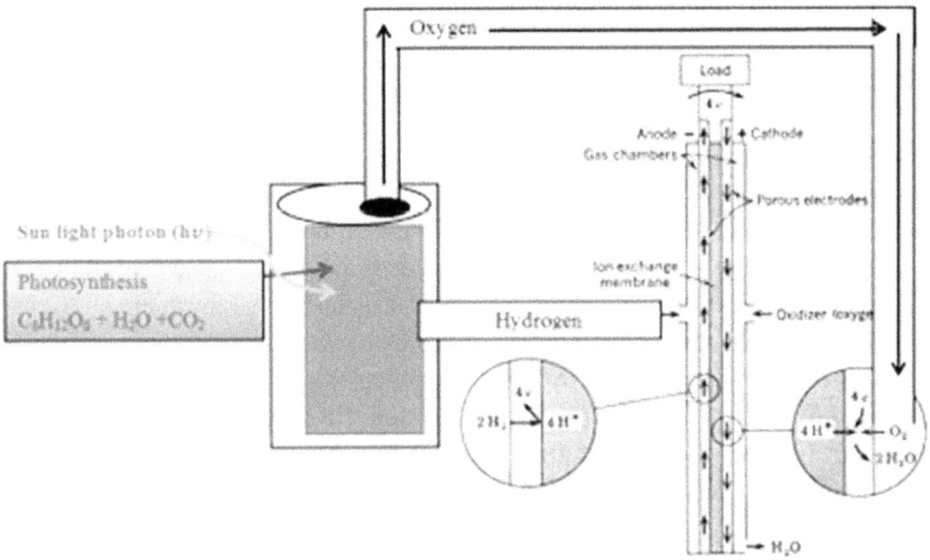

$$H_{2(g)} + ½O_{2(g)} = H_2O_{(l)}$$

or

$$H_{2(g)} + 2OH^- = 2\ H_2O\ (l) + 2e^-$$

$$½O_2\ (g) + H_2O_{(l)} + 2e^- = 2OH^-$$

$$H_{2(g)} + ½O_{2(g)} = H_2O_{(l)}$$

The standard electromotive force of hydrogen-oxygen fuel cell can calculate from the standard Gibbs free energy of formation of $H_2O_{(l)}$.

$$\Delta G^o = -\ z\ F\ E^o$$

$$-\ 56{,}690\ cal.mol^{-1} = -\ 2(23{,}060\ cal.V^{-1}\ mol^{-1})E^o$$

$$E^o = 1.229\ V$$

HYDROGEN, ENVIRONMENT AND CLIMATE CHANGE

Microalgae biomass can be used to produce biodiesel, bioethanol, biogas, bio-hydrogen, and bio-oils. Microalgae have an excessive photosynthetic efficiency and excessive growth rate. On evaluating to different lipid feedstock, seasonal and geographical constraints are ways much less of a challenge due to the availability and diversity of microalgae species over a large variety of climates and geographic areas (Aparicio et al., 2020).

Green algae as a supply of biohydrogen is a promising choice as a future electricity service given that its conversion to energy yields solely water and it has the probability to address fossil fuels issues (Show et al., 2012). Nevertheless, the quantity of hydrogen created through biological processes is insignificant compared to the quantity produced via modern chemical structures (Srirangan et al., 2011).

The lack of a large-scale approach, low yield and energy conversion efficiency, and inhibition of hydrogenase by oxygen, a byproduct of photolysis, all impede microalgae H_2 generation. Sulfur deficiency is required to avoid oxygen-induced hydrogenase inhibition. Under such conditions, oxygen evolution is lowered below that necessary for respiration, an anaerobic environment is created, and hydrogenase activity may be preserved. (Zhang et al., 2014; Zhu et al., 2014).

Combined Biorefinery System for Cost-Effective Production of Biohydrogen

Previous research has looked at the upfront costs of commercial large-scale microalgal biohydrogen production, where the nutrients and harvesting are the most expensive issues. The cost of producing biohydrogen is expected to be US$15/GJ, which is comparable to the cost of producing two-stage biohydrogen from biomass residue, which is estimated to be 19 Euro/GJ (Akkerman et al., 2002; Goswami et al., 2021). It is also stated that huge efforts should be made to recover and reuse photobioreactor building materials and growth nutrients in order to greatly minimize overall operation costs (K.-Y. Show et al., 2018). As a result, in recent years, researchers have begun to investigate the sustainable integrated fermentation refinery in order to raise the maximum biohydrogen yield by assembling hybrid hydrogen-producing systems. The operating cost of dark-fermentation-hydrogen production is reduced to some extent with this innovative approach. This system depends on producing various valuable product and using the residue biomass for fermentative H_2 production. wastewater treatment may be also incorporates. For instance, after biodiesel synthesis, lipid-extracted microalgal biomass residues (LMBRs) can be employed as a raw material for dark-fermentation-H_2 generation (Chen et al., 2016; Sengmee et al., 2017). Various multi-biorefinery systems have also been used to recover multi-bioproducts such as lipids and carotenoids, then use the residual biomass as feedstock and combine with bacteria to make biohydrogen under dark fermentation conditions (Brar et al., 2022).

ECONOMIC ADVANCE AND FUTURE VIEWPOINT

For production of hydrogen, there are a number of techniques still being used. Chemical processes including steam reforming, coal gasification, and H2O electrolysis for H2 generation are well-established systems with a wide range of industrial uses. With a cost of $ 7/GJ, steam reforming of methane is currently the most cost-effective technique of hydrogen manufacture (Kalamaras & Efstathiou, 2013). The most prevalent method for producing hydrogen is electrolysis, which costs roughly $1.09/kg when combined with geothermal as a source of power(Yilmaz et al., 2019). H_2 production cost was close to \$2.8/kg in a commercial scale facility based on photosynthetic microalgae(Melis & Happe, 2001). The cost of producing hydrogen from biomass gasification and pyrolysis is roughly \$10–14/GJ and \$8.9–15.5/GJ, depending on the equipment operated and the availability of feedstocks(Balat & Balat, 2009). Because of current constraints, biological processes such as fermentation continue to have applicability issues. The main issue is cost, which is higher due to decreased H_2 yields and process instabilities (Das & Basak, 2021a). The cost of producing hydrogen via the photofermentation process is roughly \$502.10/GJ, with the cost of operation units accounting for 90% of the cost (Bhatia et al., 2021). Solar energy is a sustainable and clean energy source, but it has a number of drawbacks, such as heating issues and fluctuating intensity. These issues can be solved by building effective photobioreactors that disperse light equally and effectively (Das & Basak, 2021b).This method has a lot of promise for future technology development to make H_2 generation more efficient and cost-effective (Brar et al., 2022). In the future, the extent of H_2 production will be determined not only the previous mentioned research, but also by economic considerations.

CONCLUSION

Economic growth, energy use, the environment, and global warming are linked, yet "securing energy and the environment at the lowest cost" is becoming more challenging. Although natural and manmade factors contribute to climate change, global warming caused by greenhouse gases, particularly CO_2, is the primary cause of anthropogenic climate change. Switching to a hydrogen economy and employing renewable energy sources would be the best option for maintaining economic growth. In the long run, hydrogen created from algal biomass and solar energy will be the most renewable energy currencies. To achieve timely implementation of educational, financial, legislative, social, and technical initiatives is required.

REFERENCES

Abanades, S., Charvin, P., Flamant, G., & Neveu, P. (2006). Screening of water-splitting thermochemical cycles potentially attractive for hydrogen production by concentrated solar energy. *Energy*, *31*(14), 2805–2822. doi:10.1016/j.energy.2005.11.002

Abomohra, A. E., & Elshobary, M. E. (2019). Biodiesel, bioethanol and biobutanol production from microalgae biomass. In Md. A. Alam & Z. Wang (Eds.), Microalgae Biotechnology for Development of Biofuel and Waste Water Treatment (pp. 293–321). Springer Singapore Press. doi:10.1007/978-981-13-2264-8

Acar, C., Dincer, I., & Zamfirescu, C. (2014). A review on selected heterogeneous photocatalysts for hydrogen production. *International Journal of Energy Research*, *38*(15), 1903–1920. doi:10.1002/er.3211

Akkerman, I., Janssen, M., Rocha, J., & Wijffels, R. H. (2002). Photobiological hydrogen production: Photochemical efficiency and bioreactor design. *International Journal of Hydrogen Energy*, *27*(11–12), 1195–1208. doi:10.1016/S0360-3199(02)00071-X

Angeli, S. D., Monteleone, G., Giaconia, A., & Lemonidou, A. A. (2014). State-of-the-art catalysts for CH4 steam reforming at low temperature. *International Journal of Hydrogen Energy*, *39*(5), 1979–1997. doi:10.1016/j.ijhydene.2013.12.001

Anwar, M., Lou, S., Chen, L., Li, H., & Hu, Z. (2019). Recent advancement and strategy on bio-hydrogen production from photosynthetic microalgae. *Bioresource Technology*, *292*(August), 121972. doi:10.1016/j.biortech.2019.121972 PMID:31444119

Aparicio, E., Rodríguez-Jasso, R. M., Lara, A., Loredo-Treviño, A., Aguilar, C. N., Kostas, E. T., & Ruiz, H. A. (2020). Biofuels production of third generation biorefinery from macroalgal biomass in the Mexican context: An overview. *Sustainable Seaweed Technologies*, 393–446.

Ashik, U. P. M., Daud, W. M. A. W., & Abbas, H. F. (2015). Production of greenhouse gas free hydrogen by thermocatalytic decomposition of methane–A review. *Renewable & Sustainable Energy Reviews*, *44*, 221–256. doi:10.1016/j.rser.2014.12.025

Ashour, M., Elshobary, M. E., El-Shenody, R., Kamil, A. W. A., & Abomohra, A. E (2019). Evaluation of a native oleaginous marine microalga Nannochloropsis oceanica for dual use in biodiesel production and aquaculture feed. *Biomass and Bioenergy*, *120*(December), 439–447. doi:10.1016/j.biombioe.2018.12.009

Balat, M., & Balat, M. (2009). Political, economic and environmental impacts of biomass-based hydrogen. *International Journal of Hydrogen Energy*, *34*(9), 3589–3603. doi:10.1016/j.ijhydene.2009.02.067

Baykara, S. Z. (2004). Hydrogen production by direct solar thermal decomposition of water, possibilities for improvement of process efficiency. *International Journal of Hydrogen Energy*, *29*(14), 1451–1458. doi:10.1016/j.ijhydene.2004.02.014

Baykara, S. Z. (2018). Hydrogen: A brief overview on its sources, production and environmental impact. *International Journal of Hydrogen Energy*, *43*(23), 10605–10614. doi:10.1016/j.ijhydene.2018.02.022

Benemann, J. R. (2000). Hydrogen production by microalgae. *Journal of Applied Phycology*, *12*(3–5), 291–300. doi:10.1023/A:1008175112704

Bhatia, S. K., Jagtap, S. S., Bedekar, A. A., Bhatia, R. K., Rajendran, K., Pugazhendhi, A., Rao, C., Atabani, A. E., Kumar, G., & Yang, Y.-H. (2021). Renewable biohydrogen production from lignocellulosic biomass using fermentation and integration of systems with other energy generation technologies. *The Science of the Total Environment*, *765*, 144429. doi:10.1016/j.scitotenv.2020.144429 PMID:33385808

Bicer, Y., & Dincer, I. (2017). Life cycle assessment of nuclear-based hydrogen and ammonia production options: A comparative evaluation. *International Journal of Hydrogen Energy*, *42*(33), 21559–21570. doi:10.1016/j.ijhydene.2017.02.002

Bourdareau, S., Mignerot, L., Heesch, S., Peters, A. F., Coelho, S. M., & Cock, J. M. (2021). Alternation of Generations in Plants and Algae. *Evolutionary Developmental Biology: A Reference Guide*, 631–644.

Brar, K. K., Cortez, A. A., Pellegrini, V. O. A., Amulya, K., Polikarpov, I., Magdouli, S., Kumar, M., Yang, Y.-H., Bhatia, S. K., & Brar, S. K. (2022). An overview on progress, advances, and future outlook for biohydrogen production technology. *International Journal of Hydrogen Energy*. Advance online publication. doi:10.1016/j.ijhydene.2022.01.156

Brentner, L. B., Peccia, J., & Zimmerman, J. B. (2010). Challenges in developing biohydrogen as a sustainable energy source: Implications for a research agenda. *Environmental Science & Technology, 44*(7), 2243–2254. doi:10.1021/es9030613 PMID:20222726

Chen, C.-Y., Chang, H.-Y., & Chang, J.-S. (2016). Producing carbohydrate-rich microalgal biomass grown under mixotrophic conditions as feedstock for biohydrogen production. *International Journal of Hydrogen Energy, 41*(7), 4413–4420. doi:10.1016/j.ijhydene.2015.05.163

Cheng, J., Yue, L., Ding, L., Li, Y.-Y., Ye, Q., Zhou, J., Cen, K., & Lin, R. (2019). Improving fermentative hydrogen and methane production from an algal bloom through hydrothermal/steam acid pretreatment. *International Journal of Hydrogen Energy, 44*(12), 5812–5820. doi:10.1016/j.ijhydene.2019.01.046

Cheng, J., Zhang, M., Song, W., Xia, A., Zhou, J., & Cen, K. (2011). Cogeneration of hydrogen and methane from Arthrospira maxima biomass with bacteria domestication and enzymatic hydrolysis. *International Journal of Hydrogen Energy, 36*(2), 1474–1481. doi:10.1016/j.ijhydene.2010.11.009

Compaore, J., & Stal, L. J. (2010). Oxygen and the light-dark cycle of nitrogenase activity in two unicellular cyanobacteria. *Environmental Microbiology, 12*(1), 54–62. doi:10.1111/j.1462-2920.2009.02034.x PMID:19691503

D'Souza, L. (2013). Thermochemical hydrogen production from water using reducible oxide materials: A critical review. *Materials for Renewable and Sustainable Energy, 2*(1), 1–12. doi:10.100740243-013-0007-0

Damen, K., van Troost, M., Faaij, A., & Turkenburg, W. (2006). A comparison of electricity and hydrogen production systems with CO2 capture and storage. Part A: Review and selection of promising conversion and capture technologies. *Progress in Energy and Combustion Science, 32*(2), 215–246. doi:10.1016/j.pecs.2005.11.005

Das, S. R., & Basak, N. (2021a). Molecular biohydrogen production by dark and photo fermentation from wastes containing starch: Recent advancement and future perspective. *Bioprocess and Biosystems Engineering, 44*(1), 1–25. doi:10.100700449-020-02422-5 PMID:32785789

Das, S. R., & Basak, N. (2021b). Molecular biohydrogen production by dark and photo fermentation from wastes containing starch: Recent advancement and future perspective. *Bioprocess and Biosystems Engineering, 44*(1), 1–25. doi:10.100700449-020-02422-5 PMID:32785789

Daud, S. N. S. S., Norddin, M. N. A. M., Jaafar, J., & Sudirman, R. (2021). Development of sulfonated poly (ether ether ketone)/polyethersulfone-crosslinked quaternary ammonium poly (ether ether ketone) bipolar membrane electrolyte via hot-press approach for hydrogen/oxygen fuel cell. *International Journal of Energy Research, 45*(6), 9210–9228. doi:10.1002/er.6453

Dincer, I., & Balta, M. T. (2011). Potential thermochemical and hybrid cycles for nuclear-based hydrogen production. *International Journal of Energy Research, 35*(2), 123–137. doi:10.1002/er.1769

Elshobary, M. E., Abo-Shady, A. M., Khairy, H. M., Essa, D., Zabed, H. M., Qi, X., & Abomohra, A. E. (2019). Influence of nutrient supplementation and starvation conditions on the biomass and lipid productivities of Micractinium reisseri grown in wastewater for biodiesel production. *Journal of Environmental Management, 250*(April), 109529. doi:10.1016/j.jenvman.2019.109529 PMID:31518792

Elshobary, M. E., El-Shenody, R. A., & Abomohra, A. E. (2021). Sequential biofuel production from seaweeds enhances the energy recovery: A case study for biodiesel and bioethanol production. *International Journal of Energy Research*, *45*(4), 6457–6467. doi:10.1002/er.6181

Elshobary, M. E., El-Shenody, R. A., Ashour, M., Zabed, H. M., & Qi, X. (2020). Antimicrobial and antioxidant characterization of bioactive components from Chlorococcum minutum. *Food Bioscience*, *35*, 100567. doi:10.1016/j.fbio.2020.100567

Elshobary, M. E., Zabed, H. M., Yun, J., Zhang, G., & Qi, X. (2021). Recent insights into microalgae-assisted microbial fuel cells for generating sustainable bioelectricity. *International Journal of Hydrogen Energy*, *46*(4), 3135–3159. doi:10.1016/j.ijhydene.2020.06.251

Eroglu, E., & Melis, A. (2016). Microalgal hydrogen production research. *International Journal of Hydrogen Energy*, *41*(30), 12772–12798. doi:10.1016/j.ijhydene.2016.05.115

Fan, H.-J., & Hall, M. B. (2001). A capable bridging ligand for Fe-only hydrogenase: Density functional calculations of a low-energy route for heterolytic cleavage and formation of dihydrogen. *Journal of the American Chemical Society*, *123*(16), 3828–3829. doi:10.1021/ja004120i PMID:11457119

Fetouh, H. A., Ismail, A. M., Hamid, H. A., & Bashier, M. O. (2019). Synthesis of promising nano-composites from an antitumer and biologically active heterocyclic compound uploaded by clay and chitosan polymers. *International Journal of Biological Macromolecules*, *137*, 1211–1220. doi:10.1016/j.ijbiomac.2019.06.187 PMID:31254576

Genç, M. S., Çelik, M., & Karasu, İ. (2012). A review on wind energy and wind–hydrogen production in Turkey: A case study of hydrogen production via electrolysis system supplied by wind energy conversion system in Central Anatolian Turkey. *Renewable & Sustainable Energy Reviews*, *16*(9), 6631–6646. doi:10.1016/j.rser.2012.08.011

Goswami, R. K., Mehariya, S., Obulisamy, P. K., & Verma, P. (2021). Advanced microalgae-based renewable biohydrogen production systems: A review. *Bioresource Technology*, *320*, 124301. doi:10.1016/j.biortech.2020.124301 PMID:33152683

Grechanik, V., Naidov, I., Bolshakov, M., & Tsygankov, A. (2021). Photoautotrophic hydrogen production by nitrogen-deprived Chlamydomonas reinhardtii cultures. *International Journal of Hydrogen Energy*, *46*(5), 3565–3575. doi:10.1016/j.ijhydene.2020.10.215

Ha, G., El-dalatony, M. M., Kurade, M. B., Salama, E., Basak, B., Kang, D., Roh, H., Lim, H., & Jeon, B. (2020). Energy-efficient pretreatments for the enhanced conversion of microalgal biomass to biofuels * Corresponding Author. *Bioresource Technology*, *123333*, 123333. Advance online publication. doi:10.1016/j.biortech.2020.123333 PMID:32305842

Hallenbeck, P. C. (2002). Biological hydrogen production; fundamentals and limiting processes. *International Journal of Hydrogen Energy*, *27*(11–12), 1185–1193. doi:10.1016/S0360-3199(02)00131-3

Happe, T., & Naber, J. D. (1993). Isolation, characterization and N-terminal amino acid sequence of hydrogenase from the green alga Chlamydomonas reinhardtii. *European Journal of Biochemistry*, *214*(2), 475–481. doi:10.1111/j.1432-1033.1993.tb17944.x PMID:8513797

Ho, S.-H. H., Huang, S.-W. W., Chen, C.-Y. Y., Hasunuma, T., Kondo, A., & Chang, J.-S. S. (2013). Bioethanol production using carbohydrate-rich microalgae biomass as feedstock. *Bioresource Technology, 135*, 191–198. doi:10.1016/j.biortech.2012.10.015 PMID:23116819

Ismail, A. A., & Bahnemann, D. W. (2014). Photochemical splitting of water for hydrogen production by photocatalysis: A review. *Solar Energy Materials and Solar Cells, 128*, 85–101. doi:10.1016/j.solmat.2014.04.037

Kalamaras, C. M., & Efstathiou, A. M. (2013). Hydrogen Production Technologies: Current State and Future Developments. *Conference Papers in Energy, 2013*, 1–9. doi:10.1155/2013/690627

Kalinci, Y., Hepbasli, A., & Dincer, I. (2009). Biomass-based hydrogen production: A review and analysis. *International Journal of Hydrogen Energy, 34*(21), 8799–8817. doi:10.1016/j.ijhydene.2009.08.078

Karthic, P., & Joseph, S. (2012). Comparison and limitations of biohydrogen production processes. *Research Journal of Biotechnology, 7*(2), 59–71.

Khairy, H. M., & El-Shafay, S. M. (2013). Seasonal variations in the biochemical composition of some common seaweed species from the coast of Abu Qir Bay, Alexandria, Egypt. *Oceanologia, 55*(2), 435–452. doi:10.5697/oc.55-2.435

Kim, D. H., & Kim, M. S. (2011). Hydrogenases for biological hydrogen production. *Bioresource Technology, 102*(18), 8423–8431. doi:10.1016/j.biortech.2011.02.113 PMID:21435869

Kosourov, S., Makarova, V., Fedorov, A. S., Tsygankov, A., Seibert, M., & Ghirardi, M. L. (2005). The Effect of Sulfur Re-Addition on H2 Photoproduction by Sulfur-Deprived Green Algae. *Photosynthesis Research, 85*(3), 295–305. doi:10.100711120-005-5105-0 PMID:16170632

Kumar, G., Bakonyi, P., Sivagurunathan, P., Kim, S.-H., Nemestóthy, N., Bélafi-Bakó, K., & Lin, C.-Y. (2015). Enhanced biohydrogen production from beverage industrial wastewater using external nitrogen sources and bioaugmentation with facultative anaerobic strains. *Journal of Bioscience and Bioengineering, 120*(2), 155–160. doi:10.1016/j.jbiosc.2014.12.011 PMID:25661265

Kumar, G., Zhen, G., Kobayashi, T., Sivagurunathan, P., Kim, S. H., & Xu, K. Q. (2016). Impact of pH control and heat pre-treatment of seed inoculum in dark H2 fermentation: A feasibility report using mixed microalgae biomass as feedstock. *International Journal of Hydrogen Energy, 41*(7), 4382–4392. doi:10.1016/j.ijhydene.2015.08.069

Kumar, R., Linga, P., Ripmeester, J. A., & Englezos, P. (2009). Two-stage clathrate hydrate/membrane process for precombustion capture of carbon dioxide and hydrogen. *Journal of Environmental Engineering, 135*(6), 411–417. doi:10.1061/(ASCE)EE.1943-7870.0000002

Kumari, J., Udawat, P., Dubey, A. K., Haque, M. I., Rathore, M. S., & Jha, B. (2017). Overexpression of SbSI-1, a nuclear protein from Salicornia brachiata confers drought and salt stress tolerance and maintains photosynthetic efficiency in transgenic tobacco. *Frontiers in Plant Science, 8*, 1215. doi:10.3389/fpls.2017.01215 PMID:28751902

Lee, D. J., Wang, Y., Zhang, Z., Lee, D. J., Zhou, X., Jing, Y., Jiang, D., Hu, J., & He, C. (2017). Photo-fermentative hydrogen production from crop residue: A mini review. *Bioresource Technology*, *229*, 222–230. doi:10.1016/j.biortech.2017.01.008 PMID:28108074

Lee, H. Y., Lee, S. E., Jung, K. H., Yeon, J. H., & Choi, W. Y. (2011). Repeated-batch operation of surface-aerated fermentor for bioethanol production from the hydrolysate of seaweed Sargassum sagamianum. *Journal of Microbiology and Biotechnology*, *21*(3), 323–331. doi:10.4014/jmb.1010.10057 PMID:21464605

Luimstra, V. M., Schuurmans, J. M., de Carvalho, C. F. M., Matthijs, H. C. P., Hellingwerf, K. J., & Huisman, J. (2019). Exploring the low photosynthetic efficiency of cyanobacteria in blue light using a mutant lacking phycobilisomes. *Photosynthesis Research*, *141*(3), 291–301. doi:10.100711120-019-00630-z PMID:30820745

Łukajtis, R., Hołowacz, I., Kucharska, K., Glinka, M., Rybarczyk, P., Przyjazny, A., & Kamiński, M. (2018). Hydrogen production from biomass using dark fermentation. *Renewable & Sustainable Energy Reviews*, *91*, 665–694. doi:10.1016/j.rser.2018.04.043

Malone, L. A., Proctor, M. S., Hitchcock, A., Hunter, C. N., & Johnson, M. P. (2021). Cytochrome b6f–Orchestrator of photosynthetic electron transfer. *Biochimica et Biophysica Acta (BBA)-. Bioenergetics*, 148380. doi:10.1016/j.bbabio.2021.148380 PMID:33460588

Mathews, J., & Wang, G. (2009). Metabolic pathway engineering for enhanced biohydrogen production. *International Journal of Hydrogen Energy*, *34*(17), 7404–7416. doi:10.1016/j.ijhydene.2009.05.078

Melis, A. (2007). Photosynthetic H 2 metabolism in Chlamydomonas reinhardtii (unicellular green algae). *Planta*, *226*(5), 1075–1086. doi:10.100700425-007-0609-9 PMID:17721788

Melis, A., & Happe, T. (2001). Hydrogen Production. Green Algae as a Source of Energy. *Plant Physiology*, *127*(3), 740–748. doi:10.1104/pp.010498 PMID:11706159

Moussa, G., Moury, R., Demirci, U. B., Şener, T., & Miele, P. (2013). Boron-based hydrides for chemical hydrogen storage. *International Journal of Energy Research*, *37*(8), 825–842. doi:10.1002/er.3027

Mubarak, M., Shaija, A., & Suchithra, T. V. (2019). Flocculation: An effective way to harvest microalgae for biodiesel production. *Journal of Environmental Chemical Engineering*, *7*(4), 103221. doi:10.1016/j.jece.2019.103221

Nagarajan, D., Lee, D. J., Kondo, A., & Chang, J. S. (2017a). Recent insights into biohydrogen production by microalgae – From biophotolysis to dark fermentation. *Bioresource Technology*, *227*(1), 373–387. doi:10.1016/j.biortech.2016.12.104 PMID:28089136

Nagarajan, D., Lee, D. J., Kondo, A., & Chang, J. S. (2017b). Recent insights into biohydrogen production by microalgae–From biophotolysis to dark fermentation. *Bioresource Technology*, *227*, 373–387. doi:10.1016/j.biortech.2016.12.104

Nicolet, Y., Lemon, B. J., Fontecilla-Camps, J. C., & Peters, J. W. (2000). A novel FeS cluster in Fe-only hydrogenases. *Trends in Biochemical Sciences*, *25*(3), 138–143. doi:10.1016/S0968-0004(99)01536-4 PMID:10694885

Nikkanen, L., Solymosi, D., Jokel, M., & Allahverdiyeva, Y. (2021). Regulatory electron transport pathways of photosynthesis in cyanobacteria and microalgae: Recent advances and biotechnological prospects. *Physiologia Plantarum, 173*(2), 514–525. doi:10.1111/ppl.13404 PMID:33764547

NOAA/ESRL. (2019). *National Oceanic and Atmospheric Administration/Earth System ResearchLaboratory*. https://www.esrl.noaa.gov/

Oey, M., Sawyer, A. L., Ross, I. L., & Hankamer, B. (2016). Challenges and opportunities for hydrogen production from microalgae. *Plant Biotechnology Journal, 14*(7), 1487–1499. doi:10.1111/pbi.12516 PMID:26801871

Osman, M. E. H., Abo-Shady, A. M., Elshobary, M. E., Abd El-Ghafar, M. O., & Abomohra, A. E.-F. F. (2020). Screening of seaweeds for sustainable biofuel recovery through sequential biodiesel and bioethanol production. *Environmental Science and Pollution Research International, 27*(26), 32481–32493. doi:10.100711356-020-09534-1 PMID:32506400

Pandey, A., Sinha, P., & Pandey, A. (2021). Hydrogen production by sequential dark and photofermentation using wet biomass hydrolysate of Spirulina platensis: Response surface methodological approach. *International Journal of Hydrogen Energy, 46*(10), 7137–7146. doi:10.1016/j.ijhydene.2020.11.205

Parthasarathy, P., & Narayanan, K. S. (2014). Hydrogen production from steam gasification of biomass: Influence of process parameters on hydrogen yield–a review. *Renewable Energy, 66*, 570–579. doi:10.1016/j.renene.2013.12.025

Patel, N., & Miotello, A. (2015). Progress in Co–B related catalyst for hydrogen production by hydrolysis of boron-hydrides: A review and the perspectives to substitute noble metals. *International Journal of Hydrogen Energy, 40*(3), 1429–1464. doi:10.1016/j.ijhydene.2014.11.052

Patel, S. K. S., Das, D., Kim, S. C., Cho, B., Kalia, V. C., & Lee, J. (2021). Integrating strategies for sustainable conversion of waste biomass into dark-fermentative hydrogen and value-added products. *Renewable & Sustainable Energy Reviews, 150*(May), 111491. doi:10.1016/j.rser.2021.111491

Qiao, M. Y., Zhang, Y.-J., Liu, L. A., Shi, L., Ma, Q.-H., Chow, W. S., & Jiang, C.-D. (2021). Do rapid photosynthetic responses protect maize leaves against photoinhibition under fluctuating light? *Photosynthesis Research, 149*(1), 57–68. doi:10.100711120-020-00780-5 PMID:32783175

Rathore, D., & Singh, A. (2013). Biohydrogen production from microalgae. In *Biofuel Technologies* (pp. 317–333). Springer. doi:10.1007/978-3-642-34519-7_13

Razu, M. H., Hossain, F., & Khan, M. (2019). Advancement of Bio-hydrogen Production from Microalgae BT - Microalgae Biotechnology for Development of Biofuel and Wastewater Treatment (Md. A. Alam & Z. Wang, Eds.). Springer Singapore. doi:10.1007/978-981-13-2264-8_17

Roblero Lucchetti, J. G. (2020). *Dry Reforming of Methane via Use of Ni, Pt and Mo over Combined Supports*. Academic Press.

Ruiz-Marin, A., Canedo-López, Y., & Chávez-Fuentes, P. (2020). Biohydrogen production by Chlorella vulgaris and Scenedesmus obliquus immobilized cultivated in artificial wastewater under different light quality. *AMB Express, 10*(1), 1–7. doi:10.118613568-020-01129-w PMID:33108605

Sallam, E. R., Khairy, H. M., Elnouby, M. S., & Fetouh, H. A. (2021). Sustainable electricity production from seawater using Spirulina platensis microbial fuel cell catalyzed by silver nanoparticles-activated carbon composite prepared by a new modified photolysis method. *Biomass and Bioenergy*, *148*, 106038. doi:10.1016/j.biombioe.2021.106038

Sengmee, D., Cheirsilp, B., Suksaroge, T. T., & Prasertsan, P. (2017). Biophotolysis-based hydrogen and lipid production by oleaginous microalgae using crude glycerol as exogenous carbon source. *International Journal of Hydrogen Energy*, *42*(4), 1970–1976. doi:10.1016/j.ijhydene.2016.10.089

Show, K.-Y., Yan, Y., Ling, M., Ye, G., Li, T., & Lee, D.-J. (2018). Hydrogen production from algal biomass – Advances, challenges and prospects. *Bioresource Technology*, *257*, 290–300. doi:10.1016/j.biortech.2018.02.105 PMID:29506887

Show, K.-Y. Y., Lee, D.-J. J., & Chang, J.-S. S. (2013). Algal biomass dehydration. *Bioresource Technology*, *135*, 720–729. doi:10.1016/j.biortech.2012.08.021 PMID:22939595

Singh, H., Rout, S., & Das, D. (2022). Dark fermentative biohydrogen production using pretreated Scenedesmus obliquus biomass under an integrated paradigm of biorefinery. *International Journal of Hydrogen Energy*, *47*(1), 102–116. doi:10.1016/j.ijhydene.2021.10.018

Singh, L., & Wahid, Z. A. (2015). Methods for enhancing bio-hydrogen production from biological process: A review. *Journal of Industrial and Engineering Chemistry*, *21*, 70–80. doi:10.1016/j.jiec.2014.05.035

Sittijunda, S., Reungsang, A., & Sompong, O. (2010). Biohydrogen production from dual digestion pretreatment of poultry slaughterhouse sludge by anaerobic self-fermentation. *International Journal of Hydrogen Energy*, *35*(24), 13427–13434. doi:10.1016/j.ijhydene.2009.11.116

Skonieczny, M. T., & Yargeau, V. (2009). Biohydrogen production from wastewater by Clostridium beijerinckii: Effect of pH and substrate concentration. *International Journal of Hydrogen Energy*, *34*(8), 3288–3294. doi:10.1016/j.ijhydene.2009.01.044

Srirangan, K., Pyne, M. E., & Chou, C. P. (2011). Biochemical and genetic engineering strategies to enhance hydrogen production in photosynthetic algae and cyanobacteria. *Bioresource Technology*, *102*(18), 8589–8604. doi:10.1016/j.biortech.2011.03.087 PMID:21514821

Steinfeld, A. (2005). Solar thermochemical production of hydrogen—a review. *Solar Energy*, *78*(5), 603–615. doi:10.1016/j.solener.2003.12.012

Torzillo, G., Scoma, A., Faraloni, C., Ena, A., & Johanningmeier, U. (2009). Increased hydrogen photoproduction by means of a sulfur-deprived Chlamydomonas reinhardtii D1 protein mutant. *International Journal of Hydrogen Energy*, *34*(10), 4529–4536. doi:10.1016/j.ijhydene.2008.07.093

Touloupakis, E., Faraloni, C., Silva Benavides, A. M., Masojídek, J., & Torzillo, G. (2021). Sustained photobiological hydrogen production by Chlorella vulgaris without nutrient starvation. *International Journal of Hydrogen Energy*, *46*(5), 3684–3694. doi:10.1016/j.ijhydene.2020.10.257

Tverskoi, D., Makarenkov, V., & Aleskerov, F. (2018). Modeling functional specialization of a cell colony under different fecundity and viability rates and resource constraint. *PLoS One*, *13*(8), e0201446. doi:10.1371/journal.pone.0201446 PMID:30089142

Uddin, M. N., Daud, W. M. A. W., & Abbas, H. F. (2013). Potential hydrogen and non-condensable gases production from biomass pyrolysis: Insights into the process variables. *Renewable & Sustainable Energy Reviews*, *27*, 204–224. doi:10.1016/j.rser.2013.06.031

Vargas, J. V. C., Mariano, A. B., Corrêa, D. O., & Ordonez, J. C. (2014). The microalgae derived hydrogen process in compact photobioreactors. *International Journal of Hydrogen Energy*, *39*(18), 9588–9598. doi:10.1016/j.ijhydene.2014.04.093

Voloshin, R. A., Rodionova, M. V., Zharmukhamedov, S. K., Veziroglu, T. N., Allakhverdiev, S. I., Nejat Veziroglu, T., & Allakhverdiev, S. I. (2016). Review: Biofuel production from plant and algal biomass. *International Journal of Hydrogen Energy*, *41*(39), 17257–17273. doi:10.1016/j.ijhydene.2016.07.084

Wang, Y., Tahir, N., Cao, W., Zhang, Q., & Lee, D.-J. (2019). Grid columnar flat panel photobioreactor with immobilized photosynthetic bacteria for continuous photofermentative hydrogen production. *Bioresource Technology*, *291*, 121806. doi:10.1016/j.biortech.2019.121806 PMID:31326683

Wanke, A., Malisic, M., Wawra, S., & Zuccaro, A. (2021). Unraveling the sugar code: The role of microbial extracellular glycans in plant–microbe interactions. *Journal of Experimental Botany*, *72*(1), 15–35. doi:10.1093/jxb/eraa414 PMID:32929496

Weaver, P. F., Lien, S., & Seibert, M. (1980). Photobiological production of hydrogen. *Solar Energy*, *24*(1), 3–45. doi:10.1016/0038-092X(80)90018-3

Xing, H., Yuan, Y., Huiyuan, Z., Yong, S., Bingxi, L., & Heping, T. (2017). Solar thermochemical hydrogen production using metallic oxides. *Energy Sources. Part A, Recovery, Utilization, and Environmental Effects*, *39*(3), 257–263. doi:10.1080/15567036.2013.878769

Yan, W., & Hoekman, S. K. (2014). Production of CO 2-free hydrogen from methane dissociation: A review. *Environmental Progress & Sustainable Energy*, *33*(1), 213–219. doi:10.1002/ep.11746

Yilmaz, C., Koyuncu, I., Alcin, M., & Tuna, M. (2019). Artificial Neural Networks based thermodynamic and economic analysis of a hydrogen production system assisted by geothermal energy on Field Programmable Gate Array. *International Journal of Hydrogen Energy*, *44*(33), 17443–17459. doi:10.1016/j.ijhydene.2019.05.049

Yin, Z., Zhu, L., Li, S., Hu, T., Chu, R., Mo, F., Hu, D., Liu, C., & Li, B. (2020). A comprehensive review on cultivation and harvesting of microalgae for biodiesel production: Environmental pollution control and future directions. *Bioresource Technology*, *301*, 122804. doi:10.1016/j.biortech.2020.122804 PMID:31982297

Zabed, H. M., Qi, X., Yun, J., & Zhang, H. (2019). Anaerobic Digestion of Microalgae Biomass for Methane Production. In *Microalgae Biotechnology for Development of Biofuel and Wastewater Treatment* (pp. 397–421). Springer. doi:10.1007/978-981-13-2264-8_16

Zhang, L., He, M., & Liu, J. (2014). The enhancement mechanism of hydrogen photoproduction in Chlorella protothecoides under nitrogen limitation and sulfur deprivation. *International Journal of Hydrogen Energy*, *39*(17), 8969–8976. doi:10.1016/j.ijhydene.2014.04.045

Zhu, L. D., Hiltunen, E., Antila, E., Zhong, J. J., Yuan, Z. H., & Wang, Z. M. (2014). Microalgal biofuels: Flexible bioenergies for sustainable development. *Renewable & Sustainable Energy Reviews*, *30*, 1035–1046. doi:10.1016/j.rser.2013.11.003

Chapter 15
Recovery of Rare Metals and Substance Production by Algae

Reda Mohamed Moghazy
Water Pollution Research Department, National Research Centre, Cairo, Egypt

ABSTRACT

Rare earth metals (REMs) are some of the most expensive materials due to global demand with a constant rise due to their critical role in advancements of technology. This drives us to find eco-friendly approaches to their efficient recovery and reuse. Many microorganism groups can develop various mechanisms to chelate metals, among these the algae that are considered to be promising an emerging solution for collecting rare metals and substances from the environments due to their high bioremediation ability through different mechanisms such as bioaccumulation, biodegradation, or biosorption inside their biomass. Hence, it is easy to recover these accumulated metals using different methodologies and reuse them in different technological aspects; on the other hand, through the bioremediation process, other substances can be produced as secondary metabolites that can be utilized as useful materials. The chapter will discuss the importance of rare metals and the effective biotechnological role of algae in order to recover them for their economic reuse in different approaches.

INTRODUCTION

In 1794, rare earth metals were discovered (Muraleedharan et al., 1994). the term "rare" does not always imply that something is uncommon in nature (Brown et al., 1990). while some are found in similar concentrations to regularly used metals in the earth's crust, others are not. because of their limited circulation and difficulty in amending, they are given this designation (Liang et al., 2014).

Rare earth metals (REMs) have an essential role in biological and medical development, as well as organometallic compounds, luminous compounds, coordination chemistry, catalysis, SOLID-STATE chemistry, environmental, and analytical chemistry (Robson C Oliveira et al., 2012). They're also used in a variety of fields and products, including audio systems, defense applications, household appliances, liquid crystal displays, automotive transducers, plasmas, glass additives, automobiles, fertilisers, sports

DOI: 10.4018/978-1-6684-2438-4.ch015

equipment, pharmaceuticals, Wind turbines, catalysts, metals, computers, lighting, and batteries (Du & Graedel, 2011).

rare metals have become one of the metals that are utilized in a variety of critical technologies which is becoming more popular over time, and the need is expected to rise in the near future (Massari & Ruberti, 2013; Moss et al., 2011). As a result, its release into the biosphere will increase day by day. As a result, there is a pressing need to address the environmental consequences of this waste. algal affinity for certain minerals may provide a significant environmental risk, or it may provide an opportunity for bioremediation of contaminated environments (Goecke, Zachleder, et al., 2015).

Rare earth metals are non-essential metals that cause both good and negative physical reactions. They are not thought to be vital in any known metabolic function, even though various studies have demonstrated that they can be advantageous under specific conditions (Goecke, Jerez, et al., 2015).

SO THAT, in this chapter, we will discuss the importance of different rare metals and the effective biotechnological role of algae in order to recover them for their economic reuse in different approaches.

RARE EARTH METALS

many essential elements are found in the rare earth element group, including scandium, yttrium, lithium, and a series of 14 lanthanides (ln) with atomic numbers ranging from 58 to 71. because they are members of group iiib in the periodic table, their physical and chemical properties are comparable, although their atomic number differs somewhat (Zhaozhou et al., 2012). Lanthanides also have electron configurations at the atomic level, which means that when the atomic number grows, the ionic radius shrinks, a phenomenon known as lanthanide contraction.

Rare metals are globally limited, in many circumstances, difficult to substitute with other metals. Rare metals are crucial for high-tech sectors such as vehicles, digital consumer electronics, and information related gadgets due to their valuable properties. As a result, rare metals are referred to as "industry vitamins," and the demand for rare metals is expanding in tandem with the development of these high-tech sectors (Kuroda & Ueda, 2010).

Furthermore, in 2014 and 2017, the European Commission identified Sc, light REMs (LREE: La, Ce, Pr, Nd, etc.) and heavy REMs (HREE: Y, Dy, Er, Yb, Ho, etc.) as critical raw materials (CRM) in their report, which was published based on the raw materials' economic importance and supply to demand ratio. REMs play an important role in the development of green, sustainable, and low-carbon technology. Furthermore, the geographical limits imposed by the viewpoint of REE resource acquisition might impose significant supply concerns on future market trends. As a result, the benefits associated with REMs, along with a significant increase in market demand, have served as a strong motivator to pique the interest of both the research and industrial sectors. (Binnemans, Jones, et al., 2013; Binnemans et al., 2015; Gao et al., 2017; X.-J. Li et al., 2016).

CHALLENGES IN THE RECOVERY OF RARE EARTH METALS

Countries like as China, the United States of America, and Australia are now the top producers of REMs, with China accounting for 90 percent of global output (Barros et al., 2019). Because they are non-biodegradable, these metals are difficult to remove from the environment, and their impact is amplified by their buildup throughout the food chain (K. Li et al., 2015). At the moment, REE are not recycled, despite the fact that they are gaining attention owing to their vital role in high-tech goods (Dodson et al., 2012) and their purification and pre-concentration have been assessed in conjunction with their rising demand (K. Li et al., 2015; Tian et al., 2012). The reuse of the REMs, in turn, reduces the environmental burden of their disposal (Sahoo et al., 2016). However, the overall REMs concentrations of these secondary wastes are often lower than those of REMs materials, and recycling yields remain extremely low, complicating efforts to construct a circular economic program. (Taggart et al., 2016).

Furthermore, the challenges associated with existing primary mining extraction tactics push the quest for efficient and sustainable alternative schemes to recover resources from secondary resources such as industrial waste streams, processing residues, and mine tailings (Binnemans, Pontikes, et al., 2013; Ramasamy et al., 2019). Acid mine drainage waste has the potential to serve as a viable secondary resource for REMs acquisition in this line (Ramasamy et al., 2019). the use of acidic solutions rich in rare earth metals (REMS), sulphur, iron, aluminum, magnesium, zinc, cobalt, nickel, and uranium allows the conventional leaching and hydrometallurgical treatment processes to be avoided (Tunsu et al., 2015).

OVERVIEW OF CONVENTIONAL RECOVERY METHODS

For the recovery of rare earth metals (REMs), many procedures have been used. solvent extraction, carbon adsorption, electrodeposition, Zinc dust cementation, and ion exchange, have all been used to recover rare metals from its solution. Reductive exchange, Precipitation, electrolytic recovery, and ion exchange, are the key recovery processes (Canda et al., 2016; Mpinga et al., 2014). The two most prevalent procedures for the preconcentration and separation of REMs from diverse matrices are solvent extraction and ion exchange (Das & Das, 2013). For the platinum group metals (PGM), solvent extraction is a regularly used approach for PGM recovery (Firmansyah et al., 2020). Solvent extraction is inefficient owing to the enormous volume of solvent used, which may cause health issues (Das & Das, 2013).

Furthermore, solvent extraction processes are typically labor-intensive and time-consuming (C.-W. Lee et al., 2017). For the recovery of PGM from solution, several methods such as ion exchange (Firmansyah et al., 2020), membrane separation, and adsorption have been developed (Ma et al., 2006). Adsorption appears to be the most appropriate approach for recovering PGM at low concentrations due to its low cost and high efficiency (Zhou et al., 2009). In an overview, (Shen & Forssberg, 2003) addressed metal recovery strategies from various slags, including incineration bottom ash.

Magnetic separation and eddy current separation are two ways mentioned to produce an iron metal fraction and a non-ferrous fraction (including Al, Cu, Zn, and other elements). They also state that using wet gravity separation resulted in 0–2 percent Ag in the heavy metal fraction. The development center for the sustainable management of recyclable waste and resources (ZAR) in Hinwil, Switzerland, is meeting the problem of ensuring efficient high material recovery in municipal solid waste incinerator (MSWI) facilities and improved environmental protection performance. (Morf et al., 2013).

Traditional approaches have some drawbacks, such as high reagent and energy consumption, limited selectivity, high operational costs, and secondary metabolite production. As a result, there is a need to create a low-cost, environmentally acceptable technology for recovering rare earth metals from watery environments (Opare et al., 2021).

Figure 1. Recovery pathway of rare earth metal by algae

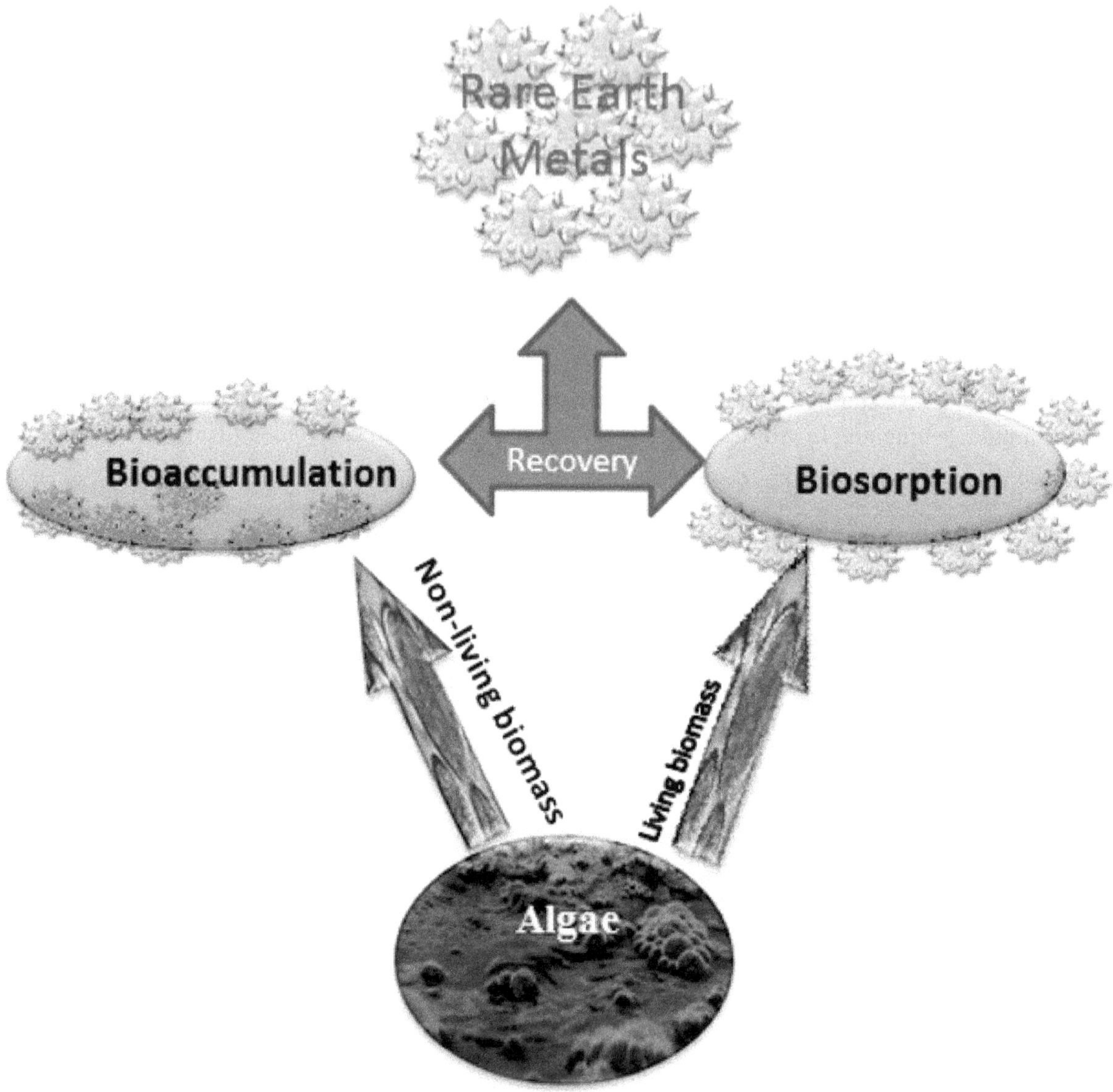

BIOREMEDIATION AS A PROMISING ECO-FRIENDLY SOLUTION FOR RARE METAL RECOVERY

Rare metals in vast quantities accumulate in high-tech goods and might be considered resources. As a result, efficient rare metal recovery from nature, wastewater, and abandoned high-tech goods is critical for both resource recovery and environmental cleaning. Nature has yielded many microbes with intrinsic biological capabilities that are ideal for use in metal recovery Fig (1) (Kuroda & Ueda, 2010).

Because of its low cost and environmental friendliness, bioadsorption of metal ions via microbial activities is gaining popularity as a new technique that can outperform conventional approaches (Eccles, 1999; Gadd & White, 1993; Lovley & Coates, 1997). Metal ion adsorption employing microorganisms is believed to offer benefits in terms of the capacity to adsorb and concentrate certain metal ions in order to recover metal ions from the hydrosphere with a mixture of metal ions at low concentrations.

Furthermore, due to their tiny size, microorganisms have a large surface area ratio, resulting in a broad contact area that can interact with metal ions in the surrounding environment. Packet-bed column techniques have been established as a successful bioadsorption procedure for practical applications, allowing for continuous bioadsorption and metal ion recovery (Vijayaraghavan et al., 2004; Volesky et al., 2003).

The cell wall matrix of algae, in particular, is constructed of a variety of polysaccharides and proteins, some of which include anionic functional groups, such as carboxylic, sulphate, and phosphate groups for biosorption.(Abdelhamid et al., 2021; Badr et al., 2016; Kandil et al., 2022; Michel et al., 2010; Moghazy & Abdo, 2018). Lanthanides are strong acids, therefore they preferentially bond to strong bases containing oxygen as the electron-donating element (Soltani et al., 2014). This indicates that REMs bind weakly to weak bases such as those with S or P as ion donors. The carboxylic and hydroxyl ionic groups are the major ligands on the algal cell surface that include at least one donor oxygen atom. (Giese, 2020). Ca^{2+} ions exchange for La^{3+}, Eu^{3+}, and Yb^{3+} binding in *Sargassum polycystum* at a 1:1 ratio, according to (Diniz & Volesky, 2005). (Vijayaraghavan et al., 2010) used energy-dispersive X-ray spectroscopy (ED-XRF) analysis to prove that Ca^{2+} peaks on the cell surface of brown algae, *Turbinaria conoides*, were decreased when new La^{3+}, Ce^{3+}, Yb^{3+}, and Eu^{3+} peaks biosorbed were present.

Macroalgae such as seaweed can biosorb 10^2–10^6 fold of lanthanides concentration higher than that found in seawater (Goecke, Jerez, et al., 2015). 1.3 µg of REMs concentration which is found in seaweed is considered a high concentration that is compared to such present in its environment (ranges between 10^{-3} and 10^{-1} µg/L) (Giese, 2020).

The brown marine macroalgae are among the algal group that researched (specially the Sargassacea family), this group have been extensively investigated as the most employed algae in REE biosorption (da Gama et al., 2014). The brown marine macroalgae have significant amounts of alginate, which is made up of numerous carboxylic groups capable of trapping cations and metallic species in solution (Amador-Castro et al., 2021). Furthermore, C. *reinhardtii* (freshwater green microalgae) can recover 1.03 mmol/g in 5 h when investigated for lanthanides biosorption,, this value was extremely close to the brown marine microalga *Turbinaria conoides* that can recover 1.11 mmol/g in 6 h. (Vijayaraghavan et al., 2010). The same observation have been done for Nd^{3+} when investigated by microalgae *Ankistrodesmus gracilis* and *Monoraphidium* sp., where the biosorptions where achieved at 0.98 and 0.94 mmol/g, respectively. (Palmieri et al., 2000), whereas the greatest result recorded for brown macroalgae *Sargassum* spp. was 0.70 mmol/g in 40 min (R C Oliveira & Garcia Jr, 2009).

Factors Affecting Biosorption of REMs

1. Effect of pH

The pH of a solution is important in the adsorption process because it controls the speciation of metal ions in the solution as well as the surface polarity of a biosorbent. Because the biosorbents studied in the literature contain one or more carboxylic, amine, hydroxyl, and thiol groups, their acid-base behavior is fully dependent on pH and hence modifies the affinity of these functional groups for other metal ions. The pH-dependent study gives important information about an adsorbent's adsorption-desorption behavior. The biosorption of REMs onto various biosorbents has been examined in the majority of research in the pH range of 1–7. (Abd El-Magied et al., 2017; Galhoum, Mafhouz, et al., 2015) because the hydrolysis of REMs ions results in the creation of insoluble hydroxides at higher pH (Ponou et al., 2014).

In certain situations, pH-dependent tests at higher pH ranges have also been conducted to illustrate the influence of precipitation on the adsorption efficacy of a biosorbent. (Oyewo et al., 2018). The most common LaIII species are: $[La_5(OH)_9]^{6+}$ (pH b 4.5); La^{3+} (4.5 b pH b 10); La $(OH)^{2+}$, $La(OH)^{2+}$ (10 b pH b 14) (Wu et al., 2011) and EuIII: Eu^{3+} (pH b 6); Eu_2 $(OH)_2$ $^{4+}$ (6 b pH b 9.5); $Eu(OH)^{2+}$, $Eu(OH)_3$ (9.5 b pH b 11); $Eu(OH)_4$ (11 b pH b 14) (Song et al., 2016) in the solution at distinct pH levels, demonstrating that different chemical species of REMs ion predominate in the aqueous solution. In several investigations, increasing the solution pH resulted in an increase in the adsorption efficacy of biosorbents for REMs (Iftekhar et al., 2017). This effect has been linked to the deprotonation of functional groups on the biosorbent surface when the pH increases, resulting in the availability of binding sites for metal ion adsorption.

Protonation of these functional groups was favored at lower pH levels, and therefore adsorption of positively charged metal ions was hampered due to electrostatic repulsion between the metal ions and protonated functionalities. As the pH rose, the protonated functionalities began to deprotonate, resulting in the formation of negatively charged functionalities with a higher affinity for positively charged REMs ions. In these circumstances, adsorption tests were carried out at lower pH levels to reduce the influence of REE precipitation as insoluble hydroxides. (Swain et al., 2018).

Some pH-dependent investigations found that an initial rise in pH was followed by a plateau (Cadogan et al., 2014). The behavior was seen for REE biosorption onto chitosan-derived biosorbents (Galhoum, Mahfouz, et al., 2015) and core–shell structured Fe3O4@carboxymethyl cellulose magnetic composite (Fe_3O_4@CMC) (pH_{zpc} ~ 1.9) (Cai et al., 2017). A change in surface polarity (from a positively charged surface at pH b pH_{zpc} to a negatively charged surface at pH N pH_{zpc}) resulted in an initial increase in adsorption performance due to the favorable interaction of the negatively charged surface with positively charged EuIII species in the case of Fe_3O_4@CMC. The complexation and precipitation caused the plateau area found in the pH range of 7.6–10.0. (Cai et al., 2017). The first increase in the adsorption capacity of chitosan-derived biosorbents was due to the binding of REMs with the hydroxyl and carboxylate/hydroxyl functionalities of chitosan. (Cadogan et al., 2015) and amino acid grafted chitosan (Galhoum, Atia, et al., 2015), respectively. The plateau area was detected at lower pH values (4–7), owing to the "buffering" action of the amino group found in both pure and modified chitosan. Clearly, the "buffering" effect minimizes the effects of pH on a biosorbent's sorption activity. (Svecova et al., 2006). A notable study on the effect of counterions (or anions) on the biosorption of LaIII onto S. *polycystum* (Diniz & Volesky, 2005) and S. *fluitans* (Palmieri et al., 2002) found that the presence of anions such as sulphate (in the solution from metal salts, and pH adjustments using H_2SO_4) reduced adsorption performance. This was owing to the solution's shift in LaIII speciation from La^{3+} to less adsorbed $[La(SO_4)]+$. Furthermore, anions such as NO_3 and Cl in the solution via pH changes or metal salts did not affect the adsorption process because La^{3+} remained the dominating species in the solution.

2. Effect of Biosorbent Dose

There is always an optimal adsorbent dose that can lead to maximal adsorption capacities by optimizing the interaction between metal ions and biosorbent binding sites. As the concentration of adsorbent increases, so does the percentage adsorption performance, which is comparable to an increase in active sites.

Most often, the adsorption percentage saturates at a given biosorbent dosage level (optimum level), beyond which the adsorption capacity rapidly declines due to the fixed concentration of REMs. (Hisada & Kawase, 2018). When the biosorbent dosage was raised from 0.01 g/ L to 0.03 g/ L, the percent adsorption of PrIII onto *T. arjuna* bark powder rose from 42% to 87%. Another 5% increase was reported in the range of 0.03–0.10 g/ L, indicating a saturation area. Adsorption efficiency (percent) was observed to decline after a given biosorbent dose level in a few experiments. (Torab-Mostaedi et al., 2015).

Adsorbent doses above the limit can make active sites unavailable to metal ions due to "sorbate-sorbate interactions," reducing adsorption capacity, a peculiar behavior known as the "Cs-effect," as reported by (Gupta & Sengupta, 2017) for the adsorption of trivalent REMs onto functionalized MWCNTs. The histogram clearly shows that the adsorbent dosage is a spread caused by differences in the functionalities present in biosorbents of various origins.

For a more cost-effective industrial use, adsorbent dosage optimization is essential, and contract duration and metal ion concentration should also be considered (Gupta & Sengupta, 2017; Iftekhar et al., 2018).

3. Effect of Ionic Strength

The ionic strength-dependent response reveals a great lot about the complexation process. For the adsorption of metal ions onto the surface of an adsorbent, two distinct complexation mechanisms, namely inner-sphere and outer-sphere complexation mechanisms, are known. If the adsorption capacity decreases as ionic strength increases, the process follows the outer-sphere (electrostatic) complexation mechanism. In contrast, for the inner-sphere (ligand exchange) complexation process, adsorption performance either rises or remains constant when electrolyte concentration increases (McBride, 1997).

There was no change in the adsorption capabilities of Fe_3O_4@CMC and aminomethylphosphonic acid grafted chitosan with increased ionic strength in the literature examined here, indicating that the adsorption process was driven by an inner-sphere complexation mechanism.(Elsalamouny et al., 2017).

An improvement in adsorption capacity with increased ionic strength was ascribed to the contraction of the electrical double layer and activity changes of the ionic species in raw coated cactus fibres (CF) and MnO2-coated cactus fibres (CF/MnO_2). It is also probable that the increasing Na^+/K^+ concentration in solution (due to ionic strength adaptation with NaCl, $NaNO_3$, $NaClO_4$, KNO_3, etc.) interrupted the hydration sphere of the REE ion and thus assisted the adsorption process (because adsorption of a metal ion onto an adsorbent includes full or partial detachment of water molecules from its hydration sphere) (Chen & Lin, 2001; Prodromou & Pashalidis, 2016).

The biosorption of REMs onto phosphorylated cactus fibers (CF PO_4), pre-treated crab shell particles (PCSP), and magnetic alginate-chitosan gel beads was inhibited by increasing ionic strength, indicating that the adsorption process was mediated by the outer-sphere complexation mechanism. The interaction of REE ions with functions on the surface of a biosorbent is shielded by increasing ionic strength. Furthermore, raising the concentration of Na^+/K^+ in the solution competes with REE ions for binding sites (Prodromou & Pashalidis, 2016; Vijayaraghavan et al., 2009).

Rare Metal and the Induction of Bioactive Substance Production

Metals in low concentrations are required for microalgae cells to grow. They are components of photosynthetic electron transport proteins (Cu, Fe) and photosynthetic water oxidizing centers (Mn), or they are vitamins (Co) (Andersen, 2005). They also act as cofactors for enzymes involved in CO_2 fixation

(Zn in carbonic anhydrase) (Moroney et al., 2001), DNA transcription (Zn in RNA polymerase), and phosphorus acquisition (Zn in alkaline phosphatase) (Sunda, 2012) or N_2 assimilation (Mo, Fe, V in nitrogenase) (Bothe et al., 2010) and nitrate reduction (Mo in nitrate and Fe in nitrite reductase) (Vega et al., 1971). However, High levels of these metals, as well as other non-essential heavy metals (Hg, As, Cd, Pb, Cr), have deleterious effects in microalgae cells (impairment of photosynthetic process, blockage of cell division, inhibition of enzyme activity) (Monteiro et al., 2012). Metals also have an impact on the shape of microalgal cells. Cadmium (Cd) accumulation in *Chlamydomonas acidophil*us cells resulted in increased cell size and polyphosphate body disintegration (Nishikawa et al., 2003). The addition of lead (Pb) in *Chlorella sorokiniana* culture caused the production of colonies of *Chlorella* cells with cytoplasmic lipid droplets and disfigured chloroplasts (Carfagna et al., 2013). When *Synechocystis* sp. cells were exposed to thallium, the thylakoid membranes were fragmented (Tl) (Aoki et al., 2013). Mitochondria in *Desmidium swartzii* cells become larger and swollen after being exposed to Zn. (Andosch et al., 2015). Aluminum (Al) and lead had a synergistic impact on *Dunaliella tertiolecta*, causing cell membrane lysis. (Saçan et al., 2007). Cell damage caused by cerium (Ce) in *Anabaena flosaquae* can also result in the release of toxins. (Yingjun et al., 2012). In *Chlamydomonas reinhardtii*, lithium (Li) can change the length and shape of the flagella. (Periz et al., 2007) or alter the structure of the polysaccharide sheath around *Ankistrodesmus gracilis* cells (Nordi et al., 2006), and can also inhibit other microalgae strains at varied dosages (Karlander & Krauss, 1972; Mota et al., 2015). Diatom cultivation In the presence of germanium (Ge), titanium (Ti), zirconium (Zr), or tin (Sn), *Synedra acus* produced changes in the form, size, and mechanical strength of silica valves in *Synedra frustules*. (Basharina et al., 2012). Although heavy metals have a usually detrimental influence on microalgae cultures, certain findings show that they can also play a positive function in microalgae production. Low quantities of lead, aluminum, and cobalt stimulated the development of *Dunaliella tertiolecta* and *Monoraphidium minutum*. (El-Sheekh et al., 2003; Saçan et al., 2007). Arsenic (As(V)) has been shown to boost the development of the cyanobacterium *Nostocminutum* (Ferrari et al., 2013) as well as the microalgae *Chlorella salina* (Karadjova et al., 2008) and *Chlorella* sp. (Knauer & Hemond, 2000). Furthermore, inorganics can help microalgae flourish in the absence of nutrients. For example, 20 g/L vanadium (VO_3) boosted the growth of *Scenedesmus obliquus* growing on iron (Fe^{3+}) deficient media by up to sixfold. Under photoautotrophic culture conditions, vanadium was nearly completely absorbed by *Scenedesmus* cells (Meisch & Bielig, 1975). In another investigation, 0.01–1 g/L vanadium (VO_3) supplementation resulted in up to 67 percent growth improvement in photoautotrophic *Chlorella pyrenoidosa* culture, even when iron (Fe^{3+}) supplementation was present in the growth media (Meisch et al., 1977). However, vanadium (VO^{3-}) at dosages greater than 1 mg/L inhibited *Chlorella pyrenoidosa* growth. (Meisch et al., 1977). Vanadium, in the forms VO_4^{3-} and V_2O_5 has also been shown to inhibit *Haematococcus lacustris* and *Scenedesmus quadricauda*. (Fargašová et al., 1999; Tran et al., 2009). Furthermore, lanthanide elements such as lanthanum (La), cerium (Ce), neodymium (Nd), europium (Eu), or gadolinium (Gd) have been reported to be a good replacement for calcium deficiency in *Desmodesmus quadricauda* culture, with Gd, La, or Nd supplementation yielding nearly the same culture dry weight as Ca supplemented media. Furthermore, a small amount of cerium added to a regular medium increased the number of *Desmodesmus* cells in culture. Lanthanide elements, on the other hand, aid in the reduction of *Desmodesmus* development when added into a manganese-deficient medium.(Goecke, Jerez, et al., 2015). Lanthanum reduced the development of *Scenedesmus quadricauda* at greater concentrations, and the inhibitory concentration of La was the same as for other lanthanides: cerium (Ce), neodymium (Nd), samarium (Sm), europium (Eu), gadolinium (Gd), terbium (Tb), dysprosium (Dy), holmium (Ho), erbium (Er), thulium (Lu) (Jin et al., 2009; Tai et al., 2010). Cerium (Ce) was stimulatory to the cyanobacterium. *Anabaena flosaquae* at low concentrations but inhibitive at high concentrations (Yingjun et al., 2012). Cd^{2+} was shown to enhance proliferation and sustain the activity of carbonic anhydrase in *Thalassiosira weissflogii* cells grown in Zn-limited media. (J. G. Lee et al., 1995). A new carbonic anhydrase using Cd^{2+} as a catalytic metal ion was recently found in *Thalassiosira weissflogii* (Alterio et al., 2015). In the presence of urea as the main nitrogen source, Ni^{2+} is required for the culture of marine diatoms such as *Phaeodactylum tricornutum*, *Cyclotella cryptica*, *Thalassiosira weissflogii*, and *Thalassiosira pseudonana* (L. Oliveira & Antia, 1984; Rees & Bekheet, 1982). Nickel is a cofactor in the enzyme urease, however, at larger doses, Ni inhibited diatom development (Egleston & Morel, 2008; L. Oliveira & Antia, 1984). Cobalt can partially compensate for a Ni deficiency (Rees & Bekheet, 1982). Metallic nanoparticles (NPs) are active against microalgae in addition to metals and metalloids. Inhibitory effects of TiO_2, ZnO, CeO_2, NiO, $BaTiO_3$, Y_2O_3, Al_2O_3, Ag, and Pt nanoparticles on numerous freshwater and marine microalgae strains were reported, and their inhibitory activity was suggested to be due to Reactive Oxygen Species (ROS) generation, or mechanical damage caused by nanoparticles themselves, but also due to metal ions released from nanoparticles, light shading effect, interactions with (Aravantinou et al., 2015; Castro-Bugallo et al., 2014; W.-M. Lee & An, 2013; Manier et al., 2013; Manzo et al., 2013; Sadiq et al., 2011; Suman et al., 2015; Xia et al., 2015). The inhibitory effect of nanoparticles is also affected by their size and the content of their aged solution or growth media (Aravantinou et al., 2015; Manier et al., 2013; Xia et al., 2015). Metal ions emitted by nanoparticles, on the other hand, can promote the development of cyanobacteria and microalgae(Pádrová et al., 2015).

CONCLUSION

Currently, It is necessity for finding out an eco-friendly approaches for recovery and reuse of REMs. Biosorption, which combines biotechnology with extractive hydrometallurgy, provides an economically viable approach for the effective removal and recovery of REMs. Among different Microorganisms groups, the group most used in REMs biosorption studies has been the algae, which can develop various mechanisms to chelate metals in living or nonliving biomass, furthermore they are considered to be promising emerging approach for re-collecting REMs from the aquatic environments due to their high ability for bioremediation. Different mechanisms of bioremediation was achieved for recovery of REMs; the most common of them are the bioaccumulation and biosorption, there are different factor affecting these mechanisms such as pH, algal dose, and ionic strength. On the other hand, the recovery process can induce a byproduct metabolites substances that can be utilized as useful materials for different aspects.

REFERENCES

Abd El-Magied, M. O., Galhoum, A. A., Atia, A. A., Tolba, A. A., Maize, M. S., Vincent, T., & Guibal, E. (2017). Cellulose and chitosan derivatives for enhanced sorption of erbium (III). *Colloids and Surfaces. A, Physicochemical and Engineering Aspects*, *529*, 580–593. doi:10.1016/j.colsurfa.2017.05.031

Abdelhamid, A. E., Labena, A., Mansor, E. S., Husien, S., & Moghazy, R. M. (2021). Highly efficient adsorptive membrane for heavy metal removal based on Ulva fasciata biomass. *Biomass Conversion and Biorefinery*. Advance online publication. doi:10.100713399-020-01250-7

Alterio, V., Langella, E., De Simone, G., & Monti, S. M. (2015). Cadmium-containing carbonic anhydrase CDCA1 in marine diatom Thalassiosira weissflogii. *Marine Drugs*, *13*(4), 1688–1697. doi:10.3390/md13041688 PMID:25815892

Amador-Castro, F., García-Cayuela, T., Alper, H. S., Rodriguez-Martinez, V., & Carrillo-Nieves, D. (2021). Valorization of pelagic sargassum biomass into sustainable applications: Current trends and challenges. *Journal of Environmental Management*, *283*, 112013. doi:10.1016/j.jenvman.2021.112013 PMID:33508553

Andersen, R. A. (2005). *Algal culturing techniques*. Elsevier.

Andosch, A., Höftberger, M., Lütz, C., & Lütz-Meindl, U. (2015). Subcellular sequestration and impact of heavy metals on the ultrastructure and physiology of the multicellular freshwater alga Desmidium swartzii. *International Journal of Molecular Sciences*, *16*(5), 10389–10410. doi:10.3390/ijms160510389 PMID:25961949

Aoki, M., Matsumoto, H., Takahashi, T., Sato, K., Kumata, H., & Fujiwara, K. (2013). Thallium Induces morphological changes in the photosynthetic apparatus of Synechocystis sp. PCC6803. In Photosynthesis Research for Food, Fuel and the Future (pp. 586–589). Springer. doi:10.1007/978-3-642-32034-7_126

Aravantinou, A. F., Tsarpali, V., Dailianis, S., & Manariotis, I. D. (2015). Effect of cultivation media on the toxicity of ZnO nanoparticles to freshwater and marine microalgae. *Ecotoxicology and Environmental Safety*, *114*, 109–116. doi:10.1016/j.ecoenv.2015.01.016 PMID:25625916

Badr, S. A., Ashmawy, A. A., El-Sherif, I. Y., & Moghazy, R. M. (2016). Non-conventional low-cost biosorbents for adsorption and desorption of heavy metals. *Research Journal of Pharmaceutical, Biological and Chemical Sciences*, *7*(4).

Barros, Ó., Costa, L., Costa, F., Lago, A., Rocha, V., Vipotnik, Z., Silva, B., & Tavares, T. (2019). Recovery of rare earth elements from wastewater towards a circular economy. *Molecules (Basel, Switzerland)*, *24*(6), 1005. doi:10.3390/molecules24061005 PMID:30871164

Basharina, T. N., Danilovtseva, E. N., Zelinskiy, S. N., Klimenkov, I. V., Likhoshway, Y. V., & Annenkov, V. V. (2012). The effect of titanium, zirconium and tin on the growth of diatom Synedra acus and morphology of its silica valves. *Silicon*, *4*(4), 239–249. doi:10.100712633-012-9119-x

Binnemans, K., Jones, P. T., Blanpain, B., Van Gerven, T., & Pontikes, Y. (2015). Towards zero-waste valorisation of rare-earth-containing industrial process residues: A critical review. *Journal of Cleaner Production*, *99*, 17–38. doi:10.1016/j.jclepro.2015.02.089

Binnemans, K., Jones, P. T., Blanpain, B., Van Gerven, T., Yang, Y., Walton, A., & Buchert, M. (2013). Recycling of rare earths: A critical review. *Journal of Cleaner Production*, *51*, 1–22. doi:10.1016/j.jclepro.2012.12.037

Binnemans, K., Pontikes, Y., Jones, P. T., Van Gerven, T., & Blanpain, B. (2013). Recovery of rare earths from industrial waste residues: a concise review. *Proceedings of the 3rd International Slag Valorisation Symposium: The Transition to Sustainable Materials Management*, 191–205.

Bothe, H., Schmitz, O., Yates, M. G., & Newton, W. E. (2010). Nitrogen fixation and hydrogen metabolism in cyanobacteria. *Microbiology and Molecular Biology Reviews*, *74*(4), 529–551. doi:10.1128/MMBR.00033-10 PMID:21119016

Brown, P. H., Rathjen, A. H., Graham, R. D., & Tribe, D. E. (1990). Rare earth elements in biological systems. *Handbook on the Physics and Chemistry of Rare Earths*, *13*, 423–452. doi:10.1016/S0168-1273(05)80135-7

Cadogan, E. I., Lee, C.-H., & Popuri, S. R. (2015). Facile synthesis of chitosan derivatives and Arthrobacter sp. biomass for the removal of europium (III) ions from aqueous solution through biosorption. *International Biodeterioration & Biodegradation*, *102*, 286–297. doi:10.1016/j.ibiod.2015.01.018

Cadogan, E. I., Lee, C.-H., Popuri, S. R., & Lin, H.-Y. (2014). Efficiencies of chitosan nanoparticles and crab shell particles in europium uptake from aqueous solutions through biosorption: Synthesis and characterization. *International Biodeterioration & Biodegradation*, *95*, 232–240. doi:10.1016/j.ibiod.2014.06.003

Cai, Y., Yuan, F., Wang, X., Sun, Z., Chen, Y., Liu, Z., Wang, X., Yang, S., & Wang, S. (2017). Synthesis of core–shell structured Fe3O4@ carboxymethyl cellulose magnetic composite for highly efficient removal of Eu (III). *Cellulose (London, England)*, *24*(1), 175–190. doi:10.100710570-016-1094-8

Canda, L., Heput, T., & Ardelean, E. (2016). Methods for recovering precious metals from industrial waste. *IOP Conference Series. Materials Science and Engineering*, *106*(1), 12020. doi:10.1088/1757-899X/106/1/012020

Carfagna, S., Lanza, N., Salbitani, G., Basile, A., Sorbo, S., & Vona, V. (2013). Physiological and morphological responses of Lead or Cadmium exposed Chlorella sorokiniana 211-8K (Chlorophyceae). *SpringerPlus*, *2*(1), 1–7. doi:10.1186/2193-1801-2-147 PMID:23641320

Castro-Bugallo, A., González-Fernández, Á., Guisande, C., & Barreiro, A. (2014). Comparative responses to metal oxide nanoparticles in marine phytoplankton. *Archives of Environmental Contamination and Toxicology*, *67*(4), 483–493. doi:10.100700244-014-0044-4 PMID:24908584

Chen, J. P., & Lin, M. (2001). Surface charge and metal ion adsorption on an H-type activated carbon: Experimental observation and modeling simulation by the surface complex formation approach. *Carbon*, *39*(10), 1491–1504. doi:10.1016/S0008-6223(00)00277-3

da Gama, B. A. P., Plouguerné, E., & Pereira, R. C. (2014). The antifouling defence mechanisms of marine macroalgae. *Advances in Botanical Research*, *71*, 413–440. doi:10.1016/B978-0-12-408062-1.00014-7

Das, N., & Das, D. (2013). Recovery of rare earth metals through biosorption: An overview. *Journal of Rare Earths*, *31*(10), 933–943. doi:10.1016/S1002-0721(13)60009-5

Diniz, V., & Volesky, B. (2005). Effect of counterions on lanthanum biosorption by Sargassum polycystum. *Water Research*, *39*(11), 2229–2236. doi:10.1016/j.watres.2005.04.004 PMID:15899510

Dodson, J. R., Hunt, A. J., Parker, H. L., Yang, Y., & Clark, J. H. (2012). Elemental sustainability: Towards the total recovery of scarce metals. *Chemical Engineering and Processing*, *51*, 69–78. doi:10.1016/j.cep.2011.09.008

Du, X., & Graedel, T. E. (2011). Global in-use stocks of the rare earth elements: A first estimate. *Environmental Science & Technology*, *45*(9), 4096–4101. doi:10.1021/es102836s PMID:21438595

Eccles, H. (1999). Treatment of metal-contaminated wastes: Why select a biological process? *Trends in Biotechnology*, *17*(12), 462–465. doi:10.1016/S0167-7799(99)01381-5 PMID:10557157

Egleston, E. S., & Morel, F. M. M. (2008). Nickel limitation and zinc toxicity in a urea-grown diatom. *Limnology and Oceanography*, *53*(6), 2462–2471. doi:10.4319/lo.2008.53.6.2462

El-Sheekh, M. M., El-Naggar, A. H., Osman, M. E. H., & El-Mazaly, E. (2003). Effect of cobalt on growth, pigments and the photosynthetic electron transport in Monoraphidium minutum and Nitzchia perminuta. *Brazilian Journal of Plant Physiology*, *15*(3), 159–166. doi:10.1590/S1677-04202003000300005

Elsalamouny, A. R., Desouky, O. A., Mohamed, S. A., Galhoum, A. A., & Guibal, E. (2017). Uranium and neodymium biosorption using novel chelating polysaccharide. *International Journal of Biological Macromolecules*, *104*, 963–968. doi:10.1016/j.ijbiomac.2017.06.081 PMID:28663149

Fargašová, A., Bumbálová, A., & Havránek, E. (1999). Ecotoxicological effects and uptake of metals (Cu+, Cu2+, Mn2+, Mo6+, Ni2+, V5+) in freshwater alga Scenedesmus quadricauda. *Chemosphere*, *38*(5), 1165–1173. doi:10.1016/S0045-6535(98)00346-4

Ferrari, S. G., Silva, P. G., González, D. M., Navoni, J. A., & Silva, H. J. (2013). Arsenic tolerance of cyanobacterial strains with potential use in biotechnology. *Revista Argentina de Microbiologia*, *45*(3), 174–179. doi:10.1016/S0325-7541(13)70021-X PMID:24165141

Firmansyah, M. L., Yoshida, W., Hanada, T., & Goto, M. (2020). Application of ionic liquids in solvent extraction of platinum group metals. *Solvent Extraction Research and Development, Japan*, *27*(1), 1–24. doi:10.15261erdj.27.1

Gadd, G. M., & White, C. (1993). Microbial treatment of metal pollution—A working biotechnology? *Trends in Biotechnology*, *11*(8), 353–359. doi:10.1016/0167-7799(93)90158-6 PMID:7764182

Galhoum, A. A., Atia, A. A., Mahfouz, M. G., Abdel-Rehem, S. T., Gomaa, N. A., Vincent, T., & Guibal, E. (2015). Dy (III) recovery from dilute solutions using magnetic-chitosan nano-based particles grafted with amino acids. *Journal of Materials Science*, *50*(7), 2832–2848. doi:10.100710853-015-8845-z

Galhoum, A. A., Mafhouz, M. G., Abdel-Rehem, S. T., Gomaa, N. A., Atia, A. A., Vincent, T., & Guibal, E. (2015). Cysteine-functionalized chitosan magnetic nano-based particles for the recovery of light and heavy rare earth metals: Uptake kinetics and sorption isotherms. *Nanomaterials (Basel, Switzerland)*, *5*(1), 154–179. doi:10.3390/nano5010154 PMID:28347004

Galhoum, A. A., Mahfouz, M. G., Abdel-Rehem, S. T., Gomaa, N. A., Atia, A. A., Vincent, T., & Guibal, E. (2015). Diethylenetriamine-functionalized chitosan magnetic nano-based particles for the sorption of rare earth metal ions [Nd (III), Dy (III) and Yb (III)]. *Cellulose (London, England)*, *22*(4), 2589–2605. doi:10.100710570-015-0677-0

Gao, Q., Xie, J.-F., Shao, Y.-T., Chen, C., Han, B., Xia, K.-S., & Zhou, C.-G. (2017). Ultrafast and high-capacity adsorption of Gd(III) onto inorganic phosphorous acid modified mesoporous SBA-15. *Chemical Engineering Journal*, *313*, 197–206. doi:10.1016/j.cej.2016.12.068

Giese, E. C. (2020). Biosorption as green technology for the recovery and separation of rare earth elements. *World Journal of Microbiology & Biotechnology*, *36*(4), 1–11. doi:10.100711274-020-02821-6 PMID:32172357

Goecke, F., Jerez, C. G., Zachleder, V., Figueroa, F. L., Bišová, K., Řezanka, T., & Vítová, M. (2015). Use of lanthanides to alleviate the effects of metal ion-deficiency in Desmodesmus quadricauda (Sphaeropleales, Chlorophyta). *Frontiers in Microbiology*, *6*, 2. doi:10.3389/fmicb.2015.00002 PMID:25674079

Goecke, F., Zachleder, V., & Vítová, M. (2015). Rare earth elements and algae: physiological effects, biorefinery and recycling. In *Algal Biorefineries* (pp. 339–363). Springer. doi:10.1007/978-3-319-20200-6_10

Gupta, N. K., & Sengupta, A. (2017). Understanding the sorption behavior of trivalent lanthanides on amide functionalized multi walled carbon nanotubes. *Hydrometallurgy*, *171*, 8–15. doi:10.1016/j.hydromet.2017.03.016

Hisada, M., & Kawase, Y. (2018). Recovery of rare-earth metal neodymium from aqueous solutions by poly-γ-glutamic acid and its sodium salt as biosorbents: Effects of solution pH on neodymium recovery mechanisms. *Journal of Rare Earths*, *36*(5), 528–536. doi:10.1016/j.jre.2018.01.001

Iftekhar, S., Ramasamy, D. L., Srivastava, V., Asif, M. B., & Sillanpää, M. (2018). Understanding the factors affecting the adsorption of Lanthanum using different adsorbents: A critical review. *Chemosphere*, *204*, 413–430. doi:10.1016/j.chemosphere.2018.04.053 PMID:29677649

Iftekhar, S., Srivastava, V., & Sillanpää, M. (2017). Synthesis and application of LDH intercalated cellulose nanocomposite for separation of rare earth elements (REEs). *Chemical Engineering Journal*, *309*, 130–139. doi:10.1016/j.cej.2016.10.028

Jin, X., Chu, Z., Yan, F., & Zeng, Q. (2009). Effects of lanthanum (III) and EDTA on the growth and competition of Microcystis aeruginosa and Scenedesmus quadricauda. *Limnologica*, *39*(1), 86–93. doi:10.1016/j.limno.2008.03.002

Kandil, H., Abdelhamid, A. E., Moghazy, R. M., & Amin, A. (2022). Functionalized PVA film with good adsorption capacity for anionic dye. *Polymer Engineering and Science*, *62*(1), 145–159. doi:10.1002/pen.25840

Karadjova, I. B., Slaveykova, V. I., & Tsalev, D. L. (2008). The biouptake and toxicity of arsenic species on the green microalga Chlorella salina in seawater. *Aquatic Toxicology (Amsterdam, Netherlands)*, *87*(4), 264–271. doi:10.1016/j.aquatox.2008.02.006 PMID:18378014

Karlander, E. P., & Krauss, R. W. (1972). Absorption and toxicity of beryllium and lithium in Chlorella vannielii Shihira and Krauss. *Chesapeake Science*, *13*(4), 245–253. doi:10.2307/1351108

Knauer, K., & Hemond, H. (2000). Accumulation and reduction of arsenate by the freshwater green alga Chlorella sp.(Chlorophyta). *Journal of Phycology*, *36*(3), 506–509. doi:10.1046/j.1529-8817.2000.99056.x PMID:29544008

Kuroda, K., & Ueda, M. (2010). Engineering of microorganisms towards recovery of rare metal ions. *Applied Microbiology and Biotechnology*, *87*(1), 53–60. doi:10.100700253-010-2581-8 PMID:20393699

Lee, C.-W., Su, H., Cai, Y.-D., Wu, M.-T., Wu, D.-C., & Shiea, J. (2017). Rapid identification of psychoactive drugs in drained gastric lavage fluid and whole blood specimens of drug overdose patients using ambient mass spectrometry. *Mass Spectrometry (Tokyo, Japan)*, *6*(2), S0056–S0056. doi:10.5702/massspectrometry.S0056 PMID:28573080

Lee, J. G., Roberts, S. B., & Morel, F. M. M. (1995). Cadmium: A nutrient for the marine diatom Thalassiosira weissflogii. *Limnology and Oceanography*, *40*(6), 1056–1063. doi:10.4319/lo.1995.40.6.1056

Lee, W.-M., & An, Y.-J. (2013). Effects of zinc oxide and titanium dioxide nanoparticles on green algae under visible, UVA, and UVB irradiations: No evidence of enhanced algal toxicity under UV pre-irradiation. *Chemosphere*, *91*(4), 536–544. doi:10.1016/j.chemosphere.2012.12.033 PMID:23357865

Li, K., Gao, Q., Yadavalli, G., Shen, X., Lei, H., Han, B., Xia, K., & Zhou, C. (2015). Selective adsorption of Gd3+ on a magnetically retrievable imprinted chitosan/carbon nanotube composite with high capacity. *ACS Applied Materials & Interfaces*, *7*(38), 21047–21055. doi:10.1021/acsami.5b07560 PMID:26355685

Li, X.-J., Yan, C.-J., Luo, W.-J., Gao, Q., Zhou, Q., Liu, C., & Zhou, S. (2016). Exceptional cerium(III) adsorption performance of poly(acrylic acid) brushes-decorated attapulgite with abundant and highly accessible binding sites. *Chemical Engineering Journal*, *284*, 333–342. doi:10.1016/j.cej.2015.09.003

Liang, T., Li, K., & Wang, L. (2014). State of rare earth elements in different environmental components in mining areas of China. *Environmental Monitoring and Assessment*, *186*(3), 1499–1513. doi:10.100710661-013-3469-8 PMID:24135922

Lovley, D. R., & Coates, J. D. (1997). Bioremediation of metal contamination. *Current Opinion in Biotechnology*, *8*(3), 285–289. doi:10.1016/S0958-1669(97)80005-5 PMID:9206008

Ma, H., Liao, X., Liu, X., & Shi, B. (2006). Recovery of platinum (IV) and palladium (II) by bayberry tannin immobilized collagen fiber membrane from water solution. *Journal of Membrane Science*, *278*(1–2), 373–380. doi:10.1016/j.memsci.2005.11.022

Manier, N., Bado-Nilles, A., Delalain, P., Aguerre-Chariol, O., & Pandard, P. (2013). Ecotoxicity of non-aged and aged CeO2 nanomaterials towards freshwater microalgae. *Environmental Pollution*, *180*, 63–70. doi:10.1016/j.envpol.2013.04.040 PMID:23727569

Manzo, S., Miglietta, M. L., Rametta, G., Buono, S., & Di Francia, G. (2013). Toxic effects of ZnO nanoparticles towards marine algae Dunaliella tertiolecta. *The Science of the Total Environment*, *445*, 371–376. doi:10.1016/j.scitotenv.2012.12.051 PMID:23361041

Massari, S., & Ruberti, M. (2013). Rare earth elements as critical raw materials: Focus on international markets and future strategies. *Resources Policy*, *38*(1), 36–43. doi:10.1016/j.resourpol.2012.07.001

McBride, M. B. (1997). A critique of diffuse double layer models applied to colloid and surface chemistry. *Clays and Clay Minerals*, *45*(4), 598–608. doi:10.1346/CCMN.1997.0450412

Meisch, H.-U., Benzschawel, H., & Bielig, H.-J. (1977). The role of vanadium in green plants. *Archives of Microbiology*, *114*(1), 67–70. doi:10.1007/BF00429632 PMID:20864

Meisch, H.-U., & Bielig, H.-J. (1975). Effect of vanadium on growth, chlorophyll formation and iron metabolism in unicellular green algae. *Archives of Microbiology*, *105*(1), 77–82. doi:10.1007/BF00447117 PMID:1190955

Michel, G., Tonon, T., Scornet, D., Cock, J. M., & Kloareg, B. (2010). The cell wall polysaccharide metabolism of the brown alga Ectocarpus siliculosus. Insights into the evolution of extracellular matrix polysaccharides in Eukaryotes. *The New Phytologist*, *188*(1), 82–97. doi:10.1111/j.1469-8137.2010.03374.x PMID:20618907

Moghazy, R. M., & Abdo, S. (2018). The efficacy of microalgal biomass collected from high rate algal pond for dyes biosorption and biofuel production. *Research Journal of Chemistry and Environment*, *22*(11), 54–60.

Monteiro, C. M., Castro, P. M. L., & Malcata, F. X. (2012). Metal uptake by microalgae: Underlying mechanisms and practical applications. *Biotechnology Progress*, *28*(2), 299–311. doi:10.1002/btpr.1504 PMID:22228490

Morf, L. S., Gloor, R., Haag, O., Haupt, M., Skutan, S., Di Lorenzo, F., & Böni, D. (2013). Precious metals and rare earth elements in municipal solid waste–sources and fate in a Swiss incineration plant. *Waste Management (New York, N.Y.)*, *33*(3), 634–644. doi:10.1016/j.wasman.2012.09.010 PMID:23085306

Moroney, J. V., Bartlett, S. G., & Samuelsson, G. (2001). Carbonic anhydrases in plants and algae. *Plant, Cell & Environment*, *24*(2), 141–153. doi:10.1111/j.1365-3040.2001.00669.x

Moss, R. L., Tzimas, E., Kara, H., Willis, P., & Kooroshy, J. (2011). Critical metals in strategic energy technologies. *Assessing Rare Metals as Supply-Chain Bottlenecks in Low-Carbon Energy Technologies, JRC Scientific and Technical Reports, EUR, 24884.*

Mota, R., Pereira, S. B., Meazzini, M., Fernandes, R., Santos, A., Evans, C. A., De Philippis, R., Wright, P. C., & Tamagnini, P. (2015). Effects of heavy metals on Cyanothece sp. CCY 0110 growth, extracellular polymeric substances (EPS) production, ultrastructure and protein profiles. *Journal of Proteomics*, *120*, 75–94. doi:10.1016/j.jprot.2015.03.004 PMID:25779463

Mpinga, C. N., Bradshaw, S. M., Akdogan, G., Snyders, C. A., & Eksteen, J. J. (2014). Evaluation of the Merrill–Crowe process for the simultaneous removal of platinum, palladium and gold from cyanide leach solutions. *Hydrometallurgy*, *142*, 36–46. doi:10.1016/j.hydromet.2013.11.004

Muraleedharan, T. R., Philip, L., Iyengar, L., & Venkobachar, C. (1994). Application studies of biosorption for monazite processing industry effluents. *Bioresource Technology*, *49*(2), 179–186. doi:10.1016/0960-8524(94)90082-5

Nishikawa, K., Yamakoshi, Y., Uemura, I., & Tominaga, N. (2003). Ultrastructural changes in Chlamydomonas acidophila (Chlorophyta) induced by heavy metals and polyphosphate metabolism. *FEMS Microbiology Ecology*, *44*(2), 253–259. doi:10.1016/S0168-6496(03)00049-7 PMID:19719642

Nordi, C. S. F., Cavagliere, T. G. W. F., Vieira, A. A. H., & Nascimento, O. R. (2006). Efeito caotrópico do íon lítio na permeabilidade da cápsula polissacarídica da microalga Ankistrodesmus gracilis (Reinsch) Korsikov (Chlorophyceae). *Acta Botanica Brasílica*, *20*(2), 449–454. doi:10.1590/S0102-33062006000200020

Oliveira, L., & Antia, N. J. (1984). Evidence of nickel ion requirement for autotrophic growth of a marine diatom with urea serving as nitrogen source. *British Phycological Journal*, *19*(2), 125–134. doi:10.1080/00071618400650131

Oliveira, R. C., & Garcia, O. Jr. (2009). Study of biosorption of rare earth metals (La, Nd, Eu, Gd) by Sargassum sp. biomass in batch systems: Physicochemical evaluation of kinetics and adsorption models. *Advanced Materials Research*, *71*, 605–608. doi:10.4028/www.scientific.net/AMR.71-73.605

Oliveira, R. C., Guibal, E., & Garcia, O. Jr. (2012). Biosorption and desorption of lanthanum (III) and neodymium (III) in fixed-bed columns with Sargassum sp.: Perspectives for separation of rare earth metals. *Biotechnology Progress*, *28*(3), 715–722. doi:10.1002/btpr.1525 PMID:22275117

Opare, E. O., Struhs, E., & Mirkouei, A. (2021). A comparative state-of-technology review and future directions for rare earth element separation. *Renewable & Sustainable Energy Reviews*, *143*, 110917. doi:10.1016/j.rser.2021.110917

Oyewo, O. A., Onyango, M. S., & Wolkersdorfer, C. (2018). Lanthanides removal from mine water using banana peels nanosorbent. *International Journal of Environmental Science and Technology*, *15*(6), 1265–1274. doi:10.100713762-017-1494-9

Pádrová, K., Lukavský, J., Nedbalová, L., Čejková, A., Cajthaml, T., Sigler, K., Vítová, M., & Řezanka, T. (2015). Trace concentrations of iron nanoparticles cause overproduction of biomass and lipids during cultivation of cyanobacteria and microalgae. *Journal of Applied Phycology*, *27*(4), 1443–1451. doi:10.100710811-014-0477-1

Palmieri, M. C., Garcia, O. Jr, & Melnikov, P. (2000). Neodymium biosorption from acidic solutions in batch system. *Process Biochemistry*, *36*(5), 441–444. doi:10.1016/S0032-9592(00)00236-3

Periz, G., Dharia, D., Miller, S. H., & Keller, L. R. (2007). Flagellar elongation and gene expression in Chlamydomonas reinhardtii. *Eukaryotic Cell*, *6*(8), 1411–1420. doi:10.1128/EC.00167-07 PMID:17573545

Ponou, J., Wang, L. P., Dodbiba, G., Okaya, K., Fujita, T., Mitsuhashi, K., Atarashi, T., Satoh, G., & Noda, M. (2014). Recovery of rare earth elements from aqueous solution obtained from Vietnamese clay minerals using dried and carbonized parachlorella. *Journal of Environmental Chemical Engineering*, *2*(2), 1070–1081. doi:10.1016/j.jece.2014.04.002

Prodromou, M., & Pashalidis, I. (2016). Europium adsorption by non-treated and chemically modified opuntia ficus indica cactus fibres in aqueous solutions. *Desalination and Water Treatment*, *57*(11), 5079–5088. doi:10.1080/19443994.2014.1002431

Ramasamy, D. L., Porada, S., & Sillanpää, M. (2019). Marine algae: A promising resource for the selective recovery of scandium and rare earth elements from aqueous systems. *Chemical Engineering Journal*, *371*, 759–768. doi:10.1016/j.cej.2019.04.106

Rees, T. A. V., & Bekheet, I. A. (1982). The role of nickel in urea assimilation by algae. *Planta*, *156*(5), 385–387. doi:10.1007/BF00393307 PMID:24272649

Saçan, M. T., Oztay, F., & Bolkent, S. (2007). Exposure of Dunaliella tertiolecta to lead and aluminum: Toxicity and effects on ultrastructure. *Biological Trace Element Research*, *120*(1), 264–272. doi:10.100712011-007-8016-4 PMID:17916979

Sadiq, I. M., Pakrashi, S., Chandrasekaran, N., & Mukherjee, A. (2011). Studies on toxicity of aluminum oxide (Al2O3) nanoparticles to microalgae species: Scenedesmus sp. and Chlorella sp. *Journal of Nanoparticle Research*, *13*(8), 3287–3299. doi:10.100711051-011-0243-0

Sahoo, P. K., Kim, K., Powell, M. A., & Equeenuddin, S. M. (2016). Recovery of metals and other beneficial products from coal fly ash: A sustainable approach for fly ash management. *International Journal of Coal Science & Technology*, *3*(3), 267–283. doi:10.100740789-016-0141-2

Shen, H., & Forssberg, E. (2003). An overview of recovery of metals from slags. *Waste Management (New York, N.Y.)*, *23*(10), 933–949. doi:10.1016/S0956-053X(02)00164-2 PMID:14614927

Soltani, N., Moore, F., Keshavarzi, B., & Sharifi, R. (2014). Geochemistry of trace metals and rare earth elements in stream water, stream sediments and acid mine drainage from Darrehzar Copper Mine, Kerman, Iran. *Water Quality, Exposure, and Health*, *6*(3), 97–114. doi:10.100712403-014-0114-x

Song, W., Liang, J., Wen, T., Wang, X., Hu, J., Hayat, T., Alsaedi, A., & Wang, X. (2016). Accumulation of Co (II) and Eu (III) by the mycelia of Aspergillus niger isolated from radionuclide-contaminated soils. *Chemical Engineering Journal*, *304*, 186–193. doi:10.1016/j.cej.2016.06.103

Suman, T. Y., Rajasree, S. R. R., & Kirubagaran, R. (2015). Evaluation of zinc oxide nanoparticles toxicity on marine algae Chlorella vulgaris through flow cytometric, cytotoxicity and oxidative stress analysis. *Ecotoxicology and Environmental Safety*, *113*, 23–30. doi:10.1016/j.ecoenv.2014.11.015 PMID:25483368

Sunda, W. (2012). Feedback interactions between trace metal nutrients and phytoplankton in the ocean. *Frontiers in Microbiology*, *3*, 204. doi:10.3389/fmicb.2012.00204 PMID:22701115

Svecova, L., Spanelova, M., Kubal, M., & Guibal, E. (2006). Cadmium, lead and mercury biosorption on waste fungal biomass issued from fermentation industry. I. Equilibrium studies. *Separation and Purification Technology*, *52*(1), 142–153. doi:10.1016/j.seppur.2006.03.024

Swain, K. K., Mishra, P. M., & Devi, A. P. (2018). Biosorption of praseodymium (III) using Terminalia arjuna bark powder in batch systems: Isotherm and kinetic studies. *Water Science and Technology*, *77*(3), 727–738. doi:10.2166/wst.2017.589 PMID:29431718

Taggart, R. K., Hower, J. C., Dwyer, G. S., & Hsu-Kim, H. (2016). Trends in the rare earth element content of US-based coal combustion fly ashes. *Environmental Science & Technology*, *50*(11), 5919–5926. doi:10.1021/acs.est.6b00085 PMID:27228215

Tai, P., Zhao, Q., Su, D., Li, P., & Stagnitti, F. (2010). Biological toxicity of lanthanide elements on algae. *Chemosphere*, *80*(9), 1031–1035. doi:10.1016/j.chemosphere.2010.05.030 PMID:20547408

Tian, M., Song, N., Wang, D., Quan, X., Jia, Q., Liao, W., & Lin, L. (2012). Applications of the binary mixture of sec-octylphenoxyacetic acid and 8-hydroxyquinoline to the extraction of rare earth elements. *Hydrometallurgy*, *111*, 109–113. doi:10.1016/j.hydromet.2011.11.002

Torab-Mostaedi, M., Asadollahzadeh, M., Hemmati, A., & Khosravi, A. (2015). Biosorption of lanthanum and cerium from aqueous solutions by grapefruit peel: Equilibrium, kinetic and thermodynamic studies. *Research on Chemical Intermediates*, *41*(2), 559–573. doi:10.100711164-013-1210-4

Tran, N.-P., Park, J.-K., Kim, Z.-H., & Lee, C.-G. (2009). Influence of sodium orthovanadate on the production of astaxanthin from green algae Haematococcus lacustris. *Biotechnology and Bioprocess Engineering; BBE*, *14*(3), 322–329. doi:10.100712257-008-0216-z

Tunsu, C., Petranikova, M., Gergorić, M., Ekberg, C., & Retegan, T. (2015). Reclaiming rare earth elements from end-of-life products: A review of the perspectives for urban mining using hydrometallurgical unit operations. *Hydrometallurgy*, *156*, 239–258. doi:10.1016/j.hydromet.2015.06.007

Vega, J. M., Herrera, J., Aparicio, P. J., Paneque, A., & Losada, M. (1971). Role of molybdenum in nitrate reduction by Chlorella. *Plant Physiology*, *48*(3), 294–299. doi:10.1104/pp.48.3.294 PMID:16657784

Vijayaraghavan, K., Jegan, J., Palanivelu, K., & Velan, M. (2004). Removal of nickel (II) ions from aqueous solution using crab shell particles in a packed bed up-flow column. *Journal of Hazardous Materials*, *113*(1–3), 223–230. doi:10.1016/j.jhazmat.2004.06.014 PMID:15363535

Vijayaraghavan, K., Mahadevan, A., Joshi, U. M., & Balasubramanian, R. (2009). An examination of the uptake of lanthanum from aqueous solution by crab shell particles. *Chemical Engineering Journal*, *152*(1), 116–121. doi:10.1016/j.cej.2009.03.040

Vijayaraghavan, K., Sathishkumar, M., & Balasubramanian, R. (2010). Biosorption of lanthanum, cerium, europium, and ytterbium by a brown marine alga, Turbinaria conoides. *Industrial & Engineering Chemistry Research*, *49*(9), 4405–4411. doi:10.1021/ie1000373

Volesky, B., Weber, J., & Park, J. M. (2003). Continuous-flow metal biosorption in a regenerable Sargassum column. *Water Research*, *37*(2), 297–306. doi:10.1016/S0043-1354(02)00282-8 PMID:12502059

Wu, D., Zhang, L., Wang, L., Zhu, B., & Fan, L. (2011). Adsorption of lanthanum by magnetic alginate-chitosan gel beads. *Journal of Chemical Technology and Biotechnology (Oxford, Oxfordshire)*, *86*(3), 345–352. doi:10.1002/jctb.2522

Xia, B., Chen, B., Sun, X., Qu, K., Ma, F., & Du, M. (2015). Interaction of TiO2 nanoparticles with the marine microalga Nitzschia closterium: Growth inhibition, oxidative stress and internalization. *The Science of the Total Environment*, *508*, 525–533. doi:10.1016/j.scitotenv.2014.11.066 PMID:25483108

Yingjun, W., Jia, L., Yun, L. Ü., Hangbiao, J., Shihuai, D., & Yunmin, Z. (2012). Effects of cerium on growth and physiological characteristics of Anabaena flosaquae. *Journal of Rare Earths*, *30*(12), 1287–1292. doi:10.1016/S1002-0721(12)60222-1

Zhaozhou, Z., Zhongliang, W., Jun, L., Yong, L. I., Zhang, Z., & Zhang, P. (2012). Distribution of rare earth elements in sewage-irrigated soil profiles in Tianjin, China. *Journal of Rare Earths*, *30*(6), 609–613. doi:10.1016/S1002-0721(12)60099-4

Zhou, L., Liu, J., & Liu, Z. (2009). Adsorption of platinum (IV) and palladium (II) from aqueous solution by thiourea-modified chitosan microspheres. *Journal of Hazardous Materials*, *172*(1), 439–446. doi:10.1016/j.jhazmat.2009.07.030 PMID:19646814

Chapter 16
Application of Algae in Food Science, Antioxidants, Animal Feed, and Aquaculture

Monika Prakash Rai
https://orcid.org/0000-0003-2167-2335
Amity Institute of Biotechnology, Amity University, Noida, India

Neelima Lal
Amity Institute of Biotechnology, Amity University, Noida, India

ABSTRACT

The commercial viability of various algal species increases with the growing demand of natural food supplements. Utilisation of algal biomass as health supplements is paving new ways in the fields of nutraceutical and pharmaceutical industries. They also play a crucial nutritional role in livestock feed and aquaculture. Various algal species are rich sources of bioactive compounds like fatty acids, essential minerals, bioactive peptides, carotenoids, vitamins, etc. and thus has the potential to compete with their synthetic counterparts in the market. The increase in demand for high value health supplements and market trends has motivated researchers and industries in developing algal novel products containing functional ingredients. Some important algae that are used as human food, antioxidants, and nutritional supplements are reviewed in this chapter. This chapter also summarizes the role of algae in animal feed industry and aquaculture. Major challenges in the application of algae as nutraceuticals and food are also discussed along with possible future directions.

INTRODUCTION

Algae consist of a group of organisms that are found both in marine as well as freshwaters. These organisms have been used as food and feed by humans for a very long time. Many researchers have demonstrated the importance of algae as human food and have revealed that they have higher nutritional value when compared to conventional crops. They are considered as a source of renewable, sustainable and economi-

DOI: 10.4018/978-1-6684-2438-4.ch016

cally feasible bioactive medicinal products and food supplements (Khan et al., 2018). The commercial value of algae is also increasing as the components of algal species such as fatty acids, carotenoids, colourants, vitamins etc are on high demand, making them a potential competitor in the market (Jibril et al., 2016). They are also used in cosmetics and play a very important role in aquaculture. Similarly, they are being used as a source of very important molecules. For instance, Polyunsaturated fatty acids are used in infant formulas and as nutritional supplements (Spolaore, et al., 2006).

BACKGROUND

Algae are known for their capability of providing bioactive compounds for the production of novel medicinal and pharmaceutical products. Hence, they are widely studied for their use in human nutrition and also for its utilization in functional foods. Microalgae are a rich source of carbon compounds and hence it can be utilized as nutraceuticals, cosmetics, pharmaceuticals, food and feed supplements (Khan et al., 2018). They are defined as a "treasure house" of various biological activities and as a source for bioactive molecules with nutraceutical and potential pharmaceutical properties. They have also been reported as rich sources of omega 3 and 6 fatty acids, bioactive peptides, vitamins and minerals (Saha and Murray, 2018). Therefore because of their sustainability and ease of cultivation both indoors and outdoors, microalgae are considered as a cutting-edge to pharmaceutical and nutraceutical industries (Saha and Murray, 2018). Macroalgae such as red, green and brown seaweeds are also rich sources of bioactive compounds with potential applications in the field of medicine with therapeutic properties, such as anticancer, antiobesity, antidiabetic, antihypertensive, antihyperlipidemic, antioxidant, anticoagulant, anti-inflammatory, immunomodulatory, antiestrogenic, thyroid stimulating, neuroprotective, antiviral, antifungal, antibacterial and tissue healing properties (Khalid et al., 2018). Generally, algal species that are commonly used as food include *Chlorella sp., Nannochloropsis sp., Tetraselmis sp., Porphyridium sp.*, Isochrysis sp., *Ascophyllum nodosum, Fucus vesiculosus, Fucus serratus, Himanthalia elongate, Porphyra umbilicalis, Palmaria palmata* and *Haematococcus sp.* (Raymundo et al., 2006) although only those foods that are formulated using *Spirulina sp.* and *Hematococus pulvalis* will be discussed in detail in the current paper.

Bioactive Compounds from Algae

Geographic distribution and seasonality play a major role in defining the nutritional composition of algal species. Some of the bioactive compounds that are of algal origin includes natural pigments (NPs), polyunsaturated fatty acids (PUFAs), lipids, proteins and polysaccharides. Similar to terrestrial plants, macroalgae has a highly flexible range of nutritional composition varying between divisions, genus and species (Wan et al., 2018). Some of the microalgal species that are currently being mass produced and their bioactive components are summarised in (Table 1).

Proteins and Peptides

Proteins and peptides are one of the major groups of compounds showing biological functionalities (Samarakoon et al.,2012). Some of the microalgae peptides have been attributed to activities such as antioxidative, antihypertensive, immunomodulatory, anticancerogenic, hepato-protective, and antico-

Figure 1. Bioactive compounds produced by algae

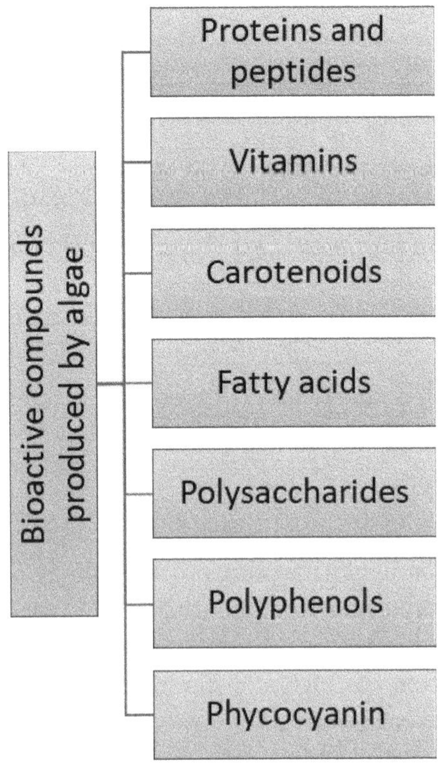

agulant activities (Suetsuna et al., 2001). A lot of important proteins and amino acids are produced by algae; for instance, proteins, peptides, and amino acids (Khan et al., 2018). Production of innovative functional food products using microalgae peptides have become of great interest (Spolaore et al., 2006; Wells et al., 2017; Raposo et al., 2016).

Vitamins and Fatty Acids

Vitamins are essential for metabolic processes as they are precursors of vital enzyme cofactors. They also show high antioxidant activity. Most of the vitamins must be orally consumed as they cannot be produced endogenously in humans (Galasso et al., 2019). Microalgae are source for various vitamins: pro-vitamin A (α- and β-carotene, apocarotenoids), vitamin C (ascorbic acid), vitamin E (tocopherols and tocotrienols) and some vitamins of the B group, such as B1 (thiamine), B2 (riboflavin), B3 (niacin) and B12 (cobalamin) (Claire et al., 1991). Microalgae have also shown the presence of Vitamin D (Brown et al., 1999; Jäpelt et al., 2013). Fatty acids can be classified as saturated (SFA), monounsaturated (MUFA) and polyunsaturated (PUFA). PUFAs are grouped into omega 3 and 6 fatty acid based on the position of first double bond with respect to the terminal methyl end of the carbon chain (Balic et al. 2020).

Table 1. Summary of referenced average nutritional compositions of the described algae expressed as g per 100 g dry weight (Bishop and Zubeck, 2012).

Component	*Chlorella vulgaris*	*Haematococcus pluvalis*	*Spirulina sp.*	*Dunaliella salina*
Protein	64.5	23.6	63	7.4
Fat	10.0	13.8	4.3	7.0
Carbohydrates	15.0	38.0	17.8	29.7
Vitamin B1(Thiamin)	0.0023	0.00047	0.001	0.0009
Vitamin B2 (Riboflavin)	0.005	0.0017	0.0045	0.0009
Vitamin B3 (Niacin)	0.025	0.0066	0.0149	0.001
Vitamin B6 (Pyridoxine)	0.0025	0.00036	0.00096	0.0004
Vitamin B12 (Cobalamine)	0.000008	0.00012	0.00016	0.000004

Carotenoids

Carotenoids are mainly involved in light harvesting process as they are the accessory pigments. They play an important role in photo-protection and photosynthetic organism. Several carotenoids such as α-carotene, β-carotene and β-crytoxanthin are precursors to the vital nutritional components like vitamin A and retinol. Various algal carotenoids such as β-carotene, zeaxanthin, antheraxantin, violaxanthin, neoxanthin and lutein are also present in terrestrial plants (Wane et al., 2018)

Polysaccharides

Principal chemical energy in algae are stored as simple sugars and polysaccharides. They also help in providing structural support for the cells (Percival, 1979). Among the macroalgal species, Kelp species (Phaeophyceae) are known to have the highest carbohydrate content ranging from 50-60% (Kraan, 2010). They are also good sources of phycocolloids and hence can be used in foods, cosmetics, pharmaceuticals and industrial products (Campo et al., 2009; Holdt and Kraan, 2011).

Algae as Human Food

The use of algae as food diet back to thousands of years when Chinese used *Nostoc* to survive during famine (Milledge, 2010). In the present scenario, food industry is one of the biggest commercially growing markets for microalgae as they are rich sources of carbohydrates, proteins, fatty acids, vitamins and minerals. Currently there is an increased worldwide demand for microalgae in the food sector (Araujo et al. 2021). To improve the human nutrition, microalgae are marketed nowadays in various forms like tablets, capsules and liquids. Pastas, biscuits, candy bars, gums and beverages with microalgae incorporated in them are also commercially available (Spolaore, et al., 2006). Macroalgae can be incorporated into flakes, flour, powder etc. The well-known nori sheets from the genus *Porphyra/Pyropia* are commonly used in sushi rolls or crispy thin snacks. Algal powders from *Spirulina sp.* containing polyunsaturated fatty acids (PUFA) are popular as food and health supplements in many countries (Venugopal, 2009; Shahidi, 2008). Other commonly available microalgae supplements in the market include single-cell protein (SCP) from *Chlorella* and *Spirulina,* which are sold in the form of tablets (Chae et al., 2006).

Table 2. Some of the most commonly used algae, their growth type, mode of cultivation and applications in various fields (Saha and Murray, 2018)

Algal species	Growth type	Cultivation	Application
Chlorella vulgaris	Photoautotrophic	Open raceway pond, tubular PBR, flat-plate photobioreactor	Whole biomass for human nutrition as tablets, powders, nectar noodles; cosmetics; aquafeed
Haematococcus pluvialis	Photoautotrophic (two phase cultivation), Mixotrophic	Open raceway pond, tubular enclosed outdoor PBR, bubble column and airlift photobioreactors, large plastic bags	Carotenoid astaxanthin, aquafeed, poultry feed, animal feed, human nutrition, cosmetics, pharmaceuticals, food-colourant, food-supplement
Arthrospira (Spirulina)	Photoautotrophic	Open raceway pond, tanks, earthen pots, basins, natural lakes	Whole biomass for human nutrition as tablets, capsules, powders; blue phycocyanin as colourant in food; source for g-linolenic acid (GLA), vitamins and minerals
Dunaliella salina	Photoautotrophic (two phase cultivation)	Unstirred open pond, lagoons, paddle wheel stirred raceway ponds, tubular photobioreactors	Carotenoid β-carotene for food and in cosmetics, human nutrition as powder, animal feed, source for proteins and glycerol
Palmaria palmata	Epiphyte on stipes of *Laminaria hyperborea* or *L. digitata*	Hatchery cultivation	Food science, immunodiagnostic, cosmetics

Some of the popular algal species, their growth type, cultivation and application are summarised in Table 2 (Saha and Murray, 2018).

Microalgae that are widely commercialized for food and nutritional supplements include *Chlorella vulgaris, Haematococcus pluvialis, Dunaliella salina* and the cyanobacteria *Spirulina maxima* (Figure 2) (Singh et al. 2005).

Spirulina as Food Supplement

Spirulina became popular when NASA successfully utilized it in space as nutritional supplements for the astronauts (Mandotra et al.,2021). It is identified as a potential antiviral, antitumor, antioxidant and antiallergic agent hence can be commercialized for pharmaceutics, food industry, agriculture, perfumery, medicine, and environmental applications (Soni et al., 2017). *Spirulina* is a multicellular, filamentous blue-green microalgae which belongs to the family, Oscillatoriaceae. They are rod-disk in shape and the multicellular cylindrical trichomes are organized in helical shaped filaments. It has high economic value and hence viable for commercial production. They possess efficient photosynthetic activity and have a doubling time of 24 hours and are also easily adaptive to wide range of environments ((Lanlan et al., 2015). Functions and nutritional benefits of *Spirulina* are given in the Table 3.

Apart from the macro and micronutrients, *Spirulina* also contains certain phytonutrients such as chlorophyll, phycocyanin and carotenoids that play an important role in metabolic processes., The nutritional profile of *Spirulina* powder, is presented in Table 4.

Figure 2. Commonly used algae and their applications

Commercial Products from Spirulina

Protein content, final product colour and consistency are the factors that decide the market price of *Spirulina* products. Higher the protein content, the higher is their commercial value. For commercial applications, the two main types of *Spirulina* being processed include *Arthrospira plantesis* and *Anthrospira maxima* (Mandotra et al., 2021; Soni et al., 2017). Some of the other commonly known *Spirulina* products are *Spirulina* Powder, *Spirulina* flakes, *Spirulina* tablets, *Spirulina* Capsules, Spiruvita-C, Dr. *Spirulina* Diavita-C, *Spirulina* drinks, *Spirulina* chocolates (Rabelo et al., 2013). For this reason, *Spirulina* is considered one of the important foods in the future. Moreover, with every passing year, novel foods formulated with *Spirulina* that are being launched into the markets are increasing. Compared to their

Table 3. Functions and nutritional benefits of Spirulina (Sudhakar et al. 2013b; Soni et al. 2017; Ishimi et al. 2006)

Benefits	Nutrients	Functions
Fights Fatigue	Vitamins, minerals, proteins	Rebuild the body's depleted nutrients
Food for brain	Nucleic acid (DNA/RNA), SOD, AFA	Produces high-quality brain tissues, increases the ability of RBC to carry oxygen to the brain Effectively balance functions of right and left brain, enhances memory
Antiaging	Nucleic acid (DNA/RNA), SOD, C/E, Beta carotene, Enzyme	Helps repair genetic elements in injured cells, slows down the aging processes.
Improves Anaemia	Iron, protein, fat, chlorophyll, B12, MVTs, minerals, proteins, pigment MVT, protein, iron	Vitamin B12 250% higher than in any food, helps to build blood components, alleviates anaemia Provides balanced elements for blood formation, prevents anaemia due to nutrient deficiency Maintains balanced nutritional intake.

Table 4. Nutritional profile of spirulina powder

Composition*	Per 100g
1. Macronutrients	
Calories, kcal	290
Water, g	4.7
Total lipids, g	7.7
Total protein, g	57.5
Carbohydrates, g	23.9
Ash, g	6.2
2. Minerals	
Calcium, mg	40.2
Iron, mg	28.5
Magnesium, mg	195.0
Phosphorous, mg	118.0
Potassium, g	1.4
Sodium, g	1.0
Zinc, mg	2.0
Copper, mg	6.1
Manganese, mg	1.9
Selenium, µg	7.2
3. Vitamins	
Vitamin A, IU	570
Vitamin K, µg	25.5
Vitamin B1, mg	2.4
Vitamin B2, mg	3.7
Vitamin B3, mg	12.8
Vitamin B6, mg	0.4
Vitamin E, mg	5.0
4. Amino acids	
Tryptophan, g	0.93
Threonine, g	2.91
Isoleucine, g	3.21
Leucine, g	4.95
Lysine, g	3.02
Methionine, g	1.15
Cysteine, g	0.66
Phenylalanine, g	2.77
Tyrosine, g	2.58
Valine, g	3.51
Arginine, g	4.15
Histidine, g	1.08
Alanine, g	4.51
Aspartic acid, g	5.79
Glutamic acid, g	8.39
Glycine, g	3.09
Proline, g	2.38
Serine, g	2.99

* Data accessed from the U.S. Department of Agriculture Food Data Central database available at https://fdc.nal.usda.gov/. Data accessed on March 2020 – FDC ID:170495.

synthetic counterparts, *Spirulina* has improved nutritional benefits that could be used to develop novel foods. *Spirulina* is an excellent source of bioactive peptides ad hence finds application in functional food industries. Moreover, they are also used as antihypertensive, antidiabetic, antiobesity, and antioxidant ingredients (Lafagra et al., 2020).

Astaxanthin: An Antioxidant Produced by Haematococcus Pluvialis

Astaxanthin, popularly known as "super antioxidant" is a commercially important compound produced by the microalga *Haematococcus pluvialis*. *Hematococcus pluvialis* is a unicellular, green, biflagellate microalga belonging to the class Chlorophyceae, order Volvocales and family Haematococcaseae (Eom et al., 2006). *H. pluvialis* is the richest source of natural astaxanthin and is thus considered as a promising source for the industrial production of astaxanthin. Compared to its synthetic counterparts, astaxanthin that are naturally produced from *H. pluvialis* is considered to have higher antioxidant capacity. Astaxanthin helps in reducing free radicals and oxidative stress and helps in maintaining a healthy human body. Astaxanthin is widely used in nutraceuticals, food and cosmetic industries. Hence it is considered to be one of the high-value microalgal products of the future (Shah et al., 2016). Astaxanthin (3,30 -dihydroxy-β, β 0 -carotene-4,40 -dione) belongs to a set of carotenoids found naturally in organisms like microalgae, fishes etc (Mularczyk et al., 2020). It is generally known for its powerful antioxidant property and its considerable potential and applications in human nutrition and health (Hussein et al., 2006).

Commercial Products from Astaxanthin

Astaxanthin is available commercially in the form of dietary supplements, oils or dried aplanospores (Zanoni et al., 2019). Studies demonstrated that increased antioxidant properties and bioavailability of astaxanthin were observed when *H. pluvialis* was administered as a dietary supplement along with olive oil (Rao et al., 2010). Furthermore, increased hypolipidemic/ hypocholesterolemic effect in plasma was observed when astaxanthin was taken along with fish oils. Commercially available astaxanthin products are sold in the forms of daily capsules, soft gels, energy drinks and powders (Ambati et al., 2014).

Algae in Animal Feed and Aquaculture

Animal Feed

Animal feed has play a major role in global food security. The consumption of algae is not only limited to human beings, but also for other animal species. Usage of seaweeds by Europeans in animal husbandry has come since the time of Romans. Seaweeds are used in domestic animal nutrition in countries like Iceland, France, and Norway. Some of the most prevalent microalgae used as animal feeds are *Schizochytrium sp., Chlorella sp., Arthrospira sp., Isochrysis sp.* and *Porphyridium sp.(* Ramaraj et al. 2019).

In conclusion, exploitation of algae as animal feed ingredients is a promising alternative to corn and soyabean and for overcoming the current world-wide competition among food-feed biofuel industries. In addition, microalgae also contribute as a sustainable alternative and helps in the protection of environment and natural resources. Therefore, addition of algae in animal feed is a very promising approach for the development and maintenance of livestock sector. The cost-effective use of microalgae as animal feed still remains as a huge challenge for the feed industry. Hence, advancement of the current technologies used for microalgal cultivation is necessary for cost effective production of microalgal biomass (Madeira et al., 2017).

Aquaculture Feed

Increasing developments in the field of aquaculture has led to rise in necessities of bulk feed materials and other resources leading to hike in its price with every passing year. Microalgae are considered as an ideal source of feed supplements as they contain bioactive components like -protein, amino acids, polysaccharides, unsaturated fatty acids and cytochrome etc. For instance, dried *Spirulina* contains 60-90% of protein and essential amino acids especially lysine and threonine. Microalgae are used as feed additives of fish, shrimp, crab, shellfish and sea cucumber (Jinghui et al., 2015). The most commonly used algal species are *Chlorella, Tetraselmis, Isochrysis, Pavlova, Phaeodactylum, Chaetoceros, Nannochloropsis, Skeletonema and Thalassiosira* (Kaparapu et al., 2018). Besides being a natural source of bioactive components, the incorporation of algae, even in very small amounts in livestock and aquaculture feeds, is known to improve the immune system, lipid metabolism, antiviral and antibacterial action, stress resistance and gut function of the aquatic organisms (Turner et al., 2002; Nakagawa, 1997; Güroy et al. 2011; Michiels et al. 2011; Nath et al. 2012; Sheikhzadeh et al. 2012).

For the use of microalgae in aquaculture they have to meet various parameters such as high nutritional value, less toxicity, digestible cell walls, easy to culture, correct cell size and shape ((Raja et al. 2004; Patil et al. 2007). Only very few microalgal species are cultured in aquaculture hatcheries based on the conditions like available strains, culture methods, physical characteristics of cell, nutritional composition, presence of toxins and digestibility (Guedes and Malcata 2012; Anon, 2010; Tredici et al., 2009; Muller-Fuega et al., 2004; Muller-Fuega et al., 2003a; Muller-Fuega et al., 2003b). A noteworthy quality which could be used in aquaculture is that bromophenols have low molecular weight and can enhance ‘sea-like’ flavours (Hansen and Lin, 2011). Some of the commonly used algal species and their applications are listed in the Table 5. Currently, studies are underway to understand on- and off- site production of microalgae, cost-effective cultivation, scaling up and improved quality control (Hemaiswarya et al., 2011).

Challenges for Using Algal Ingredients

Many studies have revealed microalgae as an optimal candidate for sources of natural antioxidants (Lee et al., 2008). Although microalgae have been used as a food and nutraceutical additive by humans since a really long time, their role in contributing for health and nutritional benefits, as well as their application in various sectors such as energy and cosmetics sectors still remains as challenges that are to be addressed (Blackburn et al., 2012). *Spirulina* is known as a “wonderful future food source” because of its exceptionally high protein content and balanced essential amino acids content. Furthermore, high consumption of microalgae in large quantities may cause increased content of nucleic acids which are metabolized to uric acids causing adverse health effects like kidney or gout stones (Gantar and Svircev, 2008). Some of the key challenges in using algal ingredients are shown in Figure 3.

FUTURE RESEARCH DIRECTIONS

Extensive studies and research activities in microalgal species will help prevent various diseases and provide more health benefits to human beings. Even in the 21st century, microalgae still remain as a group of organisms that are least explored and requires more research in the areas of bioprospecting. For example, microalgae can be used as sustainable sources of PUFAs but more research studies and

Table 5. Algal species and their application in aquaculture (Hemaiswarya et al., 2011; Kaparapu et al., 2018).

Algae/ Genus	Applications
Arthrospira platensis (Cyanophyceae); Chlorella vulgaris, C.minutissima, C.virginica, Dunaliella tertiolecta, D. salina, Haematococcus pluvialis (Chlorophyceae)	In formulated feed ingredients
Thalassiosira pseudonana (Bacillariophyta); Pavlova lutheri (Haptophyta); Isochrysis galbana Chlorella minutissima, Gomphonema sp, Isochrysis galbana, Nitzschia sp, Phaeodactylum tricornutum, Tetraselmis subcordiformis. Tetraselmis suecica, T.chui (Chlorophyceae); Chaetoceros calcitrans,C. gracilis; Skeletonema costatum.	Feed for bivalve molluscs
Crypthecodinium cohnii (dinoflagellates); Schizochytrium sp.; Ulkenia sp. Chlorella sp, Chlamydomonas sp, Nannochloris oculata, Tetraselmis tetrathele and T. chuii.	Rotifer and Artemia live prey
Tetraselmis suecica, T.chui (Chlorophyceae); Chaetoceros calcitrans, gracilis; Skeletonema costatum; Thalassiosira pseudonana (Bacillariophyta)	Feed for crustacean larvae ((shrimps, lobsters))
Nitzschia sp.; Navicula sp.; Amphora sp	Feed for gastropod molluscs and sea urchins
Isochrysis galbana; Nannochloropsis oculata	Green water" for finfish larvae
Thalassiosira weissflogii	Used in the shrimp and shellfish larviculture, considered by several hatcheries.
Chaetoceros	Used to increase vitamin levels in some shrimp hatcheries
Tetraselmis	Excellent feed for larval shrimps and contains natural amino acids
Pavlova	Used to increase the DHA/EPA levels in brood stock, oysters, clams, mussels and scallops, sterol composition so it is popular with cold water fish hatcheries (cod) for enriching rotifers

investigation of new strains are required as only few species of microalgae are being used for the production of PUFAs (Gantar and Svircev, 2008). Furthermore, attention is required in key areas such as omics technologies and systems and synthetic biology (Reijnders et al., 2014; Gudmundsson and Nogales, 2015). Regarding the safety and regulatory issues, FDA (Food and Drug Administration) regulates food produced from algal biomass. Algal biomass is categorized as "other dietary supplements" according to the classification by Center for Food Safety and Applied Nutrition. *Spirulina, Chlorella, Dunaliella, Haematococcus, Schizochytrium, P. cruentum* and *C. cohnii* are some of the algal food sources which belongs to GRAS (Generally Recognized As Safe) category. Food safety and regulations is therefore a very relevant aspect to be considered for developing microalgal products (Enzing et al., 2014).

Even though more than 60 microalgal genomes have been sequenced (David et al 2021), new omics data is very important to understand the regulatory mechanisms such as nutrient deprivation, biotic and abiotic stresses and to characterize the pathways involved in the production of nutraceuticals (Koyande et al., 2019).

Figure 3. Key challenges in the field of algal production

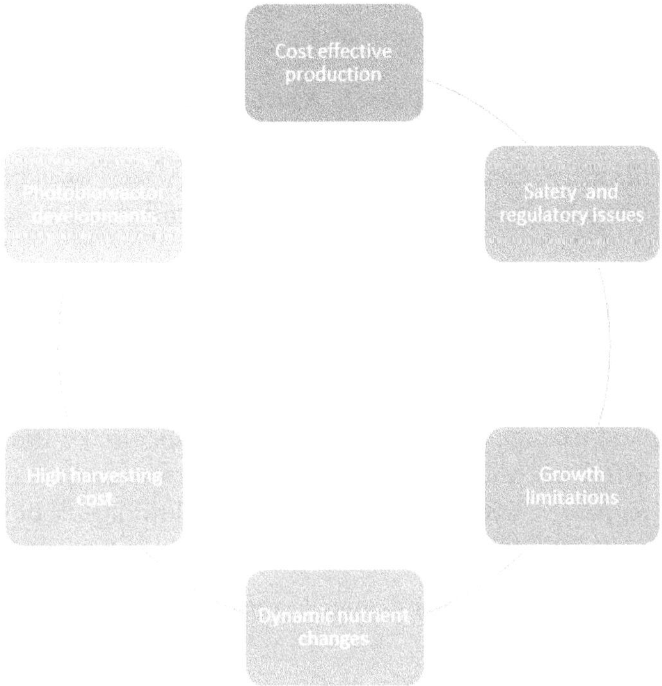

CONCLUSION

Microalgae are considered as the richest sources of renewable, sustainable and economically feasible bioactive medicinal products, food and feed supplements. Lack of incentives for microalgal-based foods production and little awareness about its health benefits are major hurdles towards success of micro-algae derived foods. Therefore, extra work should be done in genetic manipulation of microalgal species for increasing the production of target compounds. Further effort involving biotechnological treatment of the macroalgae by enzymatic degradation of algal fibres that could improve protein digestibility and, therefore, will increase their nutritional value.

ACKNOWLEDGEMENT

This work was supported by Mission innovation India unit, Department of Biotechnology, Ministry of Science & Technology, New Delhi (India) [File no. BT/PR31218/PBD/26/771/ 2019].

REFERENCES

Achankunju, J. (2016). *Bioprospecting selected microalgae for Aquaculture feed*. Academic Press.

Ambati, R. R., Phang, S. M., Ravi, S., & Aswathanarayana, R. G. (2014). Astaxanthin: Sources, extraction, stability, biological activities and its commercial applications—a review. *Marine Drugs*, *12*(1), 128–152. doi:10.3390/md12010128 PMID:24402174

Araújo, R., Vázquez Calderón, F., Sánchez López, J., Azevedo Isabel, C., Bruhn, A., Fluch, S., Garcia, T. M., Ghaderiardakani, F., Ilmjärv, T., Laurans, M., Mac Monagail, M., Mangini, S., Peteiro, C., Rebours, C., Stefansson, T., & Ullmann, J. (2021). Current Status of the Algae Production Industry in Europe: An Emerging Sector of the Blue Bioeconomy. *Frontiers in Marine Science*, *7*, 626389. doi:10.3389/fmars.2020.626389

Balić, A., Vlašić, D., Žužul, K., Marinović, B., & Mokos, Z. B. (2020). Omega-3 Versus Omega-6 Polyunsaturated Fatty Acids in the Prevention and Treatment of Inflammatory Skin Diseases. *International Journal of Molecular Sciences*, *21*(3), 741. doi:10.3390/ijms21030741 PMID:31979308

Batista, A. P., Niccolai, A., Fradinho, P., Fragoso, S., Bursic, I., Rodolfi, L., Biondi, N., Tredici, M. R., Sousa, I., & Raymundo, A. (2017). Microalgae biomass as an alternative ingredient in cookies: Sensory, physical and chemical properties, antioxidant activity and in vitro digestibility. *Algal Research*, *26*, 161–171. doi:10.1016/j.algal.2017.07.017

Becker, W. (2004). Microalgae in human and animal nutrition. In A. Richmond (Ed.), *Handbook of microalgal culture* (pp. 312–351). Blackwell.

Bernaerts, T., Gheysen, L., Kyomugasho, C., Jamsazzadeh, K., Zahra, V., & Foubert, I. (2018). Comparison of microalgal biomasses as functional food ingredients: Focus on the composition of cell wall related polysaccharides. *Algal Research*, *32*, 150–161. doi:10.1016/j.algal.2018.03.017

Bishop, W. M., & Zubeck, H. M. (2012). Evaluation of Microalgae for use as Nutraceuticals and Nutritional Supplements. *Journal of Nutrition & Food Sciences*, *2*, 147. doi:10.4172/2155-9600.1000147

Bixler, H. J., & Porse, H. (2011). A decade of change in the seaweed hydrocolloids industry. *Journal of Applied Phycology*, *23*(3), 321–335. doi:10.100710811-010-9529-3

Blackburn, S. I., & Volkman, J. K. (2012). Microalgae: A Renewable Source of Bioproducts. In N. T. Dunford (Ed.), *Food and Industrial Bioproducts and Bioprocessing* (pp. 221–241). John Wiley & Sons, Inc. doi:10.1002/9781119946083.ch9

Bonos, E., Kasapidou, E., Kargopoulos, A., Karampampas, A., Christaki, E., Florou-Paneri, P., & Nikolakakis, I. (2016). Spirulina as a functional ingredient in broiler chicken diets. *South African Journal of Animal Science*, *46*(1), 94–102. doi:10.4314ajas.v46i1.12

Boyera, N., Galey, I., & Bernard, B. A. (1998). Effect of vitamin C and its derivatives on collagen synthesis and cross-linking by normal human fibroblasts. *International Journal of Cosmetic Science*, *20*(3), 151–158. doi:10.1046/j.1467-2494.1998.171747.x PMID:18505499

Brown, M. R., Mular, M., Miller, I., Farmer, C., & Trenerry, C. (1999). The vitamin content of microalgae used in aquaculture. *Journal of Applied Phycology*, *11*(3), 247–255. doi:10.1023/A:1008075903578

Burtin, P. (2003). Nutritional value of seaweeds. *Elect J Env Agri Food Che.*, *2*, 498–503.

Campo, V. L., Kawano, D. F., Silva, D. B. Jr, & Carvalho, I. (2009). Carrageenans: Biological properties, chemical modifications and structural analysis – A review. *Carbohydrate Polymers*, *77*(2), 167–180. doi:10.1016/j.carbpol.2009.01.020

Capelli, B., Bagchi, D., & Cysewski, G. R. (2013). Synthetic astaxanthin is significantly inferior to algal-based astaxanthin as an antioxidant and may not be suitable as a human nutritional supplement. *NUTRAfoods: International Journal of Science and Marketing for Nutraceutical Actives, Raw Materials, Finish Products*, *12*(4), 145–152. doi:10.100713749-013-0051-5

Cardozo, K. H., Guaratini, T., Barros, M. P., Falcão, V. R., Tonon, A. P., Lopes, N. P., Campos, S., Torres, M. A., Souza, A. O., Colepicolo, P., & Pinto, E. (2007). Metabolites from algae with economical impact. *Comparative Biochemistry and Physiology. Toxicology & Pharmacology: CBP*, *146*(1-2), 60–78. doi:10.1016/j.cbpc.2006.05.007 PMID:16901759

Chu, W. L., Alwi, A., & Phang, S. M. (2002). Phycoerythrin production by a marine Oscillatoria (Cyanophyta). *Mal. J. Sci.*, *21*, 67–73.

Claire, C., De Roeck-Holtzhauer, Y., & Quere, I. (1991). Vitamin analysis of five planktonic microalgae and one macroalga. *Journal of Applied Phycology*, *3*(3), 259–264. doi:10.1007/BF00003584

Costa, J. A. V., & de Morais, M. G. (2014). *An Open Pond System for Microalgal Cultivation. Biofuels from Algae.* Academic Press.

Eom, S. H., Kim, Y. M., & Kim, S. K. (2012). Antimicrobial effect of phlorotannins from marine brown algae. *Food and Chemical Toxicology: An International Journal Published for the British Industrial Biological Research Association, 50*(9), 3251–3255.

Fassett, R. G., & Combes, J. S. (2011). Astaxanthin: A potential therapeutic agentin cardiovascular disease. *Marine Drugs*, *9*(3), 447–465. doi:10.3390/md9030447 PMID:21556169

Guarnieri, M. T., & Pienkos, P. T. (2015). Algal omics: Unlocking bioproduct diversity in algae cell factories. *Photosynthesis Research*, *123*(3), 255–263. doi:10.100711120-014-9989-4 PMID:24627032

Gudmundsson, S., & Nogales, J. (2015). Cyanobacteria as photosynthetic biocatalysts: A systems biology perspective. *Molecular BioSystems*, *11*(1), 60–70. doi:10.1039/C4MB00335G PMID:25382198

Guedes, A. C., & Malcata, F. X. (2012). Nutritional Value and Uses of Microalgae in Aquaculture. In Z. A. Muchlisin (Ed.), *Aquaculture* (pp. 59–78)., doi:10.5772/1516

Güroy, D., Güroy, B., & Merrifield, D. L. (2011). Effect of Dietary Ulva and Spirulina on Weight Loss and Body Composition of Rainbow Trout, Oncorhynchus mykiss (Walbaum), During a Starvation Period. Journal of Animal Physiology and Nutrition, 320–327.

Güroy, D., Güroy, B., Merrifield, D. L., Ergün, S., Tekinay, A. A., & Yiğit, M. (2011). Effect of Dietary Ulva and Spirulina on Weight Loss and Body Composition of Rainbow Trout, Oncorhynchus mykiss (Walbaum), During a Starvation Period. *Journal of Animal Physiology and Nutrition*, *95*(3), 320–327. doi:10.1111/j.1439-0396.2010.01057.x PMID:20880289

Harnedy, P. A., & FitzGerald, R. J. (2011). Bioactive proteins, peptides, and amino acids from macroalgae. *Journal of Phycology*, *47*(2), 218–232. doi:10.1111/j.1529-8817.2011.00969.x PMID:27021854

Hemaiswarya, S., Raju, R., Kumar, R., & Anbazhagan, C. (2011). Microalgae: A sustainable feed source for aquaculture. *World Journal of Microbiology & Biotechnology*, *27*(8), 1737–1746. doi:10.100711274-010-0632-z

Holdt, S. L., & Kraan, S. (2011). Bioactive compounds in seaweed: Functional food applications and legislation. *Journal of Applied Phycology*, *23*(3), 543–597. doi:10.100710811-010-9632-5

Hopkins, D. L., Clayton, E. H., Lamb, T. A., Van de Ven, R. J., Refshauge, G., Kerr, M. J., Bailes, K., Lewandowski, P., & Ponnampalam, E. N. (2014). The impact of supplementing lambs with algae on growth, meat traits and oxidative status. *Meat Science*, *98*(2), 135–141. doi:10.1016/j.meatsci.2014.05.016 PMID:24950082

Hussein, G., Sankawa, U., Goto, H., Matsumoto, K., & Watanabe, H. (2006). Astaxanthin, a carotenoid with potential in human health and nutrition. *Journal of Natural Products*, *69*(3), 443–449. doi:10.1021/np050354+ PMID:16562856

Ishimi, Y., Sugiyama, F., Ezaki, J., Fujioka, M., & Wu, J. (2006). Effects of Spirulina, a blue-green alga, on bone metabolism in overiectomized rats and hindlimb-unloaded mice. *Bioscience, Biotechnology, and Biochemistry*, *70*(2), 363–368. doi:10.1271/bbb.70.363 PMID:16495651

Jerez-Martel, I., García Poza, S., Rodríguez-Martel, G., Rico, M., Afonso-Olivares, C., Gomez, P., & Juan, L. (2017). Phenolic Profile and Antioxidant Activity of Crude Extracts from Microalgae and Cyanobacteria Strains. *Journal of Food Quality*, *2017*, 1–8. doi:10.1155/2017/2924508

Jibril, S. M., Jakada, B. H., Umar, H. Y., & Ahmad, T. A. (2016). Importance of Some Algal Species as a Source of Food and Supplement. *International Journal of Current Microbiology and Applied Sciences*, *5*(5), 186–193. doi:10.20546/ijcmas.2016.505.020

Kim, Y. J., Joo, H. W., Park, J., Kim, D. K., Jeong, K. J., & Chang, Y. K. (2015). Production of 2, 3-butanediol by Klebsiella oxytoca from various sugars in microalgal hydrolysate. *Biotechnology Progress*, *31*(6), 1669–1675. doi:10.1002/btpr.2168 PMID:26400837

Kraan, S. (2010). Mass-cultivation of carbohydrate rich macroalgae, a possible solution for sustainable biofuel production. *Mitigation and Adaptation Strategies for Global Change*, *18*(1), 27–46. doi:10.100711027-010-9275-5

Lahaye, M., & Robic, A. (2007). Structure and functional properties of ulvan, a polysaccharide from green seaweeds. *Biomacromolecules*, *8*(6), 1765–1774. doi:10.1021/bm061185q PMID:17458931

Lanlan, Z., Lin, C., Wang, J., Yu, C., Xin, G., Zhang, Z., & Tianzhong, L. (2015). Attached cultivation for improving the biomass productivity of Spirulina platensis. *Bioresource Technology*, *181*, 136–142. doi:10.1016/j.biortech.2015.01.025 PMID:25647023

Lee, S. H., Karawita, R., Affan, A., Lee, J. B., Lee, B. J., & Jeon, Y. J. (2008). Potential antioxidant activites of enzymatic digests from benthic diatoms Achnanthes longipes, Amphora coffeaeformis, and Navicula sp. (Bacillariophyceae). *J. Food Sci. Nutr.*, *13*(3), 166–175. doi:10.3746/jfn.2008.13.3.166

Li, Y.-X., Wijesekaraa, I., Li, Y., & Kim, S.-K. (2011). Phlorotannins as bioactive agents from brown algae. *Process Biochemistry*, *46*(12), 2219–2224. doi:10.1016/j.procbio.2011.09.015

Lucas, B. F., de Morais, M. G., Santos, T. D., & Costa, J. A. V. (2018). Spirulina for snack enrichment: Nutritional, physical and sensory evaluations. *Lebensmittel-Wissenschaft + Technologie*, *90*, 270–276. doi:10.1016/j.lwt.2017.12.032

Madeira, M. S., Cardoso, C., Lopes, P. A., Coelho, D., Afonso, C., Bandarra, N. M., & Prates, J. A. M. (2017). Microalgae as feed ingredients for livestock production and meat quality: A review. *Livestock Science*, *205*, 111–121. doi:10.1016/j.livsci.2017.09.020

Manach, C., Scalbert, A., Morand, C., R'em'esy, C., & Jim'enez, L. (2004). Polyphenols: Food sources andbioavailability. *The American Journal of Clinical Nutrition*, *79*(5), 727–747. doi:10.1093/ajcn/79.5.727 PMID:15113710

Marti-Quijal, F. J., Zamuz, S., Tomašević, I., Gómez, B., Rocchetti, G., Lucini, L., Remize, F., Barba, F. J., & Lorenzo, J. M. (2019). Influence of different sources of vegetable, whey and microalgae proteins on the physicochemical properties and amino acid profile of fresh pork sausages. *Lebensmittel-Wissenschaft + Technologie*, *110*, 316–323. doi:10.1016/j.lwt.2019.04.097

Masoj'ıdek, J., Kopecky, J., Giannelli, L., & Torzillo, G. ´. (2011). Productivity correlated to photobiochemical performance of Chlorella mass cultures grown outdoors in thin layer cascades. *Journal of Industrial Microbiology & Biotechnology*, *38*(2), 307–317. doi:10.100710295-010-0774-x PMID:20652359

Melis, A., & Mitra, M. (2009). Optical Properties of Microalgae for Enhanced Biofuels Production. *Optics Express*, *16*, 21807–21820. PMID:19104614

Michiels, J., Skrivanova, E., & Missotten, J. (2011). Intact Brown Seaweed (Ascophyllum nodosum) in Diets of Weaned Piglets: Effects on Performance, Gut Bacteria and Morphology and Plasma Oxidative Status. Journal of Animal Physiology and Animal Nutrition. doi:10.1111/j.1439-0396.2011.01227

Michiels, J., Skrivanova, E., & Missotten, J. (2011). Intact Brown Seaweed (Ascophyllum nodosum) in Diets of Weaned Piglets: Effects on Performance, Gut Bacteria and Morphology and Plasma Oxidative Status. Journal of Animal Physiology and Animal Nutrition. doi:10.1111/j.1439-0396.2011.01227

Milledge, J. J. (2010). The challenge of algal fuel: Economic processing of the entire algal biomass. *Condensed Matter-Materials Engineering Newsletter*, *1*(6), 4–6.

Mokronsnop, V. M., & Zolotareva, E. K. (2014). Microalgae as tocopherol producers. *Biotechnologia Acta*, *7*(2), 26–33. doi:10.15407/biotech7.02.026

Mourelle, M. L., Gómez, C. P., & Legido, J. L. (2015). Cosmética termal: Valor añadido en los centros termales. In *Proceedings of the 1st International Congress on Water Healing Spa and Quality of Life*. University of Vigo.

Mularczyk, M., Michalak, I., & Marycz, K. (2020). Astaxanthin and other Nutrients from Haematococcus pluvialis—Multifunctional Applications. *Marine Drugs*, *18*(9), 459. doi:10.3390/md18090459 PMID:32906619

Muller-Feuga, A. (2000). The role of microalgae in aquaculture: Situation and trends. *Journal of Applied Phycology*, *12*(3/5), 527–534. doi:10.1023/A:1008106304417

Muller-Feuga, A. (2004). Microalgae for Aquaculture: The Current Global Situation Future Trends In A. Richmond (Ed.), Handbook of Microalgal Culture: Biotechnology and Applied Phycology (pp. 352–364). Academic Press.

Muller-Feuga, A., Moal, J., & Kaas, R. (2003a). The Microalgae of Aquaculture. In J. G. Støttrup & L. A. McEvoy (Eds.), *Live Feeds in Marine Aquaculture* (pp. 206–252). doi:10.1002/9780470995143.ch6

Muller-Feuga, A., Robert, C., & Cahu, J. (2003b). Uses of Microalgae in Aquaculture. In J. G. Støttrup & L. A. McEvoy (Eds.), *Live Feeds in Marine Aquaculture* (pp. 253–299). doi:10.1002/9780470995143.ch7

Nagata, A., Tajima, T., & Takahashi, J. (2006). Effect of astaxanthin 5 mg on antifatigue and task performance of human. *Carotenoid Sci.*, *10*, 102–106.

Naguib, Y. M. A. (2000). Antioxidant activities of astaxanthin and related carotenoids. *Journal of Agricultural and Food Chemistry*, *48*(4), 1150–1154. doi:10.1021/jf991106k PMID:10775364

Nakagawa, H. (1997). Effect of Dietary Algae on Improvement of Lipid Metabolism in Fish. *Biomedicine and Pharmacotherapy*, *51*(8), 345–348. doi:10.1016/S0753-3322(97)88053-5 PMID:9436528

Nath, P. R., Khozin-Goldberg, I., Cohen, Z., Boussiba, S., & Zilberg, D. (2012). Dietary Supplementation with the Microalgae Parietochloris incisa Increases Survival and Stress Resistance in Guppy Poecilia reticulata Fry. *Aquaculture Nutrition*, *18*(2), 167–180. doi:10.1111/j.1365-2095.2011.00885.x

Nelson, D. R., Hazzouri, K. M., Lauersen, K. J., Jaiswal, A., Chaiboonchoe, A., Mystikou, A., Fu, W., Daakour, S., Dohai, B., Alzahmi, A., Nobles, D., Hurd, M., Sexton, J., Preston, M. J., Blanchette, J., Lomas, M. W., Amiri, K. M. A., & Salehi-Ashtiani, K. (2021). Large-scale genome sequencing reveals the driving forces of viruses in microalgal evolution. *Cell Host & Microbe*, *29*(2), 250–266.e8. doi:10.1016/j.chom.2020.12.005 PMID:33434515

Nishikawa, Y., Minenaka, Y., Ichimura, M., Tatsumi, K., Nadamoto, T., & Urabe, K. (2005). Effects of astaxanthin and vitamin C on the preventionof gastric ulcerations in stressed rats. *J. Nutr. Sci. Vit.*, *51*(3), 135–141. doi:10.3177/jnsv.51.135 PMID:16161762

Osterlie, M., Bjerkeng, B., & Liaaen-Jensen, S. (2000). Plasma appearance and distribution of astaxanthin E/Z isomers in plasma lipoproteins of after single dose administration of astaxanthin. *The Journal of Nutritional Biochemistry*, *11*(10), 482–492. doi:10.1016/S0955-2863(00)00104-2 PMID:11120445

Palanisamy, M., Töpfl, S., Berger, R. G., & Hertel, C. (2019). Physico-chemical and nutritional properties of meat analogues based on Spirulina/lupin protein mixtures. *European Food Research and Technology*, *245*(9), 1889–1898. doi:10.100700217-019-03298-w

Panis, G., & Carreon, J. R. (2016). Commercial astaxanthin production derived by green alga *Haematococcus pluvialis*: A microalgae process model and a techno-economic assessment all through production line. *Algal Research*, *18*, 175–190. doi:10.1016/j.algal.2016.06.007

Park, J. S., Chyun, J. H., Kim, Y. K., Line, L. L., & Chew, B. P. (2010). Astaxanthin decreased oxidative stress and inflammation and enhanced immune response in humans. *Nutrition & Metabolism*, *7*(1), 1–10. doi:10.1186/1743-7075-7-18 PMID:20205737

Pasquet, V., Morisset, P., Ihammouine, S., Chepied, A., Aumailley, L., Berard, J.-B., Serive, B., Kaas, R., Lanneluc, I., Thiery, V., Lafferriere, M., Piot, J.-M., Patrice, T., Cadoret, J.-P., & Picot, L. (2011). Antiproliferative Activity of Violaxanthin Isolated from Bioguided Fractionation of *Dunaliella tertiolecta* Extracts. *Marine Drugs*, *2011*(9), 819–831. doi:10.3390/md9050819 PMID:21673891

Percival, E. (1979). The polysaccharides of green, red and brown seaweeds: Their basic structure, biosynthesis and function'. *British Phycological Journal*, *14*(2), 103–117. doi:10.1080/00071617900650121

Pereira, H., Barreira, L., Figueiredo, F., Custódio, L., Vizetto-Duarte, C., Polo, C., Rešek, E., Engelen, A., & Varela, J. (2012). Polyunsaturated Fatty acids of marine macroalgae: Potential for nutritional and pharmaceutical applications. *Marine Drugs*, *10*(9), 1920–1935. doi:10.3390/md10091920 PMID:23118712

Pisal, D. S., & Lele, S. S. (2005). Carotenoid production from microalgae,Dunaliella salina. *Indian Journal of Biotechnology*, *4*, 476–483.

Priyadarshani, I., & Rath, B. (2012). Commercial and industrial applications of micro algae – A review. *Journal of Algal Biomass Utilization*, *3*(4), 89–100.

Qureshi, M. A. (1995). Immunomodulatory effects of spirulina supplementation in chickens. In *The 44th Western Poultry Disease Conference*. North Carolina State University.

Ramaraj, S., Ramalingam, R., Abeer, H., & Elsayed, F. (2019). Microalgae metabolites: A rich source for food and medicine. *Saudi Journal of Biological Sciences*, *26*(4), 709–722. doi:10.1016/j.sjbs.2017.11.003 PMID:31048995

Ranga Rao, A., Baskaran, V., Sarada, R., & Ravishankar, G. A. (2013). In vivo bioavailability and antioxidant activity of carotenoids from micro algal biomass—A repeated dose study. *Food Research International*, *54*(1), 711–717. doi:10.1016/j.foodres.2013.07.067

Ranga Rao, A., Harshvardhan Reddy, A., & Aradhya, S. M. (2010). Antibacterial properties of *Spirulina platensis, Haematococcus pluvialis, Botryococcus braunii* micro algal extracts. *Current Trends in Biotechnology and Pharmacy*, *4*, 809–819.

Ranga Rao, A., Sindhuja, H. N., Dharmesh, S. M., Sankar, K. U., Sarada, R., & Ravishankar, G. A. (2013). Effective inhibition of skin cancer, tyrosinase,and antioxidative properties by astaxanthin and astaxanthin esters from thegreen alga Haematococcus pluvialis. *Journal of Agricultural and Food Chemistry*, *61*(16), 3842–3851. doi:10.1021/jf304609j PMID:23473626

Rao, M. V., Paliyath, G., & Ormrod, D. P. (1996). Ultraviolet-B- and Ozone-Induced Biochemical Changes in Antioxidant Enzymes of Arabidopsis thaliana. *Plant Physiology*, *110*(1), 125–136. doi:10.1104/pp.110.1.125 PMID:8587977

Raposo, M. F.-D., De Morais, A. M. M. B., & De Morais, R. M. S. C. (2016). Emergent Sources of Prebiotics: Seaweeds and Microalgae. *Marine Drugs*, *14*(2), 27. doi:10.3390/md14020027 PMID:26828501

Raymundo, A., Gouveia, L., Batista, A. P., Raymundo, A., Sousa, I., & Emphis, J. (2006). Chlorella vulgaris and Haematococcus pluvialis biomass as colouring and antioxidant in food emulsions. *European Food Research and Technology*, *222*(3-4), 362–367. doi:10.100700217-005-0105-z

Reijnders, M. J., van Heck, R. G., Lum, C. M., Scaife, M. A., dos Santos, V. A., Smith, A. G., & (2014). Green genes: Bioinformatics and systems-biology innovations drive algal biotechnology. *Trends in Biotechnology*, *32*(12), 617–626. doi:10.1016/j.tibtech.2014.10.003 PMID:25457388

Richer, S., Stiles, W., Statkute, L., Pulido, J., Frankowski, J., Rudy, D., Pei, K., Tsipursky, M., & Nyland, J. (2004). Double-masked, placebo-controlled, randomized trial of lutein and antioxidant supplementation in the intervention of atrophic age-related macular degeneration: The Veterans LAST study (Lutein Antioxidant Supplementation Trial). *Optom. J. Am. Optom. Assoc.*, *2004*(75), 216–229. doi:10.1016/S1529-1839(04)70049-4 PMID:15117055

Rock, C. L. (2009). Carotenoids and Cancer. In Carotenoids (Vol. 5). Birkhäuser Verlag AG. doi:10.1007/978-3-7643-7501-0_13

Saha, S., & Murray, P. (2018). Exploitation of Microalgae Species for Nutraceutical Purposes: Cultivation Aspects. *Fermentation (Basel, Switzerland)*, *4*(2), 46. Advance online publication. doi:10.3390/fermentation4020046

Samarakoon, K., & Jeon, Y.-J. (2012). Bio-functionalities of proteins derived from marine algae–A review. *Food Research International*, *48*(2), 48. doi:10.1016/j.foodres.2012.03.013

Sánchez, M., Bernal-Castillo, J., Rozo, C., & Rodríguez, I. (2003). Spirulina (Arthrospira): An edible microorganism. A review. *Universitas Scientiarum*, *8*, 7–24.

Shahidi, F. (2009). *Marine nutraceuticals and functional foods*. CRC Press.

Sheih, I. C., Wu, T. K., & Fang, T. J. (2009, July). Antioxidant properties of a new antioxidative peptide from algae protein waste hydrolysate in different oxidation systems. *Bioresource Technology*, *100*(13), 3419–3425. doi:10.1016/j.biortech.2009.02.014 PMID:19299123

Sheikhzadeh, N., Tayefi-Nasrabadi, H., Khani Oushani, A., & Najafi Enferadi, M. H. (2012). Effects of Haematococcus pluvialis Supplementation on Antioxidant System and Metabolism in Rainbow Trout (Oncorhynchus mykiss). *Fish Physiology and Biochemistry*, *38*(1), 413–419. doi:10.100710695-011-9519-7 PMID:21695482

Sheikhzadeh, N., Tayefi-Nasrabadi, H., Khani Oushani, A., & Najafi Enferadi, M. H. (2012). Effects of Haematococcus pluvialis Supplementation on Antioxidant System and Metabolism in Rainbow Trout (Oncorhynchus mykiss). *Fish Physiology and Biochemistry*, *38*(1), 413–419. doi:10.100710695-011-9519-7 PMID:21695482

Sher, I., Tzameret, A., Peri-Chen, S., Edelshtain, V., Ioffe, M., Sayer, A., Buzhansky, L., Gazit, E., & Rotenstreich, Y. (2018). Synthetic 9-cis-beta-carotene inhibits photoreceptor degeneration in cultures of eye cups from rpe65rd12 mouse model of retinoid cycle defect. *Scientific Reports*, *2018*(8), 6130. doi:10.103841598-018-24439-3 PMID:29666392

Shi, X., Jung, K. W., Kim, H., Ahn, Y. T., & Shin, H. S. (2011). Direct fermentation of *Laminaria japonica* for biohydrogen production by anaerobic mixed cultures. *International Journal of Hydrogen Energy*, *36*(10), 5857–586. doi:10.1016/j.ijhydene.2011.01.125

Shimamatsu, H. (2004). Mass production of *Spirulina*, an edible microalga. *Hydrobiologia*, *512*(1-3), 39–44. doi:10.1023/B:HYDR.0000020364.23796.04

Simkus, A., Simkiene, A., Cernauskiene, J., Kvietkute, N., Cernauskas, A., Paleckaitis, M., & Kerziene, S. (2013). The effect of blue algae Spirulina platensis on pig growth performance and carcass and meat quality. *Vet. Med. Zoot.*, *61*, 83.

Singh, S., Bhushan, K. N., & Banerjee, U. C. (2005). Bioactive Compounds from Cyanobacteria and Microalgae: An Overview. *Critical Reviews in Biotechnology*, *25*(3), 73–95. doi:10.1080/07388550500248498 PMID:16294828

Soni, R. A., Sudhakar, K., & Rana, R. S. (2017). Spirulina–from growth to nutritional product: A review. *Trends in Food Science & Technology*, *69*, 157–171. doi:10.1016/j.tifs.2017.09.010

Spiller, G. A., & Dewell, A. (2003). Safety of an astaxanthin rich *Haemaotoccu pluvialis* algal extract: A randomized clinical trial. *Journal of Medicinal Food*, *6*(1), 51–56. doi:10.1089/109662003765184741 PMID:12804020

Spolaore, P., Joannis-Cassan, C., Duran, E., & Isambert, A. (2006). Commercial applications of microalgae. *Journal of Bioscience and Bioengineering*, *101*(2), 87–96. doi:10.1263/jbb.101.87 PMID:16569602

Spolaore, P., Joannis-Cassan, C., Duran, E., & Isambert, A. (2006). Commercial applications of microalgae. *Journal of Bioscience and Bioengineering*, *101*(2), 8796. doi:10.1263/jbb.101.87 PMID:16569602

Subba Rao, P.V., Mantri, V., Ks, G., & Kumar, S.K. (2007). *Seaweeds as a Human Diet: An Emerging Trend in the New Millennium*. . doi:10.13140/2.1.3494.7208

Sudhakar, K., & Premalatha, M. (2015). Characterization of micro algal biomass through FTIR/TGA/CHN analysis: Application to Scenedesmus sp. *Energy Sources. Part A, Recovery, Utilization, and Environmental Effects*, *37*(21), 1–8. doi:10.1080/15567036.2013.825661

Suetsuna, K., & Chen, J. (2001). Identification of Antihypertensive Peptides from Peptic Digest of Two Microalgae, Chlorella vulgaris and Spirulina platensis. *Marine Biotechnology (New York, N.Y.)*, *3*(4), 305–309. doi:10.100710126-001-0012-7 PMID:14961345

Takano, H., Takeyama, H., Nakamura, N., Sode, K., Burgess, J., Manabe, E., Hirano, M., & Matsunaga, T. (1992). CO2 removal by high-density culture of a marine cyanobacterium synechococcus sp. using an improved photobioreactor employing light-diffusing optical fibers. *Applied Biochemistry and Biotechnology*, *34/35*(1), 449–458. doi:10.1007/BF02920568

Takebe, Y., Saucedo, C. J., Lund, G., Uenishi, R., Hase, S., Tsuchiura, T., Kneteman, N., Ramessar, K., Tyrrell, D. L., Shirakura, M., Wakita, T., McMahon, J. B., & O'Keefe, B. R. (2013). Antiviral Lectins from Red and Blue-Green Algae Show Potent In Vitro and In Vivo Activity against Hepatitis C Virus. *PLoS One*, *8*(5), e64449. doi:10.1371/journal.pone.0064449 PMID:23700478

Thirumdas, R., Brnčić, M., Brnčić, S. R., Barba, F. J., Gálvez, F., Zamuz, S., Lacomba, R., & Lorenzo, J. M. (2018). Evaluating the impact of vegetal and microalgae protein sources on proximate composition, amino acid profile, and physicochemical properties of fermented Spanish "chorizo" sausages. *Journal of Food Processing and Preservation*, *42*(11), e13817. doi:10.1111/jfpp.13817

Torzillo, G., Pushparaj, B., Masojidek, J., & Vonshak, A. (2003). Biological constraints in algal biotechnology. *Biotechnology and Bioprocess Engineering; BBE*, *8*(6), 338–348. doi:10.1007/BF02949277

Tredici, M. R. (2004). Mass production of microalgae: photobioreactors. Handbook of Microalgal Cultures, Biotechnology and Applied Phycology, 178–214.

Tredici, M. R., Biondi, N., & Ponis, E. (2009). Advances in Microalgal Culture for Aquaculture Feed and Other Uses. In G. Burnell & G. Allan (Eds.), *New Technologies in Aquaculture: Improving Production Efficiency* (pp. 611–676). Quality and Environmental Management. doi:10.1533/9781845696474.3.610

Turner, J. L., Dritz, S. S., Higgins, J. J., & Minton, J. E. (2002). Effects of Ascophyllum nodosum Extract on Growth Performance and Immune Function of Young Pigs Challenged with Salmonella typhimurium. *Journal of Animal Science*, *80*(7), 1947–1953. doi:10.2527/2002.8071947x PMID:12162664

Ulcer preventive and antioxidative properties of astaxanthinfrom Haematococcus pluvialis. (2008). *Eur. J. Pharmacol.*, *590*, 387–395. doi:10.1016/j.ejphar.2008.06.042

Urala, N., & Lähteenmäki, L. (2004). Attitudes behind consumers' willingness to use functional foods. *Food Quality and Preference*, *15*(7-8), 793–803. doi:10.1016/j.foodqual.2004.02.008

Uribe-Wandurraga, Z. N., Igual, M., García-Segovia, P., & Martínez-Monzó, J. (2020). Influence of microalgae addition in formulation on colour, texture, and extrusion parameters of corn snacks. *Food Science & Technology International*, *26*(8), 685–695. doi:10.1177/1082013220924178 PMID:32375498

Venugopal, V. (2009). *Marine Products for Healthcare*. CRC Press.

Waldenstedt, L., Inborr, J., Hansson, I., & Elwinger, K. (2003). Effects of astaxanthin-rich algal meal (Haematococcus pluvialis) on growth performance, caecal campylobacter and clostridial counts and tissue astaxanthin concentration of broiler chickens. *Animal Feed Science and Technology*, *108*(1-4), 119–132. doi:10.1016/S0377-8401(03)00164-0

Wells, M. L., Potin, P., Craigie, J. S., Raven, J. A., Merchant, S. S., Helliwell, K. E., Smith, A. G., Camire, M. E., & Brawley, S. H. (2017). Algae as nutritional and functional food sources: Revisiting our understanding. *Journal of Applied Phycology*, *29*(2), 949–982. doi:10.100710811-016-0974-5 PMID:28458464

Winwood, R. (2013). Recent developments in the commercial production of DHA and EPA rich oils from micro-algae. *OCL*, *20*(6), D604. doi:10.1051/ocl/2013030

Xu, L., Weathers, P., Xiong, X.-R., & Liu, C.-Z. (2009). Microalgal Bioreactors: Challenges and Opportunities. *Engineering in Life Sciences*, *9*(3), 178–189. doi:10.1002/elsc.200800111

Zang, X., Liu, B., Liu, S., Arunakumara, K. K., & Zhang, X. (2007). Optimum conditions for transformation of *Synechocystis* sp. *PCC* 6803. *Journal of Microbiology (Seoul, Korea)*, *45*, 241–245. PMID:17618230

ADDITIONAL READING

Bagchi, D., & Nair, S. (2017). *Developing New Functional Food and Nutraceutical Products*. Academic Press.

Bilal, M., Rasheed, T., Ahmed, I., & Iqbal, H. M. N. (2017). High-value compounds from microalgae with industrial exploitability – a review. *Frontiers in Bioscience (Scholar Edition)*, *9*(3), 319–342. doi:10.2741490 PMID:28410122

Bleakley, S., & Hayes, M. (2017). Algal proteins: Extraction, application, and challenges concerning production. *Foods*, *6*(5), E33. doi:10.3390/foods6050033 PMID:28445408

Chew, K. W., Yap, J. Y., Show, P. L., Suan, N. H., Juan, J. C., Ling, T. C., Lee, D.-J., & Chang, J.-S. (2017). Microalgae biorefinery: High value products perspectives. *Bioresource Technology*, *229*, 53–62. doi:10.1016/j.biortech.2017.01.006 PMID:28107722

Gouveia, L., Batista, A. P., Sousa, I., Raymundo, A., & Bandarra, N. M. (2008). *Microalgae in novel food products. Food Chemistry Research Developments* (K. N. Papadopoulos, Ed.). Nova Science Publishers, Inc.

Nicoletti, M. (2016). Microalgae nutraceuticals. *Foods*, *5*(4), E54. doi:10.3390/foods5030054 PMID:28231149

Xu, L., Weathers, P. J., Xiong, X.-R., & Liu, C.-Z. (2009). Microalgal bioreactors: Challenges and opportunities. *Engineering in Life Sciences*, *9*(3), 178–189. doi:10.1002/elsc.200800111

KEY TERMS AND DEFINITIONS

Algae: Any of numerous groups of chlorophyll-containing, mainly aquatic eukaryotic organisms ranging from microscopic single-celled forms to multicellular forms 100 feet (30 meters) or more long, distinguished from plants by the absence of true roots, stems, and leaves and by a lack of nonreproductive cells in the reproductive structures: classified into the six phyla Euglenophyta, Crysophyta, Pyrrophyta, Chlorophyta, Phaeophyta, and Rhodophyta.

Animal Feed: Animal feed is food given to domestic animals, especially livestock, in the course of animal husbandry. There are two basic types: fodder and forage. Used alone, the word feed more often refers to fodder.

Antioxidants: A substance that inhibits oxidation, especially one used to counteract the deterioration of stored food products.

Aquaculture: The cultivation of aquatic animals and plants, especially fish, shellfish, and seaweed, in natural or controlled marine or freshwater environments; underwater agriculture.

Dietary Supplements: A product taken orally that contains one or more ingredients (such as vitamins or amino acids) that are intended to supplement one's diet and are not considered food.

Chapter 17
Algal Nanobiotechnology and Its Applications

Ahmed E. Alprol
National Institute of Oceanography and Fisheries, Egypt

ABSTRACT

Nanotechnology has been a catchphrase in recent years. Its expansion into a new field has been phenomenal. Because of their various shapes and sizes, nanoparticles differ from their conventional material. They have a larger surface area, which is necessary for many chemical methods. One of the possible solutions to the above-mentioned limitations is algae-mediated nanoparticle production. This chapter focuses on the use of algae to synthesis nanoparticles and the possible benefits of this technology over traditional methods. The creation of nanoparticles by cyanobacteria, microalgae, and macroalgae is taken into account. Metal nanoparticles derived from algae, such as gold, silver, and iron, have a wide range of applications in environmental pollution treatment, such as heavy metal removal, organic dye degradation, and antimicrobial agents, and examples of major biomedical applications of these algal-derived NPs are presented, among many others.

INTRODUCTION

In comparison to cereal-based crops, algae have been employed for a long time because of their high biomass production rate in a variety of severe habitats. Due to their numerous advantages over various agriculturally based crops, an alga is classified as a third generation biofuel. The idea of using algae to generate energy isn't new. Due to limited culture techniques, the cost of producing algal biofuels has been relatively expensive up until now. However, as time goes on, new methods are being developed for growing algae on a big scale all year, in a variety of climatic zones ranging from tropical to temperate (Batool et al., 2019). Nanotechnology is an area of science and technology that includes manipulating materials after they have been thoroughly consolidated and processed (Daniel, 2004). A variety of physical, chemical, and biological processes can produce nanoparticles. Particles are now generated in a vacuum or using other processing techniques.It uses a variety of processing methods, including atomic, molecular, and particle level processing as well as chemical modification. The main disadvantage is

DOI: 10.4018/978-1-6684-2438-4.ch017

the high expense of these approaches, as well as the low production required to use it as a material or for energy purposes. It is the consequence of a relationship between research and development and the private sector. In the near future, it is likely to have an impact on a variety of commercial industries, as well as a broad aspect of medical, defiance, and toiletries (Invernizzi, 2011). When compared to materials, the resulting (natural, engineered, or incidental) nanomaterials (e.g., nanoparticles (NPs)) have a dimension below 100 nm and unique physical–chemical–biological properties (e.g., large surface-to-volume ratio, surface functionalization, controlled targeting and release). These interesting features make them extremely appealing tools for use in a range of fields (e.g., cosmetology, pharmacy, biotechnology, chemistry, or agriculture).Algae are a large and diverse group of photosynthetic eukaryotic organisms that are both capable of photosynthesis (that is, they produce their own food using light, water, carbon dioxide, or other chemicals) and organization. It allows (that is, they produce their own food using light, water, carbon dioxide, or other chemicals). They are classed as microalgae (unicellular such as diatoms or multicellular) or macroalgae (seaweeds) based on morphological characteristics and can be found in both marine and freshwater habitats, as well as on damp rocks (Sharma et al., 2019). These water plants are non-flowering and basic. Despite the presence of chlorophyll, algae lack many structures (such as real stems, roots, leaves, and vascular tissue) that distinguish land plants (such as bryophytes and tracheophytes) (Borghans et al., 2008). Despite this, they play an important role in aquatic ecosystems (aside from the potential for toxic blooming, which can be controlled by algal-mediated NPs) and are an economically valuable biomass source for a variety of applications (including agricultural, aquacultural, pharmaceutical, cosmetical, biotechnological, energetical, and nanotechnological) (Wang et al., 2017). Green nanotechnology, particularly for algae, has recently been the subject of a lot of research because of its significant benefits over other technologies, such as high metal intake capability, low production costs, and environmental friendliness. It also reduces the need for culture maintenance because it is produced extracellularly. Algae have a diverse range of habitats (Zein & Gharib, 2014). Microalgae, which are microscopic, and macroalgae, which are macroscopic, are two types of algae. They could be eukaryotic or prokaryotic, and they can be found in a variety of environments ranging from freshwater to marine or saline water. Aside from nanoparticle manufacturing, they have the potential for use as natural colours, biofuels, and food enrichment, as well as in a number of other fields such as medicine, agriculture, and pharmaceuticals (Patel et al., 2018).Metal and non-metal NPs synthesized from plants (e.g., seaweeds) and microorganisms (e.g., bacteria) have gotten more attention than conventional physical and chemical synthesis routes, owing to their green assembly during biosynthesis, which uses clean energy processes that are naturally regulated, and therefore overcomes health and environmental toxicity .This chapter focuses on the important factors and methods involved in the algal-mediated synthesis of NPs such as gold, sulfated, palladium, silver, and many others, as well as their applicability in the real world. The fact that algae-mediated creation of magnetic nanoparticles can lead to increased growth and lipid formation in cyanobacteria and microalgae is established in another study. It also features an algal membrane bioreactor using nanoparticles, as well as (possible) biomedical uses. Finally, this chapter aims to aid in the development of safe and long-term nanotechnology generated from algal as aquatic environments.

TYPES OF NANOPARTICLE AND THEIR APPLICATIONS

There are two types of NPs that can be made using algae: organic and inorganic NPs. Poly-lysine (-PL), chitosan (CS), cationic quaternary polyelectrolytes, and quaternary ammonium compounds are examples of organic NPs. In general, organic NPs are less stable at high temperatures than inorganic NPs, hence inorganic NPs are preferred as antimicrobial polymers (LewisOscar et al., 2016). On the other hand, is a cationic L-lysine homopeptide that is highly effective against Gram-positive bacteria and spores of *Bacillus coagulans*, *Bacillus subtilis*, and *Bacillus stearothermophilus* (Patel et al., 2018). Antimicrobial action of CSNPs against viruses, bacteria, and fungi is diverse. They're harmless and biocompatible, and they can help with absorption (Iqbal et al., 2020). Acrylic and methacrylic compounds produce cationic quaternary polyelectrolytes as by-products. When parameters like hydrophobicity, surface charge, and molecular weight are changed, these compounds have a wide range of biological applications (e.g., as antimicrobials) due to their structural plasticity(Muñoz-Bonilla & Fernández-García, 2012).

Quaternary ammonium compounds are excellent disinfectants, and their antibacterial activities are based on chain length (Beyth et al., 2015). Interestingly, the synthesis of organic NPs by algae has received little attention thus far; this could be due to the organic NPs' lack of stability at high temperatures. Tiburu et al. (2017) studied the production of CSNPs in freshwater green algae by isolating deacetylated chitin (Tiburu et al., 2017). By combining chitin with caustic soda and heating it with indirect steam, the chitin was transformed into CS.

X-ray powder diffraction (XRD) examination revealed that the resultant CSNPs have an orthorhombic structure.This study focused on inorganic NPs such as Ag, zinc oxide (ZnO), copper oxide (CuO), gold, and iron oxide (magnetite Fe_3O_4 and/or its oxidised counterpart maghemite -Fe_2O_3). Because of their unique catalytic, electrical, and optical characteristics, metal NPs are the most studied nanomaterials (San & Shon, 2018).They're also easy to synthesize and modify on the surface, allowing for fine-tuning of their size and shape-dependent features (San & Shon, 2018). Because of their photocatalytic activity and propensity to produce reactive oxygen species (ROS), AgNPs have been shown to have potent antibacterial activities against a wide range of fungi, viruses, and bacteria (Garipov et al., 2019). They have anti-biofilm properties and can block UV radiation, making them a suitable coating material for medical and culinary applications (Blecher et al., 2011). CuONPs are utilised as antibacterial agents, although their efficacy is lower than that of AgNPs or ZnONPs (Ruparelia et al., 2008). Their antibacterial activity is based on their ability to break cell membranes and create reactive oxygen species (ROS) (Pelgrift &Friedman, 2013). AuNPs are deemed non-toxic and are mostly employed in gene delivery, biosensing, and cancer therapy (Ghosh et al., 2008). Iron oxide nanoparticles are widely used in magnetic resonance imaging (MRI), immunoassays, tissue healing, and as chemotherapy agents (Gupta & Gupta, 2005).

METHODS FOR SYNTHESIZING NANOPARTICLES WITH MODIFIED PROPERTIES

The techniques "bottom-up" and "top-down" can be used to synthesize NPs in two different ways. Because the initial produced NPs assemble to form a structure or cluster through chemical or biological mechanisms, the "bottom-up" methodology is also called as the self-assembly method. Suitable bulk content is reduced to pieces using chemical or physical processes in "top-down" techniques (e.g., ion etching, lithography).The "bottom-up" method (e.g., chemical/physical vapour deposition, electrochemistry) is preferred because it produces NPs with a more uniform chemical composition and fewer flaws than "top-down" procedures, which produce defective surface structures. The physical and chemical characteristics of NPs can be affected by such defective surface structures (Sudha & Rajamanickam, 2013). Although these methods produce a large quantity of NPs at a low cost, they have some disadvantages (for example, use of hazardous solvents, the processing of damaging by-products, and contamination from precursor chemicals) (Sharma et al., 2018 & Thakkar et al., 2010). Ultrasonication, electron beam, ion implantation, laser radiation, spray pyrolysis, and vapour phase have already been used in the physical synthesis of NPs (Sharma et al., 2018). These physical approaches, on the other hand, have significant limitations that make them less suited for synthesizing NPs (e.g., high cost, low production rate, and the use of a lot of energy to maintain the high temperature and pressure).

Green NPs synthesis employing biological entities (e.g., microorganisms including bacteria, fungus, and yeast, microalgae, or plant extracts including macroalgae) is gaining popularity as a way to overcome the disadvantages of physical and chemical methods used in the synthesis of NPs (Thakkar et al., 2010). Finally, the choice of preparation procedure should take into account not only the physical and chemical properties of the final product (e.g., size usually 100 nm), dispersion, shape (e.g., spherical, star-shaped, or branched NPs), chemical miscibility, and optical properties (surface plasmon resonance (SPR) effect), but also environmental factors.

FACTORS INFLUENCING THE SYNTHESIS OF NANOPARTICLES

The appropriate, effective, and optimal NPs (bio) production process is controlled by a number of chemical, physical, and biological variables. Temperature, pH, metal ion concentrations, reactant concentrations, reaction time, stirring rate, incubation time, capping agents, and the type of microbe or plant extract employed are all examples of these variables (El Nemr et al., 2010). The stability, size, and shape of the NPs can be affected by most, if not all, of these parameters. Furthermore, it is common knowledge that the toxicity of nanomaterials is mostly determined by structural characteristics such as size, shape, composition, and surface chemistry.

pH

To avoid fluctuating shapes and sizes of NPs during the synthesis reaction, the buffer strength (i.e., pH) must remain stable(Singh, M., Kalaivani, Manikandan, S., Sangeetha, 2013). At low pH, the SPR peak widens and deflects towards a longer wavelength region, resulting in a wide range of NPs (often cylindrical or triangular in shape), whereas high pH is excellent for synthesising small-sized NPs, which promotes the production of spherical NPs (Sharma et al., 2018). Using the extract of the brown macroalgae *Sargassum angustifolium* (C. Agardh). Ghaemi & Gholamipour, (2017) investigated the influence of pH by altering it from 2 to 10 during the synthesis of AgNPs. Surprisingly, the stabilising capability of these NPs varied depending on whether the microenvironment was acidic or alkaline. When compared to acidic conditions, a reasonable amount of extremely stable small-sized NPs were synthesized in alkaline conditions (e.g., pH 10). Many large-pearl-sized NPs were generated under alkaline circumstances, and they were considerably more stable than the clustered NPs formed under acidic conditions (Shou et al., 2011).

Temperature

Another important element that controls the synthesis of NPs is temperature. Various chemical processes (e.g., solvothermal, electrochemical, or templating approaches) are heavily influenced by the reaction temperature (Prathna et al., 2012). In general, the synthesis of NPs using green techniques necessitates a temperature of almost 100°C (Vincy et al., 2017). Physical processes demand temperatures greater than 350°C, but chemical synthesis requires moderate temperatures (Edison et al., 2016). At high temperatures, a rapid rate of AuNP production was found, which might be explained by a quicker reduction rate (Huang et al., 2007). However, when their conversion rate grew, their average size shrank (Zeng et al., 2011).

Reaction Time

In the synthesis of NPs, reaction time is a key problem. Indeed, modifying the reaction time in the same experiment can result in varied particle sizes. For example, Ahmad et al. (2012) biosynthesized NPs using *Ananascomosus* (L.) extract, which appeared in 2 minutes and formed spherical NPs with a mean size of 12 nm after a 5-minute process (Ahmad & Sharma, 2012). Furthermore, Aboelfetoh et al. (2017) shown that at room temperature, increasing the contact time/interaction between *Cauler paserrulata* (Forsskl) and silver ion (Ag^{+}) led to an increase in SPR peak intensity and fast synthesis of non-agglomerated AgNPs (Aboelfetoh et al.2017).

NANOPARTICLES TYPES

Nanoparticles were dividing based on two criteria: molecular base and structural base. Further investigation revealed that molecular bases may be divided into organic and inorganic groups, with carbon nanoparticles being a major component of organic nanoparticles. Fullerenes, which include nanotubes and latex balloons, are a type of organic nanoparticle. A carbon nanotube is a tube made from a one-atom thick sheet of graphite rolled into a tube. Single-walled and multi-walled tubes are the two types of tubes (San & Shon, 2018).

Magnetic, semiconductor nanoparticles, and noble metal nanoparticles are combined to form inorganic nanoparticles. Magnetic particles are made up of many parts with magnetic characteristics, such as iron, cobalt, and nickel. An external magnetic field could be used to manipulate them. Because of its numerous applications in the fields like as medicine, catalysis, magnetic particle imaging, wastewater treatment, nanofluidics, and others, it is widely utilized (Kumar et al., 2014).

We can categorise nanoparticle into three primary types' structure and properties: tube-like structures like carbon nanotubes, branching like dendrimers, and spherical structures like liposomes. Liposomes are water-soluble vesicles or tiny bubbles constructed up of phospholipids layer by layer, with the hydrophilic tail of one layer facing outward. While another layer's hydrophobic tail makes contact with the first, the vesicle is able to transport the water-soluble medication into liposomes.

ALGAL ASSISTED NANOPARTICLE BIOSYNTHESIS

Phyconanotechnology is a relatively new branch of nanoscience that involves the synthesis of nanoparticles using algae extract because they are easy to handle, could grow at low temperatures, and are less harmful in nature (Sharma et al., 2016).

In the presence of sunshine, algae are a varied group of unicellular and multicellular plants and algae that carry out photosynthesis. Microalgae are classified as microalgae based on their size, while macroalgae are defined as algae that are larger in size (El-Sherif et al., 2010). They can be found in both marine and freshwater settings, as well as terrestrial ecosystems that are damp. When in a symbiotic relationship with lichen on rocks in a dry and arid location, it aids in the absorption of moisture from the atmosphere. Chlorophyta (green algae), phaeophyta (brown algae), and rhodophyta (red algae) are the three primary divisions of algae, all of which have chlorophyll an in common. The cell walls of chlorophyta, phaeophyta, and rhodophyta include xylans and mannans, alginic acid and fucoidan, and xylan and galactans, respectively, in addition to cellulose (Davis et al., 2003).

Numerous functional groups, such as carbonyl, hydroxyl, carboxyl, sulfonate, thiol, amino, and amidic groups, are found in the stiff cell wall matrices and play an important role in bulk metal reduction and accumulation (Subramaniyam et al., 2015). Few macro and microalgal species have been investigated for the green synthesis of metal nanoparticles such as gold, silver, palladium, and iron nanoparticles in studies (Subramaniyam et al., 2015).

The production of the nanoparticle by algae is a three-step process. It all started with the creation of algal extract, which may be made by dissolving algal extract in an organic solvent or by dissolving algal extract in water. This could be done in the presence of boiling water or by heating for a fixed length of time. The next stage is to make the desired molar solution of the compound from which the nanoparticle will be generated, followed by incubation of the ionic metallic compound and maintenance of the required environment, or we might give continuous stirring. Our algae and ionic metallic compound of choice could have a significant impact on our necessary due diligence and judgments when moving through the intracellular and extracellular production processes (Subramaniyam et al., 2015). Because of the reducing agent potential of many algal aqueous extracts, extracellular nanoparticle production may occur. It has the needed potential because to the inclusion of proteins, polysaccharides, reducing sugar, pigments, and other compounds that can stimulate metal ion reduction and then precipitate as nanoparticles. While in the case of intracellular production, the ability to reduce the ionic metal compound is due to factors such as algal metabolism, which includes respiration and photosynthesis, which might be advantageous in a reduction situation. NADPH and NADPH reductase were found to be enzymatically active in a study. While in the case of intracellular production, the ability to reduce the ionic metal compound is due to factors such as algal metabolism, which includes respiration and photosynthesis, which might be advantageous in a reduction situation. It was discovered in a study that NADPH and NADPH reductase enzyme located in the electron transport system of either photosynthesis or the respiratory system had the ability to carry out metallic ion reduction. Another study discovered that nitrogenous enzyme has the ability to reduce metallic ion compounds such as Au ions to nanoparticles (Zeng et al., 2011). Algal biomass can be used to make nanoparticles through extracellular and intracellular mechanisms.

In a second investigation, microalgae may lead to the intracellular creation of nanoparticles along a dose-dependent line, followed by the subsequent reduction of metallic ion particles into nanoparticles and their subsequent release. As a result, it was proven that microalgae have evolved into a state that might lead to the synthesis of bimetallic nanoalloys in a very limited and controlled manner. pH, kind of metal, molar concentration, temperature, and duration of reducing agent have all been found to be useful in determining the shape and size of nanoparticles over time (Kumar et al., 2014).

Microalgae aided in the biosynthesis of metal nanoparticles: Silver nanoparticles have been discovered to have antifungal, anti-cancer, antibacterial, wound-healing, and catalytic properties, among other things. These features have led to possible applications in biosensors, medication delivery, ultrasonic medical imaging, and other fields. It was recommended in a study that the Ag nanoparticle be used to generate an antibacterial cotton fabric. They've also been shown to be useful in the development of biosensors that detect glucose levels. A variety of algae, as well as cyanobacteria species, have the ability to produce nanoparticles (Huang et al., 2007). Various microalgae could be used to synthesise. Sharma et al. (2016) were successful in forming Ag nanoparticles with high yield using green algae species such as *Chlorella vulgaris*.

Merin et al. (2010) used a range of species, including *Chlorella salina*, *Tetraselmisgracilis,* and different diatoms, to produce and optimise the result. AgNPs produced in this way have been proven to be harmful to *Proteus vulgaris*, *E. coli*, *Pseudomonas aeruginosa*, *Klebsiella sp*., and other bacteria (Ghaemi & Gholamipour, 2017).

Chlorella pyrenoidosa is discovered to use in the production of nanoparticles with a diameter of 5-10 nm, whereas Chlorella vulgaris is said to be viable at a concentration of AgNO3 of 10-3. In the continuously stirred environment, it also resulted in polydisperse silver nanoparticles as a result of both intracellular and extracellular processes. The particle's size ranges from 8 to 20 nanometers (Jiang et al., 2011).

MACROALGAE-AIDED NANOPARTICLE BIOSYNTHESIS

Due to the fact that gold nanoparticles have more advantageous qualities than any other material, they have always been in great demand. The features that give it an advantage over others are its ease of manufacture, capacity to be conjugated with various biocompatible compounds, light scattering, and non-toxicity, all of which contribute to its wide range of applications. It is now used in a variety of fields, including diagnosis, photonics, electronics, medicines, optics, imaging, and catalysis. It can be used to determine which analytes are present in the sensor. It has the potential to suppress cancer or tumours once activated (Ahmad &Sharma, 2012). Until date, numerous microalgae and diatoms have been employed to generate gold nanoparticles of varied shapes and sizes according to our needs. When incubated with 1 ml of $HAuCl_4$, *C. vulgaris* could produce a 90 percent yield (10 mM).The transformation of green to pink colour was utilized to validate the creation of gold nanoparticles (Prathna et al., 2012). *Tetraselmis-suecica* has been discovered to have the ability to reduce gold nanoparticles. Confirmation comes in the form of a colour change from red to yellow. When *Tetraselmiskochinensis* was exposed to 28-29°C, it produced gold nanoparticles with an average size of 15nm. Another study discovered that vigorous stirring of *C. pyrenoidosa* can result in AuNPs (Vincy et al., 2017). Some of the important study in the field of algal nanotechnology has been displayed in Table 1.

PHYCOSYNTHESIS OF METALLIC NANOPARTICLES

Gold Nanoparticles (AuNPs)

The potential of *Turbinari aconoides* and *Sargassum tenerrimum*as Brown algae to produce gold nanoparticles was investigated. The presence of hydroxyl groups in the algal extract was shown to aid the reduction of Au(III) to elemental gold and act as a capping agent for the synthesised spherical shaped gold nanoparticle of size. The presence of hydroxyl groups in the algal extract promoted the reduction of Au(III) to elemental gold and served as a capping agent for the generated spherical shaped gold nanoparticles with sizes ranging from 5-57 nm, according to the findings (Ramakrishna et al., 2016). Another brown algae species, Cystoseira baccata, was discovered to be capable of synthesising polycrystalline spherical gold nanoparticles with a size of 8.4 ± 2.2 nm.

Table 1. Lists algal species that could be used to create metal nanoparticles.

Algal species	Type of NP	Size of NP	Applications	References
Caulerpa racemosa	Ag	~25nm	Catalytic degradation Of methylene Blue	(Edison et al., 2016)
Sargasum wightii (Greville) Macroalgae	Au	8--12		(Singaravelu et al., 2007)
Chlorella vulgaris	Pd	2to15nm	Catalytic activity	(Eroglu et al.,2013)
Turbinaria conoides; Sargassum tenerrimum.	Au	27–35nm.	Catalytic reduction of nitro compounds	(Ramakrishna et al., 2016)
Codium capitatum	Ag	30nm	-	(Kannan et al.,, 2013)
Spirulina platensis	Ag	12		(Mahdieh et al.,, 2012)
Hypneam usciformis	Ag	16to 42nm	Antibacterial activity	(Vadlapudi & Amanchy, 2017)
Bifurcaria bifurcata	Cu			(Azizi et al., 2013)
Laurenciapapillosa	Au	3.5–53nm	-	(Montasser et al., 2017)
Padina gymnospora	Ag	25-40nm	Antibacterial activity	(Shiny et al., 2013)
Padina pavonica (Linn.)	Ag	10to 72nm	Microbicidal activity	(Sahayaraj et al., 2012)
Sargassum bovinum	Pd	5to10nm	Electro chemical reduction of H_2O_2	(Momeni & Nabipour, 2015)
Turbinaria conoides	A\g	2-17 nm 2-19 nm	Antimicrobial fouling Activity	(Vijayan et al., 2014)
Chlorophyta ; Caulerparacemosa	Ag			(Kathiraven et al., 2015)
Sargassum muticum	Ag	42.30-98.56nm	Insecticidal activity	(Moorthi et al., 2018)
Spirulina platensis	Ag	~5nm	Antibacterial activity	(Suganya et al., 2015)

The presence of protein acts as a capping agent and prevents the agglomeration of gold nanoparticles (González-Ballesteros et al., 2017). It was suggested that polysaccharides and polyphenols contain hydroxyl as a functional group that aids in the bioreduction of bulk material into its elemental form, and the presence of protein acts as a capping agent and prevents the agglomeration of gold nanoparticles *Pithophora Oedogonia*, a green algae, created 33nm spherical gold nanoparticles (Li & Zhang, 2016).

When aqueous extract of red seaweed *Corallinaofficinalis* was mixed with gold salt solution, El-Kassas and El-Sheekh (2014) saw a shift in the colour of the solution from brown to red, confirming the synthesis of gold nanoparticles. The nanoparticles were 14.611 nm in size and spherical in form. The hydroxyl and carbonyl groups were also found to aid in the reduction and stability of nanoparticles. The activity of the nitrate reductase enzyme was discovered to be responsible for the bioreduction of gold ion to elemental gold (Oza et al., 2012). Using different fractions of aqueous extract of brown algae *Laminaria japonica* by serial dilutions with distilled water, Ghodake and Lee (2011) reported the concentration-dependent extracellular production of gold nanoparticles within 20 minutes at 37°C.

Silver Nanoparticles (AgNPs)

Silver nanoparticles have a wide range of applications in the medical, industrial, and commercial sectors due to their unique optical, electrical, and catalytic capabilities. They are commonly employed as anti-inflammatory, antiviral, antibacterial, and antifungal drugs, as well as nano-silver coated surgical equipment and a variety of other applications (Azizi et al., 2013). Aboelfetoh et al. (2017) employed a green marine algae extract called *Caulerpaserrulata*to reduce silver ion to 10 2 nm spherical silver nanoparticles.

Selvam and Shivkumar (2015) reported the manufacture of cubic shaped silver nanoparticles of size 2- 55.8 nm from the aqueous extract of the red alga *Hypneamusci formis*, and speculated that the presence of peptides may be responsible for the reduction of silver nanoparticles. The antibacterial activity of silver nanoparticles of size 5–25 nm generated using the marine algae *Caulerparacemosa* has also been evaluated against human pathogens such as *P. mirabilis* and *S. aureus*.

Iron Nanoparticles (FeNPs): Only a few studies on the synthesis of iron metal nanoparticles from algae species have been published. For the first time, Subramaniyam et al. (2015) investigated the synthesis of iron nanoparticles from the soil microalga *Chlorococcum sp.* MM1. The produced spherical shaped iron nanoparticles were 20-50 nm in size. FTIR analysis confirmed the existence of a functional group attached to polysaccharides and glycoprotein, which is thought to act as a reducing and capping agent in the bioreduction process. Microalga, *Chlorococcum sp.* aqueous extract was allowed to react with 0.1M iron chloride solution for 48 hours. The bioreduction and capping of iron nanoparticles were caused by carbonyl and amine linkages from polysaccharides and glycoproteins found in the algal cell wall.

Gold nanoparticles were examined for antibacterial efficacy against gramme negative and gramme positive microorganisms (*Pseudomonas aeruginosa*, *Klebsiellaoxytoca*, *Enterobacterfaecalis*, *Klebsiellapneumoniae*, *Vibrio parahaemolyticus*, *Vibrio cholerae*, *Escherichia coli*, *Salmonella typhii*, *Salmonella paratyphi*, and *Proteus vulgaris*.

ALGAL-MEDIATED NANOPARTICLE SYNTHESIS OF OTHER TYPES

By reducing metal ions (Ag^+ or Au^+), algal-mediated synthesis of metallic NPs other than AgNPs and AuNPs is also conceivable. Different NPs, such as cadmium sulphide (CdS), copper oxide (CuO), iron oxide (Fe_2O_3), palladium (Pd), and zinc oxide nanoparticles (ZnONPs), can be made from microalgae, macroalgae, or cyanobacteria.

CdSNPs could be synthetized this technique by utilizing the *cyanobacterium Phormidiumtenue* (Gomont), which has unique photochemical and photophysical characteristics (Mubarak et al., 2012). The colour of the cadmium ions (Cd^{2+}) bioreduction reaction was observed until it changed from yellow to orange, indicating the synthesis of CdSNPs (Mubarak et al., 2012). These NPs were spherically formed, with a mean particle size of 5 nm, according to TEM examination.

Mandal et al. (2016) demonstrated the algal-mediated synthesis of CdSNPs using *S. platensis (Gomont)* that had been cultivated in the laboratory at room temperature (Mandal et al., 2016). The pellets were homogenised with liquid N2 in an autoclaved mortar for extraction after harvesting and shade drying in petri plates. After centrifuging the mixture for 20 minutes at 4 degrees Celsius, the supernatant was saved for further examination.

For the synthesis of CdSNPs, a solution of cadmium nitrate ($Cd(NO_3)_2$) was added to the algal mixture, and the development of yellow hue confirmed the creation of CdSNPs (Mandal et al., 2016). The synthesis was intracellular, and the NPs were spherical in shape with an average diameter of 8–12 nm, according to the TEM image (Mandal et al., 2016). Furthermore, CuONPs can be made from an extract of the brown macroalga Bifurcaria bifurcate (R. Ross), which has antibacterial, antioxidant, and antitumoral properties (Abboud, et al., 2014). After bioreduction of copper ions (Cu^{2+}) with water-soluble extracts abundantly present in the algal extract, the biosynthesized CuONPs were observed by a colour change from deep blue to dark red, which was conventionally due to the SPR phenomenon. CuONPs had a spherical morphology with particle sizes ranging from 5 to 45 nm, with a mean size of 20.66 nm, as seen in TEM micrographs.

Another work performed by El-Kassas et al. (2016) found that two brown seeweeds, *Padina pavonica* (Linnaeus) and *Sargassum acinarium* (Linnaeus), may be used to synthesise Fe_3O_4NPs. The samples were cleaned to eliminate impurities, dried at 20 degrees Celsius, and pulverised. The co-precipitation approach was then used to make Fe_3O_4NPs (El-Kassas et al., 2016). The algal extract was treated with $FeCl_3$ solution, and NPs were produced through a reduction procedure.

Application of Metal Nanoparticles in Environmental Pollution Remediation

Algae are known as "bio-nano-factories" as of their capacity to synthesize metal nanoparticles. The capacity of algae to aggregate and reduce metal particles makes them a predominant candidate for nanoparticles biosynthesis.

A wide-ranging of plant extract are operated for the biosynthesis of ZnO NPs comprising marine macroalgae (Azizi et al., 2014). Based on their broad biological applications, the massive field of nanotechnology and nanoparticles has extended beyond imagination. The compounds acquired from macroalgae are accounted for a wide scope of biological applications for example antibacterial, antifouling, and anticoagulant applications (Ramaswamy et al., 2016). Some of these recent studies have shown that marine algae are ideal for via silver, gold, and zinc nanoparticles (Ates et al., 2013 and Amirante et al., 2018).

Platinum, silver, and gold nanoparticles are widely employed in items that come into direct contact with the human body, such as shampoos, detergents, soaps, shoes, cosmetics, and toothpaste, in addition to medical and pharmaceutical applications (Singh et al., 2013).

Antimicrobial activity of silver nanoparticles of size 5–25 nm synthesized employing the marine algae *Cauler paracemosa* was tested against human pathogens such as *P. mirabilis* and *S. aureus*. The study found that silver nanoparticles not only target the cell membrane, but also enter the bacterial cell and disrupt cell division in the respiratory chain, ultimately leading to cell death (Kathiraven et al., 2015). The maximum zone of inhibition was seen against *Pseudomonas aeruginosa* (MTCC 2581) (17.2 mm), followed by Escherichia coli (MTCC 443).

Silver nanoparticles of size 34.03 nm were produced from an aqueous extract of the green alga *Pithophora oedogonia* (16.8 mm). Gold nanoparticles were investigated for antibacterial efficacy against gramme negative and gramme positive microorganisms (*Pseudomonas aeruginosa, Klebsiell aoxytoca, Enterobacter faecalis, Klebsiell apneumoniae, Vibrio parahaemolyticus, Vibrio cholerae, Escherichia coli, Salmonella typhii, Salmonella paratyphi,* and *Proteus vulgaris).*

The maximum zone of inhibition against *E. faecalis* (11 mm) was seen, while the least zone of inhibition against *K. pneumoniae*was recorded (6 mm). There was no zone of inhibition in E. coli (0 mm).

Green-synthesised nanoparticles have been proven to be more effective against gram-negative bacteria than gram-positive bacteria (Rajathi et al., 2012).

Silver nanoparticles made from micro algae *C. calcitrans*, *C. salina*, *I. galbana*, and *T. gracilis* inhibited human pathogens *Klebsiella spp*., *Proteus vulgaris*, *Pseudomonas aeruginosa*, and *E. coli* and act as a potent antibacterial agent (Merin et al., 2010).

Hassaan et al. (2022) Gelidium pulchellum algae extract was used as a bioreducing agent to synthesize zinc oxide nanoparticles (ZnO NPs) in this work.

Subramaniyam et al. (2015) employed *Chlorococcum sp*. MM11 to produce and evaluate iron nanoparticles with sizes ranging from 20 to 50 nanometers. The study found that iron nanoparticles transformed 92 percent of 4 mg L1 hexavalent chromium to Cr (III), but bulk iron only lowered 25 percent, indicating that iron nanoparticles are far more effective at eliminating contaminants from the environment. Silver nanoparticles made from the green algae species *Caulerpaserrulata* were also shown to be effective in the catalysis of Congo red (CR) dye because they had a wide surface area for electron exchange between electron source and acceptor (Aboelfetoh et al., 2017). Silver nanoparticles synthesised from freshwater microalgae *Chlorella pyrenoidosa* have been tested for photocatalytic activity against methylene blue dye, and its application in the treatment of effluent containing hazardous dye as a result of chemical processes in the industrial sectors has been suggested (Azizi et al., 2013).

MAJOR BIOAPPLICATIONS AND UNDERLYING MOLECULAR MECHANISMS OF ALGAL-MEDIATED METALLIC NANOPARTICLES SYNTHESIS

The algal-mediated production of inorganic NPs has been used in a various applications. Metallic nanoparticles such as AgNPs and AuNPs, have been widely considered for usage in biotechnological, medicinal, cosmetical, and pharmaceutical applications. Figure (1) depicts the main applications of these NPs in the realms of medicine and biotechnology.

The physicochemical reactivity of (metallic) NPs causes oxidative stress either directly or indirectly due to the formation of free radicals or reactive oxygen species (e.g., superoxide radical anions and hydroxyl radicals) as a result of the resulting formation of free radicals or reactive oxygen species (e.g., superoxide radical anions and hydroxyl radicals) (i.e., through activation of oxidative enzymatic pathways). Although some metal oxide NPs, such as titanium dioxide NPs (TiO_2NPs) used in cosmetics, food additives, and cancer therapy, are water insoluble, they can cause increased toxicity (e.g., oxidative stress, phototoxicity, genotoxicity, and immunotoxicity), especially at greater concentrations than 100 g/mL.

Because TiO_2NPs do not release toxic ions like ZnO do, their toxicity (i.e., ROS, mitochondrial depolarization, plasma membrane leakage, intracellular calcium influx, and cytokine production) could be related to their size-dependent interaction/adsorption with intracellular biomolecules (Fard et al., 2015). Additionally, because the suppression of biomolecule adsorption and the formation of hydroxyl radicals (•OH) during the photoactivation process, phototoxicity of these NPs might be reduced by coating their surfaces with CS (Fard et al., 2015).

Figure 1. Major applications of algal mediated synthesis of metallic NPs.

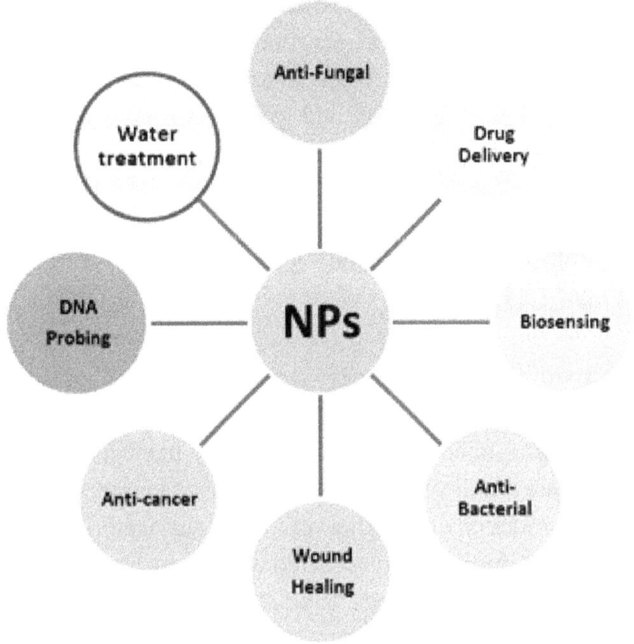

NANOPARTICLES IN AN ALGAL MEMBRANE BIOREACTOR

In terms of energy production and water cleansing, algae culture in wastewater is a promising technique. Many algal species may thrive in wastewater because micronutrients (trace metals and vitamins including cyanocobalamin and thiamin) and macronutrients (NO^{-3}, PO_4 salts with Ca, Na, K, and NH_4^+) are readily available. Solutions are created by combining these chemical salts (nutrients) with water. The materials required for algal growth are provided by nutrient solutions (together with light and carbon dioxide) (Ashour et al., 2021). As a result, nutrients from wastewater are removed, and algal biomass is used to generate electricity (Grima et al., 2003). There are a variety of strategies for collecting algal biomass, including sedimentation, air flotation, and centrifugation, in addition to chemical flocculation, however these processes are expensive on a wide scale due to their high cost (Ghoneim et al., 2014).

Membrane technology, in which high-density algae cultivation is simply done by membrane bioreactor (Hu et al., 2010), is the most advantageous strategy for algae cultivation and biomass harvesting among advanced techniques. Membrane technology has the advantage of requiring no additional chemicals such as coagulants for membrane filtration, allowing for water reuse after filtration and simplifying algal biomass separation (Rı´os et al., 2012).

Furthermore, when compared to traditional approaches, improved algal biomass recovery may be accomplished without cell injury, and algae harvesting requires less energy.

Membranes comprised of polysulfone (PSF), polyvinylidene fluoride (PVDF), and polyethersulfone (PES) are widely utilised due to their chemical and physical stability; yet, membrane fouling is a problem caused by a hydrophobic interaction between membrane materials and microbial cells (Maximous et al., 2009). Many approaches exist for improving membrane hydrophilicity and decreasing membrane fouling, including plasma treatment (Kim et al., 2011), surface coating (Madaeni & Ghaemi, 2007), and incorporation of nanomaterials (Yin et al., 2013). Nanoparticles improve membrane hydrophilicity and prevent membrane fouling, according to research. Blending carbon nanotubes with TiO_2 nanoparticles with PSF hollow fibre membranes (HFMs) improves surface modification (hydrophilicity) and antifouling properties, for example (Yin et al., 2013). Madaeni and Ghaemi (2007) coated the reverse osmosis membrane's PVA top layer with hydrophobicity and fouling can be decreased by incorporating these nanoparticles with membranes because of these properties.

ALGAL-MEDIATED SYNTHETIZED AGNPS: MEDICAL APPLICATIONS AND MECHANISMS

AgNPs have a high conductivity and are extremely sensitive to metal surface absorption. As a result, AgNPs are frequently linked to a variety of helpful theranostic applications, such as biosensing, imaging, drug administration, wound healing, cancer, and microbial (e.g., bacteria, certain fungus) therapies (Ahmad & Sharma, 2012). AgNPs are employed in a wide range of consumer products, including cosmetics, electronics, textiles, and food products, due to their effective antibacterial qualities (Fard et al., 2015). Topical ointments (e.g., creams) or implants impregnated with Ag polymers, for example, were developed to prevent infections of burned and injured areas since AgNPs have been shown to cause bacterial cell membrane lysis (Ahmad & Sharma, 2012). Furthermore, if biosynthesized NPs can bypass the immune system by overcoming biological barriers and the complicated tumour microenvironment (TME), they can be used as effective regulated and disease-targeted drug delivery systems (Lim et al., 2019).

Khalid et al. (2019)showed antibacterial, antifungal, anti-cancerous, and antiviral activities of AgNPs synthesised from an ethanolic extract of three freshwater microalgae strains, namely HM1 (DHM1), HM2 (DHM2), both obtained from Dictyosphaerium sp. (Nägeli), and HM3 (PHM3) from *Pectinodesmus sp.* These AgNPs were found to have significant efficacy against 14 bacterial strains, *Candida albicans*, the hepatocellular carcinoma (HepG2) and breast cancer (MCF7) cell lines, and the Newcastle Disease Virus (NDV) on Huh7-infected cells (Khalid et al., 2019). Unfortunately, as in several other investigations, the underlying molecular pathways were not discussed in this article. Furthermore, Venkatesan et al. (2016) demonstrated that AgNPs synthesised by extracts from the brown seaweed Ecklonia cava (Kjellman) elicit a significant anti-bacterial activity against *Escherichia coli* and *Staphylococcus aureus*, as well as an efficient antioxidant activity in vitro and an anti-cancer activity against human cervical (HeLa) cells through an apoptosis-mediated mechanism (Venkatesan et al., 2016).

MEDICAL APPLICATIONS AND MECHANISMS OF ALGAL-MEDIATED SYNTHETIZED GOLD NP

AuNPs are one of the most promising inorganic NPs that have attracted scientific and technological interests due to their ease of synthesis, chemical stability, and excellent optical and electronic properties (Fard et al., 2015).These unique properties make them appealing tools for biomedical applications in radiology as radiation enhancer, intargeted drug delivery, in biosensing for biomolecular hypersensitive detection (biosensing), in cancer diagnostics and cancer therapy by hyper thermal treatment (Khan et al., 2014).

The application of AuNPs in such fields depends primarily on the capacity to synthesize particles with regulated shape, mono dispersity, size, stability, and chemical composition.

Interestingly, they can be easily prepared by various methods (i.e., physical, chemical and the growing safe and effective biological approach) due to their exceedingly small diameter, and thereby can be easily transferred to tissues and cells just like DNA and proteins (Chugh et al., 2018). Murugesan et al. (2015), for example, demonstrated that AuNPs produced by the red macro algae *Hypneamusci formis* (Wulfen), which is also known as a highly opportunistic invader that causes enormous floating blooms, can inhibit *Aspergillus niger* and Mucorspp (Murugesan et al., 2015).

While AgNPs, which have sparked a lot of interest as antimicrobials, AuNPs have mostly contributed to a new field of research, cancer nanomedicine. This is due to the fact that, in contrast to typical anticancer medications, NPs offer a focused strategy that avoids side effects (Chugh et al., 2018). Although the majority of in vitro studies have shown that AuNPs are non-toxic to cells, their cytotoxicity is dependent on their absorption and intracellular distribution, which is dependent on the size and shape of AuNPs as well as the ligands surrounding them (Fard et al., 2015).The cytotoxicity observed in Balb/3T3 murine fibroblasts treated with 5-nm AuNPs but not 15-nm AuNPs was explained by the high quantity of small AuNPs taken up by cells compared to the bigger particle (Coradeghini et al., 2013). Due to their extensively exposed surface areas and flaws, anisotropic AuNPs (e.g., nanorods, nanourchins, and nanocages) are said to have a higher potential for oxidation than isotropic AuNPs (Kohout et al., 2018). Furthermore, some research suggests that spherical AuNPs are better suited for biomedical applications (Fard et al., 2015).

ALGAE AS A WATER CONDITIONER

Algae as a water conditioner is a technique that uses microalgae, macroalgae, or their extracts and phytochemical substances as water optimizers, resulting in better growth, pathogenic bacteria management, disease resistance, feed efficiency, and immune activation in farmed aquatic animals (Hassan et al., 2021). Green water technology is a method of rearing aquatic animals/larvae in the presence of microalgae that has been linked to higher survival and growth rates than larvae maintained in clear water (Navarro and Sarasquete, 1998 and Mansour et al., 2022). Green water technology is used to manage the aquaculture growing environment (Yang et al., 2020). Microalgae, bacteria, and zooplankton were numerous in the rearing ponds where fish larvae were reared with this system.

This technology can be based on fertilizer-stimulated natural microalgal populations, or farmed microalgae strains put into culture tanks if the system water has been pre-treated to exclude competing microorganisms (Shields and Lupatsch, 2012). The improved growth and survival rates in this technique are attributed to (1) better direct and indirect feeding of larvae, (2) lower stress levels, (3) improving environmental conditions for feeding by increasing turbidity, light, and enhanced visual contrast, (4) increased oxygenation rates, and (5) increased antibacterial properties in rearing ponds, according to several authors (Yang et al., 2020). The generation of bioactive compounds by algal cells, which have antibacterial and antioxidant components that inhibit virulence genes, is one of the mechanisms linked to the profitable and beneficial effects of green water (Kokou et al., 2012). The most common microalgal species employed for this purpose are Chlorella, N. gaditana, Nannochloropsis sp., I. galbana, Isochrysis sp., and Tetraselmis sp. (Tendencia et al., 2003).

Seaweed extract, on the other hand, has recently showed great promise as a water conditioner. On this crucial topic, just a few research have been undertaken. Previous research found that using a commercial seaweed liquid extract (TrueAlgaeMax, TAM®, made from U. lactuca, J. rubens, and P. capillacea) as an aquaculture water conditioner improved growth performance, zooplankton community and abundance, and non-specific immune responses in O. niloticus challenged with A. hydrophila (Hassan et al., 2021).

FUTURE RESEARCH DIRECTIONS

Despite every one of the research, there is still opportunity for improvement. The harvesting and lipid extraction costs are the most significant, but they can be decreased by combining traditional and modern approaches, such as the use of nanotechnology. Therefore, there is potential in elucidating the favorable effects of nanomaterial's on the biodiesel manufacturing process from algal biomass. Similarly, utilizing wastewater as a culture medium for cultivation can be considered a single solution to a range of issues. Parallel research is being conducted to improve all aspects of the potential dangers. Using algae nanohybrids in wastewater farming could be an expense way to produce biodiesel. Furthermore, because to the enormous floor area, big-scale microalgae cultivation must be done off-shore. More research is needed in this field before these environmentally friendly NPs can be safely translated into medication. Despite the successful bioreduction of metal salts to their elemental state and their application in various fields, the molecular mechanism involved during synthesis is unknown, and several other species of algal biomass remain unexplored. As a result, more research is needed to identify the most suitable and efficient biological source for nanoparticle synthesis of smaller diameters, as well as the detailed mechanism involved. Moreover, the cutting-edge capability of genome editing will be used in the future to modify the genetics of algae to express desired characteristics for viable goods.

CONCLUSION

The biological synthesis of metallic nanoparticles is an environmentally benign and cost-effective alternative to the conventional physical and chemical methods. Several scientists have looked at the biological potential of algae to produce nanoparticles of various shapes and sizes under varied settings. It was discovered that algae biomolecules act as a reducing and capping agent without the use of toxic compounds, resulting in stabilised nanoparticles. Excessive usage of hazardous organic and inorganic pollutants degrades the overall quality of the environment over time, necessitating the development of environmentally friendly techniques to reduce the negative effects of pollutants.

There has been a significant increase in the amount of attention given to nanotechnology in the last seven to eight years.

Green nanotechnology and nanotechnology's second generation are rapidly growing fields. This chapter attempts to highlight the work that has been done in the field of phyconano technology. According to the literature, algae-mediated nanotechnology could be the break we need to commercial the innovative characteristics of nanotechnology. It has stated that it will make a big impact in the near future. Algae have a diverse range of habitats, resulting in additional potential that could be used to improve our results. In comparison to cyanobacteria and green algae, less research has been done on red algae. It is also yet to be used at the cellular level.The biological approach of synthesis of metallic NPs utilising algae is discussed. AgNPs, AuNPs, and FeNPs were given special attention because of their unusual properties, which have piqued scientific and technological curiosity. For this reason, the major parameters influencing a steady and effective biosynthesis, as well as the most commonly used algae, were detailed. Some metallic NP uses in nanomedicine, as well as their underlying primary molecular mechanism-induced cytotoxicity, were discussed. Algal-mediated NPs synthesis is a quick, cost-effective, and efficient technique that has allowed nanotechnologists to synthesise desirable nanomaterials using renewable energy processes.

REFERENCES

Abboud, Y., Saffaj, T., Chagraoui, A., El Bouari, A., Brouzi, K., Tanane, O., & Ihssane, B. (2014). Biosynthesis, characterization and antimicrobial activity of copper oxide nanoparticles (CONPs) produced using brown alga extract (Bifurcaria bifurcata). *Applied Nanoscience*, *4*(5), 571–576. doi:10.100713204-013-0233-x

Aboelfetoh, E. F., El-Shenody, R. A., & Ghobara, M. M. (2017). Eco-friendly synthesis of silver nanoparticles using green algae (Caulerpaserrulata): Reaction optimization, catalytic and antibacterial activities. *Environmental Monitoring and Assessment*, *189*(7), 349. doi:10.100710661-017-6033-0 PMID:28646435

Aboelfetoh, E. F., El-Shenody, R. A., & Ghobara, M. M. (2017, July). Eco-friendly synthesis of silver nanoparticles using green algae (Caulerpa serrulata): Reaction optimization, catalytic and antibacterial activities. *Environmental Monitoring and Assessment*, *189*(7), 349. doi:10.100710661-017-6033-0 PMID:28646435

Ahmad, N., & Sharma, S. (2012). Green synthesis of silver nanoparticles using extracts of Ananas comosus. *Green Sustain. Chem.*, *2*(04), 141–147. doi:10.4236/gsc.2012.24020

Amirante, R., Demastro, G., Distaso, E., Hassaan, M. A., Mormando, A., Pantaleo, P., Tamburrano, A. M., Tedone, L., & Clodoveo, M. L. (2018). Effects of ultrasound and green synthesis ZnO nanoparticles on biogas production from olive pomace. *Energy Procedia*, (148), 940–947.

Ashour, M., Alprol, A. E., Heneash, A. M. M., Saleh, H., Abualnaja, K. M., Alhashmialameer, D., & Mansour, A. T. (2021). *Ammonia Bioremediation from Aquaculture Wastewater Effluents Using Arthrospira platensis NIOF17 / 003 : Impact of Biodiesel Residue and Potential of Ammonia-Loaded Biomass as Rotifer Feed*. Academic Press.

Ates, M., Daniels, J., Arslan, Z., Farah, I. O., & Rivera, H. F. (2013). Comparative evaluation of impact of Zn and ZnO nanoparticles on brine shrimp (Artemiasalina) larvae: Effects of particle size and solubility on toxicity. *Environmental Science. Processes & Impacts*, *15*(1), 225–233. doi:10.1039/C2EM30540B PMID:24058731

Azizi, S., Ahmad, M. B., Namvar, F., & Mohamad, R. (2014). Green biosynthesis and characterization of zinc oxide nanoparticles using brown marine macroalga Sargassum muticum aqueous extract. *Materials Letters*, *116*, 275–277. doi:10.1016/j.matlet.2013.11.038

Azizi, S., Namvar, F., Mahdavi, M., Ahmad, M. B., & Mohamad, R. (2013). Biosynthesis of silver nanoparti- cles using brown marine Macroalga, Sargassum Muticum aqueous extract. *Materials (Basel)*, *6*(12), 5942–5950. doi:10.3390/ma6125942 PMID:28788431

Azizi, S., Namvar, F., Mahdavi, M., Ahmad, M. B., & Mohamad, R. (2013). Biosynthesis of silver nanoparticles using brown marine macroalga, Sargassum muticum aqueous extract. *Materials (Basel)*, *6*(12), 5942–5950. doi:10.3390/ma6125942 PMID:28788431

Batool, A., Menaa, F., Ali, K. B., Uzair, B., & Menaa, B. (2019). Progress and prospects in translating nanobiotechnology in medical theranostics. *Current Nanoscience*, *15*, 1–23.

Beyth, N.; Houri-Haddad, Y.; Domb, A.; Khan, W.; Hazan, R. (2015). Alternative antimicrobial approach: Nano-antimicrobial materials. *Evid. Based Complementary Altern. Med.*, 246012.

Blecher, K., Nasir, A., & Friedman, A. (2011). The growing role of nanotechnology in combating infectious disease. *Virulence*, *2*(5), 395–401. doi:10.4161/viru.2.5.17035 PMID:21921677

Borghans, L., Duckworth, A. L., Heckman, J. J., & Ter Weel, B. (2008). The economics and psychology of personality traits. *The Journal of Human Resources*, *43*(4), 972–1059. doi:10.3368/jhr.43.4.972

Chugh, H., Sood, D., Chandra, I., Tomar, V., Dhawan, G., & Chandra, R. (2018). Role of gold and silver nanoparticles in cancer nano-medicine. *Artificial Cells, Nanomedicine, and Biotechnology*, *46*(sup1), 1210–1220. doi:10.1080/21691401.2018.1449118 PMID:29533101

Coradeghini, R., Gioria, S., García, C. P., Nativo, P., Franchini, F., Gilliland, D., Ponti, J., & Rossi, F. (2013). Size-dependent toxicity and cell interaction mechanisms of gold nanoparticles on mouse fibroblasts. *Toxicology Letters*, *217*(3), 205–216. doi:10.1016/j.toxlet.2012.11.022 PMID:23246733

Daniel, M. C., & Astruc, D. (2004). Gold nanoparticles: Assembly, supramolecular chemistry, quantumsize-related properties, and applications toward biology, catalysis, and nanotechnology. *Chemical Reviews*, *104*(1), 293–346. doi:10.1021/cr030698+ PMID:14719978

Davis, T. A., Volesky, B., & Mucci, A. (2003). A review of the biochemistry of heavy metal biosorption by brown algae. *Water Research*, *37*(18), 4311–4330. doi:10.1016/S0043-1354(03)00293-8 PMID:14511701

Edison, T. N. J. I., Atchudan, R., Kamal, C., & Lee, Y. R. (2016). Caulerparacemosa: A marine green alga for eco-friendly synthesis of silver nanoparticles and its catalytic degradation of methyleneblue. *Bioprocess and Biosystems Engineering*, *39*(9), 1401–1408. doi:10.100700449-016-1616-7 PMID:27129459

Zein, N., & Gharib, S. M. (2014). Spatial and temporal patterns of phytoplankton composition in Burullus Lagoon, Southern Mediterranean Coast, Egypt. *Egyptian Journal of Aquatic Research*, (April), 2009–2010. doi:10.1016/j.ejar.2014.06.004

El-Kassas, H. Y., Aly-Eldeen, M. A., & Gharib, S. M. (2016). Green synthesis of iron oxide (Fe3O4) nanoparticles using two selected brown seaweeds: Characterization and application for lead bioremediation. *Acta Oceanologica Sinica*, *35*(8), 89–98. doi:10.100713131-016-0880-3

El-Kassas, H. Y., & El-Sheekh, M. M. (2014). Cytotoxic activity of biosynthesized gold nanoparticles with an extract of the red seaweed Corallina officinalis on the MCF-7 human breast cancer cell line. *Asian Pacific Journal of Cancer Prevention*, *15*(10), 4311–4317. doi:10.7314/APJCP.2014.15.10.4311 PMID:24935390

El Nemr, A., El-sikaily, A., & Khaled, A. (2010). Modeling of adsorption isotherms of Methylene Blue onto rice husk activated carbon. *Egyptian Journal of Aquatic Research*, *36*(3), 403–425.

El-Sherif, Z. M., Nassar, M. Z., & Fahmy, M. (2010). Phytoplankton distribution in the southeastern Mediterranean Sea (Egyptian waters) in summer and winter 2005. *Egyptian Journal of Aquatic Research*, *36*, 609–621.

Eroglu, E., Chen, X., Bradshaw, M., Agarwal, V., Zou, J., Stewart, S. G., Duan, X., Lamb, R. N., Smith, S. M., Raston, C. L., & Iyer, K. S. (2013). Biogenic production of palladium nanocrystals using microalgae and their immobilization on chitosan nan- ofibers for catalytic applications. *RSC Advances*, *3*(4), 1009–1012. doi:10.1039/C2RA22402J

Fard, J. K., Jafari, S., & Eghbal, M. A. (2015). A review of molecular mechanisms involved in toxicity of nanoparticles. *Advanced Pharmaceutical Bulletin*, *5*(4), 447–454. doi:10.15171/apb.2015.061 PMID:26819915

Garipov, I. T., Khaydarov, R. R., Gapurova, O. U., Efimova, I. L., & Evgrafova, S. Y. (2019). Silver nanoparticles as a new generation of antimicrobial prophylaxis. *J. Sib. Fed. Univ. Biol.*, *12*, 266–276. doi:10.17516/1997-1389-0301

Ghaemi, M., & Gholamipour, S. (2017). Controllable synthesis and characterization of silver nanoparticles using Sargassum angostifolium. *Iran. J. Chem. Chem. Eng.*, *36*, 1–10.

Ghodake, G., & Lee, D. S. (2011). Biological Syn- thesis of Gold Nanoparticles Using the Aqueous Extract of the Brown Algae Laminaria Japonica. *Journal of Nanoelectronics and Optoelectronics*, 268–271.

Ghoneim, M. M., El-desoky, H. S., El-moselhy, K. M., Amer, A., El-naga, E. H. A., Mohamedein, L. I., & Al-prol, A. E. (2014). Removal of cadmium from aqueous solution using marine green algae, Ulva lactuca. *Egyptian Journal of Aquatic Research*, *40*(3), 235–242. doi:10.1016/j.ejar.2014.08.005

Ghosh, P., Han, G., De, M., Kim, C. K., & Rotello, V. M. (2008). Gold nanoparticles in delivery applications. *Advanced Drug Delivery Reviews*, *60*(11), 1307–1315. doi:10.1016/j.addr.2008.03.016 PMID:18555555

González-Ballesteros, N., Prado-López, S., Rodríguez-González, J. B., Lastra, M., & Rodríguez-Argüelles, M. C. (2017). Green synthesis of gold nano- particles usaing brown algae Cystoseira baccata: Its activity in colon cancer cells. *Colloids and Surfaces. B, Biointerfaces*, *153*, 190–198. doi:10.1016/j.colsurfb.2017.02.020 PMID:28242372

Grima, E. M., Belarbi, E. H., Fernandez, F. A., Medina, A. R., & Chisti, Y. (2003). Recovery of microalgal biomass and metabolites: Process options and economics. *Biotechnology Advances*, *20*(7), 491–515. doi:10.1016/S0734-9750(02)00050-2 PMID:14550018

Gupta, A. K., & Gupta, M. (2005). Synthesis and surface engineering of iron oxide nanoparticles for biomedical applications. *Biomaterials*, *26*(18), 3995–4021. doi:10.1016/j.biomaterials.2004.10.012 PMID:15626447

Hassan, S. M., Ashour, M., Sakai, N., Zhang, L., Hassanien, H. A., Gaber, A., & Ammar, G. (2021). Impact of Seaweed Liquid Extract Biostimulant on Growth, Yield, and Chemical Composition of Cucumber (Cucumis sativus). *Agriculture*, *11*(320), 1–16. doi:10.3390/agriculture11040320

Hu, H., Wang, Z., & Pan, L. (2010). Synthesis of monodisperse Fe3O4@ silica core–shell microspheres and their application for removal of heavy metal ions from water. *Journal of Alloys and Compounds*, *492*(1-2), 656–661. doi:10.1016/j.jallcom.2009.11.204

Huang, X., Jain, P. K., El-Sayed, I. H., & El-Sayed, M. A. (2007). Gold nanoparticles: Interesting optical properties and recent applications in cancer diagnostics and therapy. *Nanomedicine (London)*, *2*(5), 681–693. doi:10.2217/17435889.2.5.681 PMID:17976030

Invernizzi, N. (2011). Nanotechnology between the lab and the shop floor: What are the effects on labour? *Journal of Nanoparticle Research*, *13*(6), 2249–2268. doi:10.100711051-011-0333-z

Iqbal, H., Khan, B. A., Khan, Z. U., Razzaq, A., Khan, N. U., Menaa, B., & Menaa, F. (2020). Fabrication, physical characterizations and in vitro antibacterial activity of cefadroxil-loaded chitosan/poly (vinyl alcohol) nanofibers against staphylococcus aureus clinical isolates. *International Journal of Biological Macromolecules*, *144*, 921–931. doi:10.1016/j.ijbiomac.2019.09.169 PMID:31704336

Jiang, X., Chen, W., Chen, C., Xiong, S., & Yu, A. (2011). Role of temperature in the growth of silver nanoparticles through a synergetic reduction approach. *Nanoscale Research Letters*, *6*, 32. PMID:27502655

Kannan, R.R.R., & Staden, J. V. (2013). Synthesis of silver nanoparticles using the seaweed Codium capitatum P.C. Silva (Chlorophyceae). *South African Journal of Botany, 86*, 1–4.

Kathiraven, T., Sundaramanickam, A., Shanmugam, N., & Balasubramanian, T. (2015). Green synthesis of silver nanoparticles using marine algae Caulerpa racemosa and their antibacterial activity against some human pathogens. *Applied Nanoscience*, *5*(4), 499–504. doi:10.100713204-014-0341-2

Kathiraven, T., Sundaramanickam, A., Shanmugam, N. B. T., & Balasubramanian, T. (2015). Green synthesis of silver nanoparticles using marine algae Caulerpa racemosa and their antibacterial activity against some human pathogens. *Applied Nanoscience*, *5*(4), 499–504. doi:10.100713204-014-0341-2

Khalid, M., Khalid, N., Ahmed, I., Hanif, R., Ismail, M., & Janjua, H. A. (2019). Comparative studies of three novel freshwater microalgae strains for synthesis of silver nanoparticles: Insights of characterization, antibacterial, cytotoxicity and antiviral activities. *J. Appl. Phycol., 29*, 1851–1863.

Khan, A., Rashid, R., Murtaza, G., & Zahra, A. (2014). Gold nanoparticles: Synthesis and applications in drug delivery. *Tropical Journal of Pharmaceutical Research*, *13*(7), 1169–1177. doi:10.4314/tjpr.v13i7.23

Kim, E. S., Yu, Q., & Deng, B. (2011). Plasma surface modification of nanofiltration (NF) thin-film composite (TFC) membranes to improve anti organic fouling. *Appl. Sur. Sci.*, *257*(23), 9863–9871. doi:10.1016/j.apsusc.2011.06.059

Kohout, C., Santi, C., & Polito, L. (2018). Anisotropic gold nanoparticles in biomedical applications. *International Journal of Molecular Sciences*, *19*(11), 3385. doi:10.3390/ijms19113385 PMID:30380664

Kokou, F., Makridis, P., Kentouri, M., & Divanach, P. (2012). Antibacterial activity in microalgae cultures. *Aquaculture Research*, *43*(10), 1520–1527. doi:10.1111/j.1365-2109.2011.02955.x

Kumar, D. A., Palanichamy, V., & Roopan, S. M. (2014). Photocatalytic action of AgCl nanoparticles and its antibacterial activity. *Journal of Photochemistry and Photobiology. B, Biology*, *138*, 302–306. doi:10.1016/j.jphotobiol.2014.06.011 PMID:25022464

LewisOscar, F., Vismaya, S., Arunkumar, M., Thajuddin, N., Dhanasekaran, D., & Nithya, C. (2016). Algal nanoparticles: Synthesis and biotechnological potentials. Chapter. *Algae Org. Imminent Biotechnol., 7*, 157–182.

Li, L., & Zhang, Z. (2016). Biosynthesis of gold nanoparticles using green alga Pithophoraoedogonia with their electrochemical performance for determin- ing carbendazim in soil. *International Journal of Electrochemical Science*, 4550–4559. doi:10.20964/2016.06.13

Lim, S., Park, J., Shim, M. K., Um, W., Yoon, H. Y., Ryu, J. H., Lim, D.-K., & Kim, K. (2019). Recent advances and challenges of repurposing nanoparticle-based drug delivery systems to enhance cancer immunotherapy. *Theranostics*, *9*(25), 7906–7923. doi:10.7150/thno.38425 PMID:31695807

Madaeni, S. S., & Ghaemi, N. (2007). Characterization of self-cleaning RO membranes coated with TiO2 particles under UV irradiation. *Sci. J. Membr.*, *333*(1-2), 221–233. doi:10.1016/j.memsci.2007.07.017

Mahdieh, M., Zolanvari, A., Azimee, A. J. S. I., & Mahdieh, M. (2012). Green biosynthesis of silver nanoparticles by Spirulina platensis. *Scientia Iranica*, *19*(3), 926–929. doi:10.1016/j.scient.2012.01.010

Mandal, R. P., Sekh, S., Sarkar, N. S., Chattopadhyay, D., & De, S. (2016). Algae mediated synthesis of cadmium sulphide nanoparticles and their application in bioremediation. *Materials Research Express*, *3*(5), 055007. doi:10.1088/2053-1591/3/5/055007

Mansour, A. T., Ashour, M., Alprol, A. E., & Alsaqufi, A. S. (2022). Aquatic Plants and Aquatic Animals in the Context of Sustainability : Cultivation Techniques, Integration, and Blue Revolution. *Sustainability*, *14*(4), 1–21. doi:10.3390u14063257

Maximous, N., Nakhla, G., & Wan, W. (2009). Comparative assessment of hydrophobic and hydrophilic membrane fouling in wastewater applications. *Journal of Membrane Science*, *339*(1), 93–99. doi:10.1016/j.memsci.2009.04.034

Merin, D. D., Prakash, S., & Bhimba, B. V. (2010). Antibacterial screening of silver nanoparticles synthesized by marine micro algae. *Asian Pacific Journal of Tropical Medicine*, *3*(10), 797–799. doi:10.1016/S1995-7645(10)60191-5

Momeni, S., & Nabipour, I. (2015). A simple green synthesis of palladium nanoparticles with Sargassum algae and their electrocatalytic activities towards hydrogen peroxide. *Applied Biochemistry and Biotechnology*, *176*(7), 1937–1949. doi:10.100712010-015-1690-3 PMID:26041058

Montasser, M. S., Hegazi, M.M., Younes, A. M., Dashti, N. H., El-Sharkawey, A. E. (2017). Effect of Gold Chloride Concentration and Volume on Size and Shape of Biological Synthesized Gold Nanoparticles (AuNPs) Using Red Algae (Laurenciapapillosa). *Science of Advanced Materials, 9*(7), 1105-1113.

Moorthi, P. V., Balasubramanian, C., & Mohan, S. (2018). Algae : A potential source for nanoparticle synthesis. *Applied Biochemistry and Biotechnology*, *175*, 135. doi:10.100712010-014-1264-9 PMID:25240853

MubarakAli, D., Gopinath, V., Rameshbabu, N., & Thajuddin, N. (2012). Synthesis and characterization of CdS nanoparticles using C phycoerythrin from the marine cyanobacteria. *Materials Letters*, *74*, 8–11. doi:10.1016/j.matlet.2012.01.026

Muñoz-Bonilla, A., & Fernández García, M. (2012). Polymeric materials with antimicrobial activity. *Progress in Polymer Science*, *37*(2), 281–339. doi:10.1016/j.progpolymsci.2011.08.005

Murugesan, S., Bhuvaneswari, S., Shanthi, N., Murugakoothan, P., & Sivamurugan, V. (2015). Red alga Hypnea musciformis (Wulf) lamour mediated environmentally benign synthesis and antifungal activity of gold nano particles. *Int. J. Nanosci. Nanotech.*, *6*, 71–83.

Navarro, N., & Sarasquete, C. (1998). Use of freeze-dried microalgae for rearing gilthead seabream, Sparus aurata, larvae: I. Growth, histology and water quality. *Aquaculture (Amsterdam, Netherlands)*, *167*(3-4), 179–193. doi:10.1016/S0044-8486(98)00311-1

Navarro, N., Yúfera, M., & García-Gallego, M. (2001). Use of freeze-dried microalgae for rearing gilthead seabream, Sparus aurata L., larvae. II. Biochemical composition. *Hydrobiologia*, *452*(1/3), 69–77. doi:10.1023/A:1011971810072

Ninganagouda, S., Rathod, V., Singh, D., Hiremath, J., Singh, A. K., Mathew, J., & ul-Haq, M. (2014). Growth kinetics and mechanistic action of reactive oxygen species released by silver nanoparticles from Aspergillus niger on Escherichia coli. *BioMed Research International*, *2014*, 1–14. doi:10.1155/2014/753419 PMID:25028666

Oza, G., Pandey, S., Mewada, A., Kalita, G., & Sharon, M. (2012). Facile biosynthesis of gold nanoparticles exploiting optimum pH and temperature of freshwater alga Chlorella pyrenoidusa. *Advances in Applied Science Research*, *3*(3), 1405–1412.

Patel, J. K., Sutariya, V., Kanwar, J. R., & Pathak, Y. V. (2018). *Drug Delivery for the Retina and Posterior Segment Disease* (1st ed.). Springer. doi:10.1007/978-3-319-95807-1

Pelgrift, R. Y., & Friedman, A. J. (2013). Nanotechnology as a therapeutic tool to combat microbial resistance. *Advanced Drug Delivery Reviews*, *65*(13-14), 1803–1815. doi:10.1016/j.addr.2013.07.011 PMID:23892192

Prathna, T., Chandrasekaran, N., Raichur, A. M., & Mukherjee, A. (2012). Kinetic evolution studies of silver nanoparticles in a bio-based green synthesis process. *Colloids and Surfaces. A, Physicochemical and Engineering Aspects*, *377*(1-3), 212–216. doi:10.1016/j.colsurfa.2010.12.047

Rajathi, F. A. A., Parthiban, C., Kumar, V. G., & Anantharaman, P. (2012). Biosynthesis of antibacterial gold nanoparticles using brown alga, Stoechosper- mum marginatum (kützing). *Spectrochimica Acta Part A: Molecular and Biomolecular Spectroscopy, 99*, 166–173.

Ramakrishna, M., Babu, D. R., Gengan, R. M., & Rao, S. C. G. N. (2016). Phytoremediation using Wet- land Plants: An ecosustainable approach-An ecosustainable approach. *International Journal of Phytoremediation*, *10*(2), 133–160.

Ramaswamy, S. V. P., Narendhran, S., & Sivaraj, R. (2016). Potentiating effect of ecofriendly synthesis of copper oxide nanoparticles using brown alga: Antimicrobial and anticancer activities. *Bulletin of Materials Science*, *39*(2), 361–364. doi:10.100712034-016-1173-3

Rı'os, S. D., Salvado´, J., Farriol, X., & Torras, C. (2012). Antifouling microfiltration strategies to harvest microalgae for biofuel. *Bioresource Technology*, *119*, 406–418. doi:10.1016/j.biortech.2012.05.044 PMID:22739578

Ruparelia, J. P., Chatterjee, A. K., Duttagupta, S. P., & Mukherji, S. (2008). Strain specificity in antimicrobial activity of silver and copper nanoparticles. *Acta Biomaterialia*, *4*(3), 707–716. doi:10.1016/j.actbio.2007.11.006 PMID:18248860

Sahayaraj, K., Rajesh, S., & Rathi, J. M. (2012). Silver nanoparticles biosynthesis using marine alga Padinapavonica(Linn.) and its microbicidalactivity. *Digest Journal of Nanomaterials and Biostructures*, *7*(4), 1557–1567.

San, K. A., & Shon, Y. S. (2018). Synthesis of alkanethiolate-capped metal nanoparticles using alkyl thiosulfate ligand precursors: A method to generate promising reagents for selective catalysis. *Nanomaterials (Basel, Switzerland)*, *8*(5), 346. doi:10.3390/nano8050346 PMID:29783714

Selvam, G. G., & Sivakumar, K. (2015). Phycosyn- thesis of silver nanoparticles and photocatalytic degradation of methyl orange dye using silver (Ag) nano- particles synthesized from Hypnea musciformis (Wulfen) J.V. Lamouroux. *Applied Nanoscience, 5*, 617–622.

Sharma, A., Sharma, S., Sharma, K., Chetri, S. P. K., Vashishtha, A., Singh, P., Kumar, R., Rathi, B. A. V., & Agrawal, V. (2016). Algae as crucial organisms in advancing nanotechnology: A systematic review. *Journal of Applied Phycology*, *28*(3), 1759–1774. doi:10.100710811-015-0715-1

Sharma, D., Kanchi, S., & Bisetty, K. (2019). Biogenic synthesis of nanoparticles: A review. *Arabian Journal of Chemistry*, *12*(8), 3576–3600. doi:10.1016/j.arabjc.2015.11.002

Sharma, G., Pandey, S., Ghatak, S., Watal, G., & Rai, P. K. (2018). Potential of spectroscopic techniques in the characterization of "green nanomaterials". In *Nanomaterials in Plants, Algae, and Microorganisms* (pp. 59–77). Elsevier.

Shields, R., & Lupatsch, I. (2012). Algae for aquaculture and animal feeds. In Microalgal Biotechnology: Integration and Economy. De Gruyter. doi:10.1515/9783110298321.79

Shiny, P. J., Mukherjee, A., & Chandrasekaran, N. (2013). Marine algae mediated synthesis of the silver nanoparticles and its antibacterial efficiency. *International Journal of Pharmacy and Pharmaceutical Sciences*, *5*, 239–241.

Shou, Q., Guo, C., Yang, L., Jia, L., Liu, C., & Liu, H. (2011). Effect of pH on the single-step synthesis of gold nanoparticles using PEO–PPO–PEO triblock copolymers in aqueous media. *Journal of Colloid and Interface Science*, *363*(2), 481–489. doi:10.1016/j.jcis.2011.07.021 PMID:21855892

Singaravelu, G., Arockiamary, J., Kumar, V. G., & Govindaraju, K. (2007). Anovel extracellular synthesis of monodisperse gold nanoparticles using marine alga. *Colloids and Surfaces. B, Biointerfaces*, *57*, 97–101. doi:10.1016/j.colsurfb.2007.01.010 PMID:17350236

Singh, M., Kalaivani, R., Manikandan, S., Sangeetha, N., & Kumaraguru, A. K. (2013). Facile green synthesis of variable metallic gold nanoparticle using Padina gym- nospora, a brown marine macroalga. *Applied Nanoscience*, *3*(2), 145–151. doi:10.100713204-012-0115-7

Subramaniyam, V., Ramraj, S., Bose, S., Thavamani, P., Megharaj, M., Chen, Z. and Naidu, R. (2015) Chlorococcum sp. MM11- a novel Phyco-nanofactory for the synthesis of iron NP. *Journal of Appllied Phy- Cology, 27*(5), 1861–1869.

Sudha, S. S., & Rajamanickam, K. R. J. (2013). Microalgae mediated synthesis of silver nanoparticles and their antibacterial activity against pathogenic bacteria. *Indian Journal of Experimental Biology, 51*, 393–399. PMID:23821828

Suganya, K. S. U., Govindaraju, K., Kumar, V. G., Dhas, T. S., Karthick, V., Singaravelu, G., & Elanchezhiyan, M. (2015). Blue green alga mediated synthesis of gold nanoparticles and its antibacterial efficacy against Gram positive organisms. *Materials Science and Engineering C, 47*, 351–356. doi:10.1016/j.msec.2014.11.043 PMID:25492207

Tendencia, E. A., & dela Peña, M. (2003). Investigation of some components of the greenwater system which makes it effective in the initial control of luminous bacteria. *Aquaculture (Amsterdam, Netherlands), 218*(1-4), 115–119. doi:10.1016/S0044-8486(02)00524-0

Thakkar, K. N., Mhatre, S. S., & Parikh, R. Y. (2010). Biological synthesis of metallic nanoparticles. *Nanotechnol. Biol. Med., 6*(2), 257–262. doi:10.1016/j.nano.2009.07.002 PMID:19616126

Tiburu, E. K., Salifu, A., Aidoo, E. O., Fleischer, H. N., Manu, G., Yaya, A., Zhou, H., & Efavi, J. K. (2017). Formation of chitosan nanoparticles using deacetylated chitin isolated from freshwater algae and locally synthesized zeolite A and their influence on cancer cell growth. *Proc. J. Nano Res., 48,* 156–170. 10.4028/www.scientific.net/JNanoR.48.156

Vadlapudi, V., & Amanchy, R. (2017). Synthesis, Characterization and Antibacterial Activity of Silver Nanoparticles from Red Algae, Hypnea musciformis. *Advances in Biological Research (Faisalabad), 11*(5), 242–249.

Venkatesan, J., Kim, S.-K., & Shim, M. S. (2016). Antimicrobial, antioxidant, and anticancer activities of biosynthesized silver nanoparticles using marine algae Ecklonia cava. *Nanomaterials (Basel, Switzerland), 6*(12), 235. doi:10.3390/nano6120235 PMID:28335363

Vijayan, S. R., Santhiyagu, P., Singamuthu, M., Ahila, N. K., Jayaraman, R., & Ethiraj, K. (2014). Synthesis and characterization of silver and gold nanoparticles using aqueous extract of seaweed, Turbinariaconoides, and their antimicrofouling activity. *TheScientificWorldJournal, 2014*, 1–10. doi:10.1155/2014/938272 PMID:24672397

Vincy, W.; Mahathalana, T.J.; Sukumaran, S.; Jeeva, S. (2017). Algae as a source for synthesis of nanoparticles-a review. *Int. J. Latest Trends Eng. Technol.*, 5–9.

Wang, H.-M. D., Li, X.-C., Lee, D.-J., & Chang, J.-S. (2017). Potential biomedical applications of marine algae. *Bioresource Technology, 244*, 1407–1415. doi:10.1016/j.biortech.2017.05.198 PMID:28697977

Yang, W., Zheng, Z., Lu, K., Zheng, C., Du, Y., Wang, J., & Zhu, J. (2020). Manipulating the phytoplankton community has the potential to create a stable bacterioplankton community in a shrimp rearing environment. *Aquaculture (Amsterdam, Netherlands), 520*, 734789. doi:10.1016/j.aquaculture.2019.734789

Yin, J., Zhu, G., & Deng, B. (2013). Multi-walled carbon nanotubes (MWNTs)/polysulfone (PSU) mixed matrix hollow fiber mem- branes for enhanced water treatment. *Journal of Membrane Science*, *437*, 237–248. doi:10.1016/j.memsci.2013.03.021

Zeng, S., Yong, K.-T., Roy, I., Dinh, X.-Q., Yu, X., & Luan, F. (2011). A review on functionalized gold nanoparticles for biosensing applications. *Plasmonics*, *6*(3), 491–506. doi:10.100711468-011-9228-1

Chapter 18
Utilization of Algae in Crop Improvement and Crop Protection for a Better Agricultural System

Ashutosh Kumar
ICAR-Indian Institute of Seed Science, India

Uma Chaurasia
Jiwaji University, India

Mostafa E. Elshobary
https://orcid.org/0000-0002-2131-0679
Department of Botany, Faculty of Science, Tanta University, Egypt

Sunita Kumari
ICAR-Indian Institute of Seed Science, India

Touseef Hussain
https://orcid.org/0000-0002-3325-0423
Aligarh Muslim University, India

Akhilendra P. Bharti
ICAR-National Bureau of Agriculturally Important Microorganisms, India

Deepak Kumar Maurya
Agarkar Research Institute, India

Loknath Samanta
Mizoram University, India

Mostafa El-Sheekh
Tanta University, Egypt

ABSTRACT

In this present situation, the ongoing pressure to reduce the use of pesticides and synthetic fertilizer inputs is a major challenge for sustainable agriculture. Microbial applications are a safe and renewable mode in the maintenance of agricultural productivity. Algae are acknowledged for their wide application ranging from agriculture to industries. They play a crucial role in sustainable agriculture and are used as bio-fertilizer and soil stabilizers, decreasing the need for synthetic fertilizers. The major focus is laid on the role of algae, microalgae, and cyanobacteria in soil fertility and their beneficial roles in agriculture and the maintenance of environmental sustainability.

DOI: 10.4018/978-1-6684-2438-4.ch018

INTRODUCTION

A large, diverse group similar to plant structure found in assemblage known as algae and they are varied in size from single cells (1.0 µm) to large seaweeds up to fifty meters (Vymazal et al., 1995; Doe et al., 2016). Algae are ubiquitous, and their occurrences are everywhere as in soils, ice, snowfields, hot springs, cold spring, the deserts on earth. They are similar to plants pertinent to biochemical and physiological as they possess chlorophyll, carbohydrates, protein, and other similar products as found in higher plants. Algae produce primary organic compounds so that they play a central role in aquatic systems along with the production of oxygen necessary for the metabolism of the aquatic habitats (Lee et al., 1989; Mazard et al., 2016)

Algae are important components for soil improvement and are therefore considered as soil health indicator parameters (Bérard et al., 2005; Allen et al., 2011). Algae reduce soil erosion by maintaining the water flow in soil and play an essential function in increasing soil richness and soil reclamation (Sahu et al., 2012; Tiwari et al., 2017). Algae worthiness have been extended in agriculture widely as it helped in the biological control of agricultural pests, microbial crust formation, agricultural wastewater treatment, and recycling of treated water (Hu et al., 2003; Win et al., 2018; Xiong et al., 2018; Kumar et al., 2018). Some crucial roles performed by algae have been presented in **Fig.1**.Like other microorganisms, algae have the potential to improve the carbon contents, aeration, and texture of soil (Ibraheem et al., 2007; Sharma et al., 2012; Costa et al., 2018). Hamed (Hamed et al., 2007) also reported that algae helped in nitrogen fixation in soil. Marine algae help improve soil fertility near the coastal area farmlands by using as fertilizer, and red algae and brown algae are generally used as organic fertilizer because they are usually rich in potassium but most deficient in nitrogen and phosphorus (Alobwede et al., 2019).

In agriculture, there is an emerging concern that environmental pollution generated through a nutrient imbalance in farm fields and injudicious uses of chemical fertilizers and pesticides in agricultural practices resulted in human health problems. Hence, globally there is the inclination towards migration from chemical-based agricultural practices to conventional farming aimed at lower input in agriculture and ultimately resulted in agricultural sustainability. To meet the ever expanding demand for food production under current scenario, viable options should be considered and one such option is the use of the application of algae as fertilizers. Further, Anthropogenic climate change is one of the prime repugnant threats facing th3r4e survivability of entire humanity. Constant loss of Antarctic ice sheets and severe reduction of Greenland are catalyzing Global warming. This phenomenon is reflecting in heat waves, induction of greenhouse gas, and sea-level increase (Dawson et al., 2011; Moreira et al., 2016). Among the sustainable options, microbes of different origins have proved their worthiness in various issues which are ranging from industries to agriculture. The present chapter emphasized has been focused on the implication of algae in different fields and their versatile role in the mitigation of environmental pollutions.

VERSATILE SIGNIFICANCE OF ALGAE

Improvement of Soil Fertility

Algae are complementally applied with cyanobacteria for rice cropping systems because it is the cheapest source and easily available of natural bio-fertilizers having wetland ecosystem (Ladha et al., 2003). Algae and cyanobacteria can convert atmospheric nitrogen to ammonia. Photosynthesis supplies the reducing

Figure 1. Multiaspects of algae

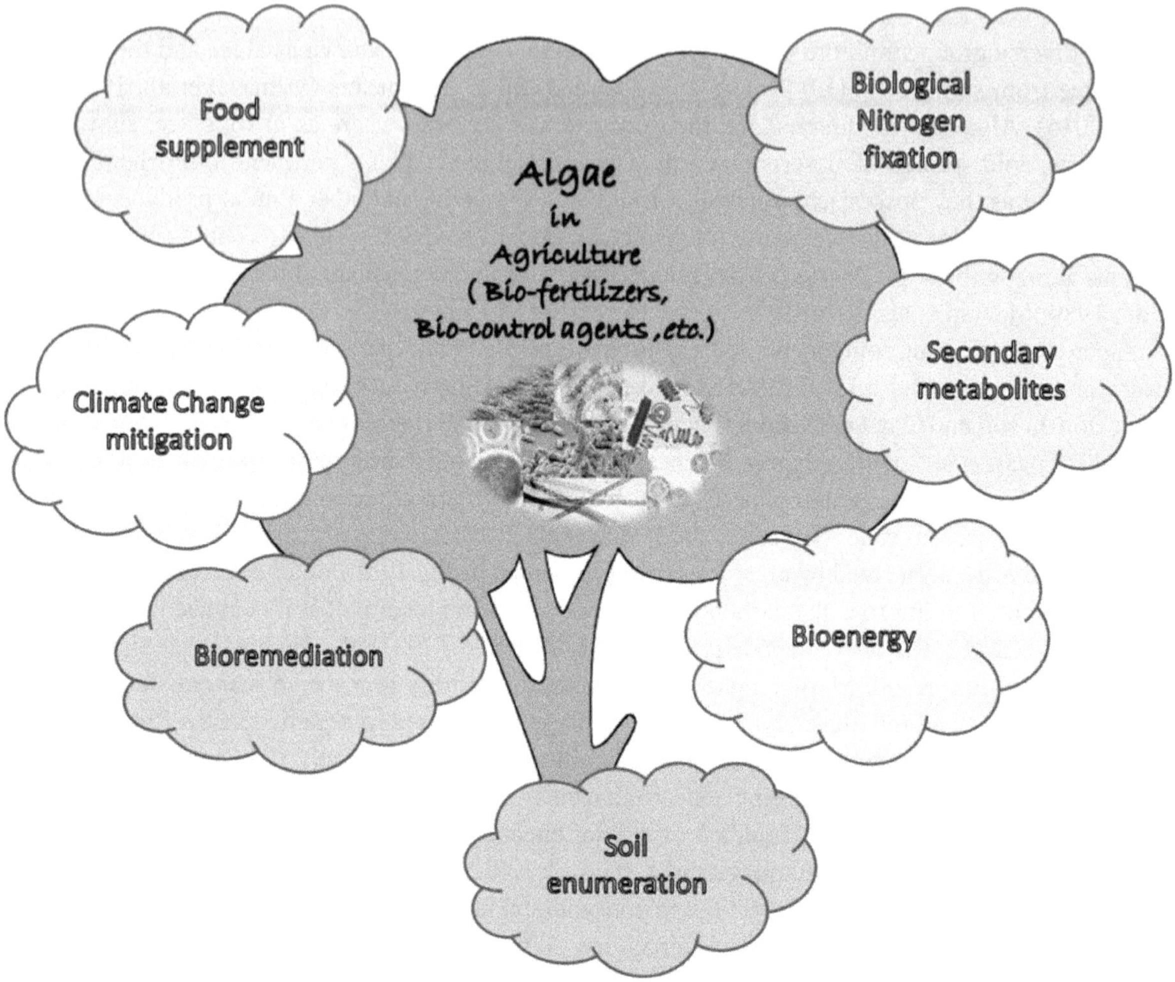

equivalent by oxidizing H_2O but evolves oxygen in the same cell which is harmful to nitrogen-fixing machinery (Zaborowska et al., 2015). Diazotrophic cyanobacteria have great potential as bio-fertilizers because, in the presence of sunlight, algae and cyanobacteria do assimilate carbon and nitrogen (Bombar et al., 2016). Symbiotic association of heterocystous cyanobacteria with water fern *Azolla* has been recognized as a fruitful area in the agricultural application (El-Zeky et al., 2005). In China, Egypt, Philippines, and India have been reported that suitable diazotrophic strains as bio-fertilizers could sustain the field-dose of herbicides by a successful biotechnological application (Tiwari et al., 1991; Singh et al., 2016). After nitrogen, phosphorus is the second indispensable supplement for plants and microbes. Some species of cyanobacteria have a slight ability to solubilize soil phosphate. Singh and Dhar (2007) have reported that algae have increased utilization efficiency and storage of micronutrients in their environments like other microorganisms; algae are also a source of organic matter in the soil (Shields et al., 1964) by decomposing of algae. Soil texture may increase after the mixing and mucilage formation in the soil so that humus content increases and easily available to other plants after few years (Li et al., 2019; Chauhan et al., 2019).

Salinity holds the vital barrier to revive soil in the region of infertile and semi-fertile areas. Metabolic and growth studies in salinity conditions give some predictable results in plants and algae (Singh et al., 2016; Chauhan et al., 2019). Some growth promoter as gibberellic acid (GA_3) could change the tolerance toward the salinity (Chauhan et al., 2019). From the stability and economic ground, prompters are expansive and impractical when it comes to large applications. Algae control a profitable role in soil remediation and increase soil fertility, which improves the plant conditions under certain environmental factor (Win et al., 2018).

Production of Extracellular Substances

Cyanobacteria expel enormously different substances that could change the physiology of plants (Rai et al., 2019). These microorganisms exert phytopathogen biocontrol and polymers, especially exopolysaccharides that support exoenzyme activity. To improve soli microbial arrangement and plant fertility, cyanobacteria secrete some metabolite as a precursor for several biomolecule counting with gibberellin, auxin vitamins, amino acids antibacterial and antifungal substances (Matson et al., 2018; Hamed et al., 2007; Wilson et al., 2016). Homeostatic levels of soil organic matter are critical to the sustainable and high production of crops. Limited variety of crop cultivation misbalance the nutrient buffer in soil and break the chain nitrogen, sulphur, carbon, and phosphorus exchange between air and soil (Singh et al., 2016; Win et al., 2018).

Heavy Metals Treatment

Metallic ion chelator is a well-studied component of microorganisms, and cyanobacteria cover a significant verity of them (Bouma-Gregson et al., 2019; Wiśniewska et al., 2019). To eliminate the heavy metal threat, cyanobacterial and microalgal follow environmental effectors, non-specific cell signalling mechanisms, and specific metal stress-protective cell regulation in reply to the contact of a toxic metal species in the cell (Ibraheem et al., 2007; Yin et al., 2018). As an extension of cellular protective mechanisms under stress conditions, cyanobacteria cover cell walls with a complex polymer. In which large percentage is branch chain polysaccharide and presence of proteoglycans develop enormous gelatine like structure, which have molecular sheave like property to hold the heavy metal and reduce their import to cell (Ahad et al., 2017; Kanamarlapudi et al., 2018).

Wastewater Treatment

Anthropogenic activity such as agricultural, civil, and industrial activities generates high concentrations of xenobiotic compounds that merge water and produce pollutant wastewater. Algae have the potential to remove the chemical from palm oil mill effluents, textile wastewater (Huisman et al., 2018; Tripathi & Hussain, 2021). It is used for the reduction of nitrogen and phosphorus from municipal wastewater. Some algae such as *Desmodesmus sp. Oscillatoria, ArthrosporaChlorella vulgaris,* and *Scenedesmusquadricauda*are have grown for wastewater treatment. *Chlorella vulgaris,* and *Scenedesmusquadricauda* were grown wastewater and found that microalgae degrade the metalaxyl, pyrimethanil, fenhexamid, iprodione, and triclopyr (Baglieri et al., 2016). Similar *Scenedesmusobliquus* and *Chlorella pyrenoidosa* have the efficiency of absorbing Pharmaceutical active compounds like ibuprofen, caffeine, carbamazepine, progesterone, and tris (2-chloroethyl) phosphate (Xiong et al., 2018).

Antimicrobial Agent

Marine microbes and algae frequently deal with complex, competitive, and hostile microbial communities. To escape from ecological pressure and to respond to such a competitive and predatory situation, they produce secondary metabolite. Marine macroalgae possess a diverse range of bioactive compound with a novel functional group; they belong to quinones, diterpenoids, phlorotannins, alkaloids, sterols, cyclic peptides, polyketides, polysaccharides, lipids, glycerols and other chemical classes(Matson et al., 2018; Hamed et al., 2018). These compounds are now an alternative to xenobiotic compounds that could be used for agricultural purposes. *Xanthomonas oryzae pv.* of *Oryzae* is a plant pathogenic bacterium whose population is inhibited by *Ulvaflexuosa*. The methanolic extract of *Padinagymnospora* has antibacterial activity against the soil-borne pathogenic bacteria *Ralstoniasolanacearum* and *Pectobacteriumcarotovora* characterized by a high proportion of plamitic acids. Acetonic extracts of the brown macroalga, *Sargassumpolyceratium* (Phaeophyceae), also showed a noticeable activity against different types of bacteria such as *Staphylococcus aureus, Erwiniacarotovora* (now known as *Pectobacteriumcarotovora*), and *Escherichia coli* (Pérez et al., 2016; Shannon et al., 2016).

Electricity and Biofuel Generation

Microbial fuel cells (MFCs) contain two reverse electrodes as the cathode (positive electrode) and a anode (negative electrode) that produce a unique class of bioelectrochemical systems (Zhao et al., 2012; Shukla et al., 2018). MFCs have great efficiency in recovering water and energy through the use as bifunctional devices. Algae have great application in MFCs to improve the overall performance. In MFCs, algae have been used for two purposes; to performing oxidation in anodophiles as a substrate in anaerobic conditions (Baicha et al., 2016) and as a useful oxygenic photosynthetic organism to generate oxygen for the cathodic reaction. Interestingly *Chlorella vulgaris* take part in MFC cathode to harvest solar energy for fixing CO_2 and generate oxygen. The main aim of utilizing algae as an oxygen producer in MFC cathode and to release back to the catholyte that generates bioelectricity followed at anode the oxidation reaction combined with successful counter reduction (Reddy et al., 2019).Microalgae Cultivation in wastewater has received great attention all around the globe for removing bioinorganic complex to clean wastewater and biodiesel production through biomass production of microalgae. Biodiesel is obtained via converting freely available nitrogen and phosphorus nutrients through the microalgae and heterotrophic bacteria symbiosis into wastewater with associated carbon dioxide (CO_2) sequestration through the photosynthesis process (Kadir et al., 2018). *Ourococcusmultisporus, Nitzschia cf. pusilla, Chlamydomonasmexicana, Scenedesmusobliquus, Chlorella vulgaris, Micractiniumreisseri, Chlorella protothecoides, Nannochloropsis* sp. and *Neochlorisole abundans* are potential candidates for treating wastewater or biodiesel and biomass production for their high lipid productivity (Madadi et al., 2016; Rincon et al., 2014). Polar and neutral lipids are preferred for biodiesel production. *Phaeodactylumtricornutum, Chlorella vulgaris, Chlorella protothecoides, Botryococcusbraunii*, and *Nannochloropsissalina* are identified to produce the fatty acids (FAs) composition of TAGs and quality (for example, iodine value (IV), cetane number (CN), FFAs (Kadir et al., 2018; Elarroussia et al., 2016).

Colouring Agent

Microalgae are used as natural colorings as red microalgae have characteristic pigment phycobiliproteins, used as natural colorants in pharmaceutical and cosmetic industries. *Spirulina* (Arthrospira) is grown commercial scale in open ponds due to a tremendous source of C-phycocyanin (Mondal et al., 2017). They are used in food products for many purposes like pasta, biscuits, puddings/gelled desserts, mayonnaise/salad dressings, chewing gum, and bread (Toker et al., 2018).

Algae Applications as Antipesticide and Antinematicide

The algae comprises of a diverse range of photosynthetic microorganism that are initially reported in freshwater and marine environments later on found in terrestrial niches via developed symbiotic relationships with other viable entities (Gartner et al., 2021; Specht et al., 2017). Seaweeds are larger eukaryotic algae also known as macroalgae, whereas tiny unicellular eukaryotes are clubbled as microalgae. Another group known as cyanobacteria (blue-green algae) exhibit evolutionary significance of prokaryotic and is highly diverse group (Specht et al., 2017). Some genera of algae showed extraordinary growth rate and generate high biomass within a lesser time (Yan et al., 2016; Rosales-Mendoza et al., 2020). Algae and their extracts potentially showed antimicrobial and nematocidal properties along with significant inhibition of agricultural important herb and insect. Hence, definite group of alage can be function as biopesticides (Kulik, 1995;Cheung et al., 2014;Hamed et al., 2018). Several reports substantiate that Alage and their derived extract have been displayed efficiently suppression of plant pathogenic such as *Pseudomonas, Xanthomonas, Erwinia, Fusarium, Verticillium,Rhizoctonia, Phytophthora, tobacco mosaic virus* (TMV) and potato virus X (PVX) (Caccamese et al., 1981;Sano, 1999; Kumar and Rengasamy,2000; Biondi et al., 2004; Nagorskaia et al., 2010; Jiménez et al., 2011;Esserti et al., 2017). The biocontrol potential of algae and its extract have been extended insects transmit diseases, effective in suppression of soil-borne nematodes, plant or fruit feeding insects (*e.g.*, fruit fly larvae) and mites (Hankins and Hockey, 1990;Sultana et al., 2012; Ali et al., 2013;Rashwan and Hammad, 2020) .

Improvement of Crop Productivity Through Alage

Phytohormones are secreted in trace amounts and accepted as signal molecules that function in the regulation of cellular processes within plants system (Lu and Xu, 2015). In the past few decades algal research particularly in microalgal genomic has depicted that signalling phytohormones are also released by diverse microalgae and cyanobacteria. (Han et al., 2018; Lu and Xu, 2015) have reported release of auxins, abscisic acid, cytokinins, ethylene and gibberellins from brown, red and green algae, cyanobacteria and diatoms. In microalgae hormones play significant role in as growth regulators for example as auxins secreted by from *C. vulgaris, D. salina, Nannochloropsis oceanica, Scenedesmus obliquus and Scenedesmus quadricauda* potentially promoted cell size enhancement and increment in chlorophyll and protein contents, along with accelerated cell division, consequence in higher biomass productiion (Mousavi et al., 2016;Kozlova et al., 2017). In cyanobacteria and microalgae, hormones molecules is efficient in the regulation of cells' response toward high salt concnetration, osmotic inbalance, oxidative stress, drought condition and nutrients' diffciency (Maršálek, et al., 1992; Yoshida et al., 2004) . Therefore, the presence of these hormones in plant overall growth and development is important. Gibberellins from plants and microalgae/cyanobacteria have at par functions and associated with stimulation of growth, pigments induction and proteins accumulation (Han et al., 2018; Lu and Xu, 2015). Gibberellins from *C. ellipsoidea, Chlorella pyrenoidosa, C. vulgaris and Microcystis aeruginosa* have been reported under beneficiary traits for increment in cell number and size and higher growth rates (Falkowska et al., 2011; Du et al., 2015 ;González-Garcinuño et al., 2016). Microalgae and cyanobacteria help in nitrogen fixation, soil stability and nutrient accession of plant through symbiotic association. Microalgae and cyanobacteria also export crucial function in the remediation and restoration of damaged soils, as micro-algae and cyanobacteria are able to: (i) modulate soils' pH; (ii) control soils' salinity and (iii) *Chlorella vulgaris, Desmodes mupleiomorphus, Desmodes pleiomorphus* and *Chlamydomonas reinhardtii* provide remedy against heavy metals contaminants in soils (Sharma et al., 2012;Prasanna, et al., 2013;Priya et al., 2014;Pan et al., 2019).

Applications of Algae as Biofertilizers

Bio-fertilization is a agriculturally sustainable practice that covers application biofertilizers to enhance nutrient contents in the soil consequence in enhanced agricultural production. The diverse species of cyanobacteria such as *Anabaena sp.*, *Nostoc* sp., and *Oscillatoria angustissima* capable of nitrogen fixing from the atmosphere. Some of the green microalgae and cyanobacteria species such as *Acutodesmus dimorphus*, *Spirulina platensis, Chlorella vulgaris*, *Scenedesmus dimorphus*, *Anabaena azolla*, and *Nostoc* sp. have been successfully applied in field as bio fertilizers to improve growth of crop. *Chlorella vulgaris* is dominant microalgae widely applied as biofertilizer (Alobwede et al., 2019). The amendment of seaweed species such as *Sargassum sp.* and *Gracilaria verrucosa* potentially resulted to chemical transformation and depicted as a indicator of soil fertility in the clay and sandy soils. The accumulation of seaweed function as conditioner to soil which improve its organic content, restored pH to neutral, and decline C/N ratio in sandy and clay soil type. Further, application of algae as biofertilization can assist the soil to improve its soil features mainly carbon content, aeration, texture, and nitrogen fixation. The accumulation of algae with other living organisms which are can minimized the soil erosion by manipulating water flow. Algae can act as role in soil reclamation, formation of microbiological crust, bio-controlling of agricultural pests, soil fertility and agricultural wastewater treatment (Abdel-Raouf et al. 2016).

Food Supplement

Increasing consumer interest in their health and the nutritional value of their food products have prompted the development of governmental approaches regarding health claims in many developed countries. Microalgae have been explored as a nutrient source with the aid of health benefits. According to the high algal photosynthetic rate, CO_2 can be fixed in different forms of sugar, fatty acid, the amino acid, in turn, used as the building blocks of polysaccharides, triacylglycerol, and polypeptides that could be utilized as a nutraceutical, food, and feed (Tang et al., 2011). Besides their production of important bioactive compounds, several pigments and phenolic compounds are already known for health benefits and pharmaceutical applications (Galasso et al., 2019).

Currently, there is increasing awareness for algal food due to the high percentage of polyunsaturated fatty acids (e.g., ω3 and ω6), essential amino acids (e.g., leucine, valine, and isoleucine) and pigments (e.g., β-carotene and lutein) and vitamins (e.g., B12, D, and E). In recent research, microalgal biomass has been expected to enrich wheat flour to produce various products or to be used as supplements to dry milk products, including baby milk and bay food, instead of soybean (Wells et al., 2017).

Microalgae are rich in proteins, peptides, and amino acids. For instance, some microalgae can contain more than half of their dry weight in terms of proteins (Khan et al., 2018). Algal peptides are now used for several beneficial properties in health care: anticoagulant, antimutagenic, antihypertensive, anticancer, anti-tyrosinase, and antioxidant (Rizzello et al., 2016). Recently, it has been shown, a glycoprotein of microalgae *Alexandriumminutum*used as an anticancer agent against lung cancer (Sansone et al., 2018). Mycosporine-like amino acids synthesized by microalgae have been demonstrated as a UV protective agent, antioxidant and anticancer compound (Lawrence et al., 2017).

In the nutritional and processed food products microalgae are representing one of the most promising sources for new product and application. *Spirulina maxima*, *Chlorella vulgaris*, *Haematococcuspluvialis*, *Diacronemavlkianum*, and *Isochrysisgalbana*are some interesting algae with potential bioactive properties, revealing a well-balanced chemical composition and a source of highly valuable molecules; these are long-chain polyunsaturated fatty acids, pigments (e.g. carotenoids, chlorophylls), sterols derivative, vitamins, hydrocolloids and other biologically active compounds (Graça et al., 2018; Balakrishnan et al., 2019). Due to high nutritive value and rich in protein content *Aphanizomenonflos-aquae, Chlorella* sp., *Dunaliellasalina, Dunaliellatertiolecta* and *Spirulinaplantensis*widely used as a human food source (Sathasivam et al., 2019).

Substance extracted from microalgae represent α- and β-carotene, apocarotenoids can be converted to vitamin A. Some isolated microalgae strains give vitamins C, D, E and some vitamins of the B group, (B1, B2, B3 and B12) (Jäpelt et al., 2013). Algae also contain various bioactive pigment classes such as chlorophylls, carotenoids and phycobiliproteins that usually have many biological activities. For example, chlorophyll-*a*, a combination of chlorophyll *a* and *b* display as antioxidant (Mishra et al., 2013), carotenoids as protective agents against UV and oxidative damage (Shaish et al., 1992) and zeaxanthin and lutein are important for the human retina (Stuart et al., 2004). Regarding PUFA's contents of algae, algae are the primary source of omega 3 and 6 in the whole marine environment. Fishes cannot produce PUFA's, but they accumulate it by feeding on microalgae (Barreira et al., 2015). PUFA's from different marine algae for instance *Nannochloropsis* sp. and *Phaeodactylum* sp. has many biological activities such as anti-inflammatory properties of EPA and DHA, improvement of cardiovascular health with vital roles in development and the growth of infants and maintaining the brain and visual systems (Janssen et al., 2014). Polysaccharides also are one of the bioactive compounds that can be accumulated to more than 50% of dry algal weight (Shunni et al., 2014). Healthy improving effects of polysaccharides are known and include improving the immune system (Bernaerts et al., 2018; De et al., 2013), or antiviral activities (Huleihel et al., 2001), or anticancer properties (Yim et al., 2005).

However, environmental stresses are the main challenges that face the full screen of bioactive substances synthesized by microalgae. Carotenoids and phycobiliproteins are well known for their responses to environmental stress (Bule et al., 2018). Others are less or partially known. Once this information is available, the beneficial effects of algal bioactive compounds and their molecular pathways should be more advanced and applicable on a large scale.

ROLE OF ALGAE IN AGRICULTURE

The ecosystem is composed of biotic as well as abiotic factors. The biotic factors constitute microorganisms that are an integral part of the ecosystem and influence plant growth by several means. The role of microorganisms in the mineralization and solubilization of different macro and micronutrients has been documented for several years (González et al., 2013). Increased environmental conditions and awareness regarding the use of agrichemicals have led to the interest in the microorganisms as well as several algae as a potential source of biopesticides, biofertilizers, and plant growth promotion (Kaushik et al., 2014; Bhooshan et al., 2018). The plant growth-promoting properties of the gibberellins suggest that the several metabolites from the microorganism can act as plant growth promoters. Similarly, the seaweed extract also acts as growth promoters. Earlier reports regarding the utilization of the seaweed extract in the growth and production of the bulb crops, fruit crops, grains, legume crops and flowers have shown advantageous (Mattner et al., 2013; Craigie et al., 2011; Rengasamy et al., 2015). They have a positive response to seed germination, root development, increasing chlorophyll content as well as resistance to the pathogen (Vinoth et al., 2015; Briceño-Domínguez et al., 2014; Elemike et al., 2019). The algae are the good sources of the growth hormones like auxins, cytokinin, gibberellins, polyamines and polyphenols etc. which can directly act as the growth promoters (González et al., 2013; Ronga et al., 2019).

Phytohormones by Algae Support the Plant Growth

The seaweed extract have the active constituents of growth hormones such as auxins, cytokinin, gibberellins, and fast migrating components (polyamines and brassinosteroids) and they all are effective at low concentrations. In addition, some major molecules such as polyphenols identified in algal extracts promote growth activity as plant growth promoters (Craigie et al., 2011; González et al., 2013; Ronga et al., 2019). Earlier auxins and their analogues were found in red-brown and green algae as well as in cyanobacteria (Stirk et al 2009; Lu et al., 2015; Žižková et al., 2016; Chung et al., 2018). The modern technologies have demonstrated the variation for the synthesis of these growth hormones varied with season and according to the developmental stages (Piotrowska-Niczyporuk et al., 2019). A number of publications show the cytokinin activity in seaweed extract (Stirk et al., 2009; Lu et al., 2015; Žižková et al., 2016). *Euglena* chloroplast has also been reported with cytokinins activity (Noble et al., 2014). There is a very little report on the production of the gibberellins in the algae. The substances with gibberellin activity have been found from *Caulerpa* and *Fucus,* extract (Stirk et al. 2009; Jacobs et al., 1993). Similarly, jasmonate and its volatile methyl ester are detected in almost all algae. This hormone has been detected in the green and red algae as well as cyanobacteria (Chung et al., 2018; Arnold et al., 2001).

The presence of polyamines, brassinosteroids as well as rhodomorphine has also been detected in several algae-like *Ulva, Chlorella, Dyctiota, Hydrodictyon, Cyanidium, Gelidium, Grateloupia, Griffithsia*and *Euglena* (Stirk et al., 2009; Marián et al., 2000; Toll et al., 2012) which can also act as growth promotion in plants. Besides these hormones, the abscisic acid has also been reported to repress the plant growth in the bioassays (Niemann et al., 1980). The component has been isolated from the green-algae *Enteromorphacompressa* and has been studied. In green microalgae (*Chlorella* sp., *Dunaliellasalina*, and *Haematococcuspluvialis*) this hormone was found and also in the thalli of brown macrophytes from the genus *Ascophyllum* and some species of *Laminaria* (Tominaga et al., 1993; Nimura et al., 2002; Schiener et al., 2015).

Table 1. Major algae division associated with the particular plant hormones.

Division	Hormones	Genera	References
Chlorophyta	IAA	*Enteromorpha, Chlorella, Cladophora,Caulerpa, Desmodesmus*	(Chung et al., 2018; Cooke et al., 2002; Stirk et al., 2003)
	Cytokines	*Protococcus,Chlorella,Scenedesmus,Chlamydomonas*	(Žižková et al., 2016; Chung et al., 2018; Ördög et al.,n2004; Stirk et al., 2003)
	Gibberellins	*Caulerpa*	(Jacobs et al., 1993)
	ABA	*Chlorella, Dunaliella, Haematococcus*	(Tominaga et al., 1993; Nimura et al., 2002; Kobayashi et al., 1997)
	Lunularic acid	*Enteromorpha*	(Niemann et al., 1980)
	Jasmonic acid	*Dunaliella, Chlorella*	(Chung et al., 2018; Arnold et al., 2001)
	Polyamines	*Ulva, Chlorella*	(Marián et al., 2000)
	Brassinosteroids	*Hydrodictyon*	(Yokota et al., 1987)
Phaeophyta	IAA	*Macrocystis, Laminaria, Fucus, Ascophyllum*	(Lu et al., 2015; Stirk et al., 2003; Polevoi et al., 200)
	Cytokinins	*Fucus, Ascophyllum, Sargassum, Macrocystis*	(Lu et al., 2015)
	Gibberellins	*Fucus*	(Lu et al., 2015)
	ABA	*Ascophyllum, Laminaria*	(Stirk et al 2009)
	Polyamines	*Dyctiota*	(Stirk et al 2009)
Rhodophyta	IAA	*Botryocladia, Porphyra*	(Sitnik et al., 2003; Labeeuw et al., 2016)
	Cytokinins	*Porphyra*	(Stirk et al., 2003)
	Jasmonic acid	*Gelidium*	(Arnold et al., 2001)
	Polyamines	*Cyanidium, Gelidium, Grateloupia*	(Marián et al., 2000; Sacramento et al., 2004)
	Rhodomorphin	*Griffithsia*	(Toll et al., 2012)
Charophyta	IAA, Cytokinins	*Chara*	(Wang et al., 2014)
Euglenophyta	Cytokinins	*Euglena*	(Noble et al., 2014)
	Jasmonic acid	*Euglena*	(Arnold et al., 2001)
	Polyamines	*Euglena*	(Marián et al.,2000)
Cyanophyta	IAA	*Oscillatoria, Chlorogloea*	(Žižková et al., 2016)
	Cytokinins	*Arthronema, Calothrix*	(Žižková et al., 2016)
	Jasmonic acid	*Spirulina*	(Han et al., 2016 ; Czerpak et al., 2006)

Secondary Metabolites and Polysaccharides from the Algae act as Growth Promoter

Polysaccharides or oligosaccharides increased germination rate, the growth rate of the root, higher yields, and favoured resistance to diseases in various crops when crud extract were spread over plant (Paulert et al., 2009; González et al., 2014). Total polysaccharide concentrations in seaweed species as *Ulvalactuca* and *Padinagymnospora* range vary between 4 and 76% of dry weight (Holdt et al., 2011). The algal dry weight hold 8 to 29% of alginates which are the major molecular representative of brown seaweed cell walls, and ulvan called Sea lettuce is the most fundamental part of green seaweed cell walls (Lahaye et al., 2007). Alginate polysaccharides posses plant growth promoting property (Hernández-Herrera et al., 2016). Alginate-derived oligosaccharides (ADO) triggered the germination, growth, and shoot elongation in different plants species by enhancing nitrogen assimilation and basal metabolism in plant (González et al., 2014). Alginate also help to hydroponic cultivation of peanut as it enhanced the growth of rice, that is obtain by treatment with γ radiation (Hien et al., 2000). Treatment with a poly-mannuronic acid fraction (Poly-Ma) of sodium alginates showed a boost in plant height over the controls in tobacco (Laporte et al., 2007). It continues the trend with common bean germination, which increase in the presence of ulvans a component of green seaweed (Paulert et al., 2009; Zou et al., 2019).

Seaweed as a Carbon Source for Plants

In principle, for the detection of the bioactive molecule, testing and assessing the efficiency of extracted seaweed product, plants could play a role of biosensors (bioassay model systems) (Rayorath et al., 2008). Tomato planets are frequently used in molecular biology and physiological studies (McCormick et al. 1986). Recently, seaweed extract is being a part of plant tissue culture experiments, different explants such as cotyledons, hypocotyl, and leaf are taken from tomatoes and used for culture.(Vikram et al., 2011). Using indole-3-butyric acid (IBA) as a reference against the seaweed extract, it elucidates the induced root formation in Mung bean cuttings (Briceño-Domínguez et al., 2014; Rengasamy et al., 2015; Lötze et al., 2015). Apparently strength of the Murashige-Skoog (MS) medium could be reduced to observe the discrete result in the presence and absence of seaweed extract in different plant tissue culture models.

Algae as a Biofertilizer

Among the yeasts, *Yarrowialipolytica* has the capacity for phosphate solubilisation. Algae, cyanobacteria and mycorrhiza have also been identified for the ability of P solubilisation (Yandigeri et al., 2011; Jia et al., 2018). Algal and bacterial growth in freshwater and sea is critically. Iron concentration is an essential parameter for growth and survival, especially in aquatic ecosystem. In sea water iron availability is scarce because of its quick transitions to insoluble ferrous form, resultant to decreasing in concentration. To dissolve this problem, marine heterotrophic bacteria, cyanobacteria and algae make use of small organic molecule called siderophores, which capture iron and bring back to cell (Årstøl et al., 2019).

Seaweeds, found in all coastal ecosystem of the world, which are multicellular marine macroalgae, play essential role for maintaining the biodiversity for the in-shore marine environment (Sangha et al., 2014). These algae have long been used in the human nutritional regime to be source of unique carbohydrates. Currently, carbohydrate polymers from seaweeds are heavily exploited commercially; three of them are: agar, carrageenans and alginates from red and brown seaweeds, respectively (Campo et al., 2009). Red algae (*Rhodophyta*) are a source of carrageenans, composed of linear polysaccharides and anionic sulfated derivatives of polysaccharides (Campo et al., 2009). Carrageenans are found in a number of commercial products, food, dairy and pharmaceutical industries (Prajapati et al., 2014; Li et al., 2014). More recently, carrageenans are tested for their potentiality on plant growth and for their roles in enhancing plant immunity.

ROLE OF ALGAE AS CLIMATE CHANGE MITIGATION

Most of the mentioned phenomena are related to anthropogenic air pollution, especially carbon dioxide (CO_2) emission. The National Oceanic and Atmospheric Administration (NOAA) have reported that CO_2 concentration has increased in the atmosphere from 391 ppm in 2012 to 409.95 ppm in 2019 (NOAA, 2019). This increase disperses the natural carbon cycle and deposites in the atmosphere. CO_2 capture is attracting scientific attention as a mitigation tool to reduce CO_2 concentration and emissions. However, CO_2 mitigation from a point source is not sufficient due to half of CO_2 emissions from the transportation sector. To achieve this mitigation goal efficiently, CO_2 capture from the atmosphere may also be essential alongside point sources. Biological CO_2 sequestration is one of the effective methods comparing to the traditional methods such as physical (membrane separation, geologic and oceanic injection) or chemical (chemical adsorption and mineral carbonation) (Zhou et al., 2017). Higher plants contribute only 3–6% reduction in total CO_2 emissions (Kao et al., 2014). However, Algalculture (microalgae and seaweed) are one of the efficient organism to capture CO_2 from atmosphere due to the following advantages (i) high photosynthetic efficacy;(ii) effectual in low-concentration CO_2; (iii) faster sequestration rate than terrestrial plants;(iv) do not compete with crops for arable land or water; (v) its biomass can be a resource of food, feed, fuel, natural bioactive compounds (Zhou et al., 2017). Accordingly, 1 kg of microalgae biomass can sequestrate approximately 1.83 kg of CO_2 (Jiang et al., 2013).

CO_2 is the key component of industrial exhaust/flue gas (10–25% v/v), besides some traces of SO_X and NO_X. Accordingly, the ideal algal strain should have high accumulation ability, highly resistant to CO_2 concentration, toxic gases, temperature instability, nutrients limitation and pH variation (Singh et al., 2013). Selecting ideal algal species and further increasing their ability to tolerate higher concentrations of CO_2 are the main pillars to face air pollution and climate changes. Several algal species have shown great potential for CO_2 sequestration, as shown in Table 2.Algal culture could be used to capture CO_2 from the atmosphere efficiently, and the resulting biomass can produce various valuable products. Notably, this mechanism is affected by different experimental conditions, such as light intensity, CO_2 concentration and cultivation methods. Manipulation of these factors could be one of the critical factors to enhance CO_2 fixation capacity and to be more tolerant of flue gas pollutants (Cheah et al., 2015).

Mutation and genetic engineering is another efficient tool to enhance microalgal cultivation, photosynthetic efficiency, as well as CO_2 fixation efficiency. γ-rays Spirulina sp. exhibiting 500% increase in biomass yield under 15% CO_2 (Cheng et al., 2017). There are numerous studies for transgene expression, and gene knock-down was developed for industrial application (Zhang et al., 2011). For instance, decreasing the dimension of the light-harvesting antenna to increase light absorption, photosynthesis rate and reduce photoinhibition (Zhang et al., 2011). Gene, for enhanced green fluorescent protein (EGFP) was successfully transformed into *chlorella vulgaris* to increase CO_2 bio-mitigation capacity (Yang et al., 2015).

On the other hand, seaweed cultivation has brought in significant environmental benefits; Chinese seaweed aquaculture has the ability to remove 75,56 t of nitrogen and 9592 t of phosphate from seawater, sequestrate 539,55t of carbon and fix 1,980,167 t of CO_2, plus releasing of 1,440,612 t of O_2 into seawater, and the production of 5809 t of iodine. Additionally, seaweed aquaculture saves chemical fertilizers about 29,313 t, pesticide 1873 t and farmland 62,492 ha in compared with terrestrial crop cultivation (Zheng et al., 2019).

Despite the perceived benefits of algae-based technologies, large-scale applications are yet to be realized since there is still a deficiency of thoughtful of the economic viability of algal

Table 2. Microalgae and cyanobacteria biomass yield, their CO_2 bio fixation rate and valuable products.

Algal groups	Microalgae species	Initial co$_2$ (%) (v/v)	Biomass (g l^{-1})	Co$_2$ fixation rate (g l^{-1} d^{-1})	Cultivation system	Valuable products	Reference
Cyanobacteria	*Anabaena* sp. CH1	10	-	1.01	Bubble column		(Chiang et al., 2011)
	Anabaena Variabilis	10	0.052	0.095	photobioreactor		(Kaur et al., 2017)
	Nostocpunctiforme	12	0.47	0.86		Proteins, lipids and carbohydrates	(García-Cubero et al., 2017)
	Spirulina sp. LEB 18	10	0.47	1.19	Erlenmeyer flask-type photobioreactors		(Duarte et al., 2017)
Microalgae	*Botryococcusbraunii*	5	3.11	0.5	Fermenter	Biofuel	(Sydney et al., 2010)
	Chlorella sp.	5	0.64	0.09	Lab-scale flask method	Organic acids and ABE solvents	(Kassim et al., 2017)
	Chlorella fusca LEB 111	10	0.91	2.55	Erlenmeyer flask-type photobioreactors		(Duarte et al., 2017)
	Chlorella minutissima	20	1.84	0.28	Lab-scale flask method	Ethanol	(Freitas et al., 2017)
	Dunaliellatertiolecta	5	2.15	0.27	Fermenter		(Sydney et al., 2010)
	Nannochloropsissp	1	0.38	0.46	Lab scale flask method	Biodiesel	(Taher et al., 2015)
	Pseudochlorococ cum sp.	1	0.31	1.182	Lab-scale flask method	Biodiesel	(Taher et al., 2015)
	*Scenedesmus*vacuolatus	12	0.63	1.15		Proteins, lipids and carbohydrates	(García-Cubero et al., 2017)
	Tetraselmis	15	0.72	0.11	Lab-scale flask method	Organic acids and ABE solvents	(Kassim et al., 2017)

cultivation (Froehlich et al., 2019; Zhou et al., 2017). Innovation and development in these areas are essential. However, there are some encounters that need to be highlighted to make commercial algal cultivation technology as the forthcoming thing for the future (Yadav et al., 2017). Algae can be a good substitute if the total running costs are reduced. Algae need a lot of water for cultivation; this may limit its potential to use it for commercial-scale CO_2 sequestration. However, using the integrated approach of using the wastewater for nutrient supply and direct use of CO_2 from a combustion process without any further costly steps should reduce the production cost; furthermore, production of by-products for bio-energy or nutritional fields may help to further reduce the total production cost (Razzak et al., 2013)

ROLE OF ALGAE IN STRESS REMOVAL

In the stress condition, the plants and crops face nutrient restriction or threat from other organisms, allowing plants to change or reseat all metabolic and regulatory systems to reduce the death trap. These stresses also influence plant growth and other fundamental processes from seedling to fruit development, and the effect can pass to the next generation. It has been reported that the application of the algal extract improves nutrient uptake and change metabolic profile to alleviate the adverse effects of the stress (Chanda et al., 2020). Algal extract and seaweed were reported to improve salinity stress tolerance in the crop. The extract increases chlorophyll concentration significantly and photosynthetic ability that helps to supply more energy to export excess salt out of the cell and recover the damaged cell (Carillo et al., 2020). The extract of *Ulva lactuca* can increase seed germination rate and growth parameter in saline pre-exposed seed and seedling. It also increases the expression of antioxidants and other axillary enzymes (Ibrahim et al., 2014). Proline, carotenoids, and exopolysaccharides are the prime molecules accumulated in halotolerant microalgae *Dunaliella salina* associated with the induction of the germination and height of coleoptiles of wheat (*Triticum aestivum*) under salt stress (El Arroussi et al., 2015).

Water inadequacy in the soil is one of the most important and established stress factors for plants and has a broad-spectrum effect on plant physiology. In the absence of water, the electron transport system and photosynthetic reaction centre become over oxidized, leading to the production of reactive oxygen species (ROS), which targets any molecule at the cellular level and inactivate or degrade active macromolecules (nucleic acid, protein, and polysaccharide). Algal extract and seaweed obtain from marine algae and microbes are the sources of the novel bioactive compounds that regulate the expression of enzymatic antioxidants, such as superoxide dismutase (SOD), catalase (CAT), and ascorbate peroxidase (APX). It has been reported that seaweed has a positive effect on growth and other physiological and morphological improvements in rice (*Oryza sativa* L.), spinach (Spinacia oleracea), and other crops (de Vasconcelos et al., 2020). High-temperature influences the chloroplast function, reduces photosynthesis and energy supply, and increases ROS and free radical load, impacting plant physiology. Similarly, low-temperature stress is also associated with a negative effect on plant growth. Microalgae and algal extracts target several pathways to improve the plant defense system and fulfil nutrient requirements to make antioxidant and critically requiring hormone in these temperature stress conditions. Treatment with seaweed extract can reduce the temperature-induced oxidative stress caused by hydrogen peroxide (H_2O_2), superoxide (O●−), and lipid peroxide (MDA). The seaweed extract showed positive effects for seed germination, seedling growth, and antioxidant capacity under heat stress in spinach (Spinacia oleracea) (Anjos et al., 2020). Similarly, the bioactive component of seaweed extract was associated with the seed germination and seedling growth of *Ceratotheca triloba* under low temperatures (Masondo et al., 2018).

CONCLUSION AND FUTURE PROSPECTS

This chapter discusses the wider applications of algae-based various applications in agriculture and other sectors along with their tentative remedies. In the modern decades, our understanding in the field of algal metabolites has improved a lot, but there are still many steps to be achieved. For the production of green energy and protein-rich foods in the future, algae are very promising for urban agriculture, as they can play a fundamental role in carbon reduction and the management of greenhouse gases. Environmental sustainability and the inadequate availability of fossil fuels have prompted the research for innovations and expansion of algal science. However, algal research only mitigated the several issues related to fossil fuel production, enhancement in agricultural production, environmental protection but also have unique properties for a wider range of applications. In extension algal liquid extract improve crop yield in adverse condition by reducing toxic stress. Further, there are certain hindrances in the commercialization of products that resulted from the algae.

ACKNOWLEDGMENT

This research received no specific grant from any funding agency in the public, commercial, or not-for-profit sectors.

REFERENCES

Abdel-Raouf, N., Al-Homaidan, A., & Ibraheem, I. 2016. Agricultural importance of algae. African J. Biotechnol. 11, 11648–11658. . v11i54 doi:10.4314/ajb

Ahad, R.I.A., Goswami, S., & Syiem, M.B. (2017). Biosorption and equilibrium isotherms study of cadmium removal by *Nostocmuscorum* Meg 1: morphological, physiological and biochemical alterations. *Biotech, 7*(2), 104.

Ali, M. Y. S., Ravikumar, S., & Beula, J. M. (2013). Mosquito Larvicidal Activity of Seaweeds Extracts against Anopheles Stephensi, Aedes aegypti and Culex Quinquefasciatus. *Asian Pacific Journal of Tropical Disease*, *3*, 196–201.

Allen, D. E., Singh, B. P., & Dalal, R. C. (2011). Soil health indicators under climate change: a review of current knowledge. In *Soil health and climate change* (pp. 25–45). Springer. doi:10.1007/978-3-642-20256-8_2

Alobwede, E., Leake, J. R., & Pandhal, J. (2019). Circular economy fertilization: Testing micro and macro algal species as soil improvers and nutrient sources for crop production in greenhouse and field conditions. *Geoderma*, *334*, 113–123. doi:10.1016/j.geoderma.2018.07.049

Anjos Neto, A.P.D., Oliveira, G.R.F., Mello, S.D.C., Silva, M.S.D., Gomes-Junior, F.G., Novembre, A. D. & Azevedo, R.A. (2020). Seed priming with seaweed extract mitigate heat stress in spinach: effect on germination, seedling growth and antioxidant capacity. *Bragantia*.

Arnold, T. M., Targett, N. M., Tanner, C. E., Hatch, W. I., & Ferrari, K. E. (2001). Evidence for Methyl Jasmonate-Induced Phlorotannin Production in *Fucus vesiculosus* (Phaeophyceae). *Journal of Phycology*, *37*(6), 1026–1029. doi:10.1046/j.1529-8817.2001.01130.x

Årstøl, E., & Hohmann-Marriott, M.F. (2019) Cyanobacterial Siderophores—Physiology, Structure, Biosynthesis, and Applications. *Marine Drugs*, *17*(5), 281.

Baglieri, A., Sidella, S., Barone, V., Fragalà, F., Silkina, A., Nègre, M., & Gennari, M. (2016). Cultivating *Chlorella vulgaris* and *Scenedesmusquadricauda* microalgae to degrade inorganic compounds and pesticides in water. *Environmental Science and Pollution Research International*, *23*(18), 18165–18174. doi:10.100711356-016-6996-3 PMID:27259964

Baicha, Z., Salar-García, M. J., Ortiz-Martínez, V. M., Hernández-Fernández, F. J., De los Ríos, A. P., Labjar, N., & Elmahi, M. (2016). A critical review on microalgae as an alternative source for bioenergy production: A promising low cost substrate for microbial fuel cells. *Fuel Processing Technology*, *154*, 104–116. doi:10.1016/j.fuproc.2016.08.017

Balakrishnan, J., Dhavamani, S., Govindan Sadasivam, S., Arumugam, M., Vellaikumar, S., Ramalingam, J., & Shanmugam, K. (2019). Omega-rich Isochrysis sp. biomass enhances brain docosahexaenoic acid levels and improves serum lipid profile and antioxidant status in Wistar rats. *Journal of the Science of Food and Agriculture*, 3.

Barreira, L., Pereira, H., Gangadhar, K. N., Custódio, L., & Varela, J. (2015). *Medicinal Effects of Microalgae-Derived Fatty Acid.* doi:10.1016/B978-0-12-800776-1.00013-3

Bérard, A., Dorigo, U., Humbert, J. F., & Martin-Laurent, F. (2005). Microalgae community structure analysis based on 18S rDNA amplification from DNA extracted directly from soil as a potential soil bioindicator. *Agronomy for Sustainable Development*, *25*(2), 285–291. doi:10.1051/agro:2005004

Bernaerts, T. M. M., Gheysen, L., Kyomugasho, C., Kermani, Z. J., Vandionant, S., Foubert, I., & Van Loey, A. M. (2018). Comparison of microalgal biomasses as functional food ingredients: Focus on the composition of cell wall related polysaccharides. *Algal Research*, *32*, 150–161. doi:10.1016/j.algal.2018.03.017

Bhooshan, N., Pabbi, S., Singh, A., Sharma, A., Jaiswal, A., & Kumar, A. (2018). Impact of blue green algae (BGA) technology: an empirical evidence from northwestern Indo-Gangetic Plains. *Biotech, 8*(8), 324.

Biondi, N., Piccardi, R., Margheri, M. C., Rodolfi, L., Smith, G. D., & Tredici, M. R. (2004). Evaluation of Nostoc Strain ATCC 53789 as a Potential Source of Natural Pesticides. *Applied and Environmental Microbiology*, *70*, 3313–3320.

Bombar, D., Paerl, R. W., & Riemann, L. (2016). Marine non-cyanobacterial diazotrophs: Moving beyond molecular detection. *Trends in Microbiology*, *24*(11), 916–927. doi:10.1016/j.tim.2016.07.002 PMID:27476748

Bouma-Gregson, K., Olm, M. R., Probst, A. J., Anantharaman, K., Power, M. E., & Banfield, J. F. (2019). Impacts of microbial assemblage and environmental conditions on the distribution of anatoxin-a producing cyanobacteria within a river network. *The ISME Journal*, *13*(6), 1618–1634. doi:10.103841396-019-0374-3 PMID:30809011

Briceño-Domínguez, D., Hernandez-Carmona, G., Moyo, M., Stirk, W., & Van Staden, J. (2014). Plant growth promoting activity of seaweed liquid extracts produced from *Macrocystis pyrifera* under different pH and temperature conditions. *Journal of Applied Phycology*, *26*(5), 2203–2210. doi:10.100710811-014-0237-2

Bule, M. H., Ahmed, I., Maqbool, F., Bilal, M., & Hmn, I. (2018). Microalgae as a source of high-value bioactive compounds. *Frontiers in Bioscience*, *10*(1), 197–216. doi:10.2741509 PMID:28930527

Caccamese, S., Azzolina, R., Furnari, G., Cormaci, M., & Grasso, S. (1981). Antimicrobial and Antiviral Activities of Some Marine Algae from Eastern Sicily. *Botanica Marina*, *24*, 365–368.

Campo, V. L., Kawano, D. F., da Silva, D. B. Jr, & Carvalho, I. (2009). Carrageenans: Biological properties, chemical modifications and structural analysis a review. *Carbohydrate Polymers*, *77*(2), 167–180. doi:10.1016/j.carbpol.2009.01.020

Carillo, P., Ciarmiello, L. F., Woodrow, P., Corrado, G., Chiaiese, P., & Rouphael, Y. (2020). Enhancing sustainability by improving plant salt tolerance through macro-and micro-algal biostimulants. *Biology (Basel)*, *9*(9), 253. doi:10.3390/biology9090253 PMID:32872247

Chanda, M.J., Benhima, R., Elmernissi, N., Kasmi, Y., & Karim, L,. Sbabou, L., & El Arroussi, H. (2020). Screening of microalgae liquid extracts for their bio stimulant properties on plant growth, nutrient uptake and metabolite profile of Solanum lycopersicum L. *Scientific Reports*, *10*(1). PMID:32071360

Chauhan, A., AbuAmarah, B. A., Kumar, A., Verma, J. S., Ghramh, H. A., Khan, K. A., & Ansari, M. J. (2019). Influence of gibberellic acid and different salt concentrations on germination percentage and physiological parameters of oat cultivars. *Saudi Journal of Biological Sciences*, *26*(6), 1298–1304. doi:10.1016/j.sjbs.2019.04.014 PMID:31516361

Cheah, W. Y., Show, P. L., Chang, J. S., Ling, T. C., & Juan, J. C. (2015). Biosequestration of atmospheric CO2 and flue gas-containing CO2 by microalgae. *Bioresource Technology*, *184*, 190–201. doi:10.1016/j.biortech.2014.11.026 PMID:25497054

Cheng, J., Lu, H., He, X., Yang, W., Zhou, J., & Cen, K. (2017). Mutation of Spirulina sp. by nuclear irradiation to improve growth rate under 15% carbon dioxide in flue gas. *Bioresource Technology*, *238*, 650–656. doi:10.1016/j.biortech.2017.04.107 PMID:28486198

Cheung, R. C. F., Wong, J. H., Pan, W. L., Chan, Y. S., Yin, C. M., Dan, X. L., Wang, H. X., Fang, E. F., Lam, S. K., Ngai, P. H. K., & … . (2014). Antifungal and Antiviral Products of Marine Organisms. *Applied Microbiology and Biotechnology*, *98*, 3475–3494.

Chiang, C. L., Lee, C. M., & Chen, P. C. (2011). Utilization of the cyanobacteria Anabaena sp. CH1 in biological carbon dioxide mitigation processes. *Bioresource Technology*, *102*(9), 5400–5405. doi:10.1016/j.biortech.2010.10.089 PMID:21232935

Chung, T.Y., Kuo, C.Y., Lin, W.J., Wang, W.L., & Chou, J.Y. (2018). Indole-3-acetic-acid-induced phenotypic plasticity in Desmodesmus algae. *Scientific Reports*, *8*(1), 10270.

Cooke, T. J., Poli, D., Sztein, A. E., & Cohen, J. D. (2002). Evolutionary Patterns in Auxin Action. *Plant Molecular Biology*, *49*(3/4), 319–338. doi:10.1023/A:1015242627321 PMID:12036257

Costa, O. Y., Raaijmakers, J. M., & Kuramae, E. E. (2018). Microbial extracellular polymeric substances: Ecological function and impact on soil aggregation. *Frontiers in Microbiology*, *9*, 9. doi:10.3389/fmicb.2018.01636 PMID:30083145

Craigie, J. S. (2011). Seaweed extract stimuli in plant science and agriculture. *Journal of Applied Phycology*, *23*(3), 371–393. doi:10.100710811-010-9560-4

Czerpak, R., Piotrowska, A., & Szulecka, K. (2006). Jasmonic acid affects changes in the growth and some components content in alga Chlorella vulgaris. *Acta Physiologiae Plantarum*, *28*(3), 195–203. doi:10.1007/BF02706531

Dawson, T. P., Jackson, S. T., House, J. I., Iain Colin, P., & Mace, G. M. (2011). Beyond predictions: Biodiversity conservation in a changing climate. *Science*, *332*(6025), 53–58. doi:10.1126cience.1200303 PMID:21454781

De Jesus Raposo, M. F., De Morais, R. M. S. C., & De Morais, A. M. M. B. (2013). Health applications of bioactive compounds from marine microalgae. *Life Sciences*, *93*(15), 479–486. doi:10.1016/j.lfs.2013.08.002 PMID:23994664

de Vasconcelos, A. C. F. (2020). Amelioration of Drought Stress on Plants under Biostimulant Sources. In *Plant Stress Physiology*. IntechOpen.

Doe, U. (2016). *National algal biofuels technology review*. Academic Press.

Du, K., Tao, H., Wen, X., Geng, Y., & Li, Y. (2015). Enhanced growth and lipid production of Chlorella pyrenoidosa by plant growth regulator GA3. *Fresenius Environmental Bulletin*, *24*, 3414–3419.

Duarte, J. H., de Morais, E. G., Radmann, E. M., & Costa, J. A. V. (2017). Biological CO_2 mitigation from coal power plant by *Chlorella fusca* and *Spirulina*sp. *Bioresource Technology*, *234*, 472–475. doi:10.1016/j.biortech.2017.03.066 PMID:28342576

El Arroussi, H., Elbaouchi, A., Benhima, R., Bendaou, N., Smouni, A., & Wahby, I. (2015). Halophilic microalgae Dunaliella salina extracts improve seed germination and seedling growth of Triticum aestivum L. under salt stress. In *II World Congress on the Use of Biostimulants in Agriculture 1148* (pp. 13-26).

El-Zeky, M. M., El-Shahat, R. M., Metwaly, G. S., & Elham, M. A. (2005). Using Cyanobacteria or *Azolla* as alternative nitrogen sources for rice production. *J. Agric. Mansoura Univ.*, *30*(9), 5567–5577.

Elarroussia, H., Elmernissia, N., Benhimaa, R., El Kadmiria, I. M., Bendaou, N., Smouni, A., & Wahbya, I. (2016). Microalgae polysaccharides a promising plant growth biostimulant. *Journal of Algal Biomass Utilization*, *7*, 55–63.

Elemike, E. E., Uzoh, I. M., Onwudiwe, D. C., & Babalola, O. O. (2019). The role of nanotechnology in the fortification of plant nutrients and improvement of crop production. *Applied Sciences (Basel, Switzerland)*, *9*(3), 499. doi:10.3390/app9030499

Esserti, S., Smaili, A., Rifai, L. A., Koussa, T., Makroum, K., Belfaiza, M., Kabil, E. M., Faize, L., Burgos, L., Alburquerque, N., & … . (2017). Protective Effect of Three Brown Seaweed Extracts against Fungal and Bacterial Diseases of Tomato. *Journal of Applied Phycology*, *29*, 1081–1093.

Falkowska, M., Pietryczuk, A., Piotrowska, A., Bajguz, A., Grygoruk, A., & Czerpak, R. (2011). The effect of gibberellic acid (GA3) on growth, metal biosorption and metabolism of the green algae Chlorella vulgaris (Chlorophyceae) Beijerinck exposed to cad-mium and lead stress. *Polish Journal of Environmental Studies*, *20*, 53–59.

Freitas, B. C. B., Cassuriaga, A. P. A., Morais, M. G., & Costa, J. A. V. (2017). Pentoses and light intensity increase the growth and carbohydrate production and alter the protein profile of *Chlorella minutissima*. *Bioresource Technology*, *28*, 248–253. doi:10.1016/j.biortech.2017.04.031 PMID:28437642

Froehlich, H. E., Afflerbach, J. C., Frazier, M., & Halpern, B. S. (2019). Blue Growth Potential to Mitigate Climate Change through Seaweed Offsetting. *Current Biology*, *29*(18), 1–7. doi:10.1016/j.cub.2019.07.041 PMID:31474532

Galasso, C., Gentile, A., Orefice, I., Ianora, A., Bruno, A., Noonan, D. M., & Brunet, C. (2019). Microalgal derivatives as potential nutraceutical and food supplements for human health: A focus on cancer prevention and interception. *Nutrients*, *11*(6), 1–22. doi:10.3390/nu11061226 PMID:31146462

García-Cubero, R., Moreno-Fernández, J., & García-González, M. (2017). Modelling growth and CO 2 fixation by *Scenedesmus vacuolatus* in continuous culture. *Algal Research*, *24*, 333–339. doi:10.1016/j.algal.2017.04.018

Gärtner, G., Stoyneva-Gärtner, M., & Uzunov, B. (2021). Algal Toxic Compounds and Their Aeroterrestrial, Airborne and Other Extremophilic Producers with Attention to Soil and Plant Contamination: A Review. *Toxins*, *13*, 322.

González, A., Contreras, R. A., & Moenne, A. (2013). Oligo-carrageenans enhance growth and contents of cellulose, essential oils and polyphenolic compounds in *Eucalyptus globulus* trees. *Molecules (Basel, Switzerland)*, *18*(8), 8740–8751. doi:10.3390/molecules18088740 PMID:23887716

González, A., Moenne, F., Gómez, M., Sáez, C. A., Contreras, R. A., & Moenne, A. (2014). Oligo-carrageenan kappa increases NADPH, ascorbate and glutathione syntheses and TRR/TRX activities enhancing photosynthesis, basal metabolism, and growth in Eucalyptus trees. *Front Plant Sci*, *5*, 512. doi:10.3389/fpls.2014.00512 PMID:25352851

González-Garcinuño, Á., Sánchez-Álvarez, J. M., Galán, M. A., & Del Valle, E. M. M. (2016). Understanding and optimizing the addition of phytohormones in the culture of microalgae for lipid production. *Biotechnology Progress*, *32*, 1203–1211.

Graça, C., Fradinho, P., Sousa, I., & Raymundo, A. (2018). Impact of Chlorella vulgaris on the rheology of wheat flour dough and bread texture. *Lebensmittel-Wissenschaft + Technologie*, *89*, 466–474. doi:10.1016/j.lwt.2017.11.024

Hamed, S. M. (2007). *Studies on nitrogen fixing Cyanobacteria* (M. Sc. Thesis). Botany Department, Faculty of Science, Beni-Suef University, Beni-Suef, Egypt.

Hamed, S. M., Abd El-Rhman, A. A., Abdel-Raouf, N., & Ibraheem, I. B. (2018). Role of marine macroalgae in plant protection & improvement for sustainable agriculture technology. *Beni-Suef University Journal of Basic and Applied Sciences*, *7*(1), 104–110. doi:10.1016/j.bjbas.2017.08.002

Han, G. Z. (2016). Evolution of jasmonate biosynthesis and signaling mechanisms. *Journal of Experimental Botany*, *68*(6), 1323–1331. doi:10.1093/jxb/erw470 PMID:28007954

Han, X., Zeng, H., Bartocci, P., Fantozzi, F., & Yan, Y. (2018). Phytohormones and Effects on Growth and Metabolites of Microalgae: A Review. *Fermentation*, *4*, 25.

Hankins, S. D., & Hockey, H. P. (1990). The Effect of a Liquid Seaweed Extract from Ascophyllum nodosum (Fucales, Phaeophyta) on the Two-Spotted Red Spider Mite Tetranychus Urticae. *Hydrobiologia*, *204*, 555–559.

Hernández-Herrera, R. M., Santacruz-Ruvalcaba, F., Zañudo-Hernández, J., & Hernández-Carmona, G. (2016). Activity of seaweed extracts and polysaccharide-enriched extracts from Ulva lactuca and Padina gymnospora as growth promoters of tomato and mung bean plants. *Journal of Applied Phycology*, *28*(4), 2549–2560. doi:10.100710811-015-0781-4

Hien, N. Q., Nagasawa, N., Tham, L. X., Yoshii, F., Dang, V. H., Mitomo, H., Makuuchi, K., & Kume, T. (2000). Growth promotion of plants with depolymerized alginates by irradiation. *Radiation Physics and Chemistry*, *59*(1), 97–101. doi:10.1016/S0969-806X(99)00522-8

Holdt, S. L., & Kraan, S. (2011). Bioactive compounds in seaweed: Functional food applications and legislation. *Journal of Applied Phycology*, *23*(3), 543–597. doi:10.100710811-010-9632-5

Hu, C., Zhang, D., Huang, Z., & Liu, Y. (2003). The vertical microdistribution of cyanobacteria and green algae within desert crusts and the development of the algal crusts. *Plant and Soil*, *257*(1), 97–111. doi:10.1023/A:1026253307432

Huisman, J., Codd, G. A., Paerl, H. W., Ibelings, B. W., Verspagen, J. M., & Visser, P. M. (2018). Cyanobacterial blooms. *Nature Reviews. Microbiology*, *16*(8), 471–483. doi:10.103841579-018-0040-1 PMID:29946124

Huleihel, M., Ishanu, V., Tal, J., & Arad, S. (2001). Antiviral effect of red microalgal polysaccharides on *Herpes simplex* and *Varicella zoster* viruses. *Journal of Applied Phycology*, *13*(2), 127–134. doi:10.1023/A:1011178225912

Ibraheem, I. B. M. (2007). Cyanobacteria as alternative biological conditioners for bioremediation of barren soil. *Egyptian J. Phycol*, *8*(1), 99–116. doi:10.21608/egyjs.2007.114548

Ibrahim, W. M., Ali, R. M., Hemida, K., & Sayed, M. A. (2014). Role of Ulva lactuca extract in alleviation of salinity stress on wheat seedlings. *TheScientificWorldJournal*, *2014*, 1–11. doi:10.1155/2014/847290 PMID:25436231

Jacobs, W. P. A. (1993). Search for Some Angiosperm Hormones and Their Metabolites in *Caulerpa paspaloides* (Chlorophyta). *Journal of Phycology*, *29*(5), 595–600. doi:10.1111/j.0022-3646.1993.00595.x

Janssen, C. I. F., & Kiliaan, A. J. (2014). Long-chain polyunsaturated fatty acids (LCPUFA) from genesis to senescence: The influence of LCPUFA on neural development, aging, and neurodegeneration. *Progress in Lipid Research*, *53*, 1–17. doi:10.1016/j.plipres.2013.10.002 PMID:24334113

Jäpelt, R. B., & Jakobsen, J. (2013). Vitamin D in plants: A review of occurrence, analysis, and biosynthesis. *Frontiers in Plant Science*, *4*(2), 136. doi:10.3389/fpls.2013.00136 PMID:23717318

Jia, R., Qu, Z., You, P., & Qu, D. (2018). Effect of biochar on photosynthetic microorganism growth and iron cycling in paddy soil under different phosphate levels. *The Science of the Total Environment*, *612*, 223–230. doi:10.1016/j.scitotenv.2017.08.126 PMID:28850841

Jiang, Y., Zhang, W., Wang, J., Chen, Y., Shen, S., & Liu, T. (2013). Utilization of simulated flue gas for cultivation of *Scenedesmus dimorphus*. *Bioresource Technology*, *128*, 359–364. doi:10.1016/j.biortech.2012.10.119 PMID:23201515

Jiménez, E., Dorta, F., Medina, C., Ramírez, A., Ramírez, I., & Peña-Cortés, H. (2011). Anti-Phytopathogenic Activities of Macro-Algae Extracts. *Marine Drugs*, *9*, 739–756.

Kadir, W. N. A., Lam, M. K., Uemura, Y., Lim, J. W., & Lee, K. T. (2018). Harvesting and pre-treatment of microalgae cultivated in wastewater for biodiesel production: A review. *Energy Conversion and Management*, *171*, 1416–1429. doi:10.1016/j.enconman.2018.06.074

Kanamarlapudi, S.L.R.K., Chintalpudi, V.K., & Muddada, S. (2018). Application of Biosorption for Removal of Heavy Metals from Wastewater. *Biosorption, 69*.

Kao, C. Y., Chen, T. Y., Chang, Y.-B., Chiu, T.-W., Lin, H.-Y., Chen, C.-D., Chang, J.-S., & Lin, C.-S. (2014). Utilization of carbon dioxide in industrial flue gases for the cultivation of microalga Chlorella sp. *Bioresource Technology*, *166*(8), 485–493. doi:10.1016/j.biortech.2014.05.094 PMID:24950094

Kassim, M. A., & Meng, T. K. (2017). Carbon dioxide (CO_2) biofixation by microalgae and its potential for biorefinery and biofuel production. *The Science of the Total Environment*, *584–585*, 1121–1129. doi:10.1016/j.scitotenv.2017.01.172 PMID:28169025

Kaur, M. (2017). *Kinetics of CO_2 sequestration by* Anabaena variabilis *in a stirred tank batch type photobioreactor*. Academic Press.

Kaushik, B. D. (2014). Developments in cyanobacterial biofertilizer. In Proc Indian Nat Sci Acad (Vol. 80, No. 2: pp. 379-388). doi:10.16943/ptinsa/2014/v80i2/55115

Khan, M. I., Jin, H. S., & Kim, J. D. (2018). The promising future of microalgae: Current status, challenges, and optimization of a sustainable and renewable industry for biofuels, feed, and other products. *Microbial Cell Factories*, *17*(1), 36. doi:10.118612934-018-0879-x PMID:29506528

Kobayashi, M., Hirai, N., Kurimura, Y., Ohigashi, H., & Tsuji, Y. (1997). Abscisic Acid-Dependent Algal Morphogenesis in the Unicellular Green Alga *Haematococcuspluvialis*. *Plant Growth Regulation*, *22*(2), 79–85. doi:10.1023/A:1005862809711

Kozlova, T. A., Hardy, B. P., Krishna, P., & Levin, D. B. (2017). Effect of phytohormones on growth and accumulation of pigments and fatty acids in the microalgae Scenedesmus quadricauda. *Algal Research*, *27*, 325–334.

Kulik, M. M. (1995). The Potential for Using Cyanobacteria (Blue-Green Algae) and Algae in the Biological Control of Plant Pathogenic Bacteria and Fungi. *European Journal of Plant Pathology*, *101*, 585–599.

Kumar, A., Agarwal, D. K., Kumar, S., Reddy, Y. M., Chintagunta, A. D., Saritha, K. V., Pal, G., & Kumar, S. P. J. (2019). Nutraceuticals derived from seed storage proteins: Implications for health wellness. *Biocatalysis and Agricultural Biotechnology*, *17*, 710–717. doi:10.1016/j.bcab.2019.01.044

Kumar, A., Imran, M., Kumar, S. P. J., Kumar, M., Singh, A. N., & Srivastava, A. K. (2018). Assessment of soil urease activity and its kinetic parameters in chilli rhizosphere under the influence of pesticides. Biopesticides. *International*, *14*, 47–57.

Kumar, K. A., & Rengasamy, R. (2000). Antibacterial Activities of Seaweed Extracts/Fractions Obtained through a TLC Profile against the Phytopathogenic Bacterium Xanthomonas Oryzae Pv. Oryzae. *Botanica Marina*, *43*, 417–421.

Labeeuw, L., Khey, J., Bramucci, A.R., Atwal, H., de la Mata, A.P., Harynuk, J., & Case, R.J. (2016). Indole-3-acetic acid is produced by *Emiliania huxleyi coccolith*-bearing cells and triggers a physiological response in bald cells. *Frontiers in Microbiology, 8*(7), 828.

Ladha, J. K., & Reddy, P. M. (2003). Nitrogen fixation in rice systems: State of knowledge and future prospects. *Plant and Soil*, *252*(1), 151–167. doi:10.1023/A:1024175307238

Lahaye, M., & Robic, A. (2007). Structure and functional properties of ulvan, a polysaccharide from green seaweeds. *Biomacromolecules*, *8*(6), 1765–1774. doi:10.1021/bm061185q PMID:17458931

Laporte, D., Vera, J., Chandía, N. P., Zuñiga, E., Matsuhiro, B., & Moenne, A. (2007). Structurally unrelated oligosaccharides obtained frommarine macroalgae differentially stimulate growth and defense against TMVin tobacco plants. *Journal of Applied Phycology*, *19*, 79–88. doi:10.100710811-006-9114-y

Lawrence, K. P., Long, P. F., & Young, A. R. (2017). Mycosporine-like Amino Acids for Skin Photoprotection. *Current Medicinal Chemistry*, 24. PMID:28554325

Lee, R. F., Valkirs, A. O., & Seligman, P. F. (1989). Importance of microalgae in the biodegradation of tributyltin in estuarine waters. *Environmental Science & Technology*, *23*(12), 1515–1518. doi:10.1021/es00070a010

Li, H., Zhao, Q., & Huang, H. (2019). Current states and challenges of salt-affected soil remediation by cyanobacteria. *The Science of the Total Environment*, *669*, 258–272. doi:10.1016/j.scitotenv.2019.03.104 PMID:30878933

Li, L., Ni, R., Shao, Y., & Mao, S. (2014). Carrageenan and its applications in drug delivery. *Carbohydrate Polymers*, *103*, 1–11. doi:10.1016/j.carbpol.2013.12.008 PMID:24528694

Lötze, E., & Hoffman, E. W. (2015). Nutrient composition and content of various biological active compounds of three South African-based commercial seaweed biostimulants. *Journal of Applied Phycology*.

Lu, Y., & Xu, J. (2015). Phytohormones in microalgae: A new opportunity for microalgal biotechnology? *Trends in Plant Science*, *20*, 273–282.

Madadi, R., Pourbabaee, A. A., Tabatabaei, M., Zahed, M. A., & Naghavi, M. R. (2016). Treatment of Petrochemical Wastewater by the Green Algae Chlorella vulgaris. *International Journal of Environmental of Research*, *10*(4), 555–560.

Marián, F. D., García-Jiménez, P., & Robain, R. R. (2000). Polyamines in Marine Macroalgae: Levels of Putrescine, Spermidine and Spermine in the Thalli and Changes in Their Concentration during Glycerol-Induced Cell Growth In Vitro. *Physiologia Plantarum*, *110*(4), 530–534. doi:10.1111/j.1399-3054.2000.1100416.x

Maršálek, B. (1992). Zahradníˇcková, H.; Hronková, M. Extracellular Abscisic Acid Produced by Cyanobacteria under Salt Stress. *Journal of Plant Physiology*, *139*, 506–508.

Masondo, N. A., Kulkarni, M. G., Finnie, J. F., & Van Staden, J. (2018). Influence of biostimulants-seed-priming on Ceratotheca triloba germination and seedling growth under low temperatures, low osmotic potential and salinity stress. *Ecotoxicology and Environmental Safety*, *147*, 43–48. doi:10.1016/j.ecoenv.2017.08.017 PMID:28826029

Matson, M. M., & Atsumi, S. (2018). Photomixotrophic chemical production in cyanobacteria. *Current Opinion in Biotechnology*, *50*, 65–71. doi:10.1016/j.copbio.2017.11.008 PMID:29179151

Mattner, S. W., Wite, D., Riches, D. A., Porter, I. J., & Arioli, T. (2013). The effect of kelp extract on seedling establishment of broccoli on contrasting soil types in southern Victoria, Australia. *Biological Agriculture and Horticulture*, *29*(4), 258–270. doi:10.1080/01448765.2013.830276

Mazard, S., Penesyan, A., Ostrowski, M., Paulsen, I. T., & Egan, S. (2016). Tiny microbes with a big impact: The role of cyanobacteria and their metabolites in shaping our future. *Marine Drugs*, *14*(5), 97. doi:10.3390/md14050097 PMID:27196915

Mishra, P., Bhoomika, K., & Dubey, R. S. (2013). Differential responses of antioxidative defense system to prolonged salinity stress in salt-tolerant and salt-sensitive Indica rice (*Oryza sativa*L.) seedlings. *Protoplasma*, *250*(1), 3–19. doi:10.100700709-011-0365-3 PMID:22194018

Mondal, M., Goswami, S., Ghosh, A., Oinam, G., Tiwari, O.N., Das, P., & Halder, G.N. (2017). Production of biodiesel from microalgae through biological carbon capture: A review. *Biotech*, *7*(2), 99.

Moreira, D., & Pires, J. C. M. (2016). Atmospheric CO_2 capture by algae: Negative carbon dioxide emission path. *Bioresource Technology*, *215*, 371–379. doi:10.1016/j.biortech.2016.03.060 PMID:27005790

Mousavi, P., Montazeri-Najafabady, N., Abolhasanzadeh, Z., Mohagheghzadeh, A., Hamidi, M., Niazi, A., Morowvat, M. H., & Ghasemi, Y. (2016). Investigating the effects of phytohormones on growth and beta-carotene production in a naturally isolates stain of Dunaliella salina. *Journal of Applied Pharmaceutical Science*, *6*, 164–171.

Nagorskaia, V. P., Reunov, A. V., Lapshina, L. A., Ermak, I. M., & Barabanova, A. O. (2010). Inhibitory effect of kappa/beta-carrageenan from red alga Tichocarpus crinitus on the development of a potato virus X infection in leaves of Datura stramonium L. *Izvestiia Akademii Nauk. Seriia Biologicheskaia*, *6*, 756–761.

Niemann, D. I., & Dorffling, K. (1980). Growth Inhibitors and Growth Promoters in *Enteromorpha compressa* (Chlorophyta). *Journal of Phycology*, *16*(3), 383–389. doi:10.1111/j.1529-8817.1980.tb03049.x

Nimura, K., & Mizuta, H. (2002). Inducible Effects of Abscisic Acid on Sporophyte Discs from *Laminaria japonica* Areschoug (*Laminariales, Phaeophyceae*). *Journal of Applied Phycology*, *14*(3), 159–163. doi:10.1023/A:1019929712633

Noble, A., Kisiala, A., Galer, A., Clysdale, D., & Emery, R. J. N. (2014). *Euglenagracilis* (Euglenophyceae) produces abscisic acid and cytokinins and responds to their exogenous application singly and in combination with other growth regulators. *European Journal of Phycology*, *49*(2), 244–254. doi:10.1080/09670262.2014.911353

Ördög, V., Stirk, W. A., van Staden, J., Novak, O., & Strnad, M. (2004). Endogenous Cytokinins in Three Genera of Microalgae from the Chlorophyta. *Journal of Phycology*, *40*(1), 88–95. doi:10.1046/j.1529-8817.2004.03046.x

Pan, S., Jeevanandam, J., & Danquah, M. K. (2019). Benefits of Algal Extracts in Sustainable Agriculture. In *Grand Challenges in Marine Biotechnology* (pp. 501–534). Springer Science and Business Media LLC.

Paulert, R., Talamini, V., Cassolato, J. E. F., Duarte, M. E. R., Noseda, M. D., Smania, A. Jr, & Stadnik, M. J. (2009). Effects of sulfated polysaccharide and alcoholic extracts from green seaweed *Ulva fasciata* on anthracnose severity and growth of common bean (*Phaseolus vulgaris* L.). *Journal of Plant Diseases and Protection*, *116*(6), 263–270. doi:10.1007/BF03356321

Pérez, M. J., Falqué, E., & Domínguez, H. (2016). Antimicrobial action of compounds from marine seaweed. *Marine Drugs*, *14*(3), 52. doi:10.3390/md14030052 PMID:27005637

Piotrowska-Niczyporuk, A., Bajguz, A., Kotowska, U., Bralska, M., & Talarek-Karwel, M. (2018). Growth, Metabolite Profile, Oxidative Status, and Phytohormone Levels in the Green Alga *Acutodesmus obliquus* Exposed to Exogenous Auxins and Cytokinins. *Journal of Plant Growth Regulation, 37*(4), 1159-74.

Polevoi, V. V., Tarakhovskaya, E. R., Maslov, Yu. I., & Polevoi, A. V. (2003). Induction of Polarity in Zygotes of *Fucusvesiculosus* L. by Auxin. *Ontogenez, 34*, 432–437. PMID:14714452

Prajapati, V. D., Maheriya, P. M., Jani, G. K., & Solanki, H. K. (2014). Carrageenan: A natural sea weed polysaccharide and its applications. *Carbohydrate Polymers, 105*, 97–112. doi:10.1016/j.carbpol.2014.01.067 PMID:24708958

Prasanna, R., Triveni, S., Bidyarani, N., Babu, S., Yadav, K., Adak, A., Khetarpal, S., Pal, M., Shivay, Y. S., & Saxena, A. K. (2013). Evaluating the efficacy of cyanobacterial formulations and biofilmed inoculants for leguminous crops. *Archives of Agronomy and Soil Science, 60*, 349–366.

Priya, M., Gurung, N., Mukherjee, K., & Bose, S. (2014). Microalgae in removal of heavy metal and organic pollutants from soil. In S. Das (Ed.), *Microbial Biodegradation and Bioremediation* (pp. 519–537). Elsevier.

Rai, A. N., Singh, A. K., & Syiem, M. B. (2019) Plant Growth-Promoting Abilities in Cyanobacteria. In Cyanobacteria (pp. 459-476). Academic Press. doi:10.1016/B978-0-12-814667-5.00023-4

Rashwan, R. S., & Hammad, D. M. (2020). Toxic Effect of Spirulina Platensis and Sargassum Vulgar as Natural Pesticides on Survival and Biological Characteristics of Cotton Leaf Worm Spodoptera littoralis. *Sci. Afr., 8*, e00323.

Rayorath, P., Khan, W., Palanisamy, R., MacKinnon, S. L., Stefanova, R., Hankins, S. D., Critchley, A. T., & Prithiviraj, B. (2008). Extracts of the brown seaweed *Ascophyllum nodosum*induce gibberellic acid (GA3)-independent amylase activity in barley. *Journal of Plant Growth Regulation, 27*(4), 370–379. doi:10.100700344-008-9063-6

Razzak, S. A., Hossain, M. M., Lucky, R. A., Bassi, A. S., & de Lasa, H. (2013). Integrated CO_2 capture, wastewater treatment and biofuel production by microalgae culturing—A review. *Renewable & Sustainable Energy Reviews, 27*, 622–653. doi:10.1016/j.rser.2013.05.063

Reddy, C. N., Nguyen, H. T., Noori, M. T., & Min, B. (2019). Potential Applications of Algae in the Cathode of Microbial Fuel Cells for Enhanced Electricity Generation with Simultaneous Nutrient Removal and Algae Biorefinery: Current Status and Future Perspectives. *Bioresource Technology, 292*, 122010. doi:10.1016/j.biortech.2019.122010 PMID:31473037

Rengasamy, K. R. R., Kulkarni, M. G., Stirk, W. A., & Van Staden, J. (2015). Eckol-a new plant growth stimulant from the brown seaweed *Ecklonia maxima. Journal of Applied Phycology, 27*(1), 581–587. doi:10.100710811-014-0337-z

Rincon, L., Jaramilo, J. J., & Cardona, C. A. (2014). Comparison of feedstocks and technologies for biodiesel production: An environmental and techno-economic evaluation. *Renewable Energy, 69*, 479–487. doi:10.1016/j.renene.2014.03.058

Rizzello, C. G., Tagliazucchi, D., Babini, E., Sefora Rutella, G., Taneyo Saa, D. L., & Gianotti, A. (2016). Bioactive peptides from vegetable food matrices: Research trends and novel biotechnologies for synthesis and recovery. *Journal of Functional Foods*, *27*, 549–569. doi:10.1016/j.jff.2016.09.023

Ronga, D., Biazzi, E., Parati, K., Carminati, D., Carminati, E., & Tava, A. (2019). Microalgal biostimulants and biofertilisers in crop productions. *Agronomy (Basel)*, *9*(4), 192. doi:10.3390/agronomy9040192

Rosales-Mendoza, S., García-Silva, I., González-Ortega, O., Sandoval-Vargas, J. M., Malla, A., & Vimolmangkang, S. (2020). The Potential of Algal Biotechnology to Produce Antiviral Compounds and Biopharmaceuticals. *Molecules (Basel, Switzerland)*, *25*, 4049.

Sacramento, A. T., García-Jiménez, P., Alcázar, R., Tiburcio, A. F., & Robaina, R. R. (2004). Influence of Polyamines on the Sporulation of *Grateloupia* (Halymeniaceae, Rhodophyta). *Journal of Phycology*, *40*(5), 887–894. doi:10.1111/j.1529-8817.2004.03183.x

Sahu, D., Priyadarshani, I., & Rath, B. (2012). Cyanobacteria–as potential biofertilizer. *CIB Tech J Microbiol*, *1*, 20–26.

Sangha, J. S., Kelloway, S., Critchley, A. T., & Prithiviraj, B. (2014). Seaweeds (Macroalgae) and their extracts as contributors of plant productivity and quality: The current status of our understanding. *Sea Plants*, *71*, 189–219. doi:10.1016/B978-0-12-408062-1.00007-X

Sano, Y. (1999). Antiviral Activity of Alginate against Infection by Tobacco Mosaic Virus. *Carbohydrate Polymers*, *38*, 183–186.

Sansone, C., Nuzzo, G., Galasso, C., Casotti, R., Fontana, A., Romano, G., & Ianora, A. (2018). The Marine Dinoflagellate *Alexandrium andersoni* Induces Cell Death in Lung and Colorectal Tumor Cell Lines. *Marine Biotechnology (New York, N.Y.)*, *20*(3), 343–352. doi:10.100710126-018-9817-5 PMID:29679250

Sathasivam,, R., Radhakrishnan,, R., & Hashem,, A., & Abd Allah E.F. (2019). Microalgae metabolites: A rich source for food and medicine. *Saudi Journal of Biological Sciences*, *26*, 709–722. doi:10.1016/j.sjbs.2017.11.003 PMID:31048995

Schiener, P., Black, K. D., Stanley, M. S., & Green, D. H. (2015). The seasonal variation in the chemical composition of the kelp species *Laminaria digitata, Laminaria hyperborea, Saccharina latissima* and *Alaria esculenta. Journal of Applied Phycology*, *27*(1), 363–373. doi:10.100710811-014-0327-1

Shaish, A., Ben-Amotz, A., & Avron, M. (1992). Biosynthesis of β-carotene in Dunaliella. *Methods in Enzymology*, *213*, 439–444. doi:10.1016/0076-6879(92)13145-N

Shannon, E., & Abu-Ghannam, N. (2016). Antibacterial derivatives of marine algae: An overview of pharmacological mechanisms and applications. *Marine Drugs*, *14*(4), 81. doi:10.3390/md14040081 PMID:27110798

Sharma, R., Khokhar, M. K., Jat, R. L., & Khandelwal, S. K. (2012). Role of algae and cyanobacteria in sustainable agriculture system. *Wudpecker Journal of Agricultural Research*, *1*(9), 381–388.

Shields, L. M., & Durrell, L. W. (1964). Algal in relation to soil fertility. *Botanical Review*, *30*(1), 92–128. doi:10.1007/BF02858614

Shukla, M., & Kumar, S. (2018). Algal growth in photosynthetic algal microbial fuel cell and its subsequent utilization for biofuels. *Renewable & Sustainable Energy Reviews*, *82*, 402–414. doi:10.1016/j.rser.2017.09.067

Shunni, Z., Yajie, W., Wei, H., Jin, X., Zhongming, W., Jingliang, X., & Zhenhong, Y. (2014). Enhanced accumulation of carbohydrate and starch in *Chlorella zofingiensis* induced by nitrogen starvation. *Applied Biochemistry and Biotechnology*, *174*(7), 2435–2445. doi:10.100712010-014-1183-9 PMID:25209553

Singh, J. S., Kumar, A., Rai, A. N., & Singh, D. P. (2016). Cyanobacteria: A precious bio-resource in agriculture, ecosystem, and environmental sustainability. *Frontiers in Microbiology*, *7*, 529. doi:10.3389/fmicb.2016.00529 PMID:27148218

Singh, N. K., & Dhar, D. W. (2007). Nitrogen and phosphorous scavenging potential in microalgae. *Indian Journal of Biotechnology*, *6*, 52–56.

Singh Bhan, U. (2013). Microalgae: A promising tool for carbon sequestration. *Mitigation and Adaptation Strategies for Global Change*, *18*(1), 73–95. doi:10.100711027-012-9393-3

Sitnik, K. M., Musatenko, L. I., Vasyuk, V. A., Vedenicheva, N. P., Generalova, V. M., Martin, G. G., & Nesterova, A. N. (2003). Gormonal'nii kompleks roslin i gribiv [Hormonal Complex in Plants and Fungi]. Kiiv: Akademperiodika.

Specht, E. A., Karunanithi, P. S., Gimpel, J. A., Ansari, W. S., & Mayfield, S. P. (2017). Host Organisms: Algae. In *Industrial Biotechnology* (pp. 605–641). John Wiley & Sons, Ltd.

Stirk, W.A., Novák, O., Hradecká, V., Pěnčík, A., Rolčík, J., Strnad, M., & Van Staden, J. (2009). Endogenous cytokinins, auxins and abscisic acid in *Ulva fasciata* (Chlorophyta) and *Dictyota humifusa* (Phaeophyta): towards understanding their biosynthesis and homoeostasis. *European Journal of Phycology, 44*(2), 231-40.

Stirk, W. A., Novak, O., Strnad, M., & van Staden, J. (2003). Cytokinins in Macroalgae. *Plant Growth Regulation*, *41*(1), 13–24. doi:10.1023/A:1027376507197

Stuart, R., William, S., Laisvyde, S., Jose, P., James, F., David, R., & Jill, N. (2004). Double-masked, placebo-controlled, randomized trial of lutein and antioxidant supplementation in the intervention of atrophic age-related macular degeneration: The Veterans LAST study (Lutein Antioxidant Supplementation Trial). *Optometry-Journal of the American Optometric Association*, *75*(4), 216–229. doi:10.1016/S1529-1839(04)70049-4 PMID:15117055

Sultana, V., Baloch, G. N., Ara, J., Ehteshamul-Haque, S., Tariq, R. M., & Athar, M. (2012). Seaweeds as an Alternative to Chemical Pesticides for the Management of Root Diseases of Sunflower and Tomato. *Journal of Applied Botany and Food Quality*, *84*, 162.

Sydney, E. B., Sturm, W., de Carvalho, J. C., Thomaz-Soccol, V., Larroche, C., Pandey, A., & Soccol, C. R. (2010). Potential carbon dioxide fixation by industrially important microalgae. *Bioresource Technology*, *101*(15), 5892–5896. doi:10.1016/j.biortech.2010.02.088 PMID:20350804

Taher, H., Al-Zuhair, S., Al-Marzouqi, A., Haik, Y., & Farid, M. (2015). Growth of microalgae using CO_2 enriched air for biodiesel production in supercritical CO2. *Renewable Energy*, *82*, 61–70. doi:10.1016/j.renene.2014.08.013

Tang, D., Han, W., Li, P., Miao, X., & Zhong, J. (2011). CO_2 biofixation and fatty acid composition of Scenedesmus obliquus and Chlorella pyrenoidosa in response to different CO2 levels. *Bioresource Technology*, *102*(3), 3071–3076. doi:10.1016/j.biortech.2010.10.047 PMID:21041075

Tiwari, D. N., Kumar, A., & Mishra, A. K. (1991). Use of cyanobacterial diazotrophic technology in rice agriculture. *Applied Biochemistry and Biotechnology*, *28/29*(1), 387–396. doi:10.1007/BF02922618

Tiwari, P. K., Misra, A. K., & Venturino, E. (2017). The role of algae in agriculture: A mathematical study. *Journal of Biological Physics*, *43*(2), 297–314. doi:10.100710867-017-9453-8 PMID:28550636

Toker, O. S. (2018). *Porphyridum Cruentum as a natural colorant in chewing gum. Food Science and Technology*. AHEAD.

Toll, A. J. (2012). *Algal Development: Molecular and Cellular Aspects*. Springer.

Tominaga, N., Takahata, M., & Tominaga, H. (1993). Effects of NaCl and KNO_3 Concentrations on the Abscisic Acid Content of *Dunaliella* sp. (Chlorophyta). *Hydrobiologya*, *267*(1-3), 163–168. doi:10.1007/BF00018798

Tripathi, S., & Hussain, T. (2021). Treatment of Industrial wastewater through new approaches using Algae biomass. In M. Shah (Ed.), *The Future of Effluent Treatment Plants-Biological Treatment Systems* (pp. 89–112). Elsevier., doi:10.1016/B978-0-12-822956-9.00006-4

Vikram, G., Srikanth, K., & Swamy, N. R. (2011). Effect of plant growth regulatorson in vitro organogenesis in cultivated tomato (*Solanum lycopersicum* L.). *J. Research in Biology*, *1*, 263–268.

Vinoth, S., Gurusaravanan, P., & Jayabalan, N. (2014). Optimization of somaticembryogenesis protocol in *Lycopersicon esculentum* L. using plant growth regulators and seaweed extracts. *Journal of Applied Phycology*, *26*(3), 1527–1537. doi:10.100710811-013-0151-z

Vymazal, J. (1995). *Algae and element cycling in wetlands*. Lewis Publishers Boca Raton.

Wang, C., Liu, Y., Li, S. S., & Han, G. Z. (2014). Origin of plant auxin biosynthesis in charophyte algae. *Trends in Plant Science*, *19*(12), 741–743. doi:10.1016/j.tplants.2014.10.004 PMID:25457112

Wells, M. L., Potin, P., Craigie, J. S., Raven, J. A., Merchant, S. S., Helliwell, K. E., & Brawley, S. H. (2017). Algae as nutritional and functional food sources: Revisiting our understanding. *Journal of Applied Phycology*, *29*(2), 949–982. doi:10.100710811-016-0974-5 PMID:28458464

Wilson, R. H., Alonso, H., & Whitney, S. M. (2016). Evolving Methanococcoides burtonii archaeal Rubisco for improved photosynthesis and plant growth. *Scientific Reports*, *6*(1), 22284. doi:10.1038rep22284 PMID:26926260

Win, T. T., Barone, G. D., Secundo, F., & Fu, P. (2018). Algal biofertilizers and plant growth stimulants for sustainable agriculture. *Industrial Biotechnology (New Rochelle, N.Y.)*, *14*(4), 203–211. doi:10.1089/ind.2018.0010

Wiśniewska, K., Lewandowska, A. U., & Śliwińska-Wilczewska, S. (2019). The importance of cyanobacteria and microalgae present in aerosols to human health and the environment–Review study. *Environment International*, *131*, 104964. doi:10.1016/j.envint.2019.104964 PMID:31351382

Xiong, J. Q., Kurade, M. B., & Jeon, B. H. (2018). Can microalgae remove pharmaceutical contaminants from water? *Trends in Biotechnology*, *36*(1), 30–44. doi:10.1016/j.tibtech.2017.09.003 PMID:28993012

Yadav, G., & Sen, R. (2017). Microalgal green refinery concept for biosequestration of carbon-dioxide vis-à-vis wastewater remediation and bioenergy production: Recent technological advances in climate research. *Journal of CO2 Utilization, 17*, 188–206. doi:10.1016/j.jcou.2016.12.006

Yan, N., Fan, C., Chen, Y., & Hu, Z. (2016). The Potential for Microalgae as Bioreactors to Produce Pharmaceuticals. *International Journal of Molecular Sciences*, *17*, 962.

Yandigeri, M. S., Yadav, A. K., Srinivasan, R., Kashyap, S., & Pabbi, S. (2011). Studies on mineral phosphate solubilization by cyanobacteria *Westiellopsis* and Anabaena. *Microbiology*, *80*(4), 558–565. doi:10.1134/S0026261711040229 PMID:22073557

Yang, B., Liu, J., Liu, B., Sun, P., Ma, X., Jiang, Y., & Chen, F. (2015). Development of a stable genetic system for *Chlorella vulgaris*—A promising green alga for CO_2 biomitigation. *Algal Research*, *12*, 134–141. doi:10.1016/j.algal.2015.08.012

Yim, J. H., Son, E., Pyo, S., & Lee, H. K. (2005). Novel sulfated polysaccharide derived from red-tide microalga *Gyrodinium impudicum* strain KG03 with immunostimulating activity in vivo. *Marine Biotechnology (New York, N.Y.)*, *7*(4), 331–338. doi:10.100710126-004-0404-6 PMID:15976942

Yin, K., Wang, Q., Lv, M., & Chen, L. (2018). Microorganism remediation strategies towards heavy metals. *Chemical Engineering Journal*.

Yokota, T., Kim, S. K., Fukui, Y., Takahashi, N., Takeuchi, Y., & Takematsu, T. (1987). Brassinosteroids and Sterols from a Green Alga, *Hydrodictyon reticulatum*: Configuration at C-24. *Phytochemistry*, *26*(2), 503–506. doi:10.1016/S0031-9422(00)81442-9

Yoshida, K., Igarashi, E., Wakatsuki, E., Miyamoto, K., & Hirata, K. (2004). Mitigation of osmotic and salt stresses by abscisic acid through reduction of stress-derived oxidative damage in Chlamydomonas reinhardtii. *Plant Science*, *167*, 1335–1341.

Zaborowska, M., Kucharski, J., & Wyszkowska, J. (2015). Using basalt flour and brown algae to improve biological properties of soil contaminated with cadmium. *Soil and Water Research*, *10*(3), 181–188. doi:10.17221/281/2014-SWR

Zhang, Y. Y., Li, H. X., Shu, W. B., Zhang, C. J., & Ye, Z. B. (2011). RNA interference of a mitochondrial APX gene improves vitamin C accumulation in tomato fruit. *Scientia Horticulturae*, *129*(2), 220–226. doi:10.1016/j.scienta.2011.03.025

Zhao, J., Li, X. F., Ren, Y. P., Wang, X. H., & Jian, C. (2012). Electricity generation from Taihu Lake cyanobacteria by sediment microbial fuel cells. *Journal of Chemical Technology and Biotechnology (Oxford, Oxfordshire)*, *87*(11), 1567–1573. doi:10.1002/jctb.3794

Zheng, Y., Jin, R., Zhang, X., Wang, Q., & Wu, J. (2019). The considerable environmental benefits of seaweed aquaculture in China. *Stochastic Environmental Research and Risk Assessment, 33*(4–6), 1203–1221. doi:10.100700477-019-01685-z

Zhou, W., Wang, J., Chen, P., Ji, C., Kang, Q., Lu, B., & Ruan, R. (2017). Bio-mitigation of carbon dioxide using microalgal systems: Advances and perspectives. *Renewable & Sustainable Energy Reviews, 76*, 1163–1175. doi:10.1016/j.rser.2017.03.065

Žižková, E., Kubeš, M., Dobrev, P.I., Přibyl, P., Šimura, J., Zahajská, L., Záveská Drábková, L., Novák, O., & Motyka, V. (2016). Control of cytokinin and auxin homeostasis in cyanobacteria and algae. *Annals of Botany, 119*(1), 151-66.

Zou, P., Lu, X., Zhao, H., Yuan, Y., Meng, L., Zhang, C., & Li, Y. (2019). Polysaccharides Derived from the Brown Algae LessoniaNigrescens Enhance Salt Stress Tolerance to Wheat Seedlings by Enhancing the Antioxidant System, and Modulating Intracellular Ion Concentration. *Frontiers in Plant Science, 10*, 48. doi:10.3389/fpls.2019.00048 PMID:30766543

Chapter 19
Microalgae as a Renewable Resource for Bioplastic Production

Sameh Ali
Botany Department, Faculty of Science, Tanta University, Egypt & Biofuels Institute, School of the Environment and Safety Engineering, Jiangsu University, China

Mudasir A. Dar
Jiangsu University, China

Mostafa El-Sheekh
Tanta University, Egypt

Tamer Elsamahy
Jiangsu University, China

Abdallah Abdelfattah
https://orcid.org/0000-0002-0795-1867
School of Environment and Safety Engineering, Jiangsu University, China & Faculty of Engineering, Tanta University, Egypt

Esraa Abdelkarim
Jiangsu University, China

Konstantinos Pispas
University of Patras, Greece

Savvas Giannis Mastropetros
University of Patras, Greece

Dimitris Zagklis
University of Patras, Greece

Michael Kornaros
https://orcid.org/0000-0001-6691-6925
University of Patras, Greece

Jianzhong Sun
Jiangsu University, China

ABSTRACT

Increase in plastic waste accumulation is considered a growing concern, resulting in white pollution. It is unavoidable that an inventive method to reduce pollution will be required. Increased recycling of plastic waste is not a practical solution. Therefore, reducing petroleum-based polymer utilization is essential for environmental sustainability. Biobased polymers are gaining appeal as a promising alternative to petroleum-based polymers. Based on several studies, biobased plastics could be produced by several microbial species, particularly algal species, rather than petroleum-based polymers. Bioplastic synthesis from microalgae is a new option that calls for further studies. Algal biorefinery that integrates bioplas-

DOI: 10.4018/978-1-6684-2438-4.ch019

tic complimentary activities will be investigated to determine its economic viability and environmental impact. Consequently, this chapter discusses the current status of bioplastic production technologies from microalgae species and different types of bioplastics produced by various algal species and the bioplastic material production methods from microalgae.

Figure 1. Plastic classification based on production origin (Reddy et al., 2013).

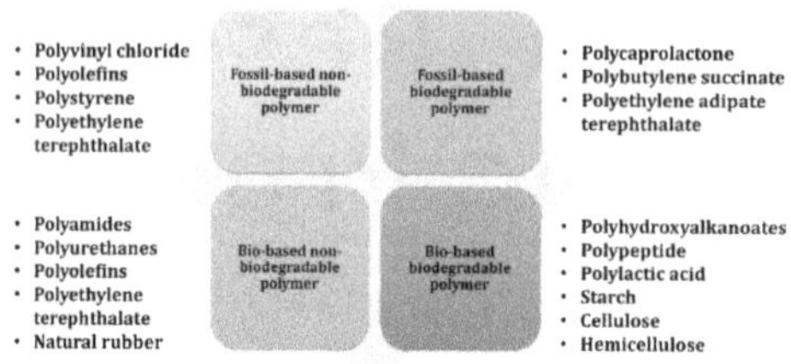

INTRODUCTION

With the increasing demand for plastic-based products and the depletion of fossil fuels production, great attention has been focused on finding sustainable resources for the production of bioplastics (Ali et al., 2019; Darwish et al., 2021). Therefore, it is urgent to decrease our reliance on crude oil for plastic production. Moreover, there is a mass need to reduce plastic accumulation in landfills. From 1950 to 2018, the total plastics production reached 8.3 billion metric tons, increasing 5% annually (Geyer et al., 2017; Ali et al., 2021a, b, c, d). At the same time, the annual production of plastics is expected to double by 2035 (800 Mt) and reach 1600 Mt by 2050. After 2020, more than 400 Mt of plastic wastes will be produced annually (Barra & Leonard, 2018). Regrettably, 76% of the overall plastic production is finally handled as wastes. Of these, 9% is recycled, 12% is incinerated, and 79% is landfilled or released to the environment (Geyer et al., 2017). Therefore, the accumulation of plastic waste showed a continuous growth of environmental pollution problems, causing severe impacts on humans and animals (Ali et al., 2021a, c, d). Plastics are classified into four types based on production origin (Fossil-based or Bio-based polymers), as shown in **Figure 1** (Reddy et al., 2013).

Imre et al. (2019) define bioplastics as biodegradable and/or derived from renewable sources, making them a viable alternative to fossil-based plastics. Bioplastic output capacity is expected to rise to 2.62 million tons by 2023 (European Bioplastics, 2020a, b). The main motivation of this output growth is the advanced innovative biobased polymeric material such as polybutylene succinate (PBS), polylactic acid (PLA), and polyhydroxyalkanoates (PHA) (European Bioplastics and nova-Institute, 2019a, b, c). However, until today, manufacturing bioplastics is significantly more expensive than those of fossil plastics (Raza et al., 2018; Tsang et al., 2019). On the other hand, using appropriate organic wastes and by-products as raw materials using different microbial strains can overcome bioplastic production costs (Jõgi & Bhat, 2020). The majority of bioplastic materials are derived from crops, and this affects the availability of food resources. Furthermore, this type of bioplastic material requires agricultural land, fertilizers, suitable conditions and water (Zeller et al., 2013). Thus, this cannot be sustained for long-term bioplastic production.

Figure 2. The current state of the art in microalgae conversion to bioplastics (Zeller et al., 2013; Khetkorn et al. 2016; Sabathini et al., 2018; Daneshvar et al., 2022).

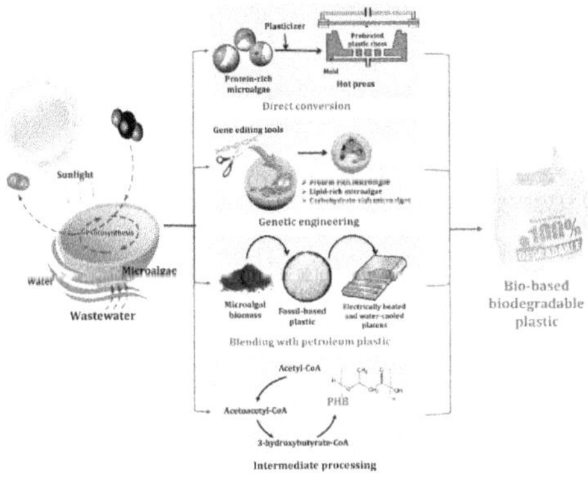

On the other hand, several studies reported that microalgae could represent an ideal and cheap source for biofuels and bioplastic production because they can grow on waste resources and do not compete with food sources. They also have a high rate of lipid accumulation (Venkata Subhash et al., 2017; Ali et al., 2020a, b, c, d). This finding is because of the ability of microalgae to utilize CO_2 during the photosynthesis process and convert it to several forms and by-products through their metabolic activities. Consequently, the microalgal cell can produce bioplastic polymers by trapping CO_2 (Daneshvar et al., 2022). At the same time, the conversion of atmospheric carbon to bioplastics is a reasonable justification for the existence of such technologies. **Figure 2** depicts the current state of the art in microalgae conversion to bioplastics. Various researches have been performed, including direct use of microalgal biomass for bioplastic production, blending microalgal biomass with fossil-based plastics, intermediates biopolymers, and genetic engineering to obtain microalgae strains that produce bioplastics.

This chapter will discuss the possibility of bioplastic production using microalgal biomass, in addition, discussing the current state of bioplastic production and the applied techniques such as the direct conversion of microalgal biomass, the blending of microalgal biomass with fossil-based plastic polymers. Furthermore, investigate the ability of genetic engineering to enhance bioplastic production from modified microalgal strains. On the other hand, discuss challenges of bioplastics production from microalgal species and the future perspectives for large-scale applications.

CURRENT STATE OF BIOPLASTIC

The current state of bioplastics showed that global bioplastic production was 2.11 million tons in 2019, according to European Bioplastics industry statistics (European Bioplastics, 2020a, b). Bio-based polypropylene (PP) and PHAs have shown the highest relative growth rate in output quantities. Since PP is a widely used plastic with various applications, the bio-based alternative is expected to expand steadily in the future continue to hold the role for the next few years; by 2023. Also, a new polymer produced from renewable resources by the polycondensation process called polyethylene furanoate (PEF) is expected to enter the bioplastics industry as an alternative to PET (Polyethylene Terephthalate) (Rosenboom et al.,

Figure 3. Bioplastic applications (European Bioplastics, 2020).

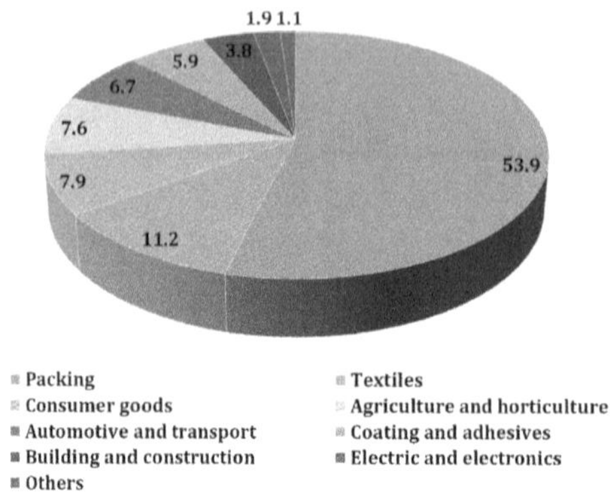

2018; European Bioplastics and nova-Institute, 2019a, b). However, some problems hinder the industrial manufacturing process, such as bioplastic deterioration and discoloration (Terzopoulou et al., 2017).

Biodegradable bioplastics as PLA, PHA, and starch blends accounted for 55.5% of all generated bioplastics in 2019. According to 2019 market results, Europe leads the world in bioplastics research and development activities. However, Asia continues to have the highest production potential with 45% of all bioplastics produced globally, and Europe produced about 25%, followed by North America (18%) and South America (12%) (European Bioplastics, 2020a, b). Bioplastics, including traditional plastics, may be used in a variety of applications, including electronics, textiles, packaging, and other consumer goods. According to European Bioplastics, the packaging application accounts for nearly 53.9% (1.14 million tons) of the overall bioplastics industry in 2019 (**Figure 3**) (European Bioplastics, 2020a, b). The portfolio of bioplastics applications is diversifying as customer and brand demand for more sustainable products continues to rise. Bioplastics are now available as alternatives to almost any form of traditional plastic. Several types of bioplastics have the same physicochemical properties as conventional plastics but with the added benefit of a lower carbon footprint and more environmentally sustainable waste disposals solutions like biodegradation or industrial composting (Ali et al., 2021a, b, c, d).

PRODUCTION OF BIOBASED AND BIODEGRADABLE PLASTICS

Bioplastic monomers can be directly generated from renewable sources or through microbial fermentation or renewable substrates such as lignocellulosic biomass. Therefore, the production of bioplastic is competed with food or feed products on land-use and represents a serious problem. However, European Bioplastics, nova-Institute (2019a, b) reported that bioplastic feedstocks are expected to be insignificant competitors with food and feed, occupying only 0.02% of total arable land today. Nevertheless, the future outlook of this production technique can represent a problem for long-term application. Alternatively, cheap non-edible renewable sources such as microalgae, lignocellulosic biomass, or food waste for bioplastics production are considered a new direction and a promising field for scientific research in this field (Ravindran & Jaiswal, 2016). Mohsenzadeh et al. (2017) reported that microbes could ferment the depolymerized starch to synthesize biopolymer for bioplastics production and/or ethylene-derived-bioethanol to produce Bio-PE.

In this sense, Suriyamongkol et al. (2007) reported that PLA's constitutive monomers, lactic acid enantiomers, can be generated by microbial fermentation. However, PHAs can be biosynthesized by living cells from a variety of renewable feedstocks, typically under unfavorable cultivation conditions such as imbalanced nutrient supply, as shown in **Table 1.** Niaounakis (2013) reported that PBS is made chemically by polycondensing 1,4-butanediol and succinic acid or its anhydride in the presence of a catalyst. Tsuji (2013) found that PLA can be produced through either direct polycondensation of lactic acid or ring-opening polymerization of lactide. Harmsen et al. (2014) stated that polytrimethylene terephthalate (PTT), polyethylene (PE), polypropylene, polyethylene terephthalate (PET), and polyamide (PA) are categorized as bioplastics if they are generated in part from renewable sources. Bio-PE is made from ethylene, which is obtained from biomass-derived ethanol through microbial fermentation, distillation, and further catalytic dehydration. Microbial fermentation is used to produce biobased succinic acid from sugarcane, cassava, or corn as a feedstock. Polybutylene adipate terephthalate (PBAT) is a biodegradable polymer made from adipic acid, butanediol, and terephthalic acid (with a biobased carbon content of up to 30%).

On the other hand, as shown in **Figure 4**, Bio-PP is created through the polymerization of propylene, which can be obtained through chemical or thermochemical processes or through integrated approaches. For example, biomass-based sugar or starch can be fermented to produce ethanol, whereas biomass, including residue from fermentation, can be gasified to produce syngas. Then, ethanol and syngas are used to synthesize propanol, which can then be catalytically dehydrated for propylene synthesis (Miranda et al., 2012). Another method is to convert glycerol from biodiesel production into propane, which can then be used to synthesize propylene (Andreeßen & Steinbüchel, 2019). There are several types of Bio-PAs that can be made from various biobased building blocks. For example, PA 4,10 is derived from 1,4-butanediamine, which is synthesized chemically from succinic acid, whereas PA 11 is derived from 11-amino-undecanoic acid, a renewable raw material derived from vegetable oils such as castor oil (Harmsen et al., 2014). Biobased PET, which is primarily used in rigid packaging, is created by combining bioethanol-derived monoethylene glycol with fossil-based terephthalic acid synthesized via conventional transesterification (Andreeßen & Steinbüchel, 2019). The biomass content of bio-PET is approximately 30% (Iwata, 2015). Finally, Bio-PTT is made from 1,3-propanediol derived from biomass and terephthalic acid derived from petroleum (Andreeßen & Steinbüchel, 2019).

MICROALGAL SPECIES POTENTIAL IN BIOPLASTIC PRODUCTION

Although the wide diversity of algal species all over the world, few microalgal species have been investigated as bioplastic producers. In this context, *Chlorella* is a kind of green algae that grows in fresh and marine water. *Chlorella* spp. contain up to 60% protein of algae wet weight. It has a stronger fracture resistance and thermal stability than *Spirulina* because of its thick cell walls (Enyidi, 2017; Dianursanti et al., 2018). In biomass–polymer blends, this species is frequently employed after comparing bioplastic manufacturing from 100% microalgae biomass with mixes incorporating additives and polymers. Zeller et al. (2013) discovered that blending is required for commercial applications. When comparing *Chlorella vulgaris* to *Spirulina* sp. in terms of product quality, tests revealed that *C. vulgaris* produced higher-quality bioplastic. Based on the physicochemical parameters of the material, *Spirulina* has better blending properties than *C. vulgaris* (Zeller et al., 2013). Dianursanti & Khalis (2018) investigated the influence of the compatibilizer ratio on the quality of the polyvinyl alcohol (PVA)-*C. vulgaris* com-

Table 1. The potential of polyhydroxyalkanoates production by microalgae.

Algal species	Bioplastic polymer	Conditions	Polymer (%Dry Weight)	References
Spirulina plantesis	P(3HB)	CO_2/acetate as carbon source	10	Jau et al. (2005)
Nostoc muscorum	P(3HB)	Phosphorus limitation, supplemented with acetate (dark incubation for 7 days)	35	Sharma & Mallick (2005)
		0.2% acetate in dark incubation for 7 days	35	
Synechococcus sp. PCC7942	PHB	Nitrogen deficiency and using acetate as carbon source	26	Nikel et al. (2006)
Nostoc muscorum	P(3HB)	0.08% propionate and 0.11% acetate at pH 8.1 for 16 days	31	Panda & Mallick (2007)
		0.4% propionate and 0.2% acetatedays at pH 8.5 for 14 days	28.2	
Synechocystis sp. PCC6803		Phosphate deficiency and using fructose/acetate as carbon source with gas-exchange limitation conditions	38	
S. platensis	PHB	Nitrogen deficiency	10	Toh et al. (2008)
Synechocystis sp. UNIWG			14	
Microalgae consortium	PHA	Wastewater as culture medium	43	Chakravarty et al. (2010)
Chlamydomonas reinhardtii	P(3HB)	Genetically engineered systems	-	Chaogang et al. (2010)
Phaeodactylum tricornutum.	PHB	Genetically engineered systems	10.6	Hempel et al. (2011)
Aulosira fertilissima CCC 444	PHB	Produced under 0.26% citrate, 0.28% acetate, and 5.58 mg/L K_2HPO_4 (incubation period of 5 days)	85	Samantaray & Mallick (2011)
	P(3HB-CO-3HV)	Phosphorus deficiency with 0.5 and 0.4% fructose and valerate, respectively.	77	
Nostoc muscorum Agardh	P(3HB-CO-3HV)	Phosphate deficiency conditions	71	Bhati & Mallick (2015)
		Nitrogen deficiency conditions	78	
Synechocystis cf. *salina*	P(3HB)	CO_2 as carbon source	7.5	Wagner et al. (2016)
Calothrix scytonemicola TISTR 8095	P(3HB)	CO_2 as carbon source and under nitrogen deficiency	25.4	Kaewbai-ngam et al. (2016)
Synechocystis sp.	PHB	Genetically engineered systems under nitrogen-free culture	35	Khetkorn et al. (2016)
Oscillatoria okeni TISTR 8549	P(3HB-CO-3HV)	Nitrogen deficiency	14.4	Taepucharoen et al. (2017)
		Nitrogen deficiency and dark condition	42.8	
		Nitrogen deficiency and supplemented with acetate in dark condition	42	
Synechocystis salina	PHA	Cultivation using Blue-Green medium (BG11) as culture medium	5.5–6.6	Kovalcik et al. (2017)
Synechococcus elongates	PHA	Phosphorus and nitrogen deficiency	17.15	Mendhulkar & Shetye (2017)
		Phosphorus deficiency	7.02	
Synechococcus subsalsus	PHA	Nitrogen deficiency	16	Costa et al. (2018)
Spirulina sp. LEB-18			12	
Aphanocapsa sp. and cf. *Chroococcidiopsis* sp.	P(3HB)	Phosphate limited conditions and permanent illumination	838 mg/L	Arias et al. (2018)
Synechocystis sp. PCC6714	PHA	UV light exposure	37	Kamravamanesh et al. (2018)

PHA; Polyhydroxyalkanoate, P(3HB-CO-3HV); Poly (3-hydroxybutyrate-co-3-hydroxyvalerate), P(3HB); Poly (3-hydroxybutyrate), PHB; Polyhydroxybutyrate.

Figure 4. Schematic diagram of polyethylene and polypropylene from biomass and microalgae (Miranda et al., 2012; Andreeßen & Steinbüchel, 2019).

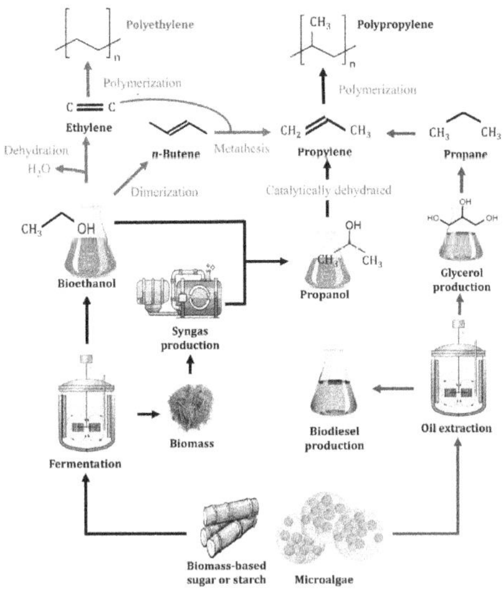

posites formed. According to the research, a compatibilizer (maleic anhydride) content of 6% in the combination produces the optimum quality. *Chlorella sorokiniana* microalgae biomass may also be used to make starch granules (1 m). The biopolymer starch is widely employed in the culinary, chemical, and bioplastic sectors. It is promising for starch-based bioplastic manufacturing due to its high gelatinization temperature (110°C) (Gifuni et al., 2018). The use of ultrasonic homogenization prior to blending can improve the homogeneity and surface characteristics of the resulting *Chlorella*–PVA blends, which can be used as a food packaging alternative (Sabathini et al., 2018). Otsuki et al. (2004) investigated the differences between *Chlorella*–PE composites with and without PE modification with maleic anhydride. The tensile strength of the composites was improved when PE was modified.

MICROALGAL BIOMASS BLENDING WITH FOSSIL-BASED PLASTICS

The blending of microalgal biomass with fossil-based plastics has the potential to extend the life of microalgal plastics while also adding mechanical value. On the other hand, the blending of synthetic plastic with microalgae as a minor component could free up unused petroleum plastic for other uses. Several studies have looked into creating hybrid materials containing varying amounts of microalgae biomass and petroleum plastics. At the same time, the overall mechanical properties of such hybrid materials are generally inferior to those of pure fossil-based plastics. The combination of microalgae biomass and synthetic plastics has advanced properties and can be applied in a wide range of applications. Zeller et al. (2013) investigated thermo-mechanical polymerization of microalgae protein to produce bioplastics and microalgae biomass blending with PE. Microalgae cells are small (~50 μm); thus, was beneficial because it could allow for better mixing when combined with fossil-based plastics. *C. vulgaris* (58% protein) and *Spirulina platensis* (57% protein) were plasticized with glycerol in various ratios in this study (0-30% wt). The best material properties were found when the algae/glycerol ratio was 80/20. When combined with polyethylene, *Spirulina*-derived plastics outperformed *Chlorella*-blended bioplastics in terms of mechanical properties. However, pure *Chlorella*-based bioplastic had better load and extension properties than pure *Spirulina*-based bioplastic, with tensile strengths of 5.7 and 3.0 MPa, respectively (**Table 2**) (Zeller et al., 2013). Sabathini et al. (2018) reported that PVA was utilized to make bioplastic films containing *Chlorella* at different solution temperatures. The bonding between blended materials was found to be weaker at higher temperatures due to its influence on the uniformity of the mixture. Therefore, ultrasonication increased the quality of the resulting blended material.

Wheat gluten has been extensively researched for use in the production of long-lasting bioplastics. Despite its brittle nature, the material's structure can be enhanced using additions and fillers. Wheat gluten's high protein concentration makes it attractive for a variety of applications (Ciapponi et al., 2019). PBS is one of the newest biopolymers, and it may be able to fulfill market demand for bioplastics. In comparison to low-density polyethylene (LDPE) and polypropylene (PP), PBS is frequently preferred (Kalia & Avérous, 2016). Biodegradable PBS can be made from either biomass or fossil fuels. Because of its ease of processing, PBS is particularly popular in the textile industry. Melt blow, multifilament, monofilament, flat, and split yarn, for example, are all made with it. It's also used to make molded products in the plastic industry (Ewa Rudnik, 2019). When PBS is mixed with other polymers, its mechanical properties improve, allowing it to be used in more applications (Xu & Guo, 2010). Blending *Spirulina* with PBS provides an economically efficient *Spirulina*-based bioplastic production (Monshupanee et al., 2016).

In addition, *Spirulina* sp. biomass was used to make bioplastic using compression molding. Wang (2014) reported that plasticization with ethylene glycol, blending with ultra-high molecular weight polyethylene, and compatibilization with *Spirulina* biomass of proper composition using polyethylene-graft-maleic anhydride (PE-g-MA) were all used to develop better performance bioplastics than those made with pure *Spirulina* biomass. In addition, activated carbon may efficiently absorb undesirable smells from algal bio-plastics. Therefore, *Spirulina* protein revealed good potentials for developing bio-plastics products. Zeller et al. (2012) investigated the stability and thermal characteristics of plasticized and blended duckweed polymers, as well as the potential of duckweed for compression molding plastics. With a 3:1 ratio of duckweed to glycerol, the best polymer stability was achieved. Ciapponi et al. (2019) found that microalgal biomass was added in the following amounts: 10, 20, and 30%. The components, including gluten, plasticizer, and microalgae, were mechanically mixed before being molded in a hot press. Microalgal filler enhanced the mechanical characteristics of plasticized gluten material. The microalgal biomass boosted the tensile modulus from 36.5 MPa to 273.1 MPa and increased the tensile strength from 3.3 MPa to 4.9 MPa in samples plasticized using 1,4-butanediol.

Moreover, additional chemicals were employed in investigations to increase the product quality and blending process efficiency. For instance, Otsuki et al. (2004) found that in the production of *Chlorella*-PP, acetone was mixed with benzoyl peroxide and Maleic anhydride to be sprayed on plastic powder to enhance blending properties. Moreover, Zhu et al. (2017) reported that sodium sulfite was also used for washing the algal biomass before blending it with PBS. PVA is well-known for its capacity to improve a product's strength, durability, and flexibility. Therefore, Dianursanti et al. (2018) stated that PVA modification with Maleic anhydride is required to increase the surface properties and stability, as well as the mechanical qualities of packing materials. **Table 2** summarizes some examples of conducted studies to produce bioplastic from blending different plastics with algal species.

Table 2. Blending microalgae biomass with different plastics and resulting mechanical properties.

Microalgal species	Blended material	Algal ratio (Algae: Blended material)	Tensile strength (Mpa)	References
Chlorella cells powder	PP	10/90	32	Zhang et al. (2000)
	PVC	20/80	30	
Nannochloropsis sp.	Corn starch/PP	14/6/80	17.1	Shi et al. (2011)
Botryococcus Braunii	PBS	20/80	21.6	Toro et al. (2013)
Nannochloropsis gaditana	PBAT	20/80	10	Torres et al. (2015)
Green algae	PLA	20/80	45	Bulota & Budtova (2015)
C. vulgaris	PVA	80/20	4.14	Dianursanti, & Khalis (2018)
Spirulina sp.	PLA	5/95	81.16	Simonic & Zemljic (2021)
		10/90	69.46	
Algal consortium biomass		5/95	62.67	
		10/90	77.23	

PBAT; Polybutylene adipate terephthalate, PBS; Poly(butylene succinate), PET; Polyethylene terephthalate, PLA, Polylactic acid, PP; Polypropylene, PS; Polystyrene, PVA; Polyvinyl alcohol, and PVC; Polyvinyl chloride.

MICROALGAL DIRECT AND INDIRECT CONVERSION TO BIOPLASTIC

The lipid, carbohydrate, or protein fractions of microalgae can be used to make bioplastics. Carbohydrate-based bioplastics are the most ancient type of bioplastic. Cellulose or starch is typically used in this form of bioplastic. In this sense, the majority of microalgae strains include a minor amount of starch. This situation further restricts the use of starch-based bioplastics in industry. However, *Chlorella* sp. consider an important source for Carbohydrate-based bioplastics because it may include 2–11% (DW) of starch (Cheng et al., 2014). Several microalgae strains have had their starch content increased by depletion of different nutrition sources such as sulfur depletion growth media, light intensity, microelement (nitrogen, phosphorus, sulfur) limitation, and the use of different agents such as cycloheximide, a particular inhibitor of cytoplasmic protein synthesis, can all help to obtain a high amount of starch. For instance, in the case of *Chlamydomonas reinhardtii* strain growth under these conditions, a starch content of up to 50% was obtained.

This suggests that this strain might be exploited as a source of starch for bioplastics. However, as the growth medium was scaled up to pilot size (Mathiot et al., 2019), the starch concentration was reduced to 25% (DW), limiting the possibility of industrial commercialization for microalgae starch-based bioplastics. Microalgae proteins have also been explored for application in the manufacturing of bioplastics. Crop protein supplies for bioplastics include soybean and sunflower seed proteins. The use of microalgae proteins rather than agricultural proteins is motivated by the desire to avert future food shortages. Microalgae's capacity to thrive in nutrient-rich wastewater makes it easier to cultivate and adds value to the environment. *Spirulina* and *Chlorella* are two microalgae strains with high protein content, more than 50%. Proteins are brittle substances; thus, plasticizers are needed to make them behave like plastic. Whole biomass may be immediately transformed into protein-based bioplastics using the hot press

Figure 5. Schematic diagram of direct and indirect conversion of protein-rich microalgae into bioplastics (Zeller et al., 2013; Kumar et al., 2014).

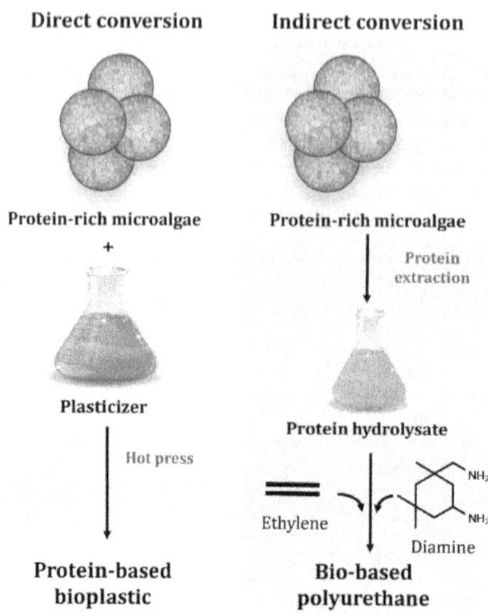

method (**Figure 5**). However, differing amino acid combinations resulted in bioplastics with varying mechanical characteristics (Zeller et al., 2013).

Microalgal protein may be utilized as raw substances for the manufacturing of thermoplastic polyurethane (PU) (Kumar et al., 2014), which is typically derived from fossil-based feedstock, in addition to direct conversion into bioplastics. The high expense of bacterial fermentation techniques, which makes bacterial bioplastic more expensive than petroleum-based polymers like polypropylene and polyethylene, is a key barrier to the commercial manufacturing of PHAs (Nagarajan et al., 2013).

GENETIC ENGINEERING

Genetic engineering has been discovered to be a potential method for modifying microalgal strains to produce biopolymers for bioplastics manufactures, such as PHB, a thermoplastic and biodegradable polyester (PS) made by specific bacterial strains (Yang et al., 2018). PHB synthesis by algae can be made possible by genetic engineering. Because the bacterial fermentation method for making PHB is expensive for bioplastics manufacture, using genetic engineering to introduce the bacterial PHB production into microalgae or macroalgae might lower production costs (Rasul et al., 2017; Yang et al., 2018).

Furthermore, when compared to genetically altered *Escherichia coli* growing on glucose as the carbon substrate, existing designed microalgae are unable to create substantial yields of plastic. However, if genetic tools become more widely available, bioplastic production may rise in the near future. Furthermore, PHB-producing microalgae might serve a dual role in a biorefinery, generating both biofuels and biomaterials from a single algal source. Advanced tools for genetic manipulation of the microalgal system have been created as our understanding of genetic systems grows (Rahman & Miller, 2017). In this sense, the biosynthesis of other bioplastic polymers as small chain length PHAs (scl PHAs) and medium-chain length PHAs (mcl PHAs) is now achievable due to the demonstration of PHA synthesis in a microalgae strain. In addition, the cost of the carbon substrate is one advantage of employing recombinant *P. tricornutum* over *E. coli*. Because of the ability of *P. tricornutum* and *C. reinhardtii* to utilize atmospheric carbon as a free carbon source. However, the cost of the carbon substrate is estimated to be roughly 38% when using the entire processing and operating costs for PHB synthesis in *E. coli*. Therefore, reducing this cost in a manufacturing environment should make bioplastic production in microalgae potentially economically viable (Rahman & Miller, 2017). As a result, genetic engineering is a potential method for modifying the genes of a specific algal strain to increase the synthesis of desired components, such as polymers or starch, in a shorter amount of time to create bioplastics (Chun et al., 2017). Plants, eukaryotic cells, and other complex creatures are genetically more complicated than algae, making genetic engineering easier (Gimpel et al., 2015; Hlavova et al., 2015).

In this context, Chaogang et al. (2010) successfully incorporated the *phb*B gene from *Ralstonia eutropha* into *C. reinhardtii*. They found that part of a native bioplastic biosynthesis pathway may be incorporated into *C. reinhardtii*. Furthermore, they were also able to incorporate *phb*B and *phb*C genes from *R. eutropha* into the nonnative host (*C. reinhardtii*), with the third enzyme *phb*A, already present in *C. reinhardtii*. This study reported that PHB production was 6 mg/g dry cell weight produced by *C. reinhardtii*. Also, Hempel et al. (2011) transferred the bacterial PHB biosynthetic pathway into the *Phaeodactylum tricornutum*'s cytosolic compartment. This method is eco-friendly and economically acceptable. It has the ability to produce PHB levels of up to 10.6% of dry algal weight by aggregating them in granule-like shape after seven days of culturing.

Noor-Mohammadi et al. (2012) developed a technique to assemble multiple-gene biosynthetic pathways in *S. cerevisiae* and then successfully transferred it into *C. reinhardtii* chloroplast. Moreover, Noor-Mohammadi et al. (2013) applied the same technique to transfer the same genes to the *C. reinhardtii* nucleus. This novel technology might help engineer microalgae for biofuel generation and other valuable products by providing a useful genetic tool for designing and integrating complicated metabolic pathways into the chloroplast genome of microalgae.

However, Rahman et al. (2013) found that the successful secretion of phasin and phasin bound PHB outside *E. coli* cells and into the culture medium. After two days of culture, 36% of the total PHB produced in the secreting strain was collected in the secreted fraction, and 64% remained in the internal fraction. Therefore, more studies should be performed to enhance bioplastic production from genetically modified organisms, where 10.6% of PHA is low when compared to similar pathways in recombinant *E. coli* systems.

PHB pathways have been successfully integrated into microalgae strains. It could be interesting to study if overexpressing the phbA gene in *C. reinhardtii* improved its capacity to accumulate PHB compared to the existing system. However, phbA is produced at natural levels. Furthermore, the purification of PHB from microalgae biomass is still considered a challenge. In this technique, algal cells are lysed, and bioplastic polymers are extracted; however, because this is a significant expense with biologically generated PHBs, other procedures must be devised to make PHB manufacturing economically viable (Kessel-Vigelius et al., 2013).

On the other hand, genetic engineering is required to investigate several mutants with specific traits and the isolation of the new desirable mutant strain. The selection and isolation of the desirable mutations take time throughout this screening phase. Another challenge with genetic modification is the stability and appropriateness of the selected algal strains for mass production under variable conditions. since genetically modified strains are generated and evaluated in a controlled laboratory environment. Variations in temperature, light intensity, pH, and contamination risk will be experienced by genetically engineered algal strains in large-scale cultures (Hlavova et al., 2015). Moreover, the genetically engineered algal strain's safety and ethical implications. It might endanger humans, animals, and plants, as well as ecosystems. If the genetically modified organism is discharged into the environment or cultivated outdoors, the risk of modified gene migration or transmitted to other organisms may occur. Despite the challenges that may develop as a result of the genetic modification procedure, genetic engineering has a lot of promise for increasing biopolymers production for bioplastic synthesis by manipulating the genes of algae strains. As a result, it is recommended that genetically modified algal strains be tested under experimental circumstances that replicate a variety of real-world scenarios.

BIOPLASTICS BIODEGRADABILITY

Biodegradation is distinguished from biodeterioration or disintegration. Biodegradation can produce long-lasting plastic molecules that are not digested or breakdown by microorganisms (Al-Tohamy et al., 2020a, b; Tosin et al., 2020). Many studies have been performed to explore the biodegradability of bioplastics in a variety of environmental settings, including soil, compost, and aquatic environments, as shown in **Table 3.**

In soil, bioplastic degradability should be discussed, and impacts in this environment should be considered as well. Soil settings have a high richness of microorganisms, making plastic biodegradation more viable than in other environments such as water or air. In this context, Boyandin et al. (2013) reported that the biodegradation of bioplastic might vary depending on the soil conditions. For example, in the soil environment of Hoa Lac, Vietnam, more than 98% of PHA films were degraded. However, the same films lost 47% of their weight in the soil of Dam Bai, Vietnam. This considerable drop in PHA film decomposition in the Dam Bai location might be attributed to the low pH (5.48), which may inhibit microbial activity. Wei et al. (2015) found that the blending of different biodegradable materials increases the biodegradability of bioplastics. It was observed that the biodegradation of PHB/potato peel waste fermentation residue (PHB/PPW-FR) biocomposite was more effective than that of a single PHB because the PPW-FR fibers decreased the crystallinity of the PHB biocomposite. Another study conducted by Harmaen et al. (2014) revealed that adding empty fruit bunch fibers accelerated PLA biocomposite biodegradation.

Because of their great microbial diversity, soil and compost were prioritized among various environmental conditions (Anstey et al., 2014). Despite the majority of plastic waste being disposed of in landfills, the biodegradation of bioplastics has received little attention. Most bioplastic wastes are disposed of by landfilling, resulting in greenhouse gas (GHG) emissions and leachate. As a result, for the recovery of plastics, other solid waste management strategies such as composting or recycling should be performed due to their advantages over landfilling or incineration and other conventional disposal techniques. Composting is a method of converting organic matter into CO_2 and humus through the action of a diverse set of microbial consortia (Kale et al., 2007).

Under certain environmental conditions, bio-based and fossil-based bioplastics, such as PHA, starch-based, PBS, PES, PLA, and PCL, are vulnerable to compost and biodegraded. The changes in household and industrial composting circumstances may result in considerable variances in bioplastic biodegradation. The biodegradation of PLA bioplastic for 11 months in home composting conditions, the biodegradation rate was very slow. This may be due to the lower temperature than the industrial-scale experiment, which might be carried out at a wide range of high-temperature range (Rudnik & Briassoulis, 2011). In order to increase the biodegradability of bioplastics in a compost environment, some research was conducted. The biodegradability of bioplastics was improved by increasing the concentration of soluble sugar in the biocomposites by adding materials with high protein content. Biofuel by-products were used in the mixture as composite bioplastics to boost the biodegradability of PBS-containing bioplastics.

When compared to pure PBS bioplastic, the introduction of a meal-based filler accelerates biodegradation, which is attributable to the high concentration of soluble sugars in meal-based fillers (Anstey et al., 2014). Because maize is a highly biodegradable substance, the incorporation of corn in PLA/corn bioplastic seems to improve compost biodegradation. As a result, bacteria were able to degrade the material and the PLA portion more quickly (Bonifer et al., 2019). Therefore, Ahn et al. (2011) found that PLA pots blended with poultry feather fibers showed a higher rate of degradation than those of the pure PLA, which might be related to the other components used in PLA pot production that inhibited the biodegradation. Wu (2014) stated that acrylic acid-grafted PHA/Rice Husk biodegradation and PHA/Rice Husk biocomposites have a direct relationship to its rice husk content. The inclusion of biomaterials in a PHA matrix improved the biocomposite qualities, including tensile strength. Although acrylic acid-grafted PHA/Rice Husk had higher tensile strength than PHA/Rice Husk, its biodegradation was somewhat slower, which was owing to the former composites' resistance to water absorption. Furthermore, Wu (2011) found that sisal fibers were disseminated in the PLA matrix, resulting in a weight loss of more than 50% after 14 weeks of soil burial. However, Jain & Tiwari (2015) reported that the use of cellulose acetate butyrate as a blending agent exacerbated PLA matrix biodegradation by making the polymer more hydrophobic.

Bioplastics may be made from renewable materials, as previously indicated. Agricultural wastes, for example, can be used to make cellulose acetate bioplastics. Mostafa et al. (2018) reported that after 14 days of composting, the biodegradation of cellulose acetate bioplastics from low-cost fiber flax and cotton linter was 44 and 35%, respectively. Some bioplastics on the market claim to be 100% biodegradable. However, their composting potential did not detect yet. The biodegradability of two distinct samples of sponge cloth bioplastics (samples A and B), which are often used for cleaning surfaces, was composted. The results revealed that sample B had a biodegradability of more than 80%, whereas sample A only decomposed minimally, implying that bioplastic biodegradability is greatly influenced by the polymeric structure the kind of environmental conditions (Vaverková & Adamcová, 2015).

Also, bioplastic wastes were discovered to be widely distributed in deep-sea habitats. Because of their semi-permanent stability in seawater, plastic wastes have the potential to cause marine pollution, which can have a negative impact on aquatic life forms (Sekiguchi, et al., 2011; Atiwesh et al., 2021). As a result, bioplastics, which are biodegradable materials, may be employed to create a sustainable environment in aquatic ecosystems. Tosin et al. (2012) investigated the biodegradability of bioplastic in six different aquatic ecosystems. They found that the deterioration in the pelagic environment was more effective than degradation in eutrophic habitat. Furthermore, the authors proposed that the maximum biodegradation might be obtained at the water-sediment contact since the environmental conditions at the interface favored the activity of plastic-degrading microorganisms. The weight loss percentage was the same for both bioplastics in static and dynamic circumstances. However, the weight loss was smaller under dynamic conditions than static conditions. This might be attributable to the fact that the dynamic setting was more realistic, with nutrient supply constraints and seawater temperature variation. Furthermore, the addition of sediments was studied to better understand its impacts on biodegradation. The sediments were researched to see if they may help with biodegradation, but no definitive association can be found (Thellen et al., 2008). The temperature of the water can also have an impact on bioplastics biodegradation. It was found that the rate of biodegradation of PHA films varied because of the change in weather temperature (Jouhara et al., 2017; Atiwesh et al., 2021). Furthermore, depending on the existing bioplastic-degrading microbes, various seawater may have had a significant influence on biodegradation. By testing the strength retention of PHB, PBS, and PCL biopolymers in three distinct seawater types, the degradation of these biopolymers was examined. The findings revealed that the strength retention varied depending on the seawater environment, which might be attributable to the various bioplastic-degrading bacteria found in these three types of seawater (Sekiguchi et al., 2011). The structure of the polymer is another factor that might influence the degree of biodegradation in marine water. Because of its increased surface area, PHA films were said to disintegrate quicker than PHA pellets. Furthermore, a bigger polymer/water contact aided in the adhesion of microbes to the polymer's surface (Volova et al., 2010). This trend was also seen in another study of PHA films in tropical soil conditions (Boyandin et al., 2013). The introduction of microscale plastics into the marine environment via wastewater discharges has recently piqued the curiosity of academics. However, no information regarding the release and/or fate of bioplastics through wastewater discharges was found in the existing literature.

CHALLENGES OF BIOPLASTICS PRODUCTION FROM MICROALGAL SPECIES

Some obstacles hamper large-scale industrialization and manufacturing despite the impressive achievements of algae-based bioplastics in-vitro.

1. The selection of suitable microalgal species is an essential concern for bioplastic production with appropriate qualities. The composition of algae biomass varies per species. For example, Johnsson & Steuer (2018) used ultrasonication and water to extract starch and PHAs from *Calothrix scytonemicola, Neochloris oleoabundans*, and *Scenedesmus almeriensis* to produce plastic films. *C. scytonemicola* was one of the promising species as PHA-rich algae that showed outstanding productivity. *S. almeriensis* starch extraction was effective, whereas *N. oleoabundans* starch extraction was unsuccessful. Nonetheless, the plastic materials made from *N. oleoabundans* were adequate, and more research might be done by synthesizing a material made up of a mixture of various polysaccharide classes, allowing for a transition from petroleum-based to biodegradable polymers. Abdo & Ali (2019) stated that a few microalgae strains could manufacture PHB. They investigated the ability of *Microcystis aeruginosa*, *Haematococcus pluvialis,* and *Chroococcus turgidus.* They found that *M. aeruginosa* exhibited the highest concentration of PHB (0.49 mg/mL), indicating that it might be utilized to produce bioplastic with strong plasticizing potential.
2. The selection of suitable polymers derived from microalgal species during the design of bioplastics is also an essential aspect in order to generate sustainable bioplastics with good qualities (polymer size, moisture content, molecular weight, brittleness, degradation rate, biodegradability, and acceptability) when compared to conventional plastics (Thakur et al., 2018). The degradation of bioplastics has been demonstrated in a number of studies to have serious environmental consequences (Folino et al., 2020). The production of GHGs during the bioplastic's degradation has become a point of contention. As a result, it is critical to address bioplastics environmental implications. Furthermore, it is also critical to make sure that bioplastics may break down in any environment without releasing any toxic gases (Rasul et al., 2017). Furthermore, Beckstrom (2019) observed the existence of disagreeable smells in bioplastics derived from algal lipid, as well as a delayed manufacturing

Table 3. Biodegradability of different types of bioplastics.

Bioplastic	Biodegradation conditions	Biodegradation rate (%)	Reference
PHB	In the seawater at 25 °C for 14 days	80	Tachibana et al. (2010)
Nylon 4 (Polyamides, Bio-based)	In the seawater at 25 °C for 21days	30	
	In the seawater at 25 °C for 25days	80	
PLA/PFF/starch (80/5/15%)	Composting at 58 °C for 60 days	53	Ahn et al. (2011)
PHB	Composting at 58 °C for 110 days	79.9	Weng et al. (2011)
PLA	In the soil with 30% moisture for 98 days	10	Wu (2011)
PLA/SF (60/40%)		>60	
PHA	In the soil /compost (90/10%), 65% humidity at 25 °C for 15 days	40–50	Arcos-Hernandez et al. (2012)
Mater-Bi bioplastic	In the seawater at 25 °C for 236 days	68.9	Tosin et al. (2012)
PHA	In the soil, 60% moisture at 20 °C for 280 days	48.5	Gómez & Michel (2013)
Starch-based	In the soil, 60% moisture at 20 °C for 110 days	14.2	
PLA	Composting (60% humidity) at 58 °C for 30 days	60	Mihai et al. (2013)
PLA/SW (70/30%)	Composting (60% humidity) at 58 °C for 30 days	40	
PLA	Synthetic material containing compost at 58 °C for 28 days	100	Arrieta et al. (2014)
PLA/PHB (75/25%)	Synthetic material containing compost at 58 °C for 35 days	100	
PHA	In the soil at 35 °C for 60 days	35	Wu (2014)
PHA/RH (60/40%)		>90	
PBS	Composting (50–55% moisture) pH 7–8 at 58–65 °C for 160 days	90	Anstey et al. (2014)
PBS/SM (75/25%), PBS/CM (75/25%), and PBS/CGM (75/25%)	Composting (50–55% moisture) pH 7–8 at 58–65 °C for 100 days	90	
PBS/switchgrass (75/25%)	Composting, 50–55% moisture, pH 7–8, at 58–65 °C for 170 days	90	
PLA/NPK (63.5/37.5%)	In the soil (80% humidity) at 30 °C for 56 days	37.4	Harmaen et al. (2014)
PLA/NPK/EFB (25/37.5/37.5%)	In the soil (80% humidity) at 30 °C for 56 days	43	
PLA	Composting (70% moisture) at 55 °C for 28 days	~70	Tabasi & Ajji (2015)
Sponge cloth (cellulose-based)	Synthetic material containing compost at 58°C for 154 days	>80	Vaverková & Adamcová (2015)
PLA (powdered)	In the soil (60% humidity) at 25 °C for 28 days	13.8	Adhikari et al. (2016)
PBS (films)		1	
PBS (powdered)		16.8	
PBS/Starch (films)		7	
PBS/Starch (powdered)		24.4	

CGM; Corn gluten meal, CM; Canola meal, EFB; Empty fruit bunch, NPK; NPK fertilizer, PBS; Polybutylene succinate, PFF; Poultry feather fibers, PHA; Polyhydroxyalkanoate, PHB; Polyhydroxybutyrate, PLA; Polylactic Acid, RH; Rice husk, SF; Sisal fiber, SM; Soy meal, and SW; Soft wood.

process owing to agglomeration tendencies caused by carbohydrates in algae biomass. Wang et al. (2016) addressed the issue of odor in research in which they blended microalgae, *Nannochloropsis* sp., and planktonic algae, catfish algae, with PE or PP. In comparison to catfish algae coupled with PP, plastic products manufactured from microalgae that included high quantities of fatty acids, especially polyunsaturated fatty acids, blended with PE had a foul-smelling odor. Absorbents such as zeolite or activated carbon will be required to eliminate or mitigate the odor since this will have a substantial impact on the commercialization of algal-based bioplastics.

3. Another hurdle to overcome before algal-based polymers can be commercialized is the algae culture system. Large-scale cultivation is required to achieve mass production of algal biomass to make polymers like starch or other chemicals. The algae may be cultivated on a large scale, either a closed photobioreactor or open pond system, each with its own set of advantages and limitations. Although open systems have cheap operating costs and are simple to maintain and scale up, they have low productivity, a significant risk of contamination, and are infeasible for mass production. Nevertheless, the closed photobioreactor promotes the development of certain algae species and is less vulnerable to contamination with a high production rate. Nonetheless, as compared to an open pond system, the scale-up costs are higher (Khoo et al., 2020; Tang et al., 2020). According to Abdul-Latif et al. (2020), the Sabah coastline, which spans 3885 km^2, may theoretically produce ~14 Mt/y of seaweed by 2050. In order to synthesize chemicals for bioplastic development and achieve a 100% reduction in CO_2 emissions, macroalgae cultivation will require more than 0.6 million hectares by 2050, based on the predicted amount of algae-based bioplastic.
4. Moreover, bioplastic waste management is critical. Composting, landfilling, incineration, and recycling are some of the ways that may be used to handle bioplastic waste. Composting is the most suited option for waste management of bioplastics among these technologies since biodegradable bioplastics decompose quickly compared to conventional plastics. PLA, for example, biodegrades in commercial composting facilities in 30-42 days on average (Chidambarampadmavathy et al., 2017).

To conclude, there are certain obstacles to bioplastics commercialization, such as algae cultivation and selecting suitable polymers or algae strains for bioplastics manufacture. Therefore, these challenges must be solved in order to avoid the ever-increasing pollution of plastic and the consumption of crude oil.

FUTURE OUTLOOK

According to a European Bioplastics nova-Institute (2019a, b), the global bioplastic manufacturing capacity can reach 2.4 million tons in 2024. Beckstrom et al. (2020) found that bioplastics might be marketed for as little as \$970/tons while simultaneously reducing GHGs emissions by 67–116%. Bioplastics' prospective market is predicted to grow and eventually replace traditional plastics. Sustainable bioplastic materials are presently being developed, and it is critical to investigate novel materials or polymers derived from renewable sources for the production of bio-based and biodegradable plastics in the long term. This is because, while certain bioplastics are bio-based, they are not totally biodegradable and may have a significant detrimental influence on the environment. Bioplastics must be created from biodegradable components that are preferably derived from natural renewable resources such as biomass, plants, trash, microalgae, and bacteria (Sidek et al., 2019). As well, bioplastic production must not compete with food resources during the long-term application. Furthermore, bioplastics must be ecological, odor-free, and have synthetic plastic characteristics, to be ideal for packaging in the food, beverage industries, and medical applications.

Through the biorefinery idea, algae are a good candidate for synthesizing polymer for sustainable bioplastics manufacturing and other co-products. Consumer-grade bioplastics may also be made using monomers or polymers produced from algal waste leftovers, which solves waste disposal issues (Kumar et al., 2020). This circular economy improves the environmental efficiency of bioplastics manufacturing processes as well as the generation of various by-products such as fuel, nutraceuticals, medicines, and cosmetics, which may be used in a variety of industries (Karan et al., 2019). Torres et al. (2015) showed the utilization of microalgal biomass and PBAT to synthesis biocomposites in the biodiesel manufacturing process. Using PBAT and microalgal biomass (20%), plasticization enhanced the mechanical characteristics, tensile modulus, and elongation of the bio-composites in the study. The finished products will disintegrate in the soil and can be utilized to enhance the agriculture quality. These results will motivate the industry to engage in bioplastics production study and development in collaboration with academics in order to increase the techniques and quality of bioplastics (Andreeßen & Steinbüchel, 2018).

The certification and legislation around the usage of bioplastics products is another concern. The certification procedure is complicated, time-consuming, and laborious because all tests are completed on the same material and by the same testing facility. This ensures that in-vitro findings do not differ from actual circumstances. There are currently no established regulations or specifications for bioplastic quality, while there are many certification schemes for bioplastic compostability (Briassoulis & Degli Innocenti, 2017). According to European Bioplastics (2020a, b), one of the certification tests for bioplastics is EN 13432 (compostability test), which ensures that the product and its constituent components are biodegradable or may be composted industrially. Chemical testing for ecotoxicity test, biodegradability, heavy metals, disintegration test, and other certification tests are included in Plastics - compostability evaluation - Test scheme and specifications (EN 13432/EN 14995). These standard criteria, policies, and regulation rules should be certified and recognized worldwide to preserve uniformity in the use of bioplastics or waste management and to provide a systematic procedure with minimal deviation for safety and quality.

Generally, the bioplastic industry's future expansion and commercial need will necessitate the discovery and development of novel biodegradable materials or polymers. This would boost manufacturing efficiency while also addressing plastic pollution and investigating new bioplastics potential regarding the quantity and quality of biodegradable bio-based plastic materials. Microalgae-based bioplastics have revealed considerable promise in terms of lowering manufacturing costs, and they are a viable supply and road to bio-based and biodegradable bioplastics.

CONCLUSION

Microalgae as a sustainable alternative to crops might be a solution to the fundamental problem of food, fuel, bio-crude, and energy imbalance. Despite the outstanding in-vitro successes of algae-based bioplastics, industrialization and large-scale production of algae-based plastics are still restricted by certain challenges. Expanding novel bioplastic materials from renewable and sustainable resources to provide sustainable alternatives to petroleum plastic is a critical problem for the plastic industry. An algae biorefinery might be a viable alternative to the old petrochemical products. Currently, there is little study on algae-based bioplastics, but it is expanding fast, and species with high purity and quantity are in high demand. It is possible to boost biopolymer synthesis through stress-induced procedures, although this typically comes at the sacrifice of productivity. To that end, genetic engineering techniques as gene editing, recombinant DNA, or blending microalgae with other plastics polymers might be advantageous to enable changed species to be commercialized. Thus, the problems of algal bioplastics have to be solved as there is already an attraction factor from the market, which finally drives the producers/users to replace the fossil-based plastics. However, there is a serious shortage of research, technical understanding, development, and investigated microalgal strains to demonstrate an alternative product for conventional plastics. It is critical to work inside and outside technical domains to generate a product at the forefront of innovation, both in terms of sustainability and acceptability, to close the material loop for bioplastics. It remains to be seen if the algal bioplastics idea will propel the algal field into the commercial world or only serve as another economic and eco-friendly by-product.

REFERENCES

Abdo, S. M., & Ali, G. H. (2019). Analysis of polyhydroxybutrate and bioplastic production from microalgae. *Bulletin of the National Research Center*, *43*(1), 97. Advance online publication. doi:10.118642269-019-0135-5

Abdul-Latif, N.-I. S., Ong, M. Y., Nomanbhay, S., Salman, B., & Show, P. L. (2020). Estimation of carbon dioxide (CO_2) reduction by utilization of algal biomass bioplastic in Malaysia using carbon emission pinch analysis (CEPA). *Bioengineered*, *11*(1), 154–164. doi:10.1080/21655979.2020.1718471 PMID:32013677

Adhikari, D., Mukai, M., Kubota, K., Kai, T., Kaneko, N., Araki, K. S., & Kubo, M. (2016). Degradation of bioplastics in soil and their degradation effects on environmental microorganisms. *Journal of Agricultural Chemistry and Environment*, *05*(01), 23–34. doi:10.4236/jacen.2016.51003

Ahn, H. K., Huda, M. S., Smith, M. C., Mulbry, W., Schmidt, W. F., & Reeves, J. B. III. (2011). Biodegradability of injection molded bioplastic pots containing polylactic acid and poultry feather fiber. *Bioresource Technology*, *102*(7), 4930–4933. doi:10.1016/j.biortech.2011.01.042 PMID:21320772

Al-Tohamy, R., Kenawy, E.-R., Sun, J., & Ali, S. S. (2020a). Performance of a newly isolated salt-tolerant yeast strain *Sterigmatomyces halophilus* SSA-1575 for azo dye decolorization and detoxification. *Frontiers in Microbiology*, *11*, 1163. doi:10.3389/fmicb.2020.01163 PMID:32595618

Al-Tohamy, R., Sun, J., Fareed, M., Kenawy, E.-R., & Ali, S. S. (2020b). Ecofriendly biodegradation of Reactive Black 5 by newly isolated *Sterigmatomyces halophilus* SSA1575, valued for textile azo dye wastewater processing and detoxification. *Scientific Reports*, *10*(1), 12370. doi:10.103841598-020-69304-4 PMID:32704008

Ali, S. S., Al-Tohamy, R., Koutra, E., El-Naggar, A. H., Kornaros, M., & Sun, J. (2020a). Valorizing lignin-like dyes and textile dyeing wastewater by a newly constructed lipid producing and lignin modifying oleaginous yeast consortium valued for biodiesel and bioremediation. *Journal of Hazardous Materials*, *403*, 123575. doi:10.1016/j.jhazmat.2020.123575 PMID:32791477

Ali, S. S., Al-Tohamy, R., Koutra, E., Kornaros, M., Khalil, K., Elsamahy, T., El-Shetehy, M., & Sun, J. (2021b). Coupling azo dye degradation and biodiesel production by manganese-dependent peroxidase producing oleaginous yeasts isolated from wood-feeding termite gut symbionts. *Biotechnology for Biofuels*, *14*(61), 1–25. doi:10.118613068-021-01906-0 PMID:33685508

Ali, S. S., Al-Tohamy, R., Manni, A., Luz, F. C., Elsamahy, T., & Sun, J. (2019). Enhanced digestion of bio-pretreated sawdust using a novel bacterial consortium: Microbial community structure and methane-producing pathways. *Fuel*, *254*, 115604. doi:10.1016/j.fuel.2019.06.012

Ali, S. S., Al-Tohamy, R., Xie, R., El-Sheekh, M. M., & Sun, J. (2020b). Construction of a new lipase- and xylanase-producing oleaginous yeast consortium capable of reactive azo dye degradation and detoxification. *Bioresource Technology*, *313*, 123631. doi:10.1016/j.biortech.2020.123631 PMID:32540694

Ali, S. S., Elsamahy, T., Al-Tohamy, R., Zhu, D., Mahmoud, Y. A.-G., Koutra, E., Metwally, M. A., Kornaros, M., & Sun, J. (2021b). Plastic wastes biodegradation: Mechanisms, challenges and future prospects. *The Science of the Total Environment*, *780*, 146590. doi:10.1016/j.scitotenv.2021.146590 PMID:34030345

Ali, S. S., Elsamahy, T., Koutra, E., Kornaros, M., El-Sheekh, M., Abdelkarim, E., Zhu, D., & Sun, J. (2021c). Degradation of conventional plastic wastes in the environment. A review on current status of knowledge and future perspectives of disposal. *The Science of the Total Environment*, *144719*, 144719. Advance online publication. doi:10.1016/j.scitotenv.2020.144719 PMID:33548729

Ali, S. S., Kornaros, M., Manni, A., Sun, J., El-Shanshoury, A. E. R., Kenawy, E.-R., & Khalil, M. A. (2020c). Enhanced anaerobic digestion performance by two artificially constructed microbial consortia capable of woody biomass degradation and chlorophenols detoxification. *Journal of Hazardous Materials*, *389*, 122076. doi:10.1016/j.jhazmat.2020.122076 PMID:32004834

Ali, S. S., Mustafa, A. M., Kornaros, M., Manni, A., Sun, J., & Khalil, M. A. (2020d). Construction of novel microbial consortia CS-5 and BC-4 valued for the degradation of catalpa sawdust and chlorophenols simultaneously with enhancing methane production. *Bioresource Technology*, *301*, 122720. doi:10.1016/j.biortech.2019.122720 PMID:31945685

Ali, S. S., Sun, J., Koutra, E., El-Zawawy, N., Elsamahy, T., & El-Shetehy, M. (2021d). Construction of a novel cold-adapted oleaginous yeast consortium valued for textile azo dye wastewater processing and biore finery. *Fuel*, *285*, 119050. doi:10.1016/j.fuel.2020.119050

Andreeßen, C., & Steinbüchel, A. (2018). Recent developments in non-biodegradable biopolymers: Precursors, production processes, and future perspectives. *Applied Microbiology and Biotechnology*, *103*(1), 143–157. doi:10.100700253-018-9483-6 PMID:30397765

Anstey, A., Muniyasamy, S., Reddy, M. M., Misra, M., & Mohanty, A. (2014). Processability and biodegradability evaluation of composites from poly(butylene succinate) (PBS) bioplastic and biofuel co-products from Ontario. *Journal of Polymers and the Environment*, *22*(2), 209–218. doi:10.100710924-013-0633-8

Arcos-Hernandez, M. V., Laycock, B., Pratt, S., Donose, B. C., Nikolić, M. A. L., Luckman, P., Werker, A., & Lant, P. A. (2012). Biodegradation in a soil environment of activated sludge derived polyhydroxy-alkanoate (PHBV). *Polymer Degradation & Stability*, *97*(11), 2301–2312. doi:10.1016/j.polymdegradstab.2012.07.035

Arias, D. M., Uggetti, E., García-Galán, M. J., & García, J. (2018). Production of polyhydroxybutyrates and carbohydrates in a mixed cyanobacterial culture: Effect of nutrients limitation and photoperiods. *New Biotechnology*, *42*, 1–11. doi:10.1016/j.nbt.2018.01.001 PMID:29306000

Arrieta, M. P., López, J., Rayón, E., & Jiménez, A. (2014). Disintegrability under composting conditions of plasticized PLA–PHB blends. *Polymer Degradation & Stability*, *108*, 307–318. doi:10.1016/j.polymdegradstab.2014.01.034

Atiwesh, G., Mikhael, A., Parrish, C. C., Banoub, J., & Le, T.-A. T. (2021). Environmental impact of bioplastic use: A review. *Heliyon*, *7*(9), e07918. doi:10.1016/j.heliyon.2021.e07918 PMID:34522811

Barra, R., & Leonard, S. A. (2018). *Plastics and the circular economy. 54th Global Environment Facility Council Meeting*, Da Nang, Viet Nam.

Beckstrom, B. D. (2019). *Bioplastic production from microalgae with fuel co-products: A techno-economic and life-cycle assessment*. Colorado State University, Libraries. https://mountainscholar.org/bitstream/handle/10217/197377/Beckstrom_colostate_0053N_15606.pdf?sequence=1

Beckstrom, B. D., Wilson, M. H., Crocker, M., & Quinn, J. C. (2020). Bioplastic feedstock production from microalgae with fuel co-products: A techno-economic and life cycle impact assessment. *Algal Research*, *46*, 101769. doi:10.1016/j.algal.2019.101769

Bhati, R., & Mallick, N. (2015). Poly(3-hydroxybutyrate-co-3-hydroxyvalerate) copolymer production by the diazotrophic cyanobacterium *Nostoc muscorum* Agardh: Process optimization and polymer characterization. *Algal Research*, *7*, 78–85. doi:10.1016/j.algal.2014.12.003

Blank, L. M., Narancic, T., Mampel, J., Tiso, T., & O'Connor, K. (2020). Biotechnological upcycling of plastic waste and other non-conventional feedstocks in a circular economy. *Current Opinion in Biotechnology*, *62*, 212–219. doi:10.1016/j.copbio.2019.11.011 PMID:31881445

Bonifer, K. S., Wen, X., Hasim, S., Phillips, E. K., Dunlap, R. N., Gann, E. R., DeBruyn, J. M., & Reynolds, T. B. (2019). *Bacillus pumilus B12* Degrades Polylactic Acid and Degradation Is Affected by Changing Nutrient Conditions. *Frontiers in Microbiology*, *10*, 2548. Advance online publication. doi:10.3389/fmicb.2019.02548 PMID:31824441

Boyandin, A. N., Prudnikova, S. V., Karpov, V. A., Ivonin, V. N., Đỗ, N. L., Nguyễn, T. H., Lê, T. M. H., Filichev, N. L., Levin, A. L., Filipenko, M. L., Volova, T. G., & Gitelson, I. I. (2013). Microbial degradation of polyhydroxyalkanoates in tropical soils. *International Biodeterioration & Biodegradation*, *83*, 77–84. doi:10.1016/j.ibiod.2013.04.014

Briassoulis, D., & Degli Innocenti, F. (2017). Standards for Soil Biodegradable Plastics. In *Green Chemistry and Sustainable Technology* (pp. 139–168). Springer Berlin Heidelberg. doi:10.1007/978-3-662-54130-2_6

Bulota, M., & Budtova, T. (2015). PLA/algae composites: Morphology and mechanical properties. *Composites. Part A, Applied Science and Manufacturing*, *73*, 109–115. doi:10.1016/j.compositesa.2015.03.001

Chakravarty, P., Mhaisalkar, V., & Chakrabarti, T. (2010). Study on poly-hydroxyalkanoate (PHA) production in pilot scale continuous mode wastewater treatment system. *Bioresource Technology*, *101*(8), 2896–2899. doi:10.1016/j.biortech.2009.11.097 PMID:20045314

Chaogang, W., Zhangli, H., Anping, L., & Baohui, J. (2010). Biosynthesis of poly-3-hydroxybutyrate (PHB) in the transgenic green algae *chlamydomonas reinhardtii*. *Journal of Phycology*, *46*(2), 396–402. doi:10.1111/j.1529-8817.2009.00789.x

Cheng, Y.-S., Labavitch, J. M., & VanderGheynst, J. S. (2014). Elevated CO_2 concentration impacts cell wall polysaccharide composition of green microalgae of the genus *Chlorella*. *Letters in Applied Microbiology*, *60*(1), 1–7. doi:10.1111/lam.12320 PMID:25163669

Chidambarampadmavathy, K., Karthikeyan, O. P., & Heimann, K. (2017). Sustainable bio-plastic production through landfill methane recycling. *Renewable & Sustainable Energy Reviews*, *71*, 555–562. doi:10.1016/j.rser.2016.12.083

Chun, Y., Mulcahy, D., Zou, L., & Kim, I. S. (2017). A short review of membrane fouling in forward osmosis processes. *Membranes*, *7*(2), 30. doi:10.3390/membranes7020030 PMID:28604649

Ciapponi, R., Turri, S., & Levi, M. (2019). Mechanical reinforcement by microalgal biofiller in novel thermoplastic biocompounds from plasticized gluten. *Materials (Basel, Switzerland)*, *12*(9), 1476. doi:10.3390/ma12091476 PMID:31067771

Costa, S. S., Miranda, A. L., Andrade, B. B., Assis, D. de J., Souza, C. O., de Morais, M. G., Costa, J. A. V., & Druzian, J. I. (2018). Influence of nitrogen on growth, biomass composition, production, and properties of polyhydroxyalkanoates (PHAs) by microalgae. *International Journal of Biological Macromolecules*, *116*, 552–562. doi:10.1016/j.ijbiomac.2018.05.064 PMID:29763703

Daneshvar, E., Wicker, R. J., Show, P.-L., & Bhatnagar, A. (2022). Biologically-mediated carbon capture and utilization by microalgae towards sustainable CO_2 biofixation and biomass valorization – A review. *Chemical Engineering Journal*, *427*, 130884. doi:10.1016/j.cej.2021.130884

Darwesh, O. M., Ali, S. S., Matter, I. A., & Elsamahy, T. (2021). Nanotextiles waste management: controlling of release and remediation of wastes. In *Nanosensors and Nanodevices for Smart Multifunctional Textiles* (pp. 267–286). Elsevier. doi:10.1016/B978-0-12-820777-2.00016-9

Dianursanti, G. M., & Noviasari, C. (2018). The effect of glycerol addition as plasticizer in *Spirulina platensis* based bioplastic. *E3S Web of Conferences, 67*, 3048. doi:10.1051/e3sconf/20186703048

Dianursanti, & Khalis, S. A. (2018). The effect of compatibilizer addition on *Chlorella vulgaris* microalgae utilization as a mixture for bioplastic. *E3S Web of Conferences, 67*, 3047. doi:10.1051/e3sconf/20186703047

Enyidi, U. (2017). *Chlorella vulgaris* as protein source in the diets of African Catfish *Clarias gariepinus*. *Fishes*, *2*(4), 17. doi:10.3390/fishes2040017

European Bioplastics. (2020a). *Certifications for bioplastics.* https://www.europeanbioplastics.org/bioplastics/standards/certification/

European Bioplastics. (2020b). *Bioplastics Market Data 2019*. Available online: https://docs.european-bioplastics.org/publications/market_data/Report_Bioplastics_Market_Data_2019.pdf

European Bioplastics and Nova-Institute. (2019a). *Bioplastics Market Update*. Available online: https://www.european-bioplastics.org/wp-content/uploads/2019/11/Report_Bioplastics-Market-Data_2019_short_version.pdf

European Bioplastics and Nova-Institute. (2019b). *Report- Bioplastics Market Data 2018*. https://www.european-bioplastics.org/wp-content/uploads/2016/02/Report_Bioplastics-Market-Data_2018.pdf

European Bioplastics, Nova-Institute, (2019c). *Bioplastics facts and figures.* https://docs.european-bioplastics.org/publications/EUBP_Facts_and_fiures.pdf

Folino, A., Karageorgiou, A., Calabrò, P. S., & Komilis, D. (2020). Biodegradation of wasted bioplastics in natural and industrial environments: A Review. *Sustainability*, *12*(15), 6030. doi:10.3390u12156030

Geyer, R., Jambeck, J. R., & Law, K. L. (2017). Production, use, and fate of all plastics ever made. *Science Advances*, *3*(7), e1700782–e1700782. doi:10.1126ciadv.1700782 PMID:28776036

Gifuni, I., Olivieri, G., Pollio, A., & Marzocchella, A. (2018). Identification of an industrial microalgal strain for starch production in biorefinery context: The effect of nitrogen and carbon concentration on starch accumulation. *New Biotechnology*, *41*, 46–54. doi:10.1016/j.nbt.2017.12.003 PMID:29237569

Gimpel, J. A., Henríquez, V., & Mayfield, S. P. (2015). In Metabolic Engineering of Eukaryotic Microalgae: Potential and Challenges Come with Great Diversity. *Frontiers in Microbiology*, *6*, 1376. doi:10.3389/fmicb.2015.01376 PMID:26696985

Gómez, E. F., & Michel, F. C. Jr. (2013). Biodegradability of conventional and bio-based plastics and natural fiber composites during composting, anaerobic digestion and long-term soil incubation. *Polymer Degradation & Stability*, *98*(12), 2583–2591. doi:10.1016/j.polymdegradstab.2013.09.018

Harmaen, A. S., Khalina, A., Azowa, I., Hassan, M. A., Tarmian, A., & Jawaid, M. (2014). Thermal and biodegradation properties of poly(lactic acid)/fertilizer/oil palm fibers blends biocomposites. *Polymer Composites*, *36*(3), 576–583. doi:10.1002/pc.22974

Harmsen, P. F. H., Hackmann, M. M., & Bos, H. L. (2014). Green building blocks for bio-based plastics. *Biofuels, Bioproducts & Biorefining*, *8*(3), 306–324. doi:10.1002/bbb.1468

Hlavova, M., Turoczy, Z., & Bisova, K. (2015). Improving microalgae for biotechnology-From genetics to synthetic biology. *Biotechnology Advances*, *33*(6), 1194–1203. doi:10.1016/j.biotechadv.2015.01.009 PMID:25656099

Imre, B., García, L., Puglia, D., & Vilaplana, F. (2019). Reactive compatibilization of plant polysaccharides and biobased polymers: Review on current strategies, expectations and reality. *Carbohydrate Polymers*, *209*, 20–37. doi:10.1016/j.carbpol.2018.12.082 PMID:30732800

Iwata, T. (2015). Biodegradable and bio-Based polymers: Future prospects of eco-friendly plastics. *Angewandte Chemie International Edition*, *54*(11), 3210–3215. doi:10.1002/anie.201410770 PMID:25583677

Jain, R., & Tiwari, A. (2015). Biosynthesis of planet friendly bioplastics using renewable carbon source. *Journal of Environmental Health Science & Engineering*, *13*(1), 11. doi:10.118640201-015-0165-3 PMID:25717378

Jau, M.-H., Yew, S.-P., Toh, P. S. Y., Chong, A. S. C., Chu, W.-L., Phang, S.-M., Najimudin, N., & Sudesh, K. (2005). Biosynthesis and mobilization of poly(3-hydroxybutyrate) [P(3HB)] by *Spirulina platensis*. *International Journal of Biological Macromolecules*, *36*(3), 144–151. doi:10.1016/j.ijbiomac.2005.05.002 PMID:16005060

Jõgi, K., & Bhat, R. (2020). Valorization of food processing wastes and by-products for bioplastic production. *Sustainable Chemistry and Pharmacy*, *18*, 100326. doi:10.1016/j.scp.2020.100326

Johnsson, N., & Steuer, F. (2018). *Bioplastic material from microalgae: Extraction of starch and pha from microalgae to create a bioplastic material, school of industrial engineering and management*. KTH Royal Institute of Technology.

Jouhara, H., Czajczyńska, D., Ghazal, H., Krzyżyńska, R., Anguilano, L., Reynolds, A. J., & Spencer, N. (2017). Municipal waste management systems for domestic use. *Energy*, *139*, 485–506. doi:10.1016/j.energy.2017.07.162

Kaewbai-ngam, A., Incharoensakdi, A., & Monshupanee, T. (2016). Increased accumulation of polyhydroxybutyrate in divergent cyanobacteria under nutrient-deprived photoautotrophy: An efficient conversion of solar energy and carbon dioxide to polyhydroxybutyrate by Calothrix scytonemicola TISTR 8095. *Bioresource Technology*, *212*, 342–347. doi:10.1016/j.biortech.2016.04.035 PMID:27130227

Kale, G., Kijchavengkul, T., Auras, R., Rubino, M., Selke, S. E., & Singh, S. P. (2007). Compostability of Bioplastic Packaging Materials: An Overview. *Macromolecular Bioscience*, *7*(3), 255–277. doi:10.1002/mabi.200600168 PMID:17370278

Kalia, S., & Avérous, L. (2016). *Biodegradable and biobased polymers for environmental and biomedical Applications*. John Wiley & Sons. doi:10.1002/9781119117360

Kamravamanesh, D., Kovacs, T., Pflügl, S., Druzhinina, I., Kroll, P., Lackner, M., & Herwig, C. (2018). Increased poly-β-hydroxybutyrate production from carbon dioxide in randomly mutated cells of cyanobacterial strain *Synechocystis* sp. PCC 6714: Mutant generation and characterization. *Bioresource Technology*, *266*, 34–44. doi:10.1016/j.biortech.2018.06.057 PMID:29957289

Karan, H., Funk, C., Grabert, M., Oey, M., & Hankamer, B. (2019). Green bioplastics as part of a circular bioeconomy. *Trends in Plant Science*, *24*(3), 237–249. doi:10.1016/j.tplants.2018.11.010 PMID:30612789

Kessel-Vigelius, S. K., Wiese, J., Schroers, M. G., Wrobel, T. J., Hahn, F., & Linka, N. (2013). An engineered plant peroxisome and its application in biotechnology. *Plant Science*, *210*, 232–240. doi:10.1016/j.plantsci.2013.06.005 PMID:23849130

Khetkorn, W., Incharoensakdi, A., Lindblad, P., & Jantaro, S. (2016). Enhancement of poly 3 hydroxy butyrate production in *Synechocystis* sp. PCC 6803 by overexpression of its native biosynthetic genes. *Bioresource Technology*, *214*, 761–768. doi:10.1016/j.biortech.2016.05.014 PMID:27213577

Khoo, K. S., Chew, K. W., Yew, G. Y., Leong, W. H., Chai, Y. H., Show, P. L., & Chen, W.-H. (2020). Recent advances in downstream processing of microalgae lipid recovery for biofuel production. *Bioresource Technology*, *304*, 122996. doi:10.1016/j.biortech.2020.122996 PMID:32115347

Kovalcik, A., Meixner, K., Mihalic, M., Zeilinger, W., Fritz, I., Fuchs, W., Kucharczyk, P., Stelzer, F., & Drosg, B. (2017). Characterization of polyhydroxyalkanoates produced by *Synechocystis salina* from digestate supernatant. *International Journal of Biological Macromolecules*, *102*, 497–504. doi:10.1016/j.ijbiomac.2017.04.054 PMID:28419829

Kumar, S., Hablot, E., Moscoso, J. L. G., Obeid, W., Hatcher, P. G., DuQuette, B. M., Graiver, D., Narayan, R., & Balan, V. (2014). Polyurethanes preparation using proteins obtained from microalgae. *Journal of Materials Science*, *49*(22), 7824–7833. doi:10.100710853-014-8493-8

Kumar, S. R., Shaiju, P. & O'Connor, K. E. (2020). Bio-based and biodegradable polymers - state-of-the-art, challenges and emerging trends. *Current Opinion in Green and Sustainable Chemistry*, *21*, 75-81.

Mathiot, C., Ponge, P., Gallard, B., Sassi, J.-F., Delrue, F., & Le Moigne, N. (2019). Microalgae starch-based bioplastics: Screening of ten strains and plasticization of unfractionated microalgae by extrusion. *Carbohydrate Polymers*, *208*, 142–151. doi:10.1016/j.carbpol.2018.12.057 PMID:30658785

Mendhulkar, V. D., & Shetye, L. A. (2017). Synthesis of Biodegradable Polymer Polyhydroxyalkanoate (PHA) in Cyanobacteria *Synechococcus elongates* Under Mixotrophic Nitrogen- and Phosphate-Mediated Stress Conditions. *Industrial Biotechnology (New Rochelle, N.Y.)*, *13*(2), 85–93. doi:10.1089/ind.2016.0021

Mihai, M., Legros, N., & Alemdar, A. (2013). Formulation-properties versatility of wood fiber bio-composites based on polylactide and polylactide/thermoplastic starch blends. *Polymer Engineering and Science*, *54*(6), 1325–1340. doi:10.1002/pen.23681

Miranda, J., Ponce, G., Arellano-Garcia, H., Maciel Filho, R., & Wolf Maciel, M. R. (2015). Syngas to higher alcohols using cu-based catalyst – a simulation approach. *Chemical Engineering Transactions*, *43*, 1519–1524. doi:10.3303/CET1543254

Mohsenzadeh, A., Zamani, A., & Taherzadeh, M. J. (2017). Bioethylene production from ethanol: A Review and techno-economical evaluation. *ChemBioEng Reviews*, *4*(2), 75–91. doi:10.1002/cben.201600025

Monshupanee, T., Nimdach, P., & Incharoensakdi, A. (2016). Two-stage (photoautotrophy and heterotrophy) cultivation enables efficient production of bioplastic poly-3-hydroxybutyrate in auto-sedimenting cyanobacterium. *Scientific Reports*, *6*(1), 37121. doi:10.1038rep37121 PMID:27845413

Mostafa, N. A., Farag, A. A., Abo-dief, H. M., & Tayeb, A. M. (2018). Production of biodegradable plastic from agricultural wastes. *Arabian Journal of Chemistry*, *11*(4), 546–553. doi:10.1016/j.arabjc.2015.04.008

Nagarajan, S., Chou, S. K., Cao, S., Wu, C., & Zhou, Z. (2013). An updated comprehensive techno-economic analysis of algae biodiesel. *Bioresource Technology*, *145*, 150–156. doi:10.1016/j.biortech.2012.11.108 PMID:23260269

Niaounakis, M. (2013). Definitions and assessment of (bio)degradation. In Biopolymers Reuse, Recycling, and Disposal (pp. 77–94). Elsevier. doi:10.1016/B978-1-4557-3145-9.00002-6

Nikel, P. I., Pettinari, M. J., Galvagno, M. A., & Méndez, B. S. (2006). Poly(3-hydroxybutyrate) synthesis by recombinant *Escherichia coli* arcA mutants in microaerobiosis. *Applied and Environmental Microbiology*, *72*(4), 2614–2620. doi:10.1128/AEM.72.4.2614-2620.2006 PMID:16597965

Noor-Mohammadi, S., Pourmir, A., & Johannes, T. W. (2012). Method to assemble and integrate biochemical pathways into the chloroplast genome of *Chlamydomonas reinhardtii. Biotechnology and Bioengineering*, *109*(11), 2896–2903. doi:10.1002/bit.24569 PMID:22674415

Noor-Mohammadi, S., Pourmir, A., & Johannes, T. W. (2013). Method for assembling and expressing multiple genes in the nucleus of microalgae. *Biotechnology Letters*, *36*(3), 561–566. doi:10.100710529-013-1378-0 PMID:24129955

Otsuki, T., Zhang, F., Kabeya, H., & Hirotsu, T. (2004). Synthesis and tensile properties of a novel composite ofChlorella and polyethylene. *Journal of Applied Polymer Science*, *92*(2), 812–816. doi:10.1002/app.13650

Panda, B., & Mallick, N. (2007). Enhanced poly-beta-hydroxybutyrate accumulation in a unicellular cyanobacterium, *Synechocystis* sp. PCC 6803. *Letters in Applied Microbiology*, *44*(2), 194–198. doi:10.1111/j.1472-765X.2006.02048.x PMID:17257260

Rahman, A., Linton, E., Hatch, A. D., Sims, R. C., & Miller, C. D. (2013). Secretion of polyhydroxybutyrate in *Escherichia coli* using a synthetic biological engineering approach. *Journal of Biological Engineering*, *7*(1), 24. doi:10.1186/1754-1611-7-24 PMID:24139229

Rahman, A., & Miller, C. D. (2017). Microalgae as a source of bioplastics. In *Algal Green Chemistry* (pp. 121–138). Elsevier. doi:10.1016/B978-0-444-63784-0.00006-0

Rasul, I., Azeem, F., Siddique, M. H., Muzammil, S., Rasul, A., Munawar, A., Afzal, M., Ali, M. A., & Nadeem, H. (2017). Algae biotechnology. In Algae Based Polymers, Blends, and Composites (pp. 301–334). Elsevier. doi:10.1016/B978-0-12-812360-7.00008-2

Ravindran, R., & Jaiswal, A. K. (2016). Exploitation of food industry waste for high-value products. *Trends in Biotechnology*, *34*(1), 58–69. doi:10.1016/j.tibtech.2015.10.008 PMID:26645658

Raza, Z. A., Abid, S., & Banat, I. M. (2018). Polyhydroxyalkanoates: Characteristics, production, recent developments and applications. *International Biodeterioration & Biodegradation*, *126*, 45–56. doi:10.1016/j.ibiod.2017.10.001

Reddy, M. M., Vivekanandhan, S., Misra, M., Bhatia, S. K., & Mohanty, A. K. (2013). Biobased plastics and bionanocomposites: Current status and future opportunities. *Progress in Polymer Science*, *38*(10–11), 1653–1689. doi:10.1016/j.progpolymsci.2013.05.006

Rosenboom, J.-G., Hohl, D. K., Fleckenstein, P., Storti, G., & Morbidelli, M. (2018). Bottle-grade polyethylene furanoate from ring-opening polymerisation of cyclic oligomers. *Nature Communications*, *9*(1), 2701. doi:10.103841467-018-05147-y PMID:30042388

Rudnik, E. (2019). Ecotoxicological assessment of compostable polymer materials. In *Compostable Polymer Materials* (pp. 293–313). Elsevier. doi:10.1016/B978-0-08-099438-3.00009-4

Rudnik, E., & Briassoulis, D. (2011). Degradation behaviour of poly(lactic acid) films and fibres in soil under Mediterranean field conditions and laboratory simulations testing. *Industrial Crops and Products*, *33*(3), 648–658. doi:10.1016/j.indcrop.2010.12.031

Sabathini, H. A., Windiani, L., Dianursanti, & Gozan, M. (2018). Mechanical physicial properties of *Chlorella*-PVA based bioplastic with ultrasonic homogenizer. *E3S Web of Conferences, 67*, 3046. doi:10.1051/e3sconf/20186703046

Samantaray, S., & Mallick, N. (2011). Production and characterization of poly-β-hydroxybutyrate (PHB) polymer from *Aulosira fertilissima. Journal of Applied Phycology*, *24*(4), 803–814. doi:10.100710811-011-9699-7

Sekiguchi, T., Saika, A., Nomura, K., Watanabe, T., Watanabe, T., Fujimoto, Y., Enoki, M., Sato, T., Kato, C., & Kanehiro, H. (2011). Biodegradation of aliphatic polyesters soaked in deep seawaters and isolation of poly(ε-caprolactone)-degrading bacteria. *Polymer Degradation & Stability*, *96*(7), 1397–1403. doi:10.1016/j.polymdegradstab.2011.03.004

Sharma, L., & Mallick, N. (2005). Accumulation of poly-β-hydroxybutyrate in *Nostoc muscorum*: Regulation by pH, light–dark cycles, N and P status and carbon sources. *Bioresource Technology*, *96*(11), 1304–1310. doi:10.1016/j.biortech.2004.10.009 PMID:15734319

Shi, B., Wideman, G., & Wang, J. H. (2011). A new approach of bioco$_2$ fixation by thermoplastic processing of microalgae. *Journal of Polymers and the Environment*, *20*(1), 124–131. doi:10.100710924-011-0329-x

Sidek, I. S., Draman, S. F. S., Abdullah, S. R. S., & Anuar, N. (2019). Current development on bioplastics and its future prospects: An introductory review. *INWASCON Technology Magazine*, 3–8. doi:10.26480/itechmag.01.2019.03.08

Simonic, M., & Zemljic, F. (2021). Production of bioplastic material from algal biomass. *Chemical Industry & Chemical Engineering Quarterly*, *27*(1), 79–84. doi:10.2298/CICEQ191024026S

Suriyamongkol, P., Weselake, R., Narine, S., Moloney, M., & Shah, S. (2007). Biotechnological approaches for the production of polyhydroxyalkanoates in microorganisms and plants— A review. *Biotechnology Advances*, *25*(2), 148–175. doi:10.1016/j.biotechadv.2006.11.007 PMID:17222526

Tabasi, R. Y., & Ajji, A. (2015). Selective degradation of biodegradable blends in simulated laboratory composting. *Polymer Degradation & Stability*, *120*, 435–442. doi:10.1016/j.polymdegradstab.2015.07.020

Tachibana, K., Hashimoto, K., Yoshikawa, M., & Okawa, H. (2010). Isolation and characterization of microorganisms degrading nylon 4 in the composted soil. *Polymer Degradation & Stability*, *95*(6), 912–917. doi:10.1016/j.polymdegradstab.2010.03.031

Taepucharoen, K., Tarawat, S., Puangcharoen, M., Incharoensakdi, A., & Monshupanee, T. (2017). Production of poly(3-hydroxybutyrate-co-3-hydroxyvalerate) under photoautotrophy and heterotrophy by non-heterocystous N_2-fixing cyanobacterium. *Bioresource Technology*, *239*, 523–527. doi:10.1016/j.biortech.2017.05.029 PMID:28533067

Tang, D. Y. Y., Yew, G. Y., Koyande, A. K., Chew, K. W., Vo, D.-V. N., & Show, P. L. (2020). Green technology for the industrial production of biofuels and bioproducts from microalgae: A review. *Environmental Chemistry Letters*, *18*(6), 1967–1985. doi:10.100710311-020-01052-3

Terzopoulou, Z., Karakatsianopoulou, E., Kasmi, N., Majdoub, M., Papageorgiou, G. Z., & Bikiaris, D. N. (2017). Effect of catalyst type on recyclability and decomposition mechanism of poly(ethylene furanoate) biobased polyester. *Journal of Analytical and Applied Pyrolysis*, *126*, 357–370. doi:10.1016/j.jaap.2017.05.010

Thakur, S., Chaudhary, J., Sharma, B., Verma, A., Tamulevicius, S., & Thakur, V. K. (2018). Sustainability of bioplastics: Opportunities and challenges. *Current Opinion in Green and Sustainable Chemistry*, *13*, 68–75. doi:10.1016/j.cogsc.2018.04.013

Thellen, C., Coyne, M., Froio, D., Auerbach, M., Wirsen, C., & Ratto, J. A. (2008). A processing, characterization and marine biodegradation study of melt-extruded polyhydroxyalkanoate (pha) films. *Journal of Polymers and the Environment*, *16*(1), 1–11. doi:10.100710924-008-0079-6

Toh, Y., Jau, P. S., Yew, M. H., Abed, S. P. M., & Sudesh, R. M. K. (2008). Comparison of polyhydroxyalkanoates biosynthesis, mobilization and the effects on cellular morphology in *Spirulina Platensis* and *Synechocystis* Sp. Uniwg. *Journal of Bio-Science*, *19*(2), 21–38. http://www.tlsr.usm.my/tlsr19022008/19022008_03.pdf

Toro, C., Reddy, M. M., Navia, R., Rivas, M., Misra, M., & Mohanty, A. K. (2013). Characterization and application in biocomposites of residual microalgal biomass generated in third generation biodiesel. *Journal of Polymers and the Environment*, *21*(4), 944–951. doi:10.100710924-013-0609-8

Torres, S., Navia, R., Campbell Murdy, R., Cooke, P., Misra, M., & Mohanty, A. K. (2015). Green composites from residual microalgae biomass and poly(butylene adipate-co-terephthalate): Processing and plasticization. *ACS Sustainable Chemistry & Engineering*, *3*(4), 614–624. doi:10.1021c500753h

Tosin, M., Barbale, M., Chinaglia, S., & Degli-Innocenti, F. (2020). Disintegration and mineralization of mulch films and leaf litter in soil. *Polymer Degradation & Stability*, *179*, 109309. doi:10.1016/j.polymdegradstab.2020.109309

Tosin, M., Weber, M., Siotto, M., Lott, C., & Degli Innocenti, F. (2012). Laboratory test methods to determine the degradation of plastics in marine environmental conditions. *Frontiers in Microbiology*, *3*, 225. doi:10.3389/fmicb.2012.00225 PMID:22737147

Tsang, Y. F., Kumar, V., Samadar, P., Yang, Y., Lee, J., Ok, Y. S., Song, H., Kim, K.-H., Kwon, E. E., & Jeon, Y. J. (2019). Production of bioplastic through food waste valorization. *Environment International*, *127*, 625–644. doi:10.1016/j.envint.2019.03.076 PMID:30991219

Tsuji, H. (2013). Poly(Lactic Acid). In Bio-Based Plastics (pp. 171–239). John Wiley & Sons Ltd. doi:10.1002/9781118676646.ch8

Vaverková, M., & Adamcová, D. (2015). Biodegrability of bioplastic materials in a controlled composting environment. *Journal of Ecological Engineering*, *16*, 155–160. doi:10.12911/22998993/2949

Venkata Subhash, G., Rajvanshi, M., Navish Kumar, B., Govindachary, S., Prasad, V., & Dasgupta, S. (2017). Carbon streaming in microalgae: Extraction and analysis methods for high value compounds. *Bioresource Technology*, *244*, 1304–1316. doi:10.1016/j.biortech.2017.07.024 PMID:28803061

Volova, T. G., Boyandin, A. N., Vasiliev, A. D., Karpov, V. A., Prudnikova, S. V., Mishukova, O. V., Boyarskikh, U. A., Filipenko, M. L., Rudnev, V. P., Bá Xuân, B., Việt Dũng, V., & Gitelson, I. I. (2010). Biodegradation of polyhydroxyalkanoates (PHAs) in tropical coastal waters and identification of PHA-degrading bacteria. *Polymer Degradation & Stability*, *95*(12), 2350–2359. doi:10.1016/j.polymdegradstab.2010.08.023

Wagner, J., Bransgrove, R., Beacham, T. A., Allen, M. J., Meixner, K., Drosg, B., Ting, V. P., & Chuck, C. J. (2016). Co-production of bio-oil and propylene through the hydrothermal liquefaction of polyhydroxybutyrate producing cyanobacteria. *Bioresource Technology*, *207*, 166–174. doi:10.1016/j.biortech.2016.01.114 PMID:26881334

Wang, K. (2014). *Bio-Plastic Potential of Spirulina Microalgae*. The University of Georgia. https://getd.libs.uga.edu/pdfs/wang_kun_201405_ms.pdf

Wang, K., Mandal, A., Ayton, E., Hunt, R., Zeller, M. A., & Sharma, S. (2016). Modification of protein rich algal-biomass to form bioplastics and odor removal. In *Protein Byproducts* (pp. 107–117). Elsevier. doi:10.1016/B978-0-12-802391-4.00006-9

Wei, L., Liang, S., & McDonald, A. G. (2015). Thermophysical properties and biodegradation behavior of green composites made from polyhydroxybutyrate and potato peel waste fermentation residue. *Industrial Crops and Products*, *69*, 91–103. doi:10.1016/j.indcrop.2015.02.011

Weng, Y.-X., Wang, X.-L., & Wang, Y.-Z. (2011). Biodegradation behavior of PHAs with different chemical structures under controlled composting conditions. *Polymer Testing*, *30*(4), 372–380. doi:10.1016/j.polymertesting.2011.02.001

Wu, C.-S. (2011). Preparation, characterization, and biodegradability of renewable resource-based composites from recycled polylactide bioplastic and sisal fibers. *Journal of Applied Polymer Science*, *123*(1), 347–355. doi:10.1002/app.34223

Wu, C.-S. (2014). Preparation and characterization of polyhydroxyalkanoate bioplastic-based green renewable composites from rice husk. *Journal of Polymers and the Environment*, *22*(3), 384–392. doi:10.100710924-014-0662-y

Xu, J., & Guo, B.-H. (2010). Poly(butylene succinate) and its copolymers: Research, development and industrialization. *Biotechnology Journal*, *5*(11), 1149–1163. doi:10.1002/biot.201000136 PMID:21058317

Yang, J. E., Park, S. J., Kim, W. J., Kim, H. J., Kim, B. J., Lee, H., Shin, J., & Lee, S. Y. (2018). One-step fermentative production of aromatic polyesters from glucose by metabolically engineered *Escherichia coli* strains. *Nature Communications*, *9*(1), 79. Advance online publication. doi:10.103841467-017-02498-w PMID:29311546

Zeller, M. A., Hunt, R., Jones, A., & Sharma, S. (2013). Bioplastics and their thermoplastic blends from *Spirulina* and *Chlorella* microalgae. *Journal of Applied Polymer Science*, *130*(5), 3263–3275. doi:10.1002/app.39559

Zeller, M. A., Hunt, R., & Sharma, S. (2012). Sustainable bioderived polymeric materials and thermoplastic blends made from floating aquatic macrophytes such as "duckweed.". *Journal of Applied Polymer Science*, *127*(1), 375–386. doi:10.1002/app.37555

Zhang, F., Endo, T., Kitagawa, R., Kabeya, H., & Hirotsu, T. (2000). Synthesis and characterization of a novel blend of polypropylene with *Chlorella*. *Journal of Materials Chemistry*, *10*(12), 2666–2672. doi:10.1039/b004489j

Zhang, F., Kabeya, H., Kitagawa, R., Hirotsu, T., Yamashita, M., & Otsuki, T. (2000). A novel polyethylene-chlorella composite. I. Characterization of *Chlorella* biologically fixing CO_2. *Journal of Applied Polymer Science*, *77*(10), 2278–2284. doi:10.1002/1097-4628(20000906)77:10<2278::AID-APP21>3.0.CO;2-Z

Zhu, N., Ye, M., Shi, D., & Chen, M. (2017). Reactive compatibilization of biodegradable poly(butylene succinate)/*Spirulina* microalgae composites. *Macromolecular Research*, *25*(2), 165–171. doi:10.100713233-017-5025-9

ABBREVIATIONS

CGM: Corn gluten meal
CM: Canola meal
EFB: Empty fruit bunch
GHG: Greenhouse gas
LDPE: Low-density polyethylene
MCL PHAs: Medium-chain length PHAs
NPK: Nitrogen, phosphorus, potassium fertilizer
P(3HB): Poly (3-hydroxybutyrate)
P(3HB-CO-3HV): Poly (3-hydroxybutyrate-co-3-hydroxyvalerate)
PA: Polyamide
PBAT: Polybutylene adipate terephthalate
PBS: Polybutylene succinate
PE: Polyethylene
PEF: Polyethylene furanoate
PE-g-MA: Polyethylene-graft-maleic anhydride
PET: Polyethylene terephthalate
PFF: Poultry feather fibers
PHA: polyhydroxyalkanoates
PHB: Polyhydroxybutyrate
PLA: Polylactic acid
PP: Polypropylene
PP: Polypropylene
PTT: Polytrimethylene terephthalate
PVA: Polyvinyl alcohol
PVC: Polyvinyl chloride
RH: Rice husk

SF: Sisal fiber
SM: Soy meal
SW: Soft wood

Chapter 20
Exploring the Potential of Algae in the Mitigation of Plastic Pollution in Aquatic Environments

Abdallah Abdelfattah
https://orcid.org/0000-0002-0795-1867
School of Environment and Safety Engineering, Jiangsu University, China & Faculty of Engineering, Tanta University, Egypt

Mudasir A. Dar
Jiangsu University, China

Hassan Ramadan
Faculty of Engineering, Tanta University, Egypt

Eslam Ibrahim El-Aswar
National Water Research Center (NWRC), Egypt

Reham Eltawab
Jiangsu University, China & Tanta University, Egypt

Esraa Abdelkarim
Jiangsu University, China

Tamer Elsamahy
Jiangsu University, China

Michael Kornaros
https://orcid.org/0000-0001-6691-6925
University of Patras, Greece

Sameh Samir Ali
Botany Department, Faculty of Science, Tanta University, Egypt & Biofuels Institute, School of the Environment and Safety Engineering, Jiangsu University, China

Liang Cheng
Jiangsu University, China

ABSTRACT

The continuous increase in global demand for plastic products caused a significant increase in plastic waste pollution. Therefore, this increase in plastic waste represents a serious problem affecting aquatic and human life because microplastics can enter the food chain and cause several diseases. Also, the convention disposal techniques appear to be ineffective strategies to mitigate plastic pollution. However, the physicochemical characteristics of plastics represent a challenge to microbial degradation. This chapter discusses the proposed eco-friendly techniques for plastic biodegradation using algae to mitigate the plastic waste crisis. Several species have been identified as excellent plastic biodegraders. However,

DOI: 10.4018/978-1-6684-2438-4.ch020

few researchers have investigated the algal role in plastic degradation. Microalgae may degrade plastic materials by employing generated toxins or enzymes. The use of algae for plastic biodegradation has been reviewed to offer new insights into various biodegradation mechanisms and contemporary bioremediation concepts for chemicals and algae-based by-products.

INTRODUCTION

Plastics are fossil-derived synthetic polymers that play a crucial role in all aspects of human activities and chemical industries. Due to all the advantages offered by plastics listed in Table 1, the demand for plastic has tripled in the past twenty-five years (Feil & Pretz, 2020). The production of plastics is expected to triple by 2050, accounting for a fifth of global oil consumption (Ali et al., 2021a, b). Moreover, COVID-19 pandemics crisis further exacerbated the situation of plastic pollution due to the intensive packaging of food items and other necessities activities. In a quest to ease livelihood, humans continue to develop recalcitrant and robust chemicals that are relatively difficult to degrade. Some substances like plastics are intentionally designed to prevent degradation while increasing their long-term usage, in other scenarios, the persistence of many industrial products is unintended (Mathews et al., 2019). Barra & Leonard (2018) found that since 2020, over 400 million tonnes of plastic wastes (PWs) will be annually produced, while plastics production is expected to double by 2035 reaching 800 million tonnes, and 1600 million tonnes by 2050.

Although plastics offer a variety of benefits, it was excessively accumulated on the Earth. Therefore, plastic pollution (white pollution) led to severe impacts on the natural environment and human health. The plastic pollution generated by massive plastic litter is wreaking havoc in the environment, affecting marine ecosystems including marine life forms and coral reefs (Chia et al., 2020). The total amount of industrial wastes resistant to natural decomposition is accumulating at an alarming rate, posing threat to local and regional economies, besides natural ecosystems, and human health (Ali et al., 2021a, b). Striving efforts to convert the plastics into natural metabolites that can mimic the biochemical reactions in living organisms is the future of plastic recycling where natural elements will be employed to mineralize plastics and return them to the biogeochemical cycles.

On the other hand, Certain PWs such as polyethylene (PE) can be degraded by some microalgae species, such as blue-green algae, diatoms, and green algae (Vimal Kumar et al., 2017). Green photosynthetic organisms, microalgae prefer to attach to the surface of plastic materials. Furthermore, blue-green microalgae are the most common microorganism found in wastewater ecosystems. According to the literature, microalgal colonies have been discovered to be dominant on the surface of discarded PE bags due to the availability of nutrients, water, and sunlight. Microalgae also can colonize the surface of polyethylene sheets, and biodegradation occurs under normal conditions (Cunha et al., 2019). Recently, a microalgae consortium was used to demonstrate positive hetero aggregation on polymers. In comparison to other techniques, algal degradation is considered a better method for treating plastic waste due to its relevance and eco-friendliness (Gnanavel et al., 2012). However, there have been few studies on the use of microalgae for plastic degradation. As a result, the purpose of this chapter is to compile the most recent advances and new knowledge about the impact of plastic waste on the aquatic environment and human health that have emerged in recent years. Plastic degradation by microorganisms or enzymes has also been discussed, with algae being a promising candidate for plastic biodegradation. Factors influencing plastic biodegradation have also been identified.

Table 1. Some of the common types of plastics and their day-to-day applications (Feil & Pretz, 2020)

Type of the plastic	Chemical formula	Composition	Applications
Acrylonitrile butadiene styrene (ABS)	$(C_8H_8C_4H_6C_3H_3N)n$	Acetonitrile, Butadiene, Styrene	Pipes, electronics, Vehicle spares, Protective head gears
Polyamide	(CO-NH)n	Amides, α, ω-amino acids, diamine, diacid	Carpets, garments, ropes, gears, seatbelts
Polyethylene	$(C_2H_4)n$	Ethylene	Wire insulations, bottles, toys, bags
Polyimide	RC(O)OC(NR)R	Dianhydride, diamine, diisocyanate	Medical tubing, high temperature adhesive
Poly-methyl methacrylate	$(C_5O_2H_8)n$	Propylene, benzene	Construction, automotive industry, medical
Polypropylene	$(C_3H_6)n$	Propylene	Automotive industry, packaging for consumer products, textiles
Polystyrene	$(C_8H_8)n$	Styrene	Food packaging, laboratory ware, electronics, automotive parts
Polytetrafluoroethylene	$(C_2F_4)n$	Tetrafluoroethylene	Cookware, buildings, dental clinics, pipes, machinery, gaskets
Polyvinyl chloride	$(C_2H_3Cl)n$	Vinyl chloride	Construction, health care, electronics, automobile
Acrylic resins	CH_2=CHCOOH	Acrylic acid, methacrylic acid, other related compounds	Decorative panels, adhesives, coatings, tiles
Epoxy resins	C_2H_4O	Epoxy monomers acidic hydroxy groups, epichlorohydrin	Adhesives, plastics, paints, coatings, sealers
Polyester resin	$(C_{10}H_8O_4)n$	Dibasic organic acids, polyhydric alcohols	Boats, flat Roofing, pond building, arts, crafts, surf boards, custom moulding
Polyurethane resin	$C_3H_8N_2O$	Polyol, isocyanate	Furniture, automotive interiors, packaging
Silicone resin	RnSiXmOy	Oligosiloxanes	Automotive, electronics, industrial molds, Packaging, artistic replications, toys
Vinyl resin	$(C_2H_3Cl)n$	Styrene, methacrylic acid, epoxy	Water pipe, window frames, gutters and downspouts, tiles, electronics

IMPACT OF PLASTIC ON THE ENVIRONMENT

Pollution due to waste plastic, particularly microplastics (plastic particles <5 mm in size), is a pervasive issue everywhere on Earth. Microplastics can be found everywhere, on land, in oceans, even in the air that we breathe (Chen et al., 2019; Ali et al., 2021a, b). In this context, Tang et al. (2021) and Chen et al. (2019) found that PWs are an ideal carrier for pathogens, chemicals, heavy metals, and organic pollutants. Furthermore, physico-chemical degradation of these wastes in the environment releases highly toxic compounds leading to deteriorating the quality of water and soil, with oceans loaded with 269 tonnes of macro-, micro-, and/or nano-plastic particles (Eriksen et al., 2014). The reactivity of plastic

Figure 1. Plastic wastes disposal techniques and their influence on the environment (Sharma et al., 2013, Ali et al., 2021a, b)

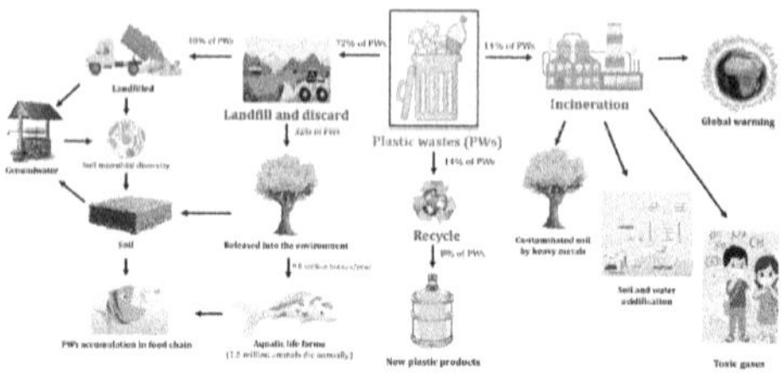

particles mainly depends on the particle size (surface area) and hydrophobicity which make plastic debris efficient sorbents. Moreover, the long-lasting accumulation of microplastic can alter the food chain, since micro and nano-plastics can be ingested by animals (Frias et al., 2010). Several plastic-type wastes have the ability to aggregate due to hydrophobic nature of these polymers (Liu et al., 2016). The marine environment provides a plethora of ecosystem services, including food security, carbon storage, waste purification, and cultural advantages (for example, recreational activities and spiritual betterment) (Liquete et al., 2013). The anthropogenic disturbances such as discharge of plastic trash into the marine ecosystem demonstrates negative impact on both different ecosystems and human health (Galloway et al., 2017). Li et al. (2020) revealed the histopathological damage in tissues of fish (*Hemiculter leucisculus*) as a result of plastic pollution and showed that the abundance of microplastics in the guts of *H. leucisculus* fluctuated from less than 1-3 mm and was estimated by ~2–16 items/g of tissue. Plastic waste discarded in various water bodies such as lakes, rivers, ponds, and other bodies of water poses a serious threat to biota. The lakes in Dhaka's megacity are one of the best examples of pollution caused by visitors discarded plastic bottles, cans, bags, and other plastic products. Plastic waste in water bodies disrupts natural flow, limits fish reproduction, and kills vital organisms, while polymers in the oceans may contribute to global warming by creating a shaded canopy that makes it difficult for plankton and other aquatic life forms to grow (Proshad et al., 2017).

Concerning air pollution due to the accumulation or disposal of these wastes, incineration of plastic waste releases a variety of pollutants into the atmosphere such as making it one of the most common sources of air pollution. For instance, particulate matter, methane, furan, nitrogen oxides, volatile organic compounds, aldehyde, metals, carbon monoxide, carbon dioxide, polyaromatic hydrocarbons, and other solid material (Sharma et al., 2013). In the case of incineration, the management of particulate matter and the temperature of the gas have an impact on the emissions generated. Most incinerators, on the other hand, operate without any additional air pollution control equipment, resulting in the emission of dangerous gases into the environment. Furthermore, additional chemical components with unknown toxicity are emitted, and the full spectrum of health consequences induced by exposure to the complete combination of emitted chemicals by plastic incineration is yet unknown (Sharma et al., 2013), as shown in Figure 1.

In oceans, microplastics are frequently consumed by marine organisms, this may influence the food chain of fish and shellfish stocks, as well as their prey, causing reduced reproduction, growth, and population levels (Galloway et al. 2017). These polluted marine creatures may eventually be ingested as seafood by humans, posing a danger to human health indirectly. As many plastic polymers are composed of toxic persistent organic pollutants (i.e., plasticizers, biocides, and flame retardants) this may alleviate the chances of toxicity to social communities (Rochman, 2015). Since aquaculture industry is very fragile, plastic pollution may have negative impacts on its productivity, viability, profitability, and safety

(Rochman, 2015). Despite that, plastic pollution immediately contributes to the economic expenditures incurred in cleaning up the environment (Science for Environment Policy, 2011).

In addition, PWs pollute the soil by settling on the surface or permeating the soil layers through a variety of channels, sludge, fertilizers, wastewater irrigation, landfilling, biosolids, and other sources, as shown in Figure 1 (Ali et al., 2021a, b; Darwesh et al., 2021). Physical degradation by Temperature and photo-oxidation cause plastics to fracture into microplastics on the soil surface. Plants and soil creatures can translocate this fragmented microplastic particles deeper into the soil polluting groundwater and worsening soil characteristics (Scheurer & Bigalke, 2018; Zhu et al., 2018).

Toxic Impact of Plastic on Algal Growth

Taipale et al. (2019) found that the mixotrophic algae (Cryptomonas sp.) exhibited growth on PE microplastics colonized by bacteria. The alga was found to sequester carbons from the polymer to produce ω-6 and ω-3 polyunsaturated fatty acids (PUFA). The prior colonization by the microbiota on PE surface resulted in enhanced growth of the microalgae as compared to control sets devoid of microbiota. However, direct contact of microalgae with the plastic polymer or its surface chemicals have a toxic impact on the growth of microalgae. Subsequently, the colonization by other microbiota eliminates the toxic chemicals covering the surface of the PE microplastics. Apart from this, the molecular weight of the polymer is also known to influence the growth of algae and ultimately degradation rate. While testing the effect of microplastic concentration, Khoironi et al. (2019) observed that higher concentration of microplastics, caused reduced growth of the microalgae. The reduced growth of microalgae could be due to two main reasons. Primarily the microplastics may cause shading effects, which limits light intensity thereby interfering with microalgal photosynthesis. Secondly, the reduced growth could be a consequence of the interactions between microalgae and microplastic, like aggregation and adsorption processes (Zhang et al., 2017). This in other words explains the impacts of microplastic on microalgae are dependent on particle size of the microplastic. According to Liu and colleagues (Liu et al., 2021), larger particles cause serious consequences by blocking light and disrupting photosynthesis, while smaller microplastics destroy the microalgal cell wall through surface absorption mechanism. In contrast, Raphidocelis subcapitata, shows higher growth rate in the media containing plastic microbeads of 63-75 mm in size (Canniff & Hoang, 2018). On a similar note, Chae et al. (2019) reported that bigger particles of microplastics (about 200 mm diameter) promoted the cell growth as well as photosynthesis in the marine microalga, Dunaliella salina. The accelerated growth was most likely owing to trace quantities of additives of microplastics such as stabilizers, phthalates, and endocrine disruptors that leached from the polymer (Chae et al. 2019). Briefly, further research is direly needed to decipher the impact of microplastics on microalgae, which serve as primary producers in ecosystems. This is also pivotal from the bioremediation viewpoint, to check the possibility of microalgae biodegrading microplastic.

Furthermore, Li et al. (2018) stated that the accumulation of synthetic PWs such as polypropylene (PP) and PE in aquatic environment eventually physically degrades into microsized plastics. However, the ability of plastics to release its toxic additives is strongly influenced by the size of their molecules (Koelmans et al., 2013). The increased number of additives released into the environment is influenced by the smaller plastic size. According to Zhu et al. (2020) and Chae et al. (2019), the size of plastics affects microorganisms' ability to adsorb additive particles that cause cell membrane damage and growth inhibition. Plastics release toxic additives added during manufacturing, such as plasticizers, polychlorinated biphenyls, dichlorodiphenyltrichloroethane, and heavy metals during degradation (Campanale et al., 2020). The release of additives or toxic chemical compounds as a result of the degradation process could be even worse for the environment (Nava & Leoni, 2021). Capolupo et al. (2020) studied at the effects of chemical additives in plastics on the microalgae Raphidocelis subcapitata and Skeletonema costatum, finding that almost all additive particles inhibited algae growth in fresh and marine environment.

Toxic Impact of Plastic on Human Health

The applied disposal technique of PWs can have a significant directly (inhalation) or indirectly (digestion) impact on human health. Microplastic persistence, in particular, can cause apoptosis, inflammation, necrosis, oxidative stress, and genotoxicity, as well as a variety of severe effects such as tissue damage, fibrosis, and carcinogenesis in the case of continued exposure (Proshad et al., 2017; Wright & Kelly, 2017). The ingestion of micro- and nano-plastics by humans and animals might allow adherent or endogenous pollutants to enter the cells (Prata et al., 2020; Wright & Kelly, 2017). PE particles (0.5–50 m), for example, can bind to nearby arteries and travel into the perivascular lymph spaces (Willert et al., 1996), eliciting a nonimmunological foreign body reaction (Doorn et al., 1996). Inhalation of particulate matter can also produce oxidative stress, which can lead to inflammation and intestinal fibrosis (Nel et al., 2006). In response to this worry, Barabad et al. (2018) investigated the PMs released by vinyl or plastic burning, finding substantial amounts of dangerous pollutants, fine and ultrafine particles, as well as acetone, benzene, and other toxic chemicals. Furthermore, all plastics contain reactive oxygen species (ROS), whose concentration can significantly increase due to interactions with light or the presence of transition metals, resulting in the formation of free radicals through the dissociation of CH bonds (Wright and Kelly, 2017). Lithner et al. (2011) have demonstrated that certain forms of plastic might produce hazardous monomers, such as mutagenic and/or carcinogenic resin monomers. Huerta Lwanga et al. (2017) investigated the transmission of low density polyethene microparticles (LDPE-MPs) via chickens and earthworms in the food chain, finding LDPE-MPs concentrations of 129.8 MPs/g faeces in chickens and 10.2 MPs/gizzard in earthworms, respectively. Finally, taking into account the mishandling of PWs as well as the worldwide repercussions of plastic on the environment, animals, and human health, innovative methods for the treatment and disposal of PWs are urgently needed.

Therefore, the relatively new phenomena of plastic invasion require concerted efforts and coordinated leadership to maximize recycling-related legislation, education programs to reduce usage. No doubt more research is needed in this area, there is concern that microplastic exposure is leading to an increase in complications in diseases like inflammation, and that microplastic accumulation in the body could have toxic effects. Although, there is a rising awareness to use environment friendly materials, the ubiquitous usage of plastics along with their undeniable benefits makes it difficult for us as individuals to minimize plastic consumption to make a collective difference globally. Even in a scenario where we can curb it, we will need to apply science to optimally mitigate the impacts of plastic pollution. In addition to inventing biodegradable alternative to plastics, scientists are continuously striving to find the natural candidates in the biodegradation process. According to Greenpeace, over 12.7 million tons of plastic waste enters the oceans annually, so having a natural ally on the seafloor will be crucial.

BIODEGRADATION OF PLASTIC WASTES

Deterioration of the plastics in the environment is a slow process, initially triggered by environmental factors like temperature, humidity, pH and UV light, etc. and finally carried out by wild microorganisms. Biodegradation is a biological process where a biological agent assimilates the organic polymer as a substrate for growth and energy, resulting in microbial biomass as the end product. The degradation of the plastics by microorganisms was first reported by Fuhs paraffin in 1961 where they used paraffin as a carbon source (Fuhs, 1961). In the environment, many microbes such as bacteria, fungus, and algae are frequently involved in plastic degradation (Rutkowska et al., 2002).

Biodegradability of polymeric materials is difficult. The slow degradation of plastics is due to their lack of water solubility as well as size of the polymer, which hinders their transport into the cells (Sivan et al., 2006). The two main properties of the polyethylene polymer are its hydrophobicity which is attributed to –CH2 groups and a high molecular weight of over 30 kDa. Despite that, biodegradation of high molecular weight polymers has been described in some microorganisms using extracellular enzymes. The extracellular enzymes degrade the main polymeric chain, resulting in intermediates of lower molecular weight with modified mechanical properties, that are readily assimilated by microorganisms (Sudhakar et al., 2008). To date, enormous number of plastic degrading bacteria and fungi have been isolated from a variety of environments which have been reviewed elsewhere (Sangale et al., 2019; Ali et al. 2021a, b; Amobonye et al., 2021) and is beyond the scope of this book chapter. However, the role of algae in bioremediation of plastics is largely ignored when compared with bacteria and fungi. Microalgae, are more preferable in comparison to bacterial systems, because bacterial endotoxins represent pollutants and need rich carbon source for growth, while algae do not contain endotoxins and being autotrophic, they require few organic carbon sources under photoautotrophic conditions (Yan et al., 2016). Furthermore, Ideonella sakaiensis, a potential bacterial model for the plastic degradation and some other PETase-producing microorganisms do not adapt well to marine environments, where the majority of plastic trash ends and accumulates (Tanasupawat et al., 2016). Therefore, the bioprospection of potential algae and their toxins might successfully break down polymeric compounds biologically (Chia et al., 2020).

Enzymes often work in moderate circumstances and have a high selectivity for the natural targets. Therefore, they are utilized in the production and hydrolysis of polymers to provide particular replacements or increase functionalization, respectively, in polymer chemistry (Sen & Puskas, 2015). To optimize the reaction's specificity, different characteristics of the two components, the polymer, and the enzyme, must be examined. Enzymes may identify polymeric chains on the surface, and functional groups can be generated. The biodegradability of degradable polymers varies depending on which enzymes are present to identify them as targets. On various polymers, enzymes have varied effects (Biundo et al., 2015). Only when the polymer chain is disrupted by hetero-atoms such as oxygen or nitrogen, or by the presence of a CC double bond, has the polymer been proved to be biodegradable. Extracellular enzymes, either constitutively produced or provoked by the presence of the material, must initiate the fragmentation of the polymer into its building components in order to begin the biodegradation of the substance. Different polymer characteristics, such as crystallinity, viscosity, and melting point, have been shown to influence these processes (Naser et al., 2021). Biodegradability is defined as the ability of living organisms or their secretory products to degrade by the activity of enzymes and/or chemical breakdown aided by abiotic events such as photodegradation, oxidation, and hydrolysis (Amass et al., 1998). The presence of polymers that hydrolyze particular linkages can stimulate enzyme expression. Specific microbes can completely degrade and remove the polymer if the building blocks are further modified and assimilate. At both the academic and industry levels, enzymatic degradation of fossil-based polymers has been extensively explored. Several investigations have looked into the ability of wild-type enzymes from various sources, as well as designer enzymes, to adsorb and attack certain man-made polymeric molecules (Austin et al., 2018; Tournier et al., 2020; Wei & Zimmermann, 2017). To increase the activity, stability, and specificity of an enzyme, it is frequently necessary to change its 3D structure (Biundo et al., 2015). Because cyanobacteria may be utilized as cell factories, they might be used to generate enzymes that can precisely breakdown polymeric structures, reducing the prevalence of micro- and nanoplastics in the environment. In vitro PET hydrolysis was explored extensively using the hydrolase PETase from Ideonella sakaiensis (Austin et al., 2018). Under a recent research, PET degradation in laboratory circumstances was much improved: over 90% of PET was destroyed in just 10 hours (Tournier et al., 2020). PETase and its variations have previously been expressed in a variety of hosts (such as Bacillus subtilis and Escherichia coli). The marine photosynthetic single-celled diatom P. tricornutum recombinantly expressed and secreted the enhanced version PETaseR280A (Moog et al., 2019). Because of its capacity to combine the benefits of a photosynthetic organism with fast development under CO2 in a saltwater-based environment, this diatom has a high potential for biotechnological applications (Hempel et al., 2011). The scientists employed a nitrate inducible promoter for the production of the enzyme in the above-mentioned study (Moog et al., 2019). The gene was altered to include the region encoding the P. tricornutum alkaline phosphatase signal peptide. This is the first instance of photosynthetic organisms being exploited as expression hosts for the synthesis of plastic-degrading enzymes that we are aware of (Moog et al., 2019). More information on the breakdown of plastic by wild-type enzymes produced from microalgae and cyanobacteria is needed, and new discoveries of this activity might lead to exciting conclusions and uses for these fascinating microorganisms. Due to quicker development than P. tricornutum, more research on the production of the wild-type enzyme PETase in photosynthetic microorganisms was done on two strains of C. reinhardtii (CC-124 and CC-503). The capacity to express the I. sakaiensis PETase was tested using C. reinhardtii strains. The strain CC-124 was shown to be able to produce the enzyme, and the activity of its cell extract was evaluated for up to 4 weeks on PET samples (Kim et al., 2020).

Factors Affecting Plastic Biodegradation

As illustrated in Figure 2, the degradation of plastic polymers is regulated by the characteristics of the polymer and its exposure to various abiotic and biotic stimuli (Ali et al., 2021a). Because its chain length and backbone composition dictate the susceptibility of a polymer to abiotic and biotic degradation, its physicochemical features play a crucial influence in its degradation. Polymeric materials, like PP, are resistant to biodegradation because of their long carbon chains. However, like with oxygen-containing polymers (such as polyurethane and polyethylene terephthalate), the insertion of heteroatoms in the carbon chain renders them sensitive to heat and bio-degradation (Singh and Sharma, 2008). Furthermore, hydrophobicity affects the efficiency of polymer degradation, with the degradation rate increasing as hydrophilicity increases. The crystallinity of polymeric polymers has an impact on their breakdown rate (Ehrenstein, 2001). Compared to amorphous polymers, the more crystalline the polymeric structure, the greater the necessity for water and oxygen that might trigger the breakdown process. A polymer's amorphous portions are also thought to be more sensitive to heat oxidation (Li et al., 2019a, b).

Furthermore, the molecular weight of a polymer might influence its breakdown rate. According to this theory, polymers with a high MW degrade more slowly due to their smaller relative surface area (Singh

Figure 2. Factors controlling the degradation rate of plastic wastes in the environment (Ali et al., 2021b)

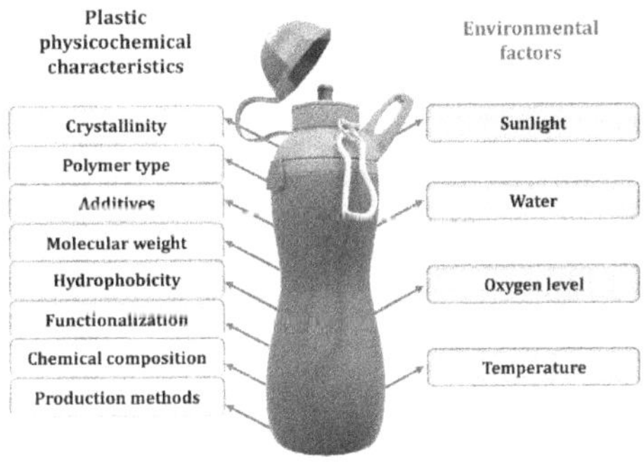

& Sharma, 2008). Plastic polymer manufacture involves various aspects, such as production procedures and additives that might impact the rate of deterioration. Nano-additives, for example, can improve the plastic polymeric properties for industrial applications. Nanoscale reinforcements increase the surface area of the contact, which improves performance. The addition of these nanoparticles to the polymeric framework intends to improve the final product's rheological, mechanical, thermal, and electrical characteristics. This allows for remarkable material qualities at extremely low nano-additives concentrations without significantly impacting polymeric properties while also making the degradation and recycling process more accessible. López de Dicastillo et al. (2020) suggest that nanoparticle additions in polymer production might help to overcome biodegradation issues. The manufacturing process may influence the degradability of the polymer. Co-polymerized PP, for example, is less photodegradable than PP made by bulk (mass) polymerization or using the Ziegler-Natta catalyst (Tang et al., 2004). Due to the presence of a functionalizing group (peroxide residue), PS polymers made by anionic polymerization are more robust against photo-oxidation than PS polymers formed via free radical polymerization (Pospíšil et al., 2006). Stabilizers are frequently employed as additives in plastic manufacture to reduce the rate of deterioration and the presence of chromophores in the polymer chain's backbone. The chemical functionalization type in a polymeric structure will irreversibly modify the material's physicochemical qualities and its degradation rate (Singh & Sharma, 2008). Because of the increased availability of photo-oxidation sites, the presence and concentration of chromophores cause the photochemical production of radicals, which begin and promote the photodegradation process.

Furthermore, the presence of metal-metal bonds in the polymer structure enhances photodegradation of the material owing to the breakage of this link during polymer irradiation (Daglen & Tyler, 2010). The morphological properties of the polymer have an impact on its degradability. In reality, as the surface roughness increases, so does the rate of deterioration, since biofilm development is favored more than on smooth surfaces owing to the increased surface area (Booth et al., 2017).

The plastic degradation process mechanisms and pace may be affected by geographical location, prevailing meteorological conditions, air pollution, and other variables (Andrady et al., 2011). For example, some studies have found that as a result of intense solar radiation, plastic sheet degradation on beaches can occur within months to a few years. However, PET bottles may remain on the bottom for more than 15 years (Fotopoulou & Karapanagioti, 2017). According to Kitamoto et al. (2011), the key factor impacting plastic degradability is sunshine intensity and the degradation rate rises with sunlight intensity due to the increase in the photo-oxidation reaction rate. Furthermore, Pischedda et al. (2019) discovered that when temperature increases, degradation rates increase as well, with the reaction rate double every 10 °C. The polymer chain becomes more mobile as the temperature rises, altering the microorganisms' enzymatic activity during biodegradation. As a result, geographic location is regarded as a critical component in the natural breakdown of plastic trash. Because of the hydrolytic cleavage of functional groups that are vulnerable to hydrolysis, resulting in polymeric chain fractionation, water availability is also an important element in the degradation process (Ali et al., 2021b). Temperature increases the hydrolysis rate of the polymeric chain, since it impacts both the oxygen transport rate and the generation of free radicals (Booth et al., 2017). The photodegradation rate in deep waters is reduced when the intensity of sunshine decreases (Booth et al., 2017).

On the other hand, due to the high humidity level at the sea surface, light degradation of polymeric chains is accelerated, resulting in the solubilization of some additives, such as photo-stabilizers (Booth et al., 2017). The rate of breakdown of plastics is influenced by the presence of oxygen (Queste et al., 2013). High oxygen levels, for example, accelerated polymeric deterioration due to the accelerated reactivity of oxygen with carbon-centered radicals created during the early degradation phases (Price & Horrocks, 2013). Because of the higher concentration of polymer alkyl radicals, these interactions resulted in enhanced scission and hence cross-linked products. As a result, we may conclude that a heated atmosphere with high oxygen and humidity levels is perfect for plastic polymer decomposition.

Algae as a Potential Candidate for Plastic Biodegradation

Micro-, and nano-plastics have shown negative impacts on aquatic ecosystems and other organisms. Bhuyar et al. (2018) found a native microalgal species that can help biodegrade plastic waste in marine environments. Microalgae synthesize enzymes that weaken the chemical bonds of the polymer, therefore requiring less energy for further breakdown. Many microalgae species have been found to devour plastic polymer (Table 2), which might help solve the pressing challenge of plastic pollution (Maurya et al., 2022).

Microalgae have been found as a diverse renewable energy resource, food, and by-products that can help alleviate the pressures caused by expanding demands (Kiran & Venkata Mohan, 2021). Microalgae cultivation can serve as a nature-based solution for plastic waste remediation with concurrent resource recoveries, such as the production of biofuels and other bio-based products (Atabani et al., 2019; Hemalatha et al., 2019; Marin-Batista et al., 2019). In a closed-loop circular bioeconomy, using microalgae production as a bio-sequestration device for absorbing pollutants and recycling nutrients can offer ecologically sustainable products. However, potential techniques should be applied to manage the various upstream and downstream bottlenecks to achieve the multifunctional potential of microalgal productions.

Many species of algae are known to grow on artificial substrata such as PE surfaces (Figure3). Microalgae that can degrade plastics are reportedly less harmful and non-toxic (Sharma et al. 2014). The adhesion of the algae to the polymer surfaces and the production of the depolymerases is the key for plastic degradation. Algal enzymes in the liquid media interact with macromolecules on the plastic surface and trigger biodegradation (Chinaglia et al., 2018). Sarmah & Rout (2019) observed higher contents of protein and carbohydrates in the algal cells when grown on the PE surface, indicating utilization of the polymer as a carbon source by algae.

Furthermore, several researchers have reported surface deterioration or disintegration of PE sheets colonized by algae (Vimal Kumar et al. 2017). Moreover, many authors have stated fouling, corrosion, hydrolysis, and penetration, as well as breakdown of leaching components and pigmentation of the polymers, during degradation of the polymer. In this context, Anabaena spiroides, a blue-green alga, shows the highest degradation (8.18%) of LDPE followed by diatom, Navicula pupula and green alga,

Table 2. List of the algal species showing degradation of the plastic waste (Maurya et al., 2022)

Algae species/type	Common name	Source	Polymer degraded	Strain type	Results	References
Uronema africanum	Cyanobacteria	Plastic bag	LDPE	Wild	microalga initiated degradation of LDPE in 30 days of incubation.	(Sanniyasi et al., 2021)
Dolichospermum spiroides	Cyanobacteria	Plastics sheets	LDPE	Wild	Microalgae can use LDPE as a substratum and the coloning on plsatic surface can decrease plastic sheet strength	(Suseela & Toppo, 2007)
Scenedesmus dimorphus	Green microalga	polyethylene dump site	LDPE HDPE	Wild	Microalgae effects on LDPE than the HDPE and the highest percentage of degradation (8.18%) was obtained from treatment using *A. spiroides*	(Vimal Kumar et al., 2017)
Anabaena spiroides	Blue-green alga					
Navicula pupula	Diatom					
Oscillatoria princeps, O. acuminate, O. subbrevis, O. willei, O. amoena, O. splendida, O. vizagapatensis, O. limnetica, O. earlei, O. peronata, O. formosa, O. okeni, O. geitleriana, O. limosa, O. chalybea, O. salina, O. rubescens, O.curviceps, O.tenuis and O. laete-virens	Blue-green alga	Domestic sewage	PE	Wild	The biodegradation end products were found to be alcohol and ester. The appearance of holes and cavities on the PE surface showed that the PE's integrity had been affected. The enzymatic activity suggested that enzymes may play a role in biodegradation.	(Sarmah & Rout, 2019)
Chlamydomonas reinhardtii	Diatom	Culture collection center	PET	Engineered	Terephthalic acid was detected as a metabolic product and morphological changes on the surface of PET film, were observed	(Kim et al., 2020)
Phormidium lucidum, Oscillatoria subbrevis, Lyngbya diguetii, Nostoc carneum, Cylindrospermum muscicola	Blue-green alga	domestic sewage	PE	Wild	Alteration in bonds of functional groups revealed the biodegradation efficiency. Also, the enhanced laccase and manganese peroxidase activity corroborated the biodegradation. Large grooves on the surface indicate the polyethylene structure disruption.	(Sarmah & Rout, 2018)
Chlorella vulgaris	Green alga	BPA	Fishery	Coculture	1-(3-methylbuthyl)-2, 3, 4, 6-tetramethylbenzene and 4-(1-hydroxy-2-methylprop-1-enyl) phenol were found to be biodegradation intermediate products	(Gulnaz and Dincer 2009)
Chlorella fusca var. vacuolata	Green alga	BPA	NA	Wild	Monohydroxybisphenol A was observed as an intermediate product. Furthermore, estrogenic activity also completely disappeared.	(Hirooka et al., 2005)
P. tricornutum	Diatom	NA	PET	Engineered	The produced metabolites from the degradation of the PET substrate were terephthalic acid and mono(2-hydroxyethyl) terephthalic acid	(Moog et al., 2019)

Scenedesmus dimorphus that exhibited [illegible]% and 3.74% degradation respectively (Vimal Kumar et al. 2017). Sarmah & Rout (2019) found that freshwater cyanobacteria (Phormidium lucidum and Oscillatoria subbrevis) which are easily available and fast growing could colonize the PE surface and degrading LDPE effectively without any pretreatment or pro-oxidant chemicals. Additionally, Gulnaz and Dincer (2009) employed a mixture of Chlorella vulgaris and Aeromonas hydrophilia for the biodegradation of bisphenol A (BPA), which is a commonly used polymer in the plastic industry. The authors further stated that BPA was destroyed quickly by algae below detection limits within 168 hours showing no estrogenic activity. Similarly, Hirooka et al. (2005) found that green alga Chlorella fusca var. vacuolata degraded BPA without estrogenic activity.

Microalgae are photoautotrophic in nature that colonize the polyethylene sheets in aquatic habitats (Sharma et al. 2014). According to Fritsch, many algae such as chaetophorales, diatoms (Bacillariophyceae), and blue-green algae (Cyanophyceae) attach to the substratum via extracellular polymeric substance (EPS). These substances (EPS) facilitate colonization of the microalgae on rocks, walls and other surfaces, causing plastic surface biodeterioration (Sanniyasi et al., 2021). The attachment of EPS occurs through surface charge, hydrophobicity, and electrostatic forces (Bhaskar & Bhosle 2004). There are very few reports on the biodegradation of LDPE by photosynthetic algae. Biodegradation of polyethylene by microalgae is an inexpensive, easy and environment friendly method when compared with conventional degradation (Vimal Kumar et al. 2017). The algal species like Oscillatoria, Phormidium, Lyngbya, Nostoc, Spirulina, Hydrocoleum, Chlorella, Pithophora, Stigeoclonium tenue, Anomoeoneis, and Nitzschia frequently colonize the polyethylene bags (Sarmah and Rout 2018, 2019). The authors further stated that the attachment of the cyanobacteria to the surface of HDPE is so strong that it could not be removed with a water jet (Sarmah and Rout 2018; 2019). The cyanobacteria, Phormidium lucidum, and Oscillatoria subbrevis have been shown to mineralize plastic in 6 weeks only (Sarmah and Rout 2019). Similarly, many microalgal species belonging to 7 orders, 9 families and 10 genera, have been enumerated from vrious polyethylene degrading locations in India (Sharma et al. 2014). In addition, the green microalga, Uronema africanum Borge, is recovered from the surface of a plastic bag collected in Kallukuttai Lake, Chennai, India. Uronema africanum has been found to initiate degradation of LDPE within 30 days of incubation (Sanniyasi et al., 2021). The same authors had earlier isolated a cyanobacterium, Dolichospermum spiroides that had colonized the surface of LDPE sheets (Suseela and Toppo 2007). These algae are responsible for biological deterioration of LDPE sheets. When the U. africanum, was incubated with LDPE sheets for 30 days, the colonized algae started to protrude greenish hair-like structures from the surfaces of LDPE sheets. Since, U. africanum is filamentous in nature, its greenish filaments adhering perpendicular to the surface of LDPE sheet were observed by light and dark-field microscopy. Concurrently, the algae caused erosion, abrasions, producing grooves and ridges on the treated LDPE sheets (Sanniyasi et al. 2021). The cyanobacteria increase the hydrophilicity of the polymer by generating carbonyl groups that are further metabolized by other microbes (Ibiene et al. 2013). Further, the authors elucidated that U. africanum produced hydrocarbons from the LDPE as revealed by GC–MS analysis (Sanniyasi et al. 2021). In addition to direct degradation and assimilation of plastics, many microalgal species can be used as expression systems to bioengineer enzymes of particular interest.

To date several conventional as well as advanced techniques have been employed to demonstrate the biodegradation of plastics by algal species. In one of the studies, FT-IR spectroscopy depicted the production of functional groups like esters, carboxylic acids, amino and nitro groups in the LDPE samples treated with algae, U. africanum (Sanniyasi et al., 2021). Similarly, Kumar and colleagues reported biodegradation of LDPE sheets by three distinct microalgae: cyanobacteria, green alga, and

Figure 3. The disposal and biodegradation of plastic waste in the oceans by microalgae.

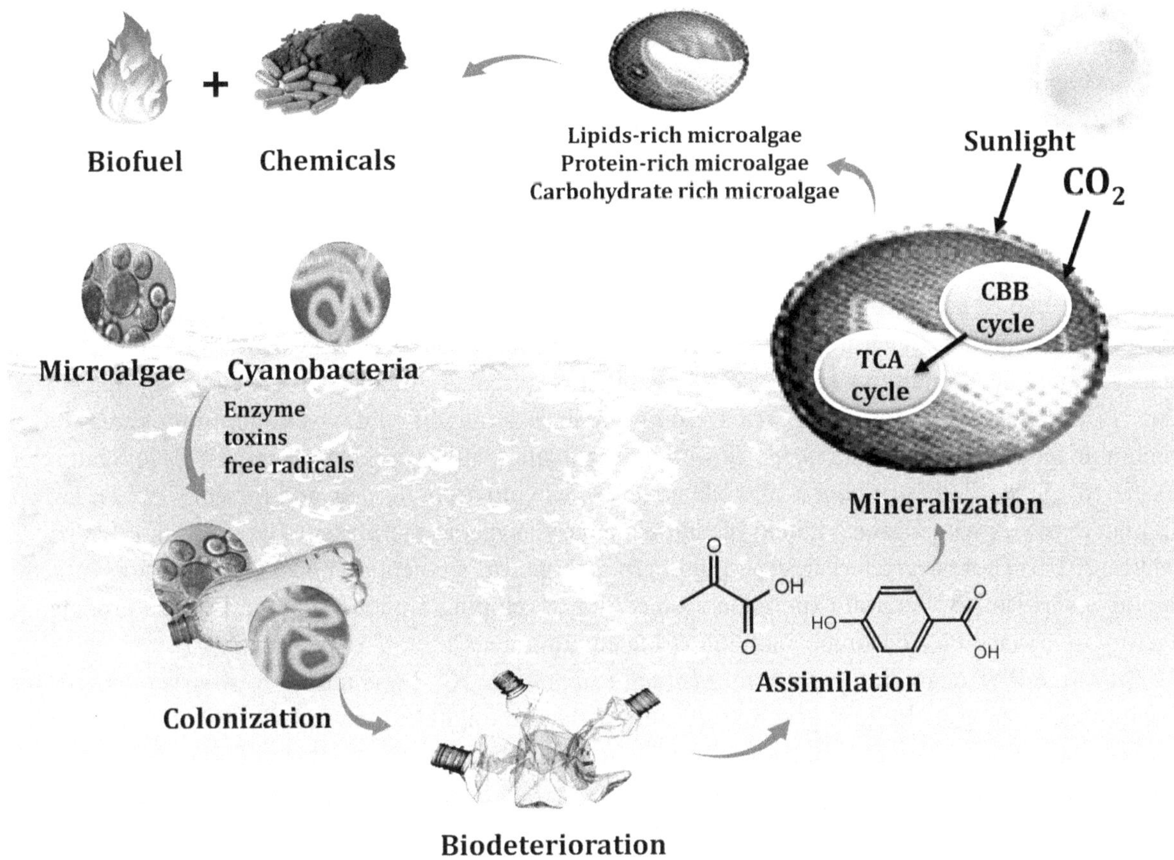

diatom (Vimal Kumar et al., 2017). The scanning electron microscopy revealed the formation of cavities on the surface of treated LDPE sheets caused due to cyanobacterium, Dolichospermum spiroides. In a similar investigation, SEM micrographs divulged cracking of the LDPE surface besides attachment of the algae (Suseela and Toppo 2007). Pit and cavity formation, along with the surface erosion are the common observations caused by cyanobacteria like Phormidium lucidum and Oscillatoria subbrevis to LDPE (Sarmah & Rout, 2018, 2019). The atomic force microscopy (AFM) was used to demonstrate the increased average area roughness (Sa), peak height (Sp), and roughness pit height (Sv) in LDPE sheets degraded by U. africanum (Sanniyasi et al., 2021). Furthermore, the arrangement of abrasions in the treated samples were similar to the 20-30 μm radial disc-like structures of algae (20–30 μm). The grooves and ridges observed in the algae (10–15 μm) were comparable to the filamentous form of the algae.

GENETIC ENGINEERING OF ALGAE FOR PLASTIC DEGRADATION

Recent studies demonstrate that many microalgae/diatoms can be modified genetically into microbial cell factories to secrete plastic-degrading enzymes. For instance, the green microalgae *Chlamydomonas reinhardtii* was transformed to produce PETase, and the transformant's cell lysate was co-incubated with

PET, causing dings and holes on the surface of PET film as well as TPA, which represents the fully degraded form of PET (Kim et al., 2020). In a similar fashion, Moog and coworkers (Moog et al. 2019) transformed *P. tricornutum* as a chassis to generate PETase, a catalytic enzyme that can break down PET and the copolymer polyethylene terephthalate glycol (PETG). These studies have led to a promising ecologically acceptable technique for biological degradation of PET using microalgae via synthetic biology.

The diatom, *Phaeodactylum tricornutum* is a single-celled marine photosynthetic eukaryote with potential biotechnological applications. Apart from being, photosynthetic organism, *P. tricornutum* is easy to culture in laboratory and grows fast in saltwater-based environments consuming CO_2. It is one of the well-known model organisms for research having a comprehensive genetic toolbox. The genome of the diatom can be easily edited or manipulated (many foreign genes can be inserted) by using standard methods of genetic engineering for the production and expression of enzymes with highest efficiency under maintained conditions (Daboussi et al., 2014; Hempel & Maier, 2016; Nymark et al., 2016). Apart from being cost-effective, *P. tricornutum* serves as a magnificent host system for expression of recombinant proteins like antibodies, as well as enzymatic pathways (Hempel et al., 2011; Slattery et al., 2018). Since *P. tricornutum* is photoautotrophic, its cultivation is easy and the cells can be grown to high densities without the addition of supplementary sugars in the presence of sunlight (Hempel & Maier, 2016). These characteristics spotlight *P. tricornutum* as a potential model organism for synthetic biology, superior to bacterial expression systems for developing a photosynthetic PETase production factory to manage PET decomposition in marine environments.

Since the PET degrading bacterium, *Ideonella sakaiensis* 201-F6 is unable to flourish under salty conditions, cloning its PETase gene into a suitable diatom might be very helpful for the bioremediation of plastics in marine environments. To this end, a group of scientists cloned PETase gene from *I. sakaiensis* 201-F6 into the marine diatom, *Phaeodactylum tricornutum* for the production of plastic degrading enzymes. To determine the PETase production, the diatom was cultured in vessels containing small particles of PET along with a copolymer polyethylene terephthalate glycol (PETG). After the test experiment, the plastic particles were observed to have grooves and holes created by the PETase (Moog et al. 2019). Besides active on PET and PETG, the authors concluded that diatom produced PETase that can degrade industrially shredded PET in a saltwater-based environment at 21 °C into terephthalic acid (TPA) and mono(2-hydroxyethyl) terephthalic acid (MHET) that are deemed as harmless products. Such striking and continuous efforts are direly needed to mitigate the plastic pollution of the oceans.

CONCLUSIONS AND FUTURE PERSPECTIVES

The current reports showed that microalgae can be used as a viable solution for management and valorization of plastic wastes. Prior knowledge of the components of various plastic polymers might help to build the potential strategies like pretreatment, size reduction, sterilization or addition of supplements, etc. for bioremediation through microalgae. Various bioreactors like biofilm-based membrane, and multistep treatment strategic technologies could improve microalgal growth on polymers. For future perspectives, microalgae cultivation with the 'sterile' polymers and cultivation in batch mode are promising for the production of value-added biomass. Moreover, the biorefinery will maximize the value creation from microalgal based valorization platforms of plastics. Using different types of plastics as a carbon source and microalgae cultivation on various support materials also require advanced research in the future. Either using the single strain of microalgae, consortium of multiple microalgae might give extraordinary biomass productivity with effective removal of various components of plastics waste from the environment. Future research on multiple biomaterial-based product development from microalgae biomass grown on plastics is highly recommended.

REFERENCES

Ali, S. S., Elsamahy, T., Al-Tohamy, R., Zhu, D., Mahmoud, Y. A.-G., Koutra, E., Metwally, M. A., Kornaros, M., & Sun, J. (2021b). Plastic wastes biodegradation: Mechanisms, challenges and future prospects. *The Science of the Total Environment*, *780*, 146590. doi:10.1016/j.scitotenv.2021.146590 PMID:34030345

Ali, S. S., Elsamahy, T., Koutra, E., Kornaros, M., El-Sheekh, M., Abdelkarim, E. A., Zhu, D., & Sun, J. (2021a). Degradation of conventional plastic wastes in the environment: A review on current status of knowledge and future perspectives of disposal. *The Science of the Total Environment*, *771*, 144719. Advance online publication. doi:10.1016/j.scitotenv.2020.144719 PMID:33548729

Amass, W., Amass, A., & Tighe, B. (1998). A review of biodegradable polymers: Uses, current developments in the synthesis and characterization of biodegradable polyesters, blends of biodegradable polymers and recent advances in biodegradation studies. *Polymer International*, *47*(2), 89–144. doi:10.1002/(SICI)1097-0126(1998100)47:2<89::AID-PI86>3.0.CO;2-F

Amobonye, A., Bhagwat, P., Singh, S., & Pillai, S. (2021). Plastic biodegradation: Frontline microbes and their enzymes. *The Science of the Total Environment*, *759*, 143536. doi:10.1016/j.scitotenv.2020.143536 PMID:33190901

Andrady, A. L., Hamid, H., & Torikai, A. (2011). Effects of solar UV and climate change on materials. *Photochemical & Photobiological Sciences*, *10*(2), 292. doi:10.1039/c0pp90038a PMID:21253664

Atabani, A. E., Shobana, S., Mohammed, M. N., Uğuz, G., Kumar, G., Arvindnarayan, S., Aslam, M., & Al-Muhtaseb, A. H. (2019). Integrated valorization of waste cooking oil and spent coffee grounds for biodiesel production: Blending with higher alcohols, FT–IR, TGA, DSC and NMR characterizations. *Fuel*, *244*, 419–430. doi:10.1016/j.fuel.2019.01.169

Austin, H. P., Allen, M. D., Donohoe, B. S., Rorrer, N. A., Kearns, F. L., Silveira, R. L., Pollard, B. C., Dominick, G., Duman, R., El Omari, K., Mykhaylyk, V., Wagner, A., Michener, W. E., Amore, A., Skaf, M. S., Crowley, M. F., Thorne, A. W., Johnson, C. W., Woodcock, H. L., . . . Beckham, G. T. (2018). Characterization and of a plastic-degrading aromatic polyesterase. *Proceedings of the National Engineering Academy of Sciences of the United States of America, 115*(19), E4350–E4357. 10.1073/pnas.1718804115

Barabad, M. L. M., Jung, W., Versoza, M. E., Kim, M., Ko, S., Park, D., & Lee, K. (2018). Emission Characteristics of Particulate Matter, Volatile Organic Compounds, and Trace Elements from the Combustion of Coals in Mongolia. *International Journal of Environmental Research and Public Health, 15*(8), 1706. doi:10.3390/ijerph15081706 PMID:30096940

Barra, R., & Leonard, S. A. (2018). *Plastics and the circular economy. 54th Global Environment Facility Council Meeting*, Da Nang, Viet Nam.

Bhaskar, P. V., & Bhosle, N. B. (2004). Microbial extracellular polysaccharides and their role in marine biogeochemical processes. *Current Science*, *88*, 45–53.

Bhuyar, P., Zagde, S., Revankar, R., Yusoff, M., Rahim, Ab. M. H., & Govindan, N. (2018). Isolation, Characterization and Partial Purification of Keratinase from Keratinolytic Bacteria. SSRN *Electronic Journal*. doi:10.2139/ssrn.3389068

Biundo, A., Hromic, A., Pavkov-Keller, T., Gruber, K., Quartinello, F., Haernvall, K., Perz, V., Arrell, M. S., Zinn, M., Ribitsch, D., & Guebitz, G. M. (2015). Characterization of a poly(butylene adipate-co-terephthalate)-hydrolyzing lipase from Pelosinus fermentans. *Applied Microbiology and Biotechnology*, *100*(4), 1753–1764. doi:10.100700253-015-7031-1 PMID:26490551

Booth, A. M., Kubowicz, S., Beegle-Krause, C., Skancke, J., Nordam, T., Landsem, E. ThroneHolst, M., & Jahren, S. (2017). *Microplastic in Global and Norwegian Marine Environments: Distributions, Degradation Mechanisms and Transport*. Norwegian Environment Agency M-918. https://www.miljodirektoratet.no/globalassets/publikasjoner/M918/M918.pdf

Campanale, C., Massarelli, C., Savino, I., Locaputo, V., & Uricchio, V. F. (2020). A Detailed Review Study on Potential Effects of Microplastics and Additives of Concern on Human Health. *International Journal of Environmental Research and Public Health*, *17*(4), 1212. doi:10.3390/ijerph17041212 PMID:32069998

Canniff, P. M., & Hoang, T. C. (2018). Microplastic ingestion by Daphnia magna and its enhancement on algal growth. *The Science of the Total Environment*, *633*, 500–507. doi:10.1016/j.scitotenv.2018.03.176 PMID:29579661

Capolupo, M., Sørensen, L., Jayasena, K. D. R., Booth, A. M., & Fabbri, E. (2020). Chemical composition and ecotoxicity of plastic and car tire rubber leachates to aquatic organisms. *Water Research*, *169*, 115270. doi:10.1016/j.watres.2019.115270 PMID:31731243

Chae, Y., Kim, D., & An, Y.-J. (2019). Effects of micro-sized polyethylene spheres on the marine microalga Dunaliella salina: Focusing on the algal cell to plastic particle size ratio. *Aquatic Toxicology (Amsterdam, Netherlands)*, *216*, 105296. doi:10.1016/j.aquatox.2019.105296 PMID:31541944

Chen, Q., Allgeier, A., Yin, D., & Hollert, H. (2019). Leaching of endocrine disrupting chemicals from marine microplastics and mesoplastics under common life stress conditions. *Environment International*, *130*, 104938. doi:10.1016/j.envint.2019.104938 PMID:31234003

Chia, W. Y., Ying Tang, D. Y., Khoo, K. S., Kay Lup, A. N., & Chew, K. W. (2020). Nature's fight against plastic pollution: Algae for plastic biodegradation and bioplastics production. *Environmental Science and Ecotechnology*, *4*, 100065. doi:10.1016/j.ese.2020.100065

Chinaglia, S., Tosin, M., & Degli-Innocenti, F. (2018). Biodegradation rate of biodegradable plastics at molecular level. *Polymer Degradation & Stability*, *147*, 237–244. doi:10.1016/j.polymdegradstab.2017.12.011

Cunha, C., Faria, M., Nogueira, N., Ferreira, A., & Cordeiro, N. (2019). Marine vs freshwater microalgae exopolymers as biosolutions to microplastics pollution. *Environmental Pollution*, *249*, 372–380. doi:10.1016/j.envpol.2019.03.046 PMID:30909130

Daboussi, F., Leduc, S., Maréchal, A., Dubois, G., Guyot, V., Perez-Michaut, C., Amato, A., Falciatore, A., Juillerat, A., Beurdeley, M., Voytas, D. F., Cavarec, L., & Duchateau, P. (2014). Genome engineering empowers the diatom Phaeodactylum tricornutum for biotechnology. *Nature Communications*, *5*(1), 3831. Advance online publication. doi:10.1038/ncomms4831 PMID:24871200

Daglen, B. C., & Tyler, D. R. (2010). Photodegradable plastics: End-of-life design principles. *Green Chemistry Letters and Reviews*, *3*(2), 69–82. doi:10.1080/17518250903506723

Darwesh, O. M., Ali, S. S., Matter, I. A., & Elsamahy, T. (2021). Nanotextiles waste management: controlling of release and remediation of wastes. In *Nanosensors and Nanodevices for Smart Multifunctional Textiles* (pp. 267–286). Elsevier. doi:10.1016/B978-0-12-820777-2.00016-9

Doorn, P. F., Campbell, P. A., & Amstutz, H. C. (1996). Metal Versus Polyethylene Wear Particles in Total Hip Replacements. *Clinical Orthopaedics and Related Research*, *329*, S206–S216. doi:10.1097/00003086-199608001-00018 PMID:8769335

Ehrenstein, G. W. (2001). Structure of Polymeric Materials. In Polymeric Materials (pp. 61–141). Carl Hanser Verlag GmbH & Co. KG. doi:10.3139/9783446434134.004

Eriksen, M., Lebreton, L. C. M., Carson, H. S., Thiel, M., Moore, C. J., Borerro, J. C., Galgani, F., Ryan, P. G., & Reisser, J. (2014). Plastic Pollution in the World's Oceans: More than 5 Trillion Plastic Pieces Weighing over 250,000 Tons Afloat at Sea. *PLoS One*, *9*(12), e111913–e111913. doi:10.1371/journal.pone.0111913 PMID:25494041

Feil, A., & Pretz, T. (2020). Mechanical recycling of packaging waste. In *Plastic Waste and Recycling* (pp. 283–319). Elsevier., doi:10.1016/B978-0-12-817880-5.00011-6

Fotopoulou, K. N., & Karapanagioti, H. K. (2017). Degradation of Various Plastics in the Environment. In *The Handbook of Environmental Chemistry* (pp. 71–92). Springer International Publishing. doi:10.1007/698_2017_11

Frias, J. P. G. L., Sobral, P., & Ferreira, A. M. (2010). Organic pollutants in microplastics from two beaches of the Portuguese coast. *Marine Pollution Bulletin*, *60*(11), 1988–1992. doi:10.1016/j.marpolbul.2010.07.030 PMID:20800853

Fuhs, G. W. (1961). Der mikrobielle Abbau von Kohlenwasserstoffen. *Archiv für Mikrobiologie*, *39*(4), 374–422. doi:10.1007/BF00411776 PMID:13702455

Galloway, T. S., Cole, M., & Lewis, C. (2017). Interactions of microplastic debris throughout the marine ecosystem. *Nature Ecology & Evolution*, *1*(5), 0116. Advance online publication. doi:10.103841559-017-0116 PMID:28812686

Gnanavel, G., Valli, V. P. M. J., & Thirumarimurugan, M. (2012). A review of biodegradation of plastics waste. *International Journal of Pharmaceutical and Chemical Sciences*, *1*(3), 670-673.

Gulnaz, O., & Dincer, S. (2019). Biodegradation of Bisphenol a by Chlorella vulgaris and Aeromonas Hydrophilia. *Journal of Applied Biological Sciences*, *3*(2), 79–84. https://www.jabsonline.org/index.php/jabs/article/view/162

Hemalatha, M., Sravan, J. S., Min, B., & Venkata Mohan, S. (2019). Microalgae biorefinery with cascading resource recovery design associated to dairy wastewater treatment. *Bioresource Technology*, *284*, 424–429. doi:10.1016/j.biortech.2019.03.106 PMID:30978645

Hempel, F., Bozarth, A. S., Lindenkamp, N., Klingl, A., Zauner, S., Linne, U., Steinbüchel, A., & Maier, U. G. (2011). Microalgae as bioreactors for bioplastic production. *Microbial Cell Factories*, *10*(1), 81. doi:10.1186/1475-2859-10-81 PMID:22004563

Hempel, F., & Maier, U. G. (2016). Microalgae as Solar-Powered Protein Factories. In *Advanced Technologies for Protein Complex Production and Characterization* (pp. 241–262). Springer International Publishing. doi:10.1007/978-3-319-27216-0_16

Hirooka, T., Nagase, H., Uchida, K., Hiroshige, Y., Ehara, Y., Nishikawa, J., Nishihara, T., Miyamoto, K., & Hirata, Z. (2005). Biodegradation of bisphenol a and disappearance of its estrogenic activity by the green alga chlorella fusca var. Vacuolata. *Environmental Toxicology and Chemistry*, *24*(8), 1896. doi:10.1897/04-259R.1 PMID:16152959

Huerta Lwanga, E., Mendoza Vega, J., Ku Quej, V., Chi, J., Sanchez del Cid, L., Chi, C., Escalona Segura, G., Gertsen, H., Salánki, T., van der Ploeg, M., Koelmans, A. A., & Geissen, V. (2017). Field evidence for transfer of plastic debris along a terrestrial food chain. *Scientific Reports*, *7*(1), 14071. doi:10.103841598-017-14588-2 PMID:29074893

Ibiene, A. A., Stanley, H. O., & Immanuel, O. M. (2013). Biodegradation of polyethylene by *Bacillus* sp. indigenous to the Niger Delta Mangrove Swamp. *Nigerian Journal of Biotechnology*, *26*, 68–79.

Khoironi, A., Anggoro, S., & Sudarno, S. (2019). Evaluation of the Interaction Among Microalgae Spirulina sp, Plastics Polyethylene Terephthalate and Polypropylene in Freshwater Environment. *Journal of Ecological Engineering*, *20*(6), 161–173. doi:10.12911/22998993/108637

Kim, J. W., Park, S.-B., Tran, Q.-G., Cho, D.-H., Choi, D.-Y., Lee, Y. J., & Kim, H.-S. (2020). Functional expression of polyethylene terephthalate-degrading enzyme (PETase) in green microalgae. *Microbial Cell Factories*, *19*(1), 97. doi:10.118612934-020-01355-8 PMID:32345276

Kiran, B. R., & Venkata Mohan, S. (2021). Microalgal Cell Biofactory-Therapeutic, Nutraceutical and Functional Food Applications. *Plants (Basel, Switzerland)*, *10*(5), 836. doi:10.3390/plants10050836 PMID:33919450

Kitamoto, H. K., Shinozaki, Y., Cao, X.-H., Morita, T., Konishi, M., Tago, K., Kajiwara, H., Koitabashi, M., Yoshida, S., Watanabe, T., Sameshima-Yamashita, Y., Nakajima-Kambe, T., & Tsushima, S. (2011). Phyllosphere yeasts rapidly break down biodegradable plastics. *AMB Express*, *1*(1), 44. doi:10.1186/2191-0855-1-44 PMID:22126328

Koelmans, A. A., Besseling, E., Wegner, A., & Foekema, E. M. (2013). Plastic as a Carrier of POPs to Aquatic Organisms: A Model Analysis. *Environmental Science & Technology*, *47*(14), 7812–7820. doi:10.1021/es401169n PMID:23758580

Li, B., Su, L., Zhang, H., Deng, H., Chen, Q., & Shi, H. (2020). Microplastics in fishes and their living environments surrounding a plastic production area. *The Science of the Total Environment*, *727*, 138662. doi:10.1016/j.scitotenv.2020.138662 PMID:32498185

Li, J., Liu, H., & Paul Chen, J. (2018). Microplastics in freshwater systems: A review on occurrence, environmental effects, and methods for microplastics detection. *Water Research*, *137*, 362–374. doi:10.1016/j.watres.2017.12.056 PMID:29580559

Li, K., Liu, Q., Fang, F., Luo, R., Lu, Q., Zhou, W., Huo, S., Cheng, P., Liu, J., Addy, M., Chen, P., Chen, D., & Ruan, R. (2019b). Microalgae-based wastewater treatment for nutrients recovery: A review. *Bioresource Technology*, *291*, 121934. doi:10.1016/j.biortech.2019.121934 PMID:31395401

Liquete, C., Piroddi, C., Drakou, E. G., Gurney, L., Katsanevakis, S., Charef, A., & Egoh, B. (2013). Current status and future prospects for the assessment of marine and coastal ecosystem services: A systematic review. *PLoS One*, *8*(7), e67737–e67737. doi:10.1371/journal.pone.0067737 PMID:23844080

Lithner, D., Larsson, Å., & Dave, G. (2011). Environmental and health hazard ranking and assessment of plastic polymers based on chemical composition. *The Science of the Total Environment*, *409*(18), 3309–3324. doi:10.1016/j.scitotenv.2011.04.038 PMID:21663944

Liu, L., Fokkink, R., & Koelmans, A. A. (2016). Sorption of polycyclic aromatic hydrocarbons to polystyrene nanoplastic. *Environmental Toxicology and Chemistry*, *35*(7), 1650–1655. doi:10.1002/etc.3311 PMID:26588181

Liu, R., Li, S., Tu, Y., & Hao, X. (2021). Capabilities and mechanisms of microalgae on removing micropollutants from wastewater: A review. *Journal of Environmental Management*, *285*, 112149. doi:10.1016/j.jenvman.2021.112149 PMID:33607565

López de Dicastillo, C., Velásquez, E., Rojas, A., Guarda, A., & Galotto, M. J. (2020). The use of nanoadditives within recycled polymers for food packaging: Properties, recyclability, and safety. *Comprehensive Reviews in Food Science and Food Safety*, *19*(4), 1760–1776. doi:10.1111/1541-4337.12575 PMID:33337105

Marin-Batista, J. D., Villamil, J. A., Rodriguez, J. J., Mohedano, A. F., & de la Rubia, M. A. (2019). Valorization of microalgal biomass by hydrothermal carbonization and anaerobic digestion. *Bioresource Technology*, *274*, 395–402. doi:10.1016/j.biortech.2018.11.103 PMID:30551042

Mathews, S. L., Epps, M. J., Blackburn, R. K., Goshe, M. B., Grunden, A. M., & Dunn, R. R. (2019). Public questions spur the discovery of new bacterial species associated with lignin bioconversion of industrial waste. *Royal Society Open Science*, *6*(3), 180748. doi:10.1098/rsos.180748 PMID:31031986

Maurya, R., Zhu, X., Valverde-Pérez, B., Ravi Kiran, B., General, T., Sharma, S., Kumar Sharma, A., Thomsen, M., Venkata Mohan, S., Mohanty, K., & Angelidaki, I. (2022). Advances in microalgal research for valorization of industrial wastewater. *Bioresource Technology*, *343*, 126128. doi:10.1016/j.biortech.2021.126128 PMID:34655786

Moog, D., Schmitt, J., Senger, J., Zarzycki, J., Rexer, K.-H., Linne, U., Erb, T., & Maier, U. G. (2019). Using a marine microalga as a chassis for polyethylene terephthalate (PET) degradation. *Microbial Cell Factories*, *18*(1), 171. doi:10.118612934-019-1220-z PMID:31601227

Naser, A. Z., Deiaba, I., & Darras, B. M. (2021). Poly(lactic acid) (PLA) and polyhydroxyalkanoates (PHAs), green alternatives to petroleum-based plastics: A review. *RSC Advances*, *2021*(11), 17151–17196. doi:10.1039/D1RA02390J PMID:35479695

Nava, V., & Leoni, B. (2021). A critical review of interactions between microplastics, microalgae and aquatic ecosystem function. *Water Research*, *188*, 116476. doi:10.1016/j.watres.2020.116476 PMID:33038716

Nel, A., Xia, T., Mädler, L., & Li, N. (2006). Toxic Potential of Materials at the Nanolevel. *Science*, *311*(5761), 622–627. doi:10.1126cience.1114397 PMID:16456071

Nymark, M., Sharma, A. K., Sparstad, T., Bones, A. M., & Winge, P. (2016). A CRISPR/Cas9 system adapted for gene editing in marine algae. *Scientific Reports*, *6*(1), 24951. doi:10.1038rep24951 PMID:27108533

Pischedda, A., Tosin, M., & Degli-Innocenti, F. (2019). Biodegradation of plastics in soil: The effect of temperature. *Polymer Degradation & Stability*, *170*, 109017. doi:10.1016/j.polymdegradstab.2019.109017

Pospíšil, J., Pilař, J., Billingham, N. C., Marek, A., Horák, Z., & Nešpůrek, S. (2006). Factors affecting accelerated testing of polymer photostability. *Polymer Degradation & Stability*, *91*(3), 417–422. doi:10.1016/j.polymdegradstab.2005.01.049

Prata, J. C., da Costa, J. P., Lopes, I., Duarte, A. C., & Rocha-Santos, T. (2020). Environmental exposure to microplastics: An overview on possible human health effects. *The Science of the Total Environment*, *702*, 134455. doi:10.1016/j.scitotenv.2019.134455 PMID:31733547

Price, D., & Horrocks, A. R. (2013). Combustion processes of textile fibres. In *Handbook of Fire Resistant Textiles* (pp. 3–25). Elsevier. doi:10.1533/9780857098931.1.3

Proshad, R., Kormoker, T., Islam, M. S., Haque, M. A., Rahman, M. M., & Mithu, M. M. R. (2017). Toxic effects of plastic on human health and environment : A consequences of health risk assessment in Bangladesh. *International Journal of Healthcare*, *6*(1), 1. doi:10.14419/ijh.v6i1.8655

Queste, B. Y., Fernand, L., Jickells, T. D., & Heywood, K. J. (2013). Spatial extent and historical context of North Sea oxygen depletion in August 2010. *Biogeochemistry*, *113*(1), 53–68. doi:10.100710533-012-9729-9 PMID:32355379

Rochman, C. M. (2015). The Complex Mixture, Fate and Toxicity of Chemicals Associated with Plastic Debris in the Marine Environment. In M. Bergmann, L. Gutow, & M. Klages (Eds.), *Marine Anthropogenic Litter*. Springer. doi:10.1007/978-3-319-16510-3_5

Rutkowska, M., Krasowska, K., Heimowska, A., Steinka, I., & Janik, H. (2002). Degradation of polyurethanes in sea water. *Polymer Degradation & Stability*, *76*(2), 233–239. doi:10.1016/S0141-3910(02)00019-8

Sangale, M. K., Shahnawaz, M., & Ade, A. B. (2019). Potential of fungi isolated from the dumping sites mangrove rhizosphere soil to degrade polythene. *Scientific Reports*, *9*(1), 5390. doi:10.103841598-019-41448-y PMID:30926843

Sanniyasi, E., Gopal, R. K., Gunasekar, D. K., & Raj, P. P. (2021). Biodegradation of low-density polyethylene (LDPE) sheet by microalga, Uronema africanum Borge. *Scientific Reports*, *11*(1), 17233. doi:10.103841598-021-96315-6 PMID:34446729

Sarmah, P., & Rout, J. (2018). Efficient biodegradation of low-density polyethylene by cyanobacteria isolated from submerged polyethylene surface in domestic sewage water. *Environmental Science and Pollution Research International*, *25*(33), 33508–33520. doi:10.100711356-018-3079-7 PMID:30267347

Sarmah, P., & Rout, J. (2019). Cyanobacterial degradation of low-density polyethylene (LDPE) by Nostoc carneum isolated from submerged polyethylene surface in domestic sewage water. *Energy, Ecology & Environment*, *4*(5), 240–252. doi:10.100740974-019-00133-6

Scheurer, M., & Bigalke, M. (2018). Microplastics in Swiss Floodplain Soils. *Environmental Science & Technology*, *52*(6), 3591–3598. doi:10.1021/acs.est.7b06003 PMID:29446629

Science for Environment Policy. (2011). *Plastic waste: ecological and human health impacts. DG environment news alert service*. European Commission. Available at https://ec.europa.eu/environment/integration/research/newsalert/pdf/IR1_en.pdf

Sen, S., & Puskas, J. E. (2015). Green polymer chemistry: Enzyme catalysis for polymer functionalization. *Molecules (Basel, Switzerland)*, *20*(5), 9358–9379. doi:10.3390/molecules20059358 PMID:26007188

Shahnawaz, M., Sangale, M. K., & Ade, A. B. (2016). Rhizosphere of Avicennia marina (Forsk.) Vierh. as a landmark for polythene degrading bacteria. *Environmental Science and Pollution Research International*, *23*(14), 14621–14635. doi:10.100711356-016-6542-3 PMID:27072028

Sharma, M., Dubey, A., & Pareek, A. (2014). Algal flora on degrading polythene waste. *CIB Tech Journal of Microbiology*, *3*, 43–47.

Sharma, R., Sharma, M., Sharma, R., & Sharma, V. (2013). The impact of incinerators on human health and environment. *Reviews on Environmental Health*, *28*(1). Advance online publication. doi:10.1515/reveh-2012-0035 PMID:23612530

Singh, B., & Sharma, N. (2008). Mechanistic implications of plastic degradation. *Polymer Degradation & Stability*, *93*(3), 561–584. doi:10.1016/j.polymdegradstab.2007.11.008

Sivan, A., Szanto, M., & Pavlov, V. (2006). Biofilm development of the polyethylene-degrading bacterium Rhodococcus ruber. *Applied Microbiology and Biotechnology*, *72*(2), 346–352. doi:10.100700253-005-0259-4 PMID:16534612

Slattery, S. S., Diamond, A., Wang, H., Therrien, J. A., Lant, J. T., Jazey, T., Lee, K., Klassen, Z., Desgagné-Penix, I., Karas, B. J., & Edgell, D. R. (2018). An Expanded Plasmid-Based Genetic Toolbox Enables Cas9 Genome Editing and Stable Maintenance of Synthetic Pathways in Phaeodactylum tricornutum. *ACS Synthetic Biology*, *7*(2), 328–338. doi:10.1021/acssynbio.7b00191 PMID:29298053

Sudhakar, M., Doble, M., Murthy, P. S., & Venkatesan, R. (2008). Marine microbe-mediated biodegradation of low- and high-density polyethylenes. *International Biodeterioration & Biodegradation*, *61*(3), 203–213. doi:10.1016/j.ibiod.2007.07.011

Suseela, M. R., & Toppo, K. (2007). Algal biofilms on polythene and its possible degradation. *Current Science*, *92*, 285–287.

Taipale, S. J., Peltomaa, E., Kukkonen, J. V. K., Kainz, M. J., Kautonen, P., & Tiirola, M. (2019). Tracing the fate of microplastic carbon in the aquatic food web by compound-specific isotope analysis. *Scientific Reports*, *9*(1), 19894. doi:10.103841598-019-55990-2 PMID:31882692

Tanasupawat, S., Takehana, T., Yoshida, S., Hiraga, K., & Oda, K. (2016). *Ideonella sakaiensis* sp. nov., isolated from a microbial consortium that degrades poly(ethylene terephthalate). *International Journal of Systematic and Evolutionary Microbiology*, *66*(8), 2813–2818. doi:10.1099/ijsem.0.001058 PMID:27045688

Tang, L., Wu, Q., & Qu, B. (2004). The effects of chemical structure and synthesis method on photodegradation of polypropylene. *Journal of Applied Polymer Science*, *95*(2), 270–279. doi:10.1002/app.21272

Tang, Y., Liu, Y., Chen, Y., Zhang, W., Zhao, J., He, S., Yang, C., Zhang, T., Tang, C., Zhang, C., & Yang, Z. (2021). A review: Research progress on microplastic pollutants in aquatic environments. *The Science of the Total Environment*, *766*, 142572. doi:10.1016/j.scitotenv.2020.142572 PMID:33183825

Tournier, V., Topham, C. M., Gilles, A., David, B., Folgoas, C., Moya-Leclair, E., Kamionka, E., Desrousseaux, M.-L., Texier, H., Gavalda, S., Cot, M., Guémard, E., Dalibey, M., Nomme, J., Cioci, G., Barbe, S., Chateau, M., André, I., Duquesne, S., & Marty, A. (2020). An engineered PET depolymerase to break down and recycle plastic bottles. *Nature*, *580*(7802), 216–219. doi:10.103841586-020-2149-4 PMID:32269349

Vimal Kumar, R., Kanna, G. R., & Elumalai, S. (2017). Biodegradation of Polyethylene by Green Photosynthetic Microalgae. *Journal of Bioremediation & Biodegradation*, *08*(01). Advance online publication. doi:10.4172/2155-6199.1000381

Wang, Y., Wang, X., & Wu, D. (2019a). Crystalline Characteristics, Mechanical Properties, Thermal Degradation Kinetics and Hydration Behavior of Biodegradable Fibers Melt-Spun from Polyoxymethylene/Poly(l-lactic acid) Blends. *Polymers*, *11*(11), 1753. Advance online publication. doi:10.3390/polym11111753 PMID:31731470

Wei, R., & Zimmermann, W. (2017). Microbial enzymes for the recycling of recalcitrant petroleum-based plastics: How far are we? *Microbial Biotechnology*, *10*(6), 1308–1322. doi:10.1111/1751-7915.12710 PMID:28371373

Willert, H. G., Semlitsch, M., & Peltier, L. F. (1996). Tissue Reactions to Plastic and Metallic Wear Products of Joint Endoprostheses. *Clinical Orthopaedics and Related Research, 333*. doi:10.1097/00003086-199612000-00002

Wright, S. L., & Kelly, F. J. (2017). Plastic and Human Health: A Micro Issue? *Environmental Science & Technology*, *51*(12), 6634–6647. doi:10.1021/acs.est.7b00423 PMID:28531345

Yan, N., Fan, C., Chen, Y., & Hu, Z. (2016). The Potential for Microalgae as Bioreactors to Produce Pharmaceuticals. *International Journal of Molecular Sciences*, *17*(6), 962. doi:10.3390/ijms17060962 PMID:27322258

Zhang, C., Chen, X., Wang, J., & Tan, L. (2017). Toxic effects of microplastic on marine microalgae Skeletonema costatum: Interactions between microplastic and algae. *Environmental Pollution*, *220*, 1282–1288. doi:10.1016/j.envpol.2016.11.005 PMID:27876228

Zhu, D., Ke, X., Christie, P., & Zhu, Y.-G. (2018). Rejoinder to "Comments on Zhu et al. (2018) Exposure of soil collembolans to microplastics perturbs their gut microbiota and alters their isotopic composition". Soil Biol. Biochem. 116 302–310". *Soil Biology & Biochemistry*, *124*, 275–276. doi:10.1016/j.soilbio.2018.05.030

Zhu, X., Zhao, W., Chen, X., Zhao, T., Tan, L., & Wang, J. (2020). Growth inhibition of the microalgae Skeletonema costatum under copper nanoparticles with microplastic exposure. *Marine Environmental Research*, *158*, 105005. doi:10.1016/j.marenvres.2020.105005 PMID:32501269

Chapter 21
Algae Biomass Conversion Technologies

Nalok Dutta
Independent Researcher, USA

Faisal Mahmood
Department of Energy Systems Engineering, Faculty of Agriculture Engineering and Technology, University of Agriculture Faisalabad, Pakistan

Abid Sarwar
Independent Researcher, Pakistan

Muhammad Sultan
Department of Agricultural Engineering, Bahauddin Zakariya University, Pakistan

Muhammad Ahmad
School of Chemistry and Chemical Engineering, Beijing Institute of Technology, Beijing, China

Muhammad Usman Khan
Independent Researcher, Pakistan

Muhammad Mubashar Omar
Faculty of Agricultural Engineering and Technology, University of Agriculture, Faisalabad, Pakistan

ABSTRACT

Biomass from algae, which is rich in proteins, carbohydrates, and lipids, could be used for the production of biofuels and chemicals. Because algal cultivation and harvesting require high energy and costs, algae-based fuel production is a challenging commercial application. At the pilot scale, this is a common bottleneck problem in algae processing for fuels or chemicals. By implementing an integrated algae biorefinery concept, the need for energy and costs can be reduced. Biopolymers, biochemicals, biofuels, and biofertilizers can all be recovered with higher economic efficiency than conventional methods. A green economy based on algae will also be more viable by reducing production costs. The purpose of this mini-review is to give information about the development of integrated biorefineries for recovery of algal-based bioproducts and their potential applications. The authors discuss the lifecycle assessment

DOI: 10.4018/978-1-6684-2438-4.ch021

and the economic aspects of an integrated algal biorefinery. A discussion of the challenges and future directions of integrated algal biorefinery is concluded.

INTRODUCTION

Environmental sustainability and energy crisis caused by overdependence on fossil fuels are the two biggest global challenges confronted by humanity today (Pietrosemoli and Rodríguez-Monroy., 2019). Fossil fuel consumption has caused global warming as a result of the huge amount of greenhouse gases (GHGs) released into the atmosphere (Martins et al. 2019). Therefore, many researchers around the world are focused on finding alternative renewable energy sources. Using first-generation edible feedstocks such as sugarcane and corn for biofuel production has caused the food versus fuel debate. Due to the expansion of biofuels, food prices may continue to rise (Gui et al., 2008). Therefore, lignocellulosic biomass feedstocks such as forest and agricultural residues have been used to produce second-generation biofuels. As a result of this complex structure, pretreatment is intensive and costly before lignocellulosic biomass can be converted into biofuels, limiting its viability (Limayem and Ricke, 2012). Hence, algae can be considered as a possible feedstock for processing third-generation biofuels, which could be a solution to these problems. The pros and cons of different generation of biomass feedstocks have been discussed in Table 1. Through the process of photosynthesis, green algae are able to survive and become the primary microorganism in aquatic environments. The algae biomass has the least utilization of nutrients when compared to biomass productivity. According to Jez et al. (2017), for rapeseed, sunflower and algae cultivation to produce 1 kg biomass per hectare, nitrogen consumption amounts to approx. 40, 45 and 0.003 mg/ha, respectively. This figure is significantly lower for algae cultivation. For rapeseed, sunflower, and algae, the phosphorus requirement is also approx. 25, 56, and 0.002 mg/ha, respectively. A wide range of commercially valuable bioproducts can be produced from algae biomass due to its diverse biochemical composition (carbohydrates, lipids, and proteins). Algal biofuels and biochemicals have been produced simultaneously from algae by integrating downstream biorefineries in recent years. Growing global energy demand could potentially be met by farming algae in aquatic environments, considering the limited agricultural land supply (Li et al., 2008). As a result of recent technological advances in photobioreactor design through integrated algae farming and biorefinery strategies, microalgae biofuel production is has come off age being more cost-effective (Brennan and Owende, 2010). Before

Table 1. The pros and cons of feedstocks across generations

Biomass generation 1	Biomass generation 2	Biomass generation 3
Source		
Edible oils	Non-edible oils	Seaweed/microalgae
Advantages		
Ecofriendly	Cheaper biomass, no competition with food crops	Higher productivity with increased oil content with the valorization of bioproducts and biofuels.
Limitations		
Competes with food supply	Complex reactions decrease process efficacy	High power requirement with additional technical hindrances

being converted to biofuel, algal biomass can be utilized for the production of various macromolecules in the form of lipids, fatty acids, and vital amines. As part of an algae biorefinery system, products of high value includes but is not limited to pharmaceuticals, biofertilizers and biomaterials (Budzianowski, 2017). Through the integration of biofuels and biochemical processing, algae biorefineries have moved closer to becoming a commercial reality.

CONVERSION TECHNOLOGIES FOR ALGAE BIOMASS

As per Barsanti et al.,2014, algae can be divided into macroalgae and microalgae based on colonization's morphology, size, and location. A major biochemical constituent of algae biomass is carbohydrates, lipids, and proteins, which can be used to make biofuel. In recent years, algae biomass has been extensively discussed as an input for biorefineries. Algal biorefineries use algae biomass as feedstock for the production of biofuels and value-added chemicals (Trivedi et al., 2015). By morphology, size and colonization of aquatic areas, Barsanti & Gualtieri (2014) categorize algae into macroalgae and microalgae. Biofuels are derived from various biochemical components of algae biomass, including carbohydrates, lipids, and proteins. Biofuels and bioproducts can be produced from algae, as shown in table 2. The process of converting algae biomass into end products involves upstream, bioreactions and downstream processing. Optimizing resource utilization, maximizing profitability, and minimizing waste are the objectives of biomass-based biorefineries. Three different processing routes can be used to recover algae biomass, while waste CO_2 and wastewater streams can be recycled to grow algae. It has been widely observed that the performance of biochemical reactions and downstream processes is dependent on the choice of algae strains during upstream cultivation, which tends to favor strains that grow quickly for accumulating biomass at high densities. During the selection of a suitable process-oriented algae strain, it is also important to consider the mechanical and physiochemical characteristics of algae cells. The algae can be maintained in the medium for multiple cycles without being significantly impacted by the surrounding environment because they are robust against shear stress and contamination induced by the cultivation system (Rodolfi et al., 2009). High biomass productivity could broaden the range of biochemical products that could be produced at the algae biorefinery. Furthermore, it is equally important to select algae that have a good ability to separate and that don't require a great deal of energy to disrupt cells (Günerken et al., 2015).

In addition, the selection of suitable algae strains may also be influenced by the type of cultivation system. A photobioreactor (PBR) and an open raceway pond (ORP) are both algae cultivation systems. Each system has its own set of targeted parameters. The survival rate of algae cells can be affected by other microorganism contamination and competition in the ORP cultivation system, for example (Cairns et al., 1972). It is therefore preferable to use a strain of algae that is resistant to bacteria and predators. The major problem encountered in PBR cultivation is the adhesion of algae to PBR walls (Zeriouh et al., 2017). One solution to this problem is to select algae species that have highly suspended characteristics. From the outset, it has been evident that the bioprocesses upstream that make up the algae cultivation process are critical in determining the downstream outputs. It is the cultivation conditions that influence the biochemical properties of algae biomass. As a result, controlling the algae cultivation parameters could provide control of the distribution of biochemical properties to meet the demands of end-products. Recently, the selection of appropriate algae strains is being conducted through pre-screening based on all

Figure 1. Potential route map of algae biorefinery

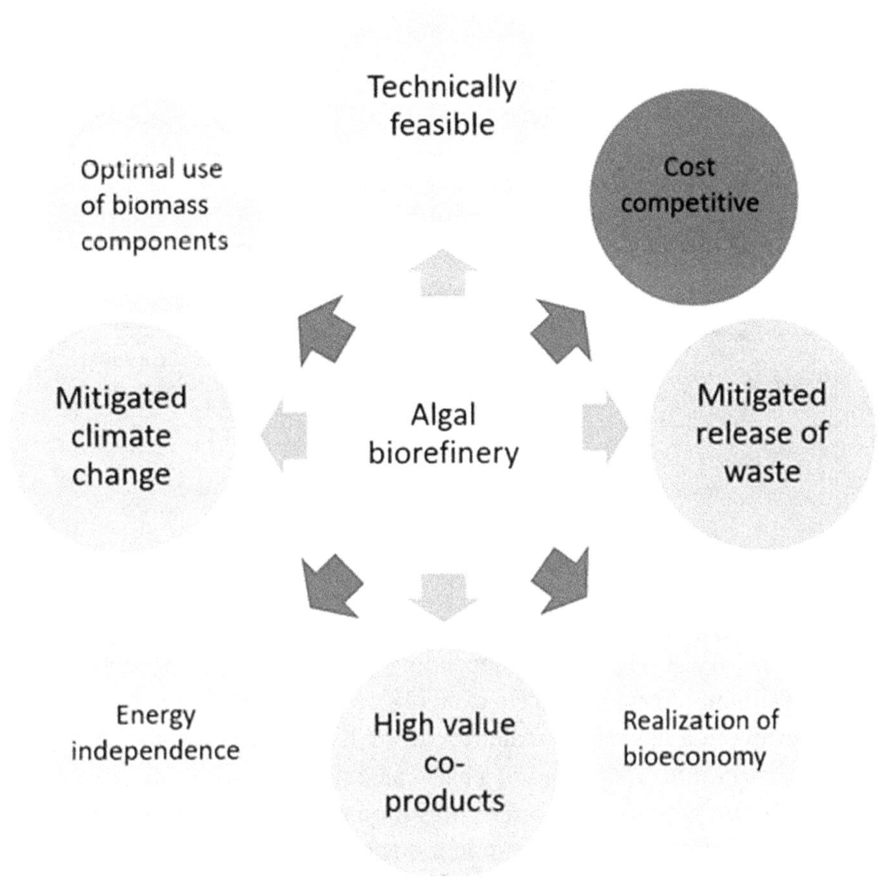

the factors mentioned above and will then be genetically modified to suit the targeted biorefinery routes. A algal biorefinery could have the following potentials if utilized to the full extent as depicted in Figure.1.

BIOCHEMICAL CONVERSION TECHNOLOGIES

Anaerobic digestion of organic biomass and fermentation of biohydrogen, bioethanol, and bioplastics through the use of bacteria and other microbes are vital steps in the biochemical conversion process. As algae have complex cell walls, enzymes have a difficult time hydrolyzing or making the substrate anaerobic, creating very low yields. To overcome this issue, a number of pretreatments and disintegration techniques have been used. Several biochemical conversion methods have been used to produce biofuels from algae biomass. An anaerobic digestion process by Banu et al. (30) generated appx.17 mL/g COD from microalgae grown on a mixed substrate (municipal wastewater). The biohydrogen at the lab-scale level was produced by Kumar et al. (31) using mixed microalgal species with a biomass concentration of ca. 2.50 g/L of algal biomass. Pilot-scale reactors have been used by some researchers to increase yields and productivity. As an example, Passos et al. (21) observed ~0.3 L of methane per gram of VS of biomethane production in high-rate raceway algal ponds with microalgal biomass. Microalgal spe-

Table 2. Bioproducts and biofuels from algae

Algae(type)	Genus	Bioenergy	Bioproducts	References
Green algae	*Scenedesmus* *Chlorella* *Nannochloropsis* *Haematococcus*	Biodiesel,bioethanol, biochar and bio-oil	Biopharmaceuticals, animal feeds, biomaterials, bio-nutrients	Biller and Ross .,2012, Chen et al. (2015), Griffiths and Harrison .,2009), John et al.,2011, Sun et al., 2018.
Diatom	*Phaeodactylu* *Cylindrotheca* *Dunaliella*	Biodiesel,bioethanol, biochar and bio-oil	Bochemicals, bionutrients	
Brown algae	*Ochromonas* *Saccharina* *Ascophyllum*	Bioethanol and biodiesel		
Blue-green algae	*Spirulina*	Bioethanol and biochar	Biopharmaceuticals, animal feeds, biomaterials, bio-nutrients	
Seaweed	*Padina* *Gelidium* *Laminaria* *Euglena* *Ulva*		Biochemicals and bionutrients	

cies will produce biofuel in varying degrees depending on the biomass concentration, the substrate, and the cultivation conditions. According to Ho et al. (48), the microalga *Chlorella vulgaris* was grown in synthetic growth media in a laboratory-scale reactor of 1 L and yielded appx.1.35g/L biomass and 3.50g/L bioethanol. *Scenedesmus dimorphus* in a lab-scale bioreactor with capacity of 2 liters produced 0.26 g/g of bioethanol when grown on the same cultivation medium as Cheng et al. (49). As shown in Figure 2, algae biorefineries combine upstream and downstream processing, incorporating upstream and downstream bioreactions. Three different processing routes can be used for algae biomass recovery, while waste CO_2 and wastewater streams can be repurposed as algae cultivation inputs.

Anaerobic Digestion

A process of anaerobic digestion is used to degrade algae biomass anaerobically, without oxygen (Ward et al., 2014). Anaerobic digestion can produce digestate from algae biomass in addition to biogas (e.g. biomethane and biohydrogen). A variety of microbial species interact with substrates (e.g. algae biomass containing lipids) during anaerobic digestion (Gonzalez-Fernandez et al., 2015). The lipid-extracted *Scenedesmus* biomass used in Yang et al. (2011) was utilized for the production of biohydrogen and biomethane in two stages. A heat-treated algae sludge was used to produce biohydrogen, and then an effluent from that process was used to produce biomethane. Deng et al. 2017 also demonstrated an algae biorefinery using Chlorella vulgaris cultivated under thermophilic conditions in pretreated anaerobic digestion swine manure to produce biofuels and to treat wastewater. Algae produced 1.0g/L of biomass, while nitrogen was removed at 99% efficiency and carbohydrates were removed at 54% efficiency. However, there are always inhibitors that stop anaerobic digestion from being efficient. Lakaniemi et al. (2011) studied whether fish algae can produce anaerobic biohydrogen and biomethane in freshwater and marine environments. Long-chain fatty acids (LCFAs) have been reported to be the main inhibitors of the anaerobic digestion process, Ma et al. (2015) reported. In the biomass of *Nannochloropsis salina*,

Figure 2. Algae biorefineries and conversion technologies

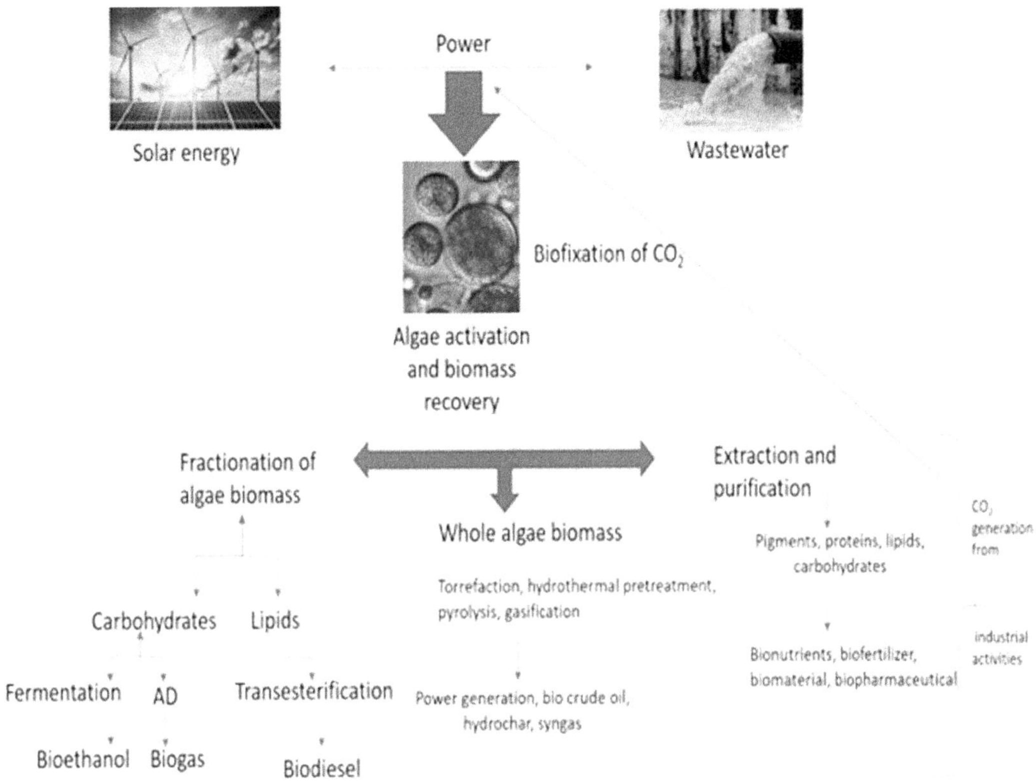

there was a negative correlation between microbiological activity and lipid concentration. Bacteria and methanogens are inhibited from producing biomethane by LCFAs because they accumulate volatile fatty acids (VFAs).

Biodiesel Production

The biorefinery route pertaining to typical chemical conversion indicates that of algae lipids into biodiesel (Chisti, 2007). Triacyglycerols and free fatty acids (FFAs) are components of algae crude lipids, which can be transesterified with alcohol to yield fatty acid alkyl esters (Demirbas, 2009b). An optimization study was conducted after the lipids from algae biomass were extracted via chemical solvents and before the conventional transesterification technique was implemented (Patil et al., 2012).

Transesterification can be achieved with catalysts such as acid catalysts (HCl, H_2SO_4), alkaline catalysts (NaOH, KOH) and other inorganic heterogeneous catalysts (ZrO_2, TiO_2). Using homogeneous acid (H_2SO_4) and alkaline (NaOH) as catalysts, Mathimani et al (2015) compared the performance of transesterification on *Chlorella sp* (22% of lipid). The results showed that acid catalyst had the highest yield ~ 60% for biodiesel conversion. The acid catalyst converted free fatty acids into esters and further converted them to esters to maximize biodiesel yield (Huang et al., 2010). As a result of the alkaline catalyst, soap was formed (unwanted side reaction) that slowed down transesterification (Park et al, 2015). According to Rahman et al. (2017), *Spirulina maxima* is normally transesterified and esterified during homogeneous acid-alkali catalysis. A catalyst (H_2SO_4) was initially used to reduce the free fatty acid (FFA) content in algae oil. The transesterification was followed by a second step using KOH as a catalyst. In this study, we confirmed that the two-step process was able to increase biodiesel yield by ~87%, compared to ~90% biodiesel from rapeseed oil.

Transesterification using supercritical fluids such as water, CO_2, or alcohol may also be another possible method of converting lipids into biodiesel, in addition to a catalytic process (Salam et al., 2016). The non-catalytic method of conversion requires high temperatures and pressures as opposed to catalytic transesterification. Supercritical transesterification offers several advantages over conventional processes, including no catalyst, shorter reaction times, higher biodiesel purity, and a lower environmental impact (Deshpande et al, 2017). Using supercritical methanol and ethanol as reactions mediums, Nan et al.,2015 investigated a noncatalytic transesterification process to produce biodiesel from *Chlorella protothecoides*. According to the results, optimal biodiesel yields were ~90% and 87%. Additionally, a continuous process for non-catalytic transesterification with supercritical CO_2 has recently been reported (Zhou et al, 2017). Chlorella sp. and Chrysophyta microalgae (~11 wt % lipid) were examined as algae with different lipid contents. The performance of continuous transesterification was compared to oil samples ~19% lipid. Chrysophyta and *Chlorella sp.* provided the highest yields of biodiesel, respectively.

Besides that, the in-situ transesterification method offers an alternative method to the conventional approach, which utilizes lipid extraction and catalytic transesterification at the same time (Ehimen et al., 2010). Supercritical in-situ transesterification (SC-IST) can produce biodiesel directly from wet algae biomass using water as the reacting medium. With supercritical water and ethanol in situ, Levine et al. (2010) produced 79.2% of fatty acid ethyl ester (FAEE) from *Chlorella Vulgaris* (53.3% lipids). Studies conducted on wet microalgae (Nannochloropsis oceanica) have demonstrated the production of biodiesel with a yield of 90% fatty acid methyl ester (FAME) (Im et al., 2014). It was also reported that Chlorella can be transesterified dry in methanol and sulphuric acid (catalyst), resulting in FAME yields of more than 90% (Viêgas et al., 2015).

After extracting lipids from algae biomass, biorefineries use it as feedstock to make bioethanol, animal feed, or pharmaceutical products. Upon comparing the dry and wet routes of in-situ transesterification, it is concluded that the wet method reduces the overall operating cost due to eliminating the drying process, which is advantageous for operating an algae biorefinery. However, supercritical conditions that require high pressures and temperatures made it less desirable than the catalytic process. Therefore, it is expected that algae biodiesel production could be facilitated via a catalytic transesterification process in algae biorefineries.

Biohydrogen Production

In a 2012 study, Miller and Ross reported that algae biomass was hydrothermally pressed to make biofuels solid hydrochar, liquid bio-oil and gaseous forms of biomethane or biohydrogen). Depending on the operating conditions, the hydrothermal process can be further categorized into carbonization, liquefaction, and gasification, where the proportions of the biofuel end products differ from one another. The physiochemical properties of algae biofuel products are believed to be influenced by hydrothermal treatment temperature and reaction time, according to Hrnčič et al (2016). Hydrothermal carbonization (HTC) is a mild reaction requiring mild temperatures and pressures (~190-275 °C, 40 bar) that produce solid hydrochar as the predominant product (Heilmann et al., 2010). Hydrothermal liquefaction (HTL) typically produces viscous bio-crude oil at temperatures ranging from 245 to 460 °C and pressure levels between 45 and 160 bar (López Barreiro et al., 2013). In contrast, hydrothermal gasification (HTG) requires thermally severe conditions at temperatures between ~400 and ~800 °C in order to produce bio-syngas, which is above the critical point of water (Yang et al., 2020). A biological hydrogen fermentation can be classified further into dark fermentation (without light source) and photo fermentation (with light source). Biological hydrogen production is usually performed by dark fermentation. Some researchers, however, recommend pretreating algae prior to dark fermentation in order to facilitate carbohydrate polymer hydrolysis. Several color light energies were used for cultivation of immobilized algae biomass and producing biohydrogen by Ruiz-Marin et al.,2020. Biomass production of *Chlorella vulgaris* and *Scenedesmus obliquus* is boosted by using urban wastewater as a growth medium. To study the impact of algal growth and biohydrogen yield, the authors used two different colors of light. Thus, blue-lit algae exhibit higher biomass concentrations than those grown in purple light. In contrast, algae cultured under purple light yield 128 mL/L of biohydrogen compared to blue light (~61 mL/L). In Table 3, a compilation of previous hydrothermal studies of algae biomass has been summarized.

Similarly, Kumar et al., 2018, used swine manure dilutions to ferment mixed algae to produce biohydrogen. A swine manure dilution ratio of 5 g/L yielded the highest biomass concentration of 2.57 g/L. Furthermore, the concentration of biomass decreased as the dilution ratio was increased for algal cultivation. Biohydrogen was then produced by fermenting the harvested biomass under dark conditions. 5 g/L of swine manure loading rate resulted in a maximum hydrogen yield of 13 mL/g algal biomass. For the production of biohydrogen, Batista et al.,2015, used urban wastewater to culture *Chlorella vulgaris* and *Scenedesmus obliquus* in tubular photobioreactors. Scientists reported that *Scenedesmus obliquus* and *Chlorella vulgaris* produced the highest yields of biohydrogen per gram of VS, coming in at ~57 mL and 41 mL respectively.

Table 3. Hydrothermal processing technologies of algal biomass in absence of water

Hydrothermal processing	Algae type	Temperature range (in °C)	Reaction duration(minutes)	End product	Yield %	References
	Chlorella vulgaris	350	60	Biocrude oil	30-35	Biller and Ross, 2011, Brown et al., 2010, Ekpo et al., 2016, Faeth and Savage, 2016, Heilmann et al., 2010, Levine et al., 2013, Li et al., 2014, Onwudili et al., 2013, Park et al., 2018, Zou et al., 2010.
	Nannochloropsis occulata	225-300	30-90	Biocrude oil	30-55	
HTL	*Dunaliella tertiolecta*	280-400	90	Biocrude oil	50-90	
	Botryococcus braunii	-	-	-	40	
	Neochloris oleoabundans	-	-	-	40	
	Nannochloropsis oculata	190-210	20-40	Hydrochar	50	
HTC	*Chlamydomonas reinhardtii*	190-220	30-120	Hydrochar	25	
	Chlorella vulgaris	150-250	60	Hydrochar	70	
	Chlorella vulgaris	500	30	Biogas	40	
HTG	*Chlorella vulgaris*	500	30	Biohydrogen	65	
	Nannochloropsis occulata	500	60	Syngas	40	

Bioethanol Production

Fermentation and anaerobic digestion are the most common processes used in algae biorefineries for converting algae biomass to bioethanol and biogas. The carbohydrates found in algae biomass can be synthesized into simple sugars via hydrolysis (Chen et al., 2013), which will then be fermented to produce bioethanol (Zhao and Su, 2014). Anaerobic digestion, on the other hand, is the degradation of biodegradable chemicals under oxygen-free conditions. Fermentation is an alternative path through which cell metabolism can gain energy, while anaerobic digestion is the decomposition of biodegradable chemicals. Rempel et al. (2019) produced bioethanol from Spirulina platensis via fermentation, and biomethane was produced via anaerobic digestion of the residue from the fermentation. An analysis of energy potential showed that fermentation coupled with anaerobic digestion produced ~13,900 kJ/kg as opposed to ~16,700 kJ/kg by direct anaerobic digestion. As a result, *Spirulina* is an interesting renewable energy resource for a variety of products. Fuels such as gasoline can be replaced by bioethanol since it is similar in physiochemical characteristics. Bioethanol yield is influenced more by the feedstock's chemical composition. Bioethanol is most commonly produced from carbohydrate-rich substrates. Among the polysaccharides found in algal biomass are starch, sugar, and cellulose. Bioethanol can therefore be produced using algal biomass as a feedstock(60).

A study of the different saccharification processes was conducted by Onay.,2019, using microalgal biomass from benthic wastewater (*Hindakia tetrachotoma*). Microalgae cultivation is carried out in a 1 L flat airlift photobioreactor with dilute distilled water used to vary the media concentration from 25% to 100%. About 0.78 g of biomass is produced per liter of medium using the lowest (25%) concentration of growth media. The biomass yield drops to 0.61 g/L when the growth media concentration is increased to 50%. This trend continues with higher concentrations. Following this, the microalgae were hydrolyzed by acidic, alkaline, and enzyme methods. A higher bioethanol yield and substrate conversion is seen in enzymatic hydrolysis with about 11.2 grams per liter. A similar pilot-scale cultivation of algae was reported by El-Mekkawi et al.,2019, in a domestic wastewater treatment plant for producing bioethanol. It is estimated that 45% of the harvested algal biomass is carbohydrates (Microcystis sp.). This algal biomass was acid hydrolyzed prior to yeast fermentation and produced ~19 grams of bioethanol per liter. In order to make bioethanol, Phwan et al.,2019, tested different levels of acids and their concentrations with microalgae. Two different acids were chosen for their study, sulfuric and acetic acid, varying in concentration from 1-9%. At 5% concentrations of sulfuric and acetic acid, respectively, the highest yield was recorded at ~0.2 and 0.281 g/g. Their study showed that sulfuric acid gave best bioethanol yield in all concentrations as compared to acetic acid for the enhancement of bioethanol production from microalgae. The results show that strong acids improve the bioethanol yield at a lower energy cost and with a low chemical requirement.

THERMOCHEMICAL CONVERSION

Biorefinery processing utilizing thermochemical upgrading utilized whole biomass for bioenergy conversion, as opposed to biological and chemical methods. Pyrolysis, gasification, and hydrothermal treatment are examples of thermochemical technologies (Chen et al., 2015). In addition to delivering excellent energy efficiency performance, hydrothermal processing of wet biomass shows excellent performance in the following processes: rapid hydrolysis, enhanced decarboxylation, and hydrodeoxygenation (Tekin et al., 2014). By utilizing the hot reactor effluents to heat the feed stream from ambient temperatures, heat integration can also be applied during hydrothermal processing, contributing up to 85% energy efficiency (Magdeldin et al., 2017). This is in contrast with a conventional conversion process that makes use of dry biomass, in which lower temperature steam is used for the reaction. The process and thermal performance assessment are therefore affected by the limited opportunities for thermal energy recovery.

Gasification

Algae biomass is converted into bio-syngas through gasification. Gases or supercritical water can be used as the medium for reaction (Mathimani et al., 2019). Hydrolysis and subsequent gasification reactions can occur in supercritical water conditions, commonly known as supercritical water gasification (SWG). Xu et al., 2011 stated that SWG occurs through the breakdown of larger biomass molecules into smaller biogas molecules. The SWG reaction was found to be endothermic at temperatures greater than ~650 °C and slightly endothermic at temperatures lower than 680 °C, according to Castello and Fiori (2011). Using supercritical water at 450-550 °C, Guan et al. (2012) gasified *Nannochloropsis sp.* to produce CO, CO_2, and CH_4 as well as intermediate products such as alkanes and aromatics. Zhao et al. (2019) have investigated the production of syngas from non-isothermal gasification of low-lipid algae - *chlorella vulgaris* and *spirulina*. According to the obtained results, the rate of H_2 generation (i.e. the increment of syngas) inversely relates to the heating rate. The results of these experiments indicate that algae biomass can be used to produce syngas on a commercial scale.

Pyrolysis

Pyrolysis is the process of heating plant material in the absence of air at a specified rate to an elevated temperature between 400 and 600 °C for the purpose of producing energy (Mathimani et al., 2019). Algae biomass pyrolysis has attracted considerable attention because of its high energy efficiency and environmental benifits. In addition to the break-down of large molecules through the pyrolysis reaction into smaller molecules, the end-products of pyrolysis are classified into three phases, somewhat similar to the hydrothermal process (Chen et al., 2015). Those include biochars produced at lower heating rates (5- 10 °C per minute) and longer reaction times (10-30 s), bio-crude oil with tars, and products produced at moderate pyrolytic conditions (400-500 °C for 2- 3-second reactions). Furthermore, they include biogases produced in a short reaction time (1–2 s) with a higher heating rate (10–600 °C/s).To produce pyrolytic bio oil, Bae et al. (2011) utilized macroalgae, which contained high nitrogen and ash contents ~6.10 wt%. At 500°C, the bio-oil yield was between 37% and 47%. Furthermore, the overall higher heating value (HHV) of bio-oil produced was similar to that of conventional terrestrial biomass, which is appx.19 MJ/kg for Undaria, 25 MJ/kg for Laminaria, and 22 MJ/kg for Porphyra. In terms of the qualitative analysis, the pyrolytic bio-oil that contains nitrogen could serve as a chemical feedstock for value-added chemical components or as biofertilizer (Chaiwong et al., 2013). In addition, a slow pyrolysis process was used to integrate bio-oil and biochar from Spirulina sp with a promising energy consumption ratio (0.49). This gave further insight into the possibility of algal biorefineries via pyrolysis.

DIRECT COMBUSTION

Most studies on dry extraction use pyrolysis and combustion techniques (Pragya and Pandey, 2016, Xu et al, 2011). These methods yield lower energy recovery and lower environmental feasibility than wet extraction techniques. The comparison study conducted by Bennion et al. (2015) showed that heating during the pyrolysis reaction and dewatering of microalgae prior to the pyrolysis step increases the demand for fossil energy substantially. GHG emissions from the dry (pyrolysis) extraction pathway increased as a result. Sun et al. (2019) reported the drying process in pyrolysis techniques causes 89% of the overall GHGs emissions. Additionally, in order to achieve high heat conversion efficiency in the in-direct combustion pathway, microalgae with a moisture content less than 50% must be dried, which requires energy intensive processing. Biomass (14–16 MJ/kg) has a lower calorific value than conventional diesel fuel, making them less competitive (Milledge et al., 2014) despite having lower global warming potential. To produce microalgae biofuels by dry extraction, the selection of an efficient and appropriate drying technology is crucial.

MAIN CHALLENGES

An algae biorefinery has gained considerable interest due to its dual functions of cost-effective wastewater treatment and algal production for biofuels, as well as using it as a raw material in various industries, including cosmetics and pharmaceuticals. Even so, there are many challenging issues that must be addressed, such as the high energy requirement to cultivate and harvest algae, and strategies to increase biomass production. ABF has higher production costs compared to PBF because of the factors listed above. Several researchers have suggested using syngas as a carbon source for algal cultivation and wastewater as a nutrient source if algal cultivation is to reduce these issues. Although algal biorefinery will reduce cultivation costs by half, it still has certain limitations when it comes to commercial application. We need to solve the biomass productivity problem (quantity and quality) and harvesting problem. The metabolic pathway and genetic material in algal biomass may be altered using genetic engineering to enhance biomass productivity. Fermenting wastewater with high nitrogen concentrations can also be used as a culture media. Using this kind of approach to produce biofuels and extract bioactive compounds will result in an increase in algal biomass productivity for commercial applications of biofuel and bioactive compounds. A specific algal species that is selected and identified for cultivation will, on the other hand, reduce the negative impact of the desired compound extraction. In order to reduce the high energy demand for ABF production, it would be beneficial to integrate upstream and downstream processes to develop a low-cost, value-added product manufacturing system. Therefore, the use of an IAB system will prove to be the best solution and the most economically viable way to scale up a process or commercialize an ABF.

CONCLUSION AND PERSPECTIVES

Due to the variety of biochemical compositions available in algae biomass, algae are extremely suitable for multi-product biorefineries. However, the most challenging aspect remained the integration of various technologies for biomass conversion in a complete algae biorefinery. More research is needed for algae biorefineries to be more sustainable and economically viable. According to the current LCA evaluation at laboratory scale, it is still feasible to realize multiple products through algae biorefinery with reduced environmental impacts. Algae biorefineries, however, can only be realized at the commercial level in the near future by counterbalancing biofuels costs with profits from bioproducts.

REFERENCES

Adesanya, V. O., Cadena, E., Scott, S. A., & Smith, A. G. (2014). Life cycle assessment on microalgal biodiesel production using a hybrid cultivation system. *Bioresource Technology*, *163*, 343–355. doi:10.1016/j.biortech.2014.04.051 PMID:24852435

Al-Saif, S. S. A. L., Abdel-Raouf, N., El-Wazanani, H. A., & Aref, I. A. (2014). Antibacterial substances from marine algae isolated from Jeddah coast of Red sea, Saudi Arabia. *Saudi Journal of Biological Sciences*, *21*(1), 57–64. doi:10.1016/j.sjbs.2013.06.001 PMID:24596500

Arenas, E. G., Rodriguez Palacio, M. C., Juantorena, A. U., Fernando, S. E. L., & Sebastian, P. J. (2017). Microalgae as a potential source for biodiesel production: Techniques, methods, and other challenges. *International Journal of Energy Research*, *41*(6), 761–789. doi:10.1002/er.3663

Ashokkumar, V., Salam, Z., Tiwari, O. N., Chinnasamy, S., Mohammed, S., & Ani, F. N. (2015). An integrated approach for biodiesel and bioethanol production from Scenedesmus bijugatus cultivated in a vertical tubular photobioreactor. *Energy Conversion and Management*, *101*, 778–786. doi:10.1016/j.enconman.2015.06.006

Badvipour, S., Eustance, E., & Sommerfeld, M. R. (2016). Process evaluation of energy requirements for feed production using dairy wastewater for algal cultivation: Theoretical approach. *Algal Research*, *19*, 207–214. doi:10.1016/j.algal.2016.08.017

Bae, Y. J., Ryu, C., Jeon, J.-K., Park, J., Suh, D. J., Suh, Y.-W., Chang, D., & Park, Y.-K. (2011). The characteristics of bio-oil produced from the pyrolysis of three marine macroalgae. *Bioresource Technology*, *102*(3), 3512–3520. doi:10.1016/j.biortech.2010.11.023 PMID:21129955

Barrera, D. J., Rosenberg, J. N., Chiu, J. G., Chang, Y.-N., Debatis, M., Ngoi, S.-M., Chang, J. T., Shoemaker, C. B., Oyler, G. A., & Mayfield, S. P. (2015). Algal chloroplast produced camelid VHH antitoxins are capable of neutralizing botulinum neurotoxin. *Plant Biotechnology Journal*, *13*(1), 117–124. doi:10.1111/pbi.12244 PMID:25229405

Barsanti, L., & Gualtieri, P. (2014). *Algae: Anatomy, Biochemistry, and Biotechnology* (2nd ed.). CRC Press.

Batan, L., Quinn, J., Willson, B., & Bradley, T. (2010). Net energy and greenhouse gas emission evaluation of biodiesel derived from microalgae. *Environmental Science & Technology*, *44*(20), 7975–7980.

Batista, A. P., Ambrosano, L., Graça, S., Sousa, C., Marques, P. A. S. S., Ribeiro, B., Botrel, E. P., Castro Neto, P., & Gouveia, L. (2015). Combining Urban Wastewater Treatment with Biohydrogen Production – An Integrated Microalgae-Based Approach. *Bioresource Technology*, *184*, 230–235. doi:10.1016/j.biortech.2014.10.064 PMID:25453433

Batista, A. P., Gouveia, L., Bandarra, N. M., Franco, J. M., & Raymundo, A. (2013). Comparison of microalgal biomass profiles as novel functional ingredient for food products. *Algal Research*, *2*(2), 164–173. doi:10.1016/j.algal.2013.01.004

Beach, E. S., Eckelman, M. J., Cui, Z., Brentner, L., & Zimmerman, J. B. (2012). Preferential technological and life cycle environmental performance of chitosan flocculation for harvesting of the green algae Neochloris oleoabundans. *Bioresource Technology*, *121*, 445–449. doi:10.1016/j.biortech.2012.06.012 PMID:22853967

Benemann, J. (2013). Microalgae for biofuels and animal feeds. *Energies*, *6*(11), 5869–5886.

Bennion, E. P., Ginosar, D. M., Moses, J., Agblevor, F., & Quinn, J. C. (2015). Lifecycle assessment of microalgae to biofuel: Comparison of thermochemical processing pathways. *Applied Energy*, *154*, 1062–1071.

Biller, P., & Ross, A. B. (2011). Potential yields and properties of oil from the hydrothermal liquefaction of microalgae with different biochemical content. *Bioresource Technology*, *102*(1), 215–225.

Biller, P., & Ross, A. B. (2012). Hydrothermal processing of algal biomass for the production of biofuels and chemicals. *Biofuels*, *3*(5), 603–623.

Brennan, L., & Owende, P. (2010). Biofuels from microalgae—A review of technologies for production, processing, and extractions of biofuels and co-products. *Renewable & Sustainable Energy Reviews*, *14*(2), 557–577. doi:10.1016/j.rser.2009.10.009

Brown, T. M., Duan, P., & Savage, P. E. (2010). Hydrothermal liquefaction and gasification of Nannochloropsis sp. *Energy & Fuels*, *24*(6), 3639–3646. doi:10.1021/ef100203u

Budzianowski, W. M. (2017). High-value low-volume bioproducts coupled to bioenergies with potential to enhance business development of sustainable biorefineries. *Renewable & Sustainable Energy Reviews*, *70*, 793–804. doi:10.1016/j.rser.2016.11.260

Byreddy, A. R., Yoganantharjah, P., Gupta, A., Gibert, Y., & Puri, M. (2019). Suitability ofnovel algal biomass as fish feed: Accumulation and distribution of omega-3 long-chain polyunsaturated fatty acid in Zebrafish. *Applied Biochemistry and Biotechnology*, *188*(1), 112–123. doi:10.100712010-018-2906-0 PMID:30345480

Cairns, J., Lanza, G. R., & Parker, B. C. (1972). Pollution related structural and functional changes in aquatic communities with emphasis on freshwater algae and protozoa. *Proceedings. Academy of Natural Sciences of Philadelphia*, *124*, 79–127.

Campbell, P. K., Beer, T., & Batten, D. (2011). Life cycle assessment of biodiesel production from microalgae in ponds. *Bioresource Technology*, *102*(1), 50–56. doi:10.1016/j.biortech.2010.06.048 PMID:20594828

Castello, D., & Fiori, L. (2011). Supercritical water gasification of biomass: Thermodynamic constraints. *Bioresource Technology*, *102*(16), 7574–7582. doi:10.1016/j.biortech.2011.05.017 PMID:21640582

Chaiwong, K., Kiatsiriroat, T., Vorayos, N., & Thararax, C. (2013). Study of bio-oil and biochar production from algae by slow pyrolysis. *Biomass and Bioenergy*, *56*, 600–606. doi:10.1016/j.biombioe.2013.05.035

Chen, C.-Y., Zhao, X.-Q., Yen, H.-W., Ho, S.-H., Cheng, C.-L., Lee, D.-J., Bai, F.-W., & Chang, J.-S. (2013). Microalgae-based carbohydrates for biofuel production. *Biochemical Engineering Journal*, *78*, 1–10.

Chen, W.-H., Lin, B.-J., Huang, M.-Y., & Chang, J.-S. (2015). Thermochemical conversion of microalgal biomass into biofuels: A review. *Bioresource Technology*, *184*, 314–327. doi:10.1016/j.biortech.2014.11.050 PubMed

Cheng, C.-H., Du, T.-B., Pi, H.-C., Jang, S.-M., Lin, Y.-H., & Lee, H.-T. (2011). Comparative study of lipid extraction from microalgae by organic solvent and supercritical CO2. *Bioresource Technology*, *102*(21), 10151–10153. doi:10.1016/j.biortech.2011.08.064 PMID:21917450

Chew, K. W., Yap, J. Y., Show, P. L., Suan, N. H., Juan, J. C., Ling, T. C., Lee, D.-J., & Chang, J.-S. (2017). Microalgae biorefinery: High value products perspectives. *Bioresource Technology*, *229*, 53–62. doi:10.1016/j.biortech.2017.01.006 PMID:28107722

Chisti, Y. (2007). Biodiesel from microalgae. *Biotechnology Advances*, *25*(3), 294–306.

Choi, S., Song, C. W., Shin, J. H., & Lee, S. Y. (2015). Biorefineries for the production of top building block chemicals and their derivatives. *Metabolic Engineering*, *28*, 223–239.

Chowdhury, R., & Franchetti, M. (2017). Life cycle energy demand from algal biofuel generated from nutrients present in the dairy waste. *Sustain. Prod. Consump.*, *9*, 22–27. doi:10.1016/j.spc.2016.07.001

Delrue, F., Li-Beisson, Y., Setier, P. A., Sahut, C., Roubaud, A., Froment, A. K., & Peltier, G. (2013). Comparison of various microalgae liquid biofuel production pathways based on energetic, economic and environmental criteria. *Bioresource Technology*, *136*, 205–212. doi:10.1016/j.biortech.2013.02.091 PMID:23567683

Delrue, F., Setier, P. A., Sahut, C., Cournac, L., Roubaud, A., Peltier, G., & Froment, A. K. (2012). An economic, sustainability, and energetic model of biodiesel production from microalgae. *Bioresource Technology*, *111*, 191–200. doi:10.1016/j.biortech.2012.02.020 PMID:22366604

Demirbas, A. (2009a). Biorefineries: Current activities and future developments. Energy Convers. Manage. 50 (11), 2782–2801. Demirbas, A., 2009b. Progress and recent trends in biodiesel fuels. *Energy Conversion and Management*, *50*(1), 14–34. doi:10.1016/j.enconman.2008.09.001

Deng, X.-Y., Gao, K., Zhang, R.-C., Addy, M., Lu, Q., Ren, H.-Y., Chen, P., Liu, Y.-H., & Ruan, R. (2017). Growing Chlorella vulgaris on thermophilic anaerobic digestion swine manure for nutrient removal and biomass production. *Bioresource Technology*, *243*, 417–425. doi:10.1016/j.biortech.2017.06.141 PMID:28688324

Deshpande, S. R., Sunol, A. K., & Philippidis, G. (2017). Status and prospects of supercritical alcohol transesterification for biodiesel production. *Wiley Interdisciplinary Reviews. Energy and Environment*, *6*(5), e252. doi:10.1002/wene.252

Ehimen, E. A., Sun, Z. F., & Carrington, C. G. (2010). Variables affecting the in situ transesterification of microalgae lipids. *Fuel*, *89*(3), 677–684. doi:10.1016/j.fuel.2009.10.011

Ekpo, U., Ross, A. B., Camargo-Valero, M. A., & Williams, P. T. (2016). A comparison of product yields and inorganic content in process streams following thermal hydrolysis and hydrothermal processing of microalgae, manure and digestate. *Bioresource Technology*, *200*, 951–960. doi:10.1016/j.biortech.2015.11.018 PMID:26615335

El-Mekkawi, S. A., Abdo, S. M., Samhan, F. A., & Ali, G. H. (2019). Optimization of Some Fermentation Conditions for Bioethanol Production from Microalgae Using Response Surface Method. *Bulletin of the National Research Center*, *43*(1), 164. doi:10.118642269-019-0205-8

Faeth, J. L., & Savage, P. E. (2016). Effects of processing conditions on biocrude yields from fast hydrothermal liquefaction of microalgae. *Bioresource Technology*, *206*, 290–293. doi:10.1016/j.biortech.2016.01.115 PMID:26879204

Fernández-Dacosta, C., Posada, J. A., Kleerebezem, R., Cuellar, M. C., & Ramirez, A. (2015). Microbial community-based polyhydroxyalkanoates (PHAs) production from wastewater: Techno-economic analysis and ex-ante environmental assessment. *Bioresource Technology*, *185*, 368–377. doi:10.1016/j.biortech.2015.03.025 PMID:25796067

Fortier, M.-O. P., Roberts, G. W., Stagg-Williams, S. M., & Sturm, B. S. M. (2014). Life cycle assessment of bio-jet fuel from hydrothermal liquefaction of microalgae. *Applied Energy*, *122*, 73–82. doi:10.1016/j.apenergy.2014.01.077

Gong, J., & You, F. (2015). Value-added chemicals from microalgae: Greener, more economical, or both? *ACS Sustainable Chemistry & Engineering*, *3*(1), 82–96. doi:10.1021c500683w

Gonzalez-Fernandez, C., Sialve, B., & Molinuevo-Salces, B. (2015). Anaerobic digestion of microalgal biomass: Challenges, opportunities and research needs. *Bioresource Technology*, *198*, 896–906. doi:10.1016/j.biortech.2015.09.095 PMID:26454349

Griffiths, M. J., & Harrison, S. T. L. (2009). Lipid productivity as a key characteristic for choosing algal species for biodiesel production. *Journal of Applied Phycology*, *21*(5), 493–507. doi:10.100710811-008-9392-7

Guan, Q., Savage, P. E., & Wei, C. (2012). Gasification of alga Nannochloropsis sp. in supercritical water. *The Journal of Supercritical Fluids*, *61*, 139–145. doi:10.1016/j.supflu.2011.09.007

Gui, M. M., Lee, K. T., & Bhatia, S. (2008). Feasibility of edible oil vs. non-edible oil vs. waste edible oil as biodiesel feedstock. *Energy*, *33*(11), 1646–1653. doi:10.1016/j.energy.2008.06.002

Günerken, E., D'Hondt, E., Eppink, M. H. M., Garcia-Gonzalez, L., Elst, K., & Wijffels, R. H. (2015). Cell disruption for microalgae biorefineries. *Biotechnology Advances*, *33*(2), 243–260. doi:10.1016/j.biotechadv.2015.01.008 PMID:25656098

Halim, R., Gladman, B., Danquah, M. K., & Webley, P. A. (2011). Oil extraction from microalgae for biodiesel production. *Bioresource Technology*, *102*(1), 178–185. doi:10.1016/j.biortech.2010.06.136 PMID:20655746

Harnedy, P. A., & FitzGerald, R. J. (2011). Bioactive proteins, peptides, and amino acids from macroalgae. *Journal of Phycology*, *47*(2), 218–232. doi:10.1111/j.1529-8817.2011.00969.x PMID:27021854

Harun, R., Yip, J. W. S., Thiruvenkadam, S., Ghani, W. A. W. A. K., Cherrington, T., & Danquah, M. K. (2014). Algal biomass conversion to bioethanol – a step-by-step assessment. *Biotechnology Journal*, *9*(1), 73–86. doi:10.1002/biot.201200353 PMID:24227697

Heilmann, S. M., Davis, H. T., Jader, L. R., Lefebvre, P. A., Sadowsky, M. J., Schendel, F. J., von Keitz, M. G., & Valentas, K. J. (2010). Hydrothermal carbonization of microalgae. *Biomass and Bioenergy*, *34*(6), 875–882. doi:10.1016/j.biombioe.2010.01.032

Heo, S.-J., Park, E.-J., Lee, K.-W., & Jeon, Y.-J. (2005). Antioxidant activities of enzymatic extracts from brown seaweeds. *Bioresource Technology*, *96*(14), 1613–1623. doi:10.1016/j.biortech.2004.07.013 PMID:15978995

Hernández, D., Riaño, B., Coca, M., & García-González, M. C. (2015). Saccharification of carbohydrates in microalgal biomass by physical, chemical and enzymatic pretreatments as a previous step for bioethanol production. *Chemical Engineering Journal*, *262*, 939–945. doi:10.1016/j.cej.2014.10.049

Ho, S.-H., Huang, S.-W., Chen, C.-Y., Hasunuma, T., Kondo, A., & Chang, J.-S. (2013). Bioethanol production using carbohydrate-rich microalgae biomass as feedstock. *Bioresource Technology*, *135*, 191–198. doi:10.1016/j.biortech.2012.10.015 PMID:23116819

Hrnčič, M. K., Kravanja, G., & Knez, Ž. (2016). Hydrothermal treatment of biomass for energy and chemicals. *Energy*, *116*, 1312–1322. doi:10.1016/j.energy.2016.06.148

Huang, G., Chen, F., Wei, D., Zhang, X., & Chen, G. (2010). Biodiesel production by microalgal biotechnology. *Applied Energy*, *87*(1), 38–46. doi:10.1016/j.apenergy.2009.06.016

Huang, Y., Li, R., Liu, H., Wang, B., Zhang, C., & Shen, Q. (2014). Novel resource utilization of refloated algal sludge to improve the quality of organic fertilizer. *Environmental Technology*, *35*(13), 1658–1667. doi:10.1080/09593330.2013.878397 PMID:24956756

Hulatt, C. J., & Thomas, D. N. (2011). Productivity, carbon dioxide uptake and net energy return of microalgal bubble column photobioreactors. *Bioresource Technology*, *102*(10), 5775–5787. doi:10.1016/j.biortech.2011.02.025 PMID:21376576

Im, H., Lee, H., Park, M. S., Yang, J.-W., & Lee, J. W. (2014). Concurrent extraction and reaction for the production of biodiesel from wet microalgae. *Bioresource Technology*, *152*, 534–537. doi:10.1016/j.biortech.2013.11.023 PMID:24291292

John, R. P., Anisha, G. S., Nampoothiri, K. M., & Pandey, A. (2011). Micro and macroalgal biomass: A renewable source for bioethanol. *Bioresource Technology*, *102*(1), 186–193.

Jorquera, O., Kiperstok, A., Sales, E. A., Embiruçu, M., & Ghirardi, M. L. (2010). Comparative energy life-cycle analyses of microalgal biomass production in open ponds and photobioreactors. *Bioresource Technology*, *101*(4), 1406–1413.

Kanda, H., Li, P., Goto, M., & Makino, H. (2015). Energy-saving lipid extraction from wet Euglena gracilis by the low-boiling-point solvent dimethyl ether. *Energies*, *8*(1), 610–620.

Khoo, H. H., Koh, C. Y., Shaik, M. S., & Sharratt, P. N. (2013). Bioenergy co-products derived from microalgae biomass via thermochemical conversion – Life cycle energy balances and CO2 emissions. *Bioresource Technology*, *143*, 298–307. doi:10.1016/j.biortech.2013.06.004 PMID:23810951

Koutra, E., Economou, C. N., Tsafrakidou, P., & Komaros, M. (2018). Bio-based products from microalgae cultivated in digestates. *Trends in Biotechnology*, *36*(8), 819–833. doi:10.1016/j.tibtech.2018.02.015 PMID:29605178

Kumar, G., Nguyen, D. D., Sivagurunathan, P., Kobayashi, T., Xu, K., & Chang, S. W. (2018). Cultivation of Microalgal Biomass Using Swine Manure for Biohydrogen Production: Impact of Dilution Ratio and Pretreatment. *Bioresource Technology*, *260*, 16–22. doi:10.1016/j.biortech.2018.03.029 PMID:29604564

Kumar, M., Olajire Oyedun, A., & Kumar, A. (2018). A review on the current status of various hydrothermal technologies on biomass feedstock. *Renewable & Sustainable Energy Reviews*, *81*, 1742–1770. doi:10.1016/j.rser.2017.05.270

Kwan, T. H., Pleissner, D., Lau, K. Y., Venus, J., Pommeret, A., & Lin, C. S. K. (2015). Technoeconomic analysis of a food waste valorization process via microalgae cultivation and co-production of plasticizer, lactic acid and animal feed from algal biomass and food waste. *Bioresource Technology*, *198*, 292–299. doi:10.1016/j.biortech.2015.09.003 PMID:26402872

Kwon, O.-M., Kim, D.-H., Kim, S.-K., & Jeong, G.-T. (2016). Production of sugars from macroalgae Gracilaria verrucosa using combined process of citric acid-catalyzed pretreatment and enzymatic hydrolysis. *Algal Research*, *13*, 293–297. doi:10.1016/j.algal.2015.12.011

Lakaniemi, A.-M., Hulatt, C. J., Thomas, D. N., Tuovinen, O. H., & Puhakka, J. A. (2011). Biogenic hydrogen and methane production from Chlorella vulgaris and Dunaliella tertiolectabiomass. *Biotechnology for Biofuels*, *4*(1), 34. doi:10.1186/1754-6834-4-34 PMID:21943287

Laurens, L. M. L., Markham, J., Templeton, D. W., Christensen, E. D., Van Wychen, S., Vadelius, E. W., Chen-Glasser, M., Dong, T., Davis, R., & Pienkos, P. T. (2017). Development of algae biorefinery concepts for biofuels and bioproducts; a perspective on process-compatible products and their impact on cost-reduction. *Energy & Environmental Science*, *10*(8), 1716–1738. doi:10.1039/C7EE01306J

Lee, R. A., & Lavoie, J.-M. (2013). From first- to third-generation biofuels: Challenges of producing a commodity from a biomass of increasing complexity. *Animal Frontiers*, *3*(2), 6–11. doi:10.2527/af.2013-0010

Levine, R. B., Pinnarat, T., & Savage, P. E. (2010). Biodiesel production from wet algal biomass through in situ lipid hydrolysis and supercritical transesterification. *Energy & Fuels*, *24*(9), 5235–5243. doi:10.1021/ef1008314

Levine, R. B., Sierra, C. O. S., Hockstad, R., Obeid, W., Hatcher, P. G., & Savage, P. E. (2013). The use of hydrothermal carbonization to recycle nutrients in algal biofuel production. *Environmental Progress & Sustainable Energy*, *32*(4), 962–975. doi:10.1002/ep.11812

Li, H., Liu, Z., Zhang, Y., Li, B., Lu, H., Duan, N., Liu, M., Zhu, Z., & Si, B. (2014). Conversion efficiency and oil quality of low-lipid high-protein and high-lipid low-protein microalgae via hydrothermal liquefaction. *Bioresource Technology*, *154*, 322–329. doi:10.1016/j.biortech.2013.12.074 PMID:24413449

Li, Y., Horsman, M., Wu, N., Lan, C. Q., & Dubois-Calero, N. (2008). Biofuels from microalgae. *Biotechnology Progress*, *24*(4), 815–820. PMID:18335954

Limayem, A., & Ricke, S. C. (2012). Lignocellulosic biomass for bioethanol production: Current perspectives, potential issues and future prospects. *Progress in Energy and Combustion Science*, *38*(4), 449–467. doi:10.1016/j.pecs.2012.03.002

López Barreiro, D., Prins, W., Ronsse, F., & Brilman, W. (2013). Hydrothermal liquefaction (HTL) of microalgae for biofuel production: State of the art review and future prospects. *Biomass and Bioenergy*, *53*, 113–127. doi:10.1016/j.biombioe.2012.12.029

Ma, J., Zhao, Q.-B., Laurens, L. L. M., Jarvis, E. E., Nagle, N. J., Chen, S., & Frear, C. S. (2015). Mechanism, kinetics and microbiology of inhibition caused by long-chain fatty acids in anaerobic digestion of algal biomass. *Biotechnology for Biofuels*, *8*(1), 141. doi:10.118613068-015-0322-z PMID:26379773

Magdeldin, M., Kohl, T., & Järvinen, M. (2017). Techno-economic assessment of the byproducts contribution from non-catalytic hydrothermal liquefaction of lignocellulose residues. *Energy*, *137*, 679–695. doi:10.1016/j.energy.2017.06.166

Martins, F., Felgueiras, C., Smithkova, M., & Caetano, N. (2019). Analysis of Fossil Fuel Energy Consumption and Environmental Impacts in European Countries. *Energies*, *12*(6), 964. doi:10.3390/en12060964

Mata, T. M., Martins, A. A., & Caetano, N. S. (2010). Microalgae for biodiesel production and other applications: A review. *Renewable & Sustainable Energy Reviews*, *14*(1), 217–232. doi:10.1016/j.rser.2009.07.020

Mathimani, T., Baldinelli, A., Rajendran, K., Prabakar, D., Matheswaran, M., Pieter van Leeuwen, R., & Pugazhendhi, A. (2019). Review on cultivation and thermochemical conversion of microalgae to fuels and chemicals: Process evaluation and knowledge gaps. *Journal of Cleaner Production*, *208*, 1053–1064. doi:10.1016/j.jclepro.2018.10.096

Mathimani, T., Uma, L., & Prabaharan, D. (2015). Homogeneous acid catalysed transesterification of marine microalga Chlorella sp. BDUG 91771 lipid – An efficient biodiesel yield and its characterization. *Renewable Energy*, *81*, 523–533. doi:10.1016/j.renene.2015.03.059

Milledge, J. J., Smith, B., Dyer, P. W., & Harvey, P. (2014). Macroalgae-derived biofuel: A review of methods of energy extraction from seaweed biomass. *Energies*, *7*(11), 7194–7222. doi:10.3390/en7117194

Miranda, J. R., Passarinho, P. C., & Gouveia, L. (2012). Pre-treatment optimization of Scenedesmus obliquus microalga for bioethanol production. *Bioresource Technology*, *104*, 342–348. doi:10.1016/j.biortech.2011.10.059 PMID:22093974

Mobin, S. M. A., Chowdhury, H., & Alam, F. (2019). Commercially important bioproducts from microalgae and their current applications - A review. *Energy Procedia*, *160*, 752–760. doi:10.1016/j.egypro.2019.02.183

Moncada, B. J., Aristizábal, M. V., & Cardona, A. C. A. (2016). Design strategies for sustainable biorefineries. *Biochemical Engineering Journal*, *116*, 122–134. doi:10.1016/j.bej.2016.06.009

Mondal, M., Goswami, S., Ghosh, A., Oinam, G., Tiwari, O. N., Das, P., Gayen, K., Mandal, M. K., & Halder, G. N. (2017). Production of biodiesel from microalgae through biological carbon capture: A review. 3. *Biotechnology*, *7*(2), 99. PMID:28560639

Mu, D., Ruan, R., Addy, M., Mack, S., Chen, P., & Zhou, Y. (2017). Life cycle assessment and nutrient analysis of various processing pathways in algal biofuel production. *Bioresource Technology*, *230*, 33–42. doi:10.1016/j.biortech.2016.12.108 PMID:28157562

Musa, M., Doshi, A., Brown, R., & Rainey, T. J. (2019). Microalgae dewatering for biofuels: A comparative techno-economic assessment using single and two-stage technologies. *Journal of Cleaner Production*, *229*, 325–336. doi:10.1016/j.jclepro.2019.05.039

Nan, Y., Liu, J., Lin, R., & Tavlarides, L. L. (2015). Production of biodiesel from microalgae oil (Chlorella protothecoides) by non-catalytic transesterification in supercritical methanol and ethanol: Process optimization. *The Journal of Supercritical Fluids*, *97*, 174–182. doi:10.1016/j.supflu.2014.08.025

Nguyen, H. S., Mäki-Arvela, P., Akhmetzyanova, U., Tišler, Z., Hachemi, I., Rudnäs, A., Smeds, A., Eränen, K., Aho, A., Kumar, N., Hemming, J., Peurla, M., & Murzin, D. Y. (2017). Direct hydrodeoxygenation of algal lipids extracted from Chlorella alga. *Journal of Chemical Technology and Biotechnology (Oxford, Oxfordshire)*, *92*(4), 741–748. doi:10.1002/jctb.5158

Norsker, N.-H., Barbosa, M. J., Vermuë, M. H., & Wijffels, R. H. (2011). Microalgal production— A close look at the economics. *Biotechnology Advances*, *29*(1), 24–27. doi:10.1016/j.biotechadv.2010.08.005 PMID:20728528

Onay, M. (2019). Bioethanol Production via Different Saccharification Strategies from H. Tetrachotoma ME03 Grown at Various Concentrations of Municipal Wastewater in a Flat-Photobioreactor. *Fuel*, *239*, 1315–1323. doi:10.1016/j.fuel.2018.11.126

Onwudili, J. A., Lea-Langton, A. R., Ross, A. B., & Williams, P. T. (2013). Catalytic hydrothermal gasification of algae for hydrogen production: Composition of reaction products and potential for nutrient recycling. *Bioresource Technology*, *127*, 72–80. doi:10.1016/j.biortech.2012.10.020 PMID:23131625

Park, J.-H., Hong, J.-Y., Jang, H. C., Oh, S. G., Kim, S.-H., Yoon, J.-J., & Kim, Y. J. (2012). Use of Gelidium amansii as a promising resource for bioethanol: A practical approach for continuous dilute-acid hydrolysis and fermentation. *Bioresource Technology*, *108*, 83–88. doi:10.1016/j.biortech.2011.12.065 PMID:22261657

Park, J.-Y., Park, M. S., Lee, Y.-C., & Yang, J.-W. (2015). Advances in direct transesterification of algal oils from wet biomass. *Bioresource Technology*, *184*, 267–275. doi:10.1016/j.biortech.2014.10.089 PMID:25466997

Park, K. Y., Lee, K., & Kim, D. (2018). Characterized hydrochar of algal biomass for producing solid fuel through hydrothermal carbonization. *Bioresource Technology*, *258*, 119–124. doi:10.1016/j.biortech.2018.03.003 PMID:29524686

Passell, H., Dhaliwal, H., Reno, M., Wu, B., Ben Amotz, A., Ivry, E., Gay, M., Czartoski, T., Laurin, L., & Ayer, N. (2013). Algae biodiesel life cycle assessment using current commercial data. *Journal of Environmental Management*, *129*, 103–111. doi:10.1016/j.jenvman.2013.06.055 PMID:23900083

Patil, P. D., Gude, V. G., Mannarswamy, A., Cooke, P., Nirmalakhandan, N., Lammers, P., & Deng, S. (2012). Comparison of direct transesterification of algal biomass under supercritical methanol and microwave irradiation conditions. *Fuel*, *97*, 822–831. doi:10.1016/j.fuel.2012.02.037

Phwan, C. K., Chew, K. W., Sebayang, A. H., Ong, H. C., Ling, T. C., Malek, M. A., Ho, Y. C., & Show, P. L. (2019). Effects of Acids Pre-Treatment on the Microbial Fermentation Process for Bioethanol Production from Microalgae. *Biotechnology for Biofuels*, *12*(1), 191. doi:10.118613068-019 1533-5 PMID:31384298

Pietrosemoli, L., & Rodríguez-Monroy, C. (2019). The Venezuelan energy crisis: Renewable energies in the transition towards sustainability. *Renew and Sustain Energ Rev*, *105*, 415-426.

Pragya, N., & Pandey, K. K. (2016). Life cycle assessment of green diesel production from microalgae. *Renewable Energy*, *86*, 623–632. doi:10.1016/j.renene.2015.08.064

Quinn, J. C., Smith, T. G., Downes, C. M., & Quinn, C. (2014). Microalgae to biofuels lifecycle assessment— Multiple pathway evaluation. *Algal Research*, *4*, 116–122. doi:10.1016/j.algal.2013.11.002

Rahman, A., Putman, R. J., Inan, K., Sal, F. A., Sathish, A., Smith, T., Nielsen, C., Sims, R. C., & Miller, C. D. (2015). Polyhydroxybutyrate production using a wastewater microalgae based media. *Algal Research*, *8*, 95–98. doi:10.1016/j.algal.2015.01.009

Rahman, M. A., Aziz, M. A., Al-khulaidi, R. A., Sakib, N., & Islam, M. (2017). Biodiesel production from microalgae Spirulina maxima by two step process: Optimization of process variable. *J. Radiat. Res. Appl. Sci.*, *10*(2), 140–147. doi:10.1016/j.jrras.2017.02.004

Rasala, B. A., Muto, M., Lee, P. A., Jager, M., Cardoso, R. M. F., Behnke, C. A., Kirk, P., Hokanson, C. A., Crea, R., Mendez, M., & Mayfield, S. P. (2010). Production of therapeutic proteins in algae, analysis of expression of seven human proteins in the chloroplast of Chlamydomonas reinhardtii. *Plant Biotechnology Journal*, *8*(6), 719–733. doi:10.1111/j.1467-7652.2010.00503.x PMID:20230484

Rempel, A., de Souza Sossella, F., Margarites, A. C., Astolfi, A. L., Steinmetz, R. L. R., Kunz, A., Treichel, H., & Colla, L. M. (2019). Bioethanol from Spirulina platensis biomass and the use of residuals to produce biomethane: An energy efficient approach. *Bioresource Technology*, *288*, 121588. doi:10.1016/j.biortech.2019.121588 PMID:31176943

Rodolfi, L., Chini Zittelli, G., Bassi, N., Padovani, G., Biondi, N., Bonini, G., & Tredici, M. R. (2009). Microalgae for oil: Strain selection, induction of lipid synthesis and outdoor mass cultivation in a low-cost photobioreactor. *Biotechnology and Bioengineering*, *102*(1), 100–112. doi:10.1002/bit.22033 PMID:18683258

Rohit, M., Chiranjeevi, P., Reddy, C. N., & Mohan, S. V. (2016). Photobioreactors for Microalgal Cultivation and influence of operational Parameters. In R. S. Singh, A. Pandey, & E. Gnansouno (Eds.), *Biofuels: Production and Future Perspectives* (p. 413). CRC Press.

Ruiz-Marin, A., Canedo-López, Y., & Chávez-Fuentes, P. (2020). Biohydrogen Production by Chlorella Vulgaris and Scenedesmus Obliquus Immobilized Cultivated in Artificial Wastewater under Different Light Quality. *AMB Express*, *10*(1), 191. doi:10.118613568-020-01129-w PMID:33108605

Salam, K. A., Velasquez-Orta, S. B., & Harvey, A. P. (2016). A sustainable integrated in situ transesterification of microalgae for biodiesel production and associated co-product-a review. *Renewable & Sustainable Energy Reviews*, *65*, 1179–1198. doi:10.1016/j.rser.2016.07.068

Samarakoon, K., & Jeon, Y.-J. (2012). Bio-functionalities of proteins derived from marine algae — A review. *Food Research International*, *48*(2), 948–960.

Samorì, C., Abbondanzi, F., Galletti, P., Giorgini, L., Mazzocchetti, L., Torri, C., & Tagliavini, E. (2015). Extraction of polyhydroxyalkanoates from mixed microbial cultures: Impact on polymer quality and recovery. *Bioresource Technology*, *189*, 195–202.

Sander, K., & Murthy, G. S. (2010). Life cycle analysis of algae biodiesel. *The International Journal of Life Cycle Assessment*, *15*(7), 704–714. doi:10.100711367-010-0194-1

Shamriz, S., & Ofoghi, H. (2018). Engineering the chloroplast of Chlamydomonas reinhardtii to express the recombinant PfCelTOS-Il2 antigen-adjuvant fusion protein. *Journal of Biotechnology*, *266*, 111–117. doi:10.1016/j.jbiotec.2017.12.015 PMID:29269249

Sharma, P., & Sharma, N. (2017). Industrial and biotechnological applications of algae: A review. *J. Adv. Plant Biol.*, *1*(1), 1–25. doi:10.14302/issn.2638-4469.japb-17-1534

Show, K.-Y., Lee, D.-J., Tay, J.-H., Lee, T.-M., & Chang, J.-S. (2015). Microalgal drying and cell disruption – Recent advances. *Bioresource Technology*, *184*, 258–266. doi:10.1016/j.biortech.2014.10.139 PMID:25465783

Silva, A. G., Carter, R., Merss, F. L. M., Corrêa, D. O., Vargas, J. V. C., Mariano, A. B., Ordonez, J. C., & Scherer, M. D. (2015). Life cycle assessment of biomass production in microalgae compact photobioreactors. *Global Change Biology. Bioenergy*, *7*(2), 184–194. doi:10.1111/gcbb.12120

Slade, R., & Bauen, A. (2013). Micro-algae cultivation for biofuels: Cost, energy balance, environmental impacts and future prospects. *Biomass and Bioenergy*, *53*, 29–38. doi:10.1016/j.biombioe.2012.12.019

Soomro, R.R., Ndikubwimana, T., Zeng, X., Lu, Y., Lin, L., & Danquah, M.K. (2016). Development of a two-stage microalgae dewatering process – A life cycle assessment approach. *Front. Plant Sci.*, *7*, 113.

Stratton, R. W., Wong, H. M., & Hileman, J. I. (2011). Quantifying variability in life cycle greenhouse gas inventories of alternative middle distillate transportation fuels. *Environmental Science & Technology*, *45*(10), 4637–4644. doi:10.1021/es102597f PMID:21513296

Subhadra, B. G., & Edwards, M. (2011). Coproduct market analysis and water footprint of simulated commercial algal biorefineries. *Applied Energy*, *88*(10), 3515–3523. doi:10.1016/j.apenergy.2010.12.051

Sun, C.-H., Fu, Q., Liao, Q., Xia, A., Huang, Y., Zhu, X., Reungsang, A., & Chang, H.-X. (2019). Life-cycle assessment of biofuel production from microalgae via various bioenergy conversion systems. *Energy*, *171*, 1033–1045. doi:10.1016/j.energy.2019.01.074

Taunt, H. N., Stoffels, L., & Purton, S. (2018). Green biologics: The algal chloroplast as a platform for making biopharmaceuticals. *Bioengineered*, *9*(1), 48–54. doi:10.1080/21655979.2017.1377867 PMID:28892417

Tekin, K., Karagöz, S., & Bektaş, S. (2014). A review of hydrothermal biomass processing. *Renewable & Sustainable Energy Reviews*, *40*, 673–687. doi:10.1016/j.rser.2014.07.216

Tredici, M. R., Bassi, N., Prussi, M., Biondi, N., Rodolfi, L., Chini Zittelli, G., & Sampietro, G. (2015). Energy balance of algal biomass production in a 1-ha "Green Wall Panel" plant: How to produce algal biomass in a closed reactor achieving a high Net Energy Ratio. *Applied Energy*, *154*, 1103–1111. doi:10.1016/j.apenergy.2015.01.086

Tredici, M. R., Rodolfi, L., Biondi, N., Bassi, N., & Sampietro, G. (2016). Techno-economic analysis of microalgal biomass production in a 1-ha Green Wall Panel (GWP®) plant. *Algal Research*, *19*, 253–263. doi:10.1016/j.algal.2016.09.005

Trivedi, J., Aila, M., Bangwal, D. P., Kaul, S., & Garg, M. O. (2015). Algae based biorefinery—How to make sense? *Renewable & Sustainable Energy Reviews*, *47*, 295–307. doi:10.1016/j.rser.2015.03.052

Udom, I., Zaribaf, B. H., Halfhide, T., Gillie, B., Dalrymple, O., Zhang, Q., & Ergas, S. J. (2013). Harvesting microalgae grown on wastewater. *Bioresource Technology*, *139*, 101–106. doi:10.1016/j.biortech.2013.04.002 PMID:23648758

Vanier, G., Stelter, S., Vanier, J., Hempel, F., Maier, U. G., Lerouge, P., Ma, J., & Bardor, M. (2018). Alga-made anti-hepatitis B antibody binds to human Fcγ receptors. *Biotechnology Journal*, *13*(4), 1700496. doi:10.1002/biot.201700496 PMID:29194986

Vasudevan, V., Stratton, R. W., Pearlson, M. N., Jersey, G. R., Beyene, A. G., Weissman, J. C., Rubino, M., & Hileman, J. I. (2012). Environmental performance of algal biofuel technology options. *Environmental Science & Technology*, *46*(4), 2451–2459. doi:10.1021/es2026399 PMID:22324757

Viêgas, C. V., Hachemi, I., Freitas, S. P., Mäki-Arvela, P., Aho, A., Hemming, J., Smeds, A., Heinmaa, I., Fontes, F. B., da Silva Pereira, D. C., Kumar, N., Aranda, D. A. G., & Murzin, D. Y. (2015). A route to produce renewable diesel from algae: Synthesis and characterization of biodiesel via in situ transesterification of Chlorella alga and its catalytic deoxygenation to renewable diesel. *Fuel*, *155*, 144–154. doi:10.1016/j.fuel.2015.03.064

Wang, B., Li, Y., Wu, N., & Lan, C. Q. (2008). CO2 bio-mitigation using microalgae. *Applied Microbiology and Biotechnology*, *79*(5), 707–718. doi:10.100700253-008-1518-y PMID:18483734

Ward, A. J., Lewis, D. M., & Green, F. B. (2014). Anaerobic digestion of algae biomass: A review. *Algal Research*, *5*, 204–214. doi:10.1016/j.algal.2014.02.001

Weschler, M. K., Barr, W. J., Harper, W. F., & Landis, A. E. (2014). Process energy comparison for the production and harvesting of algal biomass as a biofuel feedstock. *Bioresource Technology*, *153*, 108–115. doi:10.1016/j.biortech.2013.11.008 PMID:24355501

Wijffels, R. H., Barbosa, M. J., & Eppink, M. H. M. (2010). Microalgae for the production of bulk chemicals and biofuels. *Biofuels, Bioproducts & Biorefining*, *4*(3), 287–295. doi:10.1002/bbb.215

Wuang, S. C., Khin, M. C., Chua, P. Q. D., & Luo, Y. D. (2016). Use of Spirulina biomass produced from treatment of aquaculture wastewater as agricultural fertilizers. *Algal Research*, *15*, 59–64. doi:10.1016/j.algal.2016.02.009

Xu, L., Brilman, D. W. F., Withag, J. A. M., Brem, G., & Kersten, S. (2011). Assessment of a dry and a wet route for the production of biofuels from microalgae: Energy balance analysis. *Bioresource Technology*, *102*(8), 5113–5122. doi:10.1016/j.biortech.2011.01.066 PMID:21334889

Yan, N., Fan, C., Chen, Y., & Hu, Z. (2016). The Potential for Microalgae as Bioreactors to Produce Pharmaceuticals. *International Journal of Molecular Sciences*, *17*(6), 962. doi:10.3390/ijms17060962 PMID:27322258

Yang, C., Wang, S., Yang, J., Xu, D., Li, Y., Li, J., & Zhang, Y. (2020). Hydrothermal liquefaction and gasification of biomass and model compounds: A review. *Green Chemistry*, *22*(23), 8210–8232. doi:10.1039/D0GC02802A

Yang, Z., Guo, R., Xu, X., Fan, X., & Luo, S. (2011). Hydrogen and methane production from lipid-extracted microalgal biomass residues. *International Journal of Hydrogen Energy*, *36*(5), 3465–3470. doi:10.1016/j.ijhydene.2010.12.018

Yue, D., You, F., & Snyder, S. W. (2014). Biomass-to-bioenergy and biofuel supply chain optimization: Overview, key issues and challenges. *Computers & Chemical Engineering*, *66*, 36–56. doi:10.1016/j.compchemeng.2013.11.016

Zeriouh, O., Reinoso-Moreno, J. V., López-Rosales, L., Cerón-García, M. D. C., Sánchez-Mirón, A., García-Camacho, F., & Molina-Grima, E. (2017). Biofouling in photobioreactors for marine microalgae. *Critical Reviews in Biotechnology*, *37*(8), 1006–1023. doi:10.1080/07388551.2017.1299681 PMID:28427282

Zhao, B., & Su, Y. (2014). Process effect of microalgal-carbon dioxide fixation and biomass production: A review. *Renewable & Sustainable Energy Reviews*, *31*, 121–132. doi:10.1016/j.rser.2013.11.054

Zhao, M., Raheem, A., Memon, Z. M., Vuppaladadiyam, A. K., & Ji, G. (2019). Iso-conversional kinetics of low-lipid micro-algae gasification by air. *Journal of Cleaner Production*, *207*, 618–629. doi:10.1016/j.jclepro.2018.10.040

Zhou, D., Qiao, B., Li, G., Xue, S., & Yin, J. (2017). Continuous production of biodiesel from microalgae by extraction coupling with transesterification under supercritical conditions. *Bioresource Technology*, *238*, 609–615. doi:10.1016/j.biortech.2017.04.097 PMID:28482287

Zou, S., Wu, Y., Yang, M., Imdad, K., Li, C., & Tong, J. (2010). Production and characterization of bio-oil from hydrothermal liquefaction of microalgae Dunaliella tertiolecta cake. *Energy*, *35*(12), 5406–5411. doi:10.1016/j.energy.2010.07.013

Chapter 22
Towards Sustainable Use of Algae as Adsorbents for Wastewater Treatment

Victor Odhiambo Shikuku
https://orcid.org/0000-0002-2295-293X
Kaimosi Friends University College, Kenya

George Oindo Achieng'
Maseno University, Kenya

Patrick Ssebugere
Makerere University, Uganda

ABSTRACT

The occurrence of heavy metals, dyes, micronutrients, phenols, pharmaceuticals, and personal care products (PPCPs) in water resources continue to raise environmental concerns since they are known to cause detrimental effects on living organisms even at low concentrations. Conventional wastewater treatment plants have also been pointed out as point sources of loading these pollutants into the recipient surface waters. Because of the non-biodegradable nature of heavy metals and the stable structure of dyes and PPCPs, these pollutants are persistent in the environment. Studies have shown that algae (micro and macro) present an alternative source of low-cost, efficient, and sustainable biosorbent for the removal of various pollutants from water either singly or in synergy with other wastewater treatment processes. This chapter is a brief review of recent studies on the use of algae-based biosorbents for the sequestration of heavy metals, dyes, and PPCPs from wastewater. Microalgae and macroalgae are shown to be promising and sustainable materials for the biosorption of water pollutants.

INTRODUCTION

One of the primary issues in the twenty-first century is access to clean water (Al-Amshawee et al.,

DOI: 10.4018/978-1-6684-2438-4.ch022

2019). Water quality is deteriorating rapidly as a result of growing anthropogenic activities, unplanned development, and increased industrialization. Increased human population also increases the strain on access to clean water. The principal pollutants found in wastewater effluent are various types of dyes, micro- and macronutrients and metal ions, which cause harm to the aquatic environments (Varghese et al., 2019; Saravanan et al., 2021). However, in the past decade, several studies have reported the occurrence of new classes of water contaminants with no regulatory status. These pollutants are generally knowns as chemicals of emerging concern (CEC) or emerging contaminants (Shikuku, 2020). Among these contaminants are pharmaceutical and personal care products (PPCPs). The ubiquitous presence of PPCPs in the environment is of scientific concern due to their ecological toxicity such as inducing the development of drug resistant bacterial strains, mutagenicity and causing endocrine disruption. Conventional wastewater treatment plants (WWTPs) are point sources for both heavy metals (Shikuku et al., 2017) and PPCPs (Kimosop et al., 2016; Muriuki et al., 2020) loading into the recipient water bodies. The physical, chemical and biological methods of wastewater treatment have intensively been discussed in literature and textbooks, and their details will not be reiterated in this chapter. Adsorption still remains the widespread method for removal of the aforementioned contaminants from the wastewater. Various materials have been reported as possible sorbents for sequestration of water pollutants. Biochars (Ng'eno et al., 2016; Achieng' and Shikuku, 2020), clays (Shikuku and Mishra, 2021), zeolites (Ng'eno et al., 2019), and geopolymers (Tome et al., 2021; Dzoujo et al., 2021) are among the common materials, of geological, industrial, and agricultural origin, that are used in this respect. The search for alterative adsorbents is a subject of ongoing research due to the expensiveness of activated carbon. Biomass-based adsorbents are an emerging class of water purification approach due to their eco-friendliness, low-cost, and good surface characteristics. This chapter highlights the recently published information about the biosorption of PPCPs, phenols, dyes, nitrates, phosphates and heavy metals using various algal species, as well as the sorption mechanisms, and their comparative adsorptive capacities. As shown in Figure 1, the adsorption performance of the algal-based adsorbents will be dependent on the pre-treatment method of the algal biomass, which determines the type of sorbent generated, the type of pollutant and the environmental or experimental conditions.

Biosorption of Pharmaceuticals and Personal Care Products (PPCPs)

Biosorption, in regards to algae, refers to the process in which the molecules or ions of the pollutant in solution are distributed between the liquid phase and the cell wall of microalgae (Heda at al., 2021). This biosorption mechanism may entail the hydrophobic or electrostatic interactions between the pollutants and the surface functional groups of microalgae. The cell wall of the microalgae consists of xylan, mannan, alginic acid, cellulose, hemicelluloses, chitin, and lignin among other constituents, which are encapsulated by the extracellular polymeric substances (EPS) that provide binding sites for adsorption of pollutants. The extracellular polymeric substances are mainly polysaccharides, lipids, proteins, and nucleic acids either bonded to microalgal cells, or free following excretion by microalgae (Fomina and Gadd 2014). Like other pollutants, the biosorption of pharmaceutical and personal care products (PPCPs) varies depending on their chemical structures, physicochemical properties, the functional groups present, and the microalgal species. For instance, the presence of negatively charged carboxyl and hydroxyl groups on the cell wall of microalgae promotes coulombic interactions with positively charged PPCPs molecules depending on the solution pH and dissociation constants (pK_a) of the PPCP.

Figure 1. Algal biomass-based adsorbent preparation approaches (Lee et al., 2022)

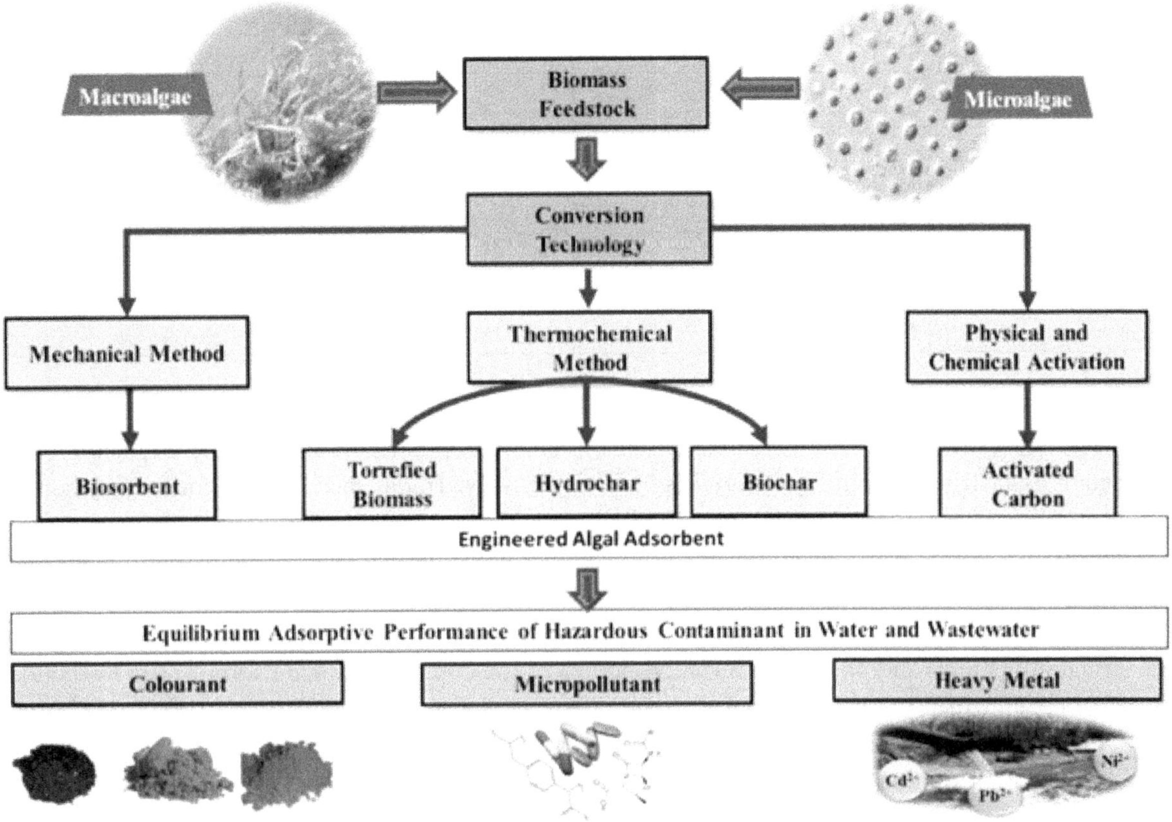

De Godos and co-workers (2012) showed that 61 ± 1% tetracycline could be removed in 43 h by the unicellular coccoid green microalga *Chlorella vulgaris* in high rate algal ponds (HRAP) and in the absence of light through biosorption mechanism. The removal of tetracycline was described by the linear sorption model with a constant of 4200 ± 200 mL/g ($r^2 = 0.99$) after 43 h. Importantly, the 98% tetracycline removal efficiency at 88 mg TSS L^{-1} of HRAP biomass indicates that efficient tetracycline removal is achievable during field wastewater treatment in HRAP under natural solar irradiation and/or under low biomass concentration operation. Wang et al. (2013) reported the biosorption of triclosan from aqueous solution by the ubiquitous green microalga *Chlorella pyrenoidosa* with a removal efficacy of 50% for an initial concentration range between 100-800 µg/L within 1 h of residence time. However, the study also reported that the triclosan treatment resulted in the disruption of the chloroplast and the release of organic material into the solution, suggesting that triclosan may affect membrane metabolism.

Guo et al. (2016) evaluated the feasibility of three freshwater green microalgae (*Chlorella sp.*, *Chlamydomonas* sp., and *Mychonastes* sp.) to remove 7-amino-cephalosporanic acid (7-ACA) from wastewater within 10 minutes. The adsorption capacities were 2.95, 3.09 ad 4.74 mg/g for *Mychonastes sp.*, *Chlamydomonas sp.*, and *Chlorella sp.*, respectively. It has been shown that the uptake of organic pollutants by microalgae is a multi-mechanistic process involving several mechanisms such as biosorption, bioaccumulation, and biodegradation all operating simultaneously (Heda at al., 2021). According to De Wilt et al. (2016), the sequestration of 62% of the mixed pharmaceutical ingredients, namely;

Table 1. Properties of PPCPs that influence biosorption

Compound	Log K_{ow}	pK_a	K_d (L/kg)	Reference
17β-Estradiol	4.01	10.4	24350	Xue et al., 2010
Clofibric acid	2.57	3.18	4.8±2.5	Ternes et al., 2004
Diclofenac	4.02	4.51	459±32	Ternes et al., 2004
Carbamazepine	2.25	7	314±205	Radjenović et al., 2009
Ibuprofen	3.79	4.91	9.5±3.1	Radjenović et al., 2009
Ketoprofen	3	4.45	226±180	Radjenović et al., 2009
Sulfamethoxazole	0.48	5.5	3.2±4.5	Radjenović et al., 2009
Erythromycin	2.48	8.88	309±271	Radjenović et al., 2009

estrone (E1), β-estradiol (E2), ethynylestradiol (EE2), carbamazepine, ibuprofen, diclofenac, metoprolol, paracetamol, and trimethoprim, by *Chlorella sorokiniana* involved multi-mechanistic sequences, whereas 34% were eliminated by biosorption. Amongst all the above-mentioned pharmaceuticals, carbamazepine was adsorbed the most adsorbed compound onto the microalgal biomass and accounted for the highest percentage (16.7%), followed by diclofenac (9.1%) and trimethoprim (7.6%), whereas the rest showed minimal contributions. The aforementioned studies underscore the fact that biosorption is dependent on the structural features and physicochemical properties of the compounds being sorbed. The molecular properties that influence adsorption include hydrophobicity (log K_{ow}), dissociation patterns (pK_a), solid-water distribution coefficient (K_d) and kinetic diameter (Shikuku et al., 2018). For example, the adsorption of trimethoprim, that predominantly exists in a cationic form in aqueous solution, onto different microalgae, namely, *Chlorella sorokiniana* and *Nannochloris sp*., has been reported to be relatively low (De Wilt et al., 2016) and not adsorbed at all (Bai and Acharya., 2016), respectively. This was attributed to its low log K_{ow} (1.4) and its ionic form. In terms of the effect of the octane-water partition coefficient (K_d) on biosorption, compounds of log K_d greater than 3.2, in general, are readily sorbed from water onto the adsorbent surface (Tadkaew et al., 2011). For instance, 26% removal efficiency of ibuprofen (K_d=3.5-4.5) by biosorption onto the freshwater diatom *Navicula* sp. was reported by Ding et al. (2017). At pH 8, the biosorption of ibuprofen onto periphyton (mixed microalgae and bacterial consortium) was reduced to less than 20% (Kang et al., 2018). At high solution pH, the ibuprofen (pK_a, 4.4) is deprotonated and thus bears a net negative charge hence electrostatic repulsion with the negatively charged surface of microalgae. This shows the interplay between log K_{ow}, K_d and pK_a parameters in driving the adsorption process. A list of these parameters for various PPCPs is presented in Table 1.

As earlier alluded to, biosorption of PPCPs onto algae biomass has also been demonstrated to be species-specific. For example, *Selenastrum capricornutum* exhibited biosorption removal efficiencies of 46% and 40% for β-estradiol (E2) and 17α-ethinylestradiol (EE2) from anaerobic sludge, respectively, while *Chlamydomonas reinhardtii* removed 86% and 71% of E2 and EE2, respectively, over the same duration (Hom-Diaz et al., 2015). Elsewhere, Zhang et al. (2014) showed that *Scenedesmus dimorphus* sorbed posted less than 10% of the maximum removal (85%) of E2 and EE2. *Cymbella sp.* could remove 97.1% naproxen while *Scenedesmus quadricauda* showed 58.8% removal under similar conditions (Ding et al., 2017). Escapa et al. (2017) showed that *Chlorella vulgaris* could remove, at least, 21% of paracetamol from the wastewater while *Tetradesmus obliquus* removed beyond 40% of the initial paracetamol concentration.

Table 2. A comparison of performance of various microalgal strains for PPCPs removal

Compound	Microalgae	Percent (%) removal	Reference
Enrofloxacin	*Chlorella vulgaris*	26	Xiong et al., 2017
	Ourococcus multisporus	18	
Carbamazepine	*Chlamydomonas mexicana*	37	Xiong et al., 2016
	Tetradesmus obliquus	30	
Diclofenac	*Chlorella vulgaris*	78	Escapa et al., 2016
	Tetradesmus obliquus	20	
Ciprofloxacin	*Tetradesmus obliquus*	79	Zhou et al., 2014
	Chlorella pyrenoidosa	75	
Norfloxacin	*Chlamydomonas reinhardtii*	53	Zhou et al., 2014
	Tetradesmus obliquus	40	
Erythromycin	*Chlamydomonas reinhardtii*	85	Zhou et al., 2014
	Chlorella vulgaris	67	

Other micropollutants reported in literature include phenols. For instance, Zheng et al. (2017) reported the biosorption of p-nitophenol by *Coelastrumsp. Pte-15*, *Chlamydomonassp.Tai-03* and *Chlorellasp. Cha-01*. The adsorption capacities were in the order 80.1, 110.7, and 82.2 mg/g, respectively. Pyrolytic conversion of the algal biomass into biochar improved both the textural properties (surface areas) and adsorption density of the adsorbent. A comparison of various algal species in the removal of PPCPs is presented in Table 2.

The species-specificity and performance may also be promoted by the presence of lipids. For example, Daneshvar et al. (2018) reported biosorption of tetracycline onto the biomass of lipids-containing *Scenedesmus quadricauda* with relatively high adsorption capacity of 295 mg/g. In a comparative study, the biomass of *Chlorella* sp. sequestered 82.7% of cefalexin before lipid extraction, and a lower amount of 71.2% after lipid extraction, confirming the lipid-mediated biosorption process (Angulo et al., 2018). It is also evident that in real wastewater, multiple mechanisms account for the overall removal of PPCPs besides biosorption. These include bioaccumulation, biodegradation and, photodegradation (Hom-Diaz et al., 2017). A brief summary of different mechanisms reported for the removal of PPCPs by various microalgae species is presented in Table 3.

Biosorption of Dyes

Relative to other biosorbents, algae-based biosorbents are appreciably suitable candidates due to their determined relatively high adsorption capacities and their high abundance in nature (Azam et al., 2020). The biosorption mechanism and efficiency depends on the physicochemical properties of the dye as well as the composition and structure of the algal cell walls. The algal cell wall consists of cellulose, hemicelluloses, chitin, proteins, and lignin among other constituents. These constituents imbue algae

Table 3. Mechanisms of PPCPs removal from water by various microalgae strains

Compound	Microalgae	Mechanism	Percent (%) removal	Reference
Sulfathiazole	*Spyrogira* sp.	Biodegradation and indirect photodegradation	36	Garcia-Rodríguez et al., 2013
Sulfapyridine			37	
Sulfamethazine			15	
Sulfamethoxazole			14	
17α-ethinylestradiol	*Desmodesmus subspicatus*	Adsorption and biodegradation	68	Maes et al., 2014
Progesterone	*Tetradesmus obliquus*	Biodegradation	95	Peng et al., 2014
Progesterone	*Chlorella pyrenoidosa*	Biodegradation	95	Peng et al., 2014
Diclofenac	*Chlorella sorokiniana*	Biodegradation, photolysis, and sorption	45	De Wilt et al., 2016
Ibuprofen			99	
Paracetamol			99	
Carbamazepine			30	
Enrofloxacin	*Tetradesmus obliquus*	Bioaccumulation, bioadsorption, and/or biodegradation	23	Xiong et al., 2017
	Ourococcus multisporus		18	

with multiple biosorption sites such as amino, amine, hydroxyl, imidazole, phosphate, and sulfate groups (Singh et al., 2020). Acid and alkali pretreatments of the biomass can be used as surface modification approaches to ameliorate the adsorption capacity of algae-based biosorbents for effective dye uptake. For example, citric acid-modified of brown algae (BA-CA) exhibited an adsorption capacity of 279.14 mg/g almost twice higher than the unmodified algae (BA) (147.77 mg/g) for the removal of crystal violet (CV) dye from water (Essekri et al., 2021). The biosorption process was strongly pH-dependent and the overall adsorption was attributed to both electrostatic attractions and electron donor–acceptor intermolecular interactions. Angelova et al. (2016) prepared a magnetically responsive iron oxide functionalized brown alga (*Sargassum horneri*) for the remotion of five water-soluble dyes of different chemical structures. The maximum sorption capacities were in the order acridine orange (193.8 mg/g)>crystal violet (167.0 mg/g)>methylene blue (158.9 mg/g)>safranin O (144.4 mg/g)>malachite green (110.4 mg/g). The biosorption processes were spontaneous, exothermic and physical. The results demonstrated the significance of chemical structures and physicochemical properties of dyes in biosorption. Gunasundari et al. (2020) fabricated ultrasonic-assisted cyanobacterium *Spirulina platensis* for the uptake of naphthol green B dye from aqueous solution with a maximum adsorption capacity of 137.9 mg/g at pH 3. The biosorption process was best described by the Freundlich isotherm and the pseudo-first order (PFO) kinetic model. Unmodified *Spirulina platensis* has been reported to have a high adsorption capacity of 482.2 mg/g for Reactive Red 120 dye with 94-99% removal efficiency (Cardoso et al., 2012), whereas *Spirulina* biomass modified with silica coated with magnetite particles had an adsorption capacity of 90.9 mg/g representing more than 97% removal efficiency (Kausar et al., 2020). The biosorption of dyes onto algae-based sorbents has been documented to be strongly pH dependent. The adsorption of cationic dyes is inhibited in acidic conditions due to competition between the dye molecules and hydrogen ions

Table 4. Biosorption capacities of various algal species for different dyes

Dye	Algae type	Adsorption capacity (mg/g)	Percent (%) removal	Reference
Methylene blue	*Chlorella* sp.	113	-	Yu et al. (2021)
Congo red	*Chlorella* sp.	164	-	Yu et al. (2021)
Remazol Black B	*Kamptonema animale*	41	99.6	Bayazıt et al. (2019)
Crystal violet	*Enteromorpha flexuosa*	119	90.3	Elgarahy et al. (2019)
Acid Red P-2BX	*Dried-Phormidium animale*	100	99.7	Gül et al. (2019)
Methylene blue	*Chlorella vulgaris*	275	83	Chin et al. (2020)
Methylene blue	*Durvillaea antarctica*	702	-	Guarín et al. (2018)

(H^+) for the same binding sites. Additionally, in strongly acidic solutions, the biosorbent surface acquire a net positive charge resulting to electrostatic repulsion with the cations. Contrarily, acidic conditions favor the uptake of anionic dyes. Table 4 represents a summary of studies on biosorption of dyes onto micro and macro-algae.

Removal of Nitrates and Phosphates

The most prevalent pollutant in groundwater resources is nitrate (Wheeler et al. 2015). The widespread use of nitrate-based fertilizers is the leading cause of nitrate contamination; as a result, agricultural areas have the greatest levels of nitrate in groundwater (Lawniczak et al. 2016). Phosphate-based fertilizer usage may likewise be linked to phosphate contamination of groundwater. Drinking water with a nitrate level higher than the maximum contaminant level (MCL) poses a health risk to newborns and, in some cases, adults. Methemoglobinemia, the outcome of nitrate conversion to nitrite in the digestive tract of newborns, is said to be common in those who have taken water with high nitrate levels (Liu et al. 2012). The use of microalgae to remove nitrate from groundwater is a long-term and cost-effective solution. Algae can absorb nitrate and convert it to biomass, which concentrates the nitrogen. *Chlorella vulgaris* and *Scenedesmus* sp. have showed a promise in removing nitrates from wastewater as immobilized and free cells (AlMomani and Örmeci 2016). Mollamohammada et al. (2020) also reported the treatment efficiency of *Chlorella sorokiniana* and *Scenedesmus* sp. immobilized in sodium alginate, for the removal of nitrates from groundwater in a batch culture. Immobilized *Scenedesmus* sp. and *Chlorella sorokiniana* cells showed 90% nitrate removal in 9 and 12 days, respectively. In real groundwater experiments, 90% of the nitrates was removed in 2 days without the use of any extra carbon source. Because they are water-insoluble, non-toxic, easy to harvest, and offer high removal effectiveness, immobilized algal beads provide a cheap alternative strategy for the removal of nitrate from water. Two microalgae strains, *Chlorella sp.* ZTY4 and *Scenedesmus* sp. LX1, were reported to effectively remove nitrogen and phosphorus with efficiencies of up to 89.8% and 92.7%, respectively (Wang et al., 2016). Other similar studies demonstrating the applicability of microalgae as a potential technology for simultaneous removal of nitrogen and phosphorus from wastewater have been reported in literature (Van Den Hende et al., 2014; Beuckels et al., 2015; Chiu et al., 2015; Ferrando and Matamoros, 2020).

Biosorption of Heavy Metals

The presence of heavy metals in water is a potential cause of serious health implications for human beings. Therefore, application of algae biomass can be considered as a possible solution for resolving current environmental problems, for example, remediation of heavy metals from water. A critique by Goswami et al. (2021) on bioremediation of heavy metals from wastewater observed that microalgae are prominent organisms that can be applied in heavy metal remotion from water. The researchers noted low heavy metal removal efficiency, which they considered to be the major downfall of microalgae. This limitation could be overcome by immobilization of cells, designing of biomass-based nanomaterials, development of consortia and maintaining of biotic and abiotic factors. However, more investigation and techniques are required to improve the removal efficiency and consumption of microalgae biomass for water purification application purposes to further contribute towards the development of a microalgae-based future.

The presence of various functional groups such as hydroxyl, carboxyl and amine groups, in all the marine macroalgae (MA) and freshwater microalgae (FA) biomass samples studied enabled efficient adsorption of Cu, Pb, Cd and Zn from industrial and synthetic water using the dried algae (Utomo et al., 2016). In this context, algae exhibit the potential for adoption in cleaning up surface water or post-treatment of wastewater; hence, minimize on the cost of eutrophication due to their biosorptive properties and fast adsorption capability. At a fixed equilibration time of 30 min, MA displayed a higher adsorption capacity for all metals compared to FA. The result ascertained that FA and MA had a higher ability in adsorbing a total metal of approximately 40 ppm from an industrial water, which is four times what was witnessed with synthetic water concentration at the same adsorbent dosage of 50 mg. Additionally, the adsorption capacities of FA for Zn were 6.87 mg/g and 7 mg/g for industrial water and synthetic water samples, correspondingly. In the case of MA, the amount of Zn adsorbed from industrial water and synthetic water samples were 7.17 mg/g and 8 mg/g, correspondingly. Notably, MA recorded higher adsorption for all the investigated metals as compared to FA. In terms of percentage adsorption, MA adsorbed 22%, 67%, 98%, and 39% of Cd, Cu, Pb and Zn, respectively, while FA adsorbed 18%, 29%, 94%, 37% of Cd, Cu, Pb and Zn, respectively.

A recent study by Ameri et al. (2020) on the application of algae as low cost and effective bio-adsorbent for removal of heavy metals from wastewater concluded that phytoremediation of wastewater and removal of metal ions from aqueous solutions using micro and macro-algae offers advantages over other methods such as coagulation/flocculation, ion exchange, membrane filtration, microfiltration, ultrafiltration among others. The study reported that properties such as low-cost raw material and cultivation, more environmentally friendly, high adsorbing capacity and metal ion uptake, high metal selectivity, no secondary pollution and especial mechanical properties for large scale production qualify algae to be a more promising and novel option for the removal of heavy metals. Reaction kinetics and isotherms were studied and in almost all cases, the pseudo second-order kinetic model whereas Langmuir and Freundlich isotherms were better harmonized with the investigational data, and the heavy metal adsorption on the algae displayed exothermic process.

Heavy metals such as Hg, Pb, Cd and Ni are considered the leading persistent inorganic pollutants. The aquatic environments are primarily contaminated with these heavy metals whose sources are mainly production and processing of metals, chemical industry, mineral industry, energy sector and management of waste and wastewater (Filote et al., 2020). Moreover, the removal of the aforementioned heavy metals from wastewater and contaminated aquatic environments requires effective solutions so that their effect on human health and the environment can be curbed in the long term. Therefore, non-conventional

Table 5. Heavy metals removal efficiencies of selected algal species

Heavy Metal	Algae type	Adsorption capacity (mg/g)	Percent (%) removal	Reference
Copper Zinc	*Chlorella* sp.	33.4 28.5	- -	Maznah et al. (2012)
Cadmium Copper Lead Zinc	Marine algae	- - - 7.17	22 67 98 39	Utomo et al. (2016)
Cadmium Copper Lead Zinc	Fresh water algae	- - - 6.87	18 29 94 37	Utomo et al. (2016)
Zinc Copper	*Chlorella vulgaris*	- -	94.1 81.7	Chan et al. (2014)
Lead Chromium Cadmium Copper	Micro-algae-Fe_2O_3	62.63 69.77 42.12 38.68	- - - -	Shen et al. (2020)

technologies based on bio-removal techniques are presently in the limelight for scientific research due to the low costs compared to the application of conventional technologies.

Ahmad et al. (2020) concluded that microalgae exhibit abundant substantial sequestering mechanism for heavy metal ions, and they, therefore, are demarcated as promising bio-sorbents. In addition, the application of microalgae demonstrated supremacy over various traditional and physiochemical methods culminating to their efficacy in large-scale remediation of wastewater.

Biomass-based adsorbents for the removal of heavy metals from wastewater has been shown to be a potential alternative water purification approach (Boukarma et al., 2022). Algae biomass-based adsorbents are alternative sources of recyclable biomass that can be adopted for adsorptive removal of heavy metal pollutants from wastewater, since they are rich in biochemical components, renewable availability, distinct properties, and high biosorption capacity. In addition, the surveyed literatures disclosed that marine algae can be applied for the abstraction of heavy metals in both dead and live forms. Nevertheless, in industrial applications, the non-living marine algae offer more practical bio-adsorbent resources for the abstraction of contaminants since toxicity of heavy metals and other contaminants do not affect dead algae biomass.

Table 5 represents a summary of studies on biosorption of heavy metals onto micro and macro-algae. This summary considers the type of algae, heavy metals removed, and the quantities of the heavy metals removed from the aqueous system.

Adsorption Mechanisms

As it has been shown, the migration and adsorption of pollutants onto algal-based adsorbents entails different mechanisms depending on the functional groups on the adsorbent surface and the textural characteristics (controlled by preparation method), and the type of pollutant. The process may be physical (physisorption), chemical (chemisorption) or both (Silva et al., 2019). In general, the adsorption process entails external diffusion of the adsorbate molecules towards the adsorbent surface against the boundary

Figure 2. Adsorption mechanisms of contaminants onto algal-based adsorbents (Lee et al., 2022)

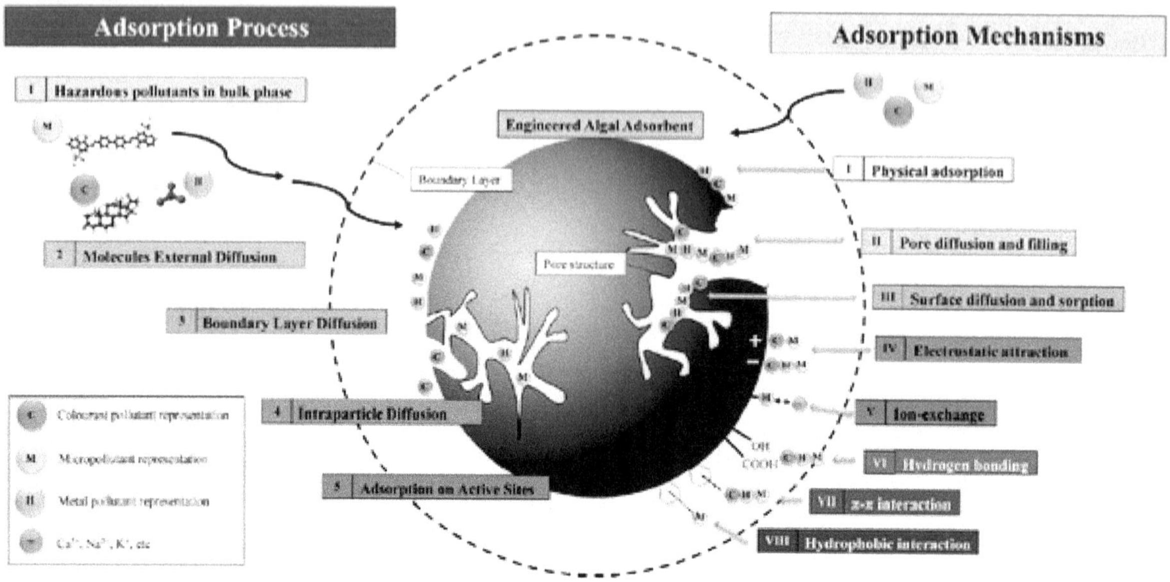

layer effect, attachment on the active sites on the adsorbent surface and pore diffusion into the inner pore structure of material. Figure 2 is a summary of the adsorption mechanisms involved in the sequestration of diverse pollutants using algal adsorbents.

CONCLUSION

Microalgae-based wastewater treatment processes are less energy-intensive, relative to conventional and advanced oxidation processes, since the energy required for algal growth is supplied through photosynthesis. The results show that microalgae offer an environmentally safe, low-cost, and efficient means for heavy metals, pharmaceuticals and personal care products (PPCPs), dyes and nutrient removal from water and can be used singly or amalgamated with other treatment methods. Additionally, the resultant byproduct, algal biomass, is a possible feedstock for biofuel production. However, there is paucity of data on life cycle assessment (LCA) of algal-based adsorbents, and adsorption of pollutants in a multi-component system. These represent possible areas of future research.

REFERENCES

Achieng', G. O., & Shikuku, V. O. (2020). Adsorption of copper ions from water onto fish scales derived biochar: Isothermal perspectives. *J. Mater. Environ. Sci.*, *11*, 1816–1827.

Ahmad, S., Pandey, A., Pathak, V. V., Tyagi, V. V., & Kothari, R. (2020). Phycoremediation: algae as eco-friendly tools for the removal of heavy metals from wastewaters. *Bioremediation of Industrial Waste for Environmental Safety,* 53-76.

Al-Amshawee, S., Yunus, M. Y. B. M., Azoddein, A. A. M., Hassell, D. G., Dakhil, I. H., & Hasan, H. A. (2020). Electrodialysis Desalination for Water and Wastewater: A Review. *Chemical Engineering Journal, 380*, 122231–122254.

Ameri, A., Tamjidi, S., Dehghankhalili, F., Farhadi, A., & Saati, M. A. (2020). Application of algae as low cost and effective bio-adsorbent for removal of heavy metals from wastewater: A review study. *Environmental Technology Reviews, 9*(1), 85–110.

Angelova, R., Baldikova, E., Pospiskova, K., Maderova, Z., Safarikova, M., & Safarik, I. (2016). Magnetically Modified *Sargassum Horneri* Biomass as an Adsorbent for Organic Dye Removal. *Journal of Cleaner Production, 137*, 189–194.

Angulo, E., Bula, L., Mercado, I., Montano, A., & Cubillan, N. (2018). Bioremediation of Cephalexin with non-living *Chlorella* sp., biomass after lipid extraction. *Bioresource Technology, 257*, 17–22.

Azam, R., Kothari, R., Singh, H. M., Ahmad, S., Ashokkumar, V., & Tyagi, V. V. (2020). Production of Algal Biomass for its Biochemical Profile Using Slaughterhouse Wastewater for Treatment under Axenic Conditions. *Bioresource Technology, 306*, 123116.

Bai, X., & Achrya, K. (2016). Removal of trimethoprim, sulfamethoxazole, and triclosan by the green alga *Nannochloris* sp. *Journal of Hazardous Materials, 315*, 70–75.

Bayazıt, G., Tastan, B. E., & Ülküye, D. G. (2019). Biosorption, Isotherm and Kinetic Properties of Common Textile Dye by *Phormidium Animale. Global NEST Journal, 21*, 1–7.

Beuckels, A., Smolders, E., & Muylaert, K. (2015). Nitrogen availability influences phosphorus removal in microalgae-based wastewater treatment. *Water Research, 77*, 98–106.

Boukarma, L., & … . (2022). Algal Biomass Valorization for the Removal of Heavy Metal Ions. In E. Lichtfouse, S. S. Muthu, & A. Khadir (Eds.), *Inorganic-Organic Composites for Water and Wastewater Treatment. Environmental Footprints and Eco-design of Products and Processes*. Springer.

Cardoso, N. F., Lima, E. C., Royer, B., Bach, M. V., Dotto, G. L., & Pinto, L. A. A. (2012). Comparison of *Spirulina Platensis* Microalgae and Commercial Activated Carbon as Adsorbents for the Removal of Reactive Red 120 Dye from Aqueous Effluents. *Journal of Hazardous Materials, 241-242*, 146–153.

Chan, A., Salsali, H., & Mcbean, E. (2014). Heavy metal removal (copper and zinc) in secondary effluent from wastewater treatment plants by microalgae. *ACS Sustainable Chemistry & Engineering, 2*, 130–137.

Chin, J. Y., Chng, L. M., Leong, S. S., Yeap, S. P., Yasin, N. H. M., & Toh, P. Y. (2020). Removal of Synthetic Dye by *Chlorella Vulgaris* Microalgae as Natural Adsorbent. *Arabian Journal for Science and Engineering, 45*, 7385–7395.

Chiu, S. Y., Kao, C. Y., Chen, T. Y., Chang, Y. B., Kuo, C. M., & Lin, C. S. (2015). Cultivation of microalgal *Chlorella* for biomass and lipid production using wastewater as nutrient resource. *Bioresource Technology, 184*, 179–189.

Daneshvar, E., Zarrinmehr, M. J., Hashtjin, A. M., Farhadian, O., & Bhatnagar, A. (2018). Versatile Applications of Freshwater and Marine Water Microalgae in Dairy Wastewater Treatment, Lipid Extraction and Tetracycline Biosorption. *Bioresource Technology, 268*, 523–530.

De Godos, I., Munoz, R., & Guieysse, B. (2012). Tetracycline removal during wastewater treatment in high-rate algal ponds. *Journal of Hazardous Materials*, *229*, 446–449.

De Wilt, A., Butkovskyi, A., Tuantet, K., Leal, L. H., Fernandes, T. V., Langenhoff, A., & Zeeman, G. (2016). Micropollutant removal in an algal treatment system fed with source separated wastewater streams. *Journal of Hazardous Materials*, *304*, 84–92.

Ding, T., Yang, M., Zhang, J., Yang, B., Lin, K., Li, J., & Gan, J. (2017). Toxicity, degradation and metabolic fate of ibuprofen on freshwater diatom *Navicula* sp. *Journal of Hazardous Materials*, *330*, 127–134.

Dzoujo, T. H., Tome, S., Shikuku, V. O., Tchuigwa, J. T., Spieß, A., Janiak, C., Etoh, A., & Dina, D. (2021). Enhanced performance of hydrogen peroxide modified pozzolan-based geopolymer for abatement of methylene blue from aqueous medium. *Silicon*. doi:10.1007/s12633-021-01264-4

Elgarahy, A. M., Elwakeel, K. Z., Elshoubaky, G. A., & Mohammad, S. H. (2019). Microwave-accelerated Sorption of Cationic Dyes onto green marine Algal Biomass. *Environmental Science and Pollution Research International*, *26*, 22704–22722.

Escapa, C., Coimbra, R. N., Paniagua, S., Garcıa, A. I., & Otero, M. (2016). Comparative assessment of diclofenac removal from water by different microalgae strains. *Algal Research*, *18*, 127–134.

Escapa, C., Coimbra, R. N., Paniagua, S., García, A. I., & Otero, M. (2017). Paracetamol and salicylic acid removal from contaminated water by microalgae. *Journal of Environmental Management*, *203*, 799–806.

Essekri, A., Hsini, A., Naciri, Y., Laabd, M., Ajmal, Z., & El Ouardi, M. (2021). Novel Citric Acid-Functionalized Brown Algae with a High Removal Efficiency of crystal Violet Dye from Colored Wastewaters: Insights into Equilibrium, Adsorption Mechanism, and Reusability. *International Journal of Phytoremediation*, *23*, 336–346.

Ferrando, L., & Matamoros, V. (2020). Attenuation of nitrates, antibiotics and pesticides from groundwater using immobilised microalgae-based systems. *The Science of the Total Environment*, *703*, 134740.

Filote, C., Roşca, M., & Hlihor, R. M. (2020). Overview of using living and non-living microorganisms for the removal of heavy metals from wastewaters. *Res. J. Agr. Sci., 52*(4).

Fomina, M., & Gadd, G. M. (2014). Biosorption: Current perspectives on concept, definition and application. *Bioresource Technology*, *160*, 3–14.

Garcia-Rodríguez, A., Matamoros, V., Fontàs, C., & Salvadó, V. (2013). The influence of light exposure, water quality and vegetation on the removal of sulfonamides and tetracyclines: A laboratory-scale study. *Chemosphere*, *90*, 2297–2302.

Goswami, R. K., Agrawal, K., Shah, M. P., & Verma, P. (2021). Bioremediation of heavy metals from wastewater: A current perspective on microalgae-based future. *Letters in Applied Microbiology*, 17.

Guarín, J. R., Moreno-Pirajan, J. C., & Giraldo, L. (2018). Kinetic Study of the Bioadsorption of Methylene Blue on the Surface of the Biomass Obtained from the Algae D. antarctica. *Journal of Chemistry*, *2018*, 1–12.

Gül, Ü. D., Taştan, B. E., & Bayazıt, G. (2019). Assessment of Algal Biomasses Having Different Cell Structures for Biosorption Properties of Acid Red P-2BX Dye. *South African Journal of Botany*, *127*, 147–152.

Gunasundari, E., Senthil Kumar, P., Rajamohan, N., & Vellaichamy, P. (2020). Feasibility of Naphthol green-b Dye Adsorption Using Microalgae: Thermodynamic and Kinetic Analysis. *Dwt*, *192*, 358–370.

Guo, W. Q., Zheng, H. S., Li, S., Du, J. S., Feng, X. C., Yin, R. L., Wu, Q. L., Rcn, N. Q., & Chang, J. S. (2016). Removal of cephalosporin antibiotics 7-ACA from wastewater during the cultivation of lipid-accumulating microalgae. *Bioresource Technology*, *221*, 284–290.

Hom-Diaz, A., Llorca, M., Rodriguez-Mozaz, S., Vicent, T., Barcelo, D., & Blanquez, P. (2015). Microalgae cultivation on wastewater digestate: Β-estradiol and 17α-ethynylestradiol degradation and transformation products identification. *Journal of Environmental Management*, *155*, 106–113.

Hom-Diaz, A., Norvill, Z. N., Blanquez, P., Vicent, T., & Guieysse, B. (2017). Ciprofloxacin removal during secondary domestic wastewater treatment in high rate algal ponds. *Chemosphere*, *180*, 33–41.

Kang, D., Zhao, Q., Wu, Y., Wu, C., & Wu, X. (2018). Removal of nutrients and pharmaceuticals and personal care products from wastewater using periphyton photobioreactors. *Bioresource Technology*, *248*, 113–119.

Kausar, R. A., & Buhaniand, S. (2020). Methylene Blue Adsorption Isotherm on *Spirulina* sp. Microalgae Biomass Coated by Silica-Magnetite. *IOP Conf. Ser. Mater. Sci. Eng.*, *857*, 012019–12028.

Kimosop, S. J., Getenga, Z. M., & Orata, F. (2016). Residue levels and discharge loads of antibiotics in wastewater treatment plants (WWTPs), hospital lagoons, and rivers within Lake Victoria Basin, Kenya. *Environmental Monitoring and Assessment*, *188*, 532.

Lee, X. J., Ong, H. C., Ooi, J., Yu, K. L., Tham, T. C., Chen, W.-H., & Ok, Y. S. (2022). Engineered macroalgal and microalgal adsorbents: Synthesis routes and adsorptive performance on hazardous water contaminants. *Journal of Hazardous Materials*, *423*, 126921.

Maes, H. M., Maletz, S. X., Ratte, H. T., Hollender, J., & Schaeffer, A. (2014). Uptake, elimination, and biotransformation of 17α-ethinylestradiol by the freshwater alga *Desmodesmus subspicatus*. *Environmental Science & Technology*, *48*, 12354–12361.

Maznah, ZBS, & Ismail, M, Halimah. (2012). Fate of thiram in an oil palm nursery during the wet season. *Journal of Oil Palm Research*, *24*, 1397–1403.

Mollamohammada, S., Aly Hassan, A., & Dahab, M. (2020). Nitrate Removal from Groundwater Using Immobilized Heterotrophic Algae. *Water, Air, and Soil Pollution*, *231*, 26.

Muriuki, C., Kairigo, P., Home, P., Ngumba, E., Raude, J., Gachanja, A., & Tuhkanen, T. (2020). Mass loading, distribution, and removal of antibiotics and antiretroviral drugs in selected wastewater treatment plants in Kenya. *The Science of the Total Environment*, *140655*(743). https://doi.org/10.1016/j.scitotenv.2020.140655

Ng'eno, E., Orata, F., Lilechi, D. B., Shikuku, V. O., & Kimosop, S. (2016). Adsorption of caffeine and ciprofloxacin onto pyrolytically derived water hyacinth biochar: Isothermal, kinetics and thermodynamics. *Journal of Chemistry and Chemical Engineering*, *10*, 185–194.

Ng'eno, E., Shikuku, V. O., Orata, F., Lilechi, D. B., & Kimosop, S. (2019). Caffeine and Ciprofloxacin Adsorption from water onto clinoptilolite: Linear isotherms, kinetics, thermodynamics and mechanistic studies. *South African Journal of Chemistry*, *72*, 136–142.

Peng, F. Q., Ying, G. G., Yang, B., Liu, S., Lai, H. J., Liu, Y. S., Chen, Z. F., & Zhou, G. J. (2014). Biotransformation of progesterone and norgestrel by two freshwater microalgae (*Scenedesmus obliquus* and *Chlorella pyrenoidosa*): Transformation kinetics and products identification. *Chemosphere*, *95*, 581–588.

Radjenović, J., Petrović, M., & Barceló, D. (2009). Fate and distribution of pharmaceuticals in wastewater and sewage sludge of the conventional activated sludge (CAS) and advanced membrane bioreactor (MBR) treatment. *Water Research*, *43*, 831–841.

Saravanan, A., Kumar, P. S., Govarthanan, M., George, C. S., Vaishnavi, S., Moulishwaran, B., & … . (2021). Adsorption Characteristics of Magnetic Nanoparticles Coated Mixed Fungal Biomass for Toxic Cr(VI) Ions in Aquatic Environment. *Chemosphere*, *267*, 129226.

Shen, L., Wang, J., Li, Z., Fan, L., Chen, R., Wu, X., Li, J., & Zeng, W. (2020). A high-efficiency Fe_2O_3@ microalgae composite for heavy metal removal from aqueous solution. *Journal of Water Process Engineering*, *33*, 101026.

Shikuku, V. O. (2020). *Effects of Emerging Chemical Contaminants on Water Resources and Environmental Health*. IGI Global Publishers.

Shikuku, V. O., & Mishra, T. (2021). Adsorption isotherm modeling for methylene blue removal onto magnetic kaolinite clay: A comparison of two-parameter isotherms. *Applied Water Science*, *11*, 103.

Shikuku, V. O., Zanella, R., Kowenje, C. O., Donato, F. F., Bandeira, N., & Prestes, O. D. (2018). Single and binary adsorption of sulfonamide antibiotics onto iron-modified clay: Linear and nonlinear isotherms, kinetics, thermodynamics, and mechanistic studies. *Applied Water Science*, *8*, 175.

Singh, S., Kumar, V., Datta, S., Dhanjal, D. S., Sharma, K., & Samuel, J. (2020). Current Advancement and Future prospect of Biosorbents for Bioremediation. *The Science of the Total Environment*, *709*, 135895.

Tadkaew, N., Hai, F. I., McDonald, J. A., Khan, S. J., & Nghiem, L. D. (2011). Removal of trace organics by MBR treatment: The role of molecular properties. *Water Research*, *45*, 2439–2451.

Ternes, T. A., Herrmann, N., Bonerz, M., Knacker, T., Siegrist, H., & Joss, A. (2004). A rapid method to measure the solid–water distribution coefficient (K_d) for pharmaceuticals and musk fragrances in sewage sludge. *Water Research*, *38*, 4075–4084.

Tome, S., Dzoujo, H., Shikuku, V., & Otieno, S. (2021). Synthesis, characterization and application of acid and alkaline activated volcanic ash-based geopolymers for adsorptive remotion of cationic and anionic dyes from water. *Ceramics International*, *47*(15), 20965–20973.

Utomo, H. D., Tan, K. X. D., Choong, Z. Y. D., Yu, J. J., Ong, J. J., & Lim, Z. B. (2016). Biosorption of heavy metal by algae biomass in surface water. *Journal of Environmental Protection*, *7*(11), 1547–1560.

Van Den Hende, S., Beelen, V., Bore, G., Boon, N., & Vervaeren, H. (2014). Up-scaling aquaculture wastewater treatment by microalgal bacterial flocs: From lab reactors to an outdoor raceway pond. *Bioresource Technology*, *159*, 342–354.

Varghese, A. G., Paul, S. A., & Latha, M. S. (2019). Remediation of Heavy Metals and Dyes from Wastewater Using Cellulose-Based Adsorbents. *Environmental Chemistry Letters*, *17*, 867–877.

Wang, S., Wang, X., Poon, K., Wang, Y., Li, S., Liu, H., Lin, S., & Cai, Z. (2013). Removal and reductive dechlorination of triclosan by *Chlorella pyrenoidosa*. *Chemosphere*, *92*, 1498–1505.

Wang, X., Wu, Y., Zhang, T., Xu, X., Dao, G., & Hu, H. (2016). Simultaneous nitrogen, phosphorous, and hardness removal from reverse osmosis concentrate by microalgae cultivation. *Water Research*, *94*, 215–224.

Xiong, J. Q., Kurade, M. B., Abou-Shanab, R. A. I., Ji, M. K., Choi, J., Kim, J. O., & Jeon, B. H. (2016). Biodegradation of carbamazepine using freshwater microalgae *Chlamydomonas mexicana* and *Scenedesmus obliquus* and the determination of its metabolic fate. *Bioresource Technology*, *205*, 183–190.

Xiong, J. Q., Kurade, M. B., & Jeon, B. H. (2017). Ecotoxicological effects of enrofloxacin and its removal by monoculture of microalgal species and their consortium. *Environ. Pollut.*, *226*, 486–493.

Xue, W., Wu, C., Xiao, K., Huang, X., Zhou, H., Tsuno, H., & Tanaka, H. (2010). Elimination and fate of selected micro-organic pollutants in a full-scale anaerobic/anoxic/aerobic process combined with membrane bioreactor for municipal wastewater reclamation. *Water Research*, *44*, 5999–6010.

Yu, K. L., Lee, X. J., Ong, H. C., Chen, W.-H., Chang, J.-S., & Lin, C.-S. (2021). Adsorptive Removal of Cationic Methylene Blue and Anionic Congo Red Dyes Using Wet-Torrefied Microalgal Biochar: Equilibrium, Kinetic and Mechanism Modeling. *Environ. Pollut.*, *272*, 115986.

Zhang, Y., Habteselassie, M. Y., Resurreccion, E. P., Mantripragada, V., Peng, S., Bauer, S., & Colosi, L. M. (2014). Evaluating removal of steroid estrogens by a model alga as a possible sustainability benefit of hypothetical integrated algae cultivation and wastewater treatment systems. *ACS Sustainable Chemistry & Engineering*, *2*, 2544–2553.

Zheng, H., Guo, W., Li, S., Chen, Y., Wu, Q., Feng, X., Yin, R., Ho, S. H., Ren, N., & Chang, J. S. (2017). Adsorption of p-nitrophenols (PNP) on microalgal biochar: Analysis of high adsorption capacity and mechanism. *Bioresource Technology*, *244*, 1456–1464.

Zhou, G. J., Ying, G. G., Liu, S., Zhou, L. J., Chen, Z. F., & Peng, F. Q. (2014). Simultaneous removal of inorganic and organic compounds in wastewater by freshwater green microalgae. *Environmental Science. Processes & Impacts*, *16*, 2018–2027.

Chapter 23
Role of Algae in Cancer

Fazrena Nadia Md Akhir
Malaysia-Japan International Institute of Technology, Universiti Teknologi Malaysia, Malaysia

Farhah Hanani Mohd Nazaruddin
Malaysia-Japan International Institute of Technology, Universiti Teknologi Malaysia, Malaysia

Nor'azizi Othman
Malaysia-Japan International Institute of Technology, Universiti Teknologi Malaysia, Malaysia

Hirofumi Hara
The University of Tokyo, Japan

Imran Ahmad
https://orcid.org/0000-0001-9774-9111
Malaysia-Japan International Institute of Technology, Universiti Teknologi Malaysia, Malaysia

ABSTRACT

Cancer is ranked as the second most common cause of death worldwide and searching new therapeutic agents for cancer treatment remains a major challenge. Despite the remarkable developments in cancer therapy in past decades, there is still an insistent necessity for innovative drugs in cancer biology, particularly in the under-explored area of marine anticancer compounds. Algae are photosynthetic organisms consisting of a total of 30,000 species that thrive in a watery environment. The identification of novel natural products and metabolites extracted from algae with anticancer potential is a major step forward in cancer therapeutic studies. Considering the huge potential for developing innovative drugs from natural compounds derived from marine algae, only a few substances have been used in cancer therapy. In this review, the authors discussed the potential antitumor effect of various species of algae for future applications in pharmaceutical industries.

INTRODUCTION

Cancer is a common medical problem that needs an effective treatment procedure. The failure of the cell

DOI: 10.4018/978-1-6684-2438-4.ch023

cycle is a familiar factor of cancer induction. The basic accessible remedial method for cancer medical care is chemotherapy and radiation. However, significant side effects have been recorded in conjunction with such therapies; for example, radiation has a bad impact on the immune system of patients. Chemotherapy is frequently the first treatment option for cancer, as these medications can kill or slow the development of cancerous cells. The noxiousness of these medications is linked, and it can change from a minor response to a serious life-threatening sickness. By offering innovative remedial treatments, these adverse effects might be reduced.

Anti-cancer properties have been found for complementary and alternative medicine substances derived from fresh or marine flora, especially micro and macroalgae. As a result, a huge microalgae diversity and a broad variety of secondary metabolites were related with biochemically and ecologically important variations. Experts are therefore continually mining bioactive components in various areas of the world from marine microalgae to evaluate pharmaceutical and medicine properties. Marine microalgae, therefore, stressed that study into their natural compounds helps to heal and alleviate human sickness.

Aquatic and photoautotrophic eukaryotes are the core characteristics of the algae. All types are photosynthesized in relation to other phototrophic eukaryotes and have a fairly simple anatomical effect. Yet, its design might range from one to multicellular to colonial. Most species of algal are single-celled. Multicellular forms are arranged in sections with different functions and have a fairly complicated structure. Algae, or photosynthetic organisms, thrive in a watery environment. Algae may grow alone or in conjunction with other species, and they can also withstand a wide range of unfavorable environmental circumstances (Barsanti et al., 2008).

Red algae, green algae, blue-green algae, and brown algae are the major types of algae, whereas macroalgae and microalgae are the two sizes of algae. Macroalgae such as seaweed are big in size multicellular algae that are visible to the human eye, whereas microalgae are tiny single-cell algae that are visible using microscopes such as prokaryotic or eukaryotic. Studies that have been conducted have been demonstrated the benefit of applying phytochemical and dietary antioxidants from various sources in alleviating chemo and radiation therapy-induced toxicities and improving therapeutic efficacy. According to Ferdous and Yusof (2021), both the micro and macro algae might be thought of as alternate natural sources of antioxidants in this perspective. Algae contain antioxidants from several groups that could be used in the field of medicines.

BACKGROUND

With a frequency of 90%, marine macroalgae, often known as seaweeds, is the most common in marine flora. The behavior of antioxidant phytochemical elements in Seaweed *Chaetomorpha sp.* extracts has gotten a lot of interest because of their potential function in illness prevention. Seaweeds also account for half of the global photosynthesis quota (Dhargalkar & Pereira, 2005). In China, seaweeds have been utilized for medicinal purposes for over 2000 years (Chengkui & Junfu, 1984). Seaweeds have captivated the interest of many researchers in the last three decades because of their capabilities to fight various diseases including gallstones, stomach ailments, eczema, cancer, renal disorders, scabies, psoriasis, asthma, arteriosclerosis, heart disease, lung diseases, and ulcers (Saadaoui et al., 2020).

For drug development, the medicinal potential of such compounds was thoroughly explored. They are a potential source of biologically active compounds, including proteins, lipids, and polyphenols with strong antibacterial properties, anti-cancer, antioxidants, antimicrobials, and antivirals (Kumar et al.,

2008; Holdt & Kraan, 2011; Ashwini et al., 2016). Hundreds of potential anti-tumor agents, in particular marine-based algae, have been isolated. In current years, several authors have been reporting the isolation of cytotoxic antitumor substances from marine organisms.

For thousands of years, nature has provided therapeutic agents, and a huge percentage of contemporary medicines can be extracted from microbes, many of which are based on their traditional usage. Plant sources have yielded a diverse range of bioactive components. Marine biotechnology is the study of using marine creatures in whole or in part to create or improve goods, improve plants or animals, or generate microorganisms for specialized applications. Some of them are presently undergoing clinical trials, preclinical studies, or more research. Marine microalgae anti-cancer chemicals have been under-examined.

Most investigations have been conducted utilizing low-resolution techniques like liquid-liquid partitioning or solid-phase extractions on microalgal extracts or fractions. Dereplication methods, fractionation based on high-performance procedures, or a thorough structural explanation of the chemicals discovered are rare. Despite the fact that aquatic compounds are outnumbered in the present pharmacopeia, the marine habitat is expected to become an essential point of supply of new compounds, since it makes up 95 percent of the biosphere (Jimeno et al., 2004).

PROBLEM IN CANCER TREATMENT

Ideally, anti-cancer medicines should operate entirely against tumor cells, however, several chemical therapies used for cancer patients now have substantial adverse effects, for example, cause bleeding, hair loss, diarrhea, and immunosuppression, on the human body (Kranz & Dobbelstein, 2012). Therefore, the identification of novel natural products and metabolites extracted from the use of tumor cells by microbes, animals, and plants without the use of toxins in normal cells is a major step forward in scientific studies.

Apoptosis has become an issue of considerable interest in cancer treatment and oncology, as well as a highly controlled programmed cell death, due to the high potential of several chemical therapies triggering apoptosis in various cancer cells. Therefore, the testing of natural compounds that may induce apoptosis in cancer cells which can be employed alone or with additional chemotherapeutic medicines is presently underway to increase therapeutic efficacy and minimize the adverse events in cancer treatment.

POTENTIAL ALGAE IN CANCER TREATMENT

Survivors of cancer frequently take supplements of vitamins or minerals, natural herbal items, and medications in order to ease the adverse effects of therapy. Vitamins, polyphenols, and carotenoids have been the most commonly prescribed antioxidants.

Microalgae might be an outstanding alternative chemical producer and it has frequently been regarded as a mother lode of pharmaceutically significant metabolites such as carotenoids, polyphenols, fatty acids, and phycobiliproteins, vitamins, resulting from microalgae' defensive mechanisms against stress conditions (Chu, 2013). These biologically active compounds have demonstrated their antioxidant capacity, including their anticancer properties both *in-vitro* and *in-vivo*. Microalgal tetraterpenoids are rich in antioxidants and have demonstrated potential anticancer action in several cell lines (Ferdous & Yusof, 2021). Furthermore, seaweeds also are an excellent source of antioxidant compounds. According to Gupta et al. (2011), fucoidans, phlorotannin, laminarin, and terpenoids are among the bioactive compounds being investigated for their antioxidant properties.

According to Ashwini et al. (2016), seaweed extracts exhibited anti-cancer activity of 24, 48, and 72 hours in which *G. corticata* chloroform extract showed higher IC_{50} levels of 341.82 μg/ml and ethanol extract used is 244.7 μg/ml for 48 hours respectively in their experimental. The two-way analysis of variance (ANOVA) detects the p-values. The structure of the treated cells differed significantly from that of the control cells. The incidence of cancer is significantly connected to oxidative stress.

Many investigations on *in vivo* and *in vitro* have shown that exogenous antioxidant treatment can inhibit free production and destruction of DNA as well as proteins, reducing cancer risk (Bonner & Arbiser, 2014; Ozben, 2015). Therefore, the in-vitro test shows that the seaweed extracts are an important source of a noble anticancer agent. The study of natural product cancer therapies has therefore steadily increased in recent years. Different cancer cell lines reported bioactive substances in marine plants (Martins et al., 2018). The prospection for the use only or with existent chemotherapy of natural antioxidants is an effective method to battle tumor development.

Red Algae

Red edible algae were seen as healthful and good nutrition. The production of red seaweed has increased substantially since the early 20th century as demand for food and industry has continued to develop. Bioactive chemicals of seaweed are produced according to the stage of algae development and capacity for interactions with changes in the environment. The principal contributions of algae in varying kinds and amounts in different species include phycobiliproteins, carotenoids, pigments, terpenes, polyphenols, phlorotannins, and polysaccharides (de Souza et al., 2012; Atashrazm et al., 2015; Lange et al., 2015). Hence, terpenes, polysaccharides, and polyphenols have an important function in the action against the disease. Porphyran is a typical *Porphyra* polysaccharide also known as red seaweed. A variety of procedures have been employed for porphyrans extraction from red seaweed, including hot water extraction, radical degradation, and ultrasonic treatment. Hypolipidemic, cancer, and anti-inflammatory porphyrans in humans have been documented (Liu et al., 2019). Table 1 presents the action of anti-cancer and potential porphyran mechanism.

Green Algae

Some researchers such as Haq et al. (2019) were aimed at characterizing and recognizing *Chaetomorpha sp.*, an active ingredient in green algae. This study also examined this macroalgae's antioxidant and cancer capabilities. In terms of its therapeutic and medical impact, the existence of silicon and oxygen as the key elements for these algae makes it rare. Oxygen-bonded silicone may be soluble and absorbed into the water, readily biologically accessible to people. Direct participation of silicone in decreased metal accumulations in Alzheimer's disease, the health of the immune system, and a reduced risk of atherosclerosis have been reported (Martin, 2007; Martin, 2013).

The presence in *Chaetomorpha sp.* of radioactive elements, like titanium and radium, is remarkable because of its medicinal ability, in particular during treatment with cancer. *Chaetomorpha sp.* is high in polyphenolic compounds and exists as the main antioxidant, helping the algae withstand oxidative stress, are hydrophilic polyphenolic chemicals such as phlorotannins that are naturally bipolar (Senthilkumar & Sudha, 2012; Farasat et al., 2013; Wu et al., 2016). From Haq et al. (2019) experiments, the total

Table 1. The activity of anti-cancer and potential porphyran mechanism.

Findings	Mechanism	References
● Red algae Porphyra yezoensis was reported to have considerable efficacy against Ehrlich cancer (53.2%). ● Four red algae demonstrated notable Meth-A fibrosarcoma antitumor activity.	-	Noda et al. (1990)
● AGS human gastric cancer cells were used to assess the apoptotic potential actions of porphyran. ● Additional porphyran of 0.1% significantly decreased DNA synthesis after a day of exposure. ● Showing that porphyran inhibited growth in cancer cells by reducing cell proliferation as well as causing apoptosis.	Negative regulation of IGF-IR phosphorylation and activation of caspase-3	Kwon & Nam (2006)
● Different quantities of pure porphyran were grown in HT-29 colon cancer cells and AGS gastric cancer cells. ● The part of polysaccharide in the porphyran preparation is the most efficient inhibition of the growth of the cell through apoptosis instead of the protein component. ● The results show a strong antiproliferative effect in vitro ($p<0.05$) of pure porphyran.	Increased activity of caspase-3.	Kwon & Nam (2007)
● Hep-2 and A375 cancer cell lines were given varying doses of phycocyanin before being irradiated with a He-Ne laser at a density of 20 J/cm2 and a wavelength of 632 nm. ● At 100 g/ml and 165 g/ml, respectively, MTT tests revealed a minimum survival rate of 29.8% (Hep-2) and 16.2% (A375). ● At the G0/G1 or G2/M checkpoints, antiproliferation and cell cycle arrest occur.	Phycocyanin binds to the tumor cells membrane with the mitogenic receptor and stimulates surface production of CD59, triggering cell apoptotic signal transduction.	Zhang et al. (2011)
● Cell cycle and antiproliferation in the G2/M phase are inhibited. ● Three porphyrans demonstrated no normal cell toxicity to humans (HL-7702) and anticancer effects were seen on Hep3B, HeLa, and MDA-MB-231. ● In relation to HeLa cells, porphyran molecular weight and solution conformation might have distinct anticancer effects.	P21 and p53 upregulation, whereas Cyclin B1 and CDK1 are negatively regulated.	He et al. (2019)

free 2, 2-Diphenyl-1-Picrylhydrazyl (DPPH) testing showed the antioxidant activity in ethanol extracts with a value of IC_{50} 9,41 ± 0,54 mg/mL and aqueous extract with an activity of IC_{50} 15,44 ± 0.98 mg/mL. These outcomes are supported by earlier research which indicated that ethanol solvents had greater antioxidant effects.

Next, a hot water extract of *Capsosiphon fulvescens* containing polysaccharides promoted apoptosis in gastric cancer cells via the PI3K/Akt pathway among green algae. According to the experiment from Kwon & Nam (2007), *C. fulvescens* PS (Cf-PS) substantially reduced the growth of cultured human cancer cells in a dose-dependent manner, according to MTS assays. Caspase-3 activation was significantly increased in Cf-PS-treated AGS cells, whereas Bcl-2 expression was decreased. Furthermore, phosphorylation of the insulin-like growth factor-I receptor (IGF-IR) was reduced in Cf-PS-treated AGS cells compared to non-treated control cells, indicating PI3-kinase (PI3K)/Akt activation. Cf-PS reduces cell proliferation and promotes apoptosis via decreasing IGF-IR signaling and the PI3K/Akt pathway, according to these findings. Dimethylsulfoniopropionate, a tertiary sulfonium metabolite present in green algae and other algal species, inhibited the growth of Ehrlich ascites carcinoma in mice.

Blue-Green Algae

Blue-green algae or also known as cyanobacteria emerged in fossil history more than a billion years past and are some of the earliest living organisms in the world (Leclerc, 2021). However, these photosynthetic microorganisms are more prevalent than their long background. One species generates a chemical molecule which is a possible new chemotherapeutic medicine for future investigation. Cyanobacteria are unicellular organisms that live all over the earth on land and in the aquatic. However, these basic beings have intricate connections to the environment around them. No claws, teeth, or a menacing appearance for defense are available in cyanobacteria. Therefore, chemical substances are used to defend against predators (Hall, 2021). They also assist the bacteria to interact with their chemical products.

Although, identifying a novel medication such as Gatorbulin-1 (GB1) and knowing about it may be quite a long process it does not take the extra time to test it then in order to make it into an authorized and safe drug. GB1 is combined with a number of other previously identified chemicals to inhibit microtubular development, several of which are already utilized as effective cancer therapies (Hall, 2021). This novel chemical will be hoped for in more effective medicines that will inhibit tumor development in people with lower side effects.

Microalgae are eukaryotic plants that account for forty percent of world production. They have naturally developed a complex system of oxygen photosynthesis with the potential to generate extremely efficient oxygen. Its benefit in the development of marine medicines is its metabolism, which can result in the creation of different biotechnology industries of diverse chemicals and potential uses. In many applications, microalgae have been utilized, including biofuels, nutrition, food, feed, bio-fertilizers, contamination control, bioremediation, and bioactive high-value chemicals that leverage their photosynthetic capacity (Chisti, 2007; Georgianna & Mayfield, 2012; Pierobon et al., 2018). They can be cultivated simply in photo-bioreactors to provide immense biomass and a renewable resource that is yet not well studied. They use solar energy and redefine carbon dioxide that mitigates the impacts of greenhouse gases and eliminates nitrogen and phosphate derivatives that may be contaminants in their respective concentration. Table 2 shows active microalgal species that have been proved to be successful in the prevention of human cancers.

According to Liu and Chen (2014), *Chlorella vulgaris* is a single alga that photosynthesis can create oxygen. Algae were previously employed as a research model to discover novel food supplies and, more recently, to boost the production of biofuels utilizing carbon dioxide as a carbon source due to their numerous sources and affordable pricing, and uniform nature. It is important to note that Chlorella is able to reduce endotoxemia and improve host protection against peritonitis without adverse effects during digestive disorders. *C. vulgaris* also has a significant amount of chlorophyll which absorbs light across a wide range of wavelengths and hence permits photosynthesis in a variety of wavelengths. This function may be utilized to create ROS under 650-nm irradiation (Song et al., 2014; Chu et al., 2014). However, it has never been discovered whether or not algae is being used as an oxygen supply in the treatment of hypoxic malignancies in conjunction with radiation and phototherapy.

Brown Algae

Moreover, the detailed finding of the raw extracts of several brown algae obtained against cancer cell lines from varied marine habitats indicates remarkable anti-cancer potential. The crude methanol extract from the sea algae obtained from the Aegean shores of Turkey showed a high inhibitory effect of 90%

Table 2. Active microalgae species to prevent human cancers.

Microalgae	Findings	Target Cells	References
Chlorella vulgaris & *Chlorella ellipsoidea*	● Both are semipurified extracts ● Inhibited cell growth HCT116 with IC_{50} values are 40.73 ± 3.71 μg/mL (*C.Vulgaris*) and 40.31 ± 4.43 μg/mL (*C. ellipsoidea*) ● *C. ellipsoidea* extracts generated 2.5 greater than *C.Vulgaris* extracts.	Colon carcinoma (HCT-116)	Cha et al. (2008)
Dunaliella tertiolecta (DT)	● The DT dichloromethane extract has demonstrated significant anti-proliferation activities in MCF-7 with violaxanthin. ● 40 μg/mL is used to observe the cytostatic activity.	Breast adenocarcinoma (MCF-7)	Pasquet et al. (2011)
Cocconeis scutellum	● More than 89.2% of fraction 3 at 1 μg/well. ● The viability of cells reduced in BT20. ● Caspase-8 and 3 activated. ● Cell cycle progression from S to G2-M phase stopped.	Breast carcinoma (BT20)	Nappo et al. (2012)
Chaetoceros calcitrans	● Inhibited growth of cells MDA-MB-231, IC_{50} is 60 μg/mL. ● B-cell leukemia/lymphoma 2(Bcl-2) is increased and regulated. Caspase-4. ● Inducing apoptosis in MDA-MB-231 cells during the arrest of the S and G2/M cell cycles produced by Bcl-2.	MCF-7 Breast adenocarcinoma (MDA-MB-231)	Goh et al. (2014)
Navicula incerta	● Cell viability of 54% with 8.25 μg/mL. ● Induction of apoptosis caspase-8, 9 have been activated.	Hepatocarcimona (HepG2)	Kim et al. (2014)
Phaeodactylum tricornutum	● The IC_{50} value was 65.15 μM. ● One of the primary apoptosis executors is caspase-3. ● The dose-dependent therapy of the anticancer drug mediates apoptosis through an apoptotic pathway p53 to control the development of HL60 cells.	Human promyelocytic leukemia (HL-60)	Samarakoon et al. (2014)

of brown algae *Padina pavonica* and *Cystoseira mediterranea* against MCF-7 and human prostate cells (Taskin et al., 2010). According to Athukorala et al. (2006), antiproliferative efficacy against murine colon cell line cancer (CT-26), human leukemia (THP-1), mouse melanoma (B-16), and human leukemia cells (U-937) were shown in *in vitro* on *Ecklonia cava* enzymatic extract that has been combined with its crude polyphenolic and polysaccharide fractions. Polyphenol extract has demonstrated the highest activity which showed the IC_{50} is 5.1 μg/mL in the nuclear staining test against cell CT-26 by apoptotic cell disappearance.

Go et al. (2010) discovered a new glycoprotein from the brown algae *Laminaria japonica* (LJGP) that inhibits the growth of HT-29 colon cancer cells. LJGP reduced the proliferation of many cancer cell lines, including AGS, HepG2, and HT-29, in a dose-dependent manner, according to MTS tests. These findings imply that LJGP suppresses HT-29 cell growth by causing apoptosis, which might be

Table 3. Anticancer activity of fucoidans isolated in brown algae.

Brown Algae	Findings	References
Undaria pinnatifida	● In P-388 tumor-bearing mice and T cell-mediated NK cells in normal mice. ● Enhanced cytolytic activity of NK cells. ● The amount of IFN-gamma increased.	Maruyama et al. (2003)
Cladosiphon novae-caledoniae	● Human uterine carcinoma HeLa cells. ● Showed anti-angiogenic activity.	Ye et al. (2005)
Cladosiphon okamuranus	● Growth inhibitory activity on stomach cancer cells. ● Inhibitory effect on anticancer agent 5-fluorouracil (5-FU).	Kawamoto et al. (2006)
Fucus vesiculosus	● In human colon carcinoma cells (HCT-15). ● Bcl-2 expression decreased, Bax expression increased, active forms of caspase-9 and caspase-3 were increased. ● Induction of apoptosis by strong activation of ERK and p38 kinase.	Hyun et al. (2009)
Fucus vesiculosus	● In human HT-29 and HCT-116 human colon cancer cells. ● Levels of caspases-8, -9, -7, and -3 were increased. ● Caspase-8 inhibitor inhibited the fucoidan-induced.	Kim et al. (2010)

mediated by a variety of mechanisms, including the Fas signaling system, the mitochondrial pathway, and cell cycle arrest. As a result, LJGP may be a potential therapeutic option for colon cancer in people.

Next, Costa et al. (2011) discovered five heterofucans from *Sargassum filipendula* using proteolytic digestion and successive acetone precipitation. Fucose, glucose, glucuronic acid, galactose, and sulfate are the primary components of these heterofucans. These findings clearly show that *S. filipendula* polysaccharides have antiproliferative and antioxidant properties effects on cervical, prostate, and liver cancer cells.

In vivo investigations on the suppressive tumor activity of brown algae have shown that the in vitro antibody potential of those algae is important for cancer treatment. The study showed a significant inhibitor effect of *S. ringgoldianum* (46.5% inhibition) *L. japonica* (57.6%), *Lessonianigrescens* (60.0%), and *Scytosiphon lomentaria* (69.8%) after oral administration of seaweed powder (1600 mg/kg body powder) on the anti-tumor activities of the powdered tissue of 21 air-dried brown algae against Ehrlich carcinoma in mice (Zorofchian Moghadamtousi et al., 2014).

Scientists are now identifying a range of components and metabolites in these different brown algae that have demonstrated extraordinary anticancer and antitumor potential. The main active metabolites of the brown algae as possible chemotherapies and chemopreventive drugs were identified as sulfated polysaccharides of fucoidan and carotenoids of fucoxanthin. Table 3 shows the anticancer activity of fucoidans that have been used from brown algae.

Active Marine-Derived Compound in Cancer Therapies

Cinachyrella sp. is the most common marine *Porifera*, having spherical bodies. Bioactive chemicals found in sponges were presented to be an effective anti-inflammatory, anti-tumor, and immunosuppressive agents. Some beneficial chemicals identified from sponges include protein kinase C inhibitors, which have been related to tumor growth and progression. Protein kinase C inhibition has been linked to the pathogenesis of joint pain, psoriasis, and tumor progression.

It was discovered that marine bacteria related to sponges *Jaspis sp.* have anticancer properties and might be used as a chemopreventive agent against cervical cancers (Utami et al., 2014). According to the findings, the antitumor activity of methanolic extracts of marine sponges *Sigmadocia pumila* and *Holothuria atra* was evaluated using both in-vitro and in-vivo methods. Human epidermoid larynx carcinoma cell line, human breast cancer cell line, African green monkey kidney normal cell line, and human cervical disease cell line were all shown to have a high degree of antitumor activity by *S. pumila* and *H. atra* (Montaser & Luesch, 2011).

The domain Protists contains a broad group of unicellular and multicellular eukaryotes known as marine algae. Protists are eukaryotes that are unicellular, colonial, filamentous, or parenchymatous in structure and lack vegetative tissue development with exception of reproduction. Several necessary elements are found in the algae range of marine water bodies, including lipids, minerals, proteins, fiber, fatty acids, polysaccharides, vitamins, and several essential amino acids (Ramasubramani et al., 2016). Many bioactive compounds found in marine algae have been linked to a variety of pharmacological characteristics, including anticancer activity (Boopathy & Kathiresan, 2010). Because of their high ROS scavenging capacity, marine algae function as powerful antioxidative agents, which may be responsible for their anticarcinogenic properties.

This research provides an overview of each chemical's active concentrations and defines a concentration range that may be utilized to determine whether or not a compound is active enough to be regarded as a medication option. In all of these examples, the active chemicals came from multicellular organisms, implying that harvesting biomass is laborious and yields limited levels of secondary metabolites, as well as a significant environmental effect from improper harvesting (Jaspars et al., 2016).

Table 4 shows the producing marine organisms with their target cancer cell lines, and the activity concentrations for the marine microalgae anti-cancer. These findings suggest that microalgae might be a viable source of cancer-fighting substances. Various substances with clinical or biotechnological potential may be found in the other fractions/extracts derived from marine microalgae, but further research is needed to prove this possibility. Various compounds derived from microalgae discussed in this analysis were investigated not just as potential anti-cancer agents, but also for other biotechnological uses.

This data indicates that microalgae might be a viable source of cancer-fighting chemicals. Various substances with clinical or biotechnological possibilities may be found in the other extracts derived from marine microalgae, but more research is needed to prove this potential. Several chemicals derived from microalgae discussed in this article have been investigated not just as potential anticancer agents, but also for other biotechnological uses.

Polyunsaturated aldehydes, for example, have been demonstrated to exhibit anti-proliferative as well as anti-bacterial properties in cancer cell lines such as human colon adenocarcinoma, lung carcinoma epithelial cells, and tumor and lymphoid, making them potential new treatment candidates (Miralto et al., 1999; Amaro et al., 2011; Paul et al., 2012; Sansone et al., 2014). A polyunsaturated fatty acid has been extensively researched as a nutraceutical or dietary supplement, with promising results in fetal development, cardiovascular disease prevention, and even cognitive function enhancement in Alzheimer's disease patients (Swanson et al., 2012).

EFFECT OF REACTIVE OXYGEN SPECIES ON CANCER CELLS

The major force for the preservation of cell metabolism and viability is oxygen, which is needed for aerobic living circumstances. Furthermore, oxygen poses a risk because of its paramagnetic properties, which will be encouraged the development of partly oxidized high reactive components defined as reac-

Table 4. Marine organism with the target cancer lines and active concentration for the active marine-derived compounds.

Marine Organism	Compound	Findings	References
Aplidium albicans	Plitidepsin (depsipeptide)	● Breast adenocarcinoma cells. ● The amounts of spermidine and spermine were significantly increased by dehydrodidemnin B. ● IC_{50}: 55.5 ng/mL	Gómez-Fabre et al. (1997)
Halichondria okadai	Eribulin mesylate (macrolide)	● Human promyelocytic leukemia cell line. ● Inhibition of tubulin polymerization *in vitro*. ● Halichondrin B-based agents as highly effective, novel anticancer drugs. ● IC_{50}: 0.657 ng/mL	Towle et al. (2001)
Dolabella auricularia	Brentuximab vedotin (antibody-drug conjugate)	● Non-Hodgkin's lymphoma cells. ● Induction of apoptosis. ● Highly potent and selective against $CD30^+$ tumor cells. ● More than 300-fold less active on antigen-negative cells. ● IC_{50}: 2.5 ng/mL	Francisco et al. (2003)
Aspergillus sp.	Plinabulin (diketopiperazine)	● Anti-multiple myeloma (MM) ● Induces apoptotic cell death in MM cell lines and tumor cells. ● IC_{50}: 2.7 - 3.375 ng/mL	Singh et al. (2011)
Cryptotheca crypta	Cytarabine (nucleoside)	● Acute myeloid leukemia cells. ● Hesperidin and silibinin showed synergistic potential and decreased the active concentration. ● IC_{50}: 272 ng/mL	Desai et al. (2015)
Halichondria okadai	Lurbinectedin (alkaloid)	● Human ovarian CCC cell lines. ● Lurbinectedin has anticancer efficacy in vitro against both chemosensitive and chemoresistant CCC cells. ● It appears to be a viable treatment for ovarian CCC. ● IC_{50}: 0.78 - 2.34 ng/mL	Takahashi et al. (2016)

tive oxygen species (ROS) (Santos-Sánchez et al., 2019). ROS levels can often be enhanced when they come into touch with exogenous or endogenous sources (Ferdous & Yusof, 2021), resulting in oxidative stress in the cell. In this stage, the ROS level approaches a harmful level and succeeds to defeat the cell's antioxidant system, escaping removal and remaining in the cell.

In cancer, ROS acts in both directions. It has the potential to be both pro-and antitumorigenic. ROS can lead to cancer formation through a variety of cancer by signaling the pathways, including MAPK/AP-1/NF-κB, which are linked to cancer spread and angiogenesis (Kashyap et al., 2019). ROS could also cause inflammation by activating the NF-κB, AP-1, HIF-1a, growth factors, inflammatory cytokines, and chemokine (Reczek & Chandel, 2017). Increased ROS levels, on the other hand, will increase oxidative stress-induced cancer cell death via initiating antitumorigenic signaling.

Role of Reactive Oxygen Species in Cell Proliferation and Survival

A rise in ROS has already been linked to increased cell growth proliferation, survival, and carcinogenesis by trying to control mitogen-activated protein kinase (MAPK), protein kinase D (PKD) through signaling the pathways, and the transcription aspects such as AP, NF-κB, HIF-1α (Galadari et al., 2017). Besides that, according to Moloney and Cotter (2018), it also can be through via the negative regulation of phosphatases and protein tyrosine phosphatase 1B, epigenetic alterations in the transcription factors and tumor controller.

Role of Reactive Oxygen Species in Genetic Instability

ROS frequently function as DNA damage mediators. When ROS accumulates in cells as a result of overproduction, they are frequently connected with DNA connection, resulting in ROS-interacting modifications such as inter-and intra-strand binds or the formation of DNA-protein crosslinks, which results in changes in gene expression. ROS induces DNA damage by oxidizing nucleoside bases and forming DNA lesions, such as 8-oxo guanine production, which results in DNA double-strand breaks (DSBs) if not repaired. ROS buildup causes mitochondrial DNA damage, strand breakage, and, eventually, degeneration (de Sá Junior et al., 2017).

Furthermore, increased ROS from oncogene activation impacts replication stress. ROS can oxidize dNTPs, which can affect polymerase activity, the breakdown of replication forks, and the creation of DSBs, all of which contribute to genomic instability (Srinivas et al., 2019). Additionally, ROS causes the activation of proteins linked with cell cycle checkpoints, resulting in cell cycle arrest. Overall, these chromosomal changes cause genetic instability, which eventually leads to carcinogenesis.

Role of Reactive Oxygen Species in Angiogenesis and Metastasis

Tumor cells are spread from the initial location to other parts of the body through blood and lymph in metastasis. According to Galadari et al. (2017), ROS can affect metastasis by activating matrix metalloproteinases (MMPs) and cathepsin expression. If the ROS levels get higher, it can activate MMP enzymes, which can stimulate or modulate a myriad of tumor progression or metastasis signaling pathways. A study by Ferdous and Yusof (2021), ROS can affect tumor migration if they are created by activated growth factor receptors, integrin assembly, and signaling kinase activation.

Furthermore, ROS induces the actin-binding protein cofilin, encouraging cell migration. However, ROS can also cause metastasis by other mechanisms, such as proteolytic degradation of glycosaminoglycan (GAG) and other ECM components. Increased ROS levels can stabilize HIFα by inhibiting prolyl hydroxylases (PHDs) and, as a result, VEGF (primary pro-angiogenic factor) activation, eventually leading to angiogenesis and cancer growth (Aggarwal et al., 2019).

Role of Reactive Oxygen Species in Chemoresistance

Chemoresistance is defined as the cancer cells' potential to avoid or tolerate the presence of medicines or therapies. It is a significant cause of the therapy's lack of effectiveness. It is also a major difficulty that oncology research is attempting to identify and solve. P-glycoprotein known as a transporter protein is a multidrug resistance protein that removes or effluxes anticancer medications from cancer cells (Galadari et al., 2017). ROS can induce the occurrence of this protein, resulting in chemoresistance and the inhibition of cell death.

Antioxidants in Cells

Next, antioxidants can be used to treat cancer when combined with chemotherapy drugs. They have the ability to increase oxidative stress in tumor cells, block transcription factors, activate apoptosis-related signaling pathways, and inhibit cell growth signaling pathways. Antioxidant supplements used during cancer therapy can minimize oxidative damage in surrounding healthy tissues, lessen side effects, and improve overall patient health and survival rate.

Antioxidants have critical roles in carcinogenesis because they regulate ROS generation. Metal exposure, for example, is said to cause ROS production and to be linked to carcinogenesis in the workplace and in the environment. According to Kongara & Karantza (2012), the Overproduction of ROS can lead to genomic instability, cellular damage, and cancer. Estrogen can potentially enhance genomic instability and accelerate breast cancer carcinogenesis by inducing ROS signaling (Okoh et al., 2011). As a result, pro-oxidant natural products from marine algae are frequently used in the development of anti-cancer medicines.

Potential Algae as a Source of Antioxidants for Use in Cancer Treatment

The antioxidant characteristics have been discovered in enzymatic extracts from a variety of marine algae species. Firstly, red algal extracts containing ethanol have antioxidant properties from the *Callophyllis japonica* and *Gracilaria tenuistipitata* algae. H_2O_2-induced cellular apoptosis was reduced, and cellular antioxidant enzymes were activated by ethanol extracts of *C. japonica* (Kang et al., 2005). The experiments with the human non-small cell lung carcinoma cell line from Yang et al. (2012), revealed that treating these cells with an aqueous extract of *G. tenuistipitata* improved their recovery from H_2O_2-induced DNA damage, inhibited cellular growth, and caused G2/M arrest.

Next, green algae-based free-radical-scavenging tests demonstrated antioxidant capabilities for sesquiterpenoids from *Ulva fasciata* Delile (Chakraborty & Paulraj, 2010). *Ulva lactuca* has a lot of flavonoids and has a lot of antioxidants. Data from animal experimental studies by Rao et al. (2004), have begun to shed light on the fact that the free radical scavenging properties of an *Ulva reticulate* hot water extract decreased hepatic oxidative stress. Antioxidant components of blue-green algae, such as *scytonemin* and *C-phycocyanin*, are also known to be effective cytotoxic agents (Abd El-Hack et al., 2019). These phytochemicals also have anti-cancer effects. Table 5 presents various records of antioxidants discovered in microalgae that have anticancer activity in vitro.

Moreover, the antioxidant properties of phlorotannins from the brown algae have been discovered using 2,2-diphenyl-1-picrylhydrazyl (DPPH)-radical scavenging assays. Methanol extracts of *Fucus vesiculosus* and *F. serratus* have been shown to protect Caco-2 cells against H_2O_2-induced DNA damage, but not from tert-butyl hydroperoxide-induced DNA damage (Shibata et al., 2007; O'Sullivan et al., 2011). *Pelvetia canaliculata* methanol extracts reduced H_2O_2-induced superoxide dismutase depletion in Caco-2 cells. Analyses of DPPH radical scavenging activity indicated that a methanol extract of the blue-green algae *Anabaena* species has antioxidant properties. Antioxidant effects of phycobiliprotein phycocyanin in a *Spirulina platensis* extract have also been demonstrated using ascorbate, iron, and H_2O_2 tests (Piñero Estrada et al., 2001). Blue-green algae's antinociceptive characteristics have received less attention thus far.

For example, the red algae, *Gracilaria corticata,* and *Sargassum oligocystum* aqueous extracts suppressed the growth of human leukemic cell lines. *Gracilaria tenuistipitata* ethanol and methanol extracts

Table 5. The antioxidants discovered in microalgae (Ferdous & Yusof, 2021).

Algae	Antioxidants	Findings	References
Dunaliella salina	β-carotene	● Human prostate cancer cell line (PC-3). ● Effective apoptosis of 32% than synthetic β-carotene. ● 27% in flow cytometry analysis. ● Inhibition rate: 79% at 50 μM ● IC_{50}: 56.1 μg/mL ● A cytochrome c and caspase test also indicated that cell death induced by algal β-carotene was higher than the β-carotene standard.	Jayappriyan et al. (2013)
Cyanophora paradoxa	β-cryptoxanthin	● Melanoma cell line (A-2058) ● Apoptosis is induced in the three lines of cancer cells. ● Inhibition rate: 90% at 100 μg/mL ● Pheophorbide a, β-cryptoxanthin, and zeaxanthin were three major pigments of strong cytotoxicity.	Baudelet et al. (2013)
Nannochloropsis oculata	Sterols	● Human promyelocytic leukemia cell line (HL-60) ● Induction of apoptosis. ● IC_{50}: 23.58 ± 0.09 μg/mL ● Revealed cytotoxic effect.	Sanjeewa et al. (2016)
Spirulina platensis	C-phycocyanin	● Human breast cancer cell line (MDA-MB-231) ● Inducing apoptosis by triggering the MAPK signaling pathways. ● IC_{50}: 189.4 μg/mL ● Protein expression levels of cyclooxygenase-2.	Jiang et al. (2018)
Tribonema sp.	Sulfated polysaccharides	● Human liver cancer cell line (HepG2) ● Induced cell apoptosis. ● Inhibition rate: 66.8% at 250 μg/mL ● M_w: 197 kDa	Chen et al. (2019)
Phaeodactylum tricornutum	Sulfated polysaccharides	● HepG2 ● Apoptosis induction without impacting the HepG2 cycle and mitosis. ● Inhibition rate: 60.37% at 250 μg/mL ● M_w: 4,810 kDa	Yang et al. (2019)

were shown to have anti-proliferative impacts on Ca9-22 oral cancer cells, as well as being implicated in cellular death, DNA damage, and oxidative stress. As a result, methanolic extracts of *Gracilaria tenuistipitata* produced from edible algae may have significant therapeutic benefits against oral squamous cell carcinoma (Yeh et al., 2012). Besides that, HT-29 colon cancer cells were used to demonstrate caspase-dependent death caused by a methanol extract of *Plocamium telfairiae* (PTE). According to Kim et al. (2007), the findings suggest that PTE can act as a chemopreventive and/or chemotherapeutic drug in colon cancer cells by reducing cell viability and inducing apoptosis.

Besides that, blue-green algae research has demonstrated the anti-cancer benefits of *Spirulina* preparations. *Spirulina*, a filamentous cyanobacterium, has a wide range of biological activity and nutritional value due to its high concentration of natural nutrients, as well as bio-modulatory and immuno-modulatory properties. *Spirulina* has also been demonstrated to have a regulatory function in lipid and carbohydrate metabolism by correcting glucose and lipid profiles in experimental animals and diabetic humans (Khan et al., 2005). They have the ability to prevent carcinogenesis due to antioxidant characteristics that preserve tissues and decrease liver, kidney, and testes damage.

FUTURE RESEARCH DIRECTIONS

The important medical and pharmaceutical capabilities of microalgae have received worldwide attention. Due to their meaningful structural variety, which implies many interactions, the mechanisms of anticancer activity by which microalgae exert their consequences are complicated. The potential of microalgae to adapt to intense adverse environments, such as ambient temperature and hydrothermal vents, makes them attractive candidates for drug development; this may be due to their unique chemicals, which are generated for survival and defense (Landsberg, 2002). Furthermore, multiple studies have been undertaken on the helpful function of microalgae metabolites in the cure of different human illnesses, with the cultivation of algae-derived chemicals receiving a great deal of attention in the pharmaceutical business (Dias et al., 2012; Shanab et al., 2012; de Morais et al., 2015). Currently, the anticancer characteristics of various algae-derived resources have been shown to modify many cellular processes including cellular cytotoxicity, tumor cell invasion, and cancer cell apoptosis.

CONCLUSION

Cancer is a significant issue, ranking second only to cardiac problems in terms of danger. Microalgae have emerged as a vast, mostly unexplored repository of common substances. It is used not just like a functional food, but it also has a historical past of use in cancer therapy in countries in Asia. According to the literature analysis, microalgae act as a useful and effective finding point with a promising outcome. It includes details about microalgae and their bioactive components, which have the possibility to treat cancer. Further investigation of isolated local algae species is required to properly study this amazing resource and emphasize the relevance of employing this organic compound to supplement current medications and offer comparable outcomes with reduced or no adverse consequences. Microalgae bioactive will offer the most likely natural alternative for chemical medicines in the future, with greater recovery rates and fewer side effects, once practical and economically possible production of marine microalga biomass is achieved.

Several kinds of biomolecules isolated and described from microalgae have shown anti-tumorigenic and anti-proliferative activity. Carotenoids, natural or sulfated polysaccharides, phenol derivatives, sterol, depsipeptides, metal-containing polyketides, and various forms of alkaloids and stypodiols were among the physiologically active byproducts found in these substances. Most of these compounds can cause cellular death by activating the caspase's apoptotic pathway, inhibiting cell microtubule nucleation, or activating intracellular caspase-independent mechanisms. Some advanced biological and molecular studies on the role and efficacy of such algal-derived anti-cancerous chemicals, as well as their characterization, are still required. Furthermore, substantial *in-vitro* and *in-vivo* studies on the active components of microalgae shall be performed to determine the potential use against various cancers.

Natural compounds derived from marine algae are high in antioxidants. Some sea algae, in fact, are able to be eaten. The facts presented in this research for crude extract fraction revealed that these compounds and their biologically active compounds had significant regulatory impacts on oxidative stress, as well as oxidative stress-related illnesses and malignancies. Several types of algae were studied for their antioxidant properties, including anti-inflammatory, antinociceptive, and anti-cancer features. These marine algae-derived materials could be employed frequently in pre-clinical investigations for medication development in the future.

Despite the enormous potential for obtaining novel formulas from marine algae organic products, only several substances can be employed in cancer therapy. There have been very few detailed publications on the ecological assessment of algae natural resources accessible. The present debate highlights the importance of huge marine water sources in the discovery of new anticancer specialties. There has not been much attention paid to the evaluation of potential molecules from the flora and fauna of marine water bodies. As a result, it is clear that the complete separation of cancer-causing chemicals and the creation of their by-products are crucial and therefore may pave the way for improved possibilities for the design and development of novel pharmacological and therapeutic treatments.

ACKNOWLEDGMENT

The authors would like to acknowledge the funding provided by Research University Grant (GUP) Tier 2 (Vote no. R.K130000.2643.18J04) for this project.

REFERENCES

Abd El-Hack, M. E., Abdelnour, S., Alagawany, M., Abdo, M., Sakr, M. A., Khafaga, A. F., Mahgoub, S. A., Elnesr, S. S., & Gebriel, M. G. (2019). Microalgae in modern cancer therapy: Current knowledge. *Biomedicine and Pharmacotherapy*, *111*, 42–50. doi:10.1016/j.biopha.2018.12.069 PMID:30576933

Aggarwal, V., Tuli, H. S., Varol, A., Thakral, F., Yerer, M. B., Sak, K., Varol, M., Jain, A., Khan, M., & Sethi, G. (2019). Role of reactive oxygen species in cancer progression: Molecular mechanisms and recent advancements. *Biomolecules*, *9*(11), 735. doi:10.3390/biom9110735 PMID:31766246

Amaro, H. M., Guedes, A. C., & Malcata, F. X. (2011). Antimicrobial activities of microalgae: an invited review. *Science against microbial pathogens: Communicating current research and technological advances, 3*, 1272-1284.

Ashwini, S., Babut, S., & Saritha, M. S. (2016). Seaweed extracts exhibit anticancer activity against HeLa cell lines. *International Journal of Current Pharmaceutical Research*, *9*(1), 114–117. doi:10.22159/ijcpr.2017v9i1.16632

Atashrazm, F., Lowenthal, R. M., Woods, G. M., Holloway, A. F., & Dickinson, J. L. (2015). Fucoidan and cancer: A multifunctional molecule with anti-tumor potential. *Marine Drugs*, *13*(4), 2327–2346. doi:10.3390/md13042327 PMID:25874926

Athukorala, Y., Kim, K. N., & Jeon, Y. J. (2006). Antiproliferative and antioxidant properties of an enzymatic hydrolysate from brown alga, Ecklonia cava. *Food and Chemical Toxicology*, *44*(7), 1065–1074. doi:10.1016/j.fct.2006.01.011 PMID:16516367

Balaji Raghavendra Rao, H., Sathivel, A., & Devaki, T. (2004). Antihepatotoxic nature of Ulva reticulata (Chlorophyceae) on acetaminophen-induced hepatoxicity in experimental rats. *Journal of Medicinal Food*, *7*(4), 495–497. doi:10.1089/jmf.2004.7.495 PMID:15671697

Barsanti, L., Coltelli, P., Evangelista, V., Frassanito, A. M., Passarelli, V., Vesentini, N., & Gualtieri, P. (2008). Oddities and curiosities in the algal world. In *Algal toxins: nature, occurrence, effect, and detection* (pp. 353–391). Springer. doi:10.1007/978-1-4020-8480-5_17

Baudelet, P. H., Gagez, A. L., Bérard, J. B., Juin, C., Bridiau, N., Kaas, R., Thiéry, V., Cadoret, J.-P., & Picot, L. (2013). Antiproliferative activity of Cyanophora paradoxa pigments in melanoma, breast, and lung cancer cells. *Marine Drugs*, *11*(11), 4390–4406. doi:10.3390/md11114390 PMID:24189278

Bonner, M. Y., & Arbiser, J. L. (2014). The antioxidant paradox: What are antioxidants and how should they be used in a therapeutic context for cancer. *Future Medicinal Chemistry*, *6*(12), 1413–1422. doi:10.4155/fmc.14.86 PMID:25329197

Boopathy, N. S., & Kathiresan, K. (2010). Anticancer drugs from marine ñora: An overview (online). *Journal of Oncology*, *2010*, 214186. PMID:21461373

Cha, K. H., Koo, S. Y., & Lee, D. U. (2008). Antiproliferative effects of carotenoids extracted from Chlorella ellipsoidea and Chlorella Vulgaris on human colon cancer cells. *Journal of Agricultural and Food Chemistry*, *56*(22), 10521–10526. doi:10.1021/jf802111x PMID:18942838

Chakraborty, K., & Paulraj, R. (2010). Sesquiterpenoids with free-radical-scavenging properties from marine macroalga Ulva fasciata Delile. *Food Chemistry*, *122*(1), 31–41. doi:10.1016/j.foodchem.2010.02.012

Chen, X., Song, L., Wang, H., Liu, S., Yu, H., Wang, X., Li, R., Liu, T., & Li, P. (2019). Partial characterization, the immune modulation and anticancer activities of sulfated polysaccharides from filamentous microalgae Tribonema sp. *Molecules (Basel, Switzerland)*, *24*(2), 322. doi:10.3390/molecules24020322 PMID:30658396

Chengkui, Z., & Junfu, Z. (1984). Chinese seaweeds in herbal medicine. In *Eleventh International Seaweed Symposium* (pp. 152-154). Springer. 10.1007/978-94-009-6560-7_24

Chisti, Y. (2007). Biodiesel from microalgae. *Biotechnology Advances*, *25*(3), 294–306. doi:10.1016/j.biotechadv.2007.02.001 PMID:17350212

Chu, M., Li, H., Wu, Q., Wo, F., & Shi, D. (2014). Pluronic-encapsulated natural chlorophyll nanocomposites for in vivo cancer imaging and photothermal/photodynamic therapies. *Biomaterials*, *35*(29), 8357–8373. doi:10.1016/j.biomaterials.2014.05.049 PMID:25002262

Chu, W. L. (2012). Biotechnological applications of microalgae. *IeJSME*, *6*(1), S24–S37.

Costa, L. S., Fidelis, G. P., Telles, C. B. S., Dantas-Santos, N., Camara, R. B. G., Cordeiro, S. L., Costa, M. S. S. P., Almeida-Lima, J., Melo-Silveira, R. F., Oliveira, R. M., Albuquerque, I. R. L., Andrade, G. P. V., & Rocha, H. A. O. (2011). Antioxidant and antiproliferative activities of heterofucans from the seaweed Sargassum filipendula. *Marine Drugs*, *9*(6), 952–966. doi:10.3390/md9060952 PMID:21747741

de Morais, M. G., Vaz, B. D. S., de Morais, E. G., & Costa, J. A. V. (2015). Biologically active metabolites synthesized by microalgae. *BioMed Research International*. PMID:26339647

de Sá, P. L. Jr, Câmara, D. A. D., Porcacchia, A. S., Fonseca, P. M. M., Jorge, S. D., Araldi, R. P., & Ferreira, A. K. (2017). The roles of ROS in cancer heterogeneity and therapy. *Oxidative Medicine and Cellular Longevity*. PMID:29123614 de Souza, L. A. R., Dore, C. M. P., Castro, A. J., de Azevedo, T. C., de Oliveira, M. T. B., Maria de Fátima, V. M., ... Leite, E. L. (2012). Galactans from the red seaweed Amansia multifida and their effects on inflammation, angiogenesis, coagulation and cell viability. *Biomedicine & Preventive Nutrition*, *2*(3), 154–162. doi:10.1016/j.bionut.2012.03.007

Desai, U. N., Shah, K. P., Mirza, S. H., Panchal, D. K., Parikh, S. K., & Rawal, R. M. (2015). Enhancement of the cytotoxic effects of Cytarabine in synergism with Hesperidine and Silibinin in Acute Myeloid Leukemia: An in-vitro approach. *Journal of Cancer Research and Therapeutics*, *11*(2), 352. doi:10.4103/0973-1482.157330 PMID:26148599

Dhargalkar, V. K., & Pereira, N. (2005). *Seaweed: Promising plant of the millennium*. Academic Press.

Dias, D. A., Urban, S., & Roessner, U. (2012). A historical overview of natural products in drug discovery. *Metabolites*, *2*(2), 303–336. doi:10.3390/metabo2020303 PMID:24957513

Farasat, M., Khavari-Nejad, R. A., Nabavi, S. M. B., & Namjooyan, F. (2013). Antioxidant properties of some filamentous green algae (Chaetomorpha Genus). *Brazilian Archives of Biology and Technology*, *56*(6), 921–927. doi:10.1590/S1516-89132013000600005

Ferdous, U. T., & Yusof, Z. N. B. (2021). Algal terpenoids: a potential source of antioxidants for cancer therapy. In *Terpenes and Terpenoids*. IntechOpen. doi:10.5772/intechopen.94122

Ferdous, U. T., & Yusof, Z. N. B. (2021). Medicinal Prospects of Antioxidants from Algal Sources in Cancer Therapy. *Frontiers in Pharmacology*, *12*, 593116. Advance online publication. doi:10.3389/fphar.2021.593116 PMID:33746748

Francisco, J. A., Cerveny, C. G., Meyer, D. L., Mixan, B. J., Klussman, K., Chace, D. F., Rejniak, S. X., Gordon, K. A., DeBlanc, R., Toki, B. E., Law, C.-L., Doronina, S. O., Siegall, C. B., Senter, P. D., & Wahl, A. F. (2003). cAC10-vcMMAE, an anti-CD30–monomethyl auristatin E conjugate with potent and selective antitumor activity. *Blood*, *102*(4), 1458–1465. doi:10.1182/blood-2003-01-0039 PMID:12714494

Galadari, S., Rahman, A., Pallichankandy, S., & Thayyullathil, F. (2017). Reactive oxygen species and cancer paradox: To promote or to suppress? *Free Radical Biology & Medicine*, *104*, 144–164. doi:10.1016/j.freeradbiomed.2017.01.004 PMID:28088622

Georgianna, D. R., & Mayfield, S. P. (2012). Exploiting diversity and synthetic biology for the production of algal biofuels. *Nature*, *488*(7411), 329–335. doi:10.1038/nature11479 PMID:22895338

Go, H., Hwang, H. J., & Nam, T. J. (2010). A glycoprotein from Laminaria japonica induces apoptosis in HT-29 colon cancer cells. *Toxicology In Vitro*, *24*(6), 1546–1553. doi:10.1016/j.tiv.2010.06.018 PMID:20615460

Goh, S. H., Alitheen, N. B. M., Yusoff, F. M., Yap, S. K., & Loh, S. P. (2014). Crude ethyl acetate extract of marine microalga, Chaetoceros calcitrans, induces Apoptosis in MDA-MB-231 breast cancer cells. *Pharmacognosy Magazine*, *10*(37), 1. doi:10.4103/0973-1296.126650 PMID:24696543

Gómez-Fabre, P. M., De Pedro, E., Medina, M. A., De Castro, I. N., & Márquez, J. (1997). Polyamine contents of human breast cancer cells treated with the cytotoxic agents chlorpheniramine and dehydrodidemnin B. *Cancer Letters*, *113*(1-2), 141–144. doi:10.1016/S0304-3835(96)04591-0 PMID:9065813

Gupta, S., & Abu-Ghannam, N. (2011). Bioactive potential and possible health effects of edible brown seaweeds. *Trends in Food Science & Technology*, *22*(6), 315–326. doi:10.1016/j.tifs.2011.03.011

Hall, D. (2021, March 8). *A Cancer Fighter from the Sea*. Retrieved August 30, 2021, from Si.edu website: https://ocean.si.edu/ocean-life/plants-algae/cancer-fighter-sea

Haq, S. H., Al-Ruwaished, G., Al-Mutlaq, M. A., Naji, S. A., Al-Mogren, M., Al-Rashed, S., Ain, Q. T., Al-Amro, A. A., & Al-Mussallam, A. (2019). Antioxidant, anticancer activity, and phytochemical analysis of green algae, Chaetomorpha collected from the Arabian Gulf. *Scientific Reports*, *9*(1), 1–7. doi:10.103841598-019-55309-1 PMID:31827196

He, D., Wu, S., Yan, L., Zuo, J., Cheng, Y., Wang, H., Liu, J., Zhang, X., Wu, M., Choi, J., & Tong, H. (2019). Antitumor bioactivity of porphyran extracted from Pyropia yezoensis Chonsoo2 on human cancer cell lines. *Journal of the Science of Food and Agriculture*, *99*(15), 6722–6730. doi:10.1002/jsfa.9954 PMID:31350864

Holdt, S. L., & Kraan, S. (2011). Bioactive compounds in seaweed: Functional food applications and legislation. *Journal of Applied Phycology*, *23*(3), 543–597. doi:10.100710811-010-9632-5

Hyun, J. H., Kim, S. C., Kang, J. I., Kim, M. K., Boo, H. J., Kwon, J. M., Koh, Y.-S., Hyun, J.-W., Park, D.-B., Yoo, E.-S., & Kang, H. K. (2009). Apoptosis inducing activity of fucoidan in HCT-15 colon carcinoma cells. *Biological & Pharmaceutical Bulletin*, *32*(10), 1760–1764. doi:10.1248/bpb.32.1760 PMID:19801840

Jaspars, M., De Pascale, D., Andersen, J. H., Reyes, F., Crawford, A. D., & Ianora, A. (2016). The marine biodiscovery pipeline and ocean medicines of tomorrow. *Journal of the Marine Biological Association of the United Kingdom*, *96*(1), 151–158. doi:10.1017/S0025315415002106

Jayappriyan, K. R., Rajkumar, R., Venkatakrishnan, V., Nagaraj, S., & Rengasamy, R. (2013). In vitro anticancer activity of natural β-carotene from Dunaliella salina EU5891199 in PC-3 cells. *Biomedicine & Preventive Nutrition*, *3*(2), 99–105. doi:10.1016/j.bionut.2012.08.003

Jiang, L., Wang, Y., Liu, G., Liu, H., Zhu, F., Ji, H., & Li, B. (2018). C-Phycocyanin exerts anti-cancer effects via the MAPK signaling pathway in MDA-MB-231 cells. *Cancer Cell International*, *18*(1), 1–14. doi:10.118612935-018-0511-5 PMID:29416441

Jimeno, J., Faircloth, G., Sousa-Faro, J. M., Scheuer, P., & Rinehart, K. (2004). New marine derived anti-cancer therapeutics— A journey from the sea to clinical trials. *Marine Drugs*, *2*(1), 14–29. doi:10.3390/md201014

Kang, K. A., Bu, H. D., Park, D. S., Go, G. M., Jee, Y., Shin, T., & Hyun, J. W. (2005). Antioxidant activity of ethanol extract of Callophyllis japonica. *Phytotherapy Research*, *19*(6), 506–510. doi:10.1002/ptr.1692 PMID:16114080

Kashyap, D., Tuli, H. S., Sak, K., Garg, V. K., Goel, N., Punia, S., & Chaudhary, A. (2019). Role of reactive oxygen species in cancer progression. *Current Pharmacology Reports*, *5*(2), 79–86. doi:10.100740495-019-00171-y

Kawamoto, H., Miki, Y., Kimura, T., Tanaka, K., Nakagawa, T., Kawamukai, M., & Matsuda, H. (2006). Effects of fucoidan from Mozuku on human stomach cell lines. *Food Science and Technology Research*, *12*(3), 218–222. doi:10.3136/fstr.12.218

Khan, Z., Bhadouria, P., & Bisen, P. S. (2005). Nutritional and therapeutic potential of Spirulina. *Current Pharmaceutical Biotechnology*, *6*(5), 373–379. doi:10.2174/138920105774370607 PMID:16248810

Kim, E. J., Park, S. Y., Lee, J. Y., & Park, J. H. Y. (2010). Fucoidan present in brown algae induces apoptosis of human colon cancer cells. *BMC Gastroenterology*, *10*(1), 1–11. doi:10.1186/1471-230X-10-96 PMID:20727207

Kim, J. Y., Yoon, M. Y., Cha, M. R., Hwang, J. H., Park, E., Choi, S. U., Park, H.-R., & Hwang, Y. I. (2007). Methanolic extracts of Plocamium telfairiae induce cytotoxicity and caspase-dependent apoptosis in HT-29 human colon carcinoma cells. *Journal of Medicinal Food*, *10*(4), 587–593. doi:10.1089/jmf.2007.002 PMID:18158827

Kim, Y. S., Li, X. F., Kang, K. H., Ryu, B., & Kim, S. K. (2014). Stigmasterol isolated from marine microalgae Navicula incerta induces apoptosis in human hepatoma HepG2 cells. *BMB Reports*, *47*(8), 433–438. doi:10.5483/BMBRep.2014.47.8.153 PMID:24286323

Kongara, S., & Karantza, V. (2012). The interplay between autophagy and ROS in tumorigenesis. *Frontiers in Oncology*, *2*, 171. doi:10.3389/fonc.2012.00171 PMID:23181220

Kranz, D., & Dobbelstein, M. (2012). A killer promoting survival: p53 as a selective means to avoid side effects of chemotherapy. *Cell Cycle (Georgetown, Tex.)*, *11*(11), 2053–2054. doi:10.4161/cc.20698 PMID:22622088

Kumar, C. S., Ganesan, P., Suresh, P. V., & Bhaskar, N. (2008). Seaweeds as a source of nutritionally beneficial compounds-a review. *Journal of Food Science and Technology*, *45*(1), 1.

Kwon, M. J., & Nam, T. J. (2006). Porphyran induces apoptosis-related signal pathways in AGS gastric cancer cell lines. *Life Sciences*, *79*(20), 1956–1962. doi:10.1016/j.lfs.2006.06.031 PMID:16876203

Kwon, M. J., & Nam, T. J. (2007). Chromatographically purified porphyran from Porphyra yezoensis effectively inhibits the proliferation of human cancer cells. *Food Science and Biotechnology*, *16*(6), 873–878.

Kwon, M. J., & Nam, T. J. (2007). A polysaccharide of the marine alga Capsosiphon fulvescens induces apoptosis in AGS gastric cancer cells via an IGF-IR-mediated PI3K/Akt pathway. *Cell Biology International*, *31*(8), 768–775. doi:10.1016/j.cellbi.2007.01.010 PMID:17344071

Landsberg, J. H. (2002). The effects of harmful algal blooms on aquatic organisms. *Reviews in Fisheries Science*, *10*(2), 113–390. doi:10.1080/20026491051695

Lange, K. W., Hauser, J., Nakamura, Y., & Kanaya, S. (2015). Dietary seaweeds and obesity. *Food Science and Human Wellness*, *4*(3), 87–96. doi:10.1016/j.fshw.2015.08.001

Leclerc, E. (2021, March 4). *Scientists Find Blue-Green Algae Chemical with Cancer Fighting Potential*. Retrieved August 30, 2021, from Smithsonianmag.com website: https://www.smithsonianmag.com/blogs/national-museum-of-natural-history/2021/03/04/scientists-find-blue-green-algae-chemical-cancer-fighting-potential/

Liu, J., & Chen, F. (2014). Biology and industrial applications of Chlorella: Advances and prospects. *Microalgae Biotechnology*, 1–35. PMID:25537445

Liu, Z., Gao, T., Yang, Y., Meng, F., Zhan, F., Jiang, Q., & Sun, X. (2019). Anti-Cancer Activity of Porphyran and Carrageenan from Red Seaweeds. *Molecules (Basel, Switzerland)*, *24*(23), 4286. doi:10.3390/molecules24234286 PMID:31775255

Martin, K. R. (2007). The chemistry of silica and its potential health benefits. *The Journal of Nutrition, Health & Aging*, *11*(2), 94. PMID:17435951

Martin, K. R. (2013). Silicon: The health benefits of a metalloid. *Interrelations between essential metal ions and human diseases*, 451-473.

Martins, R. M., Nedel, F., Guimaraes, V., Da Silva, A. F., Colepicolo, P., De Pereira, C. M., & Lund, R. G. (2018). Macroalgae extracts from Antarctica have antimicrobial and anticancer potential. *Frontiers in Microbiology*, *9*, 412. doi:10.3389/fmicb.2018.00412 PMID:29568291

Maruyama, H., Tamauchi, H., Hashimoto, M., & Nakano, T. (2003). Antitumor activity and immune response of Mekabu fucoidan extracted from Sporophyll of Undaria pinnatifida. *In Vivo (Athens, Greece)*, *17*(3), 245–249. PMID:12929574

Miralto, A., Barone, G., Romano, G., Poulet, S. A., Ianora, A., Russo, G. L., Buttino, I., Mazzarella, G., Laabir, M., Cabrini, M., & Giacobbe, M. G. (1999). The insidious effect of diatoms on copepod reproduction. *Nature*, *402*(6758), 173–176. doi:10.1038/46023

Montaser, R., & Luesch, H. (2011). Marine natural products: A new wave of drugs? *Future Medicinal Chemistry*, *3*(12), 1475–1489. doi:10.4155/fmc.11.118 PMID:21882941

Nappo, M., Berkov, S., Massucco, C., Di Maria, V., Bastida, J., Codina, C., Avila, C., Messina, P., Zupo, V., & Zupo, S. (2012). Apoptotic activity of the marine diatom Cocconeis scutellum and eicosapentaenoic acid in BT20 cells. *Pharmaceutical Biology*, *50*(4), 529–535. doi:10.3109/13880209.2011.611811 PMID:22136301

Noda, H., Amano, H., Arashima, K., & Nisizawa, K. (1990). Antitumor activity of marine algae. *Hydrobiologia*, *204*(1), 577–584. doi:10.1007/BF00040290

O'Sullivan, A. M., O'callaghan, Y. C., O'Grady, M. N., Queguineur, B., Hanniffy, D., Troy, D. J., Kerry, J. P., & O'brien, N. M. (2011). In vitro and cellular antioxidant activities of seaweed extracts prepared from five brown seaweeds harvested in spring from the west coast of Ireland. *Food Chemistry*, *126*(3), 1064–1070. doi:10.1016/j.foodchem.2010.11.127

Okoh, V., Deoraj, A., & Roy, D. (2011). Estrogen-induced reactive oxygen species-mediated signalings contribute to breast cancer. *Biochimica et Biophysica Acta (BBA)-. Revue Canadienne*, *1815*(1), 115–133. PMID:21036202

Ozben, T. (2015). Antioxidant supplementation on cancer risk and during cancer therapy: An update. *Current Topics in Medicinal Chemistry*, *15*(2), 170–178. doi:10.2174/1568026615666141209160918 PMID:25496272

Pasquet, V., Morisset, P., Ihammouine, S., Chepied, A., Aumailley, L., Berard, J. B., Serive, B., Kaas, R., Lanneluc, I., Thiery, V., Lafferriere, M., Piot, J.-M., Patrice, T., Cadoret, J.-P., & Picot, L. (2011). Antiproliferative activity of violaxanthin isolated from bioguided fractionation of Dunaliella tertiolecta extracts. *Marine Drugs*, *9*(5), 819–831. doi:10.3390/md9050819 PMID:21673891

Paul, C., Reunamo, A., Lindehoff, E., Bergkvist, J., Mausz, M. A., Larsson, H., Richter, H., Wängberg, S.-Å., Leskinen, P., Båmstedt, U., & Pohnert, G. (2012). Diatom derived polyunsaturated aldehydes do not structure the planktonic microbial community in a mesocosm study. *Marine Drugs*, *10*(4), 775–792. doi:10.3390/md10040775 PMID:22690143

Pierobon, S. C., Cheng, X., Graham, P. J., Nguyen, B., Karakolis, E. G., & Sinton, D. (2018). Emerging microalgae technology: A review. *Sustainable Energy & Fuels*, *2*(1), 13–38. doi:10.1039/C7SE00236J

Piñero Estrada, J., Bermejo Bescós, P., & Villar del Fresno, A. M. (2001). Antioxidant activity of different fractions of Spirulina platensis protean extract. *Farm*, *56*(5-7), 497–500. doi:10.1016/S0014-827X(01)01084-9 PMID:11482785

Ramasubramani, R., Praveen, R., & Sathyanarayanan, K. S. (2016). Study on the strength properties of marine algae concrete. *Rasayan Journal of Chemistry*, *4*, 706–715.

Reczek, C. R., & Chandel, N. S. (2017). *The two faces of reactive oxygen species in cancer*. Academic Press.

Saadaoui, I., Rasheed, R., Abdulrahman, N., Bounnit, T., Cherif, M., Al Jabri, H., & Mraiche, F. (2020). Algae-derived bioactive compounds with anti-lung cancer potential. *Marine Drugs*, *18*(4), 197. doi:10.3390/md18040197 PMID:32276401

Samarakoon, K. W., Ko, J. Y., Lee, J. H., Kwon, O. N., Kim, S. W., & Jeon, Y. J. (2014). Apoptotic anticancer activity of a novel fatty alcohol ester isolated from cultured marine diatom, Phaeodactylum tricornutum. *Journal of Functional Foods*, *6*, 231–240. doi:10.1016/j.jff.2013.10.011

Sanjeewa, K. K. A., Fernando, I. P. S., Samarakoon, K. W., Lakmal, H. H. C., Kim, E. A., Kwon, O. N., Dilshara, M. G., Lee, J.-B., & Jeon, Y. J. (2016). Anti-inflammatory and anti-cancer activities of sterol rich fraction of cultured marine microalga Nannochloropsis oculata. *Algae - Korean Phycological Society*, *31*(3), 277–287. doi:10.4490/algae.2016.31.6.29

Sansone, C., Braca, A., Ercolesi, E., Romano, G., Palumbo, A., Casotti, R., Francone, M., & Ianora, A. (2014). Diatom-derived polyunsaturated aldehydes activate cell death in human cancer cell lines but not normal cells. *PLoS One*, *9*(7), e101220. doi:10.1371/journal.pone.0101220 PMID:24992192

Santos-Sánchez, N. F., Salas-Coronado, R., Villanueva-Cañongo, C., & Hernández-Carlos, B. (2019). Antioxidant compounds and their antioxidant mechanism. *Antioxidants*, *10*, 1–29.

Senthilkumar, P., & Sudha, S. (2012). *Antioxidant and antibacterial properties of methanolic extract of green seaweed Chaetomorpha linum from Gulf of Mannar: Southeast coast of India*. Academic Press.

Shanab, S. M., Mostafa, S. S., Shalaby, E. A., & Mahmoud, G. I. (2012). Aqueous extracts of microalgae exhibit antioxidant and anticancer activities. *Asian Pacific Journal of Tropical Biomedicine*, *2*(8), 608–615. doi:10.1016/S2221-1691(12)60106-3 PMID:23569980

Shibata, T., Ishimaru, K., Kawaguchi, S., Yoshikawa, H., & Hama, Y. (2007). Antioxidant activities of phlorotannins isolated from Japanese Laminariaceae. In *Nineteenth International Seaweed Symposium* (pp. 255-261). Springer.

Singh, A. V., Bandi, M., Raje, N., Richardson, P., Palladino, M. A., Chauhan, D., & Anderson, K. C. (2011). A novel vascular disrupting agent plinabulin triggers JNK-mediated apoptosis and inhibits angiogenesis in multiple myeloma cells. *Blood*, *117*(21), 5692–5700. doi:10.1182/blood-2010-12-323857 PMID:21454451

Song, B. H., Lee, D. H., Kim, B. C., Ku, S. H., Park, E. J., Kwon, I. H., Kim, K. H., & Kim, K. J. (2014). Photodynamic therapy using chlorophyll-a in the treatment of acne vulgaris: A randomized, single-blind, split-face study. *Journal of the American Academy of Dermatology*, *71*(4), 764–771. doi:10.1016/j.jaad.2014.05.047 PMID:24930587

Srinivas, U. S., Tan, B. W., Vellayappan, B. A., & Jeyasekharan, A. D. (2019). ROS and the DNA damage response in cancer. *Redox Biology*, *25*, 101084. doi:10.1016/j.redox.2018.101084 PMID:30612957

Swanson, D., Block, R., & Mousa, S. A. (2012). Omega-3 fatty acids EPA and DHA: Health benefits throughout life. *Advances in Nutrition*, *3*(1), 1–7. doi:10.3945/an.111.000893 PMID:22332096

Takahashi, R., Mabuchi, S., Kawano, M., Sasano, T., Matsumoto, Y., Kuroda, H., Kozasa, K., Hashimoto, K., Sawada, K., & Kimura, T. (2016). Preclinical investigations of PM01183 (Lurbinectedin) as a single agent or in combination with other anticancer agents for clear cell carcinoma of the ovary. *PLoS One*, *11*(3), e0151050. doi:10.1371/journal.pone.0151050 PMID:26986199

Taskin, E., Caki, Z., & Ozturk, M. (2010). Assessment of in vitro antitumoral and antimicrobial activities of marine algae harvested from the eastern Mediterranean Sea. *African Journal of Biotechnology*, *9*(27), 4272–4277.

Towle, M. J., Salvato, K. A., Budrow, J., Wels, B. F., Kuznetsov, G., Aalfs, K. K., ... Littlefield, B. A. (2001). In vitro and in vivo anticancer activities of synthetic macrocyclic ketone analogues of halichondrin B. *Cancer Research*, *61*(3), 1013–1021. PMID:11221827

Utami, A. W. A., Wahyudi, A. T., & Batubara, I. (2014). Toxicity, anticancer and antioxidant activity of extracts from marine bacteria associated with sponge Jaspis sp. *International Journal of Pharma and Bio Sciences*, *5*(4), 917–923.

Wu, Q., Liu, L., Miron, A., Klímová, B., Wan, D., & Kuča, K. (2016). The antioxidant, immunomodulatory, and anti-inflammatory activities of Spirulina: An overview. *Archives of Toxicology*, *90*(8), 1817–1840. doi:10.100700204-016-1744-5 PMID:27259333

Yang, J. I., Yeh, C. C., Lee, J. C., Yi, S. C., Huang, H. W., Tseng, C. N., & Chang, H. W. (2012). Aqueous extracts of the edible Gracilaria tenuistipitata are protective against H2O2-induced DNA damage, growth inhibition, and cell cycle arrest. *Molecules (Basel, Switzerland)*, *17*(6), 7241–7254. doi:10.3390/molecules17067241 PMID:22695230

Yang, S., Wan, H., Wang, R., & Huo, D. (2019). Sulfated polysaccharides from Phaeodactylum tricornutum: Isolation, structural characteristics, and inhibiting HepG2 growth activity in vitro. *PeerJ*, *7*, e6409. doi:10.7717/peerj.6409 PMID:30809437

Ye, J., Li, Y., Teruya, K., Katakura, Y., Ichikawa, A., Eto, H., Hosoi, M., Hosoi, M., Nishimoto, S., & Shirahata, S. (2005). Enzyme-digested fucoidan extracts derived from seaweed Mozuku of Cladosiphon novae-caledoniae kylin inhibit invasion and angiogenesis of tumor cells. *Cytotechnology*, *47*(1), 117–126. doi:10.100710616-005-3761-8 PMID:19003051

Yeh, C. C., Yang, J. I., Lee, J. C., Tseng, C. N., Chan, Y. C., Hseu, Y. C., Tang, J.-Y., Chuang, L.-Y., Huang, H.-W., Chang, F.-R., & Chang, H. W. (2012). Anti-proliferative effect of methanolic extract of Gracilaria tenuistipitata on oral cancer cells involves apoptosis, DNA damage, and oxidative stress. *BMC Complementary and Alternative Medicine*, *12*(1), 1–9. doi:10.1186/1472-6882-12-142 PMID:22937998

Yu, X., Zhou, C., Yang, H., Huang, X., Ma, H., Qin, X., & Hu, J. (2015). Effect of ultrasonic treatment on the degradation and inhibition cancer cell lines of polysaccharides from Porphyra yezoensis. *Carbohydrate Polymers*, *117*, 650–656. doi:10.1016/j.carbpol.2014.09.086 PMID:25498684

Yuan, H., Song, J., Li, X., Li, N., & Dai, J. (2006). Immunomodulation and antitumor activity of κ-carrageenan oligosaccharides. *Cancer Letters*, *243*(2), 228–234. doi:10.1016/j.canlet.2005.11.032 PMID:16410037

Zhang, L. X., Cai, C. E., Guo, T. T., Gu, J. W., Xu, H. L., Zhou, Y., Wang, Y., Liu, C.-C., & He, P. M. (2011). Anti-cancer effects of polysaccharide and phycocyanin from Porphyra yezoensis. *Journal of Marine Science and Technology*, *19*(4), 377–382. doi:10.51400/2709-6998.2178

Zorofchian Moghadamtousi, S., Karimian, H., Khanabdali, R., Razavi, M., Firoozinia, M., Zandi, K., & Abdul Kadir, H. (2014). Anticancer and antitumor potential of fucoidan and fucoxanthin, two main metabolites isolated from brown algae. *TheScientificWorldJournal*, *2014*, 2014. doi:10.1155/2014/768323 PMID:24526922

KEY TERMS AND DEFINITIONS

Algae: A kind of aquatic plant that belongs to a wide group that consists of seaweeds and various single-celled forms.

Antioxidant: A compound that preserves cells from free radical damage.

Antitumor: Preventing the development of tumour.

Apoptosis: Cell death that happens as a natural and regulated element of an organism's development.

Cancer: An illness characterized by the unregulated division of aberrant cells in a body area.

Cell Line: Permanently formed cultured cells which will grow continually if provided with adequate new medium and space.

In-Vitro: A controlled environment, such as a test tube or petri dish, is used by researchers to conduct experiments outside of a living organism.

In-Vivo: A research is conducted on or inside a biological creature as a whole.

Chapter 24
Polyhydroxyalkanoates (PHAs) Production From Microalgae Cultivated in Wastewater

Savvas Giannis Mastropetros
University of Patras, Greece

Konstantinos Pispas
University of Patras, Greece

Dimitris Zagklis
University of Patras, Greece

Tamer Elsamahy
Biofuels Institute, School of the Environment and Safety Engineering, Zhenjiang, China

Sameh Ali
Botany Department, Faculty of Science, Tanta University, Egypt & Biofuels Institute, School of the Environment and Safety Engineering, Jiangsu University, China

Jianzhong Sun
Jiangsu University, China

Michael Kornaros
https://orcid.org/0000-0001-6691-6925
University of Patras, Greece

ABSTRACT

Plastic materials compose a wide range of products with small useful lifespans, leading to the production of large quantities of waste. A more easily biodegradable alternative to fossil-based plastics are bioplastics. Microalgae can produce poly (hydroxy alkanoate) esters (PHAs), which are biodegradable aliphatic polyesters. Poly (3-hydroxy butyrate) ester (PHB), belonging to the short-chain PHAs, is the most common and well characterized biopolymer. PHB compounds can be completely broken down into carbon dioxide and water under aerobic conditions and are characterized as environmentally friendly, with their thermal and mechanical properties being comparable to those of petrochemical polymers. A large number of microalgae species have been reported in literature as an alternative source of energy and carbon. In order to further mitigate the environmental footprint of microalgae cultivation for bioplastics production, a small number of published works have examined bioplastic production from microalgae cultivated in wastewater, reaching 5.5-6.5% of dry biomass weight.

DOI: 10.4018/978-1-6684-2438-4.ch024

INTRODUCTION

Since the early 1950s, the plastics commercial demand and production have increased tremendously. Plastics improve the quality of life by replacing materials such as glass, paper, metal, and wood in various applications, mainly in the field of packaging and construction (Lebreton & Andrady, 2019; Możejko-Ciesielska & Kiewisz, 2016). Conventional plastics are made of petrochemical polymers (Das et al., 2018) and qualitatively characterized by high strength, durability as well as low density and cost (Van Eygen et al., 2017). Plastic materials include polyethylene (PE), polypropylene (PP), polyvinyl chloride (PVC), polystyrene (PS), polyamides (nylons), etc., that compose a wide range of products such as automotive parts, medical equipment, agricultural tools, electrical and electronic devices (Leal Filho et al., 2019). In 2018, about 359 million tons of plastics were produced, of which 17% were produced in Europe (Plastics Europe & Conversio Market & Strategy GmbH, 2019). The industrial production of polymers is the second most important use of crude oil, after its use as an energy source (Martins et al., 2014).

Plastics consist of long-chain, high molecular weight polymers with hydrophobic properties and are difficult to decompose under ambient conditions as they exhibit significant resistance against microbial degradation (Abdo & Ali, 2019; Lee & Liew, 2020). Plastics' durability in combination with their short useful lifespan have resulted in a huge waste stream (Lee & Liew, 2020). About 40% of plastics have a usage life of less than 1 month (Achilias et al., 2007), with plastic waste released into the environment being harmful to marine, freshwater, and terrestrial ecosystems (Leal Filho et al., 2019). Some common plastic waste management strategies are both recycling and incineration, however, the recycling process has been proved a non-sustainable solution as it requires large amounts of energy (Das et al., 2018). The incineration of non-recyclable plastics is also an ineffective process, as it releases toxic compounds into the atmosphere, such as furan and dioxins (Lee & Liew, 2020). Recent researchers classify plastic wastes as hazardous, with around 1 million animals dying due to petrochemical pollution each year (Cassuriaga et al., 2018). Plastic waste disperses in all oceans causing small plastic fragments known as macro- and microplastics. The size of microplastics varies from a few μm to 500 μm, considered rarely visible to the naked eye (Andrady, 2011; Eriksen et al., 2014). It was estimated that 5.25 trillion plastic particles, weighing 268,940 t, floated at sea in 2014 (Eriksen et al., 2014). Microplastics, due to their large surface area, are able to adsorb and accumulate harmful substances, such as heavy metals and organic pollutants (Cox et al., 2019; Fang et al., 2019). Furthermore, microplastics can enter the food chain and end up in human tissues (Cox et al., 2019). Various microplastics have been found in the stomachs of three different fish species (swordfish, red and long-winged tuna) in the Mediterranean Sea, as well as in women's placenta (Ragusa et al., 2021). Obviously, plastic waste not only endangers marine creatures, but also threatens public health. In this chapter we will examine a promising alternative for plastic materials production from microalgae cultivation. These bioplastic materials are of even higher interest when produced from the cultivation of microalgae using wastewater nutrients, a fact that minimizes their economic and environmental costs.

BIOPLASTICS

The accumulation of plastic residues in the environment, the inability to manage such pollutants, the depletion of oil reserves and the increased greenhouse gas emissions (GHG emissions) have created an imperative necessity for sustainable ecological products (Cassuriaga et al., 2018). According to the

Figure 1. Classification of plastics according to the EU (European Bioplastics, 2017).

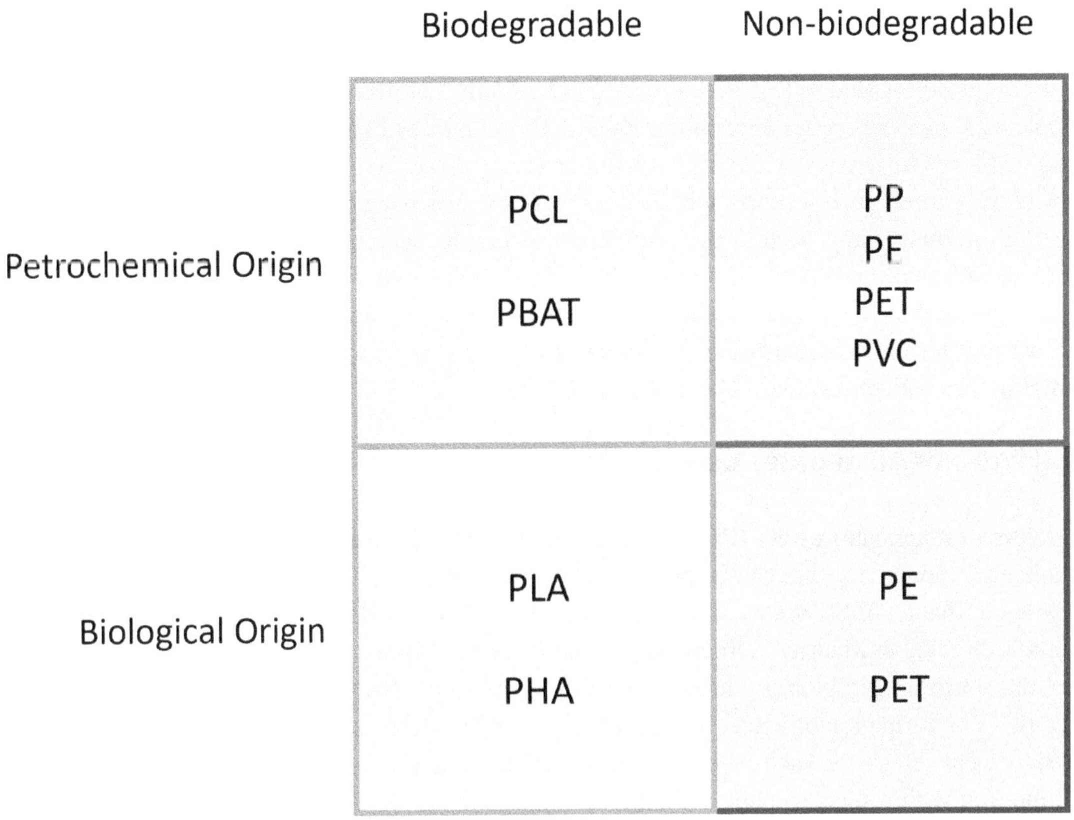

European Union, bioplastics are materials originated from biomass or biodegradable products. Thus, plastics can be categorized into four main classes according to their origin and biodegradability (European Bioplastics, 2017), (Figure 1):

1. Non-biodegradable conventional plastics of petrochemical origin, such as polypropylene (PP), polyethylene (PE), polyethylene terephthalate (PET), polyvinyl chloride (PVC) etc.
2. Non-biodegradable organically derived plastics of biomass origin, such as organic polyethylene (PE), organic polyethylene terephthalate (PET), etc.
3. Biodegradable plastics of biological origin. The most representative products of this category are some polyesters, such as polylactic acid (PLA), which is produced from lactic acid (derived mainly from the fermentation of maize starch) and polyhydroxyalkanoic esters (PHAs that are synthesized by a large number of microorganisms (Razza & Innocenti, 2012)
4. Biodegradable plastics of petrochemical origin (Polycaprolactone, PCL, Polybutylene adipate terephthalate, PBAT, etc.)

From an environmental point of view, biodegradable plastics of biological origin present a viable perspective regarding the plastic waste problem. Bioplastics are produced using renewable resources and biodegrade through anaerobic or aerobic processes (Razza & Innocenti, 2012). However, mixing bioplastics with conventional plastics of petrochemical origin in order to acquire improved properties (such as heat resistance, flexibility, durability) can lead to their non-biodegradable components ending up as microplastics in the environment, thus contributing to the problem of plastic waste accumulation.

Various life cycle assessment (LCA) studies have proven that the replacement of conventional plastics with bioplastics could reduce greenhouse gas emissions. However, the use of agricultural products for bioplastics' production not only competes with food industry but also demands massive amounts of mineral resources for crops development, as the majority of bioplastics is produced by the fermentation of agricultural goods such as corn, wheat, potatoes, rice and soy (Abdo & Ali, 2019). In contrast, an interesting alternative approach to the production of bioplastics comes from microalgae. Microalgae metabolize carbon dioxide (CO_2), in order to produce poly (3-hydroxy butyrate) esters (PHBs), part of the wider group of poly- (hydroxy alkanoate) esters (PHAs). In addition, microalgae cultivation can utilize wastewater (such as anaerobic effluents of digested wastewater) as a source of nutrients, thus minimizing the production costs (Martins et al., 2017).

Poly (Hydroxy Alkanoate) Esters (PHAs)

Poly (hydroxy alkanoate) esters (PHAs) are biodegradable aliphatic polyesters, composed from hydroxyalkanoic monomers, and can be produced by several bacteria and microalgae (cyanobacteria and eukaryotes) (Khanna & Srivastava, 2005; Martins et al., 2017). PHAs were discovered in 1920 by the French microbiologist Maurice Lemoigne, who observed the Gram-positive bacterium *Bacillus megaterium*, that form intracellular granules, consisting of mainly 3-hydroxybutyrate (PHB) (Keshavarz & Roy, 2010). The formation of PHAs from microalgae and bacteria is mostly observed under conditions of physiological cell stress, such as the lack of phosphates and nitrogenous nutrients and the light exposure limitation with respect to photosynthetic microorganisms (Możejko-Ciesielska & Kiewisz, 2016; Cassuriaga et al., 2018). PHAs are intracellularly synthesized as an alternative energy source for the cell (Kavitha et al., 2021; Możejko-Ciesielska & Kiewisz, 2016; Singh et al., 2017). The depletion of carbon source leads to depolymerization of the macromolecules so as to provide a source of carbon and energy (Możejko-Ciesielska & Kiewisz, 2016). To date, 155 PHA monomers have been found, with molecular masses in the range of 50 x 10^3 to 1 x 10^6 Daltons. Monomeric PHAs are categorized in accordance to the number of carbon atoms, into three groups: short-chain PHAs (scl-PHAs) with 3 to 5 carbon atoms, medium chain PHAs (mcl-PHAs) with 6 to 14 carbon atoms, and long chain PHAs (lcl-PHAs) with 15 or more carbon atoms (Singh et al., 2017). PHAs exhibit high crystallinity from 60% to 80% and melting temperature (T_m) between 50 °C and 180 °C (Khanna & Srivastava, 2005; Keshavarz & Roy, 2010; Rahman & Miller, 2017; Singh et al., 2017).

Poly (3-hydroxy butyrate) ester (PHB), belonging to the short-chain PHAs, is the most common and widely studied biopolymer in the group. PHB is characterized by biocompatibility, biodegradability, hydrophobicity, non-toxicity, and piezoelectric capacity. It also exhibits a melting point of 179 °C and a tensile strength of 40 MPa. These characteristics render it as a strong alternative solution against the usage of conventional petrochemical plastics (Abdo & Ali, 2019; Ansari & Fatma, 2016; Rahman & Miller, 2017; Singh et al., 2017). Growing evidence suggests that monomers of poly (3-hydroxy butyrate) (PHB) are also non-toxic (Chen & Wu, 2005). PHB compounds can be completely broken down into carbon dioxide (CO_2) and water (H_2O) under aerobic conditions, whilst they are converted into methane (CH_4) during anaerobic degradation (Cassuriaga et al., 2018). In conclusion, the above data pave the way for a wide range of applications based on PHAs, from materials packaging to industrial, agricultural, and medical applications (Keshavarz & Roy, 2010).

Biodegradation of PHAs

The term biodegradation describes a set of biochemical processes that decompose and convert organic substances to inorganic matter. Biodegradation is carried out by degraders, microorganisms (fungi, bacteria, protozoa), that grow in dead organic matter, residues produced by the ecosystems. This is a very important environmental process reducing contamination by organic wastes (Razza & Innocenti, 2012).

Polymers consist of long and bulky chains and therefore cannot penetrate the cell membrane to enter the cell, where cellular metabolism occurs. The biodegradability of polyesters of biological origin (PHAs, PLA) is in corroboration with the presence of specific enzymes, called esterases. Specifically, the process of biodegradation of PHAs begins with the hydrolysis and oxidation reactions caused by the secreted extracellular ester-specific enzymes of degraders (Razza & Innocenti, 2012). This procedure leads to the cleavage of the polymer chains in oligomers, dimers, and low molecular weight monomers (Roohi et al., 2018). These microorganisms utilize the carbon in polymer chains to produce CO_2, or other biomolecules that are necessary for their growth. Microbial biodegradation can be either aerobic or anaerobic. During the aerobic biodegradation, heterogeneous organic materials are degraded from a mixed microbial population in carbon dioxide (CO_2) and water (H_2O), with the simultaneous production of high amounts of biomass (Figure 2). This process requires aerobic conditions and a humid and warm environment. In the case of anaerobic biodegradation, the decomposition of organic matter primes the production of methane (CH_4), CO_2 and H_2O in the absence of oxygen, producing lower amounts of biomass (Figure 2) (Costa et al., 2018a; Razza & Innocenti, 2012; Roohi et al., 2018).

BIOPLASTICS PRODUCTION FROM MICROALGAE

Bioplastics have been a field of importance for both research and commercial purposes. Polyhydroxyalkanoates (PHAs), originated from biological procedures, are characterized as environmentally friendly, with their thermal and mechanical properties being comparable to those of conventional petrochemical polymers. Bacteria have been proven to be the most efficient microorganisms in PHAs production, as their productivity can reach 3.2 g L^{-1} h^{-1} (Singh et al., 2017). However, the PHA production by heterotrophic biomass requires high capital investment, due to the expensive carbon sources, the large oxygen demand, and the energy-intensive product recovery processes. In contrast, microalgae present an ecological and less expensive option for PHAs synthesis. Utilization of microalgae arises from their limited nutrient needs, as microalgal cells can photosynthesize by capturing carbon dioxide and using solar energy. In addition, microalgae can grow on different substrates, even in wastewater, exhibiting high growth rates. Wastewater treatment with simultaneous production of PHAs can be one of the future environmental strategies to reduce carbon dioxide emissions and consequently mitigate the greenhouse phenomenon in a potentially profitable way (Samantaray and Mallick, 2015; Singh et al., 2017).

Microalgae

Algae form a large group of multicellular and unicellular organisms that have the ability to photosynthesize and produce oxygen. They are distinguished into two major categories based on their size. Macroalgae are characterized by sizes ranging from a few centimeters to several meters, while microalgae are microorganisms with sizes up to a few hundreds of micrometers. Microalgae are eukaryotic photosynthetic

Figure 2. Aerobic and anaerobic degradation of PHB.

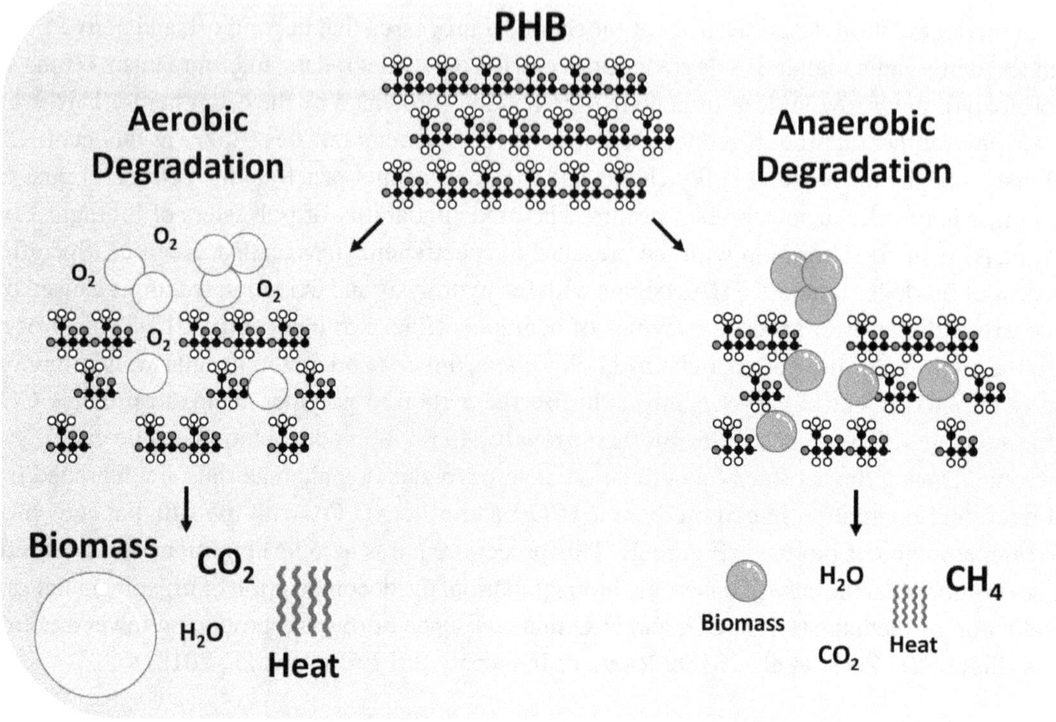

microorganisms and differ from cyanobacteria which are strictly prokaryotic cells, but this distinction is often blurry in the published literature, with cyanobacteria often being included in the microalgae group. Both eukaryotic microalgae and cyanobacteria play a dominant role in aquatic ecosystems as primary producers (Markou & Nerantzis, 2013). The oceans, which cover about 71% of the planet's surface, host more than 5000 species of microalgae which are the base of the marine food chain, responsible for 70% of earth biomass production and 50% of atmospheric oxygen (Barsanti & Gualtieri, 2006).

Microalgae are a group of phototrophic organisms that can, nevertheless, follow various alternative metabolic pathways depending on the changing habitat conditions (Rizwan et al., 2018; Singh et al., 2017). Three types of microalgal metabolism are distinguished. To grow autotrophically, cells use carbon dioxide or some other inorganic carbon source and absorb light radiation for their energy needs. Regarding the heterotrophic metabolism, organic compounds (acetic acid, maltose, glucose, etc.) are assimilated both as a carbon source and as a source of energy. By combining autotrophic and heterotrophic growth, microalgae can utilize either irradiation or organic substances as a source of energy, while carbon can be found in either inorganic or organic form (Rizwan et al., 2018; Singh et al., 2017). *Arthrospira platensis*, *Chlorella vulgaris* and *Haematococcus pluvialis* are examples of mixotrophic species (Rizwan et al., 2018).

Moreover, the wide variety of biochemical and physiological characteristics of microalgae makes them a subject of great commercial interest. Microalgae produce a number of biochemical substances, derived from the cellular metabolism, such as pigments, proteins, lipids, carbohydrates, and biopolymers (Koutra, Economou, et al., 2018). Commercial uses of microalgal biomass are relevant to the food industry, the production of organic fertilizers and the composition of pharmaceuticals and antimicrobial drugs (Koutra, Economou, et al., 2018; Rizwan et al., 2018). As previously mentioned, microalgae are also involved in environmental applications, including carbon dioxide fixation, wastewater treatment and bioremediation of heavy metals (arsenic (As), lead (Pb), mercury (Hg) etc.) (Kwon et al., 2017; Spain et al., 2021).

Cultivation of Microalgae

For the cultivation of microalgae, an adequate source of inorganic carbon (CO_2, HCO^{3-}) and light are required for photosynthesis. As already mentioned, carbon can also be supplied as sugars, acids, and alcohols. Besides carbon, nitrogen (N) is the most important nutrient that contributes to biomass production. The nitrogen content of biomass can range from 1% to 10%, or more. Nitrogen is mainly assimilated by microalgae in the form of nitrates (NO^{3-}), but ammonium (NH^{4+}) or urea are also used without adversely affecting their growth rate. Phosphorus (P) is also essential for growth and for many cellular functions such as energy transfer and nucleic acid synthesis. The preferred form by microalgae is orthophosphates (PO_4^{3-}). Sulfur (S), potassium (K), sodium (Na), iron (Fe), magnesium (Mg), calcium (Ca), and trace elements such as boron (B), copper (Cu), manganese (Mn), zinc (Zn) are equally important. In total, about 30 minerals and other organic compounds can be consumed by microalgae. In general, the biomass productivity of microalgae depends on a wide range of parameters, including the species of microalgae, nutrients, light intensity, temperature, pH level, substrate concentration and culture purity ("Handb. Microalgal Cult.," 2006).

Microalgae can grow in either open or closed systems, called bioreactors. In the case of open-type systems, microalgae cultures develop in natural or artificial open outdoor ponds and puddles. Closed systems, photobioreactors (PBRs), are made of various transparent materials, such as plastic or glass, and they differ in their design and operation characteristics. Based on their design, photoreactors are flat or tubular, horizontal, inclined, vertical, spiral, or helical ("Handb. Microalgal Cult.," 2006). Although open tanks are more durable, easier, and cheaper to construct and operate, most microalgae cannot be stored in them for long periods, as there is a high risk of contamination by fungi, bacteria, and protozoa, as well as competition against other microalgal strains that tend to dominate in the culture. In contrast, photobioreactors offer protection against contamination and more sufficient control of the conditions, ensuring the dominance of the desired strains (Rizwan et al., 2018). The choice of cultivation system depends on numerous factors, mainly on the nature of the strain, the availability of nutrients, the climate, the cultivation method, and the desirable end use of biomass (Markou & Nerantzis, 2013; Rizwan et al., 2018).

PHB Production by Microalgae

Today, about 100 strains of eukaryotic and prokaryotic (cyanobacterial) microalgae have been indicated as capable of accumulating poly (3-hydroxy butyrate) (PHB), from 0.04% to 80% of their dry mass, through photoautotrophic metabolic pathways (Cassuriaga et al., 2018; Kavitha et al., 2016). Thereafter, most studies focus on modifying cultivation conditions in order to both maximize and optimize PHB production by different microalgae strains. It has been reported that the most important factors that enhance the accumulation of PHB in microalgae biomass are:

1. The presence of organic carbon (such as acetic acid, pentoses, etc.) (Abdo & Ali, 2019; S. Zhang & Bryant, 2015).
2. Light exposure time reduction (Costa et al., 2018a).
3. Limitation of nitrogen (N) and phosphorus (P) (Krasaesueb et al., 2019; Martins et al., 2014; Samantaray & Mallick, 2015).
4. Limitation of certain heavy metals (Ni and Cu) (Samantaray & Mallick, 2015).
5. Gas transfer resistance in the culture (Samantaray & Mallick, 2015).

A large number of strains, if grown under the above conditions, can accumulate high amounts of bioplastic PHB and its copolymers. This concentration can even exceed 80% of their dry mass. Among the green microalgae, some cyanobacteria, such as *Nostoc muscorum, Spirulina platensis, Aulosira fertilissima* and *Synechocystis sp.* have been mentioned to accumulate PHB under mixotrophic growth. Polymer content of these strains, ranged from 29% to 85% (on dry biomass basis), depending on the species grown and culture medium (Ansari & Fatma, 2016). A brief overview of microalgae reported in literature as being able to produce PHAs is presented in Table 1.

WASTEWATER SUBSTRATES FOR MICROALGAE CULTIVATION

The industrial microalgae biomass production is considered unprofitable, due to the large requirements for water and nutrients. Microalgae, as well as cyanobacteria, assimilate organic carbon, inorganic components, such as nitrogen (N), phosphorus (P) and micronutrients that are abundant in wastewaters (Bhati & Mallick, 2015; Koutra et al., 2021). In addition, microalgae are able to bioremediate certain toxic heavy metals, such as arsenic (As), berilium (Be), cromiun (Cr), lead (Pb), and mercury (Hg), that are present in industrial waste (Das et al., 2018). Subsequently, the use of wastewater as a substrate for microalgae growth is a promising, alternative, ecological, and profitable solution regarding the combination of high-added-value substances production and wastewater management.

Early-stage research is conducted in the field of microalgae cultivation in wastewater aiming to bioplastic production, and more specifically, poly (hydroxy alkanoate) esters (PHAs) (Figure 3). Evidence suggests that high concentrations of PHAs with properties comparable to those of commercial plastics can be obtained. Indicatively, (Krasaesueb et al., 2019) isolated 32.48% bioplastic PHB (of biomass dry weight) in the cyanobacterium *Synechocystis sp.* PCC 6803 (ΔSphU) which was cultured on shrimp farm waste under nitrate limiting conditions. Another research reported a concentration of 65% (dry weight of biomass) of copolymer P(3HB-co-3HV) in the cyanobacterium *Nostoc muscorum* Agardh, which was grown on 10% poultry waste with CO_2 enrichment and 0.28% acetic acid, 0.30% valeric acid, and 0.38% glucose (Bhati & Mallick, 2016).

Anaerobic digestion is an important procedure regarding the renewable energy production sector as well as the field of waste management. During anaerobic digestion, organic matter is degraded by microorganisms resulting in biogas production. Biogas is a renewable energy source that consists mainly of methane (CH_4) and carbon dioxide (CO_2). The aftermath of this process is a large volume of wastewater, which is called digestate (Bjornsson et al., 2013). Digestates are considered a rich source of inorganic nutrients (ammoniacal nitrogen and phosphorus) (Ayre et al., 2017; Bjornsson et al., 2013). These characteristics render digestates as suitable substrates for microalgae cultivation. Recent research has shown that microalgae can also grow in digestates with very high concentrations of ammonium nitrogen (800-

Table 1. Brief overview of species reported in literature as able to produce PHAs.

Species	References
Spirulina LEB. 18	(Martins et al., 2014)
Spirulina sp. LEB 18, *Nostoc ellipsosporum*	(Martins et al., 2017)
Chlorella pyrenoidosa	(Das et al., 2018)
Synechocystis sp. PCC 6714	(Kamravamanesh et al., 2017)
Synechococcus sp. MA19	(Nishioka et al., 2001)
Synechocystis sp. PCC6803	(Sudesh et al., 2002)
Synechococcus elongates	(Mendhulkar & Shetye, 2017)
Synechococcus elongates	(Mendhulkar & Shetye, 2017)
Nostoc muscorum	(Mallick et al., 2007)
Nostoc muscorum	(Bhati & Mallick, 2015)
Spirulina platensis	(Toh et al., 2008)
Synechocystis sp.	(Toh et al., 2008)
Synechocystis sp.	(Panda et al., 2006)
Synechococcus subsalsus	(Costa et al., 2018b)
Chlorogloeopsis fritschii PCC 9212	(Zhang & Bryant, 2015)
Stigeoclonium sp. B23	(Mourão et al., 2020)
Scenedesmus sp.	(García et al., 2021)
Chlorella vulgaris	(Setyorini & Dianursanti, 2021)

Figure 3. Production of PHB from microalgae cultivated in wastewater.

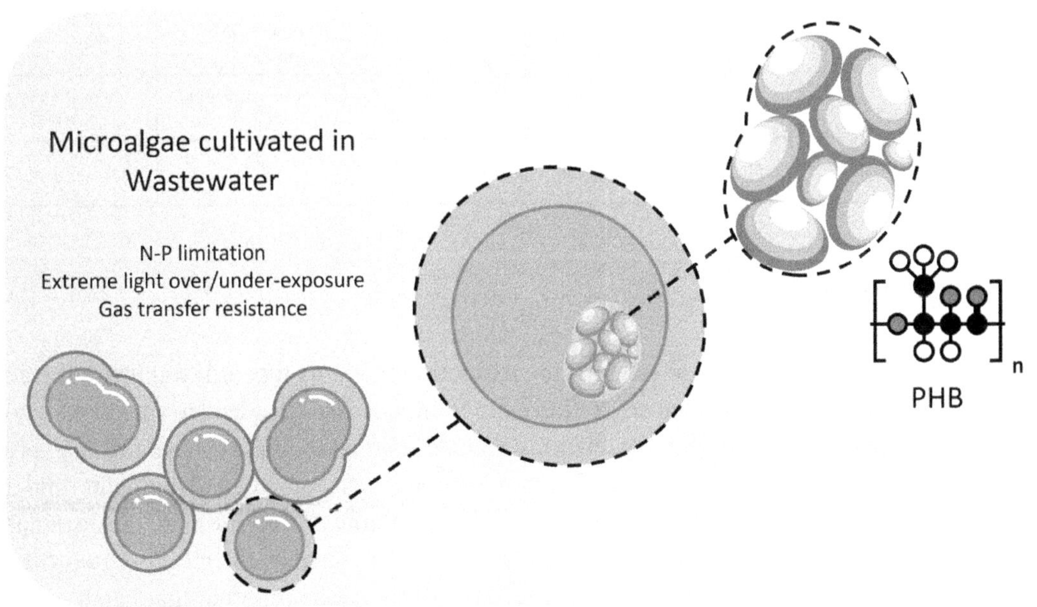

1600 mg N-NH^{+4} L^{-1}), which correspond to toxic levels for the majority of microorganisms (Ayre et al., 2017). Therefore, the use of digestates as a substrate for the cultivation of microalgae paves the way for a number of economically viable and environmentally friendly applications, such as biomethane and biofuel production as well as the production of high-added-value products, including PHA bioplastics (Koutra et al., 2018). However, limited information is available regarding the PHB production by microalgae cultivated in digestate. A recent research demonstrated the cultivation of the cyanobacterium *Synechocystis salina* in digestate and the accumulation of a 6.3% (w/w) poly (3-hydroxybutyrate) (P3HB) (Meixner et al., 2016). Nevertheless, the cultivation of microalgae in digestate substrates requires further examination, as there are obstacles that must be overcome, such as the presence of solids and potentially harmful substances (Bauer et al., 2021). Another important repressive factor against the growth of microalgae is the elevated turbidness of the digestate, due to the suspended solids, that results in the reduction of the photosynthetically active radiation (PAR) provided to the culture (Bauer et al., 2021). Table 2 summarizes the limited available published results regarding the production of bioplastics from microalgae cultivated in wastewater.

FUTURE PROSPECTS

Table 2. Published data regarding the production of bioplastics from microalgae cultivated in wastewater.

Strain	Growth media	PHB productivity	Ref.
Botryococcus braunii	60% sewage wastewater	247 mg L^{-1}	(Kavitha et al., 2016)
Aulosira fertilissima	Aquaculture wastewater	34.7% of biomass DM, 205 mg L^{-1}	(Samantaray et al., 2011)
Synechocystis sp. PCC 6803 strain ΔSphU	Shrimp wastewater	32.5% of biomass DM	(Krasaesueb et al., 2019)
Nostoc muscorum Agardh	10 g L^{-1} Poultry litter	23% of biomass DM, 144 mg L^{-1}	(Bhati & Mallick, 2016)
	10 g L^{-1} Poultry litter, 10% CO2, 0.28% acetate, 0.38% glucose, 0.30% valerate	65% of biomass DM, 774 mg L^{-1}	
Synechocystis salina	digestate supernatant	5.5% of biomass DM, 88 mg L^{-1}	(Kovalcik et al., 2017)
	1/3 diluted low-solid digestate supernatant	6.3% of biomass DM, 95 mg L^{-1}	(Meixner et al., 2016)

Biopolymers Extraction

In recent years, research has focused on establishing extraction methods that will minimize the cost of bioplastics production. It is a general concern to find extraction processes that lead to sufficient recovery of bioplastics from microalgal biomass while ensuring that the residual biomass can be further exploited. Microalgae are sources of a variety of valuable compounds (proteins, fatty acids, and pigments) and therefore their co-production alongside PHA would enhance the economic and environmental sustainability of such processes. The quantity and quality of bioproducts are determined by the biorefinery procedures that follow the biomass growth. In the work of (Fei et al., 2016). It was reported that harsh extraction

Figure 4. Steps typically followed during biopolymer extraction from biomass.

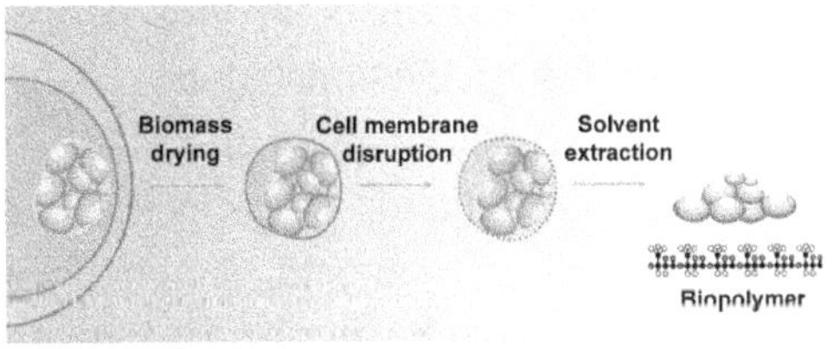

methods including high temperatures or strong acids could negatively affect the quality of the produced biopolymers. A general fact that emerges from the current literature is that extraction efficiency depends on both the methods followed and the algal strain that is under examination (Haddadi et al., 2019).

The most common extraction practices are based on organic solvents and can be divided in three steps (Figure 4). The first step is biomass drying. Removing biomass moisture can be achieved through freeze-drying, or solar drying, and is critical for the use of organic solvents in the next steps of the process. The second step of biopolymer extraction is the disruption of the cell membrane. This can be achieved by the use of organic solvents or some type of physical stress, like sonication. The second step facilitates the mass transfer of the biopolymer from inside the cell to the bulk of the extraction solution that is added during the third step of the process. Chloroform, acetone, and dichloromethane are typically used solvents due to their ability to dissolve biopolymers but not other cellular residues (Levett et al., 2016). These solvents dissolve the lipid part of the non-PHA cell mass (NCPM), which can also easily be removed prior to PHA extraction by methanol or ethanol, and dissolve both short chain length and medium chain length PHA, without dissolving any other NCPM components (Koller et al.2013). Once the dry biomass has come into contact with a suitable solvent and the extraction has been carried out, another solvent, usually methanol, is needed for the precipitation and the recovery of the crystallized biopolymer (Kosseva & Rusbandi, 2018;). Although organic solvents provide a product with a limited moisture content and do not cause a significant reduction in the molecular weight of the polymer, they come with great cost and high environmental impact (Kosseva & Rusbandi, 2018). Consequently, there is a need for more environmentally friendly processes with low cost for scaled-up bioplastic production.

A possible alternative for biopolymers recovery is biomass digestion. Cells containing PHA are hydrolyzed in an acidic or alkaline medium while the desirable product remains insoluble. On the other hand, sulfuric acid, or sodium hypochlorite, that are commonly used for pH adjustment, seem to be harmful for the recovered polymers, causing a reduction of their molecular weight and affecting the physicochemical properties of the product (Fei et al., 2016). In order to avoid such phenomena, research interest has shifted to biomass treatment through the use of enzymes capable of denaturing the cell wall without degrading PHAs (Kapritchkoff et al., 2006; Poltronieri et al., 2016).

Another idea for extraction processes free of organic solvents is the use of supercritical fluids. Supercritical carbon dioxide is mostly reported as being used for this purpose, extracting about 90% of the PHA content at purity levels that even reach 99% (Gumel et al., 2013). Moreover, supercritical carbon dioxide extraction can be applied as a secondary step to purify the biopolymers by removing oily biomass residues at 150 bar and 50 °C (Daly et al., 2018). Despite all the promising results reported in literature regarding extraction and purification using supercritical fluids, the high operational cost of such processes hinders their full-scale implementation.

The utilization of non-chlorinated solvents, such as cyclohexanone and γ-butyrolactone, show great potential in the field of non-toxic eco-friendly solvents, as they perform just as well as halogenated solvents, at a lower cost. Extraction of PHB with cyclohexanone results to sufficient recovery, but the procedure seems sensitive to temperature changes (Jiang et al., 2018). Furthermore, for the purpose of total replacement of organic solvents, the ionic liquids are proposed as extraction media. Ionic liquids are solutions consisting of ions derived from salts and act like common organic solvents due to their electrical charged anions (B. Wang et al., 2017). Either wet or dry biomass could be handled with 1-Ethyl-3-methylimidazolium diethyl phosphate, $[C_{2mim}]$ $[(C_2)_2OPO_3]$, at moderate temperatures, to recover biopolymer compounds up to 60%. It is also emphasized that the ionic liquid can be recovered enhancing the viability of the process (Dubey et al., 2018). Last but not least, Aqueous Two-Phase Extraction (ATPE) is a method based on the formation of two discrete phases containing either a pair of different polymers or a polymer and an inorganic salt that are dissolved in water (Leong, Show, et al., 2017). The temperature, the concentration of the dissolved compounds and the duration of the extraction determine the recovery rate of the bioplastics. Indicatively, ethylene oxide- propylene oxide/ sodium chloride, ethylene oxide- propylene oxide/ ammonium sulfate and polyethylene glycol/ potassium phosphate are some of the pairs of dissolved substances in an ATPE regarding PHA (Leong, Show, et al., 2017). This method is considered as technically and economically feasible both as the main extraction process and as a pretreatment step, due to its non-toxicity and scalability at limited operational cost.

In most biorefinery strategies for bioplastics production, a biomass pretreatment step is beneficial as it increases the biopolymer recovery rate and purification levels. The pretreatment techniques aim at the physicochemical disruption of the cells used and differ from each other in terms of energy requirements and efficiency (Kartik et al., 2021). The degradation of cells is possible through the application of shear stresses, which can be achieved through a bead mill or a high-pressure homogenizer. The temperature of both processes should be controlled, so that it remains close to 25 °C in order not to harm the extracted polymers (Steinbuchel et al., 2013). Thermal pretreatment is also widely applied, as the cell boundaries may be degraded due to severe temperature changes. Nonetheless, biomass heating can reduce the quality of the polymer if its temperature and duration are not properly regulated, while freezing-thawing sequences have large energy demands (Steinbuchel et al., 2013). In addition, breakage of cells can be done by sonication prior to PHA or PHB extraction. The biomass is dispersed in an ionic or non-ionic surfactant to enhance the effectivity of the ultrasounds. Sodium dodecyl sulfate (SDS), involved in sonication assay, led to extracted biopolymer with the purity of 96% (Arikawa et al., 2017). It was indicated that sample sonication pretreatment minimized the need of harmful organic solvents, as biomass treated with sodium hypochlorite presented PHB with satisfying characteristics (Martínez-Herrera et al., 2020). Non-ionic surfactants like polyoxymethylene sorbitan monolaurate could also improve PHA recovery, but the use of chloroform and acidified methanol could not be avoided in order to recover a pure product without significant molecular weight reduction (Colombo et al., 2020).

Microalgal Biomass Blends

Bio-based polymer processes are usually characterized by high cost due to extraction and purification stages that aim at a commercial product with the desirable features. Although biorefinery methods come at significant financial demands, they rarely provide a product that has not been devalued in terms of its thermal and mechanical properties (Rujnić-Sokele & Pilipović, 2017). Nowadays, the market demands high quality polymers, such as those derived from fossil fuels, and this fact conflicts with the environmental need for recyclable and biodegradable materials. Thus, efforts are made to synthesize plastics from blending biopolymers with a variety of additives or even petroleum-based polymeric compounds. Blends are considered new materials with different properties, usually more attractive, compared to those of their ingredients. However, despite the improved nature of the new material, questions are raised about their homogeneity and biodegradability (Endres, 2017).

Requirement of homogeneity arises from the need of materials without morphological abnormalities and failures. Success in blending depends not only on factors such as temperature, pressure, and processing time, but also on the compatibility of the materials to be mixed. In the case of biopolymers blends, the origin of the utilized biomass (the species of microorganisms and the growth conditions) also plays an important role. For instance, when *Chlorella vulgaris* and *Spirulina platensis* were evaluated for biopolymer production, the first species demonstrated decent bioplastic properties, while the second one was proven as more suitable for blending with polyethylene (Zeller et al., 2013).

Pre-treatment of cells containing polymers is recommended to prepare a homogeneous material from biomass, commercial plastics, and other chemical additives. Pre-treatment can be either mechanical or chemical. Biomass grinding is essential to form small polymeric particles that can easily disperse into the body of the produced blend. The better dispersion of biopolymers caused an increase in crystallinity, which affected the mechanical properties, but reduced the melting temperature of the final product (Simonic & Zemljic, 2020). Furthermore, ultra-sonication was applied before mixing Chlorella with PVA giving as a result a uniform product without surface irregularities. The sonicated biomass led to a blend with greater tensile strength and elongation capacity (Sabathini et al., 2018). PVA was also blended with Spirulina biomass residues after ultra-sonication, resulting in a material with a tensile strength of 22 MPa and 77% elongation at break. The final product exhibited water resistance which was attributed to the inorganic biomass components extracted by ultra-sonication (C. Zhang et al., 2020). Chemical methods are commonly used for biomass washes to eliminate non polymerizable compounds (Jang et al., 2013). Ethanol has been used as cell suspension medium, in order to disrupt the cell wall of *Chlorella sorokiniana* and form starch particles (Gifuni et al., 2017). In case a colorless polymeric material is required, methanol is a useful solvent, as it removes the unwanted biomass pigments (Monshupanee et al., 2016). Finally, low concentration of either base or acid degraded cellular polysaccharides with the assistance of specific enzymes, such as cellulase, have been used to produce monomers for biopolymers formation (Naresh Kumar et al., 2020).

Polymer blends are constructed inside a mold under high temperatures and escalated pressure. Both the conditions and the process duration are determined according to the preferences for the properties of the final product. Early laboratory approaches included roller mixers for blending and shaping bioplastics at high temperatures (Otsuki et al., 2004). However, the development of bioplastics production promoted the use of different methods with the prospect of being applied at a larger scale. When, internal mixers were used to synthesize bioplastic from corn starch and the biomass of microalgae, the species *Nannochloropsis gaditana* presented blends with great flexibility and permeability to oxygen (Fabra et al., 2018). In addition, solvents could be applied to cast polymeric blends provided that the correct ratio between the components and the dispersant were used. Poly Vinyl Alcohol (PVA) was combined with *Chlorella* biomass in the presence of glycerol and citric acid in distilled water and the dependence of the final material properties on the concentration of biomass was more than obvious (Sabathini et al., 2018). More complex arrangements, such as screw extrusion and injection molding, are also piloted in the field of bio-based blends. Although such applications seem unprofitable due to their energy demand, they improve blends production by limiting operating time and ensuring material homogeneity (Mathiot et al., 2019; Torres et al., 2015). During blending, the bondage between two polymeric compounds can be reinforced by the addition of substances that act as compatibilizers or plasticizers. Compatibilizers are polymeric species that can stabilize a mixture of two or more immiscible ingredients, rendering it a uniform material with improved features. More specifically, the particles of a compatibilizer are inserted into the blend bulk and reduce the interfacial tension between the immiscible components (Chen & White, 1993). On the other hand, plasticizers are involved in polymer chain mobility, or they interfere between polymer bonds. Therefore, a flexible and strong final polymeric blend can be produced (Onen Cinar et al., 2020).

Over the years, numerous blends of polymers with biomass or biomass products have been synthesized and studied in terms of their thermal and mechanical behavior. Biomass originated PHB was evaluated for its combination with Poly Propylene Carbonate (PPC). The composites were considered irreconcilable, and Poly Vinyl Acetate (PVAC) was added as a compatibilizer. As a result, the mechanical properties of the co-polymer were meliorated, but the degradation temperature decreased as well as the temperature at which the mixture crystallized (X. Wang et al., 2005). Moreover, Poly Propylene (PP) even in limited content gave PHB endurance by lowering its rigidity (Pachekoski et al., 2009). Although *Nannochloropsis* salina provided thermal stability to the blend, it had a negative impact to the mechanical properties of PVA. After the addition of Poly Diallyldimethylammonium chloride (PD) compatibilizer, the blend was homogenized and displayed a tensile strength of 20 MPa (Tran et al., 2016). The PD restored the previously low values of the elongation at break, and at the same time Young's modulus increased even in comparison with pure PVA performance (Tran et al., 2016). *Chlorella vulgaris* biomass was also examined about its blending ability with PVA. In that study, maleic anhydrate and glycerol as a plasticizer were needed in order to produce a good quality polymer blend. It was indicated that the more compatibilizer, up to 5% per weight of PVA, was added the better the results of the material in the mechanical tests were (Dianursanti et al., 2018).

Spirulina biomass is the subject of more recent research relevant to its compatibility in plastic blends. Blending *Spirulina* with non-biodegradable or slow biodegradable plastics aim potentially to increase the recyclability of the final product. Despite the devaluation of the plastic, *Spirulina* displayed remarkable miscibility with Polyethylene (PE). However, mixture elongation rate was improved by increasing PE content (Zeller et al., 2013). Although Poly Butylene Succinate (PBS) is a biodegradable polymer with a profile similar to that of PP, Spirulina was dispersed to examine new material prospects (Zhu et al., 2017). After the supplementation of maleic anhydride as a compatibilizer, both the tensile strength and the Young's modulus increased. Nevertheless, the compatibilizer was unable to fix the unsatisfactory elongation at break results. The final blend became crystalized at 73 °C and melted at 103 °C (Zhu et al., 2017). Furthermore, *Spirulina* biomass was merged with Poly Vinyl Alcohol (PVA) in several ratios to form polymeric films and glycerol was recruited for plasticizing (Shi et al., 2017). After films appraisal, it was indicated that PVA content led to high tensile strength, while glycerol was responsible for the elevated flexibility. It was also mentioned, that both PVA and glycerol contributed to the production of water-resistant recyclable films (Shi et al., 2017). Instead, water sensitivity occurred as an outcome of blending at low cost with wheat gluten, octanoic acid and butanediol despite the encouraging mechanical features (Ciapponi et al., 2019). Currently, attempts are being made to blend Spirulina biomass with Poly Lactic Acid (PLA) without the need of compatibilizers. The newly developed blend was characterized by greater crystallinity than pure PLA and satisfactory durability (Simonic & Zemljic, 2020).

Properties and Applications

Among other biodegradable plastics (such as PLA or starch-based plastics), PHAs have come to the limelight due to their variety of chemical structures and their unique material characteristics (Keshavarz & Roy, 2010). From an environmental point of view, high-density PHA fragments do not float in aquatic receptors, therefore, once discarded, they will be submerged and degraded by surface biogeochemical mechanisms (Balaji et al., 2013; Costa et al., 2018b, 2019). PHAs are completely biodegradable, and biocompatibility makes them suitable for health and medical applications (Sreenikethanam & Bajhaiya, 2021). The widespread use of PHAs in various industry sectors has been caused by the thermoplasticity that is accompanied by hydrophobicity, water insolubility and resistance to oxidative atmosphere (Balaji et al., 2013; S. S. Costa et al., 2019; Sharma et al., 2021). In addition, such bioplastics are characterized by optical purity, but good resistance to ultra-violet irradiation (compared to polypropylene) (Markl et al., 2019).

Poly-hydroxybutyrate (PHB), the most common subcategory of the PHAs group, exhibits some characteristics comparable to those of conventional plastics (such as PP, PE, PS), which could make PHB an ideal alternative to petrochemical plastics (Domínguez-Díaz et al., 2015; Khanna & Srivastava, 2005). For example, PHB has an average melting temperature of approximately 179 °C and a relatively high tensile strength of 40 MPa, whilst Polypropylene (PP) and Polystyrene (PS) show melting temperatures of 170 °C and 110 °C respectively and tensile strengths at 35 MPa and 50 MPa (Khanna & Srivastava, 2005). Furthermore, fibers of PHB homopolymer have a hard-elastic behavior. The high degree of crystallinity (60-80%), the brittleness and the very low elongation at break (5% for PHB, while 400% for PP), pose significant limitations for the utilization of PHB in a wide range of applications, such as the construction of durable building materials (Aydemir & Gardner, 2020; Bhati & Mallick, 2012; Muneer et al., 2020).

Generally, a degree of crystallinity above 50% is considered unsuitable for industrial and commercial use, as it increases fragility (Laycock et al., 2013). The brittleness of PHB is directly related to the fact that its glass transition temperature (T_g, 2-4°C) is close to room temperature (S. S. Costa et al., 2019; Markl et al., 2019; Muneer et al., 2020). Thus, the storage of PHB under ambient conditions could probably lead to further brittleness and stiffness (Domínguez-Díaz et al., 2015). A typical solution to this material failure is the addition of plasticizers, which could enhance the molecular motion and moderate the glass transition point.

Also, the temperature at which the PHB becomes thermally degraded (200–220 °C) is quite close to its melting point (T_m, 160–175 °C) often leading to failures during processing (Aydemir & Gardner, 2020; Domínguez-Díaz et al., 2015). In contrast to all of the above, medium chain length PHAs (mcl-PHAs) and copolymers seem to be more suitable for commercial and industrial applications, as they have a lower degree of crystallinity (20–40%) and lower melting temperature. This kind of PHAs is less brittle and stiff, as they exhibit increased elasticity and present much higher elongation at break (300–450%) (A Rahman & Miller, 2017).

The molecular weight and the quality of the monomers determine the physical properties of PHAs. It is important to emphasize the fact that the methods used to extract PHAs directly affect the chemical composition (monomeric sequences) and thus the physical characteristics of the polymer. Different extraction and recovery methods cause variations in molecular weight, resulting in a multiplicity of crystallinity levels and durability (Costa et al., 2018b). The chemical composition (monomeric composition) of the PHAs is also determined by the biological production process and the producing microorganism. Finally, chemical modification (by chlorination, cross-linking, epoxidation, hydroxylation, carboxylation) is suggested to alter the profile of the biologically derived PHA and adjust its features to the desired application (Koller et al., 2010).

CONCLUSION

Microalgae have been proven capable of synthesizing bioplastics like PHB as a form of energy and carbon storage, under stress. This has been validated by researchers even when microalgae were cultivated in wastewater, providing an ecofriendly solution for the necessary nutrients. This is expected to significantly reduce the carbon footprint of bioplastics and provide an eco-friendly alternative to petrochemical plastics. Research is still necessary to optimize and improve the bioplastic procedure followed in the scope of minimizing the use of organic solvents and the overall environmental footprint of the process. The downstream part of bioplastic production process can also have a strong influence on the properties of the final product, and as a result the applications for which it can be used. Through the research carried out in this field, the gap between the properties of bioplastics and fossil-based plastics has been significantly reduced, with bioplastics becoming suitable for more applications. It is certain that from a technical point of view bioplastics have a hard time competing with the well-established use of conventional plastics, but with the depletion of fossil fuels, climate change, and increased environmental awareness, times seems to be on their side.

REFERENCES

Abdo, S. M., & Ali, G. H. (2019). Analysis of polyhydroxybutrate and bioplastic production from microalgae. *Bulletin of the National Research Center*, *43*(1), 97. Advance online publication. doi:10.118642269-019-0135-5

Achilias, D. S., Roupakias, C., Megalokonomos, P., Lappas, A. A., & Antonakou, V. (2007). Chemical recycling of plastic wastes made from polyethylene (LDPE and HDPE) and polypropylene (PP). *Journal of Hazardous Materials*, *149*(3), 536–542. doi:10.1016/j.jhazmat.2007.06.076 PMID:17681427

Andrady, A. L. (2011). Microplastics in the marine environment. *Marine Pollution Bulletin*, *62*(8), 1596–1605. doi:10.1016/j.marpolbul.2011.05.030 PMID:21742351

Ansari, S., & Fatma, T. (2016). Cyanobacterial polyhydroxybutyrate (PHB): Screening, optimization and characterization. *PLoS One*, *11*(6), 1–20. doi:10.1371/journal.pone.0158168 PMID:27359097

Arikawa, H., Sato, S., Fujiki, T., & Matsumoto, K. (2017). Simple and rapid method for isolation and quantitation of polyhydroxyalkanoate by SDS-sonication treatment. *Journal of Bioscience and Bioengineering*, *124*(2), 250–254. doi:10.1016/j.jbiosc.2017.03.003 PMID:28456470

Aydemir, D., & Gardner, D. J. (2020). Biopolymer blends of polyhydroxybutyrate and polylactic acid reinforced with cellulose nanofibrils. *Carbohydrate Polymers*, *250*(July), 116867. doi:10.1016/j.carbpol.2020.116867 PMID:33049817

Ayre, J. M., Moheimani, N. R., & Borowitzka, M. A. (2017). Growth of microalgae on undiluted anaerobic digestate of piggery effluent with high ammonium concentrations. *Algal Research*, *24*, 218–226. doi:10.1016/j.algal.2017.03.023

Balaji, S., Gopi, K., & Muthuvelan, B. (2013). A review on production of poly β hydroxybutyrates from cyanobacteria for the production of bio plastics. *Algal Research*, *2*(3), 278–285. doi:10.1016/j.algal.2013.03.002

Barsanti, L., & Paolo, G. (2006). *Algae: Anatomy, Biochemistry, and Biotechnology. By. CRC Press.* Taylor & Francis Group.

Bauer, L., Ranglová, K., Masojídek, J., Drosg, B., & Meixner, K. (2021). Digestate as sustainable nutrient source for microalgae—Challenges and prospects. *Applied Sciences (Switzerland)*, *11*(3), 1–21. doi:10.3390/app11031056

Bhati, R., & Mallick, N. (2012). Production and characterization of poly(3-hydroxybutyrate-co-3-hydroxyvalerate) co-polymer by a N 2-fixing cyanobacterium, Nostoc muscorum Agardh. *Journal of Chemical Technology and Biotechnology (Oxford, Oxfordshire)*, *87*(4), 505–512. doi:10.1002/jctb.2737

Bhati, R., & Mallick, N. (2015). Poly(3-hydroxybutyrate-co-3-hydroxyvalerate) copolymer production by the diazotrophic cyanobacterium Nostoc muscorum Agardh: Process optimization and polymer characterization. *Algal Research*, *7*, 78–85. doi:10.1016/j.algal.2014.12.003

Bhati, R., & Mallick, N. (2016). Carbon dioxide and poultry waste utilization for production of polyhydroxyalkanoate biopolymers by Nostoc muscorum Agardh: Asustainable approach. *Journal of Applied Phycology*, *28*(1), 161–168. doi:10.100710811-015-0573-x

Bjornsson, W. J., Nicol, R. W., Dickinson, K. E., & McGinn, P. J. (2013). Anaerobic digestates are useful nutrient sources for microalgae cultivation: Functional coupling of energy and biomass production. *Journal of Applied Phycology*, *25*(5), 1523–1528. doi:10.100710811-012-9968-0

Cassuriaga, A. P. A., Freitas, B. C. B., Morais, M. G., & Costa, J. A. V. (2018). Innovative polyhydroxybutyrate production by Chlorella fusca grown with pentoses. *Bioresource Technology*, *265*, 456–463. doi:10.1016/j.biortech.2018.06.026 PMID:29935455

Chen, C. C., & White, J. L. (1993). Compatibilizing agents in polymer blends: Interfacial tension, phase morphology, and mechanical properties. *Polymer Engineering and Science*, *33*(14), 923–930. doi:10.1002/pen.760331409

Chen, G. Q., & Wu, Q. (2005). The application of polyhydroxyalkanoates as tissue engineering materials. *Biomaterials*, *26*(33), 6565–6578. doi:10.1016/j.biomaterials.2005.04.036 PMID:15946738

Ciapponi, R., Turri, S., & Levi, M. (2019). Mechanical Reinforcement by Microalgal Biofiller in Novel Thermoplastic Biocompounds from Plasticized Gluten. *Materials (Basel, Switzerland)*, *12*(9), 1476. doi:10.3390/ma12091476 PMID:31067771

Colombo, B., Pereira, J., Martins, M., Torres-Acosta, M. A., Dias, A. C. R. V., Lemos, P. C., Ventura, S. P. M., Eisele, G., Alekseeva, A., Adani, F., & Serafim, L. S. (2020). Recovering PHA from mixed microbial biomass: Using non-ionic surfactants as a pretreatment step. *Separation and Purification Technology*, *253*(April), 117521. Advance online publication. doi:10.1016/j.seppur.2020.117521

Costa, J. A. V., Moreira, J. B., Lucas, B. F., Da Silva Braga, V., Cassuriaga, A. P. A., & De Morais, M. G. (2018a). Recent advances and future perspectives of PHB production by cyanobacteria. *Industrial Biotechnology (New Rochelle, N.Y.)*, *14*(5), 249–256. doi:10.1089/ind.2018.0017

Costa, S. S., Miranda, A. L., Andrade, B. B., Assis, D. de J., Souza, C. O., de Morais, M. G., Costa, J. A. V., & Druzian, J. I. (2018b). Influence of nitrogen on growth, biomass composition, production, and properties of polyhydroxyalkanoates (PHAs) by microalgae. *International Journal of Biological Macromolecules*, *116*, 552–562. doi:10.1016/j.ijbiomac.2018.05.064 PMID:29763703

Costa, S. S., Miranda, A. L., de Morais, M. G., Costa, J. A. V., & Druzian, J. I. (2019). Microalgae as source of polyhydroxyalkanoates (PHAs)— A review. *International Journal of Biological Macromolecules*, *131*, 536–547. doi:10.1016/j.ijbiomac.2019.03.099 PMID:30885732

Cox, K. D., Covernton, G. A., Davies, H. L., Dower, J. F., Juanes, F., & Dudas, S. E. (2019). Human Consumption of Microplastics. *Environmental Science & Technology*, *53*(12), 7068–7074. doi:10.1021/acs.est.9b01517 PMID:31184127

Daly, S. R., Fathi, A., Bahramian, B., Manavitehrani, I., McClure, D. D., Valtchev, P., Schindeler, A., Dehghani, F., & Kavanagh, J. M. (2018, January). A green process for the purification of biodegradable poly(B-hydroxybutyrate). *The Journal of Supercritical Fluids*, *135*, 84–90. doi:10.1016/j.supflu.2018.01.007

Das, S. K., Sathish, A., & Stanley, J. (2018). Production of Biofuel and Bioplastic from Chlorella Pyrenoidosa. *Materials Today: Proceedings*, *5*(8), 16774–16781. doi:10.1016/j.matpr.2018.06.020

Dianursanti, G. M., & Noviasari, C. (2018). The effect of glycerol addition as plasticizer in Spirulina platensis based bioplastic. *E3S Web of Conferences, 67*, 3048. doi:10.1051/e3sconf/20186703048

Domínguez-Díaz, M., Meneses-Acosta, A., Romo-Uribe, A., Peña, C., Segura, D., & Espin, G. (2015). Thermo-mechanical properties, microstructure and biocompatibility in poly-β-hydroxybutyrates (PHB) produced by OP and OPN strains of Azotobacter vinelandii. *European Polymer Journal*, *63*, 101–112. doi:10.1016/j.eurpolymj.2014.12.002

Dubey, S., Bharmoria, P., Gehlot, P. S., Agrawal, V., Kumar, A., & Mishra, S. (2018). 1-Ethyl-3-methylimidazolium Diethylphosphate Based Extraction of Bioplastic "polyhydroxyalkanoates" from Bacteria: Green and Sustainable Approach. *ACS Sustainable Chemistry & Engineering*, *6*(1), 766–773. doi:10.1021/acssuschemeng.7b03096

Endres, H.-J. (2017). Large Scale Production of Stem Cells. *Advances in Biochemical Engineering/Biotechnology*, *102*(118), 1–35. doi:10.1007/10

Eriksen, M., Lebreton, L. C. M., Carson, H. S., Thiel, M., Moore, C. J., Borerro, J. C., Galgani, F., Ryan, P. G., & Reisser, J. (2014). Plastic Pollution in the World's Oceans: More than 5 Trillion Plastic Pieces Weighing over 250,000 Tons Afloat at Sea. *PLoS One*, *9*(12), e111913–e111913. doi:10.1371/journal.pone.0111913 PMID:25494041

European Bioplastics. (2017). Bioplastic Market Data 2017. *European Bioplastics*, 1–7.

Fabra, M. J., Martínez-Sanz, M., Gómez-Mascaraque, L. G., Gavara, R., & López-Rubio, A. (2018). Structural and physicochemical characterization of thermoplastic corn starch films containing microalgae. *Carbohydrate Polymers*, *186*, 184–191. doi:10.1016/j.carbpol.2018.01.039 PMID:29455977

Fang, S., Yu, W., Li, C., Liu, Y., Qiu, J., & Kong, F. (2019). Adsorption behavior of three triazole fungicides on polystyrene microplastics. *The Science of the Total Environment*, *691*, 1119–1126. doi:10.1016/j.scitotenv.2019.07.176 PMID:31466193

Fei, T., Cazeneuve, S., Wen, Z., Wu, L., & Wang, T. (2016). Effective recovery of poly-β-hydroxybutyrate (PHB) biopolymer from Cupriavidus necator using a novel and environmentally friendly solvent system. *Biotechnology Progress*, *32*(3), 678–685. doi:10.1002/btpr.2247 PMID:26871655

García, G., Sosa-Hernández, J. E., Rodas-Zuluaga, L. I., Castillo-Zacarías, C., Iqbal, H., & Parra-Saldívar, R. (2020). Accumulation of PHA in the Microalgae Scenedesmus sp. under Nutrient-Deficient Conditions. *Polymers*, *13*(1), 131. doi:10.3390/polym13010131 PMID:33396913

Gifuni, I., Olivieri, G., Krauss, I. R., D'Errico, G., Pollio, A., & Marzocchella, A. (2017). Microalgae as new sources of starch: Isolation and characterization of microalgal starch granules. *Chemical Engineering Transactions*, *57*, 1423–1428. doi:10.3303/CET1757238

Gumel, A. M., Annuar, M. S. M., & Chisti, Y. (2013). Recent Advances in the Production, Recovery and Applications of Polyhydroxyalkanoates. *Journal of Polymers and the Environment*, *21*(2), 580–605. doi:10.100710924-012-0527-1

Haddadi, M. H., Asadolahi, R., & Negahdari, B. (2019). The bioextraction of bioplastics with focus on polyhydroxybutyrate: A review. *International Journal of Environmental Science and Technology*, *16*(7), 3935–3948. doi:10.100713762-019-02352-0

Handbook of Microalgal Culture. (2006). *Handbook of Microalgal Culture.*

Jang, Y. H., Han, S. O., Sim, I. N., & Kim, H. (2013). Pretreatment effects of seaweed on the thermal and mechanical properties of seaweed/polypropylene biocomposites. *Composites. Part A, Applied Science and Manufacturing*, *47*(1), 83–90. doi:10.1016/j.compositesa.2012.11.016

Jiang, G., Johnston, B., Townrow, D. E., Radecka, I., Koller, M., Chaber, P., Adamus, G., & Kowalczuk, M. (2018). Biomass extraction using non-chlorinated solvents for biocompatibility improvement of polyhydroxyalkanoates. *Polymers*, *10*(7), 731. Advance online publication. doi:10.3390/polym10070731 PMID:30960656

Kamravamanesh, D., Pflügl, S., Nischkauer, W., Limbeck, A., Lackner, M., & Herwig, C. (2017). Photosynthetic poly-β-hydroxybutyrate accumulation in unicellular cyanobacterium Synechocystis sp. PCC 6714. *AMB Express*, *7*(1), 143. Advance online publication. doi:10.118613568-017-0443-9 PMID:28687036

Kapritchkoff, F. M., Viotti, A. P., Alli, R. C. P., Zuccolo, M., Pradella, J. G. C., Maiorano, A. E., Miranda, E. A., & Bonomi, A. (2006). Enzymatic recovery and purification of polyhydroxybutyrate produced by Ralstonia eutropha. *Journal of Biotechnology*, *122*(4), 453–462. doi:10.1016/j.jbiotec.2005.09.009 PMID:16253372

Kartik, A., Akhil, D., Lakshmi, D., Panchamoorthy Gopinath, K., Arun, J., Sivaramakrishnan, R., & Pugazhendhi, A. (2021). A critical review on production of biopolymers from algae biomass and their applications. *Bioresource Technology*, *329*, 124868. doi:10.1016/j.biortech.2021.124868 PMID:33707076

Kavitha, G., Kurinjimalar, C., Sivakumar, K., Kaarthik, M., Aravind, R., Palani, P., & Rengasamy, R. (2016). Optimization of polyhydroxybutyrate production utilizing waste water as nutrient source by Botryococcus braunii Kütz using response surface methodology. *International Journal of Biological Macromolecules*, *93*, 534–542. doi:10.1016/j.ijbiomac.2016.09.019 PMID:27612642

Kavitha, G., Kurinjimalar, C., Sivakumar, K., Kaarthik, M., Aravind, R., Palani, P., & Rengasamy, R. (2021). *International Journal of Biological Macromolecules Optimization of polyhydroxybutyrate production utilizing waste water as nutrient source by Botryococcus braunii Kütz using response surface methodology*. Academic Press.

Keshavarz, T., & Roy, I. (2010). Polyhydroxyalkanoates: Bioplastics with a green agenda. *Current Opinion in Microbiology*, *13*(3), 321–326. doi:10.1016/j.mib.2010.02.006 PMID:20227907

Khanna, S., & Srivastava, A. K. (2005). Recent advances in microbial polyhydroxyalkanoates. *Process Biochemistry*, *40*(2), 607–619. doi:10.1016/j.procbio.2004.01.053

Koller, M., Atlić, A., Dias, M., Reiterer, A., & Braunegg, G. (2010). Microbial PHA Production from Waste Raw Materials. *Microbial PHA Production from Waste Raw Materials.*, *14*, 85–119. doi:10.1007/978-3-642-03287-5_5

Koller, M., Niebelschütz, H., & Braunegg, G. (2013). Strategies for Recovery and Purification of Poly[(R)-3-Hydroxyalkanoates] (PHA) Biopolyesters from Surrounding Biomass. *Engineering in Life Sciences*, *13*(6), 549–562. doi:10.1002/elsc.201300021

Kosseva, M. R., & Rusbandi, E. (2018). Trends in the biomanufacture of polyhydroxyalkanoates with focus on downstream processing. *International Journal of Biological Macromolecules, 107*(PartA), 762–778. doi:10.1016/j.ijbiomac.2017.09.054

Koutra, E., Economou, C. N., Tsafrakidou, P., & Kornaros, M. (2018). Bio-Based Products from Microalgae Cultivated in Digestates. *Trends in Biotechnology*, *36*(8), 819–833. doi:10.1016/j.tibtech.2018.02.015 PMID:29605178

Koutra, E., Mastropetros, S. G., Ali, S. S., Tsigkou, K., & Kornaros, M. (2021). Assessing the potential of Chlorella vulgaris for valorization of liquid digestates from agro-industrial and municipal organic wastes in a biorefinery approach. *Journal of Cleaner Production*, *280*, 124352. doi:10.1016/j.jclepro.2020.124352

Kovalcik, A., Meixner, K., Mihalic, M., Zeilinger, W., Fritz, I., Fuchs, W., Kucharczyk, P., Stelzer, F., & Drosg, B. (2017). Characterization of polyhydroxyalkanoates produced by Synechocystis salina from digestate supernatant. *International Journal of Biological Macromolecules*, *102*, 497–504. doi:10.1016/j.ijbiomac.2017.04.054 PMID:28419829

Krasaesueb, N., Incharoensakdi, A., & Khetkorn, W. (2019). Utilization of shrimp wastewater for poly-β-hydroxybutyrate production by Synechocystis sp. PCC 6803 strain ΔSphU cultivated in photobioreactor. *Biotechnology Reports (Amsterdam, Netherlands)*, *23*, e00345. doi:10.1016/j.btre.2019.e00345 PMID:31193428

Kwon, H. K., Jeon, J. Y., & Oh, S. J. (2017). Potential for heavy metal (copper and zinc) removal from contaminated marine sediments using microalgae and light emitting diodes. *Ocean Science Journal*, *52*(1), 57–66. doi:10.100712601-017-0001-z

Laycock, B., Halley, P., Pratt, S., Werker, A., & Lant, P. (2013). The chemomechanical properties of microbial polyhydroxyalkanoates. *Progress in Polymer Science*, *38*(3–4), 536–583. doi:10.1016/j.progpolymsci.2012.06.003

Leal Filho, W., Saari, U., Fedoruk, M., Iital, A., Moora, H., Klöga, M., & Voronova, V. (2019). An overview of the problems posed by plastic products and the role of extended producer responsibility in Europe. *Journal of Cleaner Production*, *214*, 550–558. doi:10.1016/j.jclepro.2018.12.256

Lebreton, L., & Andrady, A. (2019). Future scenarios of global plastic waste generation and disposal. *Palgrave Communications*, *5*(1), 1–11. doi:10.105741599-018-0212-7

Lee, A., & Liew, M. S. (2020). Ecologically derived waste management of conventional plastics. *Journal of Material Cycles and Waste Management*, *22*(1), 1–10. doi:10.100710163-019-00931-4

Leong, Y. K., Show, P. L., Lan, J. C. W., Loh, H. S., Yap, Y. J., & Ling, T. C. (2017). Extraction and purification of Polyhydroxyalkanoates (PHAs): Application of Thermoseparating aqueous two-phase extraction. *Journal of Polymer Research*, *24*(10), 1–10. doi:10.100710965-017-1307-3

Levett, I., Birkett, G., Davies, N., Bell, A., Langford, A., Laycock, B., Lant, P., & Pratt, S. (2016). Techno-economic assessment of poly-3-hydroxybutyrate (PHB) production from methane - The case for thermophilic bioprocessing. *Journal of Environmental Chemical Engineering*, *4*(4), 3724–3733. doi:10.1016/j.jece.2016.07.033

Mallick, N., Gupta, S., Panda, B., & Sen, R. (2007). Process optimization for poly(3-hydroxybutyrate-co-3-hydroxyvalerate) co-polymer production by Nostoc muscorum. *Biochemical Engineering Journal*, *37*(2), 125–130. doi:10.1016/j.bej.2007.04.002

Markl, E., Grünbichler, H., & Lackner, M. (2019). Cyanobacteria for PHB Bioplastics Production: A Review. *Algae - Korean Phycological Society*, 1–10. doi:10.5772/intechopen.81536

Markou, G., & Nerantzis, E. (2013). Microalgae for high-value compounds and biofuelproduction: A review with focus on cultivation under stress conditions. *Biotechnology Advances*, *31*(8), 1532–1542. doi:10.1016/j.biotechadv.2013.07.011 PMID:23928208

Martínez-Herrera, R. E., Alemán-Huerta, M. E., Almaguer-Cantú, V., Rosas-Flores, W., Martínez-Gómez, V. J., Quintero-Zapata, I., Rivera, G., & Rutiaga-Quiñones, O. M. (2020). Efficient recovery of thermostable polyhydroxybutyrate (PHB) by a rapid and solvent-free extraction protocol assisted by ultrasound. *International Journal of Biological Macromolecules*, *164*, 771–782. doi:10.1016/j.ijbiomac.2020.07.101 PMID:32682039

Martins, R. G., Gonçalves, I. S., De Morais, M. G., & Vieira Costa, J. A. (2017). New technologies from the bioworld: Selection of biopolymer-producing microalgae. *Polímeros*, *27*(4), 285–289. doi:10.1590/0104-1428.2375

Martins, R. G., Severo Gonçalves, I., De Morais, M. G., & Costa, J. A. V. (2014). Bioprocess engineering aspects of biopolymer production by the cyanobacterium spirulina strain LEB 18. *International Journal of Polymer Science*, *2014*, 1–7. doi:10.1155/2014/895237

Mathiot, C., Ponge, P., Gallard, B., Sassi, J.-F., Delrue, F., & Le Moigne, N. (2019). Microalgae starch-based bioplastics: Screening of ten strains and plasticization of unfractionated microalgae by extrusion. *Carbohydrate Polymers*, *208*, 142–151. doi:10.1016/j.carbpol.2018.12.057 PMID:30658785

Meixner, K., Fritz, I., Daffert, C., Markl, K., Fuchs, W., & Drosg, B. (2016). Processing recommendations for using low-solids digestate as nutrient solution for poly-ß-hydroxybutyrate production with Synechocystis salina. *Journal of Biotechnology*, *240*, 61–67. doi:10.1016/j.jbiotec.2016.10.023 PMID:27984118

Mendhulkar, V. D., & Shetye, L. A. (2017). Synthesis of biodegradable polymer polyhydroxyalkanoate (PHA) in cyanobacteria Synechococcus elongates under mixotrophic nitrogen- and phosphate-mediated stress conditions. *Industrial Biotechnology (New Rochelle, N.Y.)*, *13*(2), 85–93. doi:10.1089/ind.2016.0021

Monshupanee, T., Nimdach, P., & Incharoensakdi, A. (2016). Two-stage (photoautotrophy and heterotrophy) cultivation enables efficient production of bioplastic poly-3-hydroxybutyrate in auto-sedimenting cyanobacterium. *Scientific Reports*, *6*(1), 37121. doi:10.1038rep37121 PMID:27845413

Mourão, M. M., Gradíssimo, D. G., Santos, A. V., Schneider, M. P. C., Faustino, S. M. M., Vasconcelos, V., & Xavier, L. P. (2020). Optimization of Polyhydroxybutyrate Production by Amazonian Microalga Stigeoclonium sp. B23. *Biomolecules*, *10*(12), 1628. doi:10.3390/biom10121628 PMID:33287108

Możejko-Ciesielska, J., & Kiewisz, R. (2016). Bacterial polyhydroxyalkanoates: Still fabulous? *Microbiological Research*, *192*, 271–282. doi:10.1016/j.micres.2016.07.010 PMID:27664746

Muneer, F., Rasul, I., Azeem, F., Siddique, M. H., Zubair, M., & Nadeem, H. (2020). Microbial Polyhydroxyalkanoates (PHAs): Efficient Replacement of Synthetic Polymers. *Journal of Polymers and the Environment*, *28*(9), 2301–2323. doi:10.100710924-020-01772-1

Naresh Kumar, A., Chatterjee, S., Hemalatha, M., Althuri, A., Min, B., Kim, S. H., & Venkata Mohan, S. (2020). Deoiled algal biomass derived renewable sugars for bioethanol and biopolymer production in biorefinery framework. *Bioresource Technology*, *296*(August), 122315. doi:10.1016/j.biortech.2019.122315 PMID:31706890

Nishioka, M., Nakai, K., Miyake, M., Asada, Y., & Taya, M. (2001). Production of poly-β-hydroxybutyrate by thermophilic cyanobacterium, Synechococcus sp. MA19, under phosphate-limited conditions. *Biotechnology Letters*, *23*(14), 1095–1099. doi:10.1023/A:1010551614648

Onen Cinar, S., Chong, Z. K., Kucuker, M. A., Wieczorek, N., Cengiz, U., & Kuchta, K. (2020). Bioplastic Production from Microalgae: A Review. *International Journal of Environmental Research and Public Health*, *17*(11), 3842. doi:10.3390/ijerph17113842 PMID:32481700

Otsuki, J., Zhang, F., Kabeya, H., & Hirotsu, T. (2004). Synthesis and Tensile Properties of a Novel Composite of Chlorella and Polyethylene. *Journal of Applied Polymer Science*, *92*(2), 812–816. doi:10.1002/app.13650

Pachekoski, W. M., Agnelli, J. A. M., & Belem, L. P. (2009). Thermal, mechanical and morphological properties of poly (hydroxybutyrate) and polypropylene blends after processing. *Materials Research*, *12*(2), 159–164. doi:10.1590/S1516-14392009000200008

Panda, B., Jain, P., Sharma, L., & Mallick, N. (2006). Optimization of cultural and nutritional conditions for accumulation of poly-β-hydroxybutyrate in Synechocystis sp. PCC 6803. *Bioresource Technology*, *97*(11), 1296–1301. doi:10.1016/j.biortech.2005.05.013 PMID:16046119

Plastics Europe, G. M. R., & Conversio Market & Strategy GmbH. (2019). *Plastics - the Facts 2019*. Author.

Poltronieri, P., Mezzolla, V., & D'Urso, O. F. (2016). PHB Production in Biofermentors Assisted through Biosensor Applications. *Proceedings*, *1*(2), 4. doi:10.3390/ecsa-3-E014

Ragusa, A., Svelato, A., Santacroce, C., Catalano, P., Notarstefano, V., Carnevali, O., Papa, F., Rongioletti, M. C. A., Baiocco, F., Draghi, S., D'Amore, E., Rinaldo, D., Matta, M., & Giorgini, E. (2021). Plasticenta: First evidence of microplastics in human placenta. *Environment International*, *146*, 106274. doi:10.1016/j.envint.2020.106274 PMID:33395930

Rahman, A., & Miller, C. D. (2017). Microalgae as a Source of Bioplastics. In *Algal Green Chemistry* (pp. 121–138). Elsevier. doi:10.1016/B978-0-444-63784-0.00006-0

Razza, F., & Innocenti, F. D. (2012). Bioplastics from renewable resources: The benefits of biodegradability. *Asia-Pacific Journal of Chemical Engineering*, *7*(3), S301–S309. doi:10.1002/apj.1648

Rizwan, M., Mujtaba, G., Memon, S. A., Lee, K., & Rashid, N. (2018). Exploring the potential of microalgae for new biotechnology applications and beyond: A review. *Renewable and Sustainable Energy Reviews, 92*(March), 394–404. doi:10.1016/j.rser.2018.04.034

Roja, K., Ruben Sudhakar, D., Anto, S., & Mathimani, T. (2019). Extraction and characterization of polyhydroxyalkanoates from marine green alga and cyanobacteria. *Biocatalysis and Agricultural Biotechnology*, *22*, 101358. doi:10.1016/j.bcab.2019.101358

Rujnić-Sokele, M., & Pilipović, A. (2017). Challenges and opportunities of biodegradable plastics: A mini review. *Waste Management & Research*, *35*(2), 132–140. doi:10.1177/0734242X16683272 PMID:28064843

Sabathini, H. A., Windiani, L., Dianursanti, & Gozan, M. (2018). Mechanical Physicial Properties of Chlorella-PVA based Bioplastic with Ultrasonic Homogenizer. *E3S Web of Conferences, 67*, 3046. doi:10.1051/e3sconf/20186703046

Samantaray, S., & Mallick, N. (2015). Impact of Various Stress Conditions on Poly-β-Hydroxybutyrate (PHB) Accumulation in Aulosira fertilissima CCC 444. *Current Biotechnology*, *4*(3), 366–372. doi:10.2174/2211550104666150806000642

Samantaray, S., Nayak, J. K., & Mallick, N. (2011). Wastewater utilization for poly β hydroxybutyrate production by the cyanobacterium Aulosira fertilissima in a recirculatory aquaculture system. *Applied and Environmental Microbiology*, *77*(24), 8735–8743. doi:10.1128/AEM.05275-11 PMID:21984242

Setyorini, A. I., & Dianursanti. (2021). Isolation and optimization of polyhydroxybutyrate from Chlorella vulgaris using NaClO and ultrasonication aided chemical pretreatment methods. *The 5th biomedical engineering's recent progress in biomaterials, drugs development, and medical devices: Proceedings of the 5th International Symposium of Biomedical Engineering (ISBE) 2020*. doi:10.1063/5.0047562

Sharma, V., Sehgal, R., & Gupta, R. (2021). Polyhydroxyalkanoate (PHA): Properties and Modifications. *Polymer, 212*(October), 123161. doi:10.1016/j.polymer.2020.123161

Shi, B., Liang, L., Yang, H., Zhang, L., & He, F. (2017). Glycerol-plasticized spirulina–poly(vinyl alcohol) films with improved mechanical performance. *Journal of Applied Polymer Science*, *134*(20), 1–10. doi:10.1002/app.44842

Simonic, M., & Zemljic, F. (2020). Production of bioplastic material from algal biomass. *Chemical Industry & Chemical Engineering Quarterly*, *00*, 26–26. doi:10.2298/CICEQ191024026S

Singh, A. K., Sharma, L., Mallick, N., & Mala, J. (2017). Progress and challenges in producing polyhydroxyalkanoate biopolymers from cyanobacteria. *Journal of Applied Phycology*, *29*(3), 1213–1232. doi:10.100710811-016-1006-1

Spain, O., Plöhn, M., & Funk, C. (2021). The cell wall of green microalgae and its role in heavy metal removal. *Physiologia Plantarum,* 1–10. doi:10.1111/ppl.13405

Sreenikethanam, A., & Bajhaiya, A. (2021). Algae Based Bio-Plastics: Future of Green Economy. In Biorefineries - Selected Processes. IntechOpen. doi:10.5772/intechopen.100981

Steinbuchel, A., Madkour, M. H., Heinrich, D., Alghamdi, A. M., & Shabbaj, I. I. (2013). PHA Recovery from Biomass. *Biomacromolecules*. PMID:23875914

Sudesh, K., Taguchi, K., & Doi, Y. (2002). Effect of increased PHA synthase activity on polyhydroxyalkanoates biosynthesis in Synechocystis sp. PCC6803. *International Journal of Biological Macromolecules*, *30*(2), 97–104. doi:10.1016/S0141-8130(02)00010-7 PMID:11911900

Toh, P. S., Jau, M.-H., Yew, S.-P., Abed, R. M., & Sudesh, K. (2008). Comparison of Polyhydroxyalkanoates Biosynthesis, Mobilization and the Effects on Cellular Morphology in Spirulina Platensis and Synechocystis Sp. Uniwg. *Journal of Bio-Science*, *19*(2), 21–38.

Torres, S., Navia, R., Campbell Murdy, R., Cooke, P., Misra, M., & Mohanty, A. K. (2015). Green Composites from Residual Microalgae Biomass and Poly(butylene adipate-co-terephthalate): Processing and Plasticization. *ACS Sustainable Chemistry & Engineering*, *3*(4), 614–624. doi:10.1021c500753h

Tran, D., Lee, H. R., Jung, S., Park, M. S., & Yang, J. (2016). *Lipid-extracted algal biomass based biocomposites fabrication with poly (vinyl alcohol)*. ALGAL. doi:10.1016/j.algal.2016.08.016

Van Eygen, E., Feketitsch, J., Laner, D., Rechberger, H., & Fellner, J. (2017). Comprehensive analysis and quantification of national plastic flows: The case of Austria. *Resources, Conservation and Recycling*, *117*, 183–194. doi:10.1016/j.resconrec.2016.10.017

Wang, B., Qin, L., Mu, T., Xue, Z., & Gao, G. (2017). Are Ionic Liquids Chemically Stable? *Chemical Reviews*, *117*(10), 7113–7131. doi:10.1021/acs.chemrev.6b00594 PMID:28240867

Wang, X., Peng, S., & Dong, L. (2005). Effect of poly(vinyl acetate) (PVAc) on thermal behavior and mechanical properties of poly(3-hydroxybutyrate)/poly(propylene carbonate) (PHB/ PPC) blends. *Colloid & Polymer Science*, *284*(2), 167–174. doi:10.100700396-005-1353-y

Zeller, M. A., Hunt, R., Jones, A., & Sharma, S. (2013). Bioplastics and their thermoplastic blends from Spirulina and Chlorella microalgae. *Journal of Applied Polymer Science*, *130*(5), 3263–3275. doi:10.1002/app.39559

Zhang, C., Wang, C., Cao, G., Wang, D., & Ho, S. H. (2020). A sustainable solution to plastics pollution: An eco-friendly bioplastic film production from high-salt contained Spirulina sp. residues. *Journal of Hazardous Materials*, *388*(October), 121773. doi:10.1016/j.jhazmat.2019.121773 PMID:31836373

Zhang, S., & Bryant, D. A. (2015). Biochemical validation of the glyoxylate cycle in the cyanobacterium Chlorogloeopsis fritschii strain PCC 9212. *The Journal of Biological Chemistry*, *290*(22), 14019–14030. doi:10.1074/jbc.M115.648170 PMID:25869135

Zhu, N., Ye, M., Shi, D., & Chen, M. (2017). Reactive compatibilization of biodegradable poly(butylene succinate)/Spirulina microalgae composites. *Macromolecular Research*, *25*(2), 165–171. doi:10.100713233-017-5025-9

Compilation of References

Abanades, S., Charvin, P., Flamant, G., & Neveu, P. (2006). Screening of water-splitting thermochemical cycles potentially attractive for hydrogen production by concentrated solar energy. *Energy*, *31*(14), 2805–2822. doi:10.1016/j.energy.2005.11.002

Abascal, F., Zardoya, R., & Posada, D. (2005). ProtTest: Selection of best-fit models of protein evolution. *Bioinformatics (Oxford, England)*, *21*(9), 2104–2105. doi:10.1093/bioinformatics/bti263 PMID:15647292

Abboud, Y., Saffaj, T., Chagraoui, A., El Bouari, A., Brouzi, K., Tanane, O., & Ihssane, B. (2014). Biosynthesis, characterization and antimicrobial activity of copper oxide nanoparticles (CONPs) produced using brown alga extract (Bifurcaria bifurcata). *Applied Nanoscience*, *4*(5), 571–576. doi:10.100713204-013-0233-x

Abd El-Hack, M. E., Abdelnour, S., Alagawany, M., Abdo, M., Sakr, M. A., Khafaga, A. F., Mahgoub, S. A., Elnesr, S. S., & Gebriel, M. G. (2019). Microalgae in modern cancer therapy: Current knowledge. *Biomedicine and Pharmacotherapy*, *111*, 42–50. doi:10.1016/j.biopha.2018.12.069 PMID:30576933

Abd El-Magied, M. O., Galhoum, A. A., Atia, A. A., Tolba, A. A., Maize, M. S., Vincent, T., & Guibal, E. (2017). Cellulose and chitosan derivatives for enhanced sorption of erbium (III). *Colloids and Surfaces. A, Physicochemical and Engineering Aspects*, *529*, 580–593. doi:10.1016/j.colsurfa.2017.05.031

Abdalla, A. M., Hassan, T. H., & Mansour, M. E. (2018). Performance of Wet and Dry Bagasse Combustion in Assalaya Sugar Factory - Sudan. *Innovative Energy & Research*, *7*(1). Advance online publication. doi:10.4172/2576-1463.1000179

Abdallah, N. A., Prakash, C. S., & McHughen, A. G. (2015). Genome editing for crop improvement: Challenges and opportunities. *GM Crops and Food: Biotechnology in Agriculture and the Food Chain*, *6*(4), 183–205. doi:10.1080/21645698.2015.1129937 PMID:26930114

Abdelhamid, A. E., Labena, A., Mansor, E. S., Husien, S., & Moghazy, R. M. (2021). Highly efficient adsorptive membrane for heavy metal removal based on Ulva fasciata biomass. *Biomass Conversion and Biorefinery*. Advance online publication. doi:10.100713399-020-01250-7

Abdella, B., El-Wazzan, E., El-Sersy, N. A., Sabry, S. A., & El-Helow, E. R. (2017). Pathogenicity and antibiotic susceptibility of two bacterial pathogens associated with the clam Tapes decussatus in some Egyptian fisheries. *Ege Journal of Fisheries and Aquatic Sciences*, *34*(4), 383–389. doi:10.12714/egejfas.2017.34.4.04

Abdel-Raouf, N., Al-Homaidan, A., & Ibraheem, I. 2016. Agricultural importance of algae. African J. Biotechnol. 11, 11648–11658. . v11i54 doi:10.4314/ajb

Abdel-Raouf, N., Alharbi, R. M., Al-Enazi, N. M., Alkhulaifi, M. M., & Ibraheem, I. B. M. (2018). Rapid biosynthesis of silver nanoparticles using the marine red alga Laurencia catarinensis and their characterization. *Beni-Suef University Journal of Basic and Applied Sciences*, *7*(1), 150–157. doi:10.1016/j.bjbas.2017.10.003

Abdel-Raouf, N., Al-Homaidan, A. A., & Ibraheem, I. B. M. (2012). Agricultural importance of algae. *African Journal of Biotechnology*, *11*(54), 11648–11658. doi:10.5897/AJB11.3983

Abdo, S. M., & Ali, G. H. (2019). Analysis of polyhydroxybutrate and bioplastic production from microalgae. *Bulletin of the National Research Center*, *43*(1), 97. Advance online publication. doi:10.118642269-019-0135-5

Abdullah, A.-R., Al-Radadi, N. S., Hussain, T., Faisal, S., & Ali Raza Shah, S. (2022, January). Novel biosynthesis, characterization and bio-catalytic potential of green algae (Spirogyra hyalina) mediated silver nanomaterials. *Saudi Journal of Biological Sciences*, *29*(1), 411–419. Advance online publication. doi:10.1016/j.sjbs.2021.09.013 PMID:35002436

Abdul-Latif, N.-I. S., Ong, M. Y., Nomanbhay, S., Salman, B., & Show, P. L. (2020). Estimation of carbon dioxide (CO_2) reduction by utilization of algal biomass bioplastic in Malaysia using carbon emission pinch analysis (CEPA). *Bioengineered*, *11*(1), 154–164. doi:10.1080/21655979.2020.1718471 PMID:32013677

Abidizadegan, M., Peltomaa, E., & Blomster, J. (2021). The Potential of Cryptophyte Algae in Biomedical and Pharmaceutical Applications. *Frontiers in Pharmacology*, *11*(February), 1–17. doi:10.3389/fphar.2020.618836 PMID:33603668

Aboelfetoh, E. F., El-Shenody, R. A., & Ghobara, M. M. (2017). Eco-friendly synthesis of silver nanoparticles using green algae (Caulerpaserrulata): Reaction optimization, catalytic and antibacterial activities. *Environmental Monitoring and Assessment*, *189*(7), 349. doi:10.100710661-017-6033-0 PMID:28646435

Abomohra, A. E., & Elshobary, M. E. (2019). Biodiesel, bioethanol and biobutanol production from microalgae biomass. In Md. A. Alam & Z. Wang (Eds.), Microalgae Biotechnology for Development of Biofuel and Waste Water Treatment (pp. 293–321). Springer Singapore Press. doi:10.1007/978-981-13-2264-8

Abomohra, A. E. F., El-Naggar, A. H., & Baeshen, A. A. (2018). Potential of microalgae for biodiesel production: Screening and evaluation studies. *Journal of Bioscience and Bioengineering*, *125*(2), 231–237. doi:10.1016/j.jbiosc.2017.08.020 PMID:29037768

Abomohra, A. E.-F., El-Naggar, A. H., Alaswad, S. O., Elsayed, M., Li, M., & Li, W. (2020). Enhancement of Biodiesel Yield from a Halophilic Green Microalga Isolated under Extreme Hypersaline Conditions through Stepwise Salinity Adaptation Strategy. *Bioresource Technology*, *310*, 123462. doi:10.1016/j.biortech.2020.123462 PMID:32388355

Acar, C., Dincer, I., & Zamfirescu, C. (2014). A review on selected heterogeneous photocatalysts for hydrogen production. *International Journal of Energy Research*, *38*(15), 1903–1920. doi:10.1002/er.3211

Acea, M. J., Diz, N., & Prieto-Fernandez, A. (2001). Microbial populations in heated soils inoculated with cyanobacteria. *Biology and Fertility of Soils*, *33*(2), 118–125. doi:10.1007003740000298

Achankunju, J. (2016). *Bioprospecting selected microalgae for Aquaculture feed*. Academic Press.

Achieng', G. O., & Shikuku, V. O. (2020). Adsorption of copper ions from water onto fish scales derived biochar: Isothermal perspectives. *J. Mater. Environ. Sci.*, *11*, 1816–1827.

Achilias, D. S., Roupakias, C., Megalokonomos, P., Lappas, A. A., & Antonakou, V. (2007). Chemical recycling of plastic wastes made from polyethylene (LDPE and HDPE) and polypropylene (PP). *Journal of Hazardous Materials*, *149*(3), 536–542. doi:10.1016/j.jhazmat.2007.06.076 PMID:17681427

Adesanya, V. O., Cadena, E., Scott, S. A., & Smith, A. G. (2014). Life cycle assessment on microalgal biodiesel production using a hybrid cultivation system. *Bioresource Technology*, *163*, 343–355. doi:10.1016/j.biortech.2014.04.051 PMID:24852435

Adesanya, V. O., Davey, M. P., Scott, S. A., & Smith, A. G. (2014). Kinetic modelling of growth and storage molecule production in microalgae under mixotrophic and autotrophic conditions. *Bioresource Technology*, *157*, 293–304. doi:10.1016/j.biortech.2014.01.032 PMID:24576922

Adhikari, B., & Pellegrino, J. (2015). Life-cycle assessment of five microalgae-to-biofuels processes of varying complexity. *Journal of Renewable and Sustainable Energy*, *7*(4), 043136. doi:10.1063/1.4929462

Adhikari, D., Mukai, M., Kubota, K., Kai, T., Kaneko, N., Araki, K. S., & Kubo, M. (2016). Degradation of bioplastics in soil and their degradation effects on environmental microorganisms. *Journal of Agricultural Chemistry and Environment*, *05*(01), 23–34. doi:10.4236/jacen.2016.51003

Aditya, T., & Bitu, G., & G, M. E. (2016). The Role of Algae in Pharmaceutical Development. *Special Issue: Reviews on Pharmaceutics and Nanotechnology Research & Reviews: Journal of Pharmaceutics and Nanotechnology*, *4*, 82–89.

Adl, S. M., Simpson, A. G. B., Lane, C. E., Lukeš, J., Bass, D., Bowser, S. S., ... Spiegel, F. W. (2013). Erratum: The revised classification of eukaryotes. *The Journal of Eukaryotic Microbiology*, *60*(3).

Adnan, Ong, Nomanbhay, Chew, & Show. (2019). Technologies for Biogas Upgrading to Biomethane: A Review. *Bioengineering (Basel, Switzerland)*, *6*(4), 92. doi:10.3390/bioengineering6040092

Aggarwal, V., Tuli, H. S., Varol, A., Thakral, F., Yerer, M. B., Sak, K., Varol, M., Jain, A., Khan, M., & Sethi, G. (2019). Role of reactive oxygen species in cancer progression: Molecular mechanisms and recent advancements. *Biomolecules*, *9*(11), 735. doi:10.3390/biom9110735 PMID:31766246

Aghmiuni, S. K., Siyal, S., Wang, Q., & Duan, Y. (2020). Assessment of factors affecting innovation policy in biotechnology. *Journal of Innovation and Knowledge*, *5*(3), 180–190. doi:10.1016/j.jik.2019.10.002

Agustin, M. B., Ahmmad, B., Alonzo, S. M. M., & Patriana, F. M. (2014). Bioplastic based on starch and cellulose nanocrystals from rice straw. *Journal of Reinforced Plastics and Composites*, *33*(24), 2205–2213. doi:10.1177/0731684414558325

Ahad, R.I.A., Goswami, S., & Syiem, M.B. (2017). Biosorption and equilibrium isotherms study of cadmium removal by *Nostocmuscorum* Meg 1: morphological, physiological and biochemical alterations. *Biotech, 7*(2), 104.

Ahmad, S., Pandey, A., Pathak, V. V., Tyagi, V. V., & Kothari, R. (2020). Phycoremediation: algae as eco-friendly tools for the removal of heavy metals from wastewaters. *Bioremediation of Industrial Waste for Environmental Safety,* 53-76.

Ahmad, I., & Hellebust, J. A. (1984). Osmoregulation in the extremely euryhaline marine micro-alga Chlorella autotrophica. *Plant Physiology*, *74*(4), 1010–1015. doi:10.1104/pp.74.4.1010 PMID:16663495

Ahmadi, A., Zorofchian Moghadamtousi, S., Abubakar, S., & Zandi, K. (2015). Antiviral potential of algae polysaccharides isolated from marine sources: A review. *BioMed Research International*, *2015*, 2015. doi:10.1155/2015/825203 PMID:26484353

Ahmad, N., & Sharma, S. (2012). Green synthesis of silver nanoparticles using extracts of Ananas comosus. *Green Sustain. Chem.*, *2*(04), 141–147. doi:10.4236/gsc.2012.24020

Ahn, H. K., Huda, M. S., Smith, M. C., Mulbry, W., Schmidt, W. F., & Reeves, J. B. III. (2011). Biodegradability of injection molded bioplastic pots containing polylactic acid and poultry feather fiber. *Bioresource Technology*, *102*(7), 4930–4933. doi:10.1016/j.biortech.2011.01.042 PMID:21320772

Aires, T., Serebryakova, A., Viard, F., Serrão, E. A., & Engelen, A. H. (2018). Acidification increases abundances of Vibrionales and Planctomycetia associated to a seaweed-grazer system: Potential consequences for disease and prey digestion efficiency. *PeerJ*, *6*, e4377. doi:10.7717/peerj.4377 PMID:29610702

Aiyar, P., Schaeme, D., García-Altares, M., Carrasco Flores, D., Dathe, H., Hertweck, C., Sasso, S., & Mittag, M. (2017). Antagonistic bacteria disrupt calcium homeostasis and immobilize algal cells. *Nature Communications*, *8*(1), 1–13. doi:10.103841467-017-01547-8 PMID:29170415

Ajjawi, I., Verruto, J., Aqui, M., Soriaga, L. B., Coppersmith, J., Kwok, K., ... Xu, W. (2017). Lipid production in *Nannochloropsis gaditana* is doubled by decreasing expression of a single transcriptional regulator. *Nature Biotechnology*, *35*(7), 647–652. doi:10.1038/nbt.3865 PMID:28628130

Akila, V., Manikandan, A., Sahaya Sukeetha, D., Balakrishnan, S., Ayyasamy, P. M., & Rajakumar, S. (2019). Biogas and biofertilizer production of marine macroalgae: An effective anaerobic digestion of Ulva sp. *Biocatalysis and Agricultural Biotechnology*, *18*, 101035. doi:10.1016/j.bcab.2019.101035

Akkerman, I., Janssen, M., Rocha, J., & Wijffels, R. H. (2002). Photobiological hydrogen production: Photochemical efficiency and bioreactor design. *International Journal of Hydrogen Energy*, *27*(11–12), 1195–1208. doi:10.1016/S0360-3199(02)00071-X

Akram, M. S., Cheema, M. A., Nadeem, F., Waqas, M., Bilal, M., & Saeed, M. (2020). Role of bio-fertilizers in sustainable agriculture. *Mediterranean Journal of Basic and Applied Sciences*, *4*(2), 8–23.

Al Hattab, M., Ghaly, A., & Hammouda, A. (2015). Microalgae harvesting methods for industrial production of biodiesel: Critical review and comparative analysis. *J Fundam Renew Energy Appl*, *5*(2), 1000154. doi:10.4172/2090-4541.1000154

Alam, M. A., Vandamme, D., Chun, W., Zhao, X., Foubert, I., Wang, Z., & Yuan, Z. (2016). Bioflocculation as an innovative harvesting strategy for microalgae. *Reviews in Environmental Science and Biotechnology*, *15*(4), 573–583. doi:10.100711157-016-9408-8

Alam, M. A., Wan, C., Guo, S. L., Zhao, X. Q., Huang, Z. Y., Yang, Y. L., & Bai, F. W. (2014). Characterization of the flocculating agent from the spontaneously flocculating microalga *Chlorella vulgaris* JSC-7. *Journal of Bioscience and Bioengineering*, *118*(1), 29–33. doi:10.1016/j.jbiosc.2013.12.021 PMID:24507901

Al-Amshawee, S., Yunus, M. Y. B. M., Azoddein, A. A. M., Hassell, D. G., Dakhil, I. H., & Hasan, H. A. (2020). Electrodialysis Desalination for Water and Wastewater: A Review. *Chemical Engineering Journal*, *380*, 122231–122254.

Alavijeh, R. S., Karimi, K., Wijffels, R. H., van den Berg, C., & Eppink, M. (2020). Combined Bead Milling and Enzymatic Hydrolysis for Efficient Fractionation of Lipids, Proteins, and Carbohydrates of Chlorella Vulgaris Microalgae. *Bioresource Technology*, *309*, 123321. doi:10.1016/j.biortech.2020.123321 PMID:32305840

Albertí, J., Brodhag, C., & Fullana-i-Palmer, P. (2019). First steps in life cycle assessments of cities with a sustainability perspective: A proposal for goal, function, functional unit, and reference flow. *The Science of the Total Environment*, *646*, 1516–1527. doi:10.1016/j.scitotenv.2018.07.377 PMID:30235636

Ali, M. Y. S., Ravikumar, S., & Beula, J. M. (2013). Mosquito Larvicidal Activity of Seaweeds Extracts against Anopheles Stephensi, Aedes aegypti and Culex Quinquefasciatus. *Asian Pacific Journal of Tropical Disease*, *3*, 196–201.

Ali, S. S., Al-Tohamy, R., Koutra, E., El-Naggar, A. H., Kornaros, M., & Sun, J. (2020a). Valorizing lignin-like dyes and textile dyeing wastewater by a newly constructed lipid producing and lignin modifying oleaginous yeast consortium valued for biodiesel and bioremediation. *Journal of Hazardous Materials*, *403*, 123575. doi:10.1016/j.jhazmat.2020.123575 PMID:32791477

Ali, S. S., Al-Tohamy, R., Koutra, E., Kornaros, M., Khalil, K., Elsamahy, T., El-Shetehy, M., & Sun, J. (2021b). Coupling azo dye degradation and biodiesel production by manganese-dependent peroxidase producing oleaginous yeasts isolated from wood-feeding termite gut symbionts. *Biotechnology for Biofuels*, *14*(61), 1–25. doi:10.118613068-021-01906-0 PMID:33685508

Ali, S. S., Al-Tohamy, R., Manni, A., Luz, F. C., Elsamahy, T., & Sun, J. (2019). Enhanced digestion of bio-pretreated sawdust using a novel bacterial consortium: Microbial community structure and methane-producing pathways. *Fuel*, *254*, 115604. doi:10.1016/j.fuel.2019.06.012

Ali, S. S., Al-Tohamy, R., Xie, R., El-Sheekh, M. M., & Sun, J. (2020b). Construction of a new lipase- and xylanase-producing oleaginous yeast consortium capable of reactive azo dye degradation and detoxification. *Bioresource Technology, 313*, 123631. doi:10.1016/j.biortech.2020.123631 PMID:32540694

Ali, S. S., Elsamahy, T., Al-Tohamy, R., Zhu, D., Mahmoud, Y. A.-G., Koutra, E., Metwally, M. A., Kornaros, M., & Sun, J. (2021b). Plastic wastes biodegradation: Mechanisms, challenges and future prospects. *The Science of the Total Environment*, *780*, 146590. doi:10.1016/j.scitotenv.2021.146590 PMID:34030345

Ali, S. S., Elsamahy, T., Koutra, E., Kornaros, M., El-Sheekh, M., Abdelkarim, E., Zhu, D., & Sun, J. (2021c). Degradation of conventional plastic wastes in the environment. A review on current status of knowledge and future perspectives of disposal. *The Science of the Total Environment*, *144719*, 144719. Advance online publication. doi:10.1016/j.scitotenv.2020.144719 PMID:33548729

Ali, S. S., Kornaros, M., Manni, A., Sun, J., El-Shanshoury, A. E. R., Kenawy, E.-R., & Khalil, M. A. (2020c). Enhanced anaerobic digestion performance by two artificially constructed microbial consortia capable of woody biomass degradation and chlorophenols detoxification. *Journal of Hazardous Materials*, *389*, 122076. doi:10.1016/j.jhazmat.2020.122076 PMID:32004834

Ali, S. S., Mustafa, A. M., Kornaros, M., Manni, A., Sun, J., & Khalil, M. A. (2020d). Construction of novel microbial consortia CS-5 and BC-4 valued for the degradation of catalpa sawdust and chlorophenols simultaneously with enhancing methane production. *Bioresource Technology*, *301*, 122720. doi:10.1016/j.biortech.2019.122720 PMID:31945685

Ali, S. S., Sun, J., Koutra, E., El-Zawawy, N., Elsamahy, T., & El-Shetehy, M. (2021d). Construction of a novel cold-adapted oleaginous yeast consortium valued for textile azo dye wastewater processing and biore finery. *Fuel*, *285*, 119050. doi:10.1016/j.fuel.2020.119050

Allakhverdiev, S. I., Kreslavski, V. D., Klimov, V. V., Los, D. A., Carpentier, R., & Mohanty, P. (2008). Heat stress: An overview of molecular responses in photosynthesis. *Photosynthesis Research*, *98*(1), 541–550. doi:10.100711120-008-9331-0 PMID:18649006

Allen, D. E., Singh, B. P., & Dalal, R. C. (2011). Soil health indicators under climate change: a review of current knowledge. In *Soil health and climate change* (pp. 25–45). Springer. doi:10.1007/978-3-642-20256-8_2

Almomani, F. (2020). Algal cells harvesting using cost-effective magnetic nano-particles. *The Science of the Total Environment*, *720*, 137621. doi:10.1016/j.scitotenv.2020.137621 PMID:32146403

AlNadhari, S., Al-Enazi, N. M., Alshehrei, F., & Ameen, F. (2021). A review on biogenic synthesis of metal nanoparticles using marine algae and its applications. *Environmental Research, 194*(October), 110672. doi:10.1016/j.envres.2020.110672

Alobwede, E., Leake, J. R., & Pandhal, J. (2019). Circular economy fertilization: Testing micro and macro algal species as soil improvers and nutrient sources for crop production in greenhouse and field conditions. *Geoderma*, *334*, 113–123. doi:10.1016/j.geoderma.2018.07.049

AlProl, A. E., & Elkatory, M. R. (2022). The current status of various algal industries. In Handbook of Algal Biofuels (pp. 123-147). doi:10.1016/B978-0-12-823764-9.00010-8

Al-Saif, S. S. A. L., Abdel-Raouf, N., El-Wazanani, H. A., & Aref, I. A. (2014). Antibacterial substances from marine algae isolated from Jeddah coast of Red sea, Saudi Arabia. *Saudi Journal of Biological Sciences*, *21*(1), 57–64. doi:10.1016/j.sjbs.2013.06.001 PMID:24596500

Alsenani, F., Tupally, K. R., Chua, E. T., Eltanahy, E., Alsufyani, H., Parekh, H. S., & Schenk, P. M. (2020). Evaluation of microalgae and cyanobacteria as potential sources of antimicrobial compounds. *Saudi Pharmaceutical Journal, 28*(12), 1834–1841. doi:10.1016/j.jsps.2020.11.010 PMID:33424272

Altas, D., & Bannister, T. T. (1980). Dependence of mean spectral extinction coefficient of phytoplankton on depth, water color, and species. *Limnology and Oceanography, 25*(1), 157–159. doi:10.4319/lo.1980.25.1.0157

Alterio, V., Langella, E., De Simone, G., & Monti, S. M. (2015). Cadmium-containing carbonic anhydrase CDCA1 in marine diatom Thalassiosira weissflogii. *Marine Drugs, 13*(4), 1688–1697. doi:10.3390/md13041688 PMID:25815892

Al-Tohamy, R., Kenawy, E.-R., Sun, J., & Ali, S. S. (2020a). Performance of a newly isolated salt-tolerant yeast strain *Sterigmatomyces halophilus* SSA-1575 for azo dye decolorization and detoxification. *Frontiers in Microbiology, 11*, 1163. doi:10.3389/fmicb.2020.01163 PMID:32595618

Al-Tohamy, R., Sun, J., Fareed, M., Kenawy, E.-R., & Ali, S. S. (2020b). Ecofriendly biodegradation of Reactive Black 5 by newly isolated *Sterigmatomyces halophilus* SSA1575, valued for textile azo dye wastewater processing and detoxification. *Scientific Reports, 10*(1), 12370. doi:10.103841598-020-69304-4 PMID:32704008

Alva, V., Nam, S.-Z., Söding, J., & Lupas, A. N. (2016). The MPI bioinformatics Toolkit as an integrative platform for advanced protein sequence and structure analysis. *Nucleic Acids Research, 44*(1), 410–416. doi:10.1093/nar/gkw348 PMID:27131380

Amador-Castro, F., García-Cayuela, T., Alper, H. S., Rodriguez-Martinez, V., & Carrillo-Nieves, D. (2021). Valorization of pelagic sargassum biomass into sustainable applications: Current trends and challenges. *Journal of Environmental Management, 283*, 112013. doi:10.1016/j.jenvman.2021.112013 PMID:33508553

Amaro, H. M., Guedes, A. C., & Malcata, F. X. (2011). Antimicrobial activities of microalgae: an invited review. *Science against microbial pathogens: Communicating current research and technological advances, 3*, 1272-1284.

Amass, W., Amass, A., & Tighe, B. (1998). A review of biodegradable polymers: Uses, current developments in the synthesis and characterization of biodegradable polyesters, blends of biodegradable polymers and recent advances in biodegradation studies. *Polymer International, 47*(2), 89–144. doi:10.1002/(SICI)1097-0126(1998100)47:2<89::AID-PI86>3.0.CO;2-F

Ambati, R. R., Phang, S. M., Ravi, S., & Aswathanarayana, R. G. (2014). Astaxanthin: Sources, extraction, stability, biological activities and its commercial applications—a review. *Marine Drugs, 12*(1), 128–152. doi:10.3390/md12010128 PMID:24402174

Ameri, A., Tamjidi, S., Dehghankhalili, F., Farhadi, A., & Saati, M. A. (2020). Application of algae as low cost and effective bio-adsorbent for removal of heavy metals from wastewater: A review study. *Environmental Technology Reviews, 9*(1), 85–110. doi:10.1080/21622515.2020.1831619

Amid, C., Alako, B. T. F., Balavenkataraman, K. V., Burdett, T., Burgin, J., Fan, J., Harrison, P. W., Holt, S., Hussein, A., Ivanov, E., Jayathilaka, S., Kay, S., Keane, T., Leinonen, R., Liu, X., Martinez-Villacorta, J., Milano, A., Pakseresht, A., Rahman, N., ... Cochrane, G. (2020). The European Nucleotide Archive in 2019. *Nucleic Acids Research, 48*, 70–76. doi:10.1093/nar/gkz1063 PMID:31722421

Amin, S. A., Green, D. H., Hart, M. C., Kupper, F. C., Sunda, W. G., & Carrano, C. J. (2009). Photolysis of iron-siderophore chelates promotes bacterial-algal mutualism. *Proceedings of the National Academy of Sciences of the United States of America, 106*(40), 17071–17076. doi:10.1073/pnas.0905512106 PMID:19805106

Amin, S. A., Hmelo, L. R., Van Tol, H. M., Durham, B. P., Carlson, L. T., Heal, K. R., Morales, R. L., Berthiaume, C. T., Parker, M. S., Djunaedi, B., Ingalls, A. E., Parsek, M. R., Moran, M. A., & Armbrust, E. V. (2015). Interaction and signalling between a cosmopolitan phytoplankton and associated bacteria. *Nature*, *522*(7554), 98–101. doi:10.1038/nature14488 PMID:26017307

Amirante, R., Demastro, G., Distaso, E., Hassaan, M. A., Mormando, A., Pantaleo, P., Tamburrano, A. M., Tedone, L., & Clodoveo, M. L. (2018). Effects of ultrasound and green synthesis ZnO nanoparticles on biogas production from olive pomace. *Energy Procedia*, (148), 940–947.

Amobonye, A., Bhagwat, P., Singh, S., & Pillai, S. (2021). Plastic biodegradation: Frontline microbes and their enzymes. *The Science of the Total Environment*, *759*, 143536. doi:10.1016/j.scitotenv.2020.143536 PMID:33190901

Anand, V., Gautam, R., & Vinu, R. (2017). Non-Catalytic and Catalytic Fast Pyrolysis of Schizochytrium Limacinum Microalga. *Fuel*, *205*, 1–10. doi:10.1016/j.fuel.2017.05.049

Andersen, R. A. (2005). Recipes for freshwater and seawater media. *Algal Culturing Techniques*.

Andersen, R. A. (2005). *Algal culturing techniques*. Elsevier.

Anderson, R. A. (2005). Traditional microalgae isolation techniques. In Algal Cult. Tech. Elsevier Academic Press. doi:10.1016/B978-012088426-1/50007-X

Anderson, N. L., & Anderson, N. G. (1998). Proteome and proteomics: New technologies, new concepts, and new words. *Electrophoresis*, *19*(11), 1853–1861. doi:10.1002/elps.1150191103 PMID:9740045

Andosch, A., Höftberger, M., Lütz, C., & Lütz-Meindl, U. (2015). Subcellular sequestration and impact of heavy metals on the ultrastructure and physiology of the multicellular freshwater alga Desmidium swartzii. *International Journal of Molecular Sciences*, *16*(5), 10389–10410. doi:10.3390/ijms160510389 PMID:25961949

Andrady, A. L. (2011). Microplastics in the marine environment. *Marine Pollution Bulletin*, *62*(8), 1596–1605. doi:10.1016/j.marpolbul.2011.05.030 PMID:21742351

Andrady, A. L., Hamid, H., & Torikai, A. (2011). Effects of solar UV and climate change on materials. *Photochemical & Photobiological Sciences*, *10*(2), 292. doi:10.1039/c0pp90038a PMID:21253664

Andreeßen, C., & Steinbüchel, A. (2018). Recent developments in non-biodegradable biopolymers: Precursors, production processes, and future perspectives. *Applied Microbiology and Biotechnology*, *103*(1), 143–157. doi:10.100700253-018-9483-6 PMID:30397765

Andrews, J. F. (1968). A mathematical model for the continuous culture of microorganisms utilizing inhibitory substrates. *Biotechnology and Bioengineering*, *10*(6), 707–723. doi:10.1002/bit.260100602

Angeli, S. D., Monteleone, G., Giaconia, A., & Lemonidou, A. A. (2014). State-of-the-art catalysts for CH4 steam reforming at low temperature. *International Journal of Hydrogen Energy*, *39*(5), 1979–1997. doi:10.1016/j.ijhydene.2013.12.001

Angelova, R., Baldikova, E., Pospiskova, K., Maderova, Z., Safarikova, M., & Safarik, I. (2016). Magnetically Modified *Sargassum Horneri* Biomass as an Adsorbent for Organic Dye Removal. *Journal of Cleaner Production*, *137*, 189–194.

Angulo, E., Bula, L., Mercado, I., Montano, A., & Cubillan, N. (2018). Bioremediation of Cephalexin with non-living *Chlorella* sp., biomass after lipid extraction. *Bioresource Technology*, *257*, 17–22.

An, H., Wilhelm, W. E., & Searcy, S. W. (2011). A mathematical model to design a lignocellulosic biofuel supply chain system with a case study based on a region in Central Texas. *Bioresource Technology*, *102*(17), 7860–7870. doi:10.1016/j.biortech.2011.05.060 PMID:21724388

Anjos Neto, A.P.D., Oliveira, G.R.F., Mello, S.D.C., Silva, M.S.D., Gomes-Junior, F.G., Novembre, A. D. & Azevedo, R.A. (2020). Seed priming with seaweed extract mitigate heat stress in spinach: effect on germination, seedling growth and antioxidant capacity. *Bragantia.*

Anning, T., Harris, G., & Geider, R. J. (2001). Thermal acclimation in the marine diatom *Chaetoceros calcitrans* (Bacillariophyceae). *European Journal of Phycology*, *36*(3), 233–241. doi:10.1080/09670260110001735388

Ansah, E., Wang, L., Zhang, B., & Shahbazi, A. (2018). Catalytic pyrolysis of raw and hydrothermally carbonized Chlamydomonas debaryana microalgae for denitrogenation and production of aromatic hydrocarbons. *Fuel*, *228*, 234–242. doi:10.1016/j.fuel.2018.04.163

Ansari, S., & Fatma, T. (2016). Cyanobacterial polyhydroxybutyrate (PHB): Screening, optimization and characterization. *PLoS One*, *11*(6), 1–20. doi:10.1371/journal.pone.0158168 PMID:27359097

Anstey, A., Muniyasamy, S., Reddy, M. M., Misra, M., & Mohanty, A. (2014). Processability and biodegradability evaluation of composites from poly(butylene succinate) (PBS) bioplastic and biofuel co-products from Ontario. *Journal of Polymers and the Environment*, *22*(2), 209–218. doi:10.100710924-013-0633-8

Anto, S., Mukherjee, S. S., Muthappa, R., Mathimani, T., Deviram, G., Kumar, S. S., Verma, T. N., & Pugazhendhi, A. (2020). Algae as green energy reserve: Technological outlook on biofuel production. *Chemosphere*, *242*, 125079. doi:10.1016/j.chemosphere.2019.125079 PMID:31678847

Anuradha, & Singh, J. (2021). Organic farming by biofertilizers. In *Biofertilizers*. Scrivener Publishing LLC.

Anwar, M., Lou, S., Chen, L., Li, H., & Hu, Z. (2019). Recent advancement and strategy on bio-hydrogen production from photosynthetic microalgae. *Bioresource Technology*, *292*(August), 121972. doi:10.1016/j.biortech.2019.121972 PMID:31444119

Aoki, M., Matsumoto, H., Takahashi, T., Sato, K., Kumata, H., & Fujiwara, K. (2013). Thallium Induces morphological changes in the photosynthetic apparatus of Synechocystis sp. PCC6803. In Photosynthesis Research for Food, Fuel and the Future (pp. 586–589). Springer. doi:10.1007/978-3-642-32034-7_126

Aparicio, E., Rodríguez-Jasso, R. M., Lara, A., Loredo-Treviño, A., Aguilar, C. N., Kostas, E. T., & Ruiz, H. A. (2020). Biofuels production of third generation biorefinery from macroalgal biomass in the Mexican context: An overview. *Sustainable Seaweed Technologies*, 393–446.

Apt, K. E., Grossman, A. R., & Kroth-Pancic, P. G. (1996). Stable nuclear transformation of the diatom Phaeodactylum tricornutum. *Mol. Gen. Genet. MGG*, *252*, 572–579. PMID:8914518

Arad, S., & van Moppers, D. (2013). Novel sulphated polysaccharides of red microalgae; basics and applications. In Handbook of microalgal culture: applied phycology and biotechnology. Wiley-Blackwell.

Araghian, S., Bagherzadeh, A., & Sadrabadi, R. (2015). Effect of brown algae and vermicompost application on some cherry tomato traits in hydroponic system. *Agroecology Journal*, *10*(4), 77–83.

Ara, J., Sultana, V., Qasim, R., Ehteshamul-Haque, S., & Ahmad, V. U. (2005). Biological activity of Spatoglossum asperum: A brown alga. *Phytotherapy Research*, *19*(7), 618–623. doi:10.1002/ptr.1699 PMID:16161033

Aramkitphotha, S., Tanatavikorn, H., Yenyuak, C., & Vitidsant, T. (2019). Low sulfur fuel oil from blends of microalgae pyrolysis oil and used lubricating oil: Properties and economic evaluation. *Sustainable Energy Technologies and Assessments*, *31*, 339–346. doi:10.1016/j.seta.2018.12.019

Arau'jo, E. A., Andrade, N. J., Carvalho, A. F., Ramos, A. M., Silva, C. D. S., & Silva, L. H. M. (2010). Aspectos coloidais da adesa˜o de microorganismos. *Quimica Nova*, *33*(9), 1940–1948. doi:10.1590/S0100-40422010000900022

Araújo, R., Vázquez Calderón, F., Sánchez López, J., Azevedo Isabel, C., Bruhn, A., Fluch, S., Garcia, T. M., Ghaderiardakani, F., Ilmjärv, T., Laurans, M., Mac Monagail, M., Mangini, S., Peteiro, C., Rebours, C., Stefansson, T., & Ullmann, J. (2021). Current Status of the Algae Production Industry in Europe: An Emerging Sector of the Blue Bioeconomy. *Frontiers in Marine Science*, *7*, 626389. doi:10.3389/fmars.2020.626389

Aravantinou, A. F., Tsarpali, V., Dailianis, S., & Manariotis, I. D. (2015). Effect of cultivation media on the toxicity of ZnO nanoparticles to freshwater and marine microalgae. *Ecotoxicology and Environmental Safety*, *114*, 109–116. doi:10.1016/j.ecoenv.2015.01.016 PMID:25625916

Archibald, J. M., Simpson, A. G. B., & Slamovits, C. H. (Eds.). (2017). *Handbook of the Protists* (2nd ed.). Springer.

Arcos-Hernandez, M. V., Laycock, B., Pratt, S., Donose, B. C., Nikolić, M. A. L., Luckman, P., Werker, A., & Lant, P. A. (2012). Biodegradation in a soil environment of activated sludge derived polyhydroxyalkanoate (PHBV). *Polymer Degradation & Stability*, *97*(11), 2301–2312. doi:10.1016/j.polymdegradstab.2012.07.035

Arenas, E. G., Rodriguez Palacio, M. C., Juantorena, A. U., Fernando, S. E. L., & Sebastian, P. J. (2017). Microalgae as a potential source for biodiesel production: Techniques, methods, and other challenges. *International Journal of Energy Research*, *41*(6), 761–789. doi:10.1002/er.3663

Arias, D. M., García, J., & Uggetti, E. (2020). Production of polymers by cyanobacteria grown in wastewater: Current status, challenges and future perspectives. *New Biotechnology*, *55*, 46–57. doi:10.1016/j.nbt.2019.09.001 PMID:31541716

Arias, D. M., Uggetti, E., García-Galán, M. J., & García, J. (2018). Production of polyhydroxybutyrates and carbohydrates in a mixed cyanobacterial culture: Effect of nutrients limitation and photoperiods. *New Biotechnology*, *42*, 1–11. doi:10.1016/j.nbt.2018.01.001 PMID:29306000

Arikawa, H., Sato, S., Fujiki, T., & Matsumoto, K. (2017). Simple and rapid method for isolation and quantitation of polyhydroxyalkanoate by SDS-sonication treatment. *Journal of Bioscience and Bioengineering*, *124*(2), 250–254. doi:10.1016/j.jbiosc.2017.03.003 PMID:28456470

Armbrust, E. V., Berges, J. A., Bowler, C., Green, B. R., Martinez, D., Putnam, N. H., Zhou, S., Allen, A. E., Apt, K. E., Bechner, M., Brzezinski, M. A., Chaal, B. K., Chiovitti, A., Davis, A. K., Demarest, M. S., Detter, J. C., Glavina, T., Goodstein, D., Hadi, M. Z., ... Rokhsar, D. S. (2004). The genome of the diatom Thalassiosira pseudonana: Ecology, evolution, and metabolism. *Science*, *306*(5693), 79–86. doi:10.1126cience.1101156 PMID:15459382

Arnold, T. M., Targett, N. M., Tanner, C. E., Hatch, W. I., & Ferrari, K. E. (2001). Evidence for Methyl Jasmonate-Induced Phlorotannin Production in *Fucus vesiculosus* (Phaeophyceae). *Journal of Phycology*, *37*(6), 1026–1029. doi:10.1046/j.1529-8817.2001.01130.x

Arora, K., Kaur, P., Kumar, P., Singh, A., Patel, S. K. S., Li, X., Yang, Y.-H., Bhatia, S. K., & Kulshrestha, S. (2021). Valorization of Wastewater Resources Into Biofuel and Value-Added Products Using Microalgal System. *Frontiers in Energy Research*, *9*, 646571. Advance online publication. doi:10.3389/fenrg.2021.646571

Arora, N., Yen, H.-W., & Philippidis, G. P. (2020). Harnessing the power of mutagenesis and adaptive laboratory evolution for high lipid production by oleaginous microalgae and yeasts. *Sustainability*, *12*(12), 5125. doi:10.3390u12125125

Arrieta, M. P., López, J., Rayón, E., & Jiménez, A. (2014). Disintegrability under composting conditions of plasticized PLA–PHB blends. *Polymer Degradation & Stability*, *108*, 307–318. doi:10.1016/j.polymdegradstab.2014.01.034

Arsiya, F., Sayadi, M. H., & Sobhani, S. (2017). Green synthesis of palladium nanoparticles using Chlorella vulgaris. *Materials Letters*, *186*, 113–115. doi:10.1016/j.matlet.2016.09.101

Årstøl, E., & Hohmann-Marriott, M.F. (2019) Cyanobacterial Siderophores—Physiology, Structure, Biosynthesis, and Applications. *Marine Drugs*, *17*(5), 281.

Ashik, U. P. M., Daud, W. M. A. W., & Abbas, H. F. (2015). Production of greenhouse gas free hydrogen by thermo-catalytic decomposition of methane–A review. *Renewable & Sustainable Energy Reviews*, *44*, 221–256. doi:10.1016/j.rser.2014.12.025

Ashokkumar, V., Salam, Z., Tiwari, O. N., Chinnasamy, S., Mohammed, S., & Ani, F. N. (2015). An integrated approach for biodiesel and bioethanol production from Scenedesmus bijugatus cultivated in a vertical tubular photobioreactor. *Energy Conversion and Management*, *101*, 778–786. doi:10.1016/j.enconman.2015.06.006

Ashori, A., & Nourbakhsh, A. (2010). Reinforced polypropylene composites: Effects of chemical compositions and particle size. *Bioresource Technology*, *101*(7), 2515–2519. doi:10.1016/j.biortech.2009.11.022 PMID:19948401

Ashour, M., Alprol, A. E., Heneash, A. M. M., Saleh, H., Abualnaja, K. M., Alhashmialameer, D., & Mansour, A. T. (2021). *Ammonia Bioremediation from Aquaculture Wastewater Effluents Using Arthrospira platensis NIOF17 / 003 : Impact of Biodiesel Residue and Potential of Ammonia-Loaded Biomass as Rotifer Feed.* Academic Press.

Ashour, M., Elshobary, M. E., El-Shenody, R., Kamil, A. W. A., & Abomohra, A. E (2019). Evaluation of a native oleaginous marine microalga Nannochloropsis oceanica for dual use in biodiesel production and aquaculture feed. *Biomass and Bioenergy, 120*(December), 439–447. doi:10.1016/j.biombioe.2018.12.009

Ashwini, S., Babut, S., & Saritha, M. S. (2016). Seaweed extracts exhibit anticancer activity against HeLa cell lines. *International Journal of Current Pharmaceutical Research*, *9*(1), 114–117. doi:10.22159/ijcpr.2017v9i1.16632

Atabani, A. E., Shobana, S., Mohammed, M. N., Uğuz, G., Kumar, G., Arvindnarayan, S., Aslam, M., & Al-Muhtaseb, A. H. (2019). Integrated valorization of waste cooking oil and spent coffee grounds for biodiesel production: Blending with higher alcohols, FT–IR, TGA, DSC and NMR characterizations. *Fuel*, *244*, 419–430. doi:10.1016/j.fuel.2019.01.169

Atashrazm, F., Lowenthal, R. M., Woods, G. M., Holloway, A. F., & Dickinson, J. L. (2015). Fucoidan and cancer: A multifunctional molecule with anti-tumor potential. *Marine Drugs*, *13*(4), 2327–2346. doi:10.3390/md13042327 PMID:25874926

Ates, M., Daniels, J., Arslan, Z., Farah, I. O., & Rivera, H. F. (2013). Comparative evaluation of impact of Zn and ZnO nanoparticles on brine shrimp (Artemiasalina) larvae: Effects of particle size and solubility on toxicity. *Environmental Science. Processes & Impacts*, *15*(1), 225–233. doi:10.1039/C2EM30540B PMID:24058731

Athukorala, Y., Kim, K. N., & Jeon, Y. J. (2006). Antiproliferative and antioxidant properties of an enzymatic hydrolysate from brown alga, Ecklonia cava. *Food and Chemical Toxicology*, *44*(7), 1065–1074. doi:10.1016/j.fct.2006.01.011 PMID:16516367

Atiwesh, G., Mikhael, A., Parrish, C. C., Banoub, J., & Le, T.-A. T. (2021). Environmental impact of bioplastic use: A review. *Heliyon*, *7*(9), e07918. doi:10.1016/j.heliyon.2021.e07918 PMID:34522811

Austin, H. P., Allen, M. D., Donohoe, B. S., Rorrer, N. A., Kearns, F. L., Silveira, R. L., Pollard, B. C., Dominick, G., Duman, R., El Omari, K., Mykhaylyk, V., Wagner, A., Michener, W. E., Amore, A., Skaf, M. S., Crowley, M. F., Thorne, A. W., Johnson, C. W., Woodcock, H. L., . . . Beckham, G. T. (2018). Characterization and of a plastic-degrading aromatic polyesterase. *Proceedings of the National Engineering Academy of Sciences of the United States of America, 115*(19), E4350–E4357. 10.1073/pnas.1718804115

Avagyan, A. B. (2010). New Design & Build Biological System through the Use of Microalgae Addressed to Sustainable Development. *Journal of Environmental Protection*, *01*(02), 183–200. doi:10.4236/jep.2010.12023

Awudu, I., & Zhang, J. (2013). Stochastic production planning for a biofuel supply chain under demand and price uncertainties. *Applied Energy*, *103*, 189–196. doi:10.1016/j.apenergy.2012.09.025

Aydemir, D., & Gardner, D. J. (2020). Biopolymer blends of polyhydroxybutyrate and polylactic acid reinforced with cellulose nanofibrils. *Carbohydrate Polymers*, *250*(July), 116867. doi:10.1016/j.carbpol.2020.116867 PMID:33049817

Ayre, J. M., Moheimani, N. R., & Borowitzka, M. A. (2017). Growth of microalgae on undiluted anaerobic digestate of piggery effluent with high ammonium concentrations. *Algal Research*, *24*, 218–226. doi:10.1016/j.algal.2017.03.023

Azadeh, A., Arani, H. V., & Dashti, H. (2014). A stochastic programming approach towards optimization of biofuel supply chain. *Energy*, *76*, 513–525. doi:10.1016/j.energy.2014.08.048

Azam, F. (1998). Microbial control of oceanic carbon flux: The plot thickens. *Science*, *280*(5364), 694–696. doi:10.1126cience.280.5364.694

Azam, R., Kothari, R., Singh, H. M., Ahmad, S., Ashokkumar, V., & Tyagi, V. V. (2020). Production of Algal Biomass for its Biochemical Profile Using Slaughterhouse Wastewater for Treatment under Axenic Conditions. *Bioresource Technology*, *306*, 123116.

Azeem, M., Batool, F., & Iqbal, N. (2017). Algal-based biopolymers. In Algae Based Polymers, Blends, and Composites (pp. 1-31). Elsevier. doi:10.1016/B978-0-12-812360-7.00001-X

Azizi, S., Ahmad, M. B., Ibrahim, N. A., Hussein, M. Z., & Namvar, F. (2014). Preparation and properties of poly (vinyl alcohol)/chitosan blend bio-nanocomposites reinforced by cellulose nanocrystals. *Chinese Journal of Polymer Science*, *32*(12), 1620–1627. doi:10.100710118-014-1548-0

Azizi, S., Ahmad, M. B., Namvar, F., & Mohamad, R. (2014). Green biosynthesis and characterization of zinc oxide nanoparticles using brown marine macroalga Sargassum muticum aqueous extract. *Materials Letters*, *116*, 275–277. doi:10.1016/j.matlet.2013.11.038

Azizi, S., Namvar, F., Mahdavi, M., Ahmad, M., & Mohamad, R. (2013). Biosynthesis of silver nanoparticles using brown marine macroalga, Sargassum muticum aqueous extract. *Materials (Basel)*, *6*(12), 5942–5950. doi:10.3390/ma6125942 PMID:28788431

Badr, S. A., Ashmawy, A. A., El-Sherif, I. Y., & Moghazy, R. M. (2016). Non-conventional low-cost biosorbents for adsorption and desorption of heavy metals. *Research Journal of Pharmaceutical, Biological and Chemical Sciences*, *7*(4).

Badvipour, S., Eustance, E., & Sommerfeld, M. R. (2016). Process evaluation of energy requirements for feed production using dairy wastewater for algal cultivation: Theoretical approach. *Algal Research*, *19*, 207–214. doi:10.1016/j.algal.2016.08.017

Baek, K., Kim, D. H., Jeong, J., Sim, S. J., Melis, A., Kim, J.-S., Jin, E. S., & Bae, S. (2016). DNA-free two-gene knockout in *Chlamydomonas reinhardtii* via CRISPR-Cas9 ribonucleoproteins. *Scientific Reports*, *6*(1), 30620. doi:10.1038rep30620 PMID:27466170

Baek, K., Yu, J., Jeong, J., Sim, S. J., Bae, S., & Jin, E. (2018). Photoautotrophic production of macular pigment in a Chlamydomonas reinhardtii strain generated by using DNA-free CRISPR-Cas9 RNP-mediated mutagenesis. *Biotechnology and Bioengineering*, *115*(3), 719–728. doi:10.1002/bit.26499 PMID:29150930

Bae, Y. J., Ryu, C., Jeon, J. K., Park, J., Suh, D. J., Suh, Y. W., Chang, D., & Park, Y. K. (2011). The characteristics of bio-oil produced from the pyrolysis of three marine macroalgae. *Bioresource Technology*, *102*(3), 3512–3520. doi:10.1016/j.biortech.2010.11.023 PMID:21129955

Baglieri, A., Sidella, S., Barone, V., Fragalà, F., Silkina, A., Nègre, M., & Gennari, M. (2016). Cultivating *Chlorella vulgaris* and *Scenedesmusquadricauda* microalgae to degrade inorganic compounds and pesticides in water. *Environmental Science and Pollution Research International*, *23*(18), 18165–18174. doi:10.100711356-016-6996-3 PMID:27259964

Baicha, Z., Salar-García, M. J., Ortiz-Martínez, V. M., Hernández-Fernández, F. J., De los Ríos, A. P., Labjar, N., & Elmahi, M. (2016). A critical review on microalgae as an alternative source for bioenergy production: A promising low cost substrate for microbial fuel cells. *Fuel Processing Technology*, *154*, 104–116. doi:10.1016/j.fuproc.2016.08.017

Bai, X., & Achrya, K. (2016). Removal of trimethoprim, sulfamethoxazole, and triclosan by the green alga *Nannochloris* sp. *Journal of Hazardous Materials*, *315*, 70–75.

Bakire, S., Yang, X., Ma, G., Wei, X., Yu, H., Chen, J., & Lin, H. (2018). Developing predictive models for toxicity of organic chemicals to green algae based on mode of action. *Chemosphere*, *190*, 463–470. doi:10.1016/j.chemosphere.2017.10.028 PMID:29028601

Bala, B. K. (2006). Studies on biodiesel from transformation of vegetable oils for diesel engines. *Energy Education Science and Technology*, *15*, 1–45.

Balaji Raghavendra Rao, H., Sathivel, A., & Devaki, T. (2004). Antihepatotoxic nature of Ulva reticulata (Chlorophyceae) on acetaminophen-induced hepatoxicity in experimental rats. *Journal of Medicinal Food*, *7*(4), 495–497. doi:10.1089/jmf.2004.7.495 PMID:15671697

Balaji, S., Gopi, K., & Muthuvelan, B. (2013). A review on production of poly β hydroxybutyrates from cyanobacteria for the production of bio plastics. *Algal Research*, *2*(3), 278–285. doi:10.1016/j.algal.2013.03.002

Balakrishnan, C., Jenifer, P., & Esakkilingam, M. (2013). Algal documentation and phytochemical studies of red algae Gracilaria corticata of Manapad Coast, Tamil Nadu. *Journal of Pharmacognosy and Phytochemistry*, *2*(4).

Balakrishnan, J., Dhavamani, S., Govindan Sadasivam, S., Arumugam, M., Vellaikumar, S., Ramalingam, J., & Shanmugam, K. (2019). Omega-rich Isochrysis sp. biomass enhances brain docosahexaenoic acid levels and improves serum lipid profile and antioxidant status in Wistar rats. *Journal of the Science of Food and Agriculture*, 3.

Balaman, Ş. Y., & Selim, H. (2014). A network design model for biomass to energy supply chains with anaerobic digestion systems. *Applied Energy*, *130*, 289–304. doi:10.1016/j.apenergy.2014.05.043

Balat, M., & Balat, M. (2009). Political, economic and environmental impacts of biomass-based hydrogen. *International Journal of Hydrogen Energy*, *34*(9), 3589–3603. doi:10.1016/j.ijhydene.2009.02.067

Balić, A., Vlašić, D., Žužul, K., Marinović, B., & Mokos, Z. B. (2020). Omega-3 Versus Omega-6 Polyunsaturated Fatty Acids in the Prevention and Treatment of Inflammatory Skin Diseases. *International Journal of Molecular Sciences*, *21*(3), 741. doi:10.3390/ijms21030741 PMID:31979308

Banerjee, S., Banerjee, S., Ghosh, A. K., & Das, D. (2020). Maneuvering the genetic and metabolic pathway for improving biofuel production in algae: Present status and future prospective. *Renewable & Sustainable Energy Reviews*, *133*, 1–18. doi:10.1016/j.rser.2020.110155

Barabad, M. L. M., Jung, W., Versoza, M. E., Kim, M., Ko, S., Park, D., & Lee, K. (2018). Emission Characteristics of Particulate Matter, Volatile Organic Compounds, and Trace Elements from the Combustion of Coals in Mongolia. *International Journal of Environmental Research and Public Health*, *15*(8), 1706. doi:10.3390/ijerph15081706 PMID:30096940

Barakat, K. M., El–Sayed, H. S., Khairy, H. M., El–Sheikh, M. A., Al–Rashed, S. A., Arif, I. A., & Elshobary, M. E. (2021). Effects of ocean acidification on the growth and biochemical composition of a green alga (*Ulva fasciata*) and its associated microbiota. *Saudi Journal of Biological Sciences*, *28*(9), 5106–5114. doi:10.1016/j.sjbs.2021.05.029 PMID:34466088

Barancheshme, F., & Munir, M. (2019). Development of antibiotic resistance in wastewater treatment plants. In Y. Kumar (Ed.), *Antimicrobial Resistance-A Global Threat.* IntechOpen. doi:10.5772/intechopen.81538

Barba, F. J. (2015). Current Applications and New Opportunities for the Use of Pulsed Electric Fields in Food Science and Industry. *Food Research International, 77*, 773–98.

Barbosa, M. J., Hoogakker, J., & Wijffels, R. H. (2003). Optimisation of cultivation parameters in photobioreactors for microalgae cultivation using the A-stat technique. *Biomolecular Engineering*, *20*(4-6), 115–123. doi:10.1016/S1389-0344(03)00033-9 PMID:12919788

Barbosa, M., Valentão, P., & Andrade, P. B. (2014). Bioactive compounds from macroalgae in the new millennium: Implications for neurodegenerative diseases. *Marine Drugs*, *12*(9), 4934–4972. doi:10.3390/md12094934 PMID:25257784

Barlow, J., Sims, R. C., & Quinn, J. C. (2016). Techno-economic and life-cycle assessment of an attached growth algal biorefinery. *Bioresource Technology*, *220*, 360–368. doi:10.1016/j.biortech.2016.08.091 PMID:27595701

Barnharst, T., Rajendran, A., & Hu, B. (2018). Bioremediation of synthetic intensive aquaculture wastewater by a novel feed-grade composite biofilm. *International Biodeterioration & Biodegradation*, *126*, 131–142. doi:10.1016/j.ibiod.2017.10.007

Barra, R., & Leonard, S. A. (2018). *Plastics and the circular economy. 54th Global Environment Facility Council Meeting*, Da Nang, Viet Nam.

Barreira, L., Pereira, H., Gangadhar, K. N., Custódio, L., & Varela, J. (2015). *Medicinal Effects of Microalgae-Derived Fatty Acid.* doi:10.1016/B978-0-12-800776-1.00013-3

Barrera, D. J., Rosenberg, J. N., Chiu, J. G., Chang, Y.-N., Debatis, M., Ngoi, S.-M., Chang, J. T., Shoemaker, C. B., Oyler, G. A., & Mayfield, S. P. (2015). Algal chloroplast produced camelid VHH antitoxins are capable of neutralizing botulinum neurotoxin. *Plant Biotechnology Journal*, *13*(1), 117–124. doi:10.1111/pbi.12244 PMID:25229405

Barros, A. I., Gonçalves, A. L., Simões, M., & Pires, J. C. M. (2015). Harvesting techniques applied to microalgae: A review. *Renewable & Sustainable Energy Reviews*, *41*, 1489–1500. doi:10.1016/j.rser.2014.09.037

Barros, Ó., Costa, L., Costa, F., Lago, A., Rocha, V., Vipotnik, Z., Silva, B., & Tavares, T. (2019). Recovery of rare earth elements from wastewater towards a circular economy. *Molecules (Basel, Switzerland)*, *24*(6), 1005. doi:10.3390/molecules24061005 PMID:30871164

Barsanti, L., & Gualtieri, P. (2006). Algae biotechnology. In Algae: Anatomy, Biochemistry, and Biotechnology. CRC Press.

Barsanti, L., Coltelli, P., Evangelista, V., Frassanito, A. M., Passarelli, V., Vesentini, N., & Gualtieri, P. (2008). Oddities and curiosities in the algal world. In *Algal toxins: nature, occurrence, effect, and detection* (pp. 353–391). Springer. doi:10.1007/978-1-4020-8480-5_17

Barsanti, L., & Gualtieri, P. (2014). *Algae: Anatomy, Biochemistry, and Biotechnology* (2nd ed.). CRC Press.

Barsanti, L., & Paolo, G. (2006). *Algae: Anatomy, Biochemistry, and Biotechnology. By. CRC Press.* Taylor & Francis Group.

Basharina, T. N., Danilovtseva, E. N., Zelinskiy, S. N., Klimenkov, I. V., Likhoshway, Y. V., & Annenkov, V. V. (2012). The effect of titanium, zirconium and tin on the growth of diatom Synedra acus and morphology of its silica valves. *Silicon*, *4*(4), 239–249. doi:10.100712633-012-9119-x

Batan, L., Quinn, J., Willson, B., & Bradley, T. (2010). Net energy and greenhouse gas emission evaluation of biodiesel derived from microalgae. *Environmental Science & Technology*, *44*(20), 7975–7980.

Batista, A. P., Ambrosano, L., Graça, S., Sousa, C., Marques, P. A. S. S., Ribeiro, B., Botrel, E. P., Castro Neto, P., & Gouveia, L. (2015). Combining Urban Wastewater Treatment with Biohydrogen Production – An Integrated Microalgae-Based Approach. *Bioresource Technology*, *184*, 230–235. doi:10.1016/j.biortech.2014.10.064 PMID:25453433

Batista, A. P., Gouveia, L., Bandarra, N. M., Franco, J. M., & Raymundo, A. (2013). Comparison of microalgal biomass profiles as novel functional ingredient for food products. *Algal Research*, *2*(2), 164–173. doi:10.1016/j.algal.2013.01.004

Batista, A. P., Niccolai, A., Fradinho, P., Fragoso, S., Bursic, I., Rodolfi, L., Biondi, N., Tredici, M. R., Sousa, I., & Raymundo, A. (2017). Microalgae biomass as an alternative ingredient in cookies: Sensory, physical and chemical properties, antioxidant activity and in vitro digestibility. *Algal Research*, *26*, 161–171. doi:10.1016/j.algal.2017.07.017

Batista, D., Carvalho, A. P., Costa, R., Coutinho, R., & Dobretsov, S. (2014). Extracts of macroalgae from the Brazilian coast inhibit bacterial quorum sensing. *Botanica Marina*, *57*(6), 441–447. Advance online publication. doi:10.1515/bot-2014-0031

Batool, A., Menaa, F., Ali, K. B., Uzair, B., & Menaa, B. (2019). Progress and prospects in translating nanobiotechnology in medical theranostics. *Current Nanoscience*, *15*, 1–23.

Baudelet, P. H., Gagez, A. L., Bérard, J. B., Juin, C., Bridiau, N., Kaas, R., Thiéry, V., Cadoret, J.-P., & Picot, L. (2013). Antiproliferative activity of Cyanophora paradoxa pigments in melanoma, breast, and lung cancer cells. *Marine Drugs*, *11*(11), 4390–4406. doi:10.3390/md11114390 PMID:24189278

Bauer, L., Ranglová, K., Masojídek, J., Drosg, B., & Meixner, K. (2021). Digestate as sustainable nutrient source for microalgae—Challenges and prospects. *Applied Sciences (Switzerland)*, *11*(3), 1–21. doi:10.3390/app11031056

Baweja, P., Kumar, S., & Kumar, G. (2019). Organic Fertilizer from Algae: A Novel Approach Towards Sustainable Agriculture. In B. Giri, R. Prasad, Q.-S. Wu, & A. Varma (Eds.), *Biofertilizers for Sustainable Agriculture and Environment* (pp. 353–370). Springer International Publishing. doi:10.1007/978-3-030-18933-4_16

Bayazıt, G., Tastan, B. E., & Ülküye, D. G. (2019). Biosorption, Isotherm and Kinetic Properties of Common Textile Dye by *Phormidium Animale. Global NEST Journal*, *21*, 1–7.

Baykara, S. Z. (2004). Hydrogen production by direct solar thermal decomposition of water, possibilities for improvement of process efficiency. *International Journal of Hydrogen Energy*, *29*(14), 1451–1458. doi:10.1016/j.ijhydene.2004.02.014

Baykara, S. Z. (2018). Hydrogen: A brief overview on its sources, production and environmental impact. *International Journal of Hydrogen Energy*, *43*(23), 10605–10614. doi:10.1016/j.ijhydene.2018.02.022

Beacham, T. A., Sweet, J. B., & Allen, M. J. (2017). Large scale cultivation of genetically modified microalgae: A new era for environmental risk assessment. *Algal Research*, *25*, 90–100.

Beach, E. S., Eckelman, M. J., Cui, Z., Brentner, L., & Zimmerman, J. B. (2012). Preferential technological and life cycle environmental performance of chitosan flocculation for harvesting of the green algae Neochloris oleoabundans. *Bioresource Technology*, *121*, 445–449. doi:10.1016/j.biortech.2012.06.012 PMID:22853967

Beaumont, M., Tran, R., Vera, G., Niedrist, D., Rousset, A., Pierre, R., Shastri, V. P., & Forget, A. (2021). Hydrogel-Forming Algae Polysaccharides. From Seaweed to Biomedical Applications. *Biomacromolecules*, *22*(3), 1027–1052. doi:10.1021/acs.biomac.0c01406 PMID:33577286

Béchet, Q., Shilton, A., & Guieysse, B. (2013). Modeling the effects of light and temperature on algae growth: State of the art and critical assessment for productivity prediction during outdoor cultivation. *Biotechnology Advances*, *31*(8), 1648–1663. doi:10.1016/j.biotechadv.2013.08.014 PMID:23981914

Becker, W. (2004). Microalgae in human and animal nutrition. In A. Richmond (Ed.), *Handbook of microalgal culture* (pp. 312–351). Blackwell.

Beckstrom, B. D. (2019). *Bioplastic production from microalgae with fuel co-products: A techno-economic and life-cycle assessment.* Colorado State University, Libraries. https://mountainscholar.org/bitstream/handle/10217/197377/Beckstrom_colostate_0053N_15606.pdf?sequence=1

Beckstrom, B. D., Wilson, M. H., Crocker, M., & Quinn, J. C. (2020). Bioplastic feedstock production from microalgae with fuel co-products: A techno-economic and life cycle impact assessment. *Algal Research*, *46*, 101769. doi:10.1016/j.algal.2019.101769

Behera, B., Selvanayaki, S., Jayabalan, R., & Balasubramanian, P. (2019). An in-silico Approach for enhancing the lipid productivity in microalgae by manipulating the fatty acid biosynthesis. In Soft Computing for Problem Solving, Advances in Intelligent Systems and Computing 816 (pp. 877-889). Springer Nature Singapore Pte Ltd. doi:10.1007/978-981-13-1592-3_70

Behringer, G., Ochsenkühn, M. A., Fei, C., Fanning, J., Koester, J. A., & Amin, S. A. (2018). Bacterial communities of diatoms display strong conservation across strains and time. *Frontiers in Microbiology*, *9*, 659. doi:10.3389/fmicb.2018.00659 PMID:29681892

Bekirogullari, M., Figueroa-Torres, G. M., Pittman, J. K., & Theodoropoulos, C. (2020). Models of microalgal cultivation for added-value products-A review. *Biotechnology Advances*, *44*, 107609. doi:10.1016/j.biotechadv.2020.107609 PMID:32781245

Bekirogullari, M., Pittman, J., & Theodoropoulos, C. (2018). Multi-factor kinetic modelling of microalgal biomass cultivation for optimised lipid production. *Bioresource Technology*, *269*, 417–425. doi:10.1016/j.biortech.2018.07.121 PMID:30265993

Bello, M., Ranganathan, P., & Brennan, F. (2017). Life cycle optimization for sustainable algal biofuel production using integrated nutrient recycling technology. *ACS Sustainable Chemistry & Engineering*, *5*(11), 9869–9880. doi:10.1021/acssuschemeng.7b01833

Bell, R. A. (1993). Cryptoendolithic algae of hot semiarid lands and deserts. *Journal of Phycology*, *29*(2), 133–139. doi:10.1111/j.0022-3646.1993.00133.x

Bell, T. A. S., Prithiviraj, B., Wahlen, B. D., Fields, M. W., & Peyton, B. M. (2016). A Lipid-Accumulating Alga Maintains Growth in Outdoor, Alkaliphilic Raceway Pond with Mixed Microbial Communities. *Frontiers in Microbiology*, *6*. Advance online publication. doi:10.3389/fmicb.2015.01480 PMID:26779138

Bell, W., & Mitchell, R. (1972). Chemotactic and growth responses of marine bacteria to algal extracellular products. *The Biological Bulletin*, *143*(2), 265–277. doi:10.2307/1540052

Belnap, J., & Lange, O. L. (2003). Biological Soil Crusts: Structure, Function, and Management. In I. T. Baldwin, M. M. Caldwell, G. Heldmaier, O. L. Lange, H. A. Mooney, E.-D. Schulze, & U. Sommer (Eds.), *Ecological Studies Series 150* (Vol. 150, p. 503). Springer- Verlag. doi:10.1007/978-3-642-56475-8

Benemann, J. (2013). Microalgae for biofuels and animal feeds. *Energies*, *6*(11), 5869–5886.

Benemann, J. R. (2000). Hydrogen production by microalgae. *Journal of Applied Phycology*, *12*(3–5), 291–300. doi:10.1023/A:1008175112704

Bennion, E. P., Ginosar, D. M., Moses, J., Agblevor, F., & Quinn, J. C. (2015). Lifecycle assessment of microalgae to biofuel: Comparison of thermochemical processing pathways. *Applied Energy*, *154*, 1062–1071. doi:10.1016/j.apenergy.2014.12.009

Ben-Tal, A., & Nemirovski, A. (2000). Robust solutions of linear programming problems contaminated with uncertain data. *Mathematical Programming*, *88*(3), 411–424. doi:10.1007/PL00011380

Bérard, A., Dorigo, U., Humbert, J. F., & Martin-Laurent, F. (2005). Microalgae community structure analysis based on 18S rDNA amplification from DNA extracted directly from soil as a potential soil bioindicator. *Agronomy for Sustainable Development*, *25*(2), 285–291. doi:10.1051/agro:2005004

Berjanskii, M., Liang, Y., Zhou, J., Tang, P., Stothard, P., Zhou, Y., Cruz, J., MacDonell, C., Lin, G., Lu, P., & Wishart, D. S. (2010). PROSESS: A protein structure evaluation suite and server. *Nucleic Acids Research, 38,* 633–640. . doi:10.1093/nar/gkq375

Bernaerts, T., Gheysen, L., Kyomugasho, C., Jamsazzadeh, K., Zahra, V., & Foubert, I. (2018). Comparison of microalgal biomasses as functional food ingredients: Focus on the composition of cell wall related polysaccharides. *Algal Research*, *32*, 150–161. doi:10.1016/j.algal.2018.03.017

Bernard, O., & Rémond, B. (2012). Validation of a simple model accounting for light and temperature effect on microalgal growth. *Bioresource Technology*, *123*, 520–527.

Berner, T., Dubinsky, Z., Wyman, K., & Falkowski, P. G. (1989). Photoadaptation and the "package" effect in Dunaliella tertiolecta (Chlorophyceae) 1. *Journal of Phycology*, *25*(1), 70–78. doi:10.1111/j.0022-3646.1989.00070.x

Bertsimas, D., & Sim, M. (2004). The price of robustness. *Operations Research*, *52*(1), 35–53. doi:10.1287/opre.1030.0065

Beuckels, A., Smolders, E., & Muylaert, K. (2015). Nitrogen availability influences phosphorus removal in microalgae-based wastewater treatment. *Water Research*, *77*, 98–106.

Beyaz, R., & Yildiz, M. (2017). The use of gamma irradiation in plant mutation breeding. In S. Jurić (Ed.), *Plant Engineering* (pp. 33–46). doi:10.5772/intechopen.69974

Beyth, N.; Houri-Haddad, Y.; Domb, A.; Khan, W.; Hazan, R. (2015). Alternative antimicrobial approach: Nano-antimicrobial materials. *Evid. Based Complementary Altern. Med.*, 246012.

Bhalamurugan, G. L., Valerie, O., & Mark, L. (2018). Valuable bioproducts obtained from microalgal biomass and their commercial applications: A review. *Environmental Engineering Research*, *23*(3), 229–241. doi:10.4491/eer.2017.220

Bharathiraja, B., Chakravarthy, M., Ranjith Kumar, R., Yogendran, D., Yuvaraj, D., Jayamuthunagai, J., Praveen Kumar, R., & Palani, S. (2015). Aquatic biomass (algae) as a future feed stock for bio- refineries: A review on cultivation, processing and products. *Renewable & Sustainable Energy Reviews*, *47*, 635–653. doi:10.1016/j.rser.2015.03.047

Bhaskar, P. V., & Bhosle, N. B. (2004). Microbial extracellular polysaccharides and their role in marine biogeochemical processes. *Current Science*, *88*, 45–53.

Bhatia, S. K., Jagtap, S. S., Bedekar, A. A., Bhatia, R. K., Rajendran, K., Pugazhendhi, A., Rao, C., Atabani, A. E., Kumar, G., & Yang, Y.-H. (2021). Renewable biohydrogen production from lignocellulosic biomass using fermentation and integration of systems with other energy generation technologies. *The Science of the Total Environment*, *765*, 144429. doi:10.1016/j.scitotenv.2020.144429 PMID:33385808

Bhati, R., & Mallick, N. (2012). Production and characterization of poly(3-hydroxybutyrate-co-3-hydroxyvalerate) co-polymer by a N 2-fixing cyanobacterium, Nostoc muscorum Agardh. *Journal of Chemical Technology and Biotechnology (Oxford, Oxfordshire)*, *87*(4), 505–512. doi:10.1002/jctb.2737

Bhati, R., & Mallick, N. (2015). Poly(3-hydroxybutyrate-co-3-hydroxyvalerate) copolymer production by the diazotrophic cyanobacterium *Nostoc muscorum* Agardh: Process optimization and polymer characterization. *Algal Research*, *7*, 78–85. doi:10.1016/j.algal.2014.12.003

Bhati, R., & Mallick, N. (2016). Carbon dioxide and poultry waste utilization for production of polyhydroxyalkanoate biopolymers by Nostoc muscorum Agardh: Asustainable approach. *Journal of Applied Phycology*, *28*(1), 161–168. doi:10.100710811-015-0573-x

Bhattacharjee, M. (2016). Pharmaceutically valuable bioactive compounds of algae. *Asian Journal of Pharmaceutical and Clinical Research*, *9*(6), 43–47. doi:10.22159/ajpcr.2016.v9i6.14507

Bhattacharya, D., Yoon, H. S., & Hackett, J. D. (2004). Photosynthetic eukaryotes unite: Endosymbiosis connects the dots. *BioEssays*, *26*(1), 50–60. doi:10.1002/bies.10376 PMID:14696040

Bhattacharya, M., & Goswami, S. (2020). Microalgae – A green multi-product biorefinery for future industrial prospects. *Biocatalysis and Agricultural Biotechnology*, *25*(March), 101580. doi:10.1016/j.bcab.2020.101580

Bhooshan, N., Pabbi, S., Singh, A., Sharma, A., Jaiswal, A., & Kumar, A. (2018). Impact of blue green algae (BGA) technology: an empirical evidence from northwestern Indo-Gangetic Plains. *Biotech, 8*(8), 324.

Bhuyar, P., Zagde, S., Revankar, R., Yusoff, M., Rahim, Ab. M. H., & Govindan, N. (2018). Isolation, Characterization and Partial Purification of Keratinase from Keratinolytic Bacteria. SSRN *Electronic Journal*. doi:10.2139/ssrn.3389068

Bicer, Y., & Dincer, I. (2017). Life cycle assessment of nuclear-based hydrogen and ammonia production options: A comparative evaluation. *International Journal of Hydrogen Energy*, *42*(33), 21559–21570. doi:10.1016/j.ijhydene.2017.02.002

Biggs, B. J. F., & Thomsen, H. A. (1995). Disturbance of stream periphyton by perturbations in shear stress: Time to structural failure and differences in community resistance. *Journal of Phycology*, *31*(2), 233–241. doi:10.1111/j.0022-3646.1995.00233.x

Biller, P., & Ross, A. B. (2011). Potential yields and properties of oil from the hydrothermal liquefaction of microalgae with different biochemical content. *Bioresource Technology*, *102*(1), 215–225.

Biller, P., & Ross, A. B. (2012). Hydrothermal processing of algal biomass for the production of biofuels and chemicals. *Biofuels*, *3*(5), 603–623.

Binnemans, K., Jones, P. T., Blanpain, B., Van Gerven, T., & Pontikes, Y. (2015). Towards zero-waste valorisation of rare-earth-containing industrial process residues: A critical review. *Journal of Cleaner Production*, *99*, 17–38. doi:10.1016/j.jclepro.2015.02.089

Binnemans, K., Jones, P. T., Blanpain, B., Van Gerven, T., Yang, Y., Walton, A., & Buchert, M. (2013). Recycling of rare earths: A critical review. *Journal of Cleaner Production*, *51*, 1–22. doi:10.1016/j.jclepro.2012.12.037

Binnemans, K., Pontikes, Y., Jones, P. T., Van Gerven, T., & Blanpain, B. (2013). Recovery of rare earths from industrial waste residues: a concise review. *Proceedings of the 3rd International Slag Valorisation Symposium: The Transition to Sustainable Materials Management*, 191–205.

Biondi, N., Piccardi, R., Margheri, M. C., Rodolfi, L., Smith, G. D., & Tredici, M. R. (2004). Evaluation of Nostoc Strain ATCC 53789 as a Potential Source of Natural Pesticides. *Applied and Environmental Microbiology*, *70*, 3313–3320.

Bischof, K., Hanelt, D., Tug, H., Karsten, U., Brower, P. E. M., & Wiencke, C. (1998). Acclimation of brown algal photosynthesis to ultraviolet radiation in Arctic coastal waters. (Spitsbergen, Norway). *Polar Biology*, *20*(6), 388–395. doi:10.1007003000050319

Bischof, K., Krabs, G., Wiencke, C., & Hanelt, D. (2002). Solar ultraviolet radiation affects the activity of ribulose-1, 5-bisphosphate carboxylase-oxygenase and the composition of photosynthetic and xanthophyll cycle pigments in the intertidal green alga *Ulva lactucaL. Planta*, *215*(3), 502–509. doi:10.100700425-002-0774-9 PMID:12111234

Bischof, K., & Peralta, G. (2002). Effects of solar UV-B radiation on canopy structure of *Ulva* communities from southern Spain. *Journal of Experimental Botany*, *53*(379), 2411–2421. doi:10.1093/jxb/erf091 PMID:12432033

Bishop, W. M., & Zubeck, H. M. (2012). Evaluation of Microalgae for use as Nutraceuticals and Nutritional Supplements. *Journal of Nutrition & Food Sciences*, *2*, 147. doi:10.4172/2155-9600.1000147

Biundo, A., Hromic, A., Pavkov-Keller, T., Gruber, K., Quartinello, F., Haernvall, K., Perz, V., Arrell, M. S., Zinn, M., Ribitsch, D., & Guebitz, G. M. (2015). Characterization of a poly(butylene adipate-co-terephthalate)-hydrolyzing lipase from Pelosinus fermentans. *Applied Microbiology and Biotechnology*, *100*(4), 1753–1764. doi:10.100700253-015-7031-1 PMID:26490551

Bixler, H. J., & Porse, H. (2011). A decade of change in the seaweed hydrocolloids industry. *Journal of Applied Phycology*, *23*(3), 321–335. doi:10.100710811-010-9529-3

Bjørn, A., Laurent, A., Owsianiak, M., & Olsen, S. I. (2018). Goal definition. In *Life Cycle Assessment* (pp. 67–74). Springer. doi:10.1007/978-3-319-56475-3_7

Bjornsson, W. J., Nicol, R. W., Dickinson, K. E., & McGinn, P. J. (2013). Anaerobic digestates are useful nutrient sources for microalgae cultivation: Functional coupling of energy and biomass production. *Journal of Applied Phycology*, *25*(5), 1523–1528. doi:10.100710811-012-9968-0

Blackburn, S. I., & Volkman, J. K. (2012). Microalgae: A Renewable Source of Bioproducts. In N. T. Dunford (Ed.), *Food and Industrial Bioproducts and Bioprocessing* (pp. 221–241). John Wiley & Sons, Inc. doi:10.1002/9781119946083.ch9

Blanc, G., Duncan, G., Agarkova, I., Borodovsky, M., Gurnon, J., Kuo, A., Lindquist, E., Lucas, S., Pangilinan, J., Polle, J., Salamov, A., Terry, A., Yamada, T., Dunigan, D. D., Grigoriev, I. V., Claverie, J.-M., & Van Etten, J. L. (2010). The *Chlorella variabilis* NC64A genome reveals adaptation to photosymbiosis, coevolution with viruses, and cryptic sex. *The Plant Cell*, *22*(9), 2943–2955. doi:10.1105/tpc.110.076406 PMID:20852019

Blank, L. M., Narancic, T., Mampel, J., Tiso, T., & O'Connor, K. (2020). Biotechnological upcycling of plastic waste and other non-conventional feedstocks in a circular economy. *Current Opinion in Biotechnology*, *62*, 212–219. doi:10.1016/j.copbio.2019.11.011 PMID:31881445

Bleakley, S., & Hayes, M. (2017). Algal Proteins: Extraction, Application, and Challenges Concerning Production. *Foods*, *6*(5), 1–34. doi:10.3390/foods6050033 PMID:28445408

Blecher, K., Nasir, A., & Friedman, A. (2011). The growing role of nanotechnology in combating infectious disease. *Virulence*, *2*(5), 395–401. doi:10.4161/viru.2.5.17035 PMID:21921677

Bloodgood, R. A. (2020). Prey capture in protists utilizing microtubule filled processes and surface motility. *Cytoskeleton, 77*(11), 600–611. doi:10.1002/cm.21644 PMID:33190425

Blunden, G., Morse, P. F., Mathe, I., Hohmann, J., Critchley, A. T., & Morrell, S. (2010). Betaine yields from marine algal species utilized in the preparation of seaweed extracts used in agriculture. *Natural Product Communications, 5*(4).

Boenigk, J. (2021). Evolution und Systematik der Protisten. In J. Boenigk (Ed.), *Boenigk, Biologie* (pp. 815–858). Springer Spektrum., doi:10.1007/978-3-662-61270-5_33

Boettcher, M., & McManus, M. T. (2015). Choosing the Right Tool for the Job: RNAi, TALEN, or CRISPR. *Molecular Cell, 58*(4), 575–585. doi:10.1016/j.molcel.2015.04.028 PMID:26000843

Bold, H. C., & Wynne, M. J. (1985). *Introduction to the Algae. Structure and reproduction.* Prentice Hall, Inc.

Boliko, M. C. (2019). FAO and the situation of food security and nutrition in the world. *Journal of Nutritional Science and Vitaminology, 65*(Supplement), S4–S8. doi:10.3177/jnsv.65.S4 PMID:31619643

Boller, M. L., & Carrington, E. (2006). The hydrodynamic effects of shape and size change during reconfiguration of a flexible macroalga. *The Journal of Experimental Biology, 209*(10), 1894–1903. doi:10.1242/jeb.02225 PMID:16651555

Bölling, C., & Fiehn, O. (2005). Metabolite Profiling of *Chlamydomonas reinhardtii* under Nutrient Deprivation. *Plant Physiology, 139*(4), 1995–2005. doi:10.1104/pp.105.071589 PMID:16306140

Bombar, D., Paerl, R. W., & Riemann, L. (2016). Marine non-cyanobacterial diazotrophs: Moving beyond molecular detection. *Trends in Microbiology, 24*(11), 916–927. doi:10.1016/j.tim.2016.07.002 PMID:27476748

Bonifer, K. S., Wen, X., Hasim, S., Phillips, E. K., Dunlap, R. N., Gann, E. R., DeBruyn, J. M., & Reynolds, T. B. (2019). *Bacillus pumilus B12* Degrades Polylactic Acid and Degradation Is Affected by Changing Nutrient Conditions. *Frontiers in Microbiology, 10*, 2548. Advance online publication. doi:10.3389/fmicb.2019.02548 PMID:31824441

Bonner, M. Y., & Arbiser, J. L. (2014). The antioxidant paradox: What are antioxidants and how should they be used in a therapeutic context for cancer. *Future Medicinal Chemistry, 6*(12), 1413–1422. doi:10.4155/fmc.14.86 PMID:25329197

Bonos, E., Kasapidou, E., Kargopoulos, A., Karampampas, A., Christaki, E., Florou-Paneri, P., & Nikolakakis, I. (2016). Spirulina as a functional ingredient in broiler chicken diets. *South African Journal of Animal Science, 46*(1), 94–102. doi:10.4314ajas.v46i1.12

Boopathy, N. S., & Kathiresan, K. (2010). Anticancer drugs from marine ñora: An overview (online). *Journal of Oncology, 2010*, 214186. PMID:21461373

Booth, A. M., Kubowicz, S., Beegle-Krause, C., Skancke, J., Nordam, T., Landsem, E. ThroneHolst, M., & Jahren, S. (2017). *Microplastic in Global and Norwegian Marine Environments: Distributions, Degradation Mechanisms and Transport.* Norwegian Environment Agency M-918. https://www.miljodirektoratet.no/globalassets/publikasjoner/M918/M918.pdf

Borges, L., Moron-Villarreyes, J. A., D'Oca, M. G. M., & Abreu, P. C. (2011). Effects of flocculants on lipid extraction and fatty acid composition of the microalgae Nannochloropsis oculata and Thalassiosira weissflogii. *Biomass and Bioenergy, 35*(10), 4449–4454.

Borghans, L., Duckworth, A. L., Heckman, J. J., & Ter Weel, B. (2008). The economics and psychology of personality traits. *The Journal of Human Resources, 43*(4), 972–1059. doi:10.3368/jhr.43.4.972

Bornstein, B. J., Keating, S. M., Jouraku, A., & Hucka, M. (2008). LibSBML: An API library for SBML. *Bioinformatics (Oxford, England), 24*(6), 880–881. doi:10.1093/bioinformatics/btn051 PMID:18252737

Borowitzka, M. A. (1988a). Algal growth media and sources of cultures. In M. A. Borowitzka & L. J. Borowitzka (Eds.), *Micro-algal biotechnology* (pp. 456–465). Cambridge University Press.

Borowitzka, M. A. (1997). Algae for aquaculture: Opportunities and constraints. *Journal of Applied Phycology*, *9*, 393–401.

Borowitzka, M. A. (1998). Limits to growth. In Y. S. Wong & N. F. Y. Tam (Eds.), *Wastewater treatment with algae* (pp. 203–226). Springer-Verlag.

Borowitzka, M. A. (2013). *Dunaliella*: biology, production, and markets. In A. Richmond & Q. Hu (Eds.), *Handbook of microalgal culture* (pp. 359–368). Wiley-Blackwell.

Borowitzka, M. A. (2013). High-value products from microalgae—Their development and commercialisation. *Journal of Applied Phycology*, *25*(3), 743–756.

Borowitzka, M. A., Beardall, J., & Raven, J. R. (2016). *The physiology of Microalgae*. Springer.

Borowitzka, M. A., & Moheimani, N. R. (2013). *Algae for biofuels and energy* (Vol. 5). Springer.

Borowitzka, M. A., & Moheimani, N. R. (Eds.). (2013). *Algae for Biofuels and Energy* (Vol. 5, pp. 133–152). Springer. doi:10.1007/978-94-007-5479-9_8

Borowitzka, M. A., & Vonshak, A. (2017). Scaling up microalgal cultures to commercial scale. *European Journal of Phycology*, *52*(4), 407–418.

Bothe, H., Schmitz, O., Yates, M. G., & Newton, W. E. (2010). Nitrogen fixation and hydrogen metabolism in cyanobacteria. *Microbiology and Molecular Biology Reviews*, *74*(4), 529–551. doi:10.1128/MMBR.00033-10 PMID:21119016

Bouck, G. B., & Sweeney, B. M. (1966). The fine structure and ontogeny of trichocysts in marine dinoflagellates. *Protoplasma*, *61*(1), 205–223. doi:10.1007/BF01247920 PMID:5920039

Boukarma, L., & … . (2022). Algal Biomass Valorization for the Removal of Heavy Metal Ions. In E. Lichtfouse, S. S. Muthu, & A. Khadir (Eds.), *Inorganic-Organic Composites for Water and Wastewater Treatment. Environmental Footprints and Eco-design of Products and Processes*. Springer.

Boukid, F., & Castellari, M. (2021). Food and beverages containing algae and derived ingredients launched in the market from 2015 to 2019: A front-of-pack labeling perspective with a special focus on Spain. *Foods*, *10*(1), 173. doi:10.3390/foods10010173 PMID:33467009

Bouma-Gregson, K., Olm, M. R., Probst, A. J., Anantharaman, K., Power, M. E., & Banfield, J. F. (2019). Impacts of microbial assemblage and environmental conditions on the distribution of anatoxin-a producing cyanobacteria within a river network. *The ISME Journal*, *13*(6), 1618–1634. doi:10.103841396-019-0374-3 PMID:30809011

Bourdareau, S., Mignerot, L., Heesch, S., Peters, A. F., Coelho, S. M., & Cock, J. M. (2021). Alternation of Generations in Plants and Algae. *Evolutionary Developmental Biology: A Reference Guide*, 631–644.

Boyandin, A. N., Prudnikova, S. V., Karpov, V. A., Ivonin, V. N., Đỗ, N. L., Nguyễn, T. H., Lê, T. M. H., Filichev, N. L., Levin, A. L., Filipenko, M. L., Volova, T. G., & Gitelson, I. I. (2013). Microbial degradation of polyhydroxyalkanoates in tropical soils. *International Biodeterioration & Biodegradation*, *83*, 77–84. doi:10.1016/j.ibiod.2013.04.014

Boyera, N., Galey, I., & Bernard, B. A. (1998). Effect of vitamin C and its derivatives on collagen synthesis and cross-linking by normal human fibroblasts. *International Journal of Cosmetic Science*, *20*(3), 151–158. doi:10.1046/j.1467-2494.1998.171747.x PMID:18505499

Boyer, F., Mercier, C., Bonin, A., Le Bras, Y., Taberlet, P., & Coissac, E. (2016). OBITools: A unix-inspired software package for DNA metabarcoding. *Molecular Ecology Resources*, *16*(1), 176–182. doi:10.1111/1755-0998.12428 PMID:25959493

Boyle, N. R., & Morgan, J. (2009). Flux balance analysis of primary metabolism in *Chlamydomonas reinhardtii*. *BMC Systems Biology*, *3*(4), 32–45. doi:10.1186/1752-0509-3-4 PMID:19128495

Bradley, T., Maga, D., & Antón, S. (2015). Unified approach to Life Cycle Assessment between three unique algae biofuel facilities. *Applied Energy*, *154*, 1052–1061. doi:10.1016/j.apenergy.2014.12.087

Branyikova, I., Filipenska, M., Urbanova, K., Ruzicka, M. C., Pivokonsky, M., & Branyik, T. (2011). Physicochemical approach to alkaline flocculation of Chlorella vulgaris induced by calcium phosphate precipitates. *Colloids and Surfaces. B, Biointerfaces*, *2018*(166), 54–60.

Brar, K. K., Cortez, A. A., Pellegrini, V. O. A., Amulya, K., Polikarpov, I., Magdouli, S., Kumar, M., Yang, Y.-H., Bhatia, S. K., & Brar, S. K. (2022). An overview on progress, advances, and future outlook for biohydrogen production technology. *International Journal of Hydrogen Energy*. Advance online publication. doi:10.1016/j.ijhydene.2022.01.156

Bremer, K. (1985). Summary of green plant phylogeny and classification. *Cladistics*, *1*(4), 369–385. doi:10.1111/j.1096-0031.1985.tb00434.x PMID:34965683

Brennan, B., & Regan, F. (2020). In-situ lipid and fatty acid extraction methods to recover viable products from Nannochloropsis sp. *The Science of the Total Environment*, *748*, 142464. doi:10.1016/j.scitotenv.2020.142464 PMID:33113682

Brennan, L., & Owende, P. (2010). Biofuels from microalgae a review of technologies for production, processing, and extractions of biofuels and co-products. *Renewable & Sustainable Energy Reviews*, *14*, 557–577.

Brennan, L., & Owende, P. (2010). Biofuels from microalgae—A review of technologies for production, processing, and extractions of biofuels and co-products. *Renewable & Sustainable Energy Reviews*, *14*(2), 557–577. doi:10.1016/j.rser.2009.10.009

Brenner, K., You, L., & Arnold, F. H. (2008). Engineering microbial consortia: A new frontier in synthetic biology. *Trends in Biotechnology*, *26*(9), 483–489. doi:10.1016/j.tibtech.2008.05.004 PMID:18675483

Brentner, L. B., Eckelman, M. J., & Zimmerman, J. B. (2011). Combinatorial life cycle assessment to inform process design of industrial production of algal biodiesel. *Environmental Science & Technology*, *45*(16), 7060–7067. doi:10.1021/es2006995 PMID:21662987

Brentner, L. B., Peccia, J., & Zimmerman, J. B. (2010). Challenges in developing biohydrogen as a sustainable energy source: Implications for a research agenda. *Environmental Science & Technology*, *44*(7), 2243–2254. doi:10.1021/es9030613 PMID:20222726

Briassoulis, D., & Degli Innocenti, F. (2017). Standards for Soil Biodegradable Plastics. In *Green Chemistry and Sustainable Technology* (pp. 139–168). Springer Berlin Heidelberg. doi:10.1007/978-3-662-54130-2_6

Briceño-Domínguez, D., Hernandez-Carmona, G., Moyo, M., Stirk, W., & Van Staden, J. (2014). Plant growth promoting activity of seaweed liquid extracts produced from *Macrocystis pyrifera* under different pH and temperature conditions. *Journal of Applied Phycology*, *26*(5), 2203–2210. doi:10.100710811-014-0237-2

Brito, J. C., Godinho, R., Martínez-Freiría, F., Pleguezuelos, J. M., Rebelo, H., Santos, X., Vale, C. G., Velo-Antón, G., Boratyński, Z., Carvalho, S. B., Ferreira, S., Gonçalves, D. V., Silva, T. L., Tarroso, P., Campos, J. C., Leite, J. V., Nogueira, J., Álvares, F., Sillero, N., ... Carranza, S. (2014). Unravelling biodiversity, evolution and threats to conservation in the Sahara-Sahel. *Biological Reviews of the Cambridge Philosophical Society*, *89*(1), 215–231. doi:10.1111/brv.12049 PMID:23848599

Broddrick, J. T., Du, N., Smith, S. R., Tsuji, Y., Jallet, D., Ware, M. A., Peers, G., Matsuda, Y., Dupont, C. L., Mitchell, B. G., Palsson, B. O., & Allen, A. E. (2019). Cross-compartment metabolic coupling enables flexible photoprotective mechanisms in the diatom *Phaeodactylum tricornutum*. *The New Phytologist*, *222*(3), 1364–1379. doi:10.1111/nph.15685 PMID:30636322

Brody, M., & Vatter, A. E. (1959). Observations on cellular structures of Porphyridium cruentum. *The Journal of Biophysical and Biochemical Cytology*, *5*(2), 289–294. doi:10.1083/jcb.5.2.289 PMID:13654450

Brown, M. R., Mular, M., Miller, I., Farmer, C., & Trenerry, C. (1999). The vitamin content of microalgae used in aquaculture. *Journal of Applied Phycology*, *11*(3), 247–255. doi:10.1023/A:1008075903578

Brown, P. H., Rathjen, A. H., Graham, R. D., & Tribe, D. E. (1990). Rare earth elements in biological systems. *Handbook on the Physics and Chemistry of Rare Earths*, *13*, 423–452. doi:10.1016/S0168-1273(05)80135-7

Brown, T. M., Duan, P., & Savage, P. E. (2010). Hydrothermal liquefaction and gasification of Nannochloropsis sp. *Energy & Fuels*, *24*(6), 3639–3646. doi:10.1021/ef100203u

Bruhn, A., Dahl, J., Nielsen, H. B., Nikolaisen, L., Rasmussen, M. B., Markager, S., Olesen, B., Arias, C., & Jensen, P. D. (2011). Bioenergy potential of Ulva lactuca: Biomass yield, methane production and combustion. *Bioresource Technology*, *102*(3), 2595–2604. doi:10.1016/j.biortech.2010.10.010 PMID:21044839

Buchan, A., LeCleir, G. R., Gulvik, C. A., & González, J. M. (2014). Master recyclers: Features and functions of bacteria associated with phytoplankton blooms. *Nature Reviews. Microbiology*, *12*, 686–698.

Buchmann, L., Brändle, I., Haberkorn, I., Hiestand, M., & Mathys, A. (2019). Pulsed Electric Field Based Cyclic Protein Extraction of Microalgae towards Closed-Loop Biorefinery Concepts. *Bioresource Technology*, *291*, 121870. doi:10.1016/j.biortech.2019.121870 PMID:31382092

Büdel, B. (2005). Microorganisms of biological crusts on soil surfaces. In *Microorganisms in soils: roles in genesis and functions* (pp. 307–323). Springer. doi:10.1007/3-540-26609-7_15

Büdel, B., & Friedl, T. (Eds.). (2021). *Life at Rock Surfaces. Challenged by Extreme Light, Temperature and Hydration Fluctuations*. De Gruyter. doi:10.1515/9783110646467

Budzianowski, W. M. (2017). High-value low-volume bioproducts coupled to bioenergies with potential to enhance business development of sustainable biorefineries. *Renewable & Sustainable Energy Reviews*, *70*, 793–804. doi:10.1016/j.rser.2016.11.260

Bule, M. H., Ahmed, I., Maqbool, F., Bilal, M., & Hmn, I. (2018). Microalgae as a source of high-value bioactive compounds. *Frontiers in Bioscience*, *10*(1), 197–216. doi:10.2741509 PMID:28930527

Bulota, M., & Budtova, T. (2015). PLA/algae composites: Morphology and mechanical properties. *Composites. Part A, Applied Science and Manufacturing*, *73*, 109–115. doi:10.1016/j.compositesa.2015.03.001

Burki, F., Roger, A. J., Brown, M. W., & Simpson, A. G. B. (2020). The New Tree of Eukaryotes. *Trends in Ecology & Evolution*, *35*(1), 43–55. doi:10.1016/j.tree.2019.08.008 PMID:31606140

Burlew, J. S. (1953). Current status of large-scale culture of algae. In J. S. Burlew (Ed.), *Algal culture from laboratory to pilot plant* (pp. [illegible]). Carnegie Institution.

Burtin, P. (2003). Nutritional value of seaweeds. *Elect J Env Agri Food Che.*, *2*, 498–503.

Buschmann, A. H. (2016). *Seaweed Ecology and Physiology*. Cambridge University Press.

Butler, T., & Golan, Y. (2020). Astaxanthin Production from Microalgae. In M. A. Alam, J.-L. Xu, & Z. Wang (Eds.), *Microalgae Biotechnology for Food, Health and High Value Products* (pp. 175–242). Springer Singapore. doi:10.1007/978-981-15-0169-2_6

Butterwick, C., Heaney, S. I., & Talling, J. F. (2005). Diversity in the influence of temperature on the growth rates of freshwater algae, and its ecological relevance. *Freshwater Biology*, *50*(2), 291–300.

Byreddy, A. R., Yoganantharjah, P., Gupta, A., Gibert, Y., & Puri, M. (2019). Suitability ofnovel algal biomass as fish feed: Accumulation and distribution of omega-3 long-chain polyunsaturated fatty acid in Zebrafish. *Applied Biochemistry and Biotechnology*, *188*(1), 112–123. doi:10.100712010-018-2906-0 PMID:30345480

Cabej, N. R. (2013). Epigenetics of Reproduction in Animals. In *Building the Most Complex Structure on Earth* (pp. 59–120). Elsevier. doi:10.1016/B978-0-12-401667-5.00002-X

Caccamese, S., Azzolina, R., Furnari, G., Cormaci, M., & Grasso, S. (1981). Antimicrobial and Antiviral Activities of Some Marine Algae from Eastern Sicily. *Botanica Marina*, *24*, 365–368.

Cadogan, E. I., Lee, C.-H., & Popuri, S. R. (2015). Facile synthesis of chitosan derivatives and Arthrobacter sp. biomass for the removal of europium (III) ions from aqueous solution through biosorption. *International Biodeterioration & Biodegradation*, *102*, 286–297. doi:10.1016/j.ibiod.2015.01.018

Cadogan, E. I., Lee, C.-H., Popuri, S. R., & Lin, H.-Y. (2014). Efficiencies of chitosan nanoparticles and crab shell particles in europium uptake from aqueous solutions through biosorption: Synthesis and characterization. *International Biodeterioration & Biodegradation*, *95*, 232–240. doi:10.1016/j.ibiod.2014.06.003

Cahoon, A. B., & Timko, M. P. (2003). *Biochemistry and Regulation of Chlorophyll Biosynthesis*. Springer. doi:10.1007/978-94-007-1038-2_6

Cairns, J., Lanza, G. R., & Parker, B. C. (1972). Pollution related structural and functional changes in aquatic communities with emphasis on freshwater algae and protozoa. *Proceedings. Academy of Natural Sciences of Philadelphia*, *124*, 79–127.

Cai, Y., Yuan, F., Wang, X., Sun, Z., Chen, Y., Liu, Z., Wang, X., Yang, S., & Wang, S. (2017). Synthesis of core–shell structured Fe3O4@ carboxymethyl cellulose magnetic composite for highly efficient removal of Eu (III). *Cellulose (London, England)*, *24*(1), 175–190. doi:10.100710570-016-1094-8

Caldeira, K., & Wickett, M. E. (2003). Anthropogenic carbon and ocean pH. *Nature*, *425*(6956), 365–365. doi:10.1038/425365a PMID:14508477

Callieri, C., Morabito, G., Huot, Y., Neale, P. J., & Litchman, E. (2001). Photosynthetic response of pico- and nanoplanktonic algae to UVB, UVA and PAR in a high mountain lake. *Aquatic Sciences*, *63*(3), 286–293. doi:10.1007/PL00001355

Camacho, C., Coulouris, G., Avagyan, V., Ma, N., Papadopoulos, J., Bealer, K., & Madden, T. L. (2009). BLAST+: Architecture and applications. *BMC Bioinformatics*, *10*(1), 421. doi:10.1186/1471-2105-10-421 PMID:20003500

Campanale, C., Massarelli, C., Savino, I., Locaputo, V., & Uricchio, V. F. (2020). A Detailed Review Study on Potential Effects of Microplastics and Additives of Concern on Human Health. *International Journal of Environmental Research and Public Health*, *17*(4), 1212. doi:10.3390/ijerph17041212 PMID:32069998

Campbell, M. A., Haas, B. J., Hamilton, J. P., Mount, S., & Buell, C. (2006). Comprehensive analysis of alternative splicing in rice and comparative analyses with Arabidopsis. *BMC Genomics*, *7*(1), 327–339. doi:10.1186/1471-2164-7-327 PMID:17194304

Campbell, P. K., Beer, T., & Batten, D. (2011). Life cycle assessment of biodiesel production from microalgae in ponds. *Bioresource Technology*, *102*(1), 50–56. doi:10.1016/j.biortech.2010.06.048 PMID:20594828

Campo, V. L., Kawano, D. F., Silva, D. B. Jr, & Carvalho, I. (2009). Carrageenans: Biological properties, chemical modifications and structural analysis – A review. *Carbohydrate Polymers*, *77*(2), 167–180. doi:10.1016/j.carbpol.2009.01.020

Canda, L., Heput, T., & Ardelean, E. (2016). Methods for recovering precious metals from industrial waste. *IOP Conference Series. Materials Science and Engineering*, *106*(1), 12020. doi:10.1088/1757-899X/106/1/012020

Canniff, P. M., & Hoang, T. C. (2018). Microplastic ingestion by Daphnia magna and its enhancement on algal growth. *The Science of the Total Environment*, *633*, 500–507. doi:10.1016/j.scitotenv.2018.03.176 PMID:29579661

Capelli, B., Bagchi, D., & Cysewski, G. R. (2013). Synthetic astaxanthin is significantly inferior to algal-based astaxanthin as an antioxidant and may not be suitable as a human nutritional supplement. *NUTRAfoods: International Journal of Science and Marketing for Nutraceutical Actives, Raw Materials, Finish Products*, *12*(4), 145–152. doi:10.100713749-013-0051-5

Capolupo, M., Sørensen, L., Jayasena, K. D. R., Booth, A. M., & Fabbri, E. (2020). Chemical composition and ecotoxicity of plastic and car tire rubber leachates to aquatic organisms. *Water Research*, *169*, 115270. doi:10.1016/j.watres.2019.115270 PMID:31731243

Cardon, Z. G., Gray, D. W., & Lewis, L. A. (2008). The green algal underground: Evolutionary secrets of desert cells. *Bioscience*, *58*(2), 114–122. doi:10.1641/B580206

Cardoso, N. F., Lima, E. C., Royer, B., Bach, M. V., Dotto, G. L., & Pinto, L. A. A. (2012). Comparison of *Spirulina Platensis* Microalgae and Commercial Activated Carbon as Adsorbents for the Removal of Reactive Red 120 Dye from Aqueous Effluents. *Journal of Hazardous Materials*, *241-242*, 146–153.

Cardozo, K. H. M., Guaratini, T., Barros, M. P., Falcão, V. R., Tonon, A. P., Lopes, N. P., Campos, S., Torres, M. A., Souza, A. O., Colepicolo, P., & Pinto, E. (2007). Metabolites from algae with economical impact. *Comparative Biochemistry and Physiology. Toxicology & Pharmacology: CBP*, *146*(1), 60–78. doi:10.1016/j.cbpc.2006.05.007 PMID:16901759

Carfagna, S., Lanza, N., Salbitani, G., Basile, A., Sorbo, S., & Vona, V. (2013). Physiological and morphological responses of Lead or Cadmium exposed Chlorella sorokiniana 211-8K (Chlorophyceae). *SpringerPlus*, *2*(1), 1–7. doi:10.1186/2193-1801-2-147 PMID:23641320

Carillo, P., Ciarmiello, L. F., Woodrow, P., Corrado, G., Chiaiese, P., & Rouphael, Y. (2020). Enhancing sustainability by improving plant salt tolerance through macro-and micro-algal biostimulants. *Biology (Basel)*, *9*(9), 253. doi:10.3390/biology9090253 PMID:32872247

Carneiro, M. L. N., Pradelle, F., Braga, S. L., Gomes, M. S. P., Martins, A. R. F., Turkovics, F., & Pradelle, R. N. (2017). Potential of biofuels from algae: Comparison with fossil fuels, ethanol and biodiesel in Europe and Brazil through life cycle assessment (LCA). *Renewable & Sustainable Energy Reviews*, *73*, 632–653. doi:10.1016/j.rser.2017.01.152

Carney, L. T., & Lane, T. W. (2014). Parasites in algae mass culture. *Frontiers in Microbiology*, *5*, 278.

Caron, D. A., Sanders, R. W., Lim, E. L., Marrasé, C., Amaral, L. A., Whitney, S., Aoki, R. B., & Porters, K. G. (1993). Light-dependent phagotrophy in the freshwater mixotrophic chrysophyte Dinobryon cylindricum. *Microbial Ecology*, *25*(1), 93–111. doi:10.1007/BF00182132 PMID:24189708

Carpio, R., Kuo, C.-T., de Leon, R., Schideman, L. C., & Zhang, Y. (2018). Hydrothermal liquefaction of demineralized wastewater algal biomass. *International Journal of Smart Grid and Clean Energy*, *7*(1), 13–23. doi:10.12720gce.7.1.13-23

Carvalho, A. P., Batista, D., Dobretsov, S., & Coutinho, R. (2017). Extracts of seaweeds as potential inhibitors of quorum sensing and bacterial growth. *Journal of Applied Phycology*, *29*(2), 789–797. doi:10.100710811-016-1014-1

Carvalho, A. P., Monteiro, C. M., & Malcata, F. X. (2009). Simultaneous effect of irradiance and temperature on biochemical composition of the microalga Pavlova lutheri. *Journal of Applied Phycology*, *21*(5), 543–552. doi:10.100710811-009-9415-z

Casero, M. C., Ascaso, C., Quesada, A., Mazur-Marzec, H., & Wierzchos, J. (2021). Response of Endolithic Chroococcidiopsis Strains From the Polyextreme Atacama Desert to Light Radiation. *Frontiers in Microbiology*, *11*, 3607. doi:10.3389/fmicb.2020.614875 PMID:33537015

Cassuriaga, A. P. A., Freitas, B. C. B., Morais, M. G., & Costa, J. A. V. (2018). Innovative polyhydroxybutyrate production by Chlorella fusca grown with pentoses. *Bioresource Technology*, *265*, 456–463. doi:10.1016/j.biortech.2018.06.026 PMID:29935455

Castello, D., & Fiori, L. (2011). Supercritical water gasification of biomass: Thermodynamic constraints. *Bioresource Technology*, *102*(16), 7574–7582. doi:10.1016/j.biortech.2011.05.017 PMID:21640582

Castro-Bugallo, A., González-Fernández, Á., Guisande, C., & Barreiro, A. (2014). Comparative responses to metal oxide nanoparticles in marine phytoplankton. *Archives of Environmental Contamination and Toxicology*, *67*(4), 483–493. doi:10.100700244-014-0044-4 PMID:24908584

Castro-Muñoz, R., & García-Depraect, O. (2021). Membrane-Based Harvesting Processes for Microalgae and Their Valuable-Related Molecules: A Review. *Membranes*, *11*(8), 585.

Cavalier-Smith, T. (1981). Eukaryote kingdoms: Seven or nine? *Bio Systems*, *14*(3), 461–481. doi:10.1016/0303-2647(81)90050-2 PMID:7337818

Cavalier-Smith, T. (1982). The origins of plastids. *Biological Journal of the Linnean Society. Linnean Society of London*, *17*(3), 289–306. doi:10.1111/j.1095-8312.1982.tb02023.x

Cavalier-Smith, T., & Chao, E. E. (1996). 18S rRNA sequence of Heterosigma carterae (Raphidophyceae), and the phylogeny of heterokont algae (Ochrophyta). *Phycologia*, *35*(6), 500–510. doi:10.2216/i0031-8884-35-6-500.1

Cenci, U., Bhattacharya, D., Weber, A. P., Colleoni, C., Subtil, A., & Ball, S. G. (2017). Biotic host–pathogen interactions as major drivers of plastid endosymbiosis. *Trends in Plant Science*, *22*(4), 316–328. doi:10.1016/j.tplants.2016.12.007 PMID:28089380

Chae, Y., Kim, D., & An, Y.-J. (2019). Effects of micro-sized polyethylene spheres on the marine microalga Dunaliella salina: Focusing on the algal cell to plastic particle size ratio. *Aquatic Toxicology (Amsterdam, Netherlands)*, *216*, 105296. doi:10.1016/j.aquatox.2019.105296 PMID:31541944

Chaisutyakorn, P., Praiboon, J., & Kaewsuralikhit, C. (2017). The effect of temperature on growth and lipid and fatty acid composition on marine microalgae used for biodiesel production. *Journal of Applied Phycology, 30*(1), 37–45. doi:10.1007/s10811-017-1186-3

Chai, W. S., Tan, W. G., Halimatul Munawaroh, H. S., Gupta, V. K., Ho, S. H., & Show, P. L. (2021). Multifaceted roles of microalgae in the application of wastewater biotreatment: A review. *Environmental Pollution*, *269*, 116236. doi:10.1016/j.envpol.2020.116236 PMID:33333449

Chaiwong, K., Kiatsiriroat, T., Vorayos, N., & Thararax, C. (2013). Study of bio-oil and biochar production from algae by slow pyrolysis. *Biomass and Bioenergy*, *56*, 600–606. doi:10.1016/j.biombioe.2013.05.035

Cha, K. H., Koo, S. Y., & Lee, D. U. (2008). Antiproliferative effects of carotenoids extracted from Chlorella ellipsoidea and Chlorella Vulgaris on human colon cancer cells. *Journal of Agricultural and Food Chemistry*, *56*(22), 10521–10526. doi:10.1021/jf802111x PMID:18942838

Chakdar, H., Hasan, M., Pabbi, S., Nevalainen, H., & Shukla, P. (2021). High-throughput proteomics and metabolomic studies guide re-engineering of metabolic pathways in eukaryotic microalgae: A review. *Bioresource Technology*, *321*, 124495. doi:10.1016/j.biortech.2020.124495 PMID:33307484

Chakraborty, T., & Akhtar, N. (2021). Biofertilizers: Prospectives and challenges for future. In Biofertilizers. Scrivener Publishing LLC.

Chakraborty, K., & Paulraj, R. (2010). Sesquiterpenoids with free-radical-scavenging properties from marine macroalga Ulva fasciata Delile. *Food Chemistry*, *122*(1), 31–41. doi:10.1016/j.foodchem.2010.02.012

Chakravarty, P., Mhaisalkar, V., & Chakrabarti, T. (2010). Study on poly-hydroxyalkanoate (PHA) production in pilot scale continuous mode wastewater treatment system. *Bioresource Technology*, *101*(8), 2896–2899. doi:10.1016/j.biortech.2009.11.097 PMID:20045314

Chan, A., Salsali, H., & Mcbean, E. (2014). Heavy metal removal (copper and zinc) in secondary effluent from wastewater treatment plants by microalgae. *ACS Sustainable Chemistry & Engineering*, *2*, 130–137.

Chan, C.-X., Teo, S.-S., Ho, C.-L., Othman, R. Y., & Phang, S.-M. (2004). Optimisation of RNA extraction from *Gracilaria changii* (Gracilariales, Rhodophyta). *Journal of Applied Phycology*, *16*(4), 297–301. doi:10.1023/B:JAPH.0000047782.20940.de

Chanda, M.J., Benhima, R., Elmernissi, N., Kasmi, Y., & Karim, L,. Sbabou, L., & El Arroussi, H. (2020). Screening of microalgae liquid extracts for their bio stimulant properties on plant growth, nutrient uptake and metabolite profile of Solanum lycopersicum L. *Scientific Reports*, *10*(1). PMID:32071360

Chandra, R., Das, P., Vishal, G., & Nagra, S. (2019). Factors affecting the induction of UV protectant and lipid productivity in *Lyngbya* for sequential biorefinery product recovery. *Bioresource Technology*, *278*, 303–310. doi:10.1016/j.biortech.2019.01.084 PMID:30708334

Chandra, R., Iqbal, H. M. N., Vishal, G., Lee, H.-S., & Nagra, S. (2019). Algal Biorefinery: A Sustainable Approach to Valorize Algal-Based Biomass towards Multiple Product Recovery. *Bioresource Technology*, *278*, 346–359. doi:10.1016/j.biortech.2019.01.104 PMID:30718075

Chang, K. J. L., Nichols, C. M., Blackburn, S. I., Dunstan, G. A., Koutoulis, A., & Nichols, P. D. (2014). Comparison of thraustochytrids Aurantiochytrium sp., Schizochytrium sp., Thraustochytrium sp., and Ulkenia sp. for production of biodiesel, long-chain omega-3 oils, and exopolysaccharide. *Marine Biotechnology (New York, N.Y.)*, *16*(4), 396–411. doi:10.100710126-014-9560-5 PMID:24463839

Chang, K. S., Kim, J., Park, H., Hong, S.-J., Lee, C.-G., & Jin, E. (2020). Enhanced lipid productivity in AGP knockout marine microalga *Tetraselmis* sp. using a DNA-free CRISPR-Cas9 RNP method. *Bioresource Technology*, *303*, 122932. doi:10.1016/j.biortech.2020.122932 PMID:32058903

Chaogang, W., Zhangli, H., Anping, L., & Baohui, J. (2010). Biosynthesis of poly-3-hydroxybutyrate (PHB) in the transgenic green algae *chlamydomonas reinhardtii*. *Journal of Phycology*, *46*(2), 396–402. doi:10.1111/j.1529-8817.2009.00789.x

Chatterjee, A., Singh, S., Agrawal, C., Yadav, S., Rai, R., & Rai, L. C. (2017). Role of Algae as a Biofertilizer. In R. P. Rastogi, D. Madamwar, & A. Pandey (Eds.), *Algal Green Chemistry* (pp. 189–200). Elsevier. doi:10.1016/B978-0-444-63784-0.00010-2

Chauhan, A., AbuAmarah, B. A., Kumar, A., Verma, J. S., Ghramh, H. A., Khan, K. A., & Ansari, M. J. (2019). Influence of gibberellic acid and different salt concentrations on germination percentage and physiological parameters of oat cultivars. *Saudi Journal of Biological Sciences*, *26*(6), 1298–1304. doi:10.1016/j.sjbs.2019.04.014 PMID:31516361

Chaumont, D. (1993). Biotechnology of algal biomass production: A review of systems for outdoor mass culture. *Journal of Applied Phycology*, *5*(6), 593–604. doi:10.1007/BF02184638

Chauton, M. S., Reitan, K. I., Norsker, N. H., Tveterås, R., & Kleivdal, H. T. (2015). A techno-economic analysis of industrial production of marine microalgae as a source of EPA and DHA-rich raw material for aquafeed: Research challenges and possibilities. *Aquaculture (Amsterdam, Netherlands)*, *436*, 95–103. doi:10.1016/j.aquaculture.2014.10.038

Cheah, W. Y., Show, P. L., Chang, J. S., Ling, T. C., & Juan, J. C. (2015). Biosequestration of atmospheric CO2 and flue gas-containing CO2 by microalgae. *Bioresource Technology*, *184*, 190–201. doi:10.1016/j.biortech.2014.11.026 PMID:25497054

Cheirsilp, B., Suwannarat, W., & Niyomdecha, R. (2011). Mixed culture of oleaginous yeast Rhodotorula glutinis and microalga Chlorella vulgaris for lipid production from industrial wastes and its use as biodiesel feedstock. *New Biotechnology*, *28*(4), 362–368. doi:10.1016/j.nbt.2011.01.004 PMID:21255692

Chemat, F., Abert Vian, M., Fabiano-Tixier, A. S., Nutrizio, M., Režek Jambrak, A., Munekata, P. E. S., Lorenzo, J. M., Barba, F. J., Binello, A., & Cravotto, G. (2020). A review of sustainable and intensified techniques for extraction of food and natural products. *Green Chemistry*, *22*(8), 2325–2353. doi:10.1039/C9GC03878G

Chen, J., Leng, L., Ye, C., Lu, Q., Addy, M., Wang, J., & Zhou, W. (2018). A comparative study between fungal pellet- and spore-assisted microalgae harvesting methods for algae bioflocculation. *Bioresour Technol, 259*, 181-190. .1016/j.biortech.2018.03.040 doi:https://doi.org/10

Chen, B., Lin, L., Ma, Z., Zhang, T., Chen, W., & Zou, D. (2019). Carbon and nitrogen accumulation and interspecific competition in two algae species, *Pyropia haitanensis and Ulva lactuca*, under ocean acidification conditions. *Aquaculture International*, *27*(3), 721–733. doi:10.100710499-019-00360-y

Chen, C. C., & White, J. L. (1993). Compatibilizing agents in polymer blends: Interfacial tension, phase morphology, and mechanical properties. *Polymer Engineering and Science*, *33*(14), 923–930. doi:10.1002/pen.760331409

Chen, C. Y., Ho, S. H., Liu, C. C., & Chang, J. S. (2017). Enhancing lutein production with *Chlorella sorokiniana* Mb-1 by optimizing acetate and nitrate concentrations under mixotrophic growth. *Journal of the Taiwan Institute of Chemical Engineers*, *79*, 88–96. doi:10.1016/j.jtice.2017.04.020

Chen, C. Y., Yeh, K. L., Aisyah, R., Lee, D. J., & Chang, J. S. (2011). Cultivation, photobioreactor design and harvesting of microalgae for biodiesel production: A critical review. *Bioresource Technology*, *102*, 71–81. https://doi.org/10.1016/j.biortech.2010.06.159

Chen, C. Y., Yeh, K. L., Aisyah, R., Lee, D. J., & Chang, J. S. (2011). Cultivation,photobioreactor design and harvesting of microalgae for biodiesel production: A critical review. *Bioresource Technology*, *102*, 71–81. https://doi.org/10.1016/j.biortech.2010.06.159

Chen, C.-Y., Chang, H.-Y., & Chang, J.-S. (2016). Producing carbohydrate-rich microalgal biomass grown under mixotrophic conditions as feedstock for biohydrogen production. *International Journal of Hydrogen Energy*, *41*(7), 4413–4420. doi:10.1016/j.ijhydene.2015.05.163

Chen, C.-Y., Zhao, X.-Q., Yen, H.-W., Ho, S.-H., Cheng, C.-L., Lee, D.-J., Bai, F.-W., & Chang, J.-S. (2013). Microalgae-based carbohydrates for biofuel production. *Biochemical Engineering Journal*, *78*, 1–10.

Chen, F., & Johns, M. R. (1994). Substrate inhibition of Chlamydomonas reinhardtii by acetate in heterotrophic culture. *Process Biochemistry*, *29*(4), 245–252. doi:10.1016/0032-9592(94)80064-2

Chen, G. Q., & Wu, Q. (2005). The application of polyhydroxyalkanoates as tissue engineering materials. *Biomaterials*, *26*(33), 6565–6578. doi:10.1016/j.biomaterials.2005.04.036 PMID:15946738

Chen, G.-Q., & Zhang, J. (2018). Microbial polyhydroxyalkanoates as medical implant biomaterials. *Artificial Cells, Nanomedicine, and Biotechnology*, *46*(1), 1–18. doi:10.1080/21691401.2017.1371185 PMID:28849679

Cheng, C.-H., Du, T.-B., Pi, H.-C., Jang, S.-M., Lin, Y.-H., & Lee, H.-T. (2011). Comparative study of lipid extraction from microalgae by organic solvent and supercritical CO2. *Bioresource Technology*, *102*(21), 10151–10153. doi:10.1016/j.biortech.2011.08.064 PMID:21917450

Cheng, J., Lu, H., He, X., Yang, W., Zhou, J., & Cen, K. (2017). Mutation of Spirulina sp. by nuclear irradiation to improve growth rate under 15% carbon dioxide in flue gas. *Bioresource Technology*, *238*, 650–656. doi:10.1016/j.biortech.2017.04.107 PMID:28486198

Cheng, J., Yue, L., Ding, L., Li, Y.-Y., Ye, Q., Zhou, J., Cen, K., & Lin, R. (2019). Improving fermentative hydrogen and methane production from an algal bloom through hydrothermal/steam acid pretreatment. *International Journal of Hydrogen Energy*, *44*(12), 5812–5820. doi:10.1016/j.ijhydene.2019.01.046

Cheng, J., Zhang, M., Song, W., Xia, A., Zhou, J., & Cen, K. (2011). Cogeneration of hydrogen and methane from Arthrospira maxima biomass with bacteria domestication and enzymatic hydrolysis. *International Journal of Hydrogen Energy*, *36*(2), 1474–1481. doi:10.1016/j.ijhydene.2010.11.009

Chengkui, Z., & Junfu, Z. (1984). Chinese seaweeds in herbal medicine. In *Eleventh International Seaweed Symposium* (pp. 152-154). Springer. 10.1007/978-94-009-6560-7_24

Cheng, P., Chen, D., Liu, W., Cobb, K., Zhou, N., Liu, Y., & Ruan, R. (2020). Auto-flocculation microalgae species Tribonema sp. and Synechocystis sp. with T-IPL pretreatment to improve swine wastewater nutrient removal. *The Science of the Total Environment*, *725*, 138263.

Cheng, Y.-S., Labavitch, J. M., & VanderGheynst, J. S. (2014). Elevated CO_2 concentration impacts cell wall polysaccharide composition of green microalgae of the genus *Chlorella*. *Letters in Applied Microbiology*, *60*(1), 1–7. doi:10.1111/lam.12320 PMID:25163669

Chen, J. P., & Lin, M. (2001). Surface charge and metal ion adsorption on an H-type activated carbon: Experimental observation and modeling simulation by the surface complex formation approach. *Carbon*, *39*(10), 1491–1504. doi:10.1016/S0008-6223(00)00277-3

Chen, Q., Allgeier, A., Yin, D., & Hollert, H. (2019). Leaching of endocrine disrupting chemicals from marine microplastics and mesoplastics under common life stress conditions. *Environment International*, *130*, 104938. doi:10.1016/j.envint.2019.104938 PMID:31234003

Chen, W.-H., Lin, B.-J., Huang, M.-Y., & Chang, J.-S. (2015). Thermochemical conversion of microalgal biomass into biofuels: A review. *Bioresource Technology*, *184*, 314–327. doi:10.1016/j.biortech.2014.11.050 PubMed

Chen, X., Song, L., Wang, H., Liu, S., Yu, H., Wang, X., Li, R., Liu, T., & Li, P. (2019). Partial characterization, the immune modulation and anticancer activities of sulfated polysaccharides from filamentous microalgae Tribonema sp. *Molecules (Basel, Switzerland)*, *24*(2), 322. doi:10.3390/molecules24020322 PMID:30658396

Cheung, R. C. F., Wong, J. H., Pan, W. L., Chan, Y. S., Yin, C. M., Dan, X. L., Wang, H. X., Fang, E. F., Lam, S. K., Ngai, P. H. K., & ... . (2014). Antifungal and Antiviral Products of Marine Organisms. *Applied Microbiology and Biotechnology*, *98*, 3475–3494.

Chew, K. W., Khoo, K. S., Foo, H. T., Chia, S. R., Walvekar, R., & Lim, S. S. (2021). Algae utilization and its role in the development of green cities. *Chemosphere*, *268*, 129322. Advance online publication. doi:10.1016/j.chemosphere.2020.129322 PMID:33359993

Chew, K. W., Yap, J. Y., Show, P. L., Suan, N. H., Juan, J. C., Ling, T. C., Lee, D.-J., & Chang, J.-S. (2017). Microalgae Biorefinery: High Value Products Perspectives. *Bioresource Technology*, *229*, 53–62. doi:10.1016/j.biortech.2017.01.006 PMID:28107722

Chiang, C. L., Lee, C. M., & Chen, P. C. (2011). Utilization of the cyanobacteria Anabaena sp. CH1 in biological carbon dioxide mitigation processes. *Bioresource Technology*, *102*(9), 5400–5405. doi:10.1016/j.biortech.2010.10.089 PMID:21232935

Chia, W. Y., Tang, D. Y. Y., Khoo, K. S., Lup, A. N. K., & Chew, K. W. (2020). Nature's fight against plastic pollution: Algae for plastic biodegradation and bioplastics production. *Environmental Science and Ecotechnology*, *4*, 100065. doi:10.1016/j.ese.2020.100065

Chidambarampadmavathy, K., Karthikeyan, O. P., & Heimann, K. (2017). Sustainable bio-plastic production through landfill methane recycling. *Renewable & Sustainable Energy Reviews*, *71*, 555–562. doi:10.1016/j.rser.2016.12.083

Chiellini, E., Cinelli, P., Imam, S. H., & Mao, L. (2001). Composite films based on biorelated agro-industrial waste and poly (vinyl alcohol). Preparation and mechanical properties characterization. *Biomacromolecules*, *2*(3), 1029–1037. doi:10.1021/bm010084j PMID:11710006

Chinaglia, S., Tosin, M., & Degli-Innocenti, F. (2018). Biodegradation rate of biodegradable plastics at molecular level. *Polymer Degradation & Stability*, *147*, 237–244. doi:10.1016/j.polymdegradstab.2017.12.011

Chin, J. Y., Chng, L. M., Leong, S. S., Yeap, S. P., Yasin, N. H. M., & Toh, P. Y. (2020). Removal of Synthetic Dye by *Chlorella Vulgaris* Microalgae as Natural Adsorbent. *Arabian Journal for Science and Engineering*, *45*, 7385–7395.

Chinnasamy, S., Bhatnagar, A., Hunt, R. W., & Das, K. C. (2010). Microalgae cultivation in a wastewater dominated by carpet mill effluents for biofuel applications. *Bioresource Technology*, *101*(9), 3097–3105. doi:10.1016/j.biortech.2009.12.026 PMID:20053551

Chisti, Y. (2007). Biodiesel from microalgae. *Biotechnology Advances*, *25*(3), 294–306.

Chisti, Y. (2008). Biodiesel from microalgae beats bioethanol. *Trends in Biotechnology*, *26*(3), 126–131. doi:10.1016/j.tibtech.2007.12.002 PMID:18221809

Chittora, D., Meena, M., Barupal, T., Swapnil, P., & Sharma, K. (2020). Cyanobacteria as a source of biofertilizers for sustainable agriculture. *Biochemistry and Biophysics Reports*, *22*, 100737. doi:10.1016/j.bbrep.2020.100737 PMID:32083191

Chiu, S. Y., Kao, C. Y., Chen, T. Y., Chang, Y. B., Kuo, C. M., & Lin, C. S. (2015). Cultivation of microalgal *Chlorella* for biomass and lipid production using wastewater as nutrient resource. *Bioresource Technology*, *184*, 179–189.

Chi, W., Zheng, L., He, C., Han, B., Zheng, M., Gao, W., Sun, C., Zhou, G., & Gao, X. (2017). Quorum sensing of microalgae associated marine Ponticoccus sp. PD-2 and its algicidal function regulation. *AMB Express*, *7*(1), 59. doi:10.118613568-017-0357-6 PMID:28281272

Choi, H. J. (2015). Effect of eggshells for the harvesting of microalgae species. *Biotechnology, Biotechnological Equipment*, *29*, 666–672.

Choi, S. K., Lee, J. Y., Kwon, D. Y., & Cho, K. J. (2006). Settling characteristics of problem algae in the water treatment process. *Water sci Technol: A j Int Association. Water Research*, *53*, 113–119.

Choi, S., Song, C. W., Shin, J. H., & Lee, S. Y. (2015). Biorefineries for the production of top building block chemicals and their derivatives. *Metabolic Engineering*, *28*, 223–239.

Cho, J. Y. (2012). Algicidal Activity of Marine *Alteromonas* sp. KNS-16 and Isolation of Active Compounds. *Bioscience, Biotechnology, and Biochemistry*, *76*(8), 1452–1458. doi:10.1271/bbb.120102 PMID:22878186

Chojnacka, K. (2012). Biologically Active Compounds in Seaweed Extracts - the Prospects for the Application. *The Open Conference Proceedings Journal*, *3*(1), 20–28. doi:10.2174/1876326X01203020020

Chomkhamsri, K., Wolf, M. A., & Pant, R. (2011). International reference life cycle data system (ILCD) handbook: Review schemes for life cycle assessment. *Towards Life Cycle Sustainability Management*, 107-117.

Chopin, T., & Tacon, A. G. J. (2021). 2021/04/03). Importance of Seaweeds and Extractive Species in Global Aquaculture Production. *Reviews in Fisheries Science & Aquaculture*, *29*(2), 139–148. doi:10.1080/23308249.2020.1810626

Chorus, I., & Welker, M. (2021). Toxic Cyanobacteria in Water: A Guide to Their Public Health Consequences, Monitoring and Management (2nd ed.). CRC Press. doi:10.1201/9781003081449

Chowdhury, R., & Franchetti, M. (2017). Life cycle energy demand from algal biofuel generated from nutrients present in the dairy waste. *Sustain. Prod. Consump.*, *9*, 22–27. doi:10.1016/j.spc.2016.07.001

Choy, S. Y., Prasad, K. M. N., Wu, T. Y., Raghunandan, M. E., Phang, S. M., Juan, J. C., & Ramanan, R. N. (2018). Separation of *Chlorella* biomass from culture medium by flocculation with rice starch. *Algal Research*, *30*, 162–172.

Chua, E. T., Eltanahy, E., Jung, H., Uy, M., Thomas-Hall, S. R., & Schenk, P. M. (2019). Efficient Harvesting of Nannochloropsis Microalgae via Optimized Chitosan-Mediated Flocculation. *Global Challenges*, *3*(1), 1800038.

Chugh, H., Sood, D., Chandra, I., Tomar, V., Dhawan, G., & Chandra, R. (2018). Role of gold and silver nanoparticles in cancer nano-medicine. *Artificial Cells, Nanomedicine, and Biotechnology*, *46*(sup1), 1210–1220. doi:10.1080/21691401.2018.1449118 PMID:29533101

Chu, M., Li, H., Wu, Q., Wo, F., & Shi, D. (2014). Pluronic-encapsulated natural chlorophyll nanocomposites for in vivo cancer imaging and photothermal/photodynamic therapies. *Biomaterials*, *35*(29), 8357–8373. doi:10.1016/j.biomaterials.2014.05.049 PMID:25002262

Chung, T.Y., Kuo, C.Y., Lin, W.J., Wang, W.L., & Chou, J.Y. (2018). Indole-3-acetic-acid-induced phenotypic plasticity in Desmodesmus algae. *Scientific Reports*, *8*(1), 10270.

Chungjatupornchai, W., Areerat, K., & Fa-Aroonsawat, S. (2019). Increased triacylglycerol production in oleaginous microalga Neochloris oleoabundans by overexpression of plastidial lysophosphatidic acid acyltransferase. *Microbial Cell Factories*, *18*(1), 1–11. doi:10.118612934-019-1104-2 PMID:30866936

Chungjatupornchai, W., & Fa-Aroonsawat, S. (2021). Enhanced triacylglycerol production in oleaginous microalga *Neochloris oleoabundans* by co-overexpression of lipogenic genes: Plastidial LPAAT1 and ER-located DGAT2. *Journal of Bioscience and Bioengineering*, *131*(2), 124–130. doi:10.1016/j.jbiosc.2020.09.012 PMID:33069576

Chun, Y., Mulcahy, D., Zou, L., & Kim, I. S. (2017). A short review of membrane fouling in forward osmosis processes. *Membranes*, *7*(2), 30. doi:10.3390/membranes7020030 PMID:28604649

Chu, W. L. (2012). Biotechnological applications of microalgae. *IeJSME*, *6*(1), S24–S37.

Chu, W. L., Alwi, A., & Phang, S. M. (2002). Phycoerythrin production by a marine Oscillatoria (Cyanophyta). *Mal. J. Sci., 21*, 61–[illegible].

Ciapponi, R., Turri, S., & Levi, M. (2019). Mechanical reinforcement by microalgal biofiller in novel thermoplastic biocompounds from plasticized gluten. *Materials (Basel, Switzerland), 12*(9), 1476. doi:10.3390/ma12091476 PMID:31067771

Cindana Mo'o, F. R., Wilar, G., Devkota, H. P., & Wathoni, N. (2020). Ulvan, a polysaccharide from macroalga Ulva sp.: A review of chemistry, biological activities and potential for food and biomedical applications. *Applied Sciences (Basel, Switzerland), 10*(16), 5488. doi:10.3390/app10165488

Cirri, E., & Pohnert, G. (2019). Algae-bacteria interactions that balance the planktonic microbiome. *The New Phytologist, 223*, 100–106.

Claire, C., De Roeck-Holtzhauer, Y., & Quere, I. (1991). Vitamin analysis of five planktonic microalgae and one macroalga. *Journal of Applied Phycology, 3*(3), 259–264. doi:10.1007/BF00003584

Clendennen, S. K., Zimmerman, R. C., Powers, D. A., & Alberte, R. S. (1996). Photosynthetic responses of the giant kelp Macrocystis pyrifera. Phaeophyceae to ultraviolet radiation. *Journal of Phycology, 32*(4), 614–620. doi:10.1111/j.0022-3646.1996.00614.x

Cobos, M., Paredes, J. D., Maddox, J. D., Vargas-Arana, G., Flores, L., Aguilar, C. P., Marapara, J., & Castro, J. C. (2017). Isolation and characterization of native microalgae from the Peruvian Amazon with potential for biodiesel production. *Energies, 10*(2), 224. doi:10.3390/en10020224

Cock, P. J. A., Antao, T., Chang, J. T., Chapman, B. A., Cox, C. J., Dalke, A., Friedberg, I., Hamelryck, T., Kauff, F., Wilczynski, B., & de Hoon, M. J. L. (2009). Biopython: Freely available Python tools for computational molecular biology and bioinformatics. *Bioinformatics (Oxford, England), 25*(11), 1422–1423. doi:10.1093/bioinformatics/btp163 PMID:19304878

Coelho, V. C., da Silva, C. K., Terra, A. L., Costa, J. A. V., & de Morais, M. G. (2015). Polyhydroxybutyrate production by Spirulina sp. LEB 18 grown under different nutrient concentrations. *African Journal of Microbiological Research, 9*(24), 1586–1594. doi:10.5897/AJMR2015.7530

Cole, T. M., & Buchak, E. M. (1995). *CE-QUAL-W2: A two dimensional, laterally averaged, hydrodynamic and water quality model, version 2.0: User manual.* Instruction Report EL-95-1, US Army Engineer Waterways Experiment Station.

Cole, L. A. (2016). Evolution of Chemical, Prokaryotic, and Eukaryotic Life. In *Biology of Life* (pp. 93–99). Elsevier. doi:10.1016/B978-0-12-809685-7.00013-7

Collen, J., Roeder, V., Rousvoal, S., Collin, O., Kloareg, B., & Boyen, C. (2006). An Expressed Sequence Tag Analysis of Thallus and Regenerating Protoplasts of *Chondrus Crispus* (Gigartinales, Rhodophyceae). *Journal of Phycology, 42*(1), 104–112. doi:10.1111/j.1529-8817.2006.00171.x

Colombo, B., Pereira, J., Martins, M., Torres-Acosta, M. A., Dias, A. C. R. V., Lemos, P. C., Ventura, S. P. M., Eisele, G., Alekseeva, A., Adani, F., & Serafim, L. S. (2020). Recovering PHA from mixed microbial biomass: Using non-ionic surfactants as a pretreatment step. *Separation and Purification Technology, 253*(April), 117521. Advance online publication. doi:10.1016/j.seppur.2020.117521

Compaore, J., & Stal, L. J. (2010). Oxygen and the light-dark cycle of nitrogenase activity in two unicellular cyanobacteria. *Environmental Microbiology, 12*(1), 54–62. doi:10.1111/j.1462-2920.2009.02034.x PMID:19691503

Conde, F. R., Churio, M. S., & Previtali, C. M. (2007). Experimental study of the excited-state properties and photostability of the mycosporine-like amino acid palythine in aqueous solution. *Photochemical & Photobiological Sciences*, *6*(6), 669–674. doi:10.1039/b618314j PMID:17549269

Consalvey, M., Paterson, D. M., & Underwood, G. J. C. (2004). The ups and downs of life in a benthic biofilm: Migration of benthic diatoms. *Diatom Research*, *19*(2), 181–202. doi:10.1080/0269249X.2004.9705870

Constante, A., Pillay, S., Ning, H., & Vaidya, U. K. (2015). Utilization of algae blooms as a source of natural fibers for biocomposite materials: Study of morphology and mechanical performance of Lyngbya fibers. *Algal Research*, *12*, 412–420. doi:10.1016/j.algal.2015.10.005

Cooke, T. J., Poli, D., Sztein, A. E., & Cohen, J. D. (2002). Evolutionary Patterns in Auxin Action. *Plant Molecular Biology*, *49*(3/4), 319–338. doi:10.1023/A:1015242627321 PMID:12036257

Coons, J. E., Kalb, D. M., Dale, T., & Marrone, B. L. (2014). Getting to Low-Cost Algal Biofuels: A Monograph on Conventional and Cutting-Edge Harvesting and Extraction Technologies. *Algal Research*, *6*, 250–270. doi:10.1016/j.algal.2014.08.005

Coppens, J., Grunert, O., Van Den Hende, S., Vanhoutte, I., Boon, N., Haesaert, G., & De Gelder, L. (2016). The use of microalgae as a high-value organic slow-release fertilizer results in tomatoes with increased carotenoid and sugar levels. *Journal of Applied Phycology*, *28*(4), 2367–2377. doi:10.100710811-015-0775-2

Coppola, G., Gaudio, M. T., Lopresto, C. G., Calabro, V., Curcio, S., & Chakraborty, S. (2021a). Bioplastic from renewable biomass: a facile solution for a greener environment. *Earth Systems and Environment*, 1-21.

Coppola, G., Gaudio, M. T., Lopresto, C. G., Calabro, V., Curcio, S., & Chakraborty, S. (2021b). Bioplastic from renewable biomass: A facile solution for a greener environment. *Earth Systems and Environment*, *5*(2), 231–251. doi:10.100741748-021-00208-7

Coradeghini, R., Gioria, S., García, C. P., Nativo, P., Franchini, F., Gilliland, D., Ponti, J., & Rossi, F. (2013). Size-dependent toxicity and cell interaction mechanisms of gold nanoparticles on mouse fibroblasts. *Toxicology Letters*, *217*(3), 205–216. doi:10.1016/j.toxlet.2012.11.022 PMID:23246733

Cornet, J. F., & Dussap, C. G. (2009). A simple and reliable formula for assessment of maximum volumetric productivities in photobioreactors. *Biotechnology Progress*, *25*(2), 424–435. doi:10.1002/btpr.138 PMID:19340890

Cornwall, C. E., & Hurd, C. L. (2019). Variability in the benefits of ocean acidification to photosynthetic rates of macroalgae without CO_2 concentrating mechanisms. *Marine and Freshwater Research*, *71*(3), 275–280. doi:10.1071/MF19134

Costa, J. A. V., & de Morais, M. G. (2014). *An Open Pond System for Microalgal Cultivation. Biofuels from Algae.* Academic Press.

Costa, J. A. V., Freitas, B. C. B., Cruz, C. G., Silveira, J., & Morais, M. G. (2019). Potential of microalgae as biopesticides to contribute to sustainable agriculture and environmental development. *Journal of Environmental Science and Health. Part B, Pesticides, Food Contaminants, and Agricultural Wastes*, *54*(5), 366–375. doi:10.1080/03601234.2019.1571366 PMID:30729858

Costa, J. A. V., Moreira, J. B., Lucas, B. F., Da Silva Braga, V., Cassuriaga, A. P. A., & De Morais, M. G. (2018a). Recent advances and future perspectives of PHB production by cyanobacteria. *Industrial Biotechnology (New Rochelle, N.Y.)*, *14*(5), 249–256. doi:10.1089/ind.2018.0017

Costa, L. S., Fidelis, G. P., Telles, C. B. S., Dantas-Santos, N., Camara, R. B. G., Cordeiro, S. L., Costa, M. S. S. P., Almeida Lima, J., Melo-Silveira, R. F., Oliveira, R. M., Albuquerque, I. R. L., Andrade, G. P. V., & Rocha, H. A. O. (2011). Antioxidant and antiproliferative activities of heterofucans from the seaweed Sargassum filipendula. *Marine Drugs*, *9*(6), 952–966. doi:10.3390/md9060952 PMID:21747741

Costa, L., Hemmer, J., Hillmann Wanderlind, E., Gerlach, O., Santos, A., Tamanaha, M., Bella-Cruz, A., Correa, R., Bazani, H., Radetski, C., & Almerindo, G. (2020). Green Synthesis of Gold Nanoparticles Obtained from Algae Sargassum cymosum: Optimization, Characterization and Stability. *BioNanoScience*, *10*(4), 1049–1062. Advance online publication. doi:10.100712668-020-00776-4

Costa, O. Y., Raaijmakers, J. M., & Kuramae, E. E. (2018). Microbial extracellular polymeric substances: Ecological function and impact on soil aggregation. *Frontiers in Microbiology*, *9*, 9. doi:10.3389/fmicb.2018.01636 PMID:30083145

Costa, S. S., Miranda, A. L., Andrade, B. B., Assis, D. de J., Souza, C. O., de Morais, M. G., Costa, J. A. V., & Druzian, J. I. (2018). Influence of nitrogen on growth, biomass composition, production, and properties of polyhydroxyalkanoates (PHAs) by microalgae. *International Journal of Biological Macromolecules*, *116*, 552–562. doi:10.1016/j.ijbiomac.2018.05.064 PMID:29763703

Costa, S. S., Miranda, A. L., de Morais, M. G., Costa, J. A. V., & Druzian, J. I. (2019). Microalgae as source of polyhydroxyalkanoates (PHAs)—A review. *International Journal of Biological Macromolecules*, *131*, 536–547. doi:10.1016/j.ijbiomac.2019.03.099 PMID:30885732

Couto, C., Hernández, C. P., Alves Sobrinho, R. C. M., Mendes, C. R. B., Roselet, F., & Abreu, P. C. (2021). Optimization of a low-cost fertilizer-based medium for large-scale cultivation of the coastal diatom Conticribra weissflogii using response surface methodology and its effects on biomass composition. *Journal of Applied Phycology*, *33*(5), 2767–2781.

Cox, K. D., Covernton, G. A., Davies, H. L., Dower, J. F., Juanes, F., & Dudas, S. E. (2019). Human Consumption of Microplastics. *Environmental Science & Technology*, *53*(12), 7068–7074. doi:10.1021/acs.est.9b01517 PMID:31184127

Coyer, J. A., Robertson, D. L., & Alberte, R. S. (1994). Genetic variability within a population and between diploid/haploid tissue of *Macrocystis pyrifera* (Phaeophyceae). *Journal of Phycology*, *30*(3), 545–552. doi:10.1111/j.0022-3646.1994.00545.x

Craigie, J. S. (2011). Seaweed extract stimuli in plant science and agriculture. *Journal of Applied Phycology*, *23*(3), 371–393. doi:10.100710811-010-9560-4

Craigie, J. S., Morris, E. R., Rees, D. A., & Thom, D. (1984). Alginate block structure in Phaeophyceae from Nova Scotia: Variation with species, environment and tissue-type. *Carbohydrate Polymers*, *4*(4), 237–252. doi:10.1016/0144-8617(84)90001-8

Croce, R., Cinà, F., Lombardo, A., Crispeyn, G., Cappelli, C. I., Vian, M., Maiorana, S., Benfenati, E., & Baderna, D. (2017). Aquatic toxicity of several textile dye formulations: Acute and chronic assays with *Daphnia magna* and *Raphidocelis subcapitata. Ecotoxicology and Environmental Safety*, *144*, 79–87. doi:10.1016/j.ecoenv.2017.05.046 PMID:28601520

Croft, M. T., Lawrence, A. D., Raux-Deery, E., Warren, M. J., & Smith, A. G. (2005). Algae acquire vitamin B12 through a symbiotic relationship with bacteria. *Nature*, *438*(7064), 90–93. doi:10.1038/nature04056 PMID:16267554

Cuellar-Bermudez, S. P., Aleman-Nava, G. S., Chandra, R., Garcia-Perez, J. S., Contreras-Angulo, J. R., Markou, G., Muylaert, K., Rittmann, B. E., & Parra-Saldivar, R. (2017). Nutrients utilization and contaminants removal. A review of two approaches of algae and cyanobacteria in wastewater. *Algal Research*, *24*, 438–449. doi:10.1016/j.algal.2016.08.018

Cuellar-Bermudez, S. P., Kilimtzidi, E., Devaere, J., Goiris, K., Gonzalez-Fernandez, C., Wattiez, R., & Muylaert, K. (2020). Harvesting of Arthrospira platensis with helicoidal and straight trichomes using filtration and centrifugation. *Separation Science and Technology*, *55*(13), 2381–2390.

Cuhel, R. L., Ortner, P. B., & Lean, D. R. (1984). Night synthesis of protein by algae 1. *Limnology and Oceanography*, *29*(4), 731–744. doi:10.4319/lo.1984.29.4.0731

Cunha, C., Faria, M., Nogueira, N., Ferreira, A., & Cordeiro, N. (2019). Marine vs freshwater microalgae exopolymers as biosolutions to microplastics pollution. *Environmental Pollution*, *249*, 372–380. doi:10.1016/j.envpol.2019.03.046 PMID:30909130

Curran, M. A. (2017). Overview of goal and scope definition in life cycle assessment. In *Goal and scope definition in life cycle assessment* (pp. 1–62). Springer. doi:10.1007/978-94-024-0855-3_1

Czerpak, R., Piotrowska, A., & Szulecka, K. (2006). Jasmonic acid affects changes in the growth and some components content in alga Chlorella vulgaris. *Acta Physiologiae Plantarum*, *28*(3), 195–203. doi:10.1007/BF02706531

D'Souza, L. (2013). Thermochemical hydrogen production from water using reducible oxide materials: A critical review. *Materials for Renewable and Sustainable Energy*, *2*(1), 1–12. doi:10.100740243-013-0007-0

da Gama, B. A. P., Plouguerné, E., & Pereira, R. C. (2014). The antifouling defence mechanisms of marine macroalgae. *Advances in Botanical Research*, *71*, 413–440. doi:10.1016/B978-0-12-408062-1.00014-7

Daboussi, F., Leduc, S., Maréchal, A., Dubois, G., Guyot, V., Perez-Michaut, C., Amato, A., Falciatore, A., Juillerat, A., Beurdeley, M., Voytas, D. F., Cavarec, L., & Duchateau, P. (2014). Genome engineering empowers the diatom *Phaeodactylum tricornutum* for biotechnology. *Nature Communications*, *5*(1), 3831. doi:10.1038/ncomms4831 PMID:24871200

Daglen, B. C., & Tyler, D. R. (2010). Photodegradable plastics: End-of-life design principles. *Green Chemistry Letters and Reviews*, *3*(2), 69–82. doi:10.1080/17518250903506723

Dahiya, D., Sharma, H., Rai, A. K., & Nigam, P. S. N. (2022). Application of biological systems and processes employing microbes and algae to Reduce, Recycle, Reuse (3Rs) for the sustainability of a circular bioeconomy. *AIMS Microbiology*, *8*(1), 83–102. doi:10.3934/microbiol.2022008 PMID:35496985

Dahoumane, S. A., Mechouet, M., Alvarez, F. J., Agathos, S. N., & Jeffryes, C. (2016). Microalgae: An outstanding tool in nanotechnology. *Revista Bionatura*, *1*(4), 196–201. doi:10.21931/RB/2016.01.04.7

Dal'Molin, C. G. O., Quek, L., Palfreyman, R. W., & Nielsen, L. K. (2011). AlgaGEM – a genome-scale metabolic reconstruction of algae based on the *Chlamydomonas reinhardtii* genome. *BMC Genomics*, *12*(S4), S5. doi:10.1186/1471-2164-12-S4-S5 PMID:22369158

Daly, S. R., Fathi, A., Bahramian, B., Manavitehrani, I., McClure, D. D., Valtchev, P., Schindeler, A., Dehghani, F., & Kavanagh, J. M. (2018, January). A green process for the purification of biodegradable poly(B-hydroxybutyrate). *The Journal of Supercritical Fluids*, *135*, 84–90. doi:10.1016/j.supflu.2018.01.007

Damen, K., van Troost, M., Faaij, A., & Turkenburg, W. (2006). A comparison of electricity and hydrogen production systems with CO2 capture and storage. Part A: Review and selection of promising conversion and capture technologies. *Progress in Energy and Combustion Science*, *32*(2), 215–246. doi:10.1016/j.pecs.2005.11.005

Daneshvar, E., Wicker, R. J., Show, P.-L., & Bhatnagar, A. (2022). Biologically-mediated carbon capture and utilization by microalgae towards sustainable CO_2 biofixation and biomass valorization – A review. *Chemical Engineering Journal*, *427*, 130884. doi:10.1016/j.cej.2021.130884

Daneshvar, E., Zarrinmehr, M. J., Hashtjin, A. M., Farhadian, O., & Bhatnagar, A. (2018). Versatile Applications of Freshwater and Marine Water Microalgae in Dairy Wastewater Treatment, Lipid Extraction and Tetracycline Biosorption. *Bioresource Technology*, *268*, 523–530.

Dang, Y., Holmes, D. E., Zhao, Z., Woodard, T. L., Zhang, Y., Sun, D., Wang, L.-Y., Nevin, K. P., & Lovley, D. R. (2016). Enhancing anaerobic digestion of complex organic waste with carbon-based conductive materials. *Bioresource Technology*, *220*, 516–522. doi:10.1016/j.biortech.2016.08.114 PMID:27611035

Daniel, M. C., & Astruc, D. (2004). Gold nanoparticles: Assembly, supramolecular chemistry, quantumsize-related properties, and applications toward biology, catalysis, and nanotechnology. *Chemical Reviews*, *104*(1), 293–346. doi:10.1021/cr030698+ PMID:14719978

Dao, L. H., & Beardall, J. (2016). Effects of lead on growth, photosynthetic characteristics and production of reactive oxygen species of two freshwater green algae. *Chemosphere*, *147*, 420–429. .2015.12.117 doi:10.1016/j.chemosphere

Darwesh, O. M., Ali, S. S., Matter, I. A., & Elsamahy, T. (2021). Nanotextiles waste management: controlling of release and remediation of wastes. In *Nanosensors and Nanodevices for Smart Multifunctional Textiles* (pp. 267–286). Elsevier. doi:10.1016/B978-0-12-820777-2.00016-9

Das, D. (2015). *Algal Biore nery: An Integrated Approach.* Springer International Publishing AG.

Das, N., & Das, D. (2013). Recovery of rare earth metals through biosorption: An overview. *Journal of Rare Earths*, *31*(10), 933–943. doi:10.1016/S1002-0721(13)60009-5

Das, R. N., Roy, K., & Popelier, P. L. A. (2015). Interspecies quantitative structure–toxicity–toxicity (QSTTR) relationship modeling of ionic liquids, Toxicity of ionic liquids to *V. fischeri, D. magna* and *S. vacuolatus. Ecotoxicology and Environmental Safety*, *122*, 497–520. doi:10.1016/j.ecoenv.2015.09.014 PMID:26414597

Das, S. K., Sathish, A., & Stanley, J. (2018). Production of biofuel and bioplastic from chlorella pyrenoidosa. *Materials Today: Proceedings*, *5*(8), 16774–16781. doi:10.1016/j.matpr.2018.06.020

Das, S. R., & Basak, N. (2021a). Molecular biohydrogen production by dark and photo fermentation from wastes containing starch: Recent advancement and future perspective. *Bioprocess and Biosystems Engineering*, *44*(1), 1–25. doi:10.100700449-020-02422-5 PMID:32785789

Daud, S. N. S. S., Norddin, M. N. A. M., Jaafar, J., & Sudirman, R. (2021). Development of sulfonated poly (ether ether ketone)/polyethersulfone-crosslinked quaternary ammonium poly (ether ether ketone) bipolar membrane electrolyte via hot-press approach for hydrogen/oxygen fuel cell. *International Journal of Energy Research*, *45*(6), 9210–9228. doi:10.1002/er.6453

Davidson, K., & Gurney, W. (1999). An investigation of non-steady-state algal growth. II. Mathematical modelling of co-nutrient-limited algal growth. *Journal of Plankton Research*, *21*(5), 839–858. doi:10.1093/plankt/21.5.839

Davila-Rodriguez, J. L., Escobar-Barrios, V. A., Shirai, K., & Rangel-Mendez, J. R. (2009). Synthesis of a chitin-based biocomposite for water treatment: Optimization for fluoride removal. *Journal of Fluorine Chemistry*, *130*(8), 718–726. doi:10.1016/j.jfluchem.2009.05.012

Davis, J., De Menna, F., Unger, N., Östergren, K., Loubiere, M., & Vittuari, M. (2017). *Generic strategy LCA and LCC: Guidance for LCA and LCC focused on prevention, valorisation and treatment of side flows from the food supply chain.* Academic Press.

Davis, E. M., Downs, T. D., Shi, Y., & Ajgaonkar, A. A. (1995). Recycle as an alternative to algal TSS and BOD removal from an industrial waste stabilization pond system. *Industrial Waste Conference*, *50*, 65-73.

Davison, I. R., & Pearson, G. A. (1996). Stress Tolerance in Intertidal Seaweeds. *Journal of Phycology*, *32*(2), 197–211. doi:10.1111/j.0022-3646.1996.00197.x

Davis, R., Aden, A., & Pienkos, P. T. (2011). Techno-economic analysis of autotrophic microalgae for fuel production. *Applied Energy*, *88*(10), 3524–3531. doi:10.1016/j.apenergy.2011.04.018

Davis, T. A., Volesky, B., & Mucci, A. (2003). A review of the biochemistry of heavy metal biosorption by brown algae. *Water Research*, *37*(18), 4311–4330. doi:10.1016/S0043-1354(03)00293-8 PMID:14511701

Dawson, T. P., Jackson, S. T., House, J. I., Iain Colin, P., & Mace, G. M. (2011). Beyond predictions: Biodiversity conservation in a changing climate. *Science*, *332*(6025), 53–58. doi:10.1126cience.1200303 PMID:21454781

De Bruin, A., Ibelings, B. W., Rijkeboer, M., Brehm, M., & Van Donk, E. (2004). Genetic variation in Asterionella formosa (Bacillariophyceae): Is it linked to frequent epidemics of host-specific parasitic fungi? *Journal of Phycology*, *40*, 823–830.

De Clerck, O., Guiry, M. D., Leliaert, F., Samyn, Y., & Verbruggen, H. (2013). Algal taxonomy: A road to nowhere? *Journal of Phycology*, *49*(2), 215–225. doi:10.1111/jpy.12020 PMID:27008509

De Clercq, M., Vats, A., & Biel, A. (2018). Agriculture 4.0: The Future of Farming Technology. *World Government Summit in Collaboration with Oliver Wyman*, (February), 30.

de Farias Silva, C. E., & Bertucco, A. (2016). Bioethanol from microalgae and cyanobacteria: A review and technological outlook. *Process Biochemistry*, *51*(11), 1833–1842. doi:10.1016/j.procbio.2016.02.016

De Godos, I., González, C., Becares, E., García-Encina, P. A., & Muñoz, R. (2009). Simultaneous nutrients and carbon removal during pretreated swine slurry degradation in a tubular biofilm photobioreactor. *Applied Microbiology and Biotechnology*, *82*(1), 187–194. doi:10.100700253-008-1825-3 PMID:19122998

De Godos, I., Munoz, R., & Guieysse, B. (2012). Tetracycline removal during wastewater treatment in high-rate algal ponds. *Journal of Hazardous Materials*, *229*, 446–449.

De Jesus Raposo, M. F., De Morais, R. M. S. C., & De Morais, A. M. M. B. (2013). Health applications of bioactive compounds from marine microalgae. *Life Sciences*, *93*(15), 479–486. doi:10.1016/j.lfs.2013.08.002 PMID:23994664

De Lassio, J., França, J., Espirito Santo, K., & Haddad, A. (2016). Case study: LCA methodology applied to materials management in a Brazilian residential construction site. *Journal of Engineering (Stevenage, England)*.

de Morais, M. G., Vaz, B. D. S., de Morais, E. G., & Costa, J. A. V. (2015). Biologically active metabolites synthesized by microalgae. *BioMed Research International*. PMID:26339647

de Sá, P. L. Jr, Câmara, D. A. D., Porcacchia, A. S., Fonseca, P. M. M., Jorge, S. D., Araldi, R. P., & Ferreira, A. K. (2017). The roles of ROS in cancer heterogeneity and therapy. *Oxidative Medicine and Cellular Longevity*. PMID:29123614

de Souza, L. A. R., Dore, C. M. P., Castro, A. J., de Azevedo, T. C., de Oliveira, M. T. B., Maria de Fátima, V. M., ... Leite, E. L. (2012). Galactans from the red seaweed Amansia multifida and their effects on inflammation, angiogenesis, coagulation and cell viability. *Biomedicine & Preventive Nutrition*, *2*(3), 154–162. doi:10.1016/j.bionut.2012.03.007

de Vasconcelos, A. C. F. (2020). Amelioration of Drought Stress on Plants under Biostimulant Sources. In *Plant Stress Physiology*. IntechOpen.

De Wilt, A., Butkovskyi, A., Tuantet, K., Leal, L. H., Fernandes, T. V., Langenhoff, A., & Zeeman, G. (2016). Micropollutant removal in an algal treatment system fed with source separated wastewater streams. *Journal of Hazardous Materials*, *304*, 84–92.

De-Bashan, L. E., Moreno, M., Hernandez, J. P., & Bashan, Y. (2002). Removal of ammonium and phosphorus ions from synthetic wastewater by the microalgae Chlorella vulgaris coimmobilized in alginate beads with the microalgae growth-promoting bacterium Azospirillum brasilense. *Water Research*, *36*(12), 2941–2948. doi:10.1016/S0043-1354(01)00522-X PMID:12171390

Del Campo, J. A., García-González, M., & Guerrero, M. G. (2007). Outdoor cultivation of microalgae for carotenoid production: Current state and perspectives. *Applied Microbiology and Biotechnology*, *74*(6), 1163–1174. doi:10.100700253-007-0844-9 PMID:17277962

Del Mondo, A., Smerilli, A., Sané, E., Sansone, C., & Brunet, C. (2020). Challenging microalgal vitamins for human health. *Microbial Cell Factories*, *19*(1), 1–23. doi:10.118612934-020-01459-1 PMID:33138823

del Río, P. G., Gomes-Dias, J. S., Rocha, C. M. R., Romaní, A., Garrote, G., & Domingues, L. (2020). Recent Trends on Seaweed Fractionation for Liquid Biofuels Production. *Bioresource Technology*, *299*, 122613. doi:10.1016/j.biortech.2019.122613 PMID:31870706

Dellaporta, S. L., Wood, J., & Hicks, J. B. (1983). A plant DNA minipreparation: Version II. *Plant Molecular Biology Reporter*, *1*(4), 19–21. doi:10.1007/BF02712670

Delrue, F., Li-Beisson, Y., Setier, P. A., Sahut, C., Roubaud, A., Froment, A. K., & Peltier, G. (2013). Comparison of various microalgae liquid biofuel production pathways based on energetic, economic and environmental criteria. *Bioresource Technology*, *136*, 205–212. doi:10.1016/j.biortech.2013.02.091 PMID:23567683

Delrue, F., Setier, P. A., Sahut, C., Cournac, L., Roubaud, A., Peltier, G., & Froment, A. K. (2012). An economic, sustainability, and energetic model of biodiesel production from microalgae. *Bioresource Technology*, *111*, 191–200. doi:10.1016/j.biortech.2012.02.020 PMID:22366604

Demetriou, G., Neonaki, C., Navakoudis, E., & Kotzabasis, K. (2007). Salt stress impact on the molecular structure and function of the photosynthetic apparatus—The protective role of polyamines. *Biochimica et Biophysica Acta (BBA)-Bioenergetics*, *1767*(4), 272–280. doi:10.1016/j.bbabio.2007.02.020 PMID:17408588

Demirbas, A. (2009a). Biorefineries: Current activities and future developments. Energy Convers. Manage. 50 (11), 2782–2801. Demirbas, A., 2009b. Progress and recent trends in biodiesel fuels. *Energy Conversion and Management*, *50*(1), 14–34. doi:10.1016/j.enconman.2008.09.001

Demirel, B., & Scherer, P. (2011). Trace element requirements of agricultural biogas digesters during biological conversion of renewable biomass to methane. *Biomass and Bioenergy*, *35*(3), 992–998. doi:10.1016/j.biombioe.2010.12.022

Demuez, M., Mahdy, A., Tomás-Pejó, E., González-Fernández, C., & Ballesteros, M. (2015). Enzymatic Cell Disruption of Microalgae Biomass in Biorefinery Processes. *Biotechnology and Bioengineering*, *112*(10), 1955–1966. doi:10.1002/bit.25644 PMID:25976593

Deng, X.-Y., Gao, K., Zhang, R.-C., Addy, M., Lu, Q., Ren, H.-Y., Chen, P., Liu, Y.-H., & Ruan, R. (2017). Growing Chlorella vulgaris on thermophilic anaerobic digestion swine manure for nutrient removal and biomass production. *Bioresource Technology*, *243*, 417–425. doi:10.1016/j.biortech.2017.06.141 PMID:28688324

Desai, R. K., Streefland, M., Wijffels, R. H., & Eppink, M. H. M. (2016). Novel Astaxanthin Extraction from Haematococcus Pluvialis Using Cell Permeabilising Ionic Liquids. *Green Chemistry*, *18*(5), 1261–1267. doi:10.1039/C5GC01301A

Desai, U. N., Shah, K. P., Mirza, S. H., Panchal, D. K., Parikh, S. K., & Rawal, R. M. (2015). Enhancement of the cytotoxic effects of Cytarabine in synergism with Hesperidine and Silibinin in Acute Myeloid Leukemia: An in-vitro approach. *Journal of Cancer Research and Therapeutics*, *11*(2), 352. doi:10.4103/0973-1482.157330 PMID:26148599

Deshpande, S. R., Sunol, A. K., & Philippidis, G. (2017). Status and prospects of supercritical alcohol transesterification for biodiesel production. *Wiley Interdisciplinary Reviews. Energy and Environment*, *6*(5), e252. doi:10.1002/wene.252

Després, V. R., Huffman, J. A., Burrows, S. M., Hoose, C., Safatov, A. S., Buryak, G., Fröhlich-Nowoisky, J., Elbert, W., Andreae, M. O., Pöschl, U., & Jaenicke, R. (2012). Primary biological aerosol particles in the atmosphere: A review. *Tellus. Series B, Chemical and Physical Meteorology*, *64*(1), 15598–15656. doi:10.3402/tellusb.v64i0.15598

Devadas, V. V., Khoo, K. S., Chia, W. Y., Chew, K. W., Munawaroh, H. S. H., Lam, M.-K., Lim, J.-W., Ho, Y.-C., Lee, K. T., & Show, P. L. (2021). Algae biopolymer towards sustainable circular economy. *Bioresource Technology*, *325*, 124702. doi:10.1016/j.biortech.2021.124702 PMID:33487515

Dewi, I. C., Falaise, C., Hellio, C., Bourgougnon, N., & Mouget, J. L. (2018). Anticancer, antiviral, antibacterial, and antifungal properties in microalgae. In *Microalgae in health and disease prevention* (pp. 235–261). Academic Press. doi:10.1016/B978-0-12-811405-6.00012-8

Dhandayuthapani, K., Malathy, S., Mulla, S. I., & Gupta, S. K. (2021). An Insight into the Potential Application of Microalgae in Pharmaceutical and Nutraceutical Production. In *Algae* (pp. 135–179). Springer. doi:10.1007/978-981-15-7518-1_7

Dhargalkar, V. K., & Pereira, N. (2005). *Seaweed: Promising plant of the millennium*. Academic Press.

Dhargalkar, V. K., & Pereira, N. (2005). Effect of seaweed extract on the growth, yield and nutrient uptake of soybean (*Glycine max*) under rainfed conditions. *Journal of Applied Phycology*, *1*(2), 351–355.

Dianursanti, & Khalis, S. A. (2018). The effect of compatibilizer addition on *Chlorella vulgaris* microalgae utilization as a mixture for bioplastic. *E3S Web of Conferences, 67*, 3047. doi:10.1051/e3sconf/20186703047

Dianursanti, G. M., & Noviasari, C. (2018). The effect of glycerol addition as plasticizer in *Spirulina platensis* based bioplastic. *E3S Web of Conferences, 67*, 3048. doi:10.1051/e3sconf/20186703048

Dias, D. A., Urban, S., & Roessner, U. (2012). A historical overview of natural products in drug discovery. *Metabolites*, *2*(2), 303–336. doi:10.3390/metabo2020303 PMID:24957513

Dincer, I., & Balta, M. T. (2011). Potential thermochemical and hybrid cycles for nuclear-based hydrogen production. *International Journal of Energy Research*, *35*(2), 123–137. doi:10.1002/er.1769

Ding, F., Wang, Z., Yang, X., Shi, L., Liu, J., & Chen, G. (2018). Development of classification models for predicting chronic toxicity of chemicals to *Daphnia magna* and *Pseudokirchneriella subcapitata. SAR and QSAR in Environmental Research*, *30*(1), 39–50. doi:10.1080/1062936X.2018.1545694 PMID:30477347

Ding, T., Yang, M., Zhang, J., Yang, B., Lin, K., Li, J., & Gan, J. (2017). Toxicity, degradation and metabolic fate of ibuprofen on freshwater diatom *Navicula* sp. *Journal of Hazardous Materials*, *330*, 127–134.

Diniz, V., & Volesky, B. (2005). Effect of counterions on lanthanum biosorption by Sargassum polycystum. *Water Research*, *39*(11), 2229–2236. doi:10.1016/j.watres.2005.04.004 PMID:15899510

Divya, M., & Santhanam, P. (2018). Bioremediation of wastewater using a novel method of microalgae immobilized on Twin-Layer Recirculation System (TLRS). In Basic and Applied Phytoplankton Biology (pp. 177–190). doi:10.1007/978-981-10-7938-2_8

Dmytryk, A., & Chojnacka, K. (2018). Algae as fertilizers, biostimulants, and regulators of plant growth. In *Algae Biomass: Characteristics and Applications* (pp. 115–122). Springer. doi:10.1007/978-3-319-74703-3_10

Dmytryk, A., Chojnacka, K., & Rój, E. (2018). The methods of algal biomass extraction: toward the application. In *Algae Biomass: Characterization and applications* (pp. 49–56). Springer. doi:10.1007/978-3-319-74703-3_5

Doctor, R., Palmer, A., Coleman, D., Davison, J., Hendriks, C., Kaarstad, O., & Austell, M. (2005). *Transport of CO2. IPCC Special Report on Carbon Dioxide Capture and Storage, R. Pichs-Madruga* (S. Timashev, Ed.). Intergovernmental Panel on Climate Change.

Dodds, W. K., Gudder, D. A., & Mollenhauer, D. (1995). The ecology of Nostoc. *Journal of Phycology*, *31*(1), 2–18. doi:10.1111/j.0022-3646.1995.00002.x

Dodson, J. R., Hunt, A. J., Parker, H. L., Yang, Y., & Clark, J. H. (2012). Elemental sustainability: Towards the total recovery of scarce metals. *Chemical Engineering and Processing*, *51*, 69–78. doi:10.1016/j.cep.2011.09.008

Doe, U. (2016). *National algal biofuels technology review*. Academic Press.

Domínguez-Díaz, M., Meneses-Acosta, A., Romo-Uribe, A., Peña, C., Segura, D., & Espin, G. (2015). Thermo-mechanical properties, microstructure and biocompatibility in poly-β-hydroxybutyrates (PHB) produced by OP and OPN strains of Azotobacter vinelandii. *European Polymer Journal*, *63*, 101–112. doi:10.1016/j.eurpolymj.2014.12.002

Dong, T., Knoshaug, E. P., Pienkos, P. T., & Laurens, L. M. L. (2016). Lipid Recovery from Wet Oleaginous Microbial Biomass for Biofuel Production: A Critical Review. *Applied Energy*, *177*, 879–895. doi:10.1016/j.apenergy.2016.06.002

Donoghue, P. C. J., Harrison, C. J., Paps, J., & Schneider, H. (2021). 2021/10/11/). The evolutionary emergence of land plants. *Current Biology*, *31*(19), R1281–R1298. doi:10.1016/j.cub.2021.07.038 PMID:34637740

Doorn, P. F., Campbell, P. A., & Amstutz, H. C. (1996). Metal Versus Polyethylene Wear Particles in Total Hip Replacements. *Clinical Orthopaedics and Related Research*, *329*, S206–S216. doi:10.1097/00003086-199608001-00018 PMID:8769335

Doppler, P., Kriechbaum, R., Singer, B., & Spadiut, O. (2021). Make microalgal cultures axenic again–a fast and simple workflow utilizing fluorescence-activated cell sorting. *Journal of Microbiological Methods*, *186*, 106256. doi:10.1016/j.mimet.2021.106256 PMID:34082050

Doucette, G. J., McGovern, E. R., & Babinchak, J. A. (1999). Algicidal bacteria active against Gymnodinium breve (Dinophyceae). I. Bacterial isolation and characterization of killing activity. *Journal of Phycology*, *35*(6, SUPPL.), 1447–1454. doi:10.1046/j.1529-8817.1999.3561447.x

Dow, L. (2021). How Do Quorum-Sensing Signals Mediate Algae–Bacteria Interactions? *Microorganisms*, *9*(7), 1391. doi:10.3390/microorganisms9071391 PMID:34199114

Dragosits, M., & Mattanovich, D. (2013). Adaptive laboratory evolution – principles and applications for biotechnology. *Microbial Cell Factories*, *12*(1), 64. doi:10.1186/1475-2859-12-64 PMID:23815749

Dring, J. M., Wagner, A., Boeskov, J., & Luning, K. (1996b). Sensitivity of intertidal and subtidal red algae to UVA and UVB radiation, as monitored by chlorophyll. uorescence measurements: Influence of collection depth and season, and length of irradiation. *European Journal of Phycology*, *31*(4), 293–302. doi:10.1080/09670269600651511

Dring, M. J., Makarov, V., Schoschina, E., Lorenz, M., & Luning, K. (1996a). Influence of ultraviolet–radiation on chlorophyll fluorescence and growth in different life-history stages of three species of Laminaria, Phaeophyta. *Marine Biology*, *126*, 183–191. doi:10.1007/BF00347443

Droop, M. R. (1968). Vitamin B12 and marine ecology. IV. The kinetics of uptake, growth and inhibition in Monochrysis lutheri. *Journal of the Marine Biological Association of the United Kingdom*, *48*(3), 689–733. doi:10.1017/S0025315400019238

Duarte, J. H., de Morais, E. G., Radmann, E. M., & Costa, J. A. V. (2017). Biological CO_2 mitigation from coal power plant by *Chlorella fusca* and *Spirulina*sp. *Bioresource Technology*, *234*, 472–475. doi:10.1016/j.biortech.2017.03.066 PMID:28342576

Dubey, S., Bharmoria, P., Gehlot, P. S., Agrawal, V., Kumar, A., & Mishra, S. (2018). 1-Ethyl-3-methylimidazolium Diethylphosphate Based Extraction of Bioplastic "polyhydroxyalkanoates" from Bacteria: Green and Sustainable Approach. *ACS Sustainable Chemistry & Engineering*, *6*(1), 766–773. doi:10.1021/acssuschemeng.7b03096

Dubinsky, Z., Matsukawa, R., & Karube, I. (1995). Photobiological aspects of algal mass culture. *Journal of Marine Biotechnology*, *2*, 61–65.

Du, K., Tao, H., Wen, X., Geng, Y., & Li, Y. (2015). Enhanced growth and lipid production of Chlorella pyrenoidosa by plant growth regulator GA3. *Fresenius Environmental Bulletin*, *24*, 3414–3419.

Durham, B. P., Dearth, S. P., Sharma, S., Amin, S. A., Smith, C. B., Campagna, S. R., Armbrust, E. V., & Moran, M. A. (2017). Recognition cascade and metabolite transfer in a marine bacteria-phytoplankton model system. *Environmental Microbiology*, *19*(9), 3500–3513. doi:10.1111/1462-2920.13834 PMID:28631440

Durham, B. P., Sharma, S., Luo, H., Smith, C. B., Amin, S. A., Bender, S. J., Dearth, S. P., Van Mooy, B. A., Campagna, S. R., Kujawinski, E. B., Armburst, E. V., & Moran, M. A. (2015). Cryptic carbon and sulfur cycling between surface ocean plankton. *Proceedings of the National Academy of Sciences of the United States of America*, *112*(2), 453–457. doi:10.1073/pnas.1413137112 PMID:25548163

Du, X., & Graedel, T. E. (2011). Global in-use stocks of the rare earth elements: A first estimate. *Environmental Science & Technology*, *45*(9), 4096–4101. doi:10.1021/es102836s PMID:21438595

Du, X.-P., Cai, Z.-H., Zuo, P., Meng, F.-X., Zhu, J.-M., & Zhou, J. (2020). Temporal variability of virioplankton during a Gymnodinium catenatum algal bloom. *Microorganisms*, *8*(1), 1–15. doi:10.3390/microorganisms8010107 PMID:31940944

Du, Z.-Y., Zienkiewicz, K., Vande Pol, N., Ostrom, N. E., Benning, C., & Bonito, G. M. (2019). Algal-fungal symbiosis leads to photosynthetic mycelium. *eLife*, *8*, e47815. Advance online publication. doi:10.7554/eLife.47815 PMID:31307571

Dyo, Y. M., & Purton, S. (2018). The algal chloroplast as a synthetic biology platform for production of therapeutic proteins. *Microbiology*, *164*(2), 113–121. doi:10.1099/mic.0.000599 PMID:29297850

Dzoujo, T. H., Tome, S., Shikuku, V. O., Tchuigwa, J. T., Spieß, A., Janiak, C., Etoh, A., & Dina, D. (2021). Enhanced performance of hydrogen peroxide modified pozzolan-based geopolymer for abatement of methylene blue from aqueous medium. *Silicon*. doi:10.1007/s12633-021-01264-4

Ebrahim, A., Lerman, J. A., Palsson, B. O., & Hyduke, D. R. (2013). COBRApy: COnstraints-based reconstruction and analysis for python. *BMC Systems Biology*, *7*(1), 74–79. doi:10.1186/1752-0509-7-74 PMID:23927696

Eccles, H. (1999). Treatment of metal-contaminated wastes: Why select a biological process? *Trends in Biotechnology*, *17*(12), 462–465. doi:10.1016/S0167-7799(99)01381-5 PMID:10557157

Edison, T. N. J. I., Atchudan, R., Kamal, C., & Lee, Y. R. (2016). Caulerparacemosa: A marine green alga for eco-friendly synthesis of silver nanoparticles and its catalytic degradation of methyleneblue. *Bioprocess and Biosystems Engineering*, *39*(9), 1401–1408. doi:10.100700449-016-1616-7 PMID:27129459

Edwards, M. (2010). *Algae 101 Part six: Algal classification*. Algae Industry Magazine.com.

Egan, S., Harder, T., Burke, C., Steinberg, P., Kjelleberg, S., & Thomas, T. (2013). The seaweed holobiont: Understanding seaweed-bacteria interactions. *FEMS Microbiology Reviews*, *37*(3), 462–476. doi:10.1111/1574-6976.12011 PMID:23157386

Egleston, E. S., & Morel, F. M. M. (2008). Nickel limitation and zinc toxicity in a urea-grown diatom. *Limnology and Oceanography*, *53*(6), 2462–2471. doi:10.4319/lo.2008.53.6.2462

Ehimen, E. A., Sun, Z. F., & Carrington, C. G. (2010). Variables affecting the in situ transesterification of microalgae lipids. *Fuel*, *89*(3), 677–684. doi:10.1016/j.fuel.2009.10.011

Ehrenstein, G. W. (2001). Structure of Polymeric Materials. In Polymeric Materials (pp. 61–141). Carl Hanser Verlag GmbH & Co. KG. doi:10.3139/9783446434134.004

Eikrem, W., Medlin, L. K., Henderiks, J., Rokitta, S., Rost, B., Probert, I., Throndsen, J., & Edvardsen, B. (2017). Haptophyta. In J. M. Archibald, A. G. B. Simpson, & C. H. Slamovits (Eds.), *Handbook of the Protists* (pp. 893–953). Springer International Publishing. doi:10.1007/978-3-319-28149-0_38

Eisenberg, D., Lüthy, R., & Bowie, J. U. (1997). VERIFY3D: Assessment of protein models with three-dimensional profiles. *Methods in Enzymology*, *277*, 396–404. doi:10.1016/S0076-6879(97)77022-8 PMID:9379925

Eiteman, M. A., & Ramalingam, S. (2015). Microbial production of lactic acid. *Biotechnology Letters*, *37*(5), 955–972. doi:10.100710529-015-1769-5 PMID:25604523

Ekpo, U., Ross, A. B., Camargo-Valero, M. A., & Williams, P. T. (2016). A comparison of product yields and inorganic content in process streams following thermal hydrolysis and hydrothermal processing of microalgae, manure and digestate. *Bioresource Technology*, *200*, 951–960. doi:10.1016/j.biortech.2015.11.018 PMID:26615335

El Arroussi, H., Elbaouchi, A., Benhima, R., Bendaou, N., Smouni, A., & Wahby, I. (2015). Halophilic microalgae Dunaliella salina extracts improve seed germination and seedling growth of Triticum aestivum L. under salt stress. In *II World Congress on the Use of Biostimulants in Agriculture 1148* (pp. 13-26).

El Asri, O., Afilal, M. E., Laiche, H., & Elfarh, A. (2020). Evaluation of physicochemical, microbiological, and energetic characteristics of four agricultural wastes for use in the production of green energy in Moroccan farms. *Chemical and Biological Technologies in Agriculture*, *7*(1), 1–11.

El Nemr, A., El-sikaily, A., & Khaled, A. (2010). Modeling of adsorption isotherms of Methylene Blue onto rice husk activated carbon. *Egyptian Journal of Aquatic Research*, *36*(3), 403–425.

Elarroussia, H., Elmernissia, N., Benhimaa, R., El Kadmiria, I. M., Bendaou, N., Smouni, A., & Wahbya, I. (2016). Microalgae polysaccharides a promising plant growth biostimulant. *Journal of Algal Biomass Utilization*, *7*, 55–63.

Elaya Perumal, U., & Sundararaj, R. (2020). Algae: A potential source to prevent and cure the novel coronavirus–A review. *Int. J. Emerg. Technol*, *11*(2), 479–483.

El-Barody, G. S., Moussa, M. Y., Shallan, A. M., Ali, A. M., Sabh, Z. A., & Shalaby, A. E. (2007). Contribution to the Aroma, Biological Activities, Minerals, Protein, Pigments and Lipid Contents of the Red Alga, *Asparagopsis taxiformes* (Delie) Trevisan. *Journal of Applied Sciences Research*, *3*(12), 1825–1834.

El-Din, S. M. (2015). Utilization of seaweed extracts as bio-fertilizers to stimulate the growth of wheat seedlings. *The Egyptian Journal of Experimental Biology*, *11*(1), 31–39.

Elemike, E. E., Uzoh, I. M., Onwudiwe, D. C., & Babalola, O. O. (2019). The role of nanotechnology in the fortification of plant nutrients and improvement of crop production. *Applied Sciences (Basel, Switzerland)*, *9*(3), 499. doi:10.3390/app9030499

Elena, S. F., & Lenski, R. E. (2003). Evolution experiments with microorganisms: The dynamics and genetic bases of adaptation. *Nature Reviews. Genetics*, *4*(6), 457–469. doi:10.1038/nrg1088 PMID:12776215

Elgarahy, A. M., Elwakeel, K. Z., Elshoubaky, G. A., & Mohammad, S. H. (2019). Microwave-accelerated Sorption of Cationic Dyes onto green marine Algal Biomass. *Environmental Science and Pollution Research International*, *26*, 22704–22722.

El-Gebali, S., Mistry, J., Bateman, A., Eddy, S. R., Luciani, A., Potter, S. C., Qureshi, M., Richardson, L. J., Salazar, G. A., Smart, A., Sonnhammer, E. L. L., Hirsh, L., Paladin, L., Piovesan, D., Tosatto, S. C. E., & Finn, R. D. (2019). The Pfam protein families database in 2019. *Nucleic Acids Research*, *47*(1), 427–432. doi:10.1093/nar/gky995 PMID:30357350

Elhafiz, A. A., Elhafiz, A. A., Gaur, S. S., & Hamdany, N. (2015). *Biofertilizer to grow rice, lettuce, cucumber*. doi:10.19071/rrst.2015.v7.2919

El-Kassas, H. Y., Aly-Eldeen, M. A., & Gharib, S. M. (2016). Green synthesis of iron oxide (Fe3O4) nanoparticles using two selected brown seaweeds: Characterization and application for lead bioremediation. *Acta Oceanologica Sinica*, *35*(8), 89–98. doi:10.100713131-016-0880-3

El-Kassas, H. Y., & El-Sheekh, M. M. (2014). Cytotoxic activity of biosynthesized gold nanoparticles with an extract of the red seaweed Corallina officinalis on the MCF-7 human breast cancer cell line. *Asian Pacific Journal of Cancer Prevention*, *15*(10), 4311–4317. doi:10.7314/APJCP.2014.15.10.4311 PMID:24935390

El-malek, F. A., Khairy, H., Farag, A., & Omar, S. (2020). The sustainability of microbial bioplastics, production and applications. *International Journal of Biological Macromolecules*, *157*, 319–328. doi:10.1016/j.ijbiomac.2020.04.076 PMID:32315677

El-Mekkawi, S. A., Abdo, S. M., Samhan, F. A., & Ali, G. H. (2019). Optimization of Some Fermentation Conditions for Bioethanol Production from Microalgae Using Response Surface Method. *Bulletin of the National Research Center*, *43*(1), 164. doi:10.118642269-019-0205-8

El-Mohsnawy, E., El-Sheekh, M. M., Mabrouk, M. E. M., & Zohir, W. F. (2020). Enhancing accumulation of omega 3 and 9 fatty acids in Chlorella vulgaris under mixotrophic nutrition. *Journal of Animal and Plant Sciences*, *30*(2), 485–492.

Elmowafy, E., Abdal-Hay, A., Skouras, A., Tiboni, M., Casettari, L., & Guarino, V. (2019). Polyhydroxyalkanoate (PHA): Applications in drug delivery and tissue engineering. *Expert Review of Medical Devices*, *16*(6), 467–482. doi:10.1080/17434440.2019.1615439 PMID:31058550

Elsalamouny, A. R., Desouky, O. A., Mohamed, S. A., Galhoum, A. A., & Guibal, E. (2017). Uranium and neodymium biosorption using novel chelating polysaccharide. *International Journal of Biological Macromolecules*, *104*, 963–968. doi:10.1016/j.ijbiomac.2017.06.081 PMID:28663149

El-Sheekh, M. M., El-Mohsnawy, E., Mabrouk, M. E., & Zoheir, W. F. (2020). Enhancement of biodiesel production from the green microalga Micractinium reisseri via optimization of cultivation regimes. *Journal of Taibah University for Science: JTUSCI*, *14*(1), 437–444. doi:10.1080/16583655.2020.1745505

El-Sheekh, M. M., El-Naggar, A. H., Osman, M. E. H., & El-Mazaly, E. (2003). Effect of cobalt on growth, pigments and the photosynthetic electron transport in Monoraphidium minutum and Nitzchia perminuta. *Brazilian Journal of Plant Physiology*, *15*(3), 159–166. doi:10.1590/S1677-04202003000300005

El-Sheekh, M. M., Hassan, L. H. S., & Morsi, H. H. (2021). Assessment of the in vitro anticancer activities of cyanobacteria mediated silver oxide and gold nanoparticles in human colon CaCo-2 and cervical HeLa cells. *Environmental Nanotechnology, Monitoring & Management*, *16*, 100556. doi:10.1016/j.enmm.2021.100556

El-Sherif, Z. M., Nassar, M. Z., & Fahmy, M. (2010). Phytoplankton distribution in the southeastern Mediterranean Sea (Egyptian waters) in summer and winter 2005. *Egyptian Journal of Aquatic Research*, *36*, 609–621.

Elshobary, M. E., Abo-Shady, A. M., Khairy, H. M., Essa, D., Zabed, H. M., Qi, X., & Abomohra, A. E. (2019). Influence of nutrient supplementation and starvation conditions on the biomass and lipid productivities of Micractinium reisseri grown in wastewater for biodiesel production. *Journal of Environmental Management*, *250*(April), 109529. doi:10.1016/j.jenvman.2019.109529 PMID:31518792

Elshobary, M. E., El-Shenody, R. A., & Abomohra, A. E. (2021). Sequential biofuel production from seaweeds enhances the energy recovery: A case study for biodiesel and bioethanol production. *International Journal of Energy Research*, *45*(4), 6457–6467. doi:10.1002/er.6181

Elshobary, M. E., El-Shenody, R. A., Ashour, M., Zabed, H. M., & Qi, X. (2020). Antimicrobial and antioxidant characterization of bioactive components from Chlorococcum minutum. *Food Bioscience*, *35*, 100567. doi:10.1016/j.fbio.2020.100567

Elshobary, M. E., Zabed, H. M., Yun, J., Zhang, G., & Qi, X. (2021). Recent insights into microalgae-assisted microbial fuel cells for generating sustainable bioelectricity. *International Journal of Hydrogen Energy*, *46*(4), 3135–3159. doi:10.1016/j.ijhydene.2020.06.251

Eltanahy, E., & Torky, A. (2021). *Microalgae as Cell Factories: Food and Feed-grade High-value Metabolites*. Academic Press.

Eltanahy, E., & Torky, A. (2021). Microalgae as Cell Factories: Food and Feed-Grade High-Value Metabolites. In *Microalgal Biotechnology: Recent Advances, Market Potential, and Sustainability*. Royal Society of Chemistry. doi:10.1039/9781839162473-00001

Elumalai, D., Hemavathi, M., Rekha, G. S., Pushphalatha, M., Leelavathy, R., Vignesh, A., Ashok, K., & Babu, M. (2021). Photochemical synthesizes of silver nanoparticles using Oscillatoria sancta micro algae against mosquito vectors Aedes aegypti and Anopheles stephensi. *Sensing and Bio-Sensing Research*, *100457*, 100457. Advance online publication. doi:10.1016/j.sbsr.2021.100457

El-Zabalawy, K. M., Abou-Shleel, S. M., & Abdel-Kareem, M. S. (2015). Effect of marine algae on bio-accumulation of heavy metals from polluted soil by some leafy vegetables. *Nature and Science*, *13*(3), 109–116.

El-Zeky, M. M., El-Shahat, R. M., Metwaly, G. S., & Elham, M. A. (2005). Using Cyanobacteria or *Azolla* as alternative nitrogen sources for rice production. *J. Agric. Mansoura Univ.*, *30*(9), 5567–5577.

Emblemsvåg, J., Kvadsheim, N. P., Halfdanarson, J., Koesling, M., Nystrand, B. T., Sunde, J., & Rebours, C. (2020). Strategic considerations for establishing a large-scale seaweed industry based on fish feed application: A Norwegian case study. *Journal of Applied Phycology*, *32*(6), 4159–4169.

Emparan, Q., Harun, R., & Danquah, M. K. (2019). Role of phycoremediation for nutrient removal from wastewaters: A review. *Applied Ecology and Environmental Research*, *17*(1), 889–915. doi:10.15666/aeer/1701_889915

Endres, H.-J. (2017). Large Scale Production of Stem Cells. *Advances in Biochemical Engineering/Biotechnology*, *102*(118), 1–35. doi:10.1007/10

Engesmo, A., Eikrem, W., Seoane, S., Smith, K., Edvardsen, B., Hofgaard, A., & Tomas, C. R. (2016). New insights into the morphology and phylogeny of Heterosigma akashiwo (Raphidophyceae), with the description of Heterosigma minor sp. nov. *Phycologia*, *55*(3), 279–294. doi:10.2216/15-115.1

Enyidi, U. (2017). *Chlorella vulgaris* as protein source in the diets of African Catfish *Clarias gariepinus*. *Fishes*, *2*(4), 17. doi:10.3390/fishes2040017

Eom, S. H., Kim, Y. M., & Kim, S. K. (2012). Antimicrobial effect of phlorotannins from marine brown algae. *Food and Chemical Toxicology: An International Journal Published for the British Industrial Biological Research Association, 50*(9), 3251–3255.

Eppley, R. W. (1972). Temperature and phytoplankton growth in the sea. *Fish Bulletin, 70*, 1063–1085.

Eriksen, M., Lebreton, L. C. M., Carson, H. S., Thiel, M., Moore, C. J., Borerro, J. C., Galgani, F., Ryan, P. G., & Reisser, J. (2014). Plastic Pollution in the World's Oceans: More than 5 Trillion Plastic Pieces Weighing over 250,000 Tons Afloat at Sea. *PLoS One, 9*(12), e111913–e111913. doi:10.1371/journal.pone.0111913 PMID:25494041

Eriksen, N. T., Riisgård, F. K., Gunther, W. S., & Iversen, J. J. L. (2007). On-line estimation of O 2 production, CO 2 uptake, and growth kinetics of microalgal cultures in a gas-tight photobioreactor. *Journal of Applied Phycology, 19*(2), 161–174. doi:10.100710811-006-9122-y PMID:19396354

Eroglu, E., Chen, X., Bradshaw, M., Agarwal, V., Zou, J., Stewart, S. G., Duan, X., Lamb, R. N., Smith, S. M., Raston, C. L., & Iyer, K. S. (2013). Biogenic production of palladium nanocrystals using microalgae and their immobilization on chitosan nan- ofibers for catalytic applications. *RSC Advances, 3*(4), 1009–1012. doi:10.1039/C2RA22402J

Eroglu, E., & Melis, A. (2016). Microalgal hydrogen production research. *International Journal of Hydrogen Energy, 41*(30), 12772–12798. doi:10.1016/j.ijhydene.2016.05.115

Escapa, C., Coimbra, R. N., Paniagua, S., Garcıa, A. I., & Otero, M. (2016). Comparative assessment of diclofenac removal from water by different microalgae strains. *Algal Research, 18*, 127–134.

Escapa, C., Coimbra, R. N., Paniagua, S., García, A. I., & Otero, M. (2017). Paracetamol and salicylic acid removal from contaminated water by microalgae. *Journal of Environmental Management, 203*, 799–806.

Eskandarpour, M., Dejax, P., Miemczyk, J., & Péton, O. (2015). Sustainable supply chain network design: An optimization-oriented review. *Omega, 54*, 11–32. doi:10.1016/j.omega.2015.01.006

Essekri, A., Hsini, A., Naciri, Y., Laabd, M., Ajmal, Z., & El Ouardi, M. (2021). Novel Citric Acid-Functionalized Brown Algae with a High Removal Efficiency of crystal Violet Dye from Colored Wastewaters: Insights into Equilibrium, Adsorption Mechanism, and Reusability. *International Journal of Phytoremediation, 23*, 336–346.

Esserti, S., Smaili, A., Rifai, L. A., Koussa, T., Makroum, K., Belfaiza, M., Kabil, E. M., Faize, L., Burgos, L., Alburquerque, N., & .... (2017). Protective Effect of Three Brown Seaweed Extracts against Fungal and Bacterial Diseases of Tomato. *Journal of Applied Phycology, 29*, 1081–1093.

Eswaran, K., Mairh, O. P., & Subba Rao, P. V. (2002). Inhibition of pigments and phycocolloid in a marine red alga Gracilaria edulis by ultraviolet–B radiation. *Biologia.Plantarum. Prague, 45*(1), 157–159. doi:10.1023/A:1015158204181

Etman, S. M., Elnaggar, Y. S., & Abdallah, O. Y. (2020). Fucoidan, a natural biopolymer in cancer combating: From edible algae to nanocarrier tailoring. *International Journal of Biological Macromolecules, 147*, 799–808. doi:10.1016/j.ijbiomac.2019.11.191 PMID:31770552

European Bioplastics and Nova-Institute. (2019a). *Bioplastics Market Update*. Available online: https://www.european-bioplastics.org/wp-content/uploads/2019/11/Report_Bioplastics-Market-Data_2019_short_version.pdf

European Bioplastics and Nova-Institute. (2019b). *Report- Bioplastics Market Data 2018*. https://www.european-bioplastics.org/wp-content/uploads/2016/02/Report_Bioplastics-Market-Data_2018.pdf

European Bioplastics, Nova-Institute, (2019c). *Bioplastics facts and figures*. https://docs.european-bioplastics.org/publications/EUBP_Facts_and_fiures.pdf

European Bioplastics. (2017). Bioplastic Market Data 2017. *European Bioplastics*, 1–7

European Bioplastics. (2020a). *Certifications for bioplastics*. https://www.europeanbioplastics.org/bioplastics/standards/certification/

European Bioplastics. (2020b). *Bioplastics Market Data 2019*. Available online: https://docs.european-bioplastics.org/publications/market_data/Report_Bioplastics_Market_Data_2019.pdf

Evans, J. H. (1958). The survival of freshwater algae during dry periods: Part I. An investigation of the algae of five small ponds. *Journal of Ecology*, *46*(1), 149–167. doi:10.2307/2256910

Fabra, M. J., Martínez-Sanz, M., Gómez-Mascaraque, L. G., Coll-Marqués, J. M., Martínez, J. C., & López-Rubio, A. (2017). Development and characterization of hybrid corn starch-microalgae films: Effect of ultrasound pre-treatment on structural, barrier and mechanical performance. *Algal Research*, *28*, 80–87. doi:10.1016/j.algal.2017.10.010

Fabra, M. J., Martínez-Sanz, M., Gómez-Mascaraque, L. G., Gavara, R., & López-Rubio, A. (2018). Structural and physicochemical characterization of thermoplastic corn starch films containing microalgae. *Carbohydrate Polymers*, *186*, 184–191. doi:10.1016/j.carbpol.2018.01.039 PMID:29455977

Fábregas, J., Maseda, A., Domínguez, A., & Otero, A. (2004). The cell composition of Nannochloropsis sp. changes under different irradiances in semicontinuous culture. *World Journal of Microbiology & Biotechnology*, *20*(1), 31–35. doi:10.1023/B:WIBI.0000013288.67536.ed

Fabris, M., Abbriano, R. M., Pernice, M., Sutherland, D. L., Commault, A. S., Hall, C. C., Labeeuw, L., McCauley, J. I., Kuzhiuparambil, U., Ray, P., Kahlke, T., & Ralph, P. J. (2020). Emerging Technologies in Algal Biotechnology: Toward the Establishment of a Sustainable, Algae-Based Bioeconomy. *Frontiers in Plant Science*, *11*, 279. doi:10.3389/fpls.2020.00279 PMID:32256509

Facchi, D. P., Lima, A. C., de Oliveira, J. H., Lazarin-Bidóia, D., Nakamura, C. V., Canesin, E. A., Bonafé, E. G., Monteiro, J. P., Visentainer, J. V., Muniz, E. C., & … . (2017). Polyelectrolyte complexes based on alginate/tanfloc: Optimization, characterization and medical application. *International Journal of Biological Macromolecules*, *103*, 129–138.

Faeth, J. L., & Savage, P. E. (2016). Effects of processing conditions on biocrude yields from fast hydrothermal liquefaction of microalgae. *Bioresource Technology*, *206*, 290–293. doi:10.1016/j.biortech.2016.01.115 PMID:26879204

Fain, S. R., Druehl, L. D., & Baillie, D. L. (1988). Repeat and single copy sequences are differentially conserved in the evolution of kelp chloroplast DNA. *Journal of Phycology*, *24*(3), 292–302. doi:10.1111/j.1529-8817.1988.tb04469.x

Falasco, E., Ector, L., Isaia, M., Wetzel, C. E., Hoffmann, L., & Bona, F. (2014). Diatom flora in subterranean ecosystems: A review. *International Journal of Speleology*, *43*(3), 231–251. doi:10.5038/1827-806X.43.3.1

Falcón, L. I., Magallón, S., & Castillo, A. (2010). Dating the cyanobacterial ancestor of the chloroplast. *The ISME Journal*, *4*(6), 777–783. doi:10.1038/ismej.2010.2 PMID:20200567

Falkowska, M., Pietryczuk, A., Piotrowska, A., Bajguz, A., Grygoruk, A., & Czerpak, R. (2011). The effect of gibberellic acid (GA3) on growth, metal biosorption and metabolism of the green algae Chlorella vulgaris (Chlorophyceae) Beijerinck exposed to cad-mium and lead stress. *Polish Journal of Environmental Studies*, *20*, 53–59.

Falkowski, P. G., & Raven, J. A. (2013). *Aquatic photosynthesis*. Princeton University Press.

Fang, S., Yu, W., Li, C., Liu, Y., Qiu, J., & Kong, F. (2019). Adsorption behavior of three triazole fungicides on polystyrene microplastics. *The Science of the Total Environment*, *691*, 1119–1126. doi:10.1016/j.scitotenv.2019.07.176 PMID:31466193

Fan, H.-J., & Hall, M. B. (2001). A capable bridging ligand for Fe-only hydrogenase: Density functional calculations of a low-energy route for heterolytic cleavage and formation of dihydrogen. *Journal of the American Chemical Society, 123*(16), 3828–3829. doi:10.1021/ja004120i PMID:11457119

Fan, X., Guo, R., Yuan, X., Qiu, Y., Yang, Z., Wang, F., Sun, M., & Zhao, X. (2015). Biogas production from Macrocystis pyrifera biomass in seawater system. *Bioresource Technology, 197*, 339–347. doi:10.1016/j.biortech.2015.08.128 PMID:26344241

FAO. (2014). *The state of world fisheries and aquaculture*. FAO.

Farasat, M., Khavari-Nejad, R. A., Nabavi, S. M. B., & Namjooyan, F. (2013). Antioxidant properties of some filamentous green algae (Chaetomorpha Genus). *Brazilian Archives of Biology and Technology, 56*(6), 921–927. doi:10.1590/S1516-89132013000600005

Fard, J. K., Jafari, S., & Eghbal, M. A. (2015). A review of molecular mechanisms involved in toxicity of nanoparticles. *Advanced Pharmaceutical Bulletin, 5*(4), 447–454. doi:10.15171/apb.2015.061 PMID:26819915

Fargašová, A., Bumbálová, A., & Havránek, E. (1999). Ecotoxicological effects and uptake of metals (Cu+, Cu2+, Mn2+, Mo6+, Ni2+, V5+) in freshwater alga Scenedesmus quadricauda. *Chemosphere, 38*(5), 1165–1173. doi:10.1016/S0045-6535(98)00346-4

Farooq, W., Lee, Y. C., Han, J. I., Darpito, C. H., Choi, M., & Yang, J. W. (2013). Efficient microalgae harvesting by organo-building blocks of nanoclays. *Green Chemistry, 15*, 749–755.

Faruk, O., Bledzki, A. K., Fink, H. P., & Sain, M. (2014). Progress report on natural fiber reinforced composites. *Macromolecular Materials and Engineering, 299*(1), 9–26. doi:10.1002/mame.201300008

Fassett, R. G., & Combes, J. S. (2011). Astaxanthin: A potential therapeutic agentin cardiovascular disease. *Marine Drugs, 9*(3), 447–465. doi:10.3390/md9030447 PMID:21556169

Fast, S. A., Kokabian, B., & Gude, V. G. (2014). Chitosan enhanced coagulation of algal turbid waters–comparison between rapid mix and ultrasound coagulation methods. *Chemical Engineering Journal, 244*, 403–410.

Fatemeh, S., Hennige, S., Willoughby, N., Adeloye, A., & Gutierrez, T. (2021). Integrating micro-algaemicro-algae into wastewater treatment : A review. *Science of the Total Environment, 752*(September), 142168. doi:10.1016/j.scitotenv.2020.142168

Fatima, R., Priya, M., Indurthi, L., Radhakrishnan, V., & Sudhakaran, R. (2020). Biosynthesis of silver nanoparticles using red algae Portieria hornemannii and its antibacterial activity against fish pathogens. *Microbial Pathogenesis, 138*, 103780. doi:10.1016/j.micpath.2019.103780 PMID:31622663

Fawley, M. W. (1984). Effects of light intensity and temperature interactions on growth characteristics of Phaeodactylum tricornutum (Bacillariophyceae). *Journal of Phycology, 20*, 67–72.

Fayyaz, M., Chew, K. W., Show, P. L., Ling, T. C., Ng, I.-S., & Chang, J.-S. (2020a). Genetic engineering of microalgae for enhanced biorefinery capabilities. *Biotechnology Advances, 43*, 107554. doi:10.1016/j.biotechadv.2020.107554 PMID:32437732

Feil, A., & Pretz, T. (2020). Mechanical recycling of packaging waste. In *Plastic Waste and Recycling* (pp. 283–319). Elsevier., doi:10.1016/B978-0-12-817880-5.00011-6

Fei, T., Cazeneuve, S., Wen, Z., Wu, L., & Wang, T. (2016). Effective recovery of poly-β-hydroxybutyrate (PHB) biopolymer from Cupriavidus necator using a novel and environmentally friendly solvent system. *Biotechnology Progress, 32*(3), 678–685. doi:10.1002/btpr.2247 PMID:26871655

Feng, P. Z., Deng, Z., Hu, Z., & Fan, L. (2011). Lipid accumulation and growth of Chlorella zofingiensis in flat plate photobioreactors outdoors. *Bioresource Technology*, *102*(22), 10577–10584. doi:10.1016/j.biortech.2011.08.109 PMID:21955881

Fenster, J. A., & Eckert, C. A. (2021). High-Throughput Functional Genomics for Energy Production. *Current Opinion in Biotechnology*, *67*, 7–14. doi:10.1016/j.copbio.2020.09.010 PMID:33152605

Ferdous, U. T., & Yusof, Z. N. B. (2021). Algal terpenoids: a potential source of antioxidants for cancer therapy. In *Terpenes and Terpenoids*. IntechOpen. doi:10.5772/intechopen.94122

Ferdous, U. T., & Yusof, Z. N. B. (2021). Medicinal Prospects of Antioxidants From Algal Sources in Cancer Therapy. *Frontiers in Pharmacology*, *12*(March), 1–22. doi:10.3389/fphar.2021.593116 PMID:33746748

Fernándeza, I., Vinogradovb, S. N., & Arredondo-Petera, R. (2010). Identification and *in silico* characterization of a putative ancestor to land plant non-symbiotic hemoglobins from the prasinophyceae algae *Micromonas* and *Ostreococcus*. *Global Journal of Biochemistry*, *1*(1), 18–30.

Fernández-Dacosta, C., Posada, J. A., Kleerebezem, R., Cuellar, M. C., & Ramirez, A. (2015). Microbial community-based polyhydroxyalkanoates (PHAs) production from wastewater: Techno-economic analysis and ex-ante environmental assessment. *Bioresource Technology*, *185*, 368–377. doi:10.1016/j.biortech.2015.03.025 PMID:25796067

Fernández, P. A., Roleda, M. Y., & Hurd, C. L. (2015). Effects of ocean acidification on the photosynthetic performance, carbonic anhydrase activity and growth of the giant kelp Macrocystis pyrifera. *Photosynthesis Research*, *124*(3), 293–304. doi:10.100711120-015-0138-5 PMID:25869634

Ferrando, L., & Matamoros, V. (2020). Attenuation of nitrates, antibiotics and pesticides from groundwater using immobilised microalgae-based systems. *The Science of the Total Environment*, *703*, 134740.

Ferrari, S. G., Silva, P. G., González, D. M., Navoni, J. A., & Silva, H. J. (2013). Arsenic tolerance of cyanobacterial strains with potential use in biotechnology. *Revista Argentina de Microbiologia*, *45*(3), 174–179. doi:10.1016/S0325-7541(13)70021-X PMID:24165141

Fetouh, H. A., Ismail, A. M., Hamid, H. A., & Bashier, M. O. (2019). Synthesis of promising nanocomposites from an antitumer and biologically active heterocyclic compound uploaded by clay and chitosan polymers. *International Journal of Biological Macromolecules*, *137*, 1211–1220. doi:10.1016/j.ijbiomac.2019.06.187 PMID:31254576

Ficetola, G. F., Coissac, E., Zundel, S., Riaz, T., Shehzad, W., Bessiere, J., Taberlet, P., & Pompanon, F. (2010). An *in silico* approach for the evaluation of DNA barcodes. *BMC Genomics*, *11*(1), 434–443. doi:10.1186/1471-2164-11-434 PMID:20637073

Fiehn, O. (2002). Metabolomics — the link between genotypes and phenotypes. In C. Town (Ed.), *Functional genomics* (pp. 155–171). Springer Netherlands. doi:10.1007/978-94-010-0448-0_11

Field, C. B., Behrenfeld, M. J., Randerson, J. T., & Falkowski, P. (1998). Primary production of the biosphere: Integrating terrestrial and oceanic components. *Science*, *281*(5374), 237–240. doi:10.1126cience.281.5374.237 PMID:9657713

Figueroa, F. L., Salles, S., Aguilera, J., Jimenez, C., Mercado, J., Vin. Egla, B., Flores-Moya, A., & Altamirano, M. (1997). Effects of solar radiation on photoinhibition and pigmentation in the red alga Porphyra leucostica. *Marine Ecology Progress Series*, *151*, 81–90. doi:10.3354/meps151081

Fike, J. H., Allen, V. G., Schmidt, R. E., Zhang, X., Fontenot, J. P., Bagley, C. P., Ivy, R. L., Evans, R. R., Coelho, R. W., & Wester, D. B. (2001). Tasco-Forage: I. Influence of a seaweed extract on antioxidant activity in tall fescue and in ruminants. *Journal of Animal Science*, *79*(4), 1011–1021. doi:10.2527/2001.7941011x PMID:11325175

Filote, C., Roşca, M., & Hlihor, R. M. (2020). Overview of using living and non-living microorganisms for the removal of heavy metals from wastewaters. *Res. J. Agr. Sci., 52*(4).

Firmansyah, M. L., Yoshida, W., Hanada, T., & Goto, M. (2020). Application of ionic liquids in solvent extraction of platinum group metals. *Solvent Extraction Research and Development, Japan, 27*(1), 1–24. doi:10.15261erdj.27.1

Fischer, W. W., Hemp, J., & Johnson, J. E. (2016). Evolution of Oxygenic Photosynthesis. *Annual Review of Earth and Planetary Sciences, 44*(1), 647–683. doi:10.1146/annurev-earth-060313-054810

Flechtner, V. R. (2007). North American desert microbiotic soil crust communities. In *Algae and cyanobacteria in extreme environments* (pp. 537–551). Springer. doi:10.1007/978-1-4020-6112-7_29

Flynn, K. J., Blackford, J. C., Baird, M. E., Raven, J. A., Clark, D. R., Beardall, J., Brownlee, C., Fabian, H., & Wheeler, G. L. (2012). Changes in pH at the exterior surface of plankton with ocean acidification. *Nature Climate Change, 2*(7), 510–513. doi:10.1038/nclimate1489

Flynn, K. J., Kenny, P., & Mitra, A. (2017). Minimising losses to predation during microalgae cultivation. *Journal of Applied Phycology, 29*, 1829–1840.

Folino, A., Karageorgiou, A., Calabrò, P. S., & Komilis, D. (2020). Biodegradation of wasted bioplastics in natural and industrial environments: A Review. *Sustainability, 12*(15), 6030. doi:10.3390u12156030

Fomina, M., & Gadd, G. M. (2014). Biosorption: Current perspectives on concept, definition and application. *Bioresource Technology, 160*, 3–14.

Fortier, M. O. P., Roberts, G. W., Stagg-Williams, S. M., & Sturm, B. S. (2014). Life cycle assessment of bio-jet fuel from hydrothermal liquefaction of microalgae. *Applied Energy, 122*, 73–82. doi:10.1016/j.apenergy.2014.01.077

Fotopoulou, K. N., & Karapanagioti, H. K. (2017). Degradation of Various Plastics in the Environment. In *The Handbook of Environmental Chemistry* (pp. 71–92). Springer International Publishing. doi:10.1007/698_2017_11

Fouchard, S., Pruvost, J., Degrenne, B., Titica, M., & Legrand, J. (2009). Kinetic modeling of light limitation and sulfur deprivation effects in the induction of hydrogen production with Chlamydomonas reinhardtii: Part I. Model development and parameter identification. *Biotechnology and Bioengineering, 102*(1), 232–245. doi:10.1002/bit.22034 PMID:18688816

Foucher, A., Evrard, O., Ficetola, G. F., Gielly, L., Poulain, J., Giguet-Covex, C., Laceby, J. P., Salvador-Blanes, S., Cerdan, O., & Poulenard, J. (2020). Persistence of environmental DNA in cultivated soils: implication of this memory effect for reconstructing the dynamics of land use and cover changes. *Scientific Reports, 10*(1). doi:10.1038/s41598-020-67452-1

Francisco, J. A., Cerveny, C. G., Meyer, D. L., Mixan, B. J., Klussman, K., Chace, D. F., Rejniak, S. X., Gordon, K. A., DeBlanc, R., Toki, B. E., Law, C.-L., Doronina, S. O., Siegall, C. B., Senter, P. D., & Wahl, A. F. (2003). cAC10-vcMMAE, an anti-CD30–monomethyl auristatin E conjugate with potent and selective antitumor activity. *Blood, 102*(4), 1458–1465. doi:10.1182/blood-2003-01-0039 PMID:12714494

Franklin, L. A., & Forster, R. M. (1997). The changing irradiance environment: Consequences for marine macrophyte physiology, productivity and ecology. *European Journal of Phycology, 32*(3), 207–232. doi:10.1080/09670269710001737149

Freile-Pelegrín, Y., & Robledo, D. (2013). Bioactive Phenolic Compounds from Algae. *Bioactive Compounds from Marine Foods: Plant and Animal Sources*, (May), 113–129. doi:10.1002/9781118412893.ch6

Freitas, B. C. B., Cassuriaga, A. P. A., Morais, M. G., & Costa, J. A. V. (2017). Pentoses and light intensity increase the growth and carbohydrate production and alter the protein profile of *Chlorella minutissima. Bioresource Technology, 28*, 248–253. doi:10.1016/j.biortech.2017.04.031 PMID:28437642

Frenken, T., Alacid, E., Berger, S. A., Bourne, E. C., Gerphagnon, M., Grossart, H.-P., Gsell, A. S., Ibelings, B. W., & [illegible] (2017). Integrating chytrid fungal parasites into plankton ecology: Research gaps and needs. *Environmental Microbiology*, *19*, 3802–3822.

Frias, J. P. G. L., Sobral, P., & Ferreira, A. M. (2010). Organic pollutants in microplastics from two beaches of the Portuguese coast. *Marine Pollution Bulletin*, *60*(11), 1988–1992. doi:10.1016/j.marpolbul.2010.07.030 PMID:20800853

Friedl, T. (1997). The evolution of the green algae. In *Origins of Algae and their Plastids* (pp. 87–101). Springer. doi:10.1007/978-3-7091-6542-3_4

Friedl, T., & Rybalka, N. (2012). Systematics of the green algae: A brief introduction to the current status. *Progress in Botany*, *73*, 259–280. doi:10.1007/978-3-642-22746-2_10

Friedmann, I., Lipkin, Y., & Ocampo-Paus, R. (1967). Desert algae of the Negev (Israel). *Phycologia*, *6*(4), 185–200. doi:10.2216/i0031-8884-6-4-185.1

Fritsch, F. E., & Haines, F. M. (1923). The moisture-relations of terrestrial algae. II. The changes during exposure to drought and treatment with hypertonie solutions. *Annals of Botany*, *37*(148), 683–728. doi:10.1093/oxfordjournals.aob.a089874

Fritsen, C. H., & Priscu, J. C. (1998). Cyanobacterial assemblages in permanent ice covers on Antarctic lakes: Distribution, growth rate, and temperature response of photosynthesis. *Journal of Phycology*, *34*(4), 587–597. doi:10.1046/j.1529-8817.1998.340587.x

Froehlich, H. E., Afflerbach, J. C., Frazier, M., & Halpern, B. S. (2019). Blue Growth Potential to Mitigate Climate Change through Seaweed Offsetting. *Current Biology*, *29*(18), 1–7. doi:10.1016/j.cub.2019.07.041 PMID:31474532

Fröhlich-Nowoisky, J., Kampf, C. J., Weber, B., Huffman, J. A., Pöhlker, C., Andreae, M. O., Lang-Yona, N., Burrows, S. M., Gunthe, S. S., Elbert, W., Su, H., Hoor, P., Thines, E., Hoffmann, T., Després, V. R., & Pöschla, U. (2016). Bioaerosols in the earth system: Climate, health, and ecosystem interactions. *Atmospheric Research*, *182*(15), 346–376. doi:10.1016/j.atmosres.2016.07.018

Fuentes, J. L., Garbayo, I., Cuaresma, M., Montero, Z., González-del-Valle, M., & Vílchez, C. (2016). Impact of microalgae-bacteria interactions on the production of algal biomass and associated compounds. *Marine Drugs*, *14*(5), 100. doi:10.3390/md14050100 PMID:27213407

Fuentes-Tristan, S., Parra-Saldivar, R., & Iqbal, H. M. N. (2019). Bioinspired biomolecules: Mycosporine-like amino acids and scytonemin from Lyngbya sp. with UV-protection potentialities. *Journal of Photochemistry and Photobiology. B, Biology*.

Fuhs, G. W. (1961). Der mikrobielle Abbau von Kohlenwasserstoffen. *Archiv für Mikrobiologie*, *39*(4), 374–422. doi:10.1007/BF00411776 PMID:13702455

Fujita, R. M., Wheeler, P. A., & Edwards, R. L. (1989). Assessment of macroalgal nitrogen limitation in a seasonal upwelling region. *Marine Ecology Progress Series*, *53*, 53. doi:10.3354/meps053293

Fuller, S. W., & Mathieson, A. C. (1972). Ecological studies of economic red algae. IV. Variations of carrageenan concentration and properties in *Chondrus crispus* Stackhouse. *Journal of Experimental Marine Biology and Ecology*, *10*(1), 49–58. doi:10.1016/0022-0981(72)90092-5

Fumina, S., Koki, M., & Kyosuke, N. (2020). Identification, growth, and pigment content of a spontaneous green mutant of *Pyropia kinositae* (Bangiales, Rhodophyta). *Journal of Applied Phycology*, *32*(3), 1983–1994. doi:10.100710811-020-02085-5

Furbank, R. T., & Tester, M. (2011). Phenomics–technologies to relieve the phenotyping bottleneck. *Trends in Plant Science*, *16*(12), 635–644. doi:10.1016/j.tplants.2011.09.005 PMID:22074787

Fu, S.-Y., Feng, X.-Q., Lauke, B., & Mai, Y.-W. (2008). Effects of particle size, particle/matrix interface adhesion and particle loading on mechanical properties of particulate–polymer composites. *Composites. Part B, Engineering*, *39*(6), 933–961. doi:10.1016/j.compositesb.2008.01.002

Gachelin, M., Boutoute, M., Carrier, G., Talec, A., Pruvost, E., Guihéneuf, F., Bernard, O., & Sciandra, A. (2021). Enhancing PUFA-rich polar lipids in *Tisochrysis lutea* using adaptive laboratory evolution (ALE) with oscillating thermal stress. *Applied Microbiology and Biotechnology*, *105*(1), 301–312. doi:10.100700253-020-11000-4 PMID:33201276

Gadd, G. M., & White, C. (1993). Microbial treatment of metal pollution—A working biotechnology? *Trends in Biotechnology*, *11*(8), 353–359. doi:10.1016/0167-7799(93)90158-6 PMID:7764182

Galadari, S., Rahman, A., Pallichankandy, S., & Thayyullathil, F. (2017). Reactive oxygen species and cancer paradox: To promote or to suppress? *Free Radical Biology & Medicine*, *104*, 144–164. doi:10.1016/j.freeradbiomed.2017.01.004 PMID:28088622

Galasso, C., Gentile, A., Orefice, I., Ianora, A., Bruno, A., Noonan, D. M., Sansone, C., Albini, A., & Brunet, C. (2019). Microalgal derivatives as potential nutraceutical and food supplements for human health: A focus on cancer prevention and interception. *Nutrients*, *11*(6), 1226. doi:10.3390/nu11061226 PMID:31146462

Galhoum, A. A., Atia, A. A., Mahfouz, M. G., Abdel-Rehem, S. T., Gomaa, N. A., Vincent, T., & Guibal, E. (2015). Dy (III) recovery from dilute solutions using magnetic-chitosan nano-based particles grafted with amino acids. *Journal of Materials Science*, *50*(7), 2832–2848. doi:10.100710853-015-8845-z

Galhoum, A. A., Mafhouz, M. G., Abdel-Rehem, S. T., Gomaa, N. A., Atia, A. A., Vincent, T., & Guibal, E. (2015). Cysteine-functionalized chitosan magnetic nano-based particles for the recovery of light and heavy rare earth metals: Uptake kinetics and sorption isotherms. *Nanomaterials (Basel, Switzerland)*, *5*(1), 154–179. doi:10.3390/nano5010154 PMID:28347004

Galhoum, A. A., Mahfouz, M. G., Abdel-Rehem, S. T., Gomaa, N. A., Atia, A. A., Vincent, T., & Guibal, E. (2015). Diethylenetriamine-functionalized chitosan magnetic nano-based particles for the sorption of rare earth metal ions [Nd (III), Dy (III) and Yb (III)]. *Cellulose (London, England)*, *22*(4), 2589–2605. doi:10.100710570-015-0677-0

Gallego, R., Montero, L., Cifuentes, A., Ibáñez, E., & Herrero, M. (2018). Green Extraction of Bioactive Compounds from Microalgae. *Journal of Analysis and Testing*, *2*(2), 109–123. doi:10.100741664-018-0061-9

Gallegos, C. L., & Jordan, T. E. (2002). Impact of the spring 2000 phytoplankton bloom in Chesapeake Bay on optical properties and light penetration in the Rhode River, Maryland. *Estuaries*, *25*, 508–518. doi:10.1007/BF02804886

Galloway, T. S., Cole, M., & Lewis, C. (2017). Interactions of microplastic debris throughout the marine ecosystem. *Nature Ecology & Evolution*, *1*(5), 0116. Advance online publication. doi:10.103841559-017-0116 PMID:28812686

Ganesan, A. R., Tiwari, U., & Rajauria, G. (2019). Seaweed nutraceuticals and their therapeutic role in disease prevention. *Food Science and Human Wellness*, *8*(3), 252–263. doi:10.1016/j.fshw.2019.08.001

Gan, S.-Y., Lim, P.-E., & Phang, S.-M. (2016). Genetic and Metabolic Engineering of Microalgae. In F. Bux & Y. Chisti (Eds.), *Algae Biotechnology: Products and Processes* (pp. 317–344). Springer International Publishing. doi:10.1007/978-3-319-12334-9_16

Ganter, M., Bernard, T., Moretti, S., Stelling, J., & Pagni, M. (2013). MetaNetX.org: A website and repository for accessing, analysing and manipulating metabolic networks. *Bioinformatics (Oxford, England)*, *29*(6), 815–816. doi:10.1093/bioinformatics/btt036 PMID:23357920

Gao, B., Chen, A., Zhang, W., Li, A., & Zhang, C. (2017). Co-production of lipids, eicosapentaenoic acid, fucoxanthin, and chrysolaminarin by Phaeodactylum tricornutum cultured in a flat-plate photobioreactor under varying nitrogen conditions. *Journal of Ocean University of China*, *16*(5), 916–924. doi:10.100711802-017-3174-2

Gao, G., Gao, Q., Bao, M., Xu, J., & Li, X. (2019a). Nitrogen availability modulates the Gebser, B., & Pohnert, G. (2013).Synchronized regulation of different zwitterionic metabolites in theosmoadaption of phytoplankton. *Marine Drugs*, *11*, 2168–2182.

Gao, Q., Xie, J.-F., Shao, Y.-T., Chen, C., Han, B., Xia, K.-S., & Zhou, C.-G. (2017). Ultrafast and high-capacity adsorption of Gd(III) onto inorganic phosphorous acid modified mesoporous SBA-15. *Chemical Engineering Journal*, *313*, 197–206. doi:10.1016/j.cej.2016.12.068

García-Cubero, R., Moreno-Fernández, J., & García-González, M. (2017). Modelling growth and CO 2 fixation by *Scenedesmus vacuolatus* in continuous culture. *Algal Research*, *24*, 333–339. doi:10.1016/j.algal.2017.04.018

García, D., Posadas, E., Grajeda, C., Blanco, S., Martínez-Páramo, S., Acién, G., García-Encina, P., Bolado, S., & Muñoz, R. (2017). Comparative evaluation of piggery wastewater treatment in algal-bacterial photobioreactors under indoor and outdoor conditions. *Bioresource Technology*, *245*, 483–490. doi:10.1016/j.biortech.2017.08.135 PMID:28898848

García, G., Sosa-Hernández, J. E., Rodas-Zuluaga, L. I., Castillo-Zacarías, C., Iqbal, H., & Parra-Saldívar, R. (2021). Accumulation of PHA in the microalgae Scenedesmus sp. under nutrient-deficient conditions. *Polymers*, *13*(1), 131. doi:10.3390/polym13010131 PMID:33396913

Garcia-Rodríguez, A., Matamoros, V., Fontàs, C., & Salvadó, V. (2013). The influence of light exposure, water quality and vegetation on the removal of sulfonamides and tetracyclines: A laboratory-scale study. *Chemosphere*, *90*, 2297–2302.

García-Salgado, S., Quijano, M. A., & Bonilla, M. M. (2012). Arsenic speciation in edible alga samples by microwave-assisted extraction and high performance liquid chromatography coupled to atomic fluorescence spectrometry. *Analytica Chimica Acta*, *714*, 38–46. doi:10.1016/j.aca.2011.12.001 PMID:22244135

Garcia-Vaquero, M., Sweeney, T., O'Doherty, J., & Rajauria, G. (2021). Recent Advances in the Use of Greener Extraction Technologies for the Recovery of Valuable Bioactive Compounds from Algae. In *Recent Advances in Micro and Macroalgal Processing: Food and Health Perspectives* (pp. 96–122). Wiley Online Library. doi:10.1002/9781119542650.ch4

Garipov, I. T., Khaydarov, R. R., Gapurova, O. U., Efimova, I. L., & Evgrafova, S. Y. (2019). Silver nanoparticles as a new generation of antimicrobial prophylaxis. *J. Sib. Fed. Univ. Biol.*, *12*, 266–276. doi:10.17516/1997-1389-0301

Gärtner, G., Stoyneva-Gärtner, M., & Uzunov, B. (2021). Algal Toxic Compounds and Their Aeroterrestrial, Airborne and Other Extremophilic Producers with Attention to Soil and Plant Contamination: A Review. *Toxins*, *13*, 322.

Gasiunas, G., Barrangou, R., Horvath, P., & Siksnys, V. (2012). Cas9–crRNA ribonucleoprotein complex mediates specific DNA cleavage for adaptive immunity in bacteria. *Proceedings of the National Academy of Sciences of the United States of America*, *109*(39), E2579–E2586. doi:10.1073/pnas.1208507109 PMID:22949671

Gatamaneni, B. L., Orsat, V., & Lefsrud, M. (2018). Factors Affecting Growth of Various Microalgal Species. *Environmental Engineering Science*, *35*(10), 1037–1048. doi:10.1089/ees.2017.0521

Gautam, K., Pareek, A., & Sharma, D. K. (2015). Exploiting microalgae and macroalgae for production of biofuels and biosequestration of carbon dioxide—A review. *International Journal of Green Energy*, *12*(11), 1122–1143. doi:10.1080/15435075.2014.893239

Gautam, S., & Mannan, M. A.-u. (2020). The Role of Algae in Nutraceutical and Pharmaceutical Production. In *Bioactive Natural products in Drug Discovery* (pp. 665–685). Springer. doi:10.1007/978-981-15-1394-7_24

Gavelis Gregory, S., Wakeman Kevin, C., Tillmann, U., Ripken, C., Mitarai, S., Herranz, M., Özbek, S., Holstein, T., Keeling Patrick, J., & Leander Brian, S. (2017). Microbial arms race: Ballistic "nematocysts" in dinoflagellates represent a new extreme in organelle complexity. *Science Advances*, *3*(3), e1602552. doi:10.1126ciadv.1602552 PMID:28435864

Geddie, A. W., & Hall, S. G. (2020). Development of a suitability assessment model for the cultivation of intertidal macroalgae in the United States. *The Science of the Total Environment*, *699*, 134327. doi:10.1016/j.scitotenv.2019.134327 PMID:31522045

Geider, R., MacIntyre, H., & Kana, T. (1997). Dynamic model of phytoplankton growth and acclimation: responses of the balanced growth rate and the chlorophyll a: carbon ratio to light, nutrient-limitation and temperature. *Marine Ecology Progress Series*, *148*, 187–200. doi:10.3354/meps148187

Genç, M. S., Çelik, M., & Karasu, İ. (2012). A review on wind energy and wind–hydrogen production in Turkey: A case study of hydrogen production via electrolysis system supplied by wind energy conversion system in Central Anatolian Turkey. *Renewable & Sustainable Energy Reviews*, *16*(9), 6631–6646. doi:10.1016/j.rser.2012.08.011

Georgianna, D. R., & Mayfield, S. P. (2012). Exploiting diversity and synthetic biology for the production of algal biofuels. *Nature*, *488*(7411), 329–335. doi:10.1038/nature11479 PMID:22895338

Gerardo, M. L. (2015). Harvesting of Microalgae within a Biorefinery Approach: A Review of the Developments and Case Studies from Pilot-Plants. *Algal Research, 11*, 248–62.

Gerard, V. A. (1982). In situ water motion and nutrient uptake by the giant kelp Macrocystis pyrifera. *Marine Biology*, *69*(1), 51–54. doi:10.1007/BF00396960

Ge, S., Madill, M., & Champagne, P. (2018). Use of freshwater macroalgae Spirogyra sp. for the treatment of municipal wastewaters and biomass production for biofuel applications. *Biomass and Bioenergy*, *111*, 213–223. doi:10.1016/j.biombioe.2017.03.014

Gest, H., Engelmann, T., Kamen, M., Mcmillan, C., & Barnes, C. R. (2002). History of the word photosynthesis and evolution of its definition. In Photosynthesis Research (Vol. 73). Academic Press.

Geyer, R., Jambeck, J. R., & Law, K. L. (2017). Production, use, and fate of all plastics ever made. *Science Advances*, *3*(7), e1700782–e1700782. doi:10.1126ciadv.1700782 PMID:28776036

Ghaemi, M., & Gholamipour, S. (2017). Controllable synthesis and characterization of silver nanoparticles using Sargassum angostifolium. *Iran. J. Chem. Chem. Eng.*, *36*, 1–10.

Ghodake, G., & Lee, D. S. (2011). Biological Syn- thesis of Gold Nanoparticles Using the Aqueous Ex- tract of the Brown Algae Laminaria Japonica. *Journal of Nanoelectronics and Optoelectronics*, 268–271.

Ghoneim, M. M., El-desoky, H. S., El-moselhy, K. M., Amer, A., El-naga, E. H. A., Mohamedein, L. I., & Al-prol, A. E. (2014). Removal of cadmium from aqueous solution using marine green algae, Ulva lactuca. *Egyptian Journal of Aquatic Research*, *40*(3), 235–242. doi:10.1016/j.ejar.2014.08.005

Ghosh, P., Han, G., De, M., Kim, C. K., & Rotello, V. M. (2008). Gold nanoparticles in delivery applications. *Advanced Drug Delivery Reviews*, *60*(11), 1307–1315. doi:10.1016/j.addr.2008.03.016 PMID:18555555

Giese, E. C. (2020). Biosorption as green technology for the recovery and separation of rare earth elements. *World Journal of Microbiology & Biotechnology*, *36*(4), 1–11. doi:10.100711274-020-02821-6 PMID:32172357

Gifuni, I., Olivieri, G., Krauss, I. R., D'Errico, G., Pollio, A., & Marzocchella, A. (2017). Microalgae as new sources of starch: Isolation and characterization of microalgal starch granules. *Chemical Engineering Transactions*, *57*, 1423–1428.

Gifuni, I., Olivieri, G., Pollio, A., & Marzocchella, A. (2018). Identification of an industrial microalgal strain for starch production in biorefinery context: The effect of nitrogen and carbon concentration on starch accumulation. *New Biotechnology*, *41*, 46–54. doi:10.1016/j.nbt.2017.12.003 PMID:29237569

Gimpel, J. A., Henríquez, V., & Mayfield, S. P. (2015). In Metabolic Engineering of Eukaryotic Microalgae: Potential and Challenges Come with Great Diversity. *Frontiers in Microbiology*, *6*, 1376. doi:10.3389/fmicb.2015.01376 PMID:26696985

Giordano, M., Norici, A., Ratti, S., & Raven, J. A. (2008). Role of sulfur for algae: acquisition, metabolism, ecology and evolution. *Sulfur Metabolism in Phototrophic Organisms*, 397-415.

Girard, L. (2019). Quorum sensing in *Vibrio* spp.: The complexity of multiple signalling molecules in marine and aquatic environments. *Critical Reviews in Microbiology*, *45*(4), 451–471. doi:10.1080/1040841X.2019.1624499 PMID:31241379

Gnanavel, G., Valli, V. P. M. J., & Thirumarimurugan, M. (2012). A review of biodegradation of plastics waste. *International Journal of Pharmaceutical and Chemical Sciences*, *1*(3), 670-673.

Goecke, F., Jerez, C. G., Zachleder, V., Figueroa, F. L., Bišová, K., Řezanka, T., & Vítová, M. (2015). Use of lanthanides to alleviate the effects of metal ion-deficiency in Desmodesmus quadricauda (Sphaeropleales, Chlorophyta). *Frontiers in Microbiology*, *6*, 2. doi:10.3389/fmicb.2015.00002 PMID:25674079

Goecke, F., Zachleder, V., & Vítová, M. (2015). Rare earth elements and algae: physiological effects, biorefinery and recycling. In *Algal Biorefineries* (pp. 339–363). Springer. doi:10.1007/978-3-319-20200-6_10

Goff, L. J., & Coleman, A. W. (1988). The use of plastid DNA restriction endonuclease patterns in delineating red algal species and populations. *Journal of Phycology*, *24*(3), 357–368. doi:10.1111/j.1529-8817.1988.tb04478.x

Go, H., Hwang, H. J., & Nam, T. J. (2010). A glycoprotein from Laminaria japonica induces apoptosis in HT-29 colon cancer cells. *Toxicology In Vitro*, *24*(6), 1546–1553. doi:10.1016/j.tiv.2010.06.018 PMID:20615460

Goh, S. H., Alitheen, N. B. M., Yusoff, F. M., Yap, S. K., & Loh, S. P. (2014). Crude ethyl acetate extract of marine microalga, Chaetoceros calcitrans, induces Apoptosis in MDA-MB-231 breast cancer cells. *Pharmacognosy Magazine*, *10*(37), 1. doi:10.4103/0973-1296.126650 PMID:24696543

Gómez, E. F., & Michel, F. C. Jr. (2013). Biodegradability of conventional and bio-based plastics and natural fiber composites during composting, anaerobic digestion and long-term soil incubation. *Polymer Degradation & Stability*, *98*(12), 2583–2591. doi:10.1016/j.polymdegradstab.2013.09.018

Gómez-Fabre, P. M., De Pedro, E., Medina, M. A., De Castro, I. N., & Márquez, J. (1997). Polyamine contents of human breast cancer cells treated with the cytotoxic agents chlorpheniramine and dehydrodidemnin B. *Cancer Letters*, *113*(1-2), 141–144. doi:10.1016/S0304-3835(96)04591-0 PMID:9065813

Gonçalves, A. L. (2021). The use of microalgae and cyanobacteria in the improvement of agricultural practices: A review on their bio fertilising, biostimulating and biopesticide roles. *Applied Sciences (Basel, Switzerland)*, *11*(2), 871. doi:10.3390/app11020871

Goncalves, E. C., Wilkie, A. C., Kirst, M., & Rathinasabapathi, B. (2016). Metabolic regulation of triacylglycerol accumulation in the green algae: Identification of potential targets for engineering to improve oil yield. *Plant Biotechnology Journal*, *14*(8), 1649–1660. doi:10.1111/pbi.12523 PMID:26801206

Gong, J., & You, F. (2015). Value-added chemicals from microalgae: Greener, more economical, or both? *ACS Sustainable Chemistry & Engineering*, *3*(1), 82–96. doi:10.1021c500683w

Gong, M., & Bassi, A. (2016). Carotenoids from microalgae: A review of recent developments. *Biotechnology Advances*, *34*(8), 1396–1412. doi:10.1016/j.biotechadv.2016.10.005 PMID:27816618

González, A., Contreras, R. A., & Moenne, A. (2013). Oligo-carrageenans enhance growth and contents of cellulose, essential oils and polyphenolic compounds in *Eucalyptus globulus* trees. *Molecules (Basel, Switzerland)*, *18*(8), 8740–8751. doi:10.3390/molecules18088740 PMID:23887716

González, A., Moenne, F., Gómez, M., Sáez, C. A., Contreras, R. A., & Moenne, A. (2014). Oligo-carrageenan kappa increases NADPH, ascorbate and glutathione syntheses and TRR/TRX activities enhancing photosynthesis, basal metabolism, and growth in Eucalyptus trees. *Front Plant Sci*, *5*, 512. doi:10.3389/fpls.2014.00512 PMID:25352851

González-Ballesteros, N., Prado-López, S., Rodríguez-González, J. B., Lastra, M., & Rodríguez-Argüelles, M. C. (2017). Green synthesis of gold nanoparticles using brown algae Cystoseira baccata: Its activity in colon cancer cells. *Colloids and Surfaces. B, Biointerfaces*, *153*, 190–198. doi:10.1016/j.colsurfb.2017.02.020 PMID:28242372

Gonzalez-Fernandez, C., & Ballesteros, M. (2013). Microalgae Autoflocculation: An alternative to high-energy consuming harvesting methods. *Journal of Applied Phycology*, *25*, 991–999.

Gonzalez-Fernandez, C., Sialve, B., & Molinuevo-Salces, B. (2015). Anaerobic digestion of microalgal biomass: Challenges, opportunities and research needs. *Bioresource Technology*, *198*, 896–906. doi:10.1016/j.biortech.2015.09.095 PMID:26454349

González-Garcinuño, Á., Sánchez-Álvarez, J. M., Galán, M. A., & Del Valle, E. M. M. (2016). Understanding and optimizing the addition of phytohormones in the culture of microalgae for lipid production. *Biotechnology Progress*, *32*, 1203–1211.

González-González, L. M., Eltanahy, E., & Schenk, P. M. (2019). Assessing the fertilizing potential of microalgal digestates using the marine diatom *Chaetoceros muelleri*. *Algal Research*, *41*, 101534. doi:10.1016/j.algal.2019.101534

González-González, L. M., Zhou, L., Astals, S., Thomas-Hall, S. R., Eltanahy, E., Pratt, S., Jensen, P. D., & Schenk, P. M. (2018). Biogas production coupled to repeat microalgae cultivation using a closed nutrient loop. *Bioresource Technology*, *263*, 625–630. doi:10.1016/j.biortech.2018.05.039 PMID:29800924

González-Olalla, J. M., Medina-Sánchez, J. M., Lozano, I. L., Villar-Argaiz, M., & Carrillo, P. (2018). Climate-driven shifts in algal-bacterial interaction of high-mountain lakes in two years spanning a decade. *Scientific Reports*, *8*(1), 1–12. doi:10.103841598-018-28543-2 PMID:29980756

Goodstein, D. M., Shu, S., Howson, R., Neupane, R., Hayes, R. D., Fazo, J., Mitros, T., Dirks, W., Hellsten, U., Putnam, N., & Rokhsar, D. S. (2012). Phytozome: A comparative platform for green plant genomics. *Nucleic Acids Research*, *40*(D1), 1178–1186. doi:10.1093/nar/gkr944 PMID:22110026

Gordillo, F. J., Goutx, M., Figueroa, F. L., & Niell, F. X. (1998). Effects of light intensity, CO2 and nitrogen supply on lipid class composition of Dunaliella viridis. *Journal of Applied Phycology*, *10*(2), 135–144. doi:10.1023/A:1008067022973

Gordon, R., & Brawley, S. H. (2004). Effects of water motion on propagule release from algae with complex life histories. *Marine Biology*, *145*(1), 21–29. doi:10.100700227-004-1305-y

Goswami, R. K., Agrawal, K., Shah, M. P., & Verma, P. (2021). Bioremediation of heavy metals from wastewater: A current perspective on microalgae-based future. *Letters in Applied Microbiology*, 17.

Goswami, R. K., Mehariya, S., Obulisamy, P. K., & Verma, P. (2021). Advanced microalgae-based renewable biohydrogen production systems: A review. *Bioresource Technology*, *320*, 124301. doi:10.1016/j.biortech.2020.124301 PMID:33152683

Gould, S. B., Waller, R. F., & McFadden, G. I. (2008). Plastid evolution. *Annual Review of Plant Biology*, *59*(1), 491–517. doi:10.1146/annurev.arplant.59.032607.092915 PMID:18315522

Graça, C., Fradinho, P., Sousa, I., & Raymundo, A. (2018). Impact of Chlorella vulgaris on the rheology of wheat flour dough and bread texture. *Lebensmittel-Wissenschaft + Technologie*, *89*, 466–474. doi:10.1016/j.lwt.2017.11.024

Graham, J. E., Wilcox, L. W., & Graham, L. E. (2009). *Algae*. Pearson Education, University of Wisconsin.

Graham, L. E., Graham, J. M., Wilcox, L. W. W., & Cook, M. E. (2016). *Algae (Version* (3.3 ed.). LJLM Press.

Granados, M. R., Acién, F. G., Gómez, C., Fernández-Sevilla, J. M., & Grima, E. M. (2012). Evaluation of flocculants for the recovery of freshwater microalgae. *Bioresource Technology*, *118*, 102–110.

Gray, D. W., Lewis, L. A., & Cardon, Z. G. (2007). Photosynthetic recovery following desiccation of desert green algae (Chlorophyta) and their aquatic relatives. *Plant, Cell & Environment*, *30*(10), 1240–1255. doi:10.1111/j.1365-3040.2007.01704.x PMID:17727415

Grechanik, V., Naidov, I., Bolshakov, M., & Tsygankov, A. (2021). Photoautotrophic hydrogen production by nitrogen-deprived Chlamydomonas reinhardtii cultures. *International Journal of Hydrogen Energy*, *46*(5), 3565–3575. doi:10.1016/j.ijhydene.2020.10.215

Greenwell, H. C., Laurens, L. M. L., Shields, R. J., Lovitt, R. W., & Flynn, K. J. (2010). Placing microalgae on the biofuels priority list: A review of the technological challenges. *Journal of the Royal Society, Interface*, *7*, 703–726.

Griffiths, M. J., & Harrison, S. T. L. (2009). Lipid productivity as a key characteristic for choosing algal species for biodiesel production. *Journal of Applied Phycology*, *21*(5), 493–507. doi:10.100710811-008-9392-7

Grima, E. M., Acie, F. G., Medina, A. R., & Chisti, Y. (2003). Recovery of microalgal biomass and metabolites: Process options and economics. *Biotechnology Advances*, *20*, 491–515.

Grima, E. M., Camacho, F. G., Perez, J. S., Sevilla, J. F., Fernández, F. A., & Gomez, A. C. (1994). A mathematical model of microalgal growth in light-limited chemostat culture. *Journal of Chemical Technology & Biotechnology: International Research in Process. Environmental AND Clean Technology*, *61*(2), 167–173.

Grobbelaar, J. U. (2013). Mass production of microalgae at optimal photosynthetic rates. *Photosynthesis*, 357-371.

Gröniger, A., Sinha, R. P., Klisch, M., & Häder, D. P. (2000). Photoprotective compounds in cyanobacteria, phytoplankton and macroalgae—A database. *Journal of Photochemistry and Photobiology. B, Biology*, *58*(2-3), 115–122. doi:10.1016/S1011-1344(00)00112-3 PMID:11233638

Gross, E. M. (2003). Allelopathy of aquatic autotrophs. *Critical Reviews in Plant Sciences*, *22*(3-4), 313–339. doi:10.1080/713610859

Grosser, K., Zedler, L., Schmitt, M., Dietzek, B., Popp, J., & Pohnert, G. (2012). Disruptionfree imaging by Raman spectroscopy reveals a chemical sphere with antifouling metabolites around macroalgae. *Biofouling*, *28*(7), 687–696. doi:10.1080/08927014.2012.700306 PMID:22783962

Gross, W. (2000). Ecophysiology of algae living in highly acidic environments. *Hydrobiologia*, *433*(1), 31–37. doi:10.1023/A:1004054317446

Grube, M., Seckbach, J., & Muggia, L. (Eds.). (2017). Algal and Cyanobacteria Symbioses. World Scientific. doi:10.1142/q0017

Guan, Q., Savage, P. E., & Wei, C. (2012). Gasification of alga Nannochloropsis sp. in supercritical water. *The Journal of Supercritical Fluids*, *61*, 139–145. doi:10.1016/j.supflu.2011.09.007

Guardini, Z., Dall'Osto, L., Barera, S., Jaberi, M., Cazzaniga, S., Vitulo, N., & Bassi, R. (2021). High Carotenoid Mutants of *Chlorella vulgaris* Show Enhanced Biomass Yield under High Irradiance. *Plants*, *10*(5), 911. doi:10.3390/plants10050911 PMID:34062906

Guarín, J. R., Moreno-Pirajan, J. C., & Giraldo, L. (2018). Kinetic Study of the Bioadsorption of Methylene Blue on the Surface of the Biomass Obtained from the Algae D. antarctica. *Journal of Chemistry*, *2018*, 1–12.

Guarnieri, M. T., & Pienkos, P. T. (2015). Algal omics: Unlocking bioproduct diversity in algae cell factories. *Photosynthesis Research*, *123*(3), 255–263. doi:10.100711120-014-9989-4 PMID:24627032

Gudin, C., & Chaumont, D. (1991). Cell fragility, the key problem of microalgae mass production in closed photobioreactors. *Bioresource Technology*, *38*, 141–151.

Gudmundsson, S., & Nogales, J. (2015). Cyanobacteria as photosynthetic biocatalysts: A systems biology perspective. *Molecular BioSystems*, *11*(1), 60–70. doi:10.1039/C4MB00335G PMID:25382198

Guedes, A. C., & Malcata, F. X. (2012). Nutritional Value and Uses of Microalgae in Aquaculture. In Z. A. Muchlisin (Ed.), *Aquaculture* (pp. 59–78)., doi:10.5772/1516

Guex, N., & Peitsch, M. C. (1997). SWISS-MODEL and the Swiss-PdbViewer: An environment for comparative protein modeling. *Electrophoresis*, *18*(15), 2714–2723. doi:10.1002/elps.1150181505 PMID:9504803

Gui, M. M., Lee, K. T., & Bhatia, S. (2008). Feasibility of edible oil vs. non-edible oil vs. waste edible oil as biodiesel feedstock. *Energy*, *33*(11), 1646–1653. doi:10.1016/j.energy.2008.06.002

Guiry, M. D. (2012). How many species of algae are there? *Journal of Phycology*, *48*(5), 1057–1063. doi:10.1111/j.1529-8817.2012.01222.x PMID:27011267

Guler, B. A., Deniz, I., Demirel, Z., Oncel, S. S., & Imamoglu, E. (2019). Transition from start-up to scale-up for fucoxanthin production in flat plate photobioreactor. *Journal of Applied Phycology*, *31*(3), 1525–1533.

Gulnaz, O., & Dincer, S. (2019). Biodegradation of Bisphenol a by Chlorella vulgaris and Aeromonas Hydrophilia. *Journal of Applied Biological Sciences*, *3*(2), 79–84. https://www.jabsonline.org/index.php/jabs/article/view/162

Gül, Ü. D., Taştan, B. E., & Bayazıt, G. (2019). Assessment of Algal Biomasses Having Different Cell Structures for Biosorption Properties of Acid Red P-2BX Dye. *South African Journal of Botany*, *127*, 147–152.

Gumel, A. M., Annuar, M. S. M., & Chisti, Y. (2013). Recent Advances in the Production, Recovery and Applications of Polyhydroxyalkanoates. *Journal of Polymers and the Environment*, *21*(2), 580–605. doi:10.100710924-012-0527-1

Gunasundari, E., Senthil Kumar, P., Rajamohan, N., & Vellaichamy, P. (2020). Feasibility of Naphthol green-b Dye Adsorption Using Microalgae: Thermodynamic and Kinetic Analysis. *Dwt*, *192*, 358–370.

Günerken, E., D'Hondt, E., Eppink, M. H. M., Garcia-Gonzalez, L., Elst, K., & Wijffels, R. H. (2015). Cell Disruption for Microalgae Biorefineries. *Biotechnology Advances*, *33*(2), 243–260. doi:10.1016/j.biotechadv.2015.01.008 PMID:25656098

Guo, G., Cao, W., Sun, S., Zhao, Y., & Hu, C. (2017). Nutrient removal and biogas upgrading by integrating fungal–microalgal cultivation with anaerobically digested swine wastewater treatment. *Journal of Applied Phycology*, *29*(6), 2857–2866. doi:10.100710811-017-1207-2

Guo, W. Q., Zheng, H. S., Li, S., Du, J. S., Feng, X. C., Yin, R. L., Wu, Q. L., Ren, N. Q., & Chang, J. S. (2016). Removal of cephalosporin antibiotics 7-ACA from wastewater during the cultivation of lipid-accumulating microalgae. *Bioresource Technology*, *221*, 284–290.

Gupta, A. K., & Gupta, M. (2005). Synthesis and surface engineering of iron oxide nanoparticles for biomedical applications. *Biomaterials*, *26*(18), 3995–4021. doi:10.1016/j.biomaterials.2004.10.012 PMID:15626447

Gupta, N. (2019). DNA Extraction and Polymerase Chain Reaction. *Journal of Cytology / Indian Academy of Cytologists*, *36*(2), 116–117. doi:10.4103/JOC.JOC_110_18 PMID:30992648

Gupta, N. K., & Sengupta, A. (2017). Understanding the sorption behavior of trivalent lanthanides on amide functionalized multi walled carbon nanotubes. *Hydrometallurgy*, *171*, 8–15. doi:10.1016/j.hydromet.2017.03.016

Gupta, P. L., Lee, S. M., & Choi, H. J. (2016). Integration of microalgal cultivation system for wastewater remediation and sustainable biomass production. *World Journal of Microbiology & Biotechnology*, *32*, 139–149.

Gupta, P. L., Lee, S.-M., & Choi, H.-J. (2015). A mini review: Photobioreactors for large scale algal cultivation. *World Journal of Microbiology & Biotechnology*, *31*(9), 1409–1417.

Gupta, S. K., Kumar, M., Guldhe, A., Ansari, F. A., Rawat, I., Kanney, K., & Bux, F. (2014). Design and development of polyamine polymer for harvesting microalgae for biofuels production. *Energy Conversion and Management*, *85*, 537–544.

Gupta, S., & Abu-Ghannam, N. (2011). Bioactive potential and possible health effects of edible brown seaweeds. *Trends in Food Science & Technology*, *22*(6), 315–326. doi:10.1016/j.tifs.2011.03.011

Gupta, V., Ratha, S. K., Sood, A., Chaudhary, V., & Prasanna, R. (2013). New insights into the biodiversity and applications of cyanobacteria (blue-green algae)—Prospects and challenges. *Algal Research*, *2*(2), 79–97. doi:10.1016/j.algal.2013.01.006

Güroy, D., Güroy, B., & Merrifield, D. L. (2011). Effect of Dietary Ulva and Spirulina on Weight Loss and Body Composition of Rainbow Trout, Oncorhynchus mykiss (Walbaum), During a Starvation Period. Journal of Animal Physiology and Nutrition, 320–327.

Güroy, D., Güroy, B., Merrifield, D. L., Ergün, S., Tekinay, A. A., & Yiğit, M. (2011). Effect of Dietary Ulva and Spirulina on Weight Loss and Body Composition of Rainbow Trout, Oncorhynchus mykiss (Walbaum), During a Starvation Period. *Journal of Animal Physiology and Nutrition*, *95*(3), 320–327. doi:10.1111/j.1439-0396.2010.01057.x PMID:20880289

Gutman, J., Zarka, A., & Boussiba, S. (2009). The host-range of *Paraphysoderma sedebokerensis*, a chytrid that infects *Haematococcus pluvialis. European Journal of Phycology*, *44*, 509–514.

Gutow, L., Rahman, M. M., Bartl, K., Saborowski, R., Bartsch, I., & Wiencke, C. (2014). Ocean acidification affects growth but not nutritional quality of the seaweed *Fucus vesiculosus* (Phaeophyceae, Fucales). *Journal of Experimental Marine Biology and Ecology*, *453*, 84–90. doi:10.1016/j.jembe.2014.01.005

Gwala, M., Dutta, S., & Ghosh Chaudhuri, R. (2021). Microalgae Mediated Nanomaterials Synthesis. In S. K. Mandotra, A. K. Upadhyay, & A. S. Ahluwalia (Eds.), *Algae: Multifarious Applications for a Sustainable World* (pp. 295–324). Springer Singapore. doi:10.1007/978-981-15-7518-1_13

Haberkorn, I., Buchmann, L., Häusermann, I., & Mathys, A. (2021). Nanosecond Pulsed Electric Field Processing of Microalgae Based Biorefineries Governs Growth Promotion or Selective Inactivation Based on Underlying Microbial Ecosystems. *Bioresource Technology*, *319*, 124173. doi:10.1016/j.biortech.2020.124173 PMID:33017777

Haddadi, M. H., Asadolahi, R., & Negahdari, B. (2019). The bioextraction of bioplastics with focus on polyhydroxybutyrate: A review. *International Journal of Environmental Science and Technology*, *16*(7), 3935–3948. doi:10.100713762-019-02352-0

Haddock, S. H. D., Moline, M. A., & Case, J. F. (2009). Bioluminescence in the Sea. *Annual Review of Marine Science*, *2*(1), 443–493. doi:10.1146/annurev-marine-120308-081028 PMID:21141672

Hader, D. P., Kumar, H. D., Smith, R. C., & Worrest, R. C. (1998). Effects on aquatic ecosystems. *Journal of Photochemistry and Photobiology. B, Biology*, *46*(1–3), 53–68. doi:10.1016/S1011-1344(98)00185-7

Ha, G., El-dalatony, M. M., Kurade, M. B., Salama, E., Basak, B., Kang, D., Roh, H., Lim, H., & Jeon, B. (2020). Energy-efficient pretreatments for the enhanced conversion of microalgal biomass to biofuels * Corresponding Author. *Bioresource Technology*, *123333*, 123333. Advance online publication. doi:10.1016/j.biortech.2020.123333 PMID:32305842

Halim, R. (2020). Industrial Extraction of Microalgal Pigments. In *Pigments from Microalgae Handbook* (pp. 265–308). Springer. doi:10.1007/978-3-030-50971-2_12

Halim, R., Gladman, B., Danquah, M. K., & Webley, P. A. (2011). Oil extraction from microalgae for biodiesel production. *Bioresource Technology*, *102*(1), 178–185. doi:10.1016/j.biortech.2010.06.136 PMID:20655746

Hall, D. (2021, March 8). *A Cancer Fighter from the Sea*. Retrieved August 30, 2021, from Si.edu website: https://ocean.si.edu/ocean-life/plants-algae/cancer-fighter-sea

Hallenbeck, P. C. (2002). Biological hydrogen production; fundamentals and limiting processes. *International Journal of Hydrogen Energy*, *27*(11–12), 1185–1193. doi:10.1016/S0360-3199(02)00131-3

Halligan, D. L., & Keightley, P. D. (2009). Spontaneous Mutation Accumulation Studies in Evolutionary Genetics. *Annual Review of Ecology, Evolution, and Systematics*, *40*(1), 151–172. doi:10.1146/annurev.ecolsys.39.110707.173437

Hamed, S. M. (2007). *Studies on nitrogen fixing Cyanobacteria* (M. Sc. Thesis). Botany Department, Faculty of Science, Beni-Suef University, Beni-Suef, Egypt.

Hamed, S. M., Abd El-Rhman, A. A., Abdel-Raouf, N., & Ibraheem, I. B. (2018). Role of marine macroalgae in plant protection & improvement for sustainable agriculture technology. *Beni-Suef University Journal of Basic and Applied Sciences*, *7*(1), 104–110. doi:10.1016/j.bjbas.2017.08.002

Hamida, R. S., Ali, M. A., Redhwan, A., & Bin-Meferij, M. M. (2020). Cyanobacteria – a promising platform in green nanotechnology: A review on nanoparticles fabrication and their prospective applications. *International Journal of Nanomedicine*, *15*, 6033–6066. doi:10.2147/IJN.S256134 PMID:32884261

Hamidi, M., Safarzadeh Kozani, P., Safarzadeh Kozani, P., Pierre, G., Michaud, P., & Delattre, C. (2020). Marine bacteria versus microalgae: Who is the best for biotechnological production of bioactive compounds with antioxidant properties and other biological applications? *Marine Drugs*, *18*(1), 1–38. doi:10.3390/md18010028 PMID:31905716

Hamilton, M. L., Powers, S., Napier, J. A., & Sayanova, O. (2016). Heterotrophic production of omega-3 long-chain polyunsaturated fatty acids by trophically converted marine diatom Phaeodactylum tricornutum. *Marine Drugs*, *14*(3), 53. doi:10.3390/md14030053 PMID:27005636

Handbook of Microalgal Culture. (2006). *Handbook of Microalgal Culture.*

Handbook, I. L. C. D. (2010). *International Reference Life Cycle Data System. General guide for Life Cycle Assessment. Detailed guidance. European Commission, Joint Research Centre*. Institute for Environment and Sustainability.

Hanelt, D., Melchersmann, B., Wiencke, C., & Nultsch, W. (1997a). Effects of high light stress on photosynthesis of polar macroalgae in relation to depth distribution. *Marine Ecology Progress Series*, *149*, 255–266. doi:10.3354/meps149255

Han, G. Z. (2016). Evolution of jasmonate biosynthesis and signaling mechanisms. *Journal of Experimental Botany*, *68*(6), 1323–1331. doi:10.1093/jxb/erw470 PMID:28007954

Hankins, S. D., & Hockey, H. P. (1990). The Effect of a Liquid Seaweed Extract from Ascophyllum nodosum (Fucales, Phaeophyta) on the Two-Spotted Red Spider Mite Tetranychus Urticae. *Hydrobiologia*, *204*, 555–559.

Hanschen, E. R., & Starkenburg, S. R. (2020). The state of algal genome quality and diversity. *Algal Research*, *50*, 101968. doi:10.1016/j.algal.2020.101968

Han, T. (1996a). Effect of ultraviolet-B radiation on Ulva perusa Kjellman, Chlorophyta. I. Growth and pigment content. *Algae. Kor. J. Phycol.*, *11*, 155–159.

Han, X., Zeng, H., Bartocci, P., Fantozzi, F., & Yan, Y. (2018). Phytohormones and Effects on Growth and Metabolites of Microalgae: A Review. *Fermentation*, *4*, 25.

Hao, X., Luo, L., Jouhet, J., Rébeillé, F., Maréchal, E., Hu, H., Pan, Y., Tan, X., Chen, Z., You, L., Chen, H., Wei, F., & Gong, Y. (2018). Enhanced triacylglycerol production in the diatom Phaeodactylum tricornutum by inactivation of a Hotdog-fold thioesterase gene using TALEN-based targeted mutagenesis. *Biotechnology for Biofuels*, *11*(1), 1–18. doi:10.118613068-018-1309-3 PMID:30455741

Happe, T., & Naber, J. D. (1993). Isolation, characterization and N-terminal amino acid sequence of hydrogenase from the green alga Chlamydomonas reinhardtii. *European Journal of Biochemistry*, *214*(2), 475–481. doi:10.1111/j.1432-1033.1993.tb17944.x PMID:8513797

Haq, S. H., Al-Ruwaished, G., Al-Mutlaq, M. A., Naji, S. A., Al-Mogren, M., Al-Rashed, S., Ain, Q. T., Al-Amro, A. A., & Al-Mussallam, A. (2019). Antioxidant, anticancer activity, and phytochemical analysis of green algae, Chaetomorpha collected from the Arabian Gulf. *Scientific Reports*, *9*(1), 1–7. doi:10.103841598-019-55309-1 PMID:31827196

Harith, Z. T., Yusoff, F. M., Mohammed, M. S., Shariff, M., & Din, M. (2009). Effect of different flocculants on the flocculation performance of microalgae, Chaetoceros calcitrans, cells. *African Journal of Biotechnology*, *8*, 5971–5978.

Harley, C. D. G., Anderson, K. M., Demes, K. W., Jorve, J. P., Kordas, R. L., Coyle, T. A., & Graham, M. H. (2012). Effects of climate change on global seaweed communities. *Journal of Phycology*, *48*(5), 1064–1078. doi:10.1111/j.1529-8817.2012.01224.x PMID:27011268

Harmaen, A. S., Khalina, A., Azowa, I., Hassan, M. A., Tarmian, A., & Jawaid, M. (2014). Thermal and biodegradation properties of poly(lactic acid)/fertilizer/oil palm fibers blends biocomposites. *Polymer Composites*, *36*(3), 576–583. doi:10.1002/pc.22974

Harmsen, P. F. H., Hackmann, M. M., & Bos, H. L. (2014). Green building blocks for bio-based plastics. *Biofuels, Bioproducts & Biorefining*, *8*(3), 306–324. doi:10.1002/bbb.1468

Harnedy, P. A., & FitzGerald, R. J. (2011). Bioactive proteins, peptides, and amino acids from macroalgae. *Journal of Phycology*, *47*(2), 218–232. doi:10.1111/j.1529-8817.2011.00969.x PMID:27021854

Harun, R., Yip, J. W. S., Thiruvenkadam, S., Ghani, W. A. W. A. K., Cherrington, T., & Danquah, M. K. (2014). Algal biomass conversion to bioethanol – a step-by-step assessment. *Biotechnology Journal*, *9*(1), 73–86. doi:10.1002/biot.201200353 PMID:24227697

Haslam, R. P., Hamilton, M. L., Economou, C. K., Smith, R., Hassall, K. L., Napier, J. A., & Sayanova, O. (2020). Overexpression of an endogenous type 2 diacylglycerol acyltransferase in the marine diatom Phaeodactylum tricornutum enhances lipid production and omega-3 long-chain polyunsaturated fatty acid content. *Biotechnology for Biofuels*, *13*(1), 1–17. doi:10.118613068-020-01726-8 PMID:32467729

Hassan, S. M., Ashour, M., Sakai, N., Zhang, L., Hassanien, H. A., Gaber, A., & Ammar, G. (2021). Impact of Seaweed Liquid Extract Biostimulant on Growth, Yield, and Chemical Composition of Cucumber (Cucumis sativus). *Agriculture*, *11*(320), 1–16. doi:10.3390/agriculture11040320

Hauschild, M. Z., & Bjørn, A. (2018). LCA cookbook. In *Life Cycle Assessment* (pp. 963–1048). Springer. doi:10.1007/978-3-319-56475-3_37

Heasman, M., Diemar, J., O'Connor, W., Sushames, T., & Foulkes, L. (2000). Development of extended shelf-life microalgae concentrate diets harvested by centrifugation for bivalve molluscs-a summary. *Aquaculture Research*, *31*, 637–659.

He, D., Wu, S., Yan, L., Zuo, J., Cheng, Y., Wang, H., Liu, J., Zhang, X., Wu, M., Choi, J., & Tong, H. (2019). Antitumor bioactivity of porphyran extracted from Pyropia yezoensis Chonsoo2 on human cancer cell lines. *Journal of the Science of Food and Agriculture*, *99*(15), 6722–6730. doi:10.1002/jsfa.9954 PMID:31350864

Hehenberger, E., Gast, R. J., & Keeling, P. J. (2019). A kleptoplastidic dinoflagellate and the tipping point between transient and fully integrated plastid endosymbiosis. *Proceedings of the National Academy of Sciences of the United States of America*, *116*(36), 17934–17942. doi:10.1073/pnas.1910121116 PMID:31427512

Heilmann, S. M., Davis, H. T., Jader, L. R., Lefebvre, P. A., Sadowsky, M. J., Schendel, F. J., von Keitz, M. G., & Valentas, K. J. (2010). Hydrothermal carbonization of microalgae. *Biomass and Bioenergy*, *34*(6), 875–882. doi:10.1016/j.biombioe.2010.01.032

Heirendt, L., Arreckx, S., Pfau, T., Mendoza, S. N., Richelle, A., Heinken, A., Haraldsdóttir, H. S., Wachowiak, J., Keating, S. M., Vlasov, V., Magnusdóttir, S., Ng, C. Y., Preciat, G., Žagare, A., Chan, S. H. J., Aurich, M. K., Clancy, C. M., Modamio, J., Sauls, J. T., ... Fleming, R. M. T. (2019). Creation and analysis of biochemical constraint-based models using the COBRA Toolbox v.3.0. *Nature Protocols*, *14*(3), 639–702. doi:10.103841596-018-0098-2 PMID:30787451

Heller, S. R., McNaught, A., Pletnev, I., Stein, S., & Tchekhovskoi, D. (2015). InChI, the IUPAC international chemical identifier. *Journal of Cheminformatics*, *7*(1), 23–56. doi:10.118613321-015-0068-4 PMID:26136848

Hemaiswarya, S., Raja, R., Kumar, R., & Anbazhagan, C. (2011). Microalgae: A sustainable feed source for aquaculture. *World Journal of Microbiology & Biotechnology*, *27*(8), 1737–1746. doi:10.100711274-010-0632-z

Hemalatha, M., Sravan, J. S., Min, B., & Venkata Mohan, S. (2019). Microalgae-biorefinery with cascading resource recovery design associated to dairy wastewater treatment. *Bioresource Technology*, *284*, 424–429. doi:10.1016/j.biortech.2019.03.106 PMID:30978645

Hema, R., Senthil-Kumar, M., Shivakumar, S., Chandrasekhara Reddy, P., & Udayakumar, M. (2007). Chlamydomonas reinhardtii, a model system for functional validation of abiotic stress responsive genes. *Planta*, *226*(3), 655–670. doi:10.100700425-007-0514-2 PMID:17431668

Hempel, F., Bozarth, A. S., Lindenkamp, N., Klingl, A., Zauner, S., Linne, U., Steinbüchel, A., & Maier, U. G. (2011). Microalgae as bioreactors for bioplastic production. *Microbial Cell Factories*, *10*(1), 1–6. doi:10.1186/1475-2859-10-81 PMID:22004563

Hempel, F., & Maier, U. G. (2016). Microalgae as Solar-Powered Protein Factories. In *Advanced Technologies for Protein Complex Production and Characterization* (pp. 241–262). Springer International Publishing. doi:10.1007/978-3-319-27216-0_16

Henard, C. A., Franklin, T. G., Youhenna, B., But, S., Alexander, D., Kalyuzhnaya, M. G., & Guarnieri, M. T. (2018). Biogas Biocatalysis: Methanotrophic Bacterial Cultivation, Metabolite Profiling, and Bioconversion to Lactic Acid. *Frontiers in Microbiology*, *9*, 2610. Advance online publication. doi:10.3389/fmicb.2018.02610 PMID:30429839

Hentati, F., Tounsi, L., Djomdi, D., Pierre, G., Delattre, C., Ursu, A. V., Fendri, I., Abdelkafi, S., & Michaud, P. (2020). Bioactive polysaccharides from seaweeds. *Molecules (Basel, Switzerland)*, *25*(14), 1–29. doi:10.3390/molecules25143152 PMID:32660153

Heo, S.-J., Park, E.-J., Lee, K.-W., & Jeon, Y.-J. (2005). Antioxidant activities of enzymatic extracts from brown seaweeds. *Bioresource Technology*, *96*(14), 1613–1623. doi:10.1016/j.biortech.2004.07.013 PMID:15978995

He, P. J., Mao, B., Lü, F., Shao, L. M., Lee, D. J., & Chang, J. S. (2013). The combined effect of bacteria and Chlorella vulgaris on the treatment of municipal wastewaters. *Bioresource Technology*, *146*, 562–568.

Hernandez-Carmona, G., Freile-Pelegrín, Y., & Hernández-Garibay, E. (2013). Conventional and alternative technologies for the extraction of algal polysaccharides. In *Functional ingredients from algae for foods and nutraceuticals* (pp. 475–516). Elsevier. doi:10.1533/9780857098689.3.475

Hernández, D., Riaño, B., Coca, M., & García-González, M. C. (2015). Saccharification of carbohydrates in microalgal biomass by physical, chemical and enzymatic pretreatments as a previous step for bioethanol production. *Chemical Engineering Journal*, *262*, 939–945. doi:10.1016/j.cej.2014.10.049

Hernández-Herrera, R. M., Santacruz-Ruvalcaba, F., Zañudo-Hernández, J., & Hernández-Carmona, G. (2016). Activity of seaweed extracts and polysaccharide-enriched extracts from Ulva lactuca and Padina gymnospora as growth promoters of tomato and mung bean plants. *Journal of Applied Phycology*, *28*(4), 2549–2560. doi:10.100710811-015-0781-4

Herrera-Valencia, V. A., Macario-Gonza'lez, L. A., Casais-Molina, M. L., Beltran-Aguilar, A. G., & Peraza-Echeverrı'a, S. (2012). *In Silico* cloning and characterization of the Glycerol-3-Phosphate Dehydrogenase (GPDH) gene family in the green microalga *Chlamydomonas reinhardtii*. *Current Microbiology*, *64*(5), 477–485. doi:10.100700284-012-0095-6 PMID:22358185

Hessami, M. J., Phang, S. M., Sohrabipoor, J., Zafar, F. F., & Aslanzadeh, S. (2019). The bio-methane potential of whole plant and solid residues of two species of red seaweeds: Gracilaria manilaensis and Gracilariopsis persica. *Algal Research*, *42*, 101581. doi:10.1016/j.algal.2019.101581

Hidayati, J. R., Yudiati, E., Pringgenies, D., Oktaviyanti, D. T., & Kusuma, A. P. (2020). Comparative Study on Antioxidant Activities, Total Phenolic Compound and Pigment Contents of Tropical Spirulina platensis, Gracilaria arcuata and Ulva lactuca Extracted in Different Solvents Polarity. *E3S Web of Conferences, 147*. doi:10.1051/e3sconf/202014703012

Hien, N. Q., Nagasawa, N., Tham, L. X., Yoshii, F., Dang, V. H., Mitomo, H., Makuuchi, K., & Kume, T. (2000). Growth promotion of plants with depolymerized alginates by irradiation. *Radiation Physics and Chemistry*, *59*(1), 97–101. doi:10.1016/S0969-806X(99)00522-8

Higgins, B. T., & VanderGheynst, J. S. (2014). Effects of *Escherichia coli* on Mixotrophic Growth of Chlorella minutissima and Production of Biofuel Precursors. *PLoS One*, *9*(5), e96807. doi:10.1371/journal.pone.0096807 PMID:24805253

Hilbe, J. M. (2007). STATISTICA 7: An overview. *The American Statistician*, *61*(1), 91–94. doi:10.1198/000313007X172998

Hildebrand, M., Abbriano, R. M., Polle, J. E., Traller, J. C., Trentacoste, E. M., Smith, S. R., & Davis, A. K. (2013). Metabolic and cellular organization in evolutionarily diverse microalgae as related to biofuels production. *Current Opinion in Chemical Biology*, *17*(3), 506–514. doi:10.1016/j.cbpa.2013.02.027 PMID:23538202

Hill, W. (1996). Effects of light. In Algal ecology: freshwater benthic ecosystems, Aquatic Ecology Series. Academic Press. https://doi.org/10.1016/B978-012668450-6/50034-5.

Hill, E. A., Chrisler, W. B., Beliaev, A. S., & Bernstein, H. C. (2017). A flexible microbial co-culture platform for simultaneous utilization of methane and carbon dioxide from gas feedstocks. *Bioresource Technology*, *228*, 250–256. doi:10.1016/j.biortech.2016.12.111 PMID:28092828

Hiremath, S., & Mathad, P. (2010). Impact of Salinity on the Physiological and Biochemical Traits of Chlorella vulgaris Beijerinck. *Journal of Algal Biomass Utilization*, *1*(2), 51–59.

Hirooka, T., Nagase, H., Uchida, K., Hiroshige, Y., Ehara, Y., Nishikawa, J., Nishihara, T., Miyamoto, K., & Hirata, Z. (2005). Biodegradation of bisphenol a and disappearance of its estrogenic activity by the green alga chlorella fusca var. Vacuolata. *Environmental Toxicology and Chemistry*, *24*(8), 1896. doi:10.1897/04-259R.1 PMID:16152959

Hisada, M., & Kawase, Y. (2018). Recovery of rare-earth metal neodymium from aqueous solutions by poly-γ-glutamic acid and its sodium salt as biosorbents: Effects of solution pH on neodymium recovery mechanisms. *Journal of Rare Earths*, *36*(5), 528–536. doi:10.1016/j.jre.2018.01.001

Hjuler, S. V., & Hansen, S. B. (2018). LCA of Biofuels and Biomaterials. In *Life cycle assessment* (pp. 755–782). Springer. doi:10.1007/978-3-319-56475-3_30

Hlavova, M., Turoczy, Z., & Bisova, K. (2015). Improving microalgae for biotechnology-From genetics to synthetic biology. *Biotechnology Advances*, *33*(6), 1194–1203. doi:10.1016/j.biotechadv.2015.01.009 PMID:25656099

Hmelkov, A. B., Zvyagintseva, T. N., Shevchenko, N. M., Rasin, A. B., & Ermakova, S. P. (2018). Ultrasound-assisted extraction of polysaccharides from brown alga Fucus evanescens. Structure and biological activity of the new fucoidan fractions. *Journal of Applied Phycology*, *30*(3), 2039–2046. doi:10.100710811-017-1342-9

Hoffmann, L. (1989). Algae of terrestrial habitats. *Botanical Review*, *55*(2), 77–105. doi:10.1007/BF02858529

Hofmann, L. C., Straub, S., & Bischof, K. (2013). Elevated CO2 levels affect the activity of nitrate reductase and carbonic anhydrase in the calcifying rhodophyte *Corallina officinalis*. *Journal of Experimental Botany*, *64*(4), 899–908. doi:10.1093/jxb/ers369 PMID:23314813

Höger, A., Griehl, C., & Noll, M. (2021). Infection with intracellular parasite *Amoeboaphelidium protococcarum* induces shifts in associated bacterial communities in microalgae cultures. *Journal of Applied Phycology*, *33*, 2863–2873. https://doi.org/10.1007/s10811-021-02542-9

Hoham, R. W., & Remias, D. (2020). Snow and Glacial Algae: A Review1. *Journal of Phycology*, *56*(2), 264–282. doi:10.1111/jpy.12952 PMID:31825096

Holdt, S. L., & Kraan, S. (2011). Bioactive compounds in seaweed: Functional food applications and legislation. *Journal of Applied Phycology*, *23*(3), 543–597. doi:10.100710811-010-9632-5

Holzinger, A., & Karsten, U. (2013). Desiccation stress and tolerance in green algae: Consequences for ultrastructure, physiological and molecular mechanisms. *Frontiers in Plant Science*, *4*, 327. doi:10.3389/fpls.2013.00327 PMID:23986769

Hom-Diaz, A., Llorca, M., Rodriguez-Mozaz, S., Vicent, T., Barcelo, D., & Blanquez, P. (2015). Microalgae cultivation on wastewater digestate: B-estradiol and 17α-ethynylestradiol degradation and transformation products identification. *Journal of Environmental Management*, *155*, 106–113.

Hom-Diaz, A., Norvill, Z. N., Blanquez, P., Vicent, T., & Guieysse, B. (2017). Ciprofloxacin removal during secondary domestic wastewater treatment in high rate algal ponds. *Chemosphere*, *180*, 33–41.

Hom, E. F., & Murray, A. W. (2014). Niche engineering demonstrates a latent capacity for fungal-algal mutualism. *Science*, *345*(6192), 94–98. doi:10.1126/science.1253320 PMID:24994654

Hong, Z., Lakkineni, K., Zhang, Z., & Verma, D. P. S. (2000). Removal of feedback inhibition of A1 – pyrroline - 5-carboxylate synthetase results in increased proline accumulation and protection of plants from osmotic stress. *Plant Physiology*, *122*(4), 1129–1136. doi:10.1104/pp.122.4.1129 PMID:10759508

Hopes, A., Nekrasov, V., Kamoun, S., & Mock, T. (2016). Editing of the urease gene by CRISPR-Cas in the diatom Thalassiosira pseudonana. *Plant Methods*, *12*(1), 49. doi:10.118613007-016-0148-0 PMID:27904648

Hopkins, D. L., Clayton, E. H., Lamb, T. A., Van de Ven, R. J., Refshauge, G., Kerr, M. J., Bailes, K., Lewandowski, P., & Ponnampalam, E. N. (2014). The impact of supplementing lambs with algae on growth, meat traits and oxidative status. *Meat Science*, *98*(2), 135–141. doi:10.1016/j.meatsci.2014.05.016 PMID:24950082

Hornak, K., Kasalicky, V., Simek, K., & Grossart, H. P. (2017). Strain-specific consumption and transformation of alga-derived dissolved organic matter by members of the Limnohabitans-C and Polynucleobacter-B clusters of Betaproteobacteria. *Environmental Microbiology*, *19*(11), 4519–4535. doi:10.1111/1462-2920.13900 PMID:28856804

Ho, S.-H. H., Huang, S.-W. W., Chen, C.-Y. Y., Hasunuma, T., Kondo, A., & Chang, J.-S. S. (2013). Bioethanol production using carbohydrate-rich microalgae biomass as feedstock. *Bioresource Technology*, *135*, 191–198. doi:10.1016/j.biortech.2012.10.015 PMID:23116819

Ho, S.-H., Zhang, C., Tao, F., Zhang, C., & Chen, W.-H. (2020). Microalgal Torrefaction for Solid Biofuel Production. *Trends in Biotechnology*, *38*(9), 1023–1033. doi:10.1016/j.tibtech.2020.02.009 PMID:32818442

Hossain, N., Mahlia, T. M. I., & Saidur, R. (2019). Latest development in microalgae-biofuel production with nano-additives. *Biotechnology for Biofuels*, *12*(1), 1–16. doi:10.118613068-019-1465-0 PMID:31139255

Hosseinzadeh-Bandbafha, H., Tabatabaei, M., Aghbashlo, M., Sulaiman, A., & Ghassemi, A. (2020). Life-Cycle Assessment (LCA) analysis of algal fuels. In *Biofuels from Algae* (pp. 121–151). Humana.

Houle, D., Govindaraju, D. R., & Omholt, S. (2010). Phenomics: The next challenge. *Nature Reviews. Genetics*, *11*(12), 855–866. doi:10.1038/nrg2897 PMID:21085204

Hrnčič, M. K., Kravanja, G., & Knez, Ž. (2016). Hydrothermal treatment of biomass for energy and chemicals. *Energy*, *116*, 1312–1322. doi:10.1016/j.energy.2016.06.148

Huang, G., Chen, F., Wei, D., Zhang, X., & Chen, G. (2010). Biodiesel production by microalgal biotechnology. *Applied Energy*, *87*(1), 38–46. doi:10.1016/j.apenergy.2009.06.016

Huang, L., Peng, L., & Yan, X. (2021). Multi-omics responses of red algae *Pyropia haitanensis* to intertidal desiccation during low tides. *Algal Research*, *58*, 102376. doi:10.1016/j.algal.2021.102376

Huang, X., Jain, P. K., El-Sayed, I. H., & El-Sayed, M. A. (2007). Gold nanoparticles: Interesting optical properties and recent applications in cancer diagnostics and therapy. *Nanomedicine (London)*, *2*(5), 681–693. doi:10.2217/17435889.2.5.681 PMID:17976030

Huang, Y., Li, R., Liu, H., Wang, B., Zhang, C., & Shen, Q. (2014). Novel resource utilization of refloated algal sludge to improve the quality of organic fertilizer. *Environmental Technology*, *35*(13), 1658–1667. doi:10.1080/09593330.2013.878397 PMID:24956756

Huber & Dale. (2009). Biofuels: Grassoline at the Pump. *Sci American.*, *301*(1), 52-59.

Hu, C., Zhang, D., Huang, Z., & Liu, Y. (2003). The vertical microdistribution of cyanobacteria and green algae within desert crusts and the development of the algal crusts. *Plant and Soil*, *257*(1), 97–111. doi:10.1023/A:1026253307432

Hucka, A. F. M., Sauro, H. M., Bolouri, H., Doyle, J. C., Kitano, H., Arkin, A. P., Bornstein, B. J., Bray, D., Cornish-Bowden, A., Cuellar, A. A., Dronov, S., Gilles, E. D., Ginkel, M., Gor, V., Goryanin, I. I., Hedley, W. J., Hodgman, T. C., Hofmeyr, J.-H., Hunter, P. J., ... Wang, J. (2003). The systems biology markup language (SBML): A medium for representation and exchange of biochemical network models. *Bioinformatics (Oxford, England)*, *19*(4), 524–531. doi:10.1093/bioinformatics/btg015 PMID:12611808

Huerta Lwanga, E., Mendoza Vega, J., Ku Quej, V., Chi, J., Sanchez del Cid, L., Chi, C., Escalona Segura, G., Gertsen, H., Salánki, T., van der Ploeg, M., Koelmans, A. A., & Geissen, V. (2017). Field evidence for transfer of plastic debris along a terrestrial food chain. *Scientific Reports*, *7*(1), 14071. doi:10.103841598-017-14588-2 PMID:29074893

Hug, L. A., Baker, B. J., Anantharaman, K., Brown, C. T., Probst, A. J., Castelle, C. J., Butterfield, C. N., Hernsdorf, A. W., Amano, Y., Ise, K., Suzuki, Y., Dudek, N., Relman, D. A., Finstad, K. M., Amundson, R., Thomas, B. C., & Banfield, J. F. (2016). A new view of the tree of life. *Nature Microbiology*, *1*(5), 16048. doi:10.1038/nmicrobiol.2016.48 PMID:27572647

Hu, H., Li, J. Y., Pan, X. R., Zhang, F., Ma, L. L., Wang, H. J., & Zeng, R. J. (2019). Different DHA or EPA production responses to nutrient stress in the marine microalga Tisochrysis lutea and the freshwater microalga Monodus subterraneus. *The Science of the Total Environment*, *656*, 140–149. doi:10.1016/j.scitotenv.2018.11.346 PMID:30504016

Hu, H., Wang, Z., & Pan, L. (2010). Synthesis of monodisperse Fe3O4@ silica core–shell microspheres and their application for removal of heavy metal ions from water. *Journal of Alloys and Compounds*, *492*(1-2), 656–661. doi:10.1016/j.jallcom.2009.11.204

Huisman, J., Codd, G. A., Paerl, H. W., Ibelings, B. W., Verspagen, J. M. H., & Visser, P. M. (2018). Cyanobacterial blooms. *Nature Reviews. Microbiology*, *16*(8), 471–483. doi:10.103841579-018-0040-1 PMID:29946124

Hu, J., Nagarajan, D., Zhang, Q., Chang, J.-S., & Lee, D.-J. (2018). Heterotrophic cultivation of microalgae for pigment production: A review. *Biotechnology Advances*, *36*(1), 54–67. doi:10.1016/j.biotechadv.2017.09.009 PMID:28947090

Hu, L., Feng, S., Liang, G., Du, J., Li, A., & Niu, C. (2021). CRISPR/Cas9-induced β-carotene hydroxylase mutation in *Dunaliella salina* CCAP19/18. *AMB Express*, *11*(1), 83. doi:10.118613568-021-01242-4 PMID:34097133

Hulatt, C. J., Berecz, O., Egeland, E. S., Wijffels, R. H., & Kiron, V. (2017). Polar snow algae as a valuable source of lipids? *Bioresource Technology*, *235*, 338–347. doi:10.1016/j.biortech.2017.03.130 PMID:28384586

Hulatt, C. J., & Thomas, D. N. (2011). Productivity, carbon dioxide uptake and net energy return of microalgal bubble column photobioreactors. *Bioresource Technology*, *102*(10), 5775–5787. doi:10.1016/j.biortech.2011.02.025 PMID:21376576

Huleihel, M., Ishanu, V., Tal, J., & Arad, S. (2001). Antiviral effect of red microalgal polysaccharides on *Herpes simplex* and *Varicella zoster* viruses. *Journal of Applied Phycology*, *13*(2), 127–134. doi:10.1023/A:1011178225912

Humphrey, W., Dalke, A., & Schulten, K. (1996). VMD-Visual molecular dynamics. *Journal of Molecular Graphics*, *14*(1), 33–38. doi:10.1016/0263-7855(96)00018-5 PMID:8744570

Hunter, J. D. (2007). Matplotlib: A 2D graphics environment. *Computing in Science & Engineering*, *9*(3), 90–95. doi:10.1109/MCSE.2007.55

Huntley, M. E., Johnson, Z. I., Brown, S. L., Sills, D. L., Gerber, L., Archibald, I., Machesky, S. C., Granados, J., Beal, C., & Greene, C. H. (2015). Demonstrated large-scale production of marine microalgae for fuels and feed. *Algal Research*, *10*, 249–265. doi:10.1016/j.algal.2015.04.016

Hunt, R. (2018). *Bloom Foam: Transforming Green Water Into Green Products*. Academic Press

Huo, S., Wang, Z., Zhu, S., Cui, F., Zou, B., You, W., Yuan, Z., & Dong, R. (2014). Optimization of alkaline flocculation for harvesting of *Scenedesmus quadricauda* 507 and *Chaetoceros muelleri* 862. *Energies*, *7*, 6186–6195.

Hurd, C. L. (2000). Water motion, marine macroalgal physiology, and production. *Journal of Phycology*, *36*(3), 453–472. doi:10.1046/j.1529-8817.2000.99139.x PMID:29544017

Hurd, C. L., Harrison, P. J., Bischof, K., & Lobban, C. S. (2014). *Seaweed ecology and physiology*. Cambridge University Press. doi:10.1017/CBO9781139192637

Hussein, G., Sankawa, U., Goto, H., Matsumoto, K., & Watanabe, H. (2006). Astaxanthin, a carotenoid with potential in human health and nutrition. *Journal of Natural Products*, *69*(3), 443–449. doi:10.1021/np050354+ PMID:16562856

Hu, X., Meneses, Y. E., Stratton, J., & Wang, B. (2019). Acclimation of consortium of micro-algae help removal of organic pollutants from meat processing wastewater. *Journal of Cleaner Production*, *214*, 95–102. doi:10.1016/j.jclepro.2018.12.255

Hu, Y. R., Wang, F., Wang, S. K., Liu, C. Z., & Guo, C. (2013). Efficient harvesting of marine microalgae *Nannochloropsis maritima* using magnetic nanoparticles. *Bioresource Technology*, *138*, 387–390.

Hu, Z., Zeng, X., Wang, A., Shi, C., & Duan, D. (2004). An efficient method for DNA isolation from red algae. *Journal of Applied Phycology*, *16*(3), 161–166. doi:10.1023/B:JAPH.0000048456.26639.1a

Hyun, J. H., Kim, S. C., Kang, J. I., Kim, M. K., Boo, H. J., Kwon, J. M., Koh, Y.-S., Hyun, J.-W., Park, D.-B., Yoo, E.-S., & Kang, H. K. (2009). Apoptosis inducing activity of fucoidan in HCT-15 colon carcinoma cells. *Biological & Pharmaceutical Bulletin*, *32*(10), 1760–1764. doi:10.1248/bpb.32.1760 PMID:19801840

Ianora, A., Bentley, M. G., Caldwell, G. S., Casotti, R., Cembella, A. D., Engström-Öst, J., Halsband, C., Sonnenschein, E., Legrand, C., Llewellyn, C. A., Paldavičienė, A., Pilkaityte, R., Pohnert, G., Razinkovas, A., Romano, G., Tillmann, U., & Vaiciute, D. (2011). The relevance of marine chemical ecology to plankton and ecosystem function: An emerging field. *Marine Drugs*, *9*(9), 1625–1648. doi:10.3390/md9091625 PMID:22131962

Ibiene, A. A., Stanley, H. O., & Immanuel, O. M. (2013). Biodegradation of polyethylene by *Bacillus* sp. indigenous to the Niger Delta Mangrove Swamp. *Nigerian Journal of Biotechnology*, *26*, 68–79.

Ibraheem, I. (2007). Cyanobacteria as alternative biological conditioners for bioremediation of barren soil. *Egyptian Journal of Phycology*, *8*(1), 99–117. doi:10.21608/egyjs.2007.114548

Ibrahim, W. M., Ali, R. M., Hemida, K., & Sayed, M. A. (2014). Role of Ulva lactuca extract in alleviation of salinity stress on wheat seedlings. *TheScientificWorldJournal*, *2014*, 1–11. doi:10.1155/2014/847290 PMID:25436231

Iftekhar, S., Ramasamy, D. L., Srivastava, V., Asif, M. B., & Sillanpää, M. (2018). Understanding the factors affecting the adsorption of Lanthanum using different adsorbents: A critical review. *Chemosphere*, *204*, 413–430. doi:10.1016/j.chemosphere.2018.04.053 PMID:29677649

Iftekhar, S., Srivastava, V., & Sillanpää, M. (2017). Synthesis and application of LDH intercalated cellulose nanocomposite for separation of rare earth elements (REEs). *Chemical Engineering Journal*, *309*, 130–139. doi:10.1016/j.cej.2016.10.028

Illera-Vives, M., Labandeira, S. S., Brito, L. M., López-Fabal, A., & López-Mosquera, M. E. (2015). Evaluation of compost from seaweed and fish waste as a fertilizer for horticultural use. *Scientia Horticulturae*, *186*, 101–107. doi:10.1016/j.scienta.2015.02.008

Imai, I., & Yamaguchi, M. (2012). Life cycle, physiology, ecology and red tide occurrences of the fish-killing raphidophyte Chattonella. *Harmful Algae*, *14*, 46–70. doi:10.1016/j.hal.2011.10.014

Imam, S., Schäuble, S., Valenzuela, J., López García de Lomana, A., Carter, W., Price, N. D., & Baliga, N. S. (2015). A refined genome-scale reconstruction of *Chlamydomonas* metabolism provides a platform for systems-level analyses. *The Plant Journal*, *84*(6), 1239–1256. doi:10.1111/tpj.13059 PMID:26485611

Im, H., Lee, H., Park, M. S., Yang, J.-W., & Lee, J. W. (2014). Concurrent extraction and reaction for the production of biodiesel from wet microalgae. *Bioresource Technology*, *152*, 534–537. doi:10.1016/j.biortech.2013.11.023 PMID:24291292

Imre, B., García, L., Puglia, D., & Vilaplana, F. (2019). Reactive compatibilization of plant polysaccharides and biobased polymers: Review on current strategies, expectations and reality. *Carbohydrate Polymers*, *209*, 20–37. doi:10.1016/j.carbpol.2018.12.082 PMID:30732800

Inaba, N., Trainer, V. L., Nagai, S., Kojima, S., Sakami, T., Takagi, S., & Imai, I. (2019). Dynamics of seagrass bed microbial communities in artificial Chattonella blooms: A laboratory microcosm study. *Harmful Algae*, *84*(April), 139–150. doi:10.1016/j.hal.2018.12.004 PMID:31128798

Inaba, N., Trainer, V. L., Onishi, Y., Ishii, K. I., Wyllie-Echeverria, S., & Imai, I. (2017). Algicidal and growth-inhibiting bacteria associated with seagrass and macroalgae beds in Puget Sound, WA, USA. *Harmful Algae*, *62*, 136–147. doi:10.1016/j.hal.2016.04.004 PMID:28118888

Invernizzi, N. (2011). Nanotechnology between the lab and the shop floor: What are the effects on labour? *Journal of Nanoparticle Research*, *13*(6), 2249–2268. doi:10.100711051-011-0333-z

Iqbal, M., Muhammad, G., Aslam, S., Hussain, M., Shafiq, Z., & Razzaq, H. (2021). Algal Biofertilizer. In Biofertilizers: Study and Impact. John Wiley & Sons, Inc. doi:10.1002/9781119724995.ch22

Iqbal, H., Khan, B. A., Khan, Z. U., Razzaq, A., Khan, N. U., Menaa, B., & Menaa, F. (2020). Fabrication, physical characterizations and in vitro antibacterial activity of cefadroxil-loaded chitosan/poly (vinyl alcohol) nanofibers against staphylococcus aureus clinical isolates. *International Journal of Biological Macromolecules*, *144*, 921–931. doi:10.1016/j.ijbiomac.2019.09.169 PMID:31704336

Iqbal, M., & Zafar, S. I. (1993). Effects of photon flux density, CO2, aeration rate, and inoculum density on growth and extracellular polysaccharide production by Porphyridium cruentum. *Folia Microbiologica*, *38*(6), 509–514. doi:10.1007/BF02814405

Ishimi, Y., Sugiyama, F., Ezaki, J., Fujioka, M., & Wu, J. (2006). Effects of Spirulina, a blue-green alga, on bone metabolism in overiectomized rats and hindlimb-unloaded mice. *Bioscience, Biotechnology, and Biochemistry*, *70*(2), 363–368. doi:10.1271/bbb.70.363 PMID:16495651

Ismail, A. A., & Bahnemann, D. W. (2014). Photochemical splitting of water for hydrogen production by photocatalysis: A review. *Solar Energy Materials and Solar Cells*, *128*, 85–101. doi:10.1016/j.solmat.2014.04.037

Iwamoto, H. (2004). Industrial production of microalgal cell-mass and secondary products-major industrial species-Chlorella. In Handbook of Microalgal Culture: Biotechnology and Applied Phycology. Blackwell Science.

Iwata, T. (2015). Biodegradable and bio-Based polymers: Future prospects of eco-friendly plastics. *Angewandte Chemie International Edition*, *54*(11), 3210–3215. doi:10.1002/anie.201410770 PMID:25583677

Jabeen, N., Majid, I., & Nayik, G. A. (2015). Bioplastics and food packaging: A review. *Cogent Food & Agriculture*, *1*(1), 1117749.

Jacob, A., Wiencke, C., Lehmann, H., & Kirst, G. O. (1992). *Physiology and ultrastructure of desiccation in the green alga Prasiola crispa from Antarctica*. Academic Press.

Jacobs, W. P. A. (1993). Search for Some Angiosperm Hormones and Their Metabolites in *Caulerpa paspaloides* (Chlorophyta). *Journal of Phycology*, *29*(5), 595–600. doi:10.1111/j.0022-3646.1993.00595.x

Jain, R., & Tiwari, A. (2015). Biosynthesis of planet friendly bioplastics using renewable carbon source. *Journal of Environmental Health Science & Engineering*, *13*(1), 11. doi:10.118640201-015-0165-3 PMID:25717378

Jain, S., Jain, S., Wolf, I. T., Lee, J., & Tong, Y. W. (2015). A comprehensive review on operating parameters and different pretreatment methodologies for anaerobic digestion of municipal solid waste. *Renewable & Sustainable Energy Reviews*, *52*, 142–154. doi:10.1016/j.rser.2015.07.091

Jamers, A., Blust, R., & De Coen, W. (2009). Omics in algae: Paving the way for a systems biological understanding of algal stress phenomena? *Aquatic Toxicology (Amsterdam, Netherlands)*, *92*(3), 114–121. doi:10.1016/j.aquatox.2009.02.012 PMID:19304329

James, G. O., Hocart, C. H., Hillier, W., Chen, H., Kordbacheh, F., Price, G. D., & Djordjevic, M. A. (2011). Fatty acid profiling of *Chlamydomonas reinhardtii* under nitrogen deprivation. *Bioresource Technology*, *102*(3), 3343–3351. doi:10.1016/j.biortech.2010.11.051 PMID:21146403

Jang, Y. H., Han, S. O., Sim, I. N., & Kim, H. (2013). Pretreatment effects of seaweed on the thermal and mechanical properties of seaweed/polypropylene biocomposites. *Composites. Part A, Applied Science and Manufacturing*, *47*(1), 83–90. doi:10.1016/j.compositesa.2012.11.016

Jankowska, E., Sahu, A. K., & Oleskowicz-Popiel, P. (2017). Biogas from Microalgae: Review on Microalgae's Cultivation, Harvesting and Pretreatment for Anaerobic Digestion. *Renewable & Sustainable Energy Reviews*, *75*, 692–709. doi:10.1016/j.rser.2016.11.045

Janouškovec, J., Gavelis, G. S., Burki, F., Dinh, D., Bachvaroff, T. R., Gornik, S. G., Bright, K. J., Imanian, B., Strom, S. L., Delwiche, C. F., Waller, R. F., Fensome, R. A., Leander, B. S., Rohwer, F. L., & Saldarriaga, J. F. (2017). Major transitions in dinoflagellate evolution unveiled by phylotranscriptomics. *Proceedings of the National Academy of Sciences of the United States of America*, *114*(2), E171. doi:10.1073/pnas.1614842114 PMID:28028238

Janssen, C. I. F., & Kiliaan, A. J. (2014). Long-chain polyunsaturated fatty acids (LCPUFA) from genesis to senescence: The influence of LCPUFA on neural development, aging, and neurodegeneration. *Progress in Lipid Research*, *53*, 1–17. doi:10.1016/j.plipres.2013.10.002 PMID:24334113

Janssen, M., Tramper, J., Mur, L. R., & Wijffels, R. H. (2003). Enclosed outdoor photobioreactors: Light regime, photosynthetic efficiency, scale-up, and future prospects. *Biotechnology and Bioengineering*, *81*(2), 193–210.

Jäpelt, R. B., & Jakobsen, J. (2013). Vitamin D in plants: A review of occurrence, analysis, and biosynthesis. *Frontiers in Plant Science*, *4*(2), 136. doi:10.3389/fpls.2013.00136 PMID:23717318

Jaspars, M., De Pascale, D., Andersen, J. H., Reyes, F., Crawford, A. D., & Ianora, A. (2016). The marine biodiscovery pipeline and ocean medicines of tomorrow. *Journal of the Marine Biological Association of the United Kingdom*, *96*(1), 151–158. doi:10.1017/S0025315415002106

Jau, M.-H., Yew, S.-P., Toh, P. S. Y., Chong, A. S. C., Chu, W.-L., Phang, S.-M., Najimudin, N., & Sudesh, K. (2005). Biosynthesis and mobilization of poly(3-hydroxybutyrate) [P(3HB)] by *Spirulina platensis*. *International Journal of Biological Macromolecules*, *36*(3), 144–151. doi:10.1016/j.ijbiomac.2005.05.002 PMID:16005060

Javed, F., Aslam, M., Rashid, N., Shamair, Z., Khan, A. L., Yasin, M., Fazal, T., Hafeez, A., Rehman, F., Rehman, M. S. U., Khan, Z., Iqbal, J., & Bazmi, A. A. (2019). Microalgae-Based Biofuels, Resource Recovery and Wastewater Treatment: A Pathway towards Sustainable Biorefinery. *Fuel*, *255*, 115826. doi:10.1016/j.fuel.2019.115826

Jayappriyan, K. R., Rajkumar, R., Venkatakrishnan, V., Nagaraj, S., & Rengasamy, R. (2013). In vitro anticancer activity of natural β-carotene from Dunaliella salina EU5891199 in PC-3 cells. *Biomedicine & Preventive Nutrition*, *3*(2), 99–105. doi:10.1016/j.bionut.2012.08.003

Jeffrey, C. (1967). The Origin and Differentiation of the Archegoniate Land Plants: A Second Contribution. *Kew Bulletin*, *21*(2), 335–349. doi:10.2307/4108533

Jehlee, A., Khongkliang, P., & O-Thong, S. (2017). Biogas Production from Chlorella sp. TISTR 8411 Biomass Cultivated on Biogas Effluent of Seafood Processing Wastewater. *Energy Procedia*, *138*, 853–857. doi:10.1016/j.egypro.2017.10.095

Jeong, S.-Y., & Son, H.-J. (2021). Effects of mycosubtilin homolog algicides from a marine bacterium, *Bacillus* sp. SY-1, against the harmful algal bloom species Cochlodinium polykrikoides. *Journal of Microbiology (Seoul, Korea)*, *59*(4), 389–400. doi:10.100712275-021-1086-8 PMID:33779952

Jeon, S., Lim, J.-M., Lee, H.-G., Shin, S.-E., Kang, N. K., Park, Y.-I., Oh, H.-M., Jeong, W.-J., Jeong, B., & Chang, Y. K. (2017). Current status and perspectives of genome editing technology for microalgae. *Biotechnology for Biofuels*, *10*(1), 267. doi:10.118613068-017-0957-z PMID:29163669

Jephcott, T. G., Ogtrop, F.F., Gleason, F.H., Macarthur, D.J., & Scholz, B. (2017). The ecology of chytrid and aphelid parasites of phytoplankton. In The Fungal Community. CRC Press.

Jerez-Martel, I., García Poza, S., Rodríguez-Martel, G., Rico, M., Afonso-Olivares, C., Gomez, P., & Juan, L. (2017). Phenolic Profile and Antioxidant Activity of Crude Extracts from Microalgae and Cyanobacteria Strains. *Journal of Food Quality*, *2017*, 1–8. doi:10.1155/2017/2924508

Jha, B., Kavita, K., Westphal, J., Hartmann, A., & Schmitt-Kopplin, P. (2013). Quorum sensing inhibition by asparagopsis taxiformis, a marine macro alga: Separation of the compound that interrupts bacterial communication. *Marine Drugs*, *11*(1), 253–265. doi:10.3390/md11010253 PMID:23344114

Jiang, G., Johnston, B., Townrow, D. E., Radecka, I., Koller, M., Chaber, P., Adamus, G., & Kowalczuk, M. (2018). Biomass extraction using non-chlorinated solvents for biocompatibility improvement of polyhydroxyalkanoates. *Polymers*, *10*(7), 731. Advance online publication. doi:10.3390/polym10070731 PMID:30960656

Jiang, L.-L., Zhou, J.-J., Quan, C.-S., & Xiu, Z.-L. (2017). Advances in industrial microbiome based on microbial consortium for biorefinery. *Bioresources and Bioprocessing*, *4*(1), 11. doi:10.118640643-017-0141-0 PMID:28251041

Jiang, L., Wang, Y., Liu, G., Liu, H., Zhu, F., Ji, H., & Li, B. (2018). C-Phycocyanin exerts anti-cancer effects via the MAPK signaling pathway in MDA-MB-231 cells. *Cancer Cell International*, *18*(1), 1–14. doi:10.118612935-018-0511-5 PMID:29416441

Jiang, X., Chen, W., Chen, C., Xiong, S., & Yu, A. (2011). Role of temperature in the growth of silver nanoparticles through a synergetic reduction approach. *Nanoscale Research Letters*, *6*, 32. PMID:27502655

Jiang, Y., Zhang, W., Wang, J., Chen, Y., Shen, S., & Liu, T. (2013). Utilization of simulated flue gas for cultivation of *Scenedesmus dimorphus*. *Bioresource Technology*, *128*, 359–364. doi:10.1016/j.biortech.2012.10.119 PMID:23201515

Jia, R., Qu, Z., You, P., & Qu, D. (2018). Effect of biochar on photosynthetic microorganism growth and iron cycling in paddy soil under different phosphate levels. *The Science of the Total Environment*, *612*, 223–230. doi:10.1016/j.scitotenv.2017.08.126 PMID:28850841

Jibril, S. M., Jakada, B. H., Umar, H. Y., & Ahmad, T. A. (2016). Importance of Some Algal Species as a Source of Food and Supplement. *International Journal of Current Microbiology and Applied Sciences*, *5*(5), 186–193. doi:10.20546/ijcmas.2016.505.020

Ji, H. M., Lee, H. U., Kim, E. J., Seo, S., Kim, B., Lee, G. W., Oh, Y.-K., Kim, J. Y., Huh, Y. S., & Song, H. A. (2016). Efficient harvesting of wet blue-green microalgal biomass by two-aminoclay [AC]-mixture systems. *Bioresource Technology*, *15*, 313–318.

Jiménez, E., Dorta, F., Medina, C., Ramírez, A., Ramírez, I., & Peña-Cortés, H. (2011). Anti-Phytopathogenic Activities of Macro-Algae Extracts. *Marine Drugs*, *9*, 739–756.

Jiménez-Rosado, M., Bouroudian, E., Perez-Puyana, V., Guerrero, A., & Romero, A. (2020). Evaluation of different strengthening methods in the mechanical and functional properties of soy protein-based bioplastics. *Journal of Cleaner Production*, *262*, 121517. doi:10.1016/j.jclepro.2020.121517

Jimeno, J., Faircloth, G., Sousa-Faro, J. M., Scheuer, P., & Rinehart, K. (2004). New marine derived anticancer therapeutics— A journey from the sea to clinical trials. *Marine Drugs*, *2*(1), 14–29. doi:10.3390/md201014

Jin, H.-J., Kim, J. H., Sohn, C. H., DeWreede, R. E., Choi, T. J., Towers, G. H. N., Hudson, J. B., & Hong, Y. K. (1997). Inhibition of Taq DNA polymerase by seaweed extracts from British Columbia, Canada and Korea. *Journal of Applied Phycology*, *9*(4), 383–388. doi:10.1023/A:1007925202219

Jin, X., Chu, Z., Yan, F., & Zeng, Q. (2009). Effects of lanthanum (III) and EDTA on the growth and competition of Microcystis aeruginosa and Scenedesmus quadricauda. *Limnologica*, *39*(1), 86–93. doi:10.1016/j.limno.2008.03.002

Ji, X., Jiang, M., Zhang, J., Jiang, X., & Zheng, Z. (2018). The interactions of algae-bacteria symbiotic system and its effects on nutrients removal from synthetic wastewater. *Bioresource Technology*, *247*, 44–50. doi:10.1016/j.biortech.2017.09.074 PMID:28946093

Jõgi, K., & Bhat, R. (2020). Valorization of food processing wastes and by-products for bioplastic production. *Sustainable Chemistry and Pharmacy*, *18*, 100326. doi:10.1016/j.scp.2020.100326

John, E. H., & Flynn, K. J. (2000). Modelling phosphate transport and assimilation in microalgae; how much complexity is warranted? *Ecological Modelling*, *125*(2-3), 145–157. doi:10.1016/S0304-3800(99)00178-7

John, R. P., Anisha, G. S., Nampoothiri, K. M., & Pandey, A. (2011). Micro and macroalgal biomass: A renewable source for bioethanol. *Bioresource Technology*, *102*(1), 186–193.

Johnson, W. M., Kido Soule, M. C., & Kujawinski, E. B. (2016). Evidence for quorum sensing and differential metabolite production by a marine bacterium in response to DMSP. *The ISME Journal*, *10*(9), 2304–2316. doi:10.1038/ismej.2016.6 PMID:26882264

Johnsson, N., & Steuer, F. (2018). *Bioplastic material from microalgae: Extraction of starch and pha from microalgae to create a bioplastic material, school of industrial engineering and management*. KTH Royal Institute of Technology.

Jorquera, O., Kiperstok, A., Sales, E. A., Embirucu, M., & Ghirardi, M. L. (2010). Comparative energy life-cycle analyses of microalgal biomass production in open ponds and photobioreactors. *Bioresource Technology*, *101*(4), 1406–1413. doi:10.1016/j.biortech.2009.09.038 PMID:19800784

Jose, S., & Archanaa, S. (2017). Environmental and economic sustainability of algal lipid extractions: an essential approach for the commercialization of algal biofuels. In *Algal Biofuels* (pp. 281–313). Springer. doi:10.1007/978-3-319-51010-1_14

Joubert, Y., & Fleurence, J. (2005). DNA isolation protocol for seaweeds. *Plant Molecular Biology Reporter*, *23*(2), 197–197. doi:10.1007/BF02772712

Jouhara, H., Czajczyńska, D., Ghazal, H., Krzyżyńska, R., Anguilano, L., Reynolds, A. J., & Spencer, N. (2017). Municipal waste management systems for domestic use. *Energy*, *139*, 485–506. doi:10.1016/j.energy.2017.07.162

Joung, J. K., & Sander, J. D. (2013). TALENs: A widely applicable technology for targeted genome editing. *Nature Reviews. Molecular Cell Biology*, *14*(1), 49–55. doi:10.1038/nrm3486 PMID:23169466

Juneja, A., Ceballos, R. M., & Murthy, G. S. (2013). Effects of environmental factors and nutrient availability on the biochemical composition of algae for biofuels production: A review. *Energies*, *6*(9), 4607–4638. doi:10.3390/en6094607

Juneja, A., Chaplen, F. W. R., & Murthy, G. S. (2016). Genome scale metabolic reconstruction of *Chlorella variabilis* for exploring its metabolic potential for biofuels. *Bioresource Technology*, *213*, 103–110. doi:10.1016/j.biortech.2016.02.118 PMID:26995318

K.V., S., Behera, B., & P., B. (2020). Efficacy of microalgal extracts as biostimulants through seed treatment and foliar spray for tomato cultivation. *Industrial Crops and Products*, *151*, 112453. doi:10.1016/j.indcrop.2020.112453

Kaçka, A., & Dönmez, G. (2008). Isolation of Dunaliella spp. from a hypersaline lake and their ability to accumulate glycerol. *Bioresource Technology*, *99*(17), 8348–8352. doi:10.1016/j.biortech.2008.02.042 PMID:18406610

Kadir, W. N. A., Lam, M. K., Uemura, Y., Lim, J. W., & Lee, K. T. (2018). Harvesting and pre-treatment of microalgae cultivated in wastewater for biodiesel production: A review. *Energy Conversion and Management*, *171*, 1416–1429. doi:10.1016/j.enconman.2018.06.074

Kaewbai-ngam, A., Incharoensakdi, A., & Monshupanee, T. (2016). Increased accumulation of polyhydroxybutyrate in divergent cyanobacteria under nutrient-deprived photoautotrophy: An efficient conversion of solar energy and carbon dioxide to polyhydroxybutyrate by Calothrix scytonemicola TISTR 8095. *Bioresource Technology*, *212*, 342–347. doi:10.1016/j.biortech.2016.04.035 PMID:27130227

Kalamaras, C. M., & Efstathiou, A. M. (2013). Hydrogen Production Technologies: Current State and Future Developments. *Conference Papers in Energy*, *2013*, 1–9. doi:10.1155/2013/690627

Kale, G., Kijchavengkul, T., Auras, R., Rubino, M., Selke, S. E., & Singh, S. P. (2007). Compostability of Bioplastic Packaging Materials: An Overview. *Macromolecular Bioscience*, *7*(3), 255–277. doi:10.1002/mabi.200600168 PMID:17370278

Kalia, S., & Avérous, L. (2016). *Biodegradable and biobased polymers for environmental and biomedical Applications*. John Wiley & Sons. doi:10.1002/9781119117360

Kalinci, Y., Hepbasli, A., & Dincer, I. (2009). Biomass-based hydrogen production: A review and analysis. *International Journal of Hydrogen Energy*, *34*(21), 8799–8817. doi:10.1016/j.ijhydene.2009.08.078

Källberg, M., Wang, H., Wang, S., Peng, J., Wang, Z., Lu, H., & Xu, J. (2012). Template-based protein structure modeling using the RaptorX web server. *Nature Protocols*, *7*(8), 1511–1522. doi:10.1038/nprot.2012.085 PMID:22814390

Kally, E., & Fishelson, G. (1993). *Water and peace: water resources and the Arab-Israeli peace process*. Praeger.

Kamravamanesh, D., Kovacs, T., Pflügl, S., Druzhinina, I., Kroll, P., Lackner, M., & Herwig, C. (2018). Increased poly-β-hydroxybutyrate production from carbon dioxide in randomly mutated cells of cyanobacterial strain *Synechocystis* sp. PCC 6714: Mutant generation and characterization. *Bioresource Technology*, *266*, 34–44. doi:10.1016/j.biortech.2018.06.057 PMID:29957289

Kamravamanesh, D., Pflügl, S., Nischkauer, W., Limbeck, A., Lackner, M., & Herwig, C. (2017). Photosynthetic poly-β-hydroxybutyrate accumulation in unicellular cyanobacterium Synechocystis sp. PCC 6714. *AMB Express*, *7*(1), 143. Advance online publication. doi:10.118613568-017-0443-9 PMID:28687036

Kanamarlapudi, S.L.R.K., Chintalpudi, V.K., & Muddada, S. (2018). Application of Biosorption for Removal of Heavy Metals from Wastewater. *Biosorption*, *69*.

Kanda, H., Li, P., Goto, M., & Makino, H. (2015). Energy-saving lipid extraction from wet Euglena gracilis by the low-boiling-point solvent dimethyl ether. *Energies*, *8*(1), 610–620.

Kandil, H., Abdelhamid, A. E., Moghazy, R. M., & Amin, A. (2022). Functionalized PVA film with good adsorption capacity for anionic dye. *Polymer Engineering and Science*, *62*(1), 145–159. doi:10.1002/pen.25840

Kanehisa, M., Sato, Y., Kawashima, M., Furumichi, M., & Tanabe, M. (2016). KEGG as a reference resource for gene and protein annotation. *Nucleic Acids Research*, *44*(1), 457–462. doi:10.1093/nar/gkv1070 PMID:26476454

Kang, D., Zhao, Q., Wu, Y., Wu, C., & Wu, X. (2018). Removal of nutrients and pharmaceuticals and personal care products from wastewater using periphyton photobioreactors. *Bioresource Technology*, *248*, 113–119.

Kang, K. A., Bu, H. D., Park, D. S., Go, G. M., Jee, Y., Shin, T., & Hyun, J. W. (2005). Antioxidant activity of ethanol extract of Callophyllis japonica. *Phytotherapy Research*, *19*(6), 506–510. doi:10.1002/ptr.1692 PMID:16114080

Kang, Y. H., Kang, S. K., Park, C. S., Joo, J. H., Lee, J. W., & Han, M. S. (2016). Use of lactic acid bacteria as a biological agent against the cyanobacterium Anabaena flos-aquae. *Journal of Applied Phycology*, *28*(3), 1747–1757. doi:10.100710811-015-0701-7

Kang, Y., Kim, M., Shim, C., Bae, S., & Jang, S. (2021). Potential of algae–bacteria synergistic effects on vegetable production. *Frontiers in Plant Science*, *12*, 556. doi:10.3389/fpls.2021.656662 PMID:33912211

Kannan, R.R.R., & Staden, J. V. (2013). Synthesis of silver nanoparticles using the seaweed Codium capitatum P.C. Silva (Chlorophyceae). *South African Journal of Botany*, *86*, 1–4.

Kao, P. H., & Ng, I. S. (2017). CRISPRi mediated phosphoenolpyruvate carboxylase regulation to enhance the production of lipid in *Chlamydomonas reinhardtii*. *Bioresoure Technology*, *245*(Pt B), 1527-1537. doi:10.1016/j.biortech.2017.04.111

Kao, C. Y., Chen, T. Y., Chang, Y.-B., Chiu, T.-W., Lin, H.-Y., Chen, C.-D., Chang, J.-S., & Lin, C.-S. (2014). Utilization of carbon dioxide in industrial flue gases for the cultivation of microalga Chlorella sp. *Bioresource Technology*, *166*(8), 485–493. doi:10.1016/j.biortech.2014.05.094 PMID:24950094

Kaparapu, J., & Geddada, M. N. R. (2020). *Commercial applications of algae in the field of biotechnology. In Applied Algal Biotechnology*. Nova Science Publishers, Inc.

Kapritchkoff, F. M., Viotti, A. P., Alli, R. C. P., Zuccolo, M., Pradella, J. G. C., Maiorano, A. E., Miranda, E. A., & Bonomi, A. (2006). Enzymatic recovery and purification of polyhydroxybutyrate produced by Ralstonia eutropha. *Journal of Biotechnology*, *122*(4), 453–462. doi:10.1016/j.jbiotec.2005.09.009 PMID:16253372

Karadjova, I. B., Slaveykova, V. I., & Tsalev, D. L. (2008). The biouptake and toxicity of arsenic species on the green microalga Chlorella salina in seawater. *Aquatic Toxicology (Amsterdam, Netherlands)*, *87*(4), 264–271. doi:10.1016/j.aquatox.2008.02.006 PMID:18378014

Karan, H., Funk, C., Grabert, M., Oey, M., & Hankamer, B. (2019). Green bioplastics as part of a circular bioeconomy. *Trends in Plant Science*, *24*(3), 237–249. doi:10.1016/j.tplants.2018.11.010 PMID:30612789

Karlander, E. P., & Krauss, R. W. (1972). Absorption and toxicity of beryllium and lithium in Chlorella vannielii Shihira and Krauss. *Chesapeake Science*, *13*(4), 245–253. doi:10.2307/1351108

Karsten, U. (2008). Defense Strategies of Algae and Cyanobacteria Against Solar Ultraviolet Radiation. In C. D. Amsler (Ed.), *Algal Chemical Ecology* (p. 313). Springer. doi:10.1007/978-3-540-74181-7_13

Karsten, U., Franklin, L. A., Luning, K., & Wiencke, C. (1998a). Natural ultraviolet radiation and photosynthetic active radiation induce formation of mycosporine-like amino acids in the marine macroalga *Chondrus crispus* Rhodophyta. *Planta*, *205*(2), 257–262. doi:10.1007004250050319

Karsten, U., & Holzinger, A. (2012). Light, temperature, and desiccation effects on photosynthetic activity, and drought-induced ultrastructural changes in the green alga Klebsormidium dissectum (Streptophyta) from a high alpine soil crust. *Microbial Ecology*, *63*(1), 51–63. doi:10.100700248-011-9924-6 PMID:21811791

Karsten, U., Lütz, C., & Holzinger, A. (2010). Ecophysiological performance of the aeroterrestrial green alga Klebsormidium crenulatum (Charophyceae, Streptophyta) isolated from an alpine soil crust with an emphasis on desiccation stress 1. *Journal of Phycology*, *46*(6), 1187–1197. doi:10.1111/j.1529-8817.2010.00921.x

Karsten, U., Sawall, T., Hanelt, D., Bischof, K., Figueroa, F. L., Flores Moya, A., & Wiencke, C. (1998b). An inventory of UV-absorbing mycosporine-like amino acids in macroalgae from polar to warm-temperate regions. *Botanica Marina*, *41*(1-6), 443–453. doi:10.1515/botm.1998.41.1-6.443

Karsten, U., Schumann, R., & Mostaert, A. (2007). Aeroterrestrial algae growing on man-made surfaces. In *Algae and cyanobacteria in extreme environments* (pp. 583–597). Springer. doi:10.1007/978-1-4020-6112-7_32

Karsten, U., & Wiencke, C. (1999). Factors controlling the formation of UVabsorbing mycosporine-like amino acids in the marine red alga *Palmaria palmate* from Spitzbergen (Norway). *Journal of Plant Physiology*, *155*(3), 407–415. doi:10.1016/S0176-1617(99)80124-2

Karthic, P., & Joseph, S. (2012). Comparison and limitations of biohydrogen production processes. *Research Journal of Biotechnology*, *7*(2), 59–71.

Kartik, A., Akhil, D., Lakshmi, D., Panchamoorthy Gopinath, K., Arun, J., Sivaramakrishnan, R., & Pugazhendhi, A. (2021). A critical review on production of biopolymers from algae biomass and their applications. *Bioresource Technology*, *329*, 124868. doi:10.1016/j.biortech.2021.124868 PMID:33707076

Karuppiah, T., & Azariah, V. E. (2019). Biomass Pretreatment for Enhancement of Biogas Production. In *Anaerobic Digestion*. IntechOpen. doi:10.5772/intechopen.82088

Kashyap, D., Tuli, H. S., Sak, K., Garg, V. K., Goel, N., Punia, S., & Chaudhary, A. (2019). Role of reactive oxygen species in cancer progression. *Current Pharmacology Reports*, *5*(2), 79–86. doi:10.100740495-019-00171-y

Kassim, M. A., & Meng, T. K. (2017). Carbon dioxide (CO_2) biofixation by microalgae and its potential for biorefinery and biofuel production. *The Science of the Total Environment*, *584–585*, 1121–1129. doi:10.1016/j.scitotenv.2017.01.172 PMID:28169025

Kathiraven, T., Sundaramanickam, A., Shanmugam, N., & Balasubramanian, T. (2015). Green synthesis of silver nanoparticles using marine algae Caulerpa racemosa and their antibacterial activity against some human pathogens. *Applied Nanoscience*, *5*(4), 499–504. doi:10.100713204-014-0341-2

Katju, V., & Bergthorsson, U. (2019). Old Trade, New Tricks: Insights into the Spontaneous Mutation Process from the Partnering of Classical Mutation Accumulation Experiments with High-Throughput Genomic Approaches. *Genome Biology and Evolution*, *11*(1), 136–165. doi:10.1093/gbe/evy252 PMID:30476040

Kaup, M., Trull, S., & Hom, E. F. Y. (2021). On the move: Sloths and their epibionts as model mobile ecosystems. *Biological Reviews of the Cambridge Philosophical Society*, *96*(6), 2638–2660. doi:10.1111/brv.12773 PMID:34309191

Kaur, M. (2017). *Kinetics of CO_2 sequestration by* Anabaena variabilis *in a stirred tank batch type photobioreactor*. Academic Press.

Kausar, R. A., & Buhaniand, S. (2020). Methylene Blue Adsorption Isotherm on *Spirulina* sp. Microalgae Biomass Coated by Silica Magnetite. *IOP Conf. Ser. Mater. Sci. Eng.*, *857*, 012019–12028.

Kaushik, B. D. (2014). Developments in cyanobacterial biofertilizer. In Proc Indian Nat Sci Acad (Vol. 80, No. 2: pp. 379-388). doi:10.16943/ptinsa/2014/v80i2/55115

Kavitha, G., Kurinjimalar, C., Sivakumar, K., Kaarthik, M., Aravind, R., Palani, P., & Rengasamy, R. (2021). *International Journal of Biological Macromolecules Optimization of polyhydroxybutyrate production utilizing waste water as nutrient source by Botryococcus braunii Kütz using response surface methodology*. Academic Press.

Kavitha, G., Kurinjimalar, C., Sivakumar, K., Kaarthik, M., Aravind, R., Palani, P., & Rengasamy, R. (2016). Optimization of polyhydroxybutyrate production utilizing waste water as nutrient source by Botryococcus braunii Kütz using response surface methodology. *International Journal of Biological Macromolecules*, *93*, 534–542. doi:10.1016/j.ijbiomac.2016.09.019 PMID:27612642

Kawamoto, H., Miki, Y., Kimura, T., Tanaka, K., Nakagawa, T., Kawamukai, M., & Matsuda, H. (2006). Effects of fucoidan from Mozuku on human stomach cell lines. *Food Science and Technology Research*, *12*(3), 218–222. doi:10.3136/fstr.12.218

Kazamia, E., Czesnick, H., Van Nguyen, T. T., Croft, M. T., Sherwood, E., Sasso, S., Hodson, S. J., Warren, M. J., & Smith, A. G. (2012). Mutualistic interactions between vitamin B12-dependent algae and heterotrophic bacteria exhibit regulation. *Environmental Microbiology*, *14*(6), 1466–1476. doi:10.1111/j.1462-2920.2012.02733.x PMID:22463064

Kazir, M., Abuhassira, Y., Robin, A., Nahor, O., Luo, J., Israel, A., Golberg, A., & Livney, Y. D. (2019). Extraction of proteins from two marine macroalgae, Ulva sp. and Gracilaria sp., for food application, and evaluating digestibility, amino acid composition and antioxidant properties of the protein concentrates. *Food Hydrocolloids*, *87*, 194–203. doi:10.1016/j.foodhyd.2018.07.047

Keener, K., Hoban, T., & Balasubramanian, R. (2000). Biotechnology and its applications. *The North Carolina Cooperative Extension Service, D*, 1–13. https://www.ces.ncsu.edu/depts/foodsci/ext/pubs/bioapp.html

Kent, W. J. (2002). BLAT---The BLAST-Like Alignment Tool. *Genome Research*, *12*(4), 656–664. doi:10.1101/gr.229202 PMID:11932250

Keshavarz, T., & Roy, I. (2010). Polyhydroxyalkanoates: Bioplastics with a green agenda. *Current Opinion in Microbiology*, *13*(3), 321–326. doi:10.1016/j.mib.2010.02.006 PMID:20227907

Kessel-Vigelius, S. K., Wiese, J., Schroers, M. G., Wrobel, T. J., Hahn, F., & Linka, N. (2013). An engineered plant peroxisome and its application in biotechnology. *Plant Science*, *210*, 232–240. doi:10.1016/j.plantsci.2013.06.005 PMID:23849130

Kessler, E. (1985). Upper limits of temperature for growth in *Chlorella. Plant Systematics and Evolution*, *151*, 67–71.

Ketting, R. F. (2011). The Many Faces of RNAi. *Developmental Cell*, *20*(2), 148–161. doi:10.1016/j.devcel.2011.01.012 PMID:21316584

Khairy, H. M., & El-Shafay, S. M. (2013). Seasonal variations in the biochemical composition of some common seaweed species from the coast of Abu Qir Bay, Alexandria, Egypt. *Oceanologia*, *55*(2), 435–452. doi:10.5697/oc.55-2.435

Khalid, M., Khalid, N., Ahmed, I., Hanif, R., Ismail, M., & Janjua, H. A. (2019). Comparative studies of three novel freshwater microalgae strains for synthesis of silver nanoparticles: Insights of characterization, antibacterial, cytotoxicity and antiviral activities. *J. Appl. Phycol.*, *29*, 1851–1863.

Khan, M. (2016). Heterotrophic Growth of Micro Algae. Recent Advances in Microalgal Biotechnology.

Khan, S. Siddique, R., Sajjad, W., Nabi, G., Hayat, K. M., Duan, P., & Yao, L. (2017). Biodiesel Production From Algae to Overcome the Energy Crisis. *HAYATI Journal of Biosciences, 24*(4), 163–167.

Khan, A., Rashid, R., Murtaza, G., & Zahra, A. (2014). Gold nanoparticles: Synthesis and applications in drug delivery. *Tropical Journal of Pharmaceutical Research, 13*(7), 1169–1177. doi:10.4314/tjpr.v13i7.23

Khan, K., & Ojha, P. K. (2020). Ecotoxicological modeling of organic chemicals for their acute toxicity in algae using classification and toxicophore-based approaches. *International Journal of Quantitative Structure-Property Relationships, 5*(2), 17–72. doi:10.4018/IJQSPR.2020040102

Khan, M. I., Lee, M. G., Shin, J. H., & Kim, J. D. (2017). Pretreatment optimization of the biomass of Microcystis aeruginosa for efficient bioethanol production. *AMB Express, 7*(1), 1–9. doi:10.118613568-016-0320-y PMID:28063146

Khan, M. I., Shin, J. H., & Kim, J. D. (2018). The promising future of microalgae: Current status, challenges, and optimization of a sustainable and renewable industry for biofuels, feed, and other products. *Microbial Cell Factories, 17*(1), 1–21. doi:10.118612934-018-0879-x PMID:29506528

Khanna, P., Kaur, A., & Goyal, D. (2019). Algae-based metallic nanoparticles: Synthesis, characterization and applications. *Journal of Microbiological Methods, 163*(June), 105656. doi:10.1016/j.mimet.2019.105656 PMID:31220512

Khanna, S., & Srivastava, A. K. (2005). Recent advances in microbial polyhydroxyalkanoates. *Process Biochemistry, 40*(2), 607–619. doi:10.1016/j.procbio.2004.01.053

Khanra, S., Mondal, M., Halder, G., Tiwari, O. N., Gayen, K., & Bhowmick, T. K. (2018). Downstream processing of microalgae for pigments, protein and carbohydrate in industrial application: A review. *Food and Bioproducts Processing, 110*, 60–84. doi:10.1016/j.fbp.2018.02.002

Khan, S., & Fu, P. (2020). Biotechnological perspectives on algae: A viable option for next generation biofuels. *Current Opinion in Biotechnology, 62*, 146–152. doi:10.1016/j.copbio.2019.09.020 PMID:31683048

Khan, S., Siddique, R., Sajjad, W., Nabi, G., Hayat, K. M., Duan, P., & Yao, L. (2017). Biodiesel production from algae to overcome the energy crisis. *Hayati Journal of Biosciences, 24*(4), 163–167. doi:10.1016/j.hjb.2017.10.003

Khan, W., Rayirath, U. P., Subramanian, S., Jithesh, M. N., Rayorath, P., Hodges, D. M., Critchley, A. T., Craigie, J. S., Norrie, J., & Prithiviraj, B. (2009). Seaweed extracts as biostimulants of plant growth and development. *Journal of Plant Growth Regulation, 28*(4), 386–399. doi:10.100700344-009-9103-x

Khan, Z., Bhadouria, P., & Bisen, P. S. (2005). Nutritional and therapeutic potential of Spirulina. *Current Pharmaceutical Biotechnology, 6*(5), 373–379. doi:10.2174/138920105774370607 PMID:16248810

Khasreen, M. M., Banfill, P. F., & Menzies, G. F. (2009). Life-cycle assessment and the environmental impact of buildings: A review. *Sustainability, 1*(3), 674–701. doi:10.3390u1030674

Khatami, K., Perez-Zabaleta, M., Owusu-Agyeman, I., & Cetecioglu, Z. (2020). *Waste to bioplastics: How close are we to sustainable polyhydroxyalkanoates production?* Waste Management.

Khetkorn, W., Incharoensakdi, A., Lindblad, P., & Jantaro, S. (2016). Enhancement of poly-3-hydroxybutyrate production in *Synechocystis* sp. PCC 6803 by overexpression of its native biosynthetic genes. *Bioresource Technology, 214*, 761–768. doi:10.1016/j.biortech.2016.05.014 PMID:27213577

Khoironi, A., Anggoro, S., & Sudarno, S. (2019). Evaluation of the Interaction Among Microalgae Spirulina sp, Plastics Polyethylene Terephthalate and Polypropylene in Freshwater Environment. *Journal of Ecological Engineering, 20*(6), 161–173. doi:10.12911/22998993/108637

Khona, D. K., Shirolikar, S. M., Gawde, K. K., Hom, E., Deodhar, M. A., & D'Souza, J. S. (2016). Characterization of salt stress-induced palmelloids in the green alga, Chlamydomonas reinhardtii. *Algal Research*, *16*, 434–448. doi:10.1016/j.algal.2016.03.035

Khoo, C. G., Dasan, Y. K., Lam, M. K., & Lee, K. T. (2019). Algae biorefinery: Review on a broad spectrum of downstream processes and products. *Bioresource Technology*, *292*, 121964. doi:10.1016/j.biortech.2019.121964 PMID:31451339

Khoo, H. H., Koh, C. Y., Shaik, M. S., & Sharratt, P. N. (2013). Bioenergy co-products derived from microalgae biomass via thermochemical conversion – Life cycle energy balances and CO2 emissions. *Bioresource Technology*, *143*, 298–307. doi:10.1016/j.biortech.2013.06.004 PMID:23810951

Khoo, K. S., Chew, K. W., Yew, G. Y., Leong, W. H., Chai, Y. H., Show, P. L., & Chen, W.-H. (2020). Recent advances in downstream processing of microalgae lipid recovery for biofuel production. *Bioresource Technology*, *304*, 122996. doi:10.1016/j.biortech.2020.122996 PMID:32115347

Khosravitabar, F. (2020). Microalgal biohydrogen photoproduction: Scaling up challenges and the ways forward. *Journal of Applied Phycology*, *32*(1), 277–289.

Kim, D. E., Chivian, D., & Baker, D. (2004). Protein structure prediction and analysis using the Robetta server. *Nucleic Acids Research*, *32*, 526–531. doi:10.1093/nar/gkh468 PMID:15215442

Kim, D. G., La, H. J., Ahn, C. Y., Park, Y. H., & Oh, H. M. (2011). Harvest of *Scenedesmus* sp. with bioflocculant and reuse of culture medium for subsequent high-density cultures. *Bioresource Technology*, *102*(3), 3163–3168.

Kim, D. H., & Kim, M. S. (2011). Hydrogenases for biological hydrogen production. *Bioresource Technology*, *102*(18), 8423–8431. doi:10.1016/j.biortech.2011.02.113 PMID:21435869

Kim, D. Y., Lee, K., Lee, J., Lee, Y. H., Han, J. I., Park, J. Y., & Oh, Y.-K. (2017). Acidified-flocculation process for harvesting of microalgae: Coagulant reutilization and metal-free-microalgae recovery. *Bioresource Technology*, *239*, 190–196.

Kim, D.-Y., Oh, Y.-K., Park, J.-Y., Kim, B., Choi, S.-A., & Han, J.-I. (2015). An Integrated Process for Microalgae Harvesting and Cell Disruption by the Use of Ferric Ions. *Bioresource Technology*, *191*, 469–474. doi:10.1016/j.biortech.2015.03.020 PMID:25817422

Kim, E. J., Park, S. Y., Lee, J. Y., & Park, J. H. Y. (2010). Fucoidan present in brown algae induces apoptosis of human colon cancer cells. *BMC Gastroenterology*, *10*(1), 1–11. doi:10.1186/1471-230X-10-96 PMID:20727207

Kim, E. S., Yu, Q., & Deng, B. (2011). Plasma surface modification of nanofiltration (NF) thin-film composite (TFC) membranes to improve anti organic fouling. *Appl. Sur. Sci.*, *257*(23), 9863–9871. doi:10.1016/j.apsusc.2011.06.059

Kim, G. M., Chang, W.-S., & Kim, Y.-K. (2021). Biocomposites Using Whole or Valuable Component-Extracted Microalgae Blended with Polymers: A Review. *Catalysts*, *12*(1), 25. doi:10.3390/catal12010025

Kim, H. M., Oh, C. H., & Bae, H. J. (2017). Comparison of red microalgae (Porphyridium cruentum) culture conditions for bioethanol production. *Bioresource Technology*, *233*, 44–50. doi:10.1016/j.biortech.2017.02.040 PMID:28258995

Kim, H., & Kim, J.-S. (2014). A guide to genome engineering with programmable nucleases. *Nature Reviews. Genetics*, *15*(5), 321–334. doi:10.1038/nrg3686 PMID:24690881

Kim, J. K., Um, B. H., & Kim, T. H. (2012). Bioethanol production from micro-algae, Schizocytrium sp., using hydrothermal treatment and biological conversion. *Korean Journal of Chemical Engineering*, *29*(2), 209–214. doi:10.100711814-011-0169-3

Kim, J. W., Park, S.-B., Tran, Q.-G., Cho, D.-H., Choi, D.-Y., Lee, Y. J., & Kim, H.-S. (2020). Functional expression of polyethylene terephthalate-degrading enzyme (PETase) in green microalgae. *Microbial Cell Factories*, *19*(1), 97. doi:10.118612934-020-01355-8 PMID:32345276

Kim, J. Y., Yoon, M. Y., Cha, M. R., Hwang, J. H., Park, E., Choi, S. U., Park, H.-R., & Hwang, Y. I. (2007). Methanolic extracts of Plocamium telfairiae induce cytotoxicity and caspase-dependent apoptosis in HT-29 human colon carcinoma cells. *Journal of Medicinal Food*, *10*(4), 587–593. doi:10.1089/jmf.2007.002 PMID:18158827

Kim, M. S., & Kwak, D. H. (2020). Auto/bio-flocculation conditions to separate algal particles without chemical coagulants for flotation and sedimentation processes. *Separation Science and Technology*, *55*(6), 1185–1196.

Kim, M., Ahn, J., Jeon, H., & Jin, E. (2017). Development of a *Dunaliella tertiolecta* Strain with Increased Zeaxanthin Content Using Random Mutagenesis. *Marine Drugs*, *15*(6), 189. doi:10.3390/md15060189 PMID:28635662

Kimosop, S. J., Getenga, Z. M., & Orata, F. (2016). Residue levels and discharge loads of antibiotics in wastewater treatment plants (WWTPs), hospital lagoons, and rivers within Lake Victoria Basin, Kenya. *Environmental Monitoring and Assessment*, *188*, 532.

Kim, S.-K., & Chojnacka, K. (2015). *Marine algae extracts: processes, products, and applications*. John Wiley & Sons. doi:10.1002/9783527679577

Kim, Y. J., Joo, H. W., Park, J., Kim, D. K., Jeong, K. J., & Chang, Y. K. (2015). Production of 2, 3-butanediol by Klebsiella oxytoca from various sugars in microalgal hydrolysate. *Biotechnology Progress*, *31*(6), 1669–1675. doi:10.1002/btpr.2168 PMID:26400837

Kim, Y. S., Li, X. F., Kang, K. H., Ryu, B., & Kim, S. K. (2014). Stigmasterol isolated from marine microalgae Navicula incerta induces apoptosis in human hepatoma HepG2 cells. *BMB Reports*, *47*(8), 433–438. doi:10.5483/BMBRep.2014.47.8.153 PMID:24286323

Kim, Y. S., Son, H. J., & Jeong, S. Y. (2015). Isolation of an algicide from a marine bacterium and its effects against the toxic dinoflagellate Alexandrium catenella and other harmful algal bloom species. *Journal of Microbiology (Seoul, Korea)*, *53*(8), 511–517. doi:10.100712275-015-5303-1 PMID:26224453

Kinch, J., Bagita, J., & Bate, M. (2003). *Exploring the potential for seaweed farming in Milne Bay Province*. Fisheries Newsletter-South Pacific Commission.

Kindle, K. L. (1990). High-frequency nuclear transformation of Chlamydomonas reinhardtii. *Proceedings of the National Academy of Sciences of the United States of America*, *87*(3), 1228–1232. doi:10.1073/pnas.87.3.1228 PMID:2105499

Kiple, K. F., & Ornelas, K. (2000). *The Cambridge World History of Food*. Cambridge University Press.

Kiran, B. R., & Venkata Mohan, S. (2021). Microalgal Cell Biofactory—Therapeutic, Nutraceutical and Functional Food Applications. *Plants*, *10*(5), 836. doi:10.3390/plants10050836 PMID:33919450

Kirk, J. T. O. (1994). *Light and photosynthesis in aquatic ecosystems* (2nd ed.). Cambridge University Press.

Kirst, G. O. (1990). Salinity Tolerance of Eukaryotic Marine Algae. *Annual Review of Plant Physiology and Plant Molecular Biology*, *41*(1), 21–53. doi:10.1146/annurev.pp.41.060190.000321

Kishimoto, S., & Sugiura, G. (2014). Algae as a soil conditioner. *Int Achieve Future*, *5*(4), 12–23.

Kitamoto, H. K., Shinozaki, Y., Cao, X.-H., Morita, T., Konishi, M., Tago, K., Kajiwara, H., Koitabashi, M., Yoshida, S., Watanabe, T., Sameshima-Yamashita, Y., Nakajima-Kambe, T., & Tsushima, S. (2011). Phyllosphere yeasts rapidly break down biodegradable plastics. *AMB Express*, *1*(1), 44. doi:10.1186/2191-0855-1-44 PMID:22126328

Kitaya, Y., Xiao, L., Masuda, A., Ozawa, T., Tsuda, M., & Omasa, K. (2008) Effects of temperature, photosynthetic photon flux density, photoperiod and O2 and CO2 concentrations on growth rates of the symbiotic dinoflagellate, Amphidinium sp. In *Nineteenth International Seaweed Symposium* (pp. 287–292). Springer. 10.100710811-008-9331-7

Klassen, V., Blifernez-Klassen, O., Wibberg, D., Winkler, A., Kalinowski, J., Posten, C., & Kruse, O. (2017). Highly efficient methane generation from untreated microalgae biomass. *Biotechnology for Biofuels*, *10*(1), 186. doi:10.118613068-017-0871-4 PMID:28725266

Klein, C. B., Bonomi, A., & Filho, M. R. (2018). Integration of microalgae production with industrial biofuel facilities: A critical review. *Renewable & Sustainable Energy Reviews*, *82*, 1376–1392. doi:10.1016/j.rser.2017.04.063

Knauer, K., & Hemond, H. (2000). Accumulation and reduction of arsenate by the freshwater green alga Chlorella sp.(Chlorophyta). *Journal of Phycology*, *36*(3), 506–509. doi:10.1046/j.1529-8817.2000.99056.x PMID:29544008

Knowles, E. J., & Castenholz, R. W. (2008). Effect of exogenous extracellular polysaccharides on the desiccation and freezing tolerance of rock-inhabiting phototrophic microorganisms. *FEMS Microbiology Ecology*, *66*(2), 261–270. doi:10.1111/j.1574-6941.2008.00568.x PMID:18710394

Knuckey, R. M., Brown, M. R., Robert, R., & Frampton, D. M. F. (2006). Production of microalgal concentrates by flocculation and their assessment as aquaculture feeds. *Aquacultural Engineering*, *35*, 300–313.

Kobayashi, M., Hirai, N., Kurimura, Y., Ohigashi, H., & Tsuji, Y. (1997). Abscisic Acid-Dependent Algal Morphogenesis in the Unicellular Green Alga *Haematococcuspluvialis*. *Plant Growth Regulation*, *22*(2), 79–85. doi:10.1023/A:1005862809711

Koehl, M. A. R. (1986). Seaweeds in moving water: form and mechanical function. In T. J. Givnish (Ed.), *On the economy of plant form and function* (pp. 603–634). Cambridge University Press.

Koelmans, A. A., Besseling, E., Wegner, A., & Foekema, E. M. (2013). Plastic as a Carrier of POPs to Aquatic Organisms: A Model Analysis. *Environmental Science & Technology*, *47*(14), 7812–7820. doi:10.1021/es401169n PMID:23758580

Kohout, C., Santi, C., & Polito, L. (2018). Anisotropic gold nanoparticles in biomedical applications. *International Journal of Molecular Sciences*, *19*(11), 3385. doi:10.3390/ijms19113385 PMID:30380664

Kokou, F., Makridis, P., Kentouri, M., & Divanach, P. (2012). Antibacterial activity in microalgae cultures. *Aquaculture Research*, *43*(10), 1520–1527. doi:10.1111/j.1365-2109.2011.02955.x

Koline-Sanderson. (2014). *Rotary drum vacuum filters.* Accessed on October 23, 2014 from http://www.komline.com/docs/rotary_ drum_vacuum_filter.html

Koller, M. (2019). Switching from petro-plastics to microbial polyhydroxyalkanoates (PHA): The biotechnological escape route of choice out of the plastic predicament. *EuroBiotech J*, *3*(1), 32–44. doi:10.2478/ebtj-2019-0004

Koller, M., Atlić, A., Dias, M., Reiterer, A., & Braunegg, G. (2010). Microbial PHA Production from Waste Raw Materials. *Microbial PHA Production from Waste Raw Materials.*, *14*, 85–119. doi:10.1007/978-3-642-03287-5_5

Koller, M., Muhr, A., & Braunegg, G. (2014). Microalgae as versatile cellular factories for valued products. *Algal Research*, *6*, 52–63. doi:10.1016/j.algal.2014.09.002

Koller, M., Niebelschütz, H., & Braunegg, G. (2013). Strategies for Recovery and Purification of Poly[(R)-3-Hydroxyalkanoates] (PHA) Biopolyesters from Surrounding Biomass. *Engineering in Life Sciences*, *13*(6), 549–562. doi:10.1002/elsc.201300021

Kongara, S., & Karantza, V. (2012). The interplay between autophagy and ROS in tumorigenesis. *Frontiers in Oncology*, *2*, 171. doi:10.3389/fonc.2012.00171 PMID:23181220

Koopi, H., & Buazar, F. (2018). A novel one-pot biosynthesis of pure alpha aluminum oxide nanoparticles using the macroalgae Sargassum ilicifolium: A green marine approach. *Ceramics International, 44*(8), 8940–8945. doi:10.1016/j.ceramint.2018.02.091

Kopta, T., Pavlíková, M., Sękara, A., Pokluda, R., & Maršálek, B. (2018). Effect of Bacterial-algal Biostimulant on the Yield and Internal Quality of Lettuce (Lactuca sativa L.). *Produced for Spring and Summer Crop., 46*(2), 615–621. doi:10.15835/nbha46211110

Korpinen, S., Honkanen, T., Vesakoski, O., Hemmi, A., Koivikko, R., Loponen, J., & Jormalainen, V. (2007). Macroalgal communities face the challenge of changing biotic interactions: Review with focus on the Baltic Sea. *Ambio, 36*(2/3), 203–211. doi:10.1579/0044-7447(2007)36[203:MCFTCO]2.0.CO;2 PMID:17520935

Kosourov, S., Makarova, V., Fedorov, A. S., Tsygankov, A., Seibert, M., & Ghirardi, M. L. (2005). The Effect of Sulfur Re-Addition on H2 Photoproduction by Sulfur-Deprived Green Algae. *Photosynthesis Research, 85*(3), 295–305. doi:10.100711120-005-5105-0 PMID:16170632

Kosseva, M. R., & Rusbandi, E. (2018). Trends in the biomanufacture of polyhydroxyalkanoates with focus on downstream processing. *International Journal of Biological Macromolecules, 107*(PartA), 762–778. doi:10.1016/j.ijbiomac.2017.09.054

Koutra, E., Economou, C. N., Tsafrakidou, P., & Komaros, M. (2018). Bio-based products from microalgae cultivated in digestates. *Trends in Biotechnology, 36*(8), 819–833. doi:10.1016/j.tibtech.2018.02.015 PMID:29605178

Koutra, E., Mastropetros, S. G., Ali, S. S., Tsigkou, K., & Kornaros, M. (2021). Assessing the potential of Chlorella vulgaris for valorization of liquid digestates from agro-industrial and municipal organic wastes in a biorefinery approach. *Journal of Cleaner Production, 280*, 124352. doi:10.1016/j.jclepro.2020.124352

Kouzuma, A., & Watanabe, K. (2015). Exploring the potential of algae/bacteria interactions. *Current Opinion in Biotechnology, 33*, 125–129. doi:10.1016/j.copbio.2015.02.007 PMID:25744715

Kovalcik, A., Meixner, K., Mihalic, M., Zeilinger, W., Fritz, I., Fuchs, W., Kucharczyk, P., Stelzer, F., & Drosg, B. (2017). Characterization of polyhydroxyalkanoates produced by *Synechocystis salina* from digestate supernatant. *International Journal of Biological Macromolecules, 102*, 497–504. doi:10.1016/j.ijbiomac.2017.04.054 PMID:28419829

Kozlova, T. A., Hardy, B. P., Krishna, P., & Levin, D. B. (2017). Effect of phytohormones on growth and accumulation of pigments and fatty acids in the microalgae Scenedesmus quadricauda. *Algal Research, 27*, 325–334.

Kraan, S. (2010). Mass-cultivation of carbohydrate rich macroalgae, a possible solution for sustainable biofuel production. *Mitigation and Adaptation Strategies for Global Change, 18*(1), 27–46. doi:10.100711027-010-9275-5

Kraemer, G. P., & Chapman, D. J. (1991). biomechanics and alginic acid composition during hydrodynamic adaptation by egregia menziesii (phaeophyta) juveniles. *Journal of Phycology, 27*(1), 47–53. doi:10.1111/j.0022-3646.1991.00047.x

Kraemer, S. A., Böndel, K. B., Ness, R. W., Keightley, P. D., & Colegrave, N. (2017). Fitness change in relation to mutation number in spontaneous mutation accumulation lines of *Chlamydomonas reinhardtii. Evolution, 71*(12), 2918–2929. doi:10.1111/evo.13360 PMID:28884790

Kranz, D., & Dobbelstein, M. (2012). A killer promoting survival: p53 as a selective means to avoid side effects of chemotherapy. *Cell Cycle (Georgetown, Tex.), 11*(11), 2053–2054. doi:10.4161/cc.20698 PMID:22622088

Krasaesueb, N., Incharoensakdi, A., & Khetkorn, W. (2019). Utilization of shrimp wastewater for poly-β-hydroxybutyrate production by Synechocystis sp. PCC 6803 strain ΔSphU cultivated in photobioreactor. *Biotechnology Reports (Amsterdam, Netherlands), 23*, e00345. doi:10.1016/j.btre.2019.e00345 PMID:31193428

Krasovec, M., Eyre-Walker, A., Sanchez-Ferandin, S., & Piganeau, G. (2017). Spontaneous Mutation Rate in the Smallest Photosynthetic Eukaryotes. *Molecular Biology and Evolution*, *34*(7), 1770–1779. doi:10.1093/molbev/msx119 PMID:28379581

Krasovec, M., Sanchez-Brosseau, S., Grimsley, N., & Piganeau, G. (2018). Spontaneous mutation rate as a source of diversity for improving desirable traits in cultured microalgae. *Algal Research*, *35*, 85–90. doi:10.1016/j.algal.2018.08.003

Krasovec, M., Sanchez-Brosseau, S., & Piganeau, G. (2019). First Estimation of the Spontaneous Mutation Rate in Diatoms. *Genome Biology and Evolution*, *11*(7), 1829–1837. doi:10.1093/gbe/evz130 PMID:31218358

Krassowski, M., Das, V., Sahu, S. K., & Misra, B. B. (2020). State of the Field in Multi-Omics Research: From Computational Needs to Data Mining and Sharing. *Frontiers in Genetics*, *11*(1598), 610798. Advance online publication. doi:10.3389/fgene.2020.610798 PMID:33362867

Krienitz, L. (1996). Syllabus der Boden-Luft-und Flechtenalge. Gustav Fischer Verlag.

Krienitz, L., Huss, V. A. R., & Bock, C. (2015). Chlorella: 125 years of the green survivalist. *Trends in Plant Science*, *20*(2), 67–69. doi:10.1016/j.tplants.2014.11.005 PMID:25500553

Krienitz, L., & Schagerl, M. (2016). Tiny and Tough: Microphytes of East African Soda Lakes. In M. Schagerl (Ed.), *Soda Lakes of East Africa* (pp. 149–177). Springer. doi:10.1007/978-3-319-28622-8_6

Krol, M., Maxwell, D. P., & Huner, N. P. A. (1997). Exposure of *Dunaliella salina* to low temperature mimics the high lightinduced accumulation of carotenoids and the carotenoid binding protein (Cbr). *The Plant Cell*, *38*(2), 213–216.

Kronberga, M., Karklina, D., Murniece, I., & Kruma, Z. (2011). Changes of agar-agar gel properties after replacing sucrose by inulin syrup. *Proceedings of 6th Baltic Conference on Food Science and Technology FOODBALT-2011, 'Innovations for Food Science and Production'*. Latvia University of Agriculture.

Krug, L., Erlacher, A., Markut, K., Berg, G., & Cernava, T. (2020). The microbiome of alpine snow algae shows a specific inter-kingdom connectivity and algae-bacteria interactions with supportive capacities. *The ISME Journal*, *14*(9), 2197–2210. doi:10.103841396-020-0677-4 PMID:32424246

Kudo, I., Miyamoto, M., Noiri, Y., & Maita, Y. (2000). Combined effects of temperature and iron on the growth and physiology of the marine diatom *Phaeodactylum tricornutum* (Bacillariophyceae). *Journal of Phycology*, *36*, 1096–1102.

Kuenz, A., Grimm, D., & Rahmann, G. (2021). Versatility of algae—Exploring the potential of algae for nutrient circulation. *Organic Agriculture*, *11*(2), 251–260. doi:10.100713165-020-00308-0

Kulik, M. M. (1995). The Potential for Using Cyanobacteria (Blue-Green Algae) and Algae in the Biological Control of Plant Pathogenic Bacteria and Fungi. *European Journal of Plant Pathology*, *101*, 585–599.

Kumar, S. R., Shaiju, P. & O'Connor, K. E. (2020). Bio-based and biodegradable polymers - state-of-the-art, challenges and emerging trends. *Current Opinion in Green and Sustainable Chemistry*, *21*, 75-81.

Kumar, S., Singh, N. K., Kumar, A., & Shukla, P. (2021). Next Generation Biofuel Production in the Omics Era: Potential and Prospects. In Omics Technologies for Sustainable Agriculture and Global Food Security (vol. 2, pp. 293-311). Springer.

Kumar, A. (2021). Current and Future Perspective of Microalgae for Simultaneous Wastewater Treatment and Feedstock for Biofuels Production. *Chemistry Africa*, (2), 249–275. Advance online publication. doi:10.100742250-020-00221-9

Kumar, A., Agarwal, D. K., Kumar, S., Reddy, Y. M., Chintagunta, A. D., Saritha, K. V., Pal, G., & Kumar, S. P. J. (2019). Nutraceuticals derived from seed storage proteins: Implications for health wellness. *Biocatalysis and Agricultural Biotechnology*, *17*, 710–717. doi:10.1016/j.bcab.2019.01.044

Kumar, A., Imran, M., Kumar, S. P. J., Kumar, M., Singh, A. N., & Srivastava, A. K. (2018). Assessment of soil urease activity and its kinetic parameters in chilli rhizosphere under the influence of pesticides.Biopesticides. *International, 14*, 47–57.

Kumar, B. R., Mathimani, T., Sudhakar, M. P., Rajendran, K., Nizami, A. S., Brindhadevi, K., & Pugazhendhi, A. (2021). A state of the art review on the cultivation of algae for energy and other valuable products: Application, challenges, and opportunities. *Renewable & Sustainable Energy Reviews, 138*, 110649. doi:10.1016/j.rser.2020.110649

Kumar, C. G., Joo, H. S., Kavali, R., Choi, J. W., & Chang, C. S. (2004). Characterization of an extracellular biopolymer flocculant from a haloalkalophilic Bacillus isolate. *World Journal of Microbiology & Biotechnology, 20*(8), 837–843.

Kumar, C. S., Ganesan, P., Suresh, P. V., & Bhaskar, N. (2008). Seaweeds as a source of nutritionally beneficial compounds-a review. *Journal of Food Science and Technology, 45*(1), 1.

Kumar, D. A., Palanichamy, V., & Roopan, S. M. (2014). Photocatalytic action of AgCl nanoparticles and its antibacterial activity. *Journal of Photochemistry and Photobiology. B, Biology, 138*, 302–306. doi:10.1016/j.jphotobiol.2014.06.011 PMID:25022464

Kumar, G., Bakonyi, P., Sivagurunathan, P., Kim, S.-H., Nemestóthy, N., Bélafi-Bakó, K., & Lin, C.-Y. (2015). Enhanced biohydrogen production from beverage industrial wastewater using external nitrogen sources and bioaugmentation with facultative anaerobic strains. *Journal of Bioscience and Bioengineering, 120*(2), 155–160. doi:10.1016/j.jbiosc.2014.12.011 PMID:25661265

Kumar, G., Nguyen, D. D., Sivagurunathan, P., Kobayashi, T., Xu, K., & Chang, S. W. (2018). Cultivation of Microalgal Biomass Using Swine Manure for Biohydrogen Production: Impact of Dilution Ratio and Pretreatment. *Bioresource Technology, 260*, 16–22. doi:10.1016/j.biortech.2018.03.029 PMID:29604564

Kumar, G., & Sahoo, D. (2011). Effect of seaweed liquid extract on growth and yield of *Triticum aestivum* var. Pusa Gold. *Journal of Applied Phycology, 23*(2), 251–255. doi:10.100710811-011-9660-9

Kumar, G., Shekh, A., Jakhu, S., Sharma, Y., Kapoor, R., & Sharma, T. R. (2020). Bioengineering of microalgae: Recent advances, perspectives, and regulatory challenges for industrial application. *Frontiers in Bioengineering and Biotechnology, 8*, 914. doi:10.3389/fbioe.2020.00914

Kumar, G., Zhen, G., Kobayashi, T., Sivagurunathan, P., Kim, S. H., & Xu, K. Q. (2016). Impact of pH control and heat pre-treatment of seed inoculum in dark H2 fermentation: A feasibility report using mixed microalgae biomass as feedstock. *International Journal of Hydrogen Energy, 41*(7), 4382–4392. doi:10.1016/j.ijhydene.2015.08.069

Kumari, J., Udawat, P., Dubey, A. K., Haque, M. I., Rathore, M. S., & Jha, B. (2017). Overexpression of SbSI-1, a nuclear protein from Salicornia brachiata confers drought and salt stress tolerance and maintains photosynthetic efficiency in transgenic tobacco. *Frontiers in Plant Science, 8*, 1215. doi:10.3389/fpls.2017.01215 PMID:28751902

Kumar, K. A., & Rengasamy, R. (2000). Antibacterial Activities of Seaweed Extracts/Fractions Obtained through a TLC Profile against the Phytopathogenic Bacterium Xanthomonas Oryzae Pv. Oryzae. *Botanica Marina, 43*, 417–421.

Kumar, K., & Das, D. (2012). Growth characteristics of Chlorella sorokiniana in airlift and bubble column photobioreactors. *Bioresource Technology, 116*, 307–313.

Kumar, M., Olajire Oyedun, A., & Kumar, A. (2018). A review on the current status of various hydrothermal technologies on biomass feedstock. *Renewable & Sustainable Energy Reviews, 81*, 1742–1770. doi:10.1016/j.rser.2017.05.270

Kumar, M., Sun, Y., Rathour, R., Pandey, A., Thakur, I. S., & Tsang, D. C. (2020). Algae as potential feedstock for the production of biofuels and value-added products: Opportunities and challenges. *The Science of the Total Environment*, *716*, 137116. doi:10.1016/j.scitotenv.2020.137116 PMID:32059310

Kumar, R., Linga, P., Ripmeester, J. A., & Englezos, P. (2009). Two-stage clathrate hydrate/membrane process for precombustion capture of carbon dioxide and hydrogen. *Journal of Environmental Engineering*, *135*(6), 411–417. doi:10.1061/(ASCE)EE.1943-7870.0000002

Kumar, S., Hablot, E., Moscoso, J. L. G., Obeid, W., Hatcher, P. G., DuQuette, B. M., Graiver, D., Narayan, R., & Balan, V. (2014). Polyurethanes preparation using proteins obtained from microalgae. *Journal of Materials Science*, *49*(22), 7824–7833. doi:10.100710853-014-8493-8

Kunanopparat, T., Menut, P., Morel, M.-H., & Guilbert, S. (2008). Reinforcement of plasticized wheat gluten with natural fibers: From mechanical improvement to deplasticizing effect. *Composites. Part A, Applied Science and Manufacturing*, *39*(5), 777–785. doi:10.1016/j.compositesa.2008.02.001

Kuo, C.-M., Lin, T.-H., Yang, Y.-C., Zhang, W.-X., Lai, J.-T., Wu, H.-T., Chang, J.-S., & Lin, C.-S. (2017). Ability of an alkali-tolerant mutant strain of the microalga *Chlorella* sp. AT1 to capture carbon dioxide for increasing carbon dioxide utilization efficiency. *Bioresource Technology*, *244*(Pt 1), 243–251. doi:10.1016/j.biortech.2017.07.096 PMID:28780257

Kurniawati, H. A., Ismadji, S., & Liu, J. C. (2014). Microalgae harvesting using natural saponin and chitosan. *Bioresource Technology*, *166*, 429–434.

Kuroda, K., & Ueda, M. (2010). Engineering of microorganisms towards recovery of rare metal ions. *Applied Microbiology and Biotechnology*, *87*(1), 53–60. doi:10.100700253-010-2581-8 PMID:20393699

Kusk, K. O., Christensen, A. M., & Nyholm, N. (2018). Algal growth inhibition test results of 425 organic chemical substances. *Chemosphere*, *204*, 405–412. doi:10.1016/j.chemosphere.2018.04.047 PMID:29677648

Kwan, T. H., Pleissner, D., Lau, K. Y., Venus, J., Pommeret, A., & Lin, C. S. K. (2015). Technoeconomic analysis of a food waste valorization process via microalgae cultivation and co-production of plasticizer, lactic acid and animal feed from algal biomass and food waste. *Bioresource Technology*, *198*, 292–299. doi:10.1016/j.biortech.2015.09.003 PMID:26402872

Kwon, H. K., Jeon, J. Y., & Oh, S. J. (2017). Potential for heavy metal (copper and zinc) removal from contaminated marine sediments using microalgae and light emitting diodes. *Ocean Science Journal*, *52*(1), 57–66. doi:10.100712601-017-0001-z

Kwon, H. K., Oh, S. J., & Yang, H.-S. (2013). Growth and uptake kinetics of nitrate and phosphate by benthic microalgae for phytoremediation of eutrophic coastal sediments. *Bioresource Technology*, *129*, 387–395. doi:10.1016/j.biortech.2012.11.078 PMID:23262016

Kwon, H., Lu, M., Lee, E. Y., & Lee, J. (2014). Harvesting of microalgae using flocculation combined with dissolved air flotation. *Biotechnology and Bioprocess Engineering; BBE*, *19*, 143–149.

Kwon, M. J., & Nam, T. J. (2006). Porphyran induces apoptosis-related signal pathways in AGS gastric cancer cell lines. *Life Sciences*, *79*(20), 1956–1962. doi:10.1016/j.lfs.2006.06.031 PMID:16876203

Kwon, M. J., & Nam, T. J. (2007). A polysaccharide of the marine alga Capsosiphon fulvescens induces apoptosis in AGS gastric cancer cells via an IGF-IR-mediated PI3K/Akt pathway. *Cell Biology International*, *31*(8), 768–775. doi:10.1016/j.cellbi.2007.01.010 PMID:17344071

Kwon, M. J., & Nam, T. J. (2007). Chromatographically purified porphyran from Porphyra yezoensis effectively inhibits the proliferation of human cancer cells. *Food Science and Biotechnology*, *16*(6), 873–878.

Kwon, O.-M., Kim, D.-H., Kim, S.-K., & Jeong, G.-T. (2016). Production of sugars from macroalgae Gracilaria verrucosa using combined process of citric acid-catalyzed pretreatment and enzymatic hydrolysis. *Algal Research*, *13*, 293–297. doi:10.1016/j.algal.2015.12.011

Kwon, S., Kang, N. K., Koh, H. G., Shin, S. E., Lee, B., Jeong, B. R., & Chang, Y. K. (2018). Enhancement of biomass and lipid productivity by overexpression of a bZIP transcription factor in Nannochloropsis salina. *Biotechnology and Bioengineering*, *115*(2), 331–340. doi:10.1002/bit.26465 PMID:28976541

Labeeuw, L., Khey, J., Bramucci, A.R., Atwal, H., de la Mata, A.P., Harynuk, J., & Case, R.J. (2016). Indole-3-acetic acid is produced by *Emiliania huxleyi coccolith*-bearing cells and triggers a physiological response in bald cells. *Frontiers in Microbiology, 8*(7), 828.

Ladha, J. K., & Reddy, P. M. (2003). Nitrogen fixation in rice systems: State of knowledge and future prospects. *Plant and Soil*, *252*(1), 151–167. doi:10.1023/A:1024175307238

Lahaye, M., & Robic, A. (2007). Structure and functional properties of ulvan, a polysaccharide from green seaweeds. *Biomacromolecules*, *8*(6), 1765–1774. doi:10.1021/bm061185q PMID:17458931

Lakaniemi, A.-M., Hulatt, C. J., Thomas, D. N., Tuovinen, O. H., & Puhakka, J. A. (2011). Biogenic hydrogen and methane production from Chlorella vulgaris and Dunaliella tertiolectabiomass. *Biotechnology for Biofuels*, *4*(1), 34. doi:10.1186/1754-6834-4-34 PMID:21943287

Lakatos, G., Deák, Z., Vass, I., Rétfalvi, T., Rozgonyi, S., Rákhely, G., Ördög, V., Kondorosi, É., & Maróti, G. (2014). Bacterial symbionts enhance photo-fermentative hydrogen evolution of Chlamydomonas algae. *Green Chemistry*, *16*(11), 4716–4727. doi:10.1039/C4GC00745J

Lam, M. K., & Lee, K. T. (2013). Biohydrogen Production from Algae. In Biohydrogen. Elsevier B.V. doi:10.1016/B978-0-444-59555-3.00008-8

Lam, M. K., Loy, A. C. M., Yusup, S., & Lee, K. T. (2019). Biohydrogen Production From Algae. In *Biohydrogen* (pp. 219–245). Elsevier. doi:10.1016/B978-0-444-64203-5.00009-5

Landsberg, J. H. (2002). The effects of harmful algal blooms on aquatic organisms. *Reviews in Fisheries Science*, *10*(2), 113–390. doi:10.1080/20026491051695

Lange, K. W., Hauser, J., Nakamura, Y., & Kanaya, S. (2015). Dietary seaweeds and obesity. *Food Science and Human Wellness*, *4*(3), 87–96. doi:10.1016/j.fshw.2015.08.001

Lanlan, Z., Lin, C., Wang, J., Yu, C., Xin, G., Zhang, Z., & Tianzhong, L. (2015). Attached cultivation for improving the biomass productivity of Spirulina platensis. *Bioresource Technology*, *181*, 136–142. doi:10.1016/j.biortech.2015.01.025 PMID:25647023

LaPanse, A. J., Krishnan, A., & Posewitz, M. C. (2020). Adaptive Laboratory Evolution for algal strain improvement: Methodologies and applications. *Algal Research*, 102122.

Laporte, D., Vera, J., Chandía, N. P., Zuñiga, E., Matsuhiro, B., & Moenne, A. (2007). Structurally unrelated oligosaccharides obtained frommarine macroalgae differentially stimulate growth and defense against TMVin tobacco plants. *Journal of Applied Phycology*, *19*, 79–88. doi:10.100710811-006-9114-y

Larkin, M. A., Blackshields, G., Brown, N. P., Chenna, R., McGettigan, P. A., McWilliam, H., Valentin, F., Wallace, I. M., Wilm, A., Lopez, R., Thompson, J. D., Gibson, T. J., & Higgins, D. G. (2007). Clustal W and Clustal X version 2.0. *Bioinformatics (Oxford, England)*, *23*(21), 2947–2948. doi:10.1093/bioinformatics/btm404 PMID:17846036

Larkum, A. W., Ross, I. L., Kruse, O., & Hankamer, B. (2012). Selection, breeding and engineering of microalgae for bioenergy and biofuel production. *Trends in Biotechnology*, *30*(4), 198–205. doi:10.1016/j.tibtech.2011.11.003 PMID:22178650

Laurens, L. M. L., Chen-Glasser, M., & Mcmillan, J. D. (2017). A perspective on renewable bioenergy from photosynthetic algae as feedstock for biofuels and bioproducts. *Algal Research*, *24*, 261–264. doi:10.1016/j.algal.2017.04.002

Laurens, L. M. L., Markham, J., Templeton, D. W., Christensen, E. D., Van Wychen, S., Vadelius, E. W., Chen-Glasser, M., Dong, T., Davis, R., & Pienkos, P. T. (2017). Development of algae biorefinery concepts for biofuels and bioproducts; a perspective on process-compatible products and their impact on cost-reduction. *Energy & Environmental Science*, *10*(8), 1716–1738. doi:10.1039/C7EE01306J

Lauritano, C., Ferrante, M. I., & Rogato, A. (2019). Marine Natural Products from Microalgae: An -Omics Overview. *Marine Drugs*, *17*(5), 269. doi:10.3390/md17050269 PMID:31067655

Lavoie, M., Saint-Béat, B., Strauss, J., Guérin, S., Allard, A., Hardy, S. V., Falciatore, A., & Lavaud, J. (2020). Genome-Scale Metabolic Reconstruction and *in Silico* Perturbation Analysis of the Polar Diatom *Fragilariopsis cylindrus* Predicts High Metabolic Robustness. *Biology (Basel)*, *9*(2), 30–49. doi:10.3390/biology9020030 PMID:32079178

Lawrence, K. P., Long, P. F., & Young, A. R. (2017). Mycosporine-like Amino Acids for Skin Photoprotection. *Current Medicinal Chemistry*, 24. PMID:28554325

Lawrence, K. P., Long, P. F., & Young, A. R. (2018). Mycosporine-like amino acids for skin photoprotection. *Current Medicinal Chemistry*, *25*(40), 5512–5527. doi:10.2174/0929867324666170529124237 PMID:28554325

Laycock, B., Halley, P., Pratt, S., Werker, A., & Lant, P. (2013). The chemomechanical properties of microbial polyhydroxyalkanoates. *Progress in Polymer Science*, *38*(3–4), 536–583. doi:10.1016/j.progpolymsci.2012.06.003

Laza-Martínez, A., Fernández-Marín, B., & García-Plazaola, J. I. (2019). Rapid colour changes in Euglena sanguinea (Euglenophyceae) caused by internal lipid globule migration. *European Journal of Phycology*, *54*(1), 91–101. doi:10.1080/09670262.2018.1513571

Le Lann, K., Rumin, J., Cérantola, S., Culioli, G., & Stiger-Pouvreau, V. (2014). Spatiotemporal variations of diterpene production in the brown macroalga *Bifurcaria bifurcata* from the western coasts of Brittany (France). *Journal of Applied Phycology*, *26*(2), 1207–1214. doi:10.100710811-013-0148-7

Leal Filho, W., Saari, U., Fedoruk, M., Iital, A., Moora, H., Klöga, M., & Voronova, V. (2019). An overview of the problems posed by plastic products and the role of extended producer responsibility in Europe. *Journal of Cleaner Production*, *214*, 550–558. doi:10.1016/j.jclepro.2018.12.256

Leandro, A., Pacheco, D., Cotas, J., Marques, J. C., Pereira, L., & Gonçalves, A. M. M. (2020). Seaweed's bioactive candidate compounds to food industry and global food security. In Life (Vol. 10, Issue 8). doi:10.3390/life10080140

Lean, I. J., Golder, H. M., Grant, T. M. D., & Moate, P. J. (2021). A meta-analysis of effects of dietary seaweed on beef and dairy cattle performance and methane yield. *PLoS One*, *16*(7), e0249053. doi:10.1371/journal.pone.0249053 PMID:34252114

Lebreton, L., & Andrady, A. (2019). Future scenarios of global plastic waste generation and disposal. *Palgrave Communications*, *5*(1), 1–11. doi:10.105741599-018-0212-7

Leclerc, E. (2021, March 4). *Scientists Find Blue-Green Algae Chemical with Cancer Fighting Potential.* Retrieved August 30, 2021, from Smithsonianmag.com website: https://www.smithsonianmag.com/blogs/national-museum-of-natural-history/2021/03/04/scientists-find-blue-green-algae-chemical-cancer-fighting-potential/

Lee, R. E. (2018). Phycology (5th ed.). Cambridge University Press. doi:10.1017/9781316407219

Lee, A., & Liew, M. S. (2020). Ecologically derived waste management of conventional plastics. *Journal of Material Cycles and Waste Management, 22*(1), 1–10. doi:10.100710163-019-00931-4

Lee, B. S., Koo, K. M., Ryu, J., Hong, M. J., Kim, S. H., Kwon, S. J., Kim, J.-B., Choi, J., & Ahn, J. W. (2020). Overexpression of fructose-1, 6-bisphosphate aldolase 1 enhances accumulation of fatty acids in Chlamydomonas reinhardtii. *Algal Research, 47*, 101825. doi:10.1016/j.algal.2020.101825

Lee, C.-W., Su, H., Cai, Y.-D., Wu, M.-T., Wu, D.-C., & Shiea, J. (2017). Rapid identification of psychoactive drugs in drained gastric lavage fluid and whole blood specimens of drug overdose patients using ambient mass spectrometry. *Mass Spectrometry (Tokyo, Japan), 6*(2), S0056–S0056. doi:10.5702/massspectrometry.S0056 PMID:28573080

Lee, D. J., Wang, Y., Zhang, Z., Lee, D. J., Zhou, X., Jing, Y., Jiang, D., Hu, J., & He, C. (2017). Photo-fermentative hydrogen production from crop residue: A mini review. *Bioresource Technology, 229*, 222–230. doi:10.1016/j.biortech.2017.01.008 PMID:28108074

Lee, H. Y., Lee, S. E., Jung, K. H., Yeon, J. H., & Choi, W. Y. (2011). Repeated-batch operation of surface-aerated fermentor for bioethanol production from the hydrolysate of seaweed Sargassum sagamianum. *Journal of Microbiology and Biotechnology, 21*(3), 323–331. doi:10.4014/jmb.1010.10057 PMID:21464605

Lee, H.-J., & Choi, J.-i. (2018). Isolation and characterization of a high-growth-rate strain in *Pyropia yezoensis* induced by ethyl methane sulfonate. *Journal of Applied Phycology, 30*(4), 2513–2522. doi:10.100710811-018-1426-1 PMID:30147236

Lee, H.-J., & Choi, J.-i. (2019). Enhancing temperature tolerance of *Pyropia tenera* (Bangiales) by inducing mutation. *Phycologia, 58*(5), 496–503. doi:10.1080/00318884.2019.1623547

Lee, H., Lee, H. K., An, G., & Lee, Y. K. (2007). Analysis of expressed sequence tags from the red alga *Griffithsia okiensis. Journal of Microbiology (Seoul, Korea), 45*(6), 541–546. PMID:18176538

Lee, J. G., Roberts, S. B., & Morel, F. M. M. (1995). Cadmium: A nutrient for the marine diatom Thalassiosira weissflogii. *Limnology and Oceanography, 40*(6), 1056–1063. doi:10.4319/lo.1995.40.6.1056

Lee, J., Cho, D. H., Ramanan, R., Kim, B. H., Oh, H. M., & Kim, H. S. (2013). Microalgae-associated bacteria play a key role in the flocculation of *Chlorella vulgaris. Bioresource Technology, 131*, 195–201.

Lee, M. W., Han, S. O., & Seo, Y. B. (2008). Red algae fibre/poly (butylene succinate) biocomposites: The effect of fibre content on their mechanical and thermal properties. *Composites Science and Technology, 68*(6), 1266–1272. doi:10.1016/j.compscitech.2007.12.016

Lee, R. A., & Lavoie, J.-M. (2013). From first- to third-generation biofuels: Challenges of producing a commodity from a biomass of increasing complexity. *Animal Frontiers, 3*(2), 6–11. doi:10.2527/af.2013-0010

Lee, R. F., Valkirs, A. O., & Seligman, P. F. (1989). Importance of microalgae in the biodegradation of tributyltin in estuarine waters. *Environmental Science & Technology, 23*(12), 1515–1518. doi:10.1021/es00070a010

Lee, S. H., Karawita, R., Affan, A., Lee, J. B., Lee, B. J., & Jeon, Y. J. (2008). Potential antioxidant activites of enzymatic digests from benthic diatoms Achnanthes longipes, Amphora coffeaeformis, and Navicula sp. (Bacillariophyceae). *J. Food Sci. Nutr., 13*(3), 166–175. doi:10.3746/jfn.2008.13.3.166

Lee, S.-H., & Wang, S. (2006). Biodegradable polymers/bamboo fiber biocomposite with bio-based coupling agent. *Composites. Part A, Applied Science and Manufacturing*, *37*(1), 80–91. doi:10.1016/j.compositesa.2005.04.015

Lee, W.-M., & An, Y.-J. (2013). Effects of zinc oxide and titanium dioxide nanoparticles on green algae under visible, UVA, and UVB irradiations: No evidence of enhanced algal toxicity under UV pre-irradiation. *Chemosphere*, *91*(4), 536–544. doi:10.1016/j.chemosphere.2012.12.033 PMID:23357865

Lee, X. J., Ong, H. C., Ooi, J., Yu, K. L., Tham, T. C., Chen, W.-H., & Ok, Y. S. (2022). Engineered macroalgal and microalgal adsorbents: Synthesis routes and adsorptive performance on hazardous water contaminants. *Journal of Hazardous Materials*, *423*, 126921.

Lee, Y.-C., Lee, K., & Oh, Y.-K. (2015). Recent nanoparticle engineering advances in microalgal cultivation and harvesting processes of biodiesel production: A review. *Bioresource Technology*, *184*, 63–72.

Leliaert, F., Smith, D. R., Moreau, H., Herron, M. D., Verbruggen, H., Delwiche, C. F., & De Clerck, O. (2012). Phylogeny and Molecular Evolution of the Green Algae. *Critical Reviews in Plant Sciences*, *31*(1), 1–46. doi:10.1080/07352689.2011.615705

Leong, Y. K., Chen, C. Y., Varjani, S., & Chang, J. S. (2021). Producing fucoxanthin from algae–Recent advances in cultivation strategies and downstream processing. *Bioresource Technology*, 126170. PMID:34678455

Leong, Y. K., Show, P. L., Lan, J. C. W., Loh, H. S., Yap, Y. J., & Ling, T. C. (2017). Extraction and purification of Polyhydroxyalkanoates (PHAs): Application of Thermoseparating aqueous two-phase extraction. *Journal of Polymer Research*, *24*(10), 1–10. doi:10.100710965-017-1307-3

Levasseur, M. E., Morissette, J. C., Popovic, R., & Harrison, P. (1990). Effects of long term exposure to low temperature on the photosynthetic apparatus of *Dunaliella tertiolecta* (Chlorophyceae). *Journal of Phycology*, *26*, 479–484.

Levering, J., Broddrick, J., Dupont, C. L., Peers, G., Beeri, K., Mayers, J., Gallina, A. A., Allen, A. E., Palsson, B. O., & Zengler, K. (2016). Genome-scale model reveals metabolic basis of biomass partitioning in a model diatom. *PLoS One*, *11*(5), e0155038. doi:10.1371/journal.pone.0155038 PMID:27152931

Levett, I., Birkett, G., Davies, N., Bell, A., Langford, A., Laycock, B., Lant, P., & Pratt, S. (2016). Techno-economic assessment of poly-3-hydroxybutyrate (PHB) production from methane - The case for thermophilic bioprocessing. *Journal of Environmental Chemical Engineering*, *4*(4), 3724–3733. doi:10.1016/j.jece.2016.07.033

Levine, R. B., Pinnarat, T., & Savage, P. E. (2010). Biodiesel production from wet algal biomass through in situ lipid hydrolysis and supercritical transesterification. *Energy & Fuels*, *24*(9), 5235–5243. doi:10.1021/ef1008314

Levine, R. B., Sierra, C. O. S., Hockstad, R., Obeid, W., Hatcher, P. G., & Savage, P. E. (2013). The use of hydrothermal carbonization to recycle nutrients in algal biofuel production. *Environmental Progress & Sustainable Energy*, *32*(4), 962–975. doi:10.1002/ep.11812

Levitan, O., Dinamarca, J., Zelzion, E., Lun, D. S., Guerra, L. T., Kim, M. K., Kim, J., Van Mooy, B. A. S., Bhattacharya, D., & Falkowski, P. G. (2015). Remodeling of intermediate metabolism in the diatom *Phaeodactylum tricornutum* under nitrogen stress. *Proceedings of the National Academy of Sciences of the United States of America*, *112*(2), 412–417. doi:10.1073/pnas.1419818112 PMID:25548193

Lewis, L. A. (2007). *Chlorophytaon land: independent lineages of green eukaryotes from arid lands.* In J. Seckbach (Ed.), *Algae and Cyanobacteria in extreme Environments* (pp. 571–582). Springer.

LewisOscar, F., Vismaya, S., Arunkumar, M., Thajuddin, N., Dhanasekaran, D., & Nithya, C. (2016). Algal nanoparticles: Synthesis and biotechnological potentials. Chapter. *Algae Org. Imminent Biotechnol.*, *7*, 157–182.

Leyva, L. A., Bashan, Y., Mendoza, A., & De-Bashan, L. E. (2014). Accumulation fatty acids of in Chlorella vulgaris under heterotrophic conditions in relation to activity of acetyl-CoA carboxylase, temperature, and co-immobilization with Azospirillum brasilense. *Naturwissenschaften*, *101*(10), 819–830. doi:10.100700114-014-1223-x PMID:25129521

Li, H., Watson, J., Zhang, Y., Lu, H., & Liu, Z. (2020). Environment-enhancing process for algal wastewater treatment, heavy metal control and hydrothermal biofuel production: A critical review. *Bioresource Technology, 298*(September), 122421. doi:10.1016/j.biortech.2019.122421

Li, J., Wu, X., Zhong, Y., Lu, Q., & Zhou, W. (2020). The 10th Asia-Pacific Conference on Algal Biotechnology: Thoughts and comments. *Journal of Cleaner Production, 264*. 10.1016/j.jclepro.2020.121626

Liang, T., Li, K., & Wang, L. (2014). State of rare earth elements in different environmental components in mining areas of China. *Environmental Monitoring and Assessment*, *186*(3), 1499–1513. doi:10.100710661-013-3469-8 PMID:24135922

Liang, Y. (2015). *Industrial Biorefineries and White Biotechnology. In Algal Biorefineries*. Elsevier B.V.

Li, B., Lu, F., Wei, X., & Zhao, R. (2008). Fucoidan: Structure and bioactivity. *Molecules (Basel, Switzerland)*, *13*(8), 1671–1695. doi:10.3390/molecules13081671 PMID:18794778

Li, B., Su, L., Zhang, H., Deng, H., Chen, Q., & Shi, H. (2020). Microplastics in fishes and their living environments surrounding a plastic production area. *The Science of the Total Environment*, *727*, 138662. doi:10.1016/j.scitotenv.2020.138662 PMID:32498185

Li, D., Wang, L., Zhao, Q., Wei, W., & Sun, Y. (2015). Improving high carbon dioxide tolerance and carbon dioxide fixation capability of *Chlorella* sp. by adaptive laboratory evolution. *Bioresource Technology*, *185*, 269–275. doi:10.1016/j.biortech.2015.03.011 PMID:25776894

Li, H., Handsaker, B., Wysoker, A., Fennell, T., Ruan, J., Homer, N., Marth, G., Abecasis, G., & Durbin, R. (2009). The sequence alignment/ map format and SAMtools. *Bioinformatics (Oxford, England)*, *25*(16), 2078–2079. doi:10.1093/bioinformatics/btp352 PMID:19505943

Li, H., Liu, Z., Zhang, Y., Li, B., Lu, H., Duan, N., Liu, M., Zhu, Z., & Si, B. (2014). Conversion efficiency and oil quality of low-lipid high-protein and high-lipid low-protein microalgae via hydrothermal liquefaction. *Bioresource Technology*, *154*, 322–329. doi:10.1016/j.biortech.2013.12.074 PMID:24413449

Li, H., Shen, C. R., Huang, C.-H., Sung, L.-Y., Wu, M.-Y., & Hu, Y.-C. (2016). CRISPR-Cas9 for the genome engineering of cyanobacteria and succinate production. *Metabolic Engineering*, *38*, 293–302. doi:10.1016/j.ymben.2016.09.006 PMID:27693320

Li, H., Zhao, Q., & Huang, H. (2019). Current states and challenges of salt-affected soil remediation by cyanobacteria. *The Science of the Total Environment*, *669*, 258–272. doi:10.1016/j.scitotenv.2019.03.104 PMID:30878933

Li, J., Liu, H., & Paul Chen, J. (2018). Microplastics in freshwater systems: A review on occurrence, environmental effects, and methods for microplastics detection. *Water Research*, *137*, 362–374. doi:10.1016/j.watres.2017.12.056 PMID:29580559

Li, K., Gao, Q., Yadavalli, G., Shen, X., Lei, H., Han, B., Xia, K., & Zhou, C. (2015). Selective adsorption of Gd3+ on a magnetically retrievable imprinted chitosan/carbon nanotube composite with high capacity. *ACS Applied Materials & Interfaces*, *7*(38), 21047–21055. doi:10.1021/acsami.5b07560 PMID:26355685

Li, K., Liu, Q., Fang, F., Luo, R., Lu, Q., Zhou, W., Huo, S., Cheng, P., Liu, J., Addy, M., Chen, P., Chen, D., & Ruan, R. (2019b). Microalgae-based wastewater treatment for nutrients recovery: A review. *Bioresource Technology*, *291*, 121934. doi:10.1016/j.biortech.2019.121934 PMID:31395401

Li, L., Ni, R., Shao, Y., & Mao, S. (2014). Carrageenan and its applications in drug delivery. *Carbohydrate Polymers*, *103*, 1–11. doi:10.1016/j.carbpol.2013.12.008 PMID:[illegible]

Li, L., & Zhang, Z. (2016). Biosynthesis of gold nanoparticles using green alga Pithophoraoedogonia with their electrochemical performance for determin- ing carbendazim in soil. *International Journal of Electrochemical Science*, 4550–4559. doi:10.20964/2016.06.13

Limayem, A., & Ricke, S. C. (2012). Lignocellulosic biomass for bioethanol production: Current perspectives, potential issues and future prospects. *Progress in Energy and Combustion Science*, *38*(4), 449–467. doi:10.1016/j.pecs.2012.03.002

Lim, S., Park, J., Shim, M. K., Um, W., Yoon, H. Y., Ryu, J. H., Lim, D.-K., & Kim, K. (2019). Recent advances and challenges of repurposing nanoparticle-based drug delivery systems to enhance cancer immunotherapy. *Theranostics*, *9*(25), 7906–7923. doi:10.7150/thno.38425 PMID:31695807

Lin, C. Y., & Lu, C. (2021). Development perspectives of promising lignocellulose feedstocks for production of advanced generation biofuels: A review. *Renewable and Sustainable Energy Reviews, 136*(September), 110445. doi:10.1016/j.rser.2020.110445

Lin, W.-R., Tan, S.-I., Hsiang, C.-C., Sung, P.-K., & Ng, I.-S. (2019). Challenges and opportunity of recent genome editing and multi-omics in cyanobacteria and microalgae for biorefinery. *Bioresource Technology*, *291*, 121932. doi:10.1016/j.biortech.2019.121932 PMID:31387837

Liquete, C., Piroddi, C., Drakou, E. G., Gurney, L., Katsanevakis, S., Charef, A., & Egoh, B. (2013). Current status and future prospects for the assessment of marine and coastal ecosystem services: A systematic review. *PLoS One*, *8*(7), e67737–e67737. doi:10.1371/journal.pone.0067737 PMID:23844080

Li, S. N., Wang, R., & Ho, S. H. (2021). Algae-mediated biosystems for metallic nanoparticle production: From synthetic mechanisms to aquatic environmental applications. *Journal of Hazardous Materials*, *420*(January), 126625. doi:10.1016/j.jhazmat.2021.126625 PMID:34329084

Li, S., Hu, T., Xu, Y., Wang, J., Chu, R., Yin, Z., Mo, F., & Zhu, L. (2020). A review on flocculation as an efficient method to harvest energy microalgae: Mechanisms, performances, influencing factors and perspectives. *Renewable & Sustainable Energy Reviews*, *131*, 110005. doi:10.1016/j.rser.2020.110005

Lithner, D., Larsson, Å., & Dave, G. (2011). Environmental and health hazard ranking and assessment of plastic polymers based on chemical composition. *The Science of the Total Environment*, *409*(18), 3309–3324. doi:10.1016/j.scitotenv.2011.04.038 PMID:21663944

Littler, M. M., Littler, D. S., Blair, S. M., & Norris, J. N. (1985). Deepest Known Plant Life Discovered on an Uncharted Seamount. *Science*, *227*(4682), 57–59. doi:10.1126cience.227.4682.57 PMID:17810025

Liu, H., Laws, E. A., Villareal, T. A., & Buskey, E. J. (2001). Nutrient-limited growth of Aureoumbra lagunensis (Pelagophyceae), with implications for its capability to outgrow other phytoplankton species in phosphate-limited environments. *Journal of Phycology*, *37*(4), 500–508. doi:10.1046/j.1529-8817.2001.037004500.x

Liu, J., & Chen, F. (2014). Biology and industrial applications of Chlorella: Advances and prospects. *Microalgae Biotechnology*, 1–35. PMID:25537445

Liu, J., Fu, K., Wu, C., Qin, K., Li, F., & Zhou, L. (2018). "In-Group" communication in marine *Vibrio*: A review of N-Acyl Homoserine Lactones-Driven Quorum Sensing. *Frontiers in Cellular and Infection Microbiology*, *8*(MAY), 1–17. doi:10.3389/fcimb.2018.00139 PMID:29868495

Liu, J., Pemberton, B., Lewis, J., Scales, P. J., & Martin, G. J. O. (2019). Wastewater treatment using filamentous algae – A review. In *Bioresource Technology* (Vol. 298, p. 122556). Elsevier Ltd. doi:10.1016/j.biortech.2019.122556

Liu, J., Zhu, Y., Tao, Y., Zhang, Y., Li, A., Li, T., Sang, M., & Zhang, C. (2013). Freshwater microalgae harvested via flocculation induced by pH decrease. *Biotechnology for Biofuels*, *6*, 98.

Liu, L., Fokkink, R., & Koelmans, A. A. (2016). Sorption of polycyclic aromatic hydrocarbons to polystyrene nanoplastic. *Environmental Toxicology and Chemistry*, *35*(7), 1650–1655. doi:10.1002/etc.3311 PMID:26588181

Liu, R., Li, S., Tu, Y., & Hao, X. (2021). Capabilities and mechanisms of microalgae on removing micropollutants from wastewater: A review. *Journal of Environmental Management*, *285*, 112149. doi:10.1016/j.jenvman.2021.112149 PMID:33607565

Liu, R., Peng, Y., Cao, J., & Chen, Y. (2014). Comparison on properties of lignocellulosic flour/polymer composites by using wood, cellulose, and lignin flours as fillers. *Composites Science and Technology*, *103*, 1–7. doi:10.1016/j.compscitech.2014.08.005

Liu, T., Liu, W. H., Zhao, J. S., Meng, F. Z., & Wang, H. (2017). Lutein protects against β-amyloid peptide-induced oxidative stress in cerebrovascular endothelial cells through modulation of Nrf-2 and NF-κb. *Cell Biology and Toxicology*, *33*(1), 57–67. doi:10.100710565-016-9360-y PMID:27878403

Liu, X., & Hong, Y. (2021). Microalgae-Based Wastewater Treatment and Recovery with Biomass and Value-Added Products: A Brief Review. *Current Pollution Reports*, *7*(2), 227–245. doi:10.100740726-021-00184-6

Liu, X., Zhang, D., Zhang, J., Chen, Y., Liu, X., Fan, C., Wang, R. R.-C., Hou, Y., & Hu, Z. (2021). Overexpression of the Transcription Factor AtLEC1 Significantly Improved the Lipid Content of Chlorella ellipsoidea. *Frontiers in Bioengineering and Biotechnology*, *9*, 113. doi:10.3389/fbioe.2021.626162 PMID:33681161

Liu, Z., Gao, T., Yang, Y., Meng, F., Zhan, F., Jiang, Q., & Sun, X. (2019). Anti-Cancer Activity of Porphyran and Carrageenan from Red Seaweeds. *Molecules (Basel, Switzerland)*, *24*(23), 4286. doi:10.3390/molecules24234286 PMID:31775255

Li, W. K. W. (1980). Temperature adaptation in phytoplankton: cellular and photosynthetic characteristics. In P. G. Falkowski (Ed.), *Primary productivity in the sea* (pp. 259–279). Plenum Press. doi:10.1007/978-1-4684-3890-1_15

Li, X.-J., Yan, C.-J., Luo, W.-J., Gao, Q., Zhou, Q., Liu, C., & Zhou, S. (2016). Exceptional cerium(III) adsorption performance of poly(acrylic acid) brushes-decorated attapulgite with abundant and highly accessible binding sites. *Chemical Engineering Journal*, *284*, 333–342. doi:10.1016/j.cej.2015.09.003

Li, X., Liu, B., Lao, Y., Wan, P., Mao, X., & Chen, F. (2021). Efficient magnetic harvesting of microalgae enabled by surface-initiated formation of iron nanoparticles. *Chemical Engineering Journal*, *408*, 127252.

Li, X., Yuan, Y., Cheng, D., Gao, J., Kong, L., Zhao, Q., Wei, W., & Sun, Y. (2018). Exploring stress tolerance mechanism of evolved freshwater strain *Chlorella* sp. S30 under 30 g/L salt. *Bioresource Technology*, *250*, 495–504. doi:10.1016/j.biortech.2017.11.072 PMID:29197772

Li, Y., Horsman, M., Wu, N., Lan, C. Q., & Dubois-Calero, N. (2008). Biofuels from microalgae. *Biotechnology Progress*, *24*(4), 815–820. PMID:18335954

Li, Y., Tarpeh, W. A., Nelson, K. L., & Strathmann, T. J. (2018). Quantitative Evaluation of an Integrated System for Valorization of Wastewater Algae as Bio-oil, Fuel Gas, and Fertilizer Products. *Environmental Science & Technology*, *52*(21), 12717–12727. doi:10.1021/acs.est.8b04035 PMID:30256626

Li, Y.-X., Wijesekaraa, I., Li, Y., & Kim, S.-K. (2011). Phlorotannins as bioactive agents from brown algae. *Process Biochemistry, 46*(12), 2219–2224. doi:10.1016/j.procbio.2011.08.016

Li, Y., Zhu, H., Lei, X., Zhang, H., Guan, C., Chen, Z., Zheng, W., Xu, H., Tian, Y., Yu, Z., & Zheng, T. (2015). The first evidence of deinoxanthin from Deinococcus sp. Y35 with strong algicidal effect on the toxic dinoflagellate Alexandrium tamarense. *Journal of Hazardous Materials, 290*, 87–95. doi:10.1016/j.jhazmat.2015.02.070 PMID:25746568

Li, Y., Zhu, H., Zhang, H., Chen, Z., Tian, Y., Xu, H., Zheng, T., & Zheng, W. (2014). Toxicity of algicidal extracts from Mangrovimonas yunxiaonensis strain LY01 on a HAB causing Alexandrium tamarense. *Journal of Hazardous Materials, 278*, 372–381. doi:10.1016/j.jhazmat.2014.06.032 PMID:24997253

Li, Z., Ding, R., & Floudas, C. A. (2011). A comparative theoretical and computational study on robust counterpart optimization: I. Robust linear optimization and robust mixed integer linear optimization. *Industrial & Engineering Chemistry Research, 50*(18), 10567–10603. doi:10.1021/ie200150p PMID:21935263

Löffelhardt, W., Bohnert, H. J., Bryant, D. A., & Hagemann, R. (1997). The Cyanelles of Cyanophora Paradoxa. *Critical Reviews in Plant Sciences, 16*(4), 393–413. doi:10.1080/07352689709701955

Loganathan, G. B., Orsat, V., Lefsrud, M., & Wu, B. S. (2020). A comprehensive study on the effect of light quality imparted by lightemitting diodes (LEDs) on the physiological and biochemical properties of the microalgal consortia of *Chlorella variabilis* and *Scenedesmus obliquus* cultivated in dairy wastewater. *Bioprocess and Biosystems Engineering, 43*, 1445–1455. https://doi.org/10.1007/s00449-020-02338-0

Loganathan, K., Saththasivam, J., & Sarp, S. (2018). Removal of microalgae from seawater using chitosan-alum/ferric chloride dual coagulations. *Desalination, 433*, 25–32.

Loira, N., Mendoza, S., Cortés, M. P., Rojas, N., Travisany, D., Di-Genova, A., Gajardo, N., Ehrenfeld, N., & Maass, A. (2017). Reconstruction of the microalga *Nannochloropsis salina* genome-scale metabolic model with applications to lipid production. *BMC Systems Biology, 11*(1), 66–82. doi:10.118612918-017-0441-1 PMID:28676050

Lomartire, S., Cotas, J., Pacheco, D., Marques, J. C., Pereira, L., & Gonçalves, A. M. M. (2021). Environmental impact on seaweed phenolic production and activity: An important step for compound exploitation. *Marine Drugs, 19*(5), 1–20. doi:10.3390/md19050245 PMID:33926129

López Barreiro, D., Prins, W., Ronsse, F., & Brilman, W. (2013). Hydrothermal liquefaction (HTL) of microalgae for biofuel production: State of the art review and future prospects. *Biomass and Bioenergy, 53*, 113–127. doi:10.1016/j.biombioe.2012.12.029

López de Dicastillo, C., Velásquez, E., Rojas, A., Guarda, A., & Galotto, M. J. (2020). The use of nanoadditives within recycled polymers for food packaging: Properties, recyclability, and safety. *Comprehensive Reviews in Food Science and Food Safety, 19*(4), 1760–1776. doi:10.1111/1541-4337.12575 PMID:33337105

López, A., Rico, M., Rivero, A., & de Tangil, M. S. (2011). The effects of solvents on the phenolic contents and antioxidant activity of *Stypocaulon scoparium* algae extracts. *Food Chemistry, 125*(3), 1104–1109. doi:10.1016/j.foodchem.2010.09.101

Los, D. A., Mironov, K. S., & Allakhverdiev, S. I. (2013). Regulatory role of membrane fluidity in gene expression and physiological functions. *Photosynthesis Research, 116*(2–3), 489–509. doi:10.100711120-013-9823-4 PMID:23605242

Lötze, E., & Hoffman, E. W. (2015). Nutrient composition and content of various biological active compounds of three South African-based commercial seaweed biostimulants. *Journal of Applied Phycology*.

Lovejoy, C., Vincent, W. F., Bonilla, S., Roy, S., Martineau, M. J., Terrado, R., Potvin, M., Massana, R., & Pedrós-Alió, C. (2007). distribution, phylogeny, and growth of cold-adapted picoprasinophytes in arctic seas. *Journal of Phycology*, *43*(1), 78–89. doi:10.1111/j.1529-8817.2006.00310.x

Lovley, D. R., & Coates, J. D. (1997). Bioremediation of metal contamination. *Current Opinion in Biotechnology*, *8*(3), 285–289. doi:10.1016/S0958-1669(97)80005-5 PMID:9206008

Lozano, I., Wacyk, J. M., Carrasco, J., & Cortez-San Martín, M. A. (2016). Red macroalgae *Pyropia columbina* and *Gracilaria chilensis*: Sustainable feed additive in the Salmo salar diet and the evaluation of potential antiviral activity against infectious salmon anemia virus. *Journal of Applied Phycology*, *28*(2), 1343–1351. doi:10.100710811-015-0648-8

Lucas, B. F., de Morais, M. G., Santos, T. D., & Costa, J. A. V. (2018). Spirulina for snack enrichment: Nutritional, physical and sensory evaluations. *Lebensmittel-Wissenschaft + Technologie*, *90*, 270–276. doi:10.1016/j.lwt.2017.12.032

Luengo, E., Condón-Abanto, S., Álvarez, I., & Raso, J. (2014). Effect of Pulsed Electric Field Treatments on Permeabilization and Extraction of Pigments from Chlorella Vulgaris. *The Journal of Membrane Biology*, *247*(12), 1269–1277. doi:10.100700232-014-9688-2 PMID:24880235

Lü, F., Ji, J., Shao, L., & He, P. (2013). Bacterial bioaugmentation for improving methane and hydrogen production from microalgae. *Biotechnology for Biofuels*, *6*(1), 92. doi:10.1186/1754-6834-6-92 PMID:23815806

Luimstra, V. M., Schuurmans, J. M., de Carvalho, C. F. M., Matthijs, H. C. P., Hellingwerf, K. J., & Huisman, J. (2019). Exploring the low photosynthetic efficiency of cyanobacteria in blue light using a mutant lacking phycobilisomes. *Photosynthesis Research*, *141*, 291–301. https://doi.org/10.1007/s11120-019-00630-z

Łukajtis, R., Hołowacz, I., Kucharska, K., Glinka, M., Rybarczyk, P., Przyjazny, A., & Kamiński, M. (2018). Hydrogen production from biomass using dark fermentation. *Renewable & Sustainable Energy Reviews*, *91*, 665–694. doi:10.1016/j.rser.2018.04.043

Lundquist, T. J., Woertz, I. C., Quinn, N. W. T., & Benemann, J. R. (2010). A realistic technology and engineering assessment of algae biofuel production. Energy Biosciences Institute.

Lüning, K., & Tom Dieck, I. (1989). *Environmental triggers in algal seasonality*. Academic Press.

Luning, K., & Dring, M. J. (1979). Continuous underwater light measurement near Helgoland.North Sea and its significance for characteristic light limits in the sublittoral region. *Helgoländer Wissenschaftliche Meeresuntersuchungen*, *32*(4), 403–424. doi:10.1007/BF02277985

Luo, L., Qin, B., Song, Y., & Yang, L. (2007). Seasonal and regional variations in precipitation chemistry in the Lake Taihu Basin, China. *Atmospheric Environment*, *41*(12), 2674–2679. doi:10.1016/j.atmosenv.2006.11.037

Luo, X., Li, J., Feng, J., Xie, S., & Lin, X. (2013). Evaluation of distillers grains as fillers for low density polyethylene: Mechanical, rheological and thermal characterization. *Composites Science and Technology*, *89*, 175–179. doi:10.1016/j.compscitech.2013.10.001

Lu, Q., Li, H., Zou, Y., Liu, H., & Yang, L. (2021). Astaxanthin as a microalgal metabolite for aquaculture: A review on the synthetic mechanisms, production techniques, and practical application. *Algal Research*, *54*, 102178. doi:10.1016/j.algal.2020.102178

Lutzoni, F., Nowak, M. D., Alfaro, M. E., Reeb, V., Miadlikowska, J., Krug, M., Arnold, A. E., Lewis, L. A., Swofford, D. L., Hibbett, D., Hilu, K., James, T. Y., Quandt, D., & Magallón, S. (2018). Contemporaneous radiations of fungi and plants linked to symbiosis. *Nature Communications*, *9*(1), 5451. doi:10.103841467-018-07849-9 PMID:30575731

Lutzu, G. A., & Dunford, N. T. (2018). Interactions of microalgae and other microorganisms for enhanced production of high-value compounds. [illegible]

Lu, Y., & Xu, J. (2015). Phytohormones in microalgae: A new opportunity for microalgal biotechnology? *Trends in Plant Science*, *20*, 273–282.

Lv, J., Wang, X., Liu, W., Feng, J., Liu, Q., Nan, F., Jiao, X., & Xie, S. (2018). The performance of a self-flocculating microalga *Chlorococcum* sp. GD in wastewater with different ammonia concentrations. *International Journal of Environmental Research and Public Health*, *15*, 434.

Lynch, M., Ackerman, M. S., Gout, J.-F., Long, H., Sung, W., Thomas, W. K., & Foster, P. L. (2016). Genetic drift, selection and the evolution of the mutation rate. *Nature Reviews. Genetics*, *17*(11), 704–714. doi:10.1038/nrg.2016.104 PMID:27739533

Lyons, D., Scheibling, R. E., & Van Alstyne, K. (2010). Spatial and temporal variation in DMSP content in the invasive seaweed *Codium fragile* ssp. fragile: Effects of temperature, light and grazing. *Marine Ecology Progress Series*, *417*, 51–61. doi:10.3354/meps08818

Ma, N. L., Lam, S. S., & Zaidah, R. (2015). The application of algae for cosmeceuticals in the Omics Age. *Genomics, Proteomics, Metabolomics in Nutraceuticals Functional Foods*, 476-488. doi:10.1002/9781118930458.ch37

Mackey, K. R., Morris, J. J., Morel, F. M., & Kranz, S. A. (2015). Response of photosynthesis to ocean acidification. *Oceanography (Washington, D.C.)*, *28*(2), 74–91. doi:10.5670/oceanog.2015.33

Madadi, R., Pourbabaee, A. A., Tabatabaei, M., Zahed, M. A., & Naghavi, M. R. (2016). Treatment of Petrochemical Wastewater by the Green Algae Chlorella vulgaris. *International Journal of Environmental of Research*, *10*(4), 555–560.

Madaeni, S. S., & Ghaemi, N. (2007). Characterization of self-cleaning RO membranes coated with TiO2 particles under UV irradiation. *Sci. J. Membr.*, *333*(1-2), 221–233. doi:10.1016/j.memsci.2007.07.017

Maddux, W. J., & Jones, R. F. (1964). Some interactions of temperature, light intensity, and nutrient concentration during the continuous culture of *Nitzschia closterium* and *Tetraselmis* sp. *Limnology and Oceanography*, *9*, 79–86.

Madeira, M. S., Cardoso, C., Lopes, P. A., Coelho, D., Afonso, C., Bandarra, N. M., & Prates, J. A. M. (2017). Microalgae as feed ingredients for livestock production and meat quality: A review. *Livestock Science*, *205*, 111–121. doi:10.1016/j.livsci.2017.09.020

Mader, J., Gallo, A., Schommartz, T., Handke, W., Nagel, C. H., Gunther, P., Brune, W., & Reich, K. (2016). Calcium spirulan derived from Spirulina platensis inhibits herpes simplex virus 1 attachment to human keratinocytes and protects against herpes labialis. *The Journal of Allergy and Clinical Immunology*, *137*(1), 197–203. doi:10.1016/j.jaci.2015.07.027 PMID:26341274

Maes, A., Martinez, X., Druart, K., Laurent, B., Guégan, S., Marchand, C. H., Lemaire, S. D., & Baaden, M. (2018). MinOmics, an integrative and immersive tool for multi-omics analysis. *Journal of Integrative Bioinformatics*, *15*(2), 20180006. doi:10.1515/jib-2018-0006 PMID:29927748

Maes, H. M., Maletz, S. X., Ratte, H. T., Hollender, J., & Schaeffer, A. (2014). Uptake, elimination, and biotransformation of 17α-ethinylestradiol by the freshwater alga *Desmodesmus subspicatus*. *Environmental Science & Technology*, *48*, 12354–12361.

Magdeldin, M., Kohl, T., & Järvinen, M. (2017). Techno-economic assessment of the byproducts contribution from non-catalytic hydrothermal liquefaction of lignocellulose residues. *Energy*, *137*, 679–695. doi:10.1016/j.energy.2017.06.166

Maghrabi, A. H. A., & McGuffin, L. J. (2017). ModFOLD6: An accurate web server for the global and local quality estimation of 3D protein models. *Nucleic Acids Research*, *45*(1), 416–421. doi:10.1093/nar/gkx332 PMID:28460136

Ma, H., Liao, X., Liu, X., & Shi, B. (2006). Recovery of platinum (IV) and palladium (II) by bayberry tannin immobilized collagen fiber membrane from water solution. *Journal of Membrane Science*, *278*(1–2), 373–380. doi:10.1016/j.memsci.2005.11.022

Mahdavi, M., Namvar, F., Ahmad, M., & Mohamad, R. (2013). Green biosynthesis and characterization of magnetic iron oxide (Fe 3O4) nanoparticles using seaweed (Sargassum muticum) aqueous extract. *Molecules (Basel, Switzerland)*, *18*(5), 5954–5964. doi:10.3390/molecules18055954 PMID:23698048

Mahdavi, S., Razeghi, J., Pazhouhandeh, M., Movafeghi, M., Kosari-Nasab, M., & Kianianmomeni, A. (2020). Characterization of two predicted DASH-related proteins from the green alga *Volvox carteri* provides new insights into their light-mediated transcript regulation and DNA repair activity. *Algal Research*, *52*, 102116. doi:10.1016/j.algal.2020.102116

Mahdieh, M., Zolanvari, A., Azimee, A. J. S. I., & Mahdieh, M. (2012). Green biosynthesis of silver nanoparticles by Spirulina platensis. *Scientia Iranica*, *19*(3), 926–929. doi:10.1016/j.scient.2012.01.010

Mairet, F., Bernard, O., Masci, P., Lacour, T., & Sciandra, A. (2011). Modelling neutral lipid production by the microalga Isochrysis aff. galbana under nitrogen limitation. *Bioresource Technology*, *102*(1), 142–149. doi:10.1016/j.biortech.2010.06.138 PMID:20656476

Maity, J. P., Bundschuh, J., Chen, C. Y., & Bhattacharya, P. (2014). Microalgae for third generation biofuel production, mitigation of greenhouse gas emissions and wastewater treatment: Present and future perspectives–A mini review. *Energy*, *78*, 104–113. doi:10.1016/j.energy.2014.04.003

Ma, J., Li, L., Zhao, Q., Yu, L., & Frear, C. (2021). Biomethane production from whole and extracted algae biomass: Long-term performance evaluation and microbial community dynamics. *Renewable Energy*, *170*, 38–48. doi:10.1016/j.renene.2021.01.113

Ma, J., Zhao, Q.-B., Laurens, L. L. M., Jarvis, E. E., Nagle, N. J., Chen, S., & Frear, C. S. (2015). Mechanism, kinetics and microbiology of inhibition caused by long-chain fatty acids in anaerobic digestion of algal biomass. *Biotechnology for Biofuels*, *8*(1), 141. doi:10.118613068-015-0322-z PMID:26379773

Malik, F. R., Ahmed, S., & Rizki, Y. M. (2001). Utilization of lignocellulosic waste for the preparation of nitrogenous biofertilizer. *Pakistan Journal of Biological Sciences*, *4*(10), 1217–1220. doi:10.3923/pjbs.2001.1217.1220

Mallick, N., Gupta, S., Panda, B., & Sen, R. (2007). Process optimization for poly (3-hydroxybutyrate-co-3-hydroxyvalerate) co-polymer production by Nostoc muscorum. *Biochemical Engineering Journal*, *37*(2), 125–130. doi:10.1016/j.bej.2007.04.002

Malone, L. A., Proctor, M. S., Hitchcock, A., Hunter, C. N., & Johnson, M. P. (2021). Cytochrome b6f–Orchestrator of photosynthetic electron transfer. *Biochimica et Biophysica Acta (BBA)-. Bioenergetics*, 148380. doi:10.1016/j.bbabio.2021.148380 PMID:33460588

Manach, C., Scalbert, A., Morand, C., R'em'esy, C., & Jim'enez, L. (2004). Polyphenols: Food sources andbioavailability. *The American Journal of Clinical Nutrition*, *79*(5), 727–747. doi:10.1093/ajcn/79.5.727 PMID:15113710

Mancuso, S., Briand, X., Mugnai, S., & Azzarello, E. (2006). Marine bioactive substances (IPA Extract) improve foliar ion uptake and water stress tolerance in potted" *Vitis vinifera*" plants. *Advances in Horticultural Science*, *20*(2), 1–6.

Mandal, R. P., Sekh, S., Sarkar, N. S., Chattopadhyay, D., & De, S. (2016). Algae mediated synthesis of cadmium sulphide nanoparticles and their application in bioremediation. *Materials Research Express*, *3*(5), 055007. doi:10.1088/2053-1591/3/5/055007

Manheim, D., & Nelson, Y. (2013). Settling and bioflocculation of two species of algae used in wastewater treatment and algae biomass production. *Environmental Progress & Sustainable Energy*, *32*, 946–954.

Manier, N., Bado-Nilles, A., Delalain, P., Aguerre-Chariol, O., & Pandard, P. (2013). Ecotoxicity of non-aged and aged CeO2 nanomaterials towards freshwater microalgae. *Environmental Pollution*, *180*, 63–70. doi:10.1016/j.envpol.2013.04.040 PMID:23727569

Mannina, G., Presti, D., Montiel-Jarillo, G., Carrera, J., & Suárez-Ojeda, M. E. (2020). Recovery of polyhydroxyalkanoates (PHAs) from wastewater: A review. *Bioresource Technology*, *297*, 122478. doi:10.1016/j.biortech.2019.122478 PMID:31810735

Mannweiler, K., & Hoare, M. (1992). The scale-down of an industrial disk stack centrifuge. *Bioprocess Engineering*, *8*, 19–25.

Manríquez-Hernández, J. A., Duston, J., & Garbary, D. J. (2016). Effect of irradiance on bioremediation capacity of the red algae *Chondrus crispus* and *Palmaria palmata*. *Aquaculture International*, *24*(1), 39–55. doi:10.100710499-015-9907-6

Mansori, M., Chernane, H., Latique, S., Benaliat, A., Hsissou, D., & El Kaoua, M. (2016). Effect of seaweed extract (*Ulva rigida*) on the water deficit tolerance of *Salvia officinalis* L. *Journal of Applied Phycology*, *28*(2), 1363–1370. doi:10.100710811-015-0671-9

Mansori, M., Farouk, I. A., Hsissou, D., & Kaoua, M. E. (2019). Seaweed extract treatment enhances vegetative growth and antioxidant parameters in water stressed *Salvia officinalis* L. *Journal of Materials and Environmental Sciences*, *10*(8), 756–766.

Mansour, A. T., Ashour, M., Alprol, A. E., & Alsaqufi, A. S. (2022). Aquatic Plants and Aquatic Animals in the Context of Sustainability : Cultivation Techniques, Integration, and Blue Revolution. *Sustainability*, *14*(4), 1–21. doi:10.3390u14063257

Manzo, S., Miglietta, M. L., Rametta, G., Buono, S., & Di Francia, G. (2013). Toxic effects of ZnO nanoparticles towards marine algae Dunaliella tertiolecta. *The Science of the Total Environment*, *445*, 371–376. doi:10.1016/j.scitotenv.2012.12.051 PMID:23361041

Mao, X., Chen, S. H. Y., Lu, X., Yu, J., & Liu, B. (2020). High silicate concentration facilitates fucoxanthin and eicosapentaenoic acid (EPA) production under heterotrophic condition in the marine diatom Nitzschia laevis. *Algal Research*, *52*, 102086. doi:10.1016/j.algal.2020.102086

Mao, X., Zhang, Y., Wang, X., & Liu, J. (2020). Novel insights into salinity-induced lipogenesis and carotenogenesis in the oleaginous astaxanthin-producing alga *Chromochloris zofingiensis*: A multi-omics study. *Biotechnology for Biofuels*, *13*(1), 73. doi:10.118613068-020-01714-y PMID:32322303

Mariam, A. H., Ghaly, A., & Amal, H. (2015). Microalgae harvesting methods for industrial production of biodiesel: Critical review and comparative analysis. *Fundamentals of Renewable Energy and Applications*, *5*(2), 1–26.

Marián, F. D., García-Jiménez, P., & Robain, R. R. (2000). Polyamines in Marine Macroalgae: Levels of Putrescine, Spermidine and Spermine in the Thalli and Changes in Their Concentration during Glycerol-Induced Cell Growth In Vitro. *Physiologia Plantarum*, *110*(4), 530–534. doi:10.1111/j.1399-3054.2000.1100416.x

Marin-Batista, J. D., Villamil, J. A., Rodriguez, J. J., Mohedano, A. F., & de la Rubia, M. A. (2019). Valorization of microalgal biomass by hydrothermal carbonization and anaerobic digestion. *Bioresource Technology, 274*, 395–402. doi:10.1016/j.biortech.2018.11.103 PMID:30551042

Markl, E., Grünbichler, H., & Lackner, M. (2019). Cyanobacteria for PHB Bioplastics Production: A Review. *Algae - Korean Phycological Society*, 1–10. doi:10.5772/intechopen.81536

Markou, G. (2014). Effect of various colors of light-emitting diodes (LEDs) on the biovolume composition of *Arthrospira platensis* cultivated in semi-continuous mode. *Appl Biochem Biotechnol, 172*, 2758-2768. doi:10.1007/s12010-014-0727-3

Markou, G., & Nerantzis, E. (2013). Microalgae for high value compounds and biofuels production: A review with focus on cultivation under stress conditions. *Biotechnology Advances, 31*(8), 1532–1542. doi:10.1016/j.biotechadv.2013.07.011 PMID:23928208

Marks, E. A. N., Montero, O., & Rad, C. (2019). The biostimulating effects of viable microalgal cells applied to a calcareous soil: Increases in bacterial biomass, phosphorus scavenging, and precipitation of carbonates. *The Science of the Total Environment, 692*, 784–790. doi:10.1016/j.scitotenv.2019.07.289 PMID:31539985

Maršálek, B. (1992). Zahradníˇcková, H.; Hronková, M. Extracellular Abscisic Acid Produced by Cyanobacteria under Salt Stress. *Journal of Plant Physiology, 139*, 506–508.

Martin, K. R. (2013). Silicon: The health benefits of a metalloid. *Interrelations between essential metal ions and human diseases*, 451-473.

Martínez-Herrera, R. E., Alemán-Huerta, M. E., Almaguer-Cantú, V., Rosas-Flores, W., Martínez-Gómez, V. J., Quintero-Zapata, I., Rivera, G., & Rutiaga-Quiñones, O. M. (2020). Efficient recovery of thermostable polyhydroxybutyrate (PHB) by a rapid and solvent-free extraction protocol assisted by ultrasound. *International Journal of Biological Macromolecules, 164*, 771–782. doi:10.1016/j.ijbiomac.2020.07.101 PMID:32682039

Martínez, J. M., Luengo, E., Saldaña, G., Álvarez, I., & Raso, J. (2017). C-Phycocyanin Extraction Assisted by Pulsed Electric Field from Artrosphira Platensis. *Food Research International, 99*, 1042–1047. doi:10.1016/j.foodres.2016.09.029 PMID:28865615

Martin, K. R. (2007). The chemistry of silica and its potential health benefits. *The Journal of Nutrition, Health & Aging, 11*(2), 94. PMID:17435951

Martins, F., Felgueiras, C., Smithkova, M., & Caetano, N. (2019). Analysis of Fossil Fuel Energy Consumption and Environmental Impacts in European Countries. *Energies, 12*(6), 964. doi:10.3390/en12060964

Martins, R. G., Gonçalves, I. S., De Morais, M. G., & Vieira Costa, J. A. (2017). New technologies from the bioworld: Selection of biopolymer-producing microalgae. *Polímeros, 27*(4), 285–289. doi:10.1590/0104-1428.2375

Martins, R. G., Severo Gonçalves, I., De Morais, M. G., & Costa, J. A. V. (2014). Bioprocess engineering aspects of biopolymer production by the cyanobacterium spirulina strain LEB 18. *International Journal of Polymer Science, 2014*, 1–7. doi:10.1155/2014/895237

Martins, R. M., Nedel, F., Guimaraes, V., Da Silva, A. F., Colepicolo, P., De Pereira, C. M., & Lund, R. G. (2018). Macroalgae extracts from Antarctica have antimicrobial and anticancer potential. *Frontiers in Microbiology, 9*, 412. doi:10.3389/fmicb.2018.00412 PMID:29568291

Martin, W. F., & Allen, J. F. (2018). An Algal Greening of Land. *Cell, 174*(2), 256–258. doi:10.1016/j.cell.2018.06.034 PMID:30007415

Marti-Quijal, F. J., Zamuz, S., Tomašević, I., Gómez, B., Rocchetti, G., Lucini, L., Remize, F., Barba, F. J., & Lorenzo, J. M. (2019). Influence of different sources of vegetable, whey and microalgae proteins on the physicochemical properties and amino acid profile of fresh pork sausages. *Lebensmittel-Wissenschaft + Technologie*, *110*, 316–323. doi:10.1016/j.lwt.2019.04.097

Maruyama, H., Tamauchi, H., Hashimoto, M., & Nakano, T. (2003). Antitumor activity and immune response of Mekabu fucoidan extracted from Sporophyll of Undaria pinnatifida. *In Vivo (Athens, Greece)*, *17*(3), 245–249. PMID:12929574

Masoj'ıdek, J., Kopecky, J., Giannelli, L., & Torzillo, G. ́. (2011). Productivity correlated to photobiochemical performance of Chlorella mass cultures grown outdoors in thin layer cascades. *Journal of Industrial Microbiology & Biotechnology*, *38*(2), 307–317. doi:10.100710295-010-0774-x PMID:20652359

Masondo, N. A., Kulkarni, M. G., Finnie, J. F., & Van Staden, J. (2018). Influence of biostimulants-seed-priming on Ceratotheca triloba germination and seedling growth under low temperatures, low osmotic potential and salinity stress. *Ecotoxicology and Environmental Safety*, *147*, 43–48. doi:10.1016/j.ecoenv.2017.08.017 PMID:28826029

Massari, S., & Ruberti, M. (2013). Rare earth elements as critical raw materials: Focus on international markets and future strategies. *Resources Policy*, *38*(1), 36–43. doi:10.1016/j.resourpol.2012.07.001

Mata, T. M., Martins, A. A., & Caetano, N. S. (2010). Microalgae for biodiesel production and other applications: A review. *Renewable & Sustainable Energy Reviews*, *14*, 217–232.

Mathews, J., & Wang, G. (2009). Metabolic pathway engineering for enhanced biohydrogen production. *International Journal of Hydrogen Energy*, *34*(17), 7404–7416. doi:10.1016/j.ijhydene.2009.05.078

Mathews, S. L., Epps, M. J., Blackburn, R. K., Goshe, M. B., Grunden, A. M., & Dunn, R. R. (2019). Public questions spur the discovery of new bacterial species associated with lignin bioconversion of industrial waste. *Royal Society Open Science*, *6*(3), 180748. doi:10.1098/rsos.180748 PMID:31031986

Mathimani, T., Baldinelli, A., Rajendran, K., Prabakar, D., Matheswaran, M., Pieter van Leeuwen, R., & Pugazhendhi, A. (2019). Review on cultivation and thermochemical conversion of microalgae to fuels and chemicals: Process evaluation and knowledge gaps. *Journal of Cleaner Production*, *208*, 1053–1064. doi:10.1016/j.jclepro.2018.10.096

Mathimani, T., Uma, L., & Prabaharan, D. (2015). Homogeneous acid catalysed transesterification of marine microalga Chlorella sp. BDUG 91771 lipid – An efficient biodiesel yield and its characterization. *Renewable Energy*, *81*, 523–533. doi:10.1016/j.renene.2015.03.059

Mathiot, C., Ponge, P., Gallard, B., Sassi, J.-F., Delrue, F., & Le Moigne, N. (2019). Microalgae starch-based bioplastics: Screening of ten strains and plasticization of unfractionated microalgae by extrusion. *Carbohydrate Polymers*, *208*, 142–151. doi:10.1016/j.carbpol.2018.12.057 PMID:30658785

Matson, M. M., & Atsumi, S. (2018). Photomixotrophic chemical production in cyanobacteria. *Current Opinion in Biotechnology*, *50*, 65–71. doi:10.1016/j.copbio.2017.11.008 PMID:29179151

Matsumoto, M., Yokouchi, H., Suzuki, N., Ohata, H., & Matsunaga, T. (2003). Saccharification of Marine Microalgae Using Marine Bacteria for Ethanol Production. *Applied Biochemistry and Biotechnology*, *105*(1–3), 247–254. doi:10.1385/ABAB:105:1-3:247 PMID:12721490

Matter, I. A., Bui, V. K. H., Jung, M., Seo, J. Y., Kim, Y. E., Lee, Y. C., & Oh, Y. K. (2019). Flocculation harvesting techniques for microalgae: *A review. Applied Sciences (Basel, Switzerland)*, *9*(15), 3069.

Matter, I. A., Darwesh, O. M., & Eida, M. F. (2018). Harvesting of microalgae *Scenedesmus obliquus* using chitosan-alginate dual flocculation system. *Biosci. Res*, *15*, 540–548.

Matter, I. A., Darwesh, O. M., & El-baz, F. K. (2016). Using the natural polymer chitosan in harvesting *Scenedesmus* species under di_erent concentrations and cultural pH values. *International Journal of Pharma and Bio Sciences*, *7*, 254–260.

Mattner, S. W., Wite, D., Riches, D. A., Porter, I. J., & Arioli, T. (2013). The effect of kelp extract on seedling establishment of broccoli on contrasting soil types in southern Victoria, Australia. *Biological Agriculture and Horticulture*, *29*(4), 258–270. doi:10.1080/01448765.2013.830276

Maurya, R., Zhu, X., Valverde-Pérez, B., Ravi Kiran, B., General, T., Sharma, S., Kumar Sharma, A., Thomsen, M., Venkata Mohan, S., Mohanty, K., & Angelidaki, I. (2022). Advances in microalgal research for valorization of industrial wastewater. *Bioresource Technology*, *343*, 126128. doi:10.1016/j.biortech.2021.126128 PMID:34655786

Ma, X., Zheng, H., Zhou, W., Liu, Y., Chen, P., & Ruan, R. (2016). Enhanced harvesting of *Chlorella vulgaris* using combined flocculants. *Applied Biochemistry and Biotechnology*, *180*, 791–804.

Maximous, N., Nakhla, G., & Wan, W. (2009). Comparative assessment of hydrophobic and hydrophilic membrane fouling in wastewater applications. *Journal of Membrane Science*, *339*(1), 93–99. doi:10.1016/j.memsci.2009.04.034

Maxwell, D. P., Fal, S., Trick, C. G., & Huner, N. (1994). Growth at low temperature mimics high-light acclimation in Chlorella vulgaris. *Plant Physiology*, *105*(2), 535–543.

Mayakrishnan, V., Kannappan, P., Abdullah, N., & Ahmed, A. B. A. (2013). Cardioprotective activity of polysaccharides derived from marine algae: An overview. *Trends in Food Science & Technology*, *30*(2), 98–104. doi:10.1016/j.tifs.2013.01.007

Mayo-Bean, K. E., Moran-Bruce, K., Nabholz, J. V., Meylan, W. M., & Howard, P. H. (2012). Operation manual for the ecological structure-activit,y relationship model (ECOSAR) class program estimating toxicity of industrial chemicals to aquatic organisms using the ecosar (ecological structure activity relationship) class program. ms-windows version 1.11. Office of Pollution Prevention and Toxics, US Environmental Protection Agency Syracuse Research Corporation Consortium for Environmental Risk Management.

Mazard, S., Penesyan, A., Ostrowski, M., Paulsen, I. T., & Egan, S. (2016). Tiny microbes with a big impact: The role of cyanobacteria and their metabolites in shaping our future. *Marine Drugs*, *14*(5), 97. doi:10.3390/md14050097 PMID:27196915

Maznah, ZBS, & Ismail, M, Halimah. (2012). Fate of thiram in an oil palm nursery during the wet season. *Journal of Oil Palm Research*, *24*, 1397–1403.

McBride, M. B. (1997). A critique of diffuse double layer models applied to colloid and surface chemistry. *Clays and Clay Minerals*, *45*(4), 598–608. doi:10.1346/CCMN.1997.0450412

McCoy, S. J., Krueger-Hadfield, S. A., & Mieszkowska, N. (2020). Evolutionary Phycology: Toward a Macroalgal Species Conceptual Framework. *Journal of Phycology*, *56*(6), 1404–1413. doi:10.1111/jpy.13059 PMID:32726874

McFadden, G. I. (2011). The apicoplast. *Protoplasma*, *248*(4), 641–650. doi:10.100700709-010-0250-5 PMID:21165662

Mehariya, S., Goswami, R. K., Verma, P., Lavecchia, R., & Zuorro, A. (2021). Integrated approach for wastewater treatment and biofuel production in microalgae biorefineries. *Energies*, *14*(8), 2282. Advance online publication. doi:10.3390/en14082282

Mehtani, J., Arora, N., Patel, A., Jain, P., Pruthi, P. A., Poluri, K. M., & Pruthi, V. (2017). Augmented lipid accumulation in ethyl methyl sulphonate mutants of oleaginous microalga for biodiesel production. *Bioresource Technology*, *242*, 121–127. doi:10.1016/j.biortech.2017.03.108 PMID:28366694

Meisch, H.-U., Benzschawel, H., & Bielig, H.-J. (1977). The role of vanadium in green plants. *Archives of Microbiology*, *114*(1), 67–70. doi:10.1007/BF00412983 PMID:20864

Meisch, H.-U., & Bielig, H.-J. (1975). Effect of vanadium on growth, chlorophyll formation and iron metabolism in unicellular green algae. *Archives of Microbiology*, *105*(1), 77–82. doi:10.1007/BF00447117 PMID:1190955

Meixner, K., Fritz, I., Daffert, C., Markl, K., Fuchs, W., & Drosg, B. (2016). Processing recommendations for using low-solids digestate as nutrient solution for poly-ß-hydroxybutyrate production with Synechocystis salina. *Journal of Biotechnology*, *240*, 61–67. doi:10.1016/j.jbiotec.2016.10.023 PMID:27984118

Melis, A. (2007). Photosynthetic H 2 metabolism in Chlamydomonas reinhardtii (unicellular green algae). *Planta*, *226*(5), 1075–1086. doi:10.100700425-007-0609-9 PMID:17721788

Melis, A., & Happe, T. (2001). Hydrogen Production. Green Algae as a Source of Energy. *Plant Physiology*, *127*(3), 740–748. doi:10.1104/pp.010498 PMID:11706159

Melis, A., & Mitra, M. (2009). Optical Properties of Microalgae for Enhanced Biofuels Production. *Optics Express*, *16*, 21807–21820. PMID:19104614

Mendhulkar, V. D., & Shetye, L. A. (2017). Synthesis of biodegradable polymer polyhydroxyalkanoate (PHA) in cyanobacteria Synechococcus elongates under mixotrophic nitrogen-and phosphate-mediated stress conditions. *Industrial Biotechnology (New Rochelle, N.Y.)*, *13*(2), 85–93. doi:10.1089/ind.2016.0021

Mendoza, J. L., Granados, M. R., De Godos, I., Acién, F. G., Molina, E., Banks, C., & Heaven, S. (2013). Fluid-dynamic characterization of real-scale raceway reactors for microalgae production. *Biomass and Bioenergy*, *54*, 267–275.

Merin, D. D., Prakash, S., & Bhimba, B. V. (2010). Antibacterial screening of silver nanoparticles synthe- sized by marine micro algae. *Asian Pacific Journal of Tropical Medicine*, *3*(10), 797–799. doi:10.1016/S1995-7645(10)60191-5

Metcalf, A. J., Nagygyor, A., & Boyle, N. R. (2020). Rapid Annotation of Photosynthetic Systems (RAPS): Automated algorithm to generate genome-scale metabolic networks from algal genomes. *Algal Research*, *50*(101967), 1–7. doi:10.1016/j.algal.2020.101967

Metsoviti, M. N., Papapolymerou, G., Karapanagiotidis, I. T., & Katsoulas, N. (2020). Effect of Light Intensity and Quality on Growth Rate and Composition of Chlorella vulgaris. *Plants*, *9*(1), 31. Advance online publication. doi:10.3390/plants9010031 PMID:31878279

Metting, B. (1981). The systematics and ecology of soil algae. *Botanical Review*, *47*(2), 195–312. doi:10.1007/BF02868854

Meuser, J. E., Boyd, E. S., Ananyev, G., Karns, D., Radakovits, R., Murthy, U. M. N., Ghirardi, M. L., Dismukes, G. C., Peters, J. W., & Posewitz, M. C. (2011). Evolutionary significance of an algal gene encoding an [FeFe]-hydrogenase with F-domain homology and hydrogenase activity in *Chlorella variabilis* NC64A. *Planta*, *234*(4), 829–843. doi:10.100700425-011-1431-y PMID:21643991

Meyer, J. (2007). [FeFe] hydrogenases and their evolution: A genomic perspective. *Cellular and Molecular Life Sciences*, *64*(9), 1063–1084. doi:10.100700018-007-6477-4 PMID:17353991

Meyer, N., Bigalke, A., Kaulfuss, A., & Pohnert, G. (2017). Strategies and ecological roles of algicidal bacteria. *FEMS Microbiology Reviews*, *41*(6), 880–899. doi:10.1093/femsre/fux029 PMID:28961821

Mhedhbi, E., Khelifi, N., Foladori, P., & Smaali, I. (2020). Real-Time Behavior of a Microalgae–Bacteria Consortium Treating Wastewater in a Sequencing Batch Reactor in Response to Feeding Time and Agitation Mode. *Water (Basel)*, *12*(7), 1893. doi:10.3390/w12071893

Michalak, I., Chojnacka, K., Dmytryk, A., Wilk, R., Gramza, M., & Rój, E. (2016). Evaluation of supercritical extracts of algae as biostimulants of plant growth in field trials. *Frontiers in Plant Science, 7*. doi:10.3389/fpls.2016.01591

Michel, G., Tonon, T., Scornet, D., Cock, J. M., & Kloareg, B. (2010). The cell wall polysaccharide metabolism of the brown alga Ectocarpus siliculosus. Insights into the evolution of extracellular matrix polysaccharides in Eukaryotes. *The New Phytologist, 188*(1), 82–97. doi:10.1111/j.1469-8137.2010.03374.x PMID:20618907

Michiels, J., Skrivanova, E., & Missotten, J. (2011). Intact Brown Seaweed (Ascophyllum nodosum) in Diets of Weaned Piglets: Effects on Performance, Gut Bacteria and Morphology and Plasma Oxidative Status. Journal of Animal Physiology and Animal Nutrition. doi:10.1111/j.1439-0396.2011.01227

Mihai, M., Legros, N., & Alemdar, A. (2013). Formulation-properties versatility of wood fiber biocomposites based on polylactide and polylactide/thermoplastic starch blends. *Polymer Engineering and Science, 54*(6), 1325–1340. doi:10.1002/pen.23681

Milledge, J. J. (2010). The challenge of algal fuel: Economic processing of the entire algal biomass. *Condensed Matter-Materials Engineering Newsletter, 1*(6), 4–6.

Milledge, J. J., & Heaven, S. (2013). A review of the harvesting of micro-algae for biofuel production. *Reviews in Environmental Science and Biotechnology, 12*, 165–178.

Milledge, J. J., Smith, B., Dyer, P. W., & Harvey, P. (2014). Macroalgae-derived biofuel: A review of methods of energy extraction from seaweed biomass. *Energies, 7*(11), 7194–7222. doi:10.3390/en7117194

Miralto, A., Barone, G., Romano, G., Poulet, S. A., Ianora, A., Russo, G. L., Buttino, I., Mazzarella, G., Laabir, M., Cabrini, M., & Giacobbe, M. G. (1999). The insidious effect of diatoms on copepod reproduction. *Nature, 402*(6758), 173–176. doi:10.1038/46023

Miranda, J. R., Passarinho, P. C., & Gouveia, L. (2012). Pre-treatment optimization of Scenedesmus obliquus microalga for bioethanol production. *Bioresource Technology, 104*, 342–348. doi:10.1016/j.biortech.2011.10.059 PMID:22093974

Miranda, J., Ponce, G., Arellano-Garcia, H., Maciel Filho, R., & Wolf Maciel, M. R. (2015). Syngas to higher alcohols using cu-based catalyst – a simulation approach. *Chemical Engineering Transactions, 43*, 1519–1524. doi:10.3303/CET1543254

Mishra, P., Bhoomika, K., & Dubey, R. S. (2013). Differential responses of antioxidative defense system to prolonged salinity stress in salt-tolerant and salt-sensitive Indica rice (*Oryza sativa*L.) seedlings. *Protoplasma, 250*(1), 3–19. doi:10.100700709-011-0365-3 PMID:22194018

Mobin, A., & Chowdhury, H. (2019). ScienceDirect ScienceDirect ScienceDirect Commercially important bioproducts from microalgae and their International bioproducts Commercially important microalgae and their current applications – from A review current applications – A review Assessing the. *Energy Procedia, 160*(2018), 752–760. doi:10.1016/j.egypro.2019.02.183

Mock, T., & Kroon, B. M. (2002). Photosynthetic energy conversion under extreme conditions—II: The significance of lipids under light limited growth in Antarctic sea ice diatoms. *Phytochemistry, 61*(1), 53–60. doi:10.1016/S0031-9422(02)00215-7 PMID:12165302

Mofeed, J., & Deyab, M. (2015). Monitoring for the abundance and distribution of macroalgae along Suez Canal, Egypt 1. *CATRINA, 11*(1), 81–91.

Moghazy, R. M., & Abdo, S. (2018). The efficacy of microalgal biomass collected from high rate algal pond for dyes biosorption and biofuel production. *Research Journal of Chemistry and Environment, 22*(11), 54–60.

Mohan, S. V. (2020). Algal Biorefinery Models with Self-Sustainable Closed Loop Approach: Trends and Prospective for Blue-Bioeconomy. *Bioresource Technology, 295*, 122128.

Mohan, S., Oluwafemi, O. S., Kalarikkal, N., Thomas, S., & Songca, S. P. (2016). Biopolymers–application in nanoscience and nanotechnology. *Recent Advances in Biopolymers, 1*(1), 47-66.

Mohanty, A., Misra, M., & Hinrichsen, G. (2000). Biofibres, biodegradable polymers and biocomposites: An overview. *Macromolecular Materials and Engineering, 276*(1), 1–24. doi:10.1002/(SICI)1439-2054(20000301)276:1<1::AID-MAME1>3.0.CO;2-W

Mohanty, S., Pradhan, B., Patra, S., Behera, C., Nayak, R., & Jena, M. (2020). Screening for nutritive bioactive compounds in some algal strains isolated from coastal Odisha. *Journal of Advanced Plant Sciences, 2020*(2), 1–8.

Mohn, F. H. (1980). Experiences and strategies in the recovery of biomass from mass cultures of microalgae. In G. Schelef & C. J. Soeder (Eds.), *Algae Biomass* (pp. 547–571). Elsevier.

Mohsenzadeh, A., Zamani, A., & Taherzadeh, M. J. (2017). Bioethylene production from ethanol: A Review and techno-economical evaluation. *ChemBioEng Reviews, 4*(2), 75–91. doi:10.1002/cben.201600025

Mokronsnop, V. M., & Zolotareva, E. K. (2014). Microalgae as tocopherol producers. *Biotechnologia Acta, 7*(2), 26–33. doi:10.15407/biotech7.02.026

Mollamohammada, S., Aly Hassan, A., & Dahab, M. (2020). Nitrate Removal from Groundwater Using Immobilized Heterotrophic Algae. *Water, Air, and Soil Pollution, 231*, 26.

Mollenhauer, D., Mollenhauer, R., & Kluge, M. (1996). Studies on initiation and development of the partner association in *Geosiphon pyriforme* (Kutz) v. Wettstein, a unique endocytobiotic system of a fungus (Glomales) and the cyanobacterium *Nostoc punctiforme* (Kutz) Hariot. *Protoplasma, 193*(1-4), 3–9. doi:10.1007/BF01276630

Momeni, S., & Nabipour, I. (2015). A Simple Green Synthesis of Palladium Nanoparticles with Sargassum Alga and Their Electrocatalytic Activities Towards Hydrogen Peroxide. *Applied Biochemistry and Biotechnology, 176*(7), 1937–1949. doi:10.100712010-015-1690-3 PMID:26041058

Moncada, B. J., Aristizábal, M. V., & Cardona, A. C. A. (2016). Design strategies for sustainable biorefineries. *Biochemical Engineering Journal, 116*, 122–134. doi:10.1016/j.bej.2016.06.009

Mondal, M., Goswami, S., Ghosh, A., Oinam, G., Tiwari, O.N., Das, P., & Halder, G.N. (2017). Production of biodiesel from microalgae through biological carbon capture: A review. *Biotech, 7*(2), 99.

Mondal, M., Goswami, S., Ghosh, A., Oinam, G., Tiwari, O. N., Das, P., Gayen, K., Mandal, M. K., & Halder, G. N. (2017). Production of biodiesel from microalgae through biological carbon capture: A review. 3. *Biotechnology, 7*(2), 99. PMID:28560639

Monod, J. (1949). The growth of bacterial cultures. *Annual Review of Microbiology, 3*(1), 371–394. doi:10.1146/annurev.mi.03.100149.002103

Monshupanee, T., Nimdach, P., & Incharoensakdi, A. (2016). Two-stage (photoautotrophy and heterotrophy) cultivation enables efficient production of bioplastic poly-3-hydroxybutyrate in auto-sedimenting cyanobacterium. *Scientific Reports, 6*(1), 1–9. doi:10.1038rep37121 PMID:27845413

Montaser, R., & Luesch, H. (2011). Marine natural products: A new wave of drugs? *Future Medicinal Chemistry, 3*(12), 1475–1489. doi:10.4155/fmc.11.118 PMID:21882941

Montasser, M. S., Hegazi, M.M., Younes, A. M., Dashti, N. H., El-Sharkawey, A. E. (2017). Effect of Gold Chloride Concentration and Volume on Size and Shape of Biological Synthesized Gold Nanoparti- cles (AuNPs) Using Red Algae (Laurenciapapillosa). *Science of Advanced Materials, 9*(7), 1105-1113.

Montazeri, M., Soh, L., Pérez-López, P., Zimmerman, J. B., & Eckelman, M. J. (2016). Time-dependent life cycle assessment of microalgal biorefinery co-products. *Biofuels, Bioproducts & Biorefining, 10*(4), 409–421. doi:10.1002/bbb.1649

Monteiro, C. M., Castro, P. M. L., & Malcata, F. X. (2012). Metal uptake by microalgae: Underlying mechanisms and practical applications. *Biotechnology Progress, 28*(2), 299–311. doi:10.1002/btpr.1504 PMID:22228490

Montero, O., Sobrino, C., Pares, G., & Lubian, L. M. (2002). Photoinhibition and recovery after selective short-term exposure to solar radiation of five chlorophyll c-containing marine microalgae. *Ciencias Marinas, 28*(3), 223–236. doi:10.7773/cm.v28i3.229

Montingelli, M. E., Benyounis, K. Y., Quilty, B., Stokes, J., & Olabi, A. G. (2016). Optimisation of biogas production from the macroalgae Laminaria sp. at different periods of harvesting in Ireland. *Applied Energy, 177*, 671–682. doi:10.1016/j.apenergy.2016.05.150

Moog, D., Schmitt, J., Senger, J., Zarzycki, J., Rexer, K.-H., Linne, U., Erb, T., & Maier, U. G. (2019). Using a marine microalga as a chassis for polyethylene terephthalate (PET) degradation. *Microbial Cell Factories, 18*(1), 171. doi:10.118612934-019-1220-z PMID:31601227

Moorthi, P. V., Balasubramanian, C., & Mohan, S. (2018). Algae : A potential source for nanoparticle synthesis. *Applied Biochemistry and Biotechnology, 175*, 135. doi:10.100712010-014-1264-9 PMID:25240853

Moreira, D., & Pires, J. C. M. (2016). Atmospheric CO_2 capture by algae: Negative carbon dioxide emission path. *Bioresource Technology, 215*, 371–379. doi:10.1016/j.biortech.2016.03.060 PMID:27005790

Moreno-Garcia, L., Adjallé, K., Barnabé, S., & Raghavan, G. S. V. (2017). Microalgae Biomass Production for a Biorefinery System: Recent Advances and the Way towards Sustainability. *Renewable & Sustainable Energy Reviews, 76*(March), 493–506. doi:10.1016/j.rser.2017.03.024

Morf, L. S., Gloor, R., Haag, O., Haupt, M., Skutan, S., Di Lorenzo, F., & Böni, D. (2013). Precious metals and rare earth elements in municipal solid waste–sources and fate in a Swiss incineration plant. *Waste Management (New York, N.Y.), 33*(3), 634–644. doi:10.1016/j.wasman.2012.09.010 PMID:23085306

Morillas-España, A., Lafarga, T., Acién-Fernández, F. G., Gómez-Serrano, C., & González-López, C. V. (2021). Annual production of microalgae in wastewater using pilot-scale thin-layer cascade photobioreactors. *Journal of Applied Phycology*, 1–11.

Moroney, J. V., Bartlett, S. G., & Samuelsson, G. (2001). Carbonic anhydrases in plants and algae. *Plant, Cell & Environment, 24*(2), 141–153. doi:10.1111/j.1365-3040.2001.00669.x

Morrissey, K. L. (2015). Charge-tunable Polymers as Reversible and Recyclable Flocculants for the Dewatering of Microalgae. *Biotechnology and Bioengineering, 112*(1), 74–83.

Mortain-Bertrand, A., Descolas-Gros, C., & Jupin, H. (1988). Pathway of dark inorganic carbon fixation in two species of diatoms: Influence of light regime and regulator factors on diel variations. *Journal of Plankton Research, 10*, 199–217.

Mosey, M., Douchi, D., Knoshaug, E. P., & Laurens, L. M. (2021). Methodological review of genetic engineering approaches for non-model algae. *Algal Research, 54*, 102221. doi:10.1016/j.algal.2021.102221

Moss, R. L., Tzimas, E., Kara, H., Willis, P., & Kooroshy, J. (2011). Critical metals in strategic energy technologies. *Assessing Rare Metals as Supply Chain Bottlenecks in Low Carbon Energy Technologies, JRC Scientific and Technical Reports, EUR, 24884.*

Mostafa, N. A., Farag, A. A., Abo-dief, H. M., & Tayeb, A. M. (2018). Production of biodegradable plastic from agricultural wastes. *Arabian Journal of Chemistry*, *11*(4), 546–553. doi:10.1016/j.arabjc.2015.04.008

Mota, R., Pereira, S. B., Meazzini, M., Fernandes, R., Santos, A., Evans, C. A., De Philippis, R., Wright, P. C., & Tamagnini, P. (2015). Effects of heavy metals on Cyanothece sp. CCY 0110 growth, extracellular polymeric substances (EPS) production, ultrastructure and protein profiles. *Journal of Proteomics*, *120*, 75–94. doi:10.1016/j.jprot.2015.03.004 PMID:25779463

Mourão, M. M., Gradíssimo, D. G., Santos, A. V., Schneider, M. P. C., Faustino, S. M. M., Vasconcelos, V., & Xavier, L. P. (2020). Optimization of Polyhydroxybutyrate Production by Amazonian Microalga Stigeoclonium sp. B23. *Biomolecules*, *10*(12), 1628. doi:10.3390/biom10121628 PMID:33287108

Mourelle, M. L., Gómez, C. P., & Legido, J. L. (2015). Cosmética termal: Valor añadido en los centros termales. In *Proceedings of the 1st International Congress on Water Healing Spa and Quality of Life*. University of Vigo.

Mousavi, P., Montazeri-Najafabady, N., Abolhasanzadeh, Z., Mohagheghzadeh, A., Hamidi, M., Niazi, A., Morowvat, M. H., & Ghasemi, Y. (2016). Investigating the effects of phytohormones on growth and beta-carotene production in a naturally isolates stain of Dunaliella salina. *Journal of Applied Pharmaceutical Science*, *6*, 164–171.

Moussa, G., Moury, R., Demirci, U. B., Şener, T., & Miele, P. (2013). Boron-based hydrides for chemical hydrogen storage. *International Journal of Energy Research*, *37*(8), 825–842. doi:10.1002/er.3027

Możejko-Ciesielska, J., & Kiewisz, R. (2016). Bacterial polyhydroxyalkanoates: Still fabulous? *Microbiological Research*, *192*, 271–282. doi:10.1016/j.micres.2016.07.010 PMID:27664746

Mpinga, C. N., Bradshaw, S. M., Akdogan, G., Snyders, C. A., & Eksteen, J. J. (2014). Evaluation of the Merrill–Crowe process for the simultaneous removal of platinum, palladium and gold from cyanide leach solutions. *Hydrometallurgy*, *142*, 36–46. doi:10.1016/j.hydromet.2013.11.004

Mu, D., Xin, C., & Zhou, W. (2020). Life cycle assessment and techno-economic analysis of algal biofuel production. *Microalgae Cultivation for Biofuels Production*, 281-292.

MubarakAli, D., Gopinath, V., Rameshbabu, N., & Thajuddin, N. (2012). Synthesis and characterization of CdS nanoparticles using C-phycoerythrin from the marine cyanobacteria. *Materials Letters*, *74*, 8–11. doi:10.1016/j.matlet.2012.01.026

Mubarak, M., Shaija, A., & Suchithra, T. V. (2019). Flocculation: An effective way to harvest microalgae for biodiesel production. *Journal of Environmental Chemical Engineering*, *7*(4), 103221. doi:10.1016/j.jece.2019.103221

Mu, D., Ruan, R., Addy, M., Mack, S., Chen, P., & Zhou, Y. (2017). Life cycle assessment and nutrient analysis of various processing pathways in algal biofuel production. *Bioresource Technology*, *230*, 33–42. doi:10.1016/j.biortech.2016.12.108 PMID:28157562

Mu, D., Xin, C., & Zhou, W. (2020). Life Cycle Assessment and Techno-Economic Analysis of Algal Biofuel Production. In A. Yousuf (Ed.), *Microalgae Cultivation for Biofuels Production* (pp. 281–292). Academic Press. doi:10.1016/B978-0-12-817536-1.00018-7

Mukherjee, T., & Kao, N. (2011). PLA based biopolymer reinforced with natural fibre: A review. *Journal of Polymers and the Environment*, *19*(3), 714–725. doi:10.100710924-011-0320-6

Mularczyk, M., Michalak, I., & Marycz, K. (2020). Astaxanthin and other Nutrients from Haematococcus pluvialis—Multifunctional Applications. *Marine Drugs*, *18*(9), 459. doi:10.3390/md18090459 PMID:32906619

Muller-Feuga, A. (2004). Microalgae for Aquaculture: The Current Global Situation Future Trends. In A. Richmond (Ed.), Handbook of Microalgal Culture: Biotechnology and Applied Phycology (pp. 352–364). Academic Press.

Muller-Feuga, A. (2000). The role of microalgae in aquaculture: Situation and trends. *Journal of Applied Phycology*, *12*(3/5), 527–534. doi:10.1023/A:1008106304417

Muller-Feuga, A., Moal, J., & Kaas, R. (2003a). The Microalgae of Aquaculture. In J. G. Støttrup & L. A. McEvoy (Eds.), *Live Feeds in Marine Aquaculture* (pp. 206–252). doi:10.1002/9780470995143.ch6

Muller-Feuga, A., Robert, C., & Cahu, J. (2003b). Uses of Microalgae in Aquaculture. In J. G. Støttrup & L. A. McEvoy (Eds.), *Live Feeds in Marine Aquaculture* (pp. 253–299). doi:10.1002/9780470995143.ch7

Muller-Fuega, A. (2004). Microalgae for aquaculture. The current global situation and future trends. In A. Richmond (Ed.), *Microalgal culture: biotechnology and applied phycology* (pp. 352–364). Blackwell Science.

Mullick, A., & Neogi, S. (2017). A review on acoustic methods of algal growth control by ultrasonication through existing and novel emerging technologies. *Reviews in Chemical Engineering*, *33*(5), 469–490. doi:10.1515/revce-2016-0010

Mulumba, N., & Farag, I. H. (2012). Tubular photobioreactor for microalgae biodiesel production. *International Journal of Engineering Science and Technology*, *4*(2), 703–709.

Muneer, F., Rasul, I., Azeem, F., Siddique, M. H., Zubair, M., & Nadeem, H. (2020). Microbial polyhydroxyalkanoates (PHAs): Efficient replacement of synthetic polymers. *Journal of Polymers and the Environment*, *28*(9), 2301–2323. doi:10.100710924-020-01772-1

Munir, N., Imtiaz, A., Sharif, N., & Naz, S. (2015). Optimization of growth conditions of different algal strains and determination of their lipid contents. *J Anim Plant Sci*, *25*(2), 546–553.

Munoz, ˜. C. F., de Jaeger, L., Sturme, M. H. J., Lip, K. Y. F., Olijslager, J. W. J., Springer, J., Wolbert, E. J. H., Martens, D. E., Eggink, G., Weusthuis, R. A., & Wijffels, R. H. (2018). Improved DNA/protein delivery in microalgae – A simple and reliable method for the prediction of optimal electroporation settings. *Algal Research*, *33*, 448–455. doi:10.1016/j.algal.2018.06.021

Muñoz-Bonilla, A., & Fernández-García, M. (2012). Polymeric materials with antimicrobial activity. *Progress in Polymer Science*, *37*(2), 281–339. doi:10.1016/j.progpolymsci.2011.08.005

Muñoz, C. F., Südfeld, C., Naduthodi, M. I., Weusthuis, R. A., Barbosa, M. J., Wijffels, R. H., & D'Adamo, S. (2021). Genetic engineering of microalgae for enhanced lipid production. *Biotechnology Advances*, *52*, 107836. doi:10.1016/j.biotechadv.2021.107836 PMID:34534633

Muñoz, C., Hidalgo, C., Zapata, M., Jeison, D., Riquelme, C., & Rivas, M. (2014). Use of Cellulolytic Marine Bacteria for Enzymatic Pretreatment in Microalgal Biogas Production. *Applied and Environmental Microbiology*, *80*(14), 4199–4206. doi:10.1128/AEM.00827-14 PMID:24795376

Muraleedharan, T. R., Philip, L., Iyengar, L., & Venkobachar, C. (1994). Application studies of biosorption for monazite processing industry effluents. *Bioresource Technology*, *49*(2), 179–186. doi:10.1016/0960-8524(94)90082-5

Mureed, K., Kanwal, S., Hussain, A., Noureen, S., Hussain, S., Ahmad, S., Ahmad, M., & Waqas, R. (2018). Biodiesel production from algae grown on food industry wastewater. *Environmental Monitoring and Assessment*, *190*(5), 1–11. doi:10.100710661-018-6641-3 PMID:29633020

Muriuki, C., Kairigo, P., Home, P., Ngumba, E., Raude, J., Gachanja, A., & Tuhkanen, T. (2020). Mass loading, distribution, and removal of antibiotics and antiretroviral drugs in selected wastewater treatment plants in Kenya. *The Science of the Total Environment*, *140655*(743). https://doi.org/10.1016/j.scitotenv.2020.140655

Murphy, C. F., & Allen, D. T. (2011). Energy-water nexus for mass cultivation of algae. *Environmental Science & Technology*, *45*, 5861–5868.

Murray, M. G., & Thompson, W. F. (1980). Rapid isolation of high molecular weight plant DNA. *Nucleic Acids Research*, *8*(19), 4321–4326. doi:10.1093/nar/8.19.4321 PMID:7433111

Murugesan, S., Bhuvaneswari, S., Shanthi, N., Murugakoothan, P., & Sivamurugan, V. (2015). Red alga Hypnea musciformis (Wulf) lamour mediated environmentally benign synthesis and antifungal activity of gold nano particles. *Int. J. Nanosci. Nanotech.*, *6*, 71–83.

Murwanashyaka, T., Shen, L., Yang, Z., Chang, J.-S., Manirafasha, E., Ndikubwimana, T., Chen, C., & Lu, Y. (2020). Kinetic modelling of heterotrophic microalgae culture in wastewater: Storage molecule generation and pollutants mitigation. *Biochemical Engineering Journal*, *157*, 107523. doi:10.1016/j.bej.2020.107523

Musa, M., Doshi, A., Brown, R., & Rainey, T. J. (2019). Microalgae dewatering for biofuels: A comparative techno-economic assessment using single and two-stage technologies. *Journal of Cleaner Production*, *229*, 325–336. doi:10.1016/j.jclepro.2019.05.039

Mutanda, T., Naidoo, D., Bwapwa, J. K., & Anandraj, A. (2020). Biotechnological Applications of Microalgal Oleaginous Compounds: Current Trends on Microalgal Bioprocessing of Products. *Frontiers in Energy Research*, *8*(October), 598803. Advance online publication. doi:10.3389/fenrg.2020.598803

Muthuraj, M., Selvaraj, B., Palabhanvi, B., Kumar, V., & Das, D. (2019). Enhanced lipid content in *Chlorella* sp. FC2 IITG via high energy irradiation mutagenesis. *Korean Journal of Chemical Engineering*, *36*(1), 63–70. doi:10.100711814-018-0180-z

Muto, M., Tanaka, M., Liang, Y., Yoshino, T., Matsumoto, M., & Tanaka, T. (2015). Enhancement of glycerol metabolism in the oleaginous marine diatom *Fistulifera solaris* JPCC DA0580 to improve triacylglycerol productivity. *Biotechnology for Biofuels*, *8*(1), 4. doi:10.118613068-014-0184-9 PMID:25632299

Nabti, E., Jha, B., & Hartmann, A. (2017). Impact of seaweeds on agricultural crop production as biofertilizer. *International Journal of Environmental Science and Technology*, *14*(5), 1119–1134. doi:10.100713762-016-1202-1

Nagarajan, D., Lee, D. J., Kondo, A., & Chang, J. S. (2017a). Recent insights into biohydrogen production by microalgae – From biophotolysis to dark fermentation. *Bioresource Technology*, *227*(1), 373–387. doi:10.1016/j.biortech.2016.12.104 PMID:28089136

Nagarajan, S., Chou, S. K., Cao, S., Wu, C., & Zhou, Z. (2013). An updated comprehensive techno-economic analysis of algae biodiesel. *Bioresource Technology*, *145*, 150–156. doi:10.1016/j.biortech.2012.11.108 PMID:23260269

Nagata, A., Tajima, T., & Takahashi, J. (2006). Effect of astaxanthin 5 mg on antifatigue and task performance of human. *Carotenoid Sci.*, *10*, 102–106.

Naghdi, F. G., Bai, X., Thomas-Hall, S. R., Sharma, K., & Schenk, P. M. (2016). Lipid Extraction from Wet Chaetoceros Muelleri Culture and Evaluation of Remaining Defatted Biomass. *Algal Research*, *20*, 205–212. doi:10.1016/j.algal.2016.10.011

Nagorskaia, V. P., Reunov, A. V., Lapshina, L. A., Ermak, I. M., & Barabanova, A. O. (2010). Inhibitory effect of kappa/beta-carrageenan from red alga Tichocarpus crinitus on the development of a potato virus X infection in leaves of Datura stramonium L. *Izvestiia Akademii Nauk. Seriia Biologicheskaia*, *6*, 756–761.

Naguib, Y. M. A. (2000). Antioxidant activities of astaxanthin and related carotenoids. *Journal of Agricultural and Food Chemistry*, *48*(4), 1150–1154. doi:10.1021/jf991106k PMID:10775364

Nain, L., Rana, A., Joshi, M., Jadhav, S. D., Kumar, D., Shivay, Y. S., Paul, S., & Prasanna, R. (2010). Evaluation of synergistic effects of bacterial and cyanobacterial strains as biofertilizers for wheat. *Plant and Soil*, *331*(1), 217–230. doi:10.100711104 009-0247-z

Najjar, Y. S., & Abu-Shamleh, A. (2020). Harvesting of microalgae by centrifugation for biodiesel production: A review. *Algal Research*, *51*, 102046.

Nakagawa, H. (1997). Effect of Dietary Algae on Improvement of Lipid Metabolism in Fish. *Biomedicine and Pharmacotherapy*, *51*(8), 345–348. doi:10.1016/S0753-3322(97)88053-5 PMID:9436528

Nakajima, M., Kitade, Y., Iitsuka, O., Fukuda, S., & Saga, N. (2000). Rapid extraction of high-quality genomic DNA from *Porphyra yezoensis* (Bangiales, Rhodophyta). *Phycological Research*, *48*(1), 15–17. doi:10.1046/j.1440-1835.2000.00179.x

Nakamura, K. A. Z. U. O., Bray, D. F., & Wagenaar, E. B. (1975). Ultrastructure of Chlamydomonas eugametos palmelloids induced by chloroplatinic acid treatment. *Journal of Bacteriology*, *121*(1), 338–343. doi:10.1128/jb.121.1.338-343.1975 PMID:1116991

Nalley, J. O., O'Donnell, D. R., & Litchman, E. (2018). Temperature effects on growth rates and fatty acid content in freshwater algae and cyanobacteria. *Algal Research*, *35*, 500–507. doi:10.1016/j.algal.2018.09.018

Nan, Y., Liu, J., Lin, R., & Tavlarides, L. L. (2015). Production of biodiesel from microalgae oil (Chlorella protothecoides) by non-catalytic transesterification in supercritical methanol and ethanol: Process optimization. *The Journal of Supercritical Fluids*, *97*, 174–182. doi:10.1016/j.supflu.2014.08.025

Nappo, M., Berkov, S., Massucco, C., Di Maria, V., Bastida, J., Codina, C., Avila, C., Messina, P., Zupo, V., & Zupo, S. (2012). Apoptotic activity of the marine diatom Cocconeis scutellum and eicosapentaenoic acid in BT20 cells. *Pharmaceutical Biology*, *50*(4), 529–535. doi:10.3109/13880209.2011.611811 PMID:22136301

Narancic, T., Cerrone, F., Beagan, N., & O'Connor, K. E. (2020). Recent advances in bioplastics: Application and biodegradation. *Polymers*, *12*(4), 920. doi:10.3390/polym12040920 PMID:32326661

Naresh Kumar, A., Chatterjee, S., Hemalatha, M., Althuri, A., Min, B., Kim, S. H., & Venkata Mohan, S. (2020). Deoiled algal biomass derived renewable sugars for bioethanol and biopolymer production in biorefinery framework. *Bioresource Technology*, *296*(August), 122315. doi:10.1016/j.biortech.2019.122315 PMID:31706890

Nascimento, I. A., Marques, S. S. I., Cabanelas, I. T. D., Pereira, S. A., Druzian, J. I., de Souza, C. O., Vich, D. V., de Carvalho, G. C., & Nascimento, M. A. (2013). Screening Microalgae Strains for Biodiesel Production: Lipid Productivity and Estimation of Fuel Quality Based on Fatty Acids Profiles as Selective Criteria. *BioEnergy Research*, *6*(1), 1–13. doi:10.100712155-012-9222-2

Naser, A. Z., Deiaba, I., & Darras, B. M. (2021). Poly(lactic acid) (PLA) and polyhydroxyalkanoates (PHAs), green alternatives to petroleum-based plastics: A review. *RSC Advances*, *2021*(11), 17151–17196. doi:10.1039/D1RA02390J PMID:35479695

Nath, P. R., Khozin-Goldberg, I., Cohen, Z., Boussiba, S., & Zilberg, D. (2012). Dietary Supplementation with the Microalga Parietochloris incisa Increases Survival and Stress Resistance in Guppy Poecilia reticulata Fry. *Aquaculture Nutrition*, *18*(2), 167–180. doi:10.1111/j.1365-2095.2011.00885.x

Natrah, F. M. I., Kenmegne, M. M., Wiyoto, W., Sorgeloos, P., Bossier, P., & Defoirdt, T. (2011). Effects of micro-algae commonly used in aquaculture on acyl-homoserine lactone quorum sensing. *Aquaculture (Amsterdam, Netherlands)*, *317*(1–4), 53–57. doi:10.1016/j.aquaculture.2011.04.038

Navarro, N., & Sarasquete, C. (1998). Use of freeze-dried microalgae for rearing gilthead seabream, Sparus aurata, larvae: I. Growth, histology and water quality. *Aquaculture (Amsterdam, Netherlands)*, *167*(3-4), 179–193. doi:10.1016/S0044-8486(98)00311-1

Navarro, N., Yúfera, M., & García-Gallego, M. (2001). Use of freeze-dried microalgae for rearing gilthead seabream, Sparus aurata L., larvae. II. Biochemical composition. *Hydrobiologia*, *452*(1/3), 69–77. doi:10.1023/A:1011971810072

Nava, V., & Leoni, B. (2021). A critical review of interactions between microplastics, microalgae and aquatic ecosystem function. *Water Research*, *188*, 116476. doi:10.1016/j.watres.2020.116476 PMID:33038716

Neelam, S., & Subramanyam, R. (2013). Alteration of photochemistry and protein degradation of photosystem II from Chlamydomonas reinhardtii under high salt grown cells. *Journal of Photochemistry and Photobiology. B, Biology*, *124*, 63–70. doi:10.1016/j.jphotobiol.2013.04.007 PMID:23685462

Nel, A., Xia, T., Mädler, L., & Li, N. (2006). Toxic Potential of Materials at the Nanolevel. *Science*, *311*(5761), 622–627. doi:10.1126cience.1114397 PMID:16456071

Nelson, D. R., Hazzouri, K. M., Lauersen, K. J., Jaiswal, A., Chaiboonchoe, A., Mystikou, A., Fu, W., Daakour, S., Dohai, B., Alzahmi, A., Nobles, D., Hurd, M., Sexton, J., Preston, M. J., Blanchette, J., Lomas, M. W., Amiri, K. M. A., & Salehi-Ashtiani, K. (2021). Large-scale genome sequencing reveals the driving forces of viruses in microalgal evolution. *Cell Host & Microbe*, *29*(2), 250–266.e8. doi:10.1016/j.chom.2020.12.005 PMID:33434515

Nemudryi, A. A., Valetdinova, K. R., Medvedev, S. P., & Zakian, S. M. (2014). TALEN and CRISPR/Cas Genome Editing Systems: Tools of Discovery. *Acta Naturae*, *6*(3), 19–40. doi:10.32607/20758251-2014-6-3-19-40 PMID:25349712

Ng'eno, E., Orata, F., Lilechi, D. B., Shikuku, V. O., & Kimosop, S. (2016). Adsorption of caffeine and ciprofloxacin onto pyrolytically derived water hyacinth biochar: Isothermal, kinetics and thermodynamics. *Journal of Chemistry and Chemical Engineering*, *10*, 185–194.

Ng'eno, E., Shikuku, V. O., Orata, F., Lilechi, D. B., & Kimosop, S. (2019). Caffeine and Ciprofloxacin Adsorption from water onto clinoptilolite: Linear isotherms, kinetics, thermodynamics and mechanistic studies. *South African Journal of Chemistry*, *72*, 136–142.

Ngamwonglumlert, L., & Devahastin, S. (2018). Carotenoids. Encyclopedia of Food Chemistry, 40–52.

Ng, I. S., Tan, S. I., Kao, P. H., Chang, Y. K., & Chang, J. S. (2017). Recent developments on genetic engineering of microalgae for biofuels and bio-based chemicals. *Biotechnology Journal*, *12*(10), 1600644. doi:10.1002/biot.201600644 PMID:28786539

Nguyen, H. S., Mäki-Arvela, P., Akhmetzyanova, U., Tišler, Z., Hachemi, I., Rudnäs, A., Smeds, A., Eränen, K., Aho, A., Kumar, N., Hemming, J., Peurla, M., & Murzin, D. Y. (2017). Direct hydrodeoxygenation of algal lipids extracted from Chlorella alga. *Journal of Chemical Technology and Biotechnology (Oxford, Oxfordshire)*, *92*(4), 741–748. doi:10.1002/jctb.5158

Nguyen, T. D. P., Frappart, M., Jaouen, P., Pruvost, J., & Bourseau, P. (2014). Harvesting *Chlorella vulgaris* by natural increase in pH: E_ect of medium composition. *Environmental Technology*, *35*, 1378–1388.

Nguyen, T. H. T., Park, S., Jeong, J., Shin, Y. S., Sim, S. J., & Jin, E. (2020). Enhancing lipid productivity by modulating lipid catabolism using the CRISPR-Cas9 system in *Chlamydomonas*. *Journal of Applied Phycology*, *32*(5), 2829–2840. doi:10.100710811-020-02172-7

Niaounakis, M. (2013). Definitions and assessment of (bio)degradation. In Biopolymers Reuse, Recycling, and Disposal (pp. 77–94). Elsevier. doi:10.1016/B978-1-4557-3145-9.00002-6

Nicolet, Y., Lemon, B. J., Fontecilla-Camps, J. C., & Peters, J. W. (2000). A novel FeS cluster in Fe only hydrogenases. *Trends in Biochemical Sciences*, *25*(3), 138–143. doi:10.1016/S0968-0004(99)01536-4 PMID:10694885

Niemann, D. I., & Dorffling, K. (1980). Growth Inhibitors and Growth Promoters in *Enteromorpha compressa* (Chlorophyta). *Journal of Phycology*, *16*(3), 383–389. doi:10.1111/j.1529-8817.1980.tb03049.x

Niemi & Gentili. (2021). Extraction Technologies for Functional Lipids. In *Recent Advances in Micro and Macroalgal Processing: Food and Health Perspectives*. Wiley Online Library.

Nikaido, I., Asamizu, E., Nakajima, M., Nakamura, Y., Saga, N., & Tabata, S. (2000). Generation of 10,154 Expressed Sequence Tags from a Leafy Gametophyte of a Marine Red Alga, *Porphyra yezoensis*. *DNA Research*, *7*(3), 223–227. doi:10.1093/dnares/7.3.223 PMID:10907854

Nikel, P. I., Pettinari, M. J., Galvagno, M. A., & Méndez, B. S. (2006). Poly(3-hydroxybutyrate) synthesis by recombinant *Escherichia coli* arcA mutants in microaerobiosis. *Applied and Environmental Microbiology*, *72*(4), 2614–2620. doi:10.1128/AEM.72.4.2614-2620.2006 PMID:16597965

Nikkanen, L., Solymosi, D., Jokel, M., & Allahverdiyeva, Y. (2021). Regulatory electron transport pathways of photosynthesis in cyanobacteria and microalgae: Recent advances and biotechnological prospects. *Physiologia Plantarum*, *173*(2), 514–525. doi:10.1111/ppl.13404 PMID:33764547

Nimura, K., & Mizuta, H. (2002). Inducible Effects of Abscisic Acid on Sporophyte Discs from *Laminaria japonica* Areschoug (*Laminariales, Phaeophyceae*). *Journal of Applied Phycology*, *14*(3), 159–163. doi:10.1023/A:1019929712633

Ninganagouda, S., Rathod, V., Singh, D., Hiremath, J., Singh, A. K., Mathew, J., & ul-Haq, M. (2014). Growth kinetics and mechanistic action of reactive oxygen species released by silver nanoparticles from Aspergillus niger on Escherichia coli. *BioMed Research International*, *2014*, 1–14. doi:10.1155/2014/753419 PMID:25028666

Nisha, R., Kaushik, A., & Kaushik, C. P. (2007). Effect of indigenous cyanobacterial application on structural stability and productivity of an organically poor semi-arid soil. *Geoderma*, *138*(1-2), 49–56. doi:10.1016/j.geoderma.2006.10.007

Nishikawa, K., Yamakoshi, Y., Uemura, I., & Tominaga, N. (2003). Ultrastructural changes in Chlamydomonas acidophila (Chlorophyta) induced by heavy metals and polyphosphate metabolism. *FEMS Microbiology Ecology*, *44*(2), 253–259. doi:10.1016/S0168-6496(03)00049-7 PMID:19719642

Nishikawa, Y., Minenaka, Y., Ichimura, M., Tatsumi, K., Nadamoto, T., & Urabe, K. (2005). Effects of astaxanthin and vitamin C on the preventionof gastric ulcerations in stressed rats. *J. Nutr. Sci. Vit.*, *51*(3), 135–141. doi:10.3177/jnsv.51.135 PMID:16161762

Nishioka, M., Nakai, K., Miyake, M., Asada, Y., & Taya, M. (2001). Production of poly-β-hydroxybutyrate by thermophilic cyanobacterium, Synechococcus sp. MA19, under phosphate-limited conditions. *Biotechnology Letters*, *23*(14), 1095–1099. doi:10.1023/A:1010551614648

NOAA/ESRL. (2019). *National Oceanic and Atmospheric Administration/Earth System ResearchLaboratory*. https://www.esrl.noaa.gov/

Noble, A., Kisiala, A., Galer, A., Clysdale, D., & Emery, R. J. N. (2014). *Euglenagracilis* (Euglenophyceae) produces abscisic acid and cytokinins and responds to their exogenous application singly and in combination with other growth regulators. *European Journal of Phycology*, *49*(2), 244–254. doi:10.1080/09670262.2014.911353

Noda, H., Amano, H., Arashima, K., & Nisizawa, K. (1990). Antitumor activity of marine algae. *Hydrobiologia*, *204*(1), 577–584. doi:10.1007/BF00040290

Noor-Mohammadi, S., Pourmir, A., & Johannes, T. W. (2012). Method to assemble and integrate biochemical pathways into the chloroplast genome of *Chlamydomonas reinhardtii. Biotechnology and Bioengineering*, *109*(11), 2896–2903. doi:10.1002/bit.24569 PMID:22674415

Noor-Mohammadi, S., Pourmir, A., & Johannes, T. W. (2013). Method for assembling and expressing multiple genes in the nucleus of microalgae. *Biotechnology Letters*, *36*(3), 561–566. doi:10.100710529-013-1378-0 PMID:24129955

Nordberg, H. C. M., Dusheyko, S., Hua, S., Poliakov, A., Shabalov, I., Smirnova, T., Grigoriev, I. V., & Dubchak, I. (2014). The genome portal of the Department of Energy Joint Genome Institute: 2014 updates. *Nucleic Acids Research*, *42*(1), 26–31. doi:10.1093/nar/gkt1069 PMID:24225321

Nordi, C. S. F., Cavagliere, T. G. W. F., Vieira, A. A. H., & Nascimento, O. R. (2006). Efeito caotrópico do íon lítio na permeabilidade da cápsula polissacarídica da microalga Ankistrodesmus gracilis (Reinsch) Korsikov (Chlorophyceae). *Acta Botanica Brasílica*, *20*(2), 449–454. doi:10.1590/S0102-33062006000200020

Norouzi, N. (2020). COVID-19 and management models in SMEs of the energy industry. *Majlesi Journal of Energy Management, 9*(2).

Norouzi, N., Fani, M., & Nasiri, Z. (2021). The development of a Nexus based green architecture ranking system in Iran. *Civil Engineering Beyond Limits (CEBEL), 2*(2).

Norouzi, N. (2019). An overview on Water, Energy & Environment by 2030. *International Journal of Management Perspective*, *8*(2), 11–19.

Norouzi, N. (2021a). The Pahlev Reliability Index: A measurement for the resilience of power generation technologies versus climate change. *Nuclear Engineering and Technology*, *53*(5), 1658–1663. doi:10.1016/j.net.2020.10.013

Norouzi, N. (2021b). An Overview on the renewable hydrogen market. *International Journal of Energy Studies*, *6*(1), 67–94.

Norouzi, N., & Fani, M. (2021). The prioritization and feasibility study over renewable technologies using fuzzy logic: A case study for Takestan plains. *Journal of Energy Management and Technology*, *5*(2), 12–22.

Norouzi, N., Kalantari, G., & Talebi, S. (2020). Combination of renewable energy in the refinery, with carbon emissions approach. *Biointerface Research in Applied Chemistry*, *10*(4), 5780–5786. doi:10.33263/BRIAC104.780786

Norsker, N.-H., Barbosa, M. J., Vermuë, M. H., & Wijffels, R. H. (2011). Microalgal production— A close look at the economics. *Biotechnology Advances*, *29*(1), 24–27. doi:10.1016/j.biotechadv.2010.08.005 PMID:20728528

Norsker, N.-H., Cuaresma, M., de Vree, J., Ruiz-Domínguez, M. C., García, M. C. M., Uronen, P., & Wijffels, R. (2021). *Neochloris oleoabundans* oil production in an outdoor tubular photobioreactor at pilot scale. *Journal of Applied Phycology*, *33*(3), 1327–1339.

Norton, T. A., Melkonian, M., & Andersen, R. A. (1996). Algal biodiversity. *Phycologia*, *35*(4), 308–326. doi:10.2216/i0031-8884-35-4-308.1

Nosheen, S., Ajmal, I., & Song, Y. (2021). Microbes as biofertilizers, a potential approach for sustainable crop production. *Sustainability (Switzerland)*, *13*(4), 1–20. doi:10.3390u13041868

Novoveská, L., Ross, M. E., Stanley, M. S., Pradelles, R., Wasiolek, V., & Sassi, J. F. (2019). Microalgal carotenoids: A review of production, current markets, regulations, and future direction. *Marine Drugs*, *17*(11), 640. doi:10.3390/md17110640 PMID:31766228

Nowicka-Krawczyk, P., Mühlsteinová, R., & Hauer, T. (2019). Detailed characterization of the Arthrospira type species separating commercially grown taxa into the new genus Limnospira (Cyanobacteria). *Scientific Reports*, *9*(1), 694. doi:10.103841598-018-36831-0 PMID:30679537

Nymark, M., Sharma, A. K., Sparstad, T., Bones, A. M., & Winge, P. (2016). A CRISPR/Cas9 system adapted for gene editing in marine algae. *Scientific Reports*, *6*(1), 24951. doi:10.1038rep24951 PMID:27108533

Nzayisenga, J. C., Farge, X., Groll, S. L., & Sellstedt, A. (2020). Effects of light intensity on growth and lipid production in microalgae grown in wastewater. *Biotechnology for Biofuels*, *13*(1), 1–8. doi:10.1186/s13068-019-1646-x

O'Boyle, N. M., Banck, M., James, C. A., Morley, C., Vandermeersch, T., & Hutchison, G. R. (2011). Open Babel: An open chemical toolbox. *Journal of Cheminformatics*, *3*(1), 33. doi:10.1186/1758-2946-3-33 PMID:21982300

O'Sullivan, A. M., O'callaghan, Y. C., O'Grady, M. N., Queguineur, B., Hanniffy, D., Troy, D. J., Kerry, J. P., & O'brien, N. M. (2011). In vitro and cellular antioxidant activities of seaweed extracts prepared from five brown seaweeds harvested in spring from the west coast of Ireland. *Food Chemistry*, *126*(3), 1064–1070. doi:10.1016/j.foodchem.2010.11.127

Obruca, S., Sedlacek, P., Koller, M., Kucera, D., & Pernicova, I. (2018). Involvement of polyhydroxyalkanoates in stress resistance of microbial cells: Biotechnological consequences and applications. *Biotechnology Advances*, *36*(3), 856–870. doi:10.1016/j.biotechadv.2017.12.006 PMID:29248684

Odjadjare, E. C., Mutanda, T., & Olaniran, A. O. (2017). Potential biotechnological application of microalgae: A critical review. *Critical Reviews in Biotechnology*, *37*(1), 37–52. doi:10.3109/07388551.2015.1108956 PMID:26594785

Oey, M., Sawyer, A. L., Ross, I. L., & Hankamer, B. (2016). Challenges and opportunities for hydrogen production from microalgae. *Plant Biotechnology Journal*, *14*(7), 1487–1499. doi:10.1111/pbi.12516 PMID:26801871

Okaiyeto, K., Nwodo, U. U., Okoli, S. A., Mabinya, L. V., & Okoh, A. I. (2016). Implications for public health demands alternatives to inorganic and synthetic flocculants: Bioflocculants as important candidates. *MicrobiologyOpen*, *5*(2), 177–211.

Okoh, V., Deoraj, A., & Roy, D. (2011). Estrogen-induced reactive oxygen species-mediated signalings contribute to breast cancer. *Biochimica et Biophysica Acta (BBA)-. Revue Canadienne*, *1815*(1), 115–133. PMID:21036202

Oliveira, L., & Antia, N. J. (1984). Evidence of nickel ion requirement for autotrophic growth of a marine diatom with urea serving as nitrogen source. *British Phycological Journal*, *19*(2), 125–134. doi:10.1080/00071618400650131

Oliveira, R. C., & Garcia, O. Jr. (2009). Study of biosorption of rare earth metals (La, Nd, Eu, Gd) by Sargassum sp. biomass in batch systems: Physicochemical evaluation of kinetics and adsorption models. *Advanced Materials Research*, *71*, 605–608. doi:10.4028/www.scientific.net/AMR.71-73.605

Oliveira, R. C., Guibal, E., & Garcia, O. Jr. (2012). Biosorption and desorption of lanthanum (III) and neodymium (III) in fixed-bed columns with Sargassum sp.: Perspectives for separation of rare earth metals. *Biotechnology Progress*, *28*(3), 715–722. doi:10.1002/btpr.1525 PMID:22275117

Oliver, L., Dietrich, T., Marañón, I., Villarán, M. C., & Barrio, R. J. (2020). Producing Omega-3 Polyunsaturated Fatty Acids: A Review of Sustainable Sources and Future Trends for the EPA and DHA Market. *Resources*, *9*(12), 148. doi:10.3390/resources9120148

Oliver, M. J., Jain, R., Balbuena, T. S., Agrawal, G., Gasulla, F., & Thelen, J. J. (2011). Proteome analysis of leaves of the desiccation-tolerant grass, Sporobolus stapfianus, in response to dehydration. *Phytochemistry*, *72*(10), 1273–1284. doi:10.1016/j.phytochem.2010.10.020 PMID:21109273

Onay, M. (2019). Bioethanol Production via Different Saccharification Strategies from H. Tetrachotoma ME03 Grown at Various Concentrations of Municipal Wastewater in a Flat-Photobioreactor. *Fuel*, *239*, 1315–1323. doi:10.1016/j.fuel.2018.11.126

Onen Cinar, S., Chong, Z. K., Kucuker, M. A., Wieczorek, N., Cengiz, U., & Kuchta, K. (2020). Bioplastic production from microalgae: A review. *International Journal of Environmental Research and Public Health*, *17*(11), 3842. doi:10.3390/ijerph17113842 PMID:32481700

Onwudili, J. A., Lea-Langton, A. R., Ross, A. B., & Williams, P. T. (2013). Catalytic hydrothermal gasification of algae for hydrogen production: Composition of reaction products and potential for nutrient recycling. *Bioresource Technology*, *127*, 72–80. doi:10.1016/j.biortech.2012.10.020 PMID:23131625

Ooms, M. D., Dinh, C. T., Sargent, E. H., & Sinton, D. (2016). Photon management for augmented photosynthesis. *Nature Communications*, *7*(1), 12699. doi:10.1038/ncomms12699 PMID:27581187

Opare, E. O., Struhs, E., & Mirkouei, A. (2021). A comparative state-of-technology review and future directions for rare earth element separation. *Renewable & Sustainable Energy Reviews*, *143*, 110917. doi:10.1016/j.rser.2021.110917

Ördög, V., Stirk, W. A., van Staden, J., Novak, O., & Strnad, M. (2004). Endogenous Cytokinins in Three Genera of Microalgae from the Chlorophyta. *Journal of Phycology*, *40*(1), 88–95. doi:10.1046/j.1529-8817.2004.03046.x

Orfield, N. D., Fang, A. J., Valdez, P. J., Nelson, M. C., Savage, P. E., Lin, X. N., & Keoleian, G. A. (2014). Life cycle design of an algal biorefinery featuring hydrothermal liquefaction: Effect of reaction conditions and an alternative pathway including microbial regrowth. *ACS Sustainable Chemistry & Engineering*, *2*(4), 867–874. doi:10.1021c4004983

Oslan, S. N., Shoparwe, N. F., Yusoff, A. H., Rahim, A. A., Chang, C. S., Tan, J. S., Oslan, S. N., Arumugam, K., Ariff, A. B., Sulaiman, A. Z., & Mohamed, M. S. (2021). A Review on Haematococcus pluvialis Bioprocess Optimization of Green and Red Stage Culture Conditions for the Production of Natural Astaxanthin. *Biomolecules*, *11*(2), 256. Advance online publication. doi:10.3390/biom11020256 PMID:33578851

Osmani, A., & Zhang, J. (2014). Economic and environmental optimization of a large scale sustainable dual feedstock lignocellulosic-based bioethanol supply chain in a stochastic environment. *Applied Energy*, *114*, 572–587. doi:10.1016/j.apenergy.2013.10.024

Osman, M. E. H., Abo-Shady, A. M., Elshobary, M. E., Abd El-Ghafar, M. O., & Abomohra, A. E.-F. F. (2020). Screening of seaweeds for sustainable biofuel recovery through sequential biodiesel and bioethanol production. *Environmental Science and Pollution Research International*, *27*(26), 32481–32493. doi:10.100711356-020-09534-1 PMID:32506400

Osterlie, M., Bjerkeng, B., & Liaaen-Jensen, S. (2000). Plasma appearance and distribution of astaxanthin E/Z isomers in plasma lipoproteins of after single dose administration of astaxanthin. *The Journal of Nutritional Biochemistry*, *11*(10), 482–492. doi:10.1016/S0955-2863(00)00104-2 PMID:11120445

Oswald, W. J. (1967). *Current status of microalgae from wastes (Water purification through biological oxidation of organic wastes using controlled photosynthesis, discussing algae production and harvesting)*. Academic Press.

Oswald, W. J. (1988). Large-scale algal culture systems (engineering aspects). In M. A. Borowitzka & L. J. Borowitzka (Eds.), *Micro-algal biotechnology* (pp. 357–394). Cambridge University Press.

Otero, P., López-Martínez, M. I., & García-Risco, M. R. (2019). Application of pressurized liquid extraction (PLE) to obtain bioactive fatty acids and phenols from Laminaria ochroleuca collected in Galicia (NW Spain). *Journal of Pharmaceutical and Biomedical Analysis*, *164*, 86–92. doi:10.1016/j.jpba.2018.09.057 PMID:30359842

Otsuki, T., Zhang, F., Kabeya, H., & Hirotsu, T. (2004). Synthesis and tensile properties of a novel composite ofChlorella and polyethylene. *Journal of Applied Polymer Science*, *92*(2), 812–816. doi:10.1002/app.13650

Ou, R., Xie, Y., Wolcott, M. P., Yuan, F., & Wang, Q. (2014). Effect of wood cell wall composition on the rheological properties of wood particle/high density polyethylene composites. *Composites Science and Technology*, *93*, 68–75. doi:10.1016/j.compscitech.2014.01.001

Oyewo, O. A., Onyango, M. S., & Wolkersdorfer, C. (2018). Lanthanides removal from mine water using banana peels nanosorbent. *International Journal of Environmental Science and Technology*, *15*(6), 1265–1274. doi:10.100713762-017-1494-9

Oza, G., Pandey, S., Mewada, A., Kalita, G., & Sharon, M. (2012). Facile biosynthesis of gold nano- particles exploiting optimum pH and temperature of freshwater alga Chlorella pyrenoidusa. *Advances in Applied Science Research*, *3*(3), 1405–1412.

Ozben, T. (2015). Antioxidant supplementation on cancer risk and during cancer therapy: An update. *Current Topics in Medicinal Chemistry*, *15*(2), 170–178. doi:10.2174/1568026615666141209160918 PMID:25496272

Özçimen, D., İnan, B., Morkoç, O., & Efe, A. (2017). A review on algal biopolymers. *J. Chem. Eng. Res. Updates*, *4*, 7–14. doi:10.15377/2409-983X.2017.04.2

Paajanen, P., Strauss, J., van Oosterhout, C., McMullan, M., Clark, M. D., & Mock, T. (2017). Building a locally diploid genome and transcriptome of the diatom *Fragilariopsis cylindrus*. *Scientific Data*, *4*(1), 170149. doi:10.1038data.2017.149 PMID:28994819

Pabbi, S. (2015). Blue green algae: a potential biofertilizer for rice. In *The Algae World* (pp. 449–465). Springer. doi:10.1007/978-94-017-7321-8_17

Pachekoski, W. M., Agnelli, J. A. M., & Belem, L. P. (2009). Thermal, mechanical and morphological properties of poly (hydroxybutyrate) and polypropylene blends after processing. *Materials Research*, *12*(2), 159–164. doi:10.1590/S1516-14392009000200008

Packer, A., Li, Y., Andersen, T., Hu, Q., Kuang, Y., & Sommerfeld, M. (2011). Growth and neutral lipid synthesis in green microalgae: A mathematical model. *Bioresource Technology*, *102*(1), 111–117. doi:10.1016/j.biortech.2010.06.029 PMID:20619638

Padisák, J., Soróczki-Pintér, É., & Rezner, Z. (2003). Sinking properties of some phytoplankton shapes and the relation of form resistance to morphological diversity of plankton – an experimental study. *Hydrobiologia*, *500*(1), 243–257. doi:10.1023/A:1024613001147

Pádrová, K., Lukavský, J., Nedbalová, L., Čejková, A., Cajthaml, T., Sigler, K., Vítová, M., & Řezanka, T. (2015). Trace concentrations of iron nanoparticles cause overproduction of biomass and lipids during cultivation of cyanobacteria and microalgae. *Journal of Applied Phycology*, *27*(4), 1443–1451. doi:10.100710811-014-0477-1

Paerl, R. W., Bouget, F. Y., Lozano, J. C., Verge, V., Schatt, P., Allen, E. E., Palenik, B., & Azam, F. (2017). Use of plankton-derived vitamin B1 precursors, especially thiazole-related precursor, by key marine picoeukaryotic phytoplankton. *The ISME Journal*, *11*(3), 753–765. doi:10.1038/ismej.2016.145 PMID:27935586

Pagarete, A., Ramos, A. S., Puntervoll, P., Allen, M. J., & Verdelho, V. (2021). Antiviral Potential of Algal Metabolites-A Comprehensive Review. *Marine Drugs*, *19*(2), 1–23. doi:10.3390/md19020094 PMID:33562153

Paix, B., Carriot, N., Barry-Martinet, R., Greff, S., Misson, B., Briand, J.-F., & Culioli, G. (2020). A Multi-Omics Analysis Suggests Links Between the Differentiated Surface Metabolome and Epiphytic Microbiota Along the Thallus of a Mediterranean Seaweed Holobiont. *Frontiers in Microbiology*, *11*(494), 494. doi:10.3389/fmicb.2020.00494 PMID:32269559

Palanisamy, M., Töpfl, S., Berger, R. G., & Hertel, C. (2019). Physico-chemical and nutritional properties of meat analogues based on Spirulina/lupin protein mixtures. *European Food Research and Technology*, *245*(9), 1889–1898. doi:10.100700217-019-03298-w

Palmer, J. D. (2003). The symbiotic birth and spread of plastids: How many times and whodunit? *Journal of Phycology*, *39*(1), 4–12. doi:10.1046/j.1529-8817.2003.02185.x

Palmieri, M. C., Garcia, O. Jr, & Melnikov, P. (2000). Neodymium biosorption from acidic solutions in batch system. *Process Biochemistry*, *36*(5), 441–444. doi:10.1016/S0032-9592(00)00236-3

Panahi, Y., Khosroushahi, A. Y., Sahebkar, A., & Heidari, H. R. (2019). Impact of Cultivation Condition and Media Content onChlorella vulgaris Composition. *Advanced Pharmaceutical Bulletin*, *9*(2), 182–194. doi:10.15171/apb.2019.022 PMID:31380244

Panda, B., Jain, P., Sharma, L., & Mallick, N. (2006). Optimization of cultural and nutritional conditions for accumulation of poly-β-hydroxybutyrate in Synechocystis sp. PCC 6803. *Bioresource Technology*, *97*(11), 1296–1301. doi:10.1016/j.biortech.2005.05.013 PMID:16046119

Panda, B., & Mallick, N. (2007). Enhanced poly-beta-hydroxybutyrate accumulation in a unicellular cyanobacterium, *Synechocystis* sp. PCC 6803. *Letters in Applied Microbiology*, *44*(2), 194–198. doi:10.1111/j.1472-765X.2006.02048.x PMID:17257260

Pandey, A., Sinha, P., & Pandey, A. (2021). Hydrogen production by sequential dark and photofermentation using wet biomass hydrolysate of Spirulina platensis: Response surface methodological approach. *International Journal of Hydrogen Energy*, *46*(10), 7137–7146. doi:10.1016/j.ijhydene.2020.11.205

Pandey, K. D., Shukla, P. N., Giri, D. D., & Kashyap, A. K. (2005). Cyanobacteria in alkaline soil and the effect of cyanobacteria inoculation with pyrite amendments on their reclamation. *Biology and Fertility of Soils*, *41*(6), 451–457. doi:10.100700374-005-0846-7

Panis, G., & Carreon, J. R. (2016). Commercial astaxanthin production derived by green alga Haematococcus pluvialis: A microalgae process model and a techno- economic assessment all through production line. *Algal Research*, *18*, 175–190. doi:10.1016/j.algal.2016.06.007

Pan, S., Jeevanandam, J., & Danquah, M. K. (2019). Benefits of Algal Extracts in Sustainable Agriculture. In *Grand Challenges in Marine Biotechnology* (pp. 501–534). Springer Science and Business Media LLC.

Pan, Y., Shi, B., & Zhang, Y. (2009). Research on flocculation property of bioflocculant PG.a21 Ca. *Modern Applied Science*, *3*(6), 106–112.

Papadaki, E., & Mantzouridou, F. T. (2021). Natural β-Carotene production by Blakeslea trispora cultivated in spanish-style green olive processing wastewaters. *Foods*, *10*(2), 327. doi:10.3390/foods10020327 PMID:33557096

Papenfus, H. B., Kulkarni, M. G., Stirk, W. A., Finnie, J. F., & van Staden, J. (2013). Effect of a commercial seaweed extract (Kelpak®) and polyamines on nutrient-deprived (N, P and K) okra seedlings. *Scientia Horticulturae*, *151*, 142–146. doi:10.1016/j.scienta.2012.12.022

Papone, T., Kookkhunthod, S., & Leesing, R. (2012). Microbial Oil Production by Monoculture and Mixed Cultures of Microalgae and Oleaginous Yeasts using Sugarcane Juice as Substrate. *International Journal of Biological. Life and Agricultural Sciences*, *5*(4). Advance online publication. doi:10.5281/zenodo.1332532

Pardee, K. I., Ellis, P., Bouthillier, M., Towers, G. H. N., & French, C. J. (2004). Plant virus inhibitors from marine algae. *Canadian Journal of Botany*, *82*(3), 304–309. doi:10.1139/b04-002

Parial, D., & Pal, R. (2015). Biosynthesis of monodisperse gold nanoparticles by green alga Rhizoclonium and associated biochemical changes. *Journal of Applied Phycology*, *27*(2), 975–984. doi:10.100710811-014-0355-x

Park, J. B. K., Craggs, R. J., & Shilton, A. N. (2013). Enhancing biomass energy yield from pilot-scale high rate algal ponds with recycling. *Water Research*, *47*(13), 4422–4432. doi:10.1016/j.watres.2013.04.001 PMID:23764593

Park, J. S., Chyun, J. H., Kim, Y. K., Line, L. L., & Chew, B. P. (2010). Astaxanthin decreased oxidative stress and inflammation and enhanced immune response in humans. *Nutrition & Metabolism*, *7*(1), 1–10. doi:10.1186/1743-7075-7-18 PMID:20205737

Park, J.-H., Hong, J.-Y., Jang, H. C., Oh, S. G., Kim, S.-H., Yoon, J.-J., & Kim, Y. J. (2012). Use of Gelidium amansii as a promising resource for bioethanol: A practical approach for continuous dilute-acid hydrolysis and fermentation. *Bioresource Technology*, *108*, 83–88. doi:10.1016/j.biortech.2011.12.065 PMID:22261657

Park, J., Park, B. S., Wang, P., Patidar, S. K., Kim, J. H., Kim, S.-H., & Han, M.-S. (2017). Phycospheric Native Bacteria Pelagibaca bermudensis and Stappia sp. Ameliorate Biomass Productivity of Tetraselmis striata (KCTC1432BP) in Co-cultivation System through Mutualistic Interaction. *Frontiers in Plant Science*, *8*. Advance online publication. doi:10.3389/fpls.2017.00289 PMID:28321229

Park, J.-Y., Park, M. S., Lee, Y.-C., & Yang, J.-W. (2015). Advances in direct transesterification of algal oils from wet biomass. *Bioresource Technology*, *184*, 267–275. doi:10.1016/j.biortech.2014.10.089 PMID:25466997

Park, K. Y., Lee, K., & Kim, D. (2018). Characterized hydrochar of algal biomass for producing solid fuel through hydrothermal carbonization. *Bioresource Technology*, *258*, 119–124. doi:10.1016/j.biortech.2018.03.003 PMID:29524686

Park, Y., Je, K.-W., Lee, K., Jung, S.-E., & Choi, T.-J. (2008). Growth promotion of Chlorella ellipsoidea by co-inoculation with Brevundimonas sp. isolated from the microalga. *Hydrobiologia*, *598*(1), 219–228. doi:10.100710750-007-9152-8

Parmar, A., Singh, N. K., Kaushal, A., Sonawala, S., & Madamwar, D. (2011). Purification, characterization and comparison of phycoerythrins from three different marine cyanobacterial cultures. *Bioresource Technology*, *102*(2), 1795–1802. doi:10.1016/j.biortech.2010.09.025 PMID:20889334

Parmentier, D., Manhaeghe, D., Baccini, L., Van Meirhaeghe, R., Rousseau, D. P., & Van Hulle, S. (2020). A new reactor design for harvesting algae through electrocoagulation-flotation in a continuous mode. *Algal Research*, *47*, 101828. doi:10.1016/j.algal.2020.101828

Parrot, D., Blümel, M., Utermann, C., Chianese, G., Krause, S., Kovalev, A., Gorb, S. N., & Tasdemir, D. (2019). Mapping the Surface Microbiome and Metabolome of Brown Seaweed Fucus vesiculosus by Amplicon Sequencing, Integrated Metabolomics and Imaging Techniques. *Scientific Reports*, *9*(1), 1061. doi:10.103841598-018-37914-8 PMID:30705420

Parthasarathy, P., & Narayanan, K. S. (2014). Hydrogen production from steam gasification of biomass: Influence of process parameters on hydrogen yield–a review. *Renewable Energy*, *66*, 570–579. doi:10.1016/j.renene.2013.12.025

Pascher, A. (1931). Systematische Übersicht über die mit Flagellaten in Zusammenhang stehenden Algenreihen und Versuch einer Einreihung dieser Algenstämme in die Stämme des Pflanzenreiches. *Botanisches Zentralblatt. Beihefte*, *48*, 317–332.

Pasquet, V., Morisset, P., Ihammouine, S., Chepied, A., Aumailley, L., Berard, J.-B., Serive, B., Kaas, R., Lanneluc, I., Thiery, V., Lafferriere, M., Piot, J.-M., Patrice, T., Cadoret, J.-P., & Picot, L. (2011). Antiproliferative Activity of Violaxanthin Isolated from Bioguided Fractionation of *Dunaliella tertiolecta* Extracts. *Marine Drugs*, *2011*(9), 819–831. doi:10.3390/md9050819 PMID:21673891

Passell, H., Dhaliwal, H., Reno, M., Wu, B., Ben Amotz, A., Ivry, E., Gay, M., Czartoski, T., Laurin, L., & Ayer, N. (2013). Algae biodiesel life cycle assessment using current commercial data. *Journal of Environmental Management*, *129*, 103–111. doi:10.1016/j.jenvman.2013.06.055 PMID:23900083

Passero, M., Cragin, B., Coats, E. R., McDonald, A. G., & Feris, K. (2015). Dairy wastewaters for algae cultivation, polyhydroxyalkanote reactor effluent versus anaerobic digester effluent. *BioEnergy Research*, *8*(4), 1647–1660. doi:10.100712155-015-9619-9

Patel, A. K., Singhania, R. R., Dong, C.-D., Obulisami, P. K., & Sim, S. J. (2021). 2021/12/01/). Mixotrophic biorefinery: A promising algal platform for sustainable biofuels and high value coproducts. *Renewable & Sustainable Energy Reviews*, *152*, 111669. doi:10.1016/j.rser.2021.111669

Patel, J. K., Sutariya, V., Kanwar, J. R., & Pathak, Y. V. (2018). *Drug Delivery for the Retina and Posterior Segment Disease* (1st ed.). Springer. doi:10.1007/978-3-319-95807-1

Patel, N., & Miotello, A. (2015). Progress in Co–B related catalyst for hydrogen production by hydrolysis of boronhydrides: A review and the perspectives to substitute noble metals. *International Journal of Hydrogen Energy*, *40*(3), 1429–1464. doi:10.1016/j.ijhydene.2014.11.052

Patel, S. K. S., Das, D., Kim, S. C., Cho, B., Kalia, V. C., & Lee, J. (2021). Integrating strategies for sustainable conversion of waste biomass into dark-fermentative hydrogen and value-added products. *Renewable & Sustainable Energy Reviews*, *150*(May), 111491. doi:10.1016/j.rser.2021.111491

Pate, R., Klise, G., & Wu, B. (2010). Resource demand implications for US algae biofuels production scale-up. *Applied Energy*, *88*(10), 3377–3388. doi:10.1016/j.apenergy.2011.04.023

Patil, P. D., Gude, V. G., Mannarswamy, A., Cooke, P., Nirmalakhandan, N., Lammers, P., & Deng, S. (2012). Comparison of direct transesterification of algal biomass under supercritical methanol and microwave irradiation conditions. *Fuel*, *97*, 822–831. doi:10.1016/j.fuel.2012.02.037

Patwary, Z. P., Paul, N. A., Nishitsuji, K., Campbell, A. H., Shoguchi, E., Zhao, M., & Cummins, S. F. (2021). Application of omics research in seaweeds with a focus on red seaweeds. *Briefings in Functional Genomics*, *20*(3), 148–161. doi:10.1093/bfgp/elab023 PMID:33907795

Paul, C., & Pohnert, G. (2011). Interactions of the algicidal bacterium *Kordia algicida* with diatoms: Regulated protease excretion for specific algal lysis. *PLoS One*, *6*(6), e21032. doi:10.1371/journal.pone.0021032 PMID:21695044

Paul, C., Reunamo, A., Lindehoff, E., Bergkvist, J., Mausz, M. A., Larsson, H., Richter, H., Wängberg, S.-Å., Leskinen, P., Båmstedt, U., & Pohnert, G. (2012). Diatom derived polyunsaturated aldehydes do not structure the planktonic microbial community in a mesocosm study. *Marine Drugs*, *10*(4), 775–792. doi:10.3390/md10040775 PMID:22690143

Paulert, R., Talamini, V., Cassolato, J. E. F., Duarte, M. E. R., Noseda, M. D., Smania, A. Jr, & Stadnik, M. J. (2009). Effects of sulfated polysaccharide and alcoholic extracts from green seaweed *Ulva fasciata* on anthracnose severity and growth of common bean (*Phaseolus vulgaris* L.). *Journal of Plant Diseases and Protection*, *116*(6), 263–270. doi:10.1007/BF03356321

Pelgrift, R. Y., & Friedman, A. J. (2013). Nanotechnology as a therapeutic tool to combat microbial resistance. *Advanced Drug Delivery Reviews*, *65*(13-14), 1803–1815. doi:10.1016/j.addr.2013.07.011 PMID:23892192

Peñalver, R., Lorenzo, J. M., Ros, G., Amarowicz, R., Pateiro, M., & Nieto, G. (2020). Seaweeds as a functional ingredient for a healthy diet. *Marine Drugs*, *18*(6), 1–27. doi:10.3390/md18060301 PMID:32517092

Peng, C., Li, S., Zheng, J., Huang, S., & Li, D. (2017). Harvesting microalgae with di_erent sources of starch-based cationic flocculants. *Applied Biochemistry and Biotechnology*, *181*, 112–124.

Peng, F. Q., Ying, G. G., Yang, B., Liu, S., Lai, H. J., Liu, Y. S., Chen, Z. F., & Zhou, G. J. (2014). Biotransformation of progesterone and norgestrel by two freshwater microalgae (*Scenedesmus obliquus* and *Chlorella pyrenoidosa*): Transformation kinetics and products identification. *Chemosphere*, *95*, 581–588.

Peng, Y., Xie, E., Zheng, K., Fredimoses, M., Yang, X., Zhou, X., Wang, Y., Yang, B., Lin, X., Liu, J., & Liu, Y. (2012). Nutritional and chemical composition and antiviral activity of cultivated seaweed Sargassum naozhouense Tseng et Lu. *Marine Drugs*, *11*(1), 20–32. doi:10.3390/md11010020 PMID:23271422

Peperzak, L. F., Koeman, C. R., Gieskes, W. W. C., & Joordens, J. C. A. (2003). Phytoplankton sinking rates in the Rhine region of freshwater influence. *Journal of Plankton Research*, *25*, 365–368.

Percival, E. (1979). The polysaccharides of green, red and brown seaweeds: Their basic structure, biosynthesis and function'. *British Phycological Journal*, *14*(2), 103–117. doi:10.1080/00071617900650121

Pereira, H., Barreira, L., Figueiredo, F., Custódio, L., Vizetto-Duarte, C., Polo, C., Rešek, E., Engelen, A., & Varela, J. (2012). Polyunsaturated Fatty acids of marine macroalgae: Potential for nutritional and pharmaceutical applications. *Marine Drugs*, *10*(9), 1920–1935. doi:10.3390/md10091920 PMID:23118712

Pereira, H., Schulze, P. S. C., Schüler, L. M., Santos, T., Barreira, L., & Varela, J. (2018). Fluorescence activated cell-sorting principles and applications in microalgal biotechnology. *Algal Research*, *30*, 113–120. doi:10.1016/j.algal.2017.12.013

Pereira, R. C., & Costa-Lotufo, L. V. (2012). Bioprospecting for bioactives from seaweeds:Potential, obstacles and alternatives. *Revista Brasileira de Farmacognosia*, *22*(4), 894–905. doi:10.1590/S0102-695X2012005000077

Pereira, R., & Yarish, C. (2008). Mass Production of Marine Macroalgae. In S. E. Jørgensen & B. D. Fath (Eds.), *Encyclopedia of Ecology* (pp. 2236–2247). Academic Press. doi:10.1016/B978-008045405-4.00066-5

Pérez, E. B., Pina, I. C., & Rodríguez, L. P. (2008). Kinetic model for growth of Phaeodactylum tricornutum in intensive culture photobioreactor. *Biochemical Engineering Journal*, *40*(3), 520–525. doi:10.1016/j.bej.2008.02.007

Perez-Garcia, O., de-Bashan, L. E., Hernandez, J. P., & Bashan, Y. (2010). Efficiency of growth and nutrient uptake from wastewater by heterotrophic, autotrophic, and mixotrophic cultivation of chlorella vulgaris immobilized with azospirillum brasilense. *Journal of Phycology*, *46*(4), 800–812. doi:10.1111/j.1529-8817.2010.00862.x

Pérez, M. J., Falqué, E., & Domínguez, H. (2016). Antimicrobial action of compounds from marine seaweed. *Marine Drugs*, *14*(3), 52. doi:10.3390/md14030052 PMID:27005637

Perez-Rodriguez, E., Gomez, I., Karsten, U., & Figueroa, F. L. (1998). Effects of UV-radiation on photosynthesis and excretion of UV-absorbing compounds of *Dasycladus vermicularis* Dasycladales, Chlorophyta from southern Spain. *Phycologia*, *37*(5), 379–387. doi:10.2216/i0031-8884-37-5-379.1

Perin, G., Bellan, A., Segalla, A., Meneghesso, A., Alboresi, A., & Morosinotto, T. (2015). Generation of random mutants to improve light-use efficiency of *Nannochloropsis gaditana* cultures for biofuel production. *Biotechnology for Biofuels*, *8*(1), 161. doi:10.118613068-015-0337-5 PMID:26413160

Periz, G., Dharia, D., Miller, S. H., & Keller, L. R. (2007). Flagellar elongation and gene expression in Chlamydomonas reinhardtii. *Eukaryotic Cell*, *6*(8), 1411–1420. doi:10.1128/EC.00167-07 PMID:17573545

Perotto, G., Ceseracciu, L., Simonutti, R., Paul, U. C., Guzman-Puyol, S., Tran, T.-N., Bayer, I. S., & Athanassiou, A. (2018). Bioplastics from vegetable waste via an eco-friendly water-based process. *Green Chemistry*, *20*(4), 894–902. doi:10.1039/C7GC03368K

Petsas, A. S., & Vagi, M. C. (2017). Effects on the Photosynthetic Activity of Algae after Exposure to Various Organic and Inorganic Pollutants: Review. In *Chlorophyll*. InTech., doi:10.5772/67991

Pettersen, E. F., Goddard, T. D., Huang, C. C., Couch, G. S., Greenblatt, D. M., Meng, E. C., & Ferrin, T. E. (2004). UCSF Chimera—A Visualization System for Exploratory Research and Analysis. *Journal of Computational Chemistry*, *25*(13), 1605–1612. doi:10.1002/jcc.20084 PMID:15264254

Pezzolesi, L., Mazzotti, M., Vanucci, S., & Pistocchi, R. (2019). Assimilation of inorganic nitrogen for scaling up Desmodesmus communis (Scenedesmaceae) biomass production. *Journal of Applied Phycology*, *31*(5), 2833–2844.

Pfaffinger, C. E., Schöne, D., Trunz, S., Löwe, H., & Weuster-Botz, D. (2016). Model-based optimization of microalgae areal productivity in flat-plate gas-lift photobioreactors. *Algal Research*, *20*, 153–163. doi:10.1016/j.algal.2016.10.002

Pfeifer, K., Ergal, İ., Koller, M., Basen, M., Schuster, B., & Rittmann, S. K. M. R. (2020, November). Archaea Biotechnology. *Biotechnology Advances*, *47*, 107668. Advance online publication. doi:10.1016/j.biotechadv.2020.107668 PMID:33271237

Pfeiffer, T. Z., Ivna, Š., Maronić, D. Š., & Maksimovic, I. (2017). *Regulation of Photosynthesis in Algae Under Metal Stress "Development of new technologies for modern and sustainable production of vegetables" (TR 31036) View project*. https://www.researchgate.net/publication/322096970

Pfeiffer, T. Z., Camagakevac, I. S., Maronic, D. S., & Maksimovic, I. (2017). Regulation of Photosynthesis in Algae Under Metal Stress. *Environment and Photosynthesis a Future Prospect*, (December), 261–286.

Phwan, C. K., Chew, K. W., Sebayang, A. H., Ong, H. C., Ling, T. C., Malek, M. A., Ho, Y. C., & Show, P. L. (2019). Effects of Acids Pre-Treatment on the Microbial Fermentation Process for Bioethanol Production from Microalgae. *Biotechnology for Biofuels*, *12*(1), 191. doi:10.118613068-019-1533-5 PMID:31384298

Pianosi, F., Sarrazin, F., & Wagener, T. A. (2015). A Matlab toolbox for global sensitivity analysis. *Environmental Modelling & Software*, *70*, 80–85. doi:10.1016/j.envsoft.2015.04.009

Piatek, A., Ali, Z., Baazim, H., Li, L., Abulfaraj, A., Al-Shareef, S., Aouida, M., & Mahfouz, M. M. (2015). RNA-guided transcriptional regulation in planta via synthetic dCas9-based transcription factors. *Plant Biotechnology Journal*, *13*(4), 578–589. doi:10.1111/pbi.12284 PMID:25400128

Picard, K. T., Letcher, P. M., & Powell, M. J. (2013). Evidence for a facultative mutualist nutritional relationship between the green coccoid alga *Bracteacoccus* sp. (Chlorophyceae) and the zoosporic fungus *Rhizidium phycophilum* (Chytridiomycota). *Fungal Biology*, *117*(5), 319–328. doi:10.1016/j.funbio.2013.03.003 PMID:23719219

Pierobon, S., Cheng, X., Graham, P., Nguyen, B., Karakolis, E., & Sinton, D. (2018). Emerging microalgae technology: A review. *Sustainable Energy & Fuels*, *2*(1), 13–38. doi:10.1039/C7SE00236J

Pietrosemoli, L., & Rodríguez-Monroy, C. (2019). The Venezuelan energy crisis: Renewable energies in the transition towards sustainability. *Renew and Sustain Energ Rev*, *105*, 415-426.

Piñero Estrada, J., Bermejo Bescós, P., & Villar del Fresno, A. M. (2001). Antioxidant activity of different fractions of Spirulina platensis protean extract. *Farm*, *56*(5-7), 497–500. doi:10.1016/S0014-827X(01)01084-9 PMID:11482785

Piotrowska-Niczyporuk, A., Bajguz, A., Kotowska, U., Bralska, M., & Talarek-Karwel, M. (2018). Growth, Metabolite Profile, Oxidative Status, and Phytohormone Levels in the Green Alga *Acutodesmus obliquus* Exposed to Exogenous Auxins and Cytokinins. *Journal of Plant Growth Regulation*, *37*(4), 1159-74.

Pisal, D. S., & Lele, S. S. (2005). Carotenoid production from microalgae,Dunaliella salina. *Indian Journal of Biotechnology*, *4*, 476–483.

Pischedda, A., Tosin, M., & Degli-Innocenti, F. (2019). Biodegradation of plastics in soil: The effect of temperature. *Polymer Degradation & Stability*, *170*, 109017. doi:10.1016/j.polymdegradstab.2019.109017

Piwosz, K., Vrdoljak, A., Frenken, T., González-Olalla, J. M., Šantić, D., & McKay, R. M. (2020). Light and primary production shape bacterial activity and community composition of aerobic anoxygenic phototrophic bacteria in a microcosm experiment. *mSphere 5*. doi:10.1128/mSphere.00354-20

Piwowar, A., & Harasym, J. (2020). The importance and prospects of the use of algae in agribusiness. *Sustainability*, *12*(14), 1–13. doi:10.3390u12145669

Plastics Europe, G. M. R., & Conversio Market & Strategy GmbH. (2019). *Plastics - the Facts 2019*. Author.

Plouguerné, E., Le Lann, K., Connan, S., Jechoux, G., Deslandes, E., & Stiger-Pouvreau, V. (2006). Spatial and seasonal variation in density, reproductive status, length and phenolic content of the invasive brown macroalga Sargassum muticum (Yendo) Fensholt along the coast of Western Brittany (France). *Aquatic Botany*, *85*(4), 337–344. doi:10.1016/j.aquabot.2006.06.011

Pohnert, G. (2004). Chemical defense strategies of marine organisms. *Topics in Current Chemistry*, *239*, 179–219. doi:10.1007/b95453 PMID:22160234

Polevoi, V. V., Tarakhovskaya, E. R., Maslov, Yu. I., & Polevoi, A. V. (2003). Induction of Polarity in Zygotes of *Fucusvesiculosus* L. by Auxin. *Ontogenez*, *34*, 432–437. PMID:14714452

Pollara, S. B., Becker, J. W., Nunn, B. L., Boiteau, R., Repeta, D., Mudge, M. C., Downing, G., Chase, D., Harvey, E. L., & Whalen, K. E. (2021). Bacterial Quorum-Sensing Signal Arrests Phytoplankton Cell Division and Impacts Virus-Induced Mortality. *MSphere*, *6*(3), e00009-21. Advance online publication. doi:10.1128/mSphere.00009-21 PMID:33980670

Poltronieri, P., Mezzolla, V., & D'Urso, O. F. (2016). PHB Production in Biofermentors Assisted through Biosensor Applications. *Proceedings*, *1*(2), 4. doi:10.3390/ecsa-3-E014

Ponou, J., Wang, L. P., Dodbiba, G., Okaya, K., Fujita, T., Mitsuhashi, K., Atarashi, T., Satoh, G., & Noda, M. (2014). Recovery of rare earth elements from aqueous solution obtained from Vietnamese clay minerals using dried and carbonized parachlorella. *Journal of Environmental Chemical Engineering*, *2*(2), 1070–1081. doi:10.1016/j.jece.2014.04.002

Pospíšil, J., Pilař, J., Billingham, N. C., Marek, A., Horák, Z., & Nešpůrek, S. (2006). Factors affecting accelerated testing of polymer photostability. *Polymer Degradation & Stability*, *91*(3), 417–422. doi:10.1016/j.polymdegradstab.2005.01.049

Post, A., & Larkum, A. W. D. (1993). UV-absorbing pigments, photosynthesis and UV exposure in Antarctica: Comparison of terrestrial and marine algae. *Aquatic Botany*, *45*(2-3), 231–243. doi:10.1016/0304-3770(93)90023-P

Posten, C. (2009). Design principles of photo-bioreactors for cultivation of microalgae. *Engineering in Life Sciences*, *9*(3), 163–177.

Pourkarimi, S., Hallajisani, A., Alizadehdakhel, A., & Nouralishahi, A. (2019). Biofuel production through micro- and macroalgae pyrolysis – A review of pyrolysis methods and process parameters. In *Journal of Analytical and Applied Pyrolysis* (Vol. 142, p. 104599). Elsevier B.V. doi:10.1016/j.jaap.2019.04.015

Pragya, N., & Pandey, K. K. (2016). Life cycle assessment of green diesel production from microalgae. *Renewable Energy*, *86*, 623–632. doi:10.1016/j.renene.2015.08.064

Prajapati, S. K., Malik, A., & Vijay, V. K. (2014). Comparative evaluation of biomass production and bioenergy generation potential of Chlorella spp. through anaerobic digestion. *Applied Energy*, *114*, 790–797. doi:10.1016/j.apenergy.2013.08.021

Prajapati, V. D., Maheriya, P. M., Jani, G. K., & Solanki, H. K. (2014). Carrageenan: A natural sea weed polysaccharide and its applications. *Carbohydrate Polymers*, *105*, 97–112. doi:10.1016/j.carbpol.2014.01.067 PMID:24708958

Prasad, R. C., & Prasad, B. N. (2001). Cyanobacteria as a source biofertilizer for sustainable agriculture in Nepal. *Journal in Plant Science Botanica Orientalis*, *1*, 127–133.

Prasanna, R., Babu, S., Bidyarani, N., Kumar, A., Triveni, S., Monga, D., Mukherjee, A. K., Kranthi, S., Gokte-Narkhedkar, N., Adak, A., Yadav, K., Nain, L., & Saxena, A. K. (2015). Prospecting cyanobacteria-fortified composts as plant growth promoting and biocontrol agents in cotton. *Experimental Agriculture*, *51*(1), 42–65. doi:10.1017/S0014479714000143

Prasanna, R., Kanchan, A., Ramakrishnan, B., Ranjan, K., Venkatachalam, S., Hossain, F., Shivay, Y. S., Krishnan, P., & Nain, L. (2016). Cyanobacteria-based bioinoculants influence growth and yields by modulating the microbial communities favourably in the rhizospheres of maize hybrids. *European Journal of Soil Biology*, *75*, 15–23. doi:10.1016/j.ejsobi.2016.04.001

Prasanna, R., Ramakrishnan, B., Simranjit, K., Ranjan, K., Kanchan, A., Hossain, F., & Nain, L. (2017). Cyanobacterial and rhizobial inoculation modulates the plant physiological attributes and nodule microbial communities of chickpea. *Archives of Microbiology*, *199*(9), 1311–1323. doi:10.100700203-017-1405-y PMID:28669069

Prasanna, R., Triveni, S., Bidyarani, N., Babu, S., Yadav, K., Adak, A., Khetarpal, S., Pal, M., Shivay, Y. S., & Saxena, A. K. (2013). Evaluating the efficacy of cyanobacterial formulations and biofilmed inoculants for leguminous crops. *Archives of Agronomy and Soil Science*, *60*, 349–366.

Prasath, B. B., Wang, Y., Su, Y., Zheng, W., Lin, H., & Yang, H. (2021). Coagulant plus *bacillus nitratireducens* fermentation broth technique provides a rapid algicidal effect of toxic red tide dinoflagellate. *Journal of Marine Science and Engineering*, *9*(4), 395. Advance online publication. doi:10.3390/jmse9040395

Prata, J. C., da Costa, J. P., Lopes, I., Duarte, A. C., & Rocha-Santos, T. (2020). Environmental exposure to microplastics: An overview on possible human health effects. *The Science of the Total Environment*, *702*, 134455. doi:10.1016/j.scitotenv.2019.134455 PMID:31733547

Prathima, A., & Karthikeyan, S. (2017). Characteristics of micro-algal biofuel from Botryococcus braunii. *Energy Sources. Part A, Recovery, Utilization, and Environmental Effects*, *39*(2), 206–212.

Prathna, T., Chandrasekaran, N., Raichur, A. M., & Mukherjee, A. (2012). Kinetic evolution studies of silver nanoparticles in a bio-based green synthesis process. *Colloids and Surfaces. A, Physicochemical and Engineering Aspects*, *377*(1-3), 212–216. doi:10.1016/j.colsurfa.2010.12.047

Price, D. C. (2012). Cyanophora paradoxa genome elucidates origin of photosynthesis in algae and plants (Science (843)). *Science*, *335*(6074), 1302.

Price, D. C., Chan, C. X., Yoon, H. S., Yang, E. C., Qiu, H., Weber, A. P., Schwacke, R., Gross, J., Blouin, N. A., Lane, C., Reyes-prietodion, A., Durnfordjonathan, D. G., Neilsonb, J. A. D., Franz Lang, B., Burger, G., Steiner, J. M., Löffelhardt, W., Meuser, J. E., Posewitz, M. C., ... Bhattacharya, D. (2012). *Cyanophora paradoxa* genome elucidates origin of photosynthesis in algae and plants. *Science*, *335*(6070), 843–847. doi:10.1126cience.1213561 PMID:22344442

Price, D., & Horrocks, A. R. (2013). Combustion processes of textile fibres. In *Handbook of Fire Resistant Textiles* (pp. 3–25). Elsevier. doi:10.1533/9780857098931.1.3

Priyadarshani, I., & Rath, B. (2012). Commercial and industrial applications of micro algae – A review. *Journal of Algal Biomass Utilization*, *3*(4), 89–100.

Priya, H., Prasanna, R., Ramakrishnan, B., Bidyarani, N., Babu, S., Thapa, S., & Renuka, N. (2015). Influence of cyanobacterial inoculation on the culturable microbiome and growth of rice. *Microbiological Research*, *171*, 78–89. doi:10.1016/j.micres.2014.12.011 PMID:25644956

Priya, M., Gurung, N., Mukherjee, K., & Bose, S. (2014). Microalgae in removal of heavy metal and organic pollutants from soil. In S. Das (Ed.), *Microbial Biodegradation and Bioremediation* (pp. 519–537). Elsevier.

Prodromou, M., & Pashalidis, I. (2016). Europium adsorption by non-treated and chemically modified opuntia ficus indica cactus fibres in aqueous solutions. *Desalination and Water Treatment*, *57*(11), 5079–5088. doi:10.1080/19443994.2014.1002431

Pröschold, T., & Leliaert, F. (2007). *Systematics of the green algae: Conflict of classic and modern approaches.* doi:10.1201/9780849379901.ch7

Proshad, R., Kormoker, T., Islam, M. S., Haque, M. A., Rahman, M. M., & Mithu, M. M. R. (2017). Toxic effects of plastic on human health and environment : A consequences of health risk assessment in Bangladesh. *International Journal of Healthcare*, *6*(1), 1. doi:10.14419/ijh.v6i1.8655

Prussi, M., Buffi, M., Casini, D., Chiaramonti, D., Martelli, F., Carnevale, M., & Rodolfi, L. (2014). Experimental and numerical investigations of mixing in raceway ponds for algae cultivation. *Biomass and Bioenergy*, *67*, 390–400.

Puglisi, I., Bella, E. La, Rovetto, E. I., Roberta, A., & Piero, L. (2020). *Biostimulant Effect and Biochemical Response in Lettuce Seedlings Treated with A Scenedesmus quadricauda Extract*. Academic Press.

Qian, F., Luo, Q., Yang, R., Zhu, Z., Chen, H., & Yan, X. (2015). The littoral red alga *Pyropia haitanensis* uses rapid accumulation of floridoside as the desiccation acclimation strategy. *Journal of Applied Phycology*, *27*(1), 621–632. doi:10.100710811-014-0336-0

Qiao, M. Y., Zhang, Y.-J., Liu, L. A., Shi, L., Ma, Q.-H., Chow, W. S., & Jiang, C.-D. (2021). Do rapid photosynthetic responses protect maize leaves against photoinhibition under fluctuating light? *Photosynthesis Research*, *149*(1), 57–68. doi:10.100711120-020-00780-5 PMID:32783175

Qi, F., Wu, D., Mu, R., Zhang, S., & Xu, X. (2018). Characterization of a Microalgal UV Mutant for CO_2 Biofixation and Biomass Production. *BioMed Research International*, *4375170*, 1–8. Advance online publication. doi:10.1155/2018/4375170 PMID:30671452

Qin, D., Chen, Z., Averyt, K. B., Miller, H. L., Solomon, S., Manning, M., & Tignor, M. (2007). *Ipcc, 2007: Summary for policymakers*. Academic Press.

Qiu, H., Yoon, H. S., & Bhattacharya, D. (2013). Algal endosymbionts as vectors of horizontal gene transfer in photosynthetic eukaryotes. *Frontiers in Plant Science*, *4*, 366. doi:10.3389/fpls.2013.00366 PMID:24065973

Queiroz, K., Medeiros, V., Queiroz, L., Abreu, L., Rocha, H., Ferreira, C., Jucá, M., Aoyama, H., & Leite, E. (2008). Inhibition of reverse transcriptase activity of HIV by polysaccharides of brown algae. *Biomedicine and Pharmacotherapy*, *62*(5), 303–307. doi:10.1016/j.biopha.2008.03.006 PMID:18455359

Queste, B. Y., Fernand, L., Jickells, T. D., & Heywood, K. J. (2013). Spatial extent and historical context of North Sea oxygen depletion in August 2010. *Biogeochemistry*, *113*(1), 53–68. doi:10.100710533-012-9729-9 PMID:32355379

Quinn, J. C., Smith, T. G., Downes, C. M., & Quinn, C. (2014). Microalgae to biofuels lifecycle assessment— Multiple pathway evaluation. *Algal Research*, *4*, 116–122. doi:10.1016/j.algal.2013.11.002

Qureshi, M. A. (1995). Immunomodulatory effects of spirulina supplementation in chickens. In *The 44th Western Poultry Disease Conference*. North Carolina State University.

Radjenović, J., Petrović, M., & Barceló, D. (2009). Fate and distribution of pharmaceuticals in wastewater and sewage sludge of the conventional activated sludge (CAS) and advanced membrane bioreactor (MBR) treatment. *Water Research*, *43*, 831–841.

Radulovich, R., Neori, A., Valderrama, D., Reddy, C. R. K., Cronin, H., & Forster, J. (2015). Farming of seaweeds. In B. K. Tiwari & D. J. Troy (Eds.), *Seaweed Sustainability* (pp. 27–59). Academic Press. doi:10.1016/B978-0-12-418697-2.00003-9

Ragusa, A., Svelato, A., Santacroce, C., Catalano, P., Notarstefano, V., Carnevali, O., Papa, F., Rongioletti, M. C. A., Baiocco, F., Draghi, S., D'Amore, E., Rinaldo, D., Matta, M., & Giorgini, E. (2021). Plasticenta: First evidence of microplastics in human placenta. *Environment International*, *146*, 106274. doi:10.1016/j.envint.2020.106274 PMID:33395930

Rahbari, H., Akram, A., Pazoki, M., & Aghbashlo, M. (2019). Bio-Oil Production from Sargassum Macroalgae: A Green and Healthy Source of Energy. *Jundishapur Journal of Health Sciences*. Advance online publication. doi:10.5812/jjhs.84301

Raheem, A., Prinsen, P., Vuppaladadiyam, A. K., Zhao, M., & Luque, R. (2018). A review on sustainable microalgae based biofuel and bioenergy production: Recent developments. *Journal of Cleaner Production*, *181*, 42–59. doi:10.1016/j.jclepro.2018.01.125

Raheem, A., Wan Azlina, W. A. K. G., Taufiq Yap, Y. H., Danquah, M. K., & Harun, R. (2015). Thermochemical conversion of microalgal biomass for biofuel production. In *Renewable and Sustainable Energy Reviews* (Vol. 49, pp. 990–999). Elsevier Ltd. doi:10.1016/j.rser.2015.04.186

Rahimi, M., & Jazini, M. (2021). Mixotrophic Cultivation of Chromochloris Zofingiensis on Glycerol, Acetate, and Vinasse. *Journal of Applied Phycology*, *33*(6), 3579–3590. Advance online publication. doi:10.100710811-021-02568-z

Rahman, A., Linton, E., Hatch, A. D., Sims, R. C., & Miller, C. D. (2013). Secretion of polyhydroxybutyrate in *Escherichia coli* using a synthetic biological engineering approach. *Journal of Biological Engineering*, *7*(1), 24. doi:10.1186/1754-1611-7-24 PMID:24139229

Rahman, A., & Miller, C. D. (2017). Microalgae as a source of bioplastics. In *Algal Green Chemistry* (pp. 121–138). Elsevier. doi:10.1016/B978-0-444-63784-0.00006-0

Rahman, A., Putman, R. J., Inan, K., Sal, F. A., Sathish, A., Smith, T., Nielsen, C., Sims, R. C., & Miller, C. D. (2015). Polyhydroxybutyrate production using a wastewater microalgae based media. *Algal Research*, *8*, 95–98. doi:10.1016/j.algal.2015.01.009

Rahman, M. A., Aziz, M. A., Al-khulaidi, R. A., Sakib, N., & Islam, M. (2017). Biodiesel production from microalgae Spirulina maxima by two step process: Optimization of process variable. *J. Radiat. Res. Appl. Sci.*, *10*(2), 140–147. doi:10.1016/j.jrras.2017.02.004

Rahul, R., Kumar, S., Jha, U., & Sen, G. (2015). Cationic inulin: A plant based natural biopolymer for algal biomass harvesting. *International Journal of Biological Macromolecules*, *72*, 868–874.

Rai, A. N., Singh, A. K., & Syiem, M. B. (2019) Plant Growth-Promoting Abilities in Cyanobacteria. In Cyanobacteria (pp. 459-476). Academic Press. doi:10.1016/B978-0-12-814667-5.00023-4

Rai, L. C., & Gaur, J. P. (Eds.). (2001). Algal Adaptation to Environmental Stresses Physiological, Biochemical and Molecular Mechanisms. Springer Berlin Heidelberg. doi:10.1007/978-3-642-59491-5

Rajadurai, M., Poornima, E. H., Narasimhan, S. V., Rao, V. N. R., & Venugopalan, V. P. (2005). Phytoplankton growth under temperature stress: Laboratory studies using two diatoms from a tropical coastal power station site. *Journal of Thermal Biology*, *30*(4), 299–305. doi:10.1016/j.jtherbio.2005.01.003

Rajathi, F. A. A., Parthiban, C., Kumar, V. G., & Anantharaman, P. (2012). Biosynthesis of antibacterial gold nanoparticles using brown alga, Stoechosper- mum marginatum (kützing). *Spectrochimica Acta Part A: Molecular and Biomolecular Spectroscopy, 99*, 166–173.

Raj-Kumar, P.-K., Vallon, O., & Liang, C. (2017). *In silico* analysis of the sequence features responsible for alternatively spliced introns in the model green alga *Chlamydomonas reinhardtii. Plant Molecular Biology*, *94*(3), 253–265. doi:10.100711103-017-0605-9 PMID:28364390

Raj, R., & Kumari, M. (2020). Blue green algae (BGA) and its application. *Journal of Pharmacognosy and Phytochemistry*, *9*(2), 287–296.

Rajvanshi, M., Gautam, K., Manjre, S., Kumar, G. R. K., Kumar, C., Govindachary, S., & Dasgupta, S. (2019). Stoichiometrically balanced nutrient management using a newly designed nutrient medium for large scale cultivation of Cyanobacterium aponinum. *Journal of Applied Phycology*, *31*(5), 2779–2789.

Ramachandra, T. V., & Hebbale, D. (2019, February). Bioethanol from Macroalgae: Prospects and Challenges. *Renewable & Sustainable Energy Reviews*, *117*, 109479. doi:10.1016/j.rser.2019.109479

Ramakrishna, M., Babu, D. R., Gengan, R. M., & Rao, S. C. G. N. (2016). Phytoremediation using Wet- land Plants: An ecosustainable approach-An ecosustainable approach. *International Journal of Phytoremediation*, *10*(2), 133–160.

Ramanan, R., Kim, B. H., Cho, D. H., Oh, H. M., & Kim, H. S. (2016). Algae-bacteria interactions: Evolution, ecology and emerging applications. *Biotechnology Advances*, *34*(1), 14–29. doi:10.1016/j.biotechadv.2015.12.003 PMID:26657897

Ramasamy, D. L., Porada, S., & Sillanpää, M. (2019). Marine algae: A promising resource for the selective recovery of scandium and rare earth elements from aqueous systems. *Chemical Engineering Journal*, *371*, 759–768. doi:10.1016/j.cej.2019.04.106

Ramasubramani, R., Praveen, R., & Sathyanarayanan, K. S. (2016). Study on the strength properties of marine algae concrete. *Rasayan Journal of Chemistry*, *4*, 706–715.

Ramaswamy, S. V. P., Narendhran, S., & Sivaraj, R. (2016). Potentiating effect of ecofriendly synthesis of copper oxide nanoparticles using brown alga: Antimicrobial and anticancer activities. *Bulletin of Materials Science*, *39*(2), 361–364. doi:10.100712034-016-1173-3

Ramlov, F., Souza, J., Faria, A. V., Maraschin, M., Horta, P. A., & Yokoya, N. S. (2011). Growth and accumulation of carotenoids and nitrogen compounds in Gracilaria domingensis (Kütz.) Sonder ex Dickie (Gracilariales, Rhodophyta) cultured under different irradiance and nutrient levels. *Revista Brasileira de Farmacognosia*, *21*(2), 255–261. doi:10.1590/S0102-695X2011005000081

Rana, A., Kabi, S. R., Verma, S., Adak, A., Pal, M., Shivay, Y. S., Prasanna, R., & Nain, L. (2015). Prospecting plant growth promoting bacteria and cyanobacteria as options for enrichment of macro-and micronutrients in grains in rice–wheat cropping sequence. *Cogent Food & Agriculture*, *1*(1), 1037379. doi:10.1080/23311932.2015.1037379

Ranga Rao, A., Baskaran, V., Sarada, R., & Ravishankar, G. A. (2013). In vivo bioavailability and antioxidant activity of carotenoids from micro algal biomass—A repeated dose study. *Food Research International*, *54*(1), 711–717. doi:10.1016/j.foodres.2013.07.067

Ranga Rao, A., Harshvardhan Reddy, A., & Aradhya, S. M. (2010). Antibacterial properties of *Spirulina platensis, Haematococcus pluvialis, Botryococcus braunii* micro algal extracts. *Current Trends in Biotechnology and Pharmacy*, *4*, 809–819.

Ranga Rao, A., Sindhuja, H. N., Dharmesh, S. M., Sankar, K. U., Sarada, R., & Ravishankar, G. A. (2013). Effective inhibition of skin cancer, tyrosinase,and antioxidative properties by astaxanthin and astaxanthin esters from thegreen alga Haematococcus pluvialis. *Journal of Agricultural and Food Chemistry*, *61*(16), 3842–3851. doi:10.1021/jf304609j PMID:23473626

Ranjan, K., Priya, H., Ramakrishnan, B., Prasanna, R., Venkatachalam, S., Thapa, S., Tiwari, R., Nain, L., Singh, R., & Shivay, Y. S. (2016). Cyanobacterial inoculation modifies the rhizosphere microbiome of rice planted to a tropical alluvial soil. *Applied Soil Ecology*, *108*, 195–203. doi:10.1016/j.apsoil.2016.08.010

Rao, M. G., Bharathi, P., & Akila, R. (2014). A comprehensive review on biopolymers. *Sci. Revs. Chem. Commun*, *4*(2), 61–68.

Rao, M. V., Paliyath, G., & Ormrod, D. P. (1996). Ultraviolet-B- and Ozone-Induced Biochemical Changes in Antioxidant Enzymes of Arabidopsis thaliana. *Plant Physiology*, *110*(1), 125–136. doi:10.1104/pp.110.1.125 PMID:8587977

Raposo, M. F.-D., De Morais, A. M. M. B., & De Morais, R. M. S. C. (2016). Emergent Sources of Prebiotics: Seaweeds and Microalgae. *Marine Drugs*, *14*(2), 27. doi:10.3390/md14020027 PMID:26828501

Rasala, B. A., Muto, M., Lee, P. A., Jager, M., Cardoso, R. M. F., Behnke, C. A., Kirk, P., Hokanson, C. A., Crea, R., Mendez, M., & Mayfield, S. P. (2010). Production of therapeutic proteins in algae, analysis of expression of seven human proteins in the chloroplast of Chlamydomonas reinhardtii. *Plant Biotechnology Journal*, *8*(6), 719–733. doi:10.1111/j.1467-7652.2010.00503.x PMID:20230484

Rashid, N., & Nayak, M., & Suh, W.I. (2019). Efficient microalgae removal from aqueous medium through auto-flocculation: Investigating growth-dependent role of organic matter. *Environmental Science and Pollution Research International*, *26*, 27396–27406.

Rashid, N., Park, W. K., & Selvaratnam, T. (2018). Binary culture of microalgae as an integrated approach for enhanced biomass and metabolites productivity, wastewater treatment, and bioflocculation. *Chemosphere*, *194*, 67–75.

Rashid, N., Rehman, M. S. U., Memon, S., Rahman, Z. U., Lee, K., & Han, J. I. (2013). Current status, barriers and developments in biohydrogen production by microalgae. *Renewable & Sustainable Energy Reviews*, *22*, 571–579. doi:10.1016/j.rser.2013.01.051

Rashwan, R. S., & Hammad, D. M. (2020). Toxic Effect of Spirulina Platensis and Sargassum Vulgar as Natural Pesticides on Survival and Biological Characteristics of Cotton Leaf Worm Spodoptera littoralis. *Sci. Afr.*, *8*, e00323.

Ras, M., Lardon, L., Bruno, S., Bernet, N., & Steyer, J. P. (2011). Experimental study on a coupled process of production and anaerobic digestion of *Chlorella vulgaris*. *Bioresource Technology*, *102*, 200–206.

Ras, M., Steyer, J., & Bernard, O. (2013). Temperature effect on microalgae: A crucial factor for outdoor production (A review). *Reviews in Environmental Science and Biotechnology*, *12*, 153–164. doi:10.100711157-013-9310-6

Rastogi, R. P., Sinha, R. P., Singh, S. P., & Häder, D. P. (2010). Photoprotective compounds from marine organisms. *Journal of Industrial Microbiology & Biotechnology*, *37*(6), 537–558. doi:10.100710295-010-0718-5 PMID:20401734

Rastogi, R. P., Sonani, R. R., & Madamwar, D. (2015). Cyanobacterial sunscreen scytonemin: Role in photoprotection and biomedical research. *Applied Biochemistry and Biotechnology*, *176*(6), 1551–1563. doi:10.100712010-015-1676-1 PMID:26013282

Rasul, I., Azeem, F., Siddique, M. H., Muzammil, S., Rasul, A., Munawar, A., Afzal, M., Ali, M. A., & Nadeem, H. (2017). Algae biotechnology. In Algae Based Polymers, Blends, and Composites (pp. 301–334). Elsevier. doi:10.1016/B978-0-12-812360-7.00008-2

Rathore, D., & Singh, A. (2013). Biohydrogen production from microalgae. In *Biofuel Technologies* (pp. 317–333). Springer. doi:10.1007/978-3-642-34519-7_13

Ratkowsky, D. A., Olley, J., McMeekin, T., & Ball, A. (1982). Relationship between temperature and growth rate of bacterial cultures. *Journal of Bacteriology*, *149*(1), 1–5. doi:10.1128/jb.149.1.1-5.1982 PMID:7054139

Ratnapuram, H. P., Vutukuru, S. S., & Yadavalli, R. (2018). Mixotrophic transition induced lipid productivity in Chlorella pyrenoidosa under stress conditions for biodiesel production. *Heliyon*, *4*(1), e00496. doi:10.1016/j.heliyon.2017.e00496 PMID:29322110

Raven, J. A., & Geider, R. J. (1988). Temperature and algal growth. *The New Phytologist*, *110*(4), 441–461. doi:10.1111/j.1469-8137.1988.tb00282.x

Ravindran, R., & Jaiswal, A. K. (2016). Exploitation of food industry waste for high-value products. *Trends in Biotechnology*, *34*(1), 58–69. doi:10.1016/j.tibtech.2015.10.008 PMID:26645658

Rawat, I., Kumar, R. R., Mutanda, T., & Bux, F. (2013). Biodiesel from microalgae: A critical evaluation from laboratory to large scale production. *Applied Energy*, *103*, 444–467.

Rawat, I., Kumar, R. R., Mutanda, T., & Bux, F. (2013). Biodiesel from microalgae:A critical evaluation from laboratory to large scale production. *Applied Energy*, *103*, 444–467. doi:10.1016/j.apenergy.2012.10.004

Rawat, I., Ranjith Kumar, R., Mutanda, T., & Bux, F. (2011). Dual role of microalgae: Phycoremediation of domestic wastewater and biomass production for sustainable biofuels production. *Applied Energy*, *88*(10), 3411–3424. doi:10.1016/j.apenergy.2010.11.025

Rayen, F., Behnam, T., & Dominique, P. (2019). Optimization of a raceway pond system for wastewater treatment: A review. *Critical Reviews in Biotechnology*, *39*(3), 422–435.

Raymundo, A., Gouveia, L., Batista, A. P., Raymundo, A., Sousa, I., & Emphis, J. (2006). Chlorella vulgaris and Haematococcus pluvialis biomass as colouring and antioxidant in food emulsions. *European Food Research and Technology*, *222*(3-4), 362–367. doi:10.100700217-005-0105-z

Rayorath, P., Jithesh, M. N., Farid, A., Khan, W., Palanisamy, R., Hankins, S. D., Critchley, A. T., & Prithiviraj, B. (2008). Rapid bioassays to evaluate the plant growth promoting activity of *Ascophyllum nodosum* (L.) Le Jol. using a model plant, *Arabidopsis thaliana* (L.) Heynh. *Journal of Applied Phycology*, *20*(4), 423–429. doi:10.100710811-007-9280-6

Rayorath, P., Khan, W., Palanisamy, R., MacKinnon, S. L., Stefanova, R., Hankins, S. D., Critchley, A. T., & Prithiviraj, B. (2008). Extracts of the brown seaweed *Ascophyllum nodosum*induce gibberellic acid (GA3)-independent amylase activity in barley. *Journal of Plant Growth Regulation*, *27*(4), 370–379. doi:10.100700344-008-9063-6

Raza, Z. A., Abid, S., & Banat, I. M. (2018). Polyhydroxyalkanoates: Characteristics, production, recent developments and applications. *International Biodeterioration & Biodegradation*, *126*, 45–56. doi:10.1016/j.ibiod.2017.10.001

Razu, M. H., Hossain, F., & Khan, M. (2019). Advancement of Bio-hydrogen Production from Microalgae BT - Microalgae Biotechnology for Development of Biofuel and Wastewater Treatment (Md. A. Alam & Z. Wang, Eds.). Springer Singapore. doi:10.1007/978-981-13-2264-8_17

Razza, F., & Innocenti, F. D. (2012). Bioplastics from renewable resources: The benefits of biodegradability. *Asia-Pacific Journal of Chemical Engineering*, *7*(3), S301–S309. doi:10.1002/apj.1648

Razzak, S. A., Hossain, M. M., Lucky, R. A., Bassi, A. S., & De Lasa, H. (2013). Integrated CO_2 capture, wastewater treatment and biofuel production by microalgae culturing—A review. *Renewable & Sustainable Energy Reviews*, *27*, 622–653. doi:10.1016/j.rser.2013.05.063

Reczek, C. R., & Chandel, N. S. (2017). *The two faces of reactive oxygen species in cancer*. Academic Press.

Reddy, C. N., Nguyen, H. T., Noori, M. T., & Min, B. (2019). Potential Applications of Algae in the Cathode of Microbial Fuel Cells for Enhanced Electricity Generation with Simultaneous Nutrient Removal and Algae Biorefinery: Current Status and Future Perspectives. *Bioresource Technology*, *292*, 122010. doi:10.1016/j.biortech.2019.122010 PMID:31473037

Reddy, M. M., Vivekanandhan, S., Misra, M., Bhatia, S. K., & Mohanty, A. K. (2013). Biobased plastics and bionanocomposites: Current status and future opportunities. *Progress in Polymer Science*, *38*(10–11), 1653–1689. doi:10.1016/j.progpolymsci.2013.05.006

Reed, R. H. (1984). Transient breakdown in the selective permeability of the plasma membrane of Chlorella emersonii in response to hyperosmotic shock: Implications for cell water relations and osmotic adjustment. *The Journal of Membrane Biology*, *82*(1), 83–88. doi:10.1007/BF01870734

Rees, T. A. V., & Bekheet, I. A. (1982). The role of nickel in urea assimilation by algae. *Planta*, *156*(5), 385–387. doi:10.1007/BF00393307 PMID:24272649

Reidenbach, L. B., Fernandez, P. A., Leal, P. P., Noisette, F., McGraw, C. M., Revill, A. T., Hurd, C. L., & Kübler, J. E. (2017). Growth, ammonium metabolism, and photosynthetic properties of *Ulva australis* (Chlorophyta) under decreasing pH and ammonium enrichment. *PLoS One*, *12*(11), e0188389. doi:10.1371/journal.pone.0188389 PMID:29176815

Reijnders, M. J. M. F., Van-Heck, R. G. A., Lam, C. M. C., Scaife, M. A., Dos-Santos, V. A. P. M., Smith, A. G., & Schaap, P. J. (2014). Green genes: Bioinformatics and systems-biology innovations drive algal biotechnology. *Trends in Biotechnology*, *32*(12), 617–626. doi:10.1016/j.tibtech.2014.10.003 PMID:25457388

Remelli, W., & Santabarbara, S. (2018). Excitation and emission wavelength dependence of fluorescence spectra in whole cells of the cyanobacterium Synechocystis sp. PPC6803: Influence on the estimation of Photosystem II maximal quantum efficiency. *Biochimica et Biophysica Acta. Bioenergetics*, *1859*(11), 1207–1222. doi:10.1016/j.bbabio.2018.09.366 PMID:30297025

Rempel, A., de Souza Sossella, F., Margarites, A. C., Astolfi, A. L., Steinmetz, R. L. R., Kunz, A., Treichel, H., & Colla, L. M. (2019). Bioethanol from Spirulina platensis biomass and the use of residuals to produce biomethane: An energy efficient approach. *Bioresource Technology*, *288*, 121588. doi:10.1016/j.biortech.2019.121588 PMID:31176943

Rémy, B., Mion, S., Plener, L., Elias, M., Chabrière, E., & Daudé, D. (2018). Interference in Bacterial Quorum Sensing: A Biopharmaceutical Perspective. *Frontiers in Pharmacology*, *9*, 203. Advance online publication. doi:10.3389/fphar.2018.00203 PMID:29563876

Renaud, S. M., Thinh, L. V., Lambrinidis, G., & Parry, D. L. (2002). Effect of temperature on growth, chemical composition and fatty acid composition of tropical Australian microalgae grown in batch cultures. *Aquaculture (Amsterdam, Netherlands)*, *211*(1-4), 195–214. doi:10.1016/S0044-8486(01)00875-4

Rengasamy, K. R. R., Kulkarni, M. G., Stirk, W. A., & Van Staden, J. (2015). Eckol-a new plant growth stimulant from the brown seaweed *Ecklonia maxima*. *Journal of Applied Phycology*, *27*(1), 581–587. doi:10.100710811-014-0337-z

Rengel, R., Smith, R. T., Haslam, R. P., Sayanova, O., Vila, M., & Leon, R. (2018). Overexpression of acetyl-CoA synthetase (ACS) enhances the biosynthesis of neutral lipids and starch in the green microalga Chlamydomonas reinhardtii. *Algal Research*, *31*, 183–193. doi:10.1016/j.algal.2018.02.009

Renuka, N., Guldhe, A., Prasanna, R., Singh, P., & Bux, F. (2018). Microalgae as multi-functional options in modern agriculture: Current trends, prospects and challenges. *Biotechnology Advances*, *36*(4), 1255–1273. doi:10.1016/j.biotechadv.2018.04.004 PMID:29673972

Renuka, N., Prasanna, R., Sood, A., Bansal, R., Bidyarani, N., Singh, R., Shivay, Y. S., Nain, L., & Ahluwalia, A. S. (2017). Wastewater grown microalgal biomass as inoculants for improving micronutrient availability in wheat. *Rhizosphere*, *3*, 150–159. doi:10.1016/j.rhisph.2017.04.005

Renuka, N., Sood, A., Ratha, S. K., Prasanna, R., & Ahluwalia, A. S. (2013). Evaluation of microalgal consortia for treatment of primary treated sewage effluent and biomass production. *Journal of Applied Phycology*, *25*(5), 1529–1537. doi:10.100710811-013-9982-x

Reyimu, Z., & Özçimen, D. (2017). Batch cultivation of marine microalgae Nannochloropsis oculata and Tetraselmis suecica in treated municipal wastewater toward bioethanol production. *Journal of Cleaner Production*, *150*, 40–46. doi:10.1016/j.jclepro.2017.02.189

Rhoades, D., & Hanson, A. D. (1993). Quaternary ammonium and tertiary sulphonium compounds in higher plants. *Annual Review of Plant Physiology and Plant Molecular Biology*, *44*(1), 357–384. doi:10.1146/annurev.pp.44.060193.002041

Rı'os, S. D., Salvado´, J., Farriol, X., & Torras, C. (2012). Antifouling microfiltration strategies to harvest microalgae for biofuel. *Bioresource Technology*, *119*, 406–418. doi:10.1016/j.biortech.2012.05.044 PMID:22739578

Riaño, B., Molinuevo, B., & García-González, M. C. (2011). Treatment of fish processing wastewater with microalgae-containing microbiota. *Bioresource Technology*, *102*(23), 10829–10833. doi:10.1016/j.biortech.2011.09.022 PMID:21983413

Riaz, T., Shehzad, W., Viari, A., Pompanon, F., Taberlet, P., & Coissac, E. (2011). ecoPrimers: Inference of new DNA barcode markers from whole genome sequence analysis. *Nucleic Acids Research*, *39*(21), 145–155. doi:10.1093/nar/gkr732 PMID:21930509

Richer, S., Stiles, W., Statkute, L., Pulido, J., Frankowski, J., Rudy, D., Pei, K., Tsipursky, M., & Nyland, J. (2004). Double-masked, placebo-controlled, randomized trial of lutein and antioxidant supplementation in the intervention of atrophic age-related macular degeneration: The Veterans LAST study (Lutein Antioxidant Supplementation Trial). *Optom. J. Am. Optom. Assoc.*, *2004*(75), 216–229. doi:10.1016/S1529-1839(04)70049-4 PMID:15117055

Richmond, A. (2004). Part III. Economic applications of micro algae. In Handbook of Microalgal Culture. Biotechnology Applied Phycology (pp. 255-353): Blackwell Publishing.

Richmond, A. (1996). Efficient utilization of high irradiance for production of photoautotrophic cell mass: A survey. *Journal of Applied Phycology*, *8*, 381–387.

Richmond, A. (2000). Microalgal biotechnology at the turn of the millenium: A personal view. *Journal of Applied Phycology*, *12*(3/5), 441–451. doi:10.1023/A:1008123131307

Rincon, L., Jaramilo, J. J., & Cardona, C. A. (2014). Comparison of feedstocks and technologies for biodiesel production: An environmental and techno-economic evaluation. *Renewable Energy*, *69*, 479–487. doi:10.1016/j.renene.2014.03.058

Rindi, F., & Guiry, M. D. (2004). Composition and spatial variability of terrestrial algal assemblages occurring at the bases of urban walls in Europe. *Phycologia*, *43*(3), 225–235. doi:10.2216/i0031-8884-43-3-225.1

Rioux, L.-E., & Turgeon, S. L. (2015). Seaweed carbohydrates. In *Seaweed sustainability* (pp. 141–192). Elsevier. doi:10.1016/B978-0-12-418697-2.00007-6

Rippka, R., Waterbury, J. B., & Stanier, R. Y. (1981). Isolation and purification of cyanobacteria: some general principles. The Prokaryotes, 212–220. doi:10.1007/978-3-662-13187-9_8

Rippka, R. (1988). Isolation and purification of cyanobacteria. *Methods in Enzymology*, *167*, 3–27. doi:10.1016/0076-6879(88)67004-2 PMID:3148836

Riveros, K., Sepulveda, C., Bazaes, J., Marticorena, P., Riquelme, C., & Acién, G. (2018). Overall development of a bioprocess for the outdoor production of Nannochloropsis gaditana for aquaculture. *Aquaculture Research*, *49*(1), 165–176.

Rizwan, M., Mujtaba, G., Memon, S. A., Lee, K., & Rashid, N. (2018). Exploring the potential of microalgae for new biotechnology applications and beyond: A review. *Renewable and Sustainable Energy Reviews, 92*(March), 394–404. doi:10.1016/j.rser.2018.04.034

Rizzello, C. G., Tagliazucchi, D., Babini, E., Sefora Rutella, G., Taneyo Saa, D. L., & Gianotti, A. (2016). Bioactive peptides from vegetable food matrices: Research trends and novel biotechnologies for synthesis and recovery. *Journal of Functional Foods*, *27*, 549–569. doi:10.1016/j.jff.2016.09.023

Roach, T. N., Little, M., Arts, M. G., Huckeba, J., Haas, A. F., George, E. E., ... Silveira, C. B. (2020). A Multiomic Analysis of in Situ Coral-Turf Algal Interactions. *Proceedings of the National Academy of Sciences of the United States of America*, *117*(24), 13588–13595. doi:10.1073/pnas.1915455117 PMID:32482859

Roberts, N., Hilliard, M., He, Q. P., & Wang, J. (2020). A Microalgae-Methanotroph Coculture is a Promising Platform for Fuels and Chemical Production From Wastewater. *Frontiers in Energy Research*, *8*, 563352. Advance online publication. doi:10.3389/fenrg.2020.563352

Robinson, J. T., Thorvaldsdóttir, H., Winckler, W., Guttman, M., Lander, E. S., Getz, G., & Mesirov, J. P. (2011). Integrative genomics viewer. *Nature Biotechnology*, *29*(1), 24–26. doi:10.1038/nbt.1754 PMID:21221095

Roblero Lucchetti, J. G. (2020). *Dry Reforming of Methane via Use of Ni, Pt and Mo over Combined Supports*. Academic Press.

Rochaix, J. D., & van Dillewijn, J. (1982). Transformation of the green alga *Chlamydomonas reinhardii* with yeast DNA. *Nature*, *296*(5852), 70–72. doi:10.1038/296070a0 PMID:7038510

Roche, D. B., Buenavista, M. T., & McGuffin, L. J. (2013). The FunFOLD2 server for the prediction of protein-ligand interactions. *Nucleic Acids Research*, *41*(1), 303–307. doi:10.1093/nar/gkt498 PMID:23761453

Rochman, C. M. (2015). The Complex Mixture, Fate and Toxicity of Chemicals Associated with Plastic Debris in the Marine Environment. In M. Bergmann, L. Gutow, & M. Klages (Eds.), *Marine Anthropogenic Litter*. Springer. doi:10.1007/978-3-319-16510-3_5

Rock, C. L. (2009). Carotenoids and Cancer. In Carotenoids (Vol. 5). Birkhäuser Verlag AG. doi:10.1007/978-3-7643-7501-0_13

Rodolfi, L., Chini Zittelli, G., Bassi, N., Padovani, G., Biondi, N., Bonini, G., & Tredici, M. R. (2009). Microalgae for oil: Strain selection, induction of lipid synthesis and outdoor mass cultivation in a low-cost photobioreactor. *Biotechnology and Bioengineering*, *102*(1), 100–112. doi:10.1002/bit.22033 PMID:18683258

Rodrigues, J. A. G., Eloy, Y. R. G., Vanderlei, E. D. S. O., Cavalcante, J. F., Romanos, M. T. V., & Benevides, N. M. B. (2017). An anti-dengue and anti-herpetic polysulfated fraction isolated from the coenocytic green seaweed Caulerpa cupressoides inhibits thrombin generation in vitro. *Acta Scientiarum. Biological Sciences*, *39*(2), 149–159. doi:10.4025/actascibiolsci.v39i2.28081

Roels, J. A. (1983). *Energetics and kinetics in biotechnology*. Elsevier Biomedical Press.

Rohit, M., Chiranjeevi, P., Reddy, C. N., & Mohan, S. V. (2016). Photobioreactors for Microalgal Cultivation and influence of operational Parameters. In R. S. Singh, A. Pandey, & E. Gnansouno (Eds.), *Biofuels: Production and Future Perspectives* (p. 413). CRC Press.

Roja, K., Ruben Sudhakar, D., Anto, S., & Mathimani, T. (2019). Extraction and characterization of polyhydroxyalkanoates from marine green alga and cyanobacteria. *Biocatalysis and Agricultural Biotechnology*, *22*, 101358. doi:10.1016/j.bcab.2019.101358

Roncero-Ramos, B., Román, J. R., Gómez-Serrano, C., Cantón, Y., & Acién, F. G. (2019). Production of a biocrust-cyanobacteria strain (Nostoc commune) for large-scale restoration of dryland soils. *Journal of Applied Phycology*, *31*(4), 2217–2230.

Ronga, D., Biazzi, E., Parati, K., Carminati, D., Carminati, E., & Tava, A. (2019). Microalgal biostimulants and biofertilisers in crop productions. *Agronomy (Basel)*, *9*(4), 192. doi:10.3390/agronomy9040192

Rönnberg, O., & Ruokolahti, C. (1986, January). Seasonal variation of algal epiphytes and phenolic content of Fucus vesiculosus in a northern Baltic archipelago. In *Annales Botanici Fennici* (pp. 317–323). The Finnish Botanical Publishing Board.

Roohinejad, S., Koubaa, M., Barba, F. J., Saljoughian, S., Amid, M., & Greiner, R. (2017). Application of seaweeds to develop new food products with enhanced shelf-life, quality and health-related beneficial properties. *Food Research International*, *99*, 1066–1083. doi:10.1016/j.foodres.2016.08.016 PMID:28865618

Rosales-Mendoza, S., García-Silva, I., González-Ortega, O., Sandoval-Vargas, J. M., Malla, A., & Vimolmangkang, S. (2020). The Potential of Algal Biotechnology to Produce Antiviral Compounds and Biopharmaceuticals. *Molecules (Basel, Switzerland)*, *25*, 4049.

Roselet, F., Burkert, J., & Abreu, P. C. (2016). Flocculation of nannochloropsis oculata using a tannin-based polymer: Bench scale optimization and pilot scale reproducibility. *Biomass and Bioenergy*, *87*, 55–60.

Rosenboom, J.-G., Hohl, D. K., Fleckenstein, P., Storti, G., & Morbidelli, M. (2018). Bottle-grade polyethylene furanoate from ring-opening polymerisation of cyclic oligomers. *Nature Communications*, *9*(1), 2701. doi:10.103841467-018-05147-y PMID:30042388

Rost, B., Richter, K. U., Riebesell, U., & Hansen, P. J. (2006). Inorganic carbon acquisition in red tide dinoflagellates. *Plant, Cell & Environment*, *29*(5), 810–822. doi:10.1111/j.1365-3040.2005.01450.x PMID:17087465

Roy, A., Yang, J., & Zhang, Y. (2012). COFACTOR an accurate comparative algorithm for structure-based protein function annotation. *Nucleic Acids Research*, *40*(1), 471–477. doi:10.1093/nar/gks372 PMID:22570420

Rudnik, E. (2019). Ecotoxicological assessment of compostable polymer materials. In *Compostable Polymer Materials* (pp. 293–313). Elsevier. doi:10.1016/B978-0-08-099438-3.00009-4

Rudnik, E., & Briassoulis, D. (2011). Degradation behaviour of poly(lactic acid) films and fibres in soil under Mediterranean field conditions and laboratory simulations testing. *Industrial Crops and Products*, *33*(3), 648–658. doi:10.1016/j.indcrop.2010.12.031

Ruiz-Marin, A., Canedo-López, Y., & Chávez-Fuentes, P. (2020). Biohydrogen production by Chlorella vulgaris and Scenedesmus obliquus immobilized cultivated in artificial wastewater under different light quality. *AMB Express*, *10*(1), 1–7. doi:10.118613568-020-01129-w PMID:33108605

Ruiz-Ruiz, J. C., Matus-Basto, A. J., Acereto-Escoffié, P., & Segura-Campos, M. R. (2017). Antioxidant and anti-inflammatory activities of phenolic compounds isolated from Melipona beecheii honey. *Food and Agricultural Immunology*, *28*(6), 1424–1437. doi:10.1080/09540105.2017.1347148

Rujnić-Sokele, M., & Pilipović, A. (2017). Challenges and opportunities of biodegradable plastics: A mini review. *Waste Management & Research*, *35*(2), 132–140. doi:10.1177/0734242X16683272 PMID:28064843

Rumin, J., Bonnefond, H., Saint-Jean, B., Rouxel, C., Sciandra, A., Bernard, O., Cadoret, J.-P., & Bougaran, G. (2015). The use of fluorescent Nile red and BODIPY for lipid measurement in microalgae. *Biotechnology for Biofuels*, *8*(1), 42. doi:10.118613068-015-0220-4 PMID:25788982

Rumin, J., Nicolau, E., Gonçalves de Oliveira, R. Junior, Fuentes-Grünewald, C., & Picot, L. (2020). Analysis of scientific research driving microalgae market opportunities in Europe. *Marine Drugs*, *18*(5), 264. doi:10.3390/md18050264 PMID:32443631

Ruparelia, J. P., Chatterjee, A. K., Duttagupta, S. P., & Mukherji, S. (2008). Strain specificity in antimicrobial activity of silver and copper nanoparticles. *Acta Biomaterialia*, *4*(3), 707–716. doi:10.1016/j.actbio.2007.11.006 PMID:18248860

Rutkowska, M., Krasowska, K., Heimowska, A., Steinka, I., & Janik, H. (2002). Degradation of polyurethanes in sea water. *Polymer Degradation & Stability*, *76*(2), 233–239. doi:10.1016/S0141-3910(02)00019-8

Rutten, T. P. A., Sandee, B., & Hofman, A. R. T. (2005). Phytoplankton monitoring by high performance flow cytometry: A successful approach? *Cytom. A.*, *64*(1), 16–26. doi:10.1002/cyto.a.20106 PMID:15688354

Ryu, A. J., Jeong, B., Kang, N. K., Jeon, S., Sohn, M. G., Yun, H. J., Lim, J. M., Jeong, S. W., Park, Y.-I., Jeong, W. J., Park, S., Chang, Y. K., & Jeong, K. J. (2021). Safe-Harboring based novel genetic toolkit for *Nannochloropsis salina* CCMP1776: Efficient overexpression of transgene via CRISPR/Cas9-Mediated Knock-in at the transcriptional hotspot. *Bioresource Technology*, *340*, 125676. doi:10.1016/j.biortech.2021.125676 PMID:34365302

Ryu, K. H., Kim, B., & Lee, J. H. (2019). A model-based optimization of microalgal cultivation strategies for lipid production under photoautotrophic condition. *Computers & Chemical Engineering*, *121*, 57–66. doi:10.1016/j.compchemeng.2018.10.004

Saadaoui, I., Rasheed, R., Abdulrahman, N., Bounnit, T., Cherif, M., Al Jabri, H., & Mraiche, F. (2020). Algae-derived bioactive compounds with anti-lung cancer potential. *Marine Drugs*, *18*(4), 197. doi:10.3390/md18040197 PMID:32276401

Saadaoui, I., Sedky, R., Rasheed, R., Bounnit, T., Almahmoud, A., Elshekh, A., Dalgamouni, T., al Jmal, K., Das, P., & Al Jabri, H. (2019). Assessment of the algae-based biofertilizer influence on date palm (Phoenix dactylifera L.) cultivation. *Journal of Applied Phycology*, *31*(1), 457–463. doi:10.100710811-018-1539-6

Saad, M. G., Dosoky, N. S., Zoromba, M. S., & Shafik, H. M. (2019). Algal biofuels: Current status and key challenges. *Energies*, *12*(10), 1920. Advance online publication. doi:10.3390/en12101920

Sabathini, H. A., Windiani, L., Dianursanti, & Gozan, M. (2018). Mechanical physicial properties of *Chlorella*-PVA based bioplastic with ultrasonic homogenizer. *E3S Web of Conferences, 67*, 3046. doi:10.1051/e3sconf/20186703046

Sabh, A. Z., & Shallan, A. M. (2008). Effect of organic fertilization on Bean (*Vicia faba* L.) by using different marine macroalgae in relation to the morphological, anatomical characteristics and chemical constituents of the plant. *Australian Journal of Basic and Applied Sciences*, *4*, 1076–1091.

Saçan, M. T., Oztay, F., & Bolkent, S. (2007). Exposure of Dunaliella tertiolecta to lead and aluminum: Toxicity and effects on ultrastructure. *Biological Trace Element Research*, *120*(1), 264–272. doi:10.100712011-007-8016-4 PMID:17916979

Sacramento, A. T., García-Jiménez, P., Alcázar, R., Tiburcio, A. F., & Robaina, R. R. (2004). Influence of Polyamines on the Sporulation of *Grateloupia* (Halymeniaceae, Rhodophyta). *Journal of Phycology*, *40*(5), 887–894. doi:10.1111/j.1529-8817.2004.03183.x

Sadiq, I. M., Pakrashi, S., Chandrasekaran, N., & Mukherjee, A. (2011). Studies on toxicity of aluminum oxide (Al2O3) nanoparticles to microalgae species: Scenedesmus sp. and Chlorella sp. *Journal of Nanoparticle Research*, *13*(8), 3287–3299. doi:10.100711051-011-0243-0

Sadjadi, S. J., & Omrani, H. (2008). Data envelopment analysis with uncertain data: An application for Iranian electricity distribution companies. *Energy Policy*, *36*(11), 4247–4254. doi:10.1016/j.enpol.2008.08.004

Safinaz, A. F., & Ragaa, A. H. (2013). Effect of some red marine algae as biofertilizers on growth of maize (*Zea mayz* L.) plants. *International Food Research Journal*, *20*(4), 1629–1632.

Saha, S. K., Ermis, H., & Murray, P. (2020). Marine microalgae for potential lutein production. *Applied Sciences (Basel, Switzerland)*, *10*(18), 6457. doi:10.3390/app10186457

Saha, S., & Murray, P. (2018). Exploitation of Microalgae Species for Nutraceutical Purposes: Cultivation Aspects. *Fermentation (Basel, Switzerland)*, *4*(2), 46. Advance online publication. doi:10.3390/fermentation4020046

Sahayaraj, K., Rajesh, S., & Rathi, J. M. (2012). Silver nanoparticles biosynthesis using marine alga Padinapavonica(Linn.) and its microbicidalactivity. *Digest Journal of Nanomaterials and Biostructures*, *7*(4), 1557–1567.

Sahin, M. S., Khazi, M. I., Demirel, Z., & Dalay, M. C. (2019). Variation in growth, fucoxanthin, fatty acids profile and lipid content of marine diatoms Nitzschia sp. and Nanofrustulum shiloi in response to nitrogen and iron. *Biocatalysis and Agricultural Biotechnology*, *17*, 390–398. doi:10.1016/j.bcab.2018.12.023

Sahoo, P. K., Kim, K., Powell, M. A., & Equeenuddin, S. M. (2016). Recovery of metals and other beneficial products from coal fly ash: A sustainable approach for fly ash management. *International Journal of Coal Science & Technology*, *3*(3), 267–283. doi:10.100740789-016-0141-2

Sahu, D., Priyadarshani, I., & Rath, B. (2012). Cyanobacteria–as potential biofertilizer. *CIB Tech J Microbiol*, *1*, 20–26.

Salam, K. A., Velasquez-Orta, S. B., & Harvey, A. P. (2016). A sustainable integrated in situ transesterification of microalgae for biodiesel production and associated co-product-a review. *Renewable & Sustainable Energy Reviews*, *65*, 1179–1198. doi:10.1016/j.rser.2016.07.068

Salehi-Ashtiani, K., Koussa, J., Dohai, B. S., Chaiboonchoe, A., Cai, H., Dougherty, K. A. D., Nelson, D. R., Jijakli, K., & Khraiwesh, B. (2015). Toward applications of genomics and metabolic modeling to improve algal biomass productivity. In Biomass and Biofuels from Microalgae (pp. 173-189). Springer International Publishing Switzerland. doi:10.1007/978-3-319-16640-7_10

Salem, D. M. S. A., Ismail, M. M., & Aly-Eldeen, M. A. (2019). Biogenic synthesis and antimicrobial potency of iron oxide (Fe3O4) nanoparticles using algae harvested from the Mediterranean Sea, Egypt. *Egyptian Journal of Aquatic Research*, *45*(3), 197–204. doi:10.1016/j.ejar.2019.07.002

Salim, S., Bosma, R., Vermuë, M. H., & Wijffels, R. H. (2011). Harvesting of microalgae by bio-flocculation. *Journal of Applied Phycology*, *23*, 849–855.

Sallam, E. R., Khairy, H. M., Elnouby, M. S., & Fetouh, H. A. (2021). Sustainable electricity production from seawater using Spirulina platensis microbial fuel cell catalyzed by silver nanoparticles-activated carbon composite prepared by a new modified photolysis method. *Biomass and Bioenergy*, *148*, 106038. doi:10.1016/j.biombioe.2021.106038

Samanta, P., Jang, S., Shin, S., & Kim, J. K. (2019). Effects of pH on growth and biochemical responses in Agarophyton vermiculophyllum under different temperature conditions. *Journal of Applied Phycology, 32*(1), 499–509. doi:10.1007/s10811-019-01933-3

Samantaray, S., & Mallick, N. (2011). Production and characterization of poly-β-hydroxybutyrate (PHB) polymer from *Aulosira fertilissima. Journal of Applied Phycology*, *24*(4), 803–814. doi:10.100710811-011-9699-7

Samantaray, S., & Mallick, N. (2015). Impact of Various Stress Conditions on Poly-β-Hydroxybutyrate (PHB) Accumulation in Aulosira fertilissima CCC 444. *Current Biotechnology*, *4*(3), 366–372. doi:10.2174/2211550104666150806000642

Samantaray, S., Nayak, J. K., & Mallick, N. (2011). Wastewater utilization for poly-β-hydroxybutyrate production by the cyanobacterium Aulosira fertilissima in a recirculatory aquaculture system. *Applied and Environmental Microbiology*, *77*(24), 8735–8743. doi:10.1128/AEM.05275-11 PMID:21984242

Samarakoon, K. W., Ko, J. Y., Lee, J. H., Kwon, O. N., Kim, S. W., & Jeon, Y. J. (2014). Apoptotic anticancer activity of a novel fatty alcohol ester isolated from cultured marine diatom, Phaeodactylum tricornutum. *Journal of Functional Foods*, *6*, 231–240. doi:10.1016/j.jff.2013.10.011

Samarakoon, K., & Jeon, Y.-J. (2012). Bio-functionalities of proteins derived from marine algae — A review. *Food Research International*, *48*(2), 948–960.

Samarakoon, K., & Jeon, Y.-J. (2012). Bio-functionalities of proteins derived from marine algae–A review. *Food Research International*, *48*(2), 48. doi:10.1016/j.foodres.2012.03.013

Samorì, C., Abbondanzi, F., Galletti, P., Giorgini, L., Mazzocchetti, L., Torri, C., & Tagliavini, E. (2015). Extraction of polyhydroxyalkanoates from mixed microbial cultures: Impact on polymer quality and recovery. *Bioresource Technology*, *189*, 195–202.

Sánchez-Baracaldo, P., Bianchini, G., Wilson, J. D., & Knoll, A. H. (2022). Cyanobacteria and biogeochemical cycles through Earth history. *Trends in Microbiology*, *30*(2), 143–157. doi:10.1016/j.tim.2021.05.008 PMID:34229911

Sánchez, J. F., Fernández-Sevilla, J. M., Acién, F. G., Cerón, M. C., Pérez-Parra, J., & Molina-Grima, E. (2008). Biomass and lutein productivity of Scenedesmus almeriensis: Influence of irradiance, dilution rate and temperature. *Applied Microbiology and Biotechnology*, *79*(5), 719–729. doi:10.100700253-008-1494-2 PMID:18491039

Sánchez, M., Bernal-Castillo, J., Rozo, C., & Rodríguez, I. (2003). Spirulina (Arthrospira): An edible microorganism. A review. *Universitas Scientiarum*, *8*, 7–24.

Sancho, M. M., Castillo, J. J., & El Yousfi, F. (1997). Influence of phosphorus concentration on the growth kinetics and stoichiometry of the microalga Scenedesmus obliquus. *Process Biochemistry*, *32*(8), 657–664. doi:10.1016/S0032-9592(97)00017-4

Sandepogu, M. (2018). *Seaweed Extract and Humic Acids Biostimulant to improve Growth and Post-Harvest Quality of Spinach and Lettuce*. Univ.

Sander, K., & Murthy, G. S. (2010). Life cycle analysis of algae biodiesel. *The International Journal of Life Cycle Assessment*, *15*(7), 704–714. doi:10.100711367-010-0194-1

Sandnes, J. M., Källqvist, T., Wenner, D., & Gislerød, H. R. (2005). Combined influence of light and temperature on growth rates of Nannochloropsis oceanica: Linking cellular responses to large-scale biomass production. *Journal of Applied Phycology*, *17*(6), 515–525. doi:10.100710811-005-9002-x

Sangale, M. K., Shahnawaz, M., & Ade, A. B. (2019). Potential of fungi isolated from the dumping sites mangrove rhizosphere soil to degrade polythene. *Scientific Reports*, *9*(1), 5390. doi:10.103841598-019-41448-y PMID:30926843

Sangha, J. S., Kelloway, S., Critchley, A. T., & Prithiviraj, B. (2014). Seaweeds (Macroalgae) and their extracts as contributors of plant productivity and quality: The current status of our understanding. *Sea Plants*, *71*, 189–219. doi:10.1016/B978-0-12-408062-1.00007-X

Sanjeewa, K. K. A., Fernando, I. P. S., Samarakoon, K. W., Lakmal, H. H. C., Kim, E.-A., Kwon, O.-N., Dilshara, M. G., Lee, J.-B., & Jeon, Y.-J. (2016). Anti-inflammatory and anti-cancer activities of sterol rich fraction of cultured marine microalga Nannochloropsis oculata. *Algae - Korean Phycological Society*, *31*(3), 277–287. doi:10.4490/algae.2016.31.6.29

San, K. A., & Shon, Y. S. (2018). Synthesis of alkanethiolate-capped metal nanoparticles using alkyl thiosulfate ligand precursors: A method to generate promising reagents for selective catalysis. *Nanomaterials (Basel, Switzerland)*, *8*(5), 346. doi:10.3390/nano8050346 PMID:29783714

Sankaran, R., Show, P. L., Nagarajan, D., & Chang, J. S. (2018). Exploitation and Biorefinery of Microalgae. In *Waste Biorefinery: Potential and Perspectives* (pp. 571–601). Elsevier. doi:10.1016/B978-0-444-63992-9.00019-7

Sanniyasi, E., Gopal, R. K., Gunasekar, D. K., & Raj, P. P. (2021). Biodegradation of low-density polyethylene (LDPE) sheet by microalga, Uronema africanum Borge. *Scientific Reports*, *11*(1), 17233. doi:10.103841598-021-96315-6 PMID:34446729

Sano, Y. (1999). Antiviral Activity of Alginate against Infection by Tobacco Mosaic Virus. *Carbohydrate Polymers*, *38*, 183–186.

Sansone, C., Braca, A., Ercolesi, E., Romano, G., Palumbo, A., Casotti, R., Francone, M., & Ianora, A. (2014). Diatom-derived polyunsaturated aldehydes activate cell death in human cancer cell lines but not normal cells. *PLoS One*, *9*(7), e101220. doi:10.1371/journal.pone.0101220 PMID:24992192

Sansone, C., Nuzzo, G., Galasso, C., Casotti, R., Fontana, A., Romano, G., & Ianora, A. (2018). The Marine Dinoflagellate *Alexandrium andersoni* Induces Cell Death in Lung and Colorectal Tumor Cell Lines. *Marine Biotechnology (New York, N.Y.)*, *20*(3), 343–352. doi:10.100710126-018-9817-5 PMID:29679250

Santiago, A. F., Calijuri, M. L., Assemany, P. P., Calijuri, M. D. C., & Dos Reis, A. J. D. (2013). Algal biomass production and wastewater treatment in high rate algal ponds receiving disinfected effluent. *Environmental Technology (United Kingdom)*, *34*(13–14), 1877–1885. doi:10.1080/09593330.2013.812670 PMID:24350441

Santos, C. A., Caldeira, M. L., Lopes da Silva, T., Novais, J. M., & Reis, A. (2013). Enhanced lipidic algae biomass production using gas transfer from a fermentative Rhodosporidium toruloides culture to an autotrophic Chlorella protothecoides culture. *Bioresource Technology*, *138*, 48–54. doi:10.1016/j.biortech.2013.03.135 PMID:23612161

Santos, C. A., & Reis, A. (2014). Microalgal symbiosis in biotechnology. *Applied Microbiology and Biotechnology*, *98*(13), 5839–5846. doi:10.100700253-014-5764-x PMID:24816618

Santos-Sánchez, N. F., Salas-Coronado, R., Villanueva-Cañongo, C., & Hernández-Carlos, B. (2019). Antioxidant compounds and their antioxidant mechanism. *Antioxidants*, *10*, 1–29.

Sanyano, N., Chetpattananondh, P., & Chongkhong, S. (2013). Coagulation–flocculation of marine *Chlorella* sp. for biodiesel production. *Bioresource Technology*, *147*, 471–476.

Saravanan, A., Kumar, P. S., Govarthanan, M., George, C. S., Vaishnavi, S., Moulishwaran, B., & [illegible] (2021). Adsorption Characteristics of Magnetic Nanoparticles Coated Mixed Fungal Biomass for Toxic Cr(VI) Ions in Aquatic Environment. *Chemosphere*, *267*, 129226.

Sarayloo, E., Tardu, M., Unlu, Y. S., Simsek, S., Cevahir, G., Erkey, C., & Kavakli, I. H. (2017). Understanding lipid metabolism in high-lipid-producing *Chlorella vulgaris* mutants at the genome-wide level. *Algal Research*, *28*, 244–252. doi:10.1016/j.algal.2017.11.009

Sarmah, P., & Rout, J. (2018). Efficient biodegradation of low-density polyethylene by cyanobacteria isolated from submerged polyethylene surface in domestic sewage water. *Environmental Science and Pollution Research International*, *25*(33), 33508–33520. doi:10.100711356-018-3079-7 PMID:30267347

Sarmah, P., & Rout, J. (2019). Cyanobacterial degradation of low-density polyethylene (LDPE) by Nostoc carneum isolated from submerged polyethylene surface in domestic sewage water. *Energy, Ecology & Environment*, *4*(5), 240–252. doi:10.100740974-019-00133-6

Sasi, D., Mitra, P., Vigueras, A., & Hill, G. A. (2011). Growth kinetics and lipid production using Chlorella vulgaris in a circulating loop photobioreactor. *Journal of Chemical Technology and Biotechnology (Oxford, Oxfordshire)*, *86*(6), 875–880. doi:10.1002/jctb.2603

Sathasivam, R., Radhakrishnan, R., & Hashem, A., & Abd Allah, E. F. (2019). Microalgae metabolites: A rich source for food and medicine. *Saudi Journal of Biological Sciences*, *26*(4), 709–722. doi:10.1016/j.sjbs.2017.11.003 PMID:31048995

Sathiyaraj, G., Vinosha, M., Sangeetha, D., Manikandakrishnan, M., Palanisamy, S., Sonaimuthu, M., Manikandan, R., You, S. G., & Prabhu, N. M. (2021). Bio-directed synthesis of Pt-nanoparticles from aqueous extract of red algae Halymenia dilatata and their biomedical applications. *Colloids and Surfaces. A, Physicochemical and Engineering Aspects*, *618*, 126434. doi:10.1016/j.colsurfa.2021.126434

Satthong, S., Saego, K., Kitrungloadjanaporn, P., Nuttavut, N., Amornsamankul, S., & Triampo, W. (2019). Modeling the effects of light sources on the growth of algae. *Adv Differ Equ,* 170. doi:10.1186/s13662-019-2112-6

Sayre, R. (2010). Microalgae: The Potential for Carbon Capture. *Bioscience*, *60*(9), 722–727. doi:10.1525/bio.2010.60.9.9

Schagerl, M., & Burian, A. (2016). The Ecology of African Soda Lakes: Driven by Variable and Extreme Conditions. In Soda Lakes of East Africa (pp. 295-320). Springer. doi:10.1007/978-3-319-28622-8_12

Schatz, D., & Vardi, A. (2018). Extracellular vesicles—New players in cell–cell communication in aquatic environments. *Current Opinion in Microbiology*, *43*, 148–154. doi:10.1016/j.mib.2018.01.014 PMID:29448174

Schenk, P. M., Thomas-Hall, S. R., Stephens, E., Marx, U. C., Mussgnug, J. H., Posten, C., Kruse, O., & Hankamer, B. (2008). Second generation biofuels: High-efficiency microalgae for biodiesel production. *BioEnergy Research*, *1*(1), 20–43. doi:10.100712155-008-9008-8

Scheurer, M., & Bigalke, M. (2018). Microplastics in Swiss Floodplain Soils. *Environmental Science & Technology*, *52*(6), 3591–3598. doi:10.1021/acs.est.7b06003 PMID:29446629

Schiener, P., Black, K. D., Stanley, M. S., & Green, D. H. (2015). The seasonal variation in the chemical composition of the kelp species *Laminaria digitata, Laminaria hyperborea, Saccharina latissima* and *Alaria esculenta. Journal of Applied Phycology*, *27*(1), 363–373. doi:10.100710811-014-0327-1

Schierenbeck, L., Ries, D., Rogge, K., Grewe, S., Weisshaar, B., & Kruse, O. (2015). Fast forward genetics to identify mutations causing a high light tolerant phenotype in *Chlamydomonas reinhardtii* by whole-genome-sequencing. *BMC Genomics, 16*(1), 57, 1-15. . doi:10.1186/s12864-015-1232-y

Schirrmeister, B. E., Gugger, M., & Donoghue, P. C. J. (2015). Cyanobacteria and the Great Oxidation Event: Evidence from genes and fossils. *Palaeontology*, *58*(5), 769–785. doi:10.1111/pala.12178 PMID:26924853

Schöfer, H. (2018). Cutaneous infection – would you think of algae? *Journal of the European Academy of Dermatology and Venereology*, *32*(9), 1401–1402. doi:10.1111/jdv.15202 PMID:30136764

Schussler, A., Bonfante, P., Schnepf, E., Mollenhauer, D., & Kluge, M. (1996). Characterization of the Geosiphon pyriforme symbiosome by affinity techniques: Confocal laser scanning microscopy (CLSM) and electron microscopy. *Protoplasma*, *190*(1/2), 53–67. doi:10.1007/BF01281194

Schwarz, A.-M., & Markager, S. (1999). Light absorption and photosynthesis of a benthic moss community: Importance of spectral quality of light and implications of changing light attenuation in the water column. *Freshwater Biology*, *42*, 609–623. doi:10.1046/j.1365-2427.1999.00496.x

Schwencke, J., & Carú, M. (2001). Advances in actinorhizal symbiosis: Host plant-Frankia interactions, biology, and applications in arid land reclamation. A review. *Arid Land Research and Management*, *15*(4), 285–327. doi:10.1080/153249801753127615

Schwenzfeier, A., Wierenga, P. A., & Gruppen, H. (2011). Isolation and Characterization of Soluble Protein from the Green Microalgae Tetraselmis Sp. *Bioresource Technology*, *102*(19), 9121–9127. doi:10.1016/j.biortech.2011.07.046 PMID:21831634

Science for Environment Policy. (2011). *Plastic waste: ecological and human health impacts. DG environment news alert service*. European Commission. Available at https://ec.europa.eu/environment/integration/research/newsalert/pdf/IR1_en.pdf

Scognamiglio, V., Giardi, M. T., Zappi, D., Touloupakis, E., & Antonacci, A. (2021). Photoautotrophs–Bacteria Co-Cultures: Advances, Challenges and Applications. *Materials (Basel)*, *14*(11), 3027. doi:10.3390/ma14113027 PMID:34199583

Scott, S. A., Davey, M. P., Dennis, J. S., Horst, I., Howe, C. J., Lea-Smith, D. J., & Smith, A. G. (2010). Biodiesel from algae: Challenges and prospects. *Current Opinion in Biotechnology*, *21*, 277–286.

Segev, E., Wyche, T. P., Kim, K. H., Petersen, J., Ellebrandt, C., Vlamakis, H., ... Kolter, R. (2016). Dynamic metabolic exchange governs a marine algal-bacterial interaction. *elife, 5*, e17473.

Segev, E., Wyche, T. P., Kim, K. H., Petersen, J., Ellebrandt, C., Vlamakis, H., Barteneva, N., Paulson, J. N., Chai, L., Clardy, J., & Kolter, R. (2016). Dynamic metabolic exchange governs a marine algal-bacterial interaction. *eLife*, *5*, e17473. doi:10.7554/eLife.17473 PMID:27855786

Sekiguchi, T., Saika, A., Nomura, K., Watanabe, T., Watanabe, T., Fujimoto, Y., Enoki, M., Sato, T., Kato, C., & Kanehiro, H. (2011). Biodegradation of aliphatic polyesters soaked in deep seawaters and isolation of poly(ε-caprolactone)-degrading bacteria. *Polymer Degradation & Stability*, *96*(7), 1397–1403. doi:10.1016/j.polymdegradstab.2011.03.004

Selesu, N. F., de Oliveira, V. T., Corrêa, D. O., Miyawaki, B., Mariano, A. B., Vargas, J. V., & Vieira, R. B. (2016). Maximum microalgae biomass harvesting via flocculation in large scale photobioreactor cultivation. *Canadian Journal of Chemical Engineering*, *94*, 304–309.

Selvam, G. G., & Sivakumar, K. (2015). Phycosyn- thesis of silver nanoparticles and photocatalytic deg- radation of methyl orange dye using silver (Ag) nano- particles synthesized from Hypnea musciformis (Wulfen) J.V. Lamouroux. *Applied Nanoscience, 5*, 617–622.

Semary, N. A. E. (2017). Algae and Chain Aquaculture: An Approach Towards Sustainable Agriculture. Handbook of Environmental Chemistry, 5(Part N), 1-12.

Semesi, I. S., Kangwe, J., & Björk, M. (2009). Alterations in seawater pH and CO2 affect calcification and photosynthesis in the tropical coralline alga, Hydrolithon sp.(Rhodophyta). *Estuarine, Coastal and Shelf Science*, *84*(3), 337–341. doi:10.1016/j.ecss.2009.03.038

Sengmee, D., Cheirsilp, B., Suksaroge, T. T., & Prasertsan, P. (2017). Biophotolysis-based hydrogen and lipid production by oleaginous microalgae using crude glycerol as exogenous carbon source. *International Journal of Hydrogen Energy*, *42*(4), 1970–1976. doi:10.1016/j.ijhydene.2016.10.089

Sen, S., & Puskas, J. E. (2015). Green polymer chemistry: Enzyme catalysis for polymer functionalization. *Molecules (Basel, Switzerland)*, *20*(5), 9358–9379. doi:10.3390/molecules20059358 PMID:26007188

Senthilkumar, P., & Sudha, S. (2012). *Antioxidant and antibacterial properties of methanolic extract of green seaweed Chaetomorpha linum from Gulf of Mannar: Southeast coast of India*. Academic Press.

Seo, J. Y., Lee, K., Praveenkumar, R., Kim, B., Lee, S. Y., Oh, Y.-K., & Park, S. B. (2015). Tri-functionality of Fe_3O_4-embedded carbon microparticles in microalgae harvesting. *Chemical Engineering Journal*, *2015*, 206–214.

Serrà, A., Artal, R., García-Amorós, J., Gómez, E., & Philippe, L. (2020). Circular zero-residue process using microalgae for efficient water decontamination, biofuel production, and carbon dioxide fixation. *Chemical Engineering Journal*, *388*, 124278. doi:10.1016/j.cej.2020.124278

Setyorini, A. I., & Dianursanti. (2021). Isolation and optimization of polyhydroxybutyrate from Chlorella vulgaris using NaClO and ultrasonication aided chemical pretreatment methods. *The 5th biomedical engineering's recent progress in biomaterials, drugs development, and medical devices: Proceedings of the 5th International Symposium of Biomedical Engineering (ISBE) 2020*. doi:10.1063/5.0047562

Seyedsayamdost, M. R., Case, R. J., Kolter, R., & Clardy, J. (2011). The Jekyll-and-Hyde chemistry of Phaeobacter gallaeciensis. *Nature Chemistry*, *3*(4), 331–335. doi:10.1038/nchem.1002 PMID:21430694

Seymour, J. R., & Raina, J. B. (2018). Swimming in the sea: Chemotaxis by marine bacteria. *Microbiology Australia*, *3*(1), 12–16. doi:10.1071/MA18005

Seymour, J. R., Simo, R., Ahmed, T., & Stocker, R. (2010). Chemoattraction to dimethylsulfoniopropionate throughout the marine microbial food web. *Science*, *329*(5989), 342–345. doi:10.1126cience.1188418 PMID:20647471

Shabani, N., Sowlati, T., Ouhimmou, M., & Rönnqvist, M. (2014). Tactical supply chain planning for a forest biomass power plant under supply uncertainty. *Energy*, *78*, 346–355. doi:10.1016/j.energy.2014.10.019

Shah, A. R., Ahmad, A., Srivastava, S., & Jaffar Ali, B. M. (2017). Reconstruction and analysis of a genome-scale metabolic model of *Nannochloropsis gaditana. Algal Research*, *26*, 354–364. doi:10.1016/j.algal.2017.08.014

Shahid, A., Malik, S., Zhu, H., Xu, J., Nawaz, M. Z., Nawaz, S., Asraful Alam, M., & Mehmood, M. A. (2020). Cultivating microalgae in wastewater for biomass production, pollutant removal, and atmospheric carbon mitigation; a review. *The Science of the Total Environment*, *704*, 135303. doi:10.1016/j.scitotenv.2019.135303 PMID:31818584

Shahidi, F. (2009). *Marine nutraceuticals and functional foods*. CRC Press.

Shah, M. T., Zodape, S. T., Chaudhary, D. R., Eswaran, K., & Chikara, J. (2013). Seaweed sap as an alternative liquid fertilizer for yield and quality improvement of wheat. *Journal of Plant Nutrition*, *36*(2), 192–200. doi:10.1080/01904167.2012.737886

Shahnawaz, M., Sangale, M. K., & Ade, A. B. (2016). Rhizosphere of Avicennia marina (Forsk.) Vierh. as a landmark for polythene degrading bacteria. *Environmental Science and Pollution Research International*, *23*(14), 14621–14635. doi:10.100711356-016-6542-3 PMID:27072028

Shaish, A., Ben-Amotz, A., & Avron, M. (1992). Biosynthesis of β-carotene in Dunaliella. *Methods in Enzymology, 213*, 439–444. doi:10.1016/0076-6879(92)13145-N

Sha, J., Xiong, H., Li, C., Lu, Z., Zhang, J., Zhong, H., Zhang, W., & Yan, B. (2021). 2021/07/01/). Harmful algal blooms and their eco-environmental indication. *Chemosphere, 274*, 129912. doi:10.1016/j.chemosphere.2021.129912 PMID:33979937

Shamriz, S., & Ofoghi, H. (2018). Engineering the chloroplast of Chlamydomonas reinhardtii to express the recombinant PfCelTOS-Il2 antigen-adjuvant fusion protein. *Journal of Biotechnology, 266*, 111–117. doi:10.1016/j.jbiotec.2017.12.015 PMID:29269249

Shanab, S. M., Mostafa, S. S., Shalaby, E. A., & Mahmoud, G. I. (2012). Aqueous extracts of microalgae exhibit anti-oxidant and anticancer activities. *Asian Pacific Journal of Tropical Biomedicine, 2*(8), 608–615. doi:10.1016/S2221-1691(12)60106-3 PMID:23569980

Shanmugam, Ganesan, & Rajauria. (2021). Extraction Technologies to Recover Dietary Polyphenols from Macro-and Microalgae" In Recent Advances in Micro and Macroalgal Processing: Food and Health Perspectives. Wiley Online Library.

Shannon, E., & Abu-Ghannam, N. (2016). Antibacterial derivatives of marine algae: An overview of pharmacological mechanisms and applications. *Marine Drugs, 14*(4), 81. doi:10.3390/md14040081 PMID:27110798

Shariatmadari, Z., Riahi, H., Seyed Hashtroudi, M., Ghassempour, A., & Aghashariatmadary, Z. (2013). Plant growth promoting cyanobacteria and their distribution in terrestrial habitats of Iran. *Soil Science and Plant Nutrition, 59*(4), 535–547. doi:10.1080/00380768.2013.782253

Sharma, A., Sharma, S., Sharma, K., Chetri, S. P. K., Vashishtha, A., Singh, P., Kumar, R., Rathi, B., & Agrawal, V. (2016). Algae as crucial organisms in advancing nanotechnology: a systematic review. In Journal of Applied Phycology (Vol. 28, Issue 3, pp. 1759–1774). Springer Netherlands. doi:10.100710811-015-0715-1

Sharma, V., Sehgal, R., & Gupta, R. (2021). Polyhydroxyalkanoate (PHA): Properties and Modifications. *Polymer, 212*(October), 123161. doi:10.1016/j.polymer.2020.123161

Sharma, D., Kanchi, S., & Bisetty, K. (2019). Biogenic synthesis of nanoparticles: A review. *Arabian Journal of Chemistry, 12*(8), 3576–3600. doi:10.1016/j.arabjc.2015.11.002

Sharma, G., Pandey, S., Ghatak, S., Watal, G., & Rai, P. K. (2018). Potential of spectroscopic techniques in the characterization of "green nanomaterials". In *Nanomaterials in Plants, Algae, and Microorganisms* (pp. 59–77). Elsevier.

Sharma, L., & Mallick, N. (2005). Accumulation of poly-β-hydroxybutyrate in *Nostoc muscorum*: Regulation by pH, light–dark cycles, N and P status and carbon sources. *Bioresource Technology, 96*(11), 1304–1310. doi:10.1016/j.biortech.2004.10.009 PMID:15734319

Sharma, M., Dubey, A., & Pareek, A. (2014). Algal flora on degrading polythene waste. *CIB Tech Journal of Microbiology, 3*, 43–47.

Sharma, N. K., Rai, A. K., Singh, S., & Brown, R. M. Jr. (2007). Airborne algae: Their present status and relevance. *Journal of Phycology, 43*(4), 615–627. doi:10.1111/j.1529-8817.2007.00373.x

Sharma, N. K., Rai, A. K., & Stal, L. J. (2013). *Cyanobacteria: an economic perspective*. John Wiley & Sons.

Sharma, P., & Sharma, N. (2017). Industrial and biotechnological applications of algae: A review. *J. Adv. Plant Biol., 1*(1), 1–25. doi:10.14302/issn.2638-4469.japb-17-1534

Sharma, R., Khokhar, M. K., Jat, R. L., & Khandelwal, S. K. (2012). Role of algae and cyanobacteria in sustainable agriculture system. *Wudpecker Journal of Agricultural Research*, *1*(9), 381–388.

Sharma, R., Sharma, M., Sharma, R., & Sharma, V. (2013). The impact of incinerators on human health and environment. *Reviews on Environmental Health*, *28*(1). Advance online publication. doi:10.1515/reveh-2012-0035 PMID:23612530

Sheih, I. C., Wu, T. K., & Fang, T. J. (2009, July). Antioxidant properties of a new antioxidative peptide from algae protein waste hydrolysate in different oxidation systems. *Bioresource Technology*, *100*(13), 3419–3425. doi:10.1016/j.biortech.2009.02.014 PMID:19299123

Sheikhzadeh, N., Tayefi-Nasrabadi, H., Khani Oushani, A., & Najafi Enferadi, M. H. (2012). Effects of Haematococcus pluvialis Supplementation on Antioxidant System and Metabolism in Rainbow Trout (Oncorhynchus mykiss). *Fish Physiology and Biochemistry*, *38*(1), 413–419. doi:10.100710695-011-9519-7 PMID:21695482

Shelef, G., Sukenik, A., & Green, M. (1984). *Microalgae harvesting and processing: a literature review*. Solar Energy Res Inst Golden.

Shen, H., & Forssberg, E. (2003). An overview of recovery of metals from slags. *Waste Management (New York, N.Y.)*, *23*(10), 933–949. doi:10.1016/S0956-053X(02)00164-2 PMID:14614927

Shen, L., Wang, J., Li, Z., Fan, L., Chen, R., Wu, X., Li, J., & Zeng, W. (2020). A high-efficiency Fe_2O_3@ microalgae composite for heavy metal removal from aqueous solution. *Journal of Water Process Engineering*, *33*, 101026.

Shen, M. Y., & Sali, A. (2006). Statistical potential for assessment and prediction of protein structures. *Protein Science*, *15*(11), 2507–2524. doi:10.1110/ps.062416606 PMID:17075131

Shen, Y., Cui, Y., & Yuan, W. (2013). Flocculation optimization of microalga *Nannochloropsis oculata. Applied Biochemistry and Biotechnology*, *169*, 2049–2063.

Shen, Y., Fan, Z., Chen, C., & Xu, X. (2015). An auto-flocculation strategy for Chlorella vulgaris. *Biotechnology Letters*, *37*(1), 75–80.

Shen, Y., Yuan, W., Pei, Z. J., Wu, Q., & Mao, E. (2009). Microalgae Mass Production Methods. *Transactions of the ASABE*, *52*(4), 1275–1287. doi:10.13031/2013.27771

Sher, I., Tzameret, A., Peri-Chen, S., Edelshtain, V., Ioffe, M., Sayer, A., Buzhansky, L., Gazit, E., & Rotenstreich, Y. (2018). Synthetic 9-cis-beta-carotene inhibits photoreceptor degeneration in cultures of eye cups from rpe65rd12 mouse model of retinoid cycle defect. *Scientific Reports*, *2018*(8), 6130. doi:10.103841598-018-24439-3 PMID:29666392

Shi, B., Liang, L., Yang, H., Zhang, L., & He, F. (2017). Glycerol-plasticized spirulina–poly(vinyl alcohol) films with improved mechanical performance. *Journal of Applied Polymer Science*, *134*(20), 1–10. doi:10.1002/app.44842

Shi, B., Wideman, G., & Wang, J. H. (2012). A new approach of bioCO 2 fixation by thermoplastic processing of microalgae. *Journal of Polymers and the Environment*, *20*(1), 124–131. doi:10.100710924-011-0329-x

Shibata, T., Ishimaru, K., Kawaguchi, S., Yoshikawa, H., & Hama, Y. (2007). Antioxidant activities of phlorotannins isolated from Japanese Laminariaceae. In *Nineteenth International Seaweed Symposium* (pp. 255-261). Springer.

Shields, R., & Lupatsch, I. (2012). Algae for aquaculture and animal feeds. In Microalgal Biotechnology: Integration and Economy. De Gruyter. doi:10.1515/9783110298321.79

Shields, L. M., & Durrell, L. W. (1964). Algal in relation to soil fertility. *Botanical Review*, *30*(1), 92–128. doi:10.1007/BF02858614

Shikata, T., Nukata, A., Yoshikawa, S., Matsubara, T., Yamasaki, Y., Shimasaki, Y., Oshima, Y., & Honjo, T. (2009). Effects of light quality on initiation and development of meroplanktonic diatom blooms in a eutrophic shallow sea. *Marine Biology*, *156*, 875–889.

Shikuku, V. O. (2020). *Effects of Emerging Chemical Contaminants on Water Resources and Environmental Health*. IGI Global Publishers.

Shikuku, V. O., & Mishra, T. (2021). Adsorption isotherm modeling for methylene blue removal onto magnetic kaolinite clay: A comparison of two-parameter isotherms. *Applied Water Science*, *11*, 103.

Shikuku, V. O., Zanella, R., Kowenje, C. O., Donato, F. F., Bandeira, N., & Prestes, O. D. (2018). Single and binary adsorption of sulfonamide antibiotics onto iron-modified clay: Linear and nonlinear isotherms, kinetics, thermodynamics, and mechanistic studies. *Applied Water Science*, *8*, 175.

Shimamatsu, H. (2004). Mass production of *Spirulina*, an edible microalga. *Hydrobiologia*, *512*(1-3), 39–44. doi:10.1023/B:HYDR.0000020364.23796.04

Shima, S., Matsuoka, H., Iwamoto, T., & Sakai, H. (1984). Antimicrobial action of -poly-l-lysine. *The Journal of Antibiotics*, *37*, 1449–1455.

Shin, S. G., Lee, S., Lee, C., Hwang, K., & Hwang, S. (2010). Qualitative and quantitative assessment of microbial community in batch anaerobic digestion of secondary sludge. *Bioresource Technology*, *101*(24), 9461–9470. doi:10.1016/j.biortech.2010.07.081 PMID:20705457

Shin, Y. S., Jeong, J., Nguyen, T. H. T., Kim, J. Y. H., Jin, E., & Sim, S. J. (2019). Targeted knockout of phospholipase A(2) to increase lipid productivity in *Chlamydomonas reinhardtii* for biodiesel production. *Bioresource Technology*, *271*, 368–374. doi:10.1016/j.biortech.2018.09.121 PMID:30293032

Shiny, P. J., Mukherjee, A., & Chandrasekaran, N. (2013). Marine algae mediated synthesis of the silver nanoparticles and its antibacterial efficiency. *International Journal of Pharmacy and Pharmaceutical Sciences*, *5*, 239–241.

Shi, X., Jung, K. W., Kim, H., Ahn, Y. T., & Shin, H. S. (2011). Direct fermentation of *Laminaria japonica* for biohydrogen production by anaerobic mixed cultures. *International Journal of Hydrogen Energy*, *36*(10), 5857–586. doi:10.1016/j.ijhydene.2011.01.125

Shi, X., Liu, L., Li, Y., Xiao, Y., Ding, G., Lin, S., & Chen, J. (2018). Isolation of an algicidal bacterium and its effects against the harmful-algal- bloom dinoflagellate Prorocentrum donghaiense (Dinophyceae). *Harmful Algae*, *80*(September), 72–79. doi:10.1016/j.hal.2018.09.003 PMID:30502814

Shou, Q., Guo, C., Yang, L., Jia, L., Liu, C., & Liu, H. (2011). Effect of pH on the single-step synthesis of gold nanoparticles using PEO–PPO–PEO triblock copolymers in aqueous media. *Journal of Colloid and Interface Science*, *363*(2), 481–489. doi:10.1016/j.jcis.2011.07.021 PMID:21855892

Show, K.-Y. Y., Lee, D.-J. J., & Chang, J.-S. S. (2013). Algal biomass dehydration. *Bioresource Technology*, *135*, 720–729. doi:10.1016/j.biortech.2012.08.021 PMID:22939595

Show, K.-Y., Lee, D.-J., Tay, J.-H., Lee, T.-M., & Chang, J.-S. (2015). Microalgal drying and cell disruption – Recent advances. *Bioresource Technology*, *184*, 258–266. doi:10.1016/j.biortech.2014.10.139 PMID:25465783

Show, K.-Y., Yan, Y., Ling, M., Ye, G., Li, T., & Lee, D.-J. (2018). Hydrogen production from algal biomass – Advances, challenges and prospects. *Bioresource Technology*, *257*, 290–300. doi:10.1016/j.biortech.2018.02.105 PMID:29506887

Shukla, M., & Kumar, S. (2018). Algal growth in photosynthetic algal microbial fuel cell and its subsequent utilization for biofuels. *Renewable & Sustainable Energy Reviews*, *82*, 402–414. doi:10.1016/j.rser.2017.09.067

Shumway, S. E., Burkholder, J. M., & Morton, S. L. (Eds.). (2018). *Harmful Algal Blooms: A Compendium Desk Reference*. John Wiley & Sons. doi:10.1002/9781118994672

Shunni, Z., Yajie, W., Wei, H., Jin, X., Zhongming, W., Jingliang, X., & Zhenhong, Y. (2014). Enhanced accumulation of carbohydrate and starch in *Chlorella zofingiensis* induced by nitrogen starvation. *Applied Biochemistry and Biotechnology*, *174*(7), 2435–2445. doi:10.100712010-014-1183-9 PMID:25209553

Shurin, J. B., Burkart, M. D., Mayfield, S. P., & Smith, V. H. (2016). Recent progress and future challenges in algal biofuel production. *F1000 Research*, *5*, 5. doi:10.12688/f1000research.9217.1 PMID:27781084

Shu, W.-S., & Huang, L.-N. (2022). 2022/04/01). Microbial diversity in extreme environments. *Nature Reviews. Microbiology*, *20*(4), 219–235. doi:10.103841579-021-00648-y PMID:34754082

Sial, A., Zhang, B., Zhang, A., Liu, K., Imtiaz, S. A., & Yashir, N. (2020). Microalgal–Bacterial Synergistic Interactions and Their Potential Influence in Wastewater Treatment: A Review. *BioEnergy Research*. Advance online publication. doi:10.100712155-020-10213-9

Sibbald, S. J., & Archibald, J. M. (2020). Genomic Insights into Plastid Evolution. *Genome Biology and Evolution*, *12*(7), 978–990. doi:10.1093/gbe/evaa096 PMID:32402068

Siddiqua, S., Al Mamun, A., & Enayetul Babar, S. M. (2015). Production of biodiesel from coastal macroalgae (Chara vulgaris) and optimization of process parameters using Box-Behnken design. *SpringerPlus*, *4*(1), 1–11. doi:10.118640064-015-1518-1 PMID:26636008

Siddiqui, A., Wei, Z., Boehm, M., & Ahmad, N. (2020). Engineering microalgae through chloroplast transformation to produce high-value industrial products. *Biotechnology and Applied Biochemistry*, *67*(1), 30–40. doi:10.1002/bab.1823 PMID:31538673

Sidek, I. S., Draman, S. F. S., Abdullah, S. R. S., & Anuar, N. (2019). Current development on bioplastics and its future prospects: An introductory review. *INWASCON Technology Magazine*, 3–8. doi:10.26480/itechmag.01.2019.03.08

Siegenthaler, U., & Sarmiento, J. L. (1993). Atmospheric carbon dioxide and the ocean. *Nature*, *365*(6442), 119–259. doi:10.1038/365119a0

Sigamani, S., Chinnasamy, R., Dharmaraj, R. K., Ramamurthy, D., Devarajan, N., Narayanasamy, M., & Natarajan, H. (2020). Larvicidal potency of the extracts from Chlorella sp. against Aedes aegypti. *Biocatalysis and Agricultural Biotechnology*, *27*, 101663. doi:10.1016/j.bcab.2020.101663

Sikorski, L. (2021). Effects of Sodium Chloride on Algae and Crustaceans—The Neighbouring Links of the Water Trophic Chain. *Water (Basel)*, *13*(18), 2493. doi:10.3390/w13182493

Silambarasan, S., Logeswari, P., Sivaramakrishnan, R., Incharoensakdi, A., Cornejo, P., Kamaraj, B., & Chi, N. T. L. (2021). Removal of nutrients from domestic wastewater by microalgae coupled to lipid augmentation for biodiesel production and influence of deoiled algal biomass as biofertilizer for Solanum lycopersicum cultivation. *Chemosphere*, *268*, 129323. doi:10.1016/j.chemosphere.2020.129323 PMID:33359999

Sili, C., Torzillo, G., & Vonshak, A. (2012). Arthrospira (Spirulina). In B. A. Whitton (Ed.), *Ecology of Cyanobacteria II: Their Diversity in Space and Time* (pp. 677–705). Springer Netherlands. doi:10.1007/978-94-007-3855-3_25

Silva, A. G., Carter, R., Merss, F. L. M., Corrêa, D. O., Vargas, J. V. C., Mariano, A. B., Ordonez, J. C., & Scherer, M. D. (2015). Life cycle assessment of biomass production in microalgae compact photobioreactors. *Global Change Biology. Bioenergy*, *7*(2), 184–194. doi:10.1111/gcbb.12120

Silva-Benavides, A. M., & Torzillo, G. (2012). Nitrogen and phosphorus removal through laboratory batch cultures of microalga Chlorella vulgaris and cyanobacterium Planktothrix isothrix grown as monoalgal and as co-cultures. *Journal of Applied Phycology*, *24*(2), 267–276. doi:10.100710811-011-9675-2

Silva, T. H., Alves, A., Popa, E. G., Reys, L. L., Gomes, M. E., Sousa, R. A., Silva, S. S., Mano, J. F., & Reis, R. L. (2012). Marine algae sulfated polysaccharides for tissue engineering and drug delivery approaches. *Biomatter*, *2*(4), 278–289. doi:10.4161/biom.22947 PMID:23507892

Simkus, A., Simkiene, A., Cernauskiene, J., Kvietkute, N., Cernauskas, A., Paleckaitis, M., & Kerziene, S. (2013). The effect of blue algae Spirulina platensis on pig growth performance and carcass and meat quality. *Vet. Med. Zoot.*, *61*, 83.

Simonic, M., & Zemljic, F. (2021). Production of bioplastic material from algal biomass. *Chemical Industry & Chemical Engineering Quarterly*, *27*(1), 79–84. doi:10.2298/CICEQ191024026S

Simon, J., Kósa, A., Bóka, K., Vági, P., Simon-Sarkadi, L., Mednyánszky, Z., Horváth, Á. N., Nyitrai, P., Böddi, B., & Preininger, É. (2017). Self-supporting artificial system of the green alga *Chlamydomonas reinhardtii* and the ascomycetous fungus *Alternaria infectoria*. *Symbiosis*, *71*(3), 199–209. doi:10.100713199-016-0430-y

Simpson, M. (2010). Species and Conservation in Systematics. In M. Simpson (Ed.), *Plant Systematics* (pp. 649–668). Academic Press. doi:10.1016/B978-0-12-374380-0.50019-1

Sims, R. E., Mabee, W., Saddler, J. N., & Taylor, M. (2010). An overview of second generation biofuel technologies. *Bioresource Technology*, *101*(6), 1570–1580. doi:10.1016/j.biortech.2009.11.046 PMID:19963372

Singaravelu, G., Arockiamary, J., Kumar, V. G., & Govindaraju, K. (2007). Anovel extracellular synthesis of monodisperse gold nanoparticles using marine alga. *Colloids and Surfaces. B, Biointerfaces*, *57*, 97–101. doi:10.1016/j.colsurfb.2007.01.010 PMID:17350236

Singh Bhan, U. (2013). Microalgae: A promising tool for carbon sequestration. *Mitigation and Adaptation Strategies for Global Change*, *18*(1), 73–95. doi:10.100711027-012-9393-3

Singh, A. (2021). Role of Algae in Soil Nitrogen Fixation. In C. Cruz, K. Vishwakarma, D. K. Choudhary, & A. Varma (Eds.), *Soil Nitrogen Ecology* (pp. 483–496). Springer International Publishing. doi:10.1007/978-3-030-71206-8_24

Singh, A. K., Sharma, L., Mallick, N., & Mala, J. (2017). Progress and challenges in producing polyhydroxyalkanoate biopolymers from cyanobacteria. *Journal of Applied Phycology*, *29*(3), 1213–1232. doi:10.100710811-016-1006-1

Singh, A. V., Bandi, M., Raje, N., Richardson, P., Palladino, M. A., Chauhan, D., & Anderson, K. C. (2011). A novel vascular disrupting agent plinabulin triggers JNK-mediated apoptosis and inhibits angiogenesis in multiple myeloma cells. *Blood*, *117*(21), 5692–5700. doi:10.1182/blood-2010-12-323857 PMID:21454451

Singh, A., Nigam, P. S., & Murphy, J. D. (2011). Mechanism and challenges in commercialisation of algal biofuels. *Bioresource Technology*, *102*, 26–34.

Singh, B., & Sharma, N. (2008). Mechanistic implications of plastic degradation. *Polymer Degradation & Stability*, *93*(3), 561–584. doi:10.1016/j.polymdegradstab.2007.11.008

Singh, H., Rout, S., & Das, D. (2022). Dark fermentative biohydrogen production using pretreated Scenedesmus obliquus biomass under an integrated paradigm of biorefinery. *International Journal of Hydrogen Energy*, *47*(1), 102–116. doi:10.1016/j.ijhydene.2021.10.018

Singh, J. S., Kumar, A., Rai, A. N., & Singh, D. P. (2016). Cyanobacteria: A precious bio-resource in agriculture, ecosystem, and environmental sustainability. *Frontiers in Microbiology*, *7*, 529. doi:10.3389/fmicb.2016.00529 PMID:27148218

Singh, L., & Wahid, Z. A. (2015). Methods for enhancing bio-hydrogen production from biological process: A review. *Journal of Industrial and Engineering Chemistry*, *21*, 70–80. doi:10.1016/j.jiec.2014.05.035

Singh, M., Kalaivani, R., Manikandan, S., Sangeetha, N., & Kumaraguru, A. K. (2013). Facile green synthesis of variable metallic gold nanoparticle using Padina gym- nospora, a brown marine macroalga. *Applied Nanoscience*, *3*(2), 145–151. doi:10.100713204-012-0115-7

Singh, N. K., & Dhar, D. W. (2007). Nitrogen and phosphorous scavenging potential in microalgae. *Indian Journal of Biotechnology*, *6*, 52–56.

Singh, R. N. (1941). On some phases in the life history of the terrestrial alga, Fritschiella tuberosa Iyeng., and its autecology. *The New Phytologist*, *40*(3), 170–182. doi:10.1111/j.1469-8137.1941.tb07041.x

Singh, S., Bhushan, K. N., & Banerjee, U. C. (2005). Bioactive Compounds from Cyanobacteria and Microalgae: An Overview. *Critical Reviews in Biotechnology*, *25*(3), 73–95. doi:10.1080/07388550500248498 PMID:16294828

Singh, S., Kumar, V., Datta, S., Dhanjal, D. S., Sharma, K., & Samuel, J. (2020). Current Advancement and Future prospect of Biosorbents for Bioremediation. *The Science of the Total Environment*, *709*, 135895.

Sinha, R. P. (2015). Biochemical characterization of sunscreening mycosporine-like amino acids from two Nostoc species inhabiting diverse habitats. *Protoplasma*, *252*(1), 199–208. doi:10.100700709-014-0674-4 PMID:24997122

Sinha, R. P., & Hader, D. P. (1996). Photobiology and ecophysiology of rice field cyanobacteria. *Photochemistry and Photobiology*, *64*(6), 887–896. doi:10.1111/j.1751-1097.1996.tb01852.x

Sinning, I., Michel, H., Mathis, P., & Rutherford, A. W. (1989). Characterization of four herbicide-resistant mutants of Rhodopseudomonas viridis by genetic analysis, electron paramagnetic resonance, and optical spectroscopy. *Biochemistry*, *28*(13), 5544–5553. doi:10.1021/bi00439a031 PMID:2550055

Siripornadulsil, S., Dabrowski, K., & Sayre, R. (2007). Microalgal Vaccines. In R. León, A. Galván, & E. Fernández (Eds.), *Transgenic microalgae as green cell factories* (pp. 122–128). Springer New York. doi:10.1007/978-0-387-75532-8_11

Sitnik, K. M., Musatenko, L. I., Vasyuk, V. A., Vedenicheva, N. P., Generalova, V. M., Martin, G. G., & Nesterova, A. N. (2003). Gormonal'nii kompleks roslin i gribiv [Hormonal Complex in Plants and Fungi]. Kiiv: Akademperiodika.

Sittijunda, S., Reungsang, A., & Sompong, O. (2010). Biohydrogen production from dual digestion pretreatment of poultry slaughterhouse sludge by anaerobic self-fermentation. *International Journal of Hydrogen Energy*, *35*(24), 13427–13434. doi:10.1016/j.ijhydene.2009.11.116

Sivan, A., Szanto, M., & Pavlov, V. (2006). Biofilm development of the polyethylene-degrading bacterium Rhodococcus ruber. *Applied Microbiology and Biotechnology*, *72*(2), 346–352. doi:10.100700253-005-0259-4 PMID:16534612

Sizova, I., Greiner, A., Awasthi, M., Kateriya, S., & Hegemann, P. (2013). Nuclear gene targeting in *Chlamydomonas* using engineered zinc-finger nucleases. *The Plant Journal*, *73*(5), 873–882. doi:10.1111/tpj.12066 PMID:23137232

Skonieczny, M. T., & Yargeau, V. (2009). Biohydrogen production from wastewater by Clostridium beijerinckii: Effect of pH and substrate concentration. *International Journal of Hydrogen Energy*, *34*(8), 3288–3294. doi:10.1016/j.ijhydene.2009.01.044

Slade, R., & Bauen, A. (2013). Micro-algae cultivation for biofuels: Cost, energy balance, environmental impacts and future prospects. *Biomass and Bioenergy*, *53*, 29–38. doi:10.1016/j.biombioe.2012.12.019

Slattery, S. S., Diamond, A., Wang, H., Therrien, J. A., Lant, J. T., Jazey, T., Lee, K., Klassen, Z., Desgagné-Penix, I., Karas, B. J., & Edgell, D. R. (2018). An Expanded Plasmid-Based Genetic Toolbox Enables Cas9 Genome Editing and Stable Maintenance of Synthetic Pathways in Phaeodactylum tricornutum. *ACS Synthetic Biology*, *7*(2), 328–338. doi:10.1021/acssynbio.7b00191 PMID:29298053

Smith, R. E., Cavaletto, J. F., Eadie, B. J., & Gardner, W. S. (1993). Growth and lipid composition of high Arctic ice algae during the spring bloom at Resolute, Northwest Territories, Canada. *Marine Ecology Progress Series*, *97*, 19–29. doi:10.3354/meps097019

Smriga, S., Fernandez, V. I., Mitchell, J. G., & Stocker, R. (2016). Chemotaxis toward phytoplankton drives organic matter partitioning among marine bacteria. *Proceedings of the National Academy of Sciences of the United States of America*, *113*(6), 1576–1581. doi:10.1073/pnas.1512307113 PMID:26802122

Soeder, C. J. (1981). Types of algal ponds. Wastewater for aquaculture. UFS Publication Series C, 131-135.

Sogin, E. M., Puskás, E., Dubilier, N., Liebeke, M., & Huber, J. A. (2019). Marine Metabolomics: A Method for Nontargeted Measurement of Metabolites in Seawater by Gas Chromatography-Mass Spectrometry. *mSystems*, *4*(6), e00638–e00619. doi:10.1128/mSystems.00638-19 PMID:31822601

Soh, L., Montazeri, M., Haznedaroglu, B. Z., Kelly, C., Peccia, J., Eckelman, M. J., & Zimmerman, J. B. (2014). Evaluating microalgal integrated biorefinery schemes: Empirical controlled growth studies and life cycle assessment. *Bioresource Technology*, *151*, 19–27. doi:10.1016/j.biortech.2013.10.012 PMID:24189381

Sokhansanj, S., Kumar, A., & Turhollow, A. F. (2006). Development and implementation of integrated biomass supply analysis and logistics model (IBSAL). *Biomass and Bioenergy*, *30*(10), 838–847. doi:10.1016/j.biombioe.2006.04.004

Soltani, N., Moore, F., Keshavarzi, B., & Sharifi, R. (2014). Geochemistry of trace metals and rare earth elements in stream water, stream sediments and acid mine drainage from Darrehzar Copper Mine, Kerman, Iran. *Water Quality, Exposure, and Health*, *6*(3), 97–114. doi:10.100712403-014-0114-x

Song, B. H., Lee, D. H., Kim, B. C., Ku, S. H., Park, E. J., Kwon, I. H., Kim, K. H., & Kim, K. J. (2014). Photodynamic therapy using chlorophyll-a in the treatment of acne vulgaris: A randomized, single-blind, split-face study. *Journal of the American Academy of Dermatology*, *71*(4), 764–771. doi:10.1016/j.jaad.2014.05.047 PMID:24930587

Song, T., Mårtensson, L., Eriksson, T., Zheng, W., & Rasmussen, U. (2005). Biodiversity and seasonal variation of the cyanobacterial assemblage in a rice paddy field in Fujian, China. *FEMS Microbiology Ecology*, *54*(1), 131–140. doi:10.1016/j.femsec.2005.03.008 PMID:16329979

Song, W., Liang, J., Wen, T., Wang, X., Hu, J., Hayat, T., Alsaedi, A., & Wang, X. (2016). Accumulation of Co (II) and Eu (III) by the mycelia of Aspergillus niger isolated from radionuclide-contaminated soils. *Chemical Engineering Journal*, *304*, 186–193. doi:10.1016/j.cej.2016.06.103

Soni, R. A., Sudhakar, K., & Rana, R. S. (2017). Spirulina–from growth to nutritional product: A review. *Trends in Food Science & Technology*, *69*, 157–171. doi:10.1016/j.tifs.2017.09.010

Sonnenschein, E. C., Syit, D. A., Grossart, H. P., & Ullrich, M. S. (2012). Chemotaxis of *Marinobacter adhaerens* and its impact on attachment to the diatom *Thalassiosira weissflogii*. *Applied and Environmental Microbiology*, *78*(19), 6900–6907. doi:10.1128/AEM.01790-12 PMID:22820333

Soomro, R.R., Ndikubwimana, T., Zeng, X., Lu, Y., Lin, L., & Danquah, M.K. (2016). Development of a two-stage microalgae dewatering process – A life cycle assessment approach. *Front. Plant Sci.*, *7*, 113.

Sorokin, C., & Krauss, R. W. (1958). The Effects of Light Intensity on the Growth Rates of Green Algae. *Plant Physiology, 33*(2), 109–113. doi:10.1104/pp.33.2.109 PMID:16655087

Soyster, A. L. (1973). Convex programming with set-inclusive constraints and applications to inexact linear programming. *Operations Research, 21*(5), 1154–1157. doi:10.1287/opre.21.5.1154

Spain, O., Plöhn, M., & Funk, C. (2021). The cell wall of green microalgae and its role in heavy metal removal. *Physiologia Plantarum,* 1–10. doi:10.1111/ppl.13405

Spalvins, K., Raita, S., Valters, K., & Blumberga, D. (2021). Improving single cell protein yields and amino acid profile via mutagenesis: Review of applicable amino acid inhibitors for mutant selection. *Agronomy Research (Tartu), 19*(S3). Advance online publication. doi:10.15159/ar.21.083

Sparks, T. (2012). Solid-liquid filtration: Understanding filter presses and belt filters. *Filtration & Separation, 49*, 20–24.

Specht, E. A., Karunanithi, P. S., Gimpel, J. A., Ansari, W. S., & Mayfield, S. P. (2017). Host Organisms: Algae. In *Industrial Biotechnology* (pp. 605–641). John Wiley & Sons, Ltd.

Spicer, A., & Molnar, A. (2018). Gene Editing of Microalgae: Scientific Progress and Regulatory Challenges in Europe. *Biology (Basel), 7*(1), 21. doi:10.3390/biology7010021 PMID:29509719

Spiller, G. A., & Dewell, A. (2003). Safety of an astaxanthin rich *Haemaotoccu pluvialis* algal extract: A randomized clinical trial. *Journal of Medicinal Food, 6*(1), 51–56. doi:10.1089/109662003765184741 PMID:12804020

Spinelli, F., Fiori, G., Noferini, M., Sprocatti, M., & Costa, G. (2009). Perspectives on the use of a seaweed extract to moderate the negative effects of alternate bearing in apple trees. *The Journal of Horticultural Science & Biotechnology, 84*(6), 131–137. doi:10.1080/14620316.2009.11512610

Spolaore, P., Joannis-Cassan, C., Duran, E., & Isambert, A. (2006). Commercial applications of microalgae. *Journal of Bioscience and Bioengineering, 101*(2), 87–96. doi:10.1263/jbb.101.87 PMID:16569602

Spribille, T., Tuovinen, V., Resl, P., Vanderpool, D., Wolinski, H., Aime, M. C., Schneider, K., Stabentheiner, E., Toome-Heller, M., Thor, G., Mayrhofer, H., Johannesson, H., & McCutcheon, J. P. (2016). Basidiomycete yeasts in the cortex of ascomycete macrolichens. *Science, 353*(6298), 488–492. doi:10.1126cience.aaf8287 PMID:27445309

Sproles, A. E., Fields, F. J., Smalley, T. N., Le, C. H., Badary, A., & Mayfield, S. P. (2021). Recent advancements in the genetic engineering of microalgae. *Algal Research, 53*, 102158. doi:10.1016/j.algal.2020.102158

Sreenikethanam, A., & Bajhaiya, A. (2021). Algae Based Bio-Plastics: Future of Green Economy. In Biorefineries - Selected Processes. IntechOpen. doi:10.5772/intechopen.100981

Srinivas, U. S., Tan, B. W., Vellayappan, B. A., & Jeyasekharan, A. D. (2019). ROS and the DNA damage response in cancer. *Redox Biology, 25*, 101084. doi:10.1016/j.redox.2018.101084 PMID:30612957

Srirangan, K., Pyne, M. E., & Chou, C. P. (2011). Biochemical and genetic engineering strategies to enhance hydrogen production in photosynthetic algae and cyanobacteria. *Bioresource Technology, 102*(18), 8589–8604. doi:10.1016/j.biortech.2011.03.087 PMID:21514821

St. Onge, K. R. (2018). Photosynthesis: Light Reactions. In eLS (pp. 1–8). John Wiley & Sons, Ltd. doi:10.1002/9780470015902.a0001311.pub2

Staehr, P. A., & Birkeland, M. J. (2006). Temperature acclimation of growth, photosynthesis and respiration in two mesophilic phytoplankton species. *Phycologia, 45*(6), 648–656. doi:10.2216/06-04.1

Steinbuchel, A., Madkour, M. H., Heinrich, D., Alghamdi, A. M., & Shabbaj, I. I. (2013). PHA Recovery from Biomass. *Biomacromolecules*. PMID:23875914

Steinfeld, A. (2005). Solar thermochemical production of hydrogen—a review. *Solar Energy*, *78*(5), 603–615. doi:10.1016/j.solener.2003.12.012

Stephenson, A. L., Kazamia, E., Dennis, J. S., Howe, C. J., Scott, S. A., & Smith, A. G. (2010). Life-cycle assessment of potential algal biodiesel production in the United Kingdom: A comparison of raceways and air-lift tubular bioreactors. *Energy & Fuels*, *24*(7), 4062–4077. doi:10.1021/ef1003123

Stewart, A., Rioux, D., Boyer, F., Gielly, L., Pompanon, F., Saillard, A., Thuiller, W., Valay, J.-G., Maréchal, E., & Coissac, E. (2021). Altitudinal Zonation of Green Algae Biodiversity in the French Alps. *Frontiers in Plant Science*, *12*, 679428. doi:10.3389/fpls.2021.679428 PMID:34163510

Stewart, H. L., & Carpenter, R. C. (2003). The effects of morphology and water flow on photosynthesis of marine macroalgae. *Ecology*, *84*(11), 2999–3012. doi:10.1890/02-0092

Stirk, W.A., Novák, O., Hradecká, V., Pěnčík, A., Rolčík, J., Strnad, M., & Van Staden, J. (2009). Endogenous cytokinins, auxins and abscisic acid in *Ulva fasciata* (Chlorophyta) and *Dictyota humifusa* (Phaeophyta): towards understanding their biosynthesis and homoeostasis. *European Journal of Phycology, 44*(2), 231-40.

Stirk, W. A., Bálint, P., Vambe, M., Lovász, C., Molnár, Z., van Staden, J., & Ördög, V. (2020). Effect of cell disruption methods on the extraction of bioactive metabolites from microalgal biomass. *Journal of Biotechnology*, *307*, 35–43. doi:10.1016/j.jbiotec.2019.10.012 PMID:31678206

Stirk, W. A., Novak, O., Strnad, M., & van Staden, J. (2003). Cytokinins in Macroalgae. *Plant Growth Regulation*, *41*(1), 13–24. doi:10.1023/A:1027376507197

Stirk, W. A., Ördög, V., Novák, O., Rolčík, J., Strnad, M., Bálint, P., & van Staden, J. (2013). Auxin and cytokinin relationships in 24 microalgal strains1. *Journal of Phycology*, *49*(3), 459–467. doi:10.1111/jpy.12061 PMID:27007035

Stockenreiter, M., Haupt, F., Graber, A. K., Seppälä, J., Spilling, K., Tamminen, T., & Stibor, H. (2013). Functional group richness: Implications of biodiversity for light use and lipid yield in microalgae. *Journal of Phycology*, *49*(5), 838–847. doi:10.1111/jpy.12092 PMID:27007310

Stock, F., Bilcke, G., De Decker, S., Osuna-Cruz, C. M., Van den Berge, K., Vancaester, E., De Veylder, L., Vandepoele, K., Mangelinckx, S., & Vyverman, W. (2020). Distinctive Growth and Transcriptional Changes of the Diatom Seminavis robusta in Response to Quorum Sensing Related Compounds. *Frontiers in Microbiology*, *11*(June), 1–14. doi:10.3389/fmicb.2020.01240 PMID:32582129

Stomp, M., Huisman, J., Stal, L. J., & Matthijs, H. C. P. (2007). Colorful niches of phototrophic microorganisms shaped by vibrations of the water molecule. *The ISME Journal, 1*, 271–282.

Stratton, R. W., Wong, H. M., & Hileman, J. I. (2011). Quantifying variability in life cycle greenhouse gas inventories of alternative middle distillate transportation fuels. *Environmental Science & Technology*, *45*(10), 4637–4644. doi:10.1021/es102597f PMID:21513296

Strik, W. A., Arthur, G. D., Lourens, A. F., Novok, O., Strand, M., & Van-Staden, J. (2004). Changes in seaweed concentrates when stores at an elevated temperature. *Journal of Applied Phycology*, *16*, 31–39. doi:10.1023/B:JAPH.0000019057.45363.f5

Studer, G., Rempfer, C., Waterhouse, A. M., Gumienny, R., Haas, J., & Schwede, T. (2020). QMEANDisCo – Distance Constraints Applied on Model Quality Estimation. *Bioinformatics (Oxford, England)*, *36*(6), 1765–1771. doi:10.1093/bioinformatics/btz828 PMID:31697312

Subashchandrabose, S. R., Ramakrishnan, B., Megharaj, M., Venkateswarlu, K., & Naidu, R. (2013). Mixotrophic cyanobacteria and microalgae as distinctive biological agents for organic pollutant degradation. *Environment International*, *51*, 59–72. doi:10.1016/j.envint.2012.10.007 PMID:23201778

Subba Rao, P.V., Mantri, V., Ks, G., & Kumar, S.K. (2007). *Seaweeds as a Human Diet: An Emerging Trend in the New Millennium.* . doi:10.13140/2.1.3494.7208

Subhadra, B. G., & Edwards, M. (2011). Coproduct market analysis and water footprint of simulated commercial algal biorefineries. *Applied Energy*, *88*(10), 3515–3523. doi:10.1016/j.apenergy.2010.12.051

Subramaniyam, V., Ramraj, S., Bose, S., Thavamani, P., Megharaj, M., Chen, Z. and Naidu, R. (2015). Chlorococcum sp. MM11- a novel Phyco-nanofactory for the synthesis of iron NP. *Journal of Appllied Phy- Cology, 27*(5), 1861–1869.

Sudesh, K., Taguchi, K., & Doi, Y. (2002). Effect of increased PHA synthase activity on polyhydroxyalkanoates biosynthesis in Synechocystis sp. PCC6803. *International Journal of Biological Macromolecules*, *30*(2), 97–104. doi:10.1016/S0141-8130(02)00010-7 PMID:11911900

Sudhakar, K., Mamat, R., Samykano, M., Azmi, W. H., Ishak, W. F. W., & Yusaf, T. (2018). An Overview of Marine Macroalgae as Bioresource. *Renewable & Sustainable Energy Reviews*, *91*, 165–179. doi:10.1016/j.rser.2018.03.100

Sudhakar, K., & Premalatha, M. (2015). Characterization of micro algal biomass through FTIR/TGA/CHN analysis: Application to Scenedesmus sp. *Energy Sources. Part A, Recovery, Utilization, and Environmental Effects*, *37*(21), 1–8. doi:10.1080/15567036.2013.825661

Sudhakar, M. P., & Ramesh Kumar, B. (2019). A Review on Bioenergy and Bioactive Compounds from Microalgae and Macroalgae-Sustainable Energy Perspective. *Journal of Cleaner Production*, *228*, 1320–1333. doi:10.1016/j.jclepro.2019.04.287

Sudhakar, M., Doble, M., Murthy, P. S., & Venkatesan, R. (2008). Marine microbe-mediated biodegradation of low- and high-density polyethylenes. *International Biodeterioration & Biodegradation*, *61*(3), 203–213. doi:10.1016/j.ibiod.2007.07.011

Sudha, S. S., & Rajamanickam, K. R. J. (2013). Microalgae mediated synthesis of silver nanoparticles and their antibacterial activity against pathogenic bacteria. *Indian Journal of Experimental Biology*, *51*, 393–399. PMID:23821828

Suetsuna, K., & Chen, J. (2001). Identification of Antihypertensive Peptides from Peptic Digest of Two Microalgae, Chlorella vulgaris and Spirulina platensis. *Marine Biotechnology (New York, N.Y.)*, *3*(4), 305–309. doi:10.100710126-001-0012-7 PMID:14961345

Suganya, K. S. U., Govindaraju, K., Kumar, V. G., Dhas, T. S., Karthick, V., Singaravelu, G., & Elanchezhiyan, M. (2015). Blue green alga mediated synthesis of gold nanoparticles and its antibacterial efficacy against Gram positive organisms. *Materials Science and Engineering C*, *47*, 351–356. doi:10.1016/j.msec.2014.11.043 PMID:25492207

Suganya, T., Varman, M., Masjuki, H. H., & Renganathan, S. (2016). Macroalgae and microalgae as a potential source for commercial applications along with biofuels production: A biorefinery approach. *Renewable & Sustainable Energy Reviews*, *55*, 909–941. doi:10.1016/j.rser.2015.11.026

Sukenik, A., Carmeli, Y., & Berner, T. (1989). Regulation of fatty acid composition by irradiance level in the eustigmatophyte Nannochloropsis sp. 1. *Journal of Phycology*, *25*(4), 686–692. doi:10.1111/j.0022-3646.1989.00686.x

Sultana, V., Baloch, G. N., Ara, J., Ehteshamul-Haque, S., Tariq, R. M., & Athar, M. (2012). Seaweeds as an alternative to chemical pesticides for the management of root diseases of sunflower and tomato. *Journal of Applied Botany and Food Quality*, *84*(2), 162.

Sultana, V., Baloch, G. N., Ara, J., Ehteshamul-Haque, S., Tariq, R. M., & Athar, M. (2012). Seaweeds as an Alternative to Chemical Pesticides for the Management of Root Diseases of Sunflower and Tomato. *Journal of Applied Botany and Food Quality*, *84*, 162.

Suman, T. Y., Rajasree, S. R. R., & Kirubagaran, R. (2015). Evaluation of zinc oxide nanoparticles toxicity on marine algae Chlorella vulgaris through flow cytometric, cytotoxicity and oxidative stress analysis. *Ecotoxicology and Environmental Safety*, *113*, 23–30. doi:10.1016/j.ecoenv.2014.11.015 PMID:25483368

Sumiya, N., Kawase, Y., Hayakawa, J., Matsuda, M., Nakamura, M., Era, A., Tanaka, K., Kondo, A., Hasunuma, T., Imamura, S., & Miyagishima, S. Y. (2015). Expression of cyanobacterial acyl-ACP reductase elevates the triacylglycerol level in the red alga Cyanidioschyzon merolae. *Plant & Cell Physiology*, *56*(10), 1962–1980. doi:10.1093/pcp/pcv120 PMID:26272551

Sun, C.-H., Fu, Q., Liao, Q., Xia, A., Huang, Y., Zhu, X., Reungsang, A., & Chang, H.-X. (2019). Life-cycle assessment of biofuel production from microalgae via various bioenergy conversion systems. *Energy*, *171*, 1033–1045. doi:10.1016/j.energy.2019.01.074

Sundararaman, M., Boopathi, T., & Gopinath, S. (2007). Status of Mangrove Ecosystem. In *Algae and Cyanobacteria in Extreme Environments* (pp. 209–224). Springer. doi:10.1007/978-1-4020-6112-7_11

Sunda, W. (2012). Feedback interactions between trace metal nutrients and phytoplankton in the ocean. *Frontiers in Microbiology*, *3*, 204. doi:10.3389/fmicb.2012.00204 PMID:22701115

Sung, Y. J., Choi, H. I., Lee, J. S., Hong, M. E., & Sim, S. J. (2019). Screening of oleaginous algal strains from *Chlamydomonas reinhardtii* mutant libraries via density gradient centrifugation. *Biotechnology and Bioengineering*, *116*(12), 3179–3188. doi:10.1002/bit.27149 PMID:31429923

Sun, H., Zhao, W., Mao, X., Li, Y., Wu, T., & Chen, F. (2018). High-value biomass from microalgae production platforms: Strategies and progress based on carbon metabolism and energy conversion. *Biotechnology for Biofuels*, *11*(1), 1–23. doi:10.118613068-018-1225-6 PMID:30151055

Sun, X.-M., Ren, L.-J., Bi, Z.-Q., Ji, X.-J., Zhao, Q.-Y., & Huang, H. (2018). Adaptive evolution of microalgae *Schizochytrium* sp. under high salinity stress to alleviate oxidative damage and improve lipid biosynthesis. *Bioresource Technology*, *267*, 438–444. doi:10.1016/j.biortech.2018.07.079 PMID:30032058

Surendhiran, D., & Vijay, M. (2013). Study on flocculation efficiency for harvesting *Nannochloropsis oculata* for biodiesel production. *International Journal of Chemtech Research*, *5*, 1761–1769.

Surendhiran, D., & Vijay, M. (2014). Exploration on bioflocculation of *Nannochloropsis oculata* using response surface methodology for biodiesel production. *TheScientificWorldJournal*.

Suriyamongkol, P., Weselake, R., Narine, S., Moloney, M., & Shah, S. (2007). Biotechnological approaches for the production of polyhydroxyalkanoates in microorganisms and plants— A review. *Biotechnology Advances*, *25*(2), 148–175. doi:10.1016/j.biotechadv.2006.11.007 PMID:17222526

Surzycki, R., Greenham, K., Kitayama, K., Dibal, F., Wagner, R., Rochaix, J. D., Ajam, T., & Surzycki, S. (2009). Factors effecting expression of vaccines in microalgae. *Biologicals*, *37*(3), 133–138. doi:10.1016/j.biologicals.2009.02.005 PMID:19467445

Suseela, M. R., & Toppo, K. (2007). Algal biofilms on polythene and its possible degradation. *Current Science*, *92*, 285–287.

Su, Y., Mennerich, A., & Urban, B. (2011). Municipal wastewater treatment and biomass accumulation with a wastewater-born and settleable algal-bacterial culture. *Water Research*, *45*(11), 3351–3358. doi:10.1016/j.watres.2011.03.046 PMID:21513965

Svecova, L., Spanelova, M., Kubal, M., & Guibal, E. (2006). Cadmium, lead and mercury biosorption on waste fungal biomass issued from fermentation industry. I. Equilibrium studies. *Separation and Purification Technology*, *52*(1), 142–153. doi:10.1016/j.seppur.2006.03.024

Swain, K. K., Mishra, P. M., & Devi, A. P. (2018). Biosorption of praseodymium (III) using Terminalia arjuna bark powder in batch systems: Isotherm and kinetic studies. *Water Science and Technology*, *77*(3), 727–738. doi:10.2166/wst.2017.589 PMID:29431718

Swain, S. S., Paidesetty, S. K., & Padhy, R. N. (2017). Antibacterial, antifungal and antimycobacterial compounds from cyanobacteria. *Biomedecine & Pharmacotherapie*, *90*, 760–776. doi:10.1016/j.biopha.2017.04.030 PMID:28419973

Swanson, D., Block, R., & Mousa, S. A. (2012). Omega-3 fatty acids EPA and DHA: Health benefits throughout life. *Advances in Nutrition*, *3*(1), 1–7. doi:10.3945/an.111.000893 PMID:22332096

Sydney, E. B., Sturm, W., de Carvalho, J. C., Thomaz-Soccol, V., Larroche, C., Pandey, A., & Soccol, C. R. (2010). Potential carbon dioxide fixation by industrially important microalgae. *Bioresource Technology*, *101*(15), 5892–5896. doi:10.1016/j.biortech.2010.02.088 PMID:20350804

Syed, M. S., Rafeie, M., Vandamme, D., Asadnia, M., Henderson, R., Taylor, R. A., & Warkiani, M. E. (2018). Selective separation of microalgae cells using inertial microfluidics. *Bioresource Technology*, *252*, 91–99. doi:10.1016/j.biortech.2017.12.065 PMID:29306136

Szekely, G. (2004). The role of proline in Arabidopsis thaliana osmotic stress response. *Acta Biologica Szegediensis*, *48*, 81.

Tabasi, R. Y., & Ajji, A. (2015). Selective degradation of biodegradable blends in simulated laboratory composting. *Polymer Degradation & Stability*, *120*, 435–442. doi:10.1016/j.polymdegradstab.2015.07.020

Tabatabaei, M., Tohidfar, M., Jouzani, G. S., Safarnejad, M., & Pazouki, M. (2011). Biodiesel production from genetically engineered microalgae: Future of bioenergy in Iran. *Renewable & Sustainable Energy Reviews*, *15*(4), 1918–1927. doi:10.1016/j.rser.2010.12.004

Taberlet, P., Bonin, A., Zinger, L., & Coissac, E. (2018). *Environmental DNA: For Biodiversity Research and Monitoring*. Oxford University Press. doi:10.1093/oso/9780198767220.001.0001

Taberlet, P., Prud'Homme, S. M., Campione, E., Roy, J., Miquel, C., Shehzad, W., Gielly, L., Rioux, D., Choler, P., Clément, J.-C., Melodelima, C., Pompanon, F., & Coissac, E. (2012). Soil sampling and isolation of extracellular DNA from large amount of starting material suitable for metabarcoding studies. *Molecular Ecology*, *21*(8), 1816–1820. doi:10.1111/j.1365-294X.2011.05317.x PMID:22300434

Tachibana, K., Hashimoto, K., Yoshikawa, M., & Okawa, H. (2010). Isolation and characterization of microorganisms degrading nylon 4 in the composted soil. *Polymer Degradation & Stability*, *95*(6), 912–917. doi:10.1016/j.polymdegradstab.2010.03.031

Tadkaew, N., Hai, F. I., McDonald, J. A., Khan, S. J., & Nghiem, L. D. (2011). Removal of trace organics by MBR treatment: The role of molecular properties. *Water Research*, *45*, 2439–2451.

Taepucharoen, K., Tarawat, S., Puangcharoen, M., Incharoensakdi, A., & Monshupanee, T. (2017). Production of poly(3-hydroxybutyrate-co-3-hydroxyvalerate) under photoautotrophy and heterotrophy by non-heterocystous N_2-fixing cyanobacterium. *Bioresource Technology*, *239*, 523–527. doi:10.1016/j.biortech.2017.05.029 PMID:28533067

Taggart, R. K., Hower, J. C., Dwyer, G. S., & Hsu-Kim, H. (2016). Trends in the rare earth element content of US-based coal combustion fly ashes. *Environmental Science & Technology*, *50*(11), 5919–5926. doi:10.1021/acs.est.6b00085 PMID:27228215

Taher, H., Al-Zuhair, S., Al-Marzouqi, A., Haik, Y., & Farid, M. (2015). Growth of microalgae using CO_2 enriched air for biodiesel production in supercritical CO2. *Renewable Energy*, *82*, 61–70. doi:10.1016/j.renene.2014.08.013

Tai, P., Zhao, Q., Su, D., Li, P., & Stagnitti, F. (2010). Biological toxicity of lanthanide elements on algae. *Chemosphere*, *80*(9), 1031–1035. doi:10.1016/j.chemosphere.2010.05.030 PMID:20547408

Taipale, S. J., Peltomaa, E., Kukkonen, J. V. K., Kainz, M. J., Kautonen, P., & Tiirola, M. (2019). Tracing the fate of microplastic carbon in the aquatic food web by compound-specific isotope analysis. *Scientific Reports*, *9*(1), 19894. doi:10.103841598-019-55990-2 PMID:31882692

Takahashi, R., Mabuchi, S., Kawano, M., Sasano, T., Matsumoto, Y., Kuroda, H., Kozasa, K., Hashimoto, K., Sawada, K., & Kimura, T. (2016). Preclinical investigations of PM01183 (Lurbinectedin) as a single agent or in combination with other anticancer agents for clear cell carcinoma of the ovary. *PLoS One*, *11*(3), e0151050. doi:10.1371/journal.pone.0151050 PMID:26986199

Takahashi, S., & Murata, N. (2008). How do environmental stresses accelerate photoinhibition? *Trends in Plant Science*, *13*(4), 178–182. doi:10.1016/j.tplants.2008.01.005 PMID:18328775

Takano, H., Takeyama, H., Nakamura, N., Sode, K., Burgess, J., Manabe, E., Hirano, M., & Matsunaga, T. (1992). CO2 removal by high-density culture of a marine cyanobacterium synechococcus sp. using an improved photobioreactor employing light-diffusing optical fibers. *Applied Biochemistry and Biotechnology*, *34/35*(1), 449–458. doi:10.1007/BF02920568

Takano, Y., Yamaguchi, H., Inouye, I., Moestrup, Ø., & Horiguchi, T. (2014). Phylogeny of Five Species of Nusuttodinium gen. nov. (Dinophyceae), a Genus of Unarmoured Kleptoplastidic Dinoflagellates. *Protist*, *165*(6), 759–778. doi:10.1016/j.protis.2014.09.001 PMID:25460229

Takebe, Y., Saucedo, C. J., Lund, G., Uenishi, R., Hase, S., Tsuchiura, T., Kneteman, N., Ramessar, K., Tyrrell, D. L., Shirakura, M., Wakita, T., McMahon, J. B., & O'Keefe, B. R. (2013). Antiviral Lectins from Red and Blue-Green Algae Show Potent In Vitro and In Vivo Activity against Hepatitis C Virus. *PLoS One*, *8*(5), e64449. doi:10.1371/journal.pone.0064449 PMID:23700478

Takemura, T., Imamura, S., & Tanaka, K. (2019). Identification of a chloroplast fatty acid exporter protein, CmFAX1, and triacylglycerol accumulation by its overexpression in the unicellular red alga Cyanidioschyzon merolae. *Algal Research*, *38*, 101396. doi:10.1016/j.algal.2018.101396

Talebi, A. F., Tabatabaei, M., Mohtashami, S. K., Tohidfar, M., & Moradi, F. (2013). Comparative salt stress study on intracellular ion concentration in marine and salt-adapted freshwater strains of microalgae. *Notulae Scientia Biologicae*, *5*(3), 309–315. doi:10.15835/nsb539114

TalukdarJ.DasguptaS.NagleV.BhadraB. (2020). COVID-19: Potential of microalgae-derived natural astaxanthin as adjunctive supplement in alleviating cytokine storm. *Available at* SSRN 3579738. doi:10.2139/ssrn.3579738

Talukdar, P., & Pal, R. (2017). Predictive toxicity assessment of toxins released from *Microcystis aeruginosa* kutz: An *in silico* approach. *World Journal of Pharmaceutical Research*, *6*(14), 411–423.

Tamura, K., Peterson, D., Peterson, N., Stecher, G., Nei, M., & Kumar, S. (2011). MEGA5: Molecular evolutionary genetics analysis using maximum likelihood, evolutionary distance, and maximum parsimony methods. *Molecular Biology and Evolution*, *28*(10), 2731–2739. doi:10.1093/molbev/msr121 PMID:21546353

Tanasupawat, S., Takehana, T., Yoshida, S., Hiraga, K., & Oda, K. (2016). Ideonella sakaiensis sp. nov., isolated from a microbial consortium that degrades poly(ethylene terephthalate). *International Journal of Systematic and Evolutionary Microbiology*, *66*(8), 2813–2818. doi:10.1099/ijsem.0.001058 PMID:27045688

Tang, D. Y. Y., Yew, G. Y., Koyande, A. K., Chew, K. W., Vo, D.-V. N., & Show, P. L. (2020). Green technology for the industrial production of biofuels and bioproducts from microalgae: A review. *Environmental Chemistry Letters*, *18*(6), 1967–1985. doi:10.100710311-020-01052-3

Tang, D., Han, W., Li, P., Miao, X., & Zhong, J. (2011). CO_2 biofixation and fatty acid composition of Scenedesmus obliquus and Chlorella pyrenoidosa in response to different CO2 levels. *Bioresource Technology*, *102*(3), 3071–3076. doi:10.1016/j.biortech.2010.10.047 PMID:21041075

Tang, E. P. Y., Tremblay, R., & Vincent, W. F. (1997a). Cyanobacterial dominance of polar freshwater ecosystems: Are high-latitude mat-formers adapted to low temperature? *Journal of Phycology*, *33*(2), 171–181. doi:10.1111/j.0022-3646.1997.00171.x

Tang, E. P. Y., & Vincent, W. F. (2000). Effects of daylength and temperature on the growth and photosynthesis of an Arctic cyanobacterium, *Schizothrix calcicola* (Oscillatoriaceae). *European Journal of Phycology*, *35*(3), 263–272. doi:10.1080/09670260010001735861

Tang, L., Wu, Q., & Qu, B. (2004). The effects of chemical structure and synthesis method on photodegradation of polypropylene. *Journal of Applied Polymer Science*, *95*(2), 270–279. doi:10.1002/app.21272

Tang, X., He, L. Y., Tao, X. Q., Dang, Z., Guo, C. L., Lu, G. N., & Yi, X. Y. (2010). Construction of an artificial microalgal-bacterial consortium that efficiently degrades crude oil. *Journal of Hazardous Materials*, *181*(1–3), 1158–1162. doi:10.1016/j.jhazmat.2010.05.033 PMID:20638971

Tang, Y., Liu, Y., Chen, Y., Zhang, W., Zhao, J., He, S., Yang, C., Zhang, T., Tang, C., Zhang, C., & Yang, Z. (2021). A review: Research progress on microplastic pollutants in aquatic environments. *The Science of the Total Environment*, *766*, 142572. doi:10.1016/j.scitotenv.2020.142572 PMID:33183825

Tan, J. S., Lee, S. Y., Chew, K. W., Lam, M. K., Lim, J. W., Ho, S.-H., & Show, P. L. (2020). A Review on Microalgae Cultivation and Harvesting, and Their Biomass Extraction Processing Using Ionic Liquids. *Bioengineered*, *11*(1), 116–129. doi:10.1080/21655979.2020.1711626 PMID:31909681

Tanwar, A., & Kumar, S. (2020). *Genome editing of algal species by CRISPR Cas9 for biofuels. Genome engineering via CRISPR-cas9 system*. Academic Press.

Tan, X., Zhang, D., Duan, Z., Parajuli, K., & Hu, J. (2020). Effects of light color on interspecific competition between *Microcystis aeruginosa* and *Chlorella pyrenoidosa* in batch experiment. *Environmental Science and Pollution Research International*, *27*, 344–352. https://doi.org/10.1007/s11356-019-06650-5

Tao, D. G., & Salihon, J. (2011). The Optimisation of levels of the variables pH and Concentration of Ferric chloride for Harvesting marine microalgae by flocculation. In *Proceedings of the International Conference on Food Engineering and Biotechnology (IPCBEE)*. IACSIT Press.

Tasić, M. B., Pinto, L. F. R., Klein, B. C., Veljković, V. B., & Filho, R. M. (2016). Botryococcus braunii for biodiesel production. In *Renewable and Sustainable Energy Reviews* (Vol. 64, pp. 260–270). Elsevier Ltd. doi:10.1016/j.rser.2016.06.009

Taskin, E., Caki, Z., & Ozturk, M. (2010). Assessment of in vitro antitumoral and antimicrobial activities of marine algae harvested from the eastern Mediterranean Sea. *African Journal of Biotechnology*, *9*(27), 4272–4277.

Taunt, H. N., Stoffels, L., & Purton, S. (2018). Green biologics: The algal chloroplast as a platform for making biopharmaceuticals. *Bioengineered*, *9*(1), 48–54. doi:10.1080/21655979.2017.1377867 PMID:28892417

Teeling, H., Fuchs, B. M., Bennke, C. M., Kruger, K., Chafee, M., Kappelmann, L., Reintjes, G., Waldmann, J., Quast, C., Glockner, F. O., Lucas, J., Wichels, A., Gerdts, G., Wiltshire, K. H., & Amann, R. I. (2016). Recurring patterns in bacterioplankton dynamics during coastal spring algae blooms. *eLife*, *5*, e11888. doi:10.7554/eLife.11888 PMID:27054497

Tekin, K., Karagöz, S., & Bektaş, S. (2014). A review of hydrothermal biomass processing. *Renewable & Sustainable Energy Reviews*, *40*, 673–687. doi:10.1016/j.rser.2014.07.216

Tendencia, E. A., & dela Peña, M. (2003). Investigation of some components of the greenwater system which makes it effective in the initial control of luminous bacteria. *Aquaculture (Amsterdam, Netherlands)*, *218*(1-4), 115–119. doi:10.1016/S0044-8486(02)00524-0

Teoh, M. L., Chu, W. L., & Phang, S. M. (2010). Effect of temperature change on physiology and biochemistry of algae: A review. *Malaysian Journal of Science*, *29*(2), 82–97. doi:10.22452/mjs.vol29no2.1

Terme, N. (2020). Emerging Seaweed Extraction Techniques: Enzyme-Assisted Extraction a Key Step of Seaweed Biorefinery? In *Sustainable Seaweed Technologies* (pp. 225–256). Elsevier. doi:10.1016/B978-0-12-817943-7.00009-3

Ternes, T. A., Herrmann, N., Bonerz, M., Knacker, T., Siegrist, H., & Joss, A. (2004). A rapid method to measure the solid–water distribution coefficient (K_d) for pharmaceuticals and musk fragrances in sewage sludge. *Water Research*, *38*, 4075–4084.

Terzopoulou, Z., Karakatsianopoulou, E., Kasmi, N., Majdoub, M., Papageorgiou, G. Z., & Bikiaris, D. N. (2017). Effect of catalyst type on recyclability and decomposition mechanism of poly(ethylene furanoate) biobased polyester. *Journal of Analytical and Applied Pyrolysis*, *126*, 357–370. doi:10.1016/j.jaap.2017.05.010

Tesson, S. V. M., Skjøth, C. A., Šantl-Temkiv, T., Löndahl, J., & Voordouw, G. (2016). Airborne Microalgae: Insights, Opportunities, and Challenges. *Applied and Environmental Microbiology*, *82*(7), 1978–1991. doi:10.1128/AEM.03333-15 PMID:26801574

Tevini, M., Braun, J., & Fieser, G. (1991). The protective function of the epidermal layer of rye seedlings against ultraviolet-B radiation. *Photochemistry and Photobiology*, *53*(3), 229–333. doi:10.1111/j.1751-1097.1991.tb03636.x

Thakkar, K. N., Mhatre, S. S., & Parikh, R. Y. (2010). Biological synthesis of metallic nanoparticles. *Nanotechnol. Biol. Med.*, *6*(2), 257–262. doi:10.1016/j.nano.2009.07.002 PMID:19616126

Thakur, S., Chaudhary, J., Sharma, B., Verma, A., Tamulevicius, S., & Thakur, V. K. (2018). Sustainability of bioplastics: Opportunities and challenges. *Current Opinion in Green and Sustainable Chemistry*, *13*, 68–75. doi:10.1016/j.cogsc.2018.04.013

Thamida, B. Z. N., Mandal, R., & Saiful, I. (2011). Effect of Cyanobacterial biofertilizer on the growth and yield components of two HYV of rice. *Journal of Algal Biomass Utilization*, *2*(1), 1–9.

Tharani, D., & Ananthasubramanian, M. (2020). Microalgae as sustainable producers of bioplastic. In *Microalgae biotechnology for food, health and high value products* (pp. 373–396). Springer. doi:10.1007/978-981-15-0169-2_11

Tharek, A., Yahya, A., Salleh, M. M., Jamaluddin, H., Yoshizaki, S., Hara, H., Iwamoto, K., Suzuki, I., & Mohamad, S. E. (2021). Improvement and screening of astaxanthin producing mutants of newly isolated *Coelastrum* sp. using ethyl methane sulfonate induced mutagenesis technique. *Biotechnology Reports (Amsterdam, Netherlands)*, *32*, e00673. doi:10.1016/j.btre.2021.e00673 PMID:34621628

Thellen, C., Coyne, M., Froio, D., Auerbach, M., Wirsen, C., & Ratto, J. A. (2008). A processing, characterization and marine biodegradation study of melt-extruded polyhydroxyalkanoate (pha) films. *Journal of Polymers and the Environment*, *16*(1), 1–11. doi:10.100710924-008-0079-6

Theodoropoulos, C., & Sun, C. (2019). Bioreactor Models and Modeling Approaches. In *Comprehensive Biotechnology* (pp. 663–680). Elsevier BV. doi:10.1016/B978-0-444-64046-8.00144-0

Theodoropoulos, C., & Sun, C. (2019). *Bioreactor Models and Modeling Approaches. In Comprehensive Biotechnology*. Elsevier BV.

Thiruchelvi, R., Das, A., & Sikdar, E. (2021). Bioplastics as better alternative to petro plastic. *Materials Today: Proceedings*, *37*, 1634–1639. doi:10.1016/j.matpr.2020.07.176

Thirumaran, G., Arumugam, M., Arumugam, R., & Anantharaman, P. (2009). Effect of Seaweed Liquid Fertilizer on Growth and Pigment Concentration of *Abelmoschus esculentus* (l) medikus. *American-Eurasian Journal of Agronomy*, *2*(2), 57–66.

Thirumdas, R., Brnčić, M., Brnčić, S. R., Barba, F. J., Gálvez, F., Zamuz, S., Lacomba, R., & Lorenzo, J. M. (2018). Evaluating the impact of vegetal and microalgae protein sources on proximate composition, amino acid profile, and physicochemical properties of fermented Spanish "chorizo" sausages. *Journal of Food Processing and Preservation*, *42*(11), e13817. doi:10.1111/jfpp.13817

Thomas, N. E., Balasubramanian, D. R., & Anamika, G. (2019). Microalgae as an Alternative Source for Bio-Plastic Production: A Review. *Baselius Researcher*, 116.

Thompson, J. D., Gibson, T. J., Plewniak, F., Jeanmougin, F., & Higgins, D. G. (1997). The clustal X windows interface: Flexible strategies for multiple sequence alignment aided by quality analysis tools. *Nucleic Acids Research*, *24*(24), 4876–4882. doi:10.1093/nar/25.24.4876 PMID:9396791

Thompson, P. A., Jameson, I., & Blackburn, S. I. (2009). The influence of light quality on akinete formation and germination in the toxic cyanobacterium *Anabaena circinalis*. *Harmful Algae*, *8*, 504–512. https://doi.org/10.1016/j.hal.2008.10.004

Thume, K., Gebser, B., Chen, L., Meyer, N., Kieber, D., & Pohnert, G. (2018). The metabolite dimethylsulfoxonium propionate extends the marine organosulfur cycle. *Nature*, *563*(7731), 412–415. doi:10.103841586-018-0675-0 PMID:30429546

Tian, M., Song, N., Wang, D., Quan, X., Jia, Q., Liao, W., & Lin, L. (2012). Applications of the binary mixture of sec-octylphenoxyacetic acid and 8-hydroxyquinoline to the extraction of rare earth elements. *Hydrometallurgy*, *111*, 109–113. doi:10.1016/j.hydromet.2011.11.002

Tiburu, E. K., Salifu, A., Aidoo, E. O., Fleischer, H. N., Manu, G., Yaya, A., Zhou, H., & Efavi, J. K. (2017). Formation of chitosan nanoparticles using deacetylated chitin isolated from freshwater algae and locally synthesized zeolite A and their influence on cancer cell growth. *Proc. J. Nano Res., 48,* 156–170. 10.4028/www.scientific.net/JNanoR.48.156

Titlyanov, A. E., Titlyanova, V. T., Li, X., & Huang, H. (2016). *Coral reef marine plants of Hainan island*. Academic Press.

Tiwari, D. N., Kumar, A., & Mishra, A. K. (1991). Use of cyanobacterial diazotrophic technology in rice agriculture. *Applied Biochemistry and Biotechnology*, *28/29*(1), 387–396. doi:10.1007/BF02922618

Tiwari, P. K., Misra, A. K., & Venturino, E. (2017). The role of algae in agriculture: A mathematical study. *Journal of Biological Physics*, *43*(2), 297–314. doi:10.100710867-017-9453-8 PMID:28550636

Toh, P. S., Jau, M.-H., Yew, S.-P., Abed, R. M., & Sudesh, K. (2008). Comparison of Polyhydroxyalkanoates Biosynthesis, Mobilization and the Effects on Cellular Morphology in Spirulina Platensis and Synechocystis Sp. Uniwg. *Journal of Bio-Science*, *19*(2), 21–38.

Toh, P. Y., Azenan, N. F., Wong, L., Ng, Y. S., Chng, L. M., Lim, J., & Chan, D. J. C. (2017). The role of cationic coagulant-to-cell interaction in dictating the flocculation-aided sedimentation of freshwater microalgae. *Arabian Journal for Science and Engineering*, *43*, 1–9.

Toh, Y., Jau, P. S., Yew, M. H., Abed, S. P. M., & Sudesh, R. M. K. (2008). Comparison of polyhydroxyalkanoates biosynthesis, mobilization and the effects on cellular morphology in *Spirulina Platensis* and *Synechocystis* Sp. Uniwg. *Journal of Bio-Science*, *19*(2), 21–38. http://www.tlsr.usm.my/tlsr19022008/19022008_03.pdf

Toker, O. S. (2018). *Porphyridum Cruentum as a natural colorant in chewing gum. Food Science and Technology*. AHEAD.

Tokunaga, S., Sanda, S., Uraguchi, Y., Nakagawa, S., & Sawayama, S. (2019). Overexpression of the DOF-type transcription factor enhances lipid synthesis in Chlorella vulgaris. *Applied Biochemistry and Biotechnology*, *189*(1), 116–128. doi:10.100712010-019-02990-7 PMID:30877635

Toll, A. J. (2012). *Algal Development: Molecular and Cellular Aspects*. Springer.

Tome, S., Dzoujo, H., Shikuku, V., & Otieno, S. (2021). Synthesis, characterization and application of acid and alkaline activated volcanic ash-based geopolymers for adsorptive remotion of cationic and anionic dyes from water. *Ceramics International*, *47*(15), 20965–20973.

Tominaga, N., Takahata, M., & Tominaga, H. (1993). Effects of NaCl and KNO_3 Concentrations on the Abscisic Acid Content of *Dunaliella* sp. (Chlorophyta). *Hydrobiologya*, *267*(1-3), 163–168. doi:10.1007/BF00018798

Torab-Mostaedi, M., Asadollahzadeh, M., Hemmati, A., & Khosravi, A. (2015). Biosorption of lanthanum and cerium from aqueous solutions by grapefruit peel: Equilibrium, kinetic and thermodynamic studies. *Research on Chemical Intermediates*, *41*(2), 559–573. doi:10.100711164-013-1210-4

Tork, M. B., Khalilzadeh, R., & Kouchakzadeh, H. (2017). Efficient harvesting of marine Chlorella vulgaris microalgae utilizing cationic starch nanoparticles by response surface methodology. *Bioresource Technology*, *243*, 583–588.

Toro, C., Reddy, M. M., Navia, R., Rivas, M., Misra, M., & Mohanty, A. K. (2013). Characterization and application in biocomposites of residual microalgal biomass generated in third generation biodiesel. *Journal of Polymers and the Environment*, *21*(4), 944–951. doi:10.100710924-013-0609-8

Torres, S., Navia, R., Campbell Murdy, R., Cooke, P., Misra, M., & Mohanty, A. K. (2015). Green composites from residual microalgae biomass and poly (butylene adipate-co-terephthalate): Processing and plasticization. *ACS Sustainable Chemistry & Engineering*, *3*(4), 614–624. doi:10.1021c500753h

Torzillo, G., Pushparaj, B., Masojidek, J., & Vonshak, A. (2003). Biological constraints in algal biotechnology. *Biotechnology and Bioprocess Engineering; BBE*, *8*(6), 338–348. doi:10.1007/BF02949277

Torzillo, G., Scoma, A., Faraloni, C., Ena, A., & Johanningmeier, U. (2009). Increased hydrogen photoproduction by means of a sulfur-deprived Chlamydomonas reinhardtii D1 protein mutant. *International Journal of Hydrogen Energy*, *34*(10), 4529–4536. doi:10.1016/j.ijhydene.2008.07.093

Torzillo, G., & Vonshak, A. (1994). Effect of light and temperature on the photosynthetic activity of the cyanobacterium *Spirulina plantensis*. *Biomass and Bioenergy*, *6*, 399–403.

Tosin, M., Barbale, M., Chinaglia, S., & Degli-Innocenti, F. (2020). Disintegration and mineralization of mulch films and leaf litter in soil. *Polymer Degradation & Stability*, *179*, 109309. doi:10.1016/j.polymdegradstab.2020.109309

Tosin, M., Weber, M., Siotto, M., Lott, C., & Degli Innocenti, F. (2012). Laboratory test methods to determine the degradation of plastics in marine environmental conditions. *Frontiers in Microbiology*, *3*, 225. doi:10.3389/fmicb.2012.00225 PMID:22737147

Touloupakis, E., Faraloni, C., Silva Benavides, A. M., Masojídek, J., & Torzillo, G. (2021). Sustained photobiological hydrogen production by Chlorella vulgaris without nutrient starvation. *International Journal of Hydrogen Energy*, *46*(5), 3684–3694. doi:10.1016/j.ijhydene.2020.10.257

Tournier, V., Topham, C. M., Gilles, A., David, B., Folgoas, C., Moya-Leclair, E., Kamionka, E., Desrousseaux, M.-L., Texier, H., Gavalda, S., Cot, M., Guémard, E., Dalibey, M., Nomme, J., Cioci, G., Barbe, S., Chateau, M., André, I., Duquesne, S., & Marty, A. (2020). An engineered PET depolymerase to break down and recycle plastic bottles. *Nature*, *580*(7802), 216–219. doi:10.103841586-020-2149-4 PMID:32269349

Towijit, U., Songruk, N., Lindblad, P., Incharoensakdi, A., & Jantaro, S. (2018). Co-overexpression of native phospholipid-biosynthetic genes plsX and plsC enhances lipid production in Synechocystis sp. PCC 6803. *Scientific Reports*, *8*(1), 1–13. doi:10.103841598-018-31789-5 PMID:30201972

Towle, M. J., Salvato, K. A., Budrow, J., Wels, B. F., Kuznetsov, G., Aalfs, K. K., ... Littlefield, B. A. (2001). In vitro and in vivo anticancer activities of synthetic macrocyclic ketone analogues of halichondrin B. *Cancer Research*, *61*(3), 1013–1021. PMID:11221827

Tran, D. T., Lee, H. R., Jung, S., Park, M. S., Yang, J. W., & Chang, Y. K. (2016). Preparation and characterization of poly (vinyl alcohol) biocomposites with microalgae ash. *Journal of Applied Polymer Science*, *133*(26). Advance online publication. doi:10.1002/app.43599

Tran, D.-T., Lee, H. R., Jung, S., Park, M. S., & Yang, J.-W. (2018). Lipid-extracted algal biomass based biocomposites fabrication with poly (vinyl alcohol). *Algal Research*, *31*, 525–533. doi:10.1016/j.algal.2016.08.016

Tran, N.-P., Park, J.-K., Kim, Z.-H., & Lee, C.-G. (2009). Influence of sodium orthovanadate on the production of astaxanthin from green algae Haematococcus lacustris. *Biotechnology and Bioprocess Engineering; BBE*, *14*(3), 322–329. doi:10.100712257-008-0216-z

Tredici, M. R. (2004). Mass production of microalgae: photobioreactors. Handbook of Microalgal Cultures, Biotechnology and Applied Phycology, 178–214.

Tredici, M. R., Bassi, N., Prussi, M., Biondi, N., Rodolfi, L., Chini Zittelli, G., & Sampietro, G. (2015). Energy balance of algal biomass production in a 1-ha "Green Wall Panel" plant: How to produce algal biomass in a closed reactor achieving a high Net Energy Ratio. *Applied Energy*, *154*, 1103–1111. doi:10.1016/j.apenergy.2015.01.086

Tredici, M. R., Biondi, N., & Ponis, E. (2009). Advances in Microalgal Culture for Aquaculture Feed and Other Uses. In G. Burnell & G. Allan (Eds.), *New Technologies in Aquaculture: Improving Production Efficiency* (pp. 611–676). Quality and Environmental Management. doi:10.1533/9781845696474.3.610

Tredici, M. R., Rodolfi, L., Biondi, N., Bassi, N., & Sampietro, G. (2016). Techno-economic analysis of microalgal biomass production in a 1-ha Green Wall Panel (GWP®) plant. *Algal Research*, *19*, 253–263. doi:10.1016/j.algal.2016.09.005

Trentacoste, E. M., Martinez, A. M., & Zenk, T. (2015). The place of algae in agriculture: Policies for algal biomass production. *Photosynthesis Research*, *123*(3), 305–315. doi:10.100711120-014-9985-8 PMID:24599393

Treves, H., Siemiatkowska, B., Luzarowska, U., Murik, O., Fernandez-Pozo, N., Moraes, T. A., Erban, A., Armbruster, U., Brotman, Y., Kopka, J., Rensing, S. A., Szymanski, J., & Stitt, M. (2020). Multi-omics reveals mechanisms of total resistance to extreme illumination of a desert alga. *Nature Plants*, *6*(8), 1031–1043. doi:10.103841477-020-0729-9 PMID:32719473

Triemer, R., Zimba, P., & Rowan, M. (2004). 163 Identification of Euglenoids That Produce Ichthyotoxin(S) (Euglenophyta). *Journal of Phycology*, *39*(s1), 56–57. doi:10.1111/j.0022-3646.2003.03906001_163.x

Tripathi, S., & Hussain, T. (2021). Treatment of Industrial wastewater through new approaches using Algae biomass. In M. Shah (Ed.), *The Future of Effluent Treatment Plants-Biological Treatment Systems* (pp. 89–112). Elsevier., doi:10.1016/B978-0-12-822956-9.00006-4

Trivedi, J., Aila, M., Bangwal, D. P., Kaul, S., & Garg, M. O. (2015). Algae based biorefinery—How to make sense? *Renewable & Sustainable Energy Reviews*, *47*, 295–307. doi:10.1016/j.rser.2015.03.052

Tsang, Y. F., Kumar, V., Samadar, P., Yang, Y., Lee, J., Ok, Y. S., Song, H., Kim, K.-H., Kwon, E. E., & Jeon, Y. J. (2019). Production of bioplastic through food waste valorization. *Environment International*, *127*, 625–644. doi:10.1016/j.envint.2019.03.076 PMID:30991219

Tseng, C. K. (2001). Algal biotechnology industries and research activities in China. *Journal of Applied Phycology*, *13*(4), 375–380. doi:10.1023/A:1017972812576

Tsuji, H. (2013). Poly(Lactic Acid). In Bio-Based Plastics (pp. 171–239). John Wiley & Sons Ltd. doi:10.1002/9781118676646.ch8

Tsuzuki, M., Okada, K., Isoda, H., Hirano, M., Odaka, T., Saijo, H., Aruga, R., Miyauchi, H., & Fujiwara, S. (2019). Physiological properties of photoautotrophic microalgae and cyanobacteria relevant to industrial biomass production. *Marine Biotechnology (New York, N.Y.)*, *21*(3), 406–415. doi:10.100710126-019-09890-1 PMID:30927152

Tuhy, Ł., Saeid, A., & Chojnacka, K. (2020). Algae Fertilizers. Encyclopedia of Marine Biotechnology, 1, 105-119.

Tuhy, L., Chowańska, J., & Chojnacka, K. (2013). Seaweed extracts as biostimulants of plant growth [review]. *Chemik*, *67*(7), 636–641.

Tuhy, Ł., Samoraj, M., Michalak, I., & Chojnacka, K. (2014). The application of biosorption for production of micronutrient fertilizers based on waste biomass. *Applied Biochemistry and Biotechnology*, *174*(4), 1376–1392. doi:10.100712010-014-1074-0 PMID:25108517

Tunsu, C., Petranikova, M., Gergorić, M., Ekberg, C., & Retegan, T. (2015). Reclaiming rare earth elements from end-of-life products: A review of the perspectives for urban mining using hydrometallurgical unit operations. *Hydrometallurgy*, *156*, 239–258. doi:10.1016/j.hydromet.2015.06.007

Turner, J. L., Dritz, S. S., Higgins, J. J., & Minton, J. E. (2002). Effects of Ascophyllum nodosum Extract on Growth Performance and Immune Function of Young Pigs Challenged with Salmonella typhimurium. *Journal of Animal Science*, *80*(7), 1947–1953. doi:10.2527/2002.8071947x PMID:12162664

Tverskoi, D., Makarenkov, V., & Aleskerov, F. (2018). Modeling functional specialization of a cell colony under different fecundity and viability rates and resource constraint. *PLoS One*, *13*(8), e0201446. doi:10.1371/journal.pone.0201446 PMID:30089142

Tyagi, R., Kumar, A., Tyagi, M. B., Jha, P. N., Kumar, H. D., Sinha, R. P., & Hader, D. P. (2003). Protective role of certain chemicals against UV-B-induced damage in the nitrogen-fixing cyanobacterium, *Nostoc muscorum. Journal of Basic Microbiology*, *43*(2), 137–147. doi:10.1002/jobm.200390014 PMID:12746856

U.S. EPA. (2012). *Estimation Programs Interface Suite™ for Microsoft Windows, v 4.10*. United States Environmental Protection Agency.

Ubando, A. T., Felix, C. B., & Chen, W. H. (2020). Biorefineries in circular bioeconomy: A comprehensive review. *Bioresource Technology*, *299*, 122585. doi:10.1016/j.biortech.2019.122585 PMID:31901305

Úbeda-Mínguez, P., García-Maroto, F., & Alonso, D. L. (2017). Heterologous expression of DGAT genes in the marine microalga Tetraselmis chui leads to an increase in TAG content. *Journal of Applied Phycology*, *29*(4), 1913–1926. doi:10.100710811-017-1103-9

Udani, J., & Hesslink, R. (2012). The potential use of fucoidans from brown seaweed as a dietary supplement. *Journal of Nutrition & Food Sciences*, *2*(171), 2. doi:10.4172/2155-9600.1000171

Uddin, M. N., Daud, W. M. A. W., & Abbas, H. F. (2013). Potential hydrogen and non-condensable gases production from biomass pyrolysis: Insights into the process variables. *Renewable & Sustainable Energy Reviews*, *27*, 204–224. doi:10.1016/j.rser.2013.06.031

Udom, I., Zaribaf, B. H., Halfhide, T., Gillie, B., Dalrymple, O., Zhang, Q., & Ergas, S. J. (2013). Harvesting microalgae grown on wastewater. *Bioresource Technology*, *139*, 101–106. doi:10.1016/j.biortech.2013.04.002 PMID:23648758

Uduman, N., Qi, Y., Danquah, M. K., Forde, G. M. A., & Hoadley, A. (2010). Dewatering of microalgal cultures: A major bottleneck to algae-based fuels. *Journal of Renewable and Sustainable Energy*, *2*, 012701–012715.

Ugwu, C. U., Aoyagi, H., & Uchiyama, H. (2008). Photobioreactors for mass cultivation of algae. *Bioresource Technology*, *99*(10), 4021–4028. doi:10.1016/j.biortech.2007.01.046 PMID:17379512

Ulcer preventive and antioxidative properties of astaxanthinfrom Haematococcus pluvialis. (2008). *Eur. J. Pharmacol.*, *590*, 387–395. doi:10.1016/j.ejphar.2008.06.042

Ullmann, J., & Grimm, D. (2021). Algae and their potential for a future bioeconomy, landless food production, and the socio-economic impact of an algae industry. *Organic Agriculture*, *11*(2), 261–267. doi:10.100713165-020-00337-9

Ummalyma, S. B., Gnansounou, E., Sukumaran, R. K., Sindhu, R., Pandey, A., & Sahoo, D. (2017). Bioflocculation: An alternative strategy for harvesting of microalgae—An overview. *Bioresource Technology*, *242*, 227–235.

Ummalyma, S. B., Sahoo, D., & Pandey, A. (2021). Resource recovery through bioremediation of wastewaters and waste carbon by microalgae: A circular bioeconomy approach. *Environmental Science and Pollution Research International*, *28*(42), 58837–58856. Advance online publication. doi:10.100711356-020-11645-8 PMID:33527238

Urala, N., & Lähteenmäki, L. (2004). Attitudes behind consumers' willingness to use functional foods. *Food Quality and Preference*, *15*(7-8), 793–803. doi:10.1016/j.foodqual.2004.02.008

Uribe-Wandurraga, Z. N., Igual, M., García-Segovia, P., & Martínez-Monzó, J. (2020). Influence of microalgae addition in formulation on colour, texture, and extrusion parameters of corn snacks. *Food Science & Technology International*, *26*(8), 685–695. doi:10.1177/1082013220924178 PMID:32375498

Usher, P. K., Ross, A. B., Camargo-Valero, M. A., Tomlin, A. S., & Gale, W. F. (2014). An overview of the potential environmental impacts of large-scale microalgae cultivation. *Biofuels*, *5*(3), 331–349. doi:10.1080/17597269.2014.913925

Utami, A. W. A., Wahyudi, A. T., & Batubara, I. (2014). Toxicity, anticancer and antioxidant activity of extracts from marine bacteria associated with sponge Jaspis sp. *International Journal of Pharma and Bio Sciences*, *5*(4), 917–923.

Utomo, H. D., Tan, K. X. D., Choong, Z. Y. D., Yu, J. J., Ong, J. J., & Lim, Z. B. (2016). Biosorption of heavy metal by algae biomass in surface water. *Journal of Environmental Protection*, *7*(11), 1547–1560.

Uysal, O., Uysal, F. O., & Ekinci, K. (2015). Evaluation of Microalgae as Microbial Fertilizer. *European Journal of Sustainable Development*, *4*(2), 77–82. doi:10.14207/ejsd.2015.v4n2p77

Vadlapudi, V., & Amanchy, R. (2017). Synthesis, Characterization and Antibacterial Activity of Silver Nanoparticles from Red Algae, Hypnea musciformis. *Advances in Biological Research (Faisalabad)*, *11*(5), 242–249.

Vaishampayan, A., Sinha, R. P., Hader, D. P., Dey, T., Gupta, A. K., Bhan, U., & Rao, A. L. (2001). Cyanobacterial biofertilizers in rice agriculture. *Botanical Review*, *67*(4), 453–516. doi:10.1007/BF02857893

Valdovinos-García, E. M., Barajas-Fernández, J., Olán-Acosta, M. Á., Petriz-Prieto, M. A., Guzmán-López, A., & Bravo-Sánchez, M. G. (2020). Techno-economic study of CO2 capture of a thermoelectric plant using microalgae (Chlorella vulgaris) for production of feedstock for bioenergy. *Energies*, *13*(2), 413. doi:10.3390/en13020413

Vallet, M., Strittmatter, M., Murúa, P., Lacoste, S., Dupont, J., Hubas, C., Genta-Jouve, G., Gachon, C. M. M., Kim, G. H., & Prado, S. (2018). Chemically-Mediated Interactions Between Macroalgae, Their Fungal Endophytes, and Protistan Pathogens. *Frontiers in Microbiology*, *9*, 3161. Advance online publication. doi:10.3389/fmicb.2018.03161 PMID:30627120

Van Den Hende, S., Beelen, V., Bore, G., Boon, N., & Vervaeren, H. (2014). Up-scaling aquaculture wastewater treatment by microalgal bacterial flocs: From lab reactors to an outdoor raceway pond. *Bioresource Technology*, *159*, 342–354.

van den Hoek, C., Mann, D., & Jahns, H. M. (1996). *Algae. An Introduction to Phycology*. Cambridge University Press.

Van Eygen, E., Feketitsch, J., Laner, D., Rechberger, H., & Fellner, J. (2017). Comprehensive analysis and quantification of national plastic flows: The case of Austria. *Resources, Conservation and Recycling*, *117*, 183–194. doi:10.1016/j.resconrec.2016.10.017

van Hullebusch, E. D., Yekta, S. S., Fermoso, B. C., & Fermoso, F. G. (2019). Biogeochemistry of major elements in anaerobic digesters: carbon, nitrogen, phosphorus, sulfur and iron. In *Trace Elements in Anaerobic Biotechnologies* (pp. 1–22). IWA Publishing. doi:10.2166/9781789060225_0001

Van Oosten, M. J., Pepe, O., De Pascale, S., Silletti, S., & Maggio, A. (2017). The role of biostimulants and bioeffectors as alleviators of abiotic stress in crop plants. *Chemical and Biological Technologies in Agriculture*, *4*(1), 1–12. doi:10.118640538-017-0089-5

Van Weelden, G., Bobiński, M., Okła, K., Van Weelden, W. J., Romano, A., & Pijnenborg, J. (2019). Fucoidan structure and activity in relation to anti-cancer mechanisms. *Marine Drugs*, *17*(1), 32. doi:10.3390/md17010032 PMID:30621045

Vance-Carroll, P., Uhde-Stone, C., & Allan, D. L. (2003). Phosphorus acquisition and use: Critical adaptations by plants for securing a nonrenewable resource. *The New Phytologist*, *157*(3), 423–447. doi:10.1046/j.1469-8137.2003.00695.x PMID:33873400

Vandamme, D., Beuckels, A., Markou, G., Foubert, I., & Muylaert, K. (2015). Reversible flocculation of microalgae using magnesium hydroxide. *BioEnergy Research*, *8*, 716–725.

Vandamme, D., Foubert, I., Fraeye, I., Meesschaert, B., & Muylaert, K. (2012). Flocculation of *Chlorella vulgaris* induced by high pH: Role of magnesium and calcium and practical implications. *Bioresource Technology*, *105*, 114–119.

Vandamme, D., Foubert, I., & Muylaert, K. (2013). Flocculation as a low-cost method for harvesting microalgae for bulk biomass production. *Trends in Biotechnology*, *31*, 233–239.

Vandepoele, K., Van Bel, M., Richard, G., Van Landeghem, S., Verhelst, B., Moreau, H., Van de Peer, Y., Grimsley, N., & Piganeau, G. (2013). pico-PLAZA, a genome database of microbial photosynthetic eukaryotes. *Environmental Microbiology*, *15*(8), 2147–2153. doi:10.1111/1462-2920.12174 PMID:23826978

Vanier, G., Stelter, S., Vanier, J., Hempel, F., Maier, U. G., Lerouge, P., Ma, J., & Bardor, M. (2018). Alga-made anti-hepatitis B antibody binds to human Fcγ receptors. *Biotechnology Journal*, *13*(4), 1700496. doi:10.1002/biot.201700496 PMID:29194986

Vanthoor-Koopmans, M., Wijffels, R. H., Barbosa, M. J., & Eppink, M. H. M. (2013). Biorefinery of Microalgae for Food and Fuel. *Bioresource Technology*, *135*, 142–149. doi:10.1016/j.biortech.2012.10.135 PMID:23186688

Varela Villarreal, J., Burgués, C., & Rösch, C. (2020). Acceptability of genetically engineered algae biofuels in Europe: Opinions of experts and stakeholders. *Biotechnology for Biofuels*, *13*(1), 1–21. doi:10.118613068-020-01730-y PMID:32489422

Vargas, J. V. C., Mariano, A. B., Corrêa, D. O., & Ordonez, J. C. (2014). The microalgae derived hydrogen process in compact photobioreactors. *International Journal of Hydrogen Energy*, *39*(18), 9588–9598. doi:10.1016/j.ijhydene.2014.04.093

Varghese, A. G., Paul, S. A., & Latha, M. S. (2019). Remediation of Heavy Metals and Dyes from Wastewater Using Cellulose-Based Adsorbents. *Environmental Chemistry Letters*, *17*, 867–877.

Vasudevan, V., Stratton, R. W., Pearlson, M. N., Jersey, G. R., Beyene, A. G., Weissman, J. C., Rubino, M., & Hileman, J. I. (2012). Environmental performance of algal biofuel technology options. *Environmental Science & Technology*, *46*(4), 2451–2459. doi:10.1021/es2026399 PMID:22324757

Vaverková, M., & Adamcová, D. (2015). Biodegrability of bioplastic materials in a controlled composting environment. *Journal of Ecological Engineering*, *16*, 155–160. doi:10.12911/22998993/2949

Vecchi, V., Barera, S., Bassi, R., & Dall'Osto, L. (2020). Potential and Challenges of Improving Photosynthesis in Algae. *Plants (Basel, Switzerland)*, *9*(1), 67. doi:10.3390/plants9010067 PMID:31947868

Vega, J. M., Herrera, J., Aparicio, P. J., Paneque, A., & Losada, M. (1971). Role of molybdenum in nitrate reduction by Chlorella. *Plant Physiology*, *48*(3), 294–299. doi:10.1104/pp.48.3.294 PMID:16657784

Venkata Subhash, G., Rajvanshi, M., Navish Kumar, B., Govindachary, S., Prasad, V., & Dasgupta, S. (2017). Carbon streaming in microalgae: Extraction and analysis methods for high value compounds. *Bioresource Technology*, *244*, 1304–1316. doi:10.1016/j.biortech.2017.07.024 PMID:28803061

Venkatesan, J., Kim, S.-K., & Shim, M. S. (2016). Antimicrobial, antioxidant, and anticancer activities of biosynthesized silver nanoparticles using marine algae Ecklonia cava. *Nanomaterials (Basel, Switzerland)*, *6*(12), 235. doi:10.3390/nano6120235 PMID:28335363

Venugopal, V. (2009). *Marine Products for Healthcare*. CRC Press.

Vermuë, M. H., Eppink, M. H. M., Wijffels, R. H., & Van Den Berg, C. (2018). Multi-Product Microalgae Biorefineries: From Concept towards Reality. *Trends in Biotechnology*, *36*(2), 216–227. doi:10.1016/j.tibtech.2017.10.011 PMID:29132753

Verslues, P. E., Agarwal, M., Katiyar-Agarwal, S., Zhu, J., & Zhu, J. K. (2006). Methods and concepts in quantifying resistance to drought, salt and freezing, abiotic stresses that affect plant water status. *The Plant Journal*, *45*(4), 523–539. doi:10.1111/j.1365-313X.2005.02593.x PMID:16441347

Vert, M., Doi, Y., Hellwich, K.-H., Hess, M., Hodge, P., Kubisa, P., Rinaudo, M., & Schué, F. (2012). Terminology for biorelated polymers and applications (IUPAC Recommendations 2012). *Pure and Applied Chemistry*, *84*(2), 377–410. doi:10.1351/PAC-REC-10-12-04

Viêgas, C. V., Hachemi, I., Freitas, S. P., Mäki-Arvela, P., Aho, A., Hemming, J., Smeds, A., Heinmaa, I., Fontes, F. B., da Silva Pereira, D. C., Kumar, N., Aranda, D. A. G., & Murzin, D. Y. (2015). A route to produce renewable diesel from algae: Synthesis and characterization of biodiesel via in situ transesterification of Chlorella alga and its catalytic deoxygenation to renewable diesel. *Fuel*, *155*, 144–154. doi:10.1016/j.fuel.2015.03.064

Vijayan, S. R., Santhiyagu, P., Singamuthu, M., Ahila, N. K., Jayaraman, R., & Ethiraj, K. (2014). Synthesis and characterization of silver and gold nanoparticles using aqueous extract of seaweed, Turbinariaconoides, and their antimicrofouling activity. *TheScientificWorldJournal*, *2014*, 1–10. doi:10.1155/2014/938272 PMID:24672397

Vijayaraghavan, K., Jegan, J., Palanivelu, K., & Velan, M. (2004). Removal of nickel (II) ions from aqueous solution using crab shell particles in a packed bed up-flow column. *Journal of Hazardous Materials*, *113*(1–3), 223–230. doi:10.1016/j.jhazmat.2004.06.014 PMID:15363535

Vijayaraghavan, K., Mahadevan, A., Joshi, U. M., & Balasubramanian, R. (2009). An examination of the uptake of lanthanum from aqueous solution by crab shell particles. *Chemical Engineering Journal*, *152*(1), 116–121. doi:10.1016/j.cej.2009.03.040

Vijayaraghavan, K., Sathishkumar, M., & Balasubramanian, R. (2010). Biosorption of lanthanum, cerium, europium, and ytterbium by a brown marine alga, Turbinaria conoides. *Industrial & Engineering Chemistry Research*, *49*(9), 4405–4411. doi:10.1021/ie1000373

Vikram, G., Srikanth, K., & Swamy, N. R. (2011). Effect of plant growth regulatorson in vitro organogenesis in cultivated tomato (*Solanum lycopersicum* L.). *J. Research in Biology*, *1*, 263–268.

Villain, J., Lozano, S., Halm-Lemeille, M.-P., Durrieu, G., & Bureau, R. (2014). Quantile regression model for a diverse set of chemicals: Application to acute toxicity for green algae. *Journal of Molecular Modeling*, *20*(12), 2508. Advance online publication. doi:10.100700894-014-2508-x PMID:25431186

Villanueva, R., Sousa, A., Gonçalves, M., Nilsson, M., & Hilliou, L. (2010). Production and properties of agar from the invasive marine alga, Gracilaria vermiculophylla (Gracilariales, Rhodophyta). *Journal of Applied Phycology*, *22*(2), 211–220. doi:10.100710811-009-9444-7

Vimal Kumar, R., Kanna, G. R., & Elumalai, S. (2017). Biodegradation of Polyethylene by Green Photosynthetic Microalgae. *Journal of Bioremediation & Biodegradation*, *08*(01). Advance online publication. doi:10.4172/2155-6199.1000381

Vincy, W.; Mahathalana, T.J.; Sukumaran, S.; Jeeva, S. (2017). Algae as a source for synthesis of nanoparticles-a review. *Int. J. Latest Trends Eng. Technol.*, 5–9.

Vinoth, S., Gurusaravanan, P., & Jayabalan, N. (2014). Optimization of somaticembryogenesis protocol in *Lycopersicon esculentum* L. using plant growth regulators and seaweed extracts. *Journal of Applied Phycology*, *26*(3), 1527–1537. doi:10.100710811-013-0151-z

Vishal, K. D., & Abhishek, C. (2014). Isolation and characterization of Rhizobium leguminosarum from root nodules of *Pisums sativum* L. *Journal of Academia and Industrial Research*, *2*(8), 464–467.

Volesky, B., Weber, J., & Park, J. M. (2003). Continuous-flow metal biosorption in a regenerable Sargassum column. *Water Research*, *37*(2), 297–306. doi:10.1016/S0043-1354(02)00282-8 PMID:12502059

Voleti, R. S. (2012). Experimental studies of vertical mixing in an open channel raceway for algae biofuel production (M.Sc. thesis). Utah State University.

Voloshin, R. A., Rodionova, M. V., Zharmukhamedov, S. K., Veziroglu, T. N., Allakhverdiev, S. I., Nejat Veziroglu, T., & Allakhverdiev, S. I. (2016). Review: Biofuel production from plant and algal biomass. *International Journal of Hydrogen Energy*, *41*(39), 17257–17273. doi:10.1016/j.ijhydene.2016.07.084

Volova, T. G., Boyandin, A. N., Vasiliev, A. D., Karpov, V. A., Prudnikova, S. V., Mishukova, O. V., Boyarskikh, U. A., Filipenko, M. L., Rudnev, V. P., Bá Xuân, B., Việt Dũng, V., & Gitelson, I. I. (2010). Biodegradation of polyhydroxyalkanoates (PHAs) in tropical coastal waters and identification of PHA-degrading bacteria. *Polymer Degradation & Stability*, *95*(12), 2350–2359. doi:10.1016/j.polymdegradstab.2010.08.023

Vo, T. S., Ngo, D. H., Kang, K. H., Jung, W. K., & Kim, S. K. (2015). The beneficial properties of marine polysaccharides in alleviation of allergic responses. *Molecular Nutrition & Food Research*, *59*(1), 129–138. doi:10.1002/mnfr.201400412 PMID:25379652

Vu, C. H. T., Lee, H. G., Chang, Y. K., & Oh, H. M. (2018). Axenic cultures for microalgal biotechnology: Establishment, assessment, maintenance, and applications. *Biotechnology Advances*, *36*(2), 380–396. doi:10.1016/j.biotechadv.2017.12.018 PMID:29292155

Vymazal, J. (1995). *Algae and element cycling in wetlands*. Lewis Publishers Boca Raton.

Wagner, J., Bransgrove, R., Beacham, T. A., Allen, M. J., Meixner, K., Drosg, B., Ting, V. P., & Chuck, C. J. (2016). Co-production of bio-oil and propylene through the hydrothermal liquefaction of polyhydroxybutyrate producing cyanobacteria. *Bioresource Technology*, *207*, 166–174. doi:10.1016/j.biortech.2016.01.114 PMID:26881334

Waldenstedt, L., Inborr, J., Hansson, I., & Elwinger, K. (2003). Effects of astaxanthin-rich algal meal (Haematococcus pluvialis) on growth performance, caecal campylobacter and clostridial counts and tissue astaxanthin concentration of broiler chickens. *Animal Feed Science and Technology*, *108*(1-4), 119–132. doi:10.1016/S0377-8401(03)00164-0

Wang, Y., Wang, X., & Wu, D. (2019a). Crystalline Characteristics, Mechanical Properties, Thermal Degradation Kinetics and Hydration Behavior of Biodegradable Fibers Melt-Spun from Polyoxymethylene/Poly(l-lactic acid) Blends. *Polymers*, *11*(11), 1753. Advance online publication. doi:10.3390/polym11111753 PMID:31731470

Wang, B., Li, Y., Wu, N., & Lan, C. Q. (2008). CO2 bio-mitigation using microalgae. *Applied Microbiology and Biotechnology*, *79*(5), 707–718. doi:10.100700253-008-1518-y PMID:18483734

Wang, B., Qin, L., Mu, T., Xue, Z., & Gao, G. (2017). Are Ionic Liquids Chemically Stable? *Chemical Reviews*, *117*(10), 7113–7131. doi:10.1021/acs.chemrev.6b00594 PMID:28240867

Wang, C., Liu, Y., Li, S. S., & Han, G. Z. (2014). Origin of plant auxin biosynthesis in charophyte algae. *Trends in Plant Science*, *19*(12), 741–743. doi:10.1016/j.tplants.2014.10.004 PMID:25457112

Wang, C., Li, Y., Lu, J., Deng, X., Li, H., & Hu, Z. (2018). Effect of overexpression of LPAAT and GPD1 on lipid synthesis and composition in green microalga Chlamydomonas reinhardtii. *Journal of Applied Phycology*, *30*(3), 1711–1719. doi:10.100710811-017-1349-2 PMID:29899598

Wang, H., Laughinghouse, H. D., Anderson, M. A., Chen, F., Willliams, E., Place, A. R., & Hill, R. T. (2012). Novel bacterial isolate from Permian groundwater, capable of aggregating potential biofuel-producing microalga *Nannochloropsis oceanica* IMET1. *Applied and Environmental Microbiology*, *78*, 1445–1453. https:// doi.org/10.1128/AEM.06474-11

Wang, H.-M. D., Li, X.-C., Lee, D.-J., & Chang, J.-S. (2017). Potential biomedical applications of marine algae. *Bioresource Technology*, *244*, 1407–1415. doi:10.1016/j.biortech.2017.05.198 PMID:28697977

Wang, J., & Yin, Y. (2018). Fermentative hydrogen production using pretreated microalgal biomass as feedstock. *Microbial Cell Factories*, *17*(1), 22. doi:10.118612934-018-0871-5 PMID:29444681

Wang, K. (2014). *Bio-Plastic Potential of Spirulina Microalgae*. The University of Georgia. https://getd.libs.uga.edu/pdfs/wang_kun_201405_ms.pdf

Wang, K., Mandal, A., Ayton, E., Hunt, R., Zeller, M. A., & Sharma, S. (2016). Modification of protein rich algal-biomass to form bioplastics and odor removal. In *Protein Byproducts* (pp. 107–117). Elsevier. doi:10.1016/B978-0-12-802391-4.00006-9

Wang, K., Xu, Z., & Liao, X. (2021). Bioactive compounds, health benefits, and functional food products of sea buckthorn: A review. *Critical Reviews in Food Science and Nutrition*, 1–22. doi:10.1080/10408398.2021.1905605 PMID:33783272

Wang, L., Liang, W., Yu, J., Liang, Z., Ruan, L., & Zhang, Y. (2013). Flocculation of Microcystis aeruginosa using modified larch tannin. *Environmental Science & Technology*, *47*, 5771–5777.

Wang, L., Xue, C., Wang, L., Zhao, Q., Wei, W., & Sun, Y. (2016). Strain improvement of *Chlorella* sp. for phenol biodegradation by adaptive laboratory evolution. *Bioresource Technology*, *205*, 264–268. doi:10.1016/j.biortech.2016.01.022 PMID:26803904

Wang, M., Chen, S., Zhou, W., Yuan, W., & Wang, D. (2020). Algal cell lysis by bacteria: A review and comparison to conventional methods. *Algal Research*, *46*, 101794. doi:10.1016/j.algal.2020.101794

Wang, M., Yuan, W., Chen, S., Wang, L., Zhao, S., & Li, S. (2021). Algal Lysis by Sagittula stellata for the Production of Intracellular Valuables. *Applied Biochemistry and Biotechnology*, *193*(8), 2516–2533. doi:10.100712010-021-03502-2 PMID:33779932

Wang, S., Wang, X., Poon, K., Wang, Y., Li, S., Liu, H., Lin, S., & Cai, Z. (2013). Removal and reductive dechlorination of triclosan by *Chlorella pyrenoidosa*. *Chemosphere*, *92*, 1498–1505.

Wang, X., Peng, S., & Dong, L. (2005). Effect of poly(vinyl acetate) (PVAc) on thermal behavior and mechanical properties of poly(3-hydroxybutyrate)/poly(propylene carbonate) (PHB/ PPC) blends. *Colloid & Polymer Science*, *284*(2), 167–174. doi:10.100700396-005-1353-y

Wang, X., Wu, Y., Zhang, T., Xu, X., Dao, G., & Hu, H. (2016). Simultaneous nitrogen, phosphorous, and hardness removal from reverse osmosis concentrate by microalgae cultivation. *Water Research*, *94*, 215–224.

Wang, Y., Guo, W., Yen, H. W., Ho, S. H., Lo, Y. C., Cheng, C. L., Ren, N., & Chang, J. S. (2015). Cultivation of Chlorella vulgaris JSC-6 with swine wastewater for simultaneous nutrient/COD removal and carbohydrate production. *Bioresource Technology*, *198*, 619–625.

Wang, Y., Tahir, N., Cao, W., Zhang, Q., & Lee, D.-J. (2019). Grid columnar flat panel photobioreactor with immobilized photosynthetic bacteria for continuous photofermentative hydrogen production. *Bioresource Technology*, *291*, 121806. doi:10.1016/j.biortech.2019.121806 PMID:31326683

Wanke, A., Malisic, M., Wawra, S., & Zuccaro, A. (2021). Unraveling the sugar code: The role of microbial extracellular glycans in plant–microbe interactions. *Journal of Experimental Botany*, *72*(1), 15–35. doi:10.1093/jxb/eraa414 PMID:32929496

Ward, A. J., Lewis, D. M., & Green, F. B. (2014). Anaerobic digestion of algae biomass: A review. *Algal Research*, *5*, 204–214. doi:10.1016/j.algal.2014.02.001

Ward, D. M., & Castenholz, R. W. (2000). Cyanobacteria in geothermal habitats. In B. A. Whitton & M. Potts (Eds.), *The ecology of cyanobacteria*. Kluwer Academic Publishers.

Watanabe, I., Lee, K. K., Alimagno, B. V., Sato, M., Del Rosario, D. C., & De Guzmán, M. R. (1977). *Biological nitrogen fixation in paddy field studied by in situ acetylene-reduction assays*. IRRI Research Paper Series (Philippines). no. 3.

Waterbury, J. B., & Stanier, R. Y. (1981). Isolation and growth of cyanobacteria from marine and hypersaline environments. The Prokaryotes, [illegible]. doi:10.1007/978-3-662-13187-9_9

Waterbury, J. B., & Willey, J. M. (1988). Isolation and growth of marine planktonic cyanobacteria. *Methods in Enzymology*, *167*, 100–105. doi:10.1016/0076-6879(88)67009-1

Wattier, R. A., Prodöhl, P. A., & Maggs, C. A. (2000). DNA isolation protocol for red seaweed (rhodophyta). *Plant Molecular Biology Reporter*, *18*(3), 275–281. doi:10.1007/BF02823999

Weaver, P. F., Lien, S., & Seibert, M. (1980). Photobiological production of hydrogen. *Solar Energy*, *24*(1), 3–45. doi:10.1016/0038-092X(80)90018-3

Wei, L., Jiang, Z., & Liu, B. (2022). A CRISPR/dCas9-based transcription activated system developed in marine microalga *Nannochloropsis oceanica*. *Aquaculture (Amsterdam, Netherlands)*, *546*, 737064. doi:10.1016/j.aquaculture.2021.737064

Wei, L., Liang, S., & McDonald, A. G. (2015). Thermophysical properties and biodegradation behavior of green composites made from polyhydroxybutyrate and potato peel waste fermentation residue. *Industrial Crops and Products*, *69*, 91–103. doi:10.1016/j.indcrop.2015.02.011

Wei, L., Wang, Q., Xin, Y., Lu, Y., & Xu, J. (2017). Enhancing photosynthetic biomass productivity of industrial oleaginous microalgae by overexpression of RuBisCO activase. *Algal Research*, *27*, 366–375. doi:10.1016/j.algal.2017.07.023

Wei, L., You, W., Gong, Y., El Hajjami, M., Liang, W., Xu, J., & Poetsch, A. (2020). Transcriptomic and proteomic choreography in response to light quality variation reveals key adaption mechanisms in marine *Nannochloropsis oceanica*. *The Science of the Total Environment*, *720*, 137667. https:// doi.org/10.1016/j.scitotenv.2020.137667

Wei, N., Quarterman, J., & Jin, Y.-S. (2013). Marine macroalgae: An untapped resource for producing fuels and chemicals. *Trends in Biotechnology*, *31*(2), 70–77. doi:10.1016/j.tibtech.2012.10.009 PMID:23245657

Wei, R., & Zimmermann, W. (2017). Microbial enzymes for the recycling of recalcitrant petroleum-based plastics: How far are we? *Microbial Biotechnology*, *10*(6), 1308–1322. doi:10.1111/1751-7915.12710 PMID:28371373

Wei, W., Ning, L.-W., Ye, Y.-N., Li, S.-J., Zhou, H.-Q., Huang, J., & Guo, F.-B. (2014). SMAL: A Resource of Spontaneous Mutation Accumulation Lines. *Molecular Biology and Evolution*, *31*(5), 1302–1308. doi:10.1093/molbev/msu073 PMID:24531082

Wei, Z., Wang, H., Li, X., Zhao, Q., Yin, Y., Xi, L., Ge, B., & Qin, S. (2020). Enhanced biomass and lipid production by co-cultivation of Chlorella vulgaris with Mesorhizobium sangaii under nitrogen limitation. *Journal of Applied Phycology*, *32*(1), 233–242. doi:10.100710811-019-01924-4

Wellnitz, T. A., & Ward, J. V. (2002). Does light intensity modify the effect mayfly grazers have on periphyton? *Freshwater Biology*, *39*, 135–149. https://doi.org/10.1046/j.1365-2427.1998.00270.x

Wells, M. L., Potin, P., Craigie, J. S., Raven, J. A., Merchant, S. S., Helliwell, K. E., Smith, A. G., Camire, M. E., & Brawley, S. H. (2017). Algae as Nutritional and Functional Food Sources: Revisiting Our Understanding. *Journal of Applied Phycology*, *29*(2), 949–982. doi:10.100710811-016-0974-5 PMID:28458464

Wells, M. L., Trainer, V. L., Smayda, T. J., Karlson, B. S. O., Trick, C. G., Kudela, R. M., Ishikawa, A., Bernard, S., Wulff, A., Anderson, D. M., & Cochlan, W. P. (2015). Harmful algal blooms and climate change: Learning from the past and present to forecast the future. *Harmful Algae*, *49*, 68–93. doi:10.1016/j.hal.2015.07.009 PMID:27011761

Wenderoth, K., & Rhiel, E. (2004). Influence of light quality and gassing on the vertical migration of diatoms inhabiting the Wadden Sea. *Helgoland Marine Research*, *58*, 211–215. https://doi.org/10.1007/s10152-004-0187-1

Weng, Y.-X., Wang, X.-L., & Wang, Y.-Z. (2011). Biodegradation behavior of PHAs with different chemical structures under controlled composting conditions. *Polymer Testing*, *30*(4), 372–380. doi:10.1016/j.polymertesting.2011.02.001

Wen, Z. Y., & Chen, F. (2003). Heterotrophic production of eicosapentaenoic acid by microalgae. *Biotechnology Advances*, *21*(4), 273–294. doi:10.1016/S0734-9750(03)00051-X PMID:14499126

Weschler, M. K., Barr, W. J., Harper, W. F., & Landis, A. E. (2014). Process energy comparison for the production and harvesting of algal biomass as a biofuel feedstock. *Bioresource Technology*, *153*, 108–115. doi:10.1016/j.biortech.2013.11.008 PMID:24355501

Wiederstein, M., & Sippl, M. J. (2007). ProSA-web: Interactive web service for the recognition of errors in three-dimensional structures of proteins. *Nucleic Acids Research*, *35*(Web Server), 407–410. doi:10.1093/nar/gkm290 PMID:17517781

Wieners, P. C., Mudimu, O., & Bilger, W. (2012). Desiccation-induced non-radiative dissipation in isolated green lichen algae. *Photosynthesis Research*, *113*(1), 239–247. doi:10.100711120-012-9771-4 PMID:22833109

Wienhausen, G., Noriega-Ortega, B. E., Niggemann, J., Dittmar, T., & Simon, M. (2017). The exometabolome of two model strains of the Roseobacter group: A marketplace of microbial metabolites. *Frontiers in Microbiology*, *8*, 1985. doi:10.3389/fmicb.2017.01985 PMID:29075248

Wigmosta, M. S., Coleman, A. M., Skaggs, R. J., Huesemann, M. H., & Lane, L. J. (2011). National microalgae biofuel production potential and resource demand. *Water Resources Research*, *47*(3). Advance online publication. doi:10.1029/2010WR009966

Wijesinghe, W., & Jeon, Y.-J. (2012). Biological activities and potential industrial applications of fucose rich sulfated polysaccharides and fucoidans isolated from brown seaweeds: A review. *Carbohydrate Polymers*, *88*(1), 13–20. doi:10.1016/j.carbpol.2011.12.029

Wijffels, R. H., Barbosa, M. J., & Eppink, M. H. M. (2010). Microalgae for the production of bulk chemicals and biofuels. *Biofuels, Bioproducts & Biorefining*, *4*(3), 287–295. doi:10.1002/bbb.215

Wilhelm, C., & Jakob, T. (2011, December). From photons to biomass and biofuels: Evaluation of different strategies for the improvement of algal biotechnology based on comparative energy balances. *Applied Microbiology and Biotechnology*, *92*(5), 909–919. doi:10.100700253-011-3627-2 PMID:22005740

Willert, H. G., Semlitsch, M., & Peltier, L. F. (1996). Tissue Reactions to Plastic and Metallic Wear Products of Joint Endoprostheses. *Clinical Orthopaedics and Related Research, 333*. doi:10.1097/00003086-199612000-00002

Wilson, R. H., Alonso, H., & Whitney, S. M. (2016). Evolving Methanococcoides burtonii archaeal Rubisco for improved photosynthesis and plant growth. *Scientific Reports*, *6*(1), 22284. doi:10.1038rep22284 PMID:26926260

Win, T. T., Barone, G. D., Secundo, F., & Fu, P. (2018). Algal biofertilizers and plant growth stimulants for sustainable agriculture. *Industrial Biotechnology (New Rochelle, N.Y.)*, *14*(4), 203–211. doi:10.1089/ind.2018.0010

Wintsche, B., Glaser, K., Sträuber, H., Centler, F., Liebetrau, J., Harms, H., & Kleinsteuber, S. (2016). Trace Elements Induce Predominance among Methanogenic Activity in Anaerobic Digestion. *Frontiers in Microbiology*, *7*. Advance online publication. doi:10.3389/fmicb.2016.02034 PMID:28018337

Winwood, R. (2013). Recent developments in the commercial production of DHA and EPA rich oils from micro-algae. *OCL*, *20*(6), D604. doi:10.1051/ocl/2013030

Wiśniewska, K., Lewandowska, A. U., & Śliwińska-Wilczewska, S. (2019). The importance of cyanobacteria and micro-algae present in aerosols to human health and the environment–Review study. *Environment International*, *131*(104964), 1–11. doi:10.1016/j.envint.2019.104964 PMID:31351382

Wong, T. K.-M., Ho, C.-L., Lee, W.-W., Rahim, R. A., & Phang, S.-M. (2007). Analysis of expressed sequence tags from *Sargassum binderi* (phaeophyta). *Journal of Phycology*, *43*(3), 528–534. doi:10.1111/j.1529-8817.2007.00349.x

Wood, W. F. (1987). Effect of solar ultra-violet radiation on the kelp *Ecklonia radiata. Marine Biology*, *96*(1), 143–150. doi:10.1007/BF00394848

Woodwell, G. M., Mackenzie, F. T., Houghton, R. A., Apps, M. J., Gorham, E., & Davidson, E. A. (1998). Biotic feedbacks in the warming of the Earth. *Climatic Change*, *40*(3-4), 495–518. doi:10.1023/A:1005345429236

Worden, A. Z., Follows, M. J., Giovannoni, S. J., Wilken, S., Zimmerman, A. E., & Keeling, P. J. (2015). Rethinking the marine carbon cycle: Factoring in the multifarious lifestyles of microbes. *Science*, *347*(6223), 1257594. doi:10.1126cience.1257594 PMID:25678667

World Bank. (2008). World Development Report 2008: Agriculture for Development. Washington, DC. © World Bank. *Journal of Chemical Information and Modeling*, *53*(9).

Wrede, D., Taha, M., Miranda, A. F., Kadali, K., Stevenson, T., Ball, A. S., & Mouradov, A. (2014). Co-cultivation of fungal and microalgal cells as an efficient system for harvesting microalgal cells, lipid production and wastewater treatment. *PLoS One*, *9*, e113497.

Wright, S. L., & Kelly, F. J. (2017). Plastic and Human Health: A Micro Issue? *Environmental Science & Technology*, *51*(12), 6634–6647. doi:10.1021/acs.est.7b00423 PMID:28531345

Wuang, S. C., Khin, M. C., Chua, P. Q. D., & Luo, Y. D. (2016). Use of Spirulina biomass produced from treatment of aquaculture wastewater as agricultural fertilizers. *Algal Research*, *15*, 59–64. doi:10.1016/j.algal.2016.02.009

Wu, C.-S. (2011). Preparation, characterization, and biodegradability of renewable resource-based composites from recycled polylactide bioplastic and sisal fibers. *Journal of Applied Polymer Science*, *123*(1), 347–355. doi:10.1002/app.34223

Wu, C.-S. (2014). Preparation and characterization of polyhydroxyalkanoate bioplastic-based green renewable composites from rice husk. *Journal of Polymers and the Environment*, *22*(3), 384–392. doi:10.100710924-014-0662-y

Wu, D., Zhang, L., Wang, L., Zhu, B., & Fan, L. (2011). Adsorption of lanthanum by magnetic alginate-chitosan gel beads. *Journal of Chemical Technology and Biotechnology (Oxford, Oxfordshire)*, *86*(3), 345–352. doi:10.1002/jctb.2522

Wu, G. (2010). *Assay Development: Fundamentals and Practices*. Wiley.

Wu, Q., Liu, L., Miron, A., Klímová, B., Wan, D., & Kuča, K. (2016). The antioxidant, immunomodulatory, and anti-inflammatory activities of Spirulina: An overview. *Archives of Toxicology*, *90*(8), 1817–1840. doi:10.100700204-016-1744-5 PMID:27259333

Wu, T. D., & Watanabe, C. K. (2005). GMAP: A genomic mapping and alignment program for mRNA and EST sequences. *Bioinformatics (Oxford, England)*, *21*(9), 1859–1875. doi:10.1093/bioinformatics/bti310 PMID:15728110

Wu, W., Liu, X., Wang, M., Meyer, R. S., Luo, X., Ndjiondjop, M.-N., Tan, L., Zhang, J., Wu, J., Cai, H., Sun, C., Wang, X., Wing, R. A., & Zhu, Z. (2017). A single-nucleotide polymorphism causes smaller grain size and loss of seed shattering during African rice domestication. *Nature Plants*, *3*(6), 17064. doi:10.1038/nplants.2017.64 PMID:28481332

Wyatt, N. B., Gloe, L. M., Brady, P. V., Hewson, J. C., Grillet, A. M., Hankins, M. G., & Pohl, P. I. (2012). Critical conditions for ferric chloride-induced flocculation of freshwater algae. *Biotechnology and Bioengineering*, *109*(2), 493–501.

Xia, B., Chen, B., Sun, X., Qu, K., Ma, F., & Du, M. (2015). Interaction of TiO2 nanoparticles with the marine microalga Nitzschia closterium: Growth inhibition, oxidative stress and internalization. *The Science of the Total Environment*, *508*, 525–533. doi:10.1016/j.scitotenv.2014.11.066 PMID:25483108

Xia, W., Zhu, N., Ni, Z., & Chen, M. (2017). Fabrication and characterization of biodegradable composites from poly (butylene succinate-co-butylene adipate) and Taihu Lake blue algae. *Advanced Composites Letters*, *26*(5), 096369351702600501. doi:10.1177/096369351702600501

Xing, H., Yuan, Y., Huiyuan, Z., Yong, S., Bingxi, L., & Heping, T. (2017). Solar thermochemical hydrogen production using metallic oxides. *Energy Sources. Part A, Recovery, Utilization, and Environmental Effects*, *39*(3), 257–263. doi:10.1080/15567036.2013.878769

Xin, L., Hong-Ying, H., Ke, G., & Ying-Xue, S. (2010). Effects of different nitrogen and phosphorus concentrations on the growth, nutrient uptake, and lipid accumulation of a freshwater microalga Scenedesmus sp. *Bioresource Technology*, *101*(14), 5494–5500. doi:10.1016/j.biortech.2010.02.016 PMID:20202827

Xiong, J. Q., Kurade, M. B., Abou-Shanab, R. A. I., Ji, M. K., Choi, J., Kim, J. O., & Jeon, B. H. (2016). Biodegradation of carbamazepine using freshwater microalgae *Chlamydomonas mexicana* and *Scenedesmus obliquus* and the determination of its metabolic fate. *Bioresource Technology*, *205*, 183–190.

Xiong, J. Q., Kurade, M. B., & Jeon, B. H. (2017). Ecotoxicological effects of enrofloxacin and its removal by monoculture of microalgal species and their consortium. *Environ. Pollut.*, *226*, 486–493.

Xiong, J. Q., Kurade, M. B., & Jeon, B. H. (2018). Can microalgae remove pharmaceutical contaminants from water? *Trends in Biotechnology*, *36*(1), 30–44. doi:10.1016/j.tibtech.2017.09.003 PMID:28993012

Xiu, S., & Shahbazi, A. (2012). Bio-Oil Production and Upgrading Research : A Review. *Renewable & Sustainable Energy Reviews*, *16*(7), 4406–4414. doi:10.1016/j.rser.2012.04.028

Xu, C., & Leskovar, D. I. (2015). Effects of *A. nodosum* seaweed extracts on spinach growth, physiology and nutrition value under drought stress. *Scientia Horticulturae*, *183*, 39–47. doi:10.1016/j.scienta.2014.12.004

Xue, F., Miao, J., Zhang, X., & Tan, T. (2010). A New Strategy for Lipid Production by Mix Cultivation of Spirulina platensis and Rhodotorula glutinis. *Applied Biochemistry and Biotechnology*, *160*(2), 498–503. doi:10.100712010-008-8376-z PMID:18931954

Xue, L., Zhang, Y., Zhang, T., An, L., & Wang, X. (2005). Effects of enhanced ultraviolet-B radiation on algae and cyanobacteria. *Critical Reviews in Microbiology*, *31*(2), 79–89. doi:10.1080/10408410590921727 PMID:15986833

Xue, W., Wu, C., Xiao, K., Huang, X., Zhou, H., Tsuno, H., & Tanaka, H. (2010). Elimination and fate of selected micro-organic pollutants in a full-scale anaerobic/anoxic/aerobic process combined with membrane bioreactor for municipal wastewater reclamation. *Water Research*, *44*, 5999–6010.

Xu, J., & Guo, B.-H. (2010). Poly(butylene succinate) and its copolymers: Research, development and industrialization. *Biotechnology Journal*, *5*(11), 1149–1163. doi:10.1002/biot.201000136 PMID:21058317

Xu, J., Wang, X., Sun, S., Zhao, Y., & Hu, C. (2017). Effects of influent C/N ratios and treatment technologies on integral biogas upgrading and pollutants removal from synthetic domestic sewage. *Scientific Reports*, *7*(1), 10897. doi:10.103841598-017-11207-y PMID:28883448

Xu, K., Chen, H., Wang, W., Xu, Y., Ji, D., Chen, C., & Xie, C. (2017). Responses of photosynthesis and CO2 concentrating mechanisms of marine crop *Pyropia haitanensis* thalli to large pH variations at different time scales. *Algal Research*, *28*, 200–210. doi:10.1016/j.algal.2017.10.023

Xu, L., Brilman, D. W. F., Withag, J. A. M., Brem, G., & Kersten, S. (2011). Assessment of a dry and a wet route for the production of biofuels from microalgae: Energy balance analysis. *Bioresource Technology*, *102*(8), 5113–5122. doi:10.1016/j.biortech.2011.01.066 PMID:21334889

Xu, L., Wang, F., Li, H. Z., Hu, Z. M., & Guo, C. (2010). Development of an efficient electroflocculation technology integrated with dispersed air flotation for harvesting microalgae. *Journal of Chemical Technology and Biotechnology (Oxford, Oxfordshire)*, *85*, 1504–1507.

Xu, L., Weathers, P., Xiong, X.-R., & Liu, C.-Z. (2009). Microalgal Bioreactors: Challenges and Opportunities. *Engineering in Life Sciences*, *9*(3), 178–189. doi:10.1002/elsc.200800111

Xu, R., Zhang, K., Liu, P., Khan, A., Xiong, J., Tian, F., & Li, X. (2018). A critical review on the interaction of substrate nutrient balance and microbial community structure and function in anaerobic co-digestion. *Bioresource Technology*, *247*, 1119–1127. doi:10.1016/j.biortech.2017.09.095 PMID:28958888

Xu, S., Elsayed, M., Ismail, G. A., Li, C., Wang, S., & Abomohra, A. E.-F. (2019). Evaluation of Bioethanol and Biodiesel Production from Scenedesmus Obliquus Grown in Biodiesel Waste Glycerol: A Sequential Integrated Route for Enhanced Energy Recovery. *Energy Conversion and Management*, *197*, 111907. doi:10.1016/j.enconman.2019.111907

Xu, Y., Purton, S., & Baganz, F. (2013). Chitosan flocculation to aid the harvesting of the microalga *Chlorella sorokiniana*. *Bioresource Technology*, *129*, 296–301.

Xu, Y., Wu, Q., Lei, Y., & Yao, F. (2010). Creep behavior of bagasse fiber reinforced polymer composites. *Bioresource Technology*, *101*(9), 3280–3286. doi:10.1016/j.biortech.2009.12.072 PMID:20064712

Yaakob, M. A., Mohamed, R. M. S. R., Al-Gheethi, A., Aswathnarayana Gokare, R., & Ambati, R. R. (2021). Influence of Nitrogen and Phosphorus on Microalgal Growth, Biomass, Lipid, and Fatty Acid Production: An Overview. *Cells*, *10*(2), 10–20. doi:10.3390/cells10020393 PMID:33673015

Yadav, G., & Sen, R. (2017). Microalgal green refinery concept for biosequestration of carbon-dioxide vis-à-vis wastewater remediation and bioenergy production: Recent technological advances in climate research. *Journal of CO2 Utilization*, *17*, 188–206. doi:10.1016/j.jcou.2016.12.006

Yamada, K., Suzuki, H., Takeuchi, T., Kazama, Y., Mitra, S., Abe, T., Goda, K., Suzuki, K., & Iwata, O. (2016). Efficient selective breeding of live oil-rich *Euglena gracilis* with fluorescence-activated cell sorting. *Scientific Reports*, *6*(1), 26327. doi:10.1038rep26327 PMID:27212384

Yan, C., Wang, R., Wan, J., Zhang, Q., Xue, S., Wu, X., Zhang, J., Zhang, J., Lu, Y., & Cong, W. (2016). Cellulose/microalgae composite films prepared in ionic liquids. *Algal Research*, *20*, 135–141. doi:10.1016/j.algal.2016.09.024

Yandigeri, M. S., Yadav, A. K., Srinivasan, R., Kashyap, S., & Pabbi, S. (2011). Studies on mineral phosphate solubilization by cyanobacteria *Westiellopsis* and Anabaena. *Microbiology*, *80*(4), 558–565. doi:10.1134/S0026261711040229 PMID:22073557

Yang, B., Liu, J., Liu, B., Sun, P., Ma, X., Jiang, Y., & Chen, F. (2015). Development of a stable genetic system for *Chlorella vulgaris*—A promising green alga for CO_2 biomitigation. *Algal Research*, *12*, 134–141. doi:10.1016/j.algal.2015.08.012

Yang, B., Liu, J., Ma, X., Guo, B., Liu, B., Wu, T., Jiang, Y., & Chen, F. (2017). Genetic engineering of the Calvin cycle toward enhanced photosynthetic CO 2 fixation in microalgae. *Biotechnology for Biofuels*, *10*(1), 1–13. doi:10.118613068-017-0916-8 PMID:29034004

Yang, C., Wang, S., Yang, J., Xu, D., Li, Y., Li, J., & Zhang, Y. (2020). Hydrothermal liquefaction and gasification of biomass and model compounds: A review. *Green Chemistry*, *22*(23), 8210–8232. doi:10.1039/D0GC02802A

Yang, J. E., Park, S. J., Kim, W. J., Kim, H. J., Kim, B. J., Lee, H., Shin, J., & Lee, S. Y. (2018). One-step fermentative production of aromatic polyesters from glucose by metabolically engineered *Escherichia coli* strains. *Nature Communications*, *9*(1), 79. Advance online publication. doi:10.103841467-017-02498-w PMID:29311546

Yang, J. I., Yeh, C. C., Lee, J. C., Yi, S. C., Huang, H. W., Tseng, C. N., & Chang, H. W. (2012). Aqueous extracts of the edible Gracilaria tenuistipitata are protective against H2O2-induced DNA damage, growth inhibition, and cell cycle arrest. *Molecules (Basel, Switzerland)*, *17*(6), 7241–7254. doi:10.3390/molecules17067241 PMID:22695230

Yang, J., Xu, M., Zhang, X., Hu, Q., Sommerfeld, M., & Chen, Y. (2011). Life-cycle analysis on biodiesel production from microalgae: Water footprint and nutrients balance. *Bioresource Technology*, *102*(1), 159–165. doi:10.1016/j.biortech.2010.07.017 PMID:20675125

Yang, M. D., Sykes, R. M., & Merry, C. J. (2000). Estimation of algal biological parameters using water quality modeling and SPOT satellite data. *Ecological Modelling*, *125*, 1–13.

Yang, S., Wan, H., Wang, R., & Hao, D. (2019). Sulfated polysaccharides from Phaeodactylum tricornutum: Isolation, structural characteristics, and inhibiting HepG2 growth activity in vitro. *PeerJ*, *7*, e6409. doi:10.7717/peerj.6409 PMID:30809437

Yang, W., Zheng, Z., Lu, K., Zheng, C., Du, Y., Wang, J., & Zhu, J. (2020). Manipulating the phytoplankton community has the potential to create a stable bacterioplankton community in a shrimp rearing environment. *Aquaculture (Amsterdam, Netherlands)*, *520*, 734789. doi:10.1016/j.aquaculture.2019.734789

Yang, Z., Guo, R., Xu, X., Fan, X., & Luo, S. (2011). Hydrogen and methane production from lipid-extracted microalgal biomass residues. *International Journal of Hydrogen Energy*, *36*(5), 3465–3470. doi:10.1016/j.ijhydene.2010.12.018

Yang, Z., Shang, Y., & Lu, Y. (2011). Flocculation properties of biodegradable amphoteric chitosan-based flocculants. *Chemical Engineering Journal*, *172*(1), 287–295.

Yan, N., Fan, C., Chen, Y., & Hu, Z. (2016). The potential for microalgae as bioreactors to produce pharmaceuticals. *International Journal of Molecular Sciences*, *17*(6), 962. doi:10.3390/ijms17060962 PMID:27322258

Yan, N., Fan, C., Chen, Y., & Hu, Z. (2016). The Potential for Microalgae as Bioreactors to Produce Pharmaceuticals. *International Journal of Molecular Sciences*, *17*, 962.

Yan, W., & Hoekman, S. K. (2014). Production of CO 2-free hydrogen from methane dissociation: A review. *Environmental Progress & Sustainable Energy*, *33*(1), 213–219. doi:10.1002/ep.11746

Yan, X. L., & Hino, R. (2016). *Nuclear hydrogen production handbook*. CRC press.

Yan, Y. D., & Jameson, G. J. (2004). Application of the Jameson Cell technology for algae and phosphorus removal from maturation ponds. *International Journal of Mineral Processing*, *73*, 23–28.

Yao, J., Fu, W., Wang, X., & Duan, D. (2009). Improved RNA isolation from *Laminaria japonica* Aresch (Laminariaceae, Phaeophyta). *Journal of Applied Phycology*, *21*(2), 233–238. doi:10.100710811-008-9354-0

Yao, L., Shabestary, K., Björk, S. M., Asplund-Samuelsson, J., Joensson, H. N., Jahn, M., & Hudson, E. P. (2020). Pooled CRISPRi screening of the cyanobacterium *Synechocystis* sp PCC 6803 for enhanced industrial phenotypes. *Nature Communications*, *11*(1), 1666. doi:10.103841467-020-15491-7 PMID:32245970

Yao, S., Lyu, S., An, Y., Lu, J., Gjermansen, C., & Schramm, A. (2019). Microalgae-bacteria symbiosis in microalgal growth and biofuel production: A review. *Journal of Applied Microbiology*, *126*(2), 359–368. doi:10.1111/jam.14095 PMID:30168644

Yao, Y., Zhang, M., Tian, Y., Zhao, M., Zeng, K., Zhang, B., Zhao, M., & Yin, B. (2018). Azolla biofertilizer for improving low nitrogen use efficiency in an intensive rice cropping system. *Field Crops Research*, *216*, 158–164. doi:10.1016/j.fcr.2017.11.020

Yeh, C. C., Yang, J. I., Lee, J. C., Tseng, C. N., Chan, Y. C., Hseu, Y. C., Tang, J. Y., Chuang, L.-Y., Huang, H.-W., Chang, F.-R., & Chang, H. W. (2012). Anti-proliferative effect of methanolic extract of Gracilaria tenuistipitata on oral cancer cells involves apoptosis, DNA damage, and oxidative stress. *BMC Complementary and Alternative Medicine*, *12*(1), 1–9. doi:10.1186/1472-6882-12-142 PMID:22937998

Ye, J., Li, Y., Teruya, K., Katakura, Y., Ichikawa, A., Eto, H., Hosoi, M., Hosoi, M., Nishimoto, S., & Shirahata, S. (2005). Enzyme-digested fucoidan extracts derived from seaweed Mozuku of Cladosiphon novae-caledoniae kylin inhibit invasion and angiogenesis of tumor cells. *Cytotechnology*, *47*(1), 117–126. doi:10.100710616-005-3761-8 PMID:19003051

Yılmaz Öztürk, B., Yenice Gürsu, B., & Dağ, İ. (2020). Antibiofilm and antimicrobial activities of green synthesized silver nanoparticles using marine red algae Gelidium corneum. *Process Biochemistry*, *89*, 208–219. doi:10.1016/j.procbio.2019.10.027

Yilmaz, C., Koyuncu, I., Alcin, M., & Tuna, M. (2019). Artificial Neural Networks based thermodynamic and economic analysis of a hydrogen production system assisted by geothermal energy on Field Programmable Gate Array. *International Journal of Hydrogen Energy*, *44*(33), 17443–17459. doi:10.1016/j.ijhydene.2019.05.049

Yilmaz, E., & Sönmez, M. (2017). The role of organic/bio–fertilizer amendment on aggregate stability and organic carbon content in different aggregate scales. *Soil & Tillage Research*, *168*, 118–124. doi:10.1016/j.still.2017.01.003

Yim, J. H., Son, E., Pyo, S., & Lee, H. K. (2005). Novel sulfated polysaccharide derived from red-tide microalga *Gyrodinium impudicum* strain KG03 with immunostimulating activity in vivo. *Marine Biotechnology (New York, N.Y.)*, *7*(4), 331–338. doi:10.100710126-004-0404-6 PMID:15976942

Yingjun, W., Jia, L., Yun, L. Ü., Hangbiao, J., Shihuai, D., & Yunmin, Z. (2012). Effects of cerium on growth and physiological characteristics of Anabaena flosaquae. *Journal of Rare Earths*, *30*(12), 1287–1292. doi:10.1016/S1002-0721(12)60222-1

Yin, J., Zhu, G., & Deng, B. (2013). Multi-walled carbon nanotubes (MWNTs)/polysulfone (PSU) mixed matrix hollow fiber mem- branes for enhanced water treatment. *Journal of Membrane Science*, *437*, 237–248. doi:10.1016/j.memsci.2013.03.021

Yin, K., Wang, Q., Lv, M., & Chen, L. (2018). Microorganism remediation strategies towards heavy metals. *Chemical Engineering Journal*.

Yin, Z., Zhu, L., Li, S., Hu, T., Chu, R., Mo, F., Hu, D., Liu, C., & Li, B. (2020). A comprehensive review on cultivation and harvesting of microalgae for biodiesel production: Environmental pollution control and future directions. *Bioresource Technology*, *301*, 122804. doi:10.1016/j.biortech.2020.122804 PMID:31982297

Yi, Z., Su, Y., Xu, M., Bergmann, A., Ingthorsson, S., Rolfsson, O., Salehi-Ashtiani, K., Brynjolfsson, S., & Fu, W. (2018). Chemical Mutagenesis and Fluorescence-Based High-Throughput Screening for Enhanced Accumulation of Carotenoids in a Model Marine Diatom *Phaeodactylum tricornutum. Marine Drugs*, *16*(8), 272. doi:10.3390/md16080272 PMID:30081564

Yi, Z., Xu, M., Magnusdottir, M., Zhang, Y., Brynjolfsson, S., & Fu, W. (2015). Photo-Oxidative Stress-Driven Mutagenesis and Adaptive Evolution on the Marine Diatom *Phaeodactylum tricornutum* for Enhanced Carotenoid Accumulation. *Marine Drugs*, *13*(10), 6138–6151. doi:10.3390/md13106138 PMID:26426027

Yokoi, H., Arima, T., Hirose, J., Hayashi, S., & Takasaki, Y. (1996). Flocculation properties of poly(γ-glutamic acid) produced by Bacillus subtilis. *Journal of Fermentation and Bioengineering, 82(*1), 84–87.

Yokota, T., Kim, S. K., Fukui, Y., Takahashi, N., Takeuchi, Y., & Takematsu, T. (1987). Brassinosteroids and Sterols from a Green Alga, *Hydrodictyon reticulatum*: Configuration at C-24. *Phytochemistry*, *26*(2), 503–506. doi:10.1016/S0031-9422(00)81442-9

Yoo, C., La, H. J., Kim, S. C., & Oh, H. M. (2015). Simple processes for optimized growth and harvest of Ettlia sp. by pH control using CO_2 and light irradiation. *Biotechnology and Bioengineering*, *112*, 288–296.

Yoon, H. S., Hackett, J. D., Ciniglia, C., Pinto, G., & Bhattacharya, D. (2004). A Molecular Timeline for the Origin of Photosynthetic Eukaryotes. *Molecular Biology and Evolution*, *21*(5), 809–818. doi:10.1093/molbev/msh075 PMID:14963099

Yoon, H. S., Nelson, W., Lindstrom, S. C., Boo, S. M., Pueschel, C., Qiu, H., & Bhattacharya, D. (2017). Rhodophyta. In J. M. Archibald, A. G. B. Simpson, & C. H. Slamovits (Eds.), *Handbook of the Protists* (pp. 89–133). Springer International Publishing. doi:10.1007/978-3-319-28149-0_33

Yoshida, K., Igarashi, E., Wakatsuki, E., Miyamoto, K., & Hirata, K. (2004). Mitigation of osmotic and salt stresses by abscisic acid through reduction of stress-derived oxidative damage in Chlamydomonas reinhardtii. *Plant Science*, *167*, 1335–1341.

Young, C. S., & Gobler, C. J. (2016). Ocean acidification accelerates the growth of two bloom-forming macroalgae. *PLoS One*, *11*(5), e0155152. doi:10.1371/journal.pone.0155152 PMID:27176637

You, T., & Barnett, S. M. (2004). Effect of light quality on production of extracellular polysaccharides and growth rate of Porphyridium cruentum. *Biochemical Engineering Journal*, *19*(3), 251–258. doi:10.1016/j.bej.2004.02.004

Yuan, H., Song, J., Li, X., Li, N., & Dai, J. (2006). Immunomodulation and antitumor activity of κ-carrageenan oligosaccharides. *Cancer Letters*, *243*(2), 228–234. doi:10.1016/j.canlet.2005.11.032 PMID:16410037

Yuan, S., Hu, J., Liu, Z., Hong, Y., & Wang, X. (2020). Modeling microalgae growth in continuous culture: Parameters analysis and temperature dependence. *Energy*, *195*, 116961. doi:10.1016/j.energy.2020.116961

Yue, D., You, F., & Snyder, S. W. (2014). Biomass-to-bioenergy and biofuel supply chain optimization: Overview, key issues and challenges. *Computers & Chemical Engineering*, *66*, 36–56. doi:10.1016/j.compchemeng.2013.11.016

Yugay, Y. A., Usoltseva, R. V., Silant'ev, V. E., Egorova, A. E., Karabtsov, A. A., Kumeiko, V. V., Ermakova, S. P., Bulgakov, V. P., & Shkryl, Y. N. (2020). Synthesis of bioactive silver nanoparticles using alginate, fucoidan and laminaran from brown algae as a reducing and stabilizing agent. *Carbohydrate Polymers*, *245*, 116547. doi:10.1016/j.carbpol.2020.116547 PMID:32718640

Yu, J., Hu, H., Wu, X., Wang, C., Zhou, T., Liu, Y., & Zheng, H. (2019). Continuous cultivation of Arthrospira platensis for phycocyanin production in large-scale outdoor raceway ponds using microfiltered culture medium. *Bioresource Technology*, *287*, 121420.

Yu, K. L., Lee, X. J., Ong, H. C., Chen, W.-H., Chang, J.-S., & Lin, C.-S. (2021). Adsorptive Removal of Cationic Methylene Blue and Anionic Congo Red Dyes Using Wet-Torrefied Microalgal Biochar: Equilibrium, Kinetic and Mechanism Modeling. *Environ. Pollut.*, *272*, 115986.

Yu, M., Cruz, J. M., & Li, D. M. (2019). The sustainable supply chain network competition with environmental tax policies. *International Journal of Production Economics*, *217*, 218–231. doi:10.1016/j.ijpe.2018.08.005

Yu, Q., Li, Y., Wu, B., Hu, W., He, M., & Hu, G. (2020). Novel mutagenesis and screening technologies for food microorganisms: Advances and prospects. *Applied Microbiology and Biotechnology*, *104*(4), 1517–1531. doi:10.100700253-019-10341-z PMID:31919586

Yu, X., Zhou, C., Yang, H., Huang, X., Ma, H., Qin, X., & Hu, J. (2015). Effect of ultrasonic treatment on the degradation and inhibition cancer cell lines of polysaccharides from Porphyra yezoensis. *Carbohydrate Polymers*, *117*, 650–656. doi:10.1016/j.carbpol.2014.09.086 PMID:25498684

Zabed, H. M., Qi, X., Yun, J., & Zhang, H. (2019). Anaerobic Digestion of Microalgae Biomass for Methane Production. In *Microalgae Biotechnology for Development of Biofuel and Wastewater Treatment* (pp. 397–421). Springer. doi:10.1007/978-981-13-2264-8_16

Zaborowska, M., Kucharski, J., & Wyszkowska, J. (2015). Using basalt flour and brown algae to improve biological properties of soil contaminated with cadmium. *Soil and Water Research*, *10*(3), 181–188. doi:10.17221/281/2014-SWR

Zang, X., Liu, B., Liu, S., Arunakumara, K. K., & Zhang, X. (2007). Optimum conditions for transformation of *Synechocystis* sp. *PCC* 6803. *Journal of Microbiology (Seoul, Korea)*, *45*, 241–245. PMID:17618230

Zawieja, I., & Worwąg, M. (2021). Biogas Production from Excess Sludge Oxidized with Peracetic Acid (PAA). *Energies*, *14*(12), 3434. doi:10.3390/en14123434

Zayadan, B. K., Matorin, D. N., Baimakhanova, G. B., Bolathan, K., Oraz, G. D., & Sadanov, A. K. (2014). Promising microbial consortia for producing biofertilizers for rice fields. *Microbiology*, *83*(4), 391–397. doi:10.1134/S0026261714040171 PMID:25844458

Zein, N., & Gharib, S. M. (2014). Spatial and temporal patterns of phytoplankton composition in Burullus Lagoon, Southern Mediterranean Coast, Egypt. *Egyptian Journal of Aquatic Research*, (April), 2009–2010. doi:10.1016/j.ejar.2014.06.004

Zeller, M. A., Hunt, R., Jones, A., & Sharma, S. (2013). Bioplastics and their thermoplastic blends from *Spirulina* and *Chlorella* microalgae. *Journal of Applied Polymer Science*, *130*(5), 3263–3275. doi:10.1002/app.39559

Zeller, M. A., Hunt, R., & Sharma, S. (2012). Sustainable bioderived polymeric materials and thermoplastic blends made from floating aquatic macrophytes such as "duckweed.". *Journal of Applied Polymer Science*, *127*(1), 375–386. doi:10.1002/app.37555

Zeng, S., Yong, K.-T., Roy, I., Dinh, X.-Q., Yu, X., & Luan, F. (2011). A review on functionalized gold nanoparticles for biosensing applications. *Plasmonics*, *6*(3), 491–506. doi:10.100711468-011-9228-1

Zeriouh, O., Reinoso-Moreno, J. V., López-Rosales, L., Cerón-García, M. D. C., Sánchez-Mirón, A., García-Camacho, F., & Molina-Grima, E. (2017). Biofouling in photobioreactors for marine microalgae. *Critical Reviews in Biotechnology*, *37*(8), 1006–1023. doi:10.1080/07388551.2017.1299681 PMID:28427282

Zewdie, D. T., & Ali, A. Y. (2020). Cultivation of microalgae for biofuel production: Coupling with sugarcane-processing factories. *Energy, Sustainability and Society*, *10*(1), 27. doi:10.118613705-020-00262-5

Zhang, B. (2013). New Golden Age of the Algal Technology. *Journal of Petroleum & Environmental Biotechnology*, *04*(06), 7463. doi:10.4172/2157-7463.1000e120

Zhang, C., Chen, X., Wang, J., & Tan, L. (2017). Toxic effects of microplastic on marine microalgae Skeletonema costatum: Interactions between microplastic and algae. *Environmental Pollution*, *220*, 1282–1288. doi:10.1016/j.envpol.2016.11.005 PMID:27876228

Zhang, C., Li, Q., Fu, L., Zhou, D., & Crittenden, J. C. (2018). Quorum sensing molecules in activated sludge could trigger microalgae lipid synthesis. *Bioresource Technology*, *263*(March), 576–582. doi:10.1016/j.biortech.2018.05.045 PMID:29783193

Zhang, C., Wang, C., Cao, G., Wang, D., & Ho, S.-H. (2020). A sustainable solution to plastics pollution: An eco-friendly bioplastic film production from high-salt contained Spirulina sp. residues. *Journal of Hazardous Materials*, *388*, 121773. doi:10.1016/j.jhazmat.2019.121773 PMID:31836373

Zhang, F., Endo, T., Kitagawa, R., Kabeya, H., & Hirotsu, T. (2000). Synthesis and characterization of a novel blend of polypropylene with *Chlorella*. *Journal of Materials Chemistry*, *10*(12), 2666–2672. doi:10.1039/b004489j

Zhang, F., Kabeya, H., Kitagawa, R., Hirotsu, T., Yamashita, M., & Otsuki, T. (2000). A novel polyethylene-chlorella composite. I. Characterization of *Chlorella* biologically fixing CO_2. *Journal of Applied Polymer Science*, *77*(10), 2278–2284. doi:10.1002/1097-4628(20000906)77:10<2278::AID-APP21>3.0.CO;2-Z

Zhang, F., Kabeya, H., Kitagawa, R., Hirotsu, T., Yamashita, M., & Otsuki, T. (2000). An exploratory research of PVC-Chlorella composite material (PCCM) as effective utilization of Chlorella biologically fixing CO2. *Journal of Materials Science*, *35*(10), 2603–2609. doi:10.1023/A:1004779415778

Zhang, H., Wang, H., Zheng, W., Yao, Z., Peng, Y., Zhang, S., Hu, Z., Tao, Z., & Zheng, T. (2017). Toxic effects of prodigiosin secreted by Hahella sp. KA22 on harmful alga Phaeocystis globosa. *Frontiers in Microbiology*, *8*(JUN), 1–12. doi:10.3389/fmicb.2017.00999 PMID:28634473

Zhang, J., Osmani, A., Awudu, I., & Gonela, V. (2013). An integrated optimization model for switchgrass-based bio-ethanol supply chain. *Applied Energy*, *102*, 1205–1217. doi:10.1016/j.apenergy.2012.06.054

Zhang, J., Shishatskaya, E. I., Volova, T. G., da Silva, L. F., & Chen, G.-Q. (2018). Polyhydroxyalkanoates (PHA) for therapeutic applications. *Materials Science and Engineering C*, *86*, 144–150. doi:10.1016/j.msec.2017.12.035 PMID:29525089

Zhang, K., Zhang, F., & Wu, Y. R. (2021). Emerging technologies for conversion of sustainable algal biomass into value-added products: A state-of-the-art review. *The Science of the Total Environment*, *784*, 147024. doi:10.1016/j.scitotenv.2021.147024 PMID:33895504

Zhang, L. X., Cai, C. E., Guo, T. T., Gu, J. W., Xu, H. L., Zhou, Y., Wang, Y., Liu, C.-C., & He, P. M. (2011). Anti-cancer effects of polysaccharide and phycocyanin from Porphyra yezoensis. *Journal of Marine Science and Technology*, *19*(4), 377–382. doi:10.51400/2709-6998.2178

Zhang, L., He, M., & Liu, J. (2014). The enhancement mechanism of hydrogen photoproduction in Chlorella protothecoides under nitrogen limitation and sulfur deprivation. *International Journal of Hydrogen Energy*, *39*(17), 8969–8976. doi:10.1016/j.ijhydene.2014.04.045

Zhang, S., & Bryant, D. A. (2015). Biochemical validation of the glyoxylate cycle in the cyanobacterium Chlorogloeopsis fritschii strain PCC 9212. *The Journal of Biological Chemistry*, *290*(22), 14019–14030. doi:10.1074/jbc.M115.648170 PMID:25869135

Zhang, Y. T., Jiang, J. Y., Shi, T. Q., Sun, X. M., Zhao, Q. Y., Huang, H., & Ren, L. J. (2019). Application of the CRISPR/Cas system for genome editing in microalgae. *Applied Microbiology and Biotechnology*, *103*(8), 3239–3248. doi:10.100700253-019-09726-x PMID:30877356

Zhang, Y. Y., Li, H. X., Shu, W. B., Zhang, C. J., & Ye, Z. B. (2011). RNA interference of a mitochondrial APX gene improves vitamin C accumulation in tomato fruit. *Scientia Horticulturae*, *129*(2), 220–226. doi:10.1016/j.scienta.2011.03.025

Zhang, Y., Habteselassie, M. Y., Resurreccion, E. P., Mantripragada, V., Peng, S., Bauer, S., & Colosi, L. M. (2014). Evaluating removal of steroid estrogens by a model alga as a possible sustainability benefit of hypothetical integrated algae cultivation and wastewater treatment systems. *ACS Sustainable Chemistry & Engineering*, *2*, 2544–2553.

Zhang, Y., Wang, X., Shan, T., Pang, S., & Xu, N. (2019). Transcriptome profiling of the meristem tissue of *Saccharina japonica* (Phaeophyceae, Laminariales) under severe stress of copper. *Marine Genomics*, *47*, 100671. doi:10.1016/j.margen.2019.03.006

Zhan, J., Rong, J., & Wang, Q. (2016). Mixotrophic Cultivation, a Preferable Microalgae Cultivation Mode for Biomass / Bioenergy Production, and Bioremediation, Advances and Prospect. *International Journal of Hydrogen Energy*, 1–13.

Zhao, B., & Su, Y. (2014). Process effect of microalgal-carbon dioxide fixation and biomass production: A review. *Renewable & Sustainable Energy Reviews*, *31*, 121–132. doi:10.1016/j.rser.2013.11.054

Zhao, J., Li, X. F., Ren, Y. P., Wang, X. H., & Jian, C. (2012). Electricity generation from Taihu Lake cyanobacteria by sediment microbial fuel cells. *Journal of Chemical Technology and Biotechnology (Oxford, Oxfordshire)*, *87*(11), 1567–1573. doi:10.1002/jctb.3794

Zhao, M., Raheem, A., Memon, Z. M., Vuppaladadiyam, A. K., & Ji, G. (2019). Iso-conversional kinetics of low-lipid micro-algae gasification by air. *Journal of Cleaner Production*, *207*, 618–629. doi:10.1016/j.jclepro.2018.10.040

Zhao, Z., Liu, S., Yang, X., Lei, Z., Shimizu, K., Zhang, Z., Lee, D.-J., & Adachi, Y. (2019). Stability and performance of algal-bacterial granular sludge in shaking photo-sequencing batch reactors with special focus on phosphorus accumulation. *Bioresource Technology*, *280*, 497–501. doi:10.1016/j.biortech.2019.02.071 PMID:30819417

Zhaozhou, Z., Zhongliang, W., Jun, L., Yong, L. I., Zhang, Z., & Zhang, P. (2012). Distribution of rare earth elements in sewage-irrigated soil profiles in Tianjin, China. *Journal of Rare Earths*, *30*(6), 609–613. doi:10.1016/S1002-0721(12)60099-4

Zheng, Gao, Yin, Tang, Ji, & Huang. (2012). Harvesting of microalgae by flocculation with poly (γ-glutamic acid). B*ioresource Technology,* 1*12,* 212–220.

Zheng, H., Guo, W., Li, S., Chen, Y., Wu, Q., Feng, X., Yin, R., Ho, S. H., Ren, N., & Chang, J. S. (2017). Adsorption of p-nitrophenols (PNP) on microalgal biochar: Analysis of high adsorption capacity and mechanism. *Bioresource Technology*, *244*, 1456–1464.

Zheng, Y., Jin, R., Zhang, X., Wang, Q., & Wu, J. (2019). The considerable environmental benefits of seaweed aquaculture in China. *Stochastic Environmental Research and Risk Assessment*, *33*(4–6), 1203–1221. doi:10.100700477-019-01685-z

Zheng, Y., Ye, Z., Fang, X. L., Li, Y. H., & Cai, W. M. (2008). Production and characteristics of a bioflocculant produced by Bacillus sp. F19. *Bioresource Technology*, *99*(16), 7686–7691.

Zhila, N. O., Kalacheva, G. S., & Volova, T. G. (2011). Effect of salinity on the biochemical composition of the alga Botryococcus braunii Kütz IPPAS H-252. *Journal of Applied Phycology*, *23*(1), 47–52. doi:10.100710811-010-9532-8

Zhou, D., Qiao, B., Li, G., Xue, S., & Yin, J. (2017). Continuous production of biodiesel from microalgae by extraction coupling with transesterification under supercritical conditions. *Bioresource Technology*, *238*, 609–615. doi:10.1016/j.biortech.2017.04.097 PMID:28482287

Zhou, G. J., Ying, G. G., Liu, S., Zhou, L. J., Chen, Z. F., & Peng, F. Q. (2014). Simultaneous removal of inorganic and organic compounds in wastewater by freshwater green microalgae. *Environmental Science. Processes & Impacts*, *16*, 2018–2027.

Zhou, J., Lyu, Y., Richlen, M. L., Anderson, D. M., & Cai, Z. (2016). Quorum Sensing Is a Language of Chemical Signals and Plays an Ecological Role in Algal-Bacterial Interactions. *Critical Reviews in Plant Sciences*, *35*(2), 81–105. doi:10.1080/07352689.2016.1172461 PMID:28966438

Zhou, L., Liu, J., & Liu, Z. (2009). Adsorption of platinum (IV) and palladium (II) from aqueous solution by thiourea-modified chitosan microspheres. *Journal of Hazardous Materials*, *172*(1), 439–446. doi:10.1016/j.jhazmat.2009.07.030 PMID:19646814

Zhou, W., Cheng, Y., Li, Y., Wan, Y., Liu, Y., Lin, X., & Ruan, R. (2012). Novel fungal pelletization-assisted technology for algae harvesting and wastewater treatment. *Applied Biochemistry and Biotechnology*, *167*(2), 214–228. doi:10.100712010-012-9667-y PMID:22538982

Zhou, W., Wang, J., Chen, P., Ji, C., Kang, Q., Lu, B., & Ruan, R. (2017). Bio-mitigation of carbon dioxide using microalgal systems: Advances and perspectives. *Renewable & Sustainable Energy Reviews*, *76*, 1163–1175. doi:10.1016/j.rser.2017.03.065

Zhou, W., Wang, Z., Xu, J., & Ma, L. (2018). Cultivation of microalgae Chlorella zofingiensis on municipal wastewater and biogas slurry towards bioenergy. *Journal of Bioscience and Bioengineering*, *126*(5), 644–648. doi:10.1016/j.jbiosc.2018.05.006 PMID:29801764

Zhu, C., Chen, C., & Zhao, L. (2012). Bioflocculant produced by Chlamydomonas reinhardtii. *Journal of Applied Phycology*, *24*(5), 1245–1251.

Zhu, D., Ke, X., Christie, P., & Zhu, Y.-G. (2018). Rejoinder to "Comments on Zhu et al. (2018) Exposure of soil collembolans to microplastics perturbs their gut microbiota and alters their isotopic composition". Soil Biol. Biochem. 116 302–310". *Soil Biology & Biochemistry*, *124*, 275–276. doi:10.1016/j.soilbio.2018.05.030

Zhu, L. D., Hiltunen, E., Antila, E., Zhong, J. J., Yuan, Z. H., & Wang, Z. M. (2014). Microalgal biofuels: Flexible bioenergies for sustainable development. *Renewable & Sustainable Energy Reviews*, *30*, 1035–1046. doi:10.1016/j.rser.2013.11.003

Zhu, L., Li, Z., & Hiltunen, E. (2018). Microalgae *Chlorella vulgaris* biomass harvesting by natural flocculant: Effects on biomass sedimentation, spent medium recycling and lipid extraction. *Biotechnology for Biofuels*, *11*(1), 1–10.

Zhu, N., Ye, M., Shi, D., & Chen, M. (2017). Reactive compatibilization of biodegradable poly(butylene succinate)/ *Spirulina* microalgae composites. *Macromolecular Research*, *25*(2), 165–171. doi:10.100713233-017-5025-9

Zhu, X., Zhao, W., Chen, X., Zhao, T., Tan, L., & Wang, J. (2020). Growth inhibition of the microalgae Skeletonema costatum under copper nanoparticles with microplastic exposure. *Marine Environmental Research*, *158*, 105005. doi:10.1016/j.marenvres.2020.105005 PMID:32501269

Ziesche, L., Bruns, H., Dogs, M., Wolter, L., Mann, F., Wagner-Döbler, I., Brinkhoff, T., & Schulz, S. (2015). Homoserine Lactones, Methyl Oligohydroxybutyrates, and Other Extracellular Metabolites of Macroalgae-Associated Bacteria of the Roseobacter Clade: Identification and Functions. *ChemBioChem*, *16*(14), 2094–2107. doi:10.1002/cbic.201500189 PMID:26212108

Ziganshin, A. M., Liebetrau, J., Pröter, J., & Kleinsteuber, S. (2013). Microbial community structure and dynamics during anaerobic digestion of various agricultural waste materials. *Applied Microbiology and Biotechnology*, *97*(11), 5161–5174. doi:10.100700253-013-4867-0 PMID:23624683

Zimba, P. V., Huang, I. S., Gutierrez, D., Shin, W., Bennett, M. S., & Triemer, R. E. (2017). Euglenophycin is produced in at least six species of euglenoid algae and six of seven strains of Euglena sanguinea. *Harmful Algae*, *63*, 79–84. doi:10.1016/j.hal.2017.01.010 PMID:28366403

Zingone, A., Escalera, L., Aligizaki, K., Fernández-Tejedor, M., Ismael, A., Montresor, M., Mozetič, P., Taş, S., & Totti, C. (2021). 2021/02/01/). Toxic marine microalgae and noxious blooms in the Mediterranean Sea: A contribution to the Global HAB Status Report. *Harmful Algae*, *102*, 101843. doi:10.1016/j.hal.2020.101843 PMID:33875177

Žižková, E., Kubeš, M., Dobrev, P.I., Přibyl, P., Šimura, J., Zahajská, L., Záveská Drábková, L., Novák, O., & Motyka, V. (2016). Control of cytokinin and auxin homeostasis in cyanobacteria and algae. *Annals of Botany, 119*(1), 151-66.

Zolotareva, E. K., Mokrosnop, V. M., & Stepanov, S. S. (2019). Polyphenol compounds of macroscopic and microscopic algae. *Algologia*, *29*(1), 3–29. doi:10.15407/alg29.01.003

Zorofchian Moghadamtousi, S., Karimian, H., Khanabdali, R., Razavi, M., Firoozinia, M., Zandi, K., & Abdul Kadir, H. (2014). Anticancer and antitumor potential of fucoidan and fucoxanthin, two main metabolites isolated from brown algae. *TheScientificWorldJournal*, *2014*, 2014. doi:10.1155/2014/768323 PMID:24526922

Zou, P., Lu, X., Zhao, H., Yuan, Y., Meng, L., Zhang, C., & Li, Y. (2019). Polysaccharides Derived from the Brown Algae LessoniaNigrescens Enhance Salt Stress Tolerance to Wheat Seedlings by Enhancing the Antioxidant System, and Modulating Intracellular Ion Concentration. *Frontiers in Plant Science*, *10*, 48. doi:10.3389/fpls.2019.00048 PMID:30766543

Zou, S., Wu, Y., Yang, M., Imdad, K., Li, C., & Tong, J. (2010). Production and characterization of bio-oil from hydrothermal liquefaction of microalgae Dunaliella tertiolecta cake. *Energy*, *35*(12), 5406–5411. doi:10.1016/j.energy.2010.07.013

Zou, Y., Zeng, Q., Li, H., Liu, H., & Lu, Q. (2021). Emerging technologies of algae-based wastewater remediation for bio-fertilizer production: A promising pathway to sustainable agriculture. *Journal of Chemical Technology and Biotechnology (Oxford, Oxfordshire)*, *96*(3), 551–563. doi:10.1002/jctb.6602

About the Contributors

Mostafa M. El-Sheekh has 38 years of experience in the research and teaching Dr. El-Sheekh served as the Dean Faculty of Science, Tanta University, Egypt and Cultural counselor in Egyptian Embassy in Yemen and Uzbekistan and Vice president of Tanta University for Postgraduate studies and Research. Dr. El-Sheekh holds a PhD in Phycology from Tanta University and Göttingen University, Germany. He got several postdoc fellowships in Hungary, India, Sweden, the Czech Republic, and Japan. He has authored 227 papers including 150 articles published in SCI journals (h-index=29, Scopus, h-index 35, Google scholar), 16 Book chapters, and 4 books. He also serves as editorial member and reviewer for 100 peer reviewed Journals. He is from the top 2% of the Scientists in the field of Biological Sciences as published by Stanford University depending on the Scopus database 2020 and 2021.

Norhayati Abdullah is currently the Associate Director of UTM International Kuala Lumpur and Associate Professor in Environmental Engineering at Universiti Teknologi Malaysia (UTM). She completed her tenure as a Guest Scholar at the Graduate School of Advanced Integrated Studies in Human Survivability (Shishu-kan), Kyoto University, Japan. Norhayati is recipient of the 2019/2020 ASEAN Science and Technology Fellowship focusing on the strategic recommendations for water sector transformation under the auspices of the Academy of Sciences Malaysia. Norhayati represented UTM and Malaysia at the University of Michigan Ann Arbor for the Fulbright US-ASEAN Visiting Scholar Initiatives in 2016-2017. In 2018, she received the L'Oréal-UNESCO for Women in Science Fellowship. She was Board of Director of the International Water Association (IWA) from 2014-2018, an IWA Fellow and Council Member of the Malaysian Water Association (MWA). At global level, Norhayati has been actively involved with the IWA program and activities since 2000. She received the IWA Young Water Professional (IWAYWP) Award in 2012 in recognition of her outstanding achievements in wastewater research and active involvement in various professional development programs.

Imran Ahmad is a PhD student at MJIIT, Universiti Teknologi Malaysia. Currently working on the treatment of restaurant wastewater containing FOG using microalgae. Ahmad specializes in the anaerobic treatment of industrial wastewaters, especially landfill leachate. He has published his work in peer reviewed journals, book chapters, attended and presented in many International conferences. Ahmad is also working in the editing of book to be published by Springer, Nature. Ahmad received best presentation award in 4 International conferences. He received Algae Industry Incubation Consortium (AIIC), Japan grants award and My Membrane award, Malaysia in 2021. He was also the recipient of Young Scholars grant awarded by School of Post Graduate Studies, UTM. Ahmad have teaching experience of 8 years in AICTE approved Engineering Colleges in India. He is having student membership for International Society for applied Phycology (ISAP), International Water Association (IWA), Northeast Algal Society (NEAS), Ecological Society of America (ESA), British Phycological Society, and Northeast Algal Society (NEAS).

* * *

Abdallah Abdelfattah is a teaching assistant, Department of Public Works Engineering, Faculty of Engineering, Tanta University, Egypt. A researcher specializing in wastewater treatment at the school of environment and safety engineering, Jiangsu university, Zhenjiang, China.

Bahaa Abdella got his PhD in microbiology from Sunway University, Malaysia. He is an aquatic microbiology lecturer at Kafrelsheikh University in Egypt's Faculty of Aquatic and Fisheries Sciences. His research focuses on aquatic microbiology, microbial genomics, and bioinformatics.

Samia Saad Abouelkheir is currently a Researcher of Marine Microbiology at the National Institute of Oceanography and Fisheries, Alexandria branch, Egypt. She earned her Bachelor of Science at Alexandria University in Egypt and her Ph.D. in Microbiology from the same University in 2016. In her thesis work, she pioneered the Biofabrication of gold/activated carbon/polyvinyl alcohol (Au/Ac/PVA) polymer sheet and applied this to study its application in seawater desalination. After completing her Ph.D., Samia accepted a position as an Editorial Board Member of Journal of Marine Science ISSN: 2661-3239. Journal of Marine Science - Bilingual Publishing Co is a peer-reviewed, open-access academic journal specializing in the field of marine science research. Samia's current research emphasizes the use of microorganisms as safe biological agents for the synthesis and production of promising nanomaterials providing an even vaster boost to green biotechnology as an emerging field of nanotechnology for sustainable production and cost reduction. Samia S. Abouelkheir Researcher of Marine Microbiology, Marine Environment Division, National Institute of Oceanography and Fisheries (NIOF), Alexandria, Egypt Tel. no.: 02 01122644470 ORCID: https://orcid.org/0000-0001-8559-6787.

George Achieng' is an Environmental Chemist with focus on clean environment. He is keen on ambient air quality (indoor and outdoor), stack emissions, effluent and soil. Researching in application of low-cost adsorbents in remediation of wastewater has been one of his key areas.

Ahmed E. Al Prol, is an researcher, experience in waters treatment (MSc thesis) and (Ph. D thesis); Experience in synthesis of Nano compounds ; Experience in writing article review about the natural product; Analysis of environmental samples; Analysis of environmental samples for organic pollutants; Isolation of some natural products from marine algae; Use of marine algae to remove pollutants; Use of marine algae, agriculture waste to remove pollutants; Isolation of cellulose from agriculture wastes for membrane preparation; plants extracts. Recently, his research activities focus on different adsorbents and nanomaterials production, processing and their applications in water treatment.

Sameh Ali's main research field is environmental science and engineering, and his research direction is the microbial conversion of agricultural biomass waste and the utilization of biomass energy, mainly including the biotransformation of lignocellulose, anaerobic fermentation, and organic dye wastewater bioremediation and biodegradation technology of microplastics. Especially in the fields of insect gut microbial utilization and biomimetic, through long-term cooperation with Sun Jianzhong's research team of Jiangsu University, a series of innovative and high-quality research results have been achieved in the discovery and utilization of new yeast strains. At present, more than 50 high-level SCI research papers have been published belong to the high-quality papers such as Journal of Hazardous Materials, Bioresource Technology, Science of the Total Environment, Fuel, Biotechnology for Biofuels, etc.), the comprehensive citation H factor of the article has reached 19, and its total citation rate has exceeded 900 times.

Sayed Rashad Ahmed is a researcher at the Regional center for Food and Feed, Agricultural Research Center in EGYPT. He has a Ph.D. in Natural Resources. His studies interests include wastewater treatment, waste valorization and algae research. He worked on the monitoring of water quality, algae use as fertilizer, algae for wastewater treatment and several algal research works. He has several publications in different indexed journals. He is a reviewer for many journals. He has contributed as a co-author for four chapters during this year.

Liang Cheng graduated in Environmental Engineering at Murdoch University (Australia). He worked as a postdoc at Murdoch University (Australia), Curtin University (Australia), and Nanyang Technological University (Singapore). He is now a full professor of Environmental Engineering at the School of Environment and Safety Engineering, Jiangsu University, China. His research focuses on low-energy wastewater treatment using a novel biofilm passive aeration system. Another major thematic of his group is dealing with bioelectrochemistry and extracellular electron transfer in MFC and biosensor systems. He also spent more than 10 years on biomineralization and its applications in civil engineering. He received an honorable mention in the R M Quigley Awards (2014) and Fredlund Award (2018).

Mudasir Ahmad Dar is a post-doctoral researcher, working at Biofuel Institute, School of the Environment and Safety Engineering, Jiangsu University, Zhenjiang, China. He is currently working on the biofuel production from lignocellulosic wastes by using gut bacteria. He completed his masters in Zoology from University of Pune, Maharashtra with First Division. After completing his master's, he pursued Ph.D. from the Savitribai Phule Pune University, Maharashtra, under the supervision of Prof. R. S. Pandit on the topic entitled, Bioprospection of symbiotic gut-bacteria from the snails for agro-waste disposal management. He is highly interested to work on Biofuels; Environmental biotechnology; Bioremediation, Systems biology; Gut-microbiome; Bioprospection; and modeling insects for biotechnological applications. He has presented his work at various national and international conferences and has published more than 18 papers in peer reviewed international journals of high repute related to lignocellulose bioconversion. He is serving as an Editorial and reviewer of many well reputed peer-reviewed international journals like international journal of environmental science and technology, Bioenergy research, Environmental technology and innovation, Comparative Biochemistry and Physiology-C, Environmental science and pollution research, to name a few. He qualified the prestigious exams like, CSIR-UGC-NET multiple times with AIR 41 and 101 conducted by the CSIR-HRDG Govt. of India. He is a recipient of many national awards like Maulana Azad national fellow, DST PURSE fellow, UGC-UPE phase II fellow.

Nada M. Doleib is microbiologist she received her Ph.D., M. Sc., B.S. from Khartoum Unversity. Khartoum, Sudan. she spent years as assistant professor of Microbiology at Jubba University, Sudan, and at Jeddah University, Saudi Arabia until now. Now she is the head of the Department of Biology at

Jedaah University. Teaching many courses in the field of microbiology, As research scientist at Jeddah University, she published many paper in microalgae, and applied microbiology.

Eslam El-Aswar is an assistant researcher at NWRC. He earned his BSc., MSc. and PhD in Nanochemistry from Menoufia University, Egypt in 2012, 2017 and 2017 respectively. He participated in several international conferences and workshops. In 2019, he got an internship which was offered by CAS-TWAS Centre of Excellence for Water and Environment and established in Chinese Academy of Sciences, Beijing, China. The overarching theme of his research experience related to water quality and environment. His research activities revolve around nanomaterials production, processing and their applications in water treatment and desalination. Recently, his research activities focus on removal of emerging contaminants from water bodies.

Ghadir Aly El-Chaghaby is a professor at the Regional center for Food and Feed, Agricultural Research Center in EGYPT. She has a Ph.D. in Environmental Analytical Chemistry. She has several publications in the field of environmental science especially wastewater treatment via different methods. She is the academic editor of a scientific journal and she is a reviewer for many journals. She has contributed as a co-author for four chapters during this year.

Mostafa Elshobary is an associate Professor of Phycology and a postdoc fellow at Jiangsu University, China. He has nearly 17 years of experience in research and teaching in the field of phycology and its applications. He received his PhD in Phycology from Tanta University (Egypt) and Manitoba University (Canada) as a sandwich program His research direction is microbial Biotechnology, algal biofuels (including biodiesel, bioethanol, biogas, and biohydrogen), biomass conversion, wastewater bioremediation, waste recycling for circular economy and microalgal cultivation. Notably, Dr. Elshobary has authored more than 25 manuscripts and 5 Book chapters. He also serves as an editorial member and reviewer for several international Journals".

Howida Abouel Fetouh is an assistant professor of physical chemistry, Faculty of Science, Alexandria University. She has attained B.Sc. Science (Special Chemistry), Master of Science and Ph.D. of Science at the same university, she had great experience in fields of electrochemistry, she has broad area of research interests in the production sustainable energy and fuel cells. She started from synthesis, characterization and electrochemical studies of a promising alternative ruthenium complex for dye sensitized solar cells; Production of sustainable electrical energy using aluminum fuel cells; Sustainable electricity production from seawater using Spirulina platensis microbial fuel cell catalyzed by silver nanoparticles-activated carbon composite prepared by a new modified photolysis method; Comparative Study for the Production of Sustainable Electricity from Marine Sediment Using Recyclable Low-Cost Solid Wastes Aluminum Foil and Graphite Anodes. Now she studies hard in multidisciplinary research area in the production of sustainable electricity and fuels from waste management, algal and fungi biomass for scale up these low-cost sources. She works hard for publishing in ISI journal in this research area and obtaining funds from science and Technology Development Fund and Egyptian Knowledge Bank

Shristy Gautam received the B.Sc (Zoology Honours) from University of Delhi, India in 2008 and M.Sc. (Biotechnology) from Dr. Bhim Rao Ambedkar University, Agra, India in 2010. She has done her Ph.D (Biotechnology) in 2017 from Deenbandhu ChhotuRam University of Science and Technology, Murthal, Sonipat, Haryana, India. She is currenting working as Assistant Professor at Department of Molecular Biology and Genetic Engineering, School of Bioengineering and Biosciences, Lovely Professional University, Phagwara, Punjab, India. She has also cleared CSIR-UGC NET in Life science in December 2009. Her area of research interest is related to algae. She has rich experience in culturing of algae (especially diatoms), its identification, diatom-based biofuel, dye sensitized solar cell, heavy metal detection by diatoms. She has carried out studies using mice for drowning detection, prepared a diatomatological map (D-Map) of diatoms of water bodies of Haryana, India, generated thirteen diatom axenic cultures, constructed dye sensitized solar cell (DSSC) for electricity production etc. She has published 22 research articles, three book chapters and 1 book.

Ronald Halim is an assistant professor in the School of Biosystems and Food Engineering at University College Dublin. He has multidisciplinary research interests in the sustainable production of food, fuel and biochemicals from microalgal and macroalgal biomass. After completing his PhD in microalgal cell rupture and lipid recovery from Monash University in 2014, he continued his research into scalable microalgal bioprocess engineering, developing a biorefinery system that included an autolytic-inducing step to fractionate the biomass into a range of protein, food products and biofuel precursors during his postdoctoral appointments. The passion he has for his research has taken him to various institutions around the world, including The University of Melbourne, Karlsruhe Institute of Technology and Carnegie Institution for Science (Stanford University). He has previously been successful in attracting funding from internationally competitive schemes to support his research undertakings, such as the John Stocker Postdoctoral Fellowship and the Humboldt Postdoctoral Fellowship. His group at UCD is currently carrying out research in microalgae and macroalgae biorefinery systems.

Lamiaa H.S. Hassan is a scientific researcher in algae and algal applications. She was awarded her MSc and PhD from Faculty of Science, Menoufia University, Egypt. She was interested in using microalgae for biofuel and energy production. She also studied the using of algae in nanotechnology field to develop bio-metallic nanoparticles that can be used in biological activities such as antimicrobial and anticancer agents. She participated in 8 papers, which have been published in various international journals. Currently, she is working on writing a review and a book chapter to be published by Springer.

Touseef Hussain is presently working as Chief Scientist at Matimate Agromart Pvt. Ltd., Bhavnagar, Gujarat, India, before that he was working as SRF at Div. of Plant Pathology, ICAR-IARI, New Delhi, India. He did his B.Sc. and M.Sc. from CCSU, Meerut. Ph.D. in Life Sciences from Uttarakhand Technical University, Dehradun in 2016. He has worked and have good skills and experience from Indian top reputed National Agricultural Research Institute laboratories. Later he was awarded SERB-DST-NPDF (2017-19) at Dept. of Botany, AMU, Aligarh. He is recipient of Young Scientist award and Travel grant award. He has published more than 30 research articles and 25 book Chapters in National and International Journals also Reviewer of many publishing house like, SPRINGER, ELSEVIER, WILEY, CSIR-NISCAIR. He Reviewed in many International prestigious awards like American Phytopathological Society Experiential Student Travel Awards Reviewer Committee [2019,2020], American Society Of Microbiology, International Conf. Reviewer Committee [2019, 2020] and Ohio State University Plant Sciences Symposium Virtual Poster Competition, March, 2021. He is a Lifetime member of Indian Phytopathological Society (IARI, New Delhi), American Phytopathological Society (APS) Minnesota

USA, Association of Microbiologist of India (ASM), Haryana, Indian Science Congress, Kolkata and Microbiology Society, London, UK.

Vikas Kapase is working as a Pre-Doctoral fellow at ICGEB, New Delhi in the area of algal biofuels and value-added products using metabolic engineering tools.

Hanan M. Khairy is currently a Prof. of Hydrobiology (Algae), and Head of Hydrobiology Lab., National institute of Oceanography and Fisheries, NIOF, Egypt. She earned her Bachelor of Science at Faculty of Science, Tanta University, Egypt in 1995, and her Ph.D. in Botany Department at the same University. After completing her Ph.D. in Distribution of phytoplankton in El-Dekhaila Harbor, west of Alexandria, Egypt, in relevant to Environmental condition in 2002, She continued her research in isolation, purification and identification of different algal species (marine and freshwater species) for the laboratory experiment to studying the effect of different parameters affecting their growth. Study the nutritional values of micro and macro algal species to fish larvae under. Taking a part in more than 10 projects, and supervision of different scientific theses. Reviewer in about 20 international journals, and published about 60 research papers in Scopus cited journals.

Michael Kornaros (Male) is a Professor in the Department of Chemical Engineering (DCEUP) at the University of Patras (Greece) and Director of the Laboratory of Biochemical Engineering and Environmental Technology (LBEET) in the same Department. Prof. Kornaros' main research interests are in the organic wastes valorization for the production of bio-based compounds and innovative environmental applications, anaerobic fermentation, exploitation of agro-industrial wastes and wastewaters for liquid and gaseous biofuels production, and modeling of environmental processes. He has extensive expertise in the coordination and participation in European and Cross-Border projects (more than 70). He has published more than 120 articles in peer-reviewed international scientific journals, 18 books and book chapters, and more than 150 international conference proceedings (Total citations> 4000, h-index: 37). He is an active reviewer in more than 40 international journals and conference proceedings. Throughout his career, he has chaired more than 15 sessions in National and International Conferences, while has also been a member of the Organization Committee and/or the Scientific Committee in numerous conferences.

Neelima Lal has a Master's in Biotechnology and she has experience of working in the field of biotechnology and microbiology.

Savvas Giannis Mastropetros holds a diploma in Chemical Engineering from the Department of Chemical Engineering of the University of Patras (2018). He conducted his Diploma Thesis research in the Laboratory of Biochemical Engineering and Environmental Technology, focusing on agricultural wastewater treatment through microalgal cultivation, combined with the production of high added value products. Currently, he is a PhD candidate at the same Department. He has participated in the Online Conference of Young Scientists (Mineral Resources-Environment-Chemical Engineering, February 26-28, 2021) presenting his research on the cultivation of Chlorella vulgaris in anaerobic digestion effluents with different concentrations of ammonia nitrogen. Also, Savvas Giannis Mastropetros is a co-author of the article: ''Assessing the potential of Chlorella vulgaris for valorization of liquid digestates from agro-industrial and municipal organic wastes in a biorefinery approach'' in 2020.

Fazrena Nadia Md Akhir is a senior lecturer in Malaysia-Japan International Institute of Technology (MJIIT), Universiti Teknologi Malaysia (UTM) Kuala Lumpur. She obtained her PhD in Biomedical Sciences from The University of Queensland, Australia. Her research interest are in Environmental Biotechnology, Environmental Microbiology and Cancer Immunology.

Sabrina Mohamed Moffit holds a BSc (Hons) in Biology with Psychology granted by Lancaster University (UK) and Sunway University (Malaysia). She is currently working as a research assistant while pursuing her postgraduate study under the Department of Applied Physics at Universiti Kebangsaan Malaysia (UKM). Her research is in protein structural study and its vast potential applications in numerous biotechnology industries such as in food, pharmaceutical, cosmetics, and agriculture.

Farhah Mohd Nazaruddin is a fresh graduate of the Universiti Teknologi Malaysia in the Bachelor of Chemical Process Engineering (Hons).

Aishwarya Mogal is pursuing PhD in Molecular Biology and Genetic Engineering from Lovely Professional University. The main research interests are in diatoms and their lipid biosynthesis; the nutraceutical properties of the bioactive compounds present in the diatoms; evaluation of the extracted bioactive compounds.

Prachi Nawkarkar is a Post Doctoral Researcher with the Metabolic Engineering Group working with microalgae and its genetic engineering at the International Center for Genetic Engineering and Biotechnology, New Delhi, India.

Konstantinos Pispas (Male) is a graduate student of the Materials Science Department at the University of Patras (Greece). He is conducting his diploma thesis in the Laboratory of Biochemical Engineering and Environmental Technology (LBEET) at the Department of Chemical Engineering about microalgal cultures and the production of biopolymer compounds. His research focuses on the cultivation of various microalgae and cyanobacteria species in liquid media based on anaerobic digestion effluents. The final goal is the polyesters' recovery from the microbial cells and the characterization of the final product regarding its physicochemical and mechanical properties.

Monika Prakash Rai completed her Bachelor and Master degree from Banaras Hindu University, Varanasi, India. She received her Ph.D. from School of Biochemical Engineering, IIT (BHU), Varanasi and part of her Ph.D. from Johannes Gutenberg University, Mainz, Germany in 2005. Presently, she is serving as Professor at Amity Institute of Biotechnology, Noida Campus, India. She is also a recipient of UK-India Distinguished Visiting Scientist Award in 2011. During the past 15 years, Dr. Rai has made excellent contributions to the field of algae biomass, bioremediation, and biofuels production.

Mohammadhosein Rahimi is a PhD Candidate in the School of Biosystems and Food Engineering at University College Dublin. He has multidisciplinary research interests in Algal biorefineries, Bionanotechnology, Renewable fuels and chemicals production, and Waste and biomass recovery. His research focuses on sustainable food, fuel, and biochemicals production from renewable resources like microalgae and macroalgae. Currently, he is conducting research on microalgal cultivation and biorefinery at UCD School of Biosystems and Food Engineering and UCD Conway Institute.

Shashi Kumar Rhode, is a group Leader, Metabolic Engineering group, ICGEB, New Delhi, India. his lab is working in the application of synthetic biology, metabolic engineering, genome editing, RNAi technology, chloroplast engineering approaches for enhanced lipid for sustainable algae biofuel and terpenoid drugs biosynthesis in plants . https://www.icgeb.org/metabolic-engineering/ .

Eman Ragab Sallam is currently a Researcher of Marine Chemistry at the National Institute of Oceanography and Fisheries, Alexandria branch, Egypt. She earned her Bachelor of Science at Alexandria University in Egypt. Her Ph. D. in Inorganic and Physical Chemistry, Faculty of science, Damietta University in 2017. Her thesis work involved the removal of toxic carcinogenic hexavalent chromium ion from industrial waste water in the presence of some organic nontoxic substance. After completing his Ph.D., She has multidisciplinary research interests in the sustainable bioenergy production from microalgal biomass and sea water, industrial wastewater treatment, nanotechnology and its biological applications.

Michael Schagerl is expert in algal taxonomy, physiology and cultivation, aquatic ecology and biogeochemistry. He investigates a broad range of habitats from spring ecosystems to alkaline-saline lakes, always with a focus on algae and cyanoprokaryotes. Major interests are acclimation of organisms to actual environmental conditions, biodiversity and bioindication, ecosystem functions and applied phycology, the latter related to wastewater treatment, pollution, and valuable products from algae.

Victor Shikuku holds a Ph.D. degree in Chemistry and is a Lecturer of Physical Chemistry at the Department of Physical Sciences, Kaimosi Friends University College, Kenya. Victor has published more than 20 peer-reviewed papers, 10 book chapters, and one edited University-level textbook. He has post-doctoral research experience at Trier University in Germany supported by the Alexander Humboldt Foundation. He has experience in water and wastewater treatment approaches and materials chemistry for water purification.

Patrick Ssebugere is a Senior Lecturer of Physical Chemistry at the Department of Chemistry, Makerere University. He has published more than 25 peer reviewed journals articles and 1 book chapter. He has been cited over 600 times and his h-index is 14. He is a reviewer for over 20 science journals. He has supervised over 10 MSc and 3 PhD students to completion. Dr. Ssebugere has received numerous awards including the Georg Forster Fellowship for Senior Researchers at Helmholtz Centre for Environmental Research – UFZ in Leipzig, Germany (2021), TWAS-DFG Postdoctoral Research Fellowship at Helmholtz Zentrum München, German Research Center for Environmental Health (2019). His current research grants include the COVID-19 Biomarker project supported by Presidential Scientific Initiative on Epidemics (PRESIDE), Environmental Health project funded by The World Academy of Sciences (TWAS) and Swedish International Development Cooperation Agency (SIDA), Capacity building in Environmental Sciences project supported by ICM Erasmus plus, Environmental Chemistry for Sustainable Development supported by the Austrian Partnership Programme in Higher Education and Research for Development (APPEAR), and the Development of wastewater treatment techniques using nanotechnology funded by Makerere University Research and Innovation Fund (Mak-RIF).

Mina Tajmirriahi received her M.Sc. in Chemical Engineering-Biotechnology from the department of chemical engineering, Isfahan university of technology, Isfahan, Iran, in 2019. Her master's thesis sought to explore the impacts of extractives, nonstructural components of lignocellulose, on the bioconversion of agricultural waste to biofuels (biogas and bioethanol). Based on her obtained results, she published two journal papers in 2021. She is interested in biotechnology scopes, including lignocellulose valorization, biofuels, bioproducts, biorefinery approaches, and algal biotechnology.

Dimitris Zagklis holds a Diploma in Chemical Engineering (University of Patras, 2010), an MSc in "Energy and Environment" (University of Patras, 2012) and a PhD from the Department of Chemical Engineering of the University of Patras (2015). His research activities are focused on waste and wastewater treatment with physicochemical and biological processes, and their technoeconomic and environmental assessment (SuperPro Designer and SimaPro LCA software). He has participated in 7 research projects co-funded by EU and national sources, he is a reviewer for 15 scientific journals, and guest editor of 1 special issue. Finally, he is the co-author in 21 articles published in peer-reviewed journals, 1 patent and 28 presentations in scientific conferences/workshops.

Wafaa F. Zohir is working as Pre-Doctoral Fellow, Metabolic Engineering Group, ICGEB, New Delhi, in the area of algal Biofuels and natural bioproducts from microalgae. she earned her BSc and MS from Faculty of Science, Damanhour University and has a great experience in research and teaching in the field of phycology and its applications in the Damanhour University.

Index

A

B

C

D

E

F

G

H

I

K

L

M

R

S

T

U

W

www.ingramcontent.com/pod-product-compliance
Lightning Source LLC
LaVergne TN
LVHW080357011026
849161LV00009B/2325